MAST CELLS AND BASOPHILS

MAST CELLS AND BASOPHILS

Edited by

GIANNI MARONE

Università degli Studi di Napoli Federico II, Naples, Italy

LAWRENCE M. LICHTENSTEIN

Johns Hopkins University School of Medicine, Baltimore, Maryland, U.S.A.

STEPHEN J. GALLI

Department of Pathology, Stanford University Medical Center, Stanford, California, U.S.A.

ACADEMIC PRESS

A Harcourt Science and Technology Company

San Diego San Francisco New York Boston
London Sydney Tokyo

This book is printed on acid-free paper.

Academic Press
A Harcourt Science and Technology Company
Harcourt Place, 32 Jamestown Road, London NW1 7BY, UK
http://www.academicpress.com

Academic Press
A Harcourt Science and Technology Company
525 B Street, Suite 1900, San Diego, California 92101-4495, USA
http://www.academicpress.com

ISBN 0-12-473335-2

A catalogue record for this book is available from the British Library

Typeset by Kenneth Burnley, Wirral, Cheshire
Printed in Great Britain by MPG, Bodmin, Cornwall
00 01 02 03 04 05 MP 9 8 7 6 5 4 3 2 1

To Giancarlo, Elizabeth, Joshua, Rebekah and David.
As a token of our gratitude for their love and understanding.

Contents

Section Eight
MAST CELLS IN ATHEROSCLEROSIS AND CARDIAC DISEASES

Contributors

Numbers in parentheses indicate the pages on which the authors' contributions begin. An asterisk (*) indicates the author to whom correspondence should be addressed.

S. N. ABRAHAM* (381), Department of Pathology and Microbiology, Duke University Medical Center, Durham, NC27710, U.S.A., Tel: (919) 684-3630; Fax: (919) 684-2021.

L. ALOE * (325), Institute of Neurobiology, CNR, Viale Marx 15, I-00137, Rome, Italy, Tel: 39/06/8682592; Fax: 39/06/86090370; e-mail: aloe@in.rm.cnr.it

L. ANNUNZIATO (673), Section of Pharmacology, Dept. of Neuroscience, School of Medicine, University of Naples Federico II, Via S. Pansini 5, 80131 Naples, Italy.

S. ARBILLA (625), Synthélabo Recherche, 10 rue des Carrières, 92500 Rueil Malmaison, France.

E. ARBUSTINI (455, 597), Department of Pathology, University of Pavia, Pavia, Italy.

G. C. BACCARI* (117), Stazione Zoologica 'A. Dohrn', Italy, Tel: 39-081-5665840; Fax: 39-081-5665820; e-mail: gchieffi@unina.it

P. J. BARBOSA PEREIRA (275), Departamento de Biologia Molecular y Celular Instituto de Biologia Molecular de Barcelona (IBMB) Jordi Girona, 18-26 08034 Barcelona, Spain, Tel: 34-93-400 6100, ext. 269; Fax: 34-93-204 5904; e-mail: pbpcri@ alcor.cid.csic.es

C. BELTRAME (579), Department of Internal Medicine, Section of Clinical Immunology, Allergy, and Respiratory Diseases, University of Florence, Florence, Italy.

J. BIENENSTOCK* (313), Departments of Medicine, Pathology and Molecular Medicine, McMaster University, 1200 Main Street West, HSC-3N26, Hamilton, Ontario, L8N 3Z5, Canada, Tel: 905/525-9140, ext. 22017; Fax: 905/522-4936; e-mail: bienens@ fhs.csu.mcmaster.ca

S. C. BISCHOFF* (541), Department of Gastroenterology & Hepatology, Medical School of Hannover, D-30623 Hannover, Germany, Tel: 49 511 532 3305; Fax: 49 511 532 4896; e-mail: bischoff.stephan@mh-hannover.de

N. V-BLANK (149), Unité INSERM 363, ICGM Hôpital Cochin, 27 rue du Faubourg St Jacques, 75014 Paris, France, Tel: 33 1 46 33 64 40; Fax: 33 1 46 33 92 97; e-mail: varin@cochin.inserm.fr

U. BLANK* (149), Unité d'Immuno-Allergie, Institut Pasteur, 28 rue du Dr Roux, 75724 Paris Cedex 15, Tel: 33 1 40 61 32 64; Fax: 33 140 61 33 83; e-mail: ublank@pasteur.fr

S. BONINI (325), Department of Allergy and Clinical Immunology, Second University of Naples, Italy.

J. BROWN (257), Department of Veterinary Clinical Studies, Royal (Dick) School of Veterinary Studies, The University of Edinburgh, Easter Bush Veterinary Centre, Easter Bush, Roslin, Midlothian EH25 9RG, U.K.

P. BRUHNS (185), Laboratoire d'Immunologie Cellulaire et Clinique, INSERM U.2555, Institut Curie, 26 rue d'Ulm, 75005 Paris, France.

C. CALABRESE (525), Division of Clinical Immunology and Allergy, University of Naples Federico II, Via S. Pansini 5, I-80131 Naples, Italy.

P. CASTALDO (673), Section of Pharmacology, Department of Neuroscience, School of Medicine, University of Naples Federico II, Via S. Pansini 5, 80131 Naples, Italy.

M. K. CHURCH (641), Allergy and Inflammation Division, Southampton General Hospital, Southampton, S016 6YD, U.K.

J. W. COLEMAN* (221), Department of Pharmacology, University of Liverpool, Ashton Street, Liverpool, L69 3BX, UK, Tel: (44) 151 794 5551; Fax: (44)151794 5540; e-mail: coleman@liv.ac.uk

W. R COWARD (641), Allergy and Inflammation Division, Southampton General Hospital, Southampton, S016 6YD, U.K.

M. DAËRON* (185), Laboratoire d'Immunologie Cellulaire et Clinique, INSERM U.2555, Institut Curie, 26 rue d'Ulm, 75005 Paris, France, Tel: (33)1-4432-4223; Fax: (33)1-4051-0420; e-mail: Marc.Daeroncurie.fr

C. A. DAHINDEN* (567), Institute of Immunology and Allergology Inselspital, University Hospital Bern, CH-3010 Bern, Switzerland.

F. DAL PIAZ (597), International Mass Spectrometry Facility Center, University of Naples Federico II, Via S. Pansini 5, 80131, Naples, Italy.

J. E. DAMEN (169), Terry Fox Laboratory, BC Cancer Agency, Vancouver, BC, V5Z 1L3, Canada.

G. DE CRESCENZO (455, 597), Division of Clinical Immunology and Allergy, University of Naples Federico II, Via S. Pansini 5, 80131 Naples, Italy.

A. DE PAULIS* (117, 579, 597), Division of Clinical Immunology and Allergy, University of Naples Federico II, Via S. Pansini 5, 80131 Naples, Italy, Tel: 39-081-7462218; Fax: 39-081-7462271; e-mail: depaulis@unina.it

H. DEPOORTERE (625), 31 avenue Paul Vaillant Couturier, 92220 Bagneux, France.

A. DE SANTIS (117), Stazione Zoologica 'A. Dohrn', Naples, Italy.

A. M. DVORAK* (63), Department of Pathology, East Campus, Beth Israel Deaconess Medical Center, 330 Brookline Avenue, Boston, MA 02215 , U.S.A., Tel: 617 667 3692; Fax: 617 667 2943.

M. L. ENTMAN* (507), Department of Medicine, Cardiovascular Sciences, Baylor College of Medicine, One Baylor Plaza, M/S F-602, Houston, TX 77030-3498, U.S.A.

G. FLORIO (397), Divisione di Immunologia Clinica e Allergologia, University of Naples Federico II, Via S. Pansini, 5, 80131 Naples, Italy.

V. FORTE (397), Divisione di Immunologia Clinica e Allergologia, University of Naples Federico II, Via S. Pansini, 5, 80131 Naples, Italy.

N. G. FRANGOGIANNIS (507), Department of Medicine, Cardiovascular Sciences, Baylor College of Medicine, One Baylor Plaza, M/S F-602, Houston, TX 77030-3498, U.S.A.

W. H. FRIDMAN (185), Laboratoire d'Immunologie Cellulaire et Clinique, INSERM U.2555, Institut Curie, 26 rue d'Ulm, 75005 Paris, France.

S. J. GALLI* (3, 439), Department of Pathology, Stanford University Medical Center, 300 Pasteur Drive, Stanford, CA 94305, U.S.A., Tel: (650) 723-7975; Fax: (650) 725-6902; e-mail: sgalli@leland.stanford.edu

E. W. GELFAND* (133), Division of Basic Sciences, Department of Pediatrics, National Jewish Medical and Research Center, 1400 Jackson Street, Denver, CO 80206, U.S.A.

A. GENOVESE (397, 455), Division of Clinical Immunology and Allergy, University of Naples Federico II, Via S. Pansini 5, 80131 Naples, Italy.

M. GENTILE (525, 665), Division of Clinical Immunology and Allergy, University of Naples Federico II, Via Pansini 5, 80131 Naples, Italy.

G. GIORGIO (673), Section of Pharmacology, Department of Neuroscience, School of Medicine, University of Naples Federico II, Via S. Pansini 5, 80131 Naples, Italy.

J. GOLDHILL (625), Synthélabo Recherche, 10 rue des Carrières, 92500 Rueil Malmaison, France.

F. GRANATA (455, 525, 665), Division of Clinical Immunology and Allergy, University of Naples Federico II, Via S. Pansini 5, 80131 Naples, Italy.

J.-C. GUTIERREZ-RAMOS (31), Millennium Pharmaceuticals, Inc., 4575 Sydney Street, Cambridge, MA 02139, U.S.A., Tel: 617-679-7262; Fax: 617-551-8910.

K. HARTMANN (51), Department of Dermatology, University of Cologne, Joseph-Stelzmann-Str. 9, 50931 Cologne, Germany.

C. D. HELGASON (169), Terry Fox Laboratory, BC Cancer Agency, Vancouver, BC, V5Z 1L3, Canada.

B. M. HENZ* (341), Exp. Dermatology, Charité, Campus Virchow Clinic, Augustenburger-Platz 1, D 13344 Berlin, Germany, Tel: 49-30-450- 65001; Fax: 49-30-450 65900; e-mail: magdalena.fuchs@charite.de

B. HERMES (341), Krankenhaus Neukölln, Berlin, Germany.

K. HIRAI* (209), Department of Bioregulatory Function, University of Tokyo Graduate School of Medicine, 7-3-1 Hongo, Bunkyo-ku, Tokyo 113-8655, Japan.

S. HIROTA (21), Department of Pathology, Osaka University Medical School, Yamada-oka 2-2, Suita, Osaka 565-0871, Japan.

C. M. HOGABOAM (609), University of Michigan Medical School, Department of Pathology, Ann Arbor, MI 48109-0602, U.S.A.

S. T. HOLGATE* (641), Adult Respiratory and Molecular Sciences Research, Southampton General Hospital, Southampton, S016 6YD, U.K.

M. HUBER (169), Terry Fox Laboratory, BC Cancer Agency, Vancouver, BC, V5Z 1L3, Canada.

M. HUGHES (169), Terry Fox Laboratory, BC Cancer Agency, Vancouver, BC, V5Z 1L3, Canada.

R. K, HUMPHRIES (169), Terry Fox Laboratory, BC Cancer Agency, Vancouver, BC, V5Z 1L3, Canada.

M. IIKURA (209), Department of Allergy and Rheumatology, University of Tokyo Graduate School of Medicine, 7-3-1 Hongo, Bunkyo-ku, Tokyo 113-8655, Japan.

H. H. JACOBI* (89), Allergy Unit 7511, National University Hospital, Tagensvej 20, DK 2200 Copenhagen N, Denmark.

O. JOHANSSON (89), Experimental Dermatology Unit, Department of Neuroscience, Karolinska Institute, 17177 Stockholm, Sweden.

G. L. JOHNSON (133), Division of Basic Sciences, Department of Pediatrics, National Jewish Medical and Research Center, 1400 Jackson Street, Denver, CO 80206, U.S.A.

R. KAJEKAR (355), Johns Hopkins Asthma and Allergy Center, 501 Hopkins Bayview Circle, Baltimore, MD 21224, U.S.A.

Y. KITAMURA* (21), Department of Pathology, Osaka University Medical School, Yamada-oka 2-2, Suita, Osaka 565-0871, Japan, Fax: 81-6-6879-3729; e-mail: kitamura@patho.med.osaka-u.acjp

P. A. KNIGHT (257), Department of Veterinary Clinical Studies, Royal (Dick) School of Veterinary Studies, The University of Edinburgh, Easter Bush Veterinary Centre, Easter Bush, Roslin, Midlothian EH25 9RG, U.K.

P. T. KOVANEN* (479), Wihuri Research Institute, Kalliolinnantie 4, 00140 Helsinki, Finland, Tel: 358-9-637 572; Fax: 358-9-637 476; e-mail: Petri.Kovanen@wri.fi

S. A. KRILIS* (97), The University of New South Wales, Department of Immunology, Allergy and Infectious Disease, St George Hospital, Kogarah, New South Wales, 2217, Australia, Tel: 61-2-93502955; Fax: 61-2-93503981; e-mail: s.krilis@unsw.edu.au

G. KRYSTAL* (169), Terry Fox Laboratory, BC Cancer Agency, Vancouver, BC, V5Z 1L3, Canada.

C. S. LANTZ (3, 439), Department of Biology, James Madison University, Harrisonburg, VA 22807, U.S.A.

S. LAVENS-PHILLIPS (195), Johns Hopkins Asthma and Allergy Center, 5501 Hopkins Bayview Circle, Baltimore, MD 21224, U.S.A.

L. LI (97), The University of New South Wales, Department of Immunology, Allergy and Infectious Disease, St. George Hospital, Kogarah, New South Wales, 2217, Australia.

H. LIÉNARD (185), Laboratoire d'Immunologie Cellulaire et Clinique, INSERM U.2555, Institut Curie, 26 rue d'Ulm, 75005 Paris, France.

T.-J. LIN* (419), Departments of Pathology and Microbiology & Immunology, Dalhousie University, Halifax, Nova Scotia, B3H 1E2, Canada.

N. W. LUKACS* (609), University of Michigan Medical School, Department of Pathology, Ann Arbor, MI 48109-0602, U.S.A., Tel: 734-764-51 35; Fax: 734-764-2397; e-mail: nlukacs@umich.edu

D. MACGLASHAN JR* (195), Johns Hopkins Asthma and Allergy Center, 5501 Hopkins Bayview Circle, Baltimore, MD 21224, U.S.A.

R. MALAVIYA (381), Hughes Institute, 2665 Longlake Road, St Paul, MN 55113, U.S.A.

O. MALBEC (185), Laboratoire d'Immunologie Cellulaire et Clinique, INSERM U.2555, Institut Curie, 26 rue d'Ulm, 75005 Paris, France.

G. MARONE (*397, *455, 525, 579, 597, 665), Division of Clinical Immunology and Allergy, University of Naples Federico II, Via S. Pansini 5, 80131 Naples, Italy, Tel: 39-081-7707492, Fax: 39-081-7462271; e-mail: marone@unina.it

J. S. MARSHALL (419), Departments of Pathology and Microbiology & Immunology, Dalhousie University, Halifax, Nova Scotia, B3H 1E2, Canada.

D. D. METCALFE* (51), Laboratory of Allergic Diseases, National Institute of Allergy and Infectious Diseases, National Institute of Health, Building10, Room 11C205, 10 Center Drive MSC 1881, Bethesda, MD 20892-1881, U.S.A.

A. MICERA (325), Institute of Neurobiology, CNR, Viale Marx 15, I-00137, Rome, Italy.

H. R. P. MILLER* (257), Department of Veterinary Clinical Studies, Royal (Dick) School of Veterinary Studies, The University of Edinburgh, Easter Bush Veterinary Centre, Easter Bush, Roslin, Midlothian EH25 9RG, U.K., Tel: 0131650 6102; Fax: 0131650 6588; e-mail: Hugh.Miller@ed.ac.uk

G. MINOPOLI (597), Department of Biochemistry and Medical Biotechnology, University of Naples Federico II, Naples, Italy.

S. MINUCCI (117), Dipartimento di Fisiologia Umana e Funzioni Biologiche Integrate "F. Bottazzi", Seconda Università di Napoli, Naples, Italy.

K. MIURA (195), Johns Hopkins Asthma and Allergy Center, 5501 Hopkins Bayview Circle, Baltimore, MD 21224, U.S.A.

M. MIYAMASU (209), Department of Allergy and Rheumatology, University of Tokyo Graduate School of Medicine, 7-3-1 Hongo, Bunkyo-ku, Tokyo 113-8655, Japan.

E. MORII (21), Department of Pathology, Osaka University Medical School, Yamada-oka 2-2, Suita, Osaka 565-0871, Japan.

A. C. MYERS (355), Johns Hopkins Asthma and Allergy Center, 501 Hopkins Bayview Circle, Baltimore, MD 21224, U.S.A.

T. NISHIDA (21), Department of Pathology, Osaka University Medical School, Yamada-oka 2-2, Suita, Osaka 565-0871, Japan.

S. H. P. OLIVEIRA (609), University of Michigan Medical School, Department of Pathology, Ann Arbor, MI 48109-0602, U.S.A.

A. ORIENTE (397), Divisione di Immunologia Clinica e Allergologia, Università di Napoli Federico II, Via S. Pansini 5, 80131 Naples, Italy.

C. PALUMBO (525, 665), Division of Clinical Immunology and Allergy, University of Naples Federico II, Via S. Pansini 5, 80131 Naples, Italy.

A. PANNACCIONE (673), Section of Pharmacology, Department of Neuroscience, School of Medicine, University of Naples Federico II, Via S. Pansini 5, 80131 Naples, Italy.

V. PATELLA (397, 455), Division of Clinical Immunology and Allergy, University of Naples Federico II, Via S. Pansini 5, 80131 Naples, Italy.

PH. PICHAT (625), Synthélabo Recherche, 10 rue des Carrières, 92500 Rueil Malmaison, France.

R. POLOSA (641), Adult Respiratory and Molecular Sciences Research, Southampton General Hospital, Southampton, S016 6YD, U.K.

C. POTHOULAKIS* (367), Division of Gastroenterology, Beth Israel Deaconess Medical Center, Harvard Medical School, 330 Brookline Avenue, Boston, MA 02215, U.S.A., Tel: (617) 667-1246; Fax: (617) 975-5071; e-mail: cpothoul@caregroup.harvard.edu

P. PUCCI (597), International Mass Spectrometry Facility Center, University of Naples Federico II, Via S. Pansini 5, 80131 Naples, Italy.

E. J. QUACKENBUSH* (31), Clinical Genetics, Children's Hospital, Harvard Medical School, and the Center for Blood Research, 200 Longwood Avenue, Boston, MA 02115, U.S.A., Tel: 617-278-3240; Fax: 617-278-3030.

S. W. REDDEL (97), The University of New South Wales, Department of Immunology, Allergy and Infectious Disease, St. George Hospital, Kogarah, New South Wales, 2217, Australia.

P. ROMAGNANI* (579), Department of Pathophysiology, Endocrinology Unit, University of Florence, Florence, Italy.

S. ROMAGNANI (579), Department of Internal Medicine, Section of Clinical Immunology, Allergy, and Respiratory Diseases, University of Florence, Florence, Italy.

T. RUSSO (597), Department of Biochemistry and Medical Biotechnology, University of Naples Federico II, Naples, Italy.

A. M. SANICO* (651), Department of Medicine, Division of Clinical Immunology, Johns Hopkins Asthma & Allergy Center, 5501 Hopkins Bayview Circle, Baltimore, MD 21224, U.S.A., Tel: (410) 550-2191; Fax: (410) 550-2193; e-mail: amsanico@welchjhu.edu

N. M. SCHECHTER* (275), Department of Dermatology, and Department of Biochemistry and Biophysics, University of Pennsylvania, Clinical Research Building, 415 Curie Blvd.,

Philadelphia, PA 19104-6142, U.S.A., Tel: 215-898-3680; Fax: 215-573-2033; e-mail: schechte@mail.med.upenn.edu

N. SELVE* (625), Synthélabo Recherche, 10 rue des Carrières, 92500 Rueil Malmaison, France.

G. SPADARO (397), Divisione di Immunologia Clinica e Allergologia, Università di Napoli Federico II, Via S. Pansini 5, 80131 Naples, Italy.

R. L. STEVENS* (235), Brigham and Women's Hospital, Department of Medicine, Smith Building, Room 616B, 1 Jimmy Fund Way, Boston, MA 02115 ,U.S.A., Tel: 617-525-1231; Fax: 617-525-1310; e-mail: rstevens@rics.bwh.harvard.edu

S. STROBL (275), Abteilung für Strukturforschung, Max-Planck-Institut für Biochemie, D-82152, Planegg-Martinsried, Germany, Tel: 49 89 8578 2827; Fax: 49-89-8578-3516; e-mail: strobl@biochem.mpg.de

M. TAGLIALATELA* (673), Section of Pharmacology, Department of Neuroscience, School of Medicine, University of Naples Federico II, Via S. Pansini 5, 80131 Naples, Italy, Tel: 0039-081-7463310; Fax 0039-081-7463323; e-mail: mtalial@unina.it

E. M. THORNTON (257), Department of Veterinary Clinical Studies, Royal (Dick) School of Veterinary Studies, The University of Edinburgh, Easter Bush Veterinary Centre, Easter Bush, Roslin, Midlothian EH25 9RG, U.K.

A. TOGIAS (651), Department of Medicine, Division of Respiratory and Critical Care Medicine, Johns Hopkins Asthma & Allergy Center, 5501 Hopkins Bayview Circle, Baltimore, MD 21224, U.S.A.

M. TRIGGIANI* (525, 665), Division of Clinical Immunology and Allergy, University of Naples, Federico II, Via S. Pansini 5, I-80131 Naples, Italy, Tel: 39 081 7462219; Fax: 39 0817462271; e-mail: triggian@unina.it

M. TSAI (3), Department of Pathology, Stanford University Medical Center, 300 Pasteur Drive, Stanford, CA 94305, U.S.A.

B. J. UNDEM* (355), Johns Hopkins Asthma and Allergy Center, 501 Hopkins Bayview Circle, Baltimore, MD 21224, U.S.A.

P. VALENT* (497), Department of Internal Medicine I, Division of Hematology & Hemostaseology, The University of Vienna, Währinger Gürtel 18-20, A-1090 Vienna, Austria, Tel: 43 140400 6085; Fax: 43 1402 6930.

L. VERGA (455), Department of Pathology, University of Pavia, Pavia, Italy.

A. F. WALLS* (291), Immunopharmacology Group, Mailpoint 837, Level F South Block, Southampton General Hospital, Southampton, SO16 6YD, U.K. Tel: 44 023 8079 6151; Fax: 44 023 8079 6969; e-mail: afwl@soton.ac.uk

M. WARE (169), Terry Fox Laboratory, BC Cancer Agency, Vancouver, BC, V5Z 1L3, Canada.

P. WELKER (341), D. Exp. Dermatology, Charité, Campus Virchow Clinic, Augustenburger-Platz 1, D 13344 Berlin, Germany.

B. K. WERSHIL (31), SUNY Health Sciences Center, Brooklyn, New York, NY 11203, U.S.A; Tel: 718-270-3090; Fax: 718-270-1985.

S. H. WRIGHT (257), Department of Veterinary Clinical Studies, Royal (Dick) School of Veterinary Studies, The University of Edinburgh, Easter Bush Veterinary Centre, Easter Bush, Roslin, Midlothian EH25 9RG, U.K.

M. YAMAGUCHI (209), Department of Allergy and Rheumatology, University of Tokyo Graduate School of Medicine, 7-3-1 Hongo, Bunkyo-ku, Tokyo 113-8655, Japan.

F.-G. ZHU (419), Departments of Pathology and Microbiology & Immunology, Dalhousie University, Halifax, Nova Scotia, B3H 1E2, Canada.

Preface

The histochemical characteristics of human basophils and tissue mast cells were described over a century ago by Paul Ehrlich. At that time, the basophil's and mast cell's distinguishing feature was the affinity of their specific cytoplasmic granules, now known to have distinctive ultrastructure features, for certain basic dyes. Interestingly, the cell's tinctorial properties visible by light microscopy and their granule ultrastructural characteristics have remained the key identifying features of human basophils and mast cells.

For several decades, basophils and mast cells and their mediators were considered to play mainly a proinflammatory role in various allergic disorders, and, as a result, in the 1980s mast cells and basophils fell out of favour among some immunologists. However, with the appreciation of these cells as major potential sources of multifunctional cytokines and chemokines, it became evident in the 1990s that mast cells and basophils may actually express immunoregulatory functions, as well as have roles as effector cells, in various immune disorders and protective host responses.

This is a wonderful time in mast cell and basophil research. Indeed, the last few years have witnessed unprecedented progress in our understanding of the development and function of mast cells and basophils, and of the roles played by these cells in physiological and pathological processes. To name just a few recent developments, several lines of evidence now indicate that mast cells and basophils not only express critical effector function in classic IgE-associated allergic disorders, but also play important roles in host defence against parasites, bacteria and perhaps even viruses. Indeed, it is now clear that mast cells and basophils can contribute to host defence in the context of either acquired or innate immune-responses. Moreover, these cells can be activated by different HIV-1 proteins (gp120 and Tat) and thereby represent a potentially important source of Th2 cytokines during HIV-1 infection.

Basophils and mast cells can contribute to late phase inflammatory reactions in the skin and lung. Mature basophils and mast cell precursors circulate in peripheral blood at low concentrations. However, one of their characteristics is their capacity to adhere to activated endothelial cells, and to leave the bloodstream and migrate into inflamed tissues. This probably reflects, at least in part, the expression of the chemokine receptor CCR3 on their surface. The latter observation may also help to explain why increased numbers of basophils and mast cells are present at sites of certain inflammatory responses.

The spectrum of diseases in which mast cells have been implicated has extended beyond allergic disorders to include several diseases of the cardiovascular and gastrointestinal systems and the joints, as well as the nervous system. In particular, investigations of potential anatomical and functional interactions between mast cells and the nervous system have recently attracted great interest. Mast cells and basophils are now thought to exert critical proinflammatory functions, as well as potential immunoregulatory roles, in various immune disorders through the release of such mediators as histamine, cysteinyl leukotrienes, cytokines and chemokines and neutral proteases (chymase and tryptase). Moreover, our knowledge of the intracellular signalling pathways that control the development of these cells, and the expression of basophil and mast cell functions, has also advanced rapidly, in part due to the use of powerful genetic approaches in mice to explore mast cell and basophil development and function *in vivo*.

Many of the chapters of this volume represent papers that were presented at a meeting that included these topics. This meeting, entitled 'Mast Cells and Basophils in Physiology, Pathology, and Host Defense', took place from 4th to 6th March 1999 at the Accademia Nazionale dei Lincei ['The National Academy of the Lynxes']. The Accademia Nazionale dei Lincei, which is considered the oldest secular scientific academy in Europe (Galileo became a member in 1611), is housed in the sixteenth-century Corsini Palace in the heart of Rome. This fascinating setting was an especially fitting place for this particular scientific gathering because it honoured the Nobel Laureat and Lycean, Rita Levi-Montalcini, who gave the keynote address.

The rapid advances in this field make it difficult to produce a timely reference text. Despite these difficulties, we accepted the challenge to produce what may be the first volume of the third millennium that focuses on the basic and clinically relevant aspects of mast cell and basophil biology.

The editors hope that you will enjoy this volume, and that it will convey some of the excitement that enlivens current work in basophil and mast cell biology. We would like to express our sincere appreciation to Sanofi-Synthelabo and in particular to Gianluca Visconti in Milan, whose enthusiastic support made this volume and the meeting on which it is based possible. We also thank all of the scientific contributors, as well as Jean Gilder, for their contributions to this volume.

GIANNI MARONE
LAWRENCE M. LICHTENSTEIN
STEPHEN J. GALLI

Università di Napoli Federico II

Accademia Nazionale dei Lincei

Johns Hopkins University

Acknowledgements

This volume falls within the framework of the international scientific exchange programme between the University of Naples Federico II (Italy) and the Johns Hopkins University of Baltimore (MD, USA).

I owe a debt of gratitude to Professor Fulvio Tessitore, Rector of the University of Naples Federico II, and to Edward J. Benz Jr, Chairman of the Department of Medicine of the Johns Hopkins University, both of whom encouraged and made possible this exchange programme. I am also grateful to Professor Luigi Labruna, President of the Italian National Council for the University, for his invaluable support of the scientific event on which this volume is based.

GIANNI MARONE

SECTION ONE

DEVELOPMENT OF MAST CELLS AND BASOPHILS

CHAPTER 1

Regulation of Mast Cell and Basophil Development by Stem Cell Factor and Interleukin-3

MINDY TSAI,[1] CHRIS S. LANTZ[2] and STEPHEN J. GALLI[*1]*

[1] *Department of Pathology, Stanford University Medical Center, Stanford, California and*
[2] *Department of Biology, James Madison University, Harrisonburg, Virginia, U.S.A.*

INTRODUCTION

In this chapter, we review current understanding of the effects of two cytokines, the c-kit ligand (or stem cell factor), and interleukin-3 (IL-3), on mast cell and basophil development, as well as review some of the effects of SCF on mast cell function. We will focus particularly on findings obtained in analyses of the effects of SCF or IL-3 on mast cell and basophil development and function *in vivo*. In part, this emphasis reflects the fact that recombinant human SCF (rhSCF) is already undergoing clinical testing in humans. Although rhSCF is being developed as an agent to enhance haematopoiesis, and to facilitate the harvesting of haematopoietic progenitors (1, 2), one of the effects of the repetitive treatment of human subjects with rhSCF is the induction of mast cell hyperplasia (3).

However, our focus on *in vivo* findings is also prompted by three other considerations. First, studies of mice with spontaneous or targeted mutations that affect SCF or IL-3 production, or the receptors for these cytokines, have provided information about the actual importance of endogenous SCF or IL-3 in mast cell and basophil development *in vivo*. Second, a number of studies have analysed the effects of recombinant forms of SCF or IL-3 on mast cell and basophil development in mice and other mammalian species *in vivo*. Finally, as we illustrate below, in some cases it has not been possible to use even extensive amounts of *in vitro* data to predict the results obtained when genetic or other approaches are used to analyse the effects of SCF or IL-3 on mast cell or basophil development *in vivo*.

* Corresponding author.

MAST CELLS AND BASOPHILS
ISBN 0-12-473335-2

SCF, A LIGAND FOR THE c-*kit* RECEPTOR

Mice with a double dose of mutations at either *W* or *Sl* have long been known to exhibit a hypoplastic, macrocytic anaemia, sterility and a lack of cutaneous melanocytes (reviewed in refs 4, 5). Transplantation and embryo fusion studies employing *W* or *Sl* mutant and congenic normal mice, or *in vitro* analyses employing cells or tissues derived from these animals, indicated that the deficits in the *W* mutant mice are expressed by the cells in the affected lineages, whereas those in the *Sl* mutant animals are expressed by microenvironmental cells necessary for the normal development of the affected lineages (4, 5). The complementary nature of the phenotypic abnormalities expressed by *W* or *Sl* mutant mice suggest that the *W* locus might encode a receptor expressed by haematopoietic cells, melanocytes and germ cells, whereas the *Sl* locus might encode the corresponding ligand (4).

The important observations that mutations at *W* or *Sl* also profoundly affect mast cell development were made by Kitamura *et al.* (6) and Kitamura and Go (7). These authors demonstrated that the virtual absence of mast cells in W/W^v mice, like the anaemia of these animals, reflected an abnormality intrinsic to the affected lineage (6), whereas the mast cell deficiency of Sl/Sl^d mice, which could not be corrected by bone marrow transplantation from the congenic normal (+/+) mice, reflected an abnormality in the microenvironments necessary for normal mast cell development (7). Kitamura's finding that transplantation of bone marrow cells from the congenic +/+ mice or from beige (C57BL/6-*bg*/*bg*) mice, whose mast cells can be identified unequivocally because of their giant cytoplasmic granules, repaired the mast cell deficiency of the W/W^v mice provided clear evidence that mast cells were derived from precursors that reside in the bone marrow. This work also showed that mutations at *W* had a more profound effect on the mast cell than on any other haematopoietic lineage.

Subsequently, two groups reported that the *W* gene product encodes the c-*kit* tyrosine kinase receptor (8, 9). Shortly after this discovery, three groups simultaneously reported that *Sl* encodes the corresponding ligand, which was variously named (in alphabetical order) kit ligand (KL) (10), mast cell growth factor (MGF) (11–13), steel factor (SLF or SF) (14, 15) and stem cell factor (SCF) (16–18). This chapter will use the terms SCF for the ligand and c-kit for the receptor.

The gene for SCF encodes two transmembrane proteins of 220 and 248 amino acids, which are generated by alternative splicing; both forms can be proteolytically cleaved to produce soluble forms of the molecule which retain biological activity and which spontaneously form non-covalently linked dimers in solution (reviewed in refs 19–21). While native SCF is glycosylated, the non-glycosylated, *Escherichia coli*-derived soluble recombinant forms of the extracellular ligand domain of the molecule ($rSCF^{164}$), which were used for many of the studies that we review here, have significant biological activity (reviewed in refs 20, 21).

Receptor tyrosine kinases can regulate cell survival, proliferation and differentiation by transducing extracellular signals transmitted by their cognate ligands (22). As predicted on the basis of the phenotypic abnormalities expressed by *W* or *Sl* mutant mice, SCF has been shown to promote haematopoiesis and mast cell development, as well as melanocyte survival and proliferation, and to influence the survival and proliferation of primordial germ cells (reviewed in refs 20, 21). Other findings, such as the expression of high levels of c-kit or SCF in the central nervous system, or the expression of c-kit on lymphocytes, had not been expected, because *W* or *Sl* mutant mice were not known to exhibit central nervous system abnormalities and, in general, these mutants have normal numbers of mature lymphocytes and normal B cell and T cell function (reviewed in refs 20, 21). Similarly, it has only recently become apparent that SCF–c-kit interactions have a critical role in the development of the interstitial cells of Cajal, which generate intestinal electrical pacemaker activity (23–25). Moreover, it was recently shown that mature peripheral blood eosinophils can express c-kit and that SCF stimulation of eosinophils can enhance the cells' very late antigen 4-mediated adhesion to fibronectin and the adhesion molecule VCAM-1 (26). This mechanism could, in part, account for a proposed role for SCF in influencing eosinophil recruitment during certain murine models of inflammation (27).

MULTIPLE EFFECTS OF SCF IN MAST CELL BIOLOGY

SCF can promote the *in vitro* survival of early haematopoietic progenitor cells and can act synergistically with other haematopoietic growth factors to promote the further differentiation of multiple haematopoietic lineages (1, 2, 19–21, 28–40). However, unlike most other haematopoietic lineages, mast cells retain significant expression of the SCF receptor (c-kit) into maturity, and thus exhibit responsiveness to SCF not only during their development but also, in all likelihood, throughout their mature life span (reviewed in ref. 20).

Results of *in vitro* analyses, which, in many instances, have been confirmed by *in vivo* studies, indicate that SCF can have many effects in mast cell development and function: it can maintain mast cell survival, promote chemotaxis or haptotaxis of mast cells and their precursors, promote the proliferation of immature or mature mast cells, promote the maturation of mast cell precursors or immature mast cells and alter the phenotype and mediator content of these cells, directly promote the degranulation and secretion of mediators by mast cells, enhance the mast cell's ability to secrete mediators in response to other signals, including IgE and specific antigen, and alter the expression of other receptors, including those for extracellular matrix components and neuropeptides (reviewed in refs 20, 21).

However, the specific effects of SCF on mast cell biology that are expressed under individual circumstances can be influenced significantly by other factors. For example, long-term dosing with recombinant rat SCF (rrSCF) can induce mast cell hyperplasia in multiple organs in normal rats (40). However, the pattern of expression of the mast cell-associated proteases (RMCP), RMCP I and II, by these mast cells varied according to the specific anatomical sites analysed (40). Thus, rrSCF promoted the development of mast cells that expressed predominantly RMCP I in the skin and peritoneal cavity, whereas those in the small intestinal mucosa expressed predominantly RMCP II (40). Subsequently, work in mice (28, 29), and later in rats (30), showed that IL-3 represents one of the additional cytokines that can influence the proliferation, and the serine protease phenotype, of mast cells that have been exposed to SCF.

It should be emphasized that the regulation of mast cell development in mice and humans may differ in important details (20, 31, 32). Moreover, rhSCF is now in clinical testing to determine the extent to which this cytokine, when used together with granulocyte colony-stimulating factor (G-CSF), can enhance the production and harvesting of haematopoietic progenitor cells (2). Thus, it is of particular interest to consider the effects of rhSCF on human mast cell development and function.

In vitro studies demonstrated that rhSCF can promote the development of mast cells from various sources of human haematopoietic progenitor cells (33–36). Subsequently, a Phase I study of *E. coli*-derived rhSCF showed that the administration of rhSCF (at 5–50 $\mu g\ kg^{-1}\ day^{-1}$, subcutaneously, for 14 days) to patients with advanced breast carcinoma resulted in a significant increase, by ~70%, in the numbers of cutaneous mast cells at sites that had not been directly injected with the agent (3, 37, 38). In addition, the patients exhibited increased urinary levels of the major histamine metabolite, methylhistamine (3, 37, 38), and markedly increased (by 100–1220%) serum levels of mast cell α-tryptase, as detected by an assay that can measure both the α and β forms of this protease (3). The latter finding, when taken together with the observation that, in cynomolgus monkeys, rhSCF dosing induced much higher levels of mast cell development in the liver, spleen and lymph nodes than in the skin (39), suggested that the effect of rhSCF dosing on numbers of cutaneous mast cells may have greatly underestimated the effects of the agent on mast cell populations at other anatomical sites. In any event, this work identified rhSCF as the first cytokine that can induce human mast cell hyperplasia *in vivo*, and also showed that humans may be more sensitive to this action of rhSCF than are cynomolgus monkeys (3, 37–39).

Although all of the biological effects of SCF on mast cells are of interest, none of them can be expressed unless the survival of the lineage is maintained. Work in both genetically mast cell-deficient *SCF/MGF* mutant Sl/Sl^{d} ($Mgf^{Sl}/Mgf^{Sl\text{-}d}$) mice (17, 40) and cynomolgus monkeys (39) demonstrated that rSCF can promote the survival of the mast cell lineage *in vivo*. Thus, cessation of rhSCF dosing in cynomolgus monkeys was followed by a rapid decline of tissue mast cell numbers, in some cases to nearly baseline levels (39).

Subsequently, three studies established that SCF can promote mast cell survival by suppressing apoptosis, either *in vitro* (41–43) or *in vivo* (42). Indeed, the study by Iemura *et al.* (42) indicated that apoptosis represents a mechanism which can account for striking (up to 50-fold) and rapid reductions in the size of mast cell populations *in vivo*, apparently without significant associated inflammation. The ability of SCF to protect mast cells from apoptosis is inhibited by the blockade of Ca^{2+} influx. On the other hand, the protective effects of IL-3 on mast cell apoptosis are not affected by Ca^{2+} influx inhibitors, indicating that SCF and IL-3 may maintain mast cell survival by distinct mechanisms (44).

The increased numbers of mast cells present in mast cell neoplasms or examples of naturally occurring mastocytosis may in part reflect enhanced mast cell survival. Two SCF–c-kit-dependent mechanisms which may account for increased mast cell survival in such settings *in vivo* have been described: (1) 'gain-of-function' mutations affecting c-kit itself (45–48); and (2) altered production and/or biodistribution of endogenous SCF (49–50). A third mechanism which can enhance c-kit-dependent mast cell development has been defined in recent studies of mice with various combinations of mutations at c-*kit* and *me*, which encodes the Src homology 2 domain (SH2) -containing non-transmembrane protein tyrosine phosphatase, SHP1 (51, 52). This work showed that the decrease in dermal mast cell numbers in W^{v}/W^{v} ($Kit^{W\text{-}v}/Kit^{W\text{-}v}$) mice was significantly improved by superimposition of the *me/me* genotype, which results in diminished negative regulation of c-kit signalling (51, 52).

In principle, these findings suggest that agents that can interfere with c-kit-dependent signalling in mast cells might be effective in diminishing the size of mast cell populations *in vivo*. However, c-kit is expressed on haematopoietic progenitor cells, melanocytes, germ cells and many other cell types, including certain neurons (reviewed in refs 19–21). Accordingly, the development of effective and safe approaches for manipulating the SCF receptor–ligand interaction to reduce mast cell numbers *in vivo* will require achieving either adequate target cell selectivity or clinically acceptable control of the agent's bioavailability. For example, *in vitro* studies and analyses performed *in vivo* in mice indicate that the ability of glucocorticoids to reduce mast cell populations may reflect the agent's ability to suppress local SCF production (53, 54). However, in humans, glucocorticoid treatment protocols that result in diminished numbers of dermal mast cells can also produce significant cutaneous atrophy (55).

SCF CAN REGULATE MAST CELL FUNCTION *IN VITRO* AND *IN VIVO*

Because the cell lineages that are most profoundly affected by *W* or *Sl* mutations (affecting c-kit or SCF, respectively) (reviewed in refs 19, 20) ordinarily are essentially missing in the mutant animals, it was not generally suspected that SCF might regulate the secretory function of cells that express c-kit. However, Wershil *et al.* (56) showed that SCF can induce mouse skin mast cell degranulation *in vivo* in doses as low as 140 fmol per site and that this response is c-kit-dependent, in that it occurs when dermal mast cells express the wild-type c-kit but not in phorbol 12-myristate 13-acetate (PMA) -induced dermal mast cells in W/W^v (Kit^W/Kit^{W-v}) mice that express the Kit^{W-v} mutant receptor. The receptor encoded by Kit^{W-v} has a normal extracellular ligand-binding domain, but a point mutation in the kinase domain results in markedly reduced tyrosine kinase activity upon ligand engagement (57).

Subsequently, it was shown that SCF can also induce mediator release *in vitro* from rat (58) or mouse (59) peritoneal mast cells and from human skin mast cells (60) and cultured human intestinal mast cells (61). At even lower concentrations *in vitro*, SCF can augment IgE-dependent activation of mouse peritoneal mast cells (59) or human lung (62) or skin (60) mast cells. SCF treatment can also enhance the responsiveness of mouse mast cells to the neuropeptides substance P (63) and PACAP (pituitary adenylate cyclase polypeptide) (64) *in vitro*. These findings suggest that SCF may be able to influence neuroimmune interactions by regulating the expression of neuropeptide receptors on mast cells.

The ability of SCF to promote mast cell secretion directly, and to enhance mast cell activation via $Fc_\varepsilon RI$, prompted experiments to compare the signalling pathways that were activated in mast cells that had been stimulated via c-kit as opposed to $Fc_\varepsilon RI$ (65). This work showed that patterns of activation of MAP (mitogen-activated protein) kinases and $pp90^{rsk}$ and pp70-S6 kinases were very similar in mouse mast cells that were activated through these structurally distinct receptors (65), and suggested that c-kit or $Fc_\varepsilon RI$-dependent signalling pathways may exhibit more overlap than had previously been suspected (65).

It should be emphasized, however, that the effects of SCF on mast cell secretory function are potentially complex and may vary not only according to species and type of mast cell population (reviewed in ref. 20) but also according to duration of exposure to SCF and class of mast cell mediators. For example, in purified mouse peritoneal mast cells, short-term exposure to rrSCF can both induce serotonin release directly and enhance IgE-dependent serotonin release (59). However, in immature mouse mast cells generated *in vitro*, short-term incubation with recombinant mouse SCF induces little or no mediator release (66). By contrast, longer-term incubation of such cells with recombinant mouse SCF enhances IgE-dependent prostaglandin PGD_2 generation, at least in part through effects on haematopoietic PGD_2 synthase, but simultaneously diminishes the cells' ability to release the granule-associated mediator β-hexosaminidase (66). Moreover, in studies employing rrSCF, immature mouse mast cells which had been generated in IL-3-containing medium *in vitro* secreted IL-6 and, to a lesser extent, tumour necrosis factor (TNF-α) in response to challenge with soluble rrSCF, whereas concentrations of rrSCF that were effective in inducing the release of IL-6 resulted in little or no specific release of serotonin or histamine (67). However, SCF can induce marked degranulation in immature mast cells generated from SH2-containing inositol 5′-phosphatase (SHIP) knockout mice (68). Thus, SHIP appears to play an inhibitory role in regulating the ability of SCF to initiate intracellular signalling and degranulation in mast cells *in vitro*.

Although the molecular basis for the differences in responsiveness of various mast cell populations to the effects of SCF on mediator secretion are not completely understood, the diversity of the secretory responses induced in different mast cell populations by challenge with SCF *in vitro* made it difficult to predict whether administering this cytokine to human subjects *in vivo* would provoke mast cell degranulation. Nevertheless, in a Phase I study of rhSCF (3, 37, 38), we found that subcutaneous injections of rhSCF at 5–50 $\mu g\ kg^{-1}$ induced a wheal and flare response in each of the ten subjects tested and at each rhSCF injection site, and that these reactions, when examined by transmission electron microscopy, exhibited evidence of extensive, anaphylactic-type, degranulation of dermal mast cells (3, 69). Moreover, a few subjects developed adverse events after rhSCF dosing that were consistent with the induction of systemic activation of mast cell populations (3, 37, 38). These findings strongly suggest that rhSCF can directly induce human mast cell degranulation *in vivo*, as it can *in vitro*.

It has recently been shown that chymase, a major cytoplasmic granule-associated protease of human cutaneous mast cells, can cleave human SCF at a novel site that results in the release of a soluble, but biologically active, fragment of SCF which is 7 amino acids shorter at the C-terminus than previously characterized soluble SCF (70, 71). Thus, in addition to initiating a local, mast cell-dependent inflammatory response, injection of SCF might also initiate a mast cell chymase-dependent mechanism which results in local changes in the proportion of cell membrane-associated SCF compared with soluble SCF.

The adverse effects of the mast cell activation that is induced at rhSCF injection sites *in vivo* can be largely ameliorated by pre-treatment of subjects with H_1 and H_2 antihistamines (2, 72). However, a recent study reports that it may be possible to modify SCF in a way that enhances its ability to promote haematopoiesis but does not increase its ability to enhance mast cell mediator secretion (73). Specifically, a soluble disulphide-linked dimer of mouse SCF (murine KL covalent dimer, or KL-CD), in comparison to the native, non-covalently linked KL dimer, exhibited significantly enhanced growth-promoting activity in colony-forming assays of mouse haematopoietic progenitor cells (CFU-GM) and in assays of [^{3}H]thymidine incorporation by immature, IL-3-derived mouse mast cells *in vitro*, and increased mobilization of CFU-GM in the blood and spleen of mice *in vivo*, without exhibiting a significant change in its ability to enhance either $Fc_{\varepsilon}RI$ -dependent release of hexosaminidase from IL-3-derived immature mouse mast cells *in vitro* or induce degranulation of dermal mast cells in the mouse ear *in vivo* (73). Whether the same (or other) chemical modification of rhSCF would result in an agent with an enhanced therapeutic profile (i.e. more haematopoietic cell growth-promoting activity, but unchanged or diminished ability to promote mast cell secretion) remains to be seen.

Finally, recent evidence suggests that, in some instances, increases in mast cell numbers might actually be beneficial. Mast cells can serve as key effector cells in innate immunity (reviewed in ref. 74). Studies by Maurer *et al.* (75) showed that repetitive treatment of mice with SCF can markedly improve their survival after caecal ligation and puncture, a model of acute bacterial peritonitis. Experiments using mast cell-reconstituted WBB6F$_1$-*Kit*^{W}/*Kit*$^{W\text{-}v}$ mice indicated that the enhanced survival associated with such SCF treatment reflected, at least in part, actions of SCF on mast cells (75).

INTERLEUKIN-3

Interleukin-3 is a 28-kDa glycoprotein that was first characterized as a factor that can induce the expression of 20α-hydroxysteroid dehydrogenase in the splenocytes of nude mice *in vitro* (76). It was later shown that this cytokine can promote the *in vitro* differentiation and proliferation of haematopoietic progenitor cells, leading to the generation of multipotential blast cells, mast cells, basophils, neutrophils, macrophages, eosinophils, erythrocytes, megakaryocytes and dendritic cells (77–79). Indeed, Ihle *et al.* (80) demonstrated that IL-3 represented the critical factor present in the various 'conditioned media' that were used by different groups to generate populations of immature mast cells from mouse haematopoietic cells *in vitro* (81–85).

This work, and many other *in vitro* studies, indicated that IL-3 might represent a major mast cell developmental/growth factor in the mouse, as well as a T cell-derived factor that can contribute to the enhanced development of other haematopoietic effector cells during immune responses to pathogens (86). However, studies with human haematopoietic cells indicated that IL-3 promoted the development of basophils *in vitro*, but had little if any ability, under most circumstances, to induce mast cell development (87–89).

The administration of IL-3 *in vivo* can enhance haematopoiesis in mice (90, 91) and experimental primates (89, 92) and can markedly enhance levels of circulating basophils in Rhesus macaques (89). In addition, widespread cutaneous inflammation developed at sites of recombinant human IL-3 injection in Rhesus macaques, and this was associated with a modest increase in numbers of dermal mast cells at such sites (93). However, in part because of apparently conflicting data on the extent to which human mast cells can express receptors for IL-3 (reviewed in ref. 32), it was not clear whether the ability of IL-3 to influence human mast cell development *in vitro* (88) or in Rhesus macaques *in vivo* (93) represented direct or indirect effects of the cytokine. On the other hand, both the demonstration that IL-3 can markedly enhance mast cell development in the intestines of nude mice infected with the nematode *Strongyloides ratti* (94) and the demonstration that neutralizing antibodies to IL-3 can partially suppress (by ~50%) the mast cell hyperplasia which develops in the intestines of mice infected with the nematode *Nippostrongylus brasiliensis* (95, 96) supported the hypothesis that IL-3 can contribute to the hyperplasia of mucosal mast cells which occurs in murine rodents during T cell-dependent immune responses to certain parasites.

In addition to its effects on the development of mast cells, basophils and other haematopoietic cells, many studies have shown that IL-3 can also enhance antigen presentation for T cell-dependent responses, augment macrophage cytotoxicity and adhesion, and promote the secretory function of eosinophils, basophils and mast cells

(95, 97–104). Taken together, these findings supported the hypothesis that IL-3 derived from T cells (and perhaps other sources) represents a critical link between the immune and haematopoietic systems, and may be particularly important for promoting the development, survival and effector function of tissue mast cells and blood basophils. However, the actual importance of such potential functions of IL-3 *in vivo* remained unclear. For example, mice carrying an inactivating mutation in the α chain of the heterodimeric IL-3 receptor are apparently normal, and haematopoiesis can occur *in vitro* in the absence of IL-3 (105, 106). And even though mouse T lymphocytes and mast cells (107–109) can produce IL-3 *in vitro*, the conditions in which IL-3 is expressed *in vivo*, and the sources of this cytokine in these settings, are not fully understood (110).

USING IL-3 –/– MICE TO ASSESS THE ROLE OF IL-3 IN MAST CELL AND BASOPHIL DEVELOPMENT

Mice which lack IL-3 were produced using gene targeting in embryonic stem cells (111). IL-3 –/– mice are healthy and, unlike W/W^v ($Kit^W/Kit^{W\text{-}v}$) mice, are fertile. Moreover, like mice that carry an inactivating mutation in the α chain of the heterodimeric IL-3 receptor (106) or that lack both IL-3 and the common β subunit of the receptors for IL-3, IL-5 and granulocyte–macrophage colony-stimulating factor (GM-CSF) (112), IL-3 –/– mice exhibit no detectable abnormalities in multiple aspects of haematopoiesis *in vitro* or *in vivo* (111). Thus, in comparison to mice with mutations resulting in impaired c-kit or SCF expression or function, the phenotype of IL-3 –/– mice was remarkably normal.

On the other hand, we found that IL-3 –/– mice did exhibit abnormalities in mast cell development *in vitro* and *in vivo*. In accord with previous work indicating that exogenous IL-3 can augment SCF-dependent mast cell development *in vitro* (20, 28–30, 113–115), we found that SCF induced fewer mast cells to develop *in vitro* in suspension cultures of bone marrow cells derived from IL-3 –/– mice as opposed to IL-3 +/+ mice (116). By contrast, substantially higher, and essentially equivalent, numbers of mast cells developed when bone marrow cells from either IL-3 –/– or IL-3 +/+ mice were maintained *in vitro* in exogenous SCF plus IL-3 (116).

These *in vitro* studies thus showed that endogenous IL-3 can enhance, but is not required for, mast cell development from bone marrow progenitors in the presence of exogenous SCF. To assess the role of IL-3 in mast cell development *in vivo*, we quantified mast cells in the tissues of IL-3 –/– vs. wild-type mice at baseline or after 21 daily subcutaneous injections of rrSCF (at 100 $\mu g\ kg^{-1}/day^{-1}$) or vehicle alone (116). The results of these experiments showed that endogenous IL-3 is not essential for the development of mast cells under physiological conditions *in vivo*. Indeed, in all sites examined, levels of tissue mast cells at baseline in adult IL-3 –/– mice were very similar to those in the corresponding sites in IL-3 +/+ mice. Moreover, in contrast to our observations in the *in vitro* system, we found that endogenous IL-3 was not required for rrSCF-induced mast cell hyperplasia *in vivo*. In fact, in certain tissues, mast cell levels after rrSCF treatment were significantly greater (by up to 140%) in IL-3 –/– mice than in the corresponding wild-type mice (116).

However, IL-3 contributed significantly to the increases in mast cell numbers that were observed in the intestines and spleen of mice infected with the nematode *Strongyloides venezuelensis*, and may have accounted for all of the increases in the bone marrow basophils in these animals (116). These findings strongly support the hypothesis that

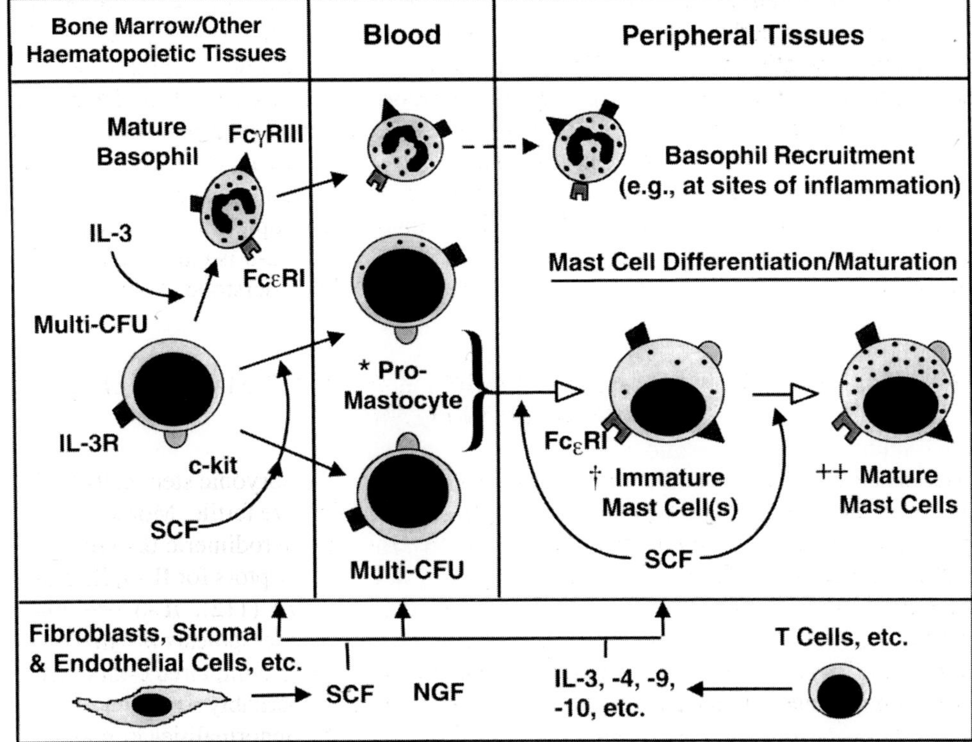

Fig. 1 A highly simplified model of mouse mast cell and basophil development. Basophils arise from c-kit⁺ multipotential haematopoietic progenitor cells, but typically mature in the bone marrow before entering the peripheral circulation. Basophils then can be recruited from the blood to sites of inflammatory or immune responses. Both human and mouse basophils generally express little or no c-kit on their surface. Mast cells also arise from multipotential haematopoietic progenitor cells, but complete major parts of their differentiation/maturation in the peripheral tissues. Unlike basophils, mast cells express high levels of c-kit on their surface throughout their development; such c-kit receptors can interact with either membrane-associated or soluble forms of SCF.

*A committed precursor of tissue mast cells, the pro-mastocyte, has been identified in mouse fetal blood. Fetal and adult mouse blood also contains multipotential haematopoietic progenitor cells which, under appropriate circumstances, can give rise to mast cells as well as other cell types. Mouse blood, haematopoietic tissues and certain other tissues may also contain unipotential mast cell progenitors that are distinct from the pro-mastocyte (see text).

† It is unclear to what extent, either in fetal or adult mouse tissues, immature mast cells that reside in the tissues are derived from pro-mastocytes as opposed to circulating multipotential haematopoietic progenitor cells.

++ Note that the phenotype of mature mast cells can vary considerably in different anatomical sites, based in part on local levels of SCF and other cytokines (e.g., IL-3, -4, -9, -10, etc.), and that the phenotypic characteristics of mast cells may vary (in some cases, reversibly) during the course of immune responses or inflammatory processes.

This figure has been modified (with permission) from one that originally appeared in Lantz, C. S. and Galli, S. J., Mast cell and basophil development. In: *Hematopoiesis: A Developmental Approach* (Zon, L. I., ed.), Oxford University Press, New York, in press.

much or all of the basophilia, and a significant fraction of the increases in intestinal and splenic mast cell populations, that occurs during Th2-type responses in mice is IL-3 driven. Our current model of the roles of SCF and IL-3 in mouse mast cell and basophil development is outlined in Fig. 1 and some explanatory notes are provided in the legend. The model specifically refers to the mouse system because the availability of mice that are defective in SCF–c-kit signalling and/or which lack IL-3 permit a direct assessment of the importance of SCF and IL-3 in mast cell and basophil development in this species *in vivo*. However, *in vitro* studies, and analyses of the effects of rhSCF *in vivo*, indicate that many of the major themes of mast cell and basophil development (as shown in Fig. 1) are very similar in mice and humans (20, 31, 32).

On the other hand, our understanding of these processes still has significant gaps. For example, the 'pro-mastocyte', which represents the earliest mast cell-committed precursor to be identified during ontogeny, has so far been identified only in the mouse (115). The mouse pro-mastocyte is defined by the phenotype Thy-1^{lo} c-Kit^{hi}, contains small numbers of cytoplasmic granules that are very similar (by ultrastructure) to those that had previously been identified in immature mouse mast cells generated in IL-3-containing medium *in vitro*, and expresses mRNAs encoding mouse mast cell-associated proteases (MC-CPA, MMCP-4 and MMCP-2) (115). However, this cell lacks expression, at the mRNA level, of $Fc_{\varepsilon}RI$. Purified pro-mastocytes can generate functionally competent mast cells at high frequencies *in vitro*, but do not exhibit developmental potential for other haematopoietic lineages.

The development of mast cells from pro-mastocytes *in vitro* occurred in cultures that had been supplemented with SCF and IL-3, but not in cultures that had been supplemented with only one of these cytokines. However, given our finding of normal levels of tissue mast cells in IL-3 –/– mice (116), it is likely that cytokines other than IL-3 can function together with SCF to regulate the maturation of pro-mastocytes *in vivo*. When transferred intraperitoneally, pro-mastocytes can reconstitute the peritoneal mast cell compartment of Kit^{W}/Kit^{W-v} mice to wild-type levels; moreover, these pro-mastocyte-derived peritoneal mast cells exhibit certain phenotypic characteristics of 'mature' peritoneal mast cells. The fetal blood pro-mastocyte population was first detected on day 14.5 of mouse gestation and, on day 15.5 of gestation, pro-mastocytes represented approximately one-fortieth of the $CD45^{+}$ leukocyte fraction in the peripheral blood of these animals. However, the numbers of pro-mastocytes in the fetal blood declined from day 15.5 until birth. The origin of pro-mastocytes is uncertain; specifically, these cells have not yet been identified in mouse fetal liver in either mid or late gestation.

Although a mast cell-committed precursor cell that was functionally or morphologically distinct from multipotent haematopoietic stem cells had not previously been purified from mouse bone marrow or blood (117, 118), prior work had established that mast cell precursor activity could be identified in day 9.5 mouse embryonic yolk sac (119) in adult mouse bone marrow (120) and in adult mouse or rat peripheral blood (118, 120–122), as well as in the mesenteric lymph nodes of *Nippostrongylus brasiliensis*-infected mice (113, 117). In humans, *in vitro* analyses indicate that mast cells can arise from a circulating c-kit^{+}, $CD34^{+}$, Ly^{-}, $CD14^{-}$, $CD17^{-}$ haematopoietic progenitor cell (123, 124) that lacks detectable expression of $Fc_{\varepsilon}RI$ (123), but not from basophils or other differentiated haematopoietic lineages (124). However, it is not known whether an equivalent to the mouse pro-mastocyte exists in humans or, indeed, to what extent pro-mastocytes are present in the blood or other tissues of adult mice.

In addition, as indicated in Fig. 1, it is not clear (even in the mouse) whether pro-mastocytes or Thy-1$^-$ c-Kit$^+$ multipotential haematopoietic progenitor cells (or some other population of mast cell progenitors) represents the most important source of the immature mast cells which are found in the peripheral tissues. For example, assays of mast cell colony formation *in vitro* indicate that unipotential precursors of mast cells (i.e. colony formation unit–mast, or CFU–Mast) may occur in the blood and other tissues of mice and rats (reviewed in refs 117, 118), but the exact relationship of these cells to the pro-mastocyte (or to the Thy1$^-$ Kit$^+$ multipotential haematopoietic progenitor cell) is not yet clear. It is possible that tissue mast cells can be derived from either pro-mastocytes or less differentiated precursors, and that the proportion of tissue mast cells which are derived from these various potential precursor populations varies at different stages of development, or in the context of different inflammatory or immune responses that are associated with changes in numbers of mast cells.

Nevertheless, it is clear, both in mice and in humans, that mature mast cells do not ordinarily circulate [mast cells may appear in the circulation after long-term dosing with rSCF (39), or in subjects with mastocytosis or mast cell leukaemia (125)]. Accordingly, in both mice and humans, much of the mast cell differentiation/maturation process occurs in the peripheral tissues; these processes are regulated by SCF and many other cytokines in the mouse and, probably, also in humans. By contrast, the weight of current evidence indicates that basophil maturation is completed (or nearly completed) before the cells are released into the peripheral circulation (20, 72, 126). And, unlike mast cells, basophils do not ordinarily reside in large numbers in peripheral tissues, but can be recruited to sites of inflammatory or immune responses (20, 72).

SUMMARY

SCF and IL-3 have distinct roles in mouse mast cell and basophil development. SCF–c-kit interactions are required for physiological mast cell development in mice, as mast cells are ordinarily essentially absent in the tissues of $Kit^W/Kit^{W\text{-}v}$ mice. Nevertheless, modest numbers of mast cells (generally, only ~3–25% of the numbers in similarly treated congenic +/+ mice) can appear in the gastrointestinal mucosal tissues or spleen of $Kit^W/Kit^{W\text{-}v}$ mice that have been infected with various nematodes, and studies in IL-3 –/– and $Kit^W/Kit^{W\text{-}v}$ IL-3 –/– mice show that IL-3 importantly contributes to this example of 'SCF-independent' mast cell hyperplasia (116). Indeed, this may represent the '*in vivo* equivalent' of the mast cell development which occurs when bone marrow cells derived from $Kit^W/Kit^{W\text{-}v}$ mice are placed in IL-3-containing medium *in vitro* (128–130).

Nevertheless, limited development of 'mucosal mast cells' occurred in *Strongyloides venezuelensis*-infected mice that were totally devoid of IL-3, indicating that other cytokines can also promote the development of such mast cells (116). Based primarily on the results of *in vitro* analyses, as well as a limited number of *in vivo* experiments, cytokines which may promote the development of certain mast cell populations in mice include IL-4 (131, 132), IL-6 (133), IL-9 (134, 135), IL-10 (136), TNF-α (133) and nerve growth factor (NGF; 137–139). Thus, while SCF may be the most important of the mast cell developmental/growth factors in mice, IL-3 (and perhaps many other cytokines) can also contribute to mast cell development in this species.

The finding of essentially normal levels of bone marrow basophils in IL-3 –/– mice indicates that IL-3 is not required for the production of this granulocyte in mice (116). However, virtually all of the basophil hyperplasia which occurred in the bone marrow of mice that were infected with *Strongyloides venezuelensis* was IL-3-dependent (116). This finding confirms a large body of evidence, derived from studies in mice, humans and other species, which indicates that IL-3 probably represents a major (if not the major) cytokine responsible for basophil hyperplasia *in vivo*.

ACKNOWLEDGEMENTS

Some of the work reviewed herein was supported by United States Public Health Service grants (AI-23990, CA-72074 and AI-33372), the Beth Israel Hospital Pathology Foundation, Inc. and AMGEN Inc. S.J.G. performed research funded by, and consults for, AMGEN Inc., under terms that are in accord with Beth Israel Deaconess Medical Center, Harvard Medical School and Stanford University conflict of interest policies.

REFERENCES

1. Morstyn, G., Brown, S., Gordon, M., *et al.* Stem cell factor is a potent synergistic factor in hematopoiesis. *Oncology* **51**:205–214, 1994.
2. Moskowitz, C. H., Stiff, P., Gordon, M. S., *et al.* Recombinant methionyl human stem cell factor and filgrastim for peripheral blood progenitor cell mobilization and transplantation in non-Hodgkin's lymphoma patients – results of a phase I/II trial. *Blood* **89**:3136–3147, 1997.
3. Costa, J. J., Demetri, G. D., Harris, T. J., *et al.* Recombinant human stem cell factor (kit ligand) promotes human mast cell and melanocyte hyperplasia and functional activation *in vivo*. *J. Exp. Med.* **183**:2681–2686, 1996.
4. Russell, E. S. Hereditary anemias of the mouse: a review for geneticists. In: *Advances in Genetics* (Caspari, E. W., ed.), Vol. 20. Academic Press, New York, 1979.
5. Silvers, W. K. Dominant spotting, patch and rump-white; steel, flexed tail, splotch and varitint waddler. In: *The Coat Colors of Mice: A Model for Gene Action and Interaction*, pp. 206–241. Springer-Verlag, New York, 1979.
6. Kitamura, Y., Go, S. and Hatanaka, K. Decrease of mast cells in *W/W^{v}* mice and their increase by bone marrow transplantation. *Blood* **52**:447–452, 1978.
7. Kitamura, Y. and Go, S. Decreased production of mast cells in *Sl/Sl^{d}* anemic mice. *Blood* **53**:492–497, 1979.
8. Chabot, B., Stephenson, D. A., Chapman, V. M., Besmer, P. and Bernstein, A. The proto-oncogene c-kit encoding a transmembrane tyrosine kinase receptor maps to the mouse W locus. *Nature* **335**:88–89, 1988.
9. Geissler, E. N., Ryan, M. A. and Housman, D. E. The dominant-white spotting (W) locus of the mouse encodes the c-kit proto-oncogene. *Cell* **55**:185–192, 1988.
10. Huang, E., Nocka, K., Beier, D. R., *et al.* The hematopoietic growth factor KL is encoded by the *Sl* locus and is the ligand of the c-kit receptor, the gene product of the *W* locus. *Cell* **63**:225–233, 1990.
11. Anderson, D. M., Lyman, S. D., Baird, A., *et al.* Molecular cloning of mast cell growth factor, a hematopoietin that is active in both membrane bound and soluble forms. *Cell* **63**:235–243, 1990.
12. Copeland, N. G., Gilbert, D. J., Cho, B. C., *et al.* Mast cell growth factor maps near the steel locus on mouse chromosome 10 and is deleted in a number of steel alleles. *Cell* **63**:175–183, 1990.
13. Williams, D. E., Eisenman, J., Baird, A., *et al.* Identification of a ligand for the c-kit proto-oncogene. Cell **63**:167–174, 1990.
14. Witte, O. N. Steel locus defines new multipotent growth factor. *Cell* **63**:5–6, 1990.
15. Williams, D. E., de Vries, P., Namen, A. E., Widmer, M. B. and Lyman, S. D. The steel factor. *Dev. Biol.* **151**:368–376, 1992.
16. Martin, T. R., Suggs, S. V., Langley, K. E., *et al.* Primary structure and functional expression of rat and human stem cell factor DNAs. *Cell* **63**:203–211, 1990.
17. Zsebo, K. M., Williams, D. A., Geissler, E. N., *et al.* Stem cell factor (SCF) is encoded at the *Sl* locus of

the mouse and is the ligand for the c-*kit* tyrosine kinase receptor. *Cell* **63**:213–224, 1990.

18. Zsebo, K. M., Wypych, J., McNiece, I. K., *et al.* Identification, purification, and biological characterization of hematopoietic stem cell factor from buffalo rat liver-conditioned medium. *Cell* **63**:195–210, 1990.
19. Besmer, P. The *kit* ligand encoded at the murine *Steel* locus: a pleiotropic growth and differentiation factor. *Curr. Opin. Cell Biol.* **3**:939–946, 1991.
20. Galli, S. J., Zsebo, K. M. and Geissler, E. N. The kit ligand, stem cell factor. *Adv. Immunol.* **55**:1–96, 1994.
21. Broudy, V. C. Stem cell factor and hematopoiesis. *Blood* **90**:1345–1364, 1997.
22. Ullrich, A. and Schlessinger, J. Signal transduction by receptors with tyrosine kinase activity. *Cell* **61**:203–212, 1990.
23. Maeda, H., Yamagata, A., Nishikawa, S., *et al.* Requirement of c-kit for development of intestinal pacemaker system. *Development* **116**:369–375, 1992.
24. Ward, S., Burns, A., Torihashi, S. and Sanders, K. Mutation in the proto-oncogene c-*kit* blocks development of interstitial cells and electrical rhythm in murine intestine. *J. Physiol.* **480**:91–97, 1994.
25. Huizinga, J. D., Thuneberg, L., Kluppel, M., Malysz, J., Mikkelsen, H. B. and Bernstein, A. *W/Kit* gene required for interstitial cells of Cajal and for intestinal pacemaker activity. *Nature* **373**:347–349, 1995.
26. Yuan, Q., Austen, K. F., Friend, D. S., Heidtman, M. and Boyce, J. A. Human peripheral blood eosinophils express a functional c-kit receptor for stem cell factor that stimulates very late antigen 4 (VLA-4) mediated cell adhesion to fibronectin and vascular cell adhesion molecule 1 (VCAM-1). *J. Exp. Med.* **186**:313–323, 1997.
27. Lukacs, N. W., Strieter, R. M., Lincoln, P. M., Brownell, E., Pullen, D. M., Schock, H. J., Chensue, S. W., Taub, D. D. and Kunkel, S. L. Stem cell factor (c-*kit* ligand) influences eosinophil recruitment and histamine levels in allergic airway inflammation. *J. Immunol.* **156**:3945–3951, 1996.
28. Tsuji, K., Zsebo, K. M. and Ogawa, M. Murine mast cell colony formation supported by IL-3, IL-4, and recombinant rat stem cell factor, ligand for c-kit. *J. Cell. Physiol.* **148**:362–369, 1991.
29. Gurish, M. F., Ghildyal, N., McNeil, H. P., Austen, K. F., Gillis, S. and Stevens, R. L. Differential expression of secretory granule proteases in mouse mast cells exposed to interleukin 3 and c-kit ligand. *J. Exp. Med.* **175**:1003–1012, 1992.
30. Haig, D. M., Huntley, J. F., Mackellar, A., *et al.* Effects of stem cell factor (kit-ligand) and interleukin-3 on the growth and serine proteinase expression of rat bone marrow-derived or serosal mast cells. *Blood* **83**:72–83, 1994.
31. Galli, S. J. Signals in the regulation of mast cell growth and development: a perspective. In: *Signal Transduction in Mast Cells and Basophils* (Razin, E., Pecht, I. and Rivera, J. eds), in press. Landes, Georgetown, 1998.
32. Li, L., Zhang, X-T. and Krilis, S. A. Factors that affect human mast cell and basophil growth. In: *Signal Transduction in Mast Cells and Basophils* (Razin E., Pecht I. and Rivera, J. eds), in press. Landes, Georgetown, 1998.
33. Irani, A-M. A., Nilsson, G., Miettinen, U., *et al.* Recombinant human stem cell factor stimulates differentiation of mast cells from dispersed human fetal liver cells. *Blood* **80**:3009–3021, 1992.
34. Kirshenbaum, A. S., Goff, J. P., Kessler, S. W., Mican, J. M., Zsebo, K. M. and Metcalfe, D. D. Effect of IL-3 and stem cell factor on the appearance of human basophils and mast cells from CD34+ pluripotent progenitor cells. *J. Immunol.* **148**:772–777, 1992.
35. Valent, P., Spanblochl, E., Sperr, W. R., *et al.* Induction of differentiation of human mast cells from bone marrow and peripheral blood mononuclear cells by recombinant human stem cell factor/kit ligand in long-term culture. *Blood* **80**:2237–2245, 1992.
36. Mitsui, H., Furitsu, T., Dvorak, A. M., *et al.* Development of human mast cells from umbilical cord blood cells by recombinant human and murine c-kit ligand. *Proc. Natl. Acad. Sci. USA* **90**:735–739, 1993.
37. Costa, J. J., Demetri, G. D., Hayes, D. F., Merica, E. A., Menchaca, D. M. and Galli, S. J. Increased skin mast cells and urine methyl histamine in patients receiving recombinant methionyl human stem cell factor. *Proc. Am. Assoc. Cancer Res.* **34**:211, 1993 (abstract).
38. Costa, J. J., Demetri, G. D., Harris, T. J., *et al.* Recombinant human stem cell factor (rhSCF) induces cutaneous mast cell activation and hyperplasia, and hyperpigmentation in humans *in vivo*. *J. Allergy Clin. Immunol.* **93**:225, 1994 (abstract).
39. Galli, S. J., Iemura, A., Garlick, D. S., Gamba-Vitalo, C., Zsebo, K. M. and Andrews, R. G. Reversible expansion of primate mast cell populations *in vivo* by stem cell factor. *J. Clin. Invest.* **91**:148–152, 1993.

40. Tsai, M., Shih, L., Newlands, G. F. J., *et al.* The rat c-kit ligand, stem cell factor, induces the development of connective tissue-type and mucosal mast cells *in vivo*. Analysis by anatomical distribution, histochemistry, and protease phenotype. *J. Exp. Med.* **174**:125–131, 1991.
41. Mekori, Y. A., Oh, C. K. and Metcalfe, D. D. IL-3-dependent murine mast cells undergo apoptosis on removal of IL-3: prevention of apoptosis by c-kit ligand. *J. Immunol.* **151**:3775–3784, 1993.
42. Iemura, A., Tsai, M., Ando, A., Wershil, B. K. and Galli, S. J. The c-kit ligand, stem cell factor, promotes mast cell survival by suppressing apoptosis. *Am. J. Pathol.* **144**:321–328, 1994.
43. Yee, N. S., Paek, I. and Besmer, P. Role of kit-ligand in proliferation and suppression of apoptosis in mast cells: basis for radiosensitivity of *White Spotting* and *Steel* mutant mice. *J. Exp. Med.* **179**:1777–1787, 1994.
44. Gommerman, J. L. and Berger, S. A. Protection from apoptosis by steel factor but not interleukin-3 is reversed through blockade of calcium influx. *Blood* **91**:1891–1900, 1998.
45. Furitsu, T., Tsujimura, T., Tono, T., *et al.* Identification of mutations in the coding sequence of the proto-oncogene c-*kit* in a human mast cell leukemia cell line causing ligand-independent activation of c-*kit* product. *J. Clin. Invest.* **92**:1736–1744, 1993.
46. Nagata, H., Worobec, A. S., Oh, C. K, *et al.* Identification of a point mutation in the catalytic domain of the protooncogene c-*kit* in peripheral blood mononuclear cells of patients who have mastocytosis with an associated hematologic disorder. *Proc. Natl. Acad. Sci. USA* **92**:10560–10564, 1995.
47. Longley Jr., B. J., Tyrrell, L., Lu, S. Z., *et al.* Somatic c-kit activating mutation in urticaria pigmentosa and aggressive mastocytosis: establishment of clonality in a human mast cell neoplasm. *Nat. Genet.* **12**:312–314, 1996.
48. London, C. A., Galli, S. J., Yuuki, T., Hu, Z-Q., Helfand, S. C. and Geissler, E. N. Spontaneous canine mast cell tumors express tandem duplications in the proto-oncogene c-*kit*. *Exp. Hematol.* **27**:689–697, 1999.
49. Longley Jr., B. J., Morganroth, G. S., Tyrrell, L., *et al.* Altered metabolism of mast-cell growth factor (c-kit ligand) in cutaneous mastocytosis. *N. Engl. J. Med.* **328**:1302–1307, 1993.
50. Otsuka, H., Kusumi, T., Kanai, S., Koyama, M., Kuno, Y. and Takizawa, R. Stem cell factor mRNA expression and production in human nasal epithelial cells: contribution to the accumulation of mast cells in the nasal epithelium of allergy. *J. Allergy Clin. Immunol.* **102**:757–764, 1998.
51. Lorenz, U., Bergemann, A. D., Steinberg, H. N., *et al.* Genetic analysis reveals cell type-specific regulation of receptor tyrosine kinase c-Kit by the protein tyrosine phosphatase SHP1. *J. Exp. Med.* **184**:1111–1126, 1996.
52. Paulson, R. F., Vesely, S., Siminovitch, K. A. and Bernstein, A. Signaling by the *W/Kit* receptor tyrosine kinase is negatively regulated *in vivo* by the protein tyrosine phosphatase SHP1. *Nat. Genet.* **13**:309–315, 1996.
53. Finotto, S., Mekori, Y. A. and Metcalfe, D. D. Glucocorticoids decrease tissue mast cell number by reducing the production of the c-kit ligand, stem cell factor, by resident cells: *in vitro* and *in vivo* evidence in murine systems. *J. Clin. Invest.* **99**:1721–1728, 1997.
54. Mekori, Y. A., Hartmann, K. and Metcalfe, D. D. Mast cell apoptosis and its regulation. In: *Signal Transduction in Mast Cells and Basophils* (Razin, E., Pecht, I. and Rivera, J., eds), in press. Landes, Georgetown, 1998.
55. Lavker, R. M. and Schechter, N. M. Cutaneous mast cell depletion results from topical corticosteroid usage. *J. Immunol.* **135**:2368–2373, 1985.
56. Wershil, B. K., Tsai, M., Geissler, E. N., Zsebo, K. M. and Galli, S. J. The rat c-*kit* ligand, stem cell factor, induces c-*kit* receptor-dependent mouse mast cell activation *in vivo*. Evidence that signaling through the c-*kit* receptor can induce expression of cellular function. *J. Exp. Med.* **175**:245–255, 1992.
57. Nocka, K., Tan, J. C., Chiu, E., *et al.* Molecular bases of dominant negative and loss of function mutations at the murine c-kit/white spotting locus: W^{37}, W^{v}, W^{41}, and W. *EMBO J.* **9**:1805–1813, 1990.
58. Nakajima, K., Hirai, K., Yamaguchi, M., *et al.* Stem cell factor has histamine releasing activity in rat connective tissue-type mast cells. *Biochem. Biophys. Res. Commun.* **183**:1076–1083, 1992.
59. Coleman, J. W., Holliday, M. R., Kimber, I., Zsebo, K. M. and Galli, S. J. Regulation of mouse peritoneal mast cell secretory function by stem cell factor, IL-3 or IL-4. *J. Immunol.* **150**:556–562, 1993.
60. Columbo, M., Horowitz, E. M., Botana, L. M., *et al.* The human recombinant c-*kit* receptor ligand, rhSCF, induces mediator release from human cutaneous mast cells and enhances IgE-dependent mediator release from both skin mast cells and peripheral blood basophils. *J. Immunol.* **149**:599–608, 1992.

61. Bischoff, S. C., Schwengberg, S., Raab, R. and Manns, M. P. Functional properties of human intestinal mast cells cultured in a new culture system: enhancement of IgE receptor-dependent mediator release and response to stem cell factor. *J. Immunol.* **159**:5560–5567, 1997.
62. Bischoff, S. C. and Dahinden, C. A. c-*kit* ligand: a unique potentiator of mediator release by human lung mast cells. *J. Exp. Med.* **175**:237–244, 1992.
63. Wershil, B. K., Lavigne, J. A. and Galli, S. J. Stem cell factor (SCF) can influence neuroimmune interactions: bone marrow-derived mast cells (BMCMC) maintained in SCF acquire the ability to release histamine and tumor necrosis factor-alpha (TNF-α) in response to substance P (SP). *FASEB J.* **8**:A742, 1994 (abstract).
64. Schmidt-Choudhury, A., Meissner, J., Seebeck, J., *et al.* Stem cell factor influences neuro-immune interactions: The response of mast cells to pituitary adenylate cyclase activiting polypeptide is altered by stem cell factor. *Regul. Pept.* **83**:73–80, 1999.
65. Tsai, M., Chen, R-H., Tam, S-Y., Blenis, J. and Galli, S. J. Activation of MAP kinases, $pp90^{rsk}$ and pp70-S6 kinases in mouse mast cells by signaling through the c-*kit* receptor tyrosine kinase or $Fc_{\varepsilon}RI$: rapamycin inhibits activation of pp70-S6 kinase and proliferation in mouse mast cells. *Eur. J. Immunol.* **23**:3286–3291, 1993.
66. Murakami, M., Matsumoto, R., Urade, Y., Austen, K. F. and Arm, J. P. c-Kit ligand mediates increased expression of cytosolic phospholipase A_2, prostaglandin endoperoxide synthase-1, and hematopoietic prostaglandin D_2 synthase and increased IgE-dependent prostaglandin D_2 generation in immature mouse mast cells. *J. Biol. Chem.* **270**:3239–3246, 1995.
67. Gagari, E., Tsai, M., Lantz, C. S., Fox, L. G. and Galli, S. J. Differential release of mast cell interleukin-6 via c-kit. *Blood* **89**:2654–2663, 1997.
68. Huber, M., Helgason, C. D., Scheid, M. P., Duronio, V., Humphries, R. K. and Krystal, G. Targeted disruption of SHIP leads to Steel factor-induced degranulation of mast cells. *EMBO J.* **17**:7311–7319, 1998.
69. Dvorak, A. M., Costa, J. J., Morgan, E. S., Monahan-Earley, R. A. and Galli, S. J. Diamine oxidase-gold ultrastructural localization of histamine in human skin biopsies containing mast cells stimulated to degranulate *in vivo* by exposure to recombinant human stem cell factor. *Blood* **90**:2893–2900, 1997.
70. Longley, B. J., Tyrrell, L., Ma, Y., *et al.* Chymase cleavage of stem cell factor yields a bioactive, soluble product. *Proc. Natl. Acad. Sci. USA* **94**:9017–9021, 1997.
71. dePaulis, A., Minopoli, G., Dal Piaz, F., Pucci, P., Russo, T. and Marone, G. Novel autocrine and paracrine loops of the stem cell factor/chymase network. *Int. Arch. Allergy Immunol.* **118**:422–425, 1999.
72. Costa, J. J. and Galli, S. J. Mast cells and basophils. In: *Clinical Immunology: Principles and Practice* (Rich, R., Fleisher, T. A., Schwartz, B. D., Shearer, W. T. and Strober, W., eds), pp. 408–430. Mosby Year Book, St. Louis, 1996.
73. Nocka, K. H., Levine, B. A., Ko, J-L., *et al.* Increased growth promoting but not mast cell degranulation potential of a covalent dimer of c-kit ligand. *Blood* **90**:3874–3883, 1997.
74. Galli, S. J., Maurer, M. and Lantz, C. S. Mast cells as sentinels of innate immunity. *Curr. Opin. Immunol.* **11**:53–59, 1999.
75. Maurer, M., Echtenacher, B., Hültner, L., Kollias, G., Männel, D. N., Langley, K. E. and Galli, S. J. The c-kit ligand, stem cell factor, can enhance innate immunity through effects on mast cells. *J. Exp. Med.* **188**:2343–2348, 1998.
76. Ihle, J. N. and Weinstein, Y. Immunological regulation of hematopoietic/lymphoid stem cell differentiation by interleukin 3. *Adv. Immunol.* **39**:1–50, 1986.
77. Rennick, D. M., Lee, F. D., Yokota, T., Arai, K. I., Cantor, H. and Nabel, G. J. A cloned MCGF cDNA encodes a multilineage hematopoietic growth factor: multiple activities of interleukin 3. *J. Immunol.* **134**:910–914, 1985.
78. Suda, T., Suda, J., Spicer, S. S. and Ogawa, M. Proliferation and differentiation in culture of mast cell progenitors derived from mast cell-deficient mice of genotype *W/W*^{v}. *J. Cell. Physiol.* **122**:187–192, 1985.
79. Caux, C., Vanbervliet, B., Massacrier, C., Durand, I. and Banchereau, J. Interleukin-3 cooperates with tumor necrosis factor-alpha for the development of human dendritic Langerhans cells from cord blood CD34(+) hematopoietic progenitor cells. *Blood* **87**:2376–2385, 1996.
80. Ihle, J. N., Keller, J., Orsozlan, S., *et al.* Biological properties of homogeneous interleukin 3. I. Demonstration of WEHI-3 growth factor activity, mast cell growth factor activity, P cell stimulating factor activity and histamine producing factor activity. *J. Immunol.* **131**:282–287, 1983.

81. Nabel, G. J., Galli, S. J., Dvorak, A. M., Dvorak, H. F. and Cantor, H. Inducer T lymphocytes synthesize a factor that stimulates proliferation of cloned mast cells. *Nature* **291**:332–334, 1981.
82. Nagao, K., Yokoro, K. and Aaronson, S. A. Continuous lines of basophil/mast cells derived from normal mouse bone marrow. *Science* **212**:333–335, 1981.
83. Razin, E., Cordon-Cardo, C. and Good, R. A. Growth of a pure population of mast cells *in vitro* with conditioned medium derived from concanavalin A-stimulated splenocytes. *Proc. Natl. Acad. Sci. USA* **78**:2559–2561, 1981.
84. Schrader, J. W., Lewis, S. J., Clark-Lewis, I. and Culvenor, J. G. The persisting (P) cell: histamine content, regulation by a T cell-derived factor, origin from a bone marrow precursor, and relationship to mast cells. *Proc. Natl. Acad. Sci. USA* **78**:323–327, 1981.
85. Tertian, G., Yung, Y. P., Guy-Grand, D. and Moore, M. A. Long-term *in vitro* culture of murine mast cells. I. Description of a growth factor-dependent culture technique. *J. Immunol.* **127**:788–794, 1981.
86. Ihle, J. N. Interleukin-3 and hematopoiesis. *Chem. Immunol.* **51**:65–106, 1992.
87. Saito, H., Hatake, K., Dvorak, A. M, *et al.* Selective differentiation and proliferation of hematopoietic cells induced by recombinant human interleukins. *Proc. Natl. Acad. Sci. USA* **85**:2288–2292, 1988.
88. Kirshenbaum, A. S., Goff, J. P., Dreskin, S. C., Irani, A. M., Schwartz, L. B. and Metcalfe, D. D. IL-3-dependent growth of basophil-like cells and mast-like cells from human bone marrow. *J. Immunol.* **142**:2424–2429, 1989.
89. Mayer, P., Valent, P., Schmidt, G., Liehl, E. and Bettelheim, P. The *in vivo* effects of recombinant human interleukin-3: demonstration of basophil differentiation factor, histamine-producing activity, and priming of GM-CSF-responsive progenitors in nonhuman primates. *Blood* **74**:613–621, 1989.
90. Kindler, V., Thorens, B., de Kossodo, S., *et al.* Stimulation of hematopoiesis *in vivo* by recombinant bacterial murine interleukin 3. *Proc. Natl. Acad. Sci. USA* **83**:1001–1005, 1986.
91. Metcalfe, D., Begley, C. G., Johnson, N. A., Nicola, N. A., Lopez, A. F. and Williamson, D. J. Effects of purified bacterially synthesized murine multi-CSF (IL-3) on hematopoiesis in normal adult mice. *Blood* **68**:46–57, 1986.
92. Donahue, R. E., Seehra, J., Metzger, M., *et al.* Human IL-3 and GM-CSF act synergistically in stimulating hematopoiesis in primates. *Science* **241**:1820–1823, 1988.
93. Volc-Platzer, B., Valent, P., Radaszkiewicz, T., Mayer, P., Bettelheim, P. and Wolff, K. Recombinant human interleukin 3 induces proliferation of inflammatory cells and keratinocytes *in vivo*. *Lab. Invest.* **64**:557–566, 1991.
94. Abe, T. and Nawa, Y. Worm expulsion and mucosal mast cell response induced by repetitive IL-3 administration in *Strongyloides ratti*-infected nude mice. *Immunology* **63**:181–185, 1988.
95. Madden, K. B., Urban, J. F., Ziltener, H. J., Schrader, J. W., Finkelman, F. D. and Katona, I. M. Antibodies to IL-3 and IL-4 suppress helminth-induced intestinal mastocytosis. *J. Immunol.* **147**:1387–1391, 1991.
96. Finkelman, F. D., Madden, K. B., Morris, S. C., *et al.* Anti-cytokine antibodies as carrier proteins. Prolongation of *in vivo* effects of exogenous cytokines by injection of cytokine–anti-cytokine antibody complexes. *J. Immunol.* **151**:1235–1244, 1993.
97. Cannistra, S. A., Vellenga, E., Groshek, P., Rambaldi, A. and Griffin, J. D. Human granulocyte–macrophage colony-stimulating factor and interleukin-3 stimulate monocyte cytotoxicity through a tumor necrosis factor-dependent mechanism. *Blood* **71**:672–676, 1988.
98. Hirai, K., Morita, Y., Misaki, Y., *et al.* Modulation of human basophil histamine release by hemopoietic growth factors. *J. Immunol.* **141**:3958–3964, 1988.
99. Kimoto, M., Kindler, V., Higaki, M., Ody, C., Izui, S. and Vassalli, P. Recombinant murine IL-3 fails to stimulate T or B lymphopoiesis *in vivo*, but enhances immune responses to T cell-dependent antigens. *J. Immunol.* **140**:1889–1894, 1988.
100. Rothenberg, M., Owen, W., Silberstein, D., *et al.* Human eosinophils have prolonged survival, enhanced functional properties, and become hypodense when exposed to human interleukin-3. *J. Clin. Invest.* **81**:1986–1992, 1988.
101. Schleimer, R. P., Derse, C. P., Friedman, B., *et al.* Regulation of human basophil mediator release by cytokines. I. Interaction with antiinflammatory steroids. *J. Immunol.* **143**:1310–1317, 1989.
102. Brunner, T., Heusser, C. H. and Dahinden, C. A. Human peripheral blood basophils primed by interleukin 3 (IL-3) produce IL-4 in response to immunoglobulin E receptor stimulation. *J. Exp. Med.* **177**:605–611, 1993.
103. MacGlashan Jr., D. W. and Schroeder, J. T. Basophils as the target and source of cytokines. In: *Biological and Molecular Aspects of Mast Cell and Basophil Differentiation and Function* (Kitamura,

Y., Yamamoto, S., Galli, S. J. and Greaves, M. W., eds), pp. 51–63. Raven Press, New York, 1995.

104. Pulaski, B. A., Yeh, K. Y., Shastri, N., *et al.* Interleukin 3 enhances cytotoxic T lymphocyte development and class I major histocompatibility complex representation of exogenous antigen by tumor-infiltrating antigen-presenting cells. *Proc. Natl. Acad. Sci. USA* **93**:3669–3674, 1996.

105. Li, C. L. and Johnson, G. R. Stimulation of multipotential, erythroid and other murine haematopoietic progenitor cells by adherent cell lines in the absence of detectable multi-CSF (IL-3). *Nature* **316**:633–636, 1985.

106. Ichihara, M., Hara, T., Takagi, M., Cho, L. C., Gorman, D. M. and Miyajima, A. Impaired interleukin-3 (IL-3) response of the A/J mouse is caused by a branch point deletion in the IL-3 receptor α subunit gene. *EMBO J.* **14**:939–950, 1995.

107. Burd, P. R., Rogers, H. W., Gordon, J. R., *et al.* Interleukin 3-dependent and -independent mast cells stimulated with IgE and antigen express multiple cytokines. *J. Exp. Med.* **170**:245–257, 1989.

108. Plaut, M., Pierce, J. H., Watson, C. J., Hanley-Hyde, J., Nordan, R. P. and Paul, W. E. Mast cell lines produce lymphokines in response to cross-linkage of FcεRI or to calcium ionophores. *Nature* **339**:64–67, 1989.

109. Wodnar-Filipowicz, A., Heusser, C. H. and Moroni, C. Production of the haemopoietic growth factors GM-CSF and interleukin-3 by mast cells in response to IgE receptor-mediated activation. *Nature* **339**:150–152, 1989.

110. Svetic, A., Madden, K. B., Zhou, X., *et al.* A primary intestinal helminthic infection rapidly induces a gut-associated elevation of Th2-associated cytokines and IL-3. *J. Immunol.* **150**:3434–3441, 1993.

111. Mach, N., Lantz, C. S., Galli, S. J., *et al.* Involvement of interleukin-3 in delayed-type hypersensitivity. *Blood* **91**:778–783, 1998.

112. Nishinakamura, R., Miyajima, A., Mee, P. J., Tybulewicz, V. L. J. and Murray, R. Hematopoiesis in mice lacking the entire granulocyte-macrophage colony-stimulating factor/interleukin-3/interleukin-5 functions. *Blood* **88**:2458–2464, 1996.

113. Lantz, C. S. and Huff, T. F. Differential responsiveness of purified mouse c-kit+ mast cells and their progenitors to IL-3 and stem cell factor. *J. Immunol.* **155**:4024–4029, 1995.

114. Rennick, D., Hunte, B., Holland, G. and Thompson-Snipes, L. Cofactors are essential for stem cell factor dependent growth and maturation of mast cell progenitors: Comparative effects of interleukin-3 (IL-3), IL-4, IL-10, and fibroblasts. *Blood* **85**:57–65, 1995.

115. Rodewald, H-R., Dessing, M., Dvorak, A. M. and Galli, S. J. Identification of a committed precursor for the mast cell lineage. *Science* **271**:818–822, 1996.

116. Lantz, C. S., Boesiger, J., Song, C. H., *et al.* Role for interleukin-3 in mast cell and basophil development and parasite immunity. *Nature* **392**:90–93, 1998.

117. Huff, T. F., Lantz, C. S., Ryan, J. J. and Leftwich, J. A. Mast cell-committed progenitors. In: *Biological and Molecular Aspects of Mast Cell Differentiation and Function* (Kitamura, Y., Yamamoto, S., Galli, S. J., Greaves, M. W., eds), pp. 105–117. Raven Press, New York, 1995.

118. Kasugai, T., Tei, H., Okada, M., *et al.* Infection with *Nippostrongylus brasiliensis* induces invasion of mast cell precursors from peripheral blood to small intestine. *Blood* **85**:1334–1340, 1995.

119. Sonoda, T., Hayashi, C. and Kitamura, Y. Presence of mast cell precursors in the yolk sac of mice. *Dev. Biol.* **97**:89–94, 1983.

120. Kitamura, Y., Hatanaka, K., Murakami, M. and Shibata, H. Presence of mast cell precursors in peripheral blood of mice demonstrated by parabiosis. *Blood* **53**:1085–1088, 1979.

121. Zucker-Franklin, D., Grusky, G., Hirayama, N. and Schnipper, E. The presence of mast cell precursors in rat peripheral blood. *Blood* **58**:544–551, 1981.

122. Sonoda, T., Ohno, T. and Kitamura, Y. Concentration of mast-cell progenitors in bone marrow, spleen, and blood of mice determined by limiting dilution analysis. *J. Cell. Physiol.* **112**:136–140, 1982.

123. Rottem, M., Okada, T., Goff, J.P. and Metcalfe, D. D. Mast cells cultured from the peripheral blood of normal donors and patients with mastocytosis originate from CD34+/ FcεRI-cell populations. *Blood* **84**:2489–2496, 1994.

124. Agis, H., Willheim, M., Sperr, W. R., *et al.* Monocytes do not make mast cells when cultured in the presence of SCF. Characterization of the circulating mast cell progenitor as a c-*kit*+, CD34+, Ly–, CD14–, CD17–, colony-forming cell. *J. Immunol.* **151**:4221–4227, 1995.

125. Metcalfe, D. D. Classification and diagnosis of mastocytosis: current status. *J. Invest. Dermatol.* **96**:2S–4S, 1991.

126. Seder, R. A., Paul, W. E., Dvorak, A. M., *et al.* Mouse splenic and bone marrow cell populations that express high-affinity Fc epsilon receptors and produce interleukin 4 are highly enriched in basophils. *Proc. Natl. Acad. Sci. USA* **88**:2835–2839, 1991.

127. Dvorak, A. M., Seder, R. A., Paul, W. E., Kissell-Rainville, S., Plaut, M. and Galli, S. J. Ultrastructural characteristics of FcεRI positive basophils in the spleen and bone marrow of mice immunized with goat anti-mouse IgD antibody. *Lab. Invest.* **68**:708–715, 1993.
128. Yung, Y.-P. and Moore, M. A. S. Long-term *in vitro* culture of murine mast cells. III. Discrimination of mast cell growth-factor and granulocyte CSF. *J. Immunol.* **129**:1256–1261, 1982.
129. Nakano, T., Sonoda, T., Hayashi, C., *et al.* Fate of bone marrow-derived cultured mast cells after intracutaneous, intraperitoneal, and intravenous transfer into genetically mast cell-deficient W/W^v mice. Evidence that cultured mast cells can give rise to both connective tissue type and mucosal mast cells. *J. Exp. Med.* **162**:1025–1043, 1985.
130. Suda, T., Suda, J., Ogawa, M. and Ihle, J. N. Permissive role of interleukin 3 (IL-3) in proliferation and differentiation of multipotential hemopoietic progenitors in culture. *J. Cell. Physiol.* **124**:182–190, 1985.
131. Mosmann, T. R., Bond, M. W., Coffman, R. L., Ohara, J. and Paul, W. E. T-cell and mast cell lines respond to B-cell stimulatory factor 1. *Proc. Natl. Acad. Sci. USA* **83**:5654–5658, 1986.
132. Smith, C. A. and Rennick, D. M. Characterization of a murine lymphokine distinct from interleukin 2 and interleukin 3 (IL-3) possessing a T-cell growth factor activity and a mast-cell growth factor activity that synergizes with IL-3. *Proc. Natl. Acad. Sci. USA* **83**:1857–1861, 1986.
133. Hu, Z-Q., Kobayashi, K., Zenda, N. and Shimamura, T. Tumor necrosis factor-α- and interleukin-6-triggered mast cell development from mouse spleen cells. *Blood* **89**:526–533, 1997.
134. Hültner, L., Druez, C., Moeller, J., *et al.* Mast cell growth-enhancing activity (MEA) is structurally related and functionally identical to the novel mouse T cell growth factor P40/TCGFIII (interleukin 9). *Eur. J. Immunol.* **20**:1413–1416, 1990.
135. Khalil, R. M., Luz, A., Mailhammer, R., *et al. Schistosoma mansoni* infection in mice augments the capacity for interleukin 3 (IL-3) and IL-9 production and concurrently enlarges progenitor pools for mast cells and granulocytes-macrophages. *Infect. Immun.* **64**:4960–4966, 1996.
136. Thompson-Snipes, L., Dhar, V., Bond, M. W., Mosmann, T. R., Moore, K. W. and Rennick, D. Interleukin-10: a novel stimulatory factor for mast cells and their progenitors. *J. Exp. Med.* **173**:507–510, 1991.
137. Aloe, L. and Levi-Montalcini, R. Mast cells increase in tissues of neonatal rats injected with the nerve growth factor. *Brain Res.* **133**:358–366, 1977.
138. Aloe, L. The effect of nerve growth factor and its antibody on mast cells *in vivo*. *J. Neuroimmunol.* **18**:1–12, 1988.
139. Matsuda, H., Kannan, Y., Ushio, H., *et al.* Nerve growth factor induces development of connective tissue-type mast cells *in vitro* from murine bone marrow cells. *J. Exp. Med.* **174**:7–14, 1991.

CHAPTER 2

Gain-of-function Mutations of c-*kit* in Human Diseases

YUKIHIKO KITAMURA,[1] SEIICHI HIROTA,[1] EIICHI MORII,[1] and TOSHIROU NISHIDA*[2]

[1]*Department of Pathology and*
[2]*Department of Surgery, Osaka University Medical School, Suita, Osaka, Japan*

INTRODUCTION

The c-*kit* is the cellular homologue of the oncogene v-*kit* of the HZ4 feline sarcoma virus and encodes a receptor tyrosine kinase that is structurally similar to the receptors of macrophage colony-stimulating factor (M-CSF), platelet-derived growth factor (PDGF) and vascular endothelial cell growth factor (VEGF) (1–4). These receptor tyrosine kinases have unique features: an extracellular domain made up of five immunoglobulin-like repeats, and a tyrosine kinase domain which is split into two domains by an insert sequence of variable length (2, 3). The structure and amino acid sequence of the c-*kit* protein are well preserved in humans, mice and rats (2, 3, 5). For many years after the discovery of v-*kit* it was not clear whether *kit* functions as a cause of human neoplasms. Recently, we demonstrated that gain-of-function mutations of c-*kit* result in the development of neoplasms of particular cell types, i.e. mast cells and interstitial cells of Cajal (ICCs).

LOSS-OF-FUNCTION MUTATION OF c-*kit*

The *W* locus of mice was demonstrated to encode the c-*kit* gene (6, 7). Many loss-of-function mutants have been reported at the *W* locus. Mice of W/W^v genotype are most frequently used. The *W* mutant allele encodes a truncated c-*kit* protein without the transmembrane domain, and the W^v mutant allele is a point mutation at the tyrosine kinase domain, resulting in a marked decrease in the kinase activity (8). Double heterozygous W/W^v mice show five abnormalities due to the loss-of-function mutations of c-*kit*: (i) anaemia due to hypoproduction of erythrocytes (9), (ii) white coat colour due to depletion

* Corresponding author.

MAST CELLS AND BASOPHILS
ISBN 0-12-473335-2

of melanocytes (9), (iii) sterility due to lack of germ cells (9), (iv) depletion of mast cells (10), (v) depletion of ICCs (11, 12).

The ligand for the c-*kit* receptor tyrosine kinase (KIT) was identified and named stem cell factor (SCF) (13–15). Since SCF is encoded by the *Sl* locus of mice, it is reasonable that homozygous or double-heterozygous mutant mice at either the *W* or *Sl* locus have the same phenotype. The most frequently used mutant mice of the *Sl* (SCF) locus are of Sl/Sl^d genotype. Sl/Sl^d mice show anaemia, white coat colour, sterility and depletion of mast cells and ICCs (9, 16, 17).

Both W/W^v and Sl/Sl^d mice are deficient in mast cells, but the mechanism is different due to the character of proteins, which are encoded by the *W* locus and the *Sl* locus, respectively. Mast cell depletion of W/W^v mice is cured by bone marrow transplantation from normal control (+/+) or Sl/Sl^d mice, but that of Sl/Sl^d mice is not (10, 16). When skin pieces were grafted from W/W^v to Sl/Sl^d mice, mast cells developed in the graft. However, mast cells did not develop in the skin pieces grafted from Sl/Sl^d to +/+ mice (16). On the other hand, the mast cell depletion of Sl/Sl^d mice is cured by injection of SCF (15). Precursors of mast cells have a defect in W/W^v mice, whereas stromal cells supporting the differentiation of mast cells have a defect in Sl/Sl^d mice.

In addition to *W* and W^v mutant alleles, many other loss-of-function mutations have been reported in the mouse c-*kit* locus (9, 18). In particular, many mutations at the tyrosine kinase domain have been reported. When such a mutant c-*kit* protein is expressed on the surface, a dominant negative phenotype is observed. This is attributed to the fact that the heterodimers composed of normal and mutant c-*kit* proteins have deficient tyrosine kinase activity (8, 19–21). For example, $W^{42}/+$ mice have an almost white coat, and the number of mast cells in the skin of $W^{42}/+$ mice is decreased to half that of +/+ mice. In contrast, the coat colour of *W*/+ mice is not diluted, and the number of mast cells in the skin of *W*/+ mice is not decreased. Since the c-*kit* protein encoded by the *W* mutant allele is not expressed on the surface due to the lack of the transmembrane domain, only normal-type c-*kit* receptors are expressed on the surface of mast cells in *W*/+ mice. Although the phenotypes of heterozygous $W^{42}/+$ and *W*/+ mice are apparently different, the phenotypes of homozygous W^{42}/W^{42} and *W*/*W* mice are comparable. Both W^{42}/W^{42} and *W*/*W* mice completely lack melanocytes and mast cells (21).

We reported first a loss-of-function mutation at the c-*kit* locus of the rat. Homozygous mutant rats at the newly found 'White spotting (*Ws*)' locus are anaemic and deficient in mast cells, melanocytes and ICCs (22, 23). Mast cells developed in the skin graft from the *Ws*/*Ws* to nude athymic rat. Since this result suggested that the *Ws* locus of the rat was comparable to the *W* locus rather than to the *Sl* locus of the mouse, Tsujimura *et al.* (5) characterized the c-*kit* gene of *Ws*/*Ws* rats. A deletion of 12 bases was found in the c-*kit* cDNA of *Ws*/*Ws* rats. Four amino acids encoded by the 12 deleted bases were located at two amino acids downstream from the tyrosine autophosphorylation site in KIT.

A loss-of-function mutant of the c-*kit* cDNA has also been found in humans. Piebaldism is known as an autosomal dominant genetic disorder characterized by congenital white patches of skin and hair from which melanocytes are depleted. Since the phenotype of human piebaldism is comparable to that of the heterozygous *W*/+ mouse, Giebel and Spritz (24) compared the c-*kit* cDNA between piebaldism and normal individuals of the same family and found a point mutation in the individuals with piebaldism. These individuals are heterozygous and do not show any other symptoms.

STOMACH LESIONS OF *W/W^v* AND *Sl/Sl^d* MICE

In addition to the discovery of mast cell deficiency in W/W^v and Sl/Sl^d mice, we found spontaneous development of forestomach papillomas and antral ulcers in W/W^v and Sl/Sl^d mice (25, 26). The stomach lesions were attributed to spontaneous bile reflux in both W/W^v and Sl/Sl^d mice (25, 26). The bile reflux was also observed in *Ws/Ws* c-*kit* mutant rats (23). At the time of the discovery, we could not understand the mechanism of the bile reflux. However, since the depletion of ICCs was identified as the fifth abnormality of W/W^v and Sl/Sl^d mice (11, 12, 17) and since ICCs regulate the rhythmical contraction of the gastrointestinal tract (11, 12, 27), the bile reflux was attributed to the depletion of ICCs.

GAIN-OF-FUNCTION MUTATION OF c-*kit* IN MAST CELL NEOPLASMS

Binding of SCF activates KIT and leads to autophosphorylation of KIT on tyrosine and to association of KIT with substrates such as phosphatidylinositol 3-kinase (PI3K) (28). In the human mast cell leukaemia cell line HMC-1, KIT was constitutively phosphorylated on tyrosine, activated, and associated with PI3K without the addition of SCF (29). The c-*kit* gene of HMC-1 cells was found to be composed of normal, wild-type allele and mutant allele with point mutations resulting in amino acid substitutions of Gly-560 for Val and Val-816 for Asp (Fig. 1). Amino acid sequences in the regions of the two mutations are completely conserved in all of mouse, rat and human KIT.

In order to determine the causal role of these mutations in the constitutive activation, mutant c-*kit* genes with Gly-560 or Val-816 were constructed and expressed in a human embryonic kidney cell line, 293T cells. In the transfected cells, KIT with either mutation was abundantly phosphorylated on tyrosine and activated in immune complex kinase reaction in the absence of SCF, whereas tyrosine phosphorylation and activation of transfected wild-type KIT was not detectable (29). Tsujimura *et al.* (30, 31) found the mutation corresponding to Val-816 of human HMC-1 cell line in the P-815 mouse mastocytoma cell line (Asp-817 to Tyr) and the RBL-2H3 rat mast cell leukaemia cell line (Asp-814 to Tyr). Both P-815 and RBL-2H3 cells show constitutive activation of KIT without the addition of SCF. There is a possibility that this mutation has induced mast cell neoplasms in the above-mentioned three species. In fact, the Asp-816 to Val mutation has been found in various types of mast cell neoplasms of human patients (32, 33).

To examine the transformation potential of the c-*kit* activation mutation, we used the murine interleukin-3 (IL-3)-dependent IC-2 mast cell line as a transfectant. The IC-2 cells did not express KIT on the surface. The Val-814 or Gly-559 murine mutant c-*kit* gene was introduced into IC-2 cells using a retroviral vector. The mutant KIT proteins expressed in IC-2 cells were constitutively phosphorylated on tyrosine and demonstrated kinase activity in the absence of SCF (34). IC-2 cells expressing either Val-814 or G-559 mutation showed factor-independent growth in suspension culture and produced tumours in nude athymic mice (34). Introduction of the murine c-*kit* gene with Val-814 or Gly-559 mutation also resulted in the malignant transformation of the IL-3-dependent Ba/F3 murine pro-B cell line (35).

Val-814 is a point mutation at the tyrosine kinase domain of the c-*kit* gene whereas

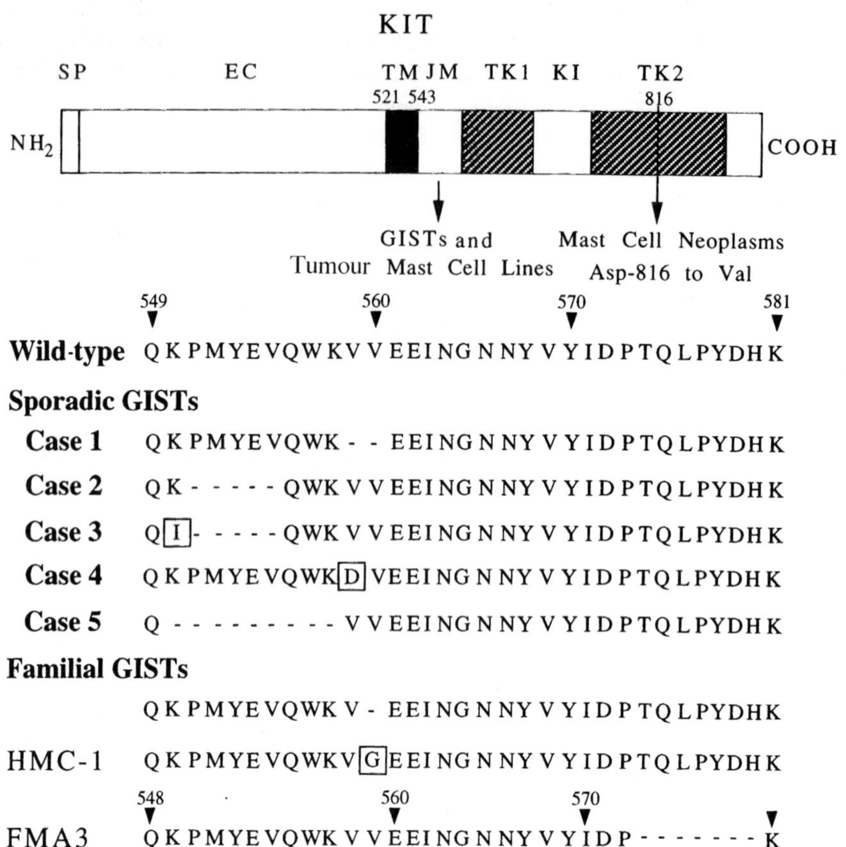

Fig. 1 Various gain-of-function mutations of c-*kit* in human mast cell neoplasms and GISTs. Deleted amino acids are shown by dashes (-) and mutated amino acids by boxes. Murine and human KIT are of different lengths, so the amino acid numbering in KIT of the FMA3 cell line is different. SP, signal peptide; EC, extracellular domain; TM, transmembrane domain; JM, juxtamembrane domain; TK1 and TK2, tyrosine kinase domains; KI, kinase insert.

Gly-559 is a point mutation between transmembrane and tyrosine kinase domains (hereafter called juxtamembrane domain) (Fig. 1). Chemical cross-linking analysis showed that a substantial fraction of the phosphorylated KIT with the Gly-559 mutation underwent dimerization even in the absence of SCF, whereas the phosphorylated c-*kit* with the Val-814 mutation did not, suggesting that distinct mechanisms resulted in the constitutive activation of c-*kit* by the Gly-559 and Val-814 mutations (35). Tsujimura *et al.* (36) found another gain-of-function mutation at the juxtamembrane domain of FMA3 murine mastocytoma cells. The c-*kit* cDNA of FMA3 cells carried an in-frame deletion of 21 base pairs (bp) (Fig. 1). The FMA3-type c-*kit* cDNA with the 21-bp deletion was introduced into the IC-2 cell line. The FMA3-type KIT was constitutively phosphorylated on tyrosine and activated (36). Moreover, the FMA3-type KIT was dimerized without stimulation by SCF. The FMA3-type KIT that spontaneously dimerized without SCF binding was not internalized even though it was activated, as is also the case with the Gly-559 KIT. IC-2 cells expressing the FMA3-type KIT grew in suspension culture without IL-3 and SCF and became leukaemic in nude athymic mice.

Although the Gly-559 mutation and the FMA3-type mutation were different in nature, a point mutation and a 21-bp deletion, respectively, their biological effects appeared comparable. Probably, the normal juxtamembrane domain may inhibit the dimerization of KIT, and mutations at the juxtamembrane domain may induce SCF-independent constitutive dimerization of KIT.

In the HMC-1 human mast cell leukaemia cell line, the c-*kit* mutations were observed in both the tyrosine kinase and juxtamembrane domains. However, only the particular point mutation (Asp-816 to Val) of the tyrosine kinase domain was detectable in mast cell tumours directly obtained from human patients (32, 33). Although mast cell tumours are rare in humans, they are the most common malignant neoplasms in dogs, representing 7–21% of all tumours. London *et al.* (37) recently found mutations of the c-*kit* gene in canine mast cell tumours. In contrast to mast cell tumours of humans, the mutations in canine mast cell tumours were located at the juxtamembrane domain. London *et al.* (37) showed that the mutation was a gain-of-function mutation in at least one of the canine mast cell tumours. The mutation site in the c-*kit* gene of mast cell tumours may be influenced by the host species.

SPORADIC GASTROINTESTINAL STROMAL TUMOURS

Since ICCs express KIT, we examined whether any human mesenchymal tumours in gastrointestinal tract expressed KIT. Authentic leiomyomas and authentic schwannomas did not express KIT, but 94% of gastrointestinal stromal tumours (GIST) expressed KIT (38). Moreover, both GIST cells and ICCs expressed CD34 (38). Although various cells, including haematopoietic stem cells, express both KIT and CD34, ICCs are the only cells that are double-positive for KIT and CD34 in normal gastrointestinal wall of humans (38). This strongly suggests that KIT and CD34 double-positive GISTs may originate from ICCs.

The complete coding region of c-*kit* was obtained from five GISTs and sequenced. Mutations were observed in the juxtamembrane domain (Fig. 1). These mutations were located within an 11-amino-acid stretch (Lys-550 to Val-560), but at non-identical sites (38). No mutations were detectable in other domains of c-*kit* cDNA, including the tyrosine kinase domain. We next examined whether the c-*kit* mutations found in the GISTs resulted in constitutive activation of the c-*kit* receptor tyrosine kinase by transient introduction of the mutant c-*kit* cDNAs into 293T human embryonic kidney cell line. The wild-type c-*kit* cDNA was introduced as a negative control, and c-*kit* mutants found in the HMC-1 mast cell leukaemia cell line were introduced as a positive control. Wild-type KIT was phosphorylated on tyrosine only when SCF was added to the culture medium. In contrast, the gain-of-function KIT mutations found in HMC-1 cells were phosphorylated on tyrosine without the addition of SCF. The magnitude of the constitutive tyrosine phosphorylation was greater in the tyrosine kinase domain mutant than in the juxtamembrane domain mutant. The c-*kit* mutants found in GISTs also showed the constitutive tyrosine phosphorylation in 293T cells without SCF. The constitutive tyrosine phosphorylation of the juxtamembrane mutant of HMC-1 cells was of similar magnitude to that of the juxtamembrane mutants of GISTs (38). In *in vitro* kinase assay, the c-*kit* mutants found in the GISTs exhibited constitutive kinase activation that was similar in magnitude to that of the juxtamembrane domain mutant of HMC-1 cells.

To investigate the biological consequences of the mutant c-*kit*, we introduced the c-*kit* mutations found in the GISTs into the mouse c-*kit* cDNA and then stably transfected into the IL-3-dependent Ba/F3 murine pro-B cell line. As a control, mouse wild-type c-*kit* cDNA was also transfected into Ba/F3 cells. Ba/F3 cells with wild-type murine c-*kit* grew in the presence of either IL-3 or SCF. Ba/F3 cells with the mutated c-*kit* grew autonomously without IL-3 and SCF. Ba/F3 cells with the mutated murine c-*kit* also grew autonomously in nude mice (38). The constitutive kinase activation of all KIT mutations found in GISTs was confirmed in Ba/F3 cells (38).

Some GISTs are typical benign tumours, and most of such GISTs are found incidentally at the time of endoscopic examination as submucosal tumours. In contrast, other GISTs metastasize to the liver and disseminate in the peritoneal cavity. Most of the latter type GISTs do not respond to radiotherapy and/or chemotherapy and ultimately kill the host. Large tumour size, the presence of intratumorous necrosis and frequent mitotic figures are considered to indicate a worse prognosis. The clinical behaviour of GISTs, however, is difficult to predict using conventional prognostic factors. Recently, Taniguchi *et al.* (39) examined whether the presence of the mutation in the juxtamembrane domain of the c-*kit* gene was important as a prognostic factor. They studied 124 GIST cases and found that the prognosis of the GISTs with the c-*kit* mutation was significantly worse than that of GISTs without the mutation.

FAMILIAL GIST

Multiple development of GISTs was found in a 60-year-old Japanese woman (case 1 in Fig. 2). A nephew of case 1 (case 3) also suffered from multiple benign GISTs. Analysis of their family pedigree revealed that many family members suffered from intestinal obstruction that may be attributable to multiple development of GISTs (Fig. 2). Case 2, a niece of case 1, received operations for benign and malignant GISTs (40).

The benign GISTs obtained from cases 1, 2 and 3 and the malignant GIST from case 2 expressed KIT. DNA was extracted from paraffin-embedded specimens of the tumours, and mutation was investigated using single-strand conformation polymorphism analysis

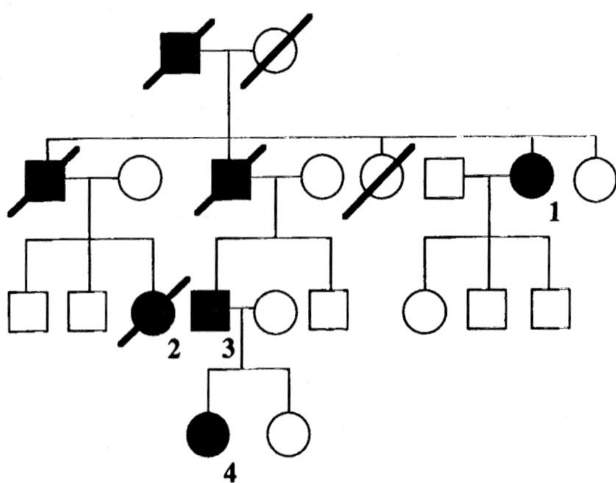

Fig. 2 A family with multiple GISTs and mutation of the c-*kit* gene. Filled symbols indicate family members with either intestinal obstruction or the c-*kit* mutation or both. Squares, males; circles, females; symbols with a dash, dead case.

(SSCP). SSCP of tumours obtained from cases 1 and 3 showed wild-type and mutant bands at exon 11 that encodes the juxtamembrane domain. Direct sequencing of the mutated bands of exon 11 revealed the deletion of one of two consecutive valines (codons 559 and 560) (Fig. 1). Unfortunately, DNA samples suitable for SSCP and direct sequencing were not obtained from tumours of case 2. Next, we obtained DNAs from peripheral leukocytes of cases 1 and 3 and their family members. Deletion of the valine residue was observed in leukocyte DNAs of cases 1, 3 and 4, but not DNAs from others. Case 4 has not reported any abdominal symptoms, probably due to the younger age.

We investigated functions of the mutated KIT by introducing it into mouse c-*kit* cDNA and then transfecting into Ba/F3 cells. Constitutive phosphorylation and kinase activation of the KIT were found. Ba/F3 cells with the mutated KIT grew autonomously both in culture and in nude mice (40).

O'Brien *et al.* (41) also reported a family with multiple GISTs without examining the c-*kit* mutation. El-Omar *et al.* (42) described a woman with leiomyomatosis and a leiomyosarcoma which arose from a leiomyoma. She also showed hyperpigmentation of the perineal skin. We consider the tumours benign and malignant GISTs. Marshall *et al.* (43) described a family with multiple gastrointestinal tumours, which appeared to be benign GISTs. The family members with multiple gastrointestinal tumours also suffered from urticaria pigmentosa or systemic mast cell disease. Since KIT plays an essential role in the development of melanocytes and mast cells, the hyperpigmentation and mast cell hyperplasia observed in these families may also be attributable to gain-of-function mutations of the c-*kit* gene. Familial GIST appears to be a cancer syndrome that might include hyperplasia of melanocytes and neoplasia of mast cells in some families.

CONCLUSION

Loss-of-function mutations of c-*kit* results in depletion of erythrocytes, melanocytes, germ cells, mast cells and ICCs. In contrast, gain-of-function mutations of c-*kit* induce neoplasms of mast cells and ICCs. The site of mutations is different between human mast cell neoplasms and human GISTs, and the mechanisms remain to be clarified. The mast cell was discovered by Ehrlich (44), and the ICC was discovered by Cajal (45). Although there are no characteristics common to mast cells and ICCs without being discovered by the great scientists of the 19th century, development and oncogenesis of these two types of cells are profoundly influenced by the SCF–KIT system.

REFERENCES

1. Besmer, P., Murphy, J. E., George, P. C., Qiu, F., Bergold, P. J., Lederman, L., Snyder, Jr. H.W., Broudeur, D., Zuckerman, E. E. and Hardy, W. D. A new acute transforming feline retrovirus and relationship of its oncogene v-*kit* with the protein kinase gene family. *Nature* **320**:415–421, 1986.
2. Qiu, F. H., Ray, H. P., Brown, Y., Barker, P. E., Jhanwar, S., Ruddle, F. H. and Besmer, P. Primary structure of c-*kit*: relationship with the CSF-1/PDGF receptor kinase family – oncogenic activation of v-*kit* involves deletion of extracellular domain and C terminus. *EMBO J.* **7**:1003–1011, 1988.
3. Yarden, Y., Kuang, W. J., Yang-Feng, T., Coussens, L., Munemitsu, S., Dull, T. J., Chen, E., Schlessinger, J., Francke, U. and Ullrich, A. Human proto-oncogene c-kit: a new cell surface receptor tyrosine kinase for an unidentified ligand. *EMBO J.* **6**:3341–3351, 1988.
4. Sibuya, M., Yamaguchi, S., Yamane, A., Ikeda, T., Tojo, A., Matsushime, H. and Sato, M. Nucleotide sequence and expression of a novel human receptor-type tyrosine kinase gene (flt) closely related to the

fms family. *Oncogene* **5**:519–524, 1990.

5. Tsujimura, T., Hirota, S., Nomura, S., Niwa, Y., Yamazaki, M., Tono, T., Morii, E., Kim, H. M., Kondo, K., Nishimune, Y. and Kitamura, Y. Characterization of *Ws* mutant allele of rats: a 12 base deletion in tyrosine kinase domain of c-*kit* gene. *Blood* **78**:1942–1946, 1991.
6. Chabot, B., Stephenson, D. A., Chapman, V. M., Besmer, P. and Bernstein, A. The proto-oncogene c-*kit* encoding a transmembrane tyrosine kinase receptor maps to the mouse *W* locus. *Nature* **335**:88–89, 1988.
7. Geissler, E. N., Ryan, M. A. and Housman, D. E. The dominant-white spotting (*W*) locus of the mouse encodes the c-*kit* proto-oncogene. *Cell* **55**:185–192, 1988.
8. Nocka, K., Tan, J. C., Chiu, E., Chu, T. Y., Ray, P., Traktman, P. and Besmer, P. Molecular bases of dominant negative and loss of function mutation at the murine c-*kit*/white spotting locus: W^{37}, W^{v}, W^{41} and *W*. *EMBO J.* **9**:1805–1813, 1990.
9. Russell, E. S. Hereditary anemias of the mouse: a review for geneticists. *Adv. Genet.* **20**:357–459, 1979.
10. Kitamura, Y., Go, S. and Hatanaka, K. Decrease of mast cells in W/W^{v} mice and their increase by bone marrow transplantation. Blood **52**:447–452, 1978.
11. Maeda, H., Yamagata, A., Nishikawa, S., Yoshinaga, K., Kobayashi, S., Nishi, K. and Nishikawa, S.-I. Requirement of c-*kit* for development of intestinal pacemaker system. *Development* **116**:369–375, 1992.
12. Huizinga, J. D., Thuneberg, L., Klüppel, M., Malysz, J., Mikkelsen, H. B. and Bernstein, A. *W/kit* gene required for interstitial cells of Cajal and for intestinal pacemaker activity. *Nature* **373**:347–349, 1995.
13. Williams, D. E., Eisenman, J., Baird, A., Rauch, C., Ness, K. V., March, C. J., Park, L. S., Martin, U., Mochizuki, D. Y., Boswell, H. S., Burgess, G. S., Cosman, D. and Lyman, S. D. Identification of a ligand for the c-*kit* proto-oncogene. *Cell* **63**:167–174, 1990.
14. Flanagan, J. G. and Leder, P. The kit ligand: a cell surface molecule altered in steel mutant fibroblasts. *Cell* **63**:185–194, 1990.
15. Zsebo, K. M., Williams, D. A., Geissler, E. N., Broudy, Y. C., Martin, F. H., Atkins, H. L., Hsu, R. Y., Birkitt, N. C., Okino, K. H., Murdock, D. C., Jacobson, F. W., Langley, K. E., Smith, K. A., Takeishi, T., Cattanach, B. M., Galli, S. J. and Suggs, S. V. Stem cell factor is encoded at the *Sl* locus of the mouse and is the ligand for the c-*kit* tyrosine kinase receptor. *Cell* **63**:213–224, 1990.
16. Kitamura, Y. and Go, S. Decreased production of mast cells in Sl/Sl^{d} anemic mice. *Blood* **53**:492–497, 1979.
17. Ward, S. M., Burns, A. J., Torihashi, S., Harney, S. C. and Sanders, K. M. Impaired development of interstitial cells and intestinal electrical rhythmicity in *steel* mutants. *Am. J. Physiol.* **269**:1577–1585, 1995.
18. Silvers, W. K. In: *The Coat Color of Mice. A Model for Mammarian Gene Action and Interaction.* Springer-Verlag, New York, 1979.
19. Nocka, K., Majumder, S., Chabot, B., Ray, P., Cervone, M., Bernstein, A. and Besmer, P. Expression of c-kit gene products in known cellular targets of *W* mutations in normal and *W* mutant mice; evidence for an impaired c-*kit* kinase in mutant mice. *Genes Dev.* **3**:816–826, 1989.
20. Reith, A. D., Rottapel, R., Giddens, E., Brady, C., Forrester, L. and Bernstein, A. *W* mutant mice with mild or severe developmental defects contain distinct point mutations in the kinase domain of the c-*kit* receptor. *Genes Dev.* **4**:390–400, 1990.
21. Tsujimura, T., Koshimizu, U., Katoh, H., Isozaki, K., Kanakura, Y., Tono, T., Adachi, S., Kasugai, T., Tei, H., Nishimune, Y., Nomura, S. and Kitamura, Y. Mast cell number in the skin of heterozygotes reflects molecular nature of c-*kit* mutation. *Blood* **81**:2530–2538, 1993.
22. Niwa, Y., Kasugai, T., Ohno, K., Morimoto, M., Yamazaki, M., Dohmae, K., Nishimune, Y., Kondo, K. and Kitamura, Y. Anemia and mast cell depletion in mutant rats that are homozygous at 'White spotting (*Ws*)' locus. *Blood* **78**:1936–1941, 1991.
23. Isozaki, K., Hirota, S., Nakama, A., Miyagawa, J.-I., Shinomura, Y., Xu, Z., Nomura, S. and Kitamura, Y. Disturbed intestinal movement, bile reflux to the stomach, and deficiency of c-*kit*-expressing cells in *Ws/Ws* mutant rats. *Gastroenterology* **109**:456–464, 1995.
24. Giebel, L. B. and Spritz, R. A. Mutation of the kit protooncogene in human piebaldism. *Proc. Natl. Acad. Sci. USA* **88**:8696–8699, 1991.
25. Kitamura, Y., Yokoyama, M., Matsuda, H. and Shimada, M. Coincidental development of forestomach papilloma and prepyloric ulcer in nontreated mutant mice of W/W^{v} *and* Sl/Sl^{d} genotypes. *Cancer Res.* **40**:3392–3397, 1980.
26. Yokoyama, M., Tatsuta, M., Baba, M. and Kitamura, Y. Bile reflux: a possible cause of stomach ulcer in

nontreated mutant mice of W/Wv genotype. *Gastroenterology* **82**:857–863, 1982.

27. Suzuki, N., Prosser, C. L. and Dahms, V. Boundary cells between longitudinal and circular layers: essential for electrical slow waves in cat intestine. *Am. J. Physiol.* **250**:G287–294, 1986.
28. Ullrich, A. and Schlessinger, J. Signal transduction by receptors with tyrosine kinase activity. *Cell* **61**:203–212, 1990.
29. Furitsu, T., Tsujimura, T., Tono, T., Ikeda, H., Kitayama, H., Koshimizu, U., Sugahara, H., Butterfield, J.H., Ashman, L.K., Kanayama, Y., Matsuzawa, Y., Kitamura, Y. and Kanakura, Y. Identification of mutations in the coding sequence of the proto-oncogene c-*kit* in a human mast cell leukemia cell line causing ligand-independent activation of c-*kit* product. *J. Clin. Invest.* **92**:1736–1744, 1993.
30. Tsujimura, T., Furitsu, T., Morimoto, M., Isozaki, K., Nomura, S., Matsuzawa, Y., Kitamura, Y. and Kanakura, Y. Ligand-independent activation of c-*kit* receptor tyrosine kinase in a murine mastocytoma cell line P-815 generated by a point mutation. *Blood* **83**:2619–2626, 1994.
31. Tsujimura, T., Furitsu, T., Morimoto, M., Kanayama, Y., Nomura, S., Matsuzawa, Y., Kitamura, Y. and Kanakura, Y. Substitution of an aspartic acid results in constitutive activation of c-*kit* receptor tyrosine kinase in a rat tumor mast cell line RBL-2H3. *Int. Arch. Allergy Immunol.* **106**:377–385, 1995.
32. Nagata, H., Worobec, A. S., Oh, C. K., Chowdhury, B. A., Tannenbaum, S., Suzuki, Y. and Metcalfe, D. D. Identification of a point mutation in the catalytic domain of the protooncogene c-*kit* in peripheral blood mononuclear cells of patients who have mastocytosis with an associated hematologic disorder. *Proc. Natl. Acad. Sci. USA* **92**:10560–10564, 1995.
33. Longley, B. J., Tyrrell, L., Lu, S.-Z., Ma, Y.-S., Langley, K., Ding, T.-G., Duffy, T., Jacobs, P., Tang, L. H. and Modlin, I. Somatic c-*kit* activating mutation in urticaria pigmentosa and aggressive mastocytosis: establishment of clonality in a human mast cell neoplasm. *Nat. Genet.* **12**:312–314, 1996.
34. Hashimoto, K., Tsujimura, T., Moriyama, Y., Yamatodani, A., Kimura, M., Tohya, K., Morimoto, M., Kitayama, H., Kanakura, Y. and Kitamura, Y. Transforming and differentiation-inducing potentials of constitutively activated c-*kit* mutant genes in the IC-2 murine interleukin-3-dependent mast cell line. *Am. J. Pathol.* **148**:189–200, 1996.
35. Kitayama, H., Kanakura, Y., Furitsu, T., Tsujimura, T., Oritani, K., Ikeda, H., Sugahara, H., Mitsui, H., Kanayama, Y., Kitamura, Y. and Matsuzawa, Y. Constitutively activating mutations of c-kit receptor tyrosine kinase confer factor-independent growth and tumorigenicity of factor-dependent hematopoietic cell lines. *Blood* **85**:790–798, 1995.
36. Tsujimura, T., Morimoto, M., Hashimoto, K., Moriyama, Y., Kitayama, H., Matsuzawa, Y., Kitamura, Y. and Kanakura, Y. Constitutive activation of c-*kit* in FMA3 murine mastocytoma cells caused by deletion of seven amino acids at the juxtamembrane domain. *Blood* **87**:273–283, 1996.
37. London, C. A., Galli, S. J., Yuuki, T., Hu, Z. Q., Helfand, S. C. and Geissler, E. N. Spontaneous canine mast cell tumors express tandem duplications in the proto-oncogene c-kit. *Exp. Hematol.* (in press).
38. Hirota, S., Isozaki, K., Moriyama, Y., Hashimoto, K., Nishida, T., Ishiguro, S., Kawano, K., Hanada, M., Kurata, A., Takeda, M., Tunio, G. M., Matsuzawa, Y., Kanakura, Y., Shinomura, Y. and Kitamura, Y. Gain-of-function mutations of c-*kit* in human gastrointestinal stromal tumors. *Science* **279**:577–580, 1998.
39. Taniguchi, M., Nishida, T., Hirota, S., Isozaki, K., Ito, T., Nomura, T., Matsuda, H. and Kitamura, Y. Effect of c-*kit* mutation on prognosis of gastrointestinal stromal tumors. *Cancer Res.* (in press).
40. Nishida, T., Hirota, S., Taniguchi, M., Hashimoto, K., Isozaki, K., Nakamura, H., Kanakura, Y., Tanaka, T., Takabayashi, A., Matsuda, H. and Kitamura, Y. Familial gastrointestinal stromal tumours with germline mutation of the KIT gene. *Nat. Genet.* **19**:323–324, 1998.
41. O'Brien, P., Kapusta, L., Dardick, I., Axler, J. and Gnidec, A. Multiple familial gastrointestinal autonomic nerve tumors and small intestinal neuronal dysplasia. *Am. J. Surg. Pathol.* **23**:198–204, 1999.
42. El-Omar, M., Davies, J., Gupta, S., Ross, H. and Thompson, R. Leiomyosarcoma in leiomyomatosis of the small intestine. *Postgrad. Med. J.* **70**:661–664, 1994.
43. Marshall, J. B., Diaz-Aris, A. A., Bochna, G. S. and Vogele, K. A. Achalasia due to diffuse esophageal leiomyomatosis and inherited as an autosomal dominant disorder; report of a family study. *Gastroenterology* **98**:1358–1365, 1990.
44. Ehrlich, P. Beitrage zur Theorie und Praxis der histologishen Farburg. *Doctoral Thesis*, University of Leipzig, 1878.
45. Cajal, S. R. Sur les ganglions et plexus nerveux de l'intestin. *C. R. Soc. Biol.* **45**:217–223, 1893.

CHAPTER 3

Modulation of Mast Cell Development from Embryonic Haematopoietic Progenitors by Eotaxin

ELIZABETH J. QUACKENBUSH,[1] BARRY K. WERSHIL[2] and JOSE-CARLOS GUTIERREZ-RAMOS[3]*

[1]*Children's Hospital, Harvard Medical School, and the Center for Blood Research, Boston, Massachusetts,*
[2]*SUNY Health Sciences Center, Brooklyn, New York, and*
[3]*Millennium Pharmaceuticals, Inc., Cambridge, Massachusetts, U.S.A.*

CHEMOKINES AND THEIR RECEPTORS PLAY DIVERSE ROLES AS MEDIATORS OF INFLAMMATION

Within the immune system, chemoattractants (also known as chemokines) are members of a large superfamily of structurally and functionally related proteins (1, 2). Chemoattractants bind to cell-surface receptors, which leads to a rapid, intracellular signalling event, cytoskeletal arrangements and cell motility, along with other cellular functions (3–5). For example, chemokine-mediated activation of integrins on circulating cells is a central event in the immune system's response to injury, inflammation or infection. Subsequently, activated cells undergo firm arrest on the vessel wall and diapedesis (6).

Chemokines are structurally classified into four subfamilies, based on highly conserved, N-terminal cysteine residues that are arranged into characteristic motifs: C, CC, CXC or CX_3C (1). New chemokines and their receptors are discovered yearly, and evidence suggests that there may be as many as 50 human chemokines (1). The functionally diverse β chemokines (containing the CC motif) make up the largest family, with nearly twenty members in humans, including macrophage inflammatory proteins (e.g. MIP-1α; 3), eotaxin (7, 8), RANTES ('regulated upon activation, normal T cell expressed and secreted'; 9) and monocyte chemotactic proteins (MCP)-1–5 (2, 10–12, 85). The C and CX_3C motifs define two chemokine subfamilies that contain only one member each. Lymphotactin has a lone C in the N-terminal domain (13) and is a potent chemoattractant for T cells, while neurotactin (also known as fractalkine) has three amino acids intervening between the first two cysteines (14, 15).

The CXC (α) chemokine subfamily is further subdivided by the presence or absence of an amino acid motif, ELR (glutamate–leucine–arginine), between the N-terminus and the

* Corresponding author.

MAST CELLS AND BASOPHILS
ISBN 0-12-473335-2

first cysteine (16). The ELR-containing CXC chemokines are uniformly potent neutrophil chemoattractants, and the prototypic human CXC chemokine, interleukin-8 (IL-8), also stimulates histamine release from basophils (17, 18).

All chemokine receptors described thus far are seven-transmembrane-spanning, G protein coupled receptors, whose antagonists include *Bordetella pertussis* toxin (3, 4). The four major classes of chemokine receptors are based on ligand specificity and have been defined as 'specific', 'shared', 'viral' and 'promiscuous'. Specific receptors only bind one ligand. Shared receptors, the largest subset, can bind more than one chemokine, but the chemokines must be members of the same family. Promiscuous receptors can bind to ligands from either family.

We now know that haematopoietic chemokines and their receptors function in processes more diverse than cell migration. For example, select chemokines can inhibit angiogenesis (19), stimulate B cell development (20) and aid human immunodeficiency virus (HIV) entry (21).

CHEMOKINES PARTICIPATE IN MULTIPLE MAST CELL AND BASOPHIL FUNCTIONS

Basophils and mast cells accumulate in allergic reactions, and their migration, differentiation and activation are required for propagation of inflammatory responses, such as allergic rhinitis, asthma and atopic dermatitis. When activated, they release multiple mediators such as histamine, IL-4, IL-13 and chemokines (22, 23). They also bear a variety of chemokine receptors that can bind multiple chemokines (Table I).

In general, less is known about the role of chemokines in mast cell and basophil development, activation and migration. Functional studies have focused on their role in the regulation of migration, cytokine expression, leukotriene formation and histamine release (Table I). Despite similar signal transduction pathways that use pertussis-sensitive G protein coupled receptors, there is remarkable variation in the ability of individual chemokines to mediate basophil and mast cell exocytosis and chemotaxis.

Chemokine receptors on basophils induce chemotaxis and mediator release

In many early studies examining chemokine-induced responses of basophils and mast cells, the ligand-binding receptor was not identified. More recently, the chemokine receptors CCR1, CCR2, CCR3, CCR5, CXCR1, CXCR2 and CXCR4 have been shown to be expressed on basophils and mast cells by mRNA and/or protein analysis (Table I).

CCR1, whose mRNA is weakly and constitutively expressed in human basophils, binds multiple CC chemokines, including RANTES, MIP-1α and MCP-3. This receptor is not expected to play a major role in basophil chemotaxis, however, as RANTES (a primary ligand) appears to induce basophil chemotaxis through CCR3 (12). Its function on basophils remains unknown, but it may mediate exocytosis rather than migration, since antibodies to CCR3 did not abrogate MIP-1α- or MCP-3-induced histamine release from human basophils.

MCP-1, which binds only to CCR2, is a weak chemoattractant of basophils but a potent stimulus of exocytosis (24). MCP-1 and MCP-3 strongly induce histamine release from human basophils, and they, along with MCP-4, readily compete for binding to CCR2 (12). The affinity of MCP-1 is approximately ten-fold higher than that of MCP-3

TABLE I
Chemokine Receptors Expressed by Basophils and Mast Cells

Chemokine receptor	Ligands	Expression pattern (receptor)	Function*
CCR1	MIP-1α; RANTES; MCP-3	Human basophils (low level mRNA)	MIP-1α: + CTX; + exocytosis
CCR2	MCP-1–4	Human basophils (low level mRNA); rat mast cells	MCP-1: + CTX; ++ exocytosis MCP-2: + CTX; ++ exocytosis MCP-3: ++ CTX; ++ exocytosis
CCR3	Exotaxin-1/2; MCP-2–4; RANTES	Human/rodent† mast cells; human/rodent basophils	Eotaxin/MCP-4: ++ CTX; + exocytosis RANTES: ++ CTX; + exocytosis
CCR5	MIP-1α; MIP-1β; RANTES	Human mast cell progenitors	Unknown
CXCR1	IL-8	Human basophils	+/– CTX; + exocytosis
CXCR2	IL-8; NAP-2	Human basophils; human mast cell progenitors	+/– CTX
CXCR4	SDF-1α	Human mast cell progenitors	+CTX

* CTX, chemotaxis; + indicates strength of activity.
† MCP-1, RANTES and MIP-1α induce histamine release from rat mast cells but not from murine or human mast cells.

and MCP-4. MCP-1 can strongly stimulate histamine release from basophils without the need for IL-3 pre-treatment (24), however, priming with IL-3 allows basophils to respond to even lower concentrations of MCP-1 (25).

Eotaxin, eotaxin-2, MCP-3 and MCP-4, all ligands for CCR3, arc potent chemoattractants for human basophils, and the activity found with each chemokine was comparable to that seen with eosinophils (12, 86). Anti-CCR3 completely abrogated eotaxin- and RANTES-induced migration, as did pertussis toxin, while MCP-3-induced migration was only minimally inhibited. Migration towards an MCP-4 gradient was also markedly inhibited, but only at MCP-4 concentrations less than 100 nm.

Binding experiments show that MCP-4 has a 40-fold higher affinity for CCR3 than MCP-3 has (12). RANTES can also bind to CCR1 and CCR4, but the nearly complete inhibition with anti-CCR3 antibodies of RANTES-induced migration suggests that this is the primary receptor on basophils through which this ligand acts. On human basophils the mRNA levels of CCR3 and CCR2 are similar, and CCR3 is strongly expressed on the surface, based upon flow cytometry data (23).

The chemokines eotaxin, RANTES, MCP-4, MCP-3 and MCP-1 rank from low to high in order of efficiency of histamine release from basophils (12). These four chemokines also induced leukotriene C_4 production in human basophils, after IL-3 priming, with the same degree of efficiency. Anti-CCR3 treatment eliminated histamine release induced by eotaxin, while decreasing the effect found with RANTES. Mediator release induced by MCP-1, MIP-1α and IL-8 was not affected by anti-CCR3 antibodies.

Hartmann *et al.* (26) also demonstrated that RANTES, MCP-1, MCP-2, MCP-3, MIP-1α and MIP-1β were able to induce histamine release from human basophils but none of them induced histamine release from human skin mast cells.

By flow cytometry, human basophils express intermediate levels of the IL-8 receptors, CXCR1 and CXCR2, in comparison to the high level of CCR3 found on the surface (23). Surprisingly, the IL-8 analogue, neutrophil-activating peptide 2 (NAP-2), which reacts with CXCR2, induced only a weak chemotactic response and had no effect on exocytosis or cytokine expression, despite a moderate level of CXCR2 surface expression (87). CXCR1 was able to induce both migration and formation of leukotriene C_4, but only from IL-3-primed basophils.

Chemokine-induced mast cell exocytosis and migration

Despite being multifunctional, secretory immune cells that originate from a common haematopoietic progenitor, basophils and mast cells respond differently to activating stimuli. Basophils respond to IL-3, IL-5 and granulocyte–macrophage colony-stimulating factor (GM-CSF), while human mast cells cannot be activated by these factors. SCF is a specific mediator of granule release from mast cells and promotes mast cell differentiation.

In contrast to basophils, human mast cells release histamine in response to stem cell factor (SCF) (27), but their response to chemokines depends upon the source of mast cells used. In one study using human mast cells isolated from lung, skin, tonsils and uterus, none of the chemokines tested (MCP-1, RANTES, MIP-1α, MIP-1β, GRO (growth-regulated oncozine) IP-10 and IL-8) induced histamine release (25). A second study by Petersen *et al.* (27) confirmed that MCP-1, RANTES and MIP-1α are unable to induce histamine release from human skin mast cells. Furthermore, eotaxin did not enhance anti-IgE-stimulated histamine release from $CCR3^+$ human lung mast cells (28). However, murine skin and peritoneal mast cells degranulated in response to MIP-1α and, to a lesser degree, to MCP-1 (29).

MCP-1 (but not MIP-1α, eotaxin, or MCP-3) directly induced pulmonary murine mast cell degranulation, *in vitro*, and caused prolonged airway hyper-reactivity and histamine release, *in vivo*, using a cockroach allergen to induce inflammation (30). MCP-1's activity was mediated through CCR2, as mice lacking CCR2 had a significantly attenuated response to the chemokine. Taub *et al.* (31) found no degranulation of murine bone marrow mast cells or mast cell lines to MIP-1α, MIP-1β, RANTES, MCP-1, IL-8, or IP-10. They also did not confirm previous results showing that MIP-1α induces histamine release from purified murine peritoneal mast cells.

Several chemokines also induce mast cell chemotaxis *in vitro*. Unstimulated murine mast cells will migrate to MCP-1 and RANTES on surfaces coated with extracellular matrix proteins (vitronectin, fibronectin and laminin), while IgE-stimulated mast cells will also migrate to RANTES, MCP-1 and MIP-1α (31). RANTES injected subcutaneously into the sole of rat paws caused mast cell accumulation that was inhibited by an anti-RANTES antibody (32).

SCF and IL-3 induce the migration of resting and activated mast cells, as well (31), and the level of response is similar to that seen with MCP-1. It is intriguing that mast cells can enhance their migratory response to certain chemokines after antigen or IgE activation. This enhancement may be unique to mast cells and crucial to optimizing migration into sites of antigen deposition in pathological situations, such as parasitic infections and allergies.

Nilsson *et al.* (33) investigated the response of the human mast cell line, HMC-1, to different chemokines by calcium flux measurements. Only CXC chemokines with the ELR tripeptide motif, such as IL-8, GRO-α, NAP-2, and epithelial cell-derived neutrophil-activating peptide-78 (ENA-78), induced calcium fluxes in HMC-1. CXCR2, but not CXCR1, mRNA was detected in HMC-1 mRNA preparations by the RNase protection assay, and flow cytometry studies confirmed these findings at the protein level (34). IL-8 induced dose-dependent chemotaxis of both HMC-1 cells and *in vitro* cultured human cord blood-derived mast cells.

CHEMOKINE SECRETION BY MAST CELLS

Several cytokines are known to be secreted by these immunoregulatory cells. For example, mast cells activated by topical dinitrofluorobenzene migrate to lymph nodes, where they become an abundant source of the T cell-recruiting chemokine, MIP-1β (35). Likewise, ENA-78 and MCP-5 are produced by mast cells during allergic airway inflammation (11, 36), and RANTES is secreted by mast cells during allergen-induced rhinitis (37).

Using HMC-1 as a source of RNA, MCP-1 and RANTES transcripts were detected at moderate and low levels in unstimulated cells, respectively, by northern blotting (38). Following phorbol ester treatment, MCP-1 and RANTES transcripts were upregulated and MIP-1α, MIP-1β and IL-8 transcripts were newly detected. MCP-1 secretion was confirmed by immunoprecipitation from supernatants of unstimulated and stimulated cells. Secretion of MIP-1α was further increased by IgE treatment (39).

Lymphotactin (Ltn) gene expression was induced in human mast cells, human basophils, HMC-1 and murine bone marrow mast cells by $Fc_{\varepsilon}RI$ aggregation (40). Transforming growth factor-β (TGF-β) enhanced $Fc_{\varepsilon}RI$-induced Ltn mRNA expression, and both MIP-1α and Ltn were upregulated in response to IL-4 following $Fc_{\varepsilon}RI$ aggregation. As with MCP-1, lymphotactin was identified in supernatants by western blotting.

Eosinophils contain eotaxin within their intracellular granules and respond to exogenous eotaxin by chemotaxis (41). Quantification of cell-associated eotaxin in different leukocyte subsets revealed that eosinophils are the principal source of this chemokine. However, murine bone marrow mast cells grown in SCF, but not in IL-3-containing conditioned medium, also secrete eotaxin when exposed to substance P (42). Eotaxin protein was not detected when these cells were stimulated by IgE and antigen. Thus, both murine mast cells and eosinophils can secrete and respond to eotaxin, likely through CCR3.

Not all sources of mast cells react accordingly, as mast cells isolated from guinea pig lung (challenged and unchallenged) did not express eotaxin by northern blot analysis (43). Airway epithelial cells appeared to be the major source of eotaxin in the guinea pig pulmonary system.

EOTAXIN

Eotaxin is a potent chemoattractant for eosinophils. It is a member of the CC chemokine family and, by northern blot analysis, is constitutively expressed only in adult intestine, thymus and skin (7, 8, 88). Eotaxin does not induce the *in vitro* chemotaxis of neutrophils, monocytes, macrophages or mast cells in adult mice (8). Human skin mast cells express CCR3 and will migrate to both eotaxin and RANTES with similar efficiency (28).

Eotaxin will also induce chemotaxis of T cells, but only after they have been stimulated by IL-2 and IL-4 to express CCR3 (44). Furthermore, eotaxin upregulated ICAM-1, CD29 and CD49a and CD49b on T cells, suggesting that T cell adhesiveness to endothelium is also promoted by eotaxin (44). Conflicting reports on mast cell responses to eotaxin may be explained, in part, by the observation that CCR3 (in humans) is preferentially associated with connective tissue mast cells rather than with mucosal mast cells (28).

The biological significance of eotaxin has been most frequently examined in animal models of inflammation. Asthma-associated airway inflammation and bronchial hyper-reactivity are characterized by massive infiltration of eosinophils, mediated in part by chemoattractants, such as eotaxin (8, 89). We and others have identified and characterized the kinetics of eotaxin expression and eosinophil infiltration in a murine model of ovalbumin-induced lung inflammation. However, the contribution of eotaxin-triggered mast cell responses (and of mast cell-released eotaxin) to lung inflammation remains poorly understood. In general, asthmatic patients have high levels of eotaxin in their bronchioalveolar fluid and increased expression of eotaxin mRNA in airway epithelial cells (45).

Eotaxin injected intraperitoneally into mice can induce a maximum dose-dependent influx of eosinophils in 1–2 h; however, in mast cell-deficient mice the influx is delayed and reduced in numbers (46). Inhibitors of lipoxygenase also caused a reduction in the number of eosinophils in peritoneal lavages. These findings suggest that leukotrienes and mast cells contribute to eotaxin-induced eosinophilia. In other studies, pre-treatment of mice with histamine or serotonin antagonists reduced eosinophil influx into the peritoneum in response to intraperitoneal eotaxin injections (47).

The role of constitutively expressed eotaxin may be to control baseline trafficking of eosinophils into non-haematopoietic tissue (88). For example, wild-type mice express eotaxin mRNA in the lamina propria of their jejunum, and eosinophils are readily detected there as well. In eotaxin-deficient mice, there is a large reduction in the number of detectable eosinophils in the lamina propria of the jejunum, as well as in the thymus. We have found by immunostaining significant amounts of eotaxin in the thymic medullary regions and along the underside of the capsule (data not shown), but the significance of this expression pattern is unknown.

A second human chemoattractant for eosinophils has been described and named eotaxin-2 (48). Despite their names, eotaxin and eotaxin-2 share only 39% identical amino acids and differ nearly completely in the N-terminal region, a region of particular importance for inducing chemotaxis. Eotaxin-2 induces chemotaxis of basophils, as well as histamine and leukotriene C_4 release, post-priming with IL-3 (48).

Given the evidence that chemokines participate in mast and basophil exocytosis and migration, we examined the function of eotaxin in mast cell differentiation. We used progenitors isolated from fetal tissues, as circulating mast cell progenitors are prevalent and actively differentiating during haematopoietic development *in utero*. Committed fetal mast cell progenitors are identified by Thy-1^{lo}c-*kit*^{hi} expression and mRNA for mast cell-associated proteases. They are present in the fetal blood at mid to late stages of gestation (49). It is presumed that these circulating cells are released from the fetal liver.

Until very recently, haematopoietic stem cells in mice were thought to originate in the yolk sac, despite an overwhelming body of evidence from other species that the earliest precursors were located within the body of the embryo proper (50). Recent evidence shows that the greatest number of murine stem cells are obtained from the aorta–gonad–mesonephros (AGM) region, an area around the endoderm of the developing gut in the abdomen (51, 90). Stem cells from this region are thought to migrate to the fetal liver and subsequently to colonize the developing thymus and bone marrow (51). In our studies we used single cell suspensions of fetal liver, fetal blood and yolk sac as sources of progenitors.

RESULTS

Eotaxin modulates myelopoiesis and mast cell development from embryonic haematopoietic progenitors

Several studies showed previously that SCF, a mast cell growth factor, acts as a potent co-mitogen for multiple types of progenitor cells when combined with other cytokines, including IL-3, IL-6 and IL-1 (52). It also acts as a chemoattractant for progenitor cells (53). Thus, we initially examined the effect of SCF combined with eotaxin on haematopoietic colony formation in methylcellulose cultures. Equal numbers of cells were plated in methylcellulose in the presence of SCF alone (20 ng ml^{-1}), eotaxin alone (50 ng ml^{-1}) and SCF combined with eotaxin. The receptor for SCF, c-*kit*, has been shown by others to be expressed on progenitor cells isolated from the yolk sac (54) and fetal liver (55). At 11 days post coitum (dpc), we found that 35% of liver cells, 3.8% of yolk sac cells and 3.2% of fetal blood cells expressed c-*kit*.

Eotaxin combined with SCF produced a synergistic response in the total number and size of colonies grown from fetal liver (Fig. 1, A), while an additive effect was seen with yolk sac and fetal blood-derived colonies (data not shown). The synergistic effect of these factors was not always represented by the colony count, but was seen when the total numbers of cells produced in methylcellulose cultures were harvested and counted. Cell counts were decreased by 52%, and colony counts by 30%, when pertussis toxin was added to fetal blood cultures induced with the combination of eotaxin and SCF (data not shown).

Macrophage and mast cell colonies are preferentially induced with SCF and eotaxin

Three types of methylcellulose colonies, CFU-erythroid (CFU-E), CFU-MIX and CFU-macrophage (CFU-M), were quantitated (Fig. 1, B, left graph). Colony type was determined by preparing cytospun preparations of individual colonies. CFU-MIX contained cells of varying size, nuclear shape and granulation, and the majority of cells were macrophages, mast cells and undefined mononuclear, weakly granulated cells. Similar patterns of CFU type were obtained with fetal liver and yolk sac. CFU-MIX are most prevalent and additively increase in number with the addition of eotaxin to SCF, as

A

Fetal Liver	*SCF*	*Eotaxin*	*SCF + Eotaxin*
Colony Count:	**187 ± 13**	**200 ± 21**	**364 ± 44**
Cell Count:			
11 dpc FL:	**67,340**	**117,580**	**298,100**
11 dpc FL:	**38,560**	**54,920**	**160,060**
12 dpc FL:	**15,750**	**40,440**	**94,410**

B

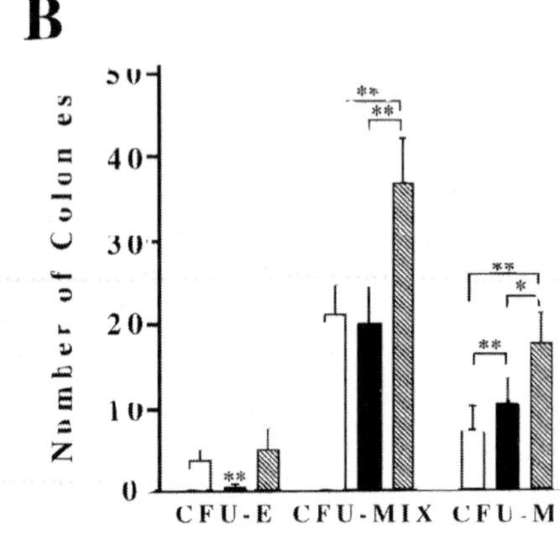

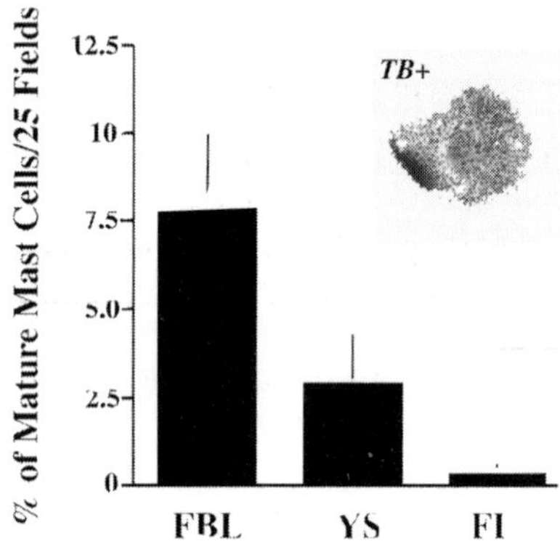

C

% OF MAC-1⁺ CELLS

	Yolk Sac			Fetal Blood			Fetal Liver		
	SCF	*EOTAXIN*	*BOTH*	*SCF*	*EOTAXIN*	*BOTH*	*SCF*	*EOTAXIN*	*BOTH*
Exp 1	12.7	12.1	15.2	2.1	6.9	14.8	17.4	24.8	34.3
Exp 2	7.7	8.3	34.2	6.8	5.8	22.0	10.5	32.4	47.2

Fig. 1 Eotaxin influences fetal haematopoietic cell differentiation. In A, the total number of fetal liver colonies and progeny cells induced (in three separate experiments) by SCF alone, eotaxin alone, or both factors combined is shown. In B (left graph), the three types of colonies induced with SCF and/or eotaxin are plotted: CFU-erythroid (CFU-E), CFU-MIX (containing cells of varying size, granulation and nuclear shape/size, with and without erythrocytes) and CFU-macrophage (CFU-M). Colony type was confirmed by preparing cytospun preparations from individual colonies with distinct morphological characteristics, after 14 days of culture. Standard error bars are shown, and statistical analysis was performed using the Friedman test: $^{**}p<0.001$, $^{*}p<0.01$. The right graph in section B shows the percentage of more differentiated mast cells identified on cytospins stained with modified Giemsa, based on granular content. The percentage of heavily granulated (mature) cells was determined by counting the number of these cells (in 25 or 100 fields at 400× magnification) on cytospun preparations made by harvesting all colony progeny induced in methylcellulose, after 14 days. Total cell counts ranged from 84 to 964. The inset in the right graph shows a mature mast cell generated after 10 days in culture with eotaxin and SCF, stained with toluidine blue (TB+). Colony progeny were harvested and immunofluorescently stained with a monoclonal antibody to Mac-1. The percentage of Mac-1^{+} cells, after background subtraction, is shown in C.

did CFU-M, while CFU-E are not induced by eotaxin alone (in any tissue type). The one exception to this pattern was seen with yolk sac, where the numbers of CFU-M were similar for each growth condition tested.

Phenotypic analysis of colony progeny was performed with the lineage markers GR-1, Mac-1, TER-119, Thy-1 and B220, and we found a synergistic increase of 1.63 times (±0.4) in the number of Mac-1^+ cells from colonies grown in the presence of both factors (Fig. 1, C). This increase was identical to the synergistic increase in the total number of cells induced by eotaxin and SCF combined. The total percentage of lineage-positive cells ranged from 15% to 51%, and progenitors (based upon replating experiments), stromal cells and mast cells were included among the lineage-negative population. Significant numbers of yolk sac- and fetal blood-derived progeny were positive for c-*kit* (data not shown).

Eotaxin influences mast cell development from embryonic progenitors

A unique subpopulation of heavily granulated cells was identified on cytospun preparations of yolk sac- and fetal blood-derived colonies induced with both factors (Fig. 1, B, right graph). These granulated cells were not obtained from cultures grown in eotaxin alone or in SCF alone at low concentrations. However, cultures of progenitor cells in SCF alone at concentrations of $\geq$ 50 ng ml^{-1} gave rise to cells with the same morphological characteristics. Acidified toluidine blue (a specific stain for mast cells) and chloroacetate esterase (an enzyme found in mast cells, but not in eosinophils or basophils) both stained the granules of these cells. The number of granules per cell, cell size and nuclear profile (both mononuclear and multinucleated cells were seen) were characteristic of mature mast cell populations derived *in vitro*. The percentage of mature mast cells obtained from 11dpc fetal blood colonies grown in SCF and eotaxin was 8 ± 3%, while cells with similar morphology obtained from yolk sac cultures represented 2.8 ± 1.7% of the total cells produced in methylcellulose after 14 days of culture (Fig. 1, B, right graph). Toluidine blue-positive cells expressing c-*kit* were identified by immunohistological staining of colony progeny derived from 11 dpc yolk sac and fetal blood (data not shown).

We also performed reverse transcriptase–polymerase chain reaction (RT–PCR) analysis for the expression of serine proteases specific for mast cells, in order to confirm that the heavily granulated cells obtained in methylcellulose cultures with SCF and eotaxin were mast cells. Mast cells derived from murine bone marrow cells cultured in spleen cell-conditioned medium are immature mast cells expressing murine mast cell protease 2 (mMCP-2) and murine mast cell carboxypeptidase A (mMC-CPA), while mast cells cultured in medium with SCF alone expressed RNA for mMCP-2, mMC-CPA and mMCP-4 (Fig. 2). Colonies grown from 11dpc fetal blood in SCF alone expressed mMCP-2 and mMC-CPA mRNA but not mMCP-4 mRNA. However, mMCP-2, mMC-CPA and mMCP-4 mRNA were expressed by progenitors grown in SCF in combination with eotaxin.

Mature mast cells are more prevalent in haematopoietic colonies derived from fetal blood than from fetal liver progenitors

When the relative number of mature mast cells in fetal blood and fetal liver cultured in SCF and eotaxin was examined, we found an increased number in fetal blood (8 ± 3%) compared with fetal liver (<1%) at 11dpc. In contrast, there are approximately ten times

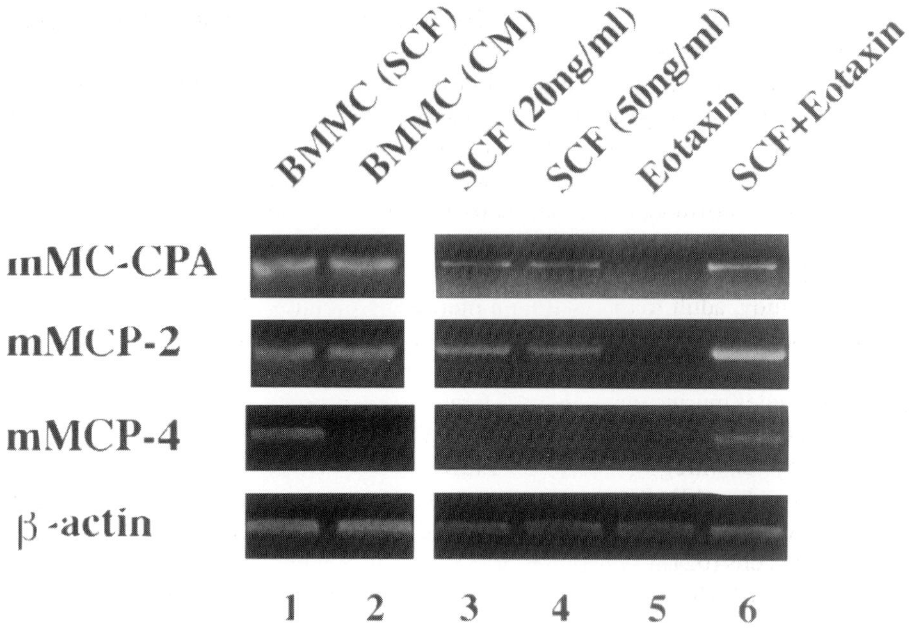

Fig. 2 Eotaxin is required for the differentiation of mast cell progenitors expressing mMCP-4. RT–PCR was used to determine the pattern of mast cell proteases (carboxypeptidase A, mMC-CPA; cellular protease 2, mMCP-2; cellular protease 4, mMCP-4) expressed in fetal blood colonies. Lanes 1 and 2 represent bone marrow mast cells (BMMC) derived from BALB/c mice, in SCF alone (lane 1) or concanavalin A stimulated conditioned medium (CM; lane 2). Lanes 3–6 represent fetal blood-derived colonies (11dpc).

more multipotent progenitors in fetal liver at 11dpc than in fetal blood. In our assay system, 2–4 times more total colonies were produced from fetal liver in the presence of SCF and eotaxin than from fetal blood and yolk sac, despite using five-fold fewer fetal liver cells to seed the cultures. The mean number of total blood and fetal liver cells per organism at 11dpc was $8.5 \pm 5.4 \times 10^5$ ($n = 8$) and $3.7 \pm 2.6 \times 10^5$ ($n = 5$), respectively, which compares favourably with reported estimates. Thus, the estimated number of fetal blood cells at 11dpc responsive to SCF in combination with eotaxin was ~500 fetal blood-derived progenitors per embryo and ~1600 fetal liver-derived progenitors per embryo, on average.

DISCUSSION

Our study showed that eotaxin in combination with SCF synergistically induced from embryonic progenitors the differentiation of granulated cells with morphological features characteristic of mature mast cells, *in vitro*. These cells were definitively identified as mast cells by their expression pattern of mast cell-specific proteases and by immunohistochemical staining. There are several possible explanations for this observation. Eotaxin, acting through CCR-3 or another receptor, could induce the secretion of additional SCF or other cytokines that accelerate mast cell proliferation, differentiation and/or survival. SCF produced by fibroblasts is known to promote mast cell survival by suppressing apoptosis (56) and to promote mast cell adherence to fibronectin and other interstitial cell types (57).

Co-incubation of mast cells with fibroblasts also induces mast cell proliferation and activation (58), and both cell populations can produce significant amounts of eotaxin (59). Hogaboam (91) demonstrated a novel role of transmembrane SCF in murine mast cell activation and eotaxin production mediated by the interaction of mast cells and fibroblasts. Bone marrow-derived mast cell–fibroblast co-cultures were established with and without direct contact, and histamine release and chemokine production were quantitated. Direct contact allows mast cells to bind both transmembrane and soluble SCF. In their study, adult mast cells were induced to produce histamine and eotaxin by transmembrane SCF. They proposed that SCF activation of mast cells could also have induced IL-4 release simultaneously, which induces eotaxin production by fibroblasts (60). IL-4 promotes murine mast cell proliferation and protease expression but is reported to have an inhibitory effect on the proliferation of human fetal liver-derived mast cell progenitors, by downregulating c-*kit* expression (61). In contrast, IL-4 and IgE synergistically induced $Fc_{\varepsilon}RI$ expression and mediator release in human umbilical cord-derived mast cells (which may represent circulating mast cells), as well as in murine peritoneal mast cells (62).

Ogasawara *et al.* (63) reported that mouse bone marrow (BM)-derived mast cells initially grown in IL-3 undergo exocytosis and cytokine expression in response to G-protein-activating polybasic compounds, after co-culture with fibroblasts and soluble c-*kit* ligand. These responses were suppressed by pertussis toxin; furthermore, they were not seen when BM-derived mast cells were cultured with soluble SCF or with fibroblasts alone. The BM mast cells co-cultured with SCF and fibroblasts expressed mMCP-4, suggesting that a fibroblast-derived mast cell maturation factor acted synergistically with SCF to mediate differentiation of BM-derived mast cells to connective tissue-type mast cells that respond to polybasic compounds. This unknown factor may be fibroblast-derived eotaxin, since it similarly aids SCF in promoting or maintaining mast cell differentiation towards an mMCP-4-expressing mast cell. This suggestion correlates well with the observation that CCR-3 is preferentially expressed on human connective tissue mast cells, which express mMCP-4 (28).

Co-operative induction of chemokine signalling by SCF has been demonstrated for haematopoietic progenitors (53). SDF-1α-mediated chemotaxis of the progenitor cell line, CTS, and of $CD34^+$ bone marrow cells was enhanced by co-incubation of cells with SCF. Treatment with SCF in combination with SDF-1α-induced phosphorylation of focal adhesion molecules, such as paxillin, and activation of p44/42 mitogen-activated kinase. While eotaxin has not been shown to induce the migration of mature (murine) mast cells, theoretically, downstream targets may also be co-operatively phosphorylated by CCR-3 and c-*kit* receptors, leading to mast cell activation, granule release and/or proliferation. Füreder *et al.* (25) reported no promotion by six chemokines of SCF-induced mast cell or IL-3-induced basophil differentiation, although eotaxin was not tested. They used healthy human peripheral blood mononuclear cells as a source of mast cells.

Very little is known about mast cell progenitor development during gestation. Murine mast cell precursors have been detected in the yolk sac as early as 9.5dpc by an *in vivo* limiting dilution method (64). At 9.5dpc, the concentration of mast cells in the yolk sac was 30 times greater than that found within the embryo body. The concentration peaked in the yolk sac at 11dpc, then rapidly declined at 13dpc. Simultaneously, the number of fetal liver mast cell precursors rose from 11dpc until 15dpc, reaching a concentration at 15dpc that was comparable to that seen in yolk sac at 11dpc.

Mast cell precursors from yolk sac at 10dpc and fetal liver at 12dpc responded better to SCF alone than to IL-3 combined with IL-1 (65), which may reflect acquisition of IL-3 and IL-1 receptors during development. In contrast, the number of mast cell colonies derived from adult bone marrow was slightly higher in the presence of IL-3 plus IL-1, compared with SCF alone. We found very low numbers of mature mast cells arising in our cultures from 11dpc fetal liver cells grown with both factors, which correlates with data that the proportion of mast cell precursors in fetal liver is low at 11dpc, despite the large number of multipotent progenitors present. We do not suggest that mast cell progenitors were absent from fetal liver at 11dpc, but rather that their responses to cytokines or chemokines may be age- and/or site-specific, as well as dependent on receptor expression and signalling patterns. Fetal liver-derived, but not yolk sac-derived, mast cell progenitors could replenish mast cell numbers in the mast cell-depleted murine strain WBB6F1-W/Wv mice, suggesting that definite (i.e. adult) mast cell stem cells were not present in the yolk sac (64).

The effect of eotaxin on the differentiation of committed human mast cell progenitors isolated from cord blood was recently reported (66). Committed progenitors were cultured with SCF, IL-6 and IL-10 for 9 weeks. After 4 weeks, cultured cells were still phenotypically committed progenitors, and they expressed four chemokine receptors: CXCR-2, CCR-3, CXCR-4 and CCR5. These receptors mediated transient calcium fluxes in response to their respective ligands; furthermore, eotaxin and SDF-1α elicited chemotaxis of the committed progenitors (as did MIP-1α and IL-8). By week 9, all progenitors had differentiated into mature mast cells and only CCR3 expression was retained. Interestingly, eotaxin now mediated a sustained calcium flux, but could no longer induce chemotaxis. In this study, eotaxin (either alone or in combination with SCF) was not mitogenic.

We also examined the effect of eotaxin on embryonic progenitors older than 11dpc and found that eotaxin's ability to induce colony formation in combination with SCF was greatest for progenitors at 11dpc but then declined sharply at 12dpc, with no or few colonies induced from progenitors at 13–14dpc (data not shown). This was true for all three tissues tested, although the decline in the number of fetal liver-derived colonies after 11dpc was not nearly as sharp or as complete as with fetal blood and yolk sac tissues. The reduced effect after 11dpc may reflect a decrease in the number of responsive progenitors, or a relative reduction secondary to a dilutional effect from the expansion of other cell types, or simply the result of migration out of a developmental compartment. Between 10 and 12dpc, the number of multipotent haematopoietic cells increases linearly in the circulation and exponentially in the fetal liver (67).

Significant levels of acidified toluidine blue-positive mast cells ($\leq$7%) are found in the circulation of fetal blood from 15dpc until birth, and by day 10, postnatally, they are undetectable on peripheral smears (B. Quackenbush, unpublished observation). Agranular precursors committed to mast cell development are also present in human cord blood mononuclear cell populations (68). Thus, there appears to be a window of time perinatally during which mast cells (presumably) acquire the proper adhesion molecule repertoire needed for transmigration from the vasculature into target tissues. This hypothesis correlates with data obtained on tissue mast cell numbers in human fetal skin and airways at 24 weeks of gestation (69). Their numbers increased gradually during gestation, and then rose dramatically in the early post-natal period. Developmental switches in the use of adhesion molecules necessary for lymphocyte trafficking have been reported to occur between birth and 48 h, postnatally (70), but perinatal changes in mast cell expression patterns have not been examined.

These observations raise the questions as to why mast cell progenitors are so prevalent in the circulation prenatally and what is their role as a circulating cell. Mast cells and basophils may play early, key roles in newborn immune defence, as significantly greater levels of spontaneous and anti-IgE-induced histamine are released from cord blood basophils than from adult blood basophils (71). In addition to their role in allergic responses, mast cells and basophils also participate in the host defence against bacterial and viral infections (72), pathogens encountered *in utero*, along with parasites. Mast cells are selectively found at sites of bacterial entry, such as skin and mucosa, and mast cell-deficient mice clear enterobacteria less efficiently than their wild-type littermates (73). Likewise, viral proteins, such as protein Fv, can trigger mediator release from human basophils and mast cells (74).

In adult mice, eotaxin in combination with SCF also functions as a granulocyte macrophage colony stimulating factor (GM-CSF) (75). Peled *et al.* (75) cultured BM progenitors for 7 days with both factors and found a large increase in $Gr\text{-}1^+$ cells. GR-1 expression is restricted to band and mature neutrophils (76). Mature neutrophils were not found in our cultures; however, day 12 yolk sac and splanchnopleural cultures could not support the growth of $GR\text{-}1^+$ cells in culture (77). The increase in Mac-1 may be due to its upregulated expression on macrophages, stem cells or committed progenitors, given that it is co-expressed with c-*kit* and CD34 on stem/progenitor cells in fetal liver and the AGM region, but not on mature mast cells. Furthermore, our time of colony analysis (at 14 days), as well as the source of progenitors, may in part explain the appearance of mature mast cells and the lack of $GR\text{-}1^+$ cells in our experiments, as mast cells would be late-appearing cells in methylcellulose cultures.

Human mast cell progenitors from cord blood retained their potential to differentiate into mature mast cells after prolonged culture (3 weeks) with GM–CSF alone. Mature mast cells emerged in culture only after subsequent transfer into SCF-containing medium (78). Fewer mature mast cells were produced compared to cultures grown with SCF alone, suggesting that at least a subset of mast cell progenitors were resistant to the lineage-determining effect of GM–CSF.

Interestingly, the adhesion molecule VLA-4 was downregulated on colony progeny induced with SCF and eotaxin, and pertussis toxin upregulated VLA-4 ($\alpha_4\beta_1$) expression when it was combined with SCF and eotaxin (data not shown). The expression pattern of VLA-4 on murine BM mast cells during differentiation has been investigated by others (79). VLA-4 levels on IL-3-supported mast cells peaked at 3 weeks and then declined to undetectable levels by 12 weeks of culture, when mature mast cells predominated. The observed inverse correlation between VLA-4 and terminal differentiation of mast cells is analogous to VLA-5 expression during keratinocyte differentiation (80).

We found red blood cells, mast cells and $Mac\text{-}1^+$ cells (primarily macrophages, based on morphology) in our methylcellulose cultures. SCF alone can induce extensive proliferation of human erythroid progenitors (81), but chemokine-induced proliferation has not been studied. In our system, eotaxin alone has minimal effect on induction of CFU-E. Thus, the progenitor we expand may be capable of multipotency or we are simultaneously expanding different committed progenitors. In human fetuses at 23–34 weeks of gestation, blood obtained by percutaneous umbilical cord sampling had a surprisingly large proportion of eosinophils (42 ± 26%) (82). Eotaxin may, thus, play a role in the expansion of multiple granulocytic subsets during fetal development, in the absence of inflammation or infection.

The yolk sac at 9–10dpc contains committed progenitors, mainly erythroid and myeloid, that migrate to the fetal liver as part of the first wave of migrating cells. We have not analysed yolk sac from 9dpc, but our peak of colony-forming activity with SCF and eotaxin (11–12dpc) correlated better temporally with a second wave of progenitors into the fetal liver. This wave includes multipotent progenitors and long-term repopulating cells from the AGM and yolk sac. The fact that adult BM progenitors lacking mature lineage markers can also be synergistically induced with SCF and eotaxin implies that a definitive progenitor is responding (75).

Two independent groups generated eotaxin-deficient mice and the number of circulating eosinophils, as well as the number of eosinophils recruited into ovalbumin-treated, inflamed airways, differed between the two groups. Rothenberg *et al.* (83) reported that the number of circulating eosinophils, as well as eosinophil recruitment into ovalbumin-treated inflamed airways, was reduced, while Yang *et al.* (84) found that the lack of eotaxin did not reduce circulating eosinophil levels or lung eosinophilia induced by aerosolized ovalbumin. The differences found may be due to the genetic backgrounds of the mice used.

In summary, our study shows that eotaxin can modulate granulopoiesis, specifically mast cell differentiation, from fetal progenitors and suggests that the study of chemokines during development is a novel and unique approach to understanding many pathological processes in the adult, as well. We propose that eotaxin production and use by activated mast cells contributes to allergic or inflammatory responses, such as antigen-induced lung inflammation, by promoting cellular activation, leukocyte recruitment and tissue damage.

REFERENCES

1. Rollins, B. J. Chemokines. *Blood* **90**:909–928, 1997.
2. Miller, M. D. and Krangel, M. S. Biology and biochemistry of the chemokines: a family of chemotactic and inflammatory cytokines. *Crit. Rev. Immunol.* **12**:17–46, 1992.
3. Bokoch, G. M. Chemoattractant signaling and leukocyte activation. *Blood* **86**:1649–1660, 1995.
4. Bargatze, R. F. and Butcher, E. C. Rapid G protein-regulated activation event involved in lymphocyte binding to high endothelial venules. *J. Exp. Med.* **178**:367–372, 1993.
5. Murphy, P. M. The molecular biology of leukocyte chemoattractant receptors. *Annu. Rev. Immunol.* **12**:593–633, 1994.
6. Springer, T. A. Traffic signals on endothelium for lymphocyte recirculation and leukocyte emigration. *Annu. Rev. Physiol.* **57**:827–872, 1995.
7. Rothenberg, M. E., Luster, A. D. and Leder, P. Murine eotaxin: an eosinophil chemoattractant inducible in endothelial cells and in interleukin 4-induced tumor suppression. *Proc. Natl. Acad. Sci. USA* **92**:8690–8694, 1995.
8. Gonzalo, J-A., Jia, G.-Q., Aguirre, V., Friend, D., Coyle, A. J., Jenkins, N. A., Lin, G.-S., Katz, H., Lichtman, A., Copeland, N., Kopf, M. and Gutiérrez-Ramos, J.-C. Mouse eotaxin expression parallels eosinophil accumulation during lung allergic inflammation but it is not restricted to a Th2-type response. *Immunity* **4**:1–14, 1996.
9. Rot, A., Krieger, M., Brunner, T., Bischoff, S. C., Schall, T. J. and Dahinden, C. A. RANTES and macrophage inflammatory protein 1 alpha induce the migration and activation of normal human eosinophil granulocytes. *J. Exp. Med.* **176**:1489–1495, 1992.

10. Van Damme, J., Proost P., Lenaerts, J. P. and Opdenakker, G. Structural and functional identification of two human, tumor-derived monocyte chemotactic proteins (MCP-2 and MCP-3) belonging to the chemokine family. *J. Exp. Med.* **176**:59–65, 1992.
11. Jia, G.-Q., Gonzalo, J. A., Lloyd, C., Kremer, L., Lu, L., Martinez-A. C., Wershil, B. K. and Gutierrez-Ramos, J.-C. Distinct expression and function of the novel mouse chemokine monocyte chemotactic protein-5 in lung allergic inflammation. *J. Exp. Med.* **184**:1939–1951, 1996.
12. Uguccioni, M., Mackay, C. R., Ochensberger, B., Loetscher, P., Rhis, S., LaRosa, G. J., Rao, P., Ponath, P. D., Baggiolini, M. and Dahinden, C. A. High expression of the chemokine receptor CCR3 in human blood basophils. Role in activation by eotaxin, MCP-4, and other chemokines. *J. Clin. Invest.* **100**:1137–1143, 1997.
13. Kelner, G. S., Kennedy, J., Bacon, K. B., Kleyensteuber, S., Largaspada, D. A., Jenkins, N. A., Copeland, N. G., Bazan, J. F., Moore, K. W., Schall, T. J. and Zlotnik, A. Lymphotactin: a cytokine that represents a new class of chemokine. *Science* **266**:1395–1399, 1994.
14. Bazan, J. F., Bacon, K. B., Hardiman, G., Wang, W., Soo, K., Rossi, D., Greaves, D. R., Zlotnik, A. and Schall, T. J. A new class of membrane-bound chemokine with a CX_3C motif. *Nature* **385**:640–644, 1997.
15. Pan, Y., Lloyd, C., Zhou, H., Dolich, S., Deeds, J., Gonzalo, J. A., Vath, J., Gosselin, M., Ma, J., Dussault, B., Woolf, E., Alperin, G., Culpepper, J., Gutierrez-Ramos, J.-C. and Gearing, D. Neurotactin: a membrane-anchored chemokine upregulated in brain inflammation. *Nature* **387**:611–617, 1997.
16. Clark-Lewis, I., Schumacher, S., Baggiolini, M. and Moser, B. Structure–activity relationships of interleukin-8 determined using chemically synthesized analogs: critical role of NH2-terminal residues and evidence for uncoupling of neutrophil chemotaxis, exocytosis, and receptor binding activities. *J. Biol. Chem.* **226**:23128–23134, 1991.
17. Dahinden, C. A., Kurimoto, Y., De Weck, A. L., Lindsey, I., Dewald, B. and Baggolini, M. The neutrophil activating peptide NAF/NAP-1 induces histamine and leukotriene release by interleukin 3-primed basophils. *J. Exp. Med.* **170**:1787–1792, 1989.
18. White, M. V., Yoshimura, T., Hook, W., Kaliner, M. and Leonard, E. J. Neutrophil attractant/activation protein (NAP-1) causes human basophil histamine release. *Immunol. Lett.* **22**:151–154, 1989.
19. Maione, T. E., Gray, G. S., Petro, J., Hunt, A. J., Donner, A. L., Bauer, S. I., Carson, H. F. and Sharpe, R. J. Inhibition of angiogenesis by recombinant human platelet factor-4 and related peptides. *Science* **247**:77–79, 1990.
20. Nagasawa, T., Kikutani, H. and Kishimoto, T. Molecular cloning and structure of a pre-B-cell growth-stimulating factor. *Proc. Natl. Acad. Sci. USA* **91**:2305–2309, 1994.
21. Bleul, C. C., Farzan, M., Choe, H., Parolin, C., Clark-Lewis, I., Sodrowski, J. and Springer, T. A. The lymphocyte chemoattractant SDF-1α is a ligand for LESTR/fusin and blocks HIV-1 entry. *Nature* **382**:829–833, 1996.
22. Gibbs, B. F., Haas, H., Falcone, F. H., Albrecht, C., Volrath, I. B., Noll, T., Wolff, H. H. and Amon, U. Purified human peripheral blood basophils release interleukin-13 and pre-formed interleukin-4 following immunological activation. *Eur. J. Immunol.* **26**:2493–2498, 1996.
23. Ochensberger, B., Tassera, L., Bifare, D., Rihs, S. and Dahinden, C. A. Regulation of cytokine expression and leukotriene formation in human basophils by growth factors, chemokines and chemotactic agents. *Eur. J. Immunol.* **29**:11–22, 1999.
24. Bischoff, S. C., Kreiger, M., Brunner, T. and Dahinden, C. A. Monocyte chemoattractant protein-1 is a potent activator of human basophils. *J. Exp. Med.* **175**:1271–1275, 1992.
25. Füreder, W., Agis, H., Semper, H., Keil, F., Maier, U., Müller, M. R., Czerwenka, K., Höfler, H., Lechner, K. and Valent, P. Differential response of human basophils and mast cells to recombinant chemokines. *Ann. Hematol.* **70**:251–258, 1995.
26. Hartmann, K., Beiglbock, F., Czarnetzki, B. M. and Zuberbier, T. Effect of CC chemokines on mediator release from human skin mast cells and basophils. *Int. Arch. Allergy Immunol.* **108**:224–230, 1995.
27. Petersen, L. J., Brasso, K., Pryds, M. and Skov, P. S. Histamine release in intact human skin by monocyte chemoattractant factor-1, RANTES, macrophage inflammatory protein-1α, stem cell factor, anti-IgE and codeine as determined by an *ex vivo* skin microdialysis technique. *J. Allergy Clin. Immunol.* **98**: 790–796, 1996.

28. Romagnani, P., De Paulis, A., Beltrame, C., Annuziato, F., Dente, V., Maggi, E., Romagnani, S. and Marone G. Tryptase–chymase double-positive human mast cells express the eotaxin receptor CCR3 and are attracted by CCR3-binding chemokines. *Am. J. Pathol.* **155**:1195–1204, 1999.
29. Alam, R., Forsythe, P. A., Stafford, S., Lett-Brown, M. A. and Grant, J. A. Macrophage inflammatory protein-1α activates basophils and mast cells. *J. Exp. Med.* **176**:781–786, 1992.
30. Campbell, E. M., Charo, I. K., Kunkel, S. L., Strieter, R. M., Boring, L., Gosling, J. and Lukacs, N. W. Monocyte chemoattractant protein-1 mediates cockroach allergen-induced bronchial hyperreactivity in normal but not CCR2–/– mice: the role of mast cells. *J. Immunol.* **163**:2160–2167, 1999.
31. Taub, D., Dastych, J., Inamura, N., Upton, J., Kelvin, D., Metcalfe, D. D. and Oppenheim, J. Bone marrow-derived murine mast cells migrate, but do not degranulate, in response to chemokines. *J. Immunol.* **154**:2393–2402, 1995.
32. Conti, P., Reale, M., Barbacane, R. C., Felaco, M., Grilli, A. and Theoharides, T. C. Mast cell recruitment after subcutaneous injection of RANTES in the sole of the rat paw. *Br. J. Haematol.* **103**:798–803, 1998.
33. Nilsson, G., Mikovits, J. A., Metcalfe, D. D. and Taub, D. D. Mast cell migratory response to interleukin-8 is mediated through interaction with chemokine receptor CXCR2/interleukin-8RB. *Blood* **93**:2791–2797, 1999.
34. Lippert, U., Artuc, M., Grutzkau, A., Moller, A., Kenderessy-Szabo, A., Schadendorf, D., Norgauer, J., Hartmann, K., Schweitzer-Stenner, R., Zuberbier, T., Henz, B. M. and Kruger-Krasagakes, S. Expression and functional activity of the IL-8 receptor type CXCR1 and CXCR2 on human mast cells. *J. Immunol.* **161**:2600–2608, 1998.
35. Wang, H. W., Tedla, N., Lloyd, A. R., Wakefield, D. and McNeil, P. H. Mast cell activation and migration to lymph nodes during induction of an immune response in mice. *J. Clin. Invest.* **102**:1617–1626, 1998.
36. Lukacs, N. W., Hogaboam, C. M., Kunkel, S. L., Chensue, S. W., Burdick, M. D., Evanoff, H. L. and Strieter, R. M. Mast cells produce ENA-78, which can function as a potent neutrophil chemoattractant during allergic airway inflammation. *J. Leukoc. Biol.* **63**:746–751, 1998.
37. Rajakulasingam, K., Hamid, Q., O'Brien, F., Shotman, E., Jose, P. J., Williams, T. J., Jacobson, M., Barkans, J. and Durham, S. R. RANTES in human allergen-induced rhinitis: cellular source and relation to tissue eosinophilia. *Am. J. Respir. Crit. Care Med.* **155**:696–703, 1997.
38. Selvan, R. S., Butterfield, J. H. and Krangel, M. S. Expression of multiple chemokine genes by a human mast cell leukemia. *J. Biol. Chem.* **269**:13893–13898, 1994.
39. Yano, K., Yamaguchi, M., de Mora, F., Lantz, C. S., Butterfield, J. H., Costa, J. J. and Galli, S. J. Production of macrophage inflammatory protein-1alpha by human mast cells: increased anti-IgE-dependent secretion after IgE-dependent enhancement of mast cell IgE-binding ability. *Lab. Invest.* **77**:185–193, 1997.
40. Ramsaeng, V., Vliagoftis, H., Oh, C. K. and Metcalfe, D. D. Lymphotactin gene expression in mast cells following Fcε receptor I aggregation. *J. Immunol.* **158**:1353–1360, 1997.
41. Nakajima, T., Yamada, H., Iikura, M., Miyamasu, M., Izumi, S., Shida, H., Ohta, K., Imai, T., Yoshie, O., Mochizuki, M., Schroder, J. M., Morita, Y., Yamamoto, K. and Hirai, K. Intracellular localization and release of eotaxin from normal eosinophils. *FEBS Lett,* **434**:226–230, 1998.
42. Lu, L. and Wershil, B. K. Substance P induces eotaxin production by murine mast cells grown in stem cell factor. *FASEB J.* **12**:A881, 1998.
43. Cook, E. B., Stahl, J. L., Lilly, C. M., Haley, K. J., Sanchez, H., Luster, A. D., Graziano, F. M. and Rothenberg, M. E. Epithelial cells are a major source of the chemokine eotaxin in the guinea pig lung. *Allergy Asthma Proc.* **19**:15–22, 1998.
44. Jinquan, T., Quan, S., Feili, G., Larsen, C. G. and Thestrup-Pedersen, K. Eotaxin activates T cells to chemotaxis and adhesion only if induced to express CCR3 by IL-2 together with IL-4. *J. Immunol.* **162**:4285–4292, 1999.
45. Lamkhioused, B., Renzi, P. M., Abi-Younes, S., Garcia-Zepada, E. A., Allakhverdi, Z., Ghaffer, O., Rothenberg, M. D., Luster, A. D. and Hamid Q. Increased expression of eotaxin in bronchioalveolar lavage and airways of asthmatics contributes to the chemotaxis of eosinophils to the site of inflammation. *J. Immunol.* **159**:4593–4601, 1997.
46. Harris, R. R., Komater, V. A., Marett, R. A., Wilcox, D. M. and Bell, R. L. Effect of mast cell deficiency

and leukotriene inhibition on the influx of eosinophils induced by eotaxin. *J. Leukoc. Biol.* **62**:688–691, 1997.

47. Das, A. M., Flower, R. J. and Perritti, M. Resident mast cells are important for eotaxin-induced eosinophil accumulation *in vivo*. *J. Leukoc. Biol.* **64**:156–162, 1998.
48. Forssmann, U., Uguccioni, M., Loetscher, P., Dahinden, C. A., Langen, H., Thelen, M. and Baggiolini, M. Eotaxin-2, a novel CC chemokine that is selective for the chemokine receptor CCR3, and acts like eotaxin on human eosinophil and basophil leukocytes. *J. Exp. Med.* **185**:2171–2176, 1997.
49. Rodewald, H.-R., Dessing, M., Dvorak, A. M. and Galli, S. J. Identification of a committed precursor for the mast cell lineage. *Science* **217**:818–822, 1997.
50. Martin, C., Beaupain, D. and Dieterlen-Lievre, F. A study of the development of the hemopoietic system using quail-chick chimeras obtained by blastoderm recombination. *Dev. Biol.* **75**:303–314, 1980.
51. Dzierzak, E. and Medvinsky, A. Mouse embryonic hematopoiesis. *Trends Genet.* **11**:359–366, 1995.
52. Ikuta, K. and Weissman, I. L. Evidence that hematopoietic stem cells express mouse c-kit but do not depend on steel factor for their generation. *Proc. Natl. Acad. Sci. USA* **89**:1502–1506, 1992.
53. Dutt, P., Wang, J. F. and Groopman, J. E. Stromal cell-derived factor-1 alpha and stem cell factor/kit ligand share signaling pathways in hemopoietic progenitors: a potential mechanism for cooperative induction of chemotaxis. *J. Immunol.* **161**:3652–3658, 1998.
54. Yoder, M. C., Hiatt, K., Dutt, P., Mukherjee, P., Bodine, D. M. and Orlic, D. Characterization of definitive lymphohematopoietic stem cells in the day 9 murine yolk sac. *Immunity* **7**:335–344, 1997.
55. Sánchez, M.-J., Holmes, A., Miles, C. and Dzierzak, E. Characterization of the first definitive hematopoietic stem cells in the AGM and liver of the embryo. *Immunity* **5**:513–525, 1996.
56. Iemura A., Tsai, M., Ando, A., Wershil, B. K. and Galli, S. J. The c-kit ligand, stem cell factor, promotes mast cell survival by suppressing apoptosis. *Am. J. Pathol.* **144**:321–328, 1994.
57. Dastych, J. and Metcalfe, D. D. Stem cell factor induces mast cell adhesion to fibronectin. *J. Immunol.* **152**:213–219, 1994.
58. Levi-Schaffer, F. and Rubinchik, E. Activated mast cells are fibrogenic for 3T3 fibroblasts. *J. Invest. Dermatol.* **104**:999–1003, 1995.
59. Bartels, J., Schluter, C., Richter, E., Noso, N., Kulke, R., Christophers, E. and Schroder, J. M. Human dermal fibroblasts express eotaxin: molecular cloning, mRNA expression and identification of eotaxin sequence variants. *Biochem. Biophys. Res. Commun.* **225**:1045–1051, 1996.
60. Mochizuki, M., Bartels, J. and Mallet, A. I. IL-4 induces eotaxin: a possible mechanism of selective eosinophil recruitment in helminth infection and atopy. *J. Immunol.* **160**:60–68, 1998.
61. Nilsson, G., Miettinen, U., Ishizaka, T., Ashman, L. K., Irani, A. M. and Schwartz, L. B. Interleukin-4 inhibits the expression of kit and tryptase during stem cell factor-dependent development of human mast cells from fetal liver cells. *Blood* **84**:1519–1527, 1994.
62. Yamaguchi, M., Sayama, K., Yano, K., Lantz, C. S., Noben-Trauth, N., Ra, C., Costa, J. J. and Galli, S. J. IgE enhances Fc epsilon receptor I expression and IgE-dependent release of histamine and lipid mediators from human umbilical cord blood-derived mast cells: synergistic effect of IL-4 and IgE on human mast cell Fc epsilon receptor I expression and mediator release. *J. Immunol.* **162**:5455–5465, 1999.
63. Ogasawara, T., Murakami, M., Suzuki-Nishimura, T., Uchida, M. K. and Kudo, I. Mouse bone marrow-derived mast cells undergo exocytosis, prostanoid generation, and cytokine expression in response to G protein-activating polybasic compounds after coculture with fibroblasts in the presence of c-kit ligand. *J. Immunol.* **158**:393–404, 1997.
64. Sonoda, T., Hayashi, C. and Kitamura, Y. Presence of mast cell precursors in the yolk sac of mice. *Dev. Biol.* 97, 89–94, 1983.
65. Keller, G., Kennedy, M., Papayannopoulou, T. and Wiles, M. V. Hematopoietic commitment during embryonic stem cell differentiation in culture. *Mol. Cell. Biol.* **13**:473–486, 1993.
66. Ochi, H., Hirani, W. M., Yuan, Q., Friend, D. S., Austen, F. K. and Boyce, J. A. T helper cell type 2 cytokine-mediated comitogenic responses and CCR3 expression during differentiation of human mast cells *in vitro*. *J. Exp. Med.* **190**:267–280, 1999.
67. Delassus, S. and Cumano, A. Circulation of hematopoietic progenitors in the mouse embryo. *Immunity* **4**:97–106, 1996.

68. Dvorak, A. M., Furitsu, T. and Ishizaka, T. Ultrastructural morphology of human mast cell progenitors in sequential cocultures of cord blood and fibroblasts. *Int. Arch. Allergy Immunol.* **100**:219–229, 1993.
69. Omi, T., Kawanami, O., Honda, M. and Akamatsu, H. Human mast cells under development of the skin and airways. *Aerugi* **40**:1407–1414, 1991.
70. Mebius, R. E., Streeter, P. R., Michie, S., Butcher, E. C. and Weissman, I. L. A developmental switch in lymphocyte homing receptor and endothelial vascular addressin expression regulates lymphocyte homing and permits $CD4^{+}CD3^{-}$ cells to colonize lymph nodes. *Proc. Natl. Acad. Sci. USA* **93**:11109–11024, 1996.
71. Damsgaard, T. E., Nielsen, B. W., Henriques, U., Hansen, B., Herlin, T. and Schiotz, P. O. Histamine releasing cells of the newborn. Mast cells from the umbilical cord matrix and basophils from cord blood. *Pediatr. Allergy Immunol.* **7**:83–90, 1996.
72. Marone, G., Casolaro, V., Patella V., Florio G. and Triggiani, M. Molecular and cellular biology of mast cells and basophils. *Int. Arch. Allergy Immunol.* **114**:207–217, 1997.
73. Echtenacher, B., Mannel, D. N. and Hultner, L. Critical protective effect of mast cells in a model of acute septic peritonitis. *Nature* **381**:75–77, 1996.
74. Patella, V., Bouvet, J.-P. and Marone, G. Protein Fv produced during viral hepatitis is a novel activator of human basophils and mast cells. *J. Immunol.* **151**:5685–5698, 1993.
75. Peled, A., Gonzalo, J.-A., Lloyd, C. and Gutierrez-Ramos, J.-C. The chemotactic cytokine eotaxin acts as a granulocyte–macrophage colony-stimulating factor during lung development. *Blood* **91**:1909–1916, 1998.
76. Fleming, T. J., Fleming, M. L. and Malek, T. R. Selective expression of Ly-6G on myeloid lineage cells in mouse bone marrow. RB6-8C5 mAB to granulocyte differentiation antigen (Gr-1) detects members of the Ly-6 family. *J. Immunol.* **151**:2399–2408, 1993.
77. Cumano, A., Dieterlen-Lievre, F. and Godin, I. Lymphoid potential, probed before circulation in mouse, is restricted to caudal intraembryonic splanchnopleura. *Cell* **86**:907–916, 1996.
78. Hjertson, M., Sundstrom, C., Nilsson, K. and Nilsson, G. The potential of human mast cell progenitors to differentiate into mature mast cells remains after prolonged culture with flt3 ligand, interleukin-3 or granulocyte-macrophage colony stimulating factor. *Br. J. Haematol.* **104**:516–522, 1999.
79. Fehlner-Gardiner, C. C., Uniyal, S., von Ballestrem, C. G. and Chan, B. M. C. Differential utilization of VLA-4 ($\alpha4\beta1$) and -5 ($\alpha5\beta1$) integrins during the development of mouse bone marrow-derived mast cells. *Differentiation* **60**:317–325, 1996.
80. Adams, J. C. and Watt, F. M. Changes in keratinocyte adhesion during terminal differentiation: reduction in fibronectin binding preceded $\alpha5\beta1$ integrin loss from the cell surface. *Cell* **63**:425–435, 1990.
81. Olweus, J., Terstappen, L. W. M. M., Thompson, P. A. and Lund-Johansen F. Expression and function of receptors for stem cell factor and erythropoietin during lineage commitment of human hematopoietic progenitors cells. *Blood* **88**:1594–1607, 1996.
82. Smith, J. B. and Tabsh, K. M. Fetal neutrophils and eosinophils express normal levels of L-selectin. *Pediatr. Res.* **34**:253–257, 1993.
83. Rothenberg, M. E., MacLean, J. A., Pearlman, E., Luster, A. D. and Leder, P. Targeted disruption of the chemokine eotaxin partially reduces antigen-induced tissue eosinophilia. *J. Exp. Med.* **185**:785–790, 1997.
84. Yang, Y., Loy, J., Ryseck, R. P., Carrasco, D. and Bravo, R. Antigen-induced eosinophilic lung inflammation develops in mice deficient in chemokine eotaxin. *Blood* **92**:3912–3923, 1998.
85. Garcia-Zepeda, E. A., Combadiere, C., Rothenberg, M. E., Sarafi, M. N., Lavigne, F., Hamid, Q., Murphy, P. M. and Luster A. D. Human monocyte chemoattractant protein (MCP)-4 is a novel CC chemokine with activities on monocytes, eosinophils, and basophils induced in allergic and nonallergic inflammation that signals through the CC chemokine receptors (CCR)-2 and -3. *J. Immunol.* **157**:5613–5626, 1996.
86. Yamada, H., Hirai, K., Miyamasu, M., Iikura, M., Misaki, Y., Shoji, S., Takaishi, T., Kasahara, T., Morita, Y. and Ito, K. Eotaxin is a potent chemokine for human basophils. *Biochem. Biophys. Res. Commun.* **231**:365–368, 1997.
87. Krieger, M., Brunner, T., Bischoff, S. C., Tschainer, V., Walz, A., Moser, B., Baggiolini, M. and Dahinden, C. A. Activation of human basophils through the interleukin receptor. *J. Immunol.*

8:2662–2667, 1992.

88. Matthews, A. N., Friend, D. S., Zimmermann, N., Sarafi, M. N., Luster, A. D., Pearlman, E., Wert, S. E., Rothenberg, M. E. Eotaxin is required for the baseline level of tissue eosinophils. *Proc. Natl. Acad. Sci. USA*. **95**:6273–6278, 1998.
89. Gonzalo, J-A., Lloyd, C. M., Wen, D., Albar, J. P., Wells, T. N., Proudfoot, A., Martinez-A. C., Dorf, M., Bjerke, T., Coyle, A. J. and Gutierrez-Ramos, J.-C. The coordinated action of CC chemokines in the lung orchestrates allergic inflammation and airway hyperresponsiveness. *J. Exp. Med.* **188**:157–167, 1998.
90. Medvinsky, A. and Dzierzak, E. Definitive hematopoiesis is autonomously initiated by the AGM region. *Cell* **86**:897–906, 1996.
91. Hogaboam, C., Kunkel, S. L., Strieter, R. M., Taub, D. D., Lincoln, P., Standiford, T. J., Lukacs, N. W. Novel role of transmembrane SCF for mast cell activation and eotaxin production in mast cell-fibroblast interactions. *J. Immunol.* **160**:6166–6171, 1998.

CHAPTER 4

Regulation and Dysregulation of Mast Cell Survival and Apoptosis

KARIN HARTMANN[1] *and DEAN D. METCALFE**[2]

[1]*Department of Dermatology, University of Cologne, Germany and*
[2]*Laboratory of Allergic Diseases, National Institute of Allergy and Infectious Diseases, Bethesda, Maryland, U.S.A.*

INTRODUCTION

Mast cells are long-living cells. In contrast to other inflammatory cells such as eosinophils and neutrophils, mast cells live for several months (1). This prolonged survival of mast cells appears to depend upon intrinsic mast cell regulators as well as upon local factors in the tissues where mast cells reside. Other than stem cell factor (SCF), which appears to be the most important survival factor for murine and human mast cells (2, 3), additional requirements for the survival of mast cells are not well characterized. It is likely that a complex interplay between mast cells and surrounding cells, mediators and extracellular matrix proteins ensures the prolonged survival of tissue mast cells (4).

Development and survival of metazoans require a balance between cell proliferation and cell death. Physiologically, most cells are terminated by a genetically programmed, active form of cell death termed 'apoptosis' (5). Apoptosis is now known to be a highly regulated, complex biological event. It is closely linked to other biological processes, including proliferation and differentiation. Apoptotic cells are morphologically characterized by vacuolization of the cytoplasm, blebbing of the cell membrane, loss of microvilli, condensation of chromatin and fragmentation of the nucleus.

Extracellular mechanisms that induce apoptosis include deprivation of growth factors (6), stimulation of so-called 'death receptors' on the cell surface, perforin, nitric oxide, bacterial toxins, ultraviolet and γ-irradiation and chemical agents. Apoptosis induced by deprivation of growth factors may be prevented by the addition of other growth factors or by adhesion to extracellular matrix proteins (7, 8). Death receptors include CD95/Fas/APO-1, two TRAIL/APO-2 receptors (TRAIL-R1/DR4 and TRAIL-R2/DR5) and one receptor for tumour necrosis factor-α (TNFα-R1/p55/CD120a) (9).

---

* Corresponding author.

ISBN 0-12-473335-2

Induction of apoptosis by the different extracellular mechanisms is tightly regulated by intracellular proteins. Of these, the Bcl-2 family of proteins has been identified as among the most important intracellular regulators of apoptosis (10, 11). Members of the Bcl-2 family include apoptosis-inhibiting, anti-apoptotic molecules such as Bcl-2, Bcl-X_L and Mcl-1, as well as apoptosis-inducing, pro-apoptotic molecules such as Bax and Bad (12, 13). Bcl-2 proteins regulate different forms of apoptosis induction downstream. These include the apoptotic pathway induced by deprivation of growth factors or by stimulation of CD95. Bcl-2 proteins do not appear to participate in all forms of apoptotic signalling. Anti-apoptotic Bcl-2 proteins that are localized in the outer membrane of mitochondria inhibit the action of caspases, another group of intracellular proteins that mediate apoptosis (14). In contrast, the pro-apoptotic cytosolic Bax forms heterodimers with Bcl-2 or Bcl-X_L that form ion channels in the outer mitochondrial membrane (15). Through these ion channels, cytochrome c and other apoptosis-inducing mitochondrial molecules may be released.

Other well-described intracellular modulators of apoptosis include the tumour suppressor p53, the oncogen c-Myc and the nuclear factor NF-κB. The gene product of p53 induces apoptosis in transformed or virus-infected cells. Accordingly, in many tumours, inactivation of p53 by point mutations has been observed. c-Myc also acts in a pro-apoptotic manner. Overexpression of Myc enhances induction of apoptosis secondary to deprivation of growth factors. In contrast, NF-κB inhibits apoptosis induction by CD95 ligand, TRAIL or TNF-α.

Phagocytotic cells recognize apoptotic cells by altered expression of cell surface molecules such as thrombospondin and phosphatidyl serine, and phagocytose these cells which become small 'apoptotic bodies'. The membrane of these apoptotic bodies is mainly intact. This assures that, during apoptosis, in contrast to necrosis, there is little if any release of mediators into the microenvironment. As recently demonstrated, mast cells that contain large amounts of cytotoxic enzymes and cytokines also do not release inflammatory mediators during apoptosis (16).

Recent research has revealed that many diseases are associated with dysregulation of apoptosis (17). Diseases with a primary deficiency of apoptosis include autoimmune diseases, graft rejection, viral infections and tumours (18). In contrast, a primary excess of apoptosis has been observed in neurodegenerative disorders and diseases of the myocardium.

REGULATION OF MAST CELL SURVIVAL AND APOPTOSIS BY GROWTH FACTORS

Apoptosis of mast cells has been demonstrated *in vivo*. In one study, an increase of mast cells in mouse skin and spleen was first induced by subcutaneous injections of SCF (19). After cessation of SCF treatment, mast cell numbers rapidly declined. Remaining mast cells showed typical morphological findings of apoptosis, thus confirming *in vitro* data that SCF-dependent mast cells undergo apoptosis after deprivation of SCF. In another series of experiments, the mechanism by which corticosteroids decrease mast cell numbers was investigated (20). Corticosteroids were shown to decrease local SCF production by fibroblasts in association with enhanced apoptosis of mast cells. Since glucocorticoids did not affect mast cell survival *in vitro* directly, these data suggest that glucocorticoids indirectly induce mast cell apoptosis in tissues by downregulating SCF production in the microenvironment.

The regulation of murine mast cell survival by growth factors has been defined *in vitro*. Murine interleukin-3 (IL-3)-dependent bone marrow-cultured mast cells (BMCMC) undergo apoptosis after deprivation of IL-3 (6, 19, 21). The addition of SCF prevents apoptosis of BMCMC after IL-3 withdrawal when added until up to 1 h after growth factor deprivation (6). Similarly, SCF-dependent BMCMC underwent apoptosis after removal of SCF, and IL-3 was then capable of protecting the SCF-starved BMCMC from apoptosis (21). Since SCF failed to inhibit apoptotic changes in BMCMC cultured from c-*kit*-mutated W/Wv mice, it was concluded that SCF prevents apoptosis via binding to c-*kit* (6). In addition, SCF prevents apoptosis directly and not through the release of other cytokines, because dexamethasone and cyclosporin A failed to inhibit the apoptosis protection by SCF. Although transforming growth factor-β (TGF-β), which is known to induce apoptosis in many other cell types, does not induce apoptosis of BMCMC, it prevents the rescue effect of SCF after IL-3 deprivation (22). This effect of TGF-β is believed to be mediated by downregulation of c-*kit* expression (23). In immature BMCMC cultured for 10–15 days in IL-3, insulin-like growth factor I (IGF-I) has additionally been found to prevent apoptosis after IL-3 removal (24). Thus, growth factor-dependent survival of murine BMCMC, as currently understood, is regulated by SCF, IL-3, TGF-β and IGF-I.

A recent study demonstrated differences between SCF- and IL-3-mediated protection from apoptosis in mouse mast cells in relation to the regulation of these protective mechanisms by calcium influx (25). Inhibition of calcium influx by econazole or ketotifen abrogated the SCF-dependent survival of BMCMC but did not affect IL-3-dependent survival. These observations suggested that calcium influx is necessary only for SCF-mediated rescue from apoptosis in BMCMC. Accordingly, the murine growth factor-independent mastocytoma cell line P815, which expresses a mutated, constitutively active c-*kit*, underwent apoptosis in response to calcium influx inhibitors without SCF being present. Experiments with cycloheximide, a known inhibitor of protein synthesis, revealed that active protein synthesis is an additional prerequisite for SCF-mediated suppression of apoptosis in BMCMC (21). Moreover, herbimycin, a known inhibitor of tyrosine kinases, has been reported to inhibit SCF-dependent protection of BMCMC from death, indicating that tyrosine kinases also take part in SCF-mediated survival (6). An interesting recent study showed that SCF-mediated survival as well as proliferation of BMCMC additionally depends on phosphatidyl-3′-kinase (PI 3-kinase) and Src kinases (26). To induce SCF-dependent proliferation, but not survival, PI 3-kinase and Src kinase pathways in turn also activated the small G protein Rac1 and protein kinase JNK (c-jun N-terminal kinase), thus dissecting different signalling pathways for SCF-induced proliferation and suppression of apoptosis.

Moreover, IL-3-mediated survival of murine mast cells has been found to depend upon Bcl-2. In SCF-dependent BMCMC, IL-3 added after deprivation of SCF significantly induced Bcl-2 gene expression (21). In contrast, inhibition of apoptosis by SCF failed to induce Bcl-2 expression in BMCMC, but correlated with an enhanced expression of Bax RNA. In agreement with this observation, another study found that apoptosis of the IL-3-dependent murine mast cell line CFTL-15 in response to IL-3 deprivation was associated with a decrease of Bcl-2 mRNA (27). In addition, transfection of CFTL-15 cells with Bcl-2 cDNA prolonged survival of CFTL-15 cells after removal of IL-3. In comparison with Bcl-2 family proteins, p53 does not seem to play a major role in the regulation of SCF-dependent survival of BMCMC. This is indicated by experiments in which p53-deficient BMCMC from p53-mutated mice showed a similar induction of apoptosis in response to

SCF withdrawal compared with wild-type BMCMC with functional p53 (21). Thus, different signalling pathways modulated by calcium influx, tyrosine kinases, PI 3-kinase, Src kinases and Bcl-2 family proteins have been implicated in the regulation of IL-3-dependent versus SCF-dependent survival as well as SCF-mediated proliferation versus survival of murine mast cells.

Nerve growth factor (NGF) has also been shown to prevent apoptosis of rat peritoneal mast cells (28). Whereas SCF also induced proliferation of rat peritoneal mast cells, NGF had no effect on proliferation. This observation suggests that SCF and NGF promote mast cell survival by different mechanisms (29). The ability of NFG to rescue mast cells from apoptosis was dependent on the density of cells (28) as well as on binding to the $p140^{trk}$ NGF receptor. Tyrosine kinase inhibitors thus blocked phosphorylation of the $p140^{trk}$ tyrosine kinase (29). NGF has also been reported to cause an increase in mRNA expression for IL-3, IL-4, IL-10, TNF-α and granulocyte–macrophage colony-stimulating factor (GM-CSF) (30). Cyclosporin A inhibited both the induction of cytokines and NGF-dependent survival, but not survival dependent on SCF or IL-3. It is therefore likely that NGF prevents apoptosis of rat mast cells indirectly in an autocrine fashion via release of cytokines that promote viability. NGF, SCF and IL-3 all induced expression of Bcl-2 mRNA in rat mast cells. However, they had no effect on the mRNA expression of Bax, Bcl-X_L, Bcl-X_S, caspase-1 and caspase-3. These observations on the intracellular regulation of apoptosis in rat peritoneal mast cells suggest that SCF-dependent survival of mouse BMCMC and rat peritoneal mast cells might be regulated in a different manner.

In contrast to murine mast cells, less is known about the regulation of human mast cell viability. Primary human mast cells cultured from cord blood mononuclear cells in SCF and IL-6 underwent apoptosis over a period of 5 days after withdrawal of SCF and IL-6 (31). SCF alone was able to fully prevent apoptosis, whereas IL-6 alone only gradually inhibited the apoptotic process. IL-3, IL-4, IL-5 and interferon-γ (INF-γ) have also been reported to enhance survival of cord blood-derived human mast cells (31, 32). In contrast to previous reports examining human lung mast cells (33), cord blood-derived human mast cells were said to express IL-3 receptors (31). The survival-promoting effect of INF-γ was comparable at different cell densities, suggesting that this effect was induced directly by INF-γ without the involvement of autocrine factors (32). In contrast, IFN-γ has been reported to inhibit mast cell development in SCF-dependent human bone marrow cell cultures (33). Other cytokines tested, such as IL-2, IL-9, IL-10, IL-11, TNFα, TGFβ, NGF and M-CSF, failed to enhance the viability of cord blood-derived mast cells (31). Taken together, SCF, IL-3, IL-4, IL-5, IL-6, and possibly INF-γ promote survival of human mast cells depending on their derivation.

REGULATION OF MAST CELL SURVIVAL AND APOPTOSIS BY DEATH RECEPTORS

Surface expression of CD95/Fas has been described on murine BMCMC and peritoneal mast cells, on the murine mast cell lines C57 and MCP-5 (34), on the murine mastocytoma line P815 and on the human mast cell line HMC-1 (35). Stimulation of CD95 with anti-CD95 monoclonal antibody (mAb) or a Fas ligand-bearing T cell line demonstrated a different sensitivity toward CD95-induced apoptosis in the different murine mast cell types, suggesting a more complex regulation of this apoptotic pathway

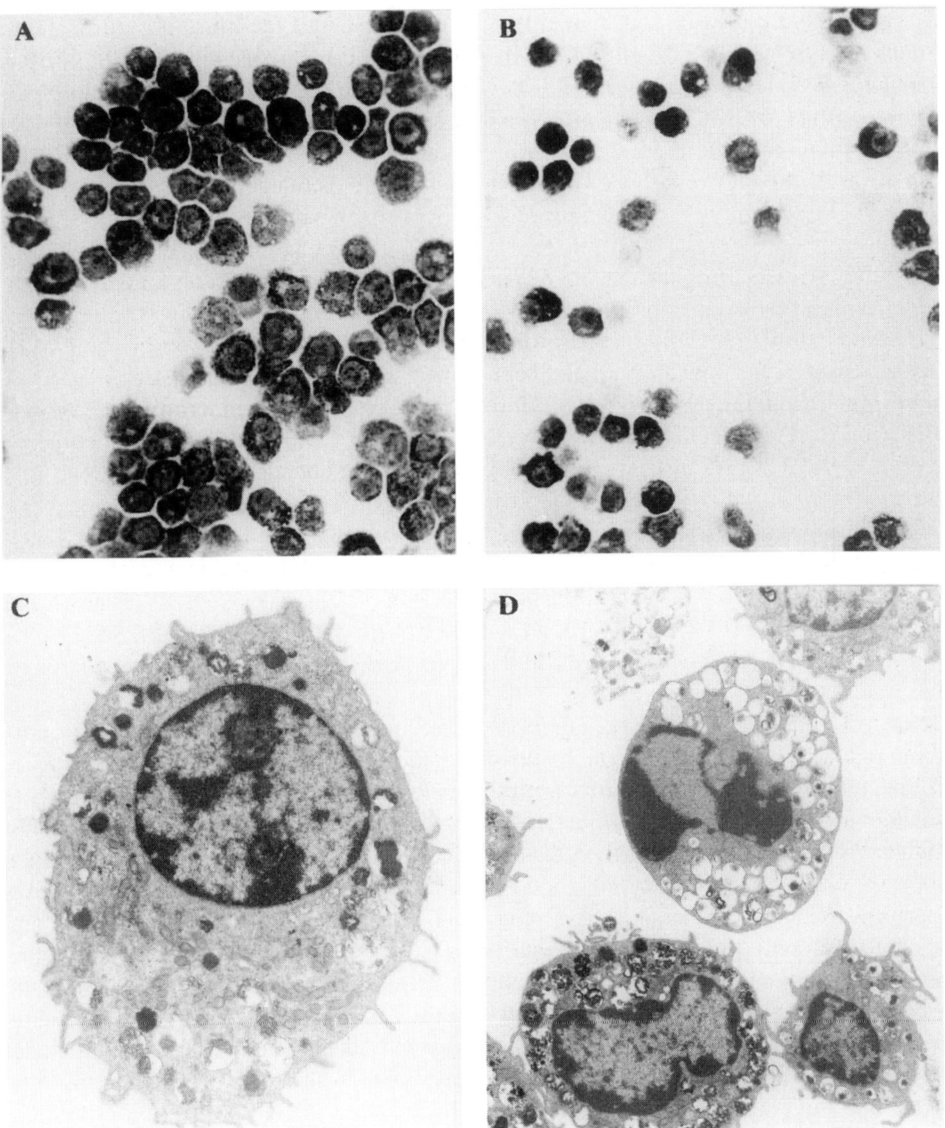

Fig. 1 Apoptosis of murine BMCMC after incubation with anti-Fas mAb and actinomycin D. (A) Giemsa staining of control BMCMB incubated with an isotype mAb. A normal morphology and dense cytoplasmatic granules are observed. (B) After treatment with anti-Fas mAb and actinomycin D for 48 h, apoptotic BMCMC are characterized by a decrease in size and an irregular shape. (C) Ultrastructure of control BMCMC incubated with an isotype mAb. A corrugated cell surface and regularly distributed nuclear chromatin are apparent. (D) After incubation with anti-Fas mAb and actinomycin D, the surface folds disappear. The cytoplasm is characterized by vacuolization, and the nuclear chromatin begins to condense.

that also appears to be independent of the amount of CD95 surface expression (34). C57 and P815 cells were highly susceptible to CD95 cross-linking and clearly underwent apoptosis after 24–48 h of stimulation with anti-CD95 mAb (34, 36). By contrast, BMCMC were at first resistant to CD95 stimulation. They underwent apoptosis only after the addition of the protein synthesis inhibitor actinomycin D (Fig. 1), thus suggesting a functional CD95 pathway which might be tightly regulated by inhibiting intracellular proteins (34). In fact, experiments in which P815 cells were transfected with Bcl-2 cDNA demonstrated that overexpression of Bcl-2 protein inhibited CD95-mediated apoptosis (36). In contrast to BMCMC, the virus-transfected cell line MCP-5 was also resistant to CD95 stimulation in the presence of actinomycin D, suggesting yet another regulatory mechanism of the CD95 pathway in transfected mast cell lines (34). In addition, SCF, TGF-β and $Fc_{\varepsilon}RI$ aggregation enhanced CD95 expression on BMCMC, and SCF as well as TGF-β was shown to partially protect BMCMC against CD95-mediated apoptosis. In P815 cells, growth arrest at the G_1 phase by thymidine or aphidicolin increased the susceptibility of these cells to CD95-mediated apoptosis (37).

A recent study found that murine BMCMC also express CD95/Fas ligand (38). However, expression of CD95 ligand was restricted to an intracellular expression, as demonstrated by polymerase chain reaction (PCR), western blotting and flow cytometry. There was no CD95 ligand expression on the cell surface. Further, BMCMC were unable to kill CD95-expressing Jurkat cells. Additional experiments are necessary to determine whether certain stimuli lead to a surface expression of CD95 ligand and whether mast cell numbers might also be controlled by autocrine CD95-mediated apoptosis.

REGULATION OF MAST CELL SURVIVAL AND APOPTOSIS BY PERFORIN, NITRIC OXIDE AND BACTERIAL TOXINS

Cytotoxic T cells kill their target cells via several independent pathways. In addition to killing induced by an interaction between death ligands, such as CD95 ligand on cytotoxic cells with CD95 receptors on target cells, cytotoxic cells also use a perforin-mediated pathway. In this apoptotic pathway, the pore-forming protein perforin acts together with the degranulation of proteases (i.e. granzymes), which finally leads to target cell death. The murine mastocytoma line P815 has been indirectly shown to be sensitive to perforin-mediated apoptosis (36). Cytotoxic T cells from CD95 ligand-defective *gld* mice, in which only the perforin pathway is believed to be functional, still reduced the viability of P815 cells. Bcl-2-transfected P815 cells were somewhat more resistant to perforin-mediated killing, thus suggesting common regulation of the CD95- and perforin-mediated apoptotic pathways regarding Bcl-2 family proteins. In contrast to CD95-mediated apoptosis, growth arrest at the G_1 phase of the cell cycle inhibited the susceptibility of P815 cells to perforin-induced lysis (37).

Another stimulus that is well known to cause apoptosis in different cell types is nitric oxide, which is synthesized by nitric oxide synthase in most cell types (39). Low concentrations of nitric oxide have homeostatic roles in the circulation and nervous system, whereas high concentrations are cytocidal and have immunomodulatory roles. One study reports that nitric oxide secreted from fibroblasts induces apoptosis of murine IL-3-dependent BMCMC when fibroblasts and BMCMC are co-cultured (40). This apoptosis was blocked by a specific nitric oxide synthase inhibitor. SCF was also able to inhibit the induction of apoptosis by nitric oxide in BMCMC.

In rat peritoneal mast cells, toxin A from *Clostridium difficile* has been reported to induce apoptosis at higher concentrations (41). After 4 h of treatment with toxin A, rat peritoneal mast cells exhibited typical features of apoptosis as assessed by electron microscopy and gel electrophoresis.

REGULATION OF MAST CELL SURVIVAL AND APOPTOSIS BY IRRADIATION AND CHEMICAL AGENTS

In murine BMCMC, γ-irradiation has been demonstrated to induce apoptosis *in vitro* (21). Irradiation with 2500 rads rapidly induced apoptosis of the majority of BMCMC within 12 h. SCF was able to protect BMCMC from irradition-induced apoptosis. SCF was still effective in suppressing apoptosis when added to the BMCMC cultures until up to 1 h after irradiation and before the cells became irreversibly committed to apoptosis. The ability of SCF to suppress apoptosis was independent of the cell cycle phase in which BMCMC were irradiated (21) and dependent on activation of both PI 3-kinase and Src kinases (26). Since SCF is known to protect mice from lethal γ-irradiation (42), it is possible that SCF exerts its radioprotective effect *in vivo* through inhibition of irradiation-induced apoptosis of stem cells.

It has been reported that the protein synthesis inhibitor cycloheximide prevents apoptosis of BMCMC induced by irradiation. This observation demonstrates that irradiation-induced apoptosis requires active protein synthesis (21). p53 has been identified as one of the proteins regulating irradiation-induced apoptosis. This is based on the observation that BMCMC from p53-mutated mice were far less sensitive to irradiation than were wild-type BMCMC.

We have recently shown that the protein synthesis inhibitor actinomycin D is effective in reducing survival of BMCMC and two murine mast cell lines (34). Concentrations of actinomycin D as low as 10 ng/ml clearly reduced the viability of BMCMC and the cell lines after 48 h. In line with this observation, another chemical agent, the protein kinase C inhibitor calphostin C, has recently been reported to induce apoptosis of the human mast cell line HMC-1 (43). Calphostin C-mediated apoptosis of HMC-1 cells was enhanced by adhesion to collagen type III and not affected by adhesion to fibronectin or laminin. This report thus demonstrates a direct influence of extracellular matrix proteins on mast cell apoptosis, as has been described for other cell types (8). Since murine BMCMC have been shown to proliferate in response to vitronectin (44), it is likely that extracellular matrix proteins may also contribute to the regulation of mast cell apoptosis.

As noted, corticosteroids failed to affect survival of murine BMCMC. This is in contrast to other cell types such as T cells that rapidly undergo apoptosis in response to corticosteroids (20). Similarly, the human mast cell line HMC-1 also appears resistant to corticosteroid-mediated apoptosis (unpublished data). In another study, the effects of cyclosporin A and FK-506 were examined on the growth of SCF-dependent human tissue mast cells (45). Unexpectedly, cyclosporin A and FK-506 not only failed to suppress, but appeared to enhance, the SCF-dependent growth of human tissue mast cells. Whether human mast cells are truly resistant to immunosuppressive drugs or whether SCF induces the resistance in these mast cells systems remains to be investigated.

PERSPECTIVE: APOPTOSIS AND MAST CELL DISEASE

Over the last several years, various diseases have been reported to be associated with a dysregulation of apoptosis (17, 18). We therefore hypothesize that dysregulation of mast cell apoptosis may also take part in the pathogenesis of mast cell disease. Thus, in several tumours and haematological diseases, mutations of different apoptosis genes as well as an altered expression of apoptosis-regulating proteins have been observed.

Tumours with mutations in the p53 gene or its regulators such as mdm2 are in general more aggressive than comparable tumours without these mutations (46). Melanoma cells have been found to escape the attack of immune effector cells by the expression of CD95 ligand. This leads to induction of apoptosis in CD95-sensitive attacking cells (47). A specific overexpression of the apoptosis-inhibiting Bcl-X_L promotes neoplastic proliferation in polycythaemia vera (48) and multiple myeloma (49).

In the majority of patients with mastocytosis, somatic mutations of c-*kit* have been identified that lead to an autonomous proliferation of mast cells (50, 51). These mutations increase aggregation of c-*kit*, and are thus referred to as activating mutations. The most common of these is Asp816Val. It could thus be speculated that this autonomous growth in mastocytosis is the result of both enhanced proliferation as well as inhibition of apoptosis.

Dysregulation of apoptosis may also play an important role in inflammatory processes (52). A recent study demonstrated that tissue eosinophilia in nasal polyps is associated with the expression of the eosinophilic growth factor IL-5 and delayed apoptosis of eosinophils (53). Inhibition of mast cell apoptosis may therefore also take part in the secondary increase of mast cells in inflammatory diseases. Thus, a better understanding of the regulation of mast cell apoptosis may also lead to new therapeutic approaches for patients with mast cell disorders.

REFERENCES

1. Fodinger, M., Fritsch, G., Winkler, K., Emminger, W., Mitterbauer, G., Gadner, H., Valent, P. and Mannhalter, C. Origin of human mast cells: development from transplanted hematopoietic stem cells after allogenic bone marrow transplantation. *Blood* **84**:2954–2959, 1994.
2. Flanagan, J. G. and Leder, P. The c-*kit* ligand: a cell surface molecule altered in steel mutant fibroblasts. *Cell* **63**:185–194, 1990.
3. Martin, F. H., Suggs, S. V. and Langley, K. E. Primary structure and functional expression of rat and human stem cell factor DNAs. *Cell* **63**:203–211, 1990.
4. Metcalfe, D. D., Baram, D. and Mekori, Y. A. Mast cells. *Physiol. Rev.* **77**:1033–1079, 1997.
5. Kerr, J. F., Wyllie, A. H. and Currie A. R. Apoptosis: a basic biological phenomenon with wide-ranging implications in tissue kinetics. *Br. J. Cancer* **26**:239–257, 1972.
6. Mekori, Y. A., Oh, C. K. and Metcalfe, D. D. IL-3-dependent murine mast cells undergo apoptosis on removal of IL-3. Prevention of apoptosis by c-kit ligand. *J. Immunol.* **151**:3775–3784, 1993.
7. Meredith, J. E., Fazeli, B. and Schwartz, M. A. The extracellular matrix as a cell survival factor. *Mol. Biol. Cell* **4**:953–961, 1993.
8. Anwar, A. R., Moqbel, R., Walsh, G. M., Kay, A. B. and Wardlaw, A. J. Adhesion to fibronectin prolongs eosinophil survival. *J. Exp. Med.* **177**, 839–843, 1993.
9. Ashkenazi, A. and Dixit, V. M. Death receptors: signaling and modulation. *Science* **281**:1305–1308, 1998.
10. Korsmeyer, S. J. Bcl-2: a repressor of lymphocyte death. *Immunol. Today* **13**:285–288, 1992.
11. Adams, J. M. and Cory, S. The Bcl-2 protein family: arbiters of cell survival. *Science* **281**:1322–1326, 1998.
12. Oltvai, Z. N., Milliman, C. L. and Korsmeyer, S. J. Bcl-2 heterodimerizes *in vivo* with a conserved homolog, Bax, that accelerates programmed cell death. *Cell* **74**:609–619, 1993.
13. Boise, L. H., Gonzalez-Garcia, M., Postema, C. E., Ding, L., Lindsten, T., Turka, L. A., Mao, X., Nunez, G. and Thompson, C. B. Bcl-x, a bcl-2-related gene that functions as a dominant regulator of apoptotic cell death. *Cell* **74**:597–608, 1993.
14. Thornberry, N. A. and Lazebnik, Y. Caspases: enemies within. *Science* **281**:1312–1316, 1998.
15. Yang, J., Liu, X., Bhalla, K., Kim, C. N., Ibrado, A. M., Cai, J., Peng, T. I., Jones, D. P. and Wang, X. Prevention of apoptosis by Bcl-2: release of cytochrome *c* from mitochondria blocked. *Science*

275:1129–1132, 1997.
16. Maurer, M. and Galli, S.J. Rapid reduction in the size of mouse cutaneous mast cell populations by apoptosis after cessation of treatment with SCF does not result in skin inflammation. *J. Invest. Dermatol.* **110**:634, 1998 (abstract).
17. Thompson, C. B. Apoptosis in the pathogenesis and treatment of disease. *Science* **267**:1456–1146, 1995.
18. Hetts, S. W. To die or not to die. An overview of apoptosis and its role in disease. *JAMA* **279**:300–307, 1998.
19. Iemura, A., Tsai, M., Ando, A., Wershil, B. K. and Galli, S. J. The c-kit ligand, stem cell factor, promotes mast cell survival by suppressing apoptosis. *Am. J. Pathol.* **144**:321–328, 1994.
20. Finotto, S., Mekori, Y. A. and Metcalfe, D. D. Glucocorticoids decrease tissue mast cell number by reducing the production of the c-kit ligand, stem cell factor, by resident cells: *in vitro* and *in vivo* evidence in murine systems. *J. Clin. Invest.* **99**:1721–1728, 1997.
21. Yee, N. S., Paek, I. and Besmer, P. Role of kit-ligand in proliferation and suppression of apoptosis in mast cells: basis for radiosensitivity of white spotting and steel mutant mice. *J. Exp. Med.* **179**:1777–1787, 1994.
22. Mekori, Y. A. and Metcalfe, D. D. Transforming growth factor-beta prevents stem cell factor-mediated rescue of mast cells from apoptosis after IL-3 deprivation. *J. Immunol.* **153**:2194–2203, 1994.
23. Dubois, C. M., Ruscetti, F. W., Stankova, J. and Keller, J. R. Transforming growth factor-β regulates c-*kit* message stability and cell-surface protein expression in hematopoietic progenitors. *Blood* **83**:3138–3145, 1994.
24. Rodriguez-Tarduchy, G., Collins, M. K. L., Garcia, I. and Lopez-Rivas, A. Insulin-like growth factor-I inhibits apoptosis in IL-3-dependent hemopoietic cells. *J. Immunol.* **149**:535–540, 1992.
25. Gommerman, J. L. and Berger S. A. Protection from apoptosis by steel factor but not interleukin-3 is reversed through blockade of calcium influx. *Blood* **91**:1891–1900, 1998.
26. Timokhina, I., Kissel, H., Stella, G. and Besmer, P. Kit signaling through PI 3-kinase and Src kinase pathways: an essential role for Rac1 and JNK activation in mast cell proliferation. *EMBO J.* **17**:6250–6262, 1998.
27. Mekori, Y. A., Oh, C. K., Dastych, J., Goff, J. P., Adachi, S., Bianchine, P. J., Worobec, A., Semere, T., Pierce, J. H. and Metcalfe, D.D. Characterization of a mast cell line that lacks the extracellular domain of membrane c-kit. *Immunology* **90**:518–525, 1997.
28. Horigome, K., Bullock, E. D. and Johnson, E. M. Effects of nerve growth factor on rat peritoneal mast cells. Survival promotion and immediate-early gene induction. *J. Biol. Chem.* **269**:2695–2702, 1994.
29. Kawamoto, K., Okada, T., Kannan, Y., Ushio, H., Matsumoto, M. and Matsuda, H. Nerve growth factor prevents apoptosis of rat peritoneal mast cells through the *trk* proto-oncogene receptor. *Blood* **86**:4638–4644, 1995.
30. Bullock, E. D. and Johnson, E. M. Nerve growth factor induces the expression of certain cytokine genes and bcl-2 in mast cells. Potential role in survival promotion. *J. Biol. Chem.* **271**:27500–27508, 1996.
31. Yanagida, M., Fukamachi, H., Ohgami, K., Kuwaki, T., Ishii, H., Uzumaki, H., Amano, K., Tokiwa, T., Mitsui, H., Saito, H., Iikura, Y., Ishizaka, T. and Nakahata, T. Effect of T-helper 2-type cytokines, interleukin-3 (IL-3), IL-4, IL-5, and IL-6 on the survival of cultured human mast cells. *Blood* **86**:3705–3714, 1995.
32. Yanagida, M., Fukamachi, H., Takei, M., Hagiwara, T., Uzumaki, H., Tokiwa, T., Saito, H., Iikura, Y. and Nakahata, T. Interferon-gamma promotes the survival and Fc epsilon RI-mediated histamine release in cultured human mast cells. *Immunology* **89**:547–552, 1996.
33. Kirshenbaum, A. S., Worobec, A. S., Davis, T. A., Ceoff, J. P., Semere, T. and Metcalfe, D. D. Inhibition of human mast cell growth and differentiation by interferon gamma-1β. *Exp. Hematol.* **26**:245–251, 1998.
34. Hartmann, K., Wagelie-Steffen, A. L., v. Stebut, E. and Metcalfe, D. D. Fas (CD95, APO-1) antigen expression and function in murine mast cells. *J. Immunol.* **159**:4006–4014, 1997.
35. Hartmann, K., Worobec, A. S., Bianchine, P. J., Mekori, Y. A. and Metcalfe, D. D. Expression of Fas antigen on a human and on murine mast cell lines: antibody to Fas antigen induces mast cell apoptosis. *J. Allergy Clin. Immunol.* **97**:261, 1996 (abstract).
36. Schroter, M., Lowin, B., Borner, C. and Tschopp, J. Regulation of Fas(Apo-1/CD95)- and perforin-mediated lytic pathways of primary cytotoxic T lymphocytes by the protooncogene *bcl*-2. *Eur. J. Immunol.* **25**:3509–3519, 1995.
37. De Leon, M., Jackson, K. M., Cavanaugh, J. R., Mbangkollo, D. and Verret, C. R. Arrest of the cell cycle reduces susceptibility of target cells to perforin-mediated lysis. *J. Cell Biochem.* **69**:425–435, 1998.
38. Wagelie-Steffen, A. L., Hartmann, K., Vliagoftis, H. and Metcalfe, D. D. Fas ligand (FasL, CD95L,

APO-1L) expression in murine mast cells. *Immunology* **94**:569–574, 1998.
39. Weller, R. Nitric oxide – a newly discovered chemical transmitter in human skin. *Br. J. Dermatol.* **137**:665–672, 1997.
40. Park, S., Jun, C., Choi, B., Lee, E., Kim, H., Cho, H. and Chung, H. Stem cell factor protects bone marrow-derived cultured mast cells (BMCMC) from cytocidal effect of nitric oxide secreted by fibroblasts in murine BMCMC–fibroblasts coculture. *Biochem. Mol. Biol. Int.* **40**:721–729, 1996.
41. Calderon, G. M., Torrez-Lopez, J., Lin, T., Chavez, B., Hernandez, M., Munoz, O., Befus, A. D. and Enciso, J. A. Effects of toxin A from *Clostridium difficile* on mast cell activation and survival. *Infect. Immun.* **66**:2755–2761, 1998.
42. Zsebo, K. M., Smith, K. A., Hartley, C. A., Greenblatt, M., Cooke, K., Rich, W. and McNiece, I. K. Radioprotection of mice by recombinant rat stem cell factor. *Proc. Natl. Acad. Sci. USA* **89**:9464–9468, 1992.
43. Brauer, G., Kruger-Krasagakes, S., v. d. Ohe M., Zhang, J., Henz, B. M. and Grutzkau, A. Extracellular matrix components modulate calphostin C-induced apoptosis in the human mast cell line-1. *J. Invest. Dermatol.* **110**:596, 1998 (abstract).
44. Bianchine, P. J., Burd, P. R. and Metcalfe, D. D. IL-3-dependent mast cells attach to plate-bound vitronectin. Demonstration of augmented proliferation in response to signals transduced via cell surface vitronectin receptors. *J. Immunol.* **149**:3665–3671, 1992.
45. Sperr, W. R., Agis, H., Czerwenka, K., Virgolini, I., Bankl, H. C., Muller, M. R., Zsebo, K., Lechner, K. and Valent, P. Effects of cyclosporin A and FK-506 on stem cell factor-induced histamine secretion and growth of human mast cells. *J. Allergy Clin. Immunol.* **98**:389–399, 1996.
46. Campbell, C., Quinn, A. G., Ro, Y. S., Angus, B. and Rees, J. L. P53 mutations are common and early events that precede tumor invasion in squamous cell neoplasia of the skin. *J. Invest. Dermatol.* **100**:746–748, 1993.
47. Hahne, M., Rimoldi, D., Schroter, M., Romero, P., Schreier, M., French, L. E., Schneider, P., Bornand, T., Fontana, A., Lienard, D., Cerottini, J. and Tschopp, J. Melanoma cell expression of Fas(Apo-1/CD95) ligand: implications for tumor immune escape. *Science* **274**:1363–1366, 1996.
48. Silva, M., Richard, C., Benito, A., Sanz, C., Olalla, I. and Fernandez-Luna, J. L. Expression of Bcl-X in erythroid precursors from patients with polycythemia vera. *N. Engl. J. Med.* **338**:564–571, 1998.
49. Tu, Y., Renner, S., Xu, F., Fleishman, A., Taylor, J., Weisz, J., Vescio, R., Rettig, M., Berenson, J., Krajewski, S., Reed, J. C. and Lichtenstein, A. Bcl-X expression in multiple myeloma: possible indicator of chemoresistance. *Cancer Res.* **58**:256–262, 1998.
50. Nagata, H., Worobec, A. S., Oh, C. K., Chowdhury, B. A., Tannenbaum, S., Suzuki, Y. and Metcalfe, D. D. Identification of a point mutation in the catalytic domain of the protooncogene c-kit in peripheral blood mononuclear cells of patients who have mastocytosis with an associated hematologic disorder. *Proc. Natl. Acad. Sci. USA* **92**:10560–10564, 1995.
51. Longley, B. J., Metcalfe, D. D., Tharp, M., Wang, X., Tyrrell, L., Lu, S., Heitjan, D. and Ma, Y. Activating and dominant inactivating c-*kit* catalytic domain mutations in distinct clinical forms of human mastocytosis. *Proc. Natl. Acad. Sci. USA*. **92**:1609–1614, 1999.
52. Simon, H. U. and Blaser, K. Inhibition of programmed eosinophil death: a key pathogenic event for eosinophilia? *Immunol. Today* **16**:53–55, 1995.
53. Simon, H. U., Yousefi, S., Schranz, C., Schapowal, A., Bachert, C. and Blaser, K. Direct demonstration of delayed eosinophil apoptosis as a mechanism causing tissue eosinophilia. *J. Immunol.* **158**:3902–3908, 1997.

SECTION TWO

PHENOTYPIC AND ULTRASTRUCTURAL FEATURES OF MAST CELLS AND BASOPHILS

CHAPTER 5

Ultrastructural Features of Human Basophil and Mast Cell Secretory Function

ANN M. DVORAK

Department of Pathology, Beth Israel Deaconess Medical Center, Boston, Massachusetts, U.S.A.

INTRODUCTION

Mast cells and basophils are critical effector cells in IgE-dependent immediate hypersensitivity responses (1–4). During such reactions, exposure to specific multivalent antigen results in the bridging of IgE molecules which are bound to $Fc_{\varepsilon}RI$ on the mast cell and/or basophil surface, initiating a series of biochemical and ultrastructural alterations, termed anaphylactic degranulation (AND), which culminate in fusion of cytoplasmic granule membranes with the plasma membrane (2, 3, 5–7). This in turn permits the release of the granule contents to the extracellular milieu. The rapid and massive release of mast cell and basophil mediators, which occurs during acute IgE-dependent reactions, leads to a number of pathophysiological consequences, prominent among which is markedly augmented vascular permeability. Persistent increased vascular permeability, which is associated with basophil and mast cell activation in many other disease processes, suggests that basophils or mast cells may be able to release mediators slowly over long periods of time, by a process distinct from AND, a process termed piecemeal degranulation (PMD), and that the extravasation of fluid and macromolecules at sites of augmented vascular permeability persists.

Using electron microscopy, we found that basophils which infiltrated certain delayed-type hypersensitivity responses in man and in experimental animals underwent a progressive loss of cytoplasmic granule contents but exhibited little or no evidence of AND (2, 8, 9). These basophils contained numerous small cytoplasmic vesicles, some of them fused to granule or plasma membranes. Similar changes were observed in tissue mast cells at sites of many disease processes or inflammatory responses, both in man and in experimental animals (2, 8, 9). These and other findings prompted us to propose a general model of basophil and mast cell degranulation that can account for varied rates of

MAST CELLS AND BASOPHILS
ISBN 0-12-473335-2

granule substance release occurring under a variety of physiological and pathological conditions (10). This model holds that loss of granule contents occurs under non-anaphylactic conditions by means of exocytotic vesicles which bud from the granule membrane, carrying with them small amounts of granule material. Upon fusion with the plasma membrane, these vesicles discharge their contents into the extracellular space. This vesicle-mediated release of granule contents is termed PMD. Because the size of cytoplasmic granules and basophils undergoing PMD appeared unchanged, we postulated that membrane lost by these cytoplasmic granules in the form of vesicles moving to the surface of the cell was replenished by a flow of vesicles from plasma membrane to cytoplasmic granules. The model proposed that the rate of vesicle flow from the plasma membrane to the granule membrane (and, in turn, from the granule to the plasma membrane) might be regulated by signals acting at the surface of basophils, including those which, at high concentrations, can also stimulate AND. According to this aspect of the model, when the rate of vesicle production induced by stimulation at the cell surface exceeds a critical level, vesicles become sufficiently numerous to promote fusion of the granule membranes with each other and then with the plasma membrane. Thus, PMD and AND might represent different points along a continuous spectrum of cellular secretory activity (11).

Earlier, we used electron-dense tracers to confirm that certain aspects of the PMD model, such as the movement of cytoplasmic vesicles from the plasma membrane to cytoplasmic granules, occurred (reviewed in ref. 2). More recently, we developed a new diamine oxidase–gold (DAOG) method (12) which localized a granule mediator, histamine, to basophil and mast cell granules and vesicles (13–23), to identify and quantify the vesicular transport of histamine in basophils and mast cells undergoing PMD (14, 15, 21, 23). DAOG (12) and another enzyme-affinity method, RNase–gold (RG) (based on the inhibition of ribonuclease by heparin), which images heparin (24), were used to assess subcellular sites of histamine and heparin in human mast cells undergoing AND (20, 25). Immunogold analyses of tumour necrosis factor-α (TNF-α), chymase, and the Charcot–Leyden crystal protein (CLC-P) in rat mast cells and human basophils, respectively, also were effective probes for determining secretory granule density and vesicular transport of these proteins in stimulated cells (26–31).

An analysis of the ultrastructural features of basophil and mast cell secretion includes the ultrastructural analysis of the basis for the development of enhanced vascular permeability stimulated by vasoactive mediators that are released from basophils and mast cells (32). These changes take place in the microcirculation. Recent ultrastructural studies have documented the role of a new endothelial cell organelle, termed the vesiculo-vacuolar organelle (VVO) (33, 34), in enhanced vascular permeability in allergic inflammation (32, 35).

The microcirculation functions to provide rapid exchange of nutrients and waste products between the blood and tissues. In normal tissues, this exchange occurs primarily across capillaries, which are the most abundant of the microvessels and which have the greatest surface area. Electron microscopy has been used to describe continuous, fenestrated and discontinuous capillaries (36, 37). While small molecules freely traverse most capillary endothelia, the passage of circulating macromolecules such as plasma proteins is severely restricted. Physiologists have postulated the existence of small and large endothelial cell pores to explain microvascular permeability (38). The anatomic basis for these pores has been much debated. It is likely that very small hydrophilic molecules (<3 nm) pass through endothelial cells, through intact inter-endothelial cell junctions, or, if lipid-soluble, exit vessels by diffusion in endothelial cell plasma membranes. Other small molecules cross the endothelial cell barrier by specific transport mechanisms. Large molecules such as plasma proteins exit most capillaries at low rates and reportedly (36, 38, 39) by two routes: (i) vesicular transport via the shuttling of 50–70-nm cytoplasmic vesicles present in endothelial cell cytoplasm (or by interconnected vesicles that form transendothelial cell channels), and (ii) inter-endothelial cell gaps. We recently identified a new endothelial cell cytoplasmic organelle, which we have termed the VVO, which provides the major route of extravasation of macromolecules at sites of augmented vascular permeability in venules associated with experimental tumours (33, 34). We also found evidence for prominent VVOs in association with extensive PMD of local mast cells in the inflamed eyelid lesions of transgenic mice which overexpress interleukin (IL)-4 (40). Using serial ultrathin sections and three-dimensional reconstructions, we found that VVOs spanned tall venular endothelium in normal vessels and that VVOs, often comprising 100 vesicular and vacuolar components, functioned as a permeability organelle in response to exposure to mast cell and basophil mediators (histamine, serotonin, VPF/VEGF) injected into their tissue vicinity (35). Morphometric (32, 35), morphologic (34) and immunocytochemical (41) studies suggest that these structures are formed by fusion of caveolae (39, 41–47) to form large structures which span the venular endothelium. Enzyme-affinity–gold studies labelled histamine within VVOs in the presence of PMD of adjacent mast cells (40), and immunoperoxidase electron microscopic labelling localized the tumour-derived cytokine vascular permeability factor/vascular endothelial growth factor (VPF/VEGF) (48–50) within VVOs in tumour vessels (51). VPF/VEGF (52) mRNA and protein have recently been identified in human mast cells by light-microscopic *in situ* hybridization and by biochemical assays of supernatants from stimulated cells (53, 54).

MECHANISM(S) OF HISTAMINE AND CLC-P SECRETION FROM HUMAN BASOPHILS, OF HISTAMINE SECRETION FROM HUMAN MAST CELLS *IN VITRO* AND OF TNF-α AND CHYMASE SECRETION FROM RAT PERITONEAL MAST CELLS

Electron microscopic morphometric studies were used to analyse the role of vesicular transport as a mechanism for effecting secretion by PMD and/or AND in human basophils stimulated with five different secretagogues [tetradecanoyl-phorbol acetate (TPA) (55), formyl-methionyl-leucyl-phenylalanine (FMLP) (11), recombinant histamine-releasing factor (rHRF) (56), monocyte chemotactic protein (MCP)-1, anti-IgE (56, 57)] and of isolated human mast cells stimulated with anti-IgE (58, 59). DAOG was used to image histamine (12–15, 17–21, 60–63) in human basophils and human mast cells and an immunogold method to image CLC-P in human basophils (15, 29–31, 61, 63).

Morphometric electron microscopic studies that imaged chymase (28) and TNF-α (26, 27) by immunogold in rat mast cells were also done. These probes allowed us to show that basophil and mast cell secretory granules and transport vesicles contained histamine (Figs 1–3) (15, 17, 18, 20, 21, 61, 63), CLC-P (Figs 2, 4) (15, 29–31, 61, 63), TNF-α (26, 27) and chymase (28). Because the probe label was a discrete gold particle of defined size, we could quantify the density of particles per unit area. These studies showed that, as the dense contents of granules decreased (as in PMD) in stimulated cells (in the

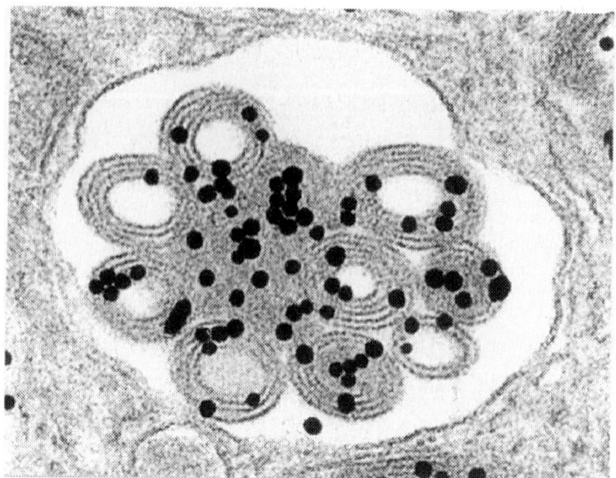

Fig. 1 Human mast cell granule prepared with DAOG shows gold-labelled histamine. ×83,000 (before enlargement). (Reproduced with permission from *J. Histochem. Cytochem.* **41**:787–800, 1993.)

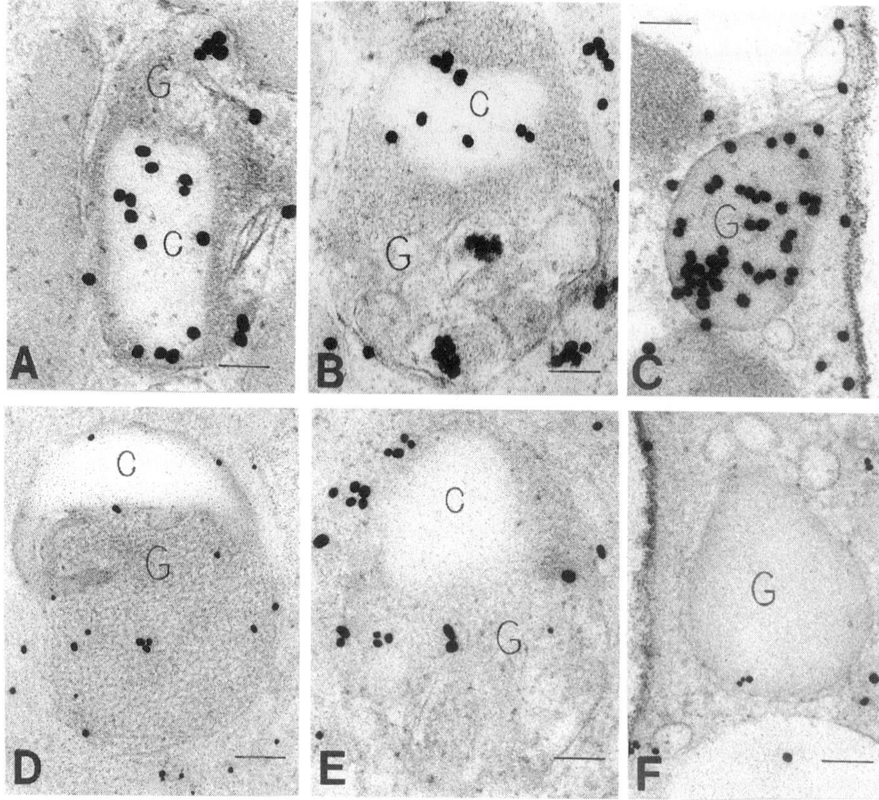

Fig. 2 CLC-P (A–C) and histamine (D–F) localization in human basophil granules. CLC-P resides in CLCs (C) and in the particulate matrix (A, B). Histamine resides in the particulate matrix only (D, E). Non-particulate primary granules are heavily labelled for CLC-P in (C) but not for histamine in (F). Bars: (A, E) 0.1 μm; (B–D, F) 0.2 μm. (Reproduced with permission from *Histol. Histopathol.* **11**:711–728, 1996.)

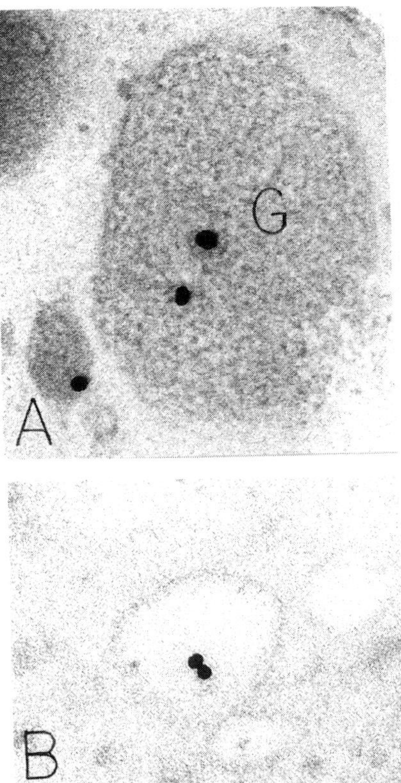

Fig. 3 TPA-stimulated human basophils prepared with DAOG after 2 min show histamine in a granule (G in A) and in electron-dense (A) and electron-lucent (B) vesicles. (A) ×117,300; (B) ×69,000 (before reduction). (Reproduced with permission from *Blood* **88**:4090–4101, 1996.)

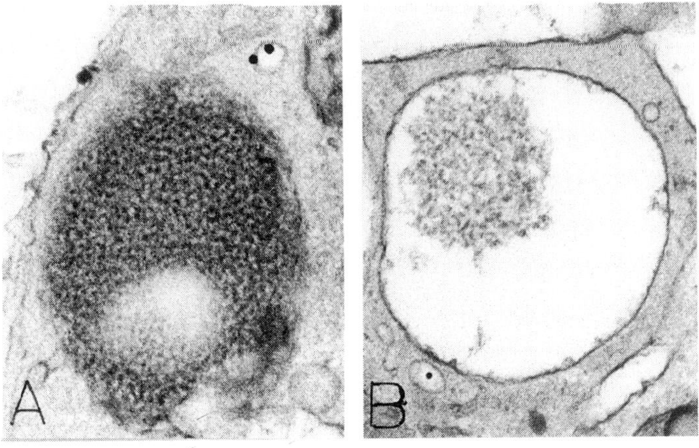

Fig. 4 FMLP-stimulated human basophil granules and vesicles prepared to detect CLC-P after 20 sec show label in electron-lucent vesicles adjacent to an unaltered granule (A) and to a partially depleted granule in basophils undergoing PMD (B). (A) ×46,000; (B) ×43,000 (before reduction). (Reproduced with permission from *Int. Arch. Allergy Immunol.* **113**:465–477, 1997.)

TABLE I
Density of DAOG Labelling in Human Basophil Granules Indicating Histamine During Secretion and Recovery

Phase	Dense granules	Altered granules	Background
Secretion	30.77* ± 16.1 (10)	4.08 ± 2.97 (10)	3.88† ± 0.9 (20)
Recovery	33.49‡ ± 4.21 (10)		5.25 ± 1.24 (10)

Density is expressed as number of gold particles per μm^2 ± SE. Numbers in parentheses indicate number of granules or Epon backgrounds counted.
* $p < 0.05$ (compared with altered granules).
† $p < 0.001$ (compared with dense granules).
‡ $p < 0.05$ (compared with altered granules or background).
Reproduced with permission from *Blood* **86**:3560–3566, 1995.

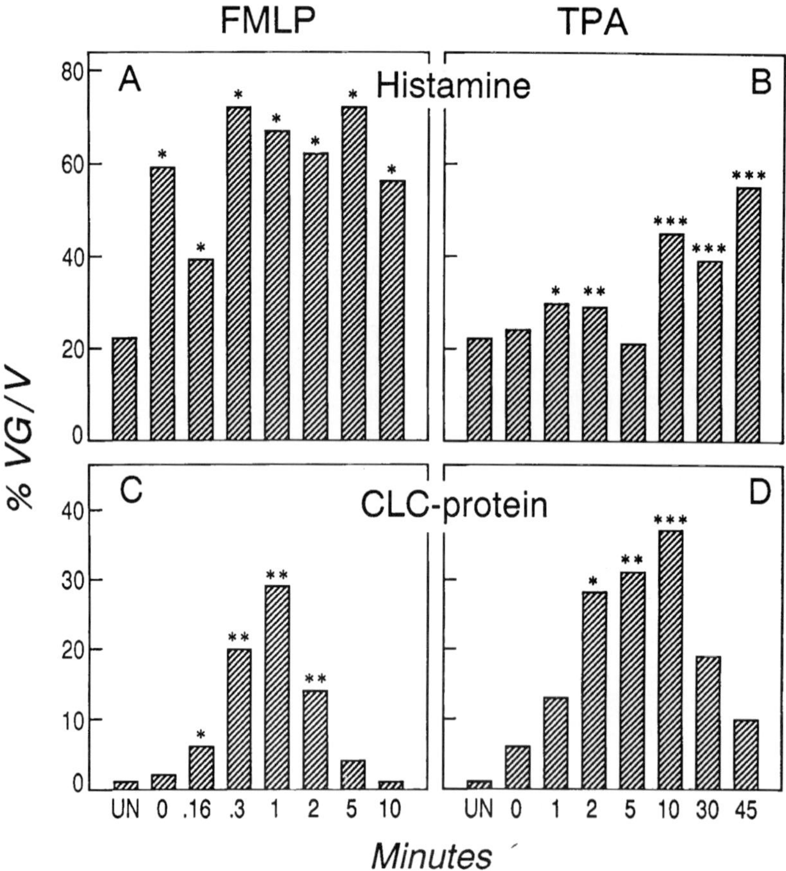

Fig. 5 %VG/V in human basophils either unstimulated (UN) or stimulated with FMLP (A, C) or TPA (B, D), recovered for electron microscopy at the indicated times and prepared to detect histamine (A, B) or CLC-P (C, D). Asterisks indicate significant values compared to UN. (Modified and reproduced with permission from *Blood* **88**:4090–4101, 1996; *Lab. Invest.* **74**:967–974, 1996; *Int. Arch. Allergy Immunol.* **113**:465–477, 1997.)

absence of exocytotic extrusion of granules), the emptying granule containers released their labelled content of histamine (18) (Table I), CLC-P (29–31), chymase (28) and TNF- α (27).

We quantified histamine and CLC-P discharge from human basophils in kinetic electron microscopic morphometric studies. This was done by calculating the proportion of gold-labelled cytoplasmic vesicles (VG) to the total vesicle (V) population (%VG/V) in unstimulated cells or cells stimulated with FMLP or TPA that were collected at appropriate time intervals to span the known kinetics of rapid (FMLP) or slow (TPA) histamine release (Fig. 5). Histamine secretion was biochemically verified in supernatants of stimulated cells (15, 21, 29, 31). Significantly larger numbers of total vesicles were gold-loaded in the FMLP-triggered cells than in the TPA-triggered cells ($p<0.001$ for 0 time and 1-, 2- and 5- min samples, and $p<0.01$ for the 10-min samples). Thus, a secretagogue that rapidly induced the anatomic continuum of PMD → AND → recovery (FMLP) also induced more histamine-loaded vesicles, compared to similar times after stimulation, than a secretagogue that slowly induced PMD, persisting to involve ~50% of the cells by 45 min and showing only minor amounts of AND and no morphological evidence of recovery (TPA).

These studies established for the first time that an important pro-inflammatory mediator, histamine, traffics from secretory granules to the extracellular milieu in small cytoplasmic vesicles in stimulated human basophils (15, 21). The association of this process with the electron microscopic release reaction defined as PMD (Table II) primarily produced by TPA and, in part, by FMLP, established vesicular transport as the

TABLE II
Human Basophil Piecemeal Degranulation

Condition/circumstance	Reference(s)
Contact allergy, skin biopsies	90–92
Delayed hypersensitivity to microbial antigens	90, 92
Basophil leukaemia, peripheral blood	93
Lymphoma	2, 94–97
Vernal conjunctivitis	98
Skin grafts	99, 100
Crohn's disease, ileum and peripheral blood	101–103
Bullous pemphigoid	104
Dialysis fluid	105
Adenocarcinoma of the lung	2, 95–97
Recombinant human IL-3-stimulated cord blood basophils	106
FMLP-stimulated peripheral blood basophils	11, 21, 30, 31
TPA-stimulated peripheral blood basophils	21, 29, 55
Ulcerative colitis, ileum	71
Sézary's syndrome	107

Reproduced with permission from Dvorak, A. M. Ultrastructural analysis of anaphylactic and piecemeal degranulation of human mast cells and basophils. In: *Immunopharmacology of Mast Cells and Basophils* (Foreman, J. C., ed.), pp. 89–113. Academic Press, London, 1993.

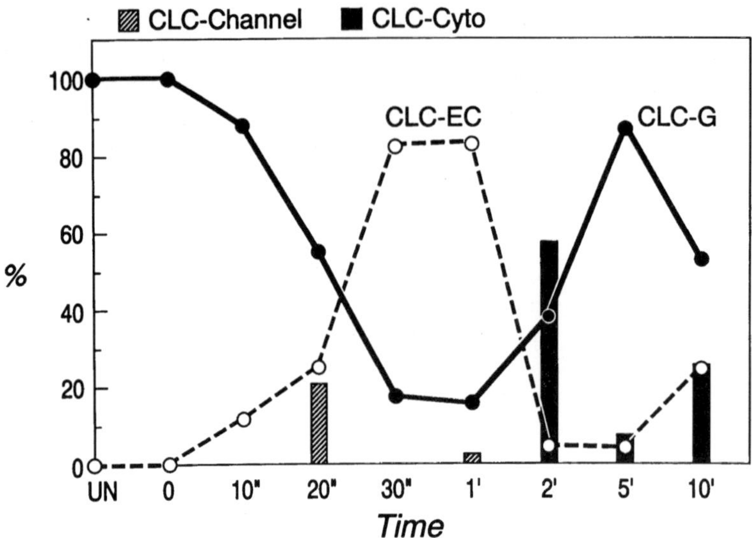

Fig. 6 %VG/V of CLC-P in formed CLCs at variable locations in human basophils stimulated with FMLP or unstimulated (UN), analysed by time. G, granule; EC, extracellular; Cyto, cytoplasmic; channel, degranulation channel. (Reproduced with permission from *Int. Arch. Allergy Immunol.* **113**:465–477, 1997.)

mechanism for effecting this type of regulated secretion. Vesicular transport of histamine was also significant in the more complex stimulated secretory and recovery model produced by exposure of human basophils to the bacterial peptide, FMLP.

Comparison of the %VG/V for histamine or CLC-P between two unique secretagogues for human basophils showed that FMLP (an extremely rapid secretagogue) induced simultaneous peaks of vesicles loaded with CLC-P or histamine – values that return to zero, in the case of CLC-P, at early recovery (10 min) (Fig. 5). This coincides with reconstitution of formed CLCs within granules at 10 minutes (30, 31) (Fig. 6). The %VG/V for histamine, however, did not return to baseline, but at 10 min was still elevated over that for unstimulated cells (Fig. 5). Thus, the behaviour of vesicle transport for these two cellular products differed considerably in FMLP-stimulated cells over a 10-min interval. The elevated %VG/V for histamine at 10 min most likely represents a combination of uptake and synthesis, since histamine release (measured biochemically) is essentially complete much earlier than 10 min post-stimulation with FMLP (64).

TPA (a slow secretagogue with morphological criteria indicating extensive and continuing PMD by 45 min, little evidence of AND, and none of recovery) achieved peak values of %VG/V for CLC-P and histamine between 10 and 15 min, also corresponding to peak histamine release times for this trigger (65, 66). As for FMLP, the %VG/V for CLC-P dropped extensively by 45 min, but the %VG/V for histamine remained high. Since there was no morphological evidence for recovery (and, therefore, synthesis) in TPA-stimulated basophils at 45 min (and little evidence of AND), the elevated vesicular traffic for histamine likely reflects ongoing PMD. These findings are consistent with electron microscopic evaluation of granule contents at 45 min after TPA. For example, granule number is not reduced, and nearly 50% of the granules are empty and represented only by their granule-sized containers; furthermore, neither particle-filled nor empty containers have residual formed CLCs. Thus, a granule source for histamine continued to exist at 45 min post-stimulation, but one for CLC-P was virtually absent.

Morphometric electron microscopic comparative studies to anti-IgE, a stimulus known to induce AND (15, 56, 57) were done using two new human basophil stimuli (Fig. 7, Table III). These studies identified important relationships between the percentage of granule–vesicle attachments (GVAs) and the incidence of PMD or AND in the stimulated samples. Thus, GVAs were found to occur soon after anti-IgE and to

TABLE III
Electron Microscopy of Human Basophils Stimulated with Anti-IgE, rHRF or MCP-1

Stimulant	Time	% GVA	% PMD	% AND
Control	30 sec	14	9	0
Control	7 min	9	18	0
Anti-IgE	2 min	40	40	13
Anti-IgE	7 min	2	12	67
rHRF	2 min	19	25	0
rHRF	7 min	21	9	16
MCP-1	5 sec	32	55	0
MCP-1	30 sec	47	28	33

Modified and reproduced with permission from *J. Allergy Clin. Immunol.* **98**:355–370, 1996.

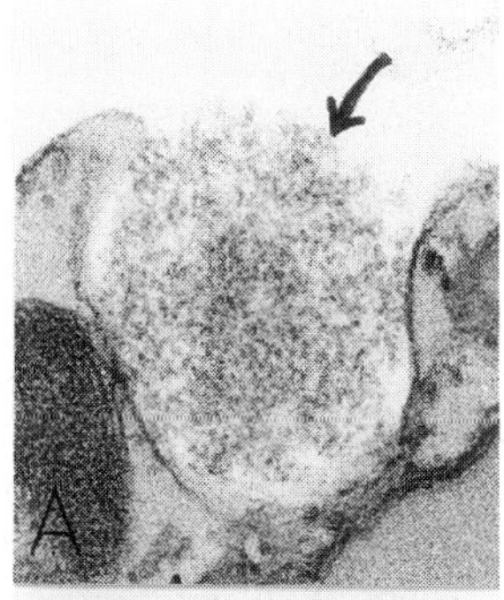

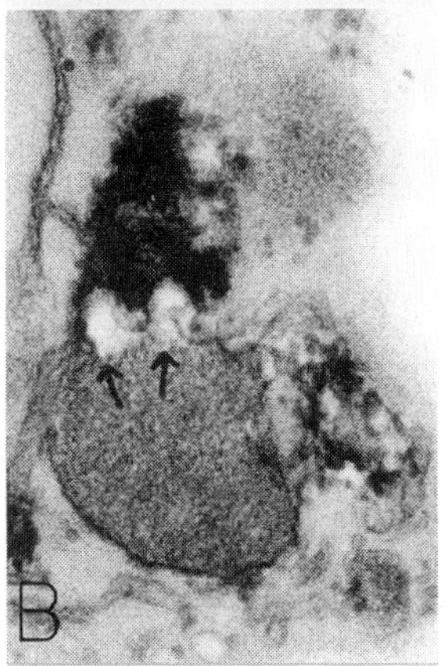

Fig. 7 Human basophils after stimulation for 2 min with anti-IgE (A) and 30 sec with MCP-1 (B) show a particle-filled granule devoid of its membrane, which has been extruded to the cell surface, in AND (curved arrow, A) and two glycogen-enveloped vesicles (arrows) attached to a granule in PMD (B). (A) ×50,000; (B) ×54,500 (before reduction). (Reproduced with permission from *J. Allergy Clin. Immunol.* **98**:355–370, 1996.)

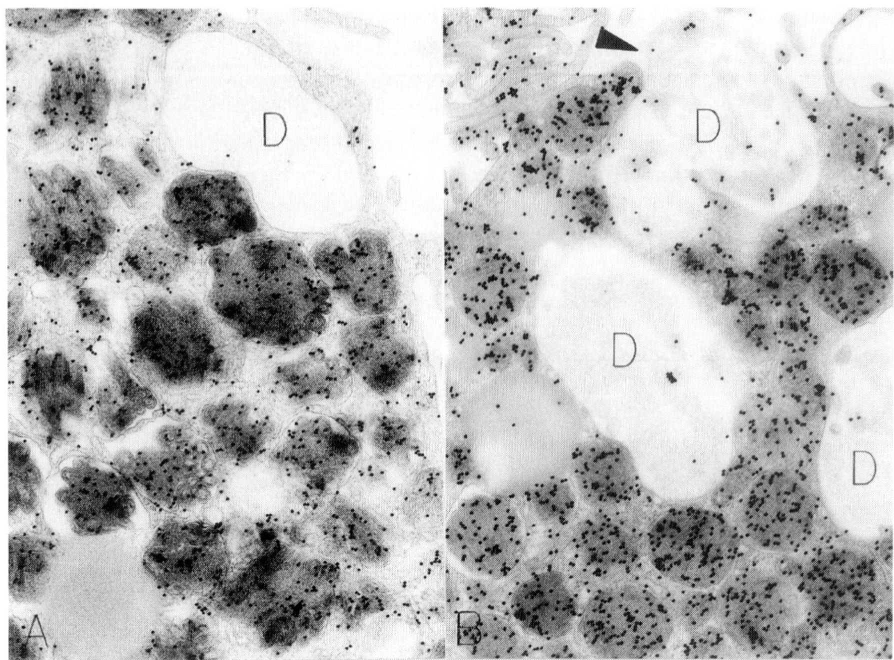

Fig. 8 Human lung mast cells at 5 min (A) and 20 min (B) post-stimulation with anti-IgE show degranulation channels (D) that either are electron-lucent and devoid of histamine (A) or contain a few residual wisps of electron-dense matrix with some DAOG label (B). The degranulation channel in (B) is open to the cell surface (arrowhead). The unaltered granules remaining in each cell are heavily labelled for histamine. (A, B) ×32,000 (before reduction). (Reproduced with permission from *J. Leukoc. Biol.* **59**:824–834, 1996.)

diminish as AND prevailed; also, a cytokine (rHRF) and a chemokine (MCP-1) each rapidly stimulated large numbers of GVAs, concomitant with both rapid histamine release and expression of PMD.

The ultrastructural trafficking of histamine in IgE-dependent AND of isolated human mast cells and recovery from secretion in short-term cultures was examined with DAOG to detect histamine (20) (Fig. 8). This study showed that electron-dense granules in unstimulated human mast cells, maintained up to 24 h in culture, and in responding anti-IgE-stimulated human mast cells, over the same time period *in vitro*, contained histamine. Alteration of granules, resulting in decreased electron density of their contents, was associated with decreased label for histamine. Electron-lucent intracytoplasmic degranulation chambers were devoid of histamine (Fig. 8). Recovering human mast cells developed new granule stores of histamine by a mixture of conservation, synthetic and endocytotic mechanisms.

HISTAMINE SECRETION FROM HUMAN MAST CELLS *IN VIVO*

PMD of human mast cells has been imaged *in vivo* by electron microscopy in a wide variety of circumstances (Table IV); AND of human mast cells *in vivo*, on the other hand, has not been so frequently captured in electron micrographs (67–69). We used the new

TABLE IV
Human Mast Cell Piecemeal Degranulation

Circumstance	Reference(s)
Urticaria pigmentosa	108
Chronic glomerulonephritis and pyelonephritis	109
Delayed hypersensitivity to microbial antigens	90–92
Contact allergy	90–92
Crohn's disease, ileum	101, 103, 110–114
Melanoma, skin	115
Bullous pemphigoid	104, 116
Whipple's disease, small intestine	117
Renal adenoma	118
Cold and mechanical stimulation, skin	119
Metastatic adenocarcinoma to lymph nodes	2, 95–97
Maxillary sinus mucosa, chronic inflammation	116
Ulcerative colitis, ileum	71
Familial polyposis coli, ileum	71
Interstitial fibrosis, lung	120
Primary and metastatic tumors	A. M. Dvorak, (unpublished data)

Reproduced with permission from Dvorak, A. M. Ultrastructural analysis of anaphylactic and piecemeal degranulation of human mast cells and basophils. In: *Immunopharmacology of Mast Cells and Basophils* (Foreman, J. C., ed.), pp. 89–113. Academic Press, London, 1993.

enzyme-affinity–gold method for histamine to document histamine stores and secretion *in vivo* by human mast cells displaying each of these secretory paths (22, 23).

Histamine Secretion from Human Mast Cells Undergoing PMD *In Vivo* in Inflammatory Bowel Disease (IBD)

Earlier, we determined that 60% of 117 biopsies, obtained from patients with IBD, exhibited evidence of human mast cell PMD; rare instances of AND also occurred (70, 71). We used DAOG to detect histamine to show that the partially empty to completely empty granule containers in PMD were also partially or completely devoid of histamine (23) (Fig. 9). This was the first demonstration of histamine secretion *in vivo* during PMD of human mast cells. Analogous studies done earlier were also the first to demonstrate histamine secretion *in vivo* during PMD of transgenic IL-4 mouse mast cells in animals with inflammatory eye disease (IED) (40).

Histamine Secretion from Human Mast Cells Undergoing Stem Cell Factor (SCF)-Induced AND *In Vivo* in Skin Biopsies

We participated in a multi-institutional study of patients with advanced breast cancer who were receiving rhSCF (67). We used DAOG to evaluate skin biopsy samples in which human mast cells were elicited and degranulated *in vivo* in these patients (22, 69). SCF has a role in the regulation of mast cell development and function (72–74) and is known

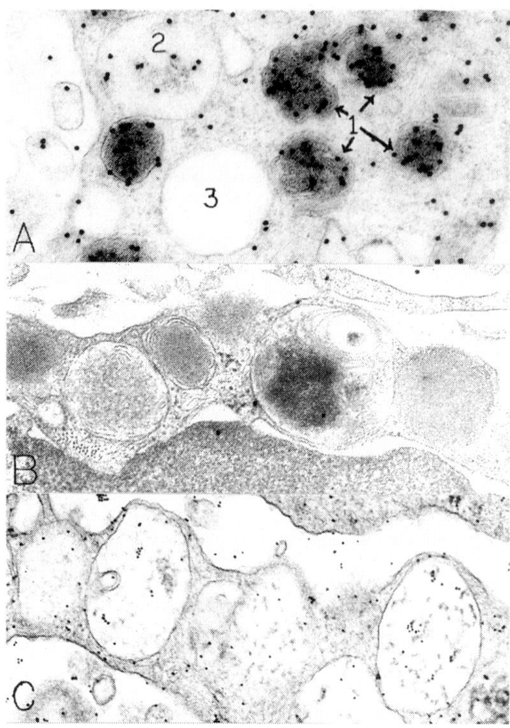

Fig. 9 (A) Secretory granules in human ileal mast cells undergoing PMD show gold label for histamine in electron-dense granules (1), gold label associated with residual electron-dense material in a partially empty granule (2), and no label in an entirely empty granule (3) in a cell undergoing PMD. (B) Absorption of DAOG reagent with histamine–agarose before staining abrogates all mast cell granule label. (C) Residual electron-dense reticular arrays contain small amounts of gold label for histamine in otherwise electron-lucent, empty, histamine-free granules of a mast cell undergoing PMD. (A) ×48,000; (B) ×39,500; (C) ×42,000 (before reduction). (Reproduced with permission from *J. Allergy Clin. Immunol.* **99**:812–820, 1997.)

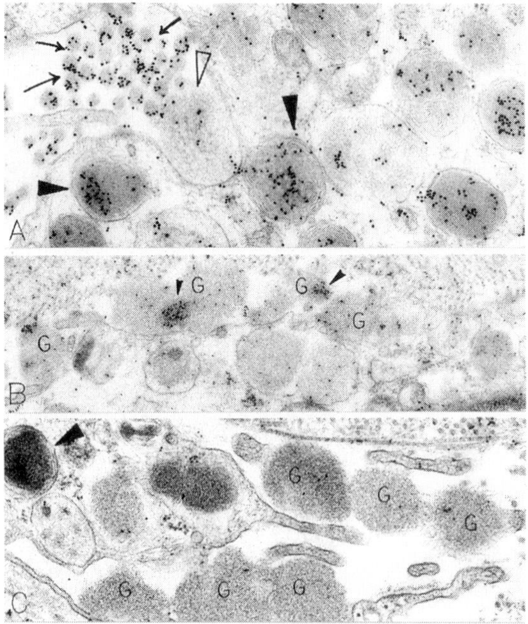

Fig. 10 Skin biopsies from sites that had been injected with rhSCF, obtained from patients who had received rhSCF for 13 days (A) or 1 day (B, C), respectively. The biopsy in (A) was obtained 1 h 45 min after the subcutaneous injection of 5 μg/kg of rhSCF, whereas the biopsy in (B, C) was obtained 1 h 40 min after the subcutaneous injection of 25 μg/kg of rhSCF. DAOG label is present on the electron-dense contents of cytoplasmic granules in the mast cell in (A) (solid arrowheads); the cell exhibits evidence of secretory activity, and an extruded, non-membrane-bound granule with altered matrix (which is located external to the plasma membrane) retains very little DAOG label for histamine (open arrowhead). However, DAOG heavily labels adjacent collagen fibres, which are seen in cross-section (arrows). In (B), multiple extruded, membrane-free mast cell granules (G), which exhibit altered matrix materials, are present in the interstitium near the degranulated mast cell; these extruded granules exhibit relatively little DAOG labelling for histamine, except in regions of the granules that contain the most electron-dense matrix material (solid arrowheads). In (C), the DAOG was absorbed with solid-phase histamine before being used to stain the specimen. Nearly all DAOG staining of intracellular mast cell granules (arrowhead), as well as of the extruded, membrane-free granules (G), was abolished by this specificity control. (A) ×52,500; (B) ×28,000; (C) ×46,500 (before reduction). (Reproduced with permission from *Blood* **90**:2893–2900, 1997.)

to promote mediator release from partially purified populations of human skin mast cells *in vitro* (75). In the new studies, we documented that rhSCF-elicited wheal and flare skin reactions contained mast cells undergoing regulated secretion by granule extrusion analogous to AND (67, 69). The DAOG method detected histamine in electron-dense granules of mast cells in control and injected skin biopsies; however, the altered matrix of membrane-free, extruded mast cell granules was largely non-reactive with DAOG (Fig. 10). These findings are the first morphological evidence of histamine secretion by classical granule exocytosis in human mast cells *in vivo* (22).

In summary, the new DAOG method for the detection of histamine in electron microscopic samples (12) allowed us to document cellular mechanisms of histamine secretion in human mast cells *in vivo* by PMD and AND for the first time.

HEPARIN SECRETION FROM HUMAN MAST CELLS *IN VITRO*

We evaluated an enzyme-affinity–gold electron microscopic method designed to identify RNA-rich structures, based on an RG probe, in human mast cells (24). As expected, the RG technique labelled RNA-containing ribosomes and nucleoli in human mast cells and in a wide variety of samples of human tissues (76). The heparin-rich secretory granules in human mast cells were also labelled (Fig. 11). Extensive studies revealed that human mast cells, isolated from lung or skin and sustained in short-term cultures, derived *de novo* in growth factor-supplemented cord blood cell cultures or present *in vivo* in multiple sites, all shared this property. We performed a large number of controls designed to examine the human mast cell granule-binding characteristics of gold alone, of irrelevant protein–gold or enzyme–gold reagents, of the role of charge and enzyme activity after various enzyme digestions, after blocking with macromolecules, after exposure to inhibitors of RNase, of heparin, or to irrelevant enzyme inhibitors, including staining of

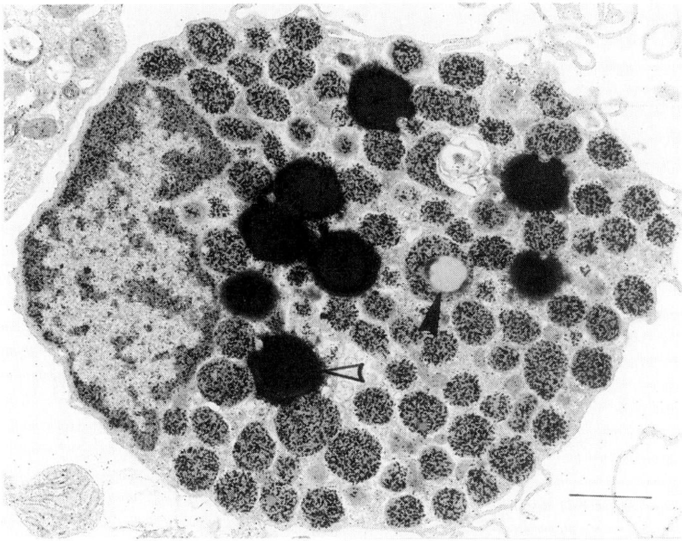

Fig. 11 Human lung mast cell *ex vivo*, stained with RG, shows labelled cytoplasmic granules and nuclei. Several large osmiophilic lipid bodies (open arrowhead) and an intragranular lipid deposit (closed arrowhead) are not labelled. Bar: 1 μm. (Reproduced with permission from *J. Histochem. Cytochem.* **46**:695–706, 1998.)

macromolecule-containing test agar blocks and a variety of combined absorption and digestion experiments of the binding of RG to human mast cell granules (24). These studies established that the RG method detected heparin in this site in conventionally prepared, well-preserved electron microscopic samples. These findings demonstrate a new use for the enzyme-affinity–gold technique in mast cell biology, based on the known property of heparin as an inhibitor of RNase (24). The RG reagent is red and is thus visible when passaged over beads to which it binds. Electron microscopic sections containing human mast cells were stained with the fluid reagent after passage over beads and the number of gold particles per granule area was calculated for each of six conditions (Fig. 12). The results show that RG did not bind to Sepharose and that it subsequently stained granules (Fig. 12A, 554 gold particles μm^2). In contrast, RG bound to heparin–agarose beads, turning them red, and the subsequent granule label was significantly reduced (Fig. 12B; 17 μm^2; $p<0.001$). A reduction in granule label that did not achieve statistical significance was seen when RG was blocked in solution with RNA and passaged over Sepharose (Fig. 12C; 55 μm^2); the beads remained unstained. When RG was blocked in solution with RNA prior to passage over heparin–agarose, the beads turned red, binding RG; granule gold label was significantly reduced (Fig. 12D; 25 μm^2; $p<0.001$). In contrast, RG, incubated in solution with heparin and then passaged over Sepharose, did not bind to the beads, leaving them white, and a significant reduction in granule gold label occurred (Fig. 12E; 18 μm^2; $p<0.001$). Finally, a double absorption of RG with heparin, first in solution, followed by heparin–agarose, resulted in red beads and abrogation of granule label (Fig. 12F; 3μm^2; $p<0.001$). Altogether, the absorption

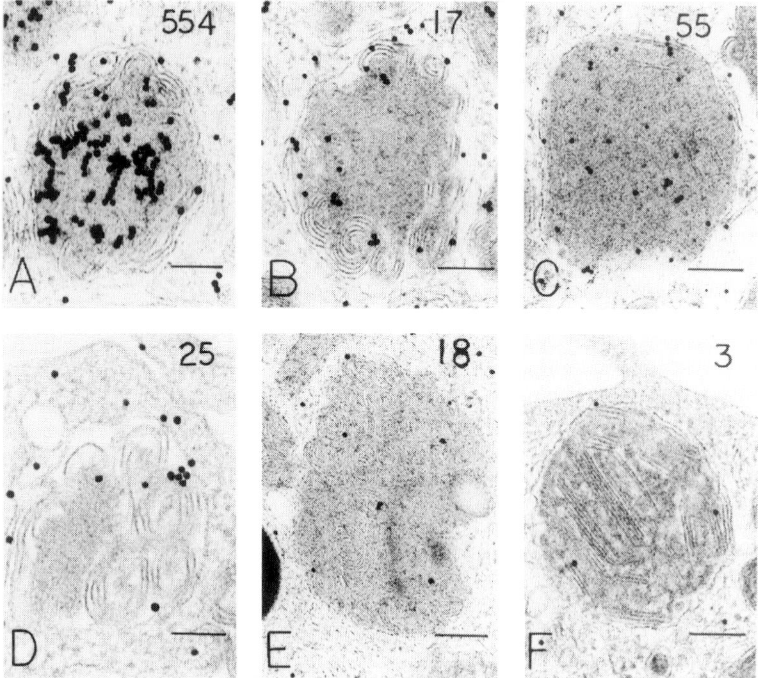

Fig. 12 Human mast cell granules labelled with RG after bead absorptions (see text). Bars: (A) 110 nm; (B, D) 130 nm; (C) 200 nm; (E) 150 nm; (F) 190 nm. (Reproduced with permission from *Clin. Exp. Allergy* **29**:1118–1128, 1999.)

experiments indicated that RG bound to human mast cell granule heparin, based on the known property of heparin as an inhibitor of the enzyme, RNase (24, 25).

We used the newly defined electron microscopic method for heparin (24) to follow the subcellular distribution of heparin in actively secreting and recovering human mast cells *in vitro* (25) (Figs 13, 14). We found that heparin was labelled by RG in electron-dense granules within non-secretory human mast cells, in electron-dense granules that persisted in secretory human mast cells at the time of maximum histamine secretion, and in electron-dense granules within recovering human mast cells. Heparin stores were absent in newly formed, electron-lucent, intracytoplasmic degranulation channels in secretory human mast cells. Electron-dense granule matrices in the process of extrusion to the cell exterior still retained heparin at the instant of cellular secretion. Non-granule heparin stores bound RG in recovering human mast cells. These locations included resolving degranulation channels (as newly emergent granules partitioned and condensed within them), and electron-dense content-containing vesicles and pro-granules within synthetic human mast cells. Ultimately, all electron microscopic patterns of human mast cell granules developed in recovering cells, and each of them contained heparin. We concluded that heparin was secreted from human mast cells which were stimulated by

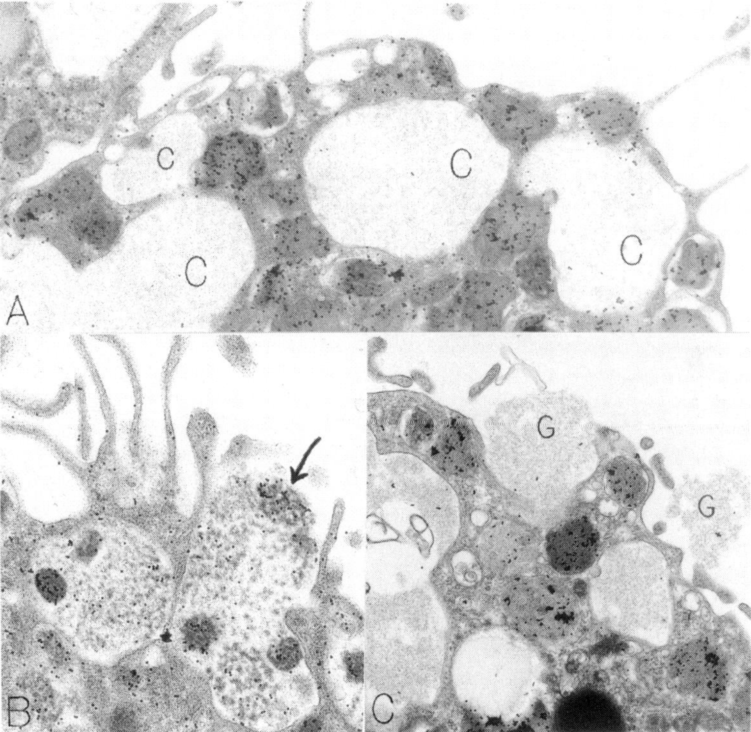

Fig. 13 Secretory mast cells, 20 min (A, B) or 6 h (C) after stimulation with anti-IgE, stained with RG, show large intracytoplasmic degranulation channels (C) devoid of electron-dense material and heparin in (A). Cytoplasmic granules that are not altered contain heparin (A, C). Extruded granules with variable electron density and heparin are present at 20 min (B) and 6 h (C) after stimulation. Note that RG label is bound to extruded, electron-dense, membrane-free granule materials at the cell surface (arrow) in the 20-min sample (B) but that two extruded granules (G) that persist in the extracellular location at 6 h are devoid of heparin. (A) ×35,000; (B) ×37,000; (C) ×24,000 (before reduction). (Reproduced with permission from *Clin. Exp. Allergy* **29**:1118–1128, 1999.)

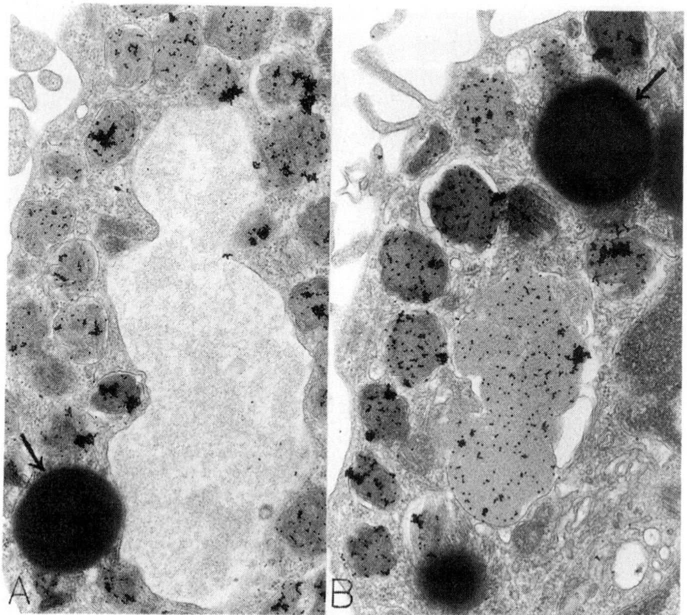

Fig. 14 Secretory mast cells, 6 h after anti-IgE stimulation and stained with RG, show degranulation channel condensation (121) as a recovery process. Unaltered granules in each cell contain heparin but lipid bodies (arrows) do not. The large degranulation channel with minor amounts of electron-dense material in (A) is devoid of heparin; the smaller, condensing degranulation channel with electron-dense granule domains in (B) contains heparin. (A) ×33,000; (B) ×36,000 (before reduction). (Reproduced with permission from *Clin. Exp. Allergy* **29**:1118–1128, 1999.)

anti-IgE, and heparin was recovered by a combination of conservative and synthetic mechanisms in human mast cells after a secretory event (25).

THE VESICULO-VACUOLAR ORGANELLE, A NEW ENDOTHELIAL PERMEABILITY STRUCTURE

VPF/VEGF was originally discovered in the late 1970s because of its capacity to increase the permeability of microvessels to plasma (77). A potent vascular permeabilizing protein in tumour culture supernatants (77) was purified and named VPF (52). Subsequently, endothelial cell mitogenic activity was found to be mediated by the same molecule, resulting in the designation of VPF/VEGF (78–81). The permeabilizing cytokine, VPF/VEGF, is among the most potent vascular permeabilizing agents known, having a potency some 50,000 times that of the mast cell- and basophil-derived mediator, histamine (49, 50, 82, 83). We now know that mast cells also synthesize and store VPF/VEGF (53, 54); thus, secretory stimuli might enhance vascular permeability by several mediators released from mast cells. We observed (33) that the predominant pathway for electron-dense tracers to leak into the subcutis from the blood vascular space in vessels in tumours secreting VPF/VEGF was by way of VVOs (Fig. 15) (34). VVOs also occurred in the venules of the normal subcutis of animals not bearing tumours, and these structures represented the predominant pathway by which tracers exited these normal vessels. However, both VVO labelling and tracer extravasation were much greater in tumour vessels than in control vessels.

Vessels associated with a mouse ovarian tumour had numbers of VVOs which were not significantly different from those of control vessels; the number of ferritin-containing vesicles and vacuoles/VVO, however, was more than twice as great in tumour vessels compared with control vessels ($p<0.001$), and the number of extravasated ferritin particles per μm of vessel perimeter was more than four times as great in tumour vessels compared with control vessels ($p<0.001$). Ferritin entered VVOs from the lumen within seconds after intravenous injection, passed from vesicle to vesicle via open stomata, then, upon reaching the abluminal surface, spilled into underlying basal lamina. Classical inter-endothelial cell gaps were not observed (Fig. 15), nor did ferritin pass between intact inter-endothelial cell junctions.

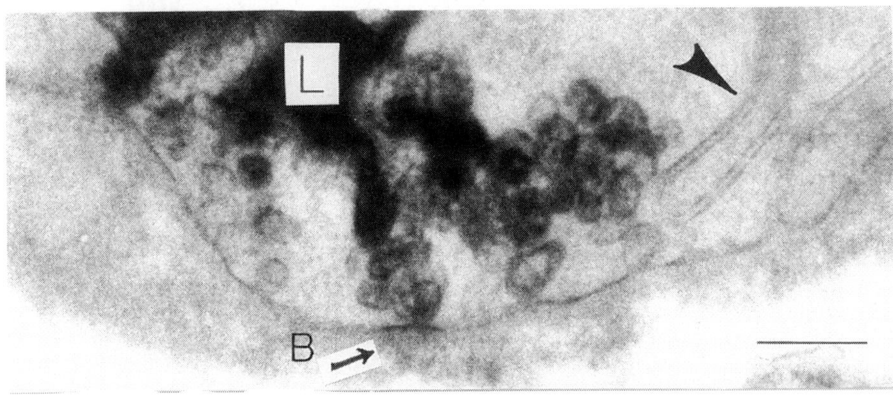

Fig. 15 Tumour vessel endothelial cell VVO at 10 sec after injecting horse radish peroxidase (HRP) intravenously. Continuity of the VVO with the vascular lumen (L) is made evident by HRP which also extends to the basal lamina (B) focally. The inter-endothelial cell junction does not contain HRP (arrowhead). Bar: 260 nm. (Reproduced with permission from *J. Leukoc. Biology.* **59**:100–115, 1996.)

Quantification of the individual vesicle and vacuoles that comprise VVOs in tumour compared with normal endothelial cells showed that the perimeters and areas of tumour vessel vesicles and vacuoles significantly exceeded that of normal endothelial cells (Fig. 16). Morphometry revealed that VVOs were enormous cytoplasmic structures (median area 0.12–0.14 μm^2 in a single electron micrograph). Specimen tilting provided evidence that individual VVO vesicles and vacuoles communicated with each other and with the plasma membrane of endothelial cells by stomata, some of which were closed by diaphragms composed of a single membrane. High magnification images with two tracers, ferritin (Fig. 17) and horseradish peroxidase (HRP) (Fig. 18), revealed that passage of macromolecules through VVOs occurred through stomata and was regulated at the level of stomatal diaphragms, thereby demonstrating a mechanism for controlling the passage of macromolecules across endothelial cells. Compared with tumour endothelial cells, little circulating ferritin and HRP entered the VVOs of normal endothelial cells because stomata joining vesicles and vacuoles to each other and to the lumen and ablumen were closed.

These findings indicate that VVOs provide a major pathway for the extravasation of circulating macromolecules across endothelial cells taller than capillary endothelial cells,

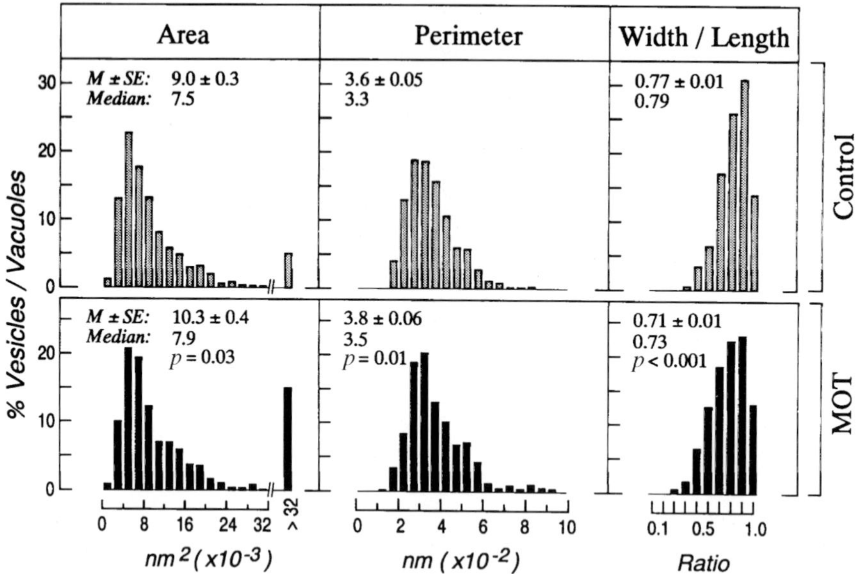

Fig. 16 A total of 521 vesicles and vacuoles from tumour endothelial cells and 487 from control endothelial cells were measured. The mean areas and perimeters of tumour endothelial cell vesicles–vacuoles were significantly larger than in normal endothelial cells.

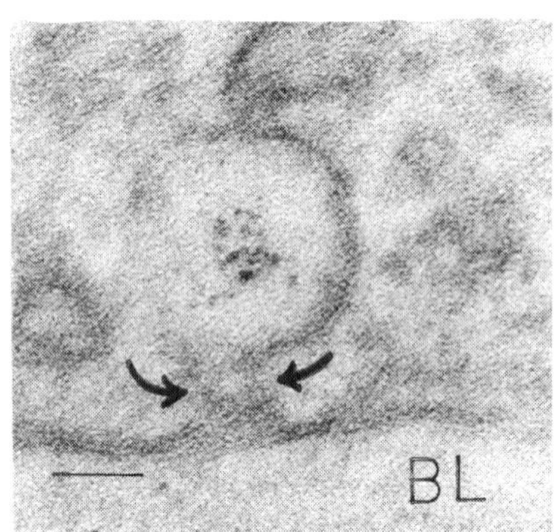

Fig. 17 Tumour vessel endothelial cell VVO vesicle shows a central stoma filled with ferritin at the basal lamina front (BL). The vesicle chamber contains little ferritin and is connected to the underlying plasma membrane by a narrow neck (arrows). Bar 63 nm. (Reproduced with permission from *J. Leukoc. Biology.* **59**:100–115, 1996.)

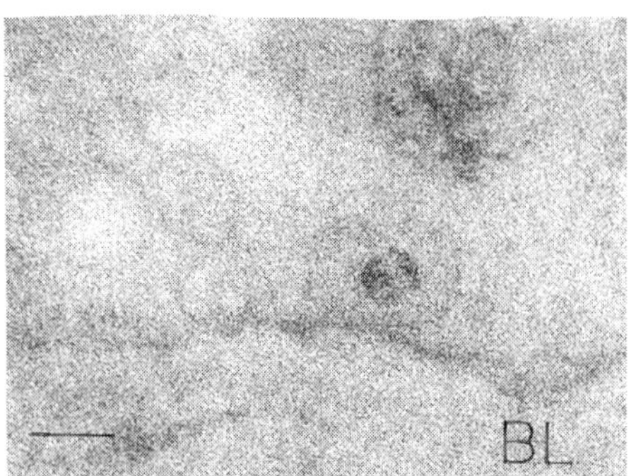

Fig. 18 Tumour endothelial cell VVO, connected to the abluminal surface, shows an open, HRP-containing stoma of a vesicle. The abluminal surface of the endothelial cell plasma membrane is focally stained with HRP beneath this vesicle; the basal lamina (BL) remains unstained. Bar: 80 nm. (Reproduced with permission from *J. Leukoc. Biology.* **59**:100–115, 1996.)

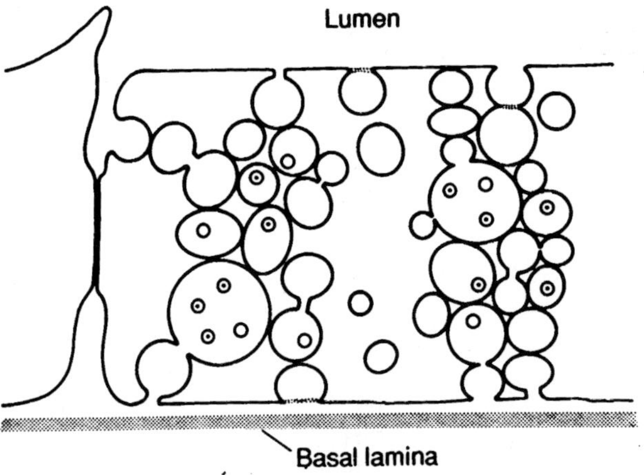

Fig. 19 Diagram showing two VVOs within an endothelial cell. The complex transcytoplasmic pathway linking the vascular lumen with the endothelial cell basal lamina is illustrated in both; one (left) opens to the inter-endothelial cell junction above its narrow mid portion. Small circular structures within vesicles and vacuoles are stomata. (Reproduced with permission from *J. Leukoc. Biology.* **59**:100–115, 1996.)

and suggest that upregulated VVO function accounts for the well-known hyperpermeability of tumour blood vessels (Fig. 19).

By morphological analysis (34), the smaller vesicles found in VVOs are similar in size and structure to the well-known plasmalemmal vesicles or caveolae that line plasma membranes in endothelial cells (84, 85). Given the obvious similarities between caveolae and VVO vesicles and vacuoles, we considered the possibility that VVOs might form from the linking together of individual caveolae. We also hypothesized that the larger vesicles and vacuoles of VVOs might form from the fusion of two or more caveola-sized vesicles. A similar fusion mechanism has been proposed in the generation of several types of cytoplasmic secretory granules in which small progranules of unit size fuse with each other in varying combinations to form larger mature granules whose volumes represent multiples of the volume of the unit pro-granule (58, 86–89). If this hypothesis is correct, then the volumes of the various VVO vesicles and vacuoles would not be expected to fall on a continuum but instead would represent multiples of the volume of smaller unit vesicles. We used a morphometric approach that has been successfully employed to demonstrate that the volumes of the cytoplasmic granules of both mast cells and pancreatic acinar cells have a periodic distribution (58, 86–89). The results make two points: (i) the size of VVO vesicles and vacuoles is heterogeneous; (ii) the volumes of individual VVO vesicles and vacuoles do not fall on a continuum but instead exhibit a modal distribution (32). The most frequent vesicle–vacuole had a unit volume of ~0.00015 μm^3; this value corresponds to a spheroid of a diameter of ~60nm, i.e. the size of typical capillary caveolae. Vacuoles corresponding to the fusion of as many as ten unit vesicles were detected. The data, therefore, are consistent with the hypothesis that larger VVO vesicles and vacuoles arise from the fusion of different numbers of caveola-sized unit vesicles. Additional evidence that VVOs are constructed from the fusion of caveolae is the presence of caveolin within VVOs (41).

VPF/VEGF is a cytokine secreted by many animal and human tumours, activated macrophages, keratinocytes, rheumatoid synovial cells, embryonic tissues and by

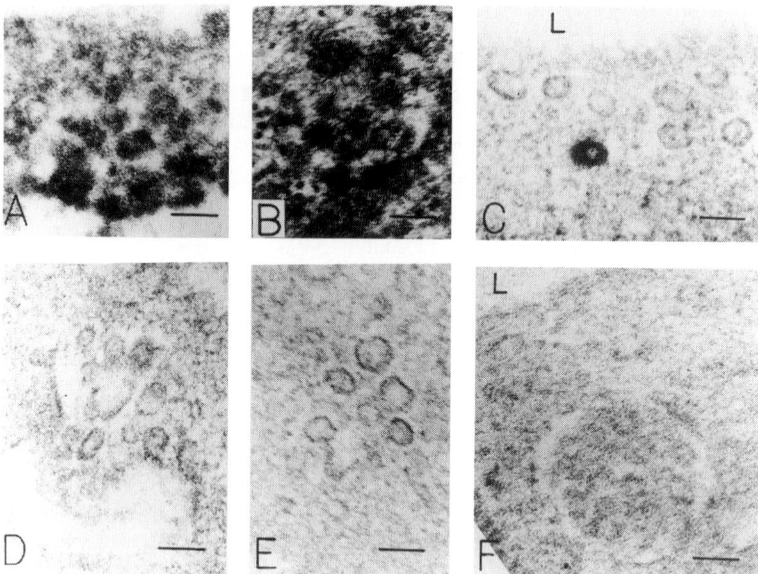

Fig. 20 Tumour endothelial cells, stained with antibody to VPF/VEGF (A–C, F), with an irrelevant antibody control (D), or with VPF/VEGF-absorbed specific antibody (E). Individual vesicle stomata within the VVO and the abluminal cell surface contain VPF in (A). (B) illustrates portions of a single VVO at mid cytoplasm. The stomata interconnecting individual vesicle components are positive for VPF; the surrounding chamber of the VVO is not. (C) One stoma of a VVO contains VPF/VEGF. The remainder of this VVO and the one adjacent to it near the luminal surface do not stain for VPF/VEGF. Note the central lucent spot within the VPF-positive stoma. This spot corresponds to a stomatal structure that has been termed a knob (84, 85), as described for caveolae. It does not stain for VPF/VEGF. Knobs are visible within the stomata of the adjacent VPF/VEGF-negative VVO. The two controls (D, E) show abluminal (D) and mid cytoplasmic (E) VVOs that are not positive. The large, round multivesicular body (F) does not contain VPF/VEGF. Bars: (A, B, E, F) 90 nm; (C) 104 nm; (D) 112 nm. (Reproduced with permission from *J. Histochem. Cytochem.* **43**:381–389, 1995.)

cultured epithelial and mesenchymal cell lines. It acts selectively on endothelial cells to increase their permeability to circulating macromolecules and to stimulate their replication. Although not detectably expressed by normal endothelial cells in the human and animal tumours we have studied, VPF/VEGF accumulates in the microvessels supplying tumours and certain inflammatory reactions in which VPF/VEGF is overexpressed. We used a pre-embedding immunocytochemical method to localize VPF/VEGF by electron microscopy in mouse tumour blood vessels. VPF/VEGF was observed on the abluminal plasma membrane and in VVOs of endothelial cells (Fig. 20).

Earlier, we determined that VVOs equivalent to those in hyperpermeable tumour vessels are present in normal venules that do not leak substantial amounts of plasma protein (33). To explain these findings, we hypothesized that VPF/VEGF increased the permeability of tumour blood vessels by increasing VVO function and that the VVOs of normal venules were relatively impermeable in the absence of exposure to VPF/VEGF. To test this hypothesis, VPF/VEGF was injected intradermally in normal animals after intravenous injection of a soluble macromolecular tracer, ferritin, whose extravasation could be followed by electron microscopy. VPF/VEGF caused normal venules to leak ferritin, and ferritin extravasated by way of VVOs, just as in hyperpermeable tumour endothelial cells (35). Like VPF/VEGF, histamine and serotonin also stimulated ferritin extravasation across venules by way of VVOs (35). Together, these data establish VVOs as the major pathway by which soluble plasma proteins exit venules in response to several mediators that increase venular hyperpermeability in allergic inflammation when mast cells undergo secretion.

ACKNOWLEDGEMENTS

Supported by NIH Grant AI-33372. I thank Peter K. Gardner for assistance in the preparation of the manuscript.

REFERENCES

1. Ishizaka, T. Mechanisms of IgE-mediated hypersensitivity. In: *Allergy. Principles and Practice*, Vol. I (Middleton, E. Jr., Reed, C. E., Ellis, E. F., Adkinson, N. F. Jr. and Yunginger, J. W., eds), pp. 71–93. C. V. Mosby, St. Louis, 1988.
2. Dvorak, A. M., ed. Basophil and Mast Cell Degranulation and Recovery. *Blood Cell Biochemistry*, Vol. 4 (Harris, J. R., ed.). Plenum Press, New York, 1991.
3. Galli, S. J. New concepts about the mast cell. *N. Engl. J. Med.* **328**:257–265, 1993.
4. Costa, J. J., Weller, P. F. and Galli, S. J. The cells of the allergic response. Mast cells, basophils and eosinophils. *JAMA* **278**:1815–1822, 1997.
5. Kinet, J.-P. The gamma-zeta dimers of Fc receptors as connectors to signal transduction. *Curr. Opin. Immunol.* **4**:43–48, 1992.
6. Beaven, M. A. and Metzger, H. Signal transduction by Fc receptor: the Fc epsilon RI case. *Immunol. Today* **14**:222–226, 1993.
7. Schwartz, L. and Huff, T. Biology of mast cells and basophils. In: *Allergy: Principles and Practice,* (Middleton, E. Jr., Reed, C. E., Ellis, E. F., Adkinson, N. F. Jr, Yunginger, J. W. and Busse, W. W., eds), pp. 135–168. C. V. Mosby, St. Louis, 1993.
8. Dvorak, A. M. Basophils and mast cells: piecemeal degranulation *in situ* and *ex vivo*: a possible mechanism for cytokine-induced function in disease. In: *Granulocyte Responses to Cytokines. Basic and Clinical Research* (Coffey, R. G., ed.), pp. 169–271. Marcel Dekker, New York, 1992.
9. Galli, S. J., Dvorak, A. M. and Dvorak, H. F. Basophils and mast cells: morphologic insights into their biology, secretory patterns, and function. In: *Mast Cell Activation and Mediator Release*, (Ishizaka, K., ed.), pp. 1–141. Karger, Basel, 1984.
10. Dvorak, H. F. and Dvorak, A. M. Basophilic leucocytes: structure, function and role in disease. *Clin. Haematol.* **4**:651–683, 1975.
11. Dvorak, A. M., Warner, J. A., Kissell, S., Lichtenstein, L. M. and MacGlashan, D. W. Jr. F-Met peptide-induced degranulation of human basophils. *Lab. Invest.* **64**:234–253, 1991.
12. Dvorak, A. M., Morgan, E. S., Schleimer, R. P. and Lichtenstein, L. M. Diamine oxidase–gold labels histamine in human mast-cell granules: a new enzyme-affinity ultrastructural method. *J. Histochem. Cytochem.* **41**:787–800, 1993.
13. Dvorak, A. M. Ultrastructural localization of histamine in human basophils and mast cells; changes associated with anaphylactic degranulation and recovery demonstrated with a diamine oxidase-gold probe. *Allergy* **52 (Suppl 34)**:14–24, 1997.
14. Dvorak, A. M. Histamine content and secretion in basophils and mast cells. In: *Progress in Histochemistry and Cytochemistry*, (Graumann, W., Bendayan, M., Bosman, F. T., Heitz, P. U., Larsson, L.-I., Ramaekers, F. C. and Schumacher, U., eds), pp. 169–350. Gustav Fischer Verlag, Jena, 1998.
15. Dvorak, A. M. Cell biology of the basophil. *Int. Rev. Cytol.* **180**:87–236, 1998.
16. Dvorak, A. M. and Morgan, E. S. Ultrastructural detection of histamine in human mast cells developing from cord blood cells cultured with human or murine recombinant c-kit ligands. *Int. Arch. Allergy Immunol.* **111**:238–244, 1996.
17. Dvorak, A. M., Morgan, E. S., Lichtenstein, L. M. and MacGlashan, D. W. Jr. Activated human basophils contain histamine in cytoplasmic vesicles. *Int. Arch. Allergy Immunol.* **105**:8–11, 1994.
18. Dvorak, A. M., MacGlashan, D. W. Jr., Morgan, E. S. and Lichtenstein, L. M. Histamine distribution in human basophil secretory granules undergoing FMLP-stimulated secretion and recovery. *Blood* **86**:3560–3566,1995.

19. Dvorak, A. M., Morgan, E. S., Monahan-Earley, R. A., Estrella, P., Schleimer, R. P., Weller, P. F., Tepper, R. I., Lichtenstein, L. M. and Galli, S. J. Analysis of mast cell activation using diamine oxidase–gold enzyme-affinity ultrastructural cytochemistry. *Int. Arch. Allergy Immunol.* **107**:87–89, 1995.
20. Dvorak, A. M., Morgan, E. S., Schleimer, R. P. and Lichtenstein, L. M. Diamine oxidase–gold ultrastructural localization of histamine in isolated human lung mast cells stimulated to undergo anaphylactic degranulation and recovery *in vitro*. *J. Leukoc. Biol.* **59**:824–834, 1996.
21. Dvorak, A. M., MacGlashan, D. W. Jr., Morgan, E. S. and Lichtenstein, L. M. Vesicular transport of histamine in stimulated human basophils. *Blood* **88**:4090–4101, 1996.
22. Dvorak, A. M., Costa, J. J., Morgan, E. S., Monahan-Earley, R. A. and Galli, S. J. Diamine oxidase–gold ultrastructural localization of histamine in human skin biopsies containing mast cells stimulated to degranulate *in vivo* by exposure to recombinant human stem cell factor. *Blood* **90**:2893–2900, 1997.
23. Dvorak, A. M. and Morgan, E. S. Diamine oxidase–gold enzyme-affinity ultrastructural demonstration that human gut mucosal mast cells secrete histamine by piecemeal degranulation *in vivo*. *J. Allergy Clin. Immunol.* **99**:812–820, 1997.
24. Dvorak, A. M. and Morgan, E. S. Ribonuclease–gold labels heparin in human mast cell granules: new use for an ultrastructural enzyme affinity technique. *J. Histochem. Cytochem.* **46**:695–706, 1998.
25. Dvorak, A. M. and Morgan, E. S. Ribonuclease–gold ultrastructural localization of heparin in isolated human lung mast cells stimulated to undergo anaphylactic degranulation and recovery *in vitro*. *Clin. Exp. Allergy* **29**:1118–1128, 1999.
26. Beil, W. J., Login, G. R., Galli, S. J. and Dvorak, A. M. Ultrastructural immunogold localization of tumor necrosis factor-α to the cytoplasmic granules of rat peritoneal mast cells with rapid microwave fixation. *J. Allergy Clin. Immunol.* **94**:531–536, 1994.
27. Beil, W. J., Login, G. R., Aoki, M., Lunardi, L. O., Morgan, E. S., Galli, S. J. and Dvorak, A. M. Tumor necrosis factor alpha immunoreactivity of rat peritoneal mast cell granules decreases during early secretion induced by 48/80: an ultrastructural immunogold morphometric analysis. *Int. Arch. Allergy Immunol.* **109**:383–389, 1996.
28. Login, G. R., Aoki, M., Yamakawa, M., Lunardi, L. O., Digenis, E. C., Tanda, N., Schwartz, L. B. and Dvorak, A. M. Immunocytochemical localization of chymase to cytoplasmic vesicles after rat peritoneal mast cell stimulation by compound 48/80. *J. Histochem. Cytochem.* **45**:1379–1391, 1997.
29. Dvorak, A. M., Ackerman, S. J., Letourneau, L., Morgan, E. S., Lichtenstein, L. M. and MacGlashan, D. W. Jr. Vesicular transport of Charcot–Leyden crystal protein in tumor-promoting phorbol diester-stimulated human basophils. *Lab. Invest.* **74**:967–974, 1996.
30. Dvorak, A. M., MacGlashan, D. W. Jr., Warner, J. A., Letourneau, L., Morgan, E. S., Lichtenstein, L. M. and Ackerman, S. J. Localization of Charcot–Leyden crystal protein in individual morphologic phenotypes of human basophils stimulated by f-Met peptide. *Clin. Exp. Allergy* **27**:452–474, 1997.
31. Dvorak, A. M., MacGlashan, D. W. Jr., Warner, J. A., Letourneau, L., Morgan, E. S., Lichtenstein, L. M. and Ackerman, S. J. Vesicular transport of Charcot–Leyden crystal protein in f-Met peptide-stimulated human basophils. *Int. Arch. Allergy Immunol.* **113**:465–477, 1997.
32. Feng, D., Nagy, J. A., Pyne, K., Hammel, I., Dvorak, H. F. and Dvorak, A. M. Pathways of macromolecular extravasation across microvascular endothelium in response to VPF/VEGF and other vasoactive mediators. *Microcirculation* **6**:23–44, 1999.
33. Kohn, S., Nagy, J. A., Dvorak, H. F. and Dvorak, A. M. Pathways of macromolecular tracer transport across venules and small veins. Structural basis for the hyperpermeability of tumor blood vessels. *Lab. Invest.* **67**:596–607, 1992.
34. Dvorak, A. M., Kohn, S., Morgan, E. S., Fox, P., Nagy, J. A. and Dvorak, H. F. The vesiculo-vacuolar organelle (VVO): a distinct endothelial cell structure that provides a transcellular pathway for macromolecular extravasation. *J. Leukoc. Biol.* **59**:100–115, 1996.
35. Feng, D., Nagy, J. A., Hipp, J., Dvorak, H. F. and Dvorak, A. M. Vesiculo-vacuolar organelles and the regulation of venule permeability to macromolecules by vascular permeability factor, histamine and serotonin. *J. Exp. Med.* **183**:1981–1986, 1996.
36. Granger, D. N. and Perry, M. A. Permeability characteristics of the microcirculation. In: *The Physiology and Pharmacology of the Microcirculation*, Vol. 1 (Martmortillaro, N. A., ed.), pp. 157–208. Academic Press, New York, 1983.
37. Simionescu, N. Cellular aspects of transcapillary exchange. *Physiol. Rev.* **63**:1536–1579, 1983.
38. Renkin, E. M. Capillary transport of macromolecules. Pores and other endothelial pathways. *J. Appl. Physiol.* **5**:315–325, 1985.
39. Schnitzer, J. E. Update on the cellular and molecular basis of capillary permeability. *Trends Cardiovasc. Med.* **3**:124–130, 1993.

40. Dvorak, A. M., Tepper, R. I., Weller, P. F., Morgan, E. S., Estrella, P., Monahan-Earley, R. A. and Galli, S. J. Piecemeal degranulation of mast cells in the inflammatory eyelid lesions of interleukin-4 transgenic mice. Evidence of mast cell histamine release *in vivo* by diamine oxidase–gold enzyme-affinity ultrastructural cytochemistry. *Blood* **83**:3600–3612, 1994.
41. Vasile, E., Qu-Hong, Dvorak, H. F. and Dvorak, A. M. Caveolae and vesiculo-vacuolar organelles in bovine capillary endothelial cells cultured with VPF/VEGF on floating matrigel–collagen gels. *J. Histochem. Cytochem.* **47**:159–167, 1999.
42. Parton, R. G. Caveolae and caveolins. *Curr. Opin. Cell Biol.* **8**:542–548, 1996.
43. Parton, R. G., Joggerst, B. and Simons, K. Regulated internalization of caveolae. *J. Cell Biol.* **127**:1199–1215, 1994.
44. Anderson, R. G. W. Plasmalemmal caveolae and GPI-anchored membrane proteins. *Curr. Opin. Cell Biol.* **5**:647–652, 1993.
45. Anderson, R. G. W. Caveolae: where incoming and outgoing messengers meet. *Proc. Natl. Acad. Sci. USA* **90**:10909–10913, 1993.
46. Schnitzer, J. E., Oh, P., Jacobson, B. S. and Dvorak, A. M. Caveolae from luminal plasmalemma of rat lung endothelium: Microdomains enriched in caveolin, Ca^{++}-ATPase, and inositol triphosphate receptor. *Proc. Natl. Acad. Sci. USA* **92**:1759–1763, 1995.
47. Schnitzer, J. E., McIntosh, D. P., Dvorak, A. M., Liu, J. and Oh, P. Separation of caveolae from associated microdomains of GPI-anchored proteins. *Science* **269**:1435–1439, 1995.
48. Dvorak, H. F., Brown, L. F., Detmar, M. and Dvorak, A. M. Vascular permeability factor/vascular endothelial cell growth factor, microvascular hyperpermeability and angiogenesis. *Am. J. Pathol.* **146**:1029–1039, 1995.
49. Dvorak, H. F., Nagy, J. A., Feng, D., Brown, L. F. and Dvorak, A. M. Vascular permeability factor/vascular endothelial growth factor and the significance of microvascular hyperpermeability in angiogenesis. *Curr. Top. Microbiol. Immunol.* **237**:97–132, 1999.
50. Brown, L. F., Detmar, M., Claffey, K., Nagy, J. A., Feng, D., Dvorak, A. M. and Dvorak, H. F. Vascular permeability factor/vascular endothelial growth factor: a multifunctional angiogenic cytokine. In: *Regulation of Angiogenesis* (Goldberg, I. D. and Rosen, E. M., eds), pp. 233–269. Birkhäuser Publishing, Basel, 1997.
51. Qu-Hong, Nagy, J. A., Senger, D. R., Dvorak, H. F. and Dvorak, A. M. Ultrastructural localization of vascular permeability factor/vascular endothelial growth factor (VPF/VEGF) to the abluminal plasma membrane and vesiculovacuolar organelles of tumor microvascular endothelium. *J. Histochem. Cytochem.* **43**:381–389, 1995.
52. Senger, D. R., Galli, S. J., Dvorak, A. M., Perruzzi, C. A., Harvey, V. S. and Dvorak, H. F. Tumor cells secrete a vascular permeability factor that promotes accumulation of ascites fluid. *Science* **219**:983–985, 1983.
53. Boesiger, J., Tsai, M., Maurer, M., Yamaguchi, M., Brown, L. F., Claffey, K. P., Dvorak, H. F. and Galli, S. J. Mast cells can secrete vascular permeability factor/vascular endothelial cell growth factor and exhibit enhanced release after immunoglobulin E-dependent upregulation of Fc_{ε} receptor I expression. *J. Exp. Med.* **188**:1135–1145, 1998.
54. Grützkau, A., Krüger-Krasagakes, S., Baumeister, H., Schwartz, C., Kögel, H., Welker, P., Lippert, U., Henz, B. M. and Moller, A. Synthesis, storage, and release of vascular endothelial cell growth factor/vascular permeability factor (VEGF/VPF) by human mast cells: implications for the biological significance of $VEGF_{206}$. *Mol. Biol. Cell* **9**:875–884, 1998.
55. Dvorak, A. M., Warner, J. A., Morgan, E., Kissell-Rainville, S., Lichtenstein, L. M. and MacGlashan, D. W. Jr. An ultrastructural analysis of tumor-promoting phorbol diester-induced degranulation of human basophils. *Am. J. Path.* **141**:1309–1322, 1992.
56. Dvorak, A. M., Schroeder, J. T., MacGlashan, D. W. Jr., Bryan, K. P., Morgan, E. S., Lichtenstein, L. M. and MacDonald, S. M. Comparative ultrastructural morphology of human basophils stimulated to release histamine by anti-IgE, recombinant IgE-dependent histamine-releasing factor, or monocyte chemotactic protein-1. *J. Allergy Clin. Immunol.* **98**:355–370, 1996.
57. Dvorak, A. M., Newball, H. H., Dvorak, H. F. and Lichtenstein, L. M. Antigen-induced IgE-mediated degranulation of human basophils. *Lab. Invest.* **43**:126–139, 1980.
58. Dvorak, A. M., Hammel, I., Schulman, E. S., Peters, S. P., MacGlashan, D. W. Jr., Schleimer, R. P., Newball, H. H., Pyne, K., Dvorak, H. F., Lichtenstein, L. M. and Galli, S.J. Differences in the behavior of cytoplasmic granules and lipid bodies during human lung mast cell degranulation. *J. Cell Biol.* **99**:1678–1687, 1984.
59. Dvorak, A. M., Schulman, E. S., Peters, S. P., MacGlashan, D. W. Jr., Newball, H. H., Schleimer, R. P.

and Lichtenstein, L. M. Immunoglobulin E-mediated degranulation of isolated human lung mast cells. *Lab. Invest.* **53**:45–56, 1985.

60. Dvorak, A. M. Ultrastructural analysis of human mast cells and basophils. In: *Human Basophils and Mast Cells. Biological Aspects*, (Marone, G., ed.), pp. 1–33. Karger, Basel, 1995.
61. Dvorak, A. M. Human basophil recovery from secretion. A review emphasizing the distribution of Charcot–Leyden crystal protein in cells stained with the electron-dense tracer, cationized ferritin. *Histol. Histopathol.* **11**:711–728, 1996.
62. Dvorak, A. M. New aspects of mast cell biology. *Int. Arch. Allergy Immunol.* **114**:1–9, 1997.
63. Dvorak, A. M. A role for vesicles in human basophil secretion. *Cell Tissue Res.* **293**:1–22, 1998.
64. Warner, J. A., Peters, S. P., Lichtenstein, L. M., Hubbard, W., Yancey, K. B., Stevenson, H. C., Miller, P. J. and MacGlashan, D. W. Jr. Differential release of mediators from human basophils: differences in arachidonic acid metabolism following activation by unrelated stimuli. *J. Leukoc. Biol.* **45**:558–571, 1989.
65. Schleimer, R. P., Gillespie, E. and Lichtenstein, L. M. Release of histamine from human leukocytes stimulated with the tumor-promoting phorbol diesters. I. Characterization of the response. *J. Immunol.* **126**:570–574, 1981.
66. Schleimer, R. P., Gillespie, E., Daiuta, R. and Lichtenstein, L. M. Release of histamine from human leukocytes stimulated with the tumor-promoting phorbol diesters. II. Interaction with other stimuli. *J. Immunol.* **128**:136–140, 1982.
67. Costa, J. J., Demetri, G. D., Harrist, T. J., Dvorak, A. M., Hayes, D. F., Merica, E. A., Menchaca, D. M., Gringeri, A. J., Schwartz, L. B. and Galli, S. J. Recombinant human stem cell factor (kit ligand) promotes human mast cell and melanocyte hyperplasia and functional activation *in vivo*. *J. Exp. Med.* **183**:2681–2686, 1996.
68. Dvorak, A. M. Human mast cells. Ultrastructural observations of *in situ, ex vivo,* and *in vitro* sites, sources, and systems. In: *The Mast Cell in Health and Disease,* (Kaliner, M. A. and Metcalfe, D. D., eds), pp. 1–90. Marcel Dekker, New York, 1992.
69. Dvorak, A. M., Costa, J. J., Monahan-Earley, R. A., Fox, P. and Galli, S. J. Ultrastructural analysis of human skin biopsy specimens from patients receiving recombinant human stem cell factor: subcutaneous injection of rhSCF induces dermal mast cell degranulation and granulocyte recruitment at the injection site. *J. Allergy Clin. Immunol.* **101**:793–806, 1998.
70. Dvorak, A. M., McLeod, R. S., Onderdonk, A. B., Monahan-Earley, R. A., Cullen, J. B., Antonioli, D. A., Morgan, E., Blair, J. E., Estrella, P., Cisneros, R. L., Cohen, Z. and Silen, W. Human gut mucosal mast cells: ultrastructural observations and anatomic variation in mast cell-nerve associations *in vivo*. *Int. Arch. Allergy Immunol.* **98**:158–168, 1992.
71. Dvorak, A. M., McLeod, R. S., Onderdonk, A., Monahan-Earley, R. A., Cullen, J. B., Antonioli, D. A., Morgan, E., Blair, J. E., Estrella, P., Cisneros, R. L., Silen, W. and Cohen, Z. Ultrastructural evidence for piecemeal and anaphylactic degranulation of human gut mucosal mast cells *in vivo*. *Int. Arch. Allergy Immunol.* **99**:74–83, 1992.
72. Galli, S. J., Zsebo, K. M. and Geissler, E. N. The kit ligand, stem cell factor. *Adv. Immunol.* **55**:1–96, 1994.
73. Mitsui, H., Furitsu, T., Dvorak, A. M., Irani, A.-M. A., Schwartz, L. B., Inagaki, N., Takei, M., Ishizaka, K., Zsebo, K. M., Gillis, S. and Ishizaka, T. Development of human mast cells from umbilical cord blood cells by recombinant human and murine c-kit ligand. *Proc. Natl. Acad. Sci. USA* **90**:735–739, 1993.
74. Furitsu, T., Saito, H., Dvorak, A. M., Schwartz, L. B., Irani, A.-M. A., Burdick, J. F., Ishizaka, K. and Ishizaka, T. Development of human mast cells *in vitro*. *Proc. Natl. Acad. Sci. USA* **86**:10039–10043, 1989.
75. Columbo, M., Horowitz, E. M., Botana, L. M., MacGlashan, Jr., D. W., Bochner, B. S., Gillis, S., Zsebo, K. M., Galli, S.J. and Lichtenstein, L.M. The human recombinant c-*kit* receptor ligand, rhSCF, induces mediator release from human cutaneous mast cells and enhances IgE-dependent mediator release from both skin mast cells and peripheral blood basophils. *J. Immunol.* **149**:599–608, 1992.
76. Dvorak, A. M. and Morgan, E. S. Ribonuclease–gold labels proteoglycan-containing cytoplasmic granules and ribonucleic acid-containing organelles – a survey. *Histol. Histopathol* **14**:597–626, 1999.
77. Dvorak, H. F., Dvorak, A. M., Manseau, E. J., Wiberg, L. and Churchill, W. H. Fibrin gel investment associated with line 1 and line 10 solid tumor growth, angiogenesis and fibroplasia in guinea pigs. Role of cellular immunity, myofibroblasts, microvascular damage, and infarction in line 1 tumor regression. *J. Nat. Cancer Inst.* **62**:1459–1472, 1979.
78. Ferrara, N. and Henzel, W. J. Pituitary follicular cells secrete a novel heparin-binding growth factor

specific for vascular endothelial cells. *Biochem. Biophys. Res. Comm.* **161**:851–858, 1989.

79. Ferrara, N., Houck, K., Jakeman, L. and Leung, D. W. Molecular and biological properties of the vascular endothelial growth factor family of proteins. *Endoc. Rev.* **13**:18–32, 1992.
80. Leung, D. W., Cachianes, G., Kuang, W.-J., Goeddel, D. V. and Ferrara, N. Vascular endothelial growth factor is a secreted angiogenic mitogen. *Science* **246**:1306–1309, 1989.
81. Gospodarowicz, D., Abraham, J. A. and Schilling, J. Isolation and characterization of a vascular endothelial cell mitogen produced by pituitary-derived folliculostellate cells. *Proc. Natl. Acad. Sci. USA* **86**:7311–7315, 1989.
82. Dvorak, H. F., Nagy, J. A., Berse, B., Brown, L. F., Yeo, K.-T., Yeo, T.-K., Dvorak, A. M., Van De Water, L., Sioussat, T. M. and Senger, D. R. Vascular permeability factor, fibrin and the pathogenesis of tumor stroma formation. *Ann. N. Y. Acad. Sci.* **667**:101–111, 1992.
83. Senger, D. R., Van De Water, L., Brown, L. F., Nagy, J. A., Yeo, K.-T., Yeo, T.-K., Berse, B., Jackman, R. W., Dvorak, A. M. and Dvorak, H. F. Vascular permeability factor (VPF, VEGF) in tumor biology. *Cancer Metastasis Rev.* **12**:303–324, 1993.
84. Palade, G. E. and Bruns, R. R. Structural modulations of plasmalemmal vesicles. *J. Cell Biol.* **37**:633–649, 1968.
85. Bruns, R. R. and Palade, G. E. Studies on blood capillaries. I. General organization of blood capillaries in muscle. *J. Cell Biol.* **37**:244-276, 1968.
86. Hammel, I., Lagunoff, D., Bauza, M. and Chi, E. Periodic, multimodal distribution of granule volumes in mast cells. *Cell Tissue Res.* **228**:51–59, 1983.
87. Hammel, I., Dvorak, A. M., Peters, S. P., Schulman, E. S., Dvorak, H. F., Lichtenstein, L. M. and Galli, S. J. Differences in the volume distributions of human lung mast cell granules and lipid bodies: evidence that the size of these organelles is regulated by distinct mechanisms. *J. Cell Biol.* **100**:1488–1492, 1985.
88. Hammel, I., Lagunoff, D. and Wysolmerski, R. Theoretical considerations on the formation of secretory granules in the rat pancreas. *Exp. Cell Res.* **204**:1–5, 1993.
89. Hammel, I., Dvorak, A. M., Fox, P., Shimoni, E. and Galli, S. J. Defective cytoplasmic granule formation. II. Differences in patterns of radiolabeling of secretory granules in beige versus normal mouse pancreatic acinar cells after [^{3}H]-glycine administration *in vivo*. *Cell Tissue Res.* **293**:445–452, 1998.
90. Dvorak, H. F., Mihm, M. C. Jr., Dvorak, A. M., Johnson, R. A., Manseau, E. J., Morgan, E. and Colvin, R. B. Morphology of delayed type hypersensitivity reactions in man. I. Quantitative description of the inflammatory response. *Lab. Invest.* **31**:111–130, 1974.
91. Dvorak, H. F., Mihm, M. C. Jr. and Dvorak, A. M. Morphology of delayed-type hypersensitivity reactions in man. *J. Invest. Dermatol.* **67**:391–401, 1976.
92. Dvorak, A. M., Mihm, M. C. Jr. and Dvorak, H. F. Morphology of delayed-type hypersensitivity reactions in man. II. Ultrastructural alterations affecting the microvasculature and the tissue mast cells. *Lab. Invest.* **34**:179–191, 1976.
93. Dvorak, A. M., Dickersin, G. R., Connell, A., Carey, R. W. and Dvorak, H. F. Degranulation mechanisms in human leukemic basophils. *Clin. Immunol. Immunopathol.* **5**:235–246, 1976.
94. Glasser, L., Corrigan, J. J. Jr. and Payne, C. Basophilic meningitis secondary to lymphoma. *Neurology* **26**:899–902, 1976.
95. Dvorak, A. M. Morphologic and immunologic characterization of human basophils, 1879 to 1985. *Riv. Immunol. Immunofarmacol.* **8**:50–83, 1988.
96. Dvorak, A. M. The fine structure of human basophils and mast cells. In: *Mast Cells, Mediators and Disease* (Holgate, S. T., ed.), pp. 29–97. Kluwer, Dordrecht, 1988.
97. Dvorak, A. M., ed. Human Mast Cells. *Advances in Anatomy, Embryology, and Cell Biology*, Vol. 114 (Beck, F., Hild, W., Kriz, W., Ortmann, R., Pauly J. E. and Schiebler, T. H., series eds). Springer-Verlag, Berlin, 1989.
98. Collin, H. B. and Allansmith, M. R. Basophils in vernal conjunctivitis in humans: an electron microscopic study. *Invest. Ophthalmol. Vis. Sci.* **16**:858–864, 1977.
99. Dvorak, H. F., Mihm, M. C. Jr., Dvorak, A. M., Barnes, B. A., Manseau, E. J. and Galli, S. J. Rejection of first-set skin allografts in man. The microvasculature is the critical target of the immune response. *J. Exp. Med.* **150**:322–337, 1979.
100. Dvorak, H. F., Mihm, M. C. Jr., Dvorak, A. M., Barnes, B. A. and Galli, S. J. The microvasculature is the critical target of the immune response in vascularized skin allograft rejection. *J. Invest. Dermatol.* **74**:280–284, 1980.
101. Dvorak, A. M., Monahan, R. A., Osage, J. E. and Dickersin, G. R. Crohn's disease: transmission electron microscopic studies. II. Immunologic inflammatory response. Alterations of mast cells, basophils, eosinophils, and the microvasculature. *Hum. Pathol.* **11**:606–619, 1980.

102. Dvorak, A. M. and Monahan, R. A. Crohn's disease. Ultrastructural studies showing basophil leukocyte granule changes and lymphocyte parallel tubular arrays in peripheral blood. *Arch. Pathol. Lab. Med.* **106**:145–149, 1982.
103. Dvorak, A. M. and Monahan, R. A. Chronic abdominal pain, malaise, constipation and diarrhea in a forty year old woman: Crohn's disease. *Norelco Rep.* **31**:18–44, 1984.
104. Dvorak, A. M., Mihm, M. C. Jr., Osage, J. E., Kwan, T. H., Austen, K. F. and Wintroub, B. U. Bullous pemphigoid, an ultrastructural study of the inflammatory response: eosinophil, basophil and mast cell granule changes in multiple biopsies from one patient. *J. Invest. Dermatol.* **78**:91–101, 1982.
105. Fox, C. C., Dvorak, A. M., MacGlashan, D. W. Jr. and Lichtenstein, L. M. Histamine-containing cells in human peritoneal fluid. *J. Immunol.* **132**:2177–2179, 1984.
106. Dvorak, A. M., Saito, H., Estrella, P., Kissell, S., Arai, N. and Ishizaka, T. Ultrastructure of eosinophils and basophils stimulated to develop in human cord blood mononuclear cell cultures containing recombinant human interleukin-5 or interleukin-3. *Lab. Invest.* **61**:116–132, 1989.
107. Dvorak, A. M. and Monahan-Earley, R. A. Case 39: Soft tissue mass and lytic lesion of the scapula in a 55-year-old man with a 16-year history of eczema. Primary T cell lymphoma of the skin (Mycosis fungoides and Sezary's syndrome). *Diagnostic Ultrastructural Pathology I. A Text-Atlas of Case Studies Illustrating the Correlative Clinical–Ultrastructural Pathologic Approach to Diagnosis*, pp. 327–346. CRC Press, Boca Raton, FL, 1992.
108. Asboe-Hansen, G. Mast cells and the skin. In *Monographs in Pathology*, Vol. 10 (Helwig, E. B. and Mostofi, F. K., eds) pp. 83–111. Williams & Wilkins, Baltimore, 1971.
109. Colvin, R. B., Dvorak, A. M. and Dvorak, H. F. Mast cells in the cortical tubular epithelium and interstitium in human renal disease. *Hum. Pathol.* **5**:315–326, 1974.
110. Dvorak, A. M. Mast-cell hyperplasia and degranulation in Crohn's disease. In: *The Mast Cell. Its Role in Health and Disease* (Pepys, J. and Edwards, A. M., eds), pp. 657–662. Pitman Medical, Kent, 1979.
111. Dvorak, A. M. Ultrastructural pathology of Crohn's disease. In: *Inflammatory Bowel Diseases – Basic Research and Clinical Implications* (Goebell, H., Peskar, B. M. and Malchow, H., eds), pp. 3–41. MTP Press, Lancaster, 1988.
112. Dvorak, A. M. and Monahan-Earley, R. A. *Diagnostic Ultrastructural Pathology. I. A Text-Atlas of Case Studies Illustrating the Correlative Clinical–Ultrastructural Pathologic Approach to Diagnosis.* CRC Press, Boca Raton, FL, 1992.
113. Dvorak, A. M. and Silen, W. Differentiation between Crohn's disease and other inflammatory conditions by electron microscopy. *Ann. Surg.* **201**:53–63, 1985.
114. Dvorak, A. M., Monahan, R. A., Osage, J. E. and Dickersin, G. R. Mast-cell degranulation in Crohn's disease. *Lancet* **i**:498, 1978 (letter).
115. Dvorak, A. M., Mihm, M. C. Jr., Osage, J. E. and Dvorak, H. F. Melanoma. An ultrastructural study of the host inflammatory and vascular responses. *J. Invest. Dermatol.* **75**:388–393, 1980.
116. Dvorak, A. M. and Kissell, S. Granule changes of human skin mast cells characteristic of piecemeal degranulation and associated with recovery during wound healing *in situ*. *J. Leukoc. Biol.* **49**:197–210, 1991.
117. Dvorak, A. M. and Monahan-Earley, R. A. Case 5: Weight loss, fatigue, diarrhea, fever and lymphadenopathy in a 48-year-old male industrial engineer. Whipple's disease. *Diagnostic Ultrastructural Pathology I. A Text-Atlas of Case Studies Illustrating the Correlative Clinical–Ultrastructural Pathologic Approach to Diagnosis*, pp. 34–45. CRC Press, Boca Raton, FL, 1992.
118. Dvorak, A. M. and Monahan, R. A. Abdominal pain, hepatomegaly and weight loss in a fifty four year old man: renal cell carcinoma and adenoma. *Norelco Rep.* **32**:61–66, 1985.
119. Murphy, G. F., Austen, K. F., Fonferko, E. and Sheffer, A. L. Morphologically distinctive forms of cutaneous mast cell degranulation induced by cold and mechanical stimuli: an ultrastructural study. *J. Allergy Clin. Immunol.* **80**:603–611, 1987.
120. Dvorak, A. M. and Monahan-Earley, R. A. Case 10: Progressive cough and shortness of breath in a 34-year-old man. Diffuse interstitial fibrosis and mast cell hyperplasia of the lung consistent with hypersensitivity pneumonitis. *Diagnostic Ultrastructural Pathology. II. A Text-Atlas of Case Studies Emphasizing Respiratory and Nervous Systems*, pp. 142–163. CRC Press, Boca Raton, FL, 1995.
121. Dvorak, A. M., Schleimer, R. P., Schulman, E. S. and Lichtenstein, L. M. Human mast cells use conservation and condensation mechanisms during recovery from degranulation. *In vitro* studies with mast cells purified from human lungs. *Lab. Invest.* **54**:663–678, 1986.

CHAPTER 6

Human Dendritic Mast Cells

HENRIK H. JACOBI[*1] *and OLLE JOHANSSON*[2]

[1]*Allergy Unit 7511, National University Hospital, Copenhagen, Denmark and*
[2]*Experimental Dermatology Unit, Department of Neuroscience, Karolinska Institute, Stockholm, Sweden*

INTRODUCTION

In recent years, it has become evident that human mast cells comprise different subtypes that display variations in their morphological (1–3), biochemical (4–7) and functional (8–12) properties. However, the identification and classification of such mast cell subtypes is still in its beginning. In this chapter we describe a subpopulation of human mast cells with dendrite-like cellular processes. The existence of such dendritic cells, side by side with ordinary round or elongated mast cells, clearly illustrates the marked heterogeneity of human mast cells.

The chapter focuses on the morphology of the human dendritic mast cells. However, we also discuss some functional aspects that may be of particular relevance to these cells.

THE MORPHOLOGY OF HUMAN DENDRITIC MAST CELLS

Mast cells are characteristic tissue cells with a single round or oval nucleus and numerous cytoplasmic granules that stain metachromatically with thiazine dyes such as toluidine blue, and also show characteristic orthochromatic staining with phthalocyanin dyes such as Alcian blue. The size and shape of mast cells show considerable variation, but human mast cells are usually described as round or elongated cells with a diameter ranging between 8 and 20 μm depending on the organ examined (1–3). The description of human mast cells as being round or elongated is, however, not entirely adequate because some mast cells may also have dendrite-like cellular processes (Fig. 1). Such dendritic mast cells have been observed in the human nasal mucosa (13), skin (14) and gastrointestinal system (unpublished data).

* Corresponding author.

MAST CELLS AND BASOPHILS
ISBN 0-12-473335-2

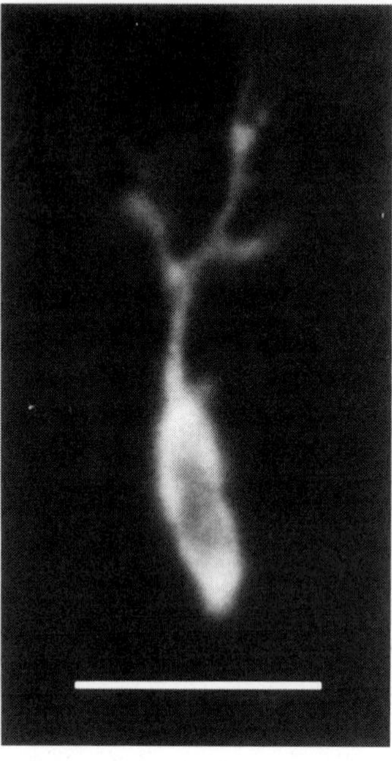

Fig. 1 Dendritic mast cell in the human skin. The cell has been labelled for histamine using the indirect immunofluorescence technique. Small second- and third-level branches are visible on the distal part of the cellular process. Bar: 25 μm. (Modified with permission from *J. Cutan. Pathol.* **25**:189–198, 1998.)

The study of human dendritic mast cells is dependent on the availability of sensitive and specific methods for the identification of mast cells in human tissues. Although conventional staining methods, using toluidine blue or Alcian blue, occasionally may reveal mast cells with an apparent dendritic morphology (15), these methods are not sensitive enough to detect the small second- or third-level branches that may be present on some of the dendrite-like cellular processes (13). In contrast, immunostaining using antibodies against mast cell mediators such as histamine, tryptase or chymase results in excellent labelling of small cellular processes; the next section will describe some of the immunohistochemical methods that we have used to study dendritic mast cells in human tissues.

Immunostaining of Mast Cells

Mast cells contain significant amounts of histamine and neutral serine proteases, and human mast cells can easily be identified in tissue sections using immunohistochemistry for histamine, tryptase or chymase (4, 16, 17). Furthermore, such immunohistochemical methods seem to be superior to conventional staining methods using toluidine blue or Alcian blue (13, 18).

To stain human mast cells for histamine, biopsies are fixed in phosphate-buffered saline (PBS) with 4% 1-ethyl-3-(3-dimethylaminopropyl)-carbodiimide (CDI) for 2 h at 4°C. After fixation, the biopsies are immersed in PBS with 10% sucrose for 24 h before sectioning in a cryostat. The sections are then stained with antihistamine antibodies raised against histamine conjugated to different carrier proteins with CDI (19). Tryptase and chymase immunohistochemistry can also be performed on CDI-fixed biopsies if the proper antitryptase and antichymase antibodies are selected (13). Alternatively, biopsies can be fixed in 4% paraformaldehyde for 2 h at 4°C and then processed as described above. The latter approach results in excellent staining with many antitryptase and antichymase antibodies (14), but completely abolishes staining with the antihistamine antibodies (17).

Although the choice of fixation procedure and primary antibodies is important for the optimal immunostaining of mast cells in tissue sections, the thickness of the tissue sections is equally important when studying mast cell subpopulations with dendrite-like cellular processes. The cellular processes bend and curve through the tissue, and it is obvious that, if the tissue sections are cut too thin, one will also reduce the likelihood that the entire length of the cellular processes will be contained within a given tissue section. We find that tissue sections with a thickness of 14–15 μm are optimal for most light-microscopy studies. For confocal microscopy, even thicker tissue sections are preferable.

Dendritic Mast Cells in the Human Nose

It is now clear that mast cells in the human nasal mucosa may have dendrite-like cellular processes (13). Such dendritic mast cells have been identified by immunostaining of nasal mucosa biopsies using the methods described above. In one study, biopsies were taken from the lower or middle nasal turbinates of healthy controls and symptom-free atopic patients. The biopsies were fixed in CDI and tissue sections were subsequently triple-stained for histamine, tryptase and chymase. In addition, tissue sections already immunostained for mast cell mediators were restained using Alcian blue. These procedures revealed both ordinary round and elongated mast cells as well as mast cells with thin, branching, dendrite-like cellular processes (13).

Figure 2 demonstrates the typical morphology of dendritic mast cells in the human nasal mucosa. The cell bodies of the dendritic mast cells are usually positive for both tryptase and chymase, demonstrating that these cells are MC_{TC} mast cells. As for the cellular processes, some of these are positive for both tryptase and chymase, whereas others are positive for histamine but seem to be negative for mast cell proteases. The dendritic mast cells are primarily localized in the subepithelial connective tissue without any obvious relationship to blood vessels or glands. Some cells have only one slender process, whereas other cells have several processes extending from different parts of the cell body. Some of the cellular processes divide into two or three terminal branches and some processes have small histamine-positive swellings along their course.

As described above, dendritic mast cells can be detected in nasal mucosa biopsies from both healthy subjects and symptom-free atopic patients. The number of cells shows considerable variation from subject to subject without any obvious differences between healthy subjects and symptom-free atopic patients. In most subjects, the dendritic mast cells are few and account for less than 5% of the mast cells in the subepithelial connective tissue (13).

Dendritic Mast Cells in the Human Skin

Dendritic mast cells in normal human skin were first detected using histamine immunohistochemistry (17, 18). In these studies, punch biopsies were taken from the palm or forearm of healthy subjects. The biopsies were fixed in CDI and tissue sections were subsequently immunostained for histamine. This procedure revealed round and elongated cells as well as dendritic histamine-positive cells. The identity of the dendritic histamine-positive cells was later confirmed using tryptase and chymase immunohistochemistry, demonstrating that these cells are MC_{TC} mast cells (14, 20).

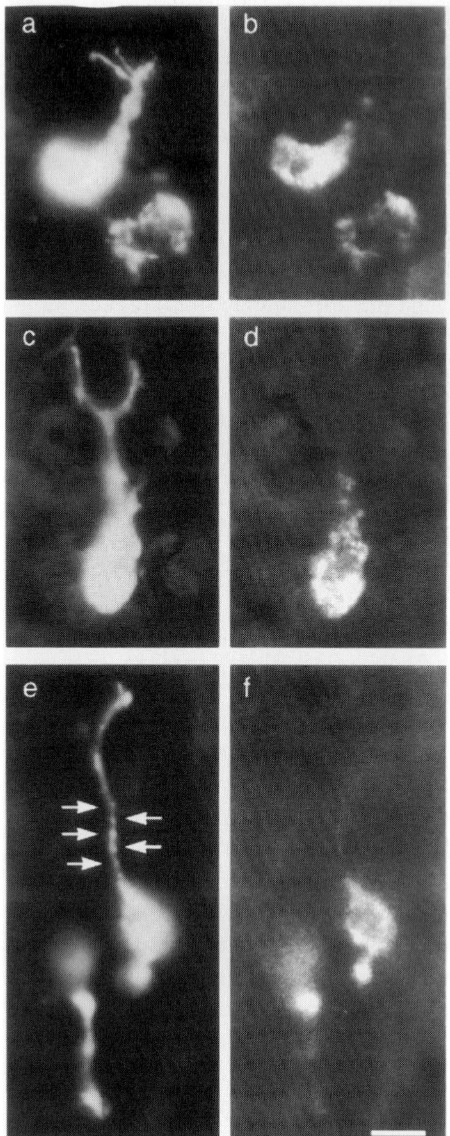

Fig. 2 Dendritic mast cells in the human nasal mucosa. (a, b) Biopsy from an allergic patient. Double-labelling for histamine (a) and tryptase (b) using the indirect immunofluorescence technique. The round cell to the right is an ordinary mast cell without cellular processes. (c, d) Biopsy from a healthy subject. Double-labelling for histamine (c) and chymase (d). (e, f) Biopsy from a healthy subject. Double-labelling for histamine (e) and tryptase (f). The arrows point to small histamine-positive swellings along the course of the cellular process. Bar: 10 μm. (Reproduced with permission from *Lab. Invest.* **78**:1179–1184, 1998.)

Although only few dendritic mast cells are present in normal skin, the relative number of these cells is increased in prurigo nodularis skin (14). Figure 3 demonstrates dendritic mast cells in prurigo nodularis skin. Dendritic mast cells are found both in the reticular and in the papillary dermis with no obvious relationship to dermal structures such as glands or vessels. The cells usually have a large cell body and the cellular processes may be quite long (>50 μm) with second- or third-level branches. At the ultrastructural level, dendritic mast cells often have a large cell body but contain relatively few granules in the cytoplasm, especially within the dendrites (14).

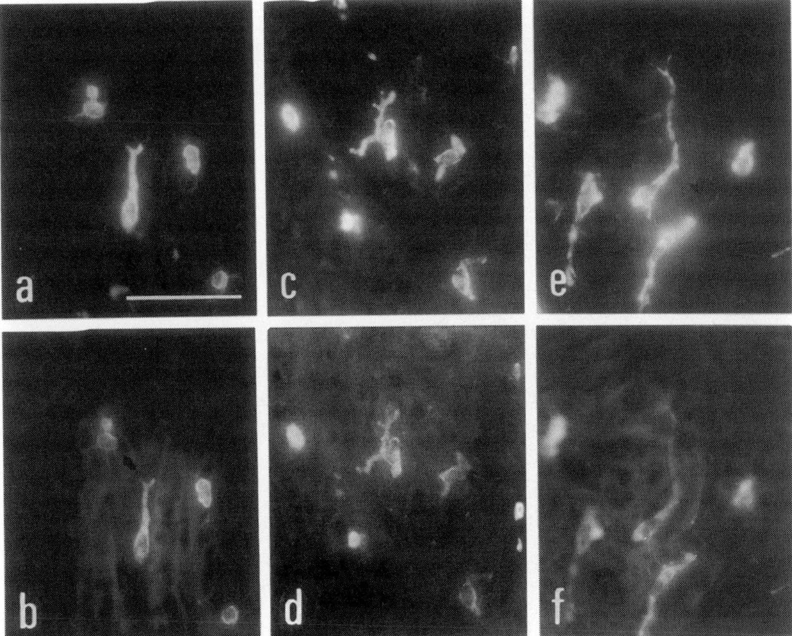

Fig. 3 Dendritic mast cells in prurigo nodularis lesional skin. Double-labelling for tryptase (a, c, e) and chymase (b, d, f) using the indirect immunofluorescence technique. Bar: 50 μm. (Reproduced with permission from *Eur. J. Dermatol.* **9**:297–299, 1999.)

FUNCTIONAL ASPECTS AND FUTURE LINES OF RESEARCH

Little is known about the function and differentiation pathways of the human dendritic mast cells. The increased number of these cells in prurigo nodularis skin (14) and at sites of allergic inflammation (unpublished data) indicate that: (i) ordinary mast cells may transform into the dendritic phenotype during an inflammatory response; and/or (ii) dendritic mast cells are selectively recruited to sites of inflammation.

The presence of dendrite-like cellular processes may simply reflect that the mast cells are migrating in the tissue. However, a similar change in shape has not been reported for other types of migrating cells. Alternatively, the cellular processes may be specialized structures with particular efferent and/or afferent functions. In this context it is interesting that, in addition to their role in immediate hypersensitivity reactions, mast cells have been implicated in conditions as diverse as antigen presentation (21, 22), host defence (23–25), tissue remodelling (26–28) and angiogenesis (29, 30). Whether the dendritic phenotype is associated with any of these functions remains to be established.

An important question to address in future studies will be whether the cellular processes are directed toward particular cell types in the tissue. This latter question can be answered by double-staining experiments using specific markers of resident tissue cells and inflammatory cells. In the human nose and skin we have not been able to detect any obvious relationship between the dendritic mast cells and blood vessels or glands. The relationship to nerve fibres has not been studied systematically but such studies are now in progress.

Other important questions will be whether the cellular processes express particular proteins and whether the cellular processes terminate as free endings in the extracellular space or make direct cell-to-cell contacts with the surrounding cells. In this context it is interesting that mast cells (at least in the murine system) may express connexins on their cytoplasmic membrane, indicating that mast cells have the potential to communicate with other cells through gap junctions (31).

CONCLUDING REMARKS

It is now clear that a subpopulation of human MC_{TC} cells have dendrite-like cellular processes. Some cells have only one slender process, whereas other cells have several processes extending from different parts of the cell body. Furthermore, the cellular processes may be quite long with second- or third-level branches, and some processes have small histamine-positive swellings along their course.

The dendritic MC_{TC} cells have been observed in the human nasal mucosa and skin as well as in other organs, and their numbers seem to increase at inflammatory sites. However, the study of these cells is only in its beginning and further studies will be necessary to reveal the functional significance of the human dendritic mast cells.

REFERENCES

1. Schulman, E. S., Kagey-Sobotka, A., MacGlashan, D. W., Adkinson, N. F., Peters, S. P., Schleimer, R. P. and Lichtenstein L. M. Heterogeneity of human mast cells. *J. Immunol.* **131**:1936–1941, 1983.
2. Galli, S. J., Dvorak, A. M. and Dvorak, H. F. Basophils and mast cells: morphologic insights into their biology, secretory patterns, and function. *Prog. Allergy* **34**:1–141, 1984.
3. Lee,T. D. G., Swieter, M., Bienenstock, J. and Befus, A. D. Heterogeneity in mast cell populations. *Clin. Immunol. Rev.* **4**:143–199, 1985.
4. Irani, A. A., Schechter, N. M., Craig, S. S., DeBlois, G. and Schwartz, L. B. Two types of human mast cells that have distinct neutral protease compositions. *Proc. Natl. Acad. Sci. USA* **83**:4464–4468, 1986.
5. Schechter, N. M., Irani, A.-M. A., Sprows, J. L., Abernethy, J., Wintroub, B. U. and Schwartz, L. B. Identification of cathepsin G-like proteinase in the MC_{TC} type of human mast cell. *J. Immunol.* **145**:2652–2661, 1990.
6. Irani, A.-M. A., Goldstein, S. M., Wintroub, B. U., Bradford, T. and Schwartz, L. B. Human mast cell carboxypeptidase. Selective localization to MC_{TC} cells. *J. Immunol.* **147**:247–253, 1991.
7. Bradding, P., Okayama, Y., Howarth, P. H., Church, M. K. and Holgate, S. T. Heterogeneity of human mast cells based on cytokine content. *J. Immunol.* **155**:297–307, 1995.
8. Church, M. K., Pao, G. J.-K. and Holgate, S. T. Characterization of histamine secretion from mechanically dispersed human lung mast cells: effects of anti-IgE, calcium ionophore A23187, compound 48/80, and basic polypeptides. *J. Immunol.* **129**:2116–2121, 1982.
9. Schulman, E. S., MacGlashan, D. W., Peters, S. P., Schleimer, R. P., Newball, H. E. I. and Lichtenstein, L. M. Human lung mast cells: purification and characterization. *J. Immunol.* **129**:2662–2667, 1982.
10. Benyon, R. C., Lowman, M. A. and Church, M. K. Human skin mast cells: their dispersion, purification, and secretory characterization. *J. Immunol.* **138**:861–867, 1987.
11. Lawrence, I. D., Warner, J. A., Cohan, V. L., Hubbard, W. C., Kagey-Sobotka, A. and Lichtenstein, L. M. Purification and characterization of human skin mast cells. Evidence for human mast cell heterogeneity. *J. Immunol.* **139**:3062–3069, 1987.
12. Patella, V., Marinò, I., Lampärter, B., Arbustini, E., Adt, M. and Marone, G. Human heart mast cells. *J. Immunol.* **154**:2855–2865, 1995.
13. Jacobi, H. H., Liang, Y., Tingsgaard, P. K., Larsen, P. L., Poulsen, L. K., Skov, P. S., Haak-Frendscho, M., Niles, A. L. and Johansson, O. Dendritic mast cells in the human nasal mucosa. *Lab. Invest.* **78**:1179–1184, 1998.

14. Liang, Y., Jacobi, H. H., Marcusson, J. A., Haak-Frendscho, M. and Johansson, O. Dendritic mast cells in prurigo nodularis skin. *Eur. J. Dermatol.* **9**:297–299, 1999.
15. Trotter, C. M., Carney, A. S. and Wilson, J. A. Mast cell distribution and morphology in human nasal turbinates following decalcification. *Rhinology* **27**:81–89, 1989.
16. Irani, A.-M. A., Bradford, T. R., Kepley, C. L., Schechter, N. M. and Schwartz, L. B. Detection of MC_T and MC_{TC} types of human mast cells by immunohistochemistry using new monoclonal anti-tryptase and anti-chymase antibodies. *J. Histochem. Cytochem.* **37**:1509–1515, 1989.
17. Johansson, O., Virtanen, M., Hilliges, M. and Yang, Q. Histamine immunohistochemistry: a new and highly sensitive method for studying cutaneous mast cells. *Histochem. J.* **24**:283–287, 1992.
18. Johansson, O., Virtanen, M., Hilliges, M. and Yang, Q. Histamine immunohistochemistry is superior to the conventional heparin-based routine staining methodology for investigations of human skin mast cells. *Histochem. J.* **26**:424–430, 1994.
19. Panula, P., Happola, O., Airaksinen, M. S., Auvinen, S. and Virkamaki, A. Carbodiimide as a tissue fixative in histamine immunohistochemistry and its application in developmental neurobiology. *J. Histochem. Cytochem.* **36**:259–269, 1988.
20. Liang, Y., Marcusson, J. A., Jacobi, H. H., Haak-Frendscho, M. and Johansson, O. Histamine-containing mast cells and their relationship to NGFr-immunoreactive nerves in pruligo nodularis: a reappraisal. *J. Cutan. Pathol.* **25**:189–198, 1998.
21. Malaviya, R., Twesten, N. J., Ross E. A., Abraham, S. N. and Pfeifer, J. D. Mast cells process bacterial Ags through a phagocytic route for class I MHC presentation to T cells. *J. Immunol.* **156**: 1490–1496, 1996.
22. Frandji, P., Tkaczyk, C., Oskeritzian, B., David, B., Desaymard, C. and Mecheri, S. Exogenous and endogenous antigens are differentially presented by mast cells to CD4+ T lymphocytes. *Eur. J. Immunol.* **26**:2517–2528, 1996.
23. Matsuda, H., Watanabe, N., Kiso, Y., Hirota, S., Ushio, H., Kannan, Y., Azuma, M., Koyama, H. and Kitamura, Y. Necessity of IgE antibodies and mast cells for manifestation of resistance against larval *Haemophysalis longicornis* ticks in mice. *J. Immunol.* **144**:259–262, 1990.
24. Echtenacher, B., Männel, D.N. and Hültner, L. Critical protective role of mast cells in a model of acute septic peritonitis. *Nature* **381**:75–77, 1996.
25. Malaviya, R., Ikeda, T., Ross, E. and Abraham, S. N. Mast cell modulation of neutrophil influx and bacterial clearance at sites of infection through TNF-α. *Nature* **381**:77–80, 1996.
26. Cairns, J. A. and Walls, A. F. Mast cell tryptase stimulates the synthesis of type I collagen in human lung fibroblasts. *J. Clin. Invest.* **99**:1313–1321, 1997.
27. Gruber, B. L., Kew, R. R., Jelaska, A., Marchese, M. J., Garlick, J., Ren, S., Schwartz L. B. and Korn, J. H. Human mast cells activate fibroblasts: tryptase is a fibrogenic factor stimulating collagen messenger ribonucleic acid synthesis and fibroblast chemotaxis. *J. Immunol.* **158**:2310–2317, 1997.
28. Kofford, M. W., Schwartz, L. B., Schechter, N. M., Yager, D. R., Diegelman, R. F. and Graham, M. F. Cleavage of type I procollagen by human mast cell chymase initiates collagen fibril formation and generates a unique carboxyl-terminal propeptide. *J. Biol. Chem.* **272**:7127–7131, 1997.
29. Marks, R. M., Roche, W. R., Czerniecki, M., Penny, R. and Nelson, D. S. Mast cell granules cause proliferation of human microvascular endothelial cells. *Lab. Invest.* **55**:289–294, 1986.
30. Blair, R. J., Meng, H., Marchese, M. J., Ren, S., Schwartz, L. B., Tonnesen, M. G. and Gruber B. L. Human mast cells stimulate vascular tube formation. Tryptase is a novel, potent angiogenic factor. *J. Clin. Invest.* **99**:2691–2700, 1997.
31. Vliagoftis, H., Hutson, A. M., Mahmudi-Azer, S., Kim, H., Rumsaeng, V., Oh, C. K., Moqbel, R. and Metcalfe, D. D. Mast cells express connexins on their cytoplasmic membrane. *J. Allergy Clin. Immunol.* **103**:656–662, 1999.

CHAPTER 7

The Phenotypic Similarities and Differences Between Human Basophils and Mast Cells

*L. LI, S. W. REDDEL and S. A. KRILIS**

Department of Medicine, The University of New South Wales, Department of Immunology, Allergy and Infectious Disease, St George Hospital, Kogarah, New South Wales, Australia

INTRODUCTION

Mast cells and basophils are well known as the primary effector cells in allergic inflammation and they represent a major source of inflammatory mediators. A number of studies support the belief that human mast cells play a unique role in a wide variety of processes, including inflammation (1), host defence (2, 3), tissue remodelling (4–6) and angiogenesis (7), whereas human basophils are important for allergic inflammation, especially in chronic disease (8). Whether mast cells and basophils are two distinct cell types or different states of a single cell lineage is not fully understood. It is apparent that mast cells and basophils share many constitutive properties but they also have differences in respect of their development, functions, morphology, secretory granular contents and cell surface molecules.

The traditional view is that mast cells arise from mast cell committed precursors in the bone marrow, circulate as agranular cells, then traverse the vascular space and enter the tissues where they complete their development. Differentiation and maturation of mast cells are most likely regulated by local micro-environmental factors (9–12). Mast cells are found in almost all of the major organs and tissues of the body, particularly in association with connective tissue structures such as blood vessels, lymphatic vessels and nerves, and in proximity to surfaces that interface the external environment such as those of the respiratory and gastrointestinal systems and the skin. In contrast, basophils differentiate and mature in the bone marrow, circulate in the blood and comprise less than 1% of total leukocytes. Normally basophils are not found in human tissue sites. However, there is increasing evidence that basophils are also involved in hypersensitivity reactions in inflamed tissue (13, 14). In this review the respective phenotypic features of basophils in blood and mast cells in tissues are examined with a discussion of their differences and similarities.

* Corresponding author.

MAST CELLS AND BASOPHILS
ISBN 0-12-473335-2

MORPHOLOGY

Mast cells and basophils were first identified by a characteristic influence of solvent and pH on the metachromatic staining reaction, which was found to be an effect on the binding of the dyes to heparin (15, 16). It is now clear that a large number of compounds stain metachromatically, in particular sulphated mucopolysaccharides and heparin, which are found in both mast cells and basophils (17, 18).

Mast cells and basophils can be distinguished on morphological grounds. It has been revealed that mast cells from most human tissue sites have a similar appearance. Analysis in a number of light- and electron-microscopic studies shows that mast cells appear as cells with a non-segmented monolobed nucleus, a surface architecture composed of narrow, elongated folds and numerous smaller and very typical cytoplasmic granules (19). Basophils, on the other hand, are typically a round cell with a few short blunt projections. The nucleus shows a polymorphonuclear morphology with marked chromatin condensation. In comparison to mast cells, basophils contain fewer and larger electron-dense granules which lack the patterns associated with mast cells (20–22).

In vitro studies have lead to a different morphological view of the nuclear profile of the mast cell. In many cases, differentiated mast cells possessing multilobular nuclei have been found occasionally in cultured mouse (23–26) and human mast cells (27–31). The findings were strongly supported by the histopathological results from rodent studies by Gurish *et al.* (32). They demonstrated clearly that mast cells with multilobular nuclei can be found occasionally in the skeletal muscle of normal mice as well as in the jejunum of helminth-infected mice and in various tissues of the V3 mastocytosis mouse (32).

The patterns of the granules in mast cells and basophils have also been examined. The results from Craig *et al.* indicate that only granules with the chymase protease exhibit a grating or lattice substructure whereas regions lacking this protease show a scroll pattern (33–35). The ultrastructural study by Weidner and Austen confirmed that mast cells from lung and bowel mucosa generally stained only for tryptase (MC_T) and that their granules were rich in scrolls; however, they emphasized the finding of significant granule morphological heterogeneity both amongst tryptose-chymase positive mast cells (MC_{TC}) found in any particular tissue site and between the individual granules of single mast cells (both MC_{TC} and MC_T) identified within a range of tissues (36). Ultrastructurally immature mast cell granules with the tryptase protease show a scroll pattern, as do their mature counterparts. Immature mast cell granules containing both tryptase and chymase proteases have an amorphous electron-dense core rather than a grating or lattice pattern (34). Basophil granules are rather uniform compared with those of the mast cell. Most granules in basophils contain electron-dense particles with neither scroll nor lattice patterns. The protein forming the Charcot–Leyden crystal, previously thought to be eosinophil-specific, has also been seen in some basophil granules (37, 38).

MEMBRANE RECEPTORS

The major aspects of mast cell activation may be initiated upon cross-linking of the high-affinity receptor $Fc_\varepsilon RI$ which is expressed on the cell surface. This occurs when multiple specific IgE antibody molecules both bind to a multivalent antigen and are also bound at their Fc region by $Fc_\varepsilon RI$. The three subunits of the IgE receptor : (α, β and γ) have now all

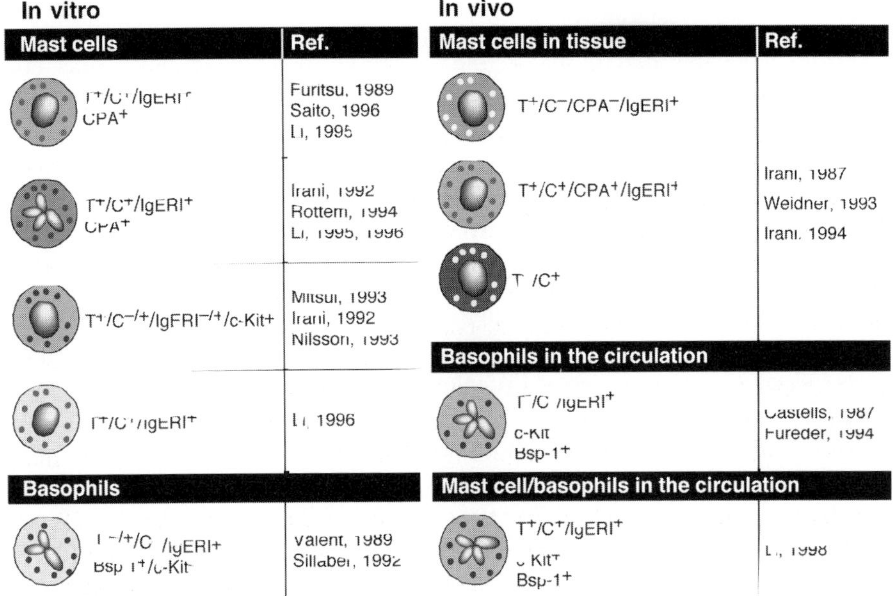

Fig. 1 Phenotypic characteristics of human mast cells and basophils. T, tryptase; C, chymase; IgERI, high-affinity IgE receptor; CPA, carboxypeptidase A; Bsp-1, basophil specific protein 1. *References*: Castells, 1987 (105); Fureder, 1994 (161); Furitsu, 1989 (192); Irani, 1987 (132); Irani, 1992 (79); Irani, 1994 (134); Li, 1995 (29); Li, 1996 (30); Li, 1996 (31); Mitsui, 1993 (82); Nilsson, 1993 (137); Rottem, 1994 (28); Saito, 1996 (193); Sillaber, 1992 (195); Valent, 1989 (194); Weidner, 1993 (133).

been cloned and sequenced (39–41). The α chain is mainly external, having only a short tail extending into the cytoplasm via one transmembrane (hydrophobic) domain. The β chain and each of the two disulphide-linked γ chains are predominantly cytoplasmic and have four and one transmembrane domains, respectively (41). The α and γ chains show substantial interspecies homology whereas the β chains are different between humans and rodents. Human basophils also express $Fc_{\varepsilon}RI$ as well as the low-affinity receptor $Fc_{\gamma}RII$ (42–44) and may also be activated by cross-linkage of IgE receptors.

In an elegant study it has been suggested that activation of basophils from atopic subjects may also occur by anti-IgG through IgG–IgE immune complexes acting on $Fc_{\varepsilon}RI$ (45). The expression of the $Fc_{\varepsilon}RI$ was once thought to be a unique characteristic of mast cells and basophils (46). However, it is also found on several potential effector cells such as Langerhan's cells in tissue (47, 48), eosinophils (49), monocytes (50) and peripheral blood dendritic cells (51). The mRNA transcripts for the α and γ chains were detected on highly purified Langerhan's cells using polymerase chain reaction (PCR) technology, but transcripts for the β chain were found only in a minority of individuals without correlation with atopic status (49). It seems that the β chain is necessary in effector cells of anaphylaxis and in cells containing pre-formed mediators packed in granules.

The cluster-of-differentiation (CD) features specific for discriminating mast cells and basophils have not yet been defined, although several studies have made the suggestion that mast cells could be isolated by using a particular pattern of differentiation antigens that are more commonly expressed on mast cells than on basophils (52, 53). It also appears that mast cells express many leukocyte antigens that are not found on basophils.

Common antigens found on both mast cells and basophils include CD45, CD55, CD59, CD63, histamine H_2 and prostaglandin D_2 receptors (54–58).

Adhesion molecules play a major role in cellular binding to other cells and to extracellular matrix glycoproteins. Integrins, which are the best known of them, are a family of non-covalently linked heterodimers of α and β subunits that mediate cell binding activity (59, 60). Both mast cells isolated by collagenase digestion from tissues and blood basophils express β_1 integrins (CD29, CD49d, CD49e) and low levels of intercellular adhesion molecule 1 (ICAM-1) (61), whereas β_2 integrins (CD18, CD11a, CD11b, CD11c), β_3 integrins (CD61/CD51) and vascular cell adhesion molecule 1 (VCAM-1) were only identified on basophils (62–64). Activation of basophils or mast cells results in enhanced expression of adhesion molecules and/or increased adherence to other cells and to extracellular matrix glycoproteins (65–69). A recent study showed that the cross-linking of surface expressed β_1 integrins on the basophils of asthmatic patients by antibody or fibronectin can trigger a significant release of histamine, but not in the case of basophils of normals or of human lung mast cells (70).

Haematopoietic cytokines play a potent role in cell differentiation and proliferation in both mouse and human. Intense studies over the years have demonstrated differences in cytokine-promoted development between mast cells and basophils, and between humans and rodents. Interleukins IL-3 and IL-5, and granulocyte–macrophage colony-stimulating factor (GM-CSF) which affect mouse mast cell growth are differentiation factors for human basophils but not for human mast cells (71, 72). It is noteworthy that human basophils, like eosinophils, express receptors for IL-3 (CD123), IL-5 (CD115) and GM-CSF (CD116) (57, 73), whereas the lack of receptors for IL-3, IL-5 and GM-CSF on tissue mast cells may explain why they do not respond to these cytokines alone. However, studies from cultured human mast cells showed that mRNAs for IL-3, IL-4, IL-5 and IL-6 receptors and binding sites for IL-3 can be detected (74).

The expression by mast cells but not basophils of c-kit (CD117), the receptor for kit ligand (variously stem cell factor, mast cell growth factor, steel factor) (75), is taken to be a key feature of distinction between the two cell types although we (see below) (31) and indirect evidence from others (76) have shown that this is not the case in all circumstances. The expression of c-kit on the mast cell surface is very important to the cell for differentiation, proliferation and survival (29, 75). Kit ligand (KL) is one of the major growth factors for mast cells and also for other haematopoietic cells (22, 29, 77–83).

The expression of urokinase receptors (CD87) on mast cells may also have implications for specific mast cell-dependent microvascular processes such as fibrinolysis, migration, or local tissue repair (84).

IL-8, a neutrophil chemotactic agent, is involved in a large number of neutrophil-driven acute and chronic inflammatory diseases. Its receptors are found on neutrophils, T lymphocytes, monocytes and keratinocytes. However, IL-8 is also involved in mast cell and basophil activation. Binding and competition studies with ^{125}I-labelled IL-8 revealed specific IL-8 receptors on basophils from normal individuals, and it was suggested that IL-8 activates human basophils by a receptor-mediated mechanism similar to that operating in neutrophils (85). Two subunits of IL-8 receptors called CXCR1 and CXCR2 have been identified (86, 87). Interestingly, in human mast cells, CXCR1 is expressed on the cell surface, whereas the CXCR2 receptor is located intracellularly in specific mast cell granules on immunoelectron microscopy (88).

One of the significant advances is the derivation of the IgM monoclonal bsp-1

antibody, which is used for purifying and isolating human basophils from blood (89, 90). It was also shown that bsp-1 is a cell surface-expressed protein and is found on the basophil but not on mast cells. Another integral membrane protein found on basophils is the 48-kDa CD40 marker which is also expressed on B cells, T cells, dendritic cells, monocytes and epithelial cells (91). Since CD40 has been clustered as a member of the nerve growth factor/tumour necrosis factor (NGF/TNF) receptor superfamily, the presence of CD40 receptor may influence cellular activation via signalling with CD40 ligand which amongst others is secreted by both basophils and mast cells.

A number of studies have searched for markers of the human mast cell progenitor. In 1991, Kirshenbaum *et al.* (92) demonstrated that both mast cell and basophil progenitors were from CD34$^+$ cells. CD34$^+$ cells cultured with recombinant human IL-3 (rhIL-3) gave rise to basophils, whereas CD34$^+$ cells co-cultured with 3T3 fibroblasts in the presence of rhIL-3 gave rise to mast cells (92). Further studies investigated additional markers such as c-kit, CD14 and CD17. The results showed that the clone which marked as CD34$^+$, c-kit$^+$, CD14$^-$ and CD17$^-$ and that separated from lymphocytes on counter flow centrifugation developed into mast cells (93). More recently, cells expressing CD34 and CD38 but without HLA-DR were found to be more likely to be human mast cell-committed progenitors (94) when single cells were cultured in a broth of various lineage promoting factors. The CD38 molecule is found on mature granulocytes, some CD34$^+$ myeloid cells and some lymphocytes. Interestingly, there were relatively few co-colonies developing from these single cells, suggesting quite an early differentiation given a stochastic model of lineage determination.

CYTOPLASMIC CONTENTS

Histamine and Heparin

One of the distinctive features of mast cells and basophils that was found in early studies is that they contain substantial amounts of histamine in their granules. Under normal conditions the content of histamine in mast cells is slightly higher than that in basophils. Exposure of human basophils to antigen or anti-IgE leads to a histamine release on average of 1 pg per cell (95), whereas lung, skin and synovial mast cells release approximately 4 pg per cell (96–98). The basophil histamine release requires 20–30 min for completion in comparison to about 10 min for mast cells (99).

Studies by Stevens *et al.* (100) and Thompson *et al.* (101) have shown that human lung mast cells contain a mixture of both heparin and chondroitin sulphate E, whereas Metcalfe *et al.* (102) showed that chondroitin sulphate A is the dominant proteoglycan in basophils. The presence of highly sulphated proteoglycans in the secretory granules of mast cells and basophils results in metachromasia which is visible by toluidine blue stain. The release of histamine and heparin could enhance neo-vascularization and endothelial cell proliferation in tissues (103, 104).

Proteolytic Enzymes

The two major proteases found in mast cell granules are chymase and tryptase. At the present time these proteases are considered to be the most selective markers for mast cells, as basophils from normal individuals contain only negligible amounts of tryptase protein and no detectable chymase protein (105). Immunohistochemical staining using antibodies to tryptase and chymase can be utilized to distinguish mast cells from basophils and other cells (106, 107).

The most predominant of the proteins comprising mast cell granules is tryptase. It has been purified from nearly all mast cells, including dispersed human lung mast cells (108), isolated skin mast cells (109) and pituitary tissue mast cells (110). It has been estimated that there is a concentration of 12–35 pg per lung or skin mast cell (111) but only 0.04 pg of tryptase per basophil which normally is not able to be detected by immunohistoreaction (105).

Four highly conserved forms of tryptase (designated tryptase I, II/β, III and α) have been described, all with a 30-amino-acid leader sequence followed by a catalytic portion of 27.4 kDa (112–115). Tryptase is bound to a proteoglycan complex in the granules that is smaller than those binding other proteases (116). The presence of histamine seems to be necessary to stabilize tryptase activity (117). According to Sakai *et al.* (118), only β-tryptase is processed to a functional form or active enzyme within the mast cells *in vivo*. However, recent findings by Stevens' group demonstrate that both α- and β-tryptase are active *in vitro* (119).

Chymase is the second major human mast cell protease (120–123) and is found at high levels in connective tissues such as the skin and submucosa of the gut and at low levels in the lung. Only one human chymase gene has been identified so far. However, the investigation carried out by McEuen *et al.* (124) showed that there are two different forms of human skin chymase by heparin–agarose chromatography. The activities of these two forms were similar in substrate specificity but the inhibitor profile was distinctly different from other chymotryptic enzymes. The study also revealed that there were different levels of expression for these two forms of chymase in the tissues (124).

Two other proteases, carboxypeptidase A (CPA) and cathepsin G, have been associated with mast cells with a chymase-expressing phenotype. Carboxypeptidase is a unique 36-kDa protease and is encoded by a single gene localized to chromosome 3 (125–127). The results of sequenced CPA cDNA sequences from both human skin and lung suggest only a single CPA gene in humans (128). Cathepsin G is a 26-kDa serine protease and a chymotryptic enzyme with a structure seemingly identical to that of neutrophil and monocyte cathepsin G (129). Cathepsin G and CPA are selectively present in the tryptase-positive/chymase-positive mast cell phenotype (130, 131).

Like T cells, mast cells exhibit substantial phenotypic and sometimes ultrastructural variation in humans. The most widely adopted classification is descriptive and based on granule protease expression. Mast cells that express tryptase along with chymase, CPA and a cathepsin G-like protease are named MC_{TC} and are found predominantly in connective tissue of skin, submucosa of stomach, submucosa of intestine, submucosa of colon and breast parenchyma, lymph nodes, conjunctiva and synovium (132–134). At human cardiac explantation the mast cells are mainly MC_{TC} and are located primarily in the appendage of the atrium (135).

Mast cells with tryptase alone are designated MC_T; these cells predominate in the intestinal mucosa, nasal mucosa and lung (132–134). In the uterus most mast cells are limited to the inner half of the myometrium with equal proportions of MC_T and MC_{TC}. There are fewer mast cells, mostly MC_{TC}, in the outer half of the myometrium and cervix, and in the endometrium, where they are mostly MC_T (136).

There is also a third type of mast cell, MC_C; these express chymase without tryptase and reside mainly in the submucosa and mucosa of the stomach, small intestinal submucosa and colonic mucosa (132–134).

Certain mast cell phenotypes can also be promoted *in vitro*. Cultures of haematopoietic progenitor cells in the presence of KL alone result in selective differentiation to mast cells of MC_T phenotype. Mast cells derived from fetal liver by culture with rhKL were predominantly of the MC_T phenotype, as were mast cells derived by co-culture of human fetal liver cells with mouse 3T3 fibroblasts (79, 137). The conditioned media obtained from a cell strain from the bone marrow of a mastocytosis patient can influence mast cell phenotypes. Li *et al.* (30) were able to show that using this conditioned media preferential MC_{TC} and MC_C phenotypes developed from unseparated bone marrow cells. Furthermore, the conditioned media with added rhKL also upregulated MC_C and MC_{TC} and downregulated MC_T phenotypes from umbilical cord blood (30).

Lipid-derived Mediators

In addition to histamine and tryptase, cross-linking of IgE on human lung mast cells also leads to the synthesis of substantial amounts of arachidonic acid metabolites, including prostaglandin D_2 (PGD_2), leukotriene C_4 (LTC_4), platelet-activating factor (PAF) and some LTB_4 (138–140). It has been noted that lung, heart and gut mast cells contain more LTC_4 and PGD_2 whereas skin mast cells synthesize more PGD_2 and much less or no LTC_4 (97, 141). In comparison, human basophils generate LTC_4 and little or no PGD_2 (139, 141). Mast cells also synthesize PGE_2, which may contribute to the inflammation which characterizes the late-phase response (141). PAF is another newly synthesized lipid-derived mediator with the ability to influence a range of different aspects of inflammation. PAF is synthesized by both human lung mast cells and basophils (142, 143). The release of PAF either from mast cells or basophils amplifies the inflammatory response, leading to the activation and recruitment of neutrophils and eosinophils.

Cytokines

In a similar fashion to that found in Th2 cells, the activation of mast cells and basophils leads to the *de novo* synthesis of several cytokines, although the mechanism has not yet been completely clarified. Basophils have a more restricted armamentarium of cytokines than mast cells do. Since cytokines play an important role in cell regulation, the recognition that mast cells produce a large number of cytokines suggests that mast cells may influence other cells within the microenvironment or play a more central role in subacute and chronic inflammatory processes.

Two cytokines have so far been shown to be produced following basophil activation: IL-4 is synthesized over a period of hours following IgE cross-linking (144, 145); the related IL-13 has also been found to be released at increasing levels over 24 h after basophil activation where it can be immunolocalized to the granules (146). It has been shown that human mast cells can secrete a greater number of cytokines than basophils. The first demonstration of mast cell cytokine production was of IL-3 and TNF-α by peritoneal mast cells (147) using immunohistostaining, but now at least nine different cytokines have been shown to be produced by tissue mast cells including TNF-α, IL-1, IL-4, IL-5, IL-6, IL-8, IL-13, NGF and KL. Human skin mast cells contain TNF-α protein and mRNA which is responsible for the upregulation of the endothelial cell adhesion molecule (ELAM-1) (148). Under allergic conditions, cutaneous mast cells also produce significant amounts of IL-1 that may contribute to lymphocytic infiltration (149).

Freshly purified human lung mast cells release IL-4, essential for the triggering of Th2 lymphocytes that themselves produce IL-4 to initiate inflammatory cell accumulation and B lymphocyte immunoglobulin class switching to IgE (150, 151). Other cytokines involved in mast cells found in normals and in the asthmatic airway are IL-5 and IL-6 which together with IL-4 and TNF-α are detected by immunostaining and would indicate that mast cells may play an important role in initiating and maintaining the inflammatory response in asthma (151). A study on human skin mast cells showed that the capability of mast cells to generate IL-8 may contribute to neutrophil recruitment (152). On activation, human lung mast cells produce IL-13 with a secretion pattern that is comparable to the IL-13 release from peripheral blood Th2 cells (153). More recently, it was found that human mast cells express NGF (154) and also IL-16, a chemoattractant factor, which is another newly found cytokine from lung and bone marrow cultured mast cells and may suggest that mast cells induce the accumulation of $CD4^+$ T cells in the inflammatory process (155).

Mast cells express the c-kit receptor which is essential for mast cell proliferation and some aspects of maturation. Recent experiments have demonstrated that mast cells from the heart (156), skin and lung (157) are also able to synthesize the mRNA of KL and secrete the protein.

In addition to IL-4 and IL-13, lung mast cells and blood basophils share the ability to express CD40 ligand. CD40 ligand shares significant amino acid homology with members of the NGF/TNF family and is viewed as a member of the NGF/TNF superfamily. The interaction between CD40 ligand as expressed by mast cells and CD40 on B cells, along with the binding of IL-4 to IL-4 receptor on B cells, satisfies the minimum requisite stimulus for immunoglobulin class switching to IgE (158) and may also have effects on B cell growth and transcriptional regulation (159). It suggests that mast cells and basophils play a key role in allergy, not only by producing inflammatory mediators, but also by directly regulating IgE production independently of T cells (158, 160).

MAST CELL PHENOTYPES IN TISSUES

Although mast cells from most human tissue sites have a similar ultrastructural appearance, they are slightly distinct in their surface markers, protease content, cytokine release and response to some external stimuli. Mast cells from adult skin and foreskin, but generally not elsewhere, express the additional marker CD88/C5aR, which also appears on the surface of the blood basophil (57, 58, 161). Uterine mast cells express CD11e/CD18 (64) and both uterine and cutaneous mast cells also exhibit CD32 marker. In patients with rheumatoid arthritis significant amounts of C5aR (CD88) was detected on synovial mast cells (162).

Using antibody to cross-link IgE to stimulate mast cells from skin, heart and lung researchers were able to demonstrate that the tryptase content in skin mast cells was much higher than that of heart and lung mast cells, whereas the release of tryptase in lung mast cells was the lowest of the three tissues examined. Heart mast cells isolated from the explanted diseased heart obtained at transplant were stimulated *in vitro*. The *de novo* synthesized mediators LTC_4 and PGD_2 were released from these mast cells, whereas PGD_2 with only a little LTC_4 was released from skin mast cells. Moreover, mast cells from skin seem to be very sensitive to stimulation, with C5a, substance P, compound 48/80 and morphine all leading to histamine release, whereas heart mast cells only reacted to C5a and compound 48/80 and lung mast cells failed to respond to C5a (163).

High chymase-like enzymatic activity was also detected in skin, uterus and stomach tissues, whereas lower levels of chymase enzymatic activity were present in lung, tonsil, colon and liver (164). In comparison to skin mast cells, uterine mast cells do not react to substance P but respond strongly with anti-IgE to release a significant amount of LTC_4 and in fact produce similar arachidonic acid metabolites to lung mast cells. Ultrastructurally, however, there are no differences between the mast cells from skin, lung and uterus (165, 166).

The results from examining the protease profile of mast cells with their differences in neutral protease composition suggest that the different mast cell phenotypes could serve differing biological and pathological roles. By examining mast cells for their specific cytokine contents, Bradding *et al.* (167) demonstrated that MC_T express IL-5, IL-6 and some IL-4, whereas MC_{TC} preferentially express IL-4 but have very little IL-5 and IL-6 in the tissues of bronchial and nasal mucosae from normal, asthmatic and allergic rhinitis patients. They found a similar predominant IL-4 pattern in skin mast cells which contain both tryptase and chymase (MC_{TC}) (167). However, there is no significant difference between MC_{TC} from skin and MC_T from lung tissue in terms of their release of KL (157).

Mast cells migrate to tissue sites with different microenvironments where they develop distinct phenotypes with different expression of serine proteases. A recent study by Longley *et al.* (12) clearly demonstrated that the microenvironment of tissues does influence mast cell phenotypic development, even if, ultrastructurally, they appear similar. Examination of mast cells of patients with urticaria pigmentosa and aggressive systemic mastocytosis whose pathological mast cells are monoclonally derived and chronically stimulated by KL, indicates that mast cells in spleen express MC_T but that in the skin the majority of mast cells are MC_{TC}. It also suggests that the KL axis does not irrevocably commit mast cells to a chymase-positive phenotype (12). The earlier observation of reduction of MC_T numbers when compared to normals in the intestinal mucosa but not MC_{TC} numbers in submucosa was obtained in patients with congenital combined immunodeficiency or acquired immunodeficiency syndrome (132). This suggests that functional T cells or their cytokines are necessary for the phenotypic development of MC_T cells in the mucosa, but that in the submucosal microenvironment T cells do not appear to be required for the development of MC_{TC}.

In the upper dermis of skin from patients with psoriasis a significantly increased number of MC_T cells are found, whereas 99% of mast cells are MC_{TC} in normal skin (168). Nasal epithelium is normally dominated by MC_{TC}; however, MC_T phenotypes were consistently increased in allergic rhinitis (169–171). The same alteration of phenotype pattern was also seen in the synovial layer in osteoarthritis and in the bladder epithelium of nephrogenic metaplasia, where in both cases the number of MC_T was significantly higher than usual (172, 173). However, in some other pathological states MC_{TC} are the predominant cell type; for example, in atherosclerosis 80–95% of mast cells in the intima of carotid arteries are MC_{TC}. There are similar findings in active vernal conjunctivitis, giant papillary conjunctivitis and allergic conjunctivitis (174, 175).

THE INVOLVEMENT OF MAST CELLS AND BASOPHILS IN ASTHMA AND DRUG REACTIONS

Allergic asthma is a complex disorder characterized by local and systemic allergic inflammation and reversible airway obstruction. There is clearly extensive evidence that mast cells and basophils act as primary effectors of allergic inflammation (176–184). Under normal conditions the level of tryptase in serum or plasma is less than 5 ng ml^{-1} (105). There are significant increases after allergen challenge in nasal lavage fluid (185) and in skin chamber fluid overlying the site of exposed dermis in an *in vivo* blister model in allergic patients for testing the direct exposure of skin mast cells to allergens (186). Redington *et al.* have stated that mast cells are more likely to be involved in the acute or early phase of the allergic response, whereas basophils are well established as the major effector cells of the chronic or late phase. One of the pieces of supportive evidence for this is that initially there is an elevation of both histamine and PGD_2 as found in mast cells, whereas there is only histamine in the second rise of the late-phase response (184).

Mast cells and basophils may be involved in drug hypersensitivity reactions. At least one of the groups clinically imitates IgE-mediated reactions, which are probably caused by non-immune mechanisms leading to the degranulation of mast cells and basophils (187, 188). The pathogenesis of the adverse reactions caused by different classes of drugs is complex and still not completely understood. *In vitro* and *in vivo* studies indicate that the release of vasoactive mediators, such as histamine and tryptase, from peripheral blood basophils and tissue mast cells plays a crucial role in determining the clinical manifestations (189).

More recently, direct evidence has revealed that human blood basophilic cells are able to store or synthesize tryptase and/or chymase and CPA, which may contribute to the level of proteases in allergic conditions (31). This study involved four groups: normals, allergy and asthma patients and those with drug allergies. Mononuclear cells from peripheral blood were prepared for the detection of antigens such as tryptase, chymase, c-kit and bsp-1. It showed that bsp-1^+ cells were increased from 0.72% to 1.4% (allergy), 1.5% (asthma) and 2.3% (drug allergy), and that nearly 50% of these basophilic cells stained for tryptase or/and chymase. Using *in situ* hybridization with chymase and α- and β-specific tryptase probes, it appeared that in normals the predominant staining was for the α-tryptase transcript with negligible staining for β-tryptase and with no detectable staining for chymase. However, both chymase and α- and β-tryptase transcripts were upregulated in basophilic cells from individuals with asthma and in those experiencing adverse drug reactions. Four out of eight patients who were exposed to allopurinol, ceftriaxone or penicillin showed significant upregulation of chymase-positive basophilic cells, whereas the other four patients experiencing amlodipidine, isoniazid or ceftriaxone drug reactions appeared to have a predominantly tryptase immunoreactivity with few chymase-positive cells. This is the first time that basophilic cells have been documented in the blood of patients with allergic conditions to show a pattern that normally is seen in mast cells in terms of the neutral protease content.

It is not clear at the present time whether mast cells and basophils can reversibly alter the expression of their neutral proteases. Nevertheless, several previous studies from rodent or human systems have shown strong support for this possibility. Friend *et al.* (190) provided evidence that mast cell subtypes can reversibly alter their expression of serine proteases, and more recently mouse mast cells were found to be able to alter their expression *in vivo* of multiple members of two distinct families of serine proteases (191). In humans it has been noted that clonal human mast cells differ in terms of their protease pattern or in their chymase and tryptase levels when these mast cells reside in varied tissue sites of a patient with systemic mastocytosis (12).

In a study by Li *et al.* (31), the peripheral blood leukocytes of patients with an allergic (atopic) disorder, asthma or drug reaction were found to have an increased proportion of metachromatic cells morphologically resembling basophils and expressing bsp-1, as well as in the majority expressing any combination of c-kit, tryptase or chymase. This study clearly demonstrated that: (i) there is an increased number of metachromatic cells in the blood of individuals with an allergic disorder; (ii) those metachromatic cells in patients with allergy, asthma or those experiencing an allergic drug reaction expressed substantial amounts of tryptase and/or chymase and CPA that are considered to be mast cell-specific proteases; (iii) interestingly most of the metachromatic tryptase- and/or chymase-positive cells also reacted to the basophil-specific marker bsp-1 which is not expressed on mast cells under normal conditions; in addition, these cells were generally of small size with multilobed nuclei which are more typical characteristics of basophils.

Therefore the presence of metachromatic cells with features of both basophils and mast cells in the peripheral blood of these patients suggests: (i) that human mast cells and basophils may under certain conditions change their contents, including their proteolytic enzymes; (ii) that these cells may accelerate their synthesis of mediators for which mRNA has been encoded; and (iii) that different phenotypes may arise from the same progenitor and the differentiated cells may not be fixed in that state but rather that it may depend upon the microenvironment.

Many of the distinctions between mast cells and basophils such as the larger size, larger uncondensed monolobed nucleus and increased cytokine expression of mast cells are not specific for these two cell classes but rather are consistent with a heightened state of cellular activation in mast cells. Likewise, the increased expression of β_2 integrins and VCAM-1 on basophils may be interpreted as reflecting a functional requirement to be able to migrate into tissues rather than a determinant of cell lineage. The metachromatic cells found in the blood of allergic patients (31) do not fit well in the mast cell/basophil dichotomy. It remains unclear whether they are best described as functional mast cells that, rather unusually, are found in the blood, or as basophils expressing supposedly mast cell-specific antigens. This research calls into question the prevailing view that mast cells and basophils are separate lineages.

ACKNOWLEDGEMENTS

This work was supported by grants from the National Health and Medical Research Council of Australia and the Clive and Vera Ramaciotti Foundation.

REFERENCES

1. Schwartz, L. B. and Austen, K. F. Structure and function of the chemical mediators of mast cells. *Prog. Allergy* **34**:271–321, 1984.
2. Echtenacher, B., Mannel, D. N. and Hultner, L. Critical protective role of mast cells in a model of acute septic peritonitis. *Nature* **381**:75–77, 1996.
3. Malaviya, R., Ikeda, T., Ross, E. and Abraham, S. N. Mast cell modulation of neutrophil influx and bacterial clearance at sites of infection through TNF-alpha. *Nature* **381**:77–80, 1996.
4. Cairns, J. A. and Walls, A. F. Mast cell tryptase is a mitogen for epithelial cells. Stimulation of IL-8 production and intercellular adhesion molecule-1 expression. *J. Immunol.* **156**:275–283, 1996.
5. Gruber, B. L., Kew, R. R., Jelaska, A., Marchese, M. J., Garlick, J., Ren, S., Schwartz, L. B. and Korn, J. H. Human mast cells activate fibroblasts: tryptase is a fibrogenic factor stimulating collagen messenger ribonucleic acid synthesis and fibroblast chemotaxis. *J. Immunol.* **158**:2310–2317, 1997.

6. Kofford, M. W., Schwartz, L. B., Schechter, N. M., Yager, D. R., Diegelmann, R. F. and Graham, M. F. Cleavage of type I procollagen by human mast cell chymase initiates collagen fibril formation and generates a unique carboxyl-terminal propeptide. *J. Biol. Chem.* **272**:7127–7131, 1997.
7. Blair, R. J., Meng, H., Marchese, M. J., Ren, S., Schwartz, L. B., Tonnesen, M. G. and Gruber, B. L. Human mast cells stimulate vascular tube formation. Tryptase is a novel, potent angiogenic factor. *J. Clin. Invest.* **99**:2691–2700, 1997.
8. Schroeder, J. T., Kagey-Sobotka, A. and Lichtenstein, L. M. The role of the basophil in allergic inflammation. *Allergy* **50**:463–472, 1995.
9. Levi-Schaffer, F., Austen, K. F., Gravallese, P. M. and Stevens, R. L. Coculture of interleukin 3-dependent mouse mast cells with fibroblasts results in a phenotypic change of the mast cells. *Proc. Natl. Acad. Sci. USA* **83**:6485–6488, 1986.
10. Ghildyal, N., Friend, D. S., Nicodemus, C. F., Austen, K. F. and Stevens, R. L. Reversible expression of mouse mast cell protease 2 mRNA and protein in cultured mast cells exposed to IL-10. *J. Immunol.* **151**:3206–3214, 1993.
11. Gurish, M. F., Pear, W. S., Stevens, R. L., Scott, M. L., Sokol, K., Ghildyal, N., Webster, M. J., Hu, X., Austen, K. F. and Baltimore, D. Tissue-regulated differentiation and maturation of a v-abl-immortalized mast cell-committed progenitor. *Immunity* **3**:175–186, 1995.
12. Longley, B. J., Tyrrell, L., Lu, S., Ma, Y., Klump, V. and Murphy, G. F. Chronically KIT-stimulated clonally-derived human mast cells show heterogeneity in different tissue microenvironments. *J. Invest. Dermatol.* **108**:792–796, 1997.
13. Liu, M. C., Hubbard, W. C., Proud, D., Stealey, B. A., Galli, S. J., Kagey-Sobotka, A., Bleecker, E. R. and Lichtenstein, L. M. Immediate and late inflammatory responses to ragweed antigen challenge of the peripheral airways in allergic asthmatics. Cellular, mediator, and permeability changes. *Am. Rev. Respir. Dis.* **144**:51–58, 1991.
14. Charlesworth, E. N., Kagey-Sobotka, A., Schleimer, R. P., Norman, P. S. and Lichtenstein, L. M. Prednisone inhibits the appearance of inflammatory mediators and the influx of eosinophils and basophils associated with the cutaneous late-phase response to allergen. *J. Immunol.* **146**:671–676, 1991.
15. Lison, L. Etudes sur la metachromasie colorants metachromatiques et substances chromotropes. *Arch. Biol.* **46**:599–668, 1935.
16. Holmgren, H. and Wilander, O. Beitrag zur Kenntris der Chemie und Funktion der Ehrlichschen Mastzellen. *Z. Mikrosk. Anat. Forsch.* **42**:242–278, 1937.
17. Kramer, H. and Windrum, G. M. The metachromatic staining reaction. *Cytochemistry* **3**:227–237, 1955.
18. Ackerman, G. A. Cytochemical properties of the blood basophilic granulocyte. *Ann. N. Y. Acad. Sci.* **103**:376–393, 1963.
19. Dvorak, A. M. Human mast cells. *Adv. Anat. Embryol. Cell Biol.* **114**:1–107, 1989.
20. Eguchi, M. Comparative electron microscopy of basophils and mast cells, *in vivo* and *in vitro*. *Electron Microsc. Rev.* **4**:293–318, 1991.
21. Dvorak, A. M. Basophil and mast cell degranulation and recovery. In: *Blood Cell Biochemistry* (Harris, R. J., ed.), Plenum Press, New York, 1991.
22. Galli, S. J. and Hammel, I. Mast cell and basophil development. *Curr. Opin. Hematol.* **1**:33–39, 1994.
23. Galli, S. J., Dvorak, A. M., Marcum, J. A., Ishizaka, T., Nabel, G., Der Simonian, H., Pyne, K., Goldin, J. M., Rosenberg, R. D., Cantor, H. and Dvorak, H. F. Mast cell clones: a model for the analysis of cellular maturation. *J. Cell Biol.* **95**:435–444, 1982.
24. Nakano, T., Sonoda, T., Hayashi, C., Yamatodani, A., Kanayama, Y., Yamamura, T., Asai, H., Yonezawa, T., Kitamura, Y. and Galli, S. J. Fate of bone marrow-derived cultured mast cells after intracutaneous, intraperitoneal, and intravenous transfer into genetically mast cell-deficient W/Wv mice. Evidence that cultured mast cells can give rise to both connective tissue type and mucosal mast cells. *J. Exp. Med.* **162**:1025–1043, 1985.
25. Dvorak, A. M., Wiberg, L., Monahan-Earley, R. A. and Galli, S. J. A simple technique to facilitate the ultrastructural analysis of cells in soft agar culture systems: demonstration of the development *in vitro* of morphologically mature mast cells and phagocytic macrophages from the bone marrow cells of genetically mast cell-deficient W/Wv or congenic normal mice. *Lab. Invest.* **62**:774–781, 1990.
26. Rottem, M., Goff, J. P., Albert, J. P. and Metcalfe, D. D. The effects of stem cell factor on the ultrastructure of Fc epsilon RI+ cells developing in IL-3-dependent murine bone marrow-derived cell cultures. *J. Immunol.* **151**:4950–4963, 1993.
27. Irani, A. M., Nilsson, G., Miettinen, U., Craig, S. S., Ashman, L. K., Ishizaka, T., Zsebo, K. M. and Schwartz, L. B. Recombinant human stem cell factor stimulates differentiation of mast cells from dispersed human fetal liver cells. *Blood* **80**:3009–3021, 1992.

28. Rottem, M., Okada, T., Goff, J. P. and Metcalfe, D. D. Mast cells cultured from the peripheral blood of normal donors and patients with mastocytosis originate from a CD34+/Fc epsilon RI– cell population. *Blood* **84**:2489–2496, 1994.
29. Li, L., Macpherson, J. J., Adelstein, S., Bunn, C. L., Atkinson, K., Phadke, K. and Krilis, S. A. Conditioned media from a cell strain derived from a patient with mastocytosis induces preferential development of cells that possess high affinity IgE receptors and the granule protease phenotype of mature cutaneous mast cells. *J. Biol. Chem.* **270**:2258–2263, 1995.
30. Li, L., Meng, X. W. and Krilis, S. A. Mast cells expressing chymase but not tryptase can be derived by culturing human progenitors in conditioned medium obtained from a human mastocytosis cell strain with c-kit ligand. *J. Immunol.* **156**: 4839–4844, 1996.
31. Li, L., Li, Y., Reddel, S. W., Cherrian, M., Friend, D. S., Stevens, R. L. and Krilis, S. A. Identification of basophilic cells in the peripheral blood of asthma, allergy, and drug-reactive patients that express mast cell granule protease. *J. Immunol.* **161**:5079–5086, 1998.
32. Gurish, M. F., Friend, D. S., Webster, M., Ghildyal, N., Nicodemus, C. F. and Stevens, R. L. Mouse mast cells that possess segmented/multi-lobular nuclei. *Blood* **90**:382–390, 1997.
33. Craig, S. S., Schechter, N. M. and Schwartz, L. B. Ultrastructural analysis of human T and TC mast cells identified by immunoelectron microscopy. *Lab. Invest.* **58**:682–691, 1988.
34. Craig, S. S., Schechter, N. M. and Schwartz, L. B. Ultrastructural analysis of maturing human T and TC mast cells *in situ*. *Lab. Invest.* **60**:147–157, 1989.
35. Craig, S. S. and Schwartz, L. B. Human MCTC type of mast cell granule: the uncommon occurrence of discrete scrolls associated with focal absence of chymase. *Lab. Invest.* **63**:581–585, 1990.
36. Weidner, N. and Austen, K. F. Ultrastructural and immunohistochemical characterization of normal mast cells at multiple body sites. *J. Invest. Dermatol.* **96**:26S–31S, 1991.
37. Ackerman, S. J., Weil, G. J. and Gleich, G. J. Formation of Charcot–Leyden crystals by human basophils. *J. Exp. Med.* **155**:1597–1609, 1982.
38. Dvorak, A. M. and Ackerman, S. J. Ultrastructural localization of the Charcot–Leyden crystal protein (lysophospholipase) to granules and intragranular crystals in mature human basophils. *Lab. Invest.* **60**:557–567, 1989.
39. Kochan, J., Pettine, L. F., Hakimi, J., Kishi, K. and Kinet, J. P. Isolation of the gene coding for the alpha subunit of the human high affinity IgE receptor. *Nucleic Acids Res.* **16**:3584, 1988.
40. Kinet, J. P., Blank, U., Ra, C., White, K., Metzger, H. and Kochan, J. Isolation and characterization of cDNAs coding for the beta subunit of the high-affinity receptor for immunoglobulin E. *Proc. Natl. Acad. Sci. USA* **85**:6483–6487, 1988.
41. Blank, U., Ra, C., Miller, L., White, K., Metzger, H. and Kinet, J. P. Complete structure and expression in transfected cells of high affinity IgE receptor. *Nature* **337**:187–189, 1989.
42. Ishizaka, T., Sterk, A. R. and Ishizaka, K. Demonstration of Fcgamma receptors on human basophil granulocytes. *J. Immunol.* **123**:578–583, 1979.
43. Anselmino, L. M., Perussia, B. and Thomas, L. L. Human basophils selectively express the Fc gamma RII (CDw32) subtype of IgG receptor. *J. Allergy Clin. Immunol.* **84**:907–914, 1989.
44. Guo, C. B., Liu, M. C., Galli, S. J., Bochner, B. S., Kagey-Sobotka, A. and Lichtenstein, L. M. Identification of IgE-bearing cells in the late-phase response to antigen in the lung as basophils. *Am. J. Respir. Cell Mol. Biol.* **10**:384–390, 1994.
45. Lichtenstein, L. M., Kagey-Sobotka, A., White, J. M. and Hamilton, R. G. Anti-human IgG causes basophil histamine release by acting on IgG–IgE complexes bound to IgE receptors. *J. Immunol.* **148**:3929–3936, 1992.
46. MacGlashan D. W. Jr, Schleimer, R. P., Peters, S. P., Schulman, E. S., Adams, G. K., Sobotka, A. K., Newball, H. H. and Lichtenstein, L. M. Comparative studies of human basophils and mast cells. *Fed. Proc.* **42**:2504–2509, 1983.
47. Bieber, T., de la Salle, H., Wollenberg, A., Hakimi, J., Chizzonite, R., Ring, J. and Hanau, D. Human epidermal Langerhans cells express the high affinity receptor for immunoglobulin E (Fc epsilon RI). *J. Exp. Med.* **175**:1285–1290, 1992.
48. Brini, A. T., Schmidt, L., Zbar, B. and Kinet, J. P. A dinucleotide repeat polymorphism in the gene for the gamma subunit of the human Fc epsilon receptors (FLER16). *Hum. Mol. Genet.* **2**:619, 1993.
49. Gounni, A. S., Lamkhioued, B., Ochiai, K., Tanaka, Y., Delaporte, E., Capron, A., Kinet, J. P. and Capron, M. High-affinity IgE receptor on eosinophils is involved in defence against parasites. *Nature* **367**:183–186, 1994.
50. Sihra, B. S., Kon, O. M., Grant, J. A. and Kay, A. B. Expression of high-affinity IgE receptors (Fc epsilon RI) on peripheral blood basophils, monocytes, and eosinophils in atopic and nonatopic subjects:

relationship to total serum IgE concentrations. *J. Allergy Clin. Immunol.* **99**:699–706, 1997.

51. Maurer, D., Fiebiger, S., Ebner, C., Reininger, B., Fischer, G. F., Wichlas, S., Jouvin, M. H., Schmitt-Egenolf, M., Kraft, D., Kinet, J. P. and Stingl, G. Peripheral blood dendritic cells express Fc epsilon RI as a complex composed of Fc epsilon RI alpha- and Fc epsilon RI gamma-chains and can use this receptor for IgE-mediated allergen presentation. *J. Immunol.* **157**:607–616, 1996.
52. Valent, P. 1995 Mack-Forster Award Lecture. Review. Mast cell differentiation antigens: expression in normal and malignant cells and use for diagnostic purposes. *Eur. J. Clin. Invest.* **25**:715–720, 1995.
53. Willheim, M., Agis, H., Sperr, W. R., Koller, M., Bankl, H. C., Kiener, H., Fritsch, G., Fureder, W., Spittler, A. and Graninger, W. Purification of human basophils and mast cells by multistep separation technique and mAb to CDw17 and CD117/c-kit. *J. Immunol. Methods* **182**:115–129, 1995.
54. Lichtenstein, L. M. and Gillespie, E. Inhibition of histamine release by histamine controlled by H2 receptor. *Nature* **244**:287–288, 1973.
55. Knol, E. F., Mul, F. P., Jansen, H., Calafat, J. and Roos, D. Monitoring human basophil activation via CD63 monoclonal antibody 435. *J. Allergy Clin. Immunol.* **88**:328–338, 1991.
56. Virgolini, I., Li, S., Sillaber, C., Majdic, O., Sinzinger, H., Lechner, K., Bettelheim, P. and Valent, P. Characterization of prostaglandin (PG)-binding sites expressed on human basophils. Evidence for a prostaglandin E1, I2, and a D2 receptor. *J. Biol. Chem.* **267**:12700–12708, 1992.
57. Fureder, W., Agis, H., Willheim, M., Bankl, H. C., Maier, U., Kishi, K., Muller, M. R., Czerwenka, K., Radaszkiewicz, T. and Butterfield, J. H. Differential expression of complement receptors on human basophils and mast cells. Evidence for mast cell heterogeneity and CD88/C5aR expression on skin mast cells. *J. Immunol.* **155**:3152–3160, 1995.
58. Ghannadan, M., Baghestanian, M., Wimazal, F., Eisenmenger, M., Latal, D., Kargul, G., Walchshofer, S., Sillaber, C., Lechner, K. and Valent, P. Phenotypic characterization of human skin mast cells by combined staining with toluidine blue and CD antibodies. *J. Invest. Dermatol.* **111**:689–695, 1998.
59. Ruoslahti, E. and Pierschbacher, M. D. New perspectives in cell adhesion: RGD and integrins. *Science* **238**:491–497, 1987.
60. Buck, C. A. and Horwitz, A. F. Integrin, a transmembrane glycoprotein complex mediating cell-substratum adhesion. *J. Cell Sci.* **Suppl 8**:231–250, 1987.
61. Bochner, B. S., Luscinskas, F. W., Gimbrone, M. A. J., Newman, W., Sterbinsky, S. A., Derse-Anthony, C. P., Klunk, D. and Schleimer, R. P. Adhesion of human basophils, eosinophils, and neutrophils to interleukin 1-activated human vascular endothelial cells: contributions of endothelial cell adhesion molecules. *J. Exp. Med.* **173**:1553–1557, 1991.
62. Bochner, B. S., MacGlashan D. W. J.r, Marcotte, G. V. and Schleimer, R. P. IgE-dependent regulation of human basophil adherence to vascular endothelium. *J. Immunol.* **142**:3180–3186, 1989.
63. Sperr, W. R., Agis, H., Czerwenka, K., Klepetko, W., Kubista, E., Boltz-Nitulescu, G., Lechner, K. and Valent, P. Differential expression of cell surface integrins on human mast cells and human basophils. *Ann. Hematol.* **65**:10–16, 1992.
64. Guo, C. B., Kagey-Sobotka, A., Lichtenstein, L. M. and Bochner, B. S. Immunophenotyping and functional analysis of purified human uterine mast cells. *Blood* **79**:708–712, 1992.
65. Bochner, B. S., Peachell, P. T., Brown, K. E. and Schleimer, R. P. Adherence of human basophils to cultured umbilical vein endothelial cells. *J. Clin. Invest.* **81**:1355–1364, 1988.
66. Thompson, H. L., Burbelo, P. D., Segui-Real, B., Yamada, Y. and Metcalfe, D. D. Laminin promotes mast cell attachment. *J. Immunol.* **143**:2323–2327, 1989.
67. Bochner, B. S., McKelvey, A. A., Sterbinsky, S. A., Hildreth, J. E., Derse, C. P., Klunk, D. A., Lichtenstein, L. M. and Schleimer, R. P. IL-3 augments adhesiveness for endothelium and CD11b expression in human basophils but not neutrophils. *J. Immunol.* **145**:1832–1837, 1990.
68. Bochner, B. S. and Sterbinsky, S. A. Altered surface expression of CD11 and Leu 8 during human basophil degranulation. *J. Immunol.* **146**:2367–2373, 1991.
69. Dastych, J., Costa, J. J., Thompson, H. L. and Metcalfe, D. D. Mast cell adhesion to fibronectin. *Immunology* **73**:478–484, 1991.
70. Lavens, S. E., Goldring, K., Thomas, L. H. and Warner, J. A. Effects of integrin clustering on human lung mast cells and basophils. *Am. J. Respir. Cell Mol. Biol.* **14**:95–103, 1996.
71. Denburg, J. A., Silver, J. E. and Abrams, J. S. Interleukin-5 is a human basophilopoietin: induction of histamine content and basophilic differentiation of HL-60 cells and of peripheral blood basophil–eosinophil progenitors. *Blood* **77**:1462–1468, 1991.
72. Tsuda, T., Wong, D., Dolovich, J., Bienenstock, J., Marshall, J. and Denburg, J. A. Synergistic effects of nerve growth factor and granulocyte–macrophage colony-stimulating factor on human basophilic cell differentiation. *Blood* **77**:971–979, 1991.

73. Stain, C., Stockinger, H., Scharf, M., Jager, U., Gossinger, H., Lechner, K. and Bettelheim, P. Human blood basophils display a unique phenotype including activation linked membrane structures. *Blood* **70**:1872–1879, 1987.
74. Yanagida, M., Fukamachi, H., Ohgami, K., Kuwaki, T., Ishii, H., Uzumaki, H., Amano, K., Tokiwa, T., Mitsui, H., Saito, H., Likura, Y., Ishizaka, T. and Nakahata, T. Effects of T-helper 2-type cytokines, interleukin-3 (IL-3), IL-4, IL-5, and IL-6 on the survival of cultured human mast cells. *Blood* **86**:3705–3714, 1995.
75. Galli, S. J., Zsebo, K. M. and Geissler, E. N. The kit ligand, stem cell factor. *Adv. Immunol.* **55**:1–96, 1994.
76. Columbo, M., Horowitz, E. M., Botana, L. M., MacGlashan, D. W. Jr, Bochner, B. S., Gillis, S., Zsebo, K. M., Galli, S. J. and Lichtenstein, L. M. The human recombinant c-kit receptor ligand, rhSCF, induces mediator release from human cutaneous mast cells and enhances IgE-dependent mediator release from both skin mast cells and peripheral blood basophils. *J. Immunol.* **149**:599–608, 1992.
77. Zsebo, K. M., Williams, D. A., Geissler, E. N., Broudy, V. C., Martin, F. H., Atkins, H. L., Hsu, R. Y., Birkett, N. C., Okino, K. H. and Murdock, D. C. Stem cell factor is encoded at the Sl locus of the mouse and is the ligand for the c-kit tyrosine kinase receptor. *Cell* **63**:213–224, 1990.
78. Lerner, N. B., Nocka, K. H., Cole, S. R., Qiu, F. H., Strife, A., Ashman, L. K. and Besmer, P. Monoclonal antibody YB5.B8 identifies the human c-kit protein product. *Blood* **77**:1876–1883, 1991.
79. Irani, A. A., Craig, S. S., Nilsson, G., Ishizaka, T. and Schwartz, L. B. Characterization of human mast cells developed *in vitro* from fetal liver cells cocultured with murine 3T3 fibroblasts. *Immunology* **77**:136–143, 1992.
80. Irani, A. M., Nilsson, G., Miettinen, U., Craig, S. S., Ashman, L. K., Ishizaka, T., Zsebo, K. M. and Schwartz, L. B. Recombinant human stem cell factor stimulates differentiation of mast cells from dispersed human fetal liver cells. *Blood* **80**:3009–3021, 1992.
81. Valent, P., Spanblochl, E., Sperr, W. R., Sillaber, C., Zsebo, K. M., Agis, H., Strobl, H., Geissler, K., Bettelheim, P. and Lechner, K. Induction of differentiation of human mast cells from bone marrow and peripheral blood mononuclear cells by recombinant human stem cell factor/kit-ligand in long-term culture. *Blood* **80**:2237–2245, 1992.
82. Mitsui, H., Furitsu, T., Dvorak, A. M., Irani, A. M., Schwartz, L. B., Inagaki, N., Takei, M., Ishizaka, K., Zsebo, K. M. and Gillis, S. Development of human mast cells from umbilical cord blood cells by recombinant human and murine c-kit ligand. *Proc. Natl. Acad. Sci. USA* **90**:735–739, 1993.
83. Nilsson, G., Miettinen, U., Ishizaka, T., Ashman, L. K., Irani, A. M. and Schwartz, L. B. Interleukin-4 inhibits the expression of Kit and tryptase during stem cell factor-dependent development of human mast cells from fetal liver cells. *Blood* **84**:1519–1527, 1994.
84. Sillaber, C., Baghestanian, M., Hofbauer, R., Virgolini, I., Bankl, H. C., Fureder, W., Agis, H., Willheim, M., Leimer, M., Scheiner, O., Binder, B. R., Kiener, H. P., Bevec, D., Fritsch, G., Majdic, O., Kress, H. G., Gadner, H., Lechner, K. and Valent, P. Molecular and functional characterization of the urokinase receptor on human mast cells. *J. Biol. Chem.* **272**:7824–7832, 1997.
85. Krieger, M., Brunner, T., Bischoff, S. C., von Tscharner, V., Walz, A., Moser, B., Baggiolini, M. and Dahinden, C. A. Activation of human basophils through the IL-8 receptor. *J. Immunol.* **149**:2662–2667, 1992.
86. Su, S. B., Mukaida, N., Wang, J., Nomura, H. and Matsushima, K. Preparation of specific polyclonal antibodies to a C-C chemokine receptor, CCR1, and determination of CCR1 expression on various types of leukocytes. *J. Leukoc. Biol.* **60**:658–666, 1996.
87. Leong, S. R., Lowman, H. B., Liu, J., Shire, S., Deforge, L. E., Gillece-Castro, B. L., McDowell, R. and Hebert, C. A. IL-8 single-chain homodimers and heterodimers: interactions with chemokine receptors CXCR1, CXCR2, and DARC. *Protein Sci.* **6**:609–617, 1997.
88. Lippert, U., Artuc, M., Grutzkau, A., Moller, A., Kenderessy-Szabo, A., Schadendorf, D., Norgauer, J., Hartmann, K., Schweitzer-Stenner, R., Zuberbier, T., Henz, B. M. and Kruger-Krasagakes, S. Expression and functional activity of the IL-8 receptor type CXCR1 and CXCR2 on human mast cells. *J. Immunol.* **161**:2600–2608, 1998.
89. Bodger, M. P. and Newton, L. A. The purification of human basophils: their immunophenotype and cytochemistry. *Br. J. Haematol.* **67**:281–284, 1987.
90. Bodger, M. P., Mounsey, G. L., Nelson, J. and Fitzgerald, P. H. A monoclonal antibody reacting with human basophils. *Blood* **69**:1414–1418, 1987.
91. Valent, P., Majdic, O., Maurer, D., Bodger, M., Muhm, M. and Bettelheim, P. Further characterization of surface membrane structures expressed on human basophils and mast cells. *Int. Arch. Allergy Appl. Immunol.* **91**:198–203, 1990.

92. Kirshenbaum, A. S., Kessler, S. W., Goff, J. P. and Metcalfe, D. D. Demonstration of the origin of human mast cells from CD34+ bone marrow progenitor cells. *J. Immunol.* **146**:1410–1415, 1991.
93. Agis, H., Willheim, M., Sperr, W. R., Wilfing, A., Kromer, E., Kabrna, E., Spanblochl, E., Strobl, H., Geissler, K., Spittler, A., Boltz-Nitulescu, G., Majdic, O., Lechner, K. and Valent, P. Monocytes do not make mast cells when cultured in the presence of SCF. Characterization of the circulating mast cell progenitor as a c-kit+, CD34+, Ly–, CD14–, CD17–, colony-forming cell. *J. Immunol.* **151**:4221–4227, 1993.
94. Kempuraj, D., Saito, H., Kaneko, A., Fukagawa, K., Nakayama, M., Toru, H., Tomikawa, M., Tachimoto, H., Ebisawa, M., Akasawa, A., Miyagi, T., Kimura, H., Nakajima, T., Tsuji, K. and Nakahata, T. Characterization of mast cell-committed progenitors present in human umbilical cord blood. *Blood* **93**:3338–3346, 1999.
95. Marone, G., Giugliano, R., Lembo, G. and Ayala, F. Human basophil releasability. II. Changes in basophil releasability in patients with atopic dermatitis. *J. Invest. Dermatol.* **87**:19–23, 1986.
96. Peters, S. P., Kagey-Sobotka, A., MacGlashan, D. W. Jr, Siegel, M. I. and Lichtenstein, L. M. The modulation of human basophil histamine release by products of the 5-lipoxygenase pathway. *J. Immunol.* **129**:797–803, 1982.
97. Lawrence, I. D., Warner, J. A., Cohan, V. L., Hubbard, W. C., Kagey-Sobotka, A. and Lichtenstein, L. M. Purification and characterization of human skin mast cells. Evidence for human mast cell heterogeneity. *J. Immunol.* **139**:3062–3069, 1987.
98. Kopicky-Burd, J. A., Kagey-Sobotka, A., Peters, S. P., Dvorak, A. M., Lennox, D. W., Lichtenstein, L. M. and Wigley, F. M. Characterization of human synovial mast cells. *J. Rheumatol.* **15**:1326–1333, 1988.
99. Peters, S. P., Schulman, E. S., Schleimer, R. P., MacGlashan, D. W. Jr, Newball, H. H. and Lichtenstein, L. M. Dispersed human lung mast cells. Pharmacologic aspects and comparison with human lung tissue fragments. *Am. Rev. Respir. Dis.* **126**:1034–1039, 1982.
100. Stevens, R. L., Fox, C. C., Lichtenstein, L. M. and Austen, K. F. Identification of chondroitin sulfate E proteoglycans and heparin proteoglycans in the secretory granules of human lung mast cells. *Proc. Natl. Acad. Sci. USA* **85**:2284–2287, 1988.
101. Thompson, H. L., Schulman, E. S. and Metcalfe, D. D. Identification of chondroitin sulfate E in human lung mast cells. *J. Immunol.* **140**:2708–2713, 1988.
102. Metcalfe, D. D., Bland, C. E. and Wasserman, S. I. Biochemical and functional characterization of proteoglycans isolated from basophils of patients with chronic myelogenous leukemia. *J. Immunol.* **132**:1943–1950, 1984.
103. Folkman, J. How is blood vessel growth regulated in normal and neoplastic tissue? G. H. A. Clowes memorial Award lecture. *Cancer Res.* **46**:467–473, 1986.
104. Norrby, K., Jakobsson, A. and Sorbo, J. Mast-cell-mediated angiogenesis: a novel experimental model using the rat mesentery. *Virchows Arch. B Cell Pathol.* **52**:195–206, 1986.
105. Castells, M. C., Irani, A. M. and Schwartz, L. B. Evaluation of human peripheral blood leukocytes for mast cell tryptase. *J. Immunol.* **138**:2184–2189, 1987.
106. Schwartz, L. B. Monoclonal antibodies against human mast cell tryptase demonstrate shared antigenic sites on subunits of tryptase and selective localization of the enzyme to mast cells. *J. Immunol.* **134**:526–531, 1985.
107. Irani, A. M., Bradford, T. R., Kepley, C. L., Schechter, N. M. and Schwartz, L. B. Detection of MCT and MCTC types of human mast cells by immunohistochemistry using new monoclonal anti-tryptase and anti-chymase antibodies. *J. Histochem. Cytochem.* **37**:1509–1515, 1989.
108. Schwartz, L. B., Lewis, R. A. and Austen, K. F. Tryptase from human pulmonary mast cells. Purification and characterization. *J. Biol. Chem.* **256**:11939–11943, 1981.
109. Harvima, I. T., Schechter, N. M., Harvima, R. J. and Fraki, J. E. Human skin tryptase: purification, partial characterization and comparison with human lung tryptase. *Biochim. Biophys. Acta* **957**:71–80, 1988.
110. Cromlish, J. A., Seidah, N. G., Marcinkiewicz, M., Hamelin, J., Johnson, D. A. and Chretien, M. Human pituitary tryptase: molecular forms, NH2-terminal sequence, immunocytochemical localization, and specificity with prohormone and fluorogenic substrates. *J. Biol. Chem.* **262**:1363–1373, 1987.
111. Schwartz, L. B., Bradford, T. R., Littman, B. H. and Wintroub, B. U. The fibrinogenolytic activity of purified tryptase from human lung mast cells. *J. Immunol.* **135**:2762–2767, 1985.
112. Miller, J. S., Westin, E. H. and Schwartz, L. B. Cloning and characterization of complementary DNA for human tryptase. *J. Clin. Invest.* **84**:1188–1195, 1989.
113. Miller, J. S., Moxley, G. and Schwartz, L. B. Cloning and characterization of a second complementary DNA for human tryptase. *J. Clin. Invest.* **86**:864–870, 1990.
114. Vanderslice, P., Ballinger, S. M., Tam, E. K., Goldstein, S. M., Craik, C. S. and Caughey, G. H. Human

mast cell tryptase: multiple cDNAs and genes reveal a multigene serine protease family. *Proc. Natl. Acad. Sci. USA* **87**:3811–3815, 1990.

115. Pallaoro, M., Fejzo, M. S., Shayesteh, L., Blount, J. L. and Caughey, G. H. Characterization of genes encoding known and novel human mast cell tryptases on chromosome 16p13.3. *J. Biol. Chem.* **274**:3355–3362, 1999.
116. Goldstein, S. M., Leong, J., Schwartz, L. B. and Cooke, D. Protease composition of exocytosed human skin mast cell protease–proteoglycan complexes. Tryptase resides in a complex distinct from chymase and carboxypeptidase. *J. Immunol.* **148**:2475–2482, 1992.
117. Schwartz, L. B. Mast cells: function and contents. *Curr. Opin. Immunol.* **6**:91–97, 1994.
118. Sakai, K., Ren, S. and Schwartz, L. B. A novel heparin-dependent processing pathway for human tryptase. Autocatalysis followed by activation with dipeptidyl peptidase I. *J. Clin. Invest.* **97**:988–995, 1996.
119. Huang, C., Li, L., Krilis, S. A., Chanasyk, K., Tang, Y., Li, Z., Hunt, J. E. and Stevens. R. L. Human tryptase α and β/II are functionally distinct due, in part, to a single amino acid difference in one of the surface loops that forms the substrate-binding cleft. *J. Biol. Chem.* **274**:19670–19676, 1999.
120. Schechter, N. M., Fraki, J. E., Geesin, J. C. and Lazarus, G. S. Human skin chymotryptic proteinase. Isolation and relation to cathepsin G and rat mast cell proteinase I. *J. Biol. Chem.* **258**:2973–2978, 1983.
121. Schechter, N. M., Choi, J. K., Slavin, D. A., Deresienski, D. T., Sayama, S., Dong, G., Lavker, R. M., Proud, D. and Lazarus, G. S. Identification of a chymotrypsin-like proteinase in human mast cells. *J. Immunol.* **137**:962–970, 1986.
122. Caughey, G. H., Zerweck, E. H. and Vanderslice, P. Structure, chromosomal assignment, and deduced amino acid sequence of a human gene for mast cell chymase. *J. Biol. Chem.* **266**:12956–12963, 1991.
123. Urata, H., Kinoshita, A., Perez, D. M., Misono, K. S., Bumpus, F. M., Graham, R. M. and Husain, A. Cloning of the gene and cDNA for human heart chymase. *J. Biol. Chem.* **266**:17173–17179, 1991.
124. McEuen, A. R., Gaca, M. D., Buckley, M. G., He, S., Gore, M. G. and Walls, A. F. Two distinct forms of human mast cell chymase – differences in affinity for heparin and in distribution in skin, heart, and other tissues. *Eur. J. Biochem.* **256**:461–470, 1998.
125. Goldstein, S. M., Kaempfer, C. E., Kealey, J. T. and Wintroub, B. U. Human mast cell carboxypeptidase. Purification and characterization. *J. Clin. Invest.* **83**:1630–1636, 1989.
126. Reynolds, D. S., Stevens, R. L., Gurley, D. S., Lane, W. S., Austen, K. F. and Serafin, W. E. Isolation and molecular cloning of mast cell carboxypeptidase A. A novel member of the carboxypeptidase gene family. *J. Biol. Chem.* **264**:20094–20099, 1989.
127. Reynolds, D. S., Gurley, D. S. and Austen, K. F. Cloning and characterization of the novel gene for mast cell carboxypeptidase A. *J. Clin. Invest.* **89**:273–282, 1992.
128. Natsuaki, M., Stewart, C. B., Vanderslice, P., Schwartz, L. B., Wintroub, B. U., Rutter, W. J. and Goldstein, S. M. Human skin mast cell carboxypeptidase: functional characterization, cDNA cloning, and genealogy. *J. Invest. Dermatol.* **99**:138–145, 1992.
129. Hohn, P. A., Popescu, N. C., Hanson, R. D., Salvesen, G. and Ley, T. J. Genomic organization and chromosomal localization of the human cathepsin G gene. *J. Biol. Chem.* **264**:13412–13419, 1989.
130. Schechter, N. M., Irani, A. M., Sprows, J. L., Abernethy, J., Wintroub, B. and Schwartz, L. B. Identification of a cathepsin G-like proteinase in the MCTC type of human mast cell. *J. Immunol.* **145**:2652–2661, 1990.
131. Irani, A. M., Goldstein, S. M., Wintroub, B. U., Bradford, T. and Schwartz, L. B. Human mast cell carboxypeptidase. Selective localization to MCTC cells. *J. Immunol.* **147**:247–253, 1991.
132. Irani, A. M., Craig, S. S., DeBlois, G., Elson, C. O., Schechter, N. M. and Schwartz, L. B. Deficiency of the tryptase-positive, chymase-negative mast cell type in gastrointestinal mucosa of patients with defective T lymphocyte function. *J. Immunol.* **138**:4381–4386, 1987.
133. Weidner, N. and Austen, K. F. Heterogeneity of mast cells at multiple body sites. Fluorescent determination of avidin binding and immunofluorescent determination of chymase, tryptase, and carboxypeptidase content. *Pathol. Res. Pract.* **189**:156–162, 1993.
134. Irani, A. M. and Schwartz, L. B. Human mast cell heterogeneity. *Allergy Proc.* **15**:303–308, 1994.
135. Sperr, W. R., Bankl, H. C., Mundigler, G., Klappacher, G., Grossschmidt, K., Agis, H., Simon, P., Laufer, P., Imhof, M. and Radaszkiewicz, T. The human cardiac mast cell: localization, isolation, phenotype, and functional characterization. *Blood* **84**:3876–3884, 1994.
136. Mori, A., Zhai, Y. L., Toki, T., Nikaido, T. and Fujii, S. Distribution and heterogeneity of mast cells in the human uterus. *Hum. Reprod.* **12**:368–372, 1997.
137. Nilsson, G., Forsberg, K., Bodger, M. P., Ashman, L. K., Zsebo, K. M., Ishizaka, T., Irani, A. M. and Schwartz, L. B. Phenotypic characterization of stem cell factor-dependent human foetal liver-derived

mast cells. *Immunology* **79**:325–330, 1993.
138. Lewis, R. A., Soter, N. A., Diamond, P. T., Austen, K. F., Oates, J. A. and Roberts, L. J. Prostaglandin D2 generation after activation of rat and human mast cells with anti-IgE. *J. Immunol.* **129**:1627–1631, 1982.
139. Warner, J. A., Peters, S. P., Lichtenstein, L. M., Hubbard, W., Yancey, K. B., Stevenson, H. C., Miller, P. J. and MacGlashan D. W. Jr. Differential release of mediators from human basophils: differences in arachidonic acid metabolism following activation by unrelated stimuli. *J. Leukoc. Biol.* **45**:558–571, 1989.
140. Harvima, I. T., Horsmanheimo, L., Naukkarinen, A. and Horsmanheimo, M. Mast cell proteinases and cytokines in skin inflammation. *Arch. Dermatol. Res.* **287**:61–67, 1994.
141. Peters, S. P., Kagey-Sobotka, A., MacGlashan, D. W. Jr and Lichtenstein, L. M. Effect of prostaglandin D2 in modulating histamine release from human basophils. *J. Pharmacol. Exp. Ther.* **228**:400–406, 1984.
142. Triggiani, M., Schleimer, R. P., Warner, J. A. and Chilton, F. H. Differential synthesis of 1-acyl-2-acetyl-*sn*-glycero-3-phosphocholine and platelet-activating factor by human inflammatory cells. *J. Immunol.* **147**:660–666, 1991.
143. Brunner, T., de Weck, A. L. and Dahinden, C. A. Platelet-activating factor induces mediator release by human basophils primed with IL-3, granulocyte–macrophage colony-stimulating factor, or IL-5. *J. Immunol.* **147**:237–242, 1991.
144. MacGlashan, D. J., White, J. M., Huang, S. K., Ono, S. J., Schroeder, J. T. and Lichtenstein, L. M. Secretion of IL-4 from human basophils. The relationship between IL-4 mRNA and protein in resting and stimulated basophils. *J. Immunol.* **152**:3006–3016, 1994.
145. Schroeder, J. T., MacGlashan D. W. Jr, Kagey-Sobotka, A., White, J. M. and Lichtenstein, L. M. IgE-dependent IL-4 secretion by human basophils. The relationship between cytokine production and histamine release in mixed leukocyte cultures. *J. Immunol.* **153**:1808–1817, 1994.
146. Li, H., Sim, T. C. and Alam, R. IL-13 released by and localized in human basophils. *J. Immunol.* **156**:4833–4838, 1996.
147. Farram, E. and Nelson, D. S. Mouse mast cells as anti-tumor effector cells. *Cell. Immunol.* **55**:294–301, 1980.
148. Walsh, L. J., Trinchieri, G., Waldorf, H. A., Whitaker, D. and Murphy, G. F. Human dermal mast cells contain and release tumor necrosis factor alpha, which induces endothelial leukocyte adhesion molecule 1. *Proc. Natl. Acad. Sci. USA* **88**:4220–4224, 1991.
149. Bochner, B. S., Charlesworth, E. N., Lichtenstein, L. M., Derse, C. P., Gillis, S., Dinarello, C. A. and Schleimer, R. P. Interleukin-1 is released at sites of human cutaneous allergic reactions. *J. Allergy Clin. Immunol.* **86**:830–839, 1990.
150. Bradding, P., Feather, I. H., Wilson, S., Bardin, P. G., Heusser, C. H., Holgate, S. T. and Howarth, P. H. Immunolocalization of cytokines in the nasal mucosa of normal and perennial rhinitic subjects. The mast cell as a source of IL-4, IL-5, and IL-6 in human allergic mucosal inflammation. *J. Immunol.* **151**:3853–3865, 1993.
151. Bradding, P., Roberts, J. A., Britten, K. M., Montefort, S., Djukanovic, R., Mueller, R., Heusser, C. H., Howarth, P. H. and Holgate, S. T. Interleukin-4, -5, and -6 and tumor necrosis factor-alpha in normal and asthmatic airways: evidence for the human mast cell as a source of these cytokines. *Am. J. Respir. Cell Mol. Biol.* **10**:471–480, 1994.
152. Moller, A., Lippert, U., Lessmann, D., Kolde, G., Hamann, K., Welker, P., Schadendorf, D., Rosenbach, T., Luger, T. and Czarnetzki, B. M. Human mast cells produce IL-8. *J. Immunol.* **151**:3261–3266, 1993.
153. Jaffe, J. S., Raible, D. G., Post, T. J., Wang, Y., Glaum, M.C., Butterfield, J.H. and Schulman, E.S. Human lung mast cell activation leads to IL-13 mRNA expression and protein release. *Am. J. Respir. Cell Mol. Biol.* **15**:473–481, 1996.
154. Nilsson, G., Forsberg-Nilsson, K., Xiang, Z., Hallbook, F., Nilsson, K. and Metcalfe, D. D. Human mast cells express functional TrkA and are a source of nerve growth factor. *Eur. J. Immunol.* **27**:2295–2301, 1997.
155. Rumsaeng, V., Cruikshank, W. W., Foster, B., Prussin, C., Kirshenbaum, A. S., Davis, T. A., Kornfeld, H., Center, D. M. and Metcalfe, D. D. Human mast cells produce the CD4+ T lymphocyte chemoattractant factor, IL-16. *J. Immunol.* **159**:2904–2910, 1997.
156. Patella, V., Marino, I., Arbustini, E., Lamparter-Schummert, B., Verga, L., Adt, M. and Marone, G. Stem cell factor in mast cells and increased mast cell density in idiopathic and ischemic cardiomyopathy. *Circulation* **97**:971–978, 1998.
157. Zhang, S., Anderson, D. F., Bradding, P., Coward, W. R., Baddeley, S. M., MacLeod, J. D., McGill, J. I., Church, M. K., Holgate, S. T. and Roche, W. R. Human mast cells express stem cell factor. *J. Pathol.*

186:59–66, 1998.

158. Gauchat, J. F., Henchoz, S., Mazzei, G., Aubry, J. P., Brunner, T., Blasey, H., Life, P., Talabot, D., Flores-Romo, L. and Thompson, J. Induction of human IgE synthesis in B cells by mast cells and basophils. *Nature* **365**:340–343, 1993.
159. Kehry, M. R. CD40-mediated signaling in B cells. Balancing cell survival, growth, and death. *J. Immunol.* **156**:2345–2348, 1996.
160. Gruss, H. J., Herrmann, F., Gattei, V., Gloghini, A., Pinto, A. and Carbone, A. CD40/CD40 ligand interactions in normal, reactive and malignant lympho-hematopoietic tissues. *Leuk. Lymphoma* **24**:393–422, 1997.
161. Fureder, W., Agis, H., Sperr, W. R., Lechner, K. and Valent, P. The surface membrane antigen phenotype of human blood basophils. *Allergy* **49**:861–865, 1994.
162. Kiener, H. P., Baghestanian, M., Dominkus, M., Walchshofer, S., Ghannadan, M., Willheim, M., Sillaber, C., Graninger, W. B., Smolen, J. S. and Valent, P. Expression of the C5a receptor (CD88) on synovial mast cells in patients with rheumatoid arthritis. *Arthritis Rheum.* **41**:233–245, 1998.
163. Patella, V., de Crescenzo, G., Ciccarelli, A., Marino, I., Adt, M. and Marone, G. Human heart mast cells: a definitive case of mast cell heterogeneity. *Int. Arch. Allergy Immunol.* **106**:386–393, 1995.
164. Urata, H., Strobel, F. and Ganten, D. Widespread tissue distribution of human chymase. *J. Hypertens.* **Suppl 12**:S17–S22, 1994.
165. Massey, W. A., Guo, C. B., Dvorak, A. M., Hubbard, W. C., Bhagavan, B. S., Cohan, V. L., Warner, J. A., Kagey-Sobotka, A. and Lichtenstein, L. M. Human uterine mast cells. Isolation, purification, characterization, ultrastructure, and pharmacology. *J. Immunol.* **147**:1621–1627, 1991.
166. Tainsh, K. R., Lau, H. Y., Liu, W. L. and Pearce, F. L. The human skin mast cell: a comparison with the human lung cell and a novel mast cell type, the uterine mast cell. *Agents Actions* **33**:16–19, 1991.
167. Bradding, P., Okayama, Y., Howarth, P. H., Church, M. K. and Holgate, S. T. Heterogeneity of human mast cells based on cytokine content. *J. Immunol.* **155**:297–307, 1995.
168. Harvima, I. T., Naukkarinen, A. Harvima, R. J., Aalto, M. L., Neittaanmaki, H. and Horsmanheimo, M. Quantitative enzyme-histochemical analysis of tryptase- and chymase-containing mast cells in psoriatic skin. *Arch. Dermatol. Res.* **282**:428–433, 1990.
169. Bentley, A. M., Jacobson, M. R., Cumberworth, V., Barkans, J. R., Moqbel, R., Schwartz, L. B., Irani, A. M., Kay, A. B. and Durham, S. R. Immunohistology of the nasal mucosa in seasonal allergic rhinitis: increases in activated eosinophils and epithelial mast cells. *J. Allergy Clin. Immunol.* **89**:877–883, 1992.
170. Otsuka, H., Inaba, M., Fujikura, T. and Kunitomo, M. Histochemical and functional characteristics of metachromatic cells in the nasal epithelium in allergic rhinitis: studies of nasal scrapings and their dispersed cells. *J. Allergy Clin. Immunol.* **96**:528–536, 1995.
171. Juliusson, S., Aldenborg, F. and Enerback, L. Proteinase content of mast cells of nasal mucosa: effects of natural allergen exposure and of local corticosteroid treatment. *Allergy* **50**:15–22, 1995.
172. Buckley, M. G., Gallagher, P. J. and Walls, A. F. Mast cell subpopulations in the synovial tissue of patients with osteoarthritis: selective increase in numbers of tryptase-positive, chymase-negative mast cells. *J. Pathol.* **186**:67–74, 1998.
173. Aldenborg, F., Peeker, R., Fall, M., Olofsson, A. and Enerback, L. Metaplastic transformation of urinary bladder epithelium: effect on mast cell recruitment, distribution, and phenotype expression. *Am. J. Pathol.* **153**:149–157, 1998.
174. Irani, A. M., Butrus, S. I., Tabbara, K. F. and Schwartz, L. B. Human conjunctival mast cells: distribution of MCT and MCTC in vernal conjunctivitis and giant papillary conjunctivitis. *J. Allergy Clin. Immunol.* **86**:34–40, 1990.
175. Jeziorska, M., McCollum, C. and Woolley, D. E. Mast cell distribution, activation, and phenotype in atherosclerotic lesions of human carotid arteries. *J. Pathol.* **182**:115–122, 1997.
176. Dvorak, H. F. and Mihm, M. C. J. Basophilic leukocytes in allergic contact dermatitis. *J. Exp. Med.* **135**:235–254, 1972.
177. Hastie, R., Heroy, J. H. and Levy, D. A. Basophil leukocytes and mast cells in human nasal secretions and scrapings studied by light microscopy. *Lab. Invest.* **40**:554–561, 1979.
178. Marone, G. The role of basophils and mast cells in the pathogenesis of pulmonary diseases. *Int. Arch. Allergy Appl. Immunol.* **76 (Suppl 1)**:70–82, 1985.
179. Bascom, R., Wachs, M., Naclerio, R. M., Pipkorn, U., Galli, S. J. and Lichtenstein, L. M. Basophil influx occurs after nasal antigen challenge: effects of topical corticosteroid pretreatment. *J. Allergy Clin. Immunol.* **81**:580–589, 1988.
180. Charlesworth, E. N., Kagey-Sobotka, A., Norman, P. S. and Lichtenstein, L. M. Effect of cetirizine on mast cell-mediator release and cellular traffic during the cutaneous late-phase reaction. *J. Allergy Clin.*

Immunol. **83**:905–912, 1989.

181. Charlesworth, E. N., Hood, A. F., Soter, N. A., Kagey-Sobotka, A., Norman, P. S. and Lichtenstein, L. M. Cutaneous late-phase response to allergen. Mediator release and inflammatory cell infiltration. *J. Clin. Invest.* **83**:1519–1526, 1989.
182. Bochner, B. S., Undem, B. J. and Lichtenstein, L. M. Immunological aspects of allergic asthma. *Annu. Rev. Immunol.* **12**:295–335, 1994.
183. Warner, J. A. and Kroegel, C. Pulmonary immune cells in health and disease: mast cells and basophils. *Eur. Respir. J.* **7**:1326–1341, 1994.
184. Redington, A. E., Polosa, R., Walls, A. F., Howarth, P. H. and Holgate, S. T. Role of mast cells and basophils in asthma. *Chem. Immunol.* **62**:22–59, 1995.
185. Baumgarten, C. R., Nichols, R. C., Naclerio, R. M., Lichtenstein, L. M., Norman, P. S. and Proud, D. Plasma kallikrein during experimentally induced allergic rhinitis: role in kinin formation and contribution to TAME-esterase activity in nasal secretions. *J. Immunol.* **137**:977–982, 1986.
186. Schwartz, L. B., Atkins, P. C., Bradford, T. R., Fleekop, P., Shalit, M. and Zweiman, B. Release of tryptase together with histamine during the immediate cutaneous response to allergen. *J. Allergy Clin. Immunol.* **80**:850–855, 1987.
187. Schlumberger, H. D. Pseudo-allergic reactions to drugs and chemicals. *Ann. Allergy* **51**:317–324, 1983.
188. Mathews, K. P. Clinical spectrum of allergic and pseudoallergic drug reactions. *J. Allergy Clin. Immunol.* **74**:558–566, 1984.
189. Genovese, A., Stellato, C., Marsella, C. V., Adt, M. and Marone, G. Role of mast cells, basophils and their mediators in adverse reactions to general anesthetics and radiocontrast media. *Int. Arch. Allergy Immunol.* **110**:13–22, 1996.
190. Friend, D. S., Ghildyal, N., Austen, K. F., Gurish, M. F., Matsumoto, R. and Stevens, R.L. Mast cells that reside at different locations in the jejunum of mice infected with *Trichinella spiralis* exhibit sequential changes in their granule ultrastructure and chymase phenotype. *J. Cell Biol.* **135**:279–290, 1996.
191. Friend, D. S., Ghildyal, N., Gurish, M. F., Hunt, J., Hu, X., Austen, K. F. and Stevens, R. L. Reversible expression of tryptases and chymases in the jejunal mast cells of mice infected with *Trichinella spiralis. J. Immunol.* **160**:5537–5545, 1998.
192. Furitsu, T., Saito, H., Dvorak, A. M., Schwartz, L. B., Irani, A. M., Burdick, J. F., Ishizaka, K. and Ishizaka, T. Development of human mast cells *in vitro. Proc. Natl. Acad. Sci. USA* **86**:10039–10043, 1989.
193. Saito, H., Ebisawa, M., Tachimoto, H., Shichijo, M., Fukagawa, K., Matsumoto, K., Iikura, Y., Awaji, T., Tsujimoto, G., Yanagida, M., Uzumaki, H., Takahashi, G., Tsuji, K. and Nakahata, T. Selective growth of human mast cells induced by Steel factor, IL-6, and prostaglandin E2 from cord blood mononuclear cells. *J. Immunol.* **157**:343–350, 1996.
194. Valent, P., Schmidt, G., Besemer, J., Mayer, P., Zenke, G., Liehl, E., Hinterberger, W., Lechner, K., Maurer, D. and Bettelheim, P. Interleukin-3 is a differentiation factor for human basophils. *Blood* **73**:1763–1769, 1989.
195. Sillaber, C., Geissler, K., Scherrer, R., Kaltenbrunner, R., Bettelheim, P., Lechner, K. and Valent, P. Type beta transforming growth factors promote interleukin-3 (IL-3)-dependent differentiation of human basophils but inhibit IL-3-dependent differentiation of human eosinophils. *Blood* **80**:634–641, 1992.

CHAPTER 8

Interactions Between Nerves and Mast Cells in Amphibians

G. CHIEFFI BACCARI,*[1] S. MINUCCI,[2] A. DE PAULIS[3] and A. DE SANTIS[1]

[1]*Stazione Zoologica 'A. Dohrn',*
[2]*Dipartimento di Fisiologia Umana e Funzioni Biologiche Integrate 'F. Bottazzi', Seconda Università di Napoli,*
[3]*Divisione di Immunologia Clinica e Allergologia, Università di Napoli 'Federico II', Naples, Italy*

INTRODUCTION

Mast cells are present in the connective tissue of the majority of vertebrates (1). Despite the extensive literature on mammalian mast cells (2–5), few studies have been devoted to the characterization of these cells in non-mammalian vertebrates. This is rather surprising, because mast cells were identified in several tissues of *Rana esculenta* (1) and *Rana pipiens* (6, 7) several decades ago and the ultrastructural and immunological characterization of mast cells in amphibians is still at a very early stage (8–11).

In this brief review, we provide up-to-date information on the characterization of mast cells that are present in various anatomical sites in the frog *Rana esculenta.* In particular, we emphasize the close association between mast cells and nerves in frog tissues.

MORPHOLOGY OF FROG MAST CELLS

Histology and Histochemistry

Mast cells are widely distributed in all frog tissues that have been examined. They are large cells with a diameter of between 8 and 30 μm. Their cytoplasm is packed with secretory granules showing the typical metachromasia originally reported by Paul Ehrlich (12). The metachromatic staining property of frog mast cells differs from that of humans in that the former are stained dark purple and the latter reddish purple (11). The mast cell density varies greatly between different tissues, being highest in the tongue (253 ± 45 mast cells mm^2). Mast cells are also present in the kidney (15.3 ± 1.4 mast cells mm^2), the testis (11, 13), the peripheral nervous system (30 ± 5.2 mm^2) and the heart (5.3 ± 0.4 mm^{-2}), as has been recently reported in humans (3, 14, 15).

* Corresponding author.

MAST CELLS AND BASOPHILS
ISBN 0-12-473335-2

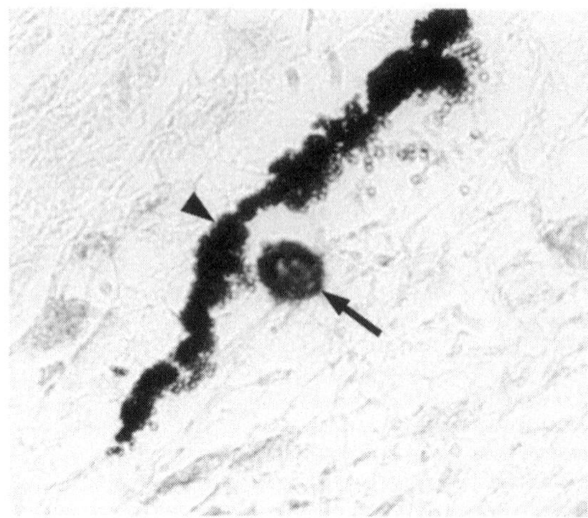

Fig. 1 Paraffin-embedded section of the heart of the frog *Rana esculenta*. A mast cell (arrow) is closely apposed to a melanocyte (arrowhead). Toluidine blue at pH 4.2. ×1250.

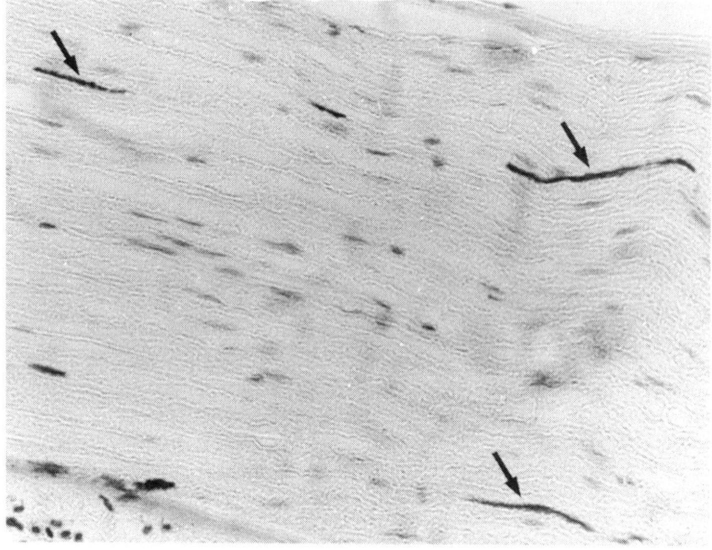

Fig. 2 Three elongated mast cells (arrows) in the sciatic nerve of the frog *Rana esculenta*. Toluidine blue at pH 4.2. ×125 (before reduction).

Interestingly, the majority of frog heart mast cells, in contrast to other anatomical sites, are usually round (Fig. 1). This is in sharp contrast to the mast cells present within nerves which are extremely elongated (Fig. 2).

The combined Alcian blue–safranin staining method reveals several subtypes of frog mast cells in different anatomical sites. In particular, blue and red granules can be detected in tongue and peritoneal mast cells, whereas heart and kidney mast cells are Alcian blue-negative/safranin-positive. These two staining patterns are characteristic of connective tissue mast cells (CTMC) in rodents and probably reflect biochemical differences in the granule content. For example, different proteoglycans and/or proteases can underlie these staining properties (16–18). The histochemical properties of mast cells revealed by the Alcian blue–safranin staining method have been used to assess maturational changes in the granules of CTMC (19). It has been postulated that the Alcian blue–safranin reaction differentiates degrees of sulphation, the more highly sulphated the polysaccharide, the greater its affinity for safranin. It was proposed that Alcian blue-positive granules contain a polysaccharide that is poor or totally lacking in *N*-sulphate, presumably a heparin precursor (19, 20). Similarly, cells that bear safranin-positive granules contain highly *N*-sulphated polysaccharide, probably heparin (19, 21). More recently, a close correlation between the presence of safranin-positive granules and the amount of rat mast cell protease I has been reported (22). Mast cells in the intestinal lamina propria of frog were Alcian blue-positive/safranin-negative, as previously reported in rodents (16–18). If the staining properties are conferred on the different frog mast cell subtypes by virtue of their proteoglycan content, it is likely that frog mast cells express a certain degree of specialization in proteoglycan synthesis.

Ultrastructure

The ultrastructural analysis of frog mast cells in different anatomical sites showed that they contained a single-lobed, central nucleus with peripheral condensation of nuclear chromatin (Fig. 3). Interestingly, the surface of frog mast cells was not adorned with folds as in human mast cells (23). A few mitochondria and free ribosomes were occasionally seen in the cytoplasm. The cytoplasm was packed with numerous membrane-bound secretory granules. These were extremely heterogeneous in shape – ovoid, fusiform or round – and displayed substructural patterns unique to the different species. Secretory

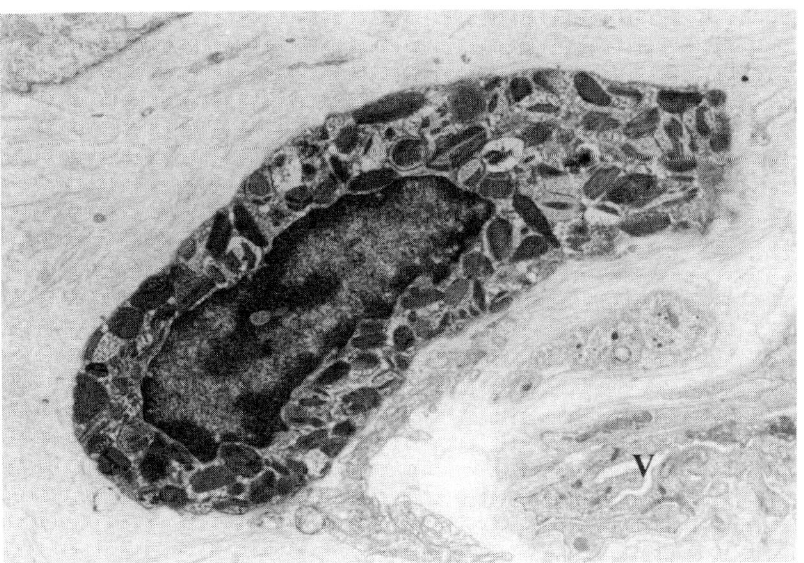

Fig. 3 Electron micrograph of a mast cell from *Rana esculenta* tongue surrounded by collagen fibres. The cytoplasm contains numerous heterogeneous, polymorphic granules. V, blood vessel. ×10,000.

granules contained an elaborate and heterogeneous substructural architecture. Some had crystal particles in hexagonal arrays, others had fusiform inclusions with a 'sandwich-like' structure and a few partial scrolls, different from those of human mast cells. Again different from human mast cells (14, 23–25), lipid bodies were never observed in frog mast cells.

Subcellular Localization of Histamine in Frog Mast Cells

The histamine content of different frog tissues varied markedly, being highest in the heart (Table I) (11). The histamine content of various frog mast cells (~0.1 pg per mast cell) was approximately 30 times lower than that of human mast cells isolated from different anatomical sites (~3 pg per cell) (14, 23).

TABLE I
Histamine content of various frog tissues

	Histamine content Wet tissue (ng g)	Protein (mg g)
Tongue	241±44.5	3.9±0.7
Heart	874±67.3	35.7±2.7
Kidney	586±146	7.4±1.8
Testis	688±76.8	12.5±1.4

A monoclonal antihistamine antibody was used to detect histamine in frog mast cells. After immunogold staining of frog tongue, low concentrations of gold particles were present over all the secretory granules in more than 95% of frog mast cells, presumably reflecting the low concentration of histamine in these cells (11).

FROG MAST CELLS AND PERIPHERAL NERVOUS SYSTEM

Occurrence of Mast Cells in Different Topographical Areas of Frog Nerves

Each vertebrate uni- and multifascicular peripheral nerve has three separate connective tissue sheets. On the outside of each nerve there is a collagenous epineurium and beneath that a perineurium which surrounds each fascicle of nerve fibres. Individual nerve fibres are embedded in the endoneurium, which completely fills the space bounded by the perineurium (26). In our study, mast cells were always detected within the epineurium, endoneurium and perineurium of the frog peripheral nerves (Fig. 4). We therefore focused our attention on the occurrence of mast cells in the large nerves of the frog, such as sciatic and brachial nerves, as well as in the glossopharyngeal and hypoglossal nerve branches which supply the tongue.

Mast cells in the sciatic and brachial nerves

In the sciatic and brachial nerves of the frog, mast cells have always been identified on the basis of their typical metachromatic secretory granules. The cytoplasm extends along the longitudinal axis of the nerve fibres (Fig. 2) and is safranin-positive with the combined Alcian blue–safranin method.

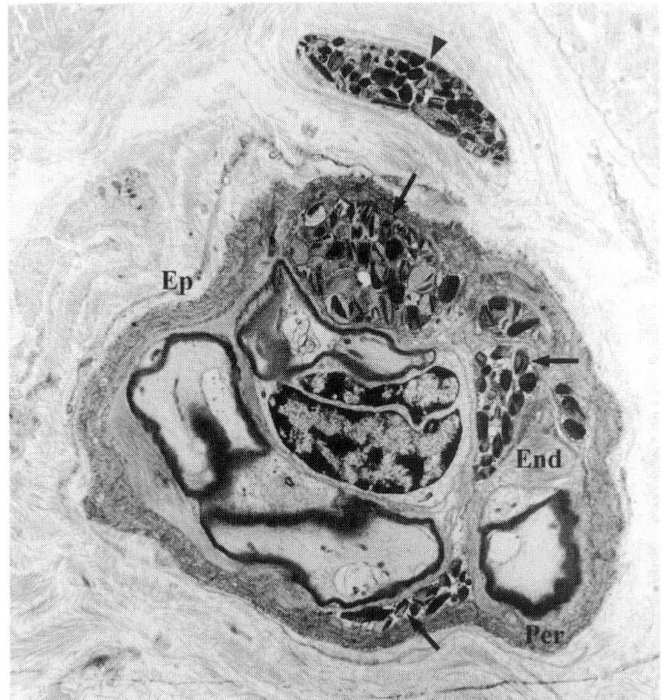

Fig. 4 Electron micrograph of a nervous fascicle from *Rana esculenta* tongue. Mast cells (arrows) are located within the endoneurium (End) and perineurium (Per). One mast cell is closely associated with the nervous fascicle (arrowhead). Ep, epineurium. ×2800 (before reduction).

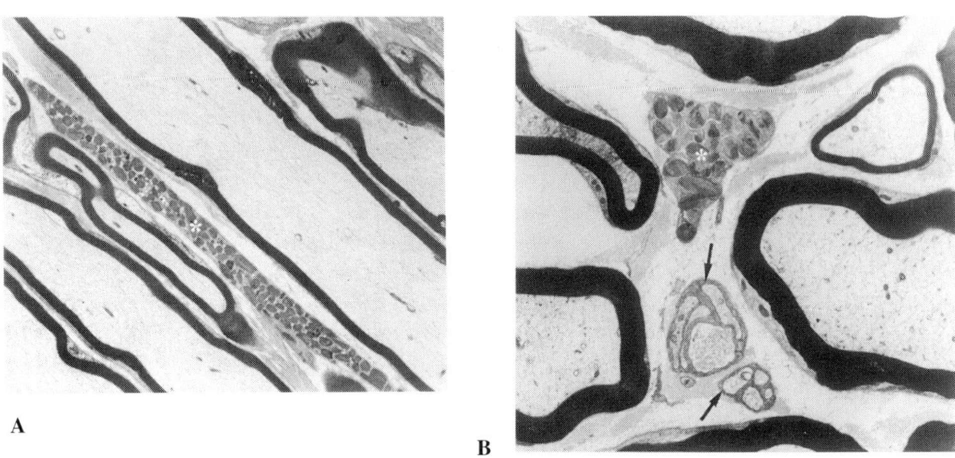

Fig. 5 (A) Electron micrograph of the longitudinal section of *Rana esculenta* sciatic nerve. One mast cell (asterisk) is located in the endoneurium ×3600 (original magnification). (B) Cross-section of *Rana esculenta* brachial nerve. An endoneurial mast cell (asterisk) surrounded by collagen fibres is present among myelinic and unmyelinic fibres (arrows). ×3600 (original magnification).

Ultrastructural examination of the sciatic and brachial nerves demonstrates that mast cells are more abundant in the epineurium and endoneurium than in the perineurium. Endoneurial mast cells surrounded by collagen fibres are in close proximity to myelinic (Fig. 5A, B) and amyelinic fibres (Fig. 5B) and are frequently present around blood vessels.

Chemical mediators, such as histamine released from mast cells, can stimulate proliferation and increase permeability in endothelial cells from mammalian tissues (27–29). Although the endothelial response to mast cell degranulation is well documented, the response of the neuron is not. However, the close anatomical association between mast cells and nerve fibres suggests the existence of reciprocal and/or feedback interactions between neurons and mast cells brought about by their metabolic products. Interestingly, *in vivo* administration of nerve growth factor (NGF) can induce mast cell hyperplasia (30). Moreover, mast cells synthesize, store and release NGF (31, 32), and NGF can cause mediator and cytokine release from mast cells (33–35). Thus, it is possible that mast cells exert a paracrine control of neurons. In addition, several neuropeptides, such as substance P, vasoactive intestinal peptide (VIP), calcitonin gene-related peptide and neurokinins, locally released from nerve fibres (36), can activate mast cells to release chemical mediators (37–39).

Mast cells in the glossopharyngeal and hypoglossal nerve branches

The frog tongue receives a dense innervation from the glossopharyngeal and hypoglossal nerves. The former is a sensorimotor nerve, the latter is a motor nerve. Numerous connective type mast cells are present around and within these nerves. Unilateral and prolonged electrical stimulations of the hypoglossal nerve in the frog cause degranulation and changes in the morphology of the mast cells on the stimulated side (Chieffi Baccari *et al.*, unpublished observations). These cells show heterochromatic nuclei and massive intracellular loss of their granules (Fig. 6).

Dimitriadu *et al.* (40) have demonstrated that prolonged or intense stimulation of the trigeminal ganglion in the rat causes degranulation and changes in the histochemical characteristics of the dura mater and tongue mast cells.

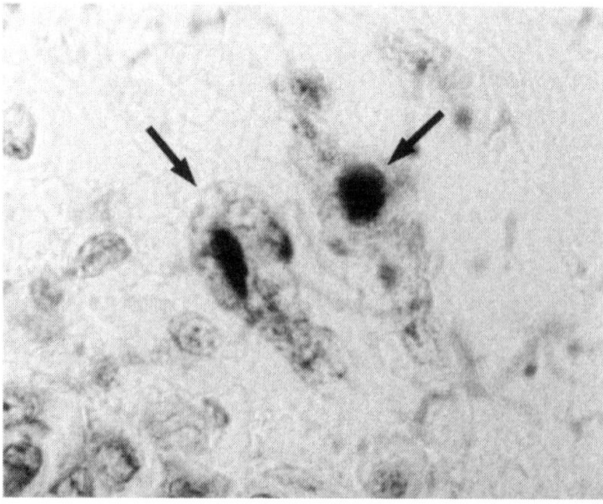

Fig. 6 Paraffin-embedded section of *Rana esculenta* tongue after electrical stimulation of hypoglossal nerve. Two degranulated mast cells (arrows) with heterochromatic nuclei can be observed in the connective tissue. Toluidine blue, pH 4.2. ×1250.

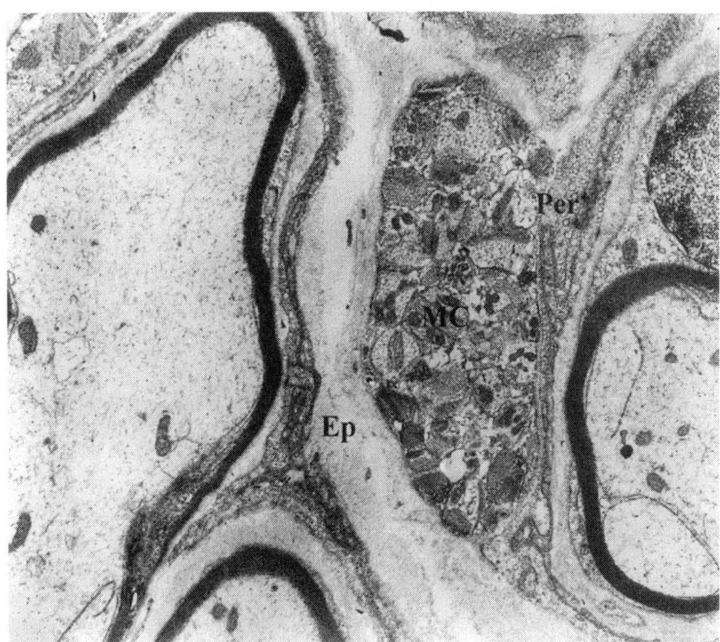

Fig. 7 Electron micrograph of a large nervous fascicle from *Rana esculenta* tongue. A mast cell (MC) is seen embedded in the epineurium (Ep) in close association to the perineurium (Per). ×8000 (before reduction).

Ultrastructural study of the large nerve fascicles in the frog tongue shows that mast cells are present in the epineurium (Fig. 7) and endoneurium (Fig. 8A) but not in the perineurium. Figure 8B shows that perineurium consists of various cellular layers linked by tight junctions. Individual perineurial cells are laden with vesicles and caveolae. The epineurial and endoneurial mast cells are always in close association to the perineurial cells (Figs 7 and 8A, B). Figure 8B shows the close anatomical association between mast cells and perineurium. The distance of the plasma membranes of the mast cell and perineural cell is less than 70 nm.

Small nerve fascicles lack epineurium, and the perineurium is limited to a few cellular layers. In these nerve fascicles, mast cells are constantly present within the lamellae of the perineurium and appear surrounded by perineurial processes (Fig. 9A, B). The plasma membranes of the two cells are at a distance of 40–100 nm and at some points they seem to be in contact.

The close anatomical association between mast cells and perineurial cells is suggestive of paracrine interactions.The perineurium is a tissue–nerve barrier which regulates the endoneurial microenvironment by limiting the passage of substances and protects axons and Schwann cells from antigens, toxins and infectious agents (26). The interaction between mast cells and perineurium may be important near body surfaces (e.g. the tongue) where fascicles of nerve fibres can be exposed to bacteria or parasites. Therefore, mast cells and their mediators could participate in the functions of the tissue–nerve barrier of the perineurium.

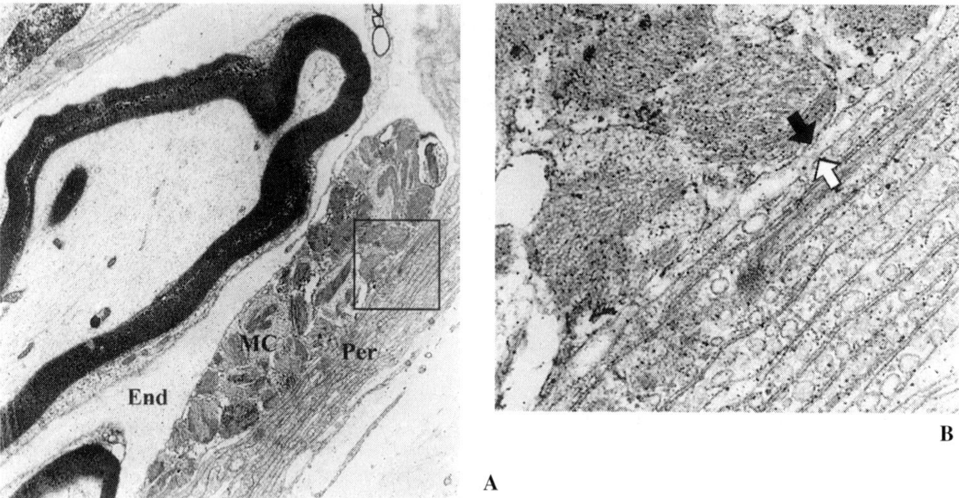

Fig. 8 (A) Electron micrograph of a nervous fascicle of *Rana esculenta* tongue. An endoneurial mast cell (MC) is closely associated to the perineurium (Per). End, endoneurium. ×6000 (original magnification). (B) High magnification of the frame indicated in (A), showing the close association between the plasma membrane of the mast cell (dark arrow) and the plasma membrane of the perineurium (pale arrow). ×22,000 (original magnification).

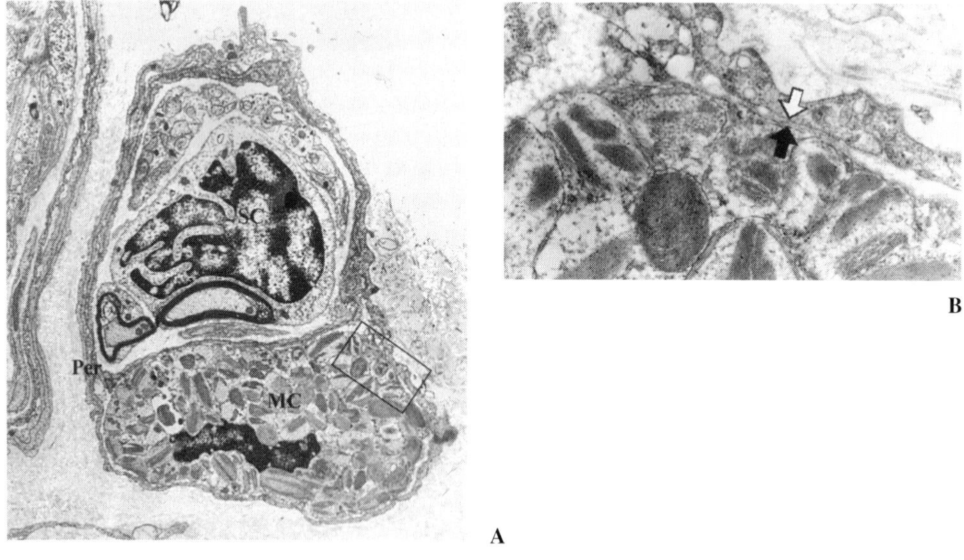

Fig. 9 (A) Electron micrograph of a small nerve fascicle from *Rana esculenta* tongue. A mast cell (MC) is embedded within the perineurial cellular layers (Per). SC, Schwann cell nucleus. ×4600 (original magnification). (B) High magnification of the frame indicated in (A), showing the close association between the plasma membrane of the mast cell (dark arrow) and the plasma membrane of the perineurial cell (pale arrow). ×28,000 (original magnification).

Development

There is little information on the ontogenesis of mast cells in the peripheral nervous system and it is unknown how and when these cells enter the nerves during embryogenesis.

Mast cells have been observed in nerves of the eye muscles of rat between days 20 and 21 of intrauterine life. Only occasionally mast cells were seen in the sciatic nerve, spinal ganglia and spinal roots at day 21 of fetal life, but during the first 2 weeks after birth the number of such cells increased considerably (41).

Our anatomical and morphological observations (42) showed that the first recognizable mast cells arised in the mesenchyme of the developing tongue (Fig. 10) in tadpole at stage 26 of Witschi's standard tables (43). The mast cells are round in shape and are distinguishable from other mesenchymal cells for the presence of few cytoplasmic granules. Their periodic acid–Schiff (PAS)-positive granules were orthochromatic when stained with toluidine blue at pH 4.2, and Alcian blue-positive when stained with Alcian blue–safranin, suggesting a 'mucosal' phenotype and the absence of heparin (21). At this larval stage no recognizable mast cells were seen within the nerves.

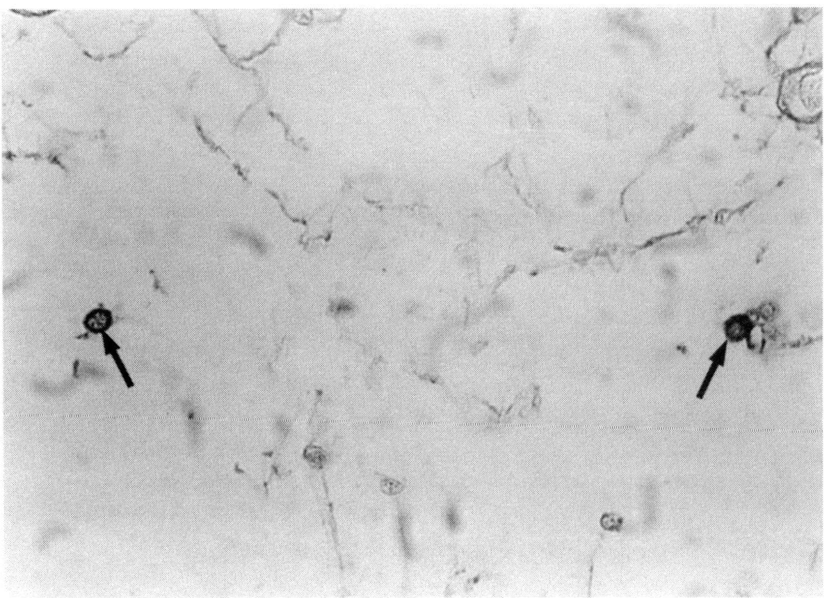

Fig. 10 Paraffin-embedded section of the tongue of *Rana esculenta* tadpole at stage 26 of the Witschi standard table (see text). Two immature mast cells (arrows), surrounded by mesenchyme, are located symmetrically on both sides of the tongue. Toluidine blue at pH 4.2. ×312.

From this stage onwards, the mast cell density significantly increased in all tissues. At stage 29 metachromatic mast cells can be observed within both tongue nerves (hypoglossal and glossopharyngeal branches) and the sciatic nerve. These cells contain granules which bind safranin. This staining suggests the presence of highly *N*-sulphated glycosaminoglycans such as heparin. These cells contain a few granules that do not yet exhibit their characteristic ultrastructural pattern. In the sciatic nerve, mast cells are

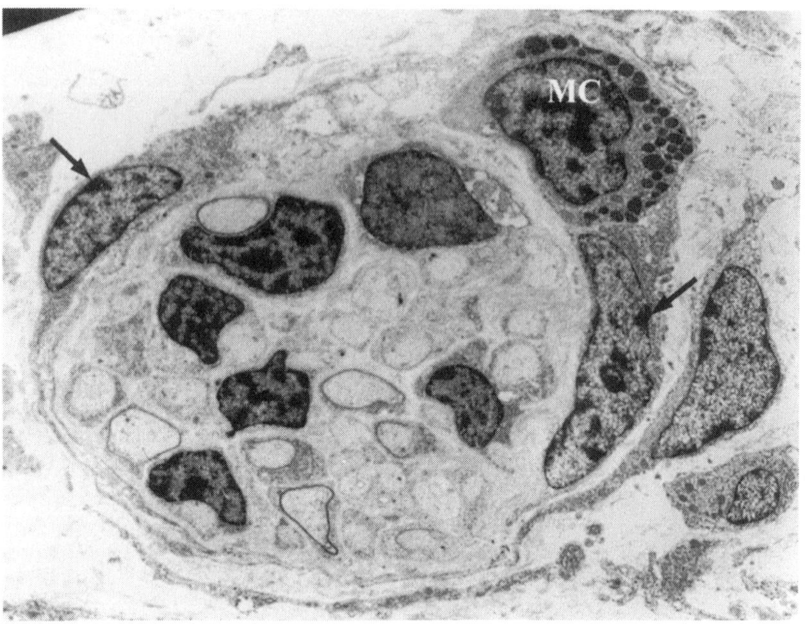

Fig. 11 Electron micrograph of a nerve fascicle from the *Rana esculenta* tadpole tongue at metamorphosis stage. A mast cell (MC) is closely associated to one of the perineurial cells (arrows). ×3600.

present in the endoneurium and epineurium predominantly in a perivascular location. At metamorphosis (stage 33) the distribution of mast cells within nerves was similar to that in the adult frog. In the sciatic nerve, elongated mast cells showing numerous metachromatic granules similar to those described for mature connective mast cells, are present in the endoneurium along the longitudinal direction of the nerve fibres. In the tongue nerve mast cells are in close association with the perineurial cells (Fig. 11).

Interestingly, the mast cell–perineurium anatomical association exists in early phases of nerve development. The perineurial cells appear early and they initially lack a basal lamina, and are not joined by the tight junctions that form a barrier in the adult. In this phase, the perineurial cells probably participate in providing and transporting nutrients to the axons and Schwann cells (44). Mast cells are already present in the first phases of nerve development and are closely associated to perineurial cells (Fig. 11). It has been proposed that factors released by the Schwann cell–axon complex during nerve development may be responsible for perineurial differentiation and organization from the surrounding mesenchyme (45–47). Therefore, it is likely that the same factors in this period are also responsible for differentiation of mast cells which share the same embryological origin with perineurial cells. In fact, today a fibroblastic, mesenchymal origin of the perineurial cells is accepted rather than a Schwann cell derivation (44, 46).

The Wallerian degeneration, during which a proliferation of perineurial cells has been shown, is also characterized by a significant perineurial mast cell hyperplasia (48). In this respect, it is likely that both perineurial and mast cells may be influenced locally by the same growth factor.

Mast cell proliferation occurs in the rat sciatic nerve undergoing Wallerian degeneration (41, 49). In our investigation, the mast cell density increases early in the distal part of the injured brachial nerve in the frog, but has still to be established whether the increasing mast cell density is due to their migration or to their multiplication *in situ*.

OCCURRENCE OF MAST CELLS IN THE CENTRAL AND PERIPHERAL NERVOUS SYSTEMS OF MAMMALS

In the last few years, the anatomical association between peripheral nerves and mast cells has been documented in a variety of mammalian tissues, including skin, diaphragm, mesentery and thymus, gastrointestinal and respiratory tracts (50–55). Mast cells have also been described in the central nervous system where they have a predominantly perivascular location (56, 57).

Previous studies (41, 58, 59) have documented that mast cells are also present within the large nerve plexa and peripheral nerve trunks of numerous mammals, including man. Their density varies significantly in different species, being numerous in the rat and guinea pig, but rare in the cat.

Furthermore, mast cells accumulate in damaged parts of human peripheral neuropathy (41, 49). They may exert important functions in inflammatory demyelinating processes. Mast cells have been localized in multiple sclerosis plaques (60–62) as well as in areas of demyelination in experimental allergic encephalomyelitis (63) or experimental allergic neuritis (64). Myelin basic protein can stimulate mast cell degranulation and induces peripheral (65) and central (66) demyelinization, and myelin proteins are degraded by mast cell proteases. In addition, during experimental allergic neuritis a delayed T cell response may depend on early release of mast cell mediators and T cell products can cause mast cell activation (64, 67, 68).

CONCLUDING REMARKS

Histochemical analysis of frog mast cells in different tissues revealed properties of the connective and mucosal types previously found in rodents (16, 17, 69). Ultrastructural study of the connective-type mast cells showed that the cytoplasmic secretory granules have unique morphological characteristics formerly unrecognized in any human or animal mast cells. The histamine content of these cells in different anatomical sites is lower than that of human mast cells (11).

The striking mast cell–nerve association documented in this study is intringuing. In the large nerves, such as sciatic and brachial nerves, mast cells are present in the endoneurium and epineurium where they are in close association with the vasa nervorum and myelinic and amyelinic fibres. In the small nerve fascicles of the tongue (body surface exposed to the external milieu), the mast cells are predominantly located between the perineurial layers, suggesting a role in the tissue–nerve barrier function of the perineurium.

Although there is some evidence to support the existence of paracrine interactions between the peripheral nervous system and mast cells, the physiological significance of these relationships remains to be established. Furthermore, the role of mast cells and their mediators in different pathological situations of the nervous system is largely unknown.

Furthermore, a better characterization of the embryonal development of mast cells in peripheral nerves could contribute to clarifying the origin of mast cells and their increase in different pathological conditions.

A comparative study of mast cells might uncover new models for investigating their pathophysiological role, particularly for their widerspread occurrence throughout the body.

REFERENCES

1. Michels, N. A. The mast cells in the lower vertebrates. *Cellule* **33**:339–461, 1923.
2. Galli, S. J. Biology of disease. New insights into 'the riddle of the mast cells': microenvironmental regulation of mast cell development and phenotypic heterogeneity. *Lab. Invest.* **62**: 5–33, 1990.
3. Marone, G., (ed.) *Human Basophils and Mast Cells: Biological Aspects.* Karger, Basel, 1995.
4. Marone, G., Spadaro, G. and Genovese, A. Biology, diagnosis and therapy of mastocytosis. In: *Human Basophils and Mast Cells: Clinical Aspects* (Marone, G., ed.), pp 1–21. Karger, Basel, 1995.
5. Metcalfe, D. D., Baram, D. and Mekori, Y. A. Mast cells. *Physiol. Rev.* **77**:1033–1079, 1997.
6. Chiu, H. and Lagunoff, D. Histochemical comparison of frog and rat mast cells. *J. Histochem. Cytochem.* **19**:369–375, 1971.
7. Chiu, H. and Lagunoff, D. Histochemical comparison of vertebrate mast cells. *Histochem. J.* **4**:135–144, 1972.
8. Setoguti, T. Electron microscopy study on the newt mast cell, especially its granule-extrusion mechanism. *J. Ultrastruct. Res.* **27**:377–395, 1969.
9. Nakao, T. and Uchinomiya, K. Fine structure of cytoplasmic granules of tadpole mast cells. *J. Electron Microsc.* **23**:57–60, 1974.
10. Vugman, I. Ultrastructure of toad (*Bufo paracnemius*) mast cells: their alteration by compound 48/80. *Anat. Anz.* **154**:425–432, 1983.
11. Chieffi Baccari, G., de Paulis, A., Di Matteo, L., Gentile, M., Marone, G. and Minucci, S. *In situ* characterization of mast cells in the frog *Rana esculenta. Cell Tissue Res.* **292**:151–162, 1998.
12. Ehrlich, P. Beiträge zur Kenntnis der granulierten Bindegewebszellen und der eosinophilen Leukocyten. *Arch. Anat. Physiol.* **3**:166–169, 1879.
13. Minucci, S., Di Matteo, L., Chieffi, P., Pierantoni, R. and Fasano, S. 17β-Estradiol effects on mast cell number and spermatogonial mitotic index in the testis of the frog, *Rana esculenta. J. Exp. Zool.* **278**:93–100, 1997.
14. Patella, V., Marinò, I., Lampärter, B., Arbustini, E., Adt, M. and Marone, G. Human mast cells. Isolation, purification, ultrastructure, and immunologic characterization. *J. Immunol.* **154**:2855–2865, 1995.
15. Patella, V., de Crescenzo, G., Marinò, I., Genovese, A., Adt, M., Gleich, G. J. and Marone G. Eosinophil granule proteins activate human heart mast cells. *J. Immunol.* **157**:1219–1225, 1996.
16. Enerbäck, L. Mast cells in rat gastrointestinal mucosa. Effects of fixation. *Acta Pathol. Microbiol. Immunol. Scand.* **66**:289–302, 1966.
17. Enerbäck, L. Mast cells in rat gastrointestinal mucosa. Dye-binding and metachromatic properties. *Acta Pathol. Microbiol. Immunol. Scand.* **66**:303–312, 1966.
18. Bienenstock, J. An update on mast cell heterogeneity. *J. Allergy Clin. Immunol.* **81**:763–769, 1988.
19. Combs, J. W., Lagunoff, D. and Benditt, E. P. Differentiation and proliferation of embryonic mast cells of the rat. *J. Cell Biol.* **25**:577–592, 1965.
20. Spicer, S. S. A correlative study of the histochemical properties of rodent acid mucopolysaccharides. *J. Histochem. Cytochem.* **8**:18–33, 1960.
21. Gaytan, F., Bellido, C., Carrera, G. and Aguilar, E. Differentiation of mast cells during postnatal development of neonatally estrogen-treated rats. *Cell Tissue Res.* **259**:25–31, 1990.
22. Koretou, O. Relationship between the staining property of mast cell granule with alcian blue-safranin O and toluidine blue O, and the content of mast cell protease I in the granule of the rat peritoneal mast cell. *Acta Histochem. Cytochem.* **21**:25–32, 1988.
23. de Paulis, A., Marinò, I., Ciccarelli, A., de Crescenzo, G., Concardi, M., Verga, L., Arbustini, E. and Marone, G. Human synovial mast cells. I. Ultrastructural *in situ* and *in vitro* immunologic characterization. *Arthritis Rheum.* **39**:1222–1233, 1996.
24. Dvorak, A. M., Hammel, I., Schulman, E. S., Peters, S. P., MacGlashan, D. W. Jr, Schleimer, R. P., Newball, H. H., Pyne, K., Dvorak, H. F., Lichtenstein, L. M. and Galli, S. J. Differences in the behavior

of cytoplasmic granules and lipid bodies during human lung mast cell degranulation. *J. Cell Biol.* **99**:1678–1687, 1984.

25. Dvorak, A. M. Basophil and mast cell degranulation and recovery. *Blood Cell Biochemistry*, Vol. 4. Plenum Press, New York, 1991.
26. Thomas, P. K. and Olsson, Y. Microscopic anatomy and function of the connective tissue components of peripheral nerve. In: *Peripheral Neuropathy* (Dyck, P. J., Thomas, P. K., Lambert, E. H. and Bunge, R., eds), pp. 168–189. W. B. Saunders, Philadelphia, 1984.
27. Powell, H. C., Myers, R. R. and Costello, M. L. Increased endoneurial fluid pressure following injection of histamine and compound 48/80 into rat peripheral nerves. *Lab. Invest.* **43**:564–572, 1980.
28. Marks, R. M., Roche, W. R., Czerniecki, M., Penny, R. and Nelson, D. S. Mast cell granules cause proliferation of human microvascular endothelial cells. *Lab. Invest.* **55**:289–294, 1986.
29. Norrby, K., Jakobsson, A. and Sorbo, J. Mast-cell-mediated angiogenesis: A novel experimental model using the rat mesentery. *Virchows Arch. B Cell Pathol.* **52**:195–206, 1986.
30. Aloe, L. and Levi-Montalcini, R. Mast cells increase in tissues of neonatal rats injected with the nerve growth factor. *Brain Res.* **133**:358–366, 1977.
31. Leon, A., Buriani, A., dal Toso, R., Fabris, M., Romanello, S., Aloe, L. and Levi-Montalcini, R. Mast cells synthesize, store, and release nerve growth factor. *Proc. Natl. Acad. Sci. USA* **91**:3739–3743, 1994.
32. Nilsson, G., Forsberg-Nilsson, K., Xiang, Z., Hallbook, F., Nilsson, K. and Metcalfe, D. D. Human mast cells express functional TrkA and are a source of nerve growth factor. *Eur. J. Immunol.* **27(9)**:2295–2301, 1997.
33. Marshall, J. S., Stead, R. H., McSharry, C., Nielsen, L. and Bienenstock, J. The role of mast cell degranulation products in mast cell hyperplasia. I. Mechanism of action of nerve growth factor. *J. Immunol.* **144**:1886–1892, 1990.
34. Horigome, K., Pryor, J. C., Bullock, E. D. and Johnson, E. M. Mediator release from mast cells by nerve growth factor. Neurothrophin specificity and receptor mediation. *J. Biol. Chem.* **268**:14881–14887, 1993.
35. Bullock, E. D. and Johnson, E. M. Nerve growth factor induces the expression of certain cytokine genes and *bcl-2* in mast cells. *J. Biol. Chem.* **271**:27500–27508, 1996.
36. Morley, J. E., Kay, N. E., Solomon, G. F. and Plotnikoff, N. P. Neuropeptides: conductors of the immune orchestra. *Life Sci.* **41**:527–544, 1987.
37. Benyon, R. C., Lowman, M. A. and Church, M. K. Human skin mast cells: their dispersion, purification and secretory characterization. *J. Immunol.* **138**:861–867, 1987.
38. Lowman, M. A., Benyon, R. C. and Church, M. K. Characterization of neuropeptide-induced histamine release from human dispersed skin mast cells. *Br. J. Pharmacol.* **95**:121–130, 1988.
39. Stellato, C., de Paulis, A., Ciccarelli, A., Cirillo, R., Patella, V., Casolaro, V. and Marone, G. Anti-imflammatory effect of cyclosporin A on human skin mast cells. *J. Invest. Dermatol.* **98**:800–804, 1992.
40. Dimitriadu, V., Buzzi, M. G., Moskowitz, M. A. and Theoharides, T. C. Trigeminal sensory fiber stimulation induces morphological changes reflecting secretion in rat dura mater mast cells. *Neuroscience* **44**:97–112, 1991.
41. Olsson, Y. Mast cells in the nervous system. Int. Rev. Cytol. **24**:27–70, 1968.
42. Chieffi Baccari, G., Rusciani, A., Di Matteo, L. and Minucci, S. Ontogenesis of the mast cells in the tongue and gastrointestinal tract of the frog, *Rana esculenta. Eur. J. Histochem.* **39(suppl 1)**:47, 1995.
43. Witschi, E. Amphibians, normal stages and fate maps. In: *Development of Vertebrates* (Witschi, E., ed.), pp 78–91. W.B. Saunders, Philadelphia, 1956.
44. du Plessis, D. G., Mouton, Y. M., Muller, C. J. F. and Geiger, D. H. An ultrastructural study of the development of the chicken perineurial sheath. *J. Anat.* **189**:631–641, 1997.
45. Vlodavsky, I., Folkman, J., Sullivan, R., Fridman, R., Ishai-Michaeli, R. and Sasse, J. Endothelial cell-derived basic fibroblast growth factor: synthesis and deposition into sub-endothelial extracellular matrix. *Proc. Natl. Acad. Sci. USA* **84**:2292–2296, 1987.
46. Bunge, M. B., Wood, P. M., Tyan, L. B., Bates, M. L. and Sanes, J. R. Perineurium originates from fibroblasts: demonstration *in vitro* with a retroviral marker. *Science.* **243**:229–231, 1989.
47. Eccleston, P. A. Regulation of Schwann cell proliferation: mechanisms involved in peripheral nerve development. *Exp. Cell Res.* **199**:1–9, 1992.
48. Latker, C. H., Wadhwani, K. C., Balbo, A. and Rapoport, S. Blood–nerve barrier in the frog during Wallerian degeneration: are axons necessary for maintenance of barrier function? *J. Comp. Neurol.* **309**:650–664, 1991.
49. Olsson, Y. and Sjostrand, J. Proliferation of mast cells in peripheral nerves during Wallerian degeneration: a radioautographic study. *Acta Neuropathol.* **13**:111–121, 1969.

50. Newson, B., Dahlstöm, A., Enerbäck, L. and Ahlman, H. Suggestive evidence for a direct innervation of mucosal mast cells. An electron microscopic study. *Neuroscience* **10**:565–570, 1983.
51. Stead, R. H., Tomioka, M., Quinonez, G., Simon, G. T., Felten, S. Y. and Bienenstock, J. Intestinal mucosal mast cells in normal and nematode-infected rat intestines are in intimate contact with peptidergic nerves. *Proc. Natl. Acad. Sci. USA* **84**:2975–2979, 1987.
52. Stead, R. H., Dixon, M. F., Bramwell, N. H., Riddell, R. H. and Bienenstock, J. Mast cells are closely apposed to nerves in the human gastrointestinal mucosa. *Gastroenterology* **97**: 575–585, 1989.
53. Arizono, N., Matsuda, S., Hattori, T., Kojima, Y., Maeda, T. and Galli, S. J. Anatomical variation in mast cell nerve associations in the rat small intestine, heart, lung, and skin. *Lab. Invest.* **62**:626–634, 1990.
54. Bienenstock, J., Stead, R. H. and Marshall, J. S. Mast cells and the Nervous System. In: *The Mast Cell in Health and Disease* (Kaliner, M. A. and Metcalfe, D. D., eds), pp. 687–698. Marcel Dekker, New York, 1993.
55. Blennerhassett, M. G. Nerve and mast cell interaction: cell conflict or information exchange? *Progr. Clin. Biol. Res.* **390**:225–241, 1994.
56. Johnson, D. and Krenger, W. Interactions of mast cells with nervous system – recent advances. *Neurochem. Res.* **17**:939–951, 1992.
57. Silver, R., Silverman, A. J., Vitkovic, L. and Lederhendler, I. I. Mast cells in the brain: evidence and functional significance. *Trends Neurosci.* **19**:25–31, 1996.
58. Enerbäck, L., Olsson, Y. and Sourander, P. Mast cells in normal and sectionated peripheral nerve. *Z. Zellforsch.* **66**:596–608, 1965.
59. Olsson, Y. Mast cells in human peripheral nerve. *Acta Neurol. Scand.* **47**:357–368, 1971.
60. Olsson, Y. Mast cells in plaques of multiple sclerosis. *Acta Neurol. Scand.* **50**:611–618, 1974.
61. Prineas J. W. and Wright R. G. Macrophages, lymphocytes, and plasma cells in the perivascular compartment in chronic multiple sclerosis. *Lab. Invest.* **38**:409–421, 1978.
62. Toms, R., Weiner, H. L. and Johnson, D. Identification of IgE-positive cells and mast cells in frozen sections of multiple sclerosis brains. *J. Neuroimmunol.* **30**:169–177, 1990.
63. Dietsch, G. N. and Hinrichs, D. J. The role of mast cells in the elicitation of experimental allergic encephalomyelitis. *J. Immunol.* **142**:1476–1481, 1989.
64. Brosnan, C. F., Lyman, W. D., Tansey, F. A. and Carter, T. H. Quantitation of mast cells in experimental allergic neuritis. *J. Neuropathol. Exp. Neurol.* **44**:196–203, 1985.
65. Johnson, D., Seeldrayers, P. A. and Weiner, H. L. The role of mast cells in demyelination. 1. Myelin proteins are degraded by mast cell proteases and myelinic basic protein and P_2 can stimulate mast cell degranulation. *Brain Res.* **444**:195–198, 1988.
66. Theoharides, T .C., Dimitriadou, V., Letourneau, R., Rozniecki, J. J., Vliagoftis, H. and Boucher, W. Synergistic action of estradiol and myelin basic protein on mast cell secretion and brain myelin changes resembling early stages of demyelination. *Neuroscience* **57**:861–871, 1993.
67. Askenase, P. W. and Van Loveren, H. Delayed-type hypersensitivity: activation of mast cells by antigen-specific-T-cell factors initiates the cascade of cellular interactions. *Immunol. Today* **4**:259–264, 1983.
68. Kaplan, A. P., Reddigari, S., Baeza, M. and Kuna, P. Histamine releasing factors and cytokine-dependent activation of basophils and mast cells. *Adv. Immunol.* **50**:237–260, 1991.
69. Tainsh K. R. and Pearce, F. L. Mast cell heterogeneity: evidence that mast cells isolated from various connective tissue locations in the rat display markedly graded phenotypes. *Int. Arch. Allergy Immunol.* **98**:26–34, 1992.

SECTION THREE

SIGNAL TRANSDUCTION IN MAST CELLS AND BASOPHILS

CHAPTER 9

Sequential Protein Kinase Activation and the Regulation of Mast Cell Cytokine Production

ERWIN W. GELFAND and GARY L. JOHNSON*

Division of Basic Sciences, Department of Pediatrics, National Jewish Medical and Research Center, Denver, Colorado, U.S.A.

INTRODUCTION

Mast cells play a central role in inflammatory and immediate allergic reactions. The binding of multivalent antigens to receptor-bound IgE and the subsequent aggregation of the high-affinity Fc receptors for IgE ($Fc_\varepsilon RI$) provide the trigger for mast cell activation. The consequences of mast cell activation are the release of pre-formed inflammatory mediators from secretory granules (degranulation) and the synthesis and secretion of a number of pro-inflammatory cytokines. These mast cell responses are regulated by intracellular signal transduction pathways triggered by $Fc_\varepsilon RI$ aggregation. In addition, mast cells express a number of other receptors which are linked to intracellular signalling pathways that are both similar and distinguishable from those triggered through $Fc_\varepsilon RI$. This review focuses on sequential protein kinase activation in mast cells which culminates in the liberation of pro-inflammatory cytokines.

SIGNALLING THROUGH MAST CELL SURFACE RECEPTORS

Among the earliest demonstrable responses to $Fc_\varepsilon RI$ aggregation is the phosphorylation of the receptor itself, priming the receptor for direct interactions with important effector molecules that activate distinct signalling pathways (1). $Fc_\varepsilon RI$ are heterotrimeric or tetrameric complexes comprising an IgE-binding α subunit, a β subunit and two γ subunits (2). The cytoplasmic domains of the β and γ subunits contain two tyrosine residues located within a conserved consensus sequence (immunoreceptor tyrosine activation motif, ITAM) (3). The presence of ITAMs in both β and γ subunits implies that they both participate in signals transduced through tetrameric $Fc_\varepsilon RI$ complexes (4, 5) and

* Corresponding author.

MAST CELLS AND BASOPHILS
ISBN 0-12-473335-2

phosphorylation of this pair on conserved tyrosines is essential for signal transduction. The extracellular domain of the α subunit is essential to generate the biochemical signals (6).

Activation through $Fc_{\varepsilon}RI$ is highlighted by the rapid phosphorylation of a set of cellular proteins on tyrosine residues, and is presumed to be mediated by the non-transmembrane protein tyrosine kinases (PTKs), Lyn and Syk (4, 5, 7, 8). Lyn, a Src family PTK, is constitutively associated with $Fc_{\varepsilon}RI$, binding to the inactive β subunit (5, 9, 10). Syk becomes activated, binding to the phosphorylated ITAMs of the $Fc_{\varepsilon}RI$ γ subunit. Lyn and Syk appear to be activated sequentially: Lyn activation is immediate and transient and appears to send the initiating signal following $Fc_{\varepsilon}RI$ engagement. Syk activation peaks after Lyn activation and may be dependent on Lyn activation. Syk activation peaks at a time that coincides with maximum tyrosine phosphorylation of the $Fc_{\varepsilon}RI$ γ subunit. The γ subunit appears essential for signalling that results in secretion (4, 7, 9), and the β subunit may be important for Ras signalling (11). The γ dimer functions as an autonomous activating complex, whereas the β chain serves as an amplifier that increases Syk activation and Ca^{2+} mobilization (9).

Aggregation of $Fc_{\varepsilon}RI$ also results in the activation of phospholipase Cγl (PLC), and is dependent on Syk activation. $Fc_{\varepsilon}RI$ cross-linking activates the lipid kinase, phosphatidylinositol 3-kinase (PI3K). A number of other proteins also undergo rapid tyrosine phosphorylation following $Fc_{\varepsilon}RI$ aggregation, including $pp60^{c\text{-}src}$, B cell tyrosine kinase (Btk) (7, 12), Spy^{75}, $p95^{vav}$, Nck (13–15) and the focal adhesion proteins, focal adhesion kinase ($pl25^{Fak}$) and paxillin (16, 17).

Studies of bone marrow-derived mast cells (BMMC) from genetically deficient mice have revealed some surprising results. Although cross-linking of $Fc_{\varepsilon}RI$ in Lyn-deficient mast cells failed to induce protein phosphorylation of various substrates and only evoked a slow Ca^{2+} response, degranulation and cytokine production proceeded normally (18). In contrast to Lyn and Syk, Btk does not physically associate with $Fc_{\varepsilon}RI$. Nonetheless, Btk plays an important role in $Fc_{\varepsilon}RI$ signalling. Btk-deficient mice demonstrate reduced anaphylactic reactions. BMMC from deficient mice also show reduced responses in terms of $Fc_{\varepsilon}RI$-induced degranulation and cytokine production under certain circumstances (19) but not in others (unpublished results). In contrast to Lyn and Btk, Syk appears essential, since, in the absence of kinase active Syk, ligation of $Fc_{\varepsilon}RI$ fails to activate a number of kinases (20), to trigger degranulation (9, 21) or to induce cytokine production (20).

The stem cell factor receptor (SCFR, c-Kit) is expressed on immature haematopoietic progenitor cells (22) and mast cells, and stem cell factor (SCF, mast cell growth factor, kit ligand or steel factor) is the ligand for the tyrosine kinase SCFR. In mice, recombinant SCF can promote mast cell hyperplasia, degranulation and mediator release (22, 23). SCFR (CD117) is a transmembrane protein that is homologous to receptors for polypeptide growth factors such as SCF. The receptor consists of an intracellular tyrosine kinase domain, a single transmembrane domain and an extracellular domain that binds SCF.

Ligation of the SCFR (by SCF) results in SCFR autophosphorylation (24) and binding of PI3K and PLC-γl to the receptor (25). Although p21Ras is activated, there is no detectable tyrosine phosphorylation of Ras GTPase-activating protein (24).

MAPK FAMILY MEMBERS

An emerging cluster of sequential protein kinase pathways regulated through surface receptors has now been identified involving members of the MAPK (mitogen-activated protein kinase) family in addition to extracellular regulated kinase-l (ERK-l) and ERK-2 (Fig. 1). These multiple linear MAPK cascades have been identified that operate through sequential activation of cytosolic protein kinases to lead signals from the plasma membrane to their cytosolic and nuclear targets, regulating nuclear transcription factors (26–29). They are characterized as serine/threonine protein kinases activated by dual phosphorylation on both a tyrosine and a threonine (30). The MAPK family includes p42/p44 MAPK (ERK-l and ERK-2), the c-Jun kinases (JNKs, which are also referred to as stress-activated protein kinases, SAPKs), and p38, the osmotic shock responsive kinase (similar to the yeast Hog 1) (31–33) (Fig. 1). All members of this family are thought to play major roles in transcriptional regulation. Several tiers in these cascades have been identified (Fig. 1). These cascades, in association with other signalling pathways, can differentially alter the phosphorylation status of the transcription machinery. The regulation of these different MAPKs involves their phosphorylation by MKKs (MAPK kinase) including the MEKs (MAPK/ERK kinases, MAPKK) (34) and JNK kinases (JNKK or SEKs, stress/ERK kinases) (35–37). The MEKs and JNKKs phosphorylate specific MAPK family members on both a tyrosine and a threonine, resulting in MAPK activation (31, 32, 35–37). Moreover, each pathway is downregulated by a variety of phosphatases, which may exhibit different specificities for the different pathways (29).

In general, the ERKs are activated by agonists for tyrosine kinase-encoded receptors

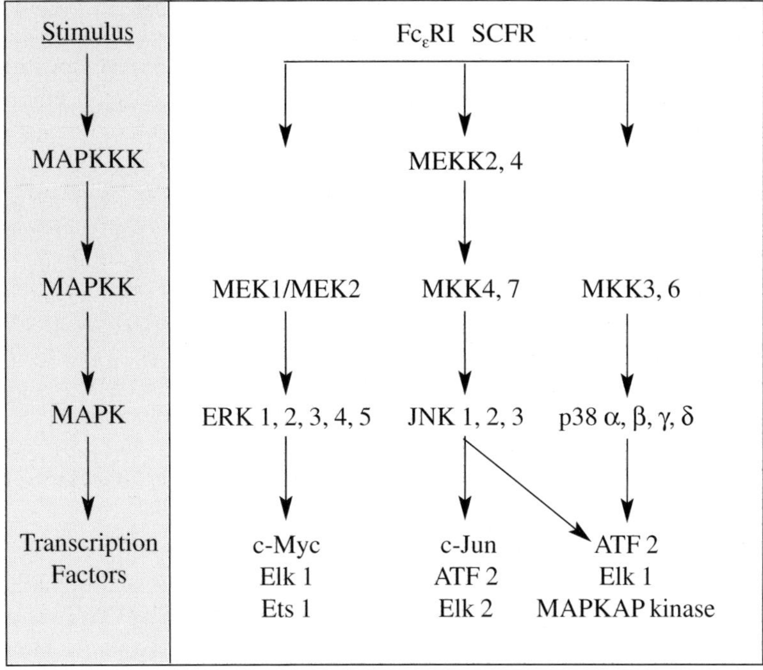

Fig. 1 Sequential activation of mitogen activated protein kinase (MAPK) pathways.

and G protein-coupled receptors, whereas the JNKs and p38 are activated by cellular stress, inflammatory cytokines and, in mast cells, $Fc_{\varepsilon}RI$ and SCF. The ERKs may induce mitogenesis and differentiation whereas the stress-activated kinases (JNK, p38) have been implicated in apoptosis, oncogenic transformation, cytokine production and inflammatory responses in different cell types.

The MKKs are in turn regulated by MAPK kinase kinases (MAPKKKs) which form a functional sequential protein kinase module (MKKK–MKK–MAPK). This group of kinases includes the Rafs (Raf-1, B-Raf, N-Raf) and the recently cloned MEKKs (MEKK 1, 2, 3, 4) (36, 38, 39). The MEKKs have been cloned and characterized (38, 40, 41). Raf kinases preferentially activate the ERK pathway, whereas the MEKKs regulate the JNK pathway. In addition to the MEKKs, germinal centre kinase (GCK), mixed lineage kinase 3 (MLK-3) and tumour progression locus 2 (Tpl-2) kinase have been shown to regulate the JNK pathway as MKKKs (42–50).

The JNKs are activated in response to many cellular stresses, including heat shock, UV irradiation (36, 51–53), CD40 ligation of B lymphocytes (54) and CD28 stimulation of T cells (55). Ten separate JNKs have been identified and are grouped into three families: JNK1, JNK2 and JNK3 (56). The JNKs function to phosphorylate c-Jun at the N-terminal regulatory sites, serine 63 and serine 73, mapping within its transactivation domain, and also result in the transcriptional activation of c-Jun (57). Other substrates of the JNKs are activating transcription factor 2 (ATF-2) (58) and Elk-1 (59). The DNA binding activity of bacterially expressed ATF-2 is increased by phosphorylation *in vitro* and *in vivo* (58). ATF-2 is a member of a group of transcription factors that bind to a similar sequence located in the promoters of many genes, including tumour necrosis factor-α (TNF-α) (60), indicating a role for the JNK pathway in the transcriptional regulation of many genes.

The stress-responsive p38 MAPK isoform has been implicated in the activation of MAPKAP-kinase 2 and the expression of pro-inflammatory cytokines (61). p38 activation is mediated by dual phosphorylation of threonine (180) and tyrosine (182) (25). Activation of p38 results in the phosphorylation and activation of ATF-2, as well as CHOP, a member of the C/EBP family of transcription factors (25, 62). JNK and p38 are often activated in parallel, but independent activation of p38 has also been observed indicating the existence of independent signalling roles for these MAP kinase cascades (36, 37). Cross-talk between the mammalian MAPK pathways also occurs. These separate pathways appear to be functionally independent and are regulated by distinct protein kinase cascades, via the activation of a unique MAPK kinase. One type, MEK1/MEK2, is a strong activator of ERK but does not phosphorylate JNK or p38 (34, 37, 63). Two other MAPK kinases, MKK3 and MKK6, activate p38 (36, 64); MKK4 (JNKK, SEKl) and MKK7 activate JNK (35–37). Upstream of the MKK is an MEKK or other MKKK and the low molecular weight GTP-binding proteins, Cdc42, Racl or Rho (64–66).

$Fc_{\varepsilon}RI$ Signalling in MC/9 Mast Cells

In order to study antigen triggering of mast cells, we generated hybridomas secreting anti-ovalbumin (OVA) IgE (67). In initial studies, we investigated signalling through $Fc_{\varepsilon}RI$ using the MC/9 murine mast cell line following addition of OVA to OVA-specific IgE-sensitized cells or DNP to DNP–IgE-sensitized cells (68). We showed that following antigen-induced aggregation of $Fc_{\varepsilon}RI$, JNK was significantly activated within 5 min and peaked at 15–20 min after addition of OVA (69). Activation by OVA was dose-dependent and specific since OVA did not induce activation of JNK in DNP–IgE-sensitized cells. Anti-mouse IgE antibody activated JNK in both DNP–IgE- and OVA–IgE-sensitized cells. In addition, ERK2 was rapidly activated within 30 sec following addition of OVA. In immunoblots using anti-ERK2 antibody, kinase activation was detected within 1 min and still observed 90 min after OVA stimulation. In these studies, MEKKl activation preceded the activation of JNK in response to triggering through $Fc_{\varepsilon}RI$. MEKKl activation was detected 30 sec after addition of OVA to IgE-sensitized cells. Maximal MEKKl activity was achieved by 3 min and was 2.5–3-fold over basal levels. MEKK1 activity decreased over the ensuing 10 min after addition of OVA.

Wortmannin is an inhibitor of PI3K with reasonable selectivity when used at concentrations below 1 μM (48). Pre-incubation of the cells (for 15 min) with wortmannin inhibited JNK activation in a dose-dependent fashion; kinase activity was almost completely eliminated in the presence of 100 nM wortmannin. In contrast, concentrations of 300 nM wortmannin failed to inhibit ERK2 activation by OVA. These findings demonstrate the role for PI3K in regulating the JNK pathway by a Src family tyrosine kinase-associated receptor. In these mast cells, the regulation of the MEKK1–JNKK–JNK pathway was dependent on the activation of PI3K. Moreover, the results clearly indicated that there was a very early separation in the signal pathways activated through $Fc_{\varepsilon}RI$ to differentially regulate JNK and ERK2 sequential protein kinase pathways.

It is also clear that the third MAPK isoform, p38, is activated by antigen cross-linking of IgE-sensitized MC/9 cells (69). The kinetics of activation are very similar to those of JNK, with detectable activity at 1 min, peaking at 5 min and gradually declining over the ensuing 60 min. In parallel to the results of wortmannin on JNK activation, the PI3K inhibitor prevented activation of p38 at similar concentrations. However, the inhibition of p38 was never as complete as the inhibition of JNK activation, suggesting that inputs in addition to PI3K are involved in differentially regulating JNK and p38 activities (69).

Mast cells have the capacity to enhance many aspects of allergic inflammation in asthma or other allergic disorders via the elaboration of a number of multifunctional cytokines. The first cytokine to be associated with mast cells was TNF-α, but mast cells represent a source of other cytokines as well, including interleukins IL-3, IL-4, IL-5, IL-6, IL-13, and granulocyte–macrophage colony-stimulating factor (GM-CSF) (6, 70). MC/9 cells are known to transcribe and produce TNF-α; other cytokines such as IL-4 or IL-6 are not detected in MC/9 cells. These cells, sensitized with anti-OVA IgE, were incubated with OVA, the supernates were harvested and levels of TNF-α assayed by ELISA. Within the first 30 min after addition of OVA, no TNF-α was detected, suggesting little pre-formed TNF-α production in MC/9 cells; TNF-α production peaked at 2.5–3 h after stimulation of $Fc_{\varepsilon}RI$ (69). Both cycloheximide and actinomycin D completely prevented TNF-α production. Addition of wortmannin inhibited TNF-α production in a dose-dependent manner (69); in contrast, the MEKl inhibitor PD 98059, at concentrations that inhibited activation of MEKl and ERK2, failed to impact TNF-α production (69).

These data indicate that signalling through $Fc_{\varepsilon}RI$ activates three members of the MAPK family: ERK, JNK and p38. Inhibition of MEKl is associated with the inhibition of ERK activation; inhibition of PI3K is associated with inhibition of JNK and p38 activation. The regulation of TNF-α production is sensitive to wortmannin but unaffected by inhibition of MEKl. These results imply that specific MAPK signalling pathways are involved in the regulation of mast cell TNF-α production triggered through $Fc_{\varepsilon}RI$.

TRANSCRIPTIONAL REGULATION OF TNF-α PRODUCTION

Little is known about the signal transduction pathways that regulate TNF-α gene expression in mast cells. In rat basophilic leukaemia (RBL) cells, Gαz, a heterotrimeric G protein, appeared to modulate the $Fc_{\varepsilon}RI$ stimulatory pathway for TNF-α synthesis (71). Some clues are also evident from the promoter for the TNF-α gene which encodes NF-κB, AP-1, AP-2, NFAT, Ets and AP-l/ATF-related regulatory elements. Within the TNF-α promoter the regulatory elements involve regulation by transcription factors that are substrates for MAPK members, including ERKs, JNKs and p38 MAPK. For example, c-Jun and ATF-2 transcription factors are strongly implicated in the control of TNF-α gene expression (72) and are substrates for JNK and p38 (57, 58). MEKK1 has been shown to also regulate the activation of NF-κB (55, 63), which has been implicated as being important in controlling TNF-α gene transcription. In addition, in monocytes, a series of pyridinylimidazole compounds have been shown to block lipopolysaccharide-induced activation of p38; inhibition of p38 MAPK inhibits TNF-α synthesis (in monocytes) primarily at the level of translation and expression (73). Thus JNK and p38 kinases appear to be major regulators of TNF-α gene expression and, in mast cells, are activated in response to ligation of $Fc_{\varepsilon}RI$.

To test these predictions, MC/9 cells were transiently transfected with the reporter plasmid, pTNF($_{-1-3111}$)Luc, having the luciferase gene regulated by the TNF-α promoter. Passively sensitized, transfected MC/9 cells responded to OVA and $Fc_{\varepsilon}RI$ aggregation with a 5–6-fold increase in luciferase expression (68). Wortmannin (100 nM) inhibited this response by 40%: in contrast, the ERK inhibitor PD 98059 actually enhanced the expression of luciferase in response to OVA. The results with wortmannin on TNF-α gene expression were similar to the effects of the inhibitor on protein expression. These results show that a component of the $Fc_{\varepsilon}RI$-mediated increase in TNF-α gene expression requires PI3K activity and this response appears largely independent of ERK stimulation.

Aggregation of $Fc_{\varepsilon}RI$ stimulated MEKK1 activity in MC/9 cells (69). Expression of an activated form of MEKK1 activated JNK; the activities of ERK and p38 were induced much less, if at all, in these cells (69). Activated MEKK1 expression in these MC/9 cells stimulated an increase in luciferase activity, resulting from activation of the TNF-α promoter. These data demonstrated that, under these conditions, activated MEKK1 is capable of stimulating TNF-α gene transcription. To further define specific control of TNF-α transcription, MC/9 cells were transfected with an inhibitory JNK2 mutant (JNK2-APF). Transfection of MC/9 cells with this mutant partially inhibited OVA-stimulated TNF-α promoter activity and inhibited MEKK1-stimulated luciferase expression (69). This partial inhibition observed with JNK2-APF may be related to it being a competitive inhibitory mutant; at present strong dominant-negative mutants for JNK or its upstream regulator, JNK kinase, have not been developed and the kinase-inactive mutants behave only as modest competitive inhibitory mutants. The findings suggested that MEKK1 stimulation of TNF-α promoter-driven luciferase expression requires JNK2. But, more importantly and relevant to subsequent findings, the limited inhibition of $Fc_{\varepsilon}RI$-mediated TNF-α promoter activity indicated that additional MEKKs were more likely to play a dominant role in these OVA-induced responses and this has been confirmed in more recent studies (see below).

Four MEKKs with homology in their kinase domains have now been cloned. All four function as MKKKs capable of activating the JNK pathway. Based on the deduced primary sequences of MEKK1, 2, 3 and 4, specific properties related to kinase activation may be predicted for each (Fig. 2 and Table I). To more directly define which MEKKs are involved in $Fc_{\varepsilon}RI$-mediated signalling of JNK, kinase-inactive inhibitory mutants (66) of the MEKKs were used. In these experiments, kinase-inactive MEKK2 and MEKK4 inhibited $Fc_{\varepsilon}RI$-stimulated JNK activity by approximately 60% and 80%, respectively, whereas inhibitory mutants of MEKK1 or MEKK3 were without effect (unpublished). Thus, although MEKK1 activation followed $Fc_{\varepsilon}RI$ aggregation and constitutively activated MEKK1 stimulated JNK, it did not appear essential for JNK activation when $Fc_{\varepsilon}RI$ was ligated.

The immunosuppressants cyclosporin A (CsA) and FK506 are potent inhibitors of TNF-α production in mast cells. Both CsA and FK506 (but not rapamycin) strongly inhibited $Fc_{\varepsilon}RI$ activation of JNK and p38 to a lesser extent (74). Cyclosporin H (CsH) which does not bind to the cyclophilins and does not have immunosuppressive activity, did not inhibit $Fc_{\varepsilon}RI$ activation of JNK or p38. In addition, incubation of MC/9 cells with CsA did not inhibit $Fc_{\varepsilon}RI$ mobilization of calcium or activation of the tyrosine kinase Syk, implying that early signal events triggered by $Fc_{\varepsilon}RI$ cross-linking are unaffected by CsA. These data indicate that the likely target for CsA via a cyclophilin complex (or FK506) is calcineurin, and identify a calcineurin-regulated JNK pathway following $Fc_{\varepsilon}RI$ aggregation of mast cells.

STEM CELL FACTOR RECEPTOR (c-KIT)

SCFR, the tyrosine kinase encoded receptor for SCF, promotes the growth and differentiation of mast cells. SCF ligation also stimulates the release of mast cell

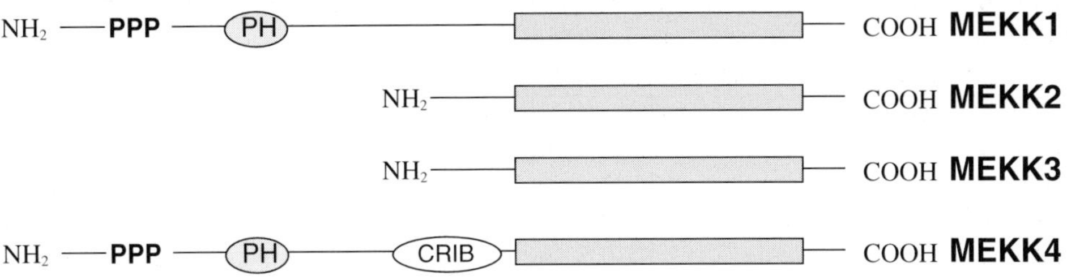

Fig. 2 Domains of the MAPKKK. Hatched bar, kinase domain; PH, pleckstrin homology domain; PPP, proline-rich domain; CRIB, Cdc42/Rac interacting binding domain.

TABLE I

Summary of demonstrated regulators of MAPK pathways

	ERK	JNK	p38
MEKK1	+	+	–
MEKK2	+	+	–
MEKK3	+	+	–
MEKK4	–	+	–

inflammatory mediators, including histamine and serotonin, and induces mast cell adherence and chemotaxis. In MC/9 cells, addition of SCF stimulated ERK, JNK and p38 in MC/9 cells (74). The activation of JNK and p38 by SCF was modest compared to signalling through $Fc_\varepsilon RI$ in these cells. Strikingly, SCFR co-stimulation with $Fc_\varepsilon RI$ markedly augmented the activation of JNK. Although SCFR activation alone did not result in measurable increases in TNF-α production in MC/9 cells, together with $Fc_\varepsilon RI$ ligation, there was a marked increase in TNF-α production; co-stimulation with SCF and OVA ($Fc_\varepsilon RI$) only weakly potentiated p38 or ERK activation, confirming that JNK was differentially regulated by $Fc_\varepsilon RI$ relative to p38 and ERK (74).

Substantiating these observations were the findings that wortmannin failed to inhibit the activation of ERK, p38 and JNK in response to SCF in MC/9 cells, in striking contrast to signalling through $Fc_\varepsilon RI$ (74). It should be noted that not every protein kinase pathway activated through $Fc_\varepsilon RI$ and SCFR is differentially inhibited by wortmannin. ERK activation through $Fc_\varepsilon RI$ was unaffected. As a corollary, AKT (PKB), which is activated in response to PI3K stimulation by either $Fc_\varepsilon RI$ or SCFR ligation, was inhibited by wortmannin.

Further support for the independence of the $Fc_\varepsilon RI$ and SCFR signalling pathways derives from studies with the immunosuppressant drugs (Table II). CsA (and FK506, but not rapamycin) strongly inhibited $Fc_\varepsilon RI$ activation of JNK and p38, but not ERK (74). In striking contrast, SCFR activation of JNK, p38 and ERK was completely unaffected by the immunosuppressants. Such results clearly document the independence of the signal pathways regulating JNK and p38 activity by $Fc_\varepsilon RI$ and SCFR. The integration of such early signalling events which differ between $Fc_\varepsilon RI$ and SCFR in their control of pathways controlling TNF-α production is likely important in overall regulation of cytokine production.

TABLE II
Functional characterization of three mast cell populations

	MC/9	BMMC	ESMC
FcεRI	+	+	+
SCFR	+	+	+
Cytokine production	TNF-α	TNF-α, IL-4	TNF-α, IL-4
Degranulation	–	+	+
Transfection	Transient/stable	–	Transient/stable
Wortmann sensitivity			
FcεRI JNK	+	+	+
FcεRI cytokine	+	+	+
SCFR JNK	–	+	+
FcεRI p38	+	+	n.d.
SCFR p38	–	–	n.d.
CsA sensitivity			
FcεRI JNK	+	+	+
FcεRI cytokine	+	+	+
SCFR JNK	–	–	–
FcεRI p38	+	+	n.d.
SCFR p38	–	–	n.d.

n.d. = not done

DIFFERENTIATION, SIGNALLING AND FUNCTION OF MAST CELLS DERIVED FROM BONE MARROW AND EMBRYONIC STEM CELLS

Mast cell development is a complex process that results in the appearance of phenotypically distinct populations of cells in different anatomical sites. Although more likely representative of cytokine-driven differences than specific attributes of the sites, connective tissue mast cells differ from mucosal mast cells primarily in their storage of large amounts of histamine. Bone marrow obtained from the femurs of mice and grown in medium containing IL-3 or IL-3–SCF rapidly differentiates into mast cells (bone marrow-derived mast cells, BMMC). $Fc_{\varepsilon}RI$ is detected within 7–10 days of culture and the cells become granulated within 3–4 weeks, confirming previous observations (75). Cells grown in IL-3–SCF take on the features of connective tissue mast cells as they are larger, contain uniformly electron-dense cytoplasmic granules, have a longer life span and stain positively with Alcian blue, safranin O, berberine sulphate and contain large amounts of histamine (unpublished observations). In contrast, the cells grown in IL-3 alone are smaller, contain cytoplasmic granules that are not uniformly electron-dense, do not stain with the various dyes, and contain less histamine. Despite these phenotypic differences, both cell types express functional $Fc_{\varepsilon}RI$: addition of OVA or DNP to passively sensitized cells (anti-OVA or anti-DNP–IgE) leads to rapid increases in cytosolic Ca^{2+}, p38, JNK and ERK activation, and TNF-α and IL-4 production (76). As in the MC/9 cells, JNK activation and cytokine production was both wortmannin-sensitive and CsA-sensitive (Table II). As the BMMC do not express FKBP12, these cells are not inhibited by FK506. Thus, these BMMC have in common most of the features of $Fc_{\varepsilon}RI$ signalling in MC/9 cells. Further, the responses to SCF are in general the same: addition of SCF augmented JNK activation and cytokine production in these cells and the SCF response was CsA-resistant (Table II).

An important feature of the BMMC is that addition of OVA to passively sensitized cells results in degranulation, measured by the release of (tritiated) serotonin or histamine (using an enzyme immunoassay) (76). Signalling through $Fc_{\varepsilon}RI$ triggers serotonin release that was inhibited by wortmannin and CsA. Moreover, the MEK1 inhibitor also blocked serotonin release.

Although BMMC are easily derived in culture and have proven useful to study the regulation of cytokine production and degranulation, genetic manipulation of these cells has proven difficult. To overcome these problems, we have derived mast cells from embryonic stem cell cultures (ESMC). The undifferentiated blast cell colonies were cultured with IL-3 or IL-3–SCF. The kinetics of mast cell differentiation from these colonies were identical to those of the BMMC, with $Fc_{\varepsilon}RI$ detected on the majority of the cells during the second week in culture and the majority of the cells exhibited cytoplasmic granules by weeks 3–4. Similar to BMMC, the phenotype of the cells was dictated by the cytokine mixture present in the culture medium; as discussed for BMMC, IL-3 alone induced cells with low histamine content (~20 nM per 10^6 cells) and IL-3–SCF induced cells with high histamine content (~600 nM per 10^6 cells) (unpublished observations). Importantly, the ESMC exhibited signalling properties similar to those described for BMMC when $Fc_{\varepsilon}RI$ or SCFR were ligated as well as for the control of cytokine production and the degranulation response (see Table II). The ability to transfect ES cells efficiently serves to make them an important genetically manipulatable system. Inducible expression and targeted disruption of specific genes as well as mast cell differentiation have been possible in this system.

IDENTIFICATION OF A CRITICAL ROLE FOR MEKK2 IN JNK ACTIVATION AND CYTOKINE PRODUCTION

As described above, multiple MEKKs have been shown to function as an MKKK in the signalling module MKKK–MKK–JNK. Initial experiments using passively sensitized MC/9 cells indicated that specific kinase-inactive inhibitory MEKKs could modulate (inhibit) OVA-induced JNK activation. In particular, kinase-inactive MEKK2 and MEKK4 strongly inhibited JNK activation in response to $Fc_{\varepsilon}RI$ ligation (unpublished observations). Importantly, inhibitory mutants of MEKK1 and MEKK3 had little or no effect in these experiments. Only kinase-inactive MEKK4 appeared to exhibit an inhibitory effect on SCFR signalling; inhibitory MEKK2 did not affect signalling through the SCFR under these conditions. These results confirmed the differential involvement of specific MEKKs in the regulation of $Fc_{\varepsilon}RI$ and SCFR signalling and ultimately cytokine production.

To further and more directly define the role of MEKK2 in regulating MAPK signalling pathways and cytokine production in mast cells, targeted disruption of the MEKK2 gene in mouse embryonic stem cells has been utilized (77). MEKK2 protein was absent in the homozygous-deficient ES cell clones and mast cells were derived from these ES cells. The loss of MEKK2 had no discernible effects on $Fc_{\varepsilon}RI$ or SCFR expression, growth and differentiation characteristics, morphology or granule content. In the MEKK2 –/– mast cells, the ability of $Fc_{\varepsilon}RI$ to activate JNK was virtually eliminated. However, activation of JNK in response to stress or UV irradiation was maintained. In contrast, MEKK1 –/– mast cells exhibited a normal JNK activation response to $Fc_{\varepsilon}RI$ aggregation. Thus, the loss of JNK activation in response to $Fc_{\varepsilon}RI$ ligation appeared specific to the loss of MEKK2 but not MEKK1. Further, in contrast to JNK activation, following ligation of $Fc_{\varepsilon}RI$, ERK and p38 activation were unaffected by the loss of MEKK2. Such findings indicate that MKKK, other than MEKK2, regulate receptor activation of ERK and p38 in mast cells and implicate an essential role for MEKK2 for JNK activation following signalling through $Fc_{\varepsilon}RI$.

The functional consequences of a deficiency of MEKK2 were also identified. Induction of mast cell degranulation following aggregation of $Fc_{\varepsilon}RI$ in MEKK2 –/– cells was indistinguishable from the response in normal (MEKK2 +/+) mast cells. However, the regulation of cytokine mRNA expression was significantly altered in MEKK2 –/– cells. In particular, the expression of mRNA for TNF-α and IL-4 was significantly reduced following $Fc_{\varepsilon}RI$ aggregation or the combination of $Fc_{\varepsilon}RI$ and SCFR signalling. These findings indicate a major role for MEKK2 in the control of JNK (but not ERK or p38) activation in mast cells and that MEKK2-dependent signalling pathways regulate cytokine production in these cells.

SUMMARY

Based on these results, a model emerges suggesting that the activation of the MEKK2 (and MEKK4) pathway and calcineurin regulation of NFATp contribute to the induction of TNF-α and other cytokine gene expression following cross-linking of $Fc_{\varepsilon}RI$ (Fig. 3). The involvement of the MEKK pathway in regulating the TNF-α promoter defines a mechanism for the previously described CRE augmentation of κ3 regulation of TNF-α

gene expression (72). The regulation of NFATp by calcineurin and transcriptional regulation via κ regulatory elements that bind NFATp in the TNF-α promoter also appears pivotal. Predictably, CsA inhibits TNF-α and IL-4 production. We have shown that this inhibition is not only the result of interference with activation of NFATp (via inhibition of the Ca^{2+}-calmodulin-dependent phosphatase calcineurin) but CsA also inhibits the activation of JNK following ligation of $Fc_{\varepsilon}RI$. In addition, the inhibition of JNK correlates with inhibition of calcineurin as the inactive cyclosporin, CsH fails to do so and FK506 (but not rapamycin) has similar effects. Thus, the immunosuppressant cyclosporins have at least two sites of action involving calcineurin in cytokine production: inhibition of the transcription factor NFAT and inhibition of the JNK pathway.

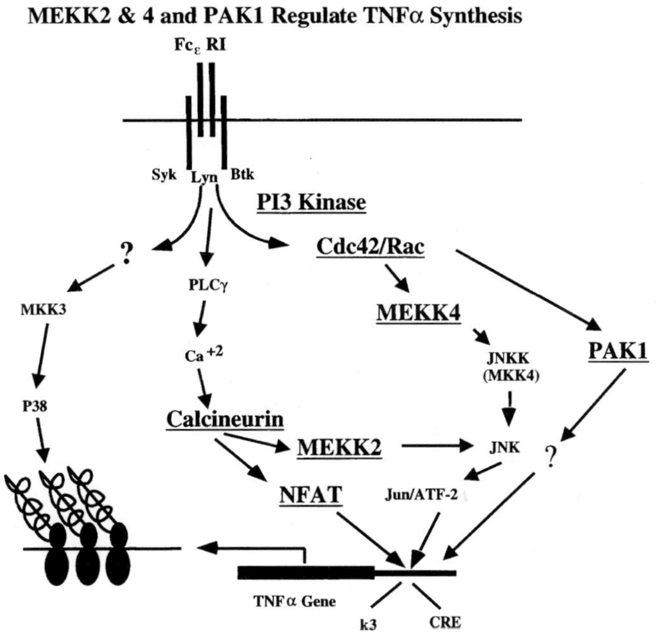

Fig. 3 Interplay of NFAT and JNK pathways. Ligation and activation of $Fc_{\varepsilon}RI$ activates the tyrosine kinases Lyn, Btk and Syk. This results in activation of PLCγ, PI3K and the MEKKs. These signalling events appear essential for the activation/regulation of p38s, JNKs, NFAT and NF-κB. The TNF-α promoter (and other cytokine promoters) is activated by NFAT and JNK. p38 may be involved in the post-transcriptional regulation of TNF-α synthesis.

ACKNOWLEDGEMENTS

We are grateful to Drs T. Ishizuka, A. Oshiba, N. Sakata, N. Terada, P. Gerwins, G. Fanger, E. Hamelmann, H. Kawasome, K. Takeda, K. Chayama, T. Garrington and T. Yujiri for their important contributions to this work and to Drs G. Keller and S. Webb for their help in growth and differentiation of mast cells. Supported by the National Institutes of Health (AI-42246).

REFERENCES

1. Paolini, R., Jouvin, M. H. and Kinet, J. P. Phosphorylation and dephosphorylation of the high affinity receptor for immunoglobulin ε immediately after receptor engagement and disengagement. *Nature* **353**:855–858, 1991.
2. Blank, U., Ra, C., Miller, L., White, K., Metzger, H. and Kinet, J. P. Complete structure and expression in transfected cells of high affinity IgE receptor. *Nature* **337**:187–189, 1989.
3. Reth, M. Antigen receptor tail clue. *Nature* **338**:383–384, 1989.
4. Jouvin, M. H., Adamczewski, M., Nemerof, R., Letourneur, O., Valle, A. and Kinet, J. P. Differential control of the tyrosine kinases Lyn and Syk by the two signaling chains of the high affinity immunoglobulin ε receptor. *J. Biol. Chem.* **269**:5918–5925, 1994.
5. Kihara, H. and Siraganian, R. P. Src homology 2 domains of Syk and Lyn bind to tyrosine-phosphorylated subunits of the high affinity IgE receptor. *J. Biol. Chem.* **269**:22427–22432, 1994.
6. Paul, W. E., Seder, R. A. and Plaut, M. Lymphokine and cytokine production by Fc epsilon RI^+ cells. *Adv. Immunol.* **53**:1–29, 1993.
7. Eiseman, E. and Bolen, J. B. Engagement of the high-affinity IgE receptor activates Src protein-related tyrosine kinases. *Nature* **355**:78–80, 1992.
8. Hutchcroft, J. E., Geahlen, R. L., Deanin, G. G. and Oliver, J. M. $Fc_{\varepsilon}RI$-mediated tyrosine phosphorylation and activation of the 72-kDa protein tyrosine kinase PTK 72 in RBL-2H3 rat tumor mast cells. *Proc. Natl. Acad. Sci. USA* **89**:9107–9111, 1992.
9. Lin, S., Cicala, C., Scharenberg, A. M. and Kinet, J. P. The $Fc_{\varepsilon}RI\beta$ subunit functions as an amplifier of $Fc_{\varepsilon}RI\gamma$-mediated cell activation signals. *Cell* **85**:985–995, 1996.
10. Shiue, L., Green, J., Green, O. M., Karus, J. L., Morgenstern, J. P., Ram, M. K., Zoller, M. J., Zydowsky, L. D., Bolen, J. B. and Brugge, J. S. Interaction of p72Syk with the gamma and beta subunits of the high-affinity receptor for immunoglobulin E, Fc epsilon RI. *Mol. Cell. Biol.* **15**:272–281, 1995.
11. Kimura, T., Kihara, H., Bhattacharyya, S., Sakamoto, H., Appella, E. and Siraganian, R. P. Downstream signaling molecules bind to different phosphorylated immunoreceptor tyrosine-based activation motif (ITAM) peptides of the high affinity IgE receptor. *J. Biol. Chem.* **270**:27962–27968, 1996.
12. Kawakami, T., Yao, L., Miura, T., Tsukada, S., Witte, O. W. and Kawakami, T. Tyrosine phosphorylation and activation of Bruton tyrosine kinase upon Fc epsilon RI cross-linking. *Mol. Cell. Biol.* **14**:5108–5113, 1994.
13. Fukamachi, H., Yamada, N., Miura, T., Kato, T., Ishikawa, M., Gulbins, E., Altman, A., Kawakami, Y. and Kawakami, T. Identification of a protein, SPY75 with repetitive helix–turn–helix motifs and an SH3 domain as a major substrate for protein tyrosine kinase(s) activated by Fc epsilon RI cross-linking. *J. Immunol.* **152**:642–652, 1994.
14. Margolis, B., Hu, P., Katzav, S., Li, W., Oliver, J. M., Ullrich, A., Weiss, A. and Schlessinger, J. Tyrosine phosphorylation of vav proto-oncogene product containing SH2 domain and transcription factor motifs. *Nature* **356**:71–74, 1992.
15. Li, W., Hu, O., Skolnik, E. Y., Ullrich, A. and Schlessinger, J. The SH2 and SH3 domain-containing Nck protein is oncogenic and a common target for phosphorylation by different surface receptors. *Mol. Cell. Biol.* **12**:5824–5833, 1992.
16. Hamawy, M. M., Mergenhagen, S. and Siraganian, R. P. Tyrosine phosphorylation of ppl25FAK by the aggregation of high affinity immunoglobulin E receptors requires cell adherence. *J. Biol. Chem.* **268**: 6851–6854, 1993.
17. Hamawy, M. M., Swaim, W. D., Minoguchi, K., de Feijter, A. W., Mergenhagen, S. E. and Siraganian, R. P. The aggregation of the high affinity IgE receptor induces tyrosine phosphorylation of paxillin, a focal adhesion protein. *J. Immunol.* **153**:4655–4662, 1994.
18. Nishizumi, H. and Yamamoto, T. Impaired tyrosine phosphorylation and Ca^{2+} mobilization, but not degranulation, in Lyn-deficient bone marrow-derived mast cells. *J. Immunol.* **158**:2350–2355, 1997.
19. Kawakami, Y., Miura, T., Bissonnette, R., Hata, D., Khan, W. N., Kitamura T., Yamamoto, M. M., Hartman, S. E., Yao, L., Alt, F. W. and Kawakami, T. Bruton's tyrosine kinase regulated apoptosis and JNK/SAPK kinase activity. *Proc. Natl. Acad. Sci. USA* **94**:3938–3942, 1997.
20. Costello, P., Downward, J. and Tybulewicz, V. L. J. Critical role for the tyrosine kinase Syk in signaling through the high affinity IgE receptor of mast cells. *Oncogene* **13**:2595–2605, 1996.
21. Zhang, J., Berenstein, E. H., Evans, R. L. and Siraganian, R. P. Transfection of Syk protein tyrosine reconstitutes high affinity IgE receptor-mediated degranulation in a Syk-negative variant of rat

basophilic leukemia RBL-2H3 cells. *J. Exp. Med.* **184**:71–79, 1996.

22. Galli, S. J., Zsebo, K. M. and Geissler, E. N. The kit ligand, stem cell factor. *Adv. Immunol.* **55**:1–96, 1994.
23. Galli, S. J., Tsai, M. and Wershil, B. K. The c-Kit receptor, stem cell factor, and mast cells. What each is teaching us about the others. *Am. J. Pathol.* **142**:965–974, 1993.
24. Rottapel, R., Reedijk, M., Williams, D. E., Lyman, S. D., Anderson, D. M., Pawson, T. and Bernstein, A. The steel/w transduction pathway: Kit autophosphorylation and its association with a unique subset of cytoplasmic signaling proteins is induced by the steel factor. *Mol. Cell. Biol.* **11**:3043–3051, 1991.
25. Davis, R. J. Transcriptional regulation by MAP kinases (review). *Mol. Reprod. Dev.* **42**:459–467, 1995.
26. Davis, R. J. The mitogen-activated protein kinase signal transduction pathway. *J. Biol. Chem.* **268**:14553–14556, 1993.
27. Johnson, G. L. and Vaillancourt, R. R. Sequential protein kinase reactions controlling cell growth and differentiation. *Curr. Opin. Cell Biol.* **6**:230–238, 1994.
28. Blumer, K. J. and Johnson, G. L. Diversity in function and regulation of MAP kinase pathways. *Trends Biochem. Sci.* **19**:236–240, 1994.
29. Waskiewicz, A. J. and Cooper, J. A. Mitogen and stress response pathways: MAP kinase cascades and phosphatase regulation in mammals and yeast. *Curr. Opin. Cell. Biol.* **7**:798–805, 1995.
30. Ray, L. B. and Sturgill, T. W. Insulin-stimulated microtubule-associated protein kinase is phosphorylated on tyrosine and threonine *in vivo*. *Proc. Natl. Acad. Sci. USA* **85**:3753–3757, 1988.
31. Boulton, T. G., Nye, S. H., Robbins, D. J., Radziejewska, N. Y. E., Morgenbesser, S. D., DePinko, R. A., Panayotatos, N., Cobb, M. H. and Yanopoulos, G. D. ERKs: a family of protein serine/threonine kinases that are activated and tyrosine phosphorylated in response to insulin and NGF. *Cell* **65**: 663–675, 1991.
32. Ahn, N. G., Weiel, J. E., Chan, C. P. and Krebs, E. G. Identification of multiple epidermal growth factor-stimulated protein serine/threonine kinases from Swiss 3T3 cells. *J. Biol. Chem.* **265**:11487–11494, 1990.
33. Ahn, N. G., Seger, R., Bratlien, R. L., Silty, C. D., Tonks, N. and Krebs, E. G. Multiple components in an epidermal growth factor-stimulated protein kinase cascade. *J. Biol. Chem.* **266**:4220–4227, 1991.
34. Crews, C. M., Alessandrini, A. and Erickson, R. L. The primary structure of MEK, a protein kinase that phosphorylates the ERK gene product. *Science* **258**:478–480, 1992.
35. Sanchez, I., Hughes, R. T., Mayer, B. J., Yee, K., Woodgett, J. R., Avruch, J., Kyriakis, J. M. and Zon, L. I. Role of the SAPK/ERK kinase-1 in the stress-activated pathway regulating transcription factor c-Jun. *Nature* **372**:794–798, 1994.
36. Lin, A., Minden, A., Marinetto, H., Lange-Carter, C. A., Johnson, G. L., Mercurio, F. and Karin, M. Identification of a dual specificity kinase that activates Jun kinases and p38–Mpk2. *Science* **268**:286–290, 1995.
37. Derijard, B., Rainglaud, J., Barrett, T., Wu, I.-H., Han, J., Ulevitch, R. J. and Davis, R. J. Independent human MAP kinase signal transduction pathways defined by MEK and MKK isoforms. *Science* **267**:682–685, 1995.
38. Blank, J. L., Gerwins, P., Elliott, E. M., Sather, S. and Johnson, G. L. Molecular cloning of MEK kinase (MEKK) 2 and 3: regulation of sequential phosphorylation pathways involving mitogen activated protein kinase and c-Jun kinase. *J. Biol. Chem.* **271**:5361–5368, 1996.
39. Wu, X., Noh, S. J., Zhou, G., Dixon, J. E. and Guan, K. L. Selective activation of MEK1 but not MEK2 by A-Raf from epidermal growth factor-stimulated Hela cells. *J. Biol. Chem.* **271**:3265–3271, 1996.
40. Gerwins, P., Blank, J. L. and Johnson, G. L. Cloning of a novel mitogen-activated protein kinase kinase kinase, MEKK4, that selectively regulates the c-Jun amino terminal kinase pathway. *J. Biol. Chem.* **272**:8288–8295, 1997.
41. Gerwins, P., Blank, J. L. and Johnson, G. L. Cloning of a novel mitogen-activated protein kinase kinase kinase, MEKK4, that selectively regulates the c-Jun amino terminal kinase pathway. *J. Biol. Chem.* **272**:8288–8295, 1997.
42. Pombo, C. M., Kehrl, H., Sanchez, P., Katz, J., Avruch, J., Zon, L. I., Woodgett, J. R., Force, T. and Kyrikis, J. M. Activation of the SAPK pathway by the human ST620 homologue germinal center kinase. *Nature* **337**:750–754, 1995.
43. Creasy, C. L. and Chernoff, J. Cloning and characterization of a member of the MST subfamily of Ste20-like kinases. *Gene* **167**:303–306, 1995.
44. Doror, D. S., Devereax, L., Dietzsh, E. and De Kretser, T. Identification of a new family of human epithelial protein kinases containing two leucine/isoleucine-zipper bearing domains. *Eur. J. Biochem.* **213**:701–710, 1983.

45. Ezoe, K., Lee, S., Strunk, K. M. and Spritz, R. A. PTK1: a novel protein kinase required for proliferation of human melanocytes. *Oncogene* **9**:935–938, 1994.
46. Holzman, L. B., Merit, S. E. and Fan, G. Identification, molecular cloning, and characterization of dual leucine-zipper bearing kinase. *J. Biol. Chem.* **269**:30808–30817, 1994.
47. Katoh, M., Hirai, M., Sugimura, T. and Terada, M. Cloning and characterization of MST, a novel (putative) serine/threonine kinase with SH3 domain. *Oncogene* **10**:1447–1451, 1995.
48. Reddy, U. R. and Pleasure, D. Cloning of a novel putative protein kinase having a leucine zipper domain from human brain. *Biochem. Biophys. Res. Commun.* **202**:613–620, 1994.
49. Patrios, C., Makris, A., Chernoff, J. and Tsichlis, P. N. Tpl-2 acts in concert with Rac and Raf-1 to activate mitogen-activated protein kinase. *Proc. Natl. Acad. Sci. USA* **91**:9755–9759, 1994.
50. Salmeron, A., Ahmad, T. B., Carlile, G. W., Pappin, D., Narsimhan, R. P. and Ley, S. C. Activation of MEK-1 and SEK-1 by Tpl-2 protoncoprotein, a novel of MAP-kinase kinase kinase. *EMBO J.* **15**:817–826, 1996.
51. Derijard, B., Hibi, M., Wu, H., Barrett, T., Su, B., Deng, T., Karin, M. and Davis, R. J. JNK1: a protein kinase stimulated by UV light and Ha-Ras that binds and phosphorylates the c-Jun activation domain. *Cell* **76**:1025–1037, 1994.
52. Kyriakis, J. M., App, H., Zhang, X.-F., Banerjie, P., Brautigan, D. L., Rapp, U. R. and Avruch, J. Raf-l activates MAP kinase-kinase. *Nature* **358**:417–421, 1992.
53. Minden, A., Lin, A., McMahon, M., Lange-Carter, C., Derijard, B., Davis, R. J., Johnson, G. L. and Karin, M. Differential activation of ERK and JNK mitogen-activated protein kinases by Raf-1 and MEKK. *Science* **266**:1719–1723, 1994.
54. Sakata, N., Patel, H. R., Terada, N., Aruffo, A., Johnson, G. L. and Gelfand, E. W. Selective activation of c-Jun kinase mitogen-activated protein kinase by CD40 on human B cells. *J. Biol. Chem.* **270**:30823–30828, 1995.
55. Su, B., Jacinto, E., Hibi, M., Kallunki, T., Karin, M. and Ben-Neriah, Y. JNK is involved in signal integration during costimulation of T lymphocytes. *Cell* **77**:727–736, 1994.
56. Gupta, S., Barrett, T., Whitmarsh, A. J., Cavanaugh, J., Sluss, H. K., Derijard, B. and Davis, R. J. Selective interaction of JNK protein kinase isoforms with transcription factors. *EMBO J.* **15**:2760–2770, 1996.
57. Pulverer, B. J., Kyriakis, J. M., Avruch, J., Nikolakaki, Z. and Woodgett, J. R. Phosphorylation of c-Jun mediated by MAP kinases. *Nature* **353**:670–674, 1991.
58. Gupta, S., Campbell, D., Derijard, B. and Davis, R. J. Transcription factor ATF2 regulation by the JNK signal transduction pathway. *Science* **267**:389–393, 1995.
59. Gille, H., Strahl, T. and Shaw, P. E. Activation of ternary complex factor Elk-1 by stress-activated protein kinases. *Curr. Biol.* **5**:1191–1200, 1995.
60. Tsai, E. Y., Jain, J., Pesavento, P. A., Rao, A. and Goldfeld, A. E. Tumor necrosis factor alpha gene regulation in activated T cells involves ATF-2/Jun and NFATp. *Mol. Cell. Biol.* **16**:459–467, 1996.
61. Beyaert, R., Cuenda, A., Berghe, W. V., Plaisance, S., Lee, J. C., Haegeman, G., Cohen, P. and Fiers, W. The p38/RK mitogen-activated protein kinase pathway regulates interleukin-6 synthesis in response to tumor necrosis factor. *EMBO J.* **15**:1914–1923, 1996.
62. Wang, X. Y. and Ron, D. Stress-induced phosphorylation and activation of the transcription factor CHOP (GADD153) by p38 MAP kinase. *Science* **272**:1347–1349, 1996.
63. Zheng, C. F. and Guan, K. L. Cloning and characterization of two distinct human extracellular signal-regulated kinase activator kinases, MEK1 and MEK2. *J. Biol. Chem.* **268**:11435–11439, 1993.
64. Han, J., Lee, J. D., Jiang, Y., Li, Z., Feng, L. and Ulevitch, R. J. Characterization of the structure and function of a novel MAP kinase kinase (MKK 6). *J. Biol. Chem.* **271**:2886–2891, 1996.
65. Minden, A., Lin A., Claret, F. X., Abo, A. and Karin, M. Selective activation of the JNK signaling cascade and c-Jun transcriptional activity by the small GTPases Rac and Cdc42Hs. *Cell* **81**:1147–1157, 1995.
66. Fanger, G. R., Johnson, N. L. and Johnson, G. L. MEK kinases are regulated by EGF and selectively interact with Rac/Cdc42. *EMBO J.* **16**:4961–4972, 1997.
67. Oshiba, A., Hamelmann, E., Takeda, K., Bradley, K., Loader, J. E., Larsen, G. L. and Gelfand, E. W. Passive transfer of immediate hypersensitivity and airway hyperresponsiveness by allergen-specific IgE and IgG1 in mice. *J. Clin. Invest.* **97**:1398–1408, 1996.
68. Ishizuka, T., Oshiba, A., Sakata, N., Terada, N., Johnson, G. L. and Gelfand, E. W. Aggregation of the $Fc_{\varepsilon}RI$ on mast cells stimulates a wortmannin-sensitive c-Jun amino-terminal kinase (JNK) activity. *J. Biol. Chem.* **271**:12762–12766, 1996.
69. Ishizuka T., Terada N., Gerwins, P., Hamelmann, E., Oshiba, A., Fanger, G. R., Johnson, G. L. and

Gelfand, E. W. Mast cell TNF-α production is regulated by MEK kinases. *Proc. Natl. Acad. Sci. USA* **94:**6358–6363, 1997.

70. Bradding, P., Roberts, J. A., Britten, K. M., Montefort, S., Djukanovic, R., Mueller, R., Heusser, C. H., Howarth, P. H. and Holgate, S. T. Interleukin-4, -5, and -6 and tumor necrosis factor-alpha in normal and asthmatic airways: evidence for the human mast cell as a source of these cytokines. *Am. J. Respir. Cell Mol. Biol.* **10**:471–480, 1994.
71. Baumgartner, R. A., Hirasawa, N., Ozawa, K., Gusovsky, F. and Beaven, M. A. Enhancement of TNF-α synthesis by overexpression of $G_{\alpha 2}$ in a mast cell line. *J. Immunol.* **157**:1625–1629, 1996.
72. Rao, A. NFATp: a transcription factor required for the co-ordinate induction of several cytokine genes. *Immunol. Today* **15**:274–281, 1994.
73. Lee, J. C., Laydon, J. T., McDonnell, P. C., Gallagher, T. F., Kumar, S., Green, D., McNulty, D., Blumenthal, M. J., Heys, J. R., Landvatter, S. W., Strickler, J. E., McLaughlin, M. M., Siemens, I. R., Fisher, S. M., Livi, G. P., White, J. R., Adams, J. L. and Youn, P. R. A protein kinase involved in the regulation of inflammatory cytokine biosynthesis. *Nature* **372**:739–746, 1994.
74. Ishizuka, I., Kawasome, H., Terada, N., Takeda, K., Gerwins, P., Keller, G. M., Johnson, G. L. and Gelfand, E. W. Stem cell factor augments $Fc_{\varepsilon}RI$-mediated TNFα production and stimulates MAP kinases via a different pathway in MC/9 mast cells. *J. Immunol.* **161**:3624–3630, 1998.
75. Thompson, H. L., Metcalfe, D. D. and Kinet, J. P. Early expression of high-affinity receptor for immunoglobulin E ($Fc_{\varepsilon}RI$) during differentiation of mouse mast cells and human basophils. *J. Clin. Invest.* **85**:1227–1233, 1990.
76. Ishizuka, T., Chayama, K., Takeda, K., Terada, N., Keller, G. M., Johnson, G. L. and Gelfand, E. W. MAP kinase activation through $Fc_{\varepsilon}RI$ and stem cell factor receptor is differentially regulated by PI3-kinase and calcineurin in mouse bone marrow-derived mast cells. *J. Immunol.* **162**: 2087–2094, 1999.
77. Garrington, T. P., Chayama, K., Webb, S., Yujiri, T., Russell, D. M., Sather, S., Gibson, S., Keller, G., Gelfand, E. W. and Johnson, G. L. MEKK2 gene disruption inhibits IgE and c-Kit ligand stimulated JNK activation and cytokine production in ES cell-derived mast cells. *Genes Dev.* (submitted 2000).

CHAPTER 10

$Fc_\varepsilon RI$-mediated Induction of TNF-α Gene Expression in Mast Cell Lines

ULRICH BLANK*[1] and NADINE VARIN-BLANK[2]

[1]Unité d'Immuno-Allergie, Institut Pasteur and [2]Unité INSERM 363, ICGM Hôpital Cochin, Paris, France

INTRODUCTION

Tumour necrosis factor-α (TNF-α) is a multifunctional cytokine that is produced by numerous cell types and plays a central role in various immune and inflammatory responses (1). TNF-α, or cachectin, was originally discovered by its ability to induce tumour regression and to cause wasting syndrome during parasite infections. The cloned cDNA (2, 3) has significant homologies to T cell-derived lymphotoxin-α (Lt-α), also called TNF-β, and lymphotoxin-β (Lt-β) (4, 5). Human TNF-α is synthesized as a 233 amino acid membrane precursor. Secretion requires a specific Zn-dependent metalloprotease which cleaves an unusually long leader peptide to give a major secreted protein of 157 amino acids (6). Secreted TNF-α forms a trimer which allows TNF-α to aggregate its receptors on target cells (7).

During the 1980s several groups described the presence of a cytotoxic activity comparable to TNF-α in normal mast cells, cytokine-dependent mast cell cultures and several interleukin (IL)-3-dependent and independent mast cell lines (8–14). Further immunological and molecular studies clearly established TNF-α as a mast cell mediator (15–17). Mast cells are the only cells known to contain significant quantities of TNF-α pre-stored in cytoplasmic granules that can be released by exocytotic discharge, providing an immediate source upon activation (16, 18, 19). They also respond to stimulation with the new synthesis of bioactive protein, which accounts for the sustained action of the cytokine at the inflammatory site (18). As mast cells are exclusively localized in tissues, the production and action of TNF-α from mast cells is thought to be essentially local. Their preferential distribution close to blood and lymphatic vessels, nerve endings and epithelia could potentialize its action by providing adequate stimuli on target cells. Mast cell-derived TNF-α has been shown to contribute importantly to host defence reactions

* Corresponding author.

ISBN 0-12-473335-2

during local infections (20, 21). It also plays a detrimental role in allergy and asthma as a mediator of late-phase inflammatory reactions (22–25). In view of its potent and pleiotropic actions, it is fundamental to maintain the biosynthesis and release of TNF-α under tight control. Evidence has accumulated from different cell types that this regulation takes place at multiple levels, including transcription, translation and secretion. In this chapter we focus on recent advances in our understanding of these mechanisms and then try to summarize what is known for mast cells.

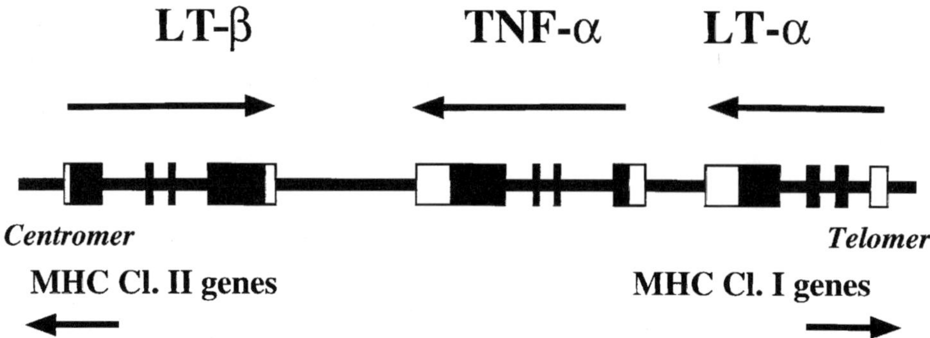

Fig. 1 Localization and structure of the genes encoding TNF-α, LT-α and LT-β on human chromosome 6. Exons and introns are marked as black and white boxes respectively. The drawing is not to scale. TNF-α is separated from LT-α and LT-β by approximately 1 and 3 kbp, respectively.

CHROMOSOMAL LOCATION AND GENE STRUCTURE OF TNF-α

In both humans and mice the TNF-α gene is located in a region between the loci encoding MHC class I and class II molecules on chromosomes 6 and 17 respectively and is closely linked to the Lt-α and Lt-β genes (Fig. 1) (26, 27). The structure of the TNF-α gene has been determined in several species and allowed extensive sequence comparison (26, 28). In all species known so far the gene presents the same structure, with the coding region consisting of four exons arranged over approximately 3 kb of DNA. As expected, the TNF-α sequence is highly conserved during evolution both in the coding region (> 60%) and in 3´ and 5´ untranslated regions (UTRs). Notably, the 3´ UTR contains multimers of a characteristic pentanucleotide, AUUUA, which is present in many cytokine mRNAs (29). Analysis of the 5´ and 3´ flanking region has also shown strong evolutionary conservation. A recent comparison between mouse and human promoter regions shows substantial homology in the whole upstream sequence extending to the Lt-α gene, the first 220 bp upstream of the transcriptional start site being the most highly conserved sequences (30). Additional high sequence conservation is also apparent further upstream (nt –350 to –800) covering particularly several NF-κB and NF-κB-related (CK-1 and CK-2) sequence elements (26, 31). Figure 2 is a schematic drawing of the promoter region of the rat gene (28) and illustrates the principal regulatory sequence elements. A sequence comparison for the highly conserved κB elements in the promoter as well as from the 3´ flanking region of the gene is separately shown in Fig. 3.

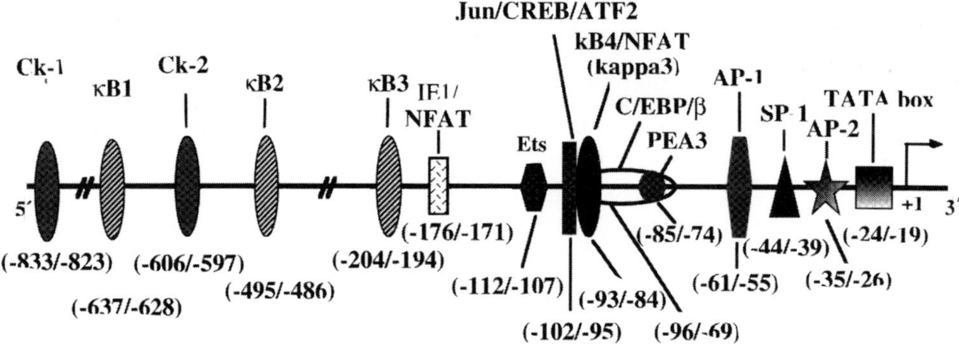

Fig. 2 A schematic representation of the rat TNF-α promoter region indicating regulatory elements potentially involved in gene transcription. The numbers indicate distance from the transcriptional start site indicated by an arrow (+1). Note that the κB2 site in rat and mouse promoters is not conserved in the human promoter. The κB4/NFAT or kappa3 element has been shown to bind both NFAT and NF-κB nuclear factors in humans.

	CK-1	**κB-1**	**CK-2**
human	**GGGGACCCCCC**	**GTGAATTCCC**	**GGGGCTGTCC**
porcine	**GGAGATACAGG**	**GTGAATTCCC**	**GGGGCTGTCC**
rabbit	**GGAGACCCCCT**	**GTGAATTCCC**	**GGGGCTGTCC**
mouse	**GGGGAATCCTT**	**GTGAATTCCC**	**GGGGCTGTCC**
rat	**GAGGAATCCTT**	**GTGAATTCCC**	**GGGGCTGTCC**

	κB-3	**CRE + κB4/NFAT (kappa3)**
human	**GGGGTATCCTT**	**TGAGCTCATGGGTTTCTGG**
porcine	**GAAGTATCCCT**	**TGAGCTCATGGGTTTCTGG**
rabbit	**GGAGTATCCTT**	**TGAGCTCATGGGTTTCTGG**
mouse	**GGAGTATCCTT**	**TGAGATCATGGTTTTCTGG**
rat	**GGAGTATCCTT**	**TGAGATCATGGTTTTCTGG**

	κB 3′ flanking
human	**GGGAATTTCC**
porcine	**GGGAATTTCC**
horse	**GGGAATTTCC**
mouse	**GGGAATTTCC**
rat	**GGGAATTTCC**

Fig. 3 Sequence alignment of κB elements conserved between different species. For the κB4/NFAT the adjacent CRE element is also shown. Differences to the human sequence are underlined. The relative positions of these elements in the rat gene can be seen in Fig. 2. The κB site in the 3′ flanking region lies 218 bp downstream of the polyadenylation signal in the rat gene.

REGULATION OF TNF-α GENE EXPRESSION IN NON-MAST CELL LINES

Regulation at the Level of Transcription

Induction of TNF-α gene expression is initiated by a number of stimuli, including various cytokines, bacterial products, immunological stimuli as well as some non-specific activators (1, 26). Following activation the increase of TNF-α mRNA steady-state levels is rapid and does not require *de novo* protein synthesis. Nuclear run-on studies performed on primary macrophages and macrophage cell lines have clearly established a transcriptional induction both by augmenting the initiation and by increasing processivity during the elongation step (32–34). Processivity was independent of protein synthesis whereas initiation of transcription was further increased in the presence of cycloheximide, suggesting the action of short-lived repressors. Several studies have allowed identification of *cis*-acting elements and nuclear factors involved in the regulation of TNF-α gene expression. Although there was initially some controversy about differences between human and murine systems (33, 35–37), a functional role for the transcription factor NF-κB is now clearly established (30, 38). All κB sites identified in the promoter region are able to bind NF-κB and functionally contribute to the induction of TNF-α gene expression. This is further confirmed by studies using pyrrolidine dithiocarbamate PDTC, a pharmacological inhibitor of NF-κB activation (39), or more recently the overexpression of the cytosolic inhibitor I-κBWT or I-κB (S32,36A), a phosphorylating-deficient mutant form (30). A κB element situated approximately 220 bp downstream of the polyadenylation signal site of the TNF-α gene (see Fig. 3) also mediates lipopolysaccharide (LPS)-induced TNF-α gene expression in murine macrophages (37). In rat astrocytes the LPS-induced TNF-α gene expression was mainly regulated by this element rather than the upstream κB sites (40). These observations indicate strong lineage- and tissue-specific regulatory mechanisms. This is further corroborated by studies in T lymphocytes where another factor, NFATp, a member of the nuclear factor of activated T cells (NFAT) family of proteins was required for the strong induction by T cell receptor ligands (41, 42). NFATp bound to the kappa3 (κB4/NFAT; see Fig. 2) site in the absence of fos/jun (AP-1) family proteins and induced transcriptional activation (42). The simultaneous binding of activating transcription factor 2 (ATF-2) and jun family members to the adjacent cyclic AMP-responsive element (CRE) significantly enhanced the effect (43). In addition to NF-κB and NFAT and depending on the stimulus and cell type, a variety of other transcription factors have also been implicated in the induction of TNF-α gene expression. These include C/EBPβ (or NF-IL6) predominantly expressed in monocytes–macrophages and binding to a site encompassing nt –74 to –100 upstream of the transcription initiation site in humans. C/EBPβ acts synergystically with c-jun in the PMA or LPS induction of the TNF-α gene in U937 cells (44). Other transcriptional activators identified include a member of the ets family that binds to the highly conserved sequence element immediately adjacent to CRE and also known as the PU.1 element (45), the Egr-1/Krox-24 transcription factor which binds to a GC-rich sequence (nt –160 and –170) of the huTNF-α gene (46), as well as the TNF-α core promoter region immediately upstream (28 bp) of the TATA box containing SP-1 and AP-2 conserved elements (47).

In conclusion, the data collected have allowed the identification of a number of transcriptional activators involved in TNF-α gene transcription. However, our understanding of this process is still sketchy as most of the studies have dealt with the independent analysis of a given promoter element under different conditions of stimulation whereas *in vivo* the co-operative occupancy of the promoter by multiple transcription factors certainly renders the situation more complex (48). So far, we know nothing about the identity of the factors that appear to repress promoter activity, as indicated in one of the studies (49).

Post-transcriptional Regulation

Post-transcriptional regulatory mechanisms have been shown to play an important role in gene expression in eukaryotic cells (50). They depend in part on specific sequences present in the 3′ UTR. One major class of these regulatory sequences is present in mRNAs encoding growth factors, oncoproteins or cytokines, including TNF-α, and consists of a conserved AU-rich sequence element (ARE) (29). This sequence is composed of several repeats of the pentanucleotide AUUUA. The ARE has been associated with both an accelerated degradation of mRNA (29) and interference with translation (51, 52). Ample evidence exists demonstrating that the production of TNF-α is regulated to a considerable extent by such cytoplasmic regulatory mechanisms of mRNA function (53, 54). TNF-α mRNA has a very short half-life (39 min) (55). Recent investigations in mice lacking the ARE sequence of the TNF-α gene have confirmed the role of this element in TNF-α message instability in the resting macrophages (56). In LPS-stimulated cells the situation was more complex, as an approximately 4-fold increase in mRNA half-life in ARE –/– mice was opposed to a stimulation-induced stabilization of TNF-α mRNA which was at least partially dependent on the ARE. At similar transcription rates the accumulation of mRNA was higher in ARE +/+ mice (90-fold) than in ARE –/– mice (46-fold). This accumulation did not seem to be due to an 'induced' splicing mechanism as recently proposed (57) since both the accumulation and disappearance of nuclear pre-mRNA was comparable in both types of mice. It is noteworthy that, even in the absence of ARE, the mRNA was still responsive to LPS-mediated stabilization, indicating the importance of other sequences than the ARE. Altogether these data suggest that in stimulated cells AREs mediate both positive and negative regulatory feedback loops. Tristetraprolin, the prototype of CCCH-zinc finger proteins, has recently been identified as an ARE binding protein able to mediate destabilization of the TNF-α message in murine macrophages (58). An example of a protein operating at the ARE during the induction/stabilization phase of the response is the ubiquitously expressed embryonic lethal abnormal visual (ELAV) RNA binding protein HuR (59). Besides mRNA stability, earlier studies have also indicated an effect of the ARE element on translation (51, 54). The phenotype of ARE –/– mice is characterized by a permissive translation of TNF-α in non-stimulated cells and the development of chronic inflammatory arthritis and bowel disease (56). In macrophages the 3′ UTR of TNF-α maintains the TNF-α mRNA in a repressed state and LPS stimulation de-represses this translational silencing (53, 60). A *trans*-dominant factor expressed constitutively in certain cell lines also overcomes the translational blockade (52). More recently, both the p38-mitogen-activated protein kinase (p38) and the c-Jun N-terminal–stress-activated protein kinase (JNK–SAPK) have been implicated in the signalling pathways leading to translational control of TNF-α. Inhibition of the LPS-mediated activation of p38 with a specific pharmacological compound, SB203580, substantially reduced TNF-α secretion in the absence of any matching effects on TNF-α mRNA levels (61). The translational block by p38 implicated the downstream MAPKAP kinase 2 (62). Furthermore, a kinase-defective mutant of SAPKβ blocked translation of TNF-α and similarly a dexamethasone-induced block of translation was overcome by overexpression of WT SAPKβ (63). Implication of both kinases was confirmed in ARE –/– mice which were shown to be no longer susceptible to translational modulation by p38 and JNK–SAPK pathways (56). However, the molecular mechanism of the translational blockade remains unclear, but TNF-α mRNA could remain unbound to the translational machinery in unstimulated cells and shift into polyribosomes upon stimulation (64). Furthermore, an inducible complex could bind to ARE sequences (65). Besides their role in translation, data obtained in a T cell line indicate that p38 and JNK–SAPK together with signal-regulated kinase (erk1/erk2) also play a role in the transcriptional control of TNF-α synthesis (66). The relative contribution of the different MAPK cascades is schematically presented in Fig. 4.

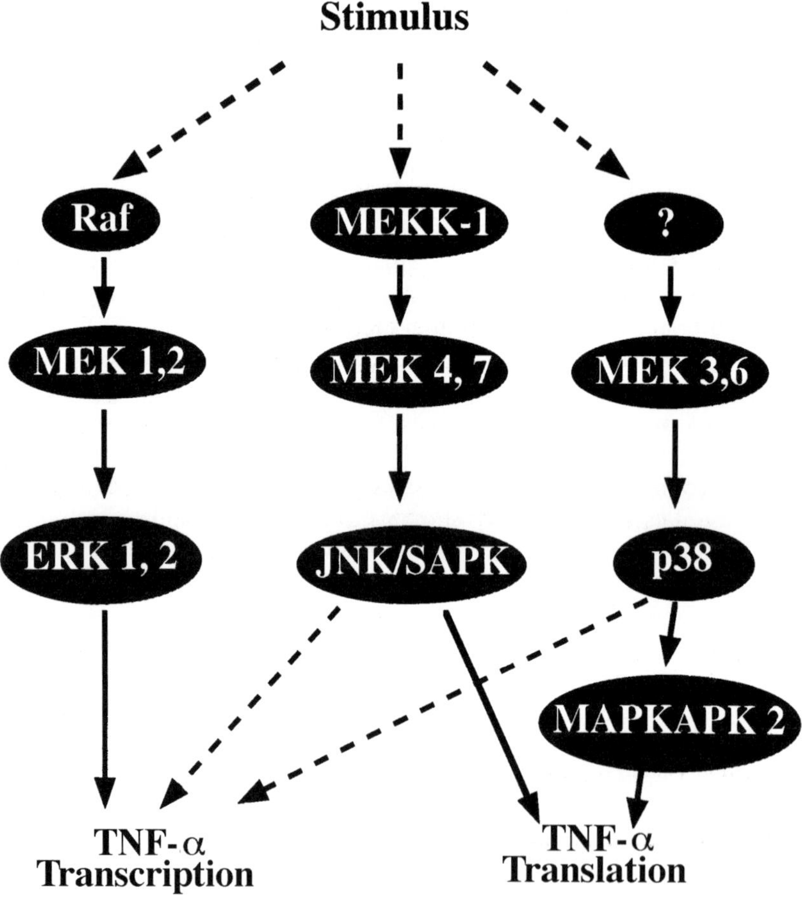

Fig. 4 Proposed signalling pathways of MAPK leading to TNF-α biosynthesis according to data obtained in macrophages and T cells (61–63). The diverse members of the MAPK are indicated. The broken arrows indicate mechanisms only shown in T cells.

Regulation at the Level of Secretion

TNF-α is a type II transmembrane protein that contains an unusually long 76 amino acid signal peptide (2). The 26-kDa transmembrane form is transported along the exocytotic pathway to the plasma membrane where it is cleaved by a highly specific membrane-bound metalloendoprotease, called TACE (6). Although a biologically active membrane-bound form of 26 kDa has been described on activated monocytes, pulse chase studies demonstrated that the membrane precursor usually decreases with a half-life of around 15–20 min to yield the 17-kDa secreted form in the medium (67). Experiments using a specific inhibitor of the TNF-α metalloprotease showed that unprocessed TNF-α became sequestered into perinuclear microsomes which co-localize with the Golgi (68). It is not completely clear to what extent secretion of TNF-α in monocytes and macrophages requires active signalling. LPS-stimulated secretion has been described to be sensitive to *Botulinum* toxin type D without affecting intracellular TNF-α levels and independently secretion could require phosphokinase C (PKC) activation (67, 69).

Fc$_\varepsilon$RI-DEPENDENT REGULATION OF TNF-α PRODUCTION

Characteristics of Fc$_\varepsilon$RI-dependent TNF-α Gene Expression in Mast Cells

Mast cells are critical effectors of IgE-dependent immune responses mediated through their high-affinity IgE receptor (Fc$_\varepsilon$RI). In these cells aggregation of receptor-bound IgE provides a potent stimulus for TNF-α production first recognized in the RBL-2H3 and PT18 mast cell lines (12, 15, 18, 70, 71). TNF-α represents a new class of mast cell mediators that are released from both pre-formed and *de novo* synthesized pools. This has been very elegantly shown in experiments using the transcriptional inhibitor actinomycin D which does not block degranulation (18). The Cl.C57.1 mast cell line, primary IL-3-dependent mast cell cultures and purified rat peritoneal mast cells (RPMC) released about 10% of the maximal TNF-α levels achieved by the Fc$_\varepsilon$RI-stimulated exocytotic discharge. In most cultured mast cell lines TNF-α is, however, not constitutively present in appreciable amounts (16). Yet, like normal mast cells, they respond to stimulation with a rapid induction of TNF-α mRNA and the secretion of newly synthesized protein (15, 72, 73). The Fc$_\varepsilon$RI-stimulated TNF-α gene expression in mast cells is tightly regulated and follows the rules of an immediate early gene with a rapid induction independent of protein synthesis (16, 73). The low levels of mRNA detected in non-stimulated cells (16, 18, 74) do not lead to a significant release of protein in short-term cultures (11, 16, 18, 74). The protein can nevertheless be accumulated in the medium of certain mast cells after 20 h (75). Upon stimulation through Fc$_\varepsilon$RI, TNF-α mRNA steady-state levels rapidly increase with maxima being achieved at around 1 h and then decline (16, 72–74). This is followed by the secretion of newly synthesized protein starting between 30 and 60 min after addition of stimulus, whereas secretion of pre-formed TNF-α is complete within 10 min (18). Nuclear run-on experiments in Cl.MC/C57.1 mast cells as well as studies using the transcriptional inhibitor actinomycin D in Cl.MC/C57.1 and RBL-2H3 cells indicate that the increase in TNF-α mRNA steady-state levels depends on *de novo* transcription (73, 76). Induction also requires a continuous aggregation of IgE receptors, steady-state mRNA levels being directly proportional to the time receptors had been aggregated (73). This demonstrates a tight and rapid coupling of receptor activation with the transcriptional machinery and was different from the regulation of the immediate early gene c-fos (77). Several transcription factors implicated in the expression of immediate early genes such as NFAT, Oct/OAP, NF-κB or an NF-κB-like factor also required continuous stimulation for their presence in the nucleus and/or their capacity to maintain gene transcription (73, 78, 79). These activities were highly dependent on the induced intracellular calcium levels as well as on the frequency of Ca^{2+} oscillations (79, 80). In contrast, the presence of AP-1 in the nucleus was not reversed after stimulus termination consistent with the data on c-fos mRNA (73, 77).

FcεRI-mediated Signalling Mechanisms Involved in Gene Expression

Although much progress has been made in the last decade in the understanding of how the aggregation of $Fc_\varepsilon RI$ is turned into a meaningful response (81), our picture regarding the $Fc_\varepsilon RI$-mediated induction of TNF-α gene expression is still far from complete due to the multiple levels and factors involved. Furthermore, growth factors present in the medium or natural environment can significantly affect $Fc_\varepsilon RI$-mediated responses. For example, while degranulation experiments in RBL-2H3 mast cells can be performed in Tyrode buffer, TNF-α production requires the presence of serum-containing medium (C. Pelletier and U. Blank, unpublished). Signalling initiated by the binding of a multivalent antigen and the consequent clustering of receptor-bound IgE launches within seconds the phosphorylation of specific tyrosine residues in the ITAM motifs of receptor β and γ subunits (81). This implicates a transphosphorylation mechanism by the Src-related kinase lyn already associated with a fraction of β chains in unstimulated receptors (82). Phosphorylated γ-ITAMs recruit the tandem SH2-containing protein tyrosine kinase syk, which is at the basis of a large number of functionally important signalling pathways implicated in gene expression (81). Activation of PLC-γ by syk leads to the generation of inositol 1,4,5-trisphosphate (IP_3) and diacylglycerol (DAG), responsible for the increase in intracellular Ca^{2+} levels and activation of PKC. Intracellular Ca^{2+} levels are also regulated via a phosphoinositide 3-kinase (PI3K) and btk-dependent pathway that leads to the sustained activation of PLC-γ (83). The sustained activation may be particularly important in late-phase anaphylactic responses including TNF-α gene expression (84). Syk activation is also necessary for the stimulation of MAPK cascades as it mediates the Shc-dependent activation of ras, an upstream regulator of erk1/erk2 (85) or the vav-dependent activation of rac-1, an upstream regulator of JNK (86). All of these events may then further lead to the activation of transcription factors regulating gene expression; however, precise mechanisms in this complex process remain to be defined.

Signalling Pathways Involved in FcεRI-induced TNF-α Production

Information on functional pathways relevant to TNF-α gene expression have been obtained through studies with pharmacological agents. Both the use of ionomycin and on reverse the Ca^{2+} chelating agent EGTA have revealed the importance of Ca^{2+} as a second messenger in TNF-α mRNA induction (72, 73). The Ca^{2+}-dependent induction of gene expression was shown to require activation of the phosphatase calcineurin. Incubation with a specific inhibitor cyclosporin A (CsA) resulted in a strong inhibition of TNF-α production, even though 50% inhibitory doses differed between various cytokines tested (4 nM, 65 nM and 130 nM for IL-1, TNF-α and IL-6, respectively) (87–90). This may reflect the complexity of the mechanisms involved as well as the data showing that calcineurin can target multiple pathways involved in cytokine gene expression including the cytoplasmic component of NFAT, IkBα and JNK family kinases (48). When mast cells were activated with bacteria expressing the fimbrial protein FimH, TNF-α production was less susceptible to blockade of calcineurin, suggesting the existence of additional albeit less potent pathways leading to TNF-α expression in mast cells (71). Secretion of bioactive TNF-α strongly depends on the activation of PKC (see below). In contrast, transcription and intracellular accumulation of TNF-α were largely independent of classical PKC isoforms (70, 89, 91). Two specific inhibitors of classical PKC isoforms as well as their depletion through long-term incubation with phorbol esters were relatively ineffective in inhibiting TNF-α production. PKC activation could nevertheless have a modulatory role as pre-incubation with PKC-activating phorbol esters further increased Fc_{ε}RI-mediated TNF-α production (89). In the CPII mast cell line evidence for a role of the atypical PKC member μ (PKC-μ) in the antigen-stimulated induction of a TNF-α reporter construct has been reported (91). While the response was insensible to the depletion of classical PKC isoforms with phorbol myristate acetate (PMA) it was still susceptible to treatment with the pharmacological agent Gö 6976, which inhibits classical PKC isoforms and the atypical form PKC-μ. Regarding PI3K activation, the results were contradictory depending on the mast cell line examined. Wortmannin, a specific inhibitor of PI3K, reduced Fc_{ε}RI-dependent TNF-α secretion in MC/9 and RBL-2H3 cell lines as well as in bone marrow-derived mast cells (BMMC). The inhibitory action on TNF-α mRNA accumulation or reporter-dependent gene expression was less pronounced, suggesting additional effects on protein secretion (89, 90, 92, 93). No inhibitory action of PI3K inhibitors on the antigen-induced TNF-α reporter or on protein expression was found in the CPII mast cell line (94). Pathways of PI3K implicate the tyrosine kinase btk acting on Ca^{2+} signalling, activation of p38 MAPK and JNK (83, 90, 93, 95, 96). The latter plays a role in TNF-α gene expression in mast cells, as overexpression of an inhibitory JNK2 mutant form suppressed the IgE-dependent expression of a TNF-α promoter-dependent reporter construct in MC/9 mast cells (92), but alternatively several growth factor receptors present on mast cells could mediate Fc_{ε}RI- and PI3K-independent signalling (93). The resistance of the CPII cell line may be explained by such a growth factor-mediated signal. The situation is equally complex regarding the role of erk1/2. Specific inhibition of MEK1 the upstream activator of erk1/2 with PD 098059 did not affect TNF-α production in MC/9 mast cells or BMMC (90, 92). However, it dose-dependently inhibited TNF-α production in the RBL-2H3 and CPII mast cells (91, 97). In the latter an even more potent effect (IC_{50} = 3 μM versus 30 μM) was noticed on TNF-α promoter-dependent reporter gene induction (91). As erk1/2 are involved in the induction of elk-1 transcription factor and c-fos which is part of the AP-1 complex the differences between cell lines remain unexplained (98). The use of selective inhibitors of p38 has revealed that this kinase, although clearly implicated in TNF-α production in T cells and macrophages (Fig. 4), is not essential in mast cells (92, 97). However, in RBL-2H3 cells inhibition of p38 and subsequent erk1/2 activity rise led to an increased TNF-α production (97). Several studies reported also inhibition of TNF-α mRNA levels and secretion by glucocorticoids, although the level of inhibition achieved differed between the various mast cells (76, 88, 99, 100). As dexamethasone inhibits the erk1/2-dependent pathway at low nanomolar concentrations (101), this may at least partly explain the inhibitory action on TNF-α production, although other mechanisms, notably on the JNK-mediated regulation of translational derepression, cannot be excluded (1). RBL-

2H3 cells became less sensitive to the dexamethasone-induced blockade of TNF-α production when a G protein, $G_{\alpha z}$, of yet unknown function was overexpressed (99).

Transcriptional Regulation

Several authors have studied the functional capacity of TNF-α promoter sequences to induce reporter gene expression (Table I). In RBL-2H3 and MC/9 mast cells the $Fc_{\varepsilon}RI$-mediated induction with full-length promoter constructs was relatively modest, in the range of 4–6-fold (73, 92). In CPII mast cells a 796-bp promoter construct containing 3′ UTR sequences was much more effective (30-fold) in inducing reporter gene expression and in these cells the minimal promoter could be resumed to the first 200 bp of the huTNF-α promoter (91, 102). A similar construct in BMMC also induced an about 5-fold activation (84). Deletions mutants at nt –128 in RBL-2H3 cells and at nt –105 and nt –86 in CPII cells abolished all inducibility (73, 91). This is in contrast to T cells, in which a deletion at nt –98 conserving the kappa3 site still retains significant inducibility, albeit reduced as compared to a longer construct (nt –142) containing the adjacent CRE binding site (43). Since in CPII mast cells addition of the missing nucleotide T at position –106 restores binding of an NFATp transcription factor, the so-called extented kappa3 site could be critical for retention of functionality (91). However, this was not verified in reporter assays. This point also needs clarification since human, mouse or rat promoters are not identical at these sites (see Fig. 3). Characterization of key regulators in TNF-α gene expression include the NFAT transcription factors first identified in RBL-2H3 cells after $Fc_{\varepsilon}RI$ and ionophore-mediated stimulation (103). Analysis using isoform-specific antibodies or molecular cloning have detected the presence of both NFAT 1 (or NFATp) and NFAT 2 (isoforms NFATc.α and NFATc.β) in mast cells (102, 104, 105). Nuclear translocation of NFAT and binding to the extended kappa3 element requires both cytosolic Ca^{2+} increase and activation of Rac-1 (103, 106). Possible synergistic and antagonistic interaction with basic leucine zipper proteins (bZip) have been reported (AP-1 proteins c-jun, junD, fosB as well as ATF-2 and Nrf1) (107). Nrf1 which requires prior NFAT activation for binding may have an impact on TNF-α production as overexpression of a transdominant-negative mutant form of Nrf1 results in the inhibition

TABLE I
Induction Levels of TNF-α Reporter Gene Constructs in Mast Cells

Mast cell line and stimulus	TNF-α promoter construct	Induction levels*	Reference
CPII ($Fc_{\varepsilon}RI$)	huTNF-α(–796)Luc; 3′UTR(TNF)	27-fold	91
	huTNF-α(–200)Luc; 3′UTR(TNF)	11-fold	
	huTNF-α(–105)Luc; 3′UTR(TNF)	No induction	
	huTNF-α(–86)Luc; 3′UTR(TNF)	No induction	
CPII (PMA/ionomycin)	huTNF-α(–796)Luc; 3′UTR(TNF)	14-fold	102
	huTNF-α(–796)Luc; 3′UTR(SV40)	4-fold	
MC/9 ($Fc_{\varepsilon}RI$)	huTNF-α(–1311)Luc	5.5-fold	92
RBL-2H3 ($Fc_{\varepsilon}RI$)	moTNF-α(–1229)CAT	2-fold	73
	moTNF-α(–128)CAT	No induction	
BMMC ($Fc_{\varepsilon}RI$)	huTNF-α(–200)Luc	4.5-fold	84

* Values are taken or estimated from cited references.

of TNF-α reporter gene expression in CPII mast cells (107). In RBL-2H3 cells evidence for the implication for another inducible factor binding to NF-κB sequences has been reported (73). Following stimulation with Fc$_\varepsilon$RI, the NF-κB-like transcription factor appeared as an upper and a lower complex in bandshift assays. Although AP-1 complexes were also induced in RBL-2H3 cells, the binding of the NF-κB-like factor was independent of AP-1 and correlated with the induction of TNF-α mRNA levels. Induction also required Ca^{2+} and activation of calcineurin, but did not depend on PKC activity (73; C. Pelletier and U. Blank, unpublished data). Pharmacological agents (PDTC and TPCK (N^{α}-tosyl-Phe-chloromethyl ketone)) inhibiting the nuclear activation of the NF-κB-like factor concomitantly inhibited TNF-α mRNA accumulation. Although this was reminiscent of NF-κB or NFAT, none of the specific antibodies raised against the various isoforms of these transcription factors was able to recognize the factor in supershift assays. Analysis of the molecular mass upon UV cross-linking revealed three bands (90, 100 and 110 kDa) that were different from NF-κB and NFAT proteins (73, 104). Interestingly, tissue-specific expression of NF-κB-like complexes of similar size have also been described in brain during development (108). However, none of these factors has yet been characterized in terms of amino acid sequence. Examination of the various κB and κB-related sequences in the rat gene showed that this factor bound with a certain variability between upper and lower complexes to all κB sites tested except κB1. The strongest binding was observed with a probe corresponding to a κB site found approximately 220 bp downstream of the polyadenylation site and already shown to play a role in the inducible TNF-α gene expression in other cells (see above). Although not yet verified, this may suggest that regions outside the promoter may also contribute to TNF-α gene expression in mast cells.

Post-transcriptional Regulatory Mechanisms

Fc$_\varepsilon$RI stimulation leads to transient TNF-α mRNA induction that cumulates at about 1 h and then declines rapidly. Incubation with cycloheximide significantly enhances TNF-α mRNA accumulation, indicating that regulatory mechanisms depend on the synthesis of new proteins (16, 73). These might act either as transcriptional suppressors or post-transcriptional. Experimental evidence revealed increased reporter gene expression in the presence of the huTNF-α 3´ UTR, suggesting the implication of a post-transcriptional regulation which involves the ARE (Table I) (102). Although p38 has been implicated in such regulation in monocytes and macrophages this has not been confirmed for mast cells (92, 97). As glucocorticoids downregulate TNF-α production without necessarily affecting TNF-α mRNA levels (22, 76), post-transcriptional regulatory mechanisms may involve JNK-activated pathways as previously described in LPS-stimulated macrophages (63) (see above). An interesting possibility for post-transcriptional regulatory mechanisms includes two CsA- and ARE-dependent pathways shown to play a role in the translational silencing of IL-3 and IL-4 but not IL-6 mRNA in autocrine IL-3-producing mast cell lines. One of these pathways requires a physical association with the translational machinery on ribosomes (109).

Regulation of Secretion

The IgE-dependent activation induces the release of TNF-α from pre-formed stores and also stimulates the release of newly synthesized proteins. It became clear that secretion of *de novo* synthesized TNF-α does not depend on the incorporation into secretory granules but requires a process analogous to the constitutive secretory pathway. Experiments conducted in the RBL-2H3 cell line which only secretes *de novo* synthesized TNF-α showed that the release required Golgi processing. Secretion could be totally blocked with the Golgi-disrupting agent brefeldin A in the absence of any effect on the release of inflammatory mediators stored in secretory granules (70). In contrast to the constitutively secreted cytokine transforming growth factor-β (TGF-β), the release of TNF-α was regulated by $Fc_{\varepsilon}RI$ and required both the activation of PKC and Ca^{2+} influx (70, 110). Dexamethasone also suppressed the release of TNF-α but had no effect on TGF-β release. An inducible component of TGF-β release was, however, suppressed to the same extent as TNF-α by dexamethasone and a PKC inhibitor, suggesting the existence of both constitutive and inducible mechanisms within the secretory pathway. Both mechanisms proceed through a Golgi compartment as they were both inhibited with brefeldin A (110). Mast cell-derived TNF-α is synthesized as a 26-kDa membrane precursor. In BMMC, pro-TNF-α can be detected following $Fc_{\varepsilon}RI$ stimulation and reaches plateau levels by 2–3 h (84). Preliminary experiments in RBL-2H3 cells using a highly specific inhibitor of the TNF-α-converting protease TACE have indicated that mast cells may use the same or a similar protease to release the secreted form (72). The mechanisms regulating the accumulation of TNF-α protein in secretory granules are less well understood but seem to depend on the differentiation status of the cell, as rapidly dividing tumour cell lines do not contain significant amounts of pre-stored TNF-α (16, 73). As mast cells do not seem to have a complete translational block observed in other cell types (60), it may be that the low levels of production of TNF-α mRNA detectable in unstimulated cells may ultimately lead to the accumulation of protein in secretory granules. The absence of activation may be the signal for sorting TNF-α into these cellular compartments. The granular storage has been confirmed by immunogold labelling using electron microscopy (19). These studies also confirm a significant reduction in TNF-α density following stimulation (111). Some release from these compartments may also occur spontaneously as long-term incubation allows the detection of significant amounts of proteins in some types of mast cells (75).

CONCLUDING REMARKS

Since the discovery of TNF-α as a mast cell-derived mediator, progress has been made in understanding the mechanisms responsible for its expression. Most of this knowledge has been brought about by studying various aspects of the $Fc_{\varepsilon}RI$-mediated TNF-α production in a variety of mast cell lines. Figure 5 shows a summary of these data. Although they have allowed certain questions to be addressed and answered, they have also revealed important differences. Strong tissue-dependent regulatory mechanisms have been shown to be operative in TNF-α gene expression. These may also apply to the different mast cell populations known to exhibit substantial structural and functional heterogeneity (112). Indeed, the heterogeneity of mast cells has recently been extended to their content in cytokines (113). Within the same biopsy, mast cells can express a heterogeneous array of cytokines (24). Thus besides $Fc_{\varepsilon}RI$, environmental factors may importantly contribute to TNF-α gene expression. A modulatory role on TNF-α expression in mast cells by several molecules involved in cell contact, cytokines or neuropeptides has been described (75, 114–116). Extensive analysis of TNF-α gene expression in numerous cells indicates a complex regulatory relationship comprising several levels. Depending on the tissue and the

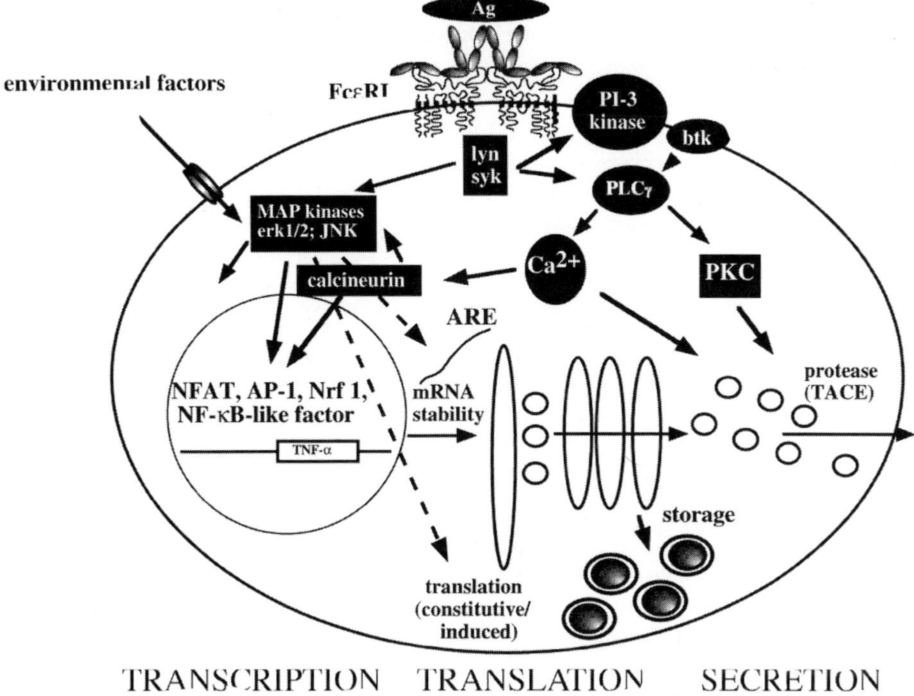

Fig. 5 Summary representation of the different mechanisms implicated in FcεRI-dependent regulation of TNF-α biosynthesis in mast cells as presented in the text.

stimulus different transcription factors can be expressed and become activated. This is exemplified by a primary activatory mode involving NF-κB in stimulated macrophages, whereas T cells and potentially mast cells have been described to function primarily through NFAT-mediated mechanisms. The latter needs clarification, however, as studies in NFAT1- and NFAT2 deficient mice have not revealed any significant effect on TNF-α production (117–119). In mast cells, a recently described NF-κB-like nuclear factor could play an important role in the tissue-specific TNF-α gene expression in mast cells (73). Differences also became apparent at the post-transcriptional level as macrophages and T cells require the activation of p38 for translational control whereas mast cells seem to be independent of p38. Mast cells, at least when fully differentiated, can accumulate substantial amounts of TNF-α in their cytoplasmic granules. Thus, the strict translational repression as observed in macrophages may not exist (60). Mast cells may express transdominant factors which can overcome translational silencing as already described for other cells (52). Much work still has to be accomplished to fully elucidate the different mechanisms that control TNF-α gene expression in mast cells. These studies may be helped by the recent generation of mice deficient in particular transcription factors as well as mice in which the 3′ UTR important for post-transcriptional mechanisms has been deleted.

REFERENCES

1. Beutler, B. *Tumor Necrosis Factors: The Molecules and Their Emerging Role in Medicine*. Raven Press, New York, 1992.
2. Pennica, D., Nedwin, G. E., Hayflick, J. S., *et al.* Human tumour necrosis factor: precursor structure, expression and homology to lymphotoxin. *Nature* **312**:724–729, 1984.
3. Beutler, B., Greenwald, D., Hulmes, J. D., *et al.* Identity of tumour necrosis factor and the macrophage-secreted factor cachectin. *Nature* **316**:552–554, 1985.
4. Gray, P. W., Aggarwal, B. B., Benton, C. V., *et al.* Cloning and expression of cDNA for human lymphotoxin, a lymphokine with tumour necrosis activity. *Nature* **312**:721–724, 1984.
5. Browning, J. L., Ngam, A., Lawton, P., *et al.* Lymphotoxin-β, a novel member of the TNF family that forms a heterodimeric complex with lymphotoxin on the cell surface. *Cell* **72**:846–847, 1993.
6. Black, R. A., Rauch, C. T., Kozlosky, C. J., *et al.* A metalloproteinase disintegrin that releases tumor necrosis factor-α from cells. *Nature* **385**: 729–733, 1997.
7. Jones, E. Y., Stuart, D. I. and Walker, N. P. Structure of tumour necrosis factor. *Nature* **338**:225–228, 1989.
8. Farram, E. and Nelson, D. S. Mouse mast cells as anti-tumor effector cells. *Cell. Immunol.* **55**:294–301, 1980.
9. Ghiara, P., Boraschi, D., Villa, L., Scapigliati, G., Taddei, C. and Tagliabue, A. *In vitro* generated mast cells express natural cytotoxicity against tumour cells. *Immunology* **55**:317–324, 1985.
10. Okuno, T., Takagaki, D., Pluznik, D. and Djeu, J. Y. Natural cytotoxic (NC) cell activity in basophilic cells: release of NC-specific cytotoxic factor by IgE receptor triggering. *J. Immunol.* **136**:4652–4658, 1986.
11. Young, J. D., Liu, C. C., Butler, G., Cohn, Z. A. and Galli, S. J. Identification, purification, and characterization of a mast cell-associated cytolytic factor related to tumor necrosis factor. *Proc. Natl. Acad. Sci. USA* **84**:9175–9179, 1987.
12. Richards, A. L., Okuno, T., Takagaki, Y. and Djieu, J. Y. Natural cytotoxic cell-specific cytotoxic factor produced by IL-3-dependent basophilic/mast cells. Relationship to TNF. *J. Immunol.* **141**:3061–3066, 1988.
13. Steffen, M., Abboud, M., Potter, G. K., Yung, Y. P. and Moore, M. A. S. Presence of tumor necrosis factor or a related factor in human basophils/mast cells. *Immunology* **66**:445–450, 1989.
14. Plaut, M., Pierce, J. H., Watson, C. J., Nordan, R. P. and Paul, W. E. Mast cells produce lymphokines in response to cross-linkage of FcεRI or to calcium ionophores. *Nature* **339**:64–67, 1989.
15. Ohno, I., Tanno, Y., Yamauchi, K. and Takishima, T. Gene expression and production of tumour necrosis factor by a rat basophilic leukaemia cell line (RBL-2H3) with IgE receptor triggering. *Immunology* **70**:88–93, 1990.
16. Gordon, J. R. and Galli, S. J. Mast cells as a source of both preformed and immunologically inducible TNF-alpha/cachectin. *Nature* **346**:274–276, 1990.
17. Walsh, L. J., Trinchieri, G., Waldorf, H. A., Whitaker, D. and Murphy, G. F. Human dermal mast cells contain and release tumor necrosis factor alpha, which induces endothelial leukocyte adhesion molecule 1. *Proc. Natl. Acad. Sci. USA* **88**:4220–4224, 1991.
18. Gordon, J. R. and Galli, S. J. Release of both preformed and newly synthesized TNF-α/cachectin by mouse mast cells stimulated via the FcεRI. A mechanism for the sustained action of mast cell-derived TNF-α during IgE-dependent biological responses. *J. Exp. Med.* **174**:103–107, 1991.
19. Beil, W. J., Login, G. R., Galli, S. J. and Dvorak, A. M. Ultrastructural immunogold localization of TNF-α to the cytoplasmic granules of rat peritoneal mast cells with rapid microwave fixation. *J. Allergy Clin. Immunol.* **94**:531–536, 1994.
20. Echtenacher, B., Mannel, D. N. and Hültner, L. Critical protective role of mast cells in a model of acute septic peritonitis. *Nature* **381**:75–77, 1996.
21. Malaviya, R., Ikeda ,T., Ross, E. and Abraham, S. N. Mast cell modulation of neutrophil influx and bacterial clearance at sites of infection through TNF-α. *Nature* **381**:77–80, 1996.
22. Wershil, B. K., Wang, Z. S., Gordon, J. R. and Galli, S. J. Recruitment of neutrophils during IgE-dependent cutaneous late phase reactions in the mouse is mast cell-dependent. Partial inhibition of the reaction with antiserum against TNF-α. *J. Clin. Invest.* **87**:446–453, 1991.
23. Broide, D. H., Lotz, M., Cuomo, A. J., Coburn, D. A., Federman, E. C. and Wasserman, S. I. Cytokines in symptomatic asthma airways. *J. Allergy Clin. Immunol.* **89**:958–967, 1992.

24. Bradding, P., Roberts, J. A., Britten, K. M., *et al.* Interleukin-4, -5, and -6 and tumor necrosis factor-alpha in normal and asthmatic airways: evidence for the human mast cell as a source of these cytokines. *Am. J. Respir. Cell Mol. Biol.* **10**:471–480, 1994.
25. Bradding, P., Mediwake, R., Feather, I. H., *et al.* TNF-α is localized to nasal mucosal mast cells and is released in acute allergic rhinitis. *Clin. Exp. Allergy* **25**:406–415, 1995.
26. Spriggs, D. R., Deutsch, S. and Kufe, D. W. Genomic structure, induction and production of TNFα. In: *Tumor Necrosis Factors. Structure, Function and Mechanism of Action* (Aggarwal, B. B. and Vilcek, J., eds), pp. 3–34. Marcel Decker, New York, 1994.
27. Lawton, P., Nelson, J., Tizard, R. and Browning, J. L. Characterisation of the mouse lymphotoxin-beta gene. *J. Immunol.* **154**:239–246, 1995.
28. Kwon, J., Chung, I. Y. and Benveniste, E. N. Cloning and sequence analysis of the rat tumor necrosis factor-encoding genes. *Gene* **132**:227–236, 1993.
29. Chen, C. Y. and Shyu, A. B. AU-rich elements: characterization and importance in mRNA degradation. *Trends Biochem. Sci.* **20**:465–470, 1995.
30. Kuprash, D. V., Udalova, I. A., Turetskaya, R. L., Kwiatkowski, D., Rice, N. A. and Nedospasov, S. A. Similarities and differences between human and murine TNF promoters in their response to lipopolysaccharide. *J. Immunol.* **162**:4045–4052, 1999.
31. Shannon, M. F., Pell, L. M., Lenardo, M. J., *et al.* A novel tumor necrosis factor-responsive transcription factor which recognizes a regulatory element in hematopoietic growth factor genes. *Mol. Cell. Biol.* **10**:2950–2959, 1990.
32. Sariban, E., Imamura, K., Luebbers, R. and Kufe, D. Transcriptional and post-transcriptional regulation of tumor necrosis factor gene expression in human monocytes. *J. Clin. Invest.* **81**:1506–1510, 1988.
33. Collart, M. A., Baeuerle, P. and Vassalli, P. Regulation of tumor necrosis factor alpha transcription in macrophages: involvement of four kappa B-like motifs and of constitutive and inducible forms of NF-kappa B. *Mol. Cell. Biol.* **10**:1498–1506, 1990.
34. Biragyn, A. and Nedospasov, S. A. Lipopolysaccharide-induced expression of TNF-α gene in the macrophage cell line ANA-1 is regulated at the level of transcription processivity. *J. Immunol.* **155**:674–683, 1995.
35. Goldfeld, A. E., Doyle, C. and Maniatis, T. Human tumor necrosis factor alpha gene regulation by virus and lipopolysaccharide. *Proc. Natl. Acad. Sci. USA* **87**:9769–9773, 1990.
36. Shakhov, A. N., Collart, M. A., Vassalli, P., Nedospasov, S. A. and Jongeneel, C. V. Kappa B-type enhancers are involved in lipopolysaccharide-mediated transcriptional activation of the tumor necrosis factor alpha gene in primary macrophages. *J. Exp. Med.* **171**:35–47, 1990.
37. Kuprash, D.V., Udalova, I. A., Turetskaya, R. L., Rice, N. R. and Nedospasov, S. A. Conserved κB element located downstream of the tumor necrosis factor-α gene: distinct NF-κB binding pattern and enhancer activity in LPS activated murine macrophages. *Oncogene* **11**:97–106, 1995.
38. Udalova, I. A., Knight, J. C., Vidal, V., Nedospasov, S. A. and Kwiatkowski, D. Complex NF-κB interactions at the distal tumor necrosis factor promoter region in human monocytes. *J. Biol. Chem.* **273**:21178–21186, 1998.
39. Löms Ziegler-Heitbrock, H. W., Sternsdorf, T., Liese, J., *et al.* Pyrrolidine dithiocarbamate inhibits NF-kappa B mobilization and TNF production in human monocytes. *J. Immunol.* **151**:6986–6993, 1993.
40. Kwon, J., Lee, S. J. and Benveniste, E. N. A 3' cis-acting element is involved in tumor necrosis factor-α gene expression in astrocytes. *J. Biol. Chem.* **271**:22383–22390, 1996.
41. Goldfeld, A. E., McCaffrey, P. G., Strominger, J. L. and Rao, A. Identification of a novel cyclosporin-sensitive element in the human tumor necrosis factor alpha gene promoter. *J. Exp. Med.* **178**:1365–1379, 1993.
42. McCaffrey, P. G., Goldfeld, A. E. and Rao, A. The role of NFATp in cyclosporin A-sensitive tumor necrosis factor-alpha gene transcription. *J. Biol. Chem.* **269**:30445–30450, 1994.
43. Tsai, E. Y., Jain, J., Pesavento, P. A., Rao, A. and Goldfeld, A. E. Tumor necrosis factor alpha gene regulation in activated T cells involves ATF-2/Jun and NFATp. *Mol. Cell. Biol.* **16**:459–467, 1996.
44. Zagariya, A., Mungre, S., Lovis, R., *et al.* Tumor necrosis factor alpha gene regulation: enhancement of C/EBPβ-induced activation by c-Jun. *Mol. Cell. Biol.* **18**:2815–2824, 1998.
45. Kramer, B., Wiegmann, K. and Kronke, M. Regulation of the human TNF promoter by the transcription factor Ets. *J. Biol. Chem.* **270**:6577–6583, 1995.
46. Kramer, B., Meichle, A., Hensel, G., Charnay, P. and Krönke, M. Characterization of an Krox-24/Egr-1-responsive element in the human tumor necrosis factor promoter. *Biochim. Biophys. Acta* **1219**:413–421, 1994.
47. Leitman, D. C., Mackow, E. R., Williams, T., Baxter, J. D. and West, B. L. The core promoter region of

the tumor necrosis factor alpha gene confers phorbol ester responsiveness to upstream transcriptional activators. *Mol. Cell. Biol.* **12**:1352–1356, 1992.

48. Jain, J., Loh, C. and Rao, A. Transcriptional regulation of the Il-2 gene. *Curr. Opin. Immunol.* **7**:333–342, 1995.
49. Rhoades, K. L., Golub, S. H. and Economou, J. S. The regulation of the human tumor necrosis factor alpha promoter region in macrophage, T cell, and B cell lines. *J. Biol. Chem.* **267**:22102–22107, 1992.
50. Wickens, M., Anderson, P. and Jackson, R. J. Life and death in the cytoplasm: messages from the 3' end. *Curr. Opin. Genet. Dev.* **7**:220–232, 1997.
51. Kruys, V., Marinx, O., Shaw, G., Deschamps, J. and Huez, G. Translational blockade imposed by cytokine-derived UA-rich sequences. *Science* **245**:852–855, 1989.
52. Kruys, V., Kemmer, K., Shakhov, A., Jongeneel, V. and Beutler, B. Constitutive activity of the tumor necrosis factor promoter is canceled by the 3' untranslated region in nonmacrophage cell lines; a trans-dominant factor overcomes this suppressive effect. *Proc. Natl. Acad. Sci. USA* **89**:673–677, 1992.
53. Beutler, B., Krochin, N., Milsark, I. W., Luedke, C. and Cerami, A. Control of cachectin (tumor necrosis factor) synthesis: mechanisms of endotoxin resistance. *Science* **232**:977–980, 1986.
54. Beutler, B., Han, J., Kruys, V. and Giroir, B. P. Coordinated regulation of TNF biosynthesis at the levels of transcription and translation. Patterns of TNF expression *in vivo.* In: *Tumor Necrosis Factors: the Molecules and their Emerging Role in Medicine* (Beutler, I. B., ed.). Raven Press, New York, 1992.
55. Han, J. H., Beutler, B. and Huez, G. Complex regulation of tumor necrosis factor mRNA turnover in lipopolysaccharide-activated macrophages. *Biochim. Biophys. Acta* **1090**:22–28, 1991.
56. Kontoyiannis, D., Pasparakis, M., Pizarro, T. T., Cominelli, F. and Kollias, G. Impaired on/off regulation of TNF biosynthesis in mice lacking TNF AU-rich elements: implications for joint and gut-associated immunopathologies. *Immunity* **10**:387–398, 1999.
57. Yang, Y., Chang, J. F., Parnes, J. R. and Fathman, C. G. T cell receptor (TCR) engagement leads to activation-induced splicing of tumor-necrosis factor (TNF) nuclear pre-mRNA. *J. Exp. Med.* **188**:247–254, 1998.
58. Carballo, E., Lai, W. S. and Blackshear, P. J. Feedback inhibition of macrophage tumor necrosis factor-α production by tristetraprolin. *Science* **281**:1001–1005, 1998.
59. Peng, S. S., Chen, C. Y., Xu, N. and Shyu, A. B. RNA stabilization by the AU-rich element binding protein, HuR, an ELAV protein. *EMBO J.* **17**:3461–3470, 1998.
60. Han, J., Brown, T. and Beutler, B. Endotoxin-responsive sequences control cachectin/tumor necrosis factor biosynthesis at the translational level. *J. Exp. Med.* **171**:465–475, 1990.
61. Lee, J. C., Laydon, J. T., McDonnell, P. C., *et al.* A protein kinase involved in the regulation of inflammatory cytokine biosynthesis. Nature **372**:739–746, 1994.
62. Kotlyarov, A., Neininger, A., Schubert, C., *et al.* MAPKAP kinase 2 is essential for LPS-induced TNF-α biosynthesis. *Nat. Cell Biol.* **1**:94–97, 1999.
63. Swantek, J. L., Cobb, M. H. and Geppert, T. D. Jun N-terminal kinase/stress-activated protein kinase (JNK/SAPK) is required for lipopolysaccharide stimulation of TNF-α translation: glucocorticoids inhibit TNF-alpha translation by blocking JNK/SAPK. *Mol. Cell. Biol.* **17**:6274–6282, 1997.
64. Prichett, W., Hand, A., Shields, J. and Dunnington, D. Mechanism of action of bicyclic imidazoles defines a translational regulatory pathway for tumor necrosis factor alpha. *J. Inflamm.* **45**:97–105, 1995.
65. Gueydan, C., Houzet, L., Marchant, A., Sels, A., Huez, G. and Kruys, V. Engagement of tumor necrosis factor mRNA by an endotoxin-inducible cytoplasmic protein. *Mol. Med.* **2**:479–488, 1996.
66. Hoffmeyer, A., Grosse-Wilde, A., Flory, E., *et al.* Different mitogen-activated protein kinase signaling pathways cooperate to regulate tumor necrosis factor α gene expression in T lymphocytes. *J. Biol. Chem.* **274**:4319–4327, 1999.
67. Pradines-Figueres, A. and Raetz, C. R. H. Processing and secretion of tumor necrosis factor α in endotoxin-treated Mono Mac 6 cells are dependent on phorbol myristate. *J. Biol. Chem.* **267**:23261–23268, 1992.
68. McGeehan, G. M., Becherer, J. D., Bast, R. C. Jr, *et al.* Regulation of tumour necrosis factor-alpha processing by a metalloproteinase inhibitor. *Nature* **370**:558–561, 1994.
69. Imamura, K., Spriggs, D., Ohno, D. and Kufe, D. Effect of botulinum toxin type D on secretion of tumor necrosis factor from human monocytes. *Mol. Cell. Biol.* **9**:2239–2243, 1989.
70. Baumgartner, R. A., Yamada, K., Deramo, V. A. and Beaven, M. A. Secretion of TNF from a rat mast cell line is a brefeldin A-sensitive and a calcium/protein kinase C-regulated process. *J. Immunol.* **153**:2609–2617, 1994.

71. Dreskin, S. and Abraham, S. Production of TNF-α by murine bone marrow derived mast cells activated by the bacterial fimbrial protein, FimH. *Clin. Immunol.* **90**:420–424, 1999.
72. Hide, I., Toriu, N., Nuibe, T., *et al.* Suppression of TNF-α secretion by azelastine in a rat mast (RBL-2H3) cell line: evidence for differential regulation of TNF-alpha release, transcription, and degranulation. *J. Immunol.* **159**:2932–2940, 1997.
73. Pelletier, C., Varin-Blank, N., Rivera, J., *et al.* Fc epsilon RI-mediated induction of TNF-alpha gene expression in the RBL-2H3 mast cell line: regulation by a novel NF-kappa B-like nuclear binding complex. *J. Immunol.* **161**:4768–4776, 1998.
74. Gurish, M. F., Ghildyal, N., Arm, J., *et al.* Cytokine mRNA are preferentially increased relative to secretory granule protein mRNA in mouse bone marrow-derived mast cells that have undergone IgE-mediated activation and degranulation. *J. Immunol.* **146**:1527–1533, 1991.
75. Bissonenette, E. Y., Chini, B. and Befus, A. D. Interferons differentially regulate histamine and TNF-α in rat intestinal mucosal mast cells. *Immunology* **86**:12–17, 1995.
76. Schmidt-Choudhury, A., Furuta, G. T., Lavigne, J. A., Galli, S. J. and Wershil, B. K. The regulation of tumor necrosis factor-alpha production in murine mast cells: pentoxifylline or dexamethasone inhibits IgE-dependent production of TNF-α by distinct mechanisms. *Cell. Immunol.* **171**:140–146, 1996.
77. Dogar, J. H., Nemeth, G. G., Durdik, J. M. and Dreskin, S. C. Fc epsilon RI-mediated expression of mRNA for c-fos in rat. *Cell. Signal.* **5**:605–613, 1993.
78. Dolmetsch, R. E., Lewis, R. S., Goodnow, C. C. and Healy, J. I. Differential activation of transcription factors induced by Ca^{2+} response amplitude and duration. *Nature* **386**:855–858, 1997.
79. Dolmetsch, R. E., Xu, K. and Lewis, R. S. Calcium oscillations increase the efficiency and specificity of gene expression. *Nature* **392**:933–936, 1998.
80. Li, W., Llopis, J., Whitney, M., Zlokarnik, G. and Tsien, R. Y. Cell-permeant caged $InsP_3$ ester shows that Ca^{2+} spike frequency can optimize gene expression. *Nature* **392**:936–941, 1998.
81. Kinet, J.-P. The high affinity IgE receptor (FcεRI): from physiology to pathology. *Annu. Rev. Immunol.* **17**:931–972, 1999.
82. Yamashita, T., Mao, S. Y. and Metzger, H. Aggregation of the high-affinity IgE receptor and enhanced activity of p53/56lyn protein-tyrosine kinase. *Proc. Natl. Acad. Sci. USA* **91**:11251–11255, 1994.
83. Scharenberg, A. M. and Kinet, J. P. PtdIns-3,4,5-P3: a regulatory nexus between tyrosine kinases and sustained calcium signals. *Cell* **94**:5–8, 1998.
84. Hata, D., Kawakami, Y., Inagaki, N., *et al.* Involvement of Bruton's tyrosine kinase in FcepsilonRI-dependent mast cell degranulation and cytokine production. *J. Exp. Med.* **187**:1235–1247, 1998.
85. Jabril-Cuenod, B., Zhang, C., Scharenberg, A., *et al.* Syk-dependent phosphorylation of Shc. A potential link between FcεRI and the ras/mitogen-activated protein kinase signaling pathway through SOS and Grb2. *J. Biol. Chem.* **271**:16268–16272, 1996.
86. Teramoto, H., Salem, P., Robbins, K. C., Bustelo, X. R. and Gutkind, J. S. Tyrosine phosphorylation of the vav protooncogene product links FcεRI to the rac-1-JNK pathway. *J. Biol. Chem.* **272**:10751–10755, 1997.
87. Kaye, R. E., Fruman, D. A., Bierer, M. W., *et al.* Effects of cyclosporin A and FK506 on Fc epsilon type I-initiated increases in cytokine mRNA in mouse bone marrow-derived progenitor mast cells: resistance to FK 506 is associated with a deficiency in FK506-binding protein FKBP12. *Proc. Natl. Acad. Sci. USA* **89**:8542–8546, 1992.
88. Wershil, B. K., Furuta, G. T., Lavigne, J. A., Choudhury, A. R., Wang, Z. S. and Galli, S. J. Dexamethasone or cyclosporin A suppress mast cell–leukocyte cytokine cascades. Multiple mechanisms of inhibition of IgE- and mast cell-dependent cutaneous inflammation in the mouse. *J. Immunol.* **154**:1391–1398, 1995.
89. Pelletier, C., Guerin-Marchand, C., Iannascoli, B., *et al.* Specific signaling pathways in the regulation of TNF-alpha mRNA synthesis and TNF-alpha secretion in RBL-2H3 mast cells stimulated through the high affinity IgE receptor. *Inflamm. Res.* **47**:493–500, 1998.
90. Ishizuka, T., Chayama, K., Takeda, K., *et al.* Mitogen-activated protein kinase activation through Fcε receptor I and stem cell factor receptor is differentially regulated by phosphatidylinositol 3-kinase and calcineurin in mouse bone marrow-derived mast cells. *J. Immunol.* **162**:2087–2094, 1999.
91. Csonga, R., Prieschl, E. E., Jaschke, D., Novotny, V. and Baumruker, T. Common and distinct signaling pathways mediate the induction of TNF-α and IL-5 in IgE plus antigen-stimulated mast cells. *J. Immunol.* **160**:273–283, 1998.
92. Ishizuka, T., Terada, N., Gerwins, N., *et al.* Mast cell tumor necrosis factor α production is regulated by MEK kinases. *Proc. Natl. Acad. Sci. USA* **94**:6358–6363, 1997.

93. Ishizuka, T., Kawasome, H., Terada, N., *et al.* Stem cell factor augments FcεRI-mediated TNF-α production and stimulates MAP kinases via a different pathway in MC/9 mast cells. *J. Immunol.* **161**:3624–3630, 1998.
94. Pendl, G. G., Prieschl, E. E., Thumb, W., Harrer, N. E., Auer, M. and Baumruker, T. Effects of phosphoinositol-3-kinase inhibitors on degranulation and gene induction in allergically triggered mouse mast cells. *Int. Arch. Allergy Immunol.* **112**:392–399, 1996.
95. Kawakami, Y., Miura, T., Bissonnette, R., *et al.* Bruton's tyrosine kinase regulates apoptosis and JNK/SAPK kinase activity. *Proc. Natl. Acad. Sci. USA* **94**:3983–3992, 1997.
96. Kawakami, Y., Hartman, S. E., Holland, P. M., Cooper, J. A. and Kawakami, T. Multiple signaling pathways for the activation of JNK in mast cells: involvement of bruton's tyrosine kinase, protein kinase C, and JNK kinases, SEK1 and MKK7. *J. Immunol.* **161**:1795–1802, 1998.
97. Zhang, C., Baumgartner, R. A., Yamada, K. and Beaven, M. A. Mitogen-activated protein (MAP) kinase regulates production of tumor necrosis factor-alpha and release of arachidonic acid in mast cells. Indications of communication between p38 and p42 MAP kinases. *J. Biol. Chem.* **272**:13397–13402, 1997.
98. Turner, H. and Cantrell, D. A. Distinct Ras effector pathways are involved in Fc epsilon R1 regulation of the transcriptional activity of Elk-1 and NFAT in mast cells. *J. Exp. Med.* **185**:43–53, 1997.
99. Baumgartner, R. A., Hirasawa, N., Ozawa, K., Gusovsky, F. and Beaven, M. A. Enhancement of TNF-α synthesis by overexpression of Gαz in a mast cell line. *J. Immunol.* **157**:1625–1629, 1996.
100. Eklund, K. K., Humphries, D. E., Xia, Z., *et al.* Glucocorticoids inhibit the cytokine-induced proliferation of mast cells, the high affinity IgE receptor-mediated expression of TNF-α, and the IL-10-induced expression of chymases. *J. Immunol.* **158**:4373–4380, 1997.
101. Rider, L. G., Hirasawa, N., Santini, F. and Beaven, M. A. Activation of the mitogen-activated protein kinase cascade is suppressed by low concentrations of dexamethasone in mast cells. *J. Immunol.* **157**:2374–2380, 1996.
102. Prieschl, E. E., Pendl, G. G., Elbe, A., *et al.* Induction of the TNF-α promoter in the murine dendritic cell line 18 and the murine mast cell line CPII is differently regulated. *J. Immunol.* **157**:2645–2653, 1996.
103. Hutchinson, L. E., McCloskey, M. A. Fc epsilon RI-mediated induction of nuclear factor of activated T-cells. *J. Biol. Chem.* **270**:16333–16338, 1995.
104. Weiss, D. L., Hural, J., Tara, D., Timmermann, L. A., Henkel, G. and Brown, M. A. Nuclear factor of activated T cells is associated with a mast cell interleukin 4 transcription complex. *Mol. Cell. Biol.* **16**:228–235, 1996.
105. Shermann, M. A., Powell, D. R., Weiss, D. L. and Brown, M. A. NFATc isoforms differentially expressed and regulated in murine T and mast cells. *J. Immunol.* **162**:2820–2828, 1999.
106. Turner, H., Gomez, M., McKenzie, E., Kirchem, A., Lennard, A. and Cantrell, D. A. Rac-1 regulates nuclear factor of activated T cells (NFAT) C1 nuclear translocation in response to FcεRI stimulation of mast cells. *J. Exp. Med.* **188**:527–537, 1998.
107. Novotny, V., Prieschl, E. E., Csonga, R., Fabjani, G. and Baumruker, T. Nrfl in a complex with fosB, c-jun, junD and ATF2 forms the AP1 component at the TNFα promoter in stimulated mast cells. *Nucleic Acids Res.* **26**:5480–5485, 1999.
108. Cauley, K. and Verma, I. M. Kappa B enhancer-binding complexes that do not contain NF-kappa B are developmentally regulated in mammalian brain. *Proc. Natl. Acad. Sci. USA* **91**:390–394, 1994.
109. Nair, A. P. K., Hirsch, H. H., Colombi, M. and Moroni, C. Cyclosporin A promotes translational silencing of autocrine interleukin-3 via a ribosome-associated deadenylation. *Mol. Cell. Biol.* **19**:889–898, 1999.
110. Baumgartner, R. A., Deramo, V. A. and Beaven, M. A. Constitutive and inducible mechanisms for synthesis and release of cytokines in immune cell lines. *J. Immunol.* **157**:4087–4093, 1996.
111. Beil, W. J., Login, G. R., Aoki, M., *et al.* TNF-α immunoreactivity of rat peritoneal mast cell granules decreases during early secretion induced by compound 48/80: an ultrastructural immunogold morphometric analysis. *Int. Arch. Allergy Immunol.* **109**:383–389, 1996.
112. Galli, S. J. Biology of disease. New insights into the 'riddle of mast cells': microenvironmental regulation of mast cell development and phenotypic heterogeneity. *Lab. Invest.* **1**:5–33, 1990.
113. Bradding, P., Okayama, Y., Howarth, P. H., Church, M. K. and Holgate, S. T. Heterogeneity of human mast cells on cytokine content. *J. Immunol.* **155**:297–307, 1995.
114. Ansel, J. C., Brown, J. R., Payan, D. G. and Brown, M. A. Substance P selectively activates TNF-alpha gene expression in mast cells. *J. Immunol.* **150**:4478–4485, 1993.

115. Arrock, M., Zuany-Amorim, C., Singer, M., Benhamou, M. and Pretolani, M. Interleukin-10 inhibits cytokine generation from mast cells. *Eur. J. Immunol.* **26**:166–170, 1996.
116. Tashiro, M., Kawakami, Y., Abe, R., *et al.* Increased secretion of TNF-α by costimulation of mast cells via CD28 and FcεRI. *J. Immunol.* **158**:2382–2389, 1997.
117. Xanthoudakis, S., Viola, J. P. B., Shaw, K. T. Y., *et al.* An enhanced response in mice lacking the transcription factor NFAT1. *Science* **272**:892–895, 1996.
118. Yoshida, H., Nishina, H., Takimoto, H., *et al.* The transcription factor NF-ATc1 regulates lymphocyte proliferation and TH2 cytokine production. *Immunity* **8**:115–124, 1998.
119. Ranger, A. M., Hodge, M. R., Gravallese, E. M. *et al.* Delayed lymphoid repopulation with defects in IL-4-driven responses produced by inactivation of NF-ATc. *Immunity* **8**:125–134, 1998.

CHAPTER 11

Regulation of Mast Cell Degranulation by SHIP

*MICHAEL HUBER, JACQUELINE E. DAMEN, MARK WARE, MICHAEL HUGHES, CHERYL D. HELGASON, R. KEITH HUMPHRIES and GERALD KRYSTAL**

Terry Fox Laboratory, BC Cancer Agency, Vancouver, Canada

INTRODUCTION

Mast cells are present at all the portals between self and non-self in humans and other mammals and play a critical role in initiating acute inflammatory responses against invading bacteria, helminthic parasites and harmless allergens (1). Upon exposure to multivalent antigens or allergens, these cells rapidly discharge pre-formed inflammatory mediators, such as histamine, proteoglycans and neutral proteases, from their many cytoplasmic granules. These mediators then act on the vasculature, smooth muscles, connective tissue, mucous glands and inflammatory cells to cause immediate-type hypersensitivity reactions (2). Understanding how this degranulation process is regulated is key to our being able to effectively treat allergies, which affect as many as 20% of the Western population, and to modulate other inflammatory responses. For example, it has been shown recently that mast cell degranulation at the site of surgical implants is responsible for the encapsulation and subsequent failure of these non-immunogenic implants via histamine-mediated chemoattraction of neutrophils and macrophages (3). Thus a more thorough understanding of thc degranulation process could have substantial clinical benefits. In this review we concentrate primarily on what is currently known about the biological properties of the recently cloned SH2-containing inositol 5′-phosphatase (SHIP) and its role in IgE-induced cell signalling and degranulation. For its role in mediating the inhibition conferred by the negative co-receptor, $Fc_{\gamma}RIIB$, we refer readers to recent articles by Scharenberg and Kinet (4), Bolland *et al.* (5) and Coggeshall (6).

* Corresponding author.

MAST CELLS AND BASOPHILS
ISBN 0-12-473335-2

STRUCTURE AND BINDING PARTNERS OF SHIP

SHIP possesses an N-terminal SH2 domain, a central inositol polyphosphate 5′-phosphatase (5-ptase) catalytic domain, two NPXY sequences, which, when phosphorylated, can bind phosphotyrosine-binding (PTB) domains, and a proline-rich C-terminus that is theoretically capable of binding to many Src-homology 3 (SH3) domain-containing proteins (reviewed in ref. 7) (see Fig. 1). Interestingly, unlike most 5-ptases, SHIP selectively hydrolyses the 5′-phosphate from phosphatidylinositol 3,4,5-trisphosphate (PI-3,4,5-P_3) and inositol 1,3,4,5-tetrakisphosphate (I-3,4,5-P_4) (8), two phosphoinositides that have been shown to play important roles in growth factor-mediated signalling (9). The human (1188 amino acids) and murine (1190 amino acids) forms of SHIP possess 87.2% identity at the amino acid level, and human SHIP maps to the long arm of chromosome 2 at the border between 2q36 and 2q37 (10).

During murine development, SHIP is first detectable in 7.5-day embryos (coincident with and dependent upon the onset of haematopoiesis) and its protein expression pattern appears restricted to haematopoietic cells (11). Interestingly, its protein expression varies considerably during haematopoiesis (11, 12), increasing substantially, for example, with T cell maturation (11) and showing a bimodal expression pattern during B cell development (Helgason *et al.*, unpublished). Also of note, SHIP is reduced in primary cells from leukaemic patients and induced expression of BCR-ABL in BA/F3 cells leads to a rapid reduction in the level of SHIP protein (13). It is thus possible that SHIP acts as a

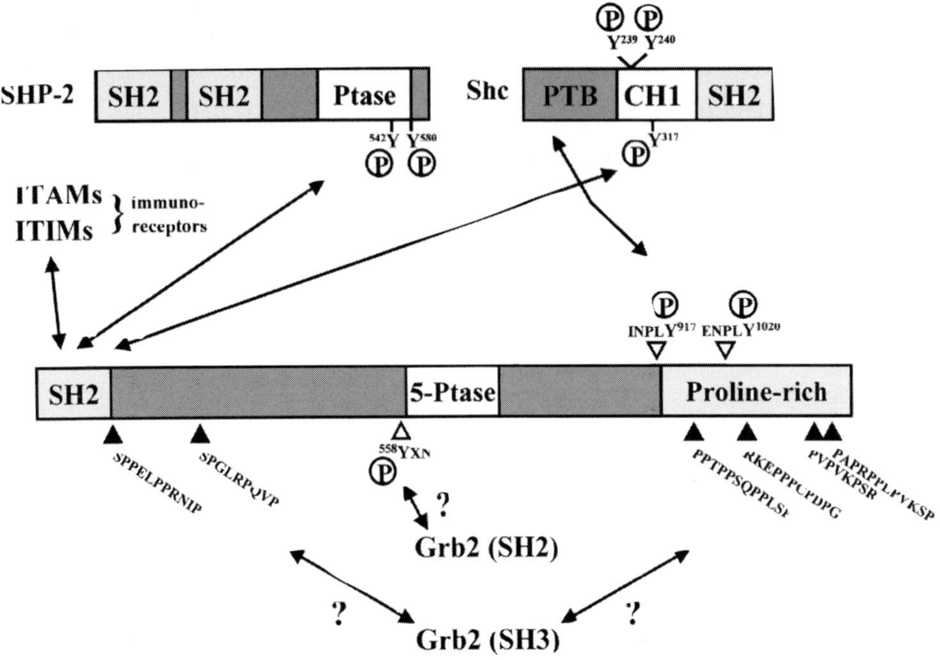

Fig. 1 The structure of SHIP, its known binding partners and its putative cleavage sites.

haematopoietic-specific tumour suppressor during myelopoiesis and its downregulation is required for the development of chronic myeloid leukaemia.

Since SHIP's 5-ptase activity does not change significantly following cytokine stimulation (8), the general consensus is that it exerts its effects via intracellular translocation to sites of synthesis of PI-3,4,5-P_3 and I-1,3,4,5-P_4 (although it has not been shown to hydrolyse I-1,3,4,5-P_4 *in vivo*). Since association with other proteins may play a significant role in this process, we and others have attempted to identify SHIP-binding partners. These studies have shown that SHIP associates, following cytokine, B or T cell receptor activation, with the adaptor protein Shc and the tyrosine phosphatase SHP-2 (7) (see Fig. 1). In addition, it binds to Grb2 in certain cell types (e.g. B cells) but not in others (7); this could be due to the activation of different kinases in different cell types which, in turn, lead to the phosphorylation of different residues in SHIP or SHIP's binding partners. Very recently, Harmer and DeFranco (14) have proposed that, in BCR-activated B cells, SHIP exists in a ternary complex with one Shc molecule and two Grb2 molecules. Specifically, they propose that the two Grb2 molecules bind via their SH2 domains to a single Shc, phosphorylated at Y^{239} and Y^{313}, and the SH3 domains of the Grb2 molecules bind the C-terminal proline-rich regions of SHIP.

SHIP has also been shown to bind via its SH2 domain, which binds preferentially to the sequence pY(Y/D)X(L/I/V) (15), to certain members of the immunoreceptor tyrosine-based inhibition motif (ITIM)-bearing family of inhibitory co-receptors. Specifically, it has been shown to bind both *in vitro* (16, 17) and *in vivo* (18) to the ITIM of Fc_γRIIB, and *in vitro* to the second ITIM of gp49B1 (19). Recently, Tridandapani *et al.* (20) have proposed, based in part on kinetic studies, that, during co-ligation of the Fc_γRIIB to the activated BCR, SHIP first transiently associates with the Fc_γRIIB to promote its tyrosine phosphorylation and then it attracts Shc and dissociates with this new binding partner from the Fc_γRIIB. As well, SHIP's SH2 domain has been demonstrated to bind *in vitro* to the tyrosine-phosphorylated immunoreceptor tyrosine-based activation motif (ITAM) within the β (21) and γ (15) subunits of the Fc_εRI as well as the ζ chain of the T cell receptor (15).

Complicating matters, SHIP can exist in at least four different molecular mass forms, depending on the cell type. The lower 135-, 125- and 110-kDa forms appear to be generated from the 145-kDa full-length protein *in vivo* by cleavage of its proline rich C-terminus (22). Although all forms become tyrosine-phosphorylated in response to cytokines, at one or both of the NPXY motifs, only the 145- and, to a lesser extent, the 135-kDa species bind Shc and only the 110-kDa form is associated with the cytoskeleton (22). These findings suggest that the different SHIP forms may actually be carrying out different functions. Interestingly, the relative proportion of the different forms has been reported to change with haematopoietic differentiation (12) and with leukaemogenesis (13, 23). Adding further complexity, Lucas and Rohrschneider (24) have recently demonstrated the presence of alternatively spliced forms of SHIP as well, consistent with earlier reports showing that, in addition to the predominant 5-kb mRNA, there are minor 6-kb, 4.5-kb and smaller SHIP mRNA species expressed in some cell types (8, 10, 12).

BIOLOGICAL PROPERTIES OF SHIP

SHIP recruitment via its SH2 domain to the tyrosine-phosphorylated ITIM of the inhibitory co-receptor Fc_γRIIB has been shown to inhibit Fc_εRI- and BCR-induced degranulation and proliferation, respectively (18, 25). More recently, SHIP has been shown, even in the absence of $Fc_{\gamma}RIIB$ co-clustering, to suppress IgE-mediated mast cell degranulation (26) and prevent Steel factor (SF)-mediated mast cell degranulation (27). It has also been shown very recently to inhibit B cell chemotaxis (Kim *et al.*, unpublished), and it is possible, given its ability to break down PI-3,4,5-P_3, that it plays a critical role in regulating many if not all PI 3-kinase (PI3K)-induced events in haematopoietic cells.

Mechanistically, these SHIP-induced effects appear to be mediated in large part if not exclusively by SHIP's ability to break down PI-3,4,5-P_3 to PI-3,4-P_2, thus reducing the ability of certain, pleckstrin homology (PH)-containing proteins [e.g, protein kinase B (PKB/Akt), phosphoinositide-dependent protein kinase 1 (PDK-1), Bruton's tyrosine kinase (Btk)] to target to the plasma membrane and be activated (5, 28–32). There is a good deal of support for this mode of action. For example, co-ligation of the $Fc_{\gamma}RIIB$ with either the $Fc_{\varepsilon}RI$ or BCR has been shown to reduce their activation by lowering their induced PI-3,4,5-P_3 levels. This, in turn, reduces the entry of extracellular calcium which is essential for downstream pathways leading to degranulation and proliferation, respectively (7, 25, 26). As well, it has been shown recently that SHIP breaks down both cytokine- and antigen-induced PI-3,4,5-P_3 to PI-3,4-P_2 in mast cells and B cells (27, 30), and this leads to a reduction in the activation of the survival-enhancing serine/threonine kinase, PKB/akt (30–32). SHIP's catalytic activity also appears to be essential for effects other than viability. For example, Vollenweider *et al.* (33) have reported that addition of catalytically active but not inactive SHIP to rat 1 fibroblasts overexpressing the insulin receptor inhibits insulin-induced GLUT4 translocation, membrane ruffling, mitogen-activated protein kinase (MAPK) phosphorylation and DNA synthesis.

However, while there is much support for SHIP exerting its biological effects in large part through its ability to hydrolyse PI-3,4,5-P_3 (and perhaps I-1,3,4,5-P_4), there are, in addition, data showing that SHIP may also act in part by competing with Grb2 for Shc and thereby reducing Ras activation (6, 34). Since the Ras pathway has been shown to be important for survival in interleukin-3 (IL-3)-stimulated cells (35), this competition may be responsible, at least in part, for the reduced viability we observed in SHIP overexpression studies in DA-ER cells (36). In fact, strong evidence for this competition playing a role in $Fc_{\gamma}RIIB$-mediated inhibition of B cell activation has been reported (6).

PROPERTIES OF THE SHIP KNOCKOUT MOUSE

Gaining insights into the biological properties of SHIP from overexpression studies has been difficult both because of difficulties inherent in overexpressing this negative regulator (36, 37) and because results obtained from these overexpression studies, especially in cells that do not normally express SHIP, do not necessarily yield accurate insights into its normal function. Specifically, overexpressed SHIP would not, by definition, be present at physiological levels nor, perhaps, at physiologically relevant subcellular locations. Thus, to gain further insight into the role that SHIP plays *in vivo* we recently generated a SHIP knockout mouse by homologous recombination in embryonic stem (ES) cells (38). Although these mice are viable and fertile, they overproduce granulocytes and macrophages and suffer from progressive splenomegaly, massive myeloid infiltration of the lungs, wasting and a shortened life span (38). Notably, granulocyte–macrophage progenitors from the bone marrow of these mice are substantially more responsive to multiple cytokines than their wild-type littermates (38), suggesting once again that SHIP is a negative regulator of haematopoietic growth factor-induced proliferation/survival. Even in the absence of added growth factors, small colonies are present in fetal calf serum (FCS)-containing methylcellulose cultures from SHIP –/– but not +/+ bone marrow, suggesting yet again that SHIP –/– progenitors have a survival advantage. This is consistent with our findings and those of Liu *et al.* (30) showing that the survival-enhancing PKB/Akt is inhibited by SHIP recruitment to the plasma membrane in mast cells (Scheid *et al.*, unpublished) and B cells (31, 32). Thus SHIP may be an important negative regulator of haematopoietic progenitor cell proliferation/survival as well as a negative regulator of end cell activation.

THE ROLE OF SHIP IN FC$_\varepsilon$RI-INDUCED DEGRANULATION OF MAST CELLS

The high-affinity receptor for IgE, Fc$_\varepsilon$RI, is composed of one α, one β and two covalently linked γ subunits (39) (see Fig. 2). The cytoplasmic domains of the β and γ subunits contain ITAMs that are critical for IgE-mediated activation (39). Since the SH2 domain of SHIP had been shown to be capable of binding to these tyrosine-phosphorylated ITAMs *in vitro* (15, 21), we investigated the role of SHIP in Fc$_\varepsilon$RI-mediated degranulation using bone marrow-derived mast cells (BMMCs) from SHIP +/+ and –/– mice. Of note, the BMMCs from both SHIP +/+ and –/– mice, following 8 weeks in culture with IL-3, were greater than 99% c-kit- and Fc$_\varepsilon$RI-positive (with similar mean fluorescences) (26). Since the presence of these two cell surface receptors is a hallmark of mature mast cells, this finding indicated that SHIP is not essential for the differentiation of BMMCs.

The standard degranulation assay, in which cells are first primed (passively presensitized) with an IgE against a specific antigen (e.g. a dinitrophenyl (DNP) group) and then exposed to a multivalent antigen [e.g. 20–30 moles of DNP covalently attached to human serum albumin (DNP–HSA)] was used to assess the degranulation potential of SHIP +/+ and –/– BMMCs. Interestingly, the SHIP –/– cells consistently released far more of their granule contents under these conditions than their +/+ counterparts. Even more remarkable, however, given that the addition of the Fc$_\varepsilon$RI-cross-linking multivalent antigen is thought to initiate the biochemical cascade that results in degranulation (39), was the finding that, when IgE alone was added (i.e. the presensitization step), there was a massive degranulation of SHIP –/– BMMCs, whereas SHIP +/+ cells, as previously reported, did not degranulate at all (26). This suggested that SHIP plays an important role in preventing inappropriate degranulation (e.g. in the absence of antigen).

We also found that the influx of extracellular calcium was substantially higher in SHIP –/– BMMCs exposed to either IgE alone or IgE plus DNP–HSA (26). Addition of EGTA to deplete extracellular calcium both prevented degranulation and markedly reduced the increase in intracellular calcium to a level that was indistinguishable in +/+ and –/– mast cells. This suggested that SHIP was acting most likely downstream of the initial release of intracellular calcium stores.

Although current dogma states that crosslinking of the IgE-pre-loaded Fc$_\varepsilon$RI by a multivalent antigen is the essential first step in triggering the signalling cascades that lead to degranulation, the calcium influx into both SHIP +/+ and –/– BMMCs and the massive degranulation in SHIP –/– BMMCs with IgE alone suggested that substantial

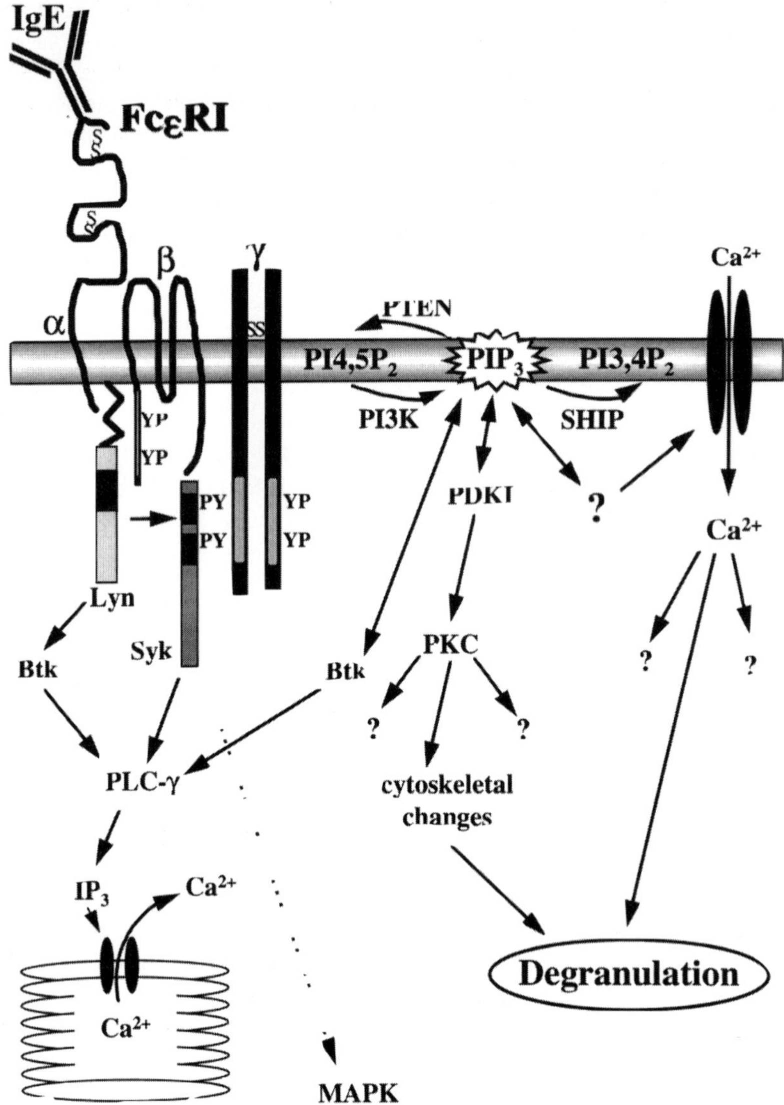

Fig. 2 Model of IgE-induced degranulation in SHIP +/+ and –/– BMMCs. The ITAMs are depicted as light grey bars in the β and γ subunits. The SH2 domains of Lyn and Syk are shown in black.

intracellular signalling was already occurring during the IgE pre-loading step (i.e. in the absence of antigen). To test this hypothesis, we explored tyrosine phosphorylation events elicited by IgE alone in SHIP +/+ and –/– BMMCs and found that both the β and γ subunits of the $Fc_{\varepsilon}RI$ were indeed tyrosine-phosphorylated in both cell types in response to IgE alone. Moreover, in +/+ cells, SHIP was constitutively tyrosine-phosphorylated and its phosphorylation level increased upon addition of IgE. Intriguingly, immunoprecipitating Shc from these cells revealed that its tyrosine phosphorylation also occurred with IgE stimulation alone but its phosphorylation was markedly reduced in SHIP –/– BMMCs. This suggested that Shc needed to bind to SHIP in order to become tyrosine-phosphorylated after $Fc_{\varepsilon}RI$ activation by IgE (26). The same result was obtained after antigen cross-linking of IgE-pre-loaded cells.

The two mitogen-activated protein kinases (MAPKs) ERK 1 and 2 were also tyrosine-phosphorylated by IgE alone in SHIP +/+ and –/– BMMCs, but in SHIP –/– cells this phosphorylation was dramatically prolonged. This is interesting for several reasons. First, since ERK activation has been shown to be dependent on Syk kinase activity (40), it suggests that IgE alone is also sufficient to activate Syk. Second, ERK activation has been shown to lead to the activation of phospholipase A_2 and the subsequent synthesis of arachidonic acid-derived metabolites (41). Thus, since ERK activation is substantially prolonged in SHIP –/– BMMCs, this suggests that SHIP may play a role in regulating the synthesis of these inflammatory mediators as well. Third, this prolonged activation of ERK in SHIP –/– cells occurs in the absence of tyrosine-phosphorylated Shc, suggesting that Shc-dependent activation of the Ras pathway may not play a significant role in the IgE-mediated activation of MAPK in these cells. Interestingly, EGTA significantly reduced IgE-induced Erk 1 and 2 phosphorylation, especially in SHIP –/– cells, suggesting that the entry of extracellular calcium is upstream of MAPK activation.

We also measured PI-3,4,5-P_3 and PI-3,4-P_2 levels in SHIP +/+ and –/– cells in response to IgE alone and found that PI-3,4,5-P_3 levels increased far higher, and PI-3,4-P_2 levels far lower, following 2 min of IgE stimulation of SHIP –/– BMMCs. This demonstrated that SHIP was the primary enzyme responsible for breaking down PI-3,4,5-P_3 *in vivo* in response to IgE-induced activation of PI3K. Of note, PI3K activity in anti-phosphotyrosine immunoprecipitates from IgE-stimulated SHIP +/+ and –/– BMMCs was identical, demonstrating that the elevation of PIP_3 levels in the –/– cells was not due to a difference in PI3K activity in the two cell types.

THE ROLE OF SHIP IN STEEL FACTOR-INDUCED SIGNALLING IN MAST CELLS

Since we found that IgE alone triggered degranulation in SHIP –/– BMMCs, we asked if SF, which is known to enhance but not trigger degranulation by itself in normal BMMCs (42), might be capable of degranulating SHIP –/– cells. Interestingly, we found SF alone did indeed stimulate massive degranulation from SHIP –/– BMMCs but, as expected, none from +/+ BMMCs (27). Worthy of note, however, is that IL-3, which stimulates the proliferation and survival of both SHIP +/+ and –/– BMMCs, did not trigger the degranulation of SHIP –/– cells. This demonstrates that the SHIP –/– BMMCs do not simply degranulate in response to any cytokine that is capable of stimulating these cells. As was found with IgE, the SF-induced degranulation was not due to higher receptor levels on SHIP –/– cells and correlated with (and was dependent upon) higher, more sustained intracellular calcium than that in SHIP +/+ cells (27). Both this influx and subsequent degranulation were completely inhibited by PI3K inhibitors, indicating that SF-induced activation of PI3K was upstream of extracellular calcium entry (27). Studies to determine if SHIP affected SF-induced extracellular calcium entry before or after the release of intracellular calcium stores revealed that, while PI3K inhibitors blocked the release of intracellular calcium, implicating PI-3,4,5-P_3 in this process, and PLC-γ2 was slightly more tyrosine-phosphorylated in SHIP –/– cells, the increases in I-1,4,5-P_3 and intracellular calcium levels were identical in –/– and +/+ BMMCs (27). This suggested that SHIP prevents SF-induced intracellular signalling in mature mast cells from progressing to degranulation by hydrolysing PI-3,4,5-P_3 and this, in turn, inhibits a crucial step between the release of calcium from intracellular stores and the influx of extracellular calcium via surface-operated calcium channels.

THE ROLE OF SHIP IN THAPSIGARGIN-INDUCED SIGNALLING IN MAST CELLS

As already mentioned, both IgE- and SF-mediated degranulation are strictly dependent on the influx of extracellular calcium (7). To study the role of SHIP in mast cell degranulation downstream of this calcium influx we used thapsigargin. This non-phorbol ester tumour promoter is a specific inhibitor of the sarcoplasmic-endoplasmic reticulum calcium-dependent ATPase which pumps calcium that leaks from the endoplasmic reticulum back into this organelle (43). Adding thapsigargin to mast cells thus results in the draining of calcium ions from the endoplasmic reticulum and PI3K-independent capacitative entry (Huber and Krystal, unpublished) of extracellular calcium through store-operated calcium surface channels, and subsequent degranulation (44). Interestingly, as was found previously with IgE and SF (7), thapsigargin stimulated far more degranulation in SHIP –/– BMMCs than in SHIP +/+ BMMCs, and this degranulation was blocked with the PI3K inhibitor, LY294002 (Huber and Krystal, unpublished). However, in contrast to IgE or SF, this heightened degranulation of SHIP –/– BMMCs was not due to a greater calcium influx into these cells, suggesting that the heightened thapsigargin-induced degranulation of SHIP –/– BMMCs was due to a PI3K-regulated pathway distinct from that regulating extracellular calcium entry. An investigation of intracellular pathways stimulated by thapsigargin in SHIP +/+ and –/– BMMCs revealed that MAPK was heavily but equally phosphorylated in both cell types. Interestingly, the pan-specific PKC inhibitor, bisindolylmaleimide, totally blocked thapsigargin-induced degranulation in both SHIP +/+ and –/– BMMCs, suggesting that a PI3K-dependent step within the mast cell degranulation process might involve a PI-3,4,5-P_3-binding protein upstream of a PKC isoform. This is interesting, given that it has been known for some time that activation of PKC is a critical event for effective degranulation to occur (45). Moreover, it is also well documented that the combination of calcium-mobilizing probes and pharmacological PKC activators leads to a synergistic increase in mast cell degranulation (45). This downstream step was not detected with PI3K inhibitors in our studies employing IgE or SF to trigger mast cell degranulation because the degranulation pathway initiated by these two agonists was blocked by PI3K inhibitors at two much earlier steps (intracellular calcium release and extracellular calcium entry) (7). Thus, thapsigargin has proven useful, by bypassing PLC-γ-induced calcium release and PI3K-mediated extracellular calcium entry, in revealing this third PI3K-regulated step in degranulation. These thapsigargin studies have also suggested that, while calcium is necessary for degranulation, it is the PI-3,4,5-P_3 rather than the intracellular calcium level that determines the extent of degranulation.

As far as the PI-3,4,5-P_3-binding protein upstream of PKC is concerned, we observed transient PI3K-mediated activation of PKB in response to thapsigargin. This is especially interesting given that PKB activation requires co-localization of the PH-containing kinase PDK-1 (which phosphorylates PKB at Thr-308) at the plasma membrane. PDK-1 has been shown very recently to phosphorylate/activate various PKC isoforms (46–48). This, coupled with our data showing that the pan-specific PKC inhibitor, bisindolylmaleimide, prevents thapsigargin-induced degranulation and that phorbol myristate acetate (PMA) does not bring thapsigargin-induced degranulation in SHIP +/+ BMMCs to levels obtained with SHIP –/– BMMCs, suggests that PDK-1 may activate a diacylglycerol (DAG)-independent PKC isotype, such as PKC-ζ (Fig. 2).

Relevant to these findings, the initiation of BMMC degranulation stimulates dramatic morphological changes in these cells. They flatten out (spread) and their cell surface transforms from a microvillous to a lamellar (ruffled) topography. This is accompanied by the polymerization of the large pool of monomeric (G) actin [approx. 50% of the total actin (49)] in these cells, and the activation of PKC has been shown to be important in this process (50, 51), perhaps by phosphorylating the actin-associated protein, myosin (45). Complicating this picture, it seems that certain PKC isoforms may enhance degranulation (52), whereas others downregulate this process (49). Preliminary studies in our laboratory suggest that certain PKC isoforms are more attracted to the membrane fraction of SHIP –/– BMMCs than +/+ BMMCs, following IgE activation (Hughes and Krystal, unpublished) and further studies are currently under way to determine their role(s) in the degranulation process.

A MODEL OF IGE-INDUCED DEGRANULATION

Pooling our results with those of other investigators, we propose a model for IgE-induced degranulation of mast cells (Fig. 2) in which binding of IgE alone to the $Fc_{\varepsilon}RI$ on normal primary mast cells activates a constitutively associated Src family member (predominantly Lyn) (53, 54), most likely via CD45 (55), to tyrosine phosphorylate the β- and γ-ITAMs. Although we can not rule out at this time that this activation by IgE alone may be due to low levels of IgE aggregates (even though we obtain the same results with HPLC-purified monomeric IgE), it is conceivable that monomeric IgE triggers signalling by causing a conformational change in the multispanning β subunit (in a fashion analogous to that proposed for seven spanner receptors). Alternatively, in keeping with the transphosphorylation model of IgE-induced signalling (which states that Lyn can only phosphorylate neighbouring receptors and not the receptor to which it is constitutively bound) (56), a small number of $Fc_{\varepsilon}RI$ aggregates may always be present even in unstimulated mast cells. Importantly, the presence of SHIP in normal mast cells ensures that this priming step does not lead to degranulation, perhaps by binding via its SH2 domain to either the C-terminal ITAM of the β chain (21) and/or the γ chain (15), or to a transmembrane protein distinct from the $Fc_{\varepsilon}RI$ that is tyrosine-phosphorylated in response to IgE binding. It may then become phosphorylated at one or both of its NPXY motifs and then attract Shc, via the latter's PTB domain, to be phosphorylated. Phosphorylated Shc may then vie with the ITAM of the $Fc_{\varepsilon}RI$ for SHIP and wrest it away. Thus Shc may serve to terminate SHIP's hydrolysis of PI-3,4,5-P_3. Inhibition of extracellular calcium entry by wortmannin or LY294002 suggests that PI3K is also translocated to the plasma membrane and activated very early in this process, but exactly how remains to be determined. Syk is attracted to the γ-ITAM (40), becomes phosphorylated by Lyn and stimulates the tyrosine phosphorylation of PLC-γ1 and γ2 (57), which in turn leads to the initial release of intracellular calcium via IP_3. This emptying of intracellular calcium triggers the entry of extracellular calcium and this entry is substantially augmented by PI3K-mediated generation of PI-3,4,5-P_3. As to how an elevation in PI-3,4,5-P_3 impacts on extracellular calcium entry, it has recently been shown that activating the BCR attracts Btk to PI-3,4,5-P_3 at the plasma membrane and this leads to its activation and subsequent phosphorylation/activation of PLC-γ2 to generate IP_3 and the sustained emptying of intracellular calcium stores that are critical for activating the store-operated calcium entry from the extracellular medium (58, 59). However, since we observe no detectable difference in the release of intracellular calcium in IgE-stimulated SHIP +/+ and –/– BMMCs (i.e. in the presence of EGTA), we hypothesize that the elevated PI-3,4,5-P_3 present in SHIP –/– cells attracts and activates Btk or another PH-containing intermediate at a step between the draining of intracellular calcium stores and extracellular calcium entry. This is in agreement with Bolland *et al.* (5), who concluded, from studying SHIP's role in mediating Fc_γRIIB inhibition of BCR activation in the chicken DT40 B cell line, that SHIP regulates extracellular calcium entry after draining of intracellular calcium stores via its ability to control Btk association with the plasma membrane. Thus SHIP appears to function as a 'gatekeeper' in normal BMMCs by keeping PI3K-generated PI-3,4,5-P_3 levels in check and this restricts extracellular calcium entry and subsequent degranulation. We also propose that SHIP regulates, via PI-3,4,5-P_3 levels, the activation of PDK-1 and downstream PKC isoforms that play a role in the cytoskeletal changes important to the degranulation process (Fig. 2).

MAJOR QUESTIONS REMAINING

One important question still to be answered is what attracts SHIP to the plasma membrane following IgE stimulation. Is it indeed the Fc_εRI itself via its β or γ subunits or a molecule analogous to the recently identified, lymphoid-specific transmembrane Cdw150 protein that is tyrosine-phosphorylated in response to BCR activation and attracts SHIP (60)?

Related to this, it is still not known what attracts PI3K to the plasma membrane following IgE-stimulated degranulation. Very recently Gupta *et al.* (61) demonstrated that SHIP binding to the Fc_γRIIB during co-clustering with the BCR induced the binding of PI3K, via its SH2 domains. It is thus possible that the binding of SHIP itself to a tyrosine-phosphorylated transmembrane protein following IgE-induced signalling plays a role in the attraction of at least some PI3K to the membrane in mast cells.

Our data suggest there are at least three PI-3,4,5-P_3-dependent pathways that must be activated by IgE for degranulation, one being required for intracellular calcium release, another for extracellular calcium entry and a third for PKC activation. It is likely that PH-containing proteins are involved in these steps and, while there is some evidence that Btk may be the one involved in the first step and PDK-1 in the third, the nature of the protein(s) involved in the second remains to be determined.

What pathways stimulated by calcium entry are critical to degranulation? Although calcium is known to trigger many intracellular events (62, 63), it is not known exactly which ones are critical for degranulation.

What is the role of the SHIP–Shc interaction in IgE-induced signalling? Does it serve to terminate the association of SHIP with the plasma membrane or does it serve to tie up Shc and thus reduce Grb2/Sos activation of Ras?

Another key question of course is which PKCs stimulate and which inhibit IgE-mediated degranulation. Once this is answered, the more difficult task of determining the pathways they activate (inactivate) to modulate degranulation can be addressed.

SUMMARY AND CONCLUSIONS

Our results both challenge the current paradigm for passive sensitization of mast cells with IgE and reveal a vital role for SHIP in both setting the threshold for and limiting degranulation by hydrolysing PI-3,4,5-P_3 and thus restricting the entry of extracellular calcium and activation of certain PKC isoforms. This raises the intriguing possibility that naturally occurring mutations in SHIP, PI3K or PTEN (64), which together determine PI-3,4,5-P_3 levels following $Fc_{\varepsilon}RI$ activation, could contribute to specific hyperallergic conditions in man. In addition, these studies suggest that the development of drugs that can specifically increase SHIP activity in mast cells might dramatically reduce allergic responses.

ACKNOWLEDGEMENTS

We would like to thank Christine Kelly for typing the manuscript. This work was supported by the NCI-C, with funds from the Terry Fox Foundation, and the MRC-C with core support from the BC Cancer Foundation and the BC Cancer Agency. M. Huber is supported by the Deutsche Forschungsgemeinschaft. M. Hughes is supported by NSERC. G. K. is a Terry Fox Cancer Research Scientist of the NCI-C supported by funds from the Canadian Cancer Society and the Terry Fox Run.

REFERENCES

1. Galli, S. J. and Wershil, B. K. The two faces of the mast cell. *Nature* **381**:21–22, 1996.
2. Schwartz, L. B. Mast cells: function and contents. *Curr. Opin. Immunol.* **6**:91–97, 1994.
3. Tang, L., Jennings, T. A. and Eaton, J. W. Mast cells mediate acute inflammatory responses to implanted biomaterials. *Proc. Natl. Acad. Sci. USA* **95**:8841–8846, 1998.
4. Scharenberg, A. M. and Kinet, J.-P. PtdIns-3,4,5-P3: a regulatory nexus between tyrosine kinases and sustained calcium signals. *Cell* **94**:5–8, 1998.
5. Bolland, S., Pearse, R. N., Kurosaki, T. and Ravetch, J. V. SHIP modulates immune receptor responses by regulating membrane association of Btk. *Immunity* **8**:509–516, 1998.
6. Coggeshall, K. M. Inhibitory signaling by B cell FcγRIIb. *Curr. Opin. Immunol.* **10**:306–312, 1998.
7. Huber, M,. Helgason, C. D., Damen, J. E., Scheid, M., Duronio, V., Liu, L., Ware, M. D., Humphries, R. K. and Krystal, G. The role of SHIP in growth factor induced signalling. In: *Progress in Biophysics & Molecular Biology*, Vol. 71 (Pawson, A. J., ed.), pp. 423–434. Elsevier Science, Amsterdam, 1999.
8. Damen, J. E., Liu, L., Rosten, P., Humphries, R. K., Jefferson, A. B., Majerus, P. W. and Krystal, G. The 145-kDa protein induced to associate with Shc by multiple cytokines is an inositol tetraphosphate and phosphatidylinositol 3,4,5-trisphosphate 5-phosphatase. *Proc. Natl. Acad. Sci. USA* **93**:1689–1693, 1996.
9. Irvine, R. Inositol phospholipids: translocation, translocation, translocation. *Curr. Biol.* **8**:R557–R559, 1998.
10. Ware, M. D., Rosten, P., Damen, J. E., Liu, L., Humphries, R. K. and Krystal, G. Cloning and characterization of human SHIP, the 145-kD inositol 5-phosphatase that associates with SHC after cytokine stimulation. *Blood* **88**:2833–2840, 1996.
11. Liu, Q., Shalaby, F., Jones, J., Bouchard, D. and Dumont, D. J. The SH2-containing inositol polyphosphate 5-phosphatase, Ship, is expressed during hematopoiesis and spermatogenesis. *Blood* **91**:2753–2759, 1998.
12. Geier, S. J., Algate, P. A., Carlberg, K., Flowers, D., Friedman, C., Trask, B. and Rohrschneider, L. R. The human SHIP gene is differentially expressed in cell lineages of the bone marrow and blood. *Blood* **89**:1876–1885, 1997.

13. Sattler, M., Salgia, R., Shrikhande, G., Verma, S., Choi, J.-L., Rohrschneider, L. R. and Griffin, J. D. The phosphatidylinositol polyphosphate 5-phosphatase SHIP and the protein tyrosine phosphatase SHP-2 form a complex in hematopoietic cells which can be regulated by BCR/ABL and growth factors. *Oncogene* **15**:2379–2384, 1997.
14. Harmer, S. L. and DeFranco, A. L. The Src homology domain 2-containing inositol phosphatase SHIP forms a ternary complex with Shc and Grb2 in antigen receptor-stimulated B lymphocytes. *J. Biol. Chem.* **274**:12183–12191, 1999.
15. Osborne, M. A., Zenner, G., Lubinus, M., Zhang, X., Songyang, Z., Cantley, L. C., Majerus, P., Burn, P. and Kochan J. P. The inositol 5′-phosphatase SHIP binds to immunoreceptor signaling motifs and responds to high affinity IgE receptor aggregation. *J. Biol. Chem.* **271**:29271–29278, 1996.
16. Vély, F., Olivero, S., Olcese, L., Moretta, A., Damen, J. E., Liu, L., Krystal, G., Cambier, J. C., Daëron, M. and Vivier, E. Differential association of phosphatases with hematopoietic co-receptors bearing immunoreceptor tyrosine-based inhibition motifs. *Eur. J. Immunol.* **27**:1994–2000, 1997.
17. Tridandapani, S., Kelley, T., Pradhan, M., Cooney, D., Justement, L. B. and Coggeshall, K. M. Recruitment and phosphorylation of SH2-containing inositol phosphatase and Shc to the B-cell Fc gamma immunoreceptor tyrosine-based inhibition motif peptide motif. *Mol. Cell. Biol.* **17**:4305–4311, 1997.
18. Ono, M., Okada, H., Bolland, S., Yanagi, S., Kurosaki, T. and Ravetch, J. V. Deletion of SHIP or SHP-1 reveals two distinct pathways for inhibitory signaling. *Cell* **90**:293–301, 1997.
19. Kuroiwa, A., Yamashita, Y., Inui, M., Yuasa, T., Ono, M., Nagabukuro, A., Matsuda, Y. and Takai, T. Association of tyrosine phosphatases SHP-1 and SHP-2, inositol 5-phosphatase SHIP with gp49B1, and chromosomal assignment of the gene. *J. Biol. Chem.* **273**:1070–1074, 1998.
20. Tridandapani, S., Pradhan, M., LaDine, J. R., Garber, S., Anderson, C. L. and Coggeshall, K. M. Protein interactions of Src homology (SH2) domain-containing inositol phosphatase (SHIP): association with Shc displaces SHIP from FcγRIIb in B cells. *J. Immunol.* **162**:1408–1414, 1999.
21. Kimura, T., Sakamoto, H., Appella, E. and Siraganian, R. P. The negative signaling molecule SH2 domain-containing inositol-polyphosphate 5-phosphatase (SHIP) binds to the tyrosine-phosphorylated beta subunit of the high affinity IgE receptor. *J. Biol. Chem.* **272**:13991–13996, 1997.
22. Damen, J. E., Liu, L., Ware, M. D., Ermolaeva, M., Majerus, P. W. and Krystal, G. Multiple forms of the SH2-containing inositol phosphatase, SHIP, are generated by C-terminal truncation. *Blood* **92**:1199–1205, 1998.
23. Odai, H., Sasaki, K., Iwamatsu, A., Nakamoto, T., Ueno, H., Yamagata, T., Mitani, K., Yazaki, Y. and Hirai, H. Purification and molecular cloning of SH2- and SH3-containing inositol polyphosphate-5-phosphatase, which is involved in the signaling pathway of granulocyte–macrophage colony-stimulating factor, erythropoietin, and Bcr-Abl. *Blood* **89**:2745–2756, 1997.
24. Lucas, D. M. and Rohrschneider, L. R. A novel spliced form of SH2-containing inositol phosphatase is expressed during myeloid development. *Blood* **93**:1922–1933, 1999.
25. Ono, M., Bolland, S., Tempst, P. and Ravetch, J. V. Role of the inositol phosphatase SHIP in negative regulation of the immune system by the receptor Fc-gamma-RIIB. *Nature* **383**:263–266, 1996.
26. Huber, M., Helgason, C. D., Damen, J. E., Liu, L., Humphries, R. K. and Krystal, G. The src homology 2-containing inositol phosphatase (SHIP) is the gatekeeper of mast cell degranulation. *Proc. Natl. Acad. Sci. USA* **95**:11330–11335, 1998.
27. Huber, M., Helgason, C. D., Scheid, M. P., Duronio, V., Humphries, R. K. and Krystal, G. Targeted disruption of SHIP leads to steel factor induced degranulation of mast cells. *EMBO J.* **27**:7311–7319, 1998.
28. Marte, B. M. and Downward, J. PKB/Akt: connecting phosphoinositide 3-kinase to cell survival and beyond. *Trends Biochem. Sci.* **22**:355–358, 1997.
29. Belham, C., Wu, S. and Avruch, J. Intracellular signalling: PDK1 – a kinase at the hub of things. *Curr. Biol.* **9**:R93–R96, 1999.
30. Liu, Q., Sasaki, T., Kozieradzki, I., Wakeham, A., Itie, A., Dumont, D. J. and Penninger, J. M. SHIP is a negative regulator of growth factor receptor-mediated PKB/Akt activation and myeloid cell survival. *Genes Dev.* **13**:786–791, 1999.
31. Aman, M. J., Lamkin, T. D., Okada, H., Kurosaki, T. and Ravichadran, K. S. The inositol phosphatase SHIP inhibits Akt/PKB activation in B cells. *J. Biol. Chem.* **273**:33922–33928, 1998.
32. Jacob, A., Cooney, D., Tridandapani, S., Kelley, T. and Coggeshall, K. M. FcγRIIb modulation of surface immunoglobulin-induced Akt activation in murine B cells. *J. Biol. Chem.* **274**:13704–13710, 1999.
33. Vollenweider, P., Clodi, M., Martin, S. S., Imamura, T., Kavanaugh, W. M. and Olefsky, J. M. An SH2 domain-containing 5′ inositolphosphatase inhibits insulin-induced GLUT4 translocation and growth

factor-induced actin filament rearrangement. *Mol. Cell. Biol.* **19**:1081–1091, 1999.

34. Liu, L., Damen, J. E., Cutler, R. L. and Krystal, G. Multiple cytokines stimulate the binding of a common 145-kilodalton protein to Shc at the Grb2 recognition site of Shc. *Mol. Cell. Biol.* **14**:6926–6935, 1994.
35. Kinoshita, T., Yokota, T., Arai, K. and Miyajima, A. Suppression of apoptotic death in hematopoietic cells by signalling through the IL-3/GM-CSF receptors. *EMBO J.* **14**:266–275, 1995.
36. Liu, L., Damen, J. E., Hughes, M. R., Babic, I., Jirik, F. R. and Krystal, G. The Src homology 2 (SH2) domain of SH2-containing inositol phosphatase (SHIP) is essential for tyrosine phosphorylation of SHIP, its association with Shc, and its induction of apoptosis. *J. Biol. Chem.* **272**:8983–8988, 1997.
37. Lioubin, M. N., Algate, P. A., Tsai, S., Carlberg, K., Aebersold, A. and Rohrschneider, L. R. p150^{Ship}, a signal transduction molecule with inositol polyphosphate-5-phosphatase activity. *Genes Dev.* **10**:1084–1095, 1996.
38. Helgason, C. D., Damen, J. E., Rosten, P., Grewal, R., Sorensen, P., Chappel, S. M., Borowski, A., Jirik, F., Krystal, G. and Humphries, R. K. Targeted disruption of SHIP leads to hemopoietic perturbations, lung pathology and a shortened lifespan. *Genes Dev.* **12**:1610–1620, 1998.
39. Alber, G., Miller, L., Jelsema, C. L., Varin-Blank, N. and Metzger, H. Structure–function relationships in the mast cell high affinity receptor for IgE. Role of the cytoplasmic domains and the beta subunit. *J. Biol. Chem.* **266**:22613–22620, 1991.
40. Hirasawa, N., Scharenberg, A., Yamamura, H., Beaven, M. A. and Kinet, J.-P. A requirement for Syk in the activation of the microtubule-associated protein kinase/phospholipase A2 pathway by Fc-epsilon RI is not shared by a G protein-coupled receptor. *J. Biol. Chem.* **270**:10960–10967, 1995.
41. Hirasawa, N. A., Santini, F. and Beaven, M. A. Activation of the mitogen-activated protein kinase/cytosolic phospholipase A_2 pathway in a rat mast cell line. Indications of different pathways for release of arachidonic acid and secretory granules. *J. Immunol.* **154**:5391–5402, 1995.
42. Vosseller, K., Stella, G., Yee, N. S. and Besmer, P. c-kit receptor signaling through its phosphatidylinositide-3′-kinase-binding site and protein kinase C: role in mast cell enhancement of degranulation, adhesion, and membrane ruffling. *Mol. Biol. Cell* **8**:909–922, 1997.
43. Thastrup, O., Dawson, A. P., Scharff, O., Foder, B., Cullen, P. J., Drobak, B. K., Bjerrum, P. J., Christensen, S. B. and Hanley, M. R. Thapsigargin, a novel molecular probe for studying intracellular calcium release and storage. *Agents Actions* **43**:187–193, 1994.
44. Putney, J. W. Jr and Bird, G. S. The signal for capacitative calcium entry. *Cell* **75**:199–201, 1993.
45. Ludowyke, R. I., Scurr, L. L. and McNally, C. M. Calcium ionophore-induced secretion from mast cells correlates with myosin light chain phosphorylation by protein kinase C. *J. Immunol.* **157**:5130–5138, 1996.
46. Le Good, J. A., Ziegler, W. H., Parekh, D. B., Alessi, D. R., Cohen, P. and Parker, P. J. Protein kinase C isotypes controlled by phosphoinositide 3-kinase through the protein kinase PDK1. *Science* **281**:2042–2045, 1998.
47. Dutil, E. M., Toker, A. and Newton, A. C. Regulation of conventional protein kinase C isozymes by phosphoinositide-dependent kinase 1 (PDK-1). *Curr. Biol.* **8**:1366–1375, 1998.
48. Chou, M. M., Hou, W., Johnson, J., Graham, L. K., Lee, M. H., Chen, C.-S., Newton, A. C., Schaffhausen, B. S. and Toker, A. Regulation of protein kinase C ζ by PI 3-kinase and PDK-1. *Curr. Biol.* **8**:1069–1077, 1998.
49. Frigeri, L. and Apgar, J. R. The role of actin microfilaments in the down-regulation of the degranulation response in RBL-2H3 mast cells. *J. Immunol.* **162**:2243–2250, 1999.
50. Pfeiffer, J. R., Seagrove, J. C., Davis, B. H., Deanin, G. G. and Oliver, J. M. Membrane and cytoskeletal changes associated with IgE-mediated serotonin release from rat basophilic leukemia cells. *J. Cell Biol.* **101**:2145–2155, 1985.
51. Apgar, J. R. Regulation of the antigen-induced F-actin response in rat basophilic leukemia cells by protein kinase C. *J. Cell Biol.* **112**:1157–1163, 1991.
52. Ozawa, K., Szallasi, Z., Kazanietz, M. G., Blumberg, P. M., Mischak, H., Mushinski, J. F. and Beaven, M. A. Ca^{2+}-dependent and Ca^{2+}-independent isozymes of protein kinase C mediate exocytosis in antigen-stimulated rat basophilic RBL-2H3 cells. Reconstitution of secretory responses with Ca^{2+} and purified isozymes in washed permeabilized cells. *J. Biol. Chem.* **268**:1749–1756, 1993.
53. Nishizumi, H. and Yamamoto, T. Impaired tyrosine phosphorylation and Ca^{2+} mobilization, but not degranulation, in lyn-deficient bone marrow-derived mast cells. *J. Immunol.* **158**:2350–2355, 1997.
54. Vonakis, B. M., Chen, H., Haleem-Smith, H. and Metzger, H. The unique domain as the site on lyn kinase for its constitutive association with the high affinity receptor for IgE. *J. Biol. Chem.* **272**:24072–24080, 1997.

55. Berger, S. A., Mak, T. W. and Paige, C. J. Leukocyte common antigen (CD45) is required for immunoglobulin E-mediated degranulation of mast cells. *J. Exp. Med.* **180**:471–476, 1994.
56. Pribluda, V. S., Pribluda, C. and Metzger, H. Transphosphorylation as the mechanism by which the high-affinity receptor for IgE is phosphorylated upon aggregation. *Proc. Natl. Acad. Sci. USA* **91**:11246–11250, 1994.
57. Zhang, J., Berenstein, E. H., Evans, R. L. and Siraganian, R. P. Transfection of Syk protein tyrosine kinase reconstitutes high affinity IgE receptor-mediated degranulation in a Syk-negative variant of rat basophilic leukemia RBL-2H3 cells. *J. Exp. Med.* **184**:71–79, 1996.
58. Scharenberg, A. M., El-Hillal, O., Fruman, D. A., Beitz, L. O., Li, Z., Lin, S., Gout, I., Cantley, L. C., Rawlings, D. J. and Kinet, J.-P. Phosphatidylinositol-3,4,5-trisphosphate (PtdIns-3,4,5-P3)/Tec kinase-dependent calcium signaling pathway: a target for SHIP-mediated inhibitory signals. *EMBO J.* **17**:1961–1972, 1998.
59. Fluckiger, A.-C., Li, Z., Kato, R. M., Wahl, M. I., Ochs, H. D., Longnecker, R., Kinet, J.-P., Witte, O. N., Scharenberg, A. M. and Rawlings, D. J. Btk/Tec kinases regulate sustained increases in intracellular Ca^{2+} following B-cell receptor activation. *EMBO J.* **17**:1973–1985, 1998.
60. Mikhalap, S., Shlapatska, L. M., Berdova, A. G., Law, C -L., Clark, E. A. and Sidorenko, S. P. CDw150 associates with Src-homology 2-containing inositol phosphatase and modulates CD95-mediated apoptosis. *J. Immunol.* **162**:5719–5727, 1999.
61. Gupta, N., Scharenberg, A. M., Fruman, D. A., Cantley, L. C., Kinet, J.-P. and Long, E. O. The SH2 domain-containing inositol 5′-phosphatase (SHIP) recruits the p85 subunit of phosphoinositide 3-kinase during FcγRIIb1-mediated inhibition of B cell receptor signaling. *J. Biol. Chem.* **274**:7489–7494, 1999.
62. Berridge, M. J. The AM and FM of calcium signalling. *Nature* **386**:759–760, 1997.
63. Putney J. W. Jr. Calcium signalling: Up, down, up, down.... What's the point? *Science* **279**:191–192, 1998.
64. Cantley, L. C. and Neel, B. G. New insights into tumor suppression: PTEN suppresses tumor formation by restraining the phosphoinositide 3-kinase/AKT pathway. *Proc. Natl. Acad. Sci. USA* **96**:4240–4245, 1999.

SECTION FOUR

REGULATION OF MAST CELL AND BASOPHIL SIGNALLING AND SECRETION

CHAPTER 12

Immunoreceptor Tyrosine-based Inhibition Motif-dependent Negative Regulation of Mast Cell Activation and Proliferation

MARC DAËRON, ODILE MALBEC, HÉLÈNE LIÉNARD, PIERRE BRUHNS and WOLF H. FRIDMAN*

Laboratoire d'Immunologie Cellulaire & Clinique, INSERM U.255, Institut Curie, Paris, France

INTRODUCTION

Immunoreceptor Tyrosine-based Inhibition Motifs (ITIMs) are molecular motifs that antagonize Immunoreceptor Tyrosine-based Activation Motifs (ITAMs) (1, 2). ITIMs are present, in variable numbers, in the intracytoplasmic domains of a large group of molecules which, when kept in close proximity to ITAM-bearing receptors, interfere with intracellular signals transduced by the latter (3). ITAM- and ITIM-bearing molecules borne by a single cell can be constitutively associated on the plasma membrane, or they can be co-aggregated by extracellular ligands. ITAMs and ITIMs are tyrosyl-phosphorylated by src family protein tyrosine kinases and, once phosphorylated, they recruit cytoplasmic molecules having Src homology 2 (SH2) domains (4, 5). Phosphorylated ITAMs recruit SH2 domain-bearing cytoplasmic kinases whereas phosphorylated ITIMs recruit SH2 domain-bearing cytoplasmic phosphatases. Two main mechanisms are used by two types of ITIM-bearing molecules whose Fc_{γ}RIIB and Killer cell Immunoglobulin-like Receptors with a Long intracytoplasmic domain (KIR-Ls) are the prototypes (6).

Fc_{γ}RIIB are low-affinity receptors for the Fc portion of IgG antibodies which exist as two or three isoforms, in humans and mice, respectively (7). They are expressed by lymphoid and myeloid cells, including mouse mast cells (8), human basophils (2) and human mast cells (our unpublished observation), and they inhibit IgE-induced mast cell (9) and basophil (2) activation when co-aggregated with Fc_{ε}RI by immune complexes. Fc_{γ}RIIB were demonstrated to negatively regulate cell activation by all ITAM-bearing receptors when co-aggregated with the latter by anti-receptor antibodies or by soluble immune complexes (2). Fc_{γ}RIIB possess a single ITIM in their intracytoplasmic domain which is tyrosyl-phosphorylated by src family protein tyrosine kinases provided by the ITAM-

* Corresponding author.

ISBN 0-12-473335-2

bearing receptors with which they are co-aggregated (10). Once phosphorylated, $Fc_{\gamma}RIIB$ recruit the SH2 domain-bearing Inositol 5-Phosphatase SHIP (11, 12). A substrate of SHIP is the membrane-anchored phosphatidylinositol 3,4,5-trisphosphate (PIP_3) (13), generated by phosphatidylinositol 3-kinase (PI3K). PIP_3 permits the membrane recruitment of molecules that possess pleckstrin homology (PH) domains and that are necessary for cell activation (14, 15). Thus, by dephosphorylating PIP_3, $Fc_{\gamma}RIIB$-associated SHIP prevents the recruitment of the Bruton tyrosine kinase (Btk) which is mandatory for a sustained Ca^{2+} mobilization (15), and extinguishes the propagation of intracellular signals generated by ITAM-bearing receptors, thereby stopping cell activation.

KIR-Ls are receptors for major histocompatibility (MHC) class I molecules. They are expressed by natural killer (NK) cells and by a subpopulation of T cells, and they inhibit cell-mediated cytotoxicity when recognizing MHC class I molecules on target cells (16). KIR-Ls were also shown to inhibit IgE-induced mast cell activation when co-aggregated with IgE receptors in a reconstitution model (17). KIR analogues are expressed by mast cells (18–20). KIR-Ls bear two ITIMS in their intracytoplasmic domain, which are phosphorylated by src family protein tyrosine kinases and recruit the two SH2 domain-bearing protein tyrosine phosphatases SHP-1 and SHP-2 (21). We found recently that the two KIR-L ITIMs differentially contribute to the recruitment of the two phosphatases (22). The substrates of SHP-1 are thought to be phosphorylated ITAMs and/or protein tyrosine kinases that phosphorylate ITAMs (21). The substrates of SHP-2 are unknown, and whether this phosphatase is actively involved in inhibition or simply removed from activating receptor complexes where it contributes to the generation of positive signals is unclear. Whatever their respective roles, protein tyrosine phosphatases recruited by KIR-Ls abort early signals transduced by ITAM-bearing receptors, thereby preventing cell activation.

Other transmembrane molecules were found to inhibit cell proliferation induced by hormone or growth factor receptors with an intrinsic protein tyrosine kinase activity (RTKs). These molecules, initially described under several names in different tissues, were understood to belong to the same multigene family. They were collectively named SIRPs, and two types, SIRP-α and SIRP-β, were recognized, differing by the presence (in SIRP-α) or the absence (in SIRP-β) of four ITIM-like motifs (23). SIRPs are widely distributed. They were found in all tissues examined in humans, including haematopoietic cells (23), in neurons and myeloid cells in rats (24), and in myeloid cells, especially macrophages, but not lymphoid cells, in mice (25). The mechanism of inhibition of RTK-dependent cell proliferation by SIRP-α is poorly understood. It was correlated with the recruitment of SHP-1 and SHP-2 by tyrosyl-phosphorylated SIRP-α (23, 26, 27), but the substrates of these phosphatases were not formally identified. Noticeably, negative regulation of cell proliferation did not require that SIRP-α and RTKs be co-aggregated by the same extracellular ligand. Indeed, the overexpression of SIRP-α was sufficient to inhibit epithelial growth factor-, platelet-derived growth factor- or insulin-induced cell proliferation of NIH-3T3 fibroblasts, and ligand-independent proliferation of the same cells infected by a retrovirus carrying an oncogenic form of RTK (23).

Because SIRP-α possess ITIM-like motifs, because SIRP-α use the same SHPs to inhibit RTK-dependent cell proliferation as KIR-Ls to inhibit ITAM-dependent cell activation, and because SIRPs are expressed by cells that also express ITAM-bearing receptors, one could wonder whether SIRP-α might also negatively regulate ITAM-bearing receptor-dependent cell activation. If so, one could hypothesize that, conversely, $Fc_{\gamma}RIIB$ might negatively regulate RTK-dependent cell proliferation. We found both hypotheses to be correct in mast cells.

SIRP-α CAN NEGATIVELY REGULATE Fc_εRI-DEPENDENT MAST CELL ACTIVATION

To investigate the hypothesis that SIRP-α might negatively regulate cell activation induced by ITAM-bearing receptors, we constructed an experimental model in mast cells (27a). Mast cells were chosen as an assay system because we found that one SIRP-α gene is expressed in human mast cells. Using oligonucleotide primers that could amplify transmembrane and intracytoplasmic sequences specific for human SIRP-α1, we indeed detected SIRP-α transcripts in RNA from both the human mast cell line HMC-1 (28) and from primary mast cells derived from normal human cord blood (29). We therefore constructed a chimeric receptor consisting of the extracellular domain of murine Fc_γRIIB and of the transmembrane and intracytoplasmic domains of a human SIRP-α. This chimera was stably expressed in the rat mast cells RBL-2H3 which constitutively express high-affinity IgE receptors (Fc_εRI) (30). The expression of the chimera affected neither the growth rate of transfectants nor IgE-induced serotonin release.

RBL transfectants were sensitized with mouse IgE and, using appropriate ligands, Fc_εRI were either aggregated or co-aggregated with the SIRP-α chimera. We found that both the release of a pre-formed mediator (serotonin) and the secretion of a newly synthesized cytokine (tumour necrosis factor-α, TNF-α) induced upon aggregation of Fc_εRI were dose-dependently inhibited when Fc_εRI were co-aggregated with the SIRP-α chimera by the same extracellular ligand. Interestingly, the simultaneous, but independent aggregation of SIRP-α and of Fc_εRI by two separate ligands did not affect mediator release. The distinct conditions required for SIRP-α to inhibit RTK-dependent cell proliferation and for the SIRP-α chimera to inhibit Fc_εRI-dependent mast cell activation may be explained by the constitutive association of SIRP-α with RTKs in macrophages (31). Whether SIRP-α constitutively expressed in mast cells can associate with Fc_εRI (possibly via their extracellular domains that were removed in the SIRP-α chimera) is not known.

When co-aggregated with Fc_εRI, the SIRP-α chimera became tyrosyl-phosphorylated, possibly by lyn provided by Fc_εRI as it was demonstrated for Fc_γRIIB (10), and recruited the two SH2 domain-bearing protein tyrosine phosphatases SHP-1 and SHP-2. Concomitantly, the phosphorylation of Fc_εRI ITAMs was reduced, possibly as a consequence of their dephosphorylation by the recruited phosphatases. The subsequent Ca^{2+} responses were attenuated and the activation of the mitogen-activated protein (MAP) kinases Erk-1 and Erk-2 was abolished. When co-aggregated with an ITAM-bearing receptor, the SIRP-α chimera therefore behaved like a typical ITIM-bearing receptor of the KIR-L type. It follows that, under appropriate conditions, SIRP-α might negatively regulate not only the proliferation of haematopoietic cells induced by RTKs, but also their many biological functions induced by ITAM-bearing receptors. These receptors include Fc receptors that associate with the ITAM-bearing transduction subunit FcRγ (32); i.e. Fc_εRI expressed by mast cells and basophils in normal individuals (33, 34), but also by eosinophils (35) and monocytes in allergic patients (36), high-affinity IgG receptors (Fc_γRI) expressed by macrophages, monocytes, dendritic cells and activated neutrophils (37, 38), high-affinity IgA receptors (Fc_αRI) expressed by monocytes and eosinophils (39, 40), and low-affinity IgG receptors (Fc_γRIIIA) expressed by mast cells, NK cells, Langerhans cells and activated monocytes (41–43). They also include the human-restricted single-chain low-affinity IgG receptors Fc_γRIIA/C expressed by B cells, monocytes, neutrophils, basophils and platelets (37), which bear one ITAM in their own intracytoplasmic domain (32). Whether SIRP-α might contribute to control IgE-dependent mast cell activation and whether a defect in such a regulation might contribute to the aetiology of allergic diseases is an attractive possibility that deserves to be considered.

Fc_γRIIB CAN NEGATIVELY REGULATE c-kit-DEPENDENT MAST CELL PROLIFERATION

Another implication of the above finding is that SIRP-α bear typical ITIMs rather than unique inhibitory motifs that would specifically negatively regulate RTK-dependent cell proliferation. As a consequence, one could hypothesize that other ITIM-bearing molecules might not only control cell activation induced by ITAM-bearing receptors, but also cell proliferation induced by RTKs. We investigated this possibility in growth factor-dependent mouse bone marrow-derived mast cells (BMMCs) that express constitutively Fc_γRIIB (8, 9) and c-kit, a typical RTK of the platelet-derived growth factor receptor family (44). When dimerized by its natural ligand, stem cell factor (SCF), c-kit autophosphorylates and recruits PI3K which generates PIP_3. PIP_3 then recruits PH domain-bearing effector molecules, including Akt, Btk and Vav (45), leading to the activation of the kinases Jnk and p38 (46). Phosphorylated c-kit also recruits adapter molecules such as Shc that lead to the activation of MAP kinases via the ras pathway (47). The combined effects of these kinases on transcription factors, such as c-Jun and c-Fos, initiate the transcription of cyclin genes whose products control the cell cycle. Ultimately, cells enter the G_1 phase, reach the M phase and divide.

We found that $F(ab')_2$ fragments of a rat anti-mouse c-kit antibody could mimic the effects of SCF and induced the proliferation of BMMCs as assessed by thymidine incorporation. Under the same conditions, intact IgG antibodies anti-c-kit failed to induce significant thymidine incorporation, unless mast cells were pre-incubated with an antibody that blocked the binding site of mouse Fc_γRIIB, or were derived from Fc_γRIIB-deficient mice. Inhibition affected not only thymidine incorporation measured 24 h following stimulation, but also the actual cell proliferation of BMMCs assessed by an increase in the number of viable cells after a 5-day culture. It did not result from an effect on cell viability, measured following staining of BMMCs with iodine propidium or with annexin V, and was correlated with an arrest of the cell cycle in G_1 (48).

When co-aggregated with c-kit, Fc_γRIIB became tyrosyl-phosphorylated. Based on the role of Fc_εRI-associated kinases demonstrated during the co-aggregation of Fc_γRIIB with Fc_εRI in lyn-deficient BMMCs (10), one can hypothesize that Fc_γRIIB were phosphorylated either by c-kit itself or by a non-receptor tyrosine kinase recruited by c-kit. Phosphorylated c-kit indeed recruits src family protein tyrosine kinases (46). Following co-aggregation with c-kit, phosphorylated Fc_γRIIB recruited SHIP, but neither SHP-1 nor SHP-2. c-kit was inducibly tyrosyl-phosphorylated when aggregated by anti-c-kit antibodies, although with a lower intensity than when dimerized by SCF. Anti-c-kit-induced phosphorylation of c-kit was not affected by the co-aggregation with Fc_γRIIB (48). A direct role of SHIP in the inhibition of c-kit-mediated proliferation remains to be demonstrated. Supporting such a role, several growth factors, including SCF, were found to induce exaggerated proliferative responses of progenitors of haematopoietic cells in SHIP-deficient mice (49).

Taken together, the above findings suggest that $Fc_{\gamma}RIIB$ may not only control the activation of cells of the immune system, but also the proliferation of haematopoietic cells, in general, and of mast cells, in particular, during ontogeny and during immune responses. The conditions under which $Fc_{\gamma}RIIB$ may have such an effect remain to be determined, and our results call for assessing the occurrence of anti-growth factor or anti-growth factor receptor antibodies in physiology and pathology.

CONCLUSION

In summary, we found that SIRP-α, which was known to negatively regulate RTK-dependent cell proliferation (23), can also negatively regulate ITAM-dependent cell activation and that inhibition is correlated with the recruitment of SHP-2 and SHP-1 by phosphorylated SIRP-α. Conversely, we found that $Fc_{\gamma}RIIB$, which was known to negatively regulate ITAM-dependent cell activation (2), can also negatively regulate RTK-dependent cell proliferation and that inhibition is correlated with the recruitment of SHIP by phosphorylated $Fc_{\gamma}RIIB$ (48). It follows that negative regulation of ITAM-dependent cell activation and negative regulation of RTK-dependent cell proliferation may be two general properties of ITIM-bearing molecules and that the same receptors may co-ordinately control the development and the functions of haematopoietic cells (Fig. 1).

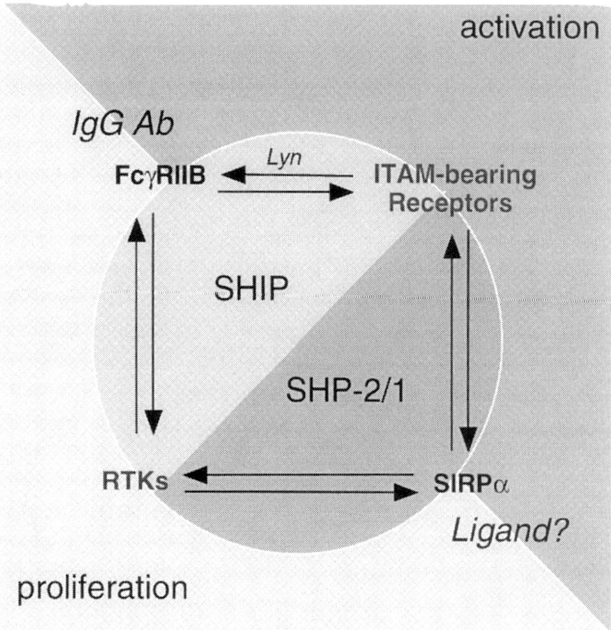

Fig. 1 Coordinate negative regulation of ITAM-dependent cell activation and of RTK-dependent cell proliferation by ITIM-bearing receptors. The figure schematizes a (mast) cell which expresses ITAM-bearing receptors (such as $Fc_{\varepsilon}RI$) and RTKs (such as c-kit) that trigger cell activation and proliferation, respectively, and ITIM-bearing receptors (such as FcγRIIB and SIRP-α) that recruit the SH2 domain-bearing phosphatases SHIP and SHP-1/2, respectively. Each ITIM-bearing receptor can negatively regulate both ITAM-dependent cell activation and RTK-dependent cell proliferation upon co-aggregation with corresponding receptors.

At the present time, the biological significance of our results is difficult to assess. The biological consequences of negative regulation of cell activation by SIRP-α will depend on whether SIRP-α expressed by mast cells are or are not constitutively associated with $Fc_{\varepsilon}RI$ as they are with RTKs. They will depend also on the nature and on the distribution of SIRP-α ligands, when they are known. One SIRP member, the neuronal adhesion molecule P84, was recently assigned an extracellular ligand. The integrin-associated protein CD47 was indeed found to bind to P84, and both molecules are co-localized in synapse-rich structures of the cerebellum and the retina, where their interactions have been proposed to control synaptic functions (50). Whether CD47 is also a ligand for other SIRP family members and/or whether other ligands bind to the polymorphic extracellular domains of SIRP-α is unknown. CD47 is widely expressed on T cells (51), on myeloid cells including neutrophils (52) and mast cells (53), on epithelial cells (52), on spleen, liver and bone marrow stromal cells (54), on neurons (24, 50) and on red blood cells (55). Because most RTKs utilize similar transduction pathways, one expects negative regulation of cell proliferation by $Fc_{\gamma}RIIB$ to affect the growth and development of a variety of cells that co-express $Fc_{\gamma}RIIB$ and RTKs, within and outside the immune system. Besides c-kit, these include platelet-derived growth factor receptors, colony-stimulating factor receptors, epithelial growth factor receptors, fibroblast growth factor receptors, nerve growth factor receptors, vascular endothelial growth factor receptors, insulin-like growth factor receptors, and insulin receptors.

Whatever the physiological relevance of their regulatory properties, $Fc_{\gamma}RIIB$ and SIRP-α can be envisioned as the potential targets for new immunotherapeutic approaches of clinical disorders associated with exaggerated mast cell responses and/or proliferation. These include allergic diseases, mastocytosis and malignant mast cell proliferations such as mastocytomas and mast cell leukaemias.

ACKNOWLEDGEMENTS

This work was supported by the Institut National de la Santé et de la Recherche Médicale (INSERM), the Institut Curie and the Association pour la Recherche sur le Cancer (ARC). H. L. is the recipient of a Rhône-Poulenc Rorer CIFRE contract, and P. B. is the recipient of a fellowship from the Ministère de l'Enseignement Supérieur et de la Recherche.

REFERENCES

1. D'Ambrosio, D., Hippen, K. H., Minskoff, S. A., Mellman, I., Pani, G., Siminovitch, K. A. and Cambier, J. C. Recruitment and activation of PTP1C in negative regulation of antigen receptor signaling by FcγRIIB1. *Science* **268**:293–296, 1995.
2. Daëron, M., Latour, S., Malbec, O., Espinosa, E., Pina, P., Pasmans, S. and Fridman, W. H. The same tyrosine-based inhibition motif, in the intracytoplasmic domain of FcγRIIB, regulates negatively BCR-, TCR- and FcR-dependent cell activation. *Immunity* **3**:635–646, 1995.
3. Vivier, E. and Daëron, M. Immunoreceptor tyrosine-based inhibition motifs. *Immunol. Today* **18**:286–291, 1997.
4. Burshtyn, D. N. and Long, E. O. Regulation through inhibitory receptors: lessons from killer cells. *Trends Cell Biol.* **7**:473–479, 1997.
5. Cambier, J. C. Antigen and Fc Receptor signaling: the awesome power of the immunoreceptor tyrosine based activation motif (ITAM). *J. Immunol.* **155**:3281–3285, 1995.

6. Daëron, M. ITIM-bearing negative coreceptors. *Immunologist* **5**:79–86, 1997.
7. Daëron, M. Fc receptor biology. *Annu. Rev. Immunol.* **15**:203–234, 1997.
8. Benhamou, M., Bonnerot, C., Fridman, W. H. and Daëron, M. Molecular heterogeneity of murine mast cell Fcγ receptors. *J. Immunol.* **144**:3071–3077, 1990.
9. Daëron, M., Malbec, O., Latour, S., Arock, M. and Fridman, W. H. Regulation of high-affinity IgE receptor-mediated mast cell activation by murine low-affinity IgG receptors. *J. Clin. Invest.* **95**:577–585, 1995.
10. Malbec, O., Fong, D., Turner, M., Tybulewicz, V. L. J., Cambier, J. C., Fridman, W. H. and Daëron, M. FcεRI-associated lyn-dependent phosphorylation of FcγRIIB during negative regulation of mast cell activation. *J. Immunol.* **160**:1647–1658, 1998.
11. Fong, D. C., Malbec, O., Arock, M., Cambier, J. C., Fridman, W. H. and Daëron, M. Selective *in vivo* recruitment of the phosphatidylinositol phosphatase SHIP by phosphorylated FcγRIIB during negative regulation of IgE-dependent mouse mast cell activation. *Immunol. Lett.* **54**:83–91, 1996.
12. Ono, M., Bolland, S., Tempst, P. and Ravetch, J. V. Role of the inositol phosphatase SHIP in negative regulation of the immune system by the receptor FcγRIIB. *Nature* **383**:263–266, 1996.
13. Damen, J. E., Liu, L., Rosten, P., Humphries, R. K., Jefferson, A. B., Majerus, P. W. and Krystal, G. The 145-kDa protein induced to associate with Shc by multiple cytokines is an inositol tetraphosphate and phosphatidylinositol 3,4,5-trisphosphate 5-phosphatase. *Proc. Natl. Acad. Sci. USA* **93**:1689–1693, 1996.
14. Bolland, S., Pearse, R. N., Kurosaki, T. and Ravetch, J. V. SHIP modulates immune receptor responses by regulating membrane association of Btk. *Immunity* **8**:509–516, 1998.
15. Scharenberg, A. M., El-Hillal, O., Fruman, D. A., Beitz, L. O., Li, Z., Lin, S., Gout, I., Cantley, L. C., Rawlings, D. J. and Kinet, J.-P. Phosphatidylinositol-3,4,5-triphosphate (PtdIns-3,4,5-P3)/Tec kinase-dependent calcium signaling pathway: a target for SHIP-mediated inhibitory signals. *EMBO J.* **17**:1961–1972, 1998.
16. Moretta, A., Bottino, C., Vitale, M., Pende, D., Biassoni, R., Mingari, M.-C. and Moretta, L. Receptors for HLA class I molecules in human natural killer cells. *Annu. Rev. Immunol.* **14**:619–648, 1996.
17. Bléry, M., Delon, J., Trautmann, A., Cambiaggi, A., Olcese, L., Biassoni, R., Moretta, L., Chavrier, P., Moretta, A., Daëron, M. and Vivier, E. Reconstituted killer-cell inhibitory receptors for MHC class I molecules control mast cell activation induced via immunoreceptor tyrosine-based activation motifs. *J. Biol. Chem.* **272**:8989–8996, 1997.
18. Katz, H. R., Vivier, E., Castells, M. C., McCormick, M. J., Chambers, J. M. and Austen, K. F. Mouse mast cell gp49B1 contains two immunoreceptor tyrosine-based inhibition motifs and suppresses mast cell activation upon coligation with FcεRI. *Proc. Natl. Acad. Sci. USA* **93**:10809–10814, 1996.
19. Long, E. O., Burshtyn, D. N., Clark, W. P., Peruzzi, M., Fajagopalan, S., Rojo, S., Wagtmann, N. and Winter, C. C. Killer cell inhibitory receptors: diversity, specificity, and function. *Immunol. Rev.* **155**:135–144, 1997.
20. Rojo, S., Burshtyn, N., Long, E. and Wagtmann, N. Type I transmembrane receptor with inhibitory function in mouse mast cells and NK cells. *J. Immunol.* **158**:9–12, 1997.
21. Binstadt, B. A., Brumbaugh, K. M., Dick, C. J., Scharenberg, A. M., Williams, B. L., Colonna, M., Lanier, L. L., Kinet, J.-P., Abraham, R. T. and Leibson, P. J. Sequential involvement of lck and SHP-1 with MHC-recognizing receptors on NK cells inhibits FcR-initiated tyrosine kinase activation. *Immunity* **5**:629–638, 1996.
22. Bruhns, P., Marchetti, P., Fridman, W. H., Vivier, E. and Daëron, M. Differential roles of N- and C-terminal ITIMs during inhibition of cell activation by killer cell inhibitory receptors. *J. Immunol.* **162**:3168–3175, 1999.
23. Kharitonenkov, A., Chen, Z., Sures, I., Wang, H., Schilling, J. and Ullrich, A. A family of proteins that inhibit signalling through tyrosine kinase receptors. *Nature* **386**:181–186, 1997.
24. Adams, S., van der Laan, L. J. V., Vernon-Wilson, E., Renardel de Lavalette, C., Döpp, E. A., Dijkstra, C. D., Simmons, D. L. and van der Berg, T. K. Signal regulatory protein is selectively expressed by myeloid and neuronal cells. *J. Immunol.* **161**:1853–1859, 1998.
25. Veillette, A., Thibaudeau, E. and Latour, S. High expression of inhibitory receptor SHPS-1 and its association with protein tyrosine phosphatase SHP-1 in macrophages. *J. Biol. Chem.* **273**:22719–22728, 1998.
26. Fujioka, Y., Matozaki, T., Noguchi, T., Iwamatsu, A., Yamao, T., Takahashi, N., Tsuda, M., Takada, T. and Kasuga, M. A novel membrane glycoprotein, SHPS-1, that binds the SH2-domain-containing protein tyrosine phosphatase SHP-2 in response to mitogens and cell adhesion. *Mol. Cell. Biol.* **16**:6887–6899, 1996.

27. Matozaki, T., Uchida, Y., Fujioka, Y. and Kasuga, M. Src kinase tyrosine phosphorylates PTP1C, a protein tyrosine phosphatase containing src homology-2 domains, that regulates cell proliferation. *Biochem. Biophys. Res. Commun.* **204**:874–881, 1994.
27a. Liénard, H., Bruhns, P., Malbec, O., Fridman, W. H. and Daëron, M. Signal regulatory proteins negatively regulate immunoreceptor-dependent cell activation. *J. Biol. Chem.* **274**:32493–32499, 1999.
28. Butterfield, J. H., Weiler, D., Dewald, G. and Gleich, G. J. Establishment of an immature mast cell line from a patient with mast cell leukemia. *Leuk. Res.* **12**:345–355, 1988.
29. Saito, H., Ebisawa, M., Tachimoto, H., Shichijo, M., Fukagawa, K., Matsumoto, K., Iikura, Y., Awaji, T., Tsujimoto, G., Yanagida, M., Uzumaki, H., Takahashi, G., Tsuji, K. and Nakahata, T. Selective growth of human mast cells induced by Steel factor, IL-6, and prostaglandin E2 from cord blood mononuclear cells. *J. Immunol.* **157**:343–350, 1996.
30. Barsumian, E. L., Isersky, C., Petrino, M. G. and Siraganian, R. P. IgE-induced histamine release from rat basophilic leukemia cell lines: isolation of releasing and nonreleasing clones. *Eur. J. Immunol.* **11**:317, 1981.
31. Timms, J. F., Carlberg, K., Gu, H., Chen, H., Kamatkar, S., Nadler, M. J., Rohrschneider, L. R. and Neel, B. G. Identification of major binding proteins and substrates for the SH2-containing protein tyrosine phosphatase SHP-1 in macrophages. *Mol. Cell. Biol.* **18**:3838–3850, 1998.
32. Reth, M. G. Antigen receptor tail clue. *Nature* **338**:383–384, 1989.
33. Blank, U., Ra, C., Miller, L., White, K., Metzger, H. and Kinet, J. P. Complete structure and expression in transfected cells of high affinity IgE receptor. *Nature* **337**:187, 1989.
34. Metzger, H., Alcaraz, G., Hohman, R., Kinet, J. P., Pribluda, V. and Quarto, R. The receptor with high affinity for immunoglobulin E. *Annu. Rev. Immunol.* **4**:419–470, 1986.
35. Gounni, A. S., Lamkhioued, B., Ochial, K., Tanaka, Y., Delaporte, E., Capron, A., Kinet, J.-P. and Capron, M. High-affinity IgE receptor on eosinophils is involved in defence against parasites. *Nature* **367**:183–186, 1994.
36. Maurer, D., Fiegiger, E., Reininger, B., Wolffwiniski, B., Jouvin, M.-H., Kilgus, O., Kinet, J.-P. and Stingl, G. Expression of a functional high affinity immunoglobulin E receptor (FcεRI) on monocytes of atopic individuals. *J. Exp. Med.* **179**:745–750, 1994.
37. Hulett, M. D. and Hogarth, P. M. Molecular basis of Fc receptor function. *Adv. Immunol.* **57**:1–127, 1994.
38. Scholl, D. A. and Geha, R. S. Physical association between the high-affinity IgG receptor and the γ subunit of the high-affinity IgE receptor (FcεRI). *Proc. Natl. Acad. Sci. USA* **90**:8847–8850, 1993.
39. Patry, C., Sibille, Y., Lehuen, A. and Monteiro, R. C. Identification of Fcα receptor (CD89) isoforms generated by alternative splicing which are differentially expressed between blood monocytes and alveolar macrophages. *J. Immunol.* **156**:4442–4448, 1996.
40. Pfefferkorn, L. C. and Yeaman, G. R. Association of IgA-Fc receptors (FcαR) with FcεRIγ2 subunits in U937 cells: aggregation induces the tyrosine phosphorylation of γ2. *J. Immunol.* **153**:3228–3236, 1994.
41. Perussia, B., Tutt, M. M., Qui, W. Q., Kuziel, W. A., Tucker, P. W., Trinchieri, G., Bennett, M., Ravetch, J. V. and Kumar, V. Murine natural killer cells express functional Fcγ receptor II encoded by the FcγRα gene. *J. Exp. Med.* **170**:73–86, 1989.
42. Ra, C., Jouvin, M. H. E., Blank, U. and Kinet, J. P. A macrophage Fcγ receptor and the mast cell receptor for IgE share an identical subunit. *Nature* **341**:752–754, 1989.
43. Weinshank, R. L., Luster, A. D. and Ravetch, J. V. Function and regulation of a murine macrophage-specific IgG Fc receptor, FcγR-α. *J. Exp. Med.* **167**:1909–1925, 1988.
44. Qiu, F., Ray, P., Brown, K., Baker, P. E., Jhanwar, S., Ruddle, F. H. and Besmer, P. Primary structure of c-kit: relationship with the CSF-1/PDGF receptor tyrosine kinase family – oncogenic activation of v-kit involves the deletion of extracellular domain and C terminus. *EMBO J.* **7**:1003–1010, 1988.
45. Salim, K., Bottomley, M. J., Querfurth, E., Zvelebil, M. J., Gout, I., Scaife, R., Margolis, R. L., Gigg, R., Smith, C. I., Driscoll, P. C., Waterfield, M. D. and Panayotou, G. Distinct specificity in the recognition of phosphoinositides by the pleckstrin homology domains of dynamin and Bruton's tyrosine kinase. *EMBO J.* **15**:6241–6250, 1996.
46. Timokhina, I., Kissel, H., Stella, G. and Besmer, P. Kit signaling through PI 3-kinase and Src kinase pathways: an essential role for Rac1 and JNK activation in mast cell proliferation. *EMBO J.* **17**:6250–6262, 1998.
47. Cutler, R. L., Liu, L., Damen, J. E. and Krystal, G. Multiple cytokines induce the tyrosine phosphorylation of Shc and its association with Grb2 in hematopoietic cells. *J. Biol. Chem.* **268**:21463–21465, 1993.
48. Malbec, O., Fridman, W. H. and Daëron, M. Negative regulation of c-kit-mediated cell proliferation by

FcγRIIB. *J. Immunol.* **162**:4424–4429, 1999.
49. Helgason, C. D., Damen, J. E., Rosten, P., Grewal, R., Sorensen, P., Chappel, S. M., Borowski, A., Jirik, F., Krystal, G. and Humphries, R. K. Targeted disruption of SHIP leads to hemopoietic perturbations, lung pathology, and shortened life span. *Genes Dev.* **12**:1610–1620, 1998.
50. Jiang, P., Lagenaur, C. F. and Narayanan, V. Integrin-associated protein is a ligand for the P84 neural adhesion molecule. *J. Biol. Chem.* **274**:559-562, 1999.
51. Reinhold, M. I., Lindberg, F. P., Kersh, G. J., Allen, P. M. and Brown, E. J. Costimulation of T cell activation by integrin-associated protein (CD47) is an adhesion-dependent, CD28-independent signaling. *J. Exp. Med.* **185**:1–11, 1997.
52. Parkos, C. A., Colgan, S. P., Liang, T. W., Nusrat, A., Bacarra, A. E., Carnes, D. K. and Madara, J. L. CD47 mediates post-adhesive events required for neutrophil migration across intestinal epithelia. *J. Cell Biol.* **132**:437–450, 1996.
53. Ghannadan, M., Baghestanian, M., Wimazal, F., Eisenmenger, M. D. L., Kargul, G., Walchshofer, S., Sillaber, C., Lechner, K. and Valent, P. Phenotypic characterization of human skin mast cells by combined staining with toluidine blue and CD antibodies. *J. Invest. Dermatol.* **111**:689–695, 1998.
54. Furuzawa, T., Yanai, N., Hara, T., Miyajima, A. and Obinata, M. Integrin-associated protein (IAP, also termed CD47) is involved in stroma-supported erythropoiesis. *J. Biochem.* **123**:101–106, 1998.
55. Cherif-Zahar, B., Matassi, G., Raynal, V., Gane, P., Delaunay, J., Arrizabalaga, B. and Cartron, J. P. Rh-deficiency of the regulator type caused by splicing mutations in the RH50 gene. *Blood* **92**:2535–2540, 1998.

CHAPTER 13

Perspectives on the Regulation of Secretion from Human Basophils and Mast Cells

*DONALD MACGLASHAN JR,**
SANDRA LAVENS-PHILLIPS and KATSUSHI MIURA

Johns Hopkins Asthma and Allergy Center, Baltimore, Maryland, U.S.A.

INTRODUCTION

As central participants in the allergic reaction, understanding the factors that regulate secretion from basophils and mast cells could suggest new therapeutic modalities for asthma and atopic disease. Studies by a large number of groups have revealed a wide variety of promising avenues for future research and therapeutic development. However, this chapter will focus on three areas that have seen some recent advances through research in this laboratory. The first of these topics will focus on the recently recognized phenomenon of $Fc_{\varepsilon}RI$ expression regulation by IgE antibody. This area of research has an immediate impact on understanding the factors that influence the successfulness of new therapies that suppress circulating IgE levels (e.g. monoclonal anti-IgE antibody). The second section will examine recent experiments which are teasing out the underlying mechanisms of the phenomenon known as cellular desensitization. The final section will examine how interleukin-3 (IL-3) modifies human basophil function.

REGULATION OF $Fc_{\varepsilon}RI$ EXPRESSION

In 1978, Malveaux and Lichtenstein reported that there was a tight correlation between the concentration of circulating IgE antibody and the number of $Fc_{\varepsilon}RI$ receptors on peripheral blood basophils (1). At that time, it was not possible to test whether IgE antibody induced changes in cell surface expression of $Fc_{\varepsilon}RI$ or whether there was a genetic linkage in the expression of the two molecules. There were a variety of studies during the 1980s and early 1990s that suggested that $Fc_{\varepsilon}RI$ expression was regulated by IgE. For example, expression of a variety of other immunoglobulin receptors were

* Corresponding author.

ISBN 0-12-473335-2

known to be regulated by their relevant antibodies (2–4). Studies in RBL cells indicated that, at a minimum, the presence of IgE could regulate the cycling of $Fc_{\varepsilon}RI$ (5). Finally, other studies in basophils indirectly suggested that IgE regulated $Fc_{\varepsilon}RI$ expression (6). It is now well known that IgE regulates $Fc_{\varepsilon}RI$ expression through *in vitro* and *in vivo* studies in both mice and man, on mast cells and basophils (7–13).

This particular piece of cell biology has relevance to therapies which suppress circulating free IgE, such as monoclonal anti-IgE antibody (8, 14, 15). The average atopic donor has circulating IgE levels greater than 400 ng ml^{-1} and basophil (and presumably mast cell) $Fc_{\varepsilon}RI$ densities greater than a quarter of a million per cell. In the absence of any regulation of $Fc_{\varepsilon}RI$ expression by IgE, and based on knowledge of basophil and mast cell sensitivity and the high affinity of IgE for $Fc_{\varepsilon}RI$, one can estimate that 2000–10,000-fold changes in circulating free IgE would have to occur to sufficiently suppress the basophil and mast cell response (8). Factoring in the regulation of $Fc_{\varepsilon}RI$ by IgE, these changes need only be 30–100-fold. In other words, as IgE levels drop, so too does expression of $Fc_{\varepsilon}RI$, and the two processes work synergistically to suppress the basophil or mast cell response (8). On the other hand, the synergism works against the therapist when IgE levels rise only slightly. Under the current protocols for anti-IgE antibody therapy, IgE levels are suppressed such that the basophil response hovers around its threshold for response to antigenic challenge. Slight increases in IgE due to mild reductions in doses of anti-IgE antibody led to modest increases in $Fc_{\varepsilon}RI$ expression that were nearly sufficient to return normal functioning to the circulating basophils (10). Thus it makes sense to develop a better understanding of the mechanism by which IgE antibody induces upregulation of $Fc_{\varepsilon}RI$ in order to develop new adjunct therapeutics which could assist the effects of IgE-suppressing therapies.

Studies in RBL cells indicated that $Fc_{\varepsilon}RI$ cycles through an internal cell compartment to return to the cell surface (5, 16, 17). The rate that the receptor is internalized appears dependent on whether the IgE is bound to the receptor, with internalization being very slow when IgE is bound. In cells incubated in the absence of IgE, approximately two-thirds of the receptor is internal. The addition of IgE then allows a 2–3-fold increased cell surface expression as IgE stabilizes the slow cycling of $Fc_{\varepsilon}RI$. However, in these cells, increases and decreases are limited to 2–3-fold changes. In human basophils and murine mast cells or basophils, the changes can range 100-fold. A hallmark of the RBL cell studies was that the total cellular mass of $Fc_{\varepsilon}RI$ did not change whereas cell surface expression increased or decreased. The addition of cycloheximide had no effect on upregulation under these conditions. In contrast, in human basophils and murine cells, mass increases or decreases along with cell surface expression (9, 11, 12). Furthermore, cycloheximide blocks upregulation in murine cells where it is possible to include this agent without killing the cells, as is true for basophils. These results indicate that these cells synthesize receptor as needed, and remove receptor when IgE is not around. An unresolved question is whether IgE induces synthesis of receptor. For human basophils, we are convinced by recent data that IgE causes a change in receptor expression by interacting with $Fc_{\varepsilon}RI$ itself (18). Although it seemed likely that this would be true, the concentration-dependence of upregulation by IgE was curious. The EC_{50} for upregulation averaged around 250 ng ml^{-1} ($\approx$1 nM), which seemed high considering the equilibrium affinity of IgE for $Fc_{\varepsilon}RI\alpha$. This affinity is not well established in human cells, but by piecing together data from forward and reverse binding experiments, it appears that the K_a for binding should be well below 0.1 nM. However, these data may also provide a clue to what is happening. A variety of experiments indicate that the forward rate constant for

IgE binding to $Fc_{\varepsilon}RI\alpha$ is quite slow. If the unbound receptor is susceptible to some kind of loss (shedding or endocytotic), then upregulation by IgE will be very dependent on the forward binding rate rather than on the equilibrium constant. It is possible to explain the somewhat high EC_{50} for upregulation by the slow forward binding rate of IgE to $Fc_{\varepsilon}RI\alpha$ on human basophils. The rate that receptor is expressed on basophils seems to be quite slow. Even at high concentrations of IgE, basophils only put 400–700 receptors on the surface per hour. This makes pulse chase experiments – to determine whether receptor is constitutively synthesized or whether its synthesis is induced – very difficult. We can, however, determine whether mRNA for $Fc_{\varepsilon}RI\alpha$ changes when the cells are exposed to IgE. An examination of the rate of cell surface upregulation indicates that it is linear in time and can be observed within 24 h of the addition of IgE (9). Thus far, a variety of experimental methodologies to measure mRNA for $Fc_{\varepsilon}RI\alpha$ have not shown any changes in this message during the first 24 h of exposure to IgE. Our current working hypothesis is that basophils constitutively synthesize this receptor and that IgE stabilizes its presence on the cell surface (Fig. 1). Whether or not this turns out to be the case, it is now apparent that the bound state of the receptor is recognized differently than the unbound state, a feature heretofore not recognized for this receptor in wild-type human or murine mast cells or basophils, although clearly suggested by the early RBL cell experiments.

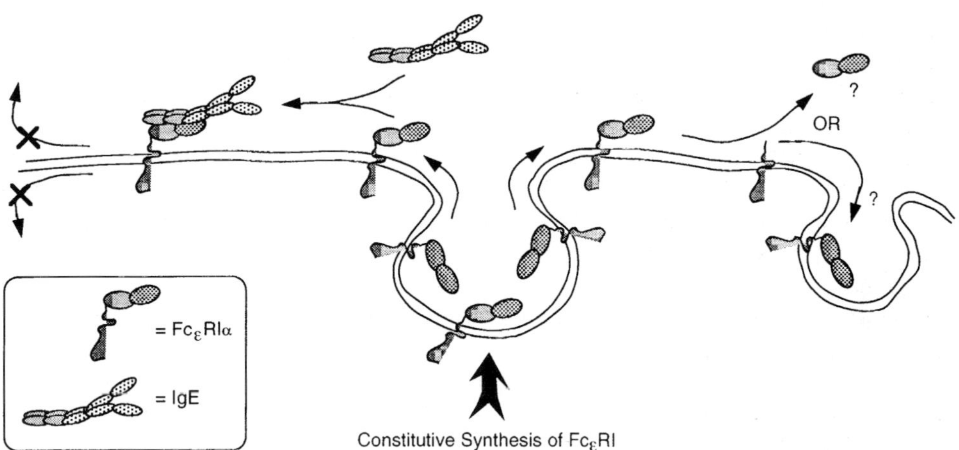

Fig. 1 Cartoon of one working hypothesis for the mechanism of IgE-mediated regulation of $Fc_{\varepsilon}RI$ expression on human basophils.

We have recently been considering an interesting implication of the upregulation process. All of the human cell studies have measured only the expression of $Fc_{\varepsilon}RI\alpha$. However, there is no *a priori* reason that upregulation should apply to all signalling components of the IgE-mediated transduction pathway. Indeed, there are indications that coordinated upregulation of all receptor subunits, let alone other signalling components, is not even necessary in human cells. It is now well recognized that $Fc_{\varepsilon}RI$ can be expressed in human cells with or without $Fc_{\varepsilon}RI\beta$ (19–21). $Fc_{\varepsilon}RI\alpha$ does need $Fc_{\varepsilon}RI\gamma$ for its cell surface expression, but this subunit is shared by other immunoreceptors and may be expressed and regulated at levels that provide little limitation for the expression of $Fc_{\varepsilon}RI\alpha$. $Fc_{\varepsilon}RI\beta$ is upregulated in murine mast cells (11) but $Fc_{\varepsilon}RI$ expression in murine cells requires coordinated expression of $Fc_{\varepsilon}RI\beta$ (22). A similar measurement has not been made using human cells. Thus, it is possible that upregulation of $Fc_{\varepsilon}RI\alpha$ is not accompanied by upregulation of $Fc_{\varepsilon}RI\beta$. Since $Fc_{\varepsilon}RI\beta$ is an important amplifier of the aggregation reaction (23), uncoordinated upregulation of these two components could have important effects on the functional consequences of $Fc_{\varepsilon}RI\alpha$ upregulation. The same reasoning applies to all other elements of the signalling cascade because the quantitative requirements for the expression of every other component remains largely unknown. One possible example of an additional rate-limiting component is lyn kinase (24). In RBL cells, there is evidence to suggest that lyn kinase activity, or its interaction with $Fc_{\varepsilon}RI$, represents a rate-limiting step in the signalling cascade. It is not clear how this occurs since lyn kinase appears to be expressed in quantities sufficient to accommodate all available receptors but the functional evidence suggests otherwise. At a minimum, the RBL cell data indicate that there may be rate-limiting components to the pathways for secretion. From this perspective, it is possible that IgE-induced upregulation of cell surface $Fc_{\varepsilon}RI\alpha$ is not necessarily accompanied by improved function. Recent experimental studies support this possibility (unpublished observations). In coordinated upregulation of the IgE-mediated signalling pathway for histamine release, the sensitivity of the cells should not change. Sensitivity is defined as the number of IgE–$Fc_{\varepsilon}RI$ aggregates required for 50% of the maximum histamine release obtainable during optimal IgE-mediated secretion. With this definition, it is possible to see that coordinated upregulation might result in a better response to a fixed amount of antigen, because the number of IgE–$Fc_{\varepsilon}RI$ complexes has increased, but the same number of complexes would be required for half-maximal release. In contrast, if uncoordinated upregulation occurs – i.e. some components are not upregulated to an extent sufficient to match the upregulation of $Fc_{\varepsilon}RI\alpha$ (think of $Fc_{\varepsilon}RI\beta$ as one example) – the sensitivity of the cell appears to decrease. This is because many formed aggregates are not actually generating the same signal as one would find with coordinated upregulation (again, view this logic from the perspective of the signal amplifier $Fc_{\varepsilon}RI\beta$). We have found that, for some preparations of basophils, uncoordinated upregulation is the norm, whereas for other preparations, coordinated upregulation is observed. This possibility raises some interesting issues about the overall regulation of the IgE-mediated response.

DESENSITIZATION

IgE-mediated histamine release is rarely found to be complete, i.e. the complete secretion of all available granule contents. Indeed, on average, it does not seem possible to induce release of more than 50% of the granules from basophils and mast cells when stimulating through $Fc_{\varepsilon}RI$ (6, 25). Simultaneous activation of these cells with other secretagogues, in addition to activation through $Fc_{\varepsilon}RI$, leads to 100% histamine release, so there is no clear limitation to secreting the entire complement of the cell's granules. Neither is this a subpopulation problem. All cells appear to participate in secretion to varying extents and all cells can respond to a variety of secretagogues (26–28). Instead, it appears that there are self-limiting features of activation. This self-limiting feature to IgE-mediated activation is not unique to this secretagogue. Most receptor-mediated events appear to have self-limited features. For example, some receptors downregulate their cell surface expression quite rapidly after activation with specific ligand (29). This is not how downregulation of the IgE-mediated response occurs (30). For many years, this self-limiting process for IgE-mediated release has been closely associated with the process called desensitization. Desensitization is an operational definition, i.e. when basophils or mast cells are challenged with aggregating antigens in the absence of extracellular calcium (which prevents degranulation), they slowly lose their responsiveness, as tested by the re-introduction of extracellular calcium. The decay in function resembles a first-order decay curve. The hypothesis is that whatever process leads to this loss of responsiveness in this type of experiment is also operational during normal secretion and serves as the self-limiting process for normal secretion. Evidence for this hypothesis comes from pharmacological studies which indicate that some drugs that inhibit desensitization enhance histamine release under normal conditions of stimulation (antigens in the presence of extracellular calcium). There are now two examples of this phenomenon. The first comes from studies using DFP (di-isopropylfluorophosphate) to regulate secretion; at low concentrations, DFP inhibits desensitization and enhances histamine release (see ref. 31 for details of these complex experiments). The second case is only recently recognized, that the lyn kinase inhibitors PP_1 and PP_2 are effective inhibitors of desensitization and, at concentrations somewhat lower than needed to inhibit normal secretion, actually enhance secretion (32). If these pharmacological studies do indeed indicate that desensitization represents the normal downregulatory process occurring in normally secreting cells, then there is a very intriguing correlation to consider. The rate that cells desensitize is inversely correlated with the maximum histamine release that can be obtained by IgE-mediated stimulation (33). In other words, the rate of desensitization varies among cell preparations from different subjects and this rate determines the ability of the cells to secrete. Two cell preparations with equal cell surface antigen-specific IgE densities and even equal early aggregation-dependent signals, might nevertheless secrete quite differently because one desensitizes more rapidly than the other. Therefore understanding the mechanisms underlying desensitization may help understand the variability in histamine release (and mediator release in general) in the population as well as provide newer tools for therapy.

As noted above, desensitization does not appear to result from downregulation of cell surface IgE–$Fc_{\varepsilon}RI$. It appears to be a process that occurs inside the cell. Indeed, it is probably not just one downregulatory process but at least two. It is possible to desensitize a cell specifically or non-specifically (34). Specific desensitization refers to desensitization being limited to the antigen used to desensitize the cell. Non-specific desensitization refers to one antigen inducing a desensitized state that effects other non-cross-reacting antigens. It should be noted that, even under conditions of non-specific desensitization, basophils and mast cells respond normally to other non-IgE-dependent secretagogues (35). Specific desensitization gradually becomes non-specific desensitization as the density of antigen-specific IgE (referring to the antigen used to desensitize the cell) increases (36). For example, at densities below 5000 antigen-specific IgEs per cell, desensitization is largely specific, whereas densities above 25,000 will cause nearly complete non-specific desensitization. These observations suggest the broad outlines of a mechanism. For desensitization to be specific, the cell need only modify in some way the aggregating receptor or closely associated components, whereas for non-specific desensitization the modification occurs to some shared components of the IgE-mediated response. This scheme, put forth 20 years ago (34, 37), may be essentially correct. Recent studies indicate that the two desensitization processes operate either before or after the activation of syk kinase. The current paradigm for IgE-mediated secretion comes from a large variety of elegant studies in RBL cells (with many similarities to studies of other immunoreceptors in other cell types) (24, 38–46). It appears that aggregation of $Fc_εRI$ shifts a steady-state phosphorylation of both the receptor and a closely associated lyn kinase (this shift in steady-state may be the result of a balance of kinases and phosphatases local to the receptor). The shift in phosphorylation is seen as an increase in phosphorylation of $Fc_εRIβ$, $Fc_εRIγ$ and lyn kinase as well as increased association of lyn kinase with $Fc_εRIβ$. Syk kinase is next recruited to the developing complex and, upon phosphorylation (much of which may be self-phosphorylation after binding of its Src-homology 2 (SH2) domain to ITAMs (immunoreceptor tyrosine-based activation motifs) in $Fc_εRIγ$), syk kinase becomes active and serves as a branch point for the phosphorylation of many downstream molecules which also become associated with the complex. In this paradigm, all downstream events lead backward to syk kinase activation. The truth of this assertion has not been adequately tested in human cells. Operating on this assumption, syk kinase is a necessary shared early resource for IgE-mediated activation. Recent studies (32) indicate that specific desensitization results from events that modify the steps preceding syk phosphorylation, i.e. after specific desensitization, syk phosphorylation no longer occurs. In contrast, following non-specific desensitization, syk phosphorylation is largely intact, suggesting that, whatever modification occurs, it occurs downstream of syk phosphorylation. Additional recent studies indicate that, during stimulation induced by anti-IgE antibody (which in human basophils induces strong non-specific desensitization), downregulation occurs between the phosphorylation of Shc (SH2 domain-containing α2 collagen-related protein) and activation of p21ras (if indeed these are sequential events). Similarly, desensitization studies indicate that a modification in the cascade must occur between the phosphorylation of syk and the elevation of cytosolic calcium. The exact location is not yet known. Thus, the state of syk phosphorylation has a relationship to the specific/non-specific desensitization observations; specific desensitization is associated with the loss of syk phosphorylation and therefore a change in events earlier than or coincident with syk activation, whereas non-specific desensitization is associated with no loss of syk phosphorylation but a loss of something downstream of syk (or strictly parallel to syk activation) because there is a loss of the calcium response and activation of p21ras (Fig. 2). Because phosphorylation of syk involves multiple tyrosines, some of which lead to inhibition of activity, it is not yet clear whether specific and non-specific desensitization involve a change in syk itself. For example, phosphorylation of tyrosine-317 might inactivate syk (47), ultimately leading to a decrease in its phosphorylation state (assuming there are local phosphatases). It

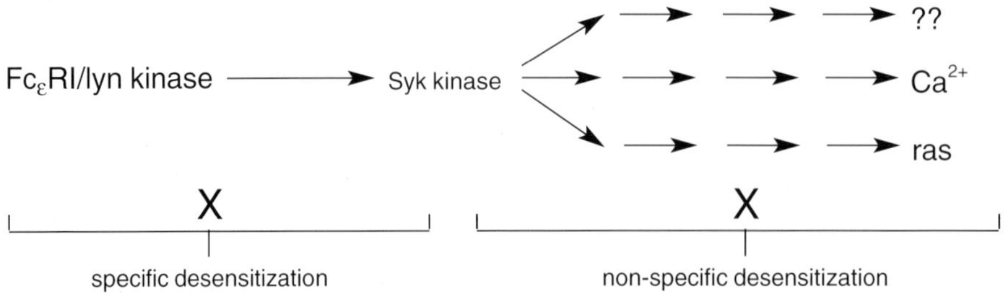

Fig. 2 Localization of the effects of desensitization to events occurring before or after the phosphorylation of syk kinase, depending on whether specific (before) or non-specific (after) desensitization has occurred.

should also be noted that we expect that specific desensitization is always occurring, but when non-specific desensitization is strong, it has the most apparent effect on function. Therefore, there should always be some loss of syk phosphorylation during non-specific desensitization, but it may account for only a small percentage of the total effect on events such as the calcium response, ras activation or mediator release.

MODIFICATION OF BASOPHIL FUNCTION BY IL-3

It is now well recognized that leukocyte function is strongly influenced by a variety of cytokines. For human basophils, the cytokine that appears to have the most significant influence is IL-3 (48–50). These cells also respond to a variety of other cytokines, causing both down- and upregulation of function; examples include interferon-γ, IL-1β, GM-CSF (granulocyte–macrophage colony-stimulating factor), IL-5, NGF (nerve growth factor), and stem cell factor (SCF) (48, 49, 51–55). However, there are marked differences in potency, efficacy and qualitative changes when compared to IL-3. Treatment of basophils with IL-3 causes a variety of functional changes that occur in different time frames. With respect to changes in secretion, we and others have observed three time domains. The first occurs rapidly, within 1–3 min of exposure to IL-3 (EC_{50} of 20 pM), and secretion of all three classes of mediators (histamine, the leukotriene LTC_4 and IL-4) is enhanced. However, enhancement is relatively modest and for LTC_4 release fades by 2 h (56). The second time domain occurs within the next 24 h. Once again, secretion of all classes of mediators is enhanced except, in this instance, enhancement is more marked, sometimes five times greater than the enhancement that occurs after several minutes. The final stage of known IL-3 effects occurs on a scale of days. This is specifically apparent for basophils of the non-releasing phenotype (i.e. basophils with IgE but without IgE-mediated secretion) (25). Treatment with IL-3 for 24 h has little or no effect on IgE-mediated release from this type of basophil, but 4 days of treatment leads to nearly normal IgE-mediated release (57).

An additional relevant feature of IL-3 effects is that secretion induced by all known secretagogues is affected, as is the secretion of all known mediators. Thus, the alterations in cell physiology induced by IL-3 appear to apply to global characteristics of secretion. This does not exclude the possibility of many specific changes to signalling cascades that are specific to various receptors, but an examination of shared events in secretion represents a starting place for understanding the mechanisms underlying the effects of IL-3. There are no indications that IL-3 changes receptor expression for secretagogues and there are no changes in the resting levels of ATP in treated cells (58). To study IL-3, we have chosen a simpler model of secretion, that of LTC_4 release induced by exposure to C5a. This secretagogue causes marked histamine release in the near absence of LTC_4 release (50, 59). Treatment of the cells with IL-3 for 5–15 min or 18–24 h allows C5a to induce LTC_4 release. Since the release before treatment is essentially zero, the change is relatively clear-cut and easier to monitor. With this as a model, we began studying the rapid upregulation by IL-3. The mechanisms underlying LTC_4 release were unknown at the time, so these studies have some ambiguity because the precise pathway to LTC_4 release remains somewhat controversial. The primary controversy for LTC_4 secretion relates to which phospholipase is responsible for generating the free arachidonic acid (AA) that is fed into the LTC_4 synthesis pathway. Although we have previously presented indirect evidence that a secretory phospholipase A_2 ($sPLA_2$; type II or type V, 14-kDa) is involved in the free AA generation (60), more recent evidence strongly supports a role for cytosolic (c) PLA_2 (61, 62). A pathway to the activation of $cPLA_2$ has been worked out from studies in a wide variety of cell types. It involves the activation of p21ras, which initiates the sequential activation of Raf-1, MEK, ERK-1/2 and finally $cPLA_2$ (63, 64). This pathway involves sequential phosphorylations with phosphorylation of $cPLA_2$ as one of its terminal steps. Inhibition of MEK with PD98059 interrupts ERK-1/2 and $cPLA_2$ phosphorylation and specifically inhibits LTC_4 release from basophils in the absence of significant effects on histamine or IL-4 secretion (62). An interesting feature of $cPLA_2$ activation is the requirement for both phosphorylation and an elevation in cytosolic calcium (64, 65). Current understanding suggests that elevated cytosolic calcium leads to the translocation of $cPLA_2$ to relevant cellular membranes, whereas phosphorylation enhances its enzymatic activity. Thus, $cPLA_2$ activity requires two simultaneous events and this simultaneity appears key to understanding why C5a does not induce LTC_4 release and why IL-3 allows its secretion. In the absence of IL-3, C5a induces a transient cytosolic calcium elevation lasting 30–45 sec. This represents the release of internal stores of calcium. An interesting contrast is stimulation with FMLP (formyl-methionyl-leucyl-phenylalanine). Like C5a, activation by FMLP is sensitive to pertussis toxin (66), indicating the common use of a GTP-binding protein, and activation is equivalently rapid. Unlike C5a, FMLP induces a two-phase calcium elevation, the initial release of internal stores (which for FMLP and C5a are of equal magnitude) as well as a strong second influx phase (59). Both secretagogues induce the phosphorylation of $cPLA_2$ (dependent on the activation of p21ras (unpublished results) and the remaining sequence of events noted above) but $cPLA_2$ phosphorylation requires greater than 30–60 sec (61). For C5a, this produces an asynchronous activation of $cPLA_2$; while cytosolic calcium is elevated, there is little or no $cPLA_2$ phosphorylation and, by the time phosphorylation occurs, cytosolic calcium has returned to resting levels. The consequence is that both $cPLA_2$ signals are not present in the same time frame and there is, therefore, little free AA generation. For FMLP, the second phase of the calcium response allows $cPLA_2$ activation and strong free AA generation and LTC_4 release.

IL-3 modifies the simultaneity of these events. IL-3 itself causes phosphorylation of $cPLA_2$ through the same pathway – i.e. activation of p21ras, etc. Like C5a and FMLP, phosphorylation of $cPLA_2$ induced by IL-3 requires greater than 1 min of exposure. In a typical C5a experiment, challenge with C5a follows exposure of the cells to IL-3 for 10–15 min. Thus, $cPLA_2$ is maximally phosphorylated during the initial rise in cytosolic calcium that follows addition of C5a. Precisely mirroring the rise and fall of the cytosolic calcium is LTC_4 release. In other words, LTC_4 stops when the cytosolic calcium signal has returned to resting within 30–45 sec. The priming by IL-3 appears to be the sole result of pre-conditioning the phosphorylation of $cPLA_2$ (61). There are no statistically significant effects of IL-3 on the C5a-induced calcium response. As expected, FMLP-induced free AA generation is also accelerated to include its generation during the initial rise in calcium (58). Thus, in large part, the fast priming effect of IL-3 on LTC_4 release is explained by its induced phosphorylation of $cPLA_2$ (Fig. 3).

The 18–24 h effects of IL-3 appear to require an alternative explanation. A closer examination of the kinetics of IL-3 priming reveal that C5a-induced LTC_4 release subsides after 2 h of IL-3 priming (56). In other words, there is little difference (i.e. no LTC_4 release) between cells treated with IL-3 for 2 h and cells not treated with IL-3 and stimulated with C5a. Phosphorylation of $cPLA_2$ has also largely returned to pre-IL-3 treatment levels. As noted above, by 24 h of IL-3 treatment, C5a again becomes a strong

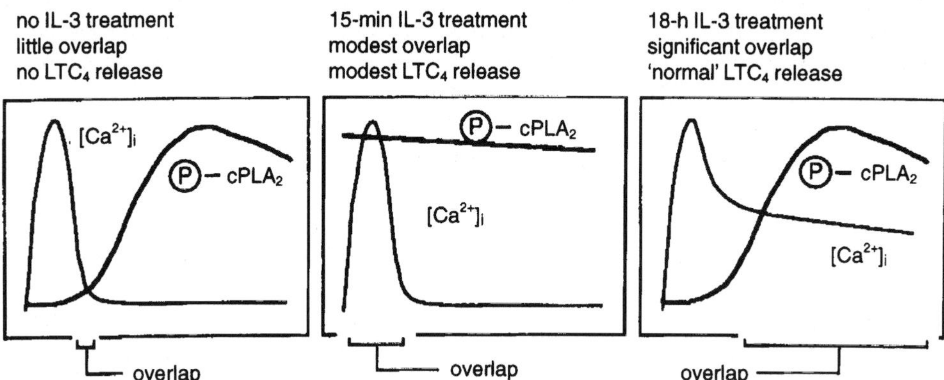

Fig. 3 Idealized kinetics of C5a-induced phosphorylation of cytosolic PLA_2 ($cPLA_2$) and the elevation of cytosolic calcium under three cytokine environments: none, 15-min pre-treatment and 18-h pre-treatment (vertical axis is either cytosolic calcium concentration or extent of $cPLA_2$ phosphorylation). The extent of overlap in the calcium elevation and $cPLA_2$ phosphorylation appears to determine the extent of free AA generation and LTC_4 release.

stimulus for LTC_4 release. Indeed, LTC_4 release is often even greater than following a short priming. An examination of $cPLA_2$ phosphorylation shows that there is no priming of this enzyme. In fact, stimulation of IL-3-treated cells with C5a looks like the untreated cells before IL-3 treatment; i.e. there is a time lag of 30–60 sec before phosphorylation of $cPLA_2$ occurs. The difference between control and IL-3-treated cells is the kinetics of the calcium response. In IL-3-treated cells there is both the first and second phase of the calcium response; i.e. the C5a-induced calcium response in IL-3-treated cells looks like the FMLP-induced calcium response in the absence of IL-3. Consequently, the overlap in calcium elevation and $cPLA_2$ phosphorylation is prolonged, apparently resulting in LTC_4 secretion that is even more similar to FMLP than in cells primed with IL-3 for only a few minutes. If EGTA is added simultaneously with FMLP (in cells not treated with IL-3), histamine release continues because it occurs within the first 30 sec and LTC_4 release is largely eliminated ($cPLA_2$ phosphorylation continues but the second phase of the calcium response is eliminated and the initial rise is intact). The same result occurs in cells treated with IL-3 for 24 h and stimulated with C5a (±EGTA). Treatment of the cells with cycloheximide and IL-3 results in responses similar to cells not treated with IL-3: the second phase calcium response (following stimulation with C5a) is nearly eliminated (the first-phase calcium is largely intact) and LTC_4 release is markedly inhibited. FMLP is not affected this way since it generates a second-phase calcium response independent of IL-3 (however, IL-3 does enhance this calcium response as well so cycloheximide also inhibits the enhanced calcium response even for FMLP). Thus, the IL-3 effect at 24 h is largely due to changes in the character of the calcium response (Fig. 3). Furthermore, these changes appear to be dependent on protein synthesis. We do not currently have an explanation for the changes in the cytosolic calcium response.

As noted above, these studies of LTC_4 release do not represent the entirety of the IL-3 effect. IL-3 affects a wide variety of basophil functions. The timing of the phases of the IL-3 effects on LTC_4 release are consistent with one being post-translational and the other being dependent on protein synthesis. We suspect that a similar dichotomy exists for other functional endpoints. Furthermore, while some changes induced by IL-3 may have global consequences, others may be specific for some *signalling* pathways. For example, there are now indications that lyn kinase may participate in the early signalling by the IL-3 receptor β subunit (67) (which is shared with the IL-5Rα subunit). Activation of lyn kinase could have side-effects or direct effects on IgE-mediated release that are distinct from the more global changes discussed above. IL-3 is a secretagogue for basophils from some donors (68). In these donors, IL-3 has secretion characteristics similar to IgE-mediated release, suggesting the hypothesis that sufficiently strong lyn kinase activation by the IL-3Rβ may initiate the signalling cascade associated with $Fc_{\varepsilon}RI$. Indeed, recent studies do indicate that IL-3 enhances early IgE-mediated signalling events. How these changes occur is a topic for future studies.

SUMMARY

This perspective has focused on only three of the many regulatory pathways controlling secretion and function in basophils and mast cells. However, there have been recent advances in our understanding of the mechanisms relevant to each of these areas of study. For each of these areas there are, of course, more questions. We have yet to determine the mechanism of $Fc_{\varepsilon}RI$ regulation by IgE or the larger implications of coordinated and uncoordinated upregulation. Although we identified a mechanistic difference for specific and non-specific desensitization, the precise nature of the changes involved in either form of desensitization remains unknown. Finally, while we now have a reasonable explanation for the effects of IL-3 on C5a-induced LTC_4 release following short- and long-term priming, we now need to understand what changes in the cell lead to enhanced cytosolic calcium responses.

REFERENCES

1. Malveaux, F. J., Conroy, M. C., Adkinson, N. F. J. and Lichtenstein, L. M. IgE receptors on human basophils. Relationship to serum IgE concentration. *J. Clin. Invest.* **62**: 176–181, 1978.
2. Lee, W. T. and Conrad, D. H. Murine B cell hybridomas bearing ligand-inducible Fc receptors for IgE. *J. Immunol.* **136**: 4573–4580, 1986.
3. Hoover, R. G., Dieckgraefe, B. K. and Lynch, R. G. T cells with Fc receptors for IgA: induction of Ta cells *in vivo* and *in vitro* by purified IgA. *J. Immunol.* **127**: 1560–1563, 1981.
4. Daeron, M., Neauport-Sautes, C., Yodoi, J., Moncuit, J. and Fridman, W. H. Receptors for immunoglobulin isotypes (FcR) on murine T cells. II. Multiple FcR induction on hybridoma T cell clones. *Eur. J. Immunol.* **15**: 668–674, 1985.
5. Furuichi, K., Rivera, J. and Isersky, C. The receptor for immunoglobulin E on rat basophilic leukemia cells: effect of ligand binding on receptor expression. *Proc. Natl. Acad. Sci. USA* **82**: 1522–1525, 1985.
6. MacGlashan, D. W. Jr. Releasability of human basophils: cellular sensitivity and maximal histamine release are independent variables. *J. Allergy Clin. Immunol.* **91**: 605–615, 1993.
7. Hsu, C. and MacGlashan, D.W. Jr. IgE antibody up-regulates high affinity IgE binding on murine bone marrow derived mast cells. *Immunol. Lett.* **52**: 129–134, 1996.
8. MacGlashan, J. D.W., Bochner, B. S., Adelman, D. C., Jardieu, P. M., Togias, A., Mckenzie-White, J., Sterbinsky, S. A., Hamilton, R. G. and Lichtenstein, L. M. Down-regulation of FcεRI expression on human basophils during *in vivo* treatment of atopic patients with anti-IgE antibody. *J. Immunol.* **158**: 1438–1445, 1997.
9. MacGlashan, D. W. Jr, White-Mckenzie, J., Chichester, K., Bochner, B. S., Davis, F. M., Schroeder, J. T. and Lichtenstein, L. M. *In vitro* regulation of FcεRIa expression on human basophils by IgE antibody. *Blood* **91**: 1633–1643, 1998.
10. Saini, S. S., MacGlashan, D. W. Jr, Sterbinsky, S. A., Togias, A., Adelman, D. C., Lichtenstein, L. M. and

Bochner, B. S. Down-regulation of human basophil IgE and FcεRIα surface densities and mediator release by anti-IgE infusions is reversible *in vitro* and *in vivo*. *J. Immunol.* **162**: 5623–5630, 1999.

11. Yamaguchi, M., Lantz, C. S., Oettgen, H. C., Katona, I. M., Fleming, T., Miyajama, I., Kinet, J. P. and Galli, S. J. IgE enhances mouse mast cell FcεRI expression *in vitro* and *in vivo*: evidence for a novel amplification mechanism in IgE-dependent reactions. *J. Exp. Med.* **185**: 663–672, 1997.
12. Lantz, C. S., Yamaguchi, M., Oettgen, H. C., Katona, I. M., Miyajama, I., Kinet, J. P. and Galli, S. J. IgE regulates mouse basophil FcεRI expression *in vivo*. *J. Immunol.* **158**: 2517–2521, 1997.
13. Yamaguchi, M., Sayama, K., Yano, K., Lantz, C. S., Noben-Trauth, N., Ra, C., Costa, J. J. and Galli, S. J. IgE enhances Fc epsilon receptor I expression and IgE-dependent release of histamine and lipid mediators from human umbilical cord blood-derived mast cells: synergistic effect of IL-4 and IgE on human mast cell Fc epsilon receptor I expression and mediator release. *J. Immunol.* **162**: 5455–5465, 1999.
14. Davis, F. M., Gossett, L. A., Pinkston, K. L., Liou, R. S., Sun, L. K., Kim, Y. W., Chang, N. T., Chang, T. W., Wagner, K., Bews, J., Brinkman, V., Towbin, H., Subramanian, N. and Heusser, C. Can anti-IgE be used to treat allergy? *Springer Semin. Immunol.* **15**: 51–73, 1993.
15. Shields, R. L., Whether, W. R., Zioncheck, K., O'Connell, L., Fendly, B., Presta, L. G., Thomas, D., Saban, R. and Jardieu, P. Inhibition of allergic reactions with antibodies to IgE. *Int. Arch. Allergy Immunol.* **107**: 308–312, 1995.
16. Furuichi, K., Rivera, J., Triche, T. and Isersky, C. The fate of IgE bound to rat basophilic leukemia cells. IV. Functional association between the receptors for IgE. *J. Immunol.* **134**: 1766–1773, 1985.
17. Furuichi, K., Rivera, J., Buonocore, L. M. and Isersky, C. Recycling of receptor-bound IgE by rat basophilic leukemia cells. *J. Immunol.* **136**: 1015–1022, 1986.
18. MacGlashan, D. W. Jr, Lichtenstein, L. M., McKenzie-White, J., Chichester, K., Henry, A. J., Sutton, B. J. and Gould, H. J. Upregulation of FcεRI on human basophils by IgE antibody is mediated by interaction of IgE with FcεRI. *J. Allergy Clin. Immunol.* (in press).
19. Miller, L., Blank, U., Metzger, H. and Kinet, J. P. Expression of high-affinity binding of human immunoglobulin E by transfected cells. *Science* **244**: 334–337, 1989.
20. Ra, C., Jouvin, M. H. and Kinet, J. P. Complete structure of the mouse mast cell receptor for IgE (Fc epsilon RI) and surface expression of chimeric receptors (rat–mouse–human) on transfected cells. *J. Biol. Chem.* **264**: 15323–15327, 1989.
21. Kuster, H., Thompson, H. and Kinet, J. P. Characterization and expression of the gene for the human Fc receptor gamma subunit. Definition of a new gene family. *J. Biol. Chem.* **265**: 6448–6452, 1990.
22. Blank, U., Ra, C., Miller, L., White, K., Metzger, H. and Kinet, J. P. Complete structure and expression in transfected cells of high affinity IgE receptor. *Nature* **337**: 187–189, 1989.
23. Lin, S., Cicaia, C., Scharenberg, A. M. and Kinet, J. P. The FcεRIβ subunit functions as an amplifier of FcεRIγ-mediated cell activation signals. *Cell* **85**: 985–995, 1996.
24. Torigoe, C., Goldstein, B., Wofsy, C. and Metzger, H. Shuttling of initiating kinase between discrete aggregates of the high affinity receptor for IgE regulates the cellular response. *Proc. Natl. Acad. Sci. USA* **94**: 1372–1377, 1997.
25. Nguyen, K. L., Gillis, S. and MacGlashan, D. W., Jr. A comparative study of releasing and nonreleasing human basophils: nonreleasing basophils lack an early component of the signal transduction pathway that follows IgE cross-linking. *J. Allergy Clin. Immunol.* **85**: 1020–1029, 1990.
26. MacGlashan, D. W. Jr and Botana, L. Biphasic Ca^{++} responses in human basophils: evidence that the initial transient elevation associated with mobilization of intracellular calcium is an insufficient signal for degranulation. *J. Immunol.* **150**: 980–991, 1993.
27. MacGlashan, D. W. Jr, Bochner, B. and Warner, J. A. Graded changes in the response of individual human basophils to stimulation: distributional behavior of early activation events. *J. Leukoc. Biol.* **55**: 13–23, 1994.
28. MacGlashan, D. W. Jr. Graded changes in the response of individual human basophils to stimulation: Distributional behavior of events temporally coincident with degranulation. *J. Leukoc. Biol.* **58**: 177–188, 1995.
29. Neidel, J. E., Kahone, I. and Cuatracasas, P. Receptor-mediated internalization of fluorescent chemotactic peptide in human neutrophils. *Science* **205**: 1412, 1979.
30. MacGlashan, D. W. Jr, Mogowski, M. and Lichtenstein, L. M. Studies of antigen binding on human basophils. II. Continued expression of antigen-specific IgE during antigen-induced desensitization. *J. Immunol.* **130**: 2337–2342, 1983.
31. Kazimierczak, W., Meier, H. L., MacGlashan, D. W. Jr and Lichtenstein, L. M. An antigen-activated

DFP-inhibitable enzyme controls basophil desensitization. *J. Immunol.* **132**: 399–405, 1984.
32. Lavens-Phillips, S. E. and MacGlashan, D. W., Jr. Pharmacology of human basophil desensitization. *FASEB J.* **13**: A338, 1999.
33. MacGlashan, D. W. Jr, Laven-Phillips, S. and Miura, K. IgE-mediated desensitization in human basophils and mast cells. *Front. Biosci.* webpage: http://www.bioscience.org/1998/v3/d/macglash/list.htm
34. Sobotka, A. K., Dembo, M., Goldstein, B. and Lichtenstein, L. M. Antigen-specific desensitization of human basophils. *J. Immunol.* **122**: 511–517, 1979.
35. Lichtenstein, L. M. and De Bernardo, R. IgE mediated histamine release: *in vitro* separation into two phases. *Int. Arch. Allergy Appl. Immunol.* **41**: 56–71, 1971.
36. MacGlashan, D. W. Jr and Lichtenstein, L. M. The transition from specific to nonspecific desensitization in human basophils. *J. Immunol.* **127**: 2410–2414, 1981.
37. Dembo, M. and Goldstein, B. A model of cell activation and desensitization by surface immunoglobin: the case of histamine release from human basophils. *Cell 59–67,* 1980.
38. Jouvin, M. H., Adamczewski, M., Numerof, R., Letourneur, O., Valle, A. and Kinet, J. P. Differential control of the tyrosine kinase lyn and syk by the two signaling chains of the high affinity immunoglobulin E receptor. *J. Biol. Chem.* **269**: 5918–5925, 1994.
39. Yamashita, T., Mao, S. Y. and Metzger, H. Aggregation of the high-affinity IgE receptor and enhanced activity of p53/56lyn protein-tyrosine kinase. *Proc. Natl. Acad. Sci. USA* **91**: 11251–11255, 1994.
40. Pribluda, V. S., Pribluda, C. and Metzger, H. Biochemical evidence that the phosphorylated tyrosines, serines and threonines on the aggregated high affinity receptor for IgE are in the immunoreceptor tyrosine-based activation motifs. *J. Biol. Chem.* **272**: 11185–11192, 1997.
41. Mao, S. Y. and Metzger, H. Characterization of protein-tyrosine phosphatases that dephosphorylate the high affinity IgE receptor. *J. Biol. Chem.* **272**: 14067–14073, 1997.
42. Benhamou, M., Stephan, V., Gutkind, S. J., Robbins, K. C. and Siraganian, R. P. Protein tyrosine phosphorylation in the degranulation step of RBL-2H3 cells. *FASEB J.* **5**: A1007, 1991.
43. Stephan, V., Benhamou, M., Gutkind, J. S., Robbins, K. C. and Siraganian, R. P. Fc epsilon RI-induced protein tyrosine phosphorylation of pp72 in rat basophilic leukemia cells (RBL-2H3). Evidence for a novel signal transduction pathway unrelated to G protein activation and phosphatidylinositol hydrolysis. *J. Biol. Chem.* **267**: 5434–5441, 1992.
44. Kihara, H. and Siraganian, R. P. Src homology 2 domains of Syk and Lyn bind to tyrosine-phosphorylated subunits of the high affinity IgE receptor. *J. Biol. Chem.* **269**: 22427–22432, 1994.
45. Minoguchi, K., Swaim, W. D., Berenstein, E. H. and Siraganian, R. P. Src family tyrosine kinase p53/56lyn, a serine kinase and Fc epsilon RI associate with alpha-galactosyl derivatives of ganglioside GD1b in rat basophilic leukemia RBL-2H3 cells. *J. Biol. Chem.* **269**: 5249–5254, 1994.
46. Hutchcroft, J. E., Geahlen, R. L., Deanin, G. G. and Oliver, J. M. Fc epsilon RI-mediated tyrosine phosphorylation and activation of the 72-kDa protein-tyrosine kinase, PTK72, in RBL-2H3 rat tumor mast cells. *Proc. Natl. Acad. Sci. USA* **89**: 9107–9111, 1992.
47. Keshvara, L., Isaacson, C. C., Yankee, T. M., Sarac, R., Harrison, M. L. and Geahlen, R. L. Syk- and Lyn-dependent phosphorylation of syk on multiple tyrosines following B cell activation includes a site that negatively regulates signaling. *J. Immunol.* **161**: 5276–5283, 1998.
48. Hirai, K., Morita, Y., Misaki, Y., Ohta, K., Suzuki, S., Motoyoshi, K. and Miyamoto, T. Modulation of human basophil histamine release by hematopoetic growth factors. *J. Immunol.* **141**: 3957–3961, 1988.
49. Schleimer, R. P., Derse, C. P., Friedman, B., Gillis, S., Plaut, M., Lichtenstein, L. M. and MacGlashan, D. W. Jr. Regulation of human basophil mediator release by cytokines. I. Interaction with antiinflammatory steroids. *J. Immunol.* **143**: 1310–1317, 1989.
50. Kurimoto, Y., de Weck, A. L. and Dahinden, C. A. Interleukin 3-dependent mediator release in basophils triggered by C5a. *J. Exp. Med.* **170**: 467–479, 1989.
51. Bischoff, S. C., Brunner, T., De, W. A. L. and Dahinden, C. A. Interleukin 5 modifies histamine release and leukotriene generation by human basophils in response to diverse agonists. *J. Exp. Med.* **172**: 1577–1582, 1990.
52. Bischoff, S. C. and Dahinden, C. A. Effect of nerve growth factor on the release of inflammatory mediators by mature human basophils. *Blood* **79**: 2662–2669, 1992.
53. Hirai, K., Yamaguchi, M., Misaki, Y., Takaishi, T., Ohta, K., Morita, Y., Ito, K. and Miyamoto, T. Enhancement of human basophil histamine release by interleukin 5. *J. Exp. Med.* **172**: 1525–1528, 1990.
54. Massey, W. A., Randall, T. C., Kagey, S. A., Warner, J. A., MacDonald, S. M., Gillis, S., Allison, A. C. and Lichtenstein, L. M. Recombinant human IL-1 alpha and -1 beta potentiate IgE-mediated histamine

release from human basophils. *J. Immunol.* **143**: 1875–1880, 1989.

55. Columbo, M., Horowitz, E. M., Botana, L. M., MacGlashan, D. W. Jr, Bochner, B. S., Gillis, S., Zsebo, K. M., Galli, S. J. and Lichtenstein, L. M. The human recombinant c-kit receptor ligand, rhSCF, induces mediator release from human cutaneous mast cells and enhances IgE-dependent mediator release from both skin mast cells and peripheral blood basophils. *J. Immunol.* **149**: 599–608, 1992.
56. Miura, K. and MacGlashan, D. W. Jr. Dual phase priming by interleukin-3 for leukotriene C4 generation in human basophils. *FASEB J.* **13**: A326, 1999.
57. Yamaguchi, M., Hirai, K., Ohta, K., Suzuki, K., Kitani, S., Takaishi, T., Ito, K., Ra, C. and Morita, Y. Culturing in the presence of IL-3 converts anti-IgE nonresponding basophils into responding basophils. *J. Allergy Clin. Immunol.* **97**: 1279–1287, 1996.
58. MacGlashan, D. W. Jr and Hubbard, W. C. Interleukin-3 alters free arachidonic acid generation in C5a-stimulated human basophils. *J. Immunol.* **151**: 6358–6369, 1993.
59. MacGlashan, D. W. Jr and Warner, J. A. Stimulus-dependent leukotriene release from human basophils: a comparative study of C5a and Fmet-leu-phe. *J. Leukoc. Biol.* **49**: 29–40, 1991.
60. Hundley, T. R., Marshall, L., Hubbard, W. C. and MacGlashan, D. W. Jr. Characteristics of arachidonic acid generation in human basophils: relationship between the effects of inhibitors of secretory phospholipase A2 activity and leukotriene C4 release. *J. Pharmacol. Exp. Ther.* **284**: 847–857, 1998.
61. Miura, K., Hubbard, W. C. and MacGlashan, D. W. Jr. Phosphorylation of cytosolic PLA2 (cPLA2) by interleukin-3 (IL-3) is associated with increased free arachidonic acid and LTC4 release in human basophils. *J. Allergy Clin. Immunol.* **102**: 512–520, 1998.
62. Miura, K., Schroeder, J. T., Hubbard, W. C. and MacGlashan, D. W. Jr. Extracellular signal-regulated kinases regulate leukotriene C4 generation, but not histamine release or IL-4 production from human basophils. *J. Immunol.* **162**: 4198–4206, 1999.
63. Seger, R. and Krebs, E. G. The MAPK signaling cascade. *FASEB J.* **9**: 726, 1995.
64. Lih-Ling, L., Wartmann, M., Lin, A. Y., Knopf, J. L., Seth, A. and Davis, R. J. cPLA2 is phosphorylated and activated by MAP kinase. *Cell* **72**: 269–278, 1993.
65. Wijkander, J. and Dundler, R. Macrophage arachidonate-mobilizing phospholipase A2: role Ca^{2+} for membrane binding but not for catalytic activity. *Biochem. Biophys. Res. Commun.* **184**: 118–124, 1992.
66. Warner, J. A., Yancey, K. B. and MacGlashan, D. W. Jr. The effect of pertussis toxin on mediator release from human basophils. *J. Immunol.* **139**: 161–165, 1987.
67. Alam, R., Pazdrak, K., Stafford, S. and Forsythe, P. The interleukin-5/receptor interaction activates Lyn and Jak2 tyrosine kinases and propagates signals via the Ras-Raf-1–MAP kinase and the Jak–STAT pathways in eosinophils. *Int. Arch. Allergy Appl. Immunol.* **107**: 226–227, 1995.
68. MacDonald, S. M., Schleimer, R. P., Kagey, S. A., Gillis, S. and Lichtenstein, L. M. Recombinant IL-3 induces histamine release from human basophils. *J. Immunol.* **142**: 3527–3532, 1989.

CHAPTER 14

Interactions between Secretory IgA and Human Basophils

KOICHI HIRAI,*[1] MOTOYASU IIKURA[2], MISATO MIYAMASU[2] and MASAO YAMAGUCHI[2]

[1]*Department of Bioregulatory Function and*
[2]*Department of Allergy and Rheumatology, University of Tokyo Graduate School of Medicine, Tokyo, Japan*

INTRODUCTION

Basophilic leukocytes were first identified by Paul Ehrlich based on their metachromatic staining properties with aniline dye. The granules of basophils contain vasoactive amines such as histamine, and cross-linking of IgE bound to high-affinity receptors for IgE ($Fc_{\varepsilon}RI$) with specific antigens leads to the release of these mediators. These characteristics of basophilic leukocytes (i.e. metachromasia, $Fc_{\varepsilon}RI$ and histamine content) are also shared by tissue mast cells. Hence, it had long been believed that basophils represent a precursor form of tissue mast cells, and that migrated basophils convert their phenotype to mast cells in extravascular tissues. This hypothesis was indeed attractive and seemed plausible, but it has now been clearly demonstrated that these cell types belong to completely different lineages. Growth of both cell types is controlled by different factors. The distinguishing features include their *in vivo* location, surface structures, mediator contents and responses to various secretagogues. On the other hand, increasing evidence has demonstrated that eosinophils are the most closely related cells to basophils. Although eosinophils lack the metachromatic staining property and histamine, basophils and eosinophils share a number of surface structures, such as integrins and cytokine receptors (1). Even expression of $Fc_{\varepsilon}RI$ is seen in eosinophils under certain circumstances (2). Furthermore, the proliferation, differentiation and activation of both types of cells are regulated by the same growth factors, i.e. interleukins IL-3 and IL-5 and granulocyte–macrophage colony-stimulating factor (GM-CSF). Moreover, CC chemokine receptor 3 (CCR3) is fully responsible for the migration of both cell types (3, 4). It has been well established that accumulation of eosinophils is prominent at the inflammatory sites of allergic diseases, and that this cell type plays the role of an effector in the pathogenesis of these disorders by virtue of the release of diverse

* Corresponding author.

MAST CELLS AND BASOPHILS
ISBN 0-12-473335-2

toxic mediators. The same situation has also been observed for human basophils. Like eosinophils, basophilic leukocytes are clearly involved in allergic disorders, especially in the late-phase allergic reactions. We and others have demonstrated the presence of basophils in the inflamed tissues of allergic diseases such as bronchial asthma (5–7). In addition, studies on experimentally induced allergic reactions have revealed the influx of basophils as well as eosinophils to the sites of inflammation several hours after antigen exposure (8). Finally, analyses of chemical mediators at the sites of late-phase allergic reactions indicate that rather than mast cells, basophils are active participants in these reactions (9, 10).

Although the pivotal role of IgE in the activation of basophils has been well documented, secretory IgA (sIgA) is the most abundant immunoglobulin isotype at the inflammatory sites of allergic disorders, such as the upper and lower airways, and the gastrointestinal tract. This review will discuss the stimulating effects of sIgA on human basophils. We first review the structure and metabolism of sIgA and then sIgA-mediated basophil activation. The underlying mechanisms responsible for sIgA-mediated basophil activation are also discussed. Given the close relation between basophilic and eosinophilic leukocytes, we will include comparisons of both cells whenever possible.

STRUCTURE AND METABOLISM OF sIgA

sIgA is a 390-kDa immunoglobulin protein, consisting of dimeric IgA molecules (two heavy chains, αα, and two light chains, κκ or λλ) connected by disulphide bonds and associated with a joining (J) chain and a secretory component (SC) (11,12). The human IgA heavy chain α is composed of one variable domain and three constant domains. The J chain interacts with the Cα3 domain through covalent bonds, and the SC forms disulphide bonds with the Cα2 domain (Fig. 1). To date, two IgA subclasses (IgA1 and IgA2) have been identified; 80–90% of serum IgA consists of the IgA1 subclass, whereas

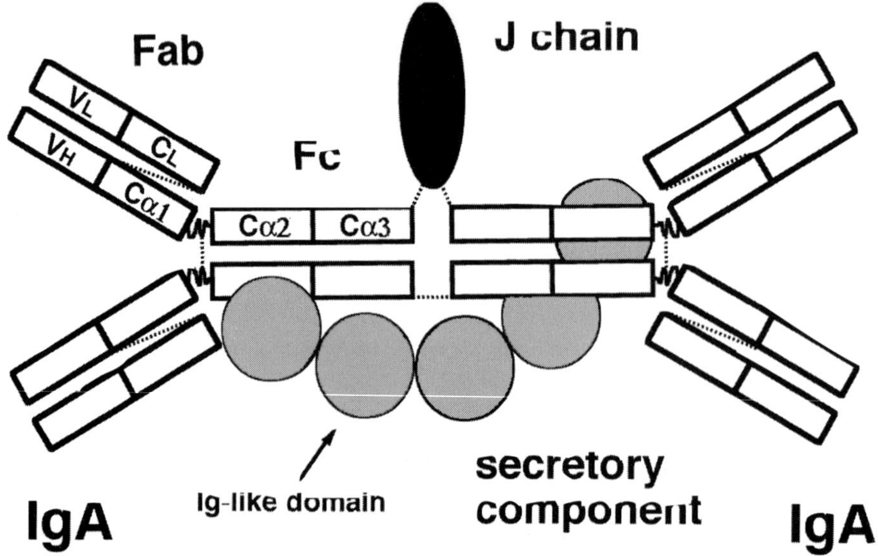

Fig. 1 Structure of sIgA.

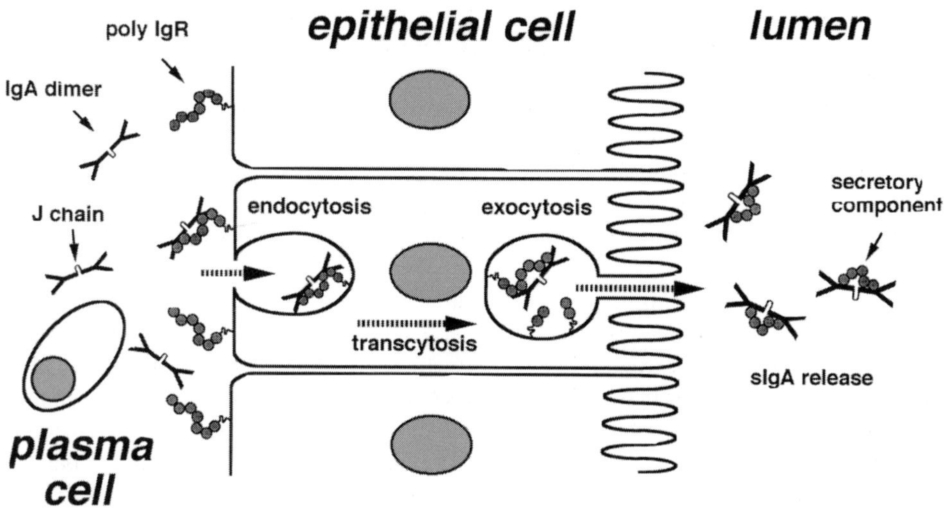

Fig. 2 The metabolism of sIgA. IgA dimers including a single J chain are produced by plasma cells in the lamina propria and bind to the poly (p) IgR, polypeptides containing seven Ig-like domains, on the basolateral surface of epithelial cells. The IgA–pIgR complexes are taken up by receptor-mediated endocytosis, undergo transcytosis in cytoplasmic vesicles and are finally secreted into the lumen on the opposite side. In the lumen, pIgR is cleaved by proteases, and the fragment attached to the Fc region of the dimeric IgA is called the 'secretory component'.

26–50% of IgA in external secretions consists of the IgA2 subclass (13). The most remarkable structural difference between both subclasses exists in the hinge region of the α heavy chain. The hinge region of the α chain of IgA1 possesses specific sites which can be cleaved by several IgA proteases, derived from microorganisms such as *Haemophilus influenzae*, *Streptococcus pneumoniae* and *Neisseria meningitidis* (12). On the other hand, a deletion of the corresponding segment of the α chain is observed in IgA2, making IgA2 resistant to cleavage by these bacterial IgA proteases. Although no functional differences, such as in the binding capacity, have been identified between the IgA1 and IgA2 subclasses, IgA2's resistance to bacterial IgA proteases may provide an advantage to sIgA for preventing against bacterial infections of mucosal tissues.

sIgA is produced by plasma cells in the lamina propria of mucosal tissues. Metabolism of sIgA is illustrated in Fig. 2. Antigen contacting the mucosal membrane penetrates into the deeper layers and stimulates differentiation of B lymphocytes into IgA-producing plasma cells. IgA-producing plasma cells in the lamina propria synthesize dimeric IgA containing a J chain, which regulates the stability of dimeric IgA. The J chain has also been shown to be essential for stable association of the IgA dimer with the SC (12). After binding to a polymeric Ig receptor (pIgR) expressed on the basolateral surface of epithelial cells, the dimeric IgA–pIgR complex is transported across the epithelium and is finally released from the apical membrane into the epithelial lining fluid. pIgR is indispensable for selective transport of IgA into external secretions (11). pIgR consists of seven Ig-like domains, and cleavage of pIgR by proteases is important in liberating sIgA from epithelial cells. The extracellular domain of pIgR is the SC itself (12). The SC protects dimeric IgA from proteolytic degradation in the mucosal environment (14). Free SC is also produced by epithelial cells, but unbound form of SC is highly susceptible to cleavage by proteolytic enzymes (15).

PATHOPHYSIOLOGY OF sIgA

The mucosal surface of the airways and digestive tract represents the primary barrier against invasion by microorganisms. In the external secretions of mucosal tissues, IgA is the most abundant immunoglobulin, and the majority of this isotype consists of sIgA (16). Because of their quantitative predominance in the mucosal tissues, both IgA and sIgA contribute to host defences against invading microorganisms. IgA and potentially sIgA play protective roles against microorganisms invading via the mucosal surfaces by virtue of toxin neutralization and prevention of bacterial adherence to the mucosal surfaces (17, 18). In fact, individuals with IgA deficiency are apt to experience recurrent or unusual types of infections (18). *In vitro* studies demonstrate that neutrophils and macrophages can remove IgA-containing immune complexes. Both types of cells also show antibody-dependent cell-mediated cytotoxicity (ADCC) and phagocytosis via $Fc_{\alpha}R$. A recent report revealed that sIgA inhibits human immunodeficiency virus (HIV) infection *in vitro* through its capacity of either neutralization or ADCC (17).

The mucosal surface is also the site of inflammation associated with allergic diseases, such as bronchial asthma, allergic rhinitis and food allergy. In addition to their well-established protective roles against bacterial and viral infections, several *in vivo* as well as *in vitro* studies have indicated that IgA and sIgA contribute to the pathogenesis of allergic inflammation. In asthmatics, the levels of IgA in the sputum and bronchoalveolar lavage fluid are significantly higher than those in normal controls (16, 19). The level of mite-specific IgA in sputum is also significantly higher in mite-sensitive asthmatics than in mite-insensitive asthmatics (20). In ragweed-specific allergic rhinitics, the levels of ragweed-specific IgA as well as the eosinophil count and eosinophil-derived neurotoxin (EDN) level increased in the nasal lavage fluid during the pollen season (21). The eosinophil cationic protein (ECP) level, eosinophil count and sIgA level in the nasal lavage of allergic rhinitis patients increased in the late phase of antigen challenge (22). Finally, studies of asthmatics and allergic rhinitics demonstrated that the increases in the levels of ECP and EDN correlated with elevation of the IgA concentration (19, 23), suggesting a possible link between sIgA and cellular activation of eosinophils *in vivo*. On the other hand, Burrows and Cooper reported inconclusive results (18). In IgA deficiency patients, allergic disease (i.e. allergic rhinitis, asthma and eczema) occurs at higher incidence than in normal individuals (18, 24). The reason for increased allergy in these patients remained unclear.

In vitro studies have demonstrated that both sIgA and IgA have potent ability to cause degranulation of eosinophils (25, 26). Eosinophil-directed haematopoietins (i.e. IL-5, IL-3 and GM-CSF) enhance both IgA- and sIgA-induced eosinophil degranulation, whereas interferon-γ (IFN-γ) downregulates this degranulation (26). Immobilized sIgA and IgA both induce the production of several cytokines in eosinophils (27). Furthermore, pre-coating of polycarbonate filters with sIgA resulted in a significant increase in the migratory response to IL-8 of eosinophils from atopic donors (28).

sIgA-MEDIATED BASOPHIL ACTIVATION

The role of IgE as reaginic antibodies on human basophils is well established. Several reports have demonstrated that IgG4 antibodies also induce basophil degranulation. IgA antibodies have been shown to lack the ability to release histamine from basophils. On the other hand, there have been no reports of a possible role for sIgA in basophil activation. By using sIgA-conjugated Sepharose beads, we recently demonstrated that sIgA is capable of activating basophils (29). sIgA-mediated basophil activation is completely dependent on priming with a basophil-directed haematopoietin, such as IL-3 (29).

IL-3 has been demonstrated to be a growth and differentiation factor for murine mast cells. In humans, however, a substantial body of evidence indicates that IL-3 is not involved in the proliferation of mast cells, but rather is responsible for the growth and differentiation of basophils. In general, growth factors have dual properties: they act not only on progenitors to stimulate their proliferation/differentiation but also prime mature cells to increase their biological functions. The same is true for human basophils (30): the results from our and other laboratories have clearly indicated that IL-3 is a potent activator of peripheral basophils. In addition to such activities as induction of migration (31), adherence (32) and prolongation of survival (33), IL-3 is capable of priming basophils for increased releasability (34, 35). Although IL-3 induces direct histamine release from a small and selected subset of atopic patients, it potentiates basophil releasability in almost all normal as well as atopic individuals.

When freshly isolated basophils were incubated with sIgA conjugated to Sepharose 4B beads, no significant degranulation was induced. However, when basophils were pre-treated with IL-3 for short time periods, these cells liberated significant amounts of histamine in response to the immobilized sIgA (29) (Fig. 3). Short-term exposure to IL-3

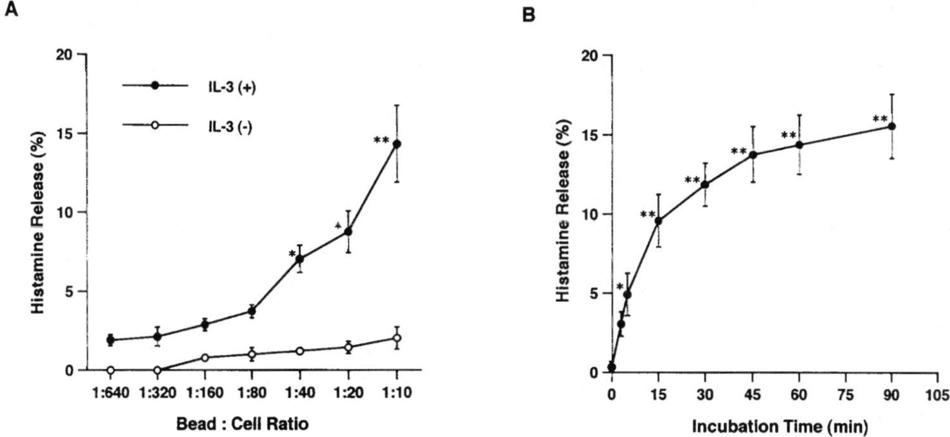

Fig. 3 (A) Induction of histamine release from basophils by sIgA. Semipurified basophils were pre-incubated with (closed circles) or without (open circles) IL-3 at 5 ng ml^{-1} for 30 min at 37°C and then mixed with a buffer containing various amounts of sIgA-coated beads (the relative numbers of beads to basophils are indicated by the bead:cell ratio). After incubation for 45 min at 37°C, the supernatants were collected. Histamine release is expressed as a percentage of the total cellular histamine content after subtracting the spontaneous release in the presence of ovalbumin (OVA)-coated beads instead of sIgA-coated beads. Bars represent the SEM (n=5). *p<0.05, **p<0.01 vs. OVA-induced release from corresponding cells. (B) Time course of sIgA-induced histamine release. IL-3-primed basophils were stimulated with sIgA at a 1:10 bead:cell ratio for various time periods. Bars represent the SEM (n=5). *p<0.05, **p<0.01, vs. percentage of histamine release at 0 min. (Reproduced from *J. Immunol.* **161**:1510–1515, 1998. Reprinted by permission of The Journal of Immunology. Copyright 1998. The American Association of Immunologists.)

is sufficient to induce sIgA-mediated degranulation in basophils: degranulation was maximized by 5 min of pre-treatment with IL-3. The order of addition of IL-3 and sIgA seems critical: prior treatment with IL-3 was essential for sIgA-mediated basophil histamine release (29). These results indicate that IL-3 and sIgA act differently on basophils – i.e. IL-3 as a priming factor and sIgA as a secretagogue. As is observed in C3a (36) and platelet-activating factor (37), stimulation with sIgA is unable to transduce signals sufficient to induce degranulation, and IL-3 renders basophils susceptible to stimulation with sIgA.

In general, the priming effects of IL-3 are considered to be exerted via IL-3R, and the β chain of IL-3 receptors is important in transducing the signals leading to the activation of basophils. The same situation is also observed in sIgA-mediated basophil degranulation. IL-3 exhibited a significant priming effect on sIgA-mediated basophil degranulation even at picomolar concentrations, indicating that this effect was exerted via high-affinity IL-3 receptors. In addition, IL-5 and GM-CSF, whose receptors have the same β chain, also induced sIgA-mediated basophil degranulation (29), indicating that intracellular signalling is mediated via the promiscuous β chain of the receptors.

Besides pre-formed mediators such as histamine, basophils generate two other classes of mediators: *de novo* synthesized lipid mediators such as leukotriene (LT)C_4 and late synthesized proteins such as cytokines. Immobilized sIgA also stimulated basophils to generate newly synthesized lipid mediators. In this case, priming with IL-3 may not be essential in some donors: non-primed basophils from such donors released comparable amounts of LTC_4 in response to sIgA (29). Although basophils are able to synthesize and release multiple cytokines such as IL-4 in response to $Fc_\varepsilon RI$-dependent and -independent activation (38, 39), the ability of sIgA to induce cytokine synthesis remains undefined.

RECEPTORS FOR sIgA

As mentioned above, sIgA consists of dimeric IgA, a J chain and an SC. Although digestion experiments clearly indicated that the $(Fc)_2$–J–SC portion but not the Fab portion of sIgA was essential for sIgA-mediated basophil activation (29) (Fig. 4), it seems unlikely that sIgA introduces activation signals via $Fc_\alpha R$. Although expression of $Fc_\alpha RI$ (CD89) has been identified in many types of leukocytes (40), there has been no evidence for expression of CD89 in human basophils. Flow cytometric analyses by Valent and colleagues revealed that human basophils lack CD89 expression (41). By using monoclonal antibodies (mAbs) against CD89 (donated by Dr H. Kubagawa, University of Alabama), we observed that treatment of basophils with IL-3 did not induce the expression of CD89 (M. Iikura *et al.*, unpublished observation). Furthermore, basophil degranulation by sIgA was not antagonized by an excess of the fluid phase of IgA (29). These notions along with the fact that monomeric IgA in immobilized form is devoid of degranulation-inducing capacity collectively indicate that $Fc_\alpha RI$ is not involved in basophil degranulation elicited by sIgA. Recently, a novel alternatively spliced form of $Fc_\alpha R$ was identified in eosinophils (42, 43). This receptor lacks the entire second extracellular domain of $Fc_\alpha RI$ and binds both sIgA and dimeric IgA, but not monomeric IgA such as serum IgA. It has not been determined whether basophils express this type of $Fc_\alpha R$. However, it also seems very unlikely that this type of $Fc_\alpha R$ is responsible for sIgA-mediated basophil degranulation, since immobilized dimeric IgA also had no effect on histamine release from IL-3-primed basophils (M. Iikura *et al.*, unpublished observation).

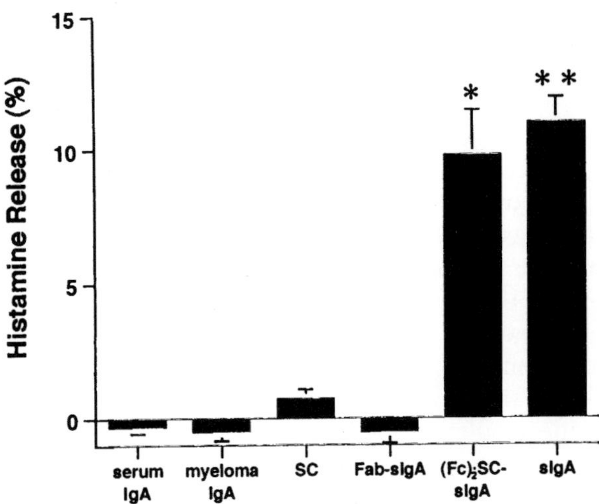

Fig. 4 Degranulation of basophils by various IgA preparations or SC. IL-3-primed basophils (n=4) were incubated with beads coated with whole IgA molecules (serum IgA, myeloma IgA, or sIgA) or sIgA-derived fragments (SC, Fab fragment, or $(Fc)_2$·SC fragment) at a 1:10 bead:cell ratio for 45 min at 37°C, and histamine in the supernatant was analysed. Bars represent the SEM. *$p<0.05$, **$p<0.01$, vs. OVA-induced release from corresponding cells. (Reproduced from *J. Immunol.* **161**:1510–1515, 1998. Reprinted by permission of The Journal of Immunology. Copyright 1998. The American Association of Immunologists.)

On the other hand, flow cytometric analyses have demonstrated constitutive expression of $Fc_{\alpha}RI$ (CD89) on the surface of human eosinophils (40). sIgA as well as IgA in immobilized form induces strong degranulation (25) and superoxide production (44) in eosinophils. $Fc_{\alpha}RI$ is apparently involved in these processes, but sIgA-mediated eosinophil activation is not mediated merely via binding to $Fc_{\alpha}RI$. sIgA always induces a higher magnitude of degranulation than monomeric IgA (25, 26). Eosinophil superoxide production initiated by sIgA was only partially inhibited by treatment with anti-$Fc_{\alpha}RI$ antibody, whereas IgA-mediated superoxide production was completely abolished (45). These observations strongly indicate that sIgA transduccs intracellular signals in eosinophils via two different pathways: $Fc_{\alpha}R$ and an undetermined binding site(s) for sIgA. Concerning the ligand for the undetermined binding site for sIgA in eosinophils, the SC has been considered to be a candidate. To date, we have not observed that SC fully compensates for the functional ability of sIgA on basophil histamine release. Immobilized preparations of SC induced weak, but statistically not significant, degranulation of IL-3-primed basophils. Because of the difficulty in obtaining a large number of purified basophils, we have not succeeded in finding direct binding of SC to human basophils to date. On the other hand, several reports have demonstrated direct binding of the SC to eosinophils. Lamkhioued *et al.* reported that a 15-kDa protein is responsible for specific binding of the SC to eosinophils (46). They also showed a mean number of 11×10^4 receptors per cell, with a K_a of 2.76×10^8 M^{-1} by Scatchard analysis (47). More recently, Motegi and Kita also demonstrated direct binding of radiolabelled SC to eosinophils (44). Both groups have also shown that SC by itself is capable of inducing biological functions in eosinophils. Free SC induces the release of eosinophil peroxidase (EPO) and EDN from eosinophils (44, 46). Motegi and Kita also reported that immobilized SC in combination with IL-5 induces superoxide generation. Importantly, they observed that the binding sites of SC on eosinophils are functionally related to β_2 integrins: treatment with anti-CD18 mAb completely abolished SC-induced eosinophil superoxide production, albeit direct binding of SC to β_2 integrins was not determined.

Several reports have clearly revealed that β_2 integrins mediate the effector function of human eosinophils: β_2 integrin is shown to be essential for the degranulation induced by several secretagogues. Eosinophils incubated in albumin-coated polystyrene plates degranulate in response to soluble stimuli such as PAF (48). The degranulation induced by these soluble agonists could not be observed in non-adherent eosinophils. In this system β_2 integrin seems to work as a counter-receptor: blocking of CD18 completely, while blocking of CD11b moderately, attenuated this adhesion-dependent eosinophil degranulation. Furthermore, CD11b/CD18 also seems to be essential for eosinophil degranulation by immobilized IgG (49). β_2 integrins CD11b/CD18 potentially introduce essential co-stimulatory signals in these cases. In contrast to eosinophils, the effects of β_2 integrin on basophil degranulation remain poorly defined. However, the results of several reports have suggested the possibility that β_2 integrins also mediate the effector function of human basophils. Zymosan activates an alternative pathway of complement, and C_3b*i*-coupled zymosan particles transduce intracellular signals mainly via binding to CD11b/CD18. Serum-treated zymosan particles have been shown to enhance IgE-mediated histamine release from basophils *in vitro* (50). In addition, an *in vivo* study demonstrated that infection by rhinovirus, which utilizes β_2 integrin for entry, increased the plasma histamine content after antigen provocation in allergic subjects (51). Although not proved, it is strongly postulated that ligation of β_2 integrins can introduce enough signals to activate the biological functions of human basophils.

In contrast to the immobilized form of sIgA, the non-immobilized form of sIgA was unable to induce degranulation. When IL-3-primed basophils were treated with sIgA and subsequently stimulated with polyclonal anti-IgA or anti-sIgA antibodies, we could not observe any significant histamine release. Treatment with aggregated sIgA also failed to elicit basophil degranulation (K. Hirai *et al.*, unpublished observation). These findings suggest that cell adhesion to a solid surface is critical for sIgA-mediated basophil degranulation. Furthermore, adhesion to β_2 integrin is known to be dependent on both Ca^{2+} and Mg^{2+}. In contrast to IgE-mediated release, which is dependent only on extracellular Ca^{2+}, sIgA-mediated basophil degranulation is dependent on both Ca^{2+} and Mg^{2+}. Finally, expression of β_2 integrin on basophils is shown to be upregulated by the treatment with IL-3 (32). Although further investigations are required, these notions raise the possibility that, as in eosinophils, adhesion via β_2 integrin is also important in sIgA-mediated basophil activation.

CLINICAL RELEVANCE OF sIgA-MEDIATED BASOPHIL ACTIVATION

The *in vitro* elaboration of mediators from basophils by sIgA may indicate the existence of *in vivo* mechanisms of activation of basophils by sIgA in allergic inflammation. Peripheral basophils do not respond to sIgA, and sIgA-mediated basophil degranulation is totally dependent on priming with basophil-active cytokines such as IL-3 and IL-5. However, these cytokines are locally produced by Th2 cells which have infiltrated sites of ongoing allergic reactions, and basophils which migrate to inflamed tissues are potentially primed by these cytokines *in vivo*. Furthermore, the magnitude of histamine release evoked by sIgA is as strong as that by anti-IgE antibodies. As shown in the present study, sIgA conjugated to Sepharose beads consistently induced histamine release, but the release was rather low (~15% of total histamine) compared with IgE-mediated degranulation. We recently established a new assay system by using sIgA immobilized on a plastic surface in place of Sepharose beads. Under this condition, sIgA induced strong histamine release from basophils (~50% of total histamine content), which is comparable to that induced by anti-IgE (M. Iikura *et al.*, unpublished observation). These findings, along with the fact that sIgA is the most abundant immunoglobulin isotype in the secretions from mucosal tissues, strongly suggest that sIgA-mediated activation of basophils represents an important mechanism in the pathogenesis of allergic inflammation of mucosal surfaces.

CONCLUDING REMARKS

An increasing body of evidence has demonstrated that basophils represent a class of allergic inflammatory cells. Along with the progression of allergic reactions, basophils migrate from the circulation to the sites of inflammation and contribute to the pathogenesis of the late-phase reaction through the release of various mediators. Although sIgA-mediated eosinophil activation has been well established, basophils are also capable of being stimulated by sIgA. In contrast to eosinophils, $Fc_{\alpha}RI$ does not seem to be involved in this process. The binding sites on basophils as well as the molecules responsible for the ligation have not been identified, but β_2 integrin and the SC might play essential roles in sIgA-mediated basophil activation. Since sIgA is the most abundant immunoglobulin isotype at the inflammatory sites of allergic disorders and basophils are active participants in allergic late-phase reactions, sIgA is potentially involved in exacerbation of the inflammation associated with allergic disorders by stimulating both basophils and eosinophils.

ACKNOWLEDGEMENTS

We thank Ms S. Takeyama for her excellent secretarial help. This work was supported by a grant from the Manabe Medical Foundation, grants-in-aid from the Ministry of Health and Welfare of Japan (to K.H. and M.Y.), grants-in-aid from the Ministry of Education, Science, Sports and Culture of Japan (to K.H. and M.Y.), and the Japan Society for the Promotion of Science Research Fellowships for Young Scientists (to M.I. and M.M.).

REFERENCES

1. Hirai, K., Miyamasu, M., Takaishi, T. and Morita, Y. Regulation of the function of eosinophils and basophils. *Crit. Rev. Immunol.* **17**:325–352, 1997.
2. Gounni, A. S., Lamkhioued, B., Ochiai, K., *et al.* High-affinity IgE receptor on eosinophils is involved in defence against parasites. *Nature* **367**:183–186, 1994.
3. Yamada, H., Hirai, K., Miyamasu, M., *et al.* Eotaxin is a potent chemotaxin for human basophils. *Biochem. Biophys. Res. Commun.* **231**:365–368, 1997.
4. Uguccioni, M., Mackay, C. R., Ochensberger, B., *et al.* High expression of the chemokine receptor CCR3 in human blood basophils. Role in activation by eotaxin, MCP-4, and other chemokines. *J. Clin. Invest.* **100**:1137–1143, 1997.
5. Iliopoulos, O., Baroody, F. M., Naclerio, R. M., Bochner, B. S., Kagey, S. A. and Lichtenstein, L. M.

Histamine-containing cells obtained from the nose hours after antigen challenge have functional and phenotypic characteristics of basophils. *J. Immunol.* **148**:2223–2228, 1992.

6. Koshino, T., Teshima, S., Fukushima, N., *et al.* Identification of basophils by immunohistochemistry in the airways of post-mortem cases of fatal asthma. *Clin. Exp. Allergy* **23**:919–925, 1993.
7. Guo, C. B., Liu, M. C., Galli, S. J., Bochner, B. S., Kagey-Sobotka, A. and Lichtenstein, L. M. Identification of IgE-bearing cells in the late-phase response to antigen in the lung as basophils. *Am. J. Respir. Cell Mol. Biol.* **10**:384–390, 1994.
8. Charlesworth, E. N. Role of basophils and mast cells in acute and late reactions in the skin. *Chem. Immunol.* **62**:84–107, 1995.
9. Naclerio, R. M., Proud, D., Togias, A. G., *et al.* Inflammatory mediators in late antigen-induced rhinitis. *N. Engl. J. Med.* **313**:65–70, 1985.
10. Naclerio, R. M., Hubbard, W., Lichtenstein, L. M., Kagey-Sobotka, A. and Proud, D. Origin of late phase histamine release. *J. Allergy Clin. Immunol.* **98**:721–723, 1996.
11. Brandtzaeg, P. and Prydz, H. Direct evidence for an integrated function of J chain and secretory component in epithelial transport of immunoglobulins. *Nature* **311**:71–73, 1984.
12. Mestecky, J. and McGhee, J. R. Immunoglobulin A (IgA): molecular and cellular interactions involved in IgA biosynthesis and immune response. *Adv. Immunol.* **40**:153–245, 1987.
13. Delacroix, D. L., Dive, C., Rambaud, J. C. and Vaerman, J. P. IgA subclasses in various secretions and in serum. *Immunology* **47**:383–385, 1982.
14. Lindh, E. Increased resistance of immunoglobulin A dimers to proteolytic degradation after binding of secretory component. *J. Immunol.* **114**:284–286, 1975.
15. Brandtzaeg, P. Human secretory component – IV. Aggregation and fragmentation of free secretory component. *Immunochemistry* **12**:877–881, 1975.
16. Peebles, R. S. Jr, Liu, M. C., Lichtenstein, L. M. and Hamilton, R. G. IgA, IgG and IgM quantification in bronchoalveolar lavage fluids from allergic rhinitics, allergic asthmatics, and normal subjects by monoclonal antibody-based immunoenzymetric assays. *J. Immunol. Methods* **179**:77–86, 1995.
17. Bukawa, H., Sekigawa, K., Hamajima, K., *et al.* Neutralization of HIV-1 by secretory IgA induced by oral immunization with a new macromolecular multicomponent peptide vaccine candidate. *Nat. Med.* **1**:681–685, 1995.
18. Burrows, P. D. and Cooper, M. D. IgA deficiency. *Adv. Immunol.* **65**:245–276, 1997.
19. Nahm, D. H. and Park, H. S. Correlation between IgA antibody and eosinophil cationic protein levels in induced sputum from asthmatic patients. *Clin. Exp. Allergy* **27**:676–681, 1997.
20. Kitani, S., Ito, K. and Miyamoto, T. IgG, IgA, and IgM antibodies to mite in sera and sputa from asthmatic patients. *Ann. Allergy* **55**:612–620, 1985.
21. Reed, C. E., Bubak, M., Dunnette, S., *et al.* Ragweed-specific IgA in nasal lavage fluid of ragweed-sensitive allergic rhinitis patients: increase during the pollen season. *Int. Arch. Allergy Appl. Immunol.* **94**:275–277, 1991.
22. Terada, N., Terada, Y., Shirotori, K., Ishikawa, K., Togawa, K. and Konno, A. Immunoglobulin as an eosinophil degranulation factor: change in immunoglobulin level in nasal lavage fluid after antigen challenge. *Acta Otolaryngol.* **116**:863–867, 1996.
23. Fujisawa, T., Uchida, Y., Kamiya, H. and Sakurai, M. Allergen-specific IgA and eosinophil in asthma. *J. Allergy Clin. Immunol.* **85**:283, 1990.
24. Ostergaard, P. A. Clinical and immunological features of transient IgA deficiency in children. *Clin. Exp. Immunol.* **40**:561–565, 1980.
25. Abu-Ghazaleh, R. I., Fujisawa, T., Mestecky, J., Kyle, R. A. and Gleich, G. J. IgA-induced eosinophil degranulation. *J. Immunol.* **142**:2393–2400, 1989.
26. Fujisawa, T., Abu-Ghazaleh, R., Kita, H., Sanderson, C. J. and Gleich, G. J. Regulatory effect of cytokines on eosinophil degranulation. *J. Immunol.* **144**:642–646, 1990.
27. Nakajima, H., Gleich, G. J. and Kita, H. Constitutive production of IL-4 and IL-10 and stimulated production of IL-8 by normal peripheral blood eosinophils. *J. Immunol.* **156**:4859–4866, 1996.
28. Shute, J. K., Lindley, I., Peichl, P., Holgate, S. T., Church, M. K. and Djukanovic, R. Mucosal IgA is an important moderator of eosinophil responses to tissue-derived chemoattractants. *Int. Arch. Allergy. Immunol.* **107**:340–341, 1995.
29. Iikura, M., Yamaguchi, M., Fujisawa, T., *et al.* Secretory IgA induces degranulation of IL-3-primed basophils. *J. Immunol.* **161**:1510–1515, 1998.
30. Hirai, K., Morita, Y. and Miyamoto, T. Hemopoietic growth factors regulate basophil function and viability. *Immunol. Ser.* **57**:587–600, 1992.

31. Yamaguchi, M., Hirai, K., Shoji, S., *et al.* Haemopoietic growth factors induce human basophil migration *in vitro. Clin. Exp. Allergy* **22**:379–383, 1992.
32. Bochner, B. S., McKelvey, A. A., Sterbinsky, S. A., *et al.* IL-3 augments adhesiveness for endothelium and CD11b expression in human basophils but not neutrophils. *J. Immunol.* **145**:1832–1837, 1990.
33. Yamaguchi, M., Hirai, K., Morita, Y., *et al.* Hemopoietic growth factors regulate the survival of human basophils *in vitro. Int. Arch. Allergy Immunol.* **97**:322–329, 1992.
34. Hirai, K., Morita, Y., Misaki, Y., *et al.* Modulation of human basophil histamine release by hemopoietic growth factors. *J. Immunol.* **141**:3958–3964, 1988.
35. Kurimoto, Y., de Weck, A. L. and Dahinden, C. A. Interleukin 3-dependent mediator release in basophils triggered by C5a. *J. Exp. Med.* **170**:467–479, 1989.
36. Bischoff, S. C., de Weck, A. L. and Dahinden, C. A. Interleukin 3 and granulocyte/macrophage-colony-stimulating factor render human basophils responsive to low concentrations of complement component C3a. *Proc. Natl. Acad. Sci. USA* **87**:6813–6817, 1990.
37. Brunner, T., de Weck, A. L. and Dahinden, C. A. Platelet-activating factor induces mediator release by human basophils primed with IL-3, granulocyte–macrophage colony-stimulating factor, or IL-5. *J. Immunol.* **147**:237–242, 1991.
38. Brunner, T., Heusser, C. H. and Dahinden, C. A. Human peripheral blood basophils primed by interleukin 3 (IL-3) produce IL-4 in response to immunoglobulin E receptor stimulation. *J. Exp. Med.* **177**:605–611, 1993.
39. MacGlashan, D. J., White, J. M., Huang, S. K., Ono, S. J., Schroeder, J. T. and Lichtenstein, L. M. Secretion of IL-4 from human basophils. The relationship between IL-4 mRNA and protein in resting and stimulated basophils. *J. Immunol.* **152**:3006–3016, 1994.
40. Morton, H. C., van Egmond, M. and van de Winkel, J. G. Structure and function of human IgA Fc receptors (Fc alpha R). *Crit. Rev. Immunol.* **16**:423–440, 1996.
41. Füreder, W., Agis, H., Sperr, W. R., Lechner, K. and Valent, P. The surface membrane antigen phenotype of human blood basophils. *Allergy* **49**:861–865, 1994.
42. Pleass, R. J., Andrews, P. D., Kerr, M. A. and Woof, J. M. Alternative splicing of the human IgA Fc receptor CD89 in neutrophils and eosinophils. *Biochem. J.* **318**:771–777, 1996.
43. van Dijk, T. B., Bracke, M., Caldenhoven, E., *et al.* Cloning and characterization of Fc alpha Rb, a novel Fc alpha receptor (CD89) isoform expressed in eosinophils and neutrophils. *Blood* **88**:4229–4238, 1996.
44. Motegi, Y. and Kita, H. Interaction with secretory component stimulates effector functions of human eosinophils but not of neutrophils. *J. Immunol.* **161**:4340–4346, 1998.
45. Motegi, Y., Gleich, G. and Kita, H. Eosinophils, but not neutrophils, are preferentially activated by secretory IgA (sIgA) compared with serum IgA: potential roles of secretory component (SC). *J. Allergy Clin. Immunol.* **99**:120, 1997.
46. Lamkhioued, B., Gounni, A. S., Gruart, V., Pierce, A., Capron, A. and Capron, M. Human eosinophils express a receptor for secretory component. Role in secretory IgA-dependent activation. *Eur. J. Immunol.* **25**:117–125, 1995.
47. Capron, M., Gruart, V., Broussolle, A. and Capron, A. Binding site for secretory component on human eosinophils. *FASEB J.* **5**:640, 1991.
48. Horie, S. and Kita, H. CD11b/CD18 (Mac-1) is required for degranulation of human eosinophils induced by human recombinant granulocyte–macrophage colony-stimulating factor and platelet-activating factor. *J. Immunol.* **152**:5457–5467, 1994.
49. Kaneko, M., Horie, S., Kato, M., Gleich, G. J. and Kita, H. A crucial role for beta 2 integrin in the activation of eosinophils stimulated by IgG. *J. Immunol.* **155**:2631–2641, 1995.
50. Thomas, L. L. and Lichtenstein, L. M. Augmentation of antigen-stimulated histamine release from human basophils by serum-treated zymosan particles. I. Characteristics of enhancement. *J. Immunol.* **123**:1462–1467, 1979.
51. Calhoun, W. J., Swenson, C. A., Dick, E. C., Schwartz, L. B., Lemanske, R. F. Jr and Busse W. W. Experimental rhinovirus 16 infection potentiates histamine release after antigen bronchoprovocation in allergic subjects. *Am. Rev. Respir. Dis.* **144**:1267–1273, 1991.

CHAPTER 15

Regulation of Mast Cell Secretion by Interferon-γ and Nitric Oxide

JOHN W. COLEMAN

Department of Pharmacology, University of Liverpool, Liverpool, U.K.

INTRODUCTION

In this chapter the following will be reviewed: the effects of interferon (IFN)-γ and other interferons on mast cells; the role of nitric oxide in regulation of mast cell function *in vitro* and mast cell-dependent processes *in vivo*; mast cells as a possible source of nitric oxide; the mechanism of action of nitric oxide on mast cells; the clinical implications of nitric oxide regulation of mast cells.

INTERFERON AND MAST CELLS

The interferons are a family of cytokines of 17–20 kDa and primary polypeptide chain length of 143–166 amino acids, that share several biological properties, including antiviral, growth-inhibitory and immune-regulatory activity, but with different potencies. There are three classes of interferon. IFN-α and IFN-β show limited sequence homology but share a common receptor; they are induced in virally infected cells and are principally antiviral in activity. IFN-γ shows no homology to the other classes, is produced by natural killer (NK) and Th1 cells in response to antigen or cytokines, has its own receptor, and is primarily immune-regulatory in function (1, 2). IFN-γ is the interferon that has been studied most extensively for effects on mast cells; it is also the most potent interferon with the most diverse effects on mast cells.

Several reports have shown regulatory effects of interferons on IgE-mediated degranulation of mast cells. Swieter *et al.* (3) showed that IFN-α/β inhibited antigen-induced histamine release from rat peritoneal mast cells. IFN-α/β gave maximum inhibition of release at a concentration of approximately 250 ng ml^{-1}. At about the same time it was reported (4) that mitogen-activated mouse spleen cells produced a protein

MAST CELLS AND BASOPHILS
ISBN 0-12-473335-2

factor that inhibited IgE/antigen-induced histamine and serotonin release from mouse peritoneal mast cells. This factor was subsequently shown to be IFN-γ, which proved to be 10–100 times more potent than IFN-α/β, giving significant inhibition of degranulation at concentrations as low as 0.2 ng ml^{-1} (5). IFN-α/β (3) and IFN-γ (5) inhibited degranulation induced by IgE/antigen far more readily than that induced by compound 48/80, suggesting selectivity of the cytokine for signalling through the IgE receptor. In both the above systems (rat mast cells with IFN-α/β, mouse mast cells with IFN-γ) the cells required culture with the cytokine for 24 h for its effects to be fully manifested. This time dependence suggests that the action of these cytokines on mast cells involves a possible induction or inhibition of regulatory pathways, rather than a direct inhibition of early signalling events.

IFN-γ inhibits not only release of mast cell granule-associated mediators such as histamine, but also arachidonate (6). Both IFN-α/β and IFN-γ suppress the cytotoxicity of rat peritoneal mast cells towards target tumour cells by inhibition of tumour necrosis factor (TNF)-α release (7). Inhibition of TNF-α release by IFN-α/β and -γ is also seen for rat intestinal mucosal mast cells, but histamine release is inhibited only in peritoneal mast cells (8). Thus, it seems that intestinal mast cell degranulation and cytokine production are differentially controlled by IFN-α/β and -γ. This regulation of mast cell TNF-α production is at the level of mRNA (9).

The effects of IFN-γ on human basophils are disputed. In contrast to a report that IFN-γ enhances anti-IgE-induced histamine release from human basophils (10), a more recent study (11) shows that immunotherapy with wasp venom, which elevates Th1 cell activity and IFN-γ production, decreases allergen-induced basophil release of histamine and sulphidoleukotrienes. Release of these mediators was also inhibited by IFN-γ *in vitro* (11).

IFN-γ regulates the growth of mast cells. It inhibits the IL-3/IL-4-dependent proliferation of mouse peritoneal mast cells (12) and the IL-3-dependent growth of mouse bone marrow-derived mast cells (BMMC) (13). It also inhibits the stem cell factor-induced growth of human mast cells from $CD34^+$ bone marrow precursors, and proliferation of human HMC-1 mast cells (14). IFN-γ did not influence the secretory responsiveness of human BMMC (14). In apparent contradiction, although using different sources of mast cells, IFN-γ has been reported to be important for survival and differentiation of mouse splenic mast cell precursors (15), and for survival and IgE-mediated degranulation of cultured human cord blood-derived mast cells (16). Very little information is available on the effects of interferons on the activation of mature human tissue mast cells. In line with its effects on macrophages (1, 2), IFN-γ upregulates the expression of MHC class II molecules on rat peritoneal mast cells (17), mouse BMMC (18) and human HMC-1 mast cells (19, 20). The upregulation of MHC class II expression on mouse BMMC does not appear to relate to antigen presentation to T cells. In fact, IFN-γ inhibits antigen presentation by this type of mast cell (18), and this correlates to inhibition of expression of the co-stimulatory molecules CD80 and CD86 (21).

Overall, IFN-γ, and in some cases also IFN-α/β, inhibit mast cell activation, mediator release and growth in rodent models. The mediators that are regulated appear to vary between mast cell phenotypes, but IgE-dependent signalling appears to be preferentially the targeted process. As mentioned above, the extended *in vitro* incubation times required for interferons to inhibit IgE-mediated degranulation of rat and mouse mast cells (3, 5) suggest their action may be via induction of regulatory proteins or other factors. Furthermore, the observation that IFN-γ is considerably more potent as an inhibitor of

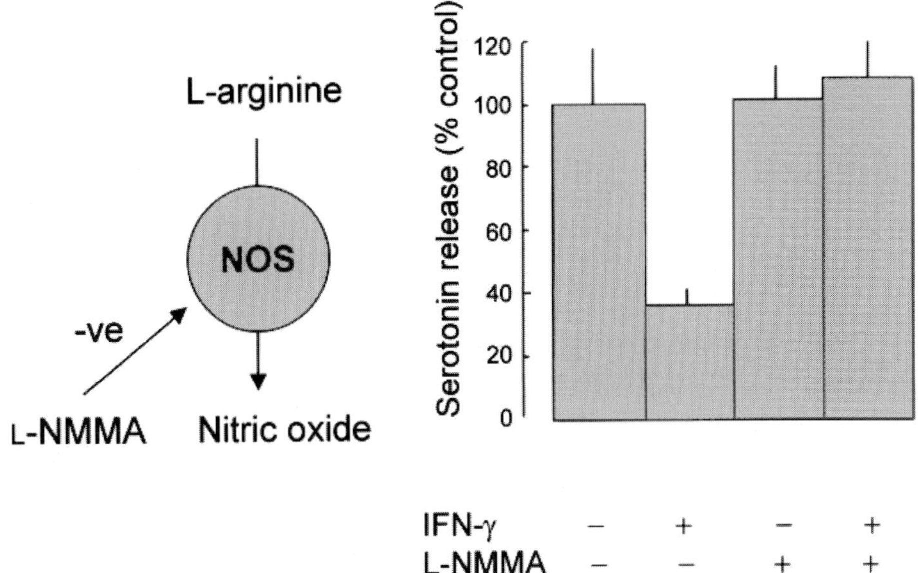

Fig. 1 Reversal of the mast cell inhibitory effect of IFN-γ by the nitric oxide synthase (NOS) inhibitor L-NMMA. (Modified from *J. Immunol.* **159**: 1444–1450, 1997, with permission of the authors and publishers, copyright 1997, The American Association of Immunologists.)

antigen-induced serotonin release from mast cells in unfractionated peritoneal populations, compared to purified peritoneal mast cells, implicates induction of an intermediate factor (22). IFN-γ induces production of nitrite (a stable metabolite of nitric oxide) by mouse peritoneal cells, and the inhibitory effects of IFN-γ are blocked by the nitric oxide synthase (NOS) inhibitor *N*-monomethyl-L-arginine (L-NMMA) (Fig. 1) or by culture of the cells in medium lacking arginine, the substrate for nitric oxide synthesis (22). These results show that the action of IFN-γ on mouse mast cells in mixed peritoneal populations is by induction of nitric oxide synthesis. In contrast to mouse peritoneal cells, those from the rat spontaneously synthesize nitric oxide at high levels and this corresponds to low-level responsiveness of antigen-triggered mast cells (23). In the rat system, IFN-γ has little effect on nitric oxide synthesis since this is already at a high level, but again L-NMMA enhances antigen-induced serotonin release, demonstrating that secretory responsiveness is suppressed by nitric oxide (23). Also, interleukin (IL)-4 enhances mast cell degranulation by inhibiting nitric oxide synthesis by rat peritoneal cells (23).

It would be of interest to investigate the role of nitric oxide in the reported effects of IFN-γ on other aspects of mast cell function, such as TNF-α production, inhibition of growth and expression of MHC class II. It is unlikely that the effects of IFN-α or IFN-β would be through nitric oxide since these are very different molecules to IFN-γ, and are not recognized inducers of NOS. Certainly there is more to be learned about the differences in action of IFN-γ compared to IFN-α/β on mast cells, and the role of nitric oxide as a more universally important intermediate in regulation of mast cells.

DO MAST CELLS PRODUCE INTERFERON-γ?

There is some controversy and rather incomplete information relating to whether mast cells produce IFN-γ and under which conditions. Williams and Coleman (24) reported anti-IgE-induced expression of IFN-γ mRNA and mRNA for several other cytokines in purified rat peritoneal mast cells, and Burd *et al.* (25) reported IFN-γ mRNA expression in mouse mast cell lines. However, Gupta *et al.* (26) found that IL-12 but not anti-IgE induced the release of IFN-γ protein from rat peritoneal mast cells. Interestingly, the IL-12-induced IFN-γ release was not accompanied by histamine release. IFN-γ immunoreactivity can be detected in mast cell granules in human skin, and mast cell positivity for IFN-γ is dramatically elevated in lesional skin samples from patients with psoriasis. Furthermore, human HMC-1 mast cells express IFN-γ mRNA and release IFN-γ protein when stimulated with phorbol myristate acetate (27).

NITRIC OXIDE AND NITRIC OXIDE SYNTHASES

Nitric oxide is a simple small molecule consisting of a single atom each of nitrogen and oxygen. It is uncharged but has an unpaired outer electron, and thus belongs to the class of highly reactive molecules known as radicals. It has a biological half-life of only a few seconds, and in tissues has an activation range of only a few micrometers. It enters cells freely by virtue of its small size and neutrality, and its selectivity is based on the close proximity of target cells and the availability of nitric oxide-regulated molecules in those cells (28, 29).

Nitric oxide exerts a range of physiological and toxic effects. It is the formerly named endothelium-derived relaxing factor released by the blood vessel walls to relax underlying smooth muscle cells, and it also acts as a neurotransmitter in the brain (28, 29). At higher concentrations, as produced by the inducible form of NOS, nitric oxide exerts cytotoxic and immune regulatory activities (30, 31). For example, nitric oxide has been claimed to influence the Th1/Th2 cell balance by selectively inhibiting IFN-γ expression (31), and mice genetically deleted of NOS II (see below) have depleted Th2-mediated immunity (32).

Nitric oxide is synthesized from arginine and molecular oxygen by a family of NOS. These enzymes are large and structurally and functionally complex, requiring several co-factors for activity. There are three isotypes of NOS. Two of these are constitutively expressed: type I or neuronal NOS found mainly in the brain, and type III, or endothelial NOS, found mainly in endothelial cells but also elsewhere. These forms of NOS are activated by physiological stimuli to produce transient low levels of nitric oxide via elevation of cellular calcium. Type II, or inducible NOS, is not present in resting cells but its expression is induced by cytokines or bacterial lipopolysaccharide acting on macrophages, neutrophils, or epithelial cells. Induction of NOS II leads to sustained synthesis of nitric oxide at high levels, and this is calcium-independent (29).

Nitric oxide is a unique intracellular messenger in that it is not cell-specific; it has no cell surface receptor but enters all cells equally by virtue of its lack of charge and small size. Its biological effects depend on its concentration, the proximity of target cells and the availability of intracellular molecular targets. At low concentrations, as produced by constitutive forms of NOS, it targets haem enzymes such as guanylate cyclase to induce synthesis of cyclic guanosine monophosphate (cGMP). At higher concentrations, as produced by type II NOS, nitric oxide exerts cytostatic, apoptotic or immune regulatory effects by poorly defined mechanisms, which may involve nitrosylation of proteins or modification of DNA, or interference with mitochondrial enzymes (29).

NITRIC OXIDE EFFECTS ON MAST CELLS

Salvemini and co-workers reported that inhibition of synthesis of a nitric oxide-like factor from rat peritoneal mast cells led to enhanced histamine release in response to a range of activators, including compound 48/80, calcium ionophore and lipopolysaccharide (33, 34). Release of the nitric oxide-like factor was less from mast cells from spontaneously hypertensive rats compared with normal rats, and this was associated with increased histamine release from isolated mast cells (35). Thus, both *in vitro* and *in vivo*, elevated synthesis of nitric oxide is associated with reduced mast cell secretory responsiveness. Subsequent studies confirmed that nitric oxide generated by mouse or rat peritoneal cells, either after induction by IFN-γ or spontaneously, is an effective inhibitor of IgE-mediated mast cell degranulation (22, 23). Nitric oxide donors such as sodium nitroprusside and *S*-nitrosoglutathione inhibit IgE-mediated serotonin release from purified mouse and rat peritoneal mast cells (22, 23) (Fig. 2) and β-hexosaminidase and TNF-α release from mouse BMMC (36). Thus, nitric oxide targets mast cells directly. The inhibitory effect of sodium nitroprusside on anti-IgE or calcium ionophore-induced histamine release from rat mast cells or human basophils is potentiated by *N*-acetylcysteine, a thiol that increases the bioavailability of the radical (37).

Nitric oxide also regulates growth of cultured mast cells and mast cell lines: when released from fibroblasts it inhibits the growth of mouse BMMC (38) and induces apoptosis in a murine mastocytoma cell (39).

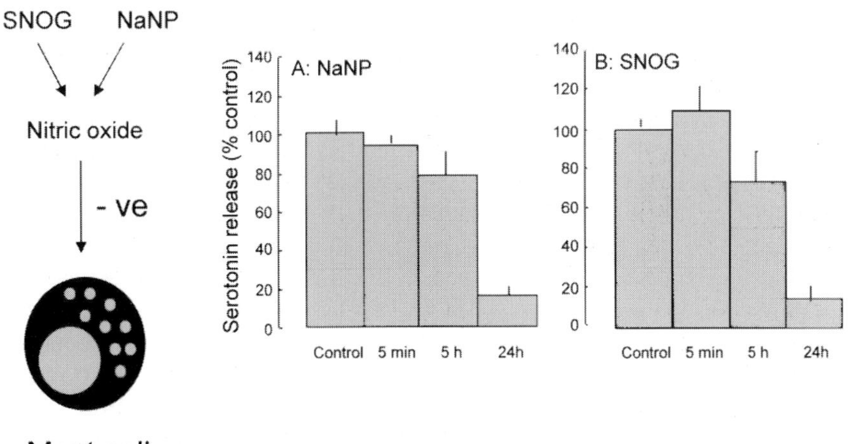

Fig. 2 Time dependence of the mast cell inhibitory effects of the nitric oxide donors sodium nitroprusside (NaNP) and *S*-nitrosoglutathione (SNOG). (Modified from *J. Immunol.* **159**: 1444–1450, 1997 with permission of the authors and publishers, copyright 1997, The American Association of Immunologists.)

DO MAST CELLS PRODUCE NITRIC OXIDE?

At the beginning of the 1990s two groups claimed that mast cells produce nitric oxide. Salvemini *et al.* (33, 40) reported that unstimulated rat serosal mast cells generated a nitric oxide-like factor that inhibited thrombin-induced aggregation of platelets. Generation of the factor was blocked by a NOS inhibitor and this effect was overcome by L-arginine. Agitation of the mast cells by stirring or stimulation with lipopolysaccharide elevated cellular cGMP – a physiological marker of nitric oxide activity. At around the same time Befus and colleagues reported likewise that resting rat peritoneal mast cells released a nitric oxide-like factor that inhibited thrombin-induced platelet aggregation; again the effect was reversed by NOS inhibitors (41). In the same study, addition of NOS inhibitors or the nitric oxide scavenger haemoglobin to the mast cells partially diminished their TNF-α-mediated cytotoxicity towards WEHI target tumour cells but did not inhibit the direct toxic effect of recombinant TNF-α. The authors concluded that mast cells released nitric oxide alongside TNF-α and this led to augmentation of the cytotoxic effects of the cytokine (41). Release of the nitric oxide-like platelet-dispersing activity and of nitrite from the rat peritoneal cells was enhanced by IL-1β (42). More recently, Bidri *et al.* (43) reported induced protein and mRNA expression for type II NOS, and production of nitrite, after IgE/antigen-induced activation of mouse BMMC.

In our laboratory, we have consistently been unable to obtain evidence that purified rat or mouse peritoneal mast cells, or mouse mast cell lines, stimulated with IFN-γ or IgE/antigen, produce functionally significant levels of nitric oxide. Accumulation of cell supernatant nitrite at micromolar levels (corresponding to several nanomoles per million cells) can be detected readily by the Griess assay from IFN-γ-stimulated unfractionated mouse peritoneal cells or unstimulated rat peritoneal cells, but resting or IFN-γ-stimulated purified peritoneal mast cells do not produce detectable nitrite (22, 23). We have also consistently failed to detect nitrite production from antigen or anti-IgE-activated peritoneal mast cells or mast cell lines (unpublished observations). Unfractionated mouse and rat peritoneal cells produce 30–40 times more IFN-γ-stimulated nitric oxide, measured as nitrite accumulation, than purified peritoneal mast cells, and the low amounts of nitrite generated by the purified mast cells can be accounted for fully by the small proportion (2–3%) of contaminating non-mast cells. Interestingly, purification of rat mast cells from peritoneal cells, that spontaneously produce high levels of nitric oxide, leads to a marked increase in secretory responsiveness to anti-IgE, indicating that removal of the source of nitric oxide-producing cells (non-mast cells) removes the suppressive influence of nitric oxide (23). Also revealing is the observation that the NOS inhibitor L-NMMA enhances IgE-mediated serotonin release from IFN-γ-stimulated mouse or resting rat mast cells in unfractionated peritoneal populations, but has no effect on responses of purified peritoneal mast cells (22, 23). Thus, even if the purified mast cells are producing nitric oxide at low levels, this is insufficient to inhibit mast cell degranulation. These findings appear to contradict those of Salvemini *et al.* (33, 40), who claimed that mast cell-derived nitric oxide was sufficient to inhibit mast cell degranulation. However, their mast cell preparations contained 10–15% contaminating non-mast cells, presumably mainly macrophages that spontaneously generate nitric oxide (23), and these preparations were stimulated by physical agitation or lipopolysaccharide – agents that are not selective for mast cells and, in fact, at least in the latter case, are more likely to stimulate macrophages.

Overall, it appears that rodent peritoneal mast cells probably do not express the inducible form of NOS that generates high levels of nitric oxide. Because inhibition of degranulation requires these higher levels of nitric oxide, mast cells almost certainly do not self-inhibit by nitric oxide synthesis. On the other hand, neighbouring cells, such as macrophages, neutrophils or epithelial cells, that express type II NOS, are readily able to inhibit mast cell degranulation by high-level synthesis of nitric oxide. Production of the platelet-dispersing nitric oxide-like activity from resting or agitated rat mast cells (33, 40, 41) may represent low-level production by constitutive forms of NOS (NOS I or III). This would be consistent with observations that human skin and nasal mast cells express type I NOS (44, 45). The constitutive NOS isotypes produce approximately 1000 times less nitric oxide than produced by NOS II, but this is the sufficient and appropriate level for its physiological actions through guanylate cyclase, such as smooth muscle relaxation, neurotransmission and platelet dispersion (29). This said, cells that produce nitric oxide by constitutive NOS, which is calcium-activated, do not respond to nitric oxide by guanylate cyclase activation, since this enzyme is calcium-inhibited. This mechanism seems to have evolved to ensure that cells producing low physiological levels of nitric oxide (e.g. endothelial cells) do not self-target, but rather aim their nitric oxide at neighbouring cells such as smooth muscle cells (29). Thus it seems highly unlikely that nitric oxide produced by constitutive NOS in mast cells can self-inhibit secretory responses.

HOW DOES NITRIC OXIDE REGULATE MAST CELLS?

Some studies have claimed that nitric oxide regulates mast cell cGMP. For example, Salvemini *et al.* (33, 40) reported that mast cell-derived nitric oxide elevated mast cell cGMP, and Bidri *et al.* (36) reported that sodium nitroprusside induced a transient (single time point) increase in cGMP in mouse BMMC. Based on two types of experiment, we believe that cGMP is not involved in the mast cell inhibitory activity of nitric oxide. Firstly, incubation of purified rat peritoneal mast cells with 8-bromo-cGMP, a cell-permeable analogue of cGMP, at a range of concentrations and for periods of up to 24 h, failed to influence anti-IgE-induced serotonin release. Secondly, addition of the guanylate cyclase inhibitor 1*H*-[1,2,4]oxadiazolo[4,3-*a*]quinoxalin-1-one (ODQ) to rat peritoneal cells spontaneously generating nitric oxide again failed to modify IgE-mediated secretion, even though the ODQ was effective under the same conditions at blocking the relaxing effect of nitric oxide donors on vascular smooth muscle (23). We also found that the nitric oxide donors sodium nitroprusside and *S*-nitrosoglutathione required 24 h of incubation with mouse peritoneal mast cells to give optimal inhibition of degranulation (Fig. 2). This suggests a time-dependent effect which is unlikely to be through cGMP, and more likely involves interference with protein or enzyme function, or induction or regulation of gene expression.

Nitric oxide can interact with superoxide to generate the highly reactive peroxynitrite species (46). However, culture of purified rat peritoneal mast cells with superoxide dismutase, which breaks down superoxide, or the peroxynitrite scavenger uric acid, fails to influence the nitric oxide-dependent inhibition of degranulation (23). Also, superoxide is generated by peritoneal macrophages only in the absence of extracellular L-arginine (47). Thus we believe the action of nitric oxide on mast cells does not involve superoxide. The molecular targets for nitric oxide in mast cells remain to be elucidated, but a prime candidate must be thiolated proteins that could be nitrosylated to disrupt their function (48, 49).

NITRIC OXIDE AND MAST CELLS *IN VIVO* – CLINICAL IMPLICATIONS

Studies in rats have thrown light on the regulatory role of nitric oxide in mast cell-dependent processes *in vivo*. Administration of L-NMMA enhanced mucosal mast cell degranulation, seen as elevated plasma rat mast cell protease II, and at the same time enhanced gut epithelial permeability (50). Intravital microscopy has been used to study nitric oxide regulation of mast cell activation and inflammatory cell adherence in the microvasculature of rats *in vivo*. The nitric oxide donor spermine-NO inhibited mast cell degranulation, granulocyte adhesion to blood vessels and rolling, and microvascular leakage of albumin (51). Also, intravenous administration of the nitric oxide synthase inhibitor nitro-L-arginine methyl ester (L-NAME) enhanced degranulation of perivascular mast cells and this was associated with increased adherence of leukocytes to the vascular endothelium (52). These results show that, in the rat, nitric oxide inhibits mast cell activation and subsequent leukocyte recruitment *in vivo*. In contrast, antigen-induced microvascular leakage in guinea pig lungs was suppressed by intravenous injection of L-NAME, suggesting that antigen-induced nitric oxide induces or enhances microvascular leakage in this species (53).

Nitric oxide may inhibit mast cell-dependent cardiac anaphylaxis. Addition of the NOS inhibitor to isolated guinea pig hearts increased antigen-induced histamine release and prolonged antigen-induced arrhythmias, whereas the nitric oxide donor sodium nitroprusside reduced these effects (54).

Studies of nitric oxide effects on mast cell function *in vivo* in humans are of course more difficult, and there is also debate about the cellular sources of nitric oxide, and the abundance of NOS II in humans. Although human macrophages are not a rich source of nitric oxide, nitric oxide is produced by many other human cell types. For example, in human skin, epidermal keratinocytes express type II and type III NOS, endothelial cells, fibroblasts and secretory glands express type III NOS, and mast cells express type I NOS (44). Type I NOS has also been localized to human nasal mucosal mast cells (45). Human epithelial cells are a rich source of nitric oxide produced by type II NOS in response to IFN-γ and other cytokines (55, 56) and therefore these cells are well placed to regulate mast cells. Nitric oxide is generated at high levels by the lungs of asthmatic patients, such that it can be detected readily in exhaled breath (57). Therefore there would appear to be sufficient sources of nitric oxide, presumably generated by type II NOS, to potentially regulate mast cell degranulation and mediator release in lung tissue. Considering the demonstrated *in vivo* mast cell inhibitory role of nitric oxide in animal models, and the high levels of nitric oxide generated in human asthmatic lung, there could be potential for exploiting nitric oxide donors in controlling mast cell-dependent inflammation in lung and other tissues in human inflammatory diseases.

CONCLUSION: MAST CELLS, INTERFERON-γ AND NITRIC OXIDE IN CELL INTERACTIONS AND DISEASE

In conclusion, the relationship between mast cells, IFN-γ and nitric oxide is becoming clearer. IFN-γ activates the synthesis of nitric oxide by type II NOS in rodent macrophages and human epithelial cells and the nitric oxide then directly inhibits mast cell secretory responses. Thus levels of IFN-γ, reflecting NK cell or Th1 cell activation (58), may indirectly influence mast cell-dependent physiological and pathological processes. It is possible that mast cell secretory responsiveness is determined by the intensity and nature of the ongoing immune response. Nitric oxide inhibits mast cell secretory responses directly; therefore nitric oxide from any source could inhibit mast cell activation, provided it is produced at sufficiently high levels. Mast cells themselves do not represent a sufficiently powerful source of nitric oxide, as would be produced by type II NOS, to exert autocrine inhibition of secretory responses, or regulation of other immune or inflammatory cell types. However, low-level nitric oxide synthesis by mast cells, perhaps by type I NOS, could lead to physiological effects, such as dispersion of platelets or relaxation of smooth muscle cells. Figure 3 outlines diagrammatically the roles of IFN-γ and nitric oxide in mast cell interactions with other cells.

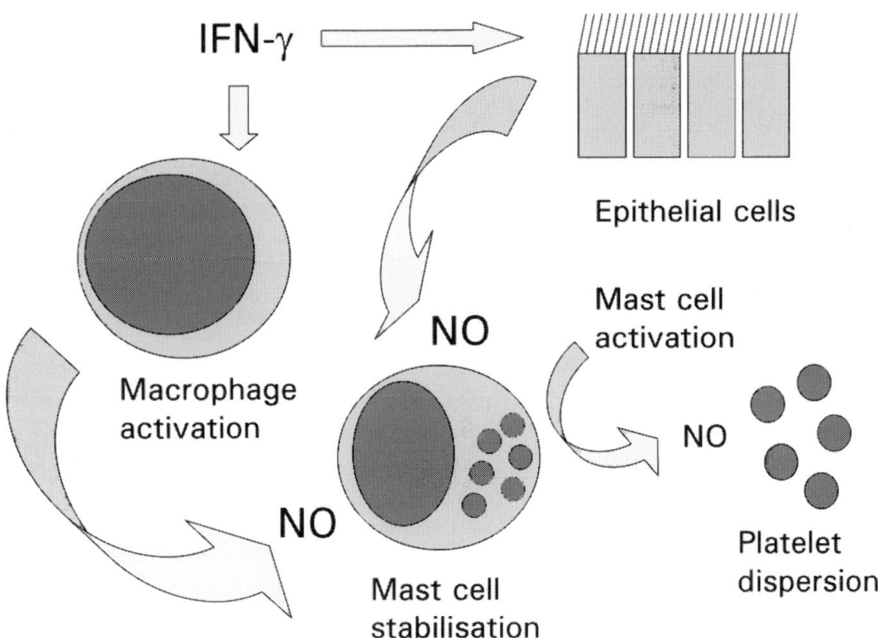

Fig. 3 Schematic representation of the roles of IFN-γ and nitric oxide in mast cell stabilization, and possible physiological effect of mast cell-derived nitric oxide on platelet dispersion. Large font NO as produced at high levels by inducible (type II) NOS; small font NO as produced at low levels by constitutive (type I/III) NOS.

Many important questions remain to be answered:

- Does nitric oxide inhibit mast cell cytokine expression, and, if so, of which cytokines?
- Is the effect of nitric oxide on mast cells selective, or are other inflammatory and immune cells equally readily suppressed?
- Is the nitric oxide effect dependent on mast cell phenotype or microenvironmental factors?
- What is the biochemical mechanism of action of nitric oxide on mast cells?
- Can nitric oxide, produced in human tissues, suppress mast cell activation and mast cell-dependent processes, and, if so, is there any therapeutic potential for nitric oxide in control of inflammatory disease?

ACKNOWLEDGEMENTS

The work of the author is supported by the Medical Research Council and The Wellcome Trust.

REFERENCES

1. Farrar M. A. and Schreiber, R. D. The molecular cell biology of interferon-γ and its receptor. *Annu. Rev. Immunol.* **11**:571–611, 1993.
2. Billiau, A. Interferon-γ. Biology and role in pathogenesis. *Adv. Immunol.* **62**:61–130, 1996.
3. Swieter, M., Ghali, W. A., Rimmer, C. and Befus, A. D. Interferon-α/β inhibits IgE-dependent histamine release from rat mast cells. *Immunology* **66**:606–610, 1989.
4. Coleman, J. W. Conditioned medium from concanavalin A-stimulated spleen cells inhibits the IgE-dependent sensitization of murine peritoneal mast cells *in vitro*. *Immunology* **69**: 150–154, 1990.
5. Coleman, J. W., Buckley, M. G., Holliday, M. R. and Morris, A. G. Interferon-γ inhibits serotonin release from mouse peritoneal mast cells. *Eur. J. Immunol.* **21**:2559–2564, 1991.
6. Holliday, M. R., Banks, E. M. S., Dearman, R. J., Kimber, I. and Coleman, J. W. Interactions of IFN-γ with IL-3 and IL-4 in the regulation of serotonin and arachidonate release from mouse peritoneal mast cells. *Immunology* **82**:70–74, 1994.
7. Bissonnette, E. Y. and Befus, A. D. Inhibition of mast cell-mediated cytotoxicity by IFN-α/β and -γ. *J. Immunol.* **145**:3385–3390, 1990.
8. Bissonnette, E. Y., Chin, B. and Befus, A. D. Interferons differentially regulate histamine and TNF-α in rat intestinal mucosal mast cells. *Immunology* **86**:12–17, 1995.
9. Enciso, J. A., Bissonnette, E. Y. and Befus, A. D. Regulation of messenger-RNA levels of TNF-α and the alpha-chain of the high-affinity receptor for IgE in mast-cells by IFN-γ and α/β. *Int. Arch. Allergy Immunol.* **110**:114–123, 1996.
10. Schleimer, R. P., Derse, C. P., Friedman, B. Gillis, S., Plaut, M., Lichtenstein, L.M. and MacGlashan, D. W. Regulation of human basophil mediator release by cytokines. I. Interaction with antiinflammatory steroids. *J. Immunol.* **143**:1310–1317, 1989.
11. Pierkes, M., Bellinghausen, I., Hultsch, T., Metz, G., Knop, J. and Saloga, J. Decreased release of histamine and sulfidoleukotrienes by human peripheral blood leukocytes after wasp venom immunotherapy is partially due to induction of IL-10 and IFN-γ production of T cells. *J. Allergy Clin. Immunol.* **103**:326–332, 1999.
12. Takagi, M., Koike, K. and Nakahata, T. Antiproliferative effect of IFN-γ on proliferation of mouse connective tissue-type mast-cells. *J. Immunol.* **145**:1880–1884, 1990.
13. Nafziger, J., Arock, M., Guillosson, J.-J. and Wietzerbin, J. Specific high-affinity receptors for interferon-γ on mouse bone marrow-derived mast cells: inhibitory effect of interferon-γ on mast cell precursors. *Eur. J. Immunol.* **20**:113–117, 1990.
14. Kirshenbaum, A. S., Worobec, A. S., Davis, T. A., Goff, J. P., Semere, T. and Metcalfe, D. D. Inhibition

of human mast cell growth and differentiation by interferon-γ1b. *Exp. Hematol.* **26**:245–251, 1998.
15. Hu, Z. Q., Zenda, N. and Shimamura, T. Down-regulation by IL-4 and up-regulation by IFN-γ of mast cell induction from mouse spleen cells. *J. Immunol.* **156**:3925–3931, 1996.
16. Yanagida, M., Fukamachi, H., Takei, M., Hagiwara, T., Uzumaki, H., Tokiwa, T., Saito, H., Iikura, Y. and Nakahata, T. Interferon-γ promotes the survival and FcεRI-mediated histamine release in cultured human mast cells. *Immunology* **89**:547–552, 1996.
17. Warbrick, E. V., Taylor, A. M., Botchkarev, V. A. and Coleman, J. W. Rat connective tissue-type mast-cells express MHC class-II – up-regulation by IFN-γ. *Cell. Immunol.* **163**:222–228, 1995.
18. Frandji, P., Tkaczyk, C., Oskeritzian, C., Lapeyre, J., Peronet, R., David, B., Guillet, J. G. and Mecheri, S. Presentation of soluble-antigens by mast-cells: upregulation by interleukin-4 and granulocyte/macrophage colony-stimulating factor and downregulation by interferon-γ. *Cell. Immunol.* **163**:37–46, 1995.
19. Love, K. S., Lakshmanan, R. R., Butterfield, J. H. and Fox, C. C. IFN-γ-stimulated enhancement of MHC class-II antigen expression by the human mast-cell line HMC-1. *Cell. Immunol.* **170**:85–90, 1996.
20. Grabbe, J., Karau, L., Welker, P., Ziegler, A. and Henz, B. M. Induction of MHC class II antigen expression on human HMC-1 mast cells. *J. Dermatol. Sci.* **16**:67–73, 1997.
21. Frandji, P., Tkaczyk, C., Oskeritzian, C., Desaymard, C. and Mecheri, S. Exogenous and endogenous antigens are differentially presented by mast cells to CD4+ T lymphocytes. *Eur. J. Immunol.* **26**:2517–2528, 1996.
22. Eastmond, N. C., Banks, E. M. S. and Coleman, J. W. Nitric oxide inhibits IgE-mediated degranulation of mast cells and is the principal intermediate in IFN-γ-induced suppression of exocytosis. *J. Immunol.* **159**:1444–1450, 1997.
23. deSchoolmeester, M. L., Eastmond, N. C., Dearman, R. J., Kimber, I., Basketter, D. A. and Coleman, J. W. Reciprocal effects of interleukin-4 and interferon-γ on immunoglobulin E-mediated mast cell degranulation: a role for nitric oxide but not peroxynitrite or cyclic guanosine monophosphate. *Immunology* **96**:138–144, 1999.
24. Williams, C. M. M. and Coleman, J. W. Induced expression of messenger-RNA for IL-5, IL-6, TNF-α, MIP-2 and IFN-γ in immunologically activated rat peritoneal mast cells – inhibition by dexamethasone and cyclosporin-A. *Immunology* **86**:244–249, 1995.
25. Burd, P. R., Rogers, H. W., Gordon, J. R., Martin, C. A., Jayaraman, S., Wilson, S. D., Dvorak, A. M., Galli, S. J. and Dorf, M. E. Interleukin-3-dependent and interleukin-3-independent mast-cells stimulated with IgE and antigen express multiple cytokines. *J. Exp. Med.* **170**:245–257, 1989.
26. Gupta, A. A., Leal-Berumen, I., Croitoru, K. and Marshall, J. S. Rat peritoneal mast cells produce IFN-γ following IL-12 treatment but not in response to IgE-mediated activation. *J. Immunol.* **157**:2123–2128, 1996.
27. Ackermann, L., Harvima, I. T., Pelkonen, J., Ritamaki-Salo, V., Naukkarinen, A., Harvima, R. J. and Horsmanheimo, M. Mast cells in psoriatic skin are strongly positive for interferon-γ. *Br. J. Dermatol.* **140**:624–633, 1999.
28. Moncada, S., Palmer, R. M. J. and Higgs, E. A. Nitric oxide: physiology, pathophysiology and pharmacology. *Pharmacol. Rev.* **43**:109–142, 1991.
29. Lincoln, J., Hoyle, C. H. V. and Burnstock, G. *Nitric Oxide in Health and Disease.* Cambridge University Press, Cambridge, 1997.
30. Langrehr, J. M., Hoffman, R. A., Lancaster, J. R. and Simmons, R. L. Nitric oxide – a new endogenous immunomodulator. *Transplantation* **55**:1205–1212, 1993.
31. Kolb, H. and Kolb-Bachofen, V. Nitric oxide in autoimmune disease: cytotoxic or regulatory mediator? *Immunol. Today* **19**:556–561, 1998.
32. Wei, X. Q., Charles, I. G., Smith, A., Ure, J., Feng, C. J., Huang, F. P., Xu, D. M., Muller, W., Moncada, S. and Liew, F. Y. Altered immune-responses in mice lacking inducible nitric-oxide synthase. *Nature* **375**:408–411, 1995.
33. Salvemini, D., Masini, E., Pistelli, A., Mannaioni, P. F. and Vane, J. Nitric-oxide – a regulatory mediator of mast-cell reactivity. *J. Cardiovasc. Pharmacol.* **17**:S258–S264, 1991.
34. Masini, E., Salvemini, D., Pistelli, A., Mannaioni, P. F. and Vane, J. R. Rat mast cells synthesize a nitric oxide like-factor which modulates the release of histamine. *Agents Actions* **33**:61–63, 1991.
35. Masini, E., Mannaioni, P. F., Pistelli, A., Salvemini, D. and Vane, J. Impairment of the L-arginine–nitric oxide pathway in mast cells from spontaneously hypertensive rats. *Biochem. Biophys. Res. Commun.* **28**:1178–1182, 1991.
36. Bidri, M., Becherel, P.-A., Le Goff, L., Pieroni, L., Guillosson, J.-J., Debre, P. and Arock, M.

Involvement of cyclic nucleotides in the immunomodulatory effects of nitric oxide on murine mast cells. *Biochem. Biophys. Res. Commun.* **210**:507–517, 1995.

37. Iikura, M., Takaishi, T., Hirai, K., Yamada, H., Iida, M., Koshino, T. and Morita, Y. Exogenous nitric oxide regulates the degranulation of human basophils and rat peritoneal mast cells. *Int. Arch. Allergy Immunol.* **115**:129–136, 1998.
38. Kim, H. M. Nitric oxide released from Swiss 3T3 fibroblasts acts as a cytostatic agent for cultured mast cells. *Pharmacology* **55**:285–291, 1997.
39. Kitajima, I., Kawahara, K., Nakajima, T., Soejima, Y., Matsuyama, T. and Maruyama, I. Nitric oxide-mediated apoptosis in murine mastocytoma. *Biochem. Biophys. Res. Commun.* **204**:244–251, 1994.
40. Salvemini, D., Masini, E., Anggard, E., Mannaioni, P. F. and Vane, J. Synthesis of a nitric oxide-like factor from L-arginine by rat serosal mast cells: stimulation of guanylate cyclase and inhibition of platelet aggregation. *Biochem. Biophys. Res. Commun.* **169**:596–601, 1990.
41. Bissonnette, E. Y., Hogaboam, C. M., Wallace, J. L. and Befus, A. D. Potentiation of tumor necrosis factor-α-mediated cytotoxicity of mast cells by their production of nitric oxide. *J. Immunol.* **147**:3060–3065, 1991.
42. Hogaboam, C. M., Befus, A. D. and Wallace, J. L. Modulation of rat mast cell reactivity by IL-1β. Divergent effects on nitric oxide and platelet-activating factor release. *J. Immunol.* **151**:3767–3774, 1993.
43. Bidri, M., Ktorza, S., Vouldoukis, I., LeGoff, L., Debre, P., Guillosson, J. J. and Arock, M. Nitric oxide pathway is induced by FcεRI and up-regulated by stem cell factor in mouse mast cells. *Eur. J. Immunol.* **27**:2907–2913, 1997.
44. Shimizu, Y., Sakai, M., Umemura, Y. and Ueda, H. Immunohistochemical localization of nitric oxide synthase in normal human skin: expression of endothelial-type and inducible-type nitric oxide synthase in keratinocytes. *J. Dermatol.* **24**:80–87, 1997.
45. Bacci. S., Rucci, L., Riccardi-Arbi, R. and Borghi-Cirri, M. B. Colocalization of tumor necrosis factor-α and nitric oxide-synthase immunoreactivity in mast cell granules of nasal mucosa. *Histol. Histopathol.* **13**:1011–1014, 1998.
46. Wink, D. A., Hanbauer, I., Grisham, M. B., Laval, F., Nims, R. W., Laval, R., Cook, J., Pacelli, R., Liebman, J., Krishna, M., Ford, P. C. and Mitchell, J. B. Chemical biology of nitric oxide: regulation and protective and toxic mechanisms. *Curr. Top. Cell. Regul.* **34**:159–187, 1996.
47. Xia, Y. and Zweier, J. L. Superoxide and peroxynitrite generation from inducible nitric oxide synthase in macrophages. *Proc. Natl. Acad. Sci. USA* **94**:6954–6958, 1997.
48. Stamler, J. S., Simon, D. I., Osborne, J. A., Mullins, M. E., Jaraki, O., Michel, T., Singel, D. J. and Loscalzo, J. *S*-nitrosylation of proteins with nitric oxide: synthesis and characterisation of biologically active compounds. *Proc. Natl. Acad. Sci. USA* **89**:444–448, 1992.
49. Stamler, J.S. Redox signalling: nitrosylation and related target interactions of nitric oxide. *Cell* **78**:931–936, 1994.
50. Kanwar, S., Wallace, J. L., Befus, D. and Kubes, P. Nitric oxide synthesis inhibition increases epithelial permeability via mast cells. *Am. J. Physiol.* **266**:G222–G229, 1994.
51. Gaboury, J. P., Niu, X. F. and Kubes, P. Nitric oxide inhibits numerous features of mast cell-induced inflammation. *Circulation* **15**:318–326, 1996.
52. Kimura, M., Mitani, H., Bandoh, T., Totsuka, T. and Hayashi, S. Mast cell degranulation in rat mesenteric venule: effects of L-NAME, methylene blue, and ketotifen. *Pharmacol. Res.* **39**:397–402, 1999.
53. Miura, M., Ichinose, M., Kageyama, N., Tomaki, M., Takahashi, T., Ishikawa, J., Ohuchi, Y., Oyake, T., Endoh, N. and Shirato, K. Endogenous nitric oxide modifies antigen-induced microvascular leakage in sensitized guinea pig airways. *J. Allergy Clin. Immunol.* **98**:144–151, 1996.
54. Masini, E., Gambassi, F., Di Bello, M.G., Mugnai, L., Raspanti, S. and Mannaioni, P. F. Nitric-oxide modulates cardiac and mast-cell anaphylaxis. *Agents Actions* **41**:C89–C90, 1994.
55. Asano, K., Chee, C. B., Gaston, B., Lilly, C. M., Gerard, C., Drazen, J. M. and Stamler, J. S. Constitutive and inducible nitric oxide synthase gene expression, regulation, and activity in human lung epithelial cells. *Proc. Natl. Acad. Sci. USA* **91**:10089–10093, 1994.
56. Robbins, R. A., Barnes, P. J., Springall, D. R., Warren, J. B., Kwon, O. J., Buttery, L. D., Wilson, A. J., Geller, D. A. and Polak, J. M. Expression of inducible nitric oxide synthase in human lung epithelial cells. *Biochem. Biophys. Res. Commun.* **203**:209–218, 1994.
57. Kharitonov, S. A., Yates, D., Robbins, R. A., Logan-Sinclair, R., Shinebourne, E. A. and Barnes, P. J. Increased nitric oxide in exhaled air of asthmatic patients. *Lancet* **343**:133–135, 1994.
58. Mosmann, T. R. and Coffman, R. L. Th1-cell and Th2-cell – different patterns of lymphokine secretion lead to different functional properties. *Annu. Rev. Immunol.* **7**:145–173, 1989.

SECTION FIVE

STRUCTURE AND FUNCTION OF MAST CELL PROTEASES

CHAPTER 16

Human and Mouse Mast Cell Tryptases

RICHARD L. STEVENS

Harvard Medical School and Brigham and Women's Hospital, Boston, Massachusetts, U.S.A.

INTRODUCTION

Tryptases are major constituents of the secretory granules of mast cells (MCs) and five of the eight and three of the 14 neutral proteases which have been cloned from human and mouse MCs, respectively, are members of this family of serine proteases. Tryptase-specific antibodies and cDNA probes have been invaluable reagents for investigators interested in understanding how MC-committed progenitors home to tissues, proliferate, differentiate and mature. Although substantial progress has been made in our understanding of the factors and mechanisms that regulate tryptase expression, the physiological substrates of these proteases are just beginning to be identified.

Tryptases have been found in every mammalian species that has been examined to date. Thus, their genes evolved relatively early in the evolution of mammals. The observation that humans express more tryptases than mice also indicates that there was strong evolutionary pressure to increase the number of these neutral proteases because of their importance in inflammatory responses. Although mouse MC protease (mMCP)-6, mMCP-7 and mMCP-11/transmembrane tryptase (mTMT) are highly homologous to one another, these three tryptases are functionally distinct.

The amino acid sequences of human tryptases α and II/β are 93% identical, yet, even in this instance, the residues which comprise their respective substrate-binding clefts differ substantially. As the number of tryptase genes increased in humans, there was strong evolutionary pressure for the newly generated genes to mutate certain residues in the seven loops that form the substrate-binding cleft of each protease to create a panel of functionally distinct tryptases. The recent discovery that the $Fc_{\varepsilon}RI^{+}$/metachromatic cells in the blood of patients with varied allergic disorders have abnormally high levels of multiple tryptases highlight the need to understand the function of these granule proteases in humans and mice.

ISBN 0-12-473335-2

IDENTIFICATION AND CLONING OF HUMAN TRYPTASE GENES

Using a histochemical approach, Glenner and Cohen (1) concluded that most human MCs contain tryptic-like neutral proteases. The subsequent observation that IgE/antigen-activated human lung MCs exocytose serine proteases which can cleave tosyl-Arg methyl ester led to the conclusion that these tryptases are preferentially stored in the cell's secretory granules (2). Although in retrospect the purified tryptase was almost certainly a complex mixture of highly homologous proteases, Schwartz *et al.* (3) isolated a 140-kDa tryptase from dissociated human lung MCs which consisted of four ~35-kDa subunits. A monoclonal antibody (4) generated against this tryptase preparation was used to isolate the human tryptase α cDNA from a lung expression cDNA library (5). Re-screening of the library resulted in the isolation of the human tryptase β cDNA (6). Because the two cDNAs were 93% identical, it was not clear in 1990 whether the homologous cDNAs originated from different genes or from two alleles of the same gene. Polymerase chain reaction analysis of human/hamster somatic hybrids, however, did show at that time that the tryptase(s) gene resided on human chromosome 16 (6).

Using a canine tryptase cDNA (7), Vanderslice *et al.* (8) screened a human skin cDNA library in 1990 to isolate the cDNAs that encode the three nearly identical human tryptases, designated I, II and III. Analysis of the nucleotide sequences of the resulting five cDNAs from the two groups revealed that human tryptases β and II are identical. Thus, to give equal credit to both groups, this neutral protease is now designated here as human tryptase II/β. Assuming the patient used to prepare the skin cDNA library did not have polyploid chromosomes, the Vanderslice *et al.* (8) nucleotide sequence and genomic blot data were the first indicators that human chromosome 16 possesses multiple genes which encode nearly identical tryptases.

The first human tryptase gene to be sequenced in its entirety was the tryptase I gene (8). Although this gene consists of six exons, the gene is only 1.8 kilobase (kb) in size. Its first exon encodes the 5′ untranslated region (UTR), its second exon encodes the hydrophobic signal peptide, and its remaining exons encode the mature portion of the translated protease. The codons which correspond to the catalytic triad of the mature enzyme reside on exons 3, 4 and 6, respectively. The most striking difference between the tryptase I and trypsin genes is the additional intron in the former gene which separates the portion of the transcript which corresponds to the 5′ UTR and its translated signal peptide. Subsequent mapping and sequencing analyses of human chromosome 16p13.3 finally established that the varied human tryptase transcripts are derived from distinct genes (9, 10).

As noted below, a less-related tryptase (designated mTMT) has been cloned from mouse MCs (10). Screening of genomic and cDNA libraries with a mTMT probe ultimately resulted in the isolation and characterization of the 3.1-kb hTMT gene. Pallaoro *et al.* (9) identified two additional mMCP-7-like genes (or two different alleles of a single mMCP-7-like gene) in the human genome which may be pseudogenes because at least one of them encodes a truncated protein. Although azurocidin is unable to cleave proteins due to mutations in two of the three residues in its catalytic triad, this neutrophil-specific serine protease homolog exhibits potent antimicrobial activity (11). Thus, a truncated tryptase which is unable to cleave a protein still might be functionally important. If the mMCP-7-like genes are expressed in human MCs and if they encode functional proteins, they would represent the sixth and seventh human tryptases.

Although allelic variation of a gene does not occur in an inbred animal such as the BALB/c mouse, allelic differences can be considerable in the human genome. Thus, the actual number of human tryptase genes cannot be conclusively determined until the nucleotide sequence of the entire human genome is deduced.

IDENTIFICATION AND CLONING OF MOUSE TRYPTASE GENES

In 1981, Razin *et al.* (12) and others noted that a relative homogeneous population of MCs (designated mBMMCs) developed when mouse bone marrow cells were cultured for 3 weeks in the presence of T cell-conditioned media. While these interleukin-(IL)-3-dependent (13) MCs were relatively immature in terms of their granulation, DuBuske *et al.* (14) noted that they expressed at least four serine proteases. Based on their ability to cleave the trypsin-susceptible substrates tosyl-Arg methyl ester and *N*-CBz-Lys thiobenzyl ester, it was predicted that two or more of the different-sized proteases were tryptases.

The derivation of retrovirus-immortalized mouse MC lines (15) and the development of highly sensitive amino acid sequencing techniques for SDS–PAGE-resolved proteins (16) were essential technological advances for obtaining the N-terminal amino acid sequence data (17) which enabled the cloning of the cDNAs and genes that encode many of the mouse MC proteases. The first five serine proteases identified in mouse MCs possessed chymotrypsin-like features. However, the sixth protease (mMCP-6) found in the lysates of serosal MCs and KiSV-MCs was a 32-kDa protein with features similar to those of pancreatic trypsin. Based on the obtained amino acid sequence data (17), a redundant oligonucleotide was used to isolate the 1.8-kb mMCP-6 gene and its 1-kb transcript (18). Chromosome mapping (19) and nucleotide sequencing data (18) revealed that the mMCP-6 gene consists of six exons.

Re-screening of mouse cDNA and genomic libraries with a full-length mMCP-6 cDNA under conditions of moderate stringency resulted in the isolation of the homologous 2.3-kb mMCP-7 gene and its 1.2-kb cDNA (20). Comparison of the 5′ end of the mMCP-7 transcript with its gene revealed that the region of the gene which corresponds to the first intron in the mMCP-6 and human tryptase I genes is not removed during the post-transcriptional processing of the mMCP-7 transcript because of a point mutation. This splice-site mutation causes the processed mMCP-7 transcript to have a 5′ UTR of one hundred and ninety five nucleotides. Although its 5′ UTR is longer than that of any other known MC-specific protease transcript, the mMCP-7 transcript is translated (21, 22) and the processed protein is a functional tryptase (23).

The mMCP-6 and mMCP-7 genes reside ~1.2 centiMorgans apart on the syntenic region of chromosome 17 (19, 24). The observation that serine protease genes in large superfamilies often reside within 7 kb of one another on their respective chromosomes (25, 26), coupled with the observation that there are at least five human tryptase genes (8–10), raised the possibility that other mouse tryptase genes are located on the chromosome in between the mMCP-6 and mMCP-7 genes. A chromosome walk-approach was therefore carried out to isolate the 3.7-kb mMCP-11/mTMT gene and its 1.2-kb transcript (10). The mTMT gene resides 2.3 kb 3′ of the mMCP-6 gene, and both genes are orientated in the same direction on the chromosome. Like the mMCP-6 gene, the mTMT gene contains five exons. Three to five genomic fragments were obtained when a mouse genomic DNA blot was probed under conditions of moderate stringency with a mTMT cDNA. Because these weaker hybridizing fragments did not correspond to the mMCP-6 or mMCP-7 genes, there appear to be at least two additional mTMT-like genes in the mouse genome which have yet to be sequenced.

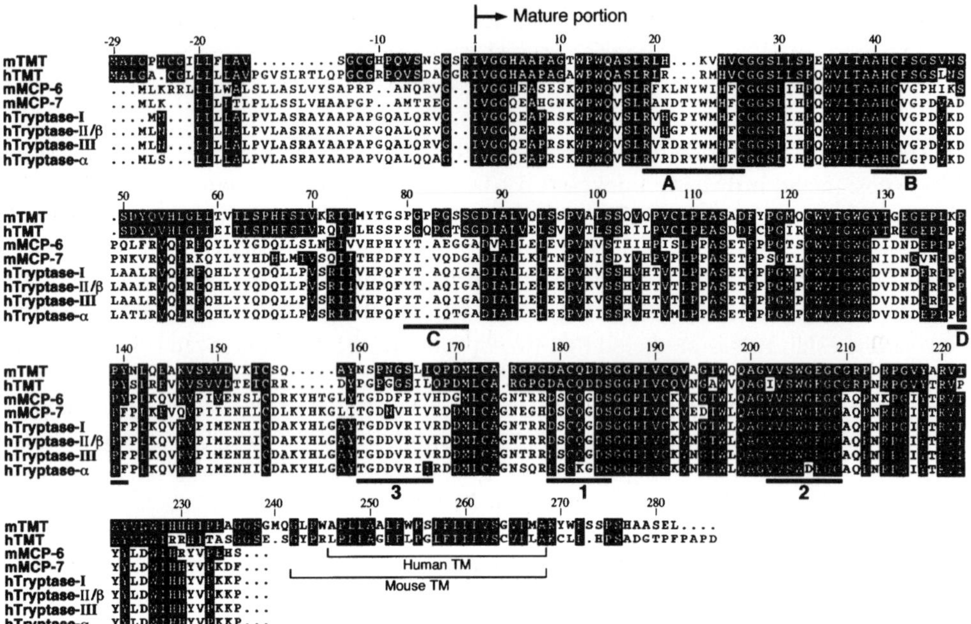

Fig. 1 Comparison of the amino acid sequences of the three known mouse tryptases and five known human tryptases. The amino acid sequences of the mouse tryptases mMCP-6, mMCP-7 and mTMT, and of the human tryptases I, II/β, III and hTMT are compared. Identical amino acids in the sequences are shaded. Numbering begins at the first residue in the mature portion of the tryptase. The transmembrane segments of mTMT and hTMT are bracketed. The seven putative loops (designated A–D and 1–3) that form the substrate-binding pockets of these tryptases are underlined. (Reproduced from *J. Biol. Chem.* **274**: 30784–30793, 1999.)

BIOCHEMICAL FEATURES OF HUMAN AND MOUSE TRYPTASES

Like all other serine proteases, the mouse and human tryptases are initially translated as 'pre-pro' zymogens (Fig. 1). mTMT, for example, is predicted to be translated as an inactive precursor consisting of a signal peptide of 19 residues, a propeptide of ten residues, and a mature domain of 282 residues (10). Most mouse and human tryptases do not have the membrane-spanning segment which is present in mTMT and hTMT (Fig. 1). Thus, their mature domains consist of approximately 245 residues. While pro-mMCP-6 and pro-mMCP-7 also possess ten-residue pro-peptides, the amino acid sequences of their activation peptides differ considerably from that of pro-mTMT. All mouse MC chymases have a Glu/Gly-Glu propeptide (26–30). Because no mouse or human tryptase has a comparable pro-peptide, it is unlikely that human and mouse tryptases are activated in MCs by the same dipeptidylpeptidase I-dependent pathway which activates the chymase family of serine proteases whose genes reside on chromosome 14 (31).

Sakai *et al.* (32) were unable to generate enzymatically active human tryptase α in insect cells. Except for residue –3, the pro-peptides of human tryptases α, I, II/β and III are 100% identical (5, 6, 8). Residue –3 is a Gln in tryptase α but is an Arg in tryptases I,

II/β, and III. Although residue –3 is Gly, Gly and Glu in mTMT (10), hTMT (10) and a gerbil tryptase (33), respectively, Sakai *et al.* (32) proposed that tryptases undergo an unusual multistep activation process in MCs which is exquisitely dependent on heparin and Arg $^{-3}$. In order to evaluate experimentally the postulated importance of Arg $^{-3}$ in tryptase expression, Huang *et al.* (34) induced insect cells to express a Gln $^{-3}$ derivative of human tryptase II/β. Because the mutated tryptase was secreted into the conditioned media and because it could be activated *in vitro*, it now appears that Arg $^{-3}$ is not needed for the proper folding of tryptases or the alignment of their disulphide bonds.

The perceived importance of heparin in the maturation process of tryptase zymogens was experimentally addressed by Humphries *et al.* (35) in their analysis of a transgenic mouse which is unable to express heparin due to targeted disruption of the *N*-deacetylase/*N*-sulphotransferase-2 (NDST-2) gene. Although the mBMMCs developed from NDST-2 null mice were unable to store carboxypeptidase A and the chymase mMCP-5 in their granules, these MCs were able to store enzymatically active tryptases. Confirmation that heparin is not essential for maturation of tryptase zymogens was obtained by Mirza *et al.* (36) when they discovered that it was possible to generate enzymatically active human tryptases α and II/β in a mammalian cell which is unable to produce heparin. While the Mirza (36) and Humphries (35) studies now suggest that pro-tryptase α is converted into an enzymatically active tryptase in human MCs, confirmation is needed. Whether or not tryptase α is a functional protein is a clinically important question because the levels of tryptase α are elevated in the blood of patients with systemic mastocytosis (37) and in the circulating metachromatic cells in the blood of patients with varied allergic disorders (38). Immunoreactive tryptase α also has been detected in the synovial fluid of some arthritis patients (39).

The mature forms of all human and mouse tryptases have an N-terminal amino acid sequence of Ile-Val-Gly-Gly. Although all MC chymases have an N-terminal amino acid sequence of Ile-Ile-Gly-Gly, the relative importance of the second residue in these two families of proteases has not been determined experimentally. The mature domains of all mouse and human tryptases have at least one potential *N*-linked glycosylation site. SDS–PAGE/immunoblot analysis of mBMMCs lysates before and after *N*-glycanase treatment revealed that mMCP-7 is glycosylated (21). Whether or not mature mMCP-6 or mTMT contain *N*-linked glycans has not been determined. Nevertheless, at least two human tryptases are glycosylated (40, 41). The observation that those human MC tryptases which form tetramers have an *N*-linked glycosylation site near the conserved Trp-rich domain (5, 6, 8) which is required for tetramer formation (42) raised the possibility that mannose-type oligosaccharides might be needed for tetramer formation. Thus, both Asn^{21} and Asn^{102} in mMCP-7 were mutated to Gln to obtain a carbohydrate-deficient mutant of this tryptase (42). Analysis of the resulting mutant indicated that *N*-linked glycans are not essential for enzymatic activity or tetramer formation. However, they contribute substantially to the overall thermal stability of the tryptase.

Most tryptases have specific Tyr-, Pro- and Trp-rich domains which are needed for tetramer formation (42–44). Because these domains are not present in hTMT and mTMT, it is unlikely that these two tryptases are able to form a similar tetramer structure. All mouse and human tryptases have eight conserved Cys residues which are needed for the formation of four disulphide bonds in each folded protease. mTMT and hTMT have one and four additional Cys residues, respectively. A tryptase-like protease has been cloned from dog mastocytoma cells (7) which has four more Cys residues than mMCP-6. Because this dog tryptase forms a tetramer bearing disulphide-linked monomers (45), it is possible that the additional Cys residues in mTMT and hTMT allow theses tryptases to form dimers with themselves or with other proteins.

The amino acid sequences of mature mMCP-6 and mMCP-7 are 71% identical, whereas mTMT is only 45% and 46% identical to mMCP-6 and mMCP-7, respectively. If the dissimilar pre-pro-peptide and C-terminal extension peptide of mTMT are taken into account, the extent of homology of mTMT with the other two members of its family is considerably lower. At pH 5.5, the pH of the secretory granule, mMCP-6 and mMCP-7 have net charges of +6 and +4, respectively, and are ionically bound to serglycin proteoglycans like the granule chymases. Models of the three-dimensional structures of mMCP-6 and mMCP-7 have revealed that each properly folded tryptase has at least one positively charged region on its surface which is located far from its substrate-binding cleft (22, 46). Site-directed mutagenesis of this region in pro-mMCP-7 confirmed that it is the face which binds to serglycin proteoglycans (46). Because this face is His-rich in mMCP-7 (46), this tryptase is able to selectively dissociate from its proteoglycan when the mMCP-7/proteoglycan complex is exocytosed into a neutral pH environment (22). At pH >6.5, mature mTMT has an overall charge of –6 (10). However, at pH <6.5, mature mTMT has an overall charge of +4 and these positively charged residues are aligned predominantly on one face. Thus, it is possible that mTMT and hTMT also interact with negatively charged molecules during their biosynthesis and/or catabolism.

Because trypsin is enzymatically active in its ~30-kDa monomer state, it was a surprise when Schwartz *et al.* (3) and others discovered that certain human MC tryptases exist as tetramers. While it has been proposed that heparin is needed for human lung- and skin-derived tryptases to form stable, enzymatically active tetramers (47–49), Mirza *et al.* (36) were able to generate enzymatically active human tryptases α and II/β in heparin-negative COS cells. Pereira *et al.* (44) also were able to crystallize human tryptase II/β in the absence of heparin. More conclusive evidence that tetramer formation is not exclusively dependent on heparin was obtained in the mouse. When the V3 mastocytosis mouse was induced to undergo passive systemic anaphylaxis *in vivo*, the exocytosed mMCP-7 was able to circulate in the blood for longer than 1 h as an enzymatically active, homotypic tetramer free of proteoglycan (22). As noted above, the transgenic mice which are unable to express fully sulphated heparin are able to produce and store enzymatically active tryptases in their secretory granules (35). When recombinant mMCP-6 and mMCP-7 zymogens are initially expressed in insect cells, they are secreted from the cells in their inactive monomer states. However, when their bioengineered pro-peptides are removed, the tryptases spontaneously undergo a structural change to form the enzymatically active tetramers (42). The enzymatic activity of mMCP-7 tetramer is substantially greater than that of the sum of four mMCP-7 monomers. Nevertheless, because enzymatically active, recombinant mMCP-6 and mMCP-7 can be generated in the absence of a glycosaminoglycan (23, 50), the proteoglycan-binding domain on the surface of each tryptase is not the most important structural determinant for either inducing or maintaining the tetramer state.

Analysis of the MCs in NDST-2 null mice revealed that disruption of heparin biosynthesis does not impact tryptase expression as long as the MCs contain chondroitin sulphate E (35). All MCs express serglycin proteoglycans (also known as secretory granule proteoglycans) (51–53), but the nature of the glycosaminoglycan attached to serglycin varies considerably depending on the MC's tissue microenvironment. Chondroitin sulphate/heparin hybrid proteoglycans have been found in some cultured MCs (54). However, in most instances, only one type of glycosaminoglycan is attached to a newly expressed serglycin peptide core. The MCs which reside in the peritoneal cavity, lymph node and jejunal epithelium of rodents contain serglycin proteoglycans with predominantly heparin (55), chondroitin sulphate D (56) and chondroitin sulphate E/diB (57), respectively. mBMMCs developed with IL-3 synthesize predominantly chondroitin sulphate E proteoglycans (13, 58), but these MCs will produce heparin proteoglycans if they are subsequently exposed to fibroblasts (59) or c-*kit* ligand/stem cell factor (60). The MCs residing in human lung produce heparin and chondroitin sulphate E proteoglycans in an approximately 2:1 molar ratio (61–63).

Thirty enzymes or more are required to synthesize the varied types of glycosaminoglycans attached to serglycin. The reason why MCs expend so much energy to produce diverse types of serglycin proteoglycans has perplexed investigators for more than a decade. Nevertheless, the fact that the mechanisms which control the post-translational modification of serglycin have been conserved for at least 80 million years indicates their importance. Cutaneous MCs which cannot produce heparin are unable to package carboxypeptidase A and mMCP-5 in their granules (35). However, these same MCs are able to store normal levels of mMCP-7 and at least some mMCP-6 in their granules. In addition, the chondroitin sulphate diB/E population of MCs which resides in the jejunal epithelium of these transgenic mice are able to store normal levels of the chymase mMCP-2. These data suggest that the type of glycosaminoglycan attached to serglycin is of critical importance for controlling which specific cassettes of proteases are allowed to be packaged in the cell's secretory granule and in what molar ratio. The reason why human lung MCs produce so much more chondroitin sulphate E than heparin now can be explained by the requirement of this cell to package more tryptases in its granules relative to the heparin/chymase-rich MCs which reside in the mouse peritoneal cavity. Every functional study on mouse and human tryptases has focused on the cofactor role of heparin. The data from the NDST-2 null mouse now suggest that the more relevant cofactor for mMCP-6 probably is chondroitin sulphate E. Heparin restricts the substrate specificities of mMCP-6 (50) and rMCP-I (64). Because chondroitin sulphate E and heparin are so different structurally, it is highly unlikely that these two glycosaminoglycans influence the substrate specificities of their associated tryptases in identical manners. Thus, the cofactor role of varied types of serglycin proteoglycans in tryptase function needs to be re-evaluated.

Because other serine proteases do not form tetramers, the functional significance of the multimeric unit remains to be determined. Pereira *et al.* (44) recently described the crystal structure of human tryptase II/β complexed to 4-amidinophenylpyruvic acid. Assuming the deduced structure of the tryptase/inhibitor complex is representative of the native tryptase in solution, the most surprising feature of the tetramer unit is the arrangement of the monomers with their substrate-binding clefts facing inside the 50×30 Å central pore. This observation suggests that the tetramer unit is a novel mechanism the MC uses to physically restrict the substrate specificities of its tryptases. mMCP-7 also forms a tetramer (22) and one of its physiological substrates is the α chain of fibrinogen (23). Fibrinogen is a 340-kDa dimer with each monomer consisting of three distinct chains (65). Because the crystal structure of the human tryptase II/β tetramer predicts that the substrate-binding domains of the mMCP-7 monomers are buried inside the tetramer, it is unclear how mMCP-7 can specifically cleave an internal site in fibrinogen's α chain.

Recent studies carried out with recombinant mMCP-6 and mMCP-7 have revealed that these two tryptases can form heterotypic complexes *in vitro* (42). This new finding offers an explanation as to why a portion of exocytosed mMCP-7 is retained for an extended period of time in connective tissues whereas homotypic tetramer complexes of mMCP-7 are able to rapidly exit inflammatory sites (22). Because the proteoglycan domain of mMCP-6 is rich in Lys and Arg residues, exocytosed mMCP-6/proteoglycan complexes cannot easily dissociate in the extracellular matrix. Thus, mMCP-6/mMCP-7 heterotypic tetramers would be retained in tissues if the tetramers are ionically bound to the proteoglycan via their mMCP-6 monomers rather than their mMCP-7 monomers. Assuming human tryptases α, I, II/β and III also can form heterotypic tetramers, the unusual structural unit might have evolved simply to allow the assembly of different cassettes of functionally distinct neutral proteases so that the MC can vary its proteolytic attack on a multidomain protein.

EXPRESSION OF MOUSE AND HUMAN TRYPTASES IN CELLS AND TISSUES

RNA blot analyses revealed that the mMCP-6 and mMCP-7 genes are transcribed in IL-3-developed BALB/c mBMMCs (18, 20) and in those *in vivo*-differentiated MCs which reside in the ear, heart, tongue and skin of the BALB/c mouse (66). Although neither tryptase transcript is present in the expanded population of MCs which reside in the jejunal epithelium during helminth infection, many of the MCs residing in the jejunum during the recovery phase of helminth infection express both tryptases (67). While these data initially suggested that the mMCP-6 and mMCP-7 transcripts are coordinately expressed in all populations of MCs, it was subsequently discovered that the mMCP-6$^+$ MCs residing in the peritoneal cavity of the BALB/c mouse do not express mMCP-7 (66). Kinetic analysis of tryptase expression in mBMMCs (20) and adoptive transfer experiments carried out in BALB/c mice with the V3-MC line (68) confirmed that mouse MCs can differentially express their tryptases.

Transcription of the mMCP-6 and mMCP-7 genes occurs within 1 week after bone marrow cells are cultured in the presence of IL-3 (20). The chymase mMCP-5 is also expressed early during the differentiation of committed progenitors into MCs (29). However, unlike the mMCP-5 transcript whose level remains relatively constant in these cultured cells, the levels of the two tryptase transcripts decrease progressively the longer the MCs are exposed to IL-3. The observation that the cellular level of the mMCP-7 transcript decreases at a much faster rate than the mMCP-6 transcript (20) was the first indicator that tryptase expression at the mRNA level is differentially regulated in MCs by a transcription and/or post-transcriptional mechanism.

The V3-MC line contains mMCP-6 mRNA but not mMCP-7 mRNA (68). Nevertheless, when this v-*abl*-immortalized cell line is adoptively transferred into BALB/c mice, those immortalized V3-MCs which home to the spleen express both tryptases. RNA blot analysis of varied mouse tissues indicated that mTMT is also expressed in a tissue-dependent manner (10). Because the mTMT transcript is not abundant in skeletal muscle which is rich in mMCP-6$^+$/mMCP-7$^+$ MCs, mTMT is not coordinately expressed in tissues with mMCP-6 or mMCP-7. Of those analysed tissues, the intestine contains the highest level of mTMT mRNA.

The chymases mMCP-2 and mMCP-4 are expressed in a strain-dependent manner (66, 69). Thus, varied mouse strains were evaluated to determine whether or not they differ in their tryptase expression. These studies ultimately revealed that the MCs in the skin of the

C57BL/6 mouse lacked mMCP-7 protein (21) because of a point mutation at the gene's exon 2/intron 2 splice site (70). Of the four categories of splice site mutations which have been identified in genome databases, 51%, 32%, 11% and 6% of the mutations result in exon skipping, activation of a cryptic splice site, creation of a pseudo exon within an intron, and intron retention, respectively. The mMCP-7 gene in the C57BL/6 strain is the only mammalian gene that has been found so far that has undergone two splice site mutations which cause the retention of an intron and the activation of a cryptic splice site at two different positions in the gene.

mBMMCs developed from W/W^v and C57BL/6 mice contain high levels of mTMT mRNA and this tryptase transcript is expressed early during the differentiation of uncommitted progenitors into MCs (10). Based on these data, it was a surprise to discover that the steady-state level of the mTMT transcript is below detection in BALB/c and 129/Sv mBMMCs. Unlike the mMCP-7 gene in C57BL/6 mice (70), sequence analysis failed to reveal an obvious defect in the mTMT gene in BALB/c and 129/Sv mice. The level of mTMT mRNA in the intestine of the BALB/c mouse is substantially lower than that in the intestine of the C57BL/6 mouse. Because its presence indicates that the mTMT gene is transcribed *in vivo* in the BALB/c mouse, there may be a strain-dependent genetic defect in the gene's promoter or an associated transcription factor.

Most mTMT transcripts in the intestine of the C57BL/6 mouse and its mBMMCs are ~1.2 kb in size. However, larger sized transcripts are occasionally detected. The mTMT and hTMT cDNAs lack classical AATAAA polyadenylylation regulatory sites in their respective 3′ UTRs. Thus, it is likely that some mTMT and hTMT transcripts have longer 3′ UTRs due to inefficient use of the first available polyadenylylation site after the translation-termination codon. While it was initially thought that the mMCP-2 and mMCP-4 genes were not transcribed in BALB/c mBMMCs, subsequent studies revealed that these chymase transcripts are expressed but often are rapidly degraded in this mouse strain by a novel post-transcriptional mechanism which depends on glucocorticoids and certain cytokines (71, 72). The expression of mMCP-7 mRNA in BALB/c mBMMCs is also regulated, in part, by a IL-10-dependent post-transcriptional mechanism (Stevens, unpublished data). Repetitive motifs residing in the 3′ UTR often regulate the stability of transcripts that encode important immunoregulatory proteins (73). The finding that the 3′UTR of the mTMT transcript has cytosine-rich motifs which resemble those repetitive motifs in the mMCP-2 and mMCP-4 transcripts (71) now raises the possibility that the strain-dependent expression of mTMT mRNA is regulated, in part, by a post-transcriptional mechanism.

The DNA-binding proteins which enhance or suppress transcription of the different tryptase genes are just beginning to be identified. MCs express the related transcription factors GATA-1, GATA-2 and GATA-3 (74). MCs can be developed from GATA-1 null mice (75), but not from GATA-2 null mice (76). GATA-2 is present in mBMMCs (74) and *in situ* hybridization studies have revealed that MC-committed progenitors express GATA-2 as they differentiate in the skin into recognizable mMCP-6$^+$/mMCP-7$^+$ MCs (77). While these data suggest that GATA-2 is essential for MC development, cutaneous MCs decrease their expression of GATA-2 as they undergo their final stages of granule maturation. Thus, tryptase expression must be controlled by additional transcription factors. *mi/mi* mice are MC-deficient (78) due to a genetic defect in the *mi* transcription factor (79). Because MCs generated from *mi/mi* mice lack mMCP-6 mRNA and do not recognize the mMCP-6 promoter, it has been proposed that the *mi* transcription factor directly participates in the transcription of the mMCP-6 gene (80). mMCP-6$^+$ MCs express polyomavirus enhancer binding protein 2 (PEBP2) (81). Overlapping PEBP2- and *mi*-binding motifs are present in the 5′ flanking region of the mMCP-6 gene and data have been obtained which suggest that these two transcription factors act in synergy to regulate the transcription of the mMCP-6 gene (81). Recent studies carried out by Singh and coworkers have revealed that the PU.1 transcription factor also is of critical importance in MC development. PU.1 null mouse embryos lack granulated MCs in their skin (Walsh *et al.*, unpublished observation). Although MC-committed progenitors can be generated from PU.1 null mice, the resulting cells express low levels of mMCP-6 and other granule proteases.

Because it has become apparent only recently that the tryptases α, I, II/β, III and hTMT transcripts are derived from distinct genes, less is known about the differential expression of these five tryptases at the mRNA or protein level relative to the three mouse tryptases. In normal humans, tryptase expression appears to be highly restricted to MCs. Chymase-positive/tryptase-negative MCs have been found *in vivo* (82) and *in vitro* (83), but most human MCs express one or more tryptases (84). It had been proposed that the protease phenotype of a human MC is irreversibly pre-determined before its progenitor leaves the bone marrow. Nevertheless, it is now known that, like that in the mouse, the cell's current and previous tissue microenvironments determine which panel of proteases a human MC will store in its secretory granules (85, 86).

Although its clinical significance remains to be determined, certain human transformed non-MC lines contain tryptase transcripts. For example, the Mono Mac 6 and U-937 cell lines contain tryptase α and I mRNA, respectively (87–89). The basophils which reside in the peripheral blood of normal subjects do not contain tryptase I, II/β or III, but these $Fc_{\varepsilon}RI^{+}$/metachromatic cells do contain a small amount of tryptase α protein (90) and mRNA (91). Cutaneous and jejunal human MCs contain hTMT mRNA and protein (10). However, the tissue expression of this human tryptase appears to be less restricted than mTMT. The level of hTMT mRNA is below detection in adult spleen, brain and skeletal muscle, but most adult human tissues contain measurable levels of hTMT mRNA. That hTMT mRNA is present in fetal spleen indicates that hTMT expression is developmentally controlled in certain human tissues. The observation that many human adenocarcinomas contain hTMT mRNA also raises the possibility that this tryptase contributes to the cancerous state of certain transformed non-MCs.

FUNCTION OF HUMAN AND MOUSE TRYPTASES

Until recently, all functional studies on MC proteases were carried out using preparations that had been purified to varying degrees from tissues. All mouse MCs (and probably all human MCs) contain multiple proteases in their granules that are similar in their biochemical properties. Because of the nearly impossible task of purifying these proteases to homogeneity, the majority of functional studies on tryptases contained multiple proteases in the analysed preparations. Recent functional studies on mMCP-6 and mMCP-7 have indicated that these two tryptases evolved to carry out different functions (23, 50). To minimize the contamination issue in future functional studies on human and mouse tryptases, greater emphasis will have to be placed on the use of recombinant material.

The discovery that exocytosed mMCP-7 is able to make its way into blood of the V3 mastocytosis mouse where it circulates for more than 1 h as an enzymatically active

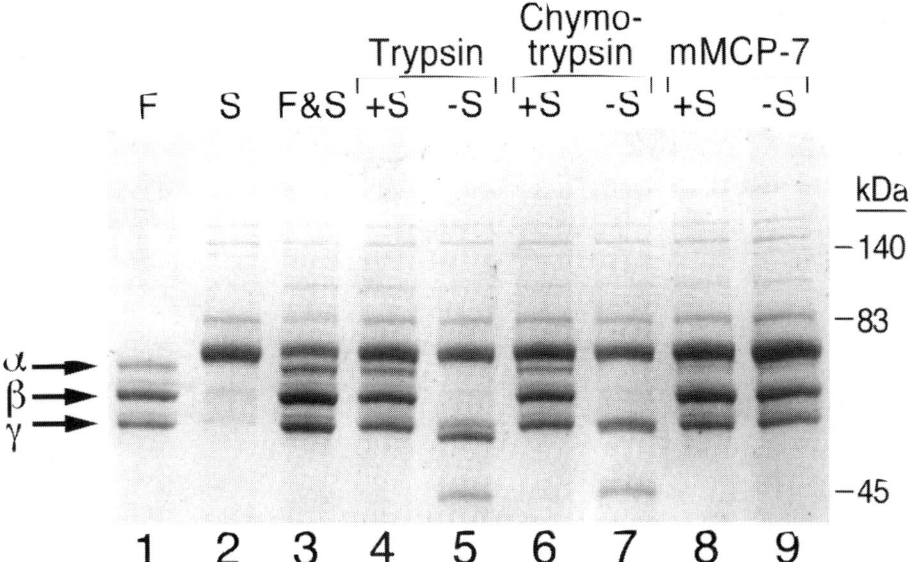

Fig. 2 Relative susceptibility of fibrinogen to trypsin, chymotrypsin and mMCP-7. Fibrinogen was incubated with trypsin (lanes 4 and 5), chymotrypsin (lanes 6 and 7), or recombinant mMCP-7 (lanes 8 and 9) before (lanes 5, 7, and 9) and after (lanes 4, 6, and 8) exposure to plasma. Fibrinogen alone, plasma alone and the combination of fibrinogen and plasma are depicted in lanes 1, 2 and 3, respectively. The migration positions of three molecular weight standards are shown on the right. The arrows on the left indicate the α, β and γ chains of fibrinogen. The major Coomassie blue-stained protein in mouse plasma (lane 2) that is slightly larger than the α chain of fibrinogen is albumin. The depicted SDS–PAGE gel was run to optimally resolve albumin and the α chain of fibrinogen. Thus, the 34-kDa fragment of the α chain of fibrinogen which is specifically formed after mMCP-7 treatment is not detected because it has run off the gel. (Reproduced from *J. Biol. Chem.* **272**: 31885–31893, 1997.)

homotypic tetramer (22) raised the possibility that one of its physiological substrates is a protein which normally resides in blood. SDS–PAGE analysis of the plasma of a V3 mastocytosis mouse undergoing passive systemic anaphylaxis revealed the presence of large amounts of fibrinogen α chain fragments (23). Mouse blood is rich in protease inhibitors. Although many proteases can degrade fibrinogen if the digestion reaction is carried out in the absence of serum, recombinant mMCP-7 is rather unique in its ability to selectively degrade the α chain of fibrinogen in the presence of serum (Fig. 2). Screening of a phage-display peptide library revealed that mMCP-7 has a very restricted substrate specificity and prefers a sequence which resides in the C-terminal domain of the α chain of fibrinogen. The concentration of fibrinogen is approximately 3 mg ml^{-1}. Even though oedema is considerable during a MC inflammatory response, one does not see the deposition of large amounts of cross-linked fibrin in the inflamed tissue. Based on the *in vivo* and *in vitro* data, it now appears that mMCP-7 helps prevent clot formation in the mouse during a MC-mediated inflammatory response by rapidly inactivating fibrinogen before it can be converted into fibrin by thrombin at the blood–endothelium barrier.

Although the primary amino acid sequences of mMCP-6 and mMCP-7 are very similar, subsequent *in vivo* and *in vitro* studies revealed that one of the major functions of mMCP-6 is to regulate neutrophil extravasation into tissues (50). When recombinant mMCP-6 was injected into the peritoneal cavities of various mouse strains, the number of neutrophils in this tissue site increased greater than 50-fold. Unlike most forms of acute

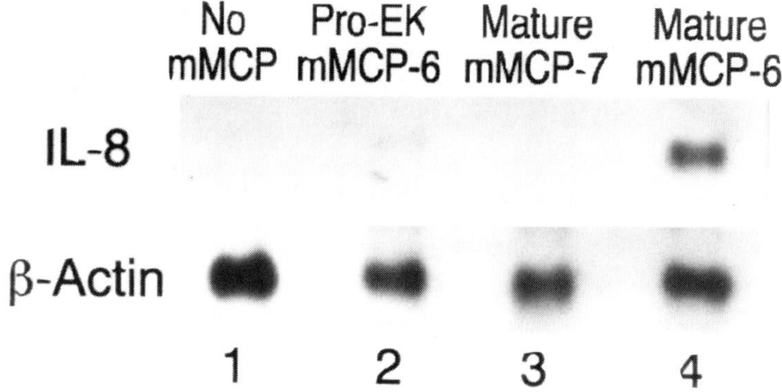

Fig. 3 IL-8 mRNA levels in tryptase-treated human endothelial cells. Blots containing total RNA isolated from human endothelial cells exposed *in vitro* for 2 h to no mMCP (lane 1), pro-EK-mMCP-6 (lane 2), mature mMCP-7 (lane 3), or mature mMCP-6 (lane 4) were probed with the relevant cDNA to compare the levels of IL-8 mRNA in the four populations of endothelial cells relative to that of the β-actin transcript. (Reproduced from *J. Immunol.* **160**: 1910–1919, 1998.)

inflammation, the mMCP-6-mediated peritonitis was relatively long-lasting and neutrophil-specific. Recombinant mMCP-6 did not induce cultured human umbilical vein endothelial cells to express tumour necrosis factor-α, RANTES, IL-1α, or IL-6, but this tryptase did induce endothelial cells to express large amounts of IL-8 (Fig. 3). Interestingly, enzymatically active mMCP-7 was unable to induce neutrophil extravasation into the peritoneal cavity and was unable to induce endothelial cells to express IL-8. The accumulated data now suggest that mMCP-6 and mMCP-7 work in concert to modulate inflammatory responses.

While the preferred physiological substrates of the various human tryptases remain to be elucidated, several potential roles have been suggested based on findings from *in vitro* studies carried out with different tryptase preparations. Because granule proteases from mBMMCs can directly degrade fibronectin (14) and because a human tryptase can activate pro-collagenases (92, 93), human tryptases have been implicated in the catabolism of extracellular matrix proteins. A lung-derived human tryptase can degrade high molecular weight kininogen (94) and activate single-chain urinary-type plasminogen activator (95). Thus, it also has been proposed that certain human tryptases participate in the regulation of coagulation by a mechanism which is different from that of mMCP-7. A lung-derived human tryptase can cleave the human complement component C3 to generate C3a (96) which has potent smooth muscle cell-contracting and vascular permeability activity.

Increased numbers of MCs are often found in healing wounds. In various lung and skin fibrotic diseases, these MCs tend to be in close proximity to the proliferating fibroblasts. That fibroblasts lose their contact inhibition and multiply when they are co-cultured with some, but not all, populations of MCs led to the concept that MC heterogeneity and/or activation is important in fibrosis (97). MCs produce so many cytokines which regulate fibroblast growth that relatively little attention has been placed on understanding the potential role of tryptases in fibrosis. Nevertheless, at least one human tryptase is mitogenic for fibroblasts (98–100), epithelial cells (101), and endothelial cells (102), and can induced dermal fibroblasts to slightly increase their levels of pro-collagen α1 mRNA (99, 103). It has been reported that human tryptases α and II/β induce the proliferation of cells by specifically cleaving the activation peptide of protease-activated receptor (PAR) 2 (36). However, the findings are controversial because it also has been reported that the tryptase-mediated effect is very small relative to that of trypsin (104) and because other domains in PAR-2 are tryptase-susceptible and ultimately cause inactivation of the receptor (105). Moreover, others (34) have reported that recombinant human tryptases α and II/β are functionally distinct. The concept that a human tryptase stimulates both the production and degradation of collagen at the same time is also difficult to comprehend at this moment in time.

In terms of the role of MC tryptases in lung function, the levels of immunoreactive tryptase are increased in the bronchial airways of allergen-challenged individuals (106). Low molecular weight inhibitors of tryptic enzymes block antigen-induced airway constriction and tissue inflammatory responses in sheep (107) and guinea pigs (108). Furthermore, a canine tryptase is able to enhance the histamine-mediated contractile response of smooth muscle *in vitro* (109). While these findings suggest that tryptases promote bronchoconstriction via a PAR-2-dependent mechanism, it has been reported recently that PAR-2 actually plays a protective role in bronchoconstriction by relaxing airways (110). When a small amount of recombinant human tryptase II/β or mMCP-6 was injected into the trachea of mice, a selective and long-lasting extravasation of neutrophils occurred in the lungs of the recipients (111). Despite the tryptase II/β-mediated airway neutrophilia, there was no change in the airway responsiveness of these animals to methylcholine. The ability of human tryptase II/β and mMCP-6 to selectively recruit neutrophils into the alveolar airspaces in a manner that does not compromise lung function now suggests that certain MC tryptases play an important protective role in respiratory infections. The observation that recombinant human tryptase α was unable to induce neutrophil extravasation indicates that some of the conflicting data on the functional role of human tryptase in the lung is a consequence of the tryptase preparation used in the analysis.

TRYPTASE METABOLISM

Because recombinant MC tryptases are just beginning to become available, little is known about how these neutral proteases are metabolized *in vivo*. The amount of storage space inside a MC is limited. Because certain populations of MCs are long-lived (112), the MC must continually catabolize its older granules if the cell is not induced to degranulate. It is well known that MCs contain small amounts of classical lysosomal enzymes in their granules (2, 113) and *in vitro* studies have revealed that rat serosal MCs continuously degrade their granule proteoglycans (114). Nevertheless, virtually nothing is known about how this cell slowly catabolizes the tryptases in its older granules.

When exocytosed from $Fc_{\varepsilon}RI$-activated MCs, some protease/proteoglycan macromolecular complexes are endocytosed by eosinophils, macrophages, fibroblasts and other cell types (115–117). Thus, one way of removing and inactivating an exocytosed tryptase is for an adjacent cell to endocytose the tryptase/proteoglycan complex and degrade it in one of its own lysosomes. However, the finding that the concentration of mMCP-7 can get very high in the peripheral blood of the V3 mastocytosis mouse 20 min after the induction of systemic anaphylaxis (22) indicates that at least some exocytosed MC tryptase is able to make its way into the blood stream.

At least one human tryptase is resistant to a number of protease inhibitors (118). As noted above, most MC tryptases exist as tetramers, and loss of activity rapidly occurs when the tetramer dissociates into four monomers (47–49). Because a rat tryptase has been found in the liver complexed to α_1-macroglobulin (119), it has been assumed that physical constraints prevent the α-macroglobulin family of protease inhibitors from efficient entrapment and inactivation of most tryptases until they dissociate into monomers. In the case of mMCP-7, the amino acid sequence of the peptide which it prefers to cleave (23) is not present in the bait region of any known protease inhibitor. Thus, the restricted specificities of certain tryptases is a contributing factor in their ability to escape rapid inactivation. As noted above, mMCP-6 induces the selective extravasation of neutrophils into tissues (50). The *in vitro* findings that the neutrophil granule proteins lactoferrin (120) and myeloperoxidase (121) are potent inhibitors of at least one tryptase suggest that neutrophils have the capacity to dampen certain tryptase-mediated responses in tissues. Inter-α-trypsin inhibitor is readily endocytosed by human MCs and is able to enter the cell's secretory granules (122). This observation is clinically relevant because the endocytosis of inter-α-trypsin inhibitor by rat serosal MCs results in the inactivation of rMCP-6 (123). Bovine pancreatic trypsin inhibitor also has been found in bovine liver MCs tightly bound to one of its tryptases (124). Finally, the observation that secretory leukocyte protease inhibitor dampens tryptase-mediated hyper-responsiveness in isolated guinea pig bronchi (108) raises the possibility that this protease inhibitor counteracts tryptase activity in airways.

One of the most distinctive features of hTMT and mTMT is the additional transmembrane segment located at its C-terminus (10). Because this segment does not have a Tyr residue, the C-terminus of hTMT and mTMT cannot undergo Tyr phosphorylation. However, because the cytosolic domains have a conserved Ser residue, it is possible that this residue undergoes phosphorylation during the metabolism of mTMT and hTMT. Although the mechanism by which mTMT and hTMT are removed from the surface of an activated MC has not been deduced, it now appears that the varied human and mouse tryptases are inactivated in different manners.

CONCLUDING REMARKS

Although substantial progress has been made, many questions remain unanswered concerning the regulation and function of the various mouse and human tryptases. The discovery of a new tryptase nearly every year during the last decade raises the possibility that not every human and mouse tryptase gene has been identified. Thus, it is anticipated that the nucleotide sequence data from the human and mouse genome initiatives finally will allow us to understand the degree of complexity of the tryptase loci on mouse chromosome 17 and human chromosome 16. The generation of recombinant human and mouse tryptases and the development of transgenic mice which vary in their expression of a particular tryptase should prove invaluable for the deduction of the functions of these neutral proteases. Once it is understood why human and mouse MCs express so many tryptases, attention will shift to the complex transcriptional, post-transcriptional and post-translational mechanisms which regulate tryptase expression in normal and diseased humans and mice. In this regard, the positive and negative *cis*-acting elements in the promoters of each tryptase gene must be identified, as well as their associated DNA-binding proteins. The post-transcriptional mechanisms which counteract high levels of tryptase transcripts in cells also need to be better understood. The importance of the pro-peptide in the activation mechanism must be elucidated, as well as the co-factor role of the varied glycosaminoglycans which are attached to serglycin proteoglycans. While it is now clear that tryptases are important in MC-mediated inflammatory responses, the release of too much tryptase has potentially catastrophic consequences. Thus, eventually investigators will need to better understand the varied mechanisms the body uses to counteract each tryptase once its level gets too high in a tissue.

REFERENCES

1. Glenner, G. G. and Cohen, L. A. Histochemical demonstration of species-specific trypsin-like enzyme in mast cells. *Nature* **105**:846–847, 1960.
2. Schwartz, L. B., Lewis, R. A., Seldin, D. and Austen, K. F. Acid hydrolases and tryptase from secretory granules of dispersed human lung mast cells. *J. Immunol.* **126**:1290–1294, 1981.
3. Schwartz, L. B., Lewis, R. A. and Austen, K. F. Tryptase from human pulmonary mast cells: purification and characterization. *J. Biol. Chem.* **256**:11939–11943, 1981.
4. Schwartz, L. B. Monoclonal antibodies against human mast cell tryptase demonstrate shared antigenic sites on subunits of tryptase and selective localization of the enzyme to mast cells. *J. Immunol.* **134**:526–531, 1985.
5. Miller, J. S., Westin, E. H. and Schwartz, L. B. Cloning and characterization of complementary DNA for human tryptase. *J. Clin. Invest.* **84**:1188–1195, 1989.
6. Miller, J. S., Moxley, G. and Schwartz, L. B. Cloning and characterization of a second complementary DNA for human tryptase. *J. Clin. Invest.* **86**:864–870, 1990.
7. Vanderslice, P., Craik, C. S., Nadel, J. A. and Caughey, G. H. Molecular cloning of dog mast cell tryptase and a related protease: structural evidence of a unique mode of serine protease activation. *Biochemistry* **28**:4148–4155, 1989.
8. Vanderslice, P., Ballinger, S. M., Tam, E. K., Goldstein, S. M., Craik, C. S. and Caughey, G. H. Human mast cell tryptase: multiple cDNAs and genes reveal a multigene serine protease family. *Proc. Natl. Acad. Sci. USA* **87**:3811–3815, 1990.
9. Pallaoro, M., Fejzo, M. S., Shayesteh, L., Blount, J. L. and Caughey, G. H. Characterization of genes encoding known and novel human mast cell tryptases on chromosome 16p13.3. *J. Biol. Chem.* **274**:3355–3362, 1999.
10. Wong, G. W., Tang, Y., Feyfant, E., Sali, A., Li, L., Li, Y., Huang, C., Friend, D. S., Krilis, S. A. and Stevens, R. L. Identification of a new member of the tryptase family of mouse and human mast cell proteases which possesses a novel C-terminal, membrane-spanning hydrophobic extension. *J. Biol. Chem.* **274**:30784–30793, 1999.
11. Campanelli, D., Detmers, P. A., Nathan, C. F. and Gabay, J. E. Azurocidin and a homologous serine protease from neutrophils. Differential antimicrobial and proteolytic properties. *J. Clin. Invest.* **85**:904–915, 1990.
12. Razin, E., Cordon-Cardo, C. and Good, R. A. Growth of a pure population of mouse mast cells *in vitro* with conditioned medium derived from concanavalin A-stimulated splenocytes. *Proc. Natl. Acad. Sci. USA* **78**:2559–2561, 1981.
13. Razin, E., Ihle, J. N., Seldin, D., Mencia-Huerta, J.-M., Katz, H. R., LeBlanc, P. A., Hein, A., Caulfield, J. P., Austen, K. F. and Stevens, R. L. Interleukin 3: a differentiation and growth factor for the mouse mast cell that contains chondroitin sulfate E proteoglycan. *J. Immunol.* **132**:1479–1486, 1984.
14. DuBuske, L., Austen, K. F., Czop, J. and Stevens, R. L. Granule-associated serine neutral proteases of the mouse bone marrow-derived mast cell that degrade fibronectin: their increase after sodium butyrate treatment of the cells. *J. Immunol.* **133**:1535–1541, 1984.
15. Reynolds, D. S., Serafin, W. E., Faller, D. V., Wall, D. A., Abbas, A. K., Dvorak, A. M., Austen, K. F. and Stevens, R. L. Immortalization of murine connective tissue-type mast cells at multiple stages of their differentiation by coculture of splenocytes with fibroblasts that produce Kirsten sarcoma virus. *J. Biol. Chem.* **263**:12783–12791, 1988.
16. Matsudaira, P. Sequence from picomole quantities of proteins electroblotted onto polyvinylidene difluoride membranes. *J. Biol. Chem.* **262**:10035–10038, 1987.
17. Reynolds, D. S., Stevens, R. L., Lane, W. S., Carr, M. H., Austen, K. F. and Serafin, W. E. Different

mouse mast cell populations express various combinations of at least six distinct mast cell serine proteases. *Proc. Natl. Acad. Sci. USA* **87**:3230–3234, 1990.

18. Reynolds, D. S., Gurley, D. S., Austen, K. F. and Serafin, W. E. Cloning of the cDNA and gene of mouse mast cell protease-6. Transcription by progenitor mast cells and mast cells of the connective tissue subclass. *J. Biol. Chem.* **266**:3847–3853, 1991.
19. Gurish, M. F., Nadeau, J. H., Johnson, K. R., McNeil, H. P., Grattan, K. M., Austen, K. F. and Stevens, R. L. A closely linked complex of mouse mast cell-specific chymase genes on chromosome 14. *J. Biol. Chem.* **268**:11372–11379, 1993.
20. McNeil, H. P., Reynolds, D. S., Schiller, V., Ghildyal, N., Gurley, D. S., Austen, K. F. and Stevens, R. L. Isolation, characterization, and transcription of the gene encoding mouse mast cell protease 7. *Proc. Natl. Acad. Sci. USA* **89**:11174–11178, 1992.
21. Ghildyal, N., Friend, D. S., Freelund, R., Austen, K. F., McNeil, H. P., Schiller, V. and Stevens, R. L. Lack of expression of the tryptase, mMCP-7, in C57BL/6J mouse mast cells. *J. Immunol.* **153**:2624–2630, 1994.
22. Ghildyal, N., Friend, D. S., Stevens, R. L., Austen, K. F., Huang, C., Penrose, J. F., Sali, A. and Gurish, M. F. Fate of two mast cell tryptases in V3 mastocytosis and normal BALB/c mice undergoing passive systemic anaphylaxis. Prolonged retention of exocytosed mMCP-6 in connective tissues and rapid accumulation of enzymatically active mMCP-7 in the blood. *J. Exp. Med.* **184**:1061–1073, 1996.
23. Huang, C., Wong, G. W., Ghildyal, N., Gurish, M. F., Sali, A., Matsumoto, R., Qiu, W.-T. and Stevens, R. L. The tryptase, mouse mast cell protease 7, exhibits anticoagulant activity *in vivo* and *in vitro* due to its ability to degrade fibrinogen in the presence of the diverse array of protease inhibitors in plasma. *J. Biol. Chem.* **272**:31885–31893, 1997.
24. Gurish, M. F., Johnson, K. R., Webster, M. J., Stevens, R. L. and Nadeau, J. H. Location of the mouse mast cell protease 7 gene (Mcpt7) to chromosome 17. *Mamm. Genome* **5**:656–657, 1994.
25. Rowen, L., Koop, B. F. and Hood, L. The complete 685-kilobase DNA sequence of the human β T cell receptor. *Science* **272**:1755–1762, 1996.
26. Hunt, J. E., Friend, D. S., Gurish, M. F., Feyfant, E., Sali, A., Huang, C., Ghildyal, N., Stechschulte, S., Austen, K. F. and Stevens, R. L. Mouse mast cell protease 9, a novel member of the chromosome 14 family of serine proteases that is selectively expressed in uterine mast cells. *J. Biol. Chem.* **272**:29158–29166, 1997.
27. Serafin, W. E., Reynolds, D. S., Rogelj, S., Lane, W. S., Conder, G. A., Johnson, S. S., Austen, K. F. and Stevens, R. L. Identification and molecular cloning of a novel mouse mucosal mast cell serine protease. *J. Biol. Chem.* **265**:423–429, 1990.
28. Serafin, W. E., Sullivan, T. P., Conder, G. A., Ebrahimi, A., Marcham, P., Johnson, S. S., Austen, K. F. and Reynolds, D. S. Cloning of the cDNA and gene for mouse mast cell protease-4. Demonstration of its late transcription in mast cell subclasses and analysis of its homology to subclass-specific neutral proteases of the mouse and rat. *J. Biol. Chem.* **266**:1934–1941, 1991.
29. McNeil, H. P., Austen, K. F., Somerville, L. L., Gurish, M. F. and Stevens, R. L. Molecular cloning of the mouse mast cell protease-5 gene. A novel secretory granule protease expressed early in the differentiation of serosal mast cells. *J. Biol. Chem.* **266**:20316–20322, 1991.
30. Huang, R., Blom, T. and Hellman, L. Cloning and structural analysis of mMCP-1, mMCP-4, and mMCP-5, three mouse mast cell-specific serine proteases. *Eur. J. Immunol.* **21**:1611–1621, 1991.
31. McGuire, M. J., Lipsky, P. E. and Thiele, D. L. Generation of active myeloid and lymphoid granule serine proteases requires processing by the granule thiol protease dipeptidyl peptidase I. *J. Biol. Chem.* **268**:2458–2467, 1993.
32. Sakai, K., Ren, S. and Schwartz, L. B. A novel heparin-dependent processing pathway for human tryptase, autocatalysis followed by activation with dipeptidyl peptidase I. *J. Clin. Invest.* **97**:988–995, 1996.
33. Murakumo, Y., Ide, H., Itoh, H., Tomita, M., Kobayashi, T., Maruyama, H., Horii, Y. and Nawa, Y. Cloning of the cDNA encoding mast cell tryptase of Mongolian gerbil, Meriones unguiculatus, and its preferential expression in the intestinal mucosa. *Biochem. J.* **309**:921–926, 1995.
34. Huang, C., Li, L., Krilis, S. A., Chanasyk, K., Tang, Y., Li, Z., Hunt, J. E. and Stevens, R. L. Human tryptases α and β/II are functionally distinct due, in part, to a single amino acid difference in one of the surface loops that forms the substrate-binding cleft. *J. Biol. Chem.* **274**:19670–19676, 1999.
35. Humphries, D. E., Wong, G. W., Friend, D. S., Gurish, M. F., Qiu, W.-T., Huang, C., Sharpe, A. H. and Stevens, R. L. Heparin is essential for the storage of specific granule proteases in mast cells. *Nature* **400**:769–772, 1999.

36. Mirza, H., Schmidt, V. A., Derian, C. K., Jesty, J. and Bahou, W. F. Mitogenic responses mediated through the proteinase-activated receptor 2 are induced by expressed forms of mast cell α or β tryptases. *Blood* **90**:3914–3922, 1997.
37. Schwartz, L. B., Sakai, K., Bradford, T. R., Ren, S., Zweiman, B., Worobec, A. S. and Metcalfe, D. D. The α form of human tryptase is the predominant type present in blood at baseline in normal subjects and is elevated in those with systemic mastocytosis. *J. Clin. Invest.* **96**:2702–2710, 1995.
38. Li, L., Li, Y., Reddel, S. W., Cherrian, M., Friend, D. S., Stevens, R. L. and Krilis, S. A. Identification of basophilic cells that express mast cell granule proteases in the peripheral blood of asthma, allergy, and drug-reactive patients. *J. Immunol.* **161**:5079–5086, 1998.
39. Buckley, M. G., Walters, C., Wong, W. M., Cawley, M. I., Ren, S., Schwartz, L. B. and Walls, A. F. Mast cell activation in arthritis, detection of α- and β-tryptase, histamine and eosinophil cationic protein in synovial fluid. *Clin. Sci.* **93**:363–370, 1997.
40. Cromlish, J.A., Seidah, N. G., Marcinkiewicz, M., Hamelin, J., Johnson, D. A. and Chrétien, M. Human pituitary tryptase: molecular forms, NH_2-terminal sequence, immunocytochemical localization, and specificity with prohormone and fluorogenic substrates. *J. Biol. Chem.* **262**:1363–1373, 1987.
41. Little, S. S. and Johnson, D. A. Human mast cell tryptase isoforms: separation and examination of substrate-specificity differences. *Biochem. J.* **307**:341–346, 1995.
42. Huang, C., Morales, G., Vagi, A., Chanasyk, K., Ferrazzi, M., Burklow, C., Qiu, W.-T., Feyfant, E., Sali, A. and Stevens, R. L. Formation of enzymatically active, homotypic and heterotypic tetramers of mouse mast cell tryptases: dependence on a conserved Trp-rich domain on the surface. *J. Biol. Chem.* **275**:351–358, 2000.
43. Johnson, D. A. and Barton, G. J. Mast cell tryptases: examination of unusual characteristics by multiple sequence alignment and molecular modeling. *Protein Sci.* **1**:370–377, 1992.
44. Pereira, P. J. B., Bergner, A., Macedo-Ribeiro, S., Huber, R., Matschiner, G., Fritz, H., Sommerhoff, C. P. and Bode, W. Human β tryptase is a ring-like tetramer with active sites facing a central pore. *Nature* **392**:306–311, 1998.
45. Raymond, W. W., Tam, E. K., Blount, J. L. and Caughey, G. H. Purification and characterization of dog mast cell protease-3, an oligomeric relative of tryptases. *J. Biol. Chem.* **270**:13164–13170, 1995.
46. Matsumoto, R., Sali, A., Ghildyal, N., Karplus, M. and Stevens, R. L. Packaging of proteases and proteoglycans in the granules of mast cells and other hematopoietic cells: a cluster of histidines on mouse mast cell protease 7 regulates its binding to heparin serglycin proteoglycans. *J. Biol. Chem.* **270**:19524–19539, 1995.
47. Schwartz, L. B. and Bradford, T. R. Regulation of tryptase from human lung mast cells by heparin. Stabilization of the active tetramer. *J. Biol. Chem.* **261**:7372–7379, 1986.
48. Schechter, N. M. Eng, G. Y., and McCaslin, D. R. Human skin tryptase: kinetic characterization of its spontaneous inactivation. *Biochemistry* **32**:2617–2625, 1993.
49. Addington, A. K. and Johnson, D. A. Inactivation of human lung tryptase: evidence for a re-activatable tetrameric intermediate and active monomers. *Biochemistry* **35**:13511–13518, 1996.
50. Huang C., Friend, D. S., Qiu, W-T., Wong, G. W., Morales, G., Hunt, J. and Stevens, R. L. Induction of a selective and persistent extravasation of neutrophils into the peritoneal cavity by the tryptase mouse mast cell protease 6. *J. Immunol.* **160**:1910–1919, 1998.
51. Bourdon, M. A., Shiga, M. and Ruoslahti, E. Identification from cDNA of the precursor form of a chondroitin sulfate proteoglycan core protein. *J. Biol. Chem.* **261**:12534–12537, 1986.
52. Tantravahi, R. V., Stevens, R. L., Austen, K. F. and Weis, J. H. A single gene in mast cells encodes the core peptides of heparin and chondroitin sulfate proteoglycans. *Proc. Natl. Acad. Sci. USA* **83**:9207–9210, 1986.
53. Avraham, S., Austen, K. F., Nicodemus, C. F., Gartner, M. C. and Stevens, R. L. Cloning and characterization of the mouse gene that encodes the peptide core of secretory granule proteoglycans and expression of this gene in transfected rat-1 fibroblasts. *J. Biol. Chem.* **264**:16719–16726, 1989.
54. Seldin, D. C., Austen, K. F. and Stevens, R. L. Purification and characterization of protease-resistant secretory granule proteoglycans containing chondroitin sulfate di-B and heparin-like glycosaminoglycans from rat basophilic leukemia cells. *J. Biol. Chem.* **260**:11131–11139, 1985.
55. Yurt, R. W., Leid, R. W. Jr, Austen, K. F. and Silbert, J. E. Native heparin from rat peritoneal mast cells. *J. Biol. Chem.* **252**:518–521, 1977.
56. Davidson, S., Gilead, L., Amira, M., Ginsburg, H. and Razin, E. Synthesis of chondroitin sulfate D and heparin proteoglycans in murine lymph node-derived mast cells. *J. Biol. Chem.* **265**:12324–12330, 1990.

57. Stevens, R. L., Lee, T. D. G., Seldin, D. C., Austen, K. F., Befus, A. D. and Bienenstock, J. Intestinal mucosal mast cells from rats infected with *Nippostrongylus brasiliensis* contain protease-resistant chondroitin sulfate di-B proteoglycans. *J. Immunol.* **137**:291–295, 1986.
58. Razin, E., Stevens, R. L., Akiyama, F., Schmid, K. and Austen, K. F. Culture from mouse bone marrow of a subclass of mast cells possessing a distinct chondroitin sulfate proteoglycan with glycosaminoglycans rich in *N*-acetylgalactosamine-4,6-disulfate. *J. Biol. Chem.* **257**:7229–7236, 1982.
59. Levi-Schaffer, F., Austen, K. F., Gravallese, P. and Stevens, R. L. Coculture of interleukin 3-dependent mouse mast cells with fibroblasts results in a phenotypic change of the mast cells. *Proc. Natl. Acad. Sci. USA* **83**:6485–6488, 1986.
60. Gurish, M. F., Ghildyal, N., McNeil, H. P., Austen, K. F., Gillis, S. and Stevens, R. L. Differential expression of secretory granule proteases in mouse mast cells exposed to interleukin 3 and c-*kit* ligand. *J. Exp. Med.* **175**:1003–1012, 1992.
61. Metcalfe, D. D., Lewis, R. A., Silbert, J. E., Rosenberg, R. D., Wasserman, S. I. and Austen, K. F. Isolation and characterization of heparin from human lung. *J. Clin. Invest.* **64**:1537–1543, 1979.
62. Stevens, R. L., Fox, C. C., Lichtenstein, L. M. and Austen, K. F. Identification of chondroitin sulfate E proteoglycans and heparin proteoglycans in the secretory granules of human lung mast cells. *Proc. Natl. Acad. Sci. USA* **85**:2284–2287, 1988.
63. Thompson, H. L., Schulman, E. S. and Metcalfe, D. D. Identification of chondroitin sulfate E in human lung mast cells. *J. Immunol.* **140**:2708–2713, 1988.
64. Le Trong, H., Neurath, H. and Woodbury, R. G. Substrate specificity of the chymotrypsin-like protease in secretory granules isolated from rat mast cells. *Proc. Natl. Acad. Sci. USA* **84**:364–367, 1987.
65. Blombäck, B. Fibrinogen and fibrin – proteins with complex roles in hemostasis and thrombosis. *Thromb. Res.* **83**:1–75, 1996.
66. Stevens, R. L., Friend, D. S., McNeil, H. P., Schiller, V., Ghildyal, N. and Austen, K. F. Strain-specific and tissue-specific expression of mouse mast cell secretory granule proteases. *Proc. Natl. Acad. Sci. USA* **91**:128–132, 1993.
67. Friend, D. S., Ghildyal, N., Gurish, M. F., Hunt, J., Hu, X., Austen, K. F. and Stevens, R. L. Reversible expression of tryptases and chymases in the jejunal mast cells of mice infected with *Trichinella spiralis*. *J. Immunol.* **160**:5537–5545, 1998.
68. Gurish, M. F., Pear, W. S., Stevens, R. L., Scott, M. L., Sokol, K., Ghildyal, N., Webster, M. J., Hu, X., Austen, K. F., Baltimore, D. and Friend, D. S. Tissue-regulated differentiation and maturation of a v-*abl*-immortalized mast cell-committed progenitor. *Immunity* **3**:175–186, 1995.
69. Eklund, K. K., Ghildyal, N., Austen, K. F., Friend, D. S., Schiller, V. and Stevens, R. L. Mouse bone marrow-derived mast cells (mBMMC) obtained *in vitro* from mice that are mast cell-deficient *in vivo* express the same panel of granule proteases as mBMMC and serosal mast cells from their normal littermates. *J. Exp. Med.* **180**:67–73, 1994.
70. Hunt, J. E., Stevens, R. L., Austen, K. F., Zhang, J., Xia, Z. and Ghildyal, N. Natural disruption of the mouse mast cell protease 7 gene in the C57BL/6 mouse. *J. Biol. Chem.* **271**:2851–2855, 1996.
71. Xia, Z., Ghildyal, N., Austen, K. F. and Stevens, R. L. Post-transcriptional regulation of chymase expression in mast cells. A cytokine-dependent mechanism for controlling the expression of granule neutral proteases of hematopoietic cells. *J. Biol. Chem.* **271**:8747–8753, 1996.
72. Eklund, K. K., Humphries, D. E., Xia, Z., Ghildyal, N., Friend, D. S., Gross, V. and Stevens, R. L. Glucocorticoids inhibit the cytokine-induced proliferation of mast cells, the high-affinity IgE receptor-mediated expression of TNF-α, and the IL-10-induced expression of chymases. *J. Immunol.* **158**:4373–4380, 1997.
73. Shaw, G. and Kamen, R. A conserved AU sequence from the 3' untranslated region of GM-CSF mRNA mediates selective mRNA degradation. *Cell* **46**:659–667, 1986.
74. Zon, L. I., Gurish, M. F., Stevens, R. L., Mather, C., Reynolds, D. S., Austen, K. F. and Orkin, S. H. GATA-binding transcription factors in mast cells regulate the promoter of the mast cell carboxypeptidase A gene. *J. Biol. Chem.* **266**:22948–22953, 1991.
75. Pevny, L., Lin, C.-S., D'Agati, V., Simon, M. C., Orkin, S. H. and Costantini, F. Development of hematopoietic cell lacking transcription factor GATA-1. *Development* **121**:163–172, 1995.
76. Tsai, F.-Y. and Orkin, S. H. Transcription factor GATA-2 is required for proliferation/survival of early hematopoietic cells and mast cell formation, but not for erythroid and myeloid terminal differentiation. *Blood* **89**:3636–3643, 1997.
77. Jippo, T., Mizuno, H., Xu, Z., Nomura, S., Yamamoto, M. and Kitamura, Y. Abundant expression of transcription factor GATA-2 in proliferating but not in differentiated mast cells in tissues of mice:

demonstration by *in situ* hybridization. *Blood* **87**:993–998, 1996.

78. Stechschulte, D. J., Sharma, R., Dileepan, K. N., Simpson, K. M., Aggarwal, N., Clancy, J. Jr and Jilka, R. L. Effect of the *mi* allele on mast cells, basophils, natural killer cells, and osteoclasts in C57BL/6J mice. *J. Cell. Physiol.* **132**:565–570, 1987.
79. Hughes, M. J., Lingrel, J. B., Krakowsky, J. M. and Anderson, K. P. A helix–loop–helix transcription factor-like gene is located at the *mi* locus. *J. Biol. Chem.* **268**:20687–20690, 1993.
80. Morii, E., Tsujimura, T., Jippo, T., Hashimoto, K., Takebayashi, K., Tsujino, K., Nomura, S., Yamamoto, M. and Kitamura, Y. Regulation of mouse mast cell protease 6 gene expression by transcription factor encoded by the *mi* locus. *Blood* **88**:2488–2494, 1996.
81. Ogihara, H., Kanno, T, Morii, E., Kim, D-K., Lee, Y-M., Sato, M., Kim, W.-Y., Nomura, S., Ito, Y. and Kitamura, Y. Synergy of PEBP2/CBF with mi transcription factor (MITF) for transactivation of mouse mast cell protease 6 gene. *Oncogene* **18**:4632–4639, 1999.
82. Weidner, N. and Austen, K. F. Heterogeneity of mast cells at multiple body sites. *Pathol. Res. Pract.* **189**:156–162, 1993.
83. Li, L., Meng, X. W. and Krilis, S. A. Mast cells expressing chymase but not tryptase can be derived by culturing human progenitors in conditioned medium obtained from a human mastocytosis cell strain with c-*kit* ligand. *J. Immunol.* **156**:4839–4844, 1996.
84. Schwartz, L. B., Irani, A. M. A., Roller, K., Castells, M. C. and Schechter, N. M. Quantitation of histamine, tryptase, and chymase in dispersed human T and TC mast cells. *J. Immunol.* **138**:2611–2615, 1987.
85. Longley, J., Tyrrell, L., Lu, S., Ma, Y., Klump, V. and Murphy, G. F. Chronically *kit*-stimulated clonally-derived human mast cells show heterogeneity in different tissue microenvironments. *J. Invest. Dermatol.* **108**:792–796, 1997.
86. Toru, H., Eguchi, M., Matsumoto, R., Yanagida, M., Yata, J. and Nakahata, T. Interleukin-4 promotes the development of tryptase and chymase double-positive human mast cells accompanied by cell maturation. *Blood* **91**:187–195, 1998.
87. Huang, R., Abrink, M., Gobl, A. E., Nilsson, G., Aveskogh, M., Larsson, L. G., Nilsson, K. and Hellman, L. Expression of a mast cell tryptase in the human monocytic cell lines U-937 and Mono Mac 6. *Scand. J. Immunol.* **38**:359–367, 1993.
88. Blom, T. and Hellman, L. Characterization of a tryptase mRNA expressed in the human basophil cell line KU812. *Scand. J. Immunol.* **37**:203–208, 1993.
89. Abrink, M., Gobl, A. E., Huang, R., Nilsson, K. and Hellman, L. Human cell lines U-937, THP-1, and Mono Mac 6 represent relatively immature cells of the monocyte–macrophage cell lineage. *Leukemia* **8**:1579–1584, 1994.
90. Castells, M. C., Irani, A. M. and Schwartz, L. B. Evaluation of human peripheral blood leukocytes for mast cell tryptase. *J. Immunol.* **138**:2184–2189, 1987.
91. Xia, H. Z., Kepley, C. L., Sakai, K., Chelliah, J., Irani, A. M. and Schwartz, L. B. Quantitation of tryptase, chymase, FcεRI-α, and FcεRI-γ gamma mRNAs in human mast cells and basophils by competitive reverse transcription–polymerase chain reaction. *J. Immunol.* **154**:5472–5480, 1995.
92. Gruber, B. L., Marchese, M. J., Suzuki, K., Schwartz, L. B., Okada, Y., Nagase, H. and Ramamurthy, N. S. Synovial procollagenase activation by human mast cell tryptase. Dependence upon matrix metalloproteinase 3 activation. *J. Clin. Invest.* **84**:1657–1662, 1989.
93. Lohi, J., Harvima, I. and Keski-Oja, J. Pericellular substrates of human mast cell tryptase: 72,000 dalton gelatinase and fibronectin. *J. Cell. Biochem.* **50**:337–349, 1992.
94. Maier, M., Spragg, J. and Schwartz, L. B. Inactivation of human high molecular weight kininogen by human mast cell tryptase. *J. Immunol.* **130**:2352–2356, 1983.
95. Stack, M. S. and Johnson, D. A. Human mast cell tryptase activates single-chain urinary-type plasminogen activator (pro-urokinase). *J. Biol. Chem.* **269**:9416–9419, 1994.
96. Schwartz, L. B., Kawahara, M. S., Hugli, T. E., Vik, D., Fearon, D. T. and Austen, K. F. Generation of C3a anaphylatoxin from human C3 by mast cell tryptase. *J. Immunol.* **130**:1891–1895, 1983.
97. Dayton, E. T., Caulfield, J. P., Hein, A., Austen, K. F. and Stevens, R. L. Regulation of growth rate of mouse fibroblasts by IL-3-activated mouse bone marrow-derived mast cells. *J. Immunol.* **142**:4307–4313, 1989.
98. Hartmann, T., Ruoss, S. J., Raymond, W. W., Seuwen, K. and Caughey, G. H. Human tryptase as a potent, cell-specific mitogen: role of signaling pathways in synergistic responses. *Am. J. Physiol.* **262**:L528–L534, 1992.
99. Cairns, J. A. and Walls, A. F. Mast cell tryptase stimulates the synthesis of type I collagen in human lung fibroblasts. *J. Clin. Invest.* **99**:1313–1321, 1997.

100. Abe, M., Kurosawa, M., Ishikawa, O., Miyachi, Y. and Kido, H. Mast cell tryptase stimulates both human dermal fibroblast proliferation and type I collagen production. *Clin. Exp. Allergy* **28**:1509–1517, 1998.
101. Cairns, J. A. and Walls, A. F. Mast cell tryptase is a mitogen for epithelial cells: stimulation of IL-8 production and intercellular adhesion molecule-1 expression. *J. Immunol.* **156**:275–283, 1996.
102. Compton, S. J., Cairns, J. A., Holgate, S. T. and Walls, A. F. The role of mast cell tryptase in regulating endothelial cell proliferation, cytokine release, and adhesion molecule expression: tryptase induces expression of mRNA for IL-1β and IL-8 and stimulates the selective release of IL-8 from human umbilical vein endothelial cells. *J. Immunol.* **161**:1939–1946, 1998.
103. Gruber, B. L., Kew, R. R., Jelaska, A., Marchese, M. J., Garlick, J., Ren, S., Schwartz, L. B. and Korn, J. H. Human mast cells activate fibroblasts: tryptase is a fibrogenic factor stimulating collagen messenger ribonucleic acid synthesis and fibroblast chemotaxis. *J. Immunol.* **158**:2310–2317, 1997.
104. Schechter, N. M., Brass, L. F., Lavker, R. M. and Jensen, P. J. Reaction of mast cell proteases tryptase and chymase with protease activated receptors (PARs) on keratinocytes and fibroblasts. *J. Cell. Physiol.* **176**:365–373, 1998.
105. Molino, M., Barnathan, E. S., Numerof, R., Clark, J., Dreyer, M., Cumashi, A., Hoxie, J. A, Schechter, N., Woolkalis, M. and Brass, L. F. Interactions of mast cell tryptase with thrombin receptors and PAR-2. *J. Biol. Chem.* **272**:4043–4049, 1997.
106. Svensson, C., Grönneberg, R., Andersson, M., Alkner, U., Andersson, O., Billing, B., Gilljam, H., Greiff, L. and Persson, C. G. Allergen challenge-induced entry of α2-macroglobulin and tryptase into human nasal and bronchial airways. *J. Allergy Clin. Immunol.* **96**:239–246, 1995.
107. Clark, J. M., Abraham, W. M., Fishman, C. E., Forteza, R., Ahmed, A., Cortes, A., Warne, R. L., Moore, W. R. and Tanaka, R. D. Tryptase inhibitors block allergen-induced airway and inflammatory responses in allergic sheep. *Am. J. Respir. Crit. Care Med.* **152**:2076–2083, 1995.
108. Barrios, V. E., Middleton, S. C., Kashem, M. A., Havill, A. M., Toombs, C. F. and Wright, C. D. Tryptase mediates hyperresponsiveness in isolated guinea pig bronchi. *Life Sci.* **63**:2295–2303, 1998.
109. Sekizawa, K., Caughey, G. H., Lazarus, S. C., Gold, W. M. and Nadel, J. A. Mast cell tryptase causes airway smooth muscle hyperreponsiveness in dogs. *J. Clin. Invest.* **83**:175–179, 1989.
110. Cocks, T. M., Fong, B., Chow, J. M., Anderson, G. P., Frauman, A. G., Goldie, R. G., Henry, P. J., Carr, M. J., Hamilton, J. R. and Moffatt, J. D. A protective role for protease-activated receptors in the airways. *Nature* **398**:156–160, 1999.
111. Huang, C., De Sanctis, G. T., Jiao, A., Friend, D. S., Drazen, J. M. and Stevens, R. L. Human mast cell tryptase II/β selectively recruits neutrophils into the airways of mice without altering airway reactivity. *FASEB J.* **13**:A325, 1999 (abstract).
112. Padawer, J. Mast cells: extended lifespan and lack of granule turnover under normal *in vivo* conditions. *Exp. Mol. Pathol.* **20**:269–280, 1974.
113. Schwartz, L. B. and Austen, K. F. Enzymes of the mast cell granule. *J. Immunol.* **74**:349–353, 1980.
114. Stevens, R. L. and Austen, K. F. Effect of *p*-nitrophenyl-D-xyloside on proteoglycan and glycosaminoglycan biosynthesis in rat serosal mast cell cultures. *J. Biol. Chem.* **257**:253–259, 1982.
115. Welsh, R. A. and Geer, J. C. Phagocytosis of mast cell granules by the eosinophilic leukocyte in the rat. *Am. J. Pathol.* **35**:103–111, 1959.
116. Fabian, I., Bleiberg, I. and Aronson, M. Increased uptake and desulphation of heparin by mouse macrophages in the presence of polycations. *Biochim. Biophys. Acta* **544**:69–76, 1978.
117. Subba-Rao, P. V., Friedman, M. M., Atkins, F. M. and Metcalfe, D. D. Phagocytosis of mast cell granules by cultured fibroblasts. *J. Immunol.* **130**:341–349, 1983.
118. Alter, S. C., Kramps, J. A., Janoff, A. and Schwartz, L. B. Interactions of human mast cell tryptase with biological protease inhibitors. *Arch. Biochem. Biophys.* **276**:26–31, 1990.
119. Tsuji, A., Akamatsu, T., Nagamune, H. and Matsuda, Y. Identification of targeting proteinase for rat alpha 1-macroglobulin *in vivo*. Mast-cell tryptase is a major component of the α1-macroglobulin-proteinase complex endocytosed into rat liver lysosomes. *Biochem. J.* **298**:79–85, 1994.
120. Elrod, K. C., Moore, W. R., Abraham, W. M. and Tanaka, R. D. Lactoferrin, a potent tryptase inhibitor, abolishes late-phase airway responses in allergic sheep. *Am. J. Respir. Crit. Care Med.* **156**:375–381, 1997.
121. Cregar, L., Elrod, K. C., Putnam, D. and Moore, W. R. Neutrophil myeloperoxidase is a potent and selective inhibitor of mast cell tryptase. *Arch. Biochem. Biophys.* **366**:125–130, 1999.
122. Ide, H., Itoh, H., Yoshida, E., Kobayashi, T., Tomita, M., Maruyama, H., Osada, Y., Nakahata, T. and Nawa, Y. Immunohistochemical demonstration of inter-α-trypsin inhibitor light chain (bikunin) in human mast cells. *Cell Tissue Res.* **297**:149–154, 1999.

123. Itoh, H., Ide, H., Ishikawa, N. and Nawa, Y. Mast cell protease inhibitor, trypstatin, is a fragment of inter-α-trypsin inhibitor light chain. *J. Biol. Chem.* **269**:3818–3822, 1994.
124. Fiorucci, L., Erba, F., Falasca, L., Dini, L. and Ascoli, F. Localization and interaction of bovine pancreatic trypsin inhibitor and tryptase in the granules of bovine mast cells. *Biochim. Biophys. Acta* **1243**:407–413, 1995.

CHAPTER 17

Expression, Function and Regulation of Mast Cell Granule Chymases During Mucosal Allergic Responses

*PAMELA A. KNIGHT, STEVEN H. WRIGHT, ELISABETH M. THORNTON, JEREMY BROWN and HUGH R. P. MILLER**

Royal (Dick) School of Veterinary Studies, University of Edinburgh, Midlothian, U.K.

INTRODUCTION

Intestinal mucosal mast cells (IMMC) are involved in gut hypersensitivities (1), in gastric inflammation (2) and in protective responses against gastrointestinal nematodes (3). Intestinal nematode infection is associated with extensive IMMC hyperplasia in primates, domestic animals and rodents (4). In mice, substantial IMMC recruitment occurs within 4–7 days of nematode establishment in the gut, (5) and over 90% of the IMMC differentiate intraepithelially (6, 7). In mice, therefore, the interaction between the surface receptors of IMMC and of enterocytes may have a significant influence on the differentiation and function of both cell types.

In rodents, the β-chymases mouse mast cell protease 1 (mMCP-1) and rat mast cell protease 2 (rMCP-2) are highly soluble with restricted chymotryptic specificity, and are predominantly expressed at mucosal surfaces, where they are substantially upregulated during nematode infections (8). mMCP-1 is strictly confined to IMMC, whereas rMCP-2 is expressed by mast cells with a wider tissue distribution (8). Since gastrointestinal nematode infection and expulsion results in accumulation of predominantly intraepithelial mMCP-1^+ IMMC and secretion of high levels of this chymase both into the blood and the gut lumen (7, 8), it seems likely that this chymase has a functional role in the immune response to gut nematodes. In order to investigate the *in vivo* role of mMCP-1 directly, we have generated mice with a homozygous null mutation of the mMCP-1 gene and investigated their response to nematode infection (9). The preliminary results of this work suggest that the response to infection with certain nematodes may be compromised and that the kinetics of mast cell hyperplasia, as well as mucosal pathology, are affected by the absence of expression of mMCP-1 (9). These findings are discussed below.

* Corresponding author.

MAST CELLS AND BASOPHILS
ISBN 0-12-473335-2

In addition to investigating the functional role of mMCP-1 in nematode expulsion, we have investigated the cytokines that regulate the expression of mMCP-1. *In vivo* studies have shown that the most critical cytokine essential for the generation and survival of proliferating IMMC populations is stem cell factor (SCF) (c-*kit* ligand or Steel factor) (10), which is associated with the mucosal epithelium (11). SCF interacts with its tyrosine kinase receptor encoded by the c-*kit* proto-oncogene (10) expressed on the surface of IMMC. The second mechanism that influences IMMC proliferation during intestinal nematode infection is mediated by T cells in the mucosal environment. The T cell-derived cytokine, interleukin-3 (IL-3) induces IMMC hyperplasia (12), and antibodies to IL-3 and IL-4 suppress IMMC proliferation in nematode-induced hyperplasia (13). Studies with transgenic models have established that IL-9 also induces IMMC hyperplasia *in vivo* (14).

The expression of the IMMC-specific β-chymase mMCP-1 is induced *in vitro* in cultures of mouse bone marrow-derived mast cells (mBMMC) when the cells are exposed to T cell conditioned medium (15), or to the cytokines IL-9 or IL-10 together with SCF (16, 17). It is possible, given that IL-9 transgenic mice express high levels of mMCP-1 in the blood (14), that IL-9 is the main stimulus for mMCP-1 expression in parasitized mice. However, we have recently shown that addition of the multifunctional cytokine, transforming growth factor-β1 (TGF-β1), to mBMMC cultures induces a 100–1000-fold increase in the expression of mMCP-1 (18). This increase is associated with morphological changes to the cells and to the extracellular release of microgram quantities of mMCP-1 into the culture supernatant (18). Additional experiments suggested that the limited expression of mMCP-1 in the presence of SCF/IL-9 may similarly be strictly regulated by endogenously secreted TGF-β1 (18). Some of the experiments showing these data are reviewed below.

INVESTIGATING THE RESPONSE OF mMCP-1 –/– MICE TO GASTROINTESTINAL NEMATODES

mMCP-1 –/– Mice Express other Chymases Normally

The targeted deletion of the mMCP-1 gene and generation of mMCP-1 –/– mice has been described by us in detail previously (9). The targeting of the mMCP-1 gene was shown to be successful by genomic analysis, and this was further confirmed by the absence of mMCP-1 transcripts in both uninfected and parasitized mMCP-1 –/– mice (9). Immunohistochemistry and immunoassay established that mature mMCP-1 protein was undetectable in jejunal tissue in uninfected and parasitized mice (9). The genetic and tissue analyses, therefore, provided convincing evidence that the mMCP-1 gene had been successfully targeted. It was also shown that transcripts for mMCP-2, which is also expressed by IMMC, and for mMCP-4 and mMCP-5, typical of connective tissue mast cells, were expressed at normal levels in mMCP-1 –/– mice (9). In conclusion, the targeting was specific for mMCP-1 and did not affect transcription of the other mast cell chymases found in the 850-kb gene cluster on chromosome 14 (19).

Response of mMCP-1 –/– Mice to *Nippostrongylus brasiliensis* Infection

Infection kinetics

Initial experiments in which mMCP-1 –/– mice (MF-1/S-129) and random-bred and +/+ MF-1 controls were infected subcutaneously with 500 mouse-adapted strain *Nippostrongylus brasiliensis* L_3 have already been described (9). *N. brasiliensis* infection resulted in substantially altered mast cell kinetics (see below), and a marginal, but statistically significant, delay in worm expulsion (9). A second, similar experiment in which mMCP-1–/– mice and random-bred +/+ MF-1 controls were infected with 400 mouse-adapted strain *N. brasiliensis* L_3 focused on whether the targeted deletion of mMCP-1 may also result in altered gut pathology. Faecal egg counts (f.e.c.) were monitored daily, and at the first indication of a drop in f.e.c. on day 8 of infection all mice were killed and worm burdens assessed. Jejunal samples were fixed in Carnoy's fluid and in 4% paraformaldeyde for toluidine blue and esterase staining for mast cells, respectively (9), and in 4% paraformaldehyde for morphometric analysis of the mucosa (see below). In this experiment, there was no indication of a direct effect on parasite kinetics; f.e.c. values dropped more rapidly in this experiment (data not shown) and no difference in worm burden was detected on day 8 (Table I), although this may in part be due to the lower level of the initial larval infection.

TABLE I
Worm Kinetics in mMCP-1 –/– Mice

Parasite species	Days post-infection	Worm burden (mean ± SE)		
		mMCP-1–/–	mMCP-1+/+	*P* value
*N. brasiliensis**	8	73±14 (*n* =10)	40±17 (*n* =10)	0.05
N. brasiliensis	8	42.4±19.1 (*n* = 10)	20.1±13.4 (*n* = 9)	NS
T. spiralis	18	29.1±5.6	8.3±2.1	0.05

Data represent adult worm burdens recovered from the small intestine (modified Baerman's) at a timepoint during period of expulsion from mMCP-1 +/+ and –/– mice infected with *N. brasiliensis* (two separate experiments) and *T. spiralis*. See text for details.
* Data from Wastling *et al.* (9).

Altered mast cell kinetics

Toluidine blue-stained mast cells in Carnoy's-fixed jejunum were significantly ($p < 0.03$, Mann–Whitney) more abundant in mMCP-1 –/– mice (median 17.1 (range 5.5–72.8) IMMC per villus/crypt unit (VCU); $n = 10$) than in control mMCP-1 +/+ mice (median 9.8 (range 2.7–13.6) IMMC per VCU; $n = 9$). In contrast, esterase-positive mast cells in paraformaldehyde-fixed jejunum were generally less abundant in mMCP-1 –/– mice (median 7.05 (range 1.6–12.9) IMMC per VCU; $n = 10$) than in control mMCP-1 +/+ mice (median 10.5 (range 4.9–17.6) IMMC per VCU; $n = 9$).

When toluidine blue-positive mast cells were counted per mm^2 in the crypts and basal lamina propria (in order to minimize errors due to intervillous spaces), mean values of 1505 (SE ± 257) IMMC in –/– mice and of 542 (SE ± 97) /mm^2 ($p < 0.004$,

TABLE II

Summary of the Results (mean ± SE) from mMCP-1 –/– (*n* = 10) and mMCP-1 +/+ (*n* = 9) Mice for Samples Taken on Day 8 Following Infection with 400 Mouse-adapted *N. brasiliensis* L_3

	mMCP–/–	mMCP+/+	% difference –/– vs +/+	*P* value
Mast cells (mm^{-2})	1505 ± 257	542 ± 97	280	0.004
mMCP-1 ($\mu g\ m^{-1}$)	0	220 ± 20	NA	NA
Villus height (μm)	261 ± 23	352 ± 26	35	0.03
Crypt depth (μm)	177 ± 12	127 ± 28	28	0.005
Worm burden	42 ± 19	20 ± 13	NA	NS
RT-PCR				
mMCP-1	–	+++		
mMCP-2	++	++		
mMCP-4	+	+		
mMCP-5	+	+		

Egg counts had fallen from a maximum 11,000 (day 6) to 0 e.p.g. (day 8). Mast cells are predominantly (>90%) intraepithelial in both groups, but in the mMCP-1 –/– group the granules are smaller, stain weakly for esterase and not at all for mMCP-1; the mast cells are concentrated in the crypts. By contrast, the mMCP-1 +/+ mast cells are intensely esterase-positive and mMCP-1^+ with larger granules, and are located predominantly at the crypt/villus junctions.

Mann–Whitney) in +/+ mice were obtained (Table II). In both groups, the cells were predominantly (> 95%) intraepithelial, and a higher proportion of the IMMC in –/– mice were located basally in the crypts when compared with +/+ controls, where the cells were more abundant at the crypt/villus junction. In uninfected control mice, IMMC were so rare that it was not feasible to count per unit area, and even when counted per VCU median values of 0.03 IMMC per VCU were obtained for both mMCP-1 –/– (range 0–0.07) and mMCP-1+/+ (range 0–0.04) groups of mice (*n* = 8 for both groups). The few cells that were detected were again intraepithelial in both groups, but it was not possible to determine whether there was the distinctive pattern of localization in the infected groups. These values are approximately 570- and 303-fold lower than the medians for infected –/– and +/+ groups, respectively. These data are consistent with published data (9), which showed that, at the time of worm expulsion, toluidine blue-positive IMMC were significantly more abundant in mMCP-1 –/– mice than in controls.

Other features that were consistent with previous work (9) included the smaller granule size and lack of esterase staining of IMMC in parasitized –/– mice when compared with +/+ controls, although the expression of other chymases (mMCP-2, -4 and -5) in the gut mucosa was comparable in both groups of infected mice as judged by semiquantitative RT-PCR (9) (Table II). At the ultrastructural level, the granules in the –/– mice lacked the stellate outlines and internal crystalline structures (Fig. 1B) that are typically present in wild-type IMMC (Fig. 1A) (20, 21). Instead, the –/– granules were oval, and a proportion (~30%) had unusual intragranular divisions with dense, unstructured cores separated into several segments (Fig. 1B) as described previously (9). These data suggest that the mMCP-1 –/– mice generated IMMC as efficiently as wild-type mice but that there were substantial effects on granule morphology and biochemistry.

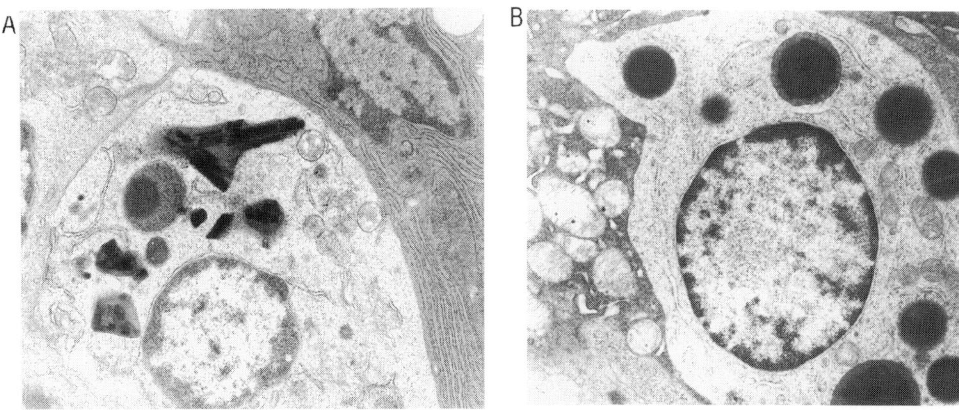

Fig. 1 Electron micrographs showing typical IMMC from +/+ (A) and –/– (B) jejunal sections, from day 8 post-infection. The crystalline bodies, clearly visible in A, are only found in +/+ IMMC, whereas granules containing internal divisions (B) were exclusive to the –/– IMMC. Original magnifications: ×16,000 (A) and ×14,000 (B).

One of the more interesting observations described in Table II, and confirmed in the study by Wastling *et al.* (9), was the increased numbers of IMMC in the parasitized gut of mMCP-1–/– mice when compared with +/+ controls. It is possible that this observation is due to lack of mMCP-1 in the granules and the failure of the IMMC to secrete this highly soluble chymase into the lateral spaces between the epithelial cells. Previous evidence from studies with RMCP-II indicate that soluble β-chymases may alter epithelial permeability by disrupting the epithelial tight junctions (22, 23) (see discussion section). We have speculated that this could allow IMMC to migrate into the gut lumen (9), for which there is a precedent in parasitized sheep (24).

Mucosal pathology is more severe in parasitized –/– mice

For an initial assessment of mucosal pathology, paraformaldehyde-fixed tissues were pinned on to thick cardboard and fixed for 6 h in 4% paraformaldehyde before being transferred to 75% ethanol (15). After processing into paraffin wax, 4-μm sections were cut along the longitudinal axis of the intestine and at right angles to the mucosa so that the villi appeared as finger-like projections. The lengths of at least 20 villi and 20 crypts per sample of jejunum (n = 8–9) were measured using a ×25 objective lens and a ×10 eyepiece containing a graticule calibrated against a micrometer. The data shown in Table II indicate that, in addition to the altered mast cell kinetics in –/– mice, there was significant elongation of the crypts and shortening of the villi in the infected jejunum of –/– mice compared with +/+ controls.

Response of mMCP-1 –/– Mice to Infection by other Gastrointestinal Nematodes

Current evidence suggests that expulsion mechanisms operate selectively according to both host and parasite species (25). There is considerable doubt that the immune rejection of *N. brasiliensis* involves IMMC in mice (26), which may be more influenced by mucins secreted by goblet cells in the gut (25). In contrast, IMMC involvement has been more clearly demonstrated in the immune rejection of the intestinal nematodes *Strongyloides ratti* and *Trichinella spiralis* (12, 14). Variation in effector mechanisms may reflect variation in the niche occupied in the intestinal environment by adult worms from different species. In contrast to *N. brasiliensis*, where adult worms are located on the surface of the mucosa (27), *T. spiralis* adults are found within the epithelium itself, whereas *S. ratti* adult worms are subepithelial (27, 28). It is possible that the close association of *T. spiralis* and *S. ratti* with the epithelium brings these parasites into close proximity to the IMMC where they may be more influenced by IMMC-derived products and epithelial turnover. Therefore we undertook to investigate the immune responses of mMCP-1 –/– mice to these nematode species.

In order to reduce the genetic differences within experimental groups and to investigate mast cell responses on a more familiar immunological background, we have been backcrossing MF1/129/mMCP-1 +/– mice onto a Balb/C background. We have been using Balb/C F6 mMCP-1 –/– and mMCP-1 +/+ mice (backcrossed five times) in initial experiments to assess responses to *S. ratti* and *T. spiralis*. Two groups of five Balb/C F6 mMCP-1 –/– and Balb/C F6 mMCP-1 +/+ mice were infected subcutaneously with 500 *S. ratti* L_3 and individual faecal larval counts were monitored for both groups. There was a trend towards higher larval counts in the faeces from the mMCP-1 –/– mice, which was significant on days 6 and 8–11 post-infection (Fig. 2).

In a second experiment, Balb/C F6 mMCP-1 –/– and mMCP-1 +/+ littermates were infected with 500 *T. spiralis* infective stage larvae, and groups of six killed on days 6 and 18 post-infection (Table I). Worm establishment on day 6 was similar in both groups (mean worm burden 186.5 (SE 36.5) and 204 (SE 52.6) from mMCP-1 +/+ and –/– mice,

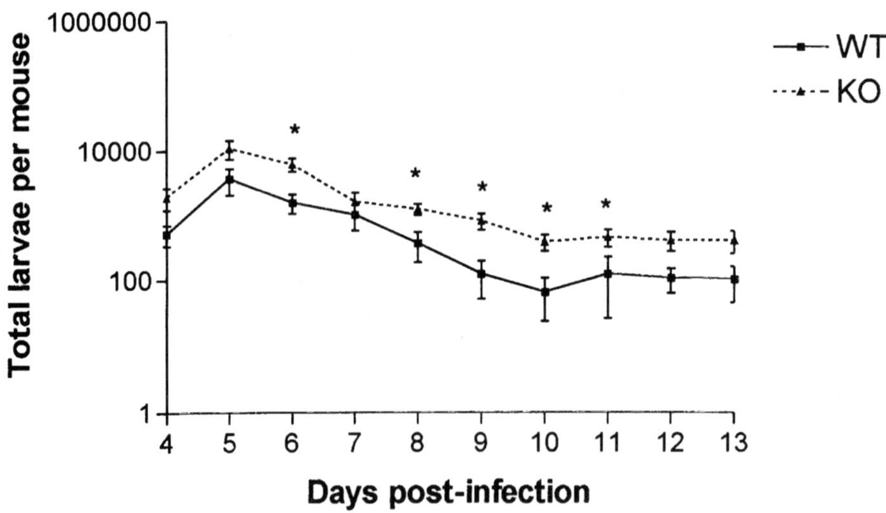

Fig. 2 Graph to show kinetics of infection with *S. ratti*. Two groups of five Balb/C F6 mMCP-1 –/– and Balb/C F6 mMCP-1 +/+ mice were infected subcutaneously with 500 *S. ratti* L_3 and individual faecal larval counts were monitored for both groups. There was a trend towards higher larval counts in the faeces from the mMCP-1 –/– mice, which was significant on days 6 and 8–11 post-infection.

respectively), but worm burdens were significantly higher in mMCP-1 –/– mice on day 18 post-infection (Table I). These preliminary experiments suggest that the absence of mMCP-1 in Balb/C mice results in a small but significant delay in the expulsion of these nematode species.

Further experiments are currently underway to determine the full kinetics of worm expulsion and to confirm whether mast cell hyperplasia differs between congenic mMCP-1 –/– and +/+ mice, and at what stage of infection this occurs. Similarly, more accurate microdissection techniques (29) will be used to monitor the kinetics of mucosal pathology in both groups. Future work will focus on exploration of the influence of mMCP-1 on vascular and epithelial permeability and the rates of turnover of IMMC and of epithelium using our *in vivo* knockout model. However, a more immediate goal will be to determine the relationship between the IMMC and the epithelial cells themselves using *in vitro*-derived IMMC (see below).

REGULATION OF THE EXPRESSION OF mMCP-1 IN mBMMC

Past attempts to generate IMMC-like cells, using mBMMC, have not been successful. Initially, mBMMC grown in the presence of IL-3 were considered to share many features in common with IMMC (30), although they expressed little or none of the IMMC-specific chymases (15, 31). In contrast, mBMMC supported by T cell-derived conditioned medium (15), or by a combination of SCF/IL-9 or SCF/IL-10, expressed the IMMC-specific chymase, mMCP-1 (16, 17). It is clear from *in vivo* studies using mice transgenic for overexpression of IL-9 (14) and from *in vitro* experiments with mBMMC, that IL-9 is involved in the expression of mMCP-1 and mMCP-2 (16).

In parasitized mice, additional factors such as SCF (11), that are associated with the mucosal epithelium may be particularly important because >95% of IMMC differentiate intraepithelially in mouse intestine (6, 7, 32). Another candidate is TGF-β1 since it is expressed by enterocytes and by several other cell types in the gut (33, 34). This cytokine has a number of immunoregulatory roles, including that of promoting the expression of the surface integrin α_E on mBMMC (35, 36). TGF-β1 modulates expression of other genes in mast cells, including an $Fc_\varepsilon RI$-induced member of the chemokine family, lymphotactin (37). The expression of SCF (11) and of TGF-β1 (33) by enterocytes provides a potential mechanism for regulating the differentiation of IMMC which are intimately associated with gut epithelium. The hypothesis that TGF-β1 regulates the differentiation and hyperplasia of IMMC was explored by culturing mBMMC in the presence or absence of this cytokine. The aim of the study was to determine whether exposure of mBMMC to TGF-β1 altered their protease phenotype.

TGF-β1 Promotes the Enhanced Expression and Secretion of mMCP-1

In order to test the hypothesis that TGF-β1 regulates the expression of IMMC-specific β-chymases, bone marrow cells were cultured in the presence of WEHI (IL-3-enriched medium) (15%), rrSCF (50 ng ml^{-1}), and rmIL-9 (5 ng ml^{-1}) for 7 days to produce >95% mast cells that were > 90% c-kit$^+$ and 20±4% mMCP-1$^+$ (18). The cells were split into separate flasks and supplemented with WEH/SCF/IL-9 to which was added either vehicle alone, or TGF-β1 at a final concentration of 1 ng ml^{-1} of culture supernatant.

After adding TGF-β1, the proportion of mMCP-1$^+$ mBMMC increased to 99% within

7 days, whereas in control flasks, lacking TGF-β1, a maximum of 30% mMCP-1$^+$ mBMMC were detected at 7 days. The differences between the values for control and TGF-β1-supplemented flasks were highly significant (18) on days 4 and 7 after addition of TGF-β1. A 500-fold increase in the level of mMCP-1 in the culture supernatants to 6000 ng ml^{-1} was observed in the presence of TGF-β1 and, although there was a gradual increase in the controls, maximum values were 44 ng ml^{-1} on day 7 (18). Similarly, the mMCP-1 content of cell pellets on day 7 was substantially higher in the TGF-β1-supplemented mBMMC than in controls (18).

In the culture supplemented with WEHI/SCF/IL-9, transcripts for the chymases mMCP-1, -2, -4 and -5 were detected at similar levels (18); this is in agreement with previous work (16). When the culture was supplemented with TGF-β1, the transcription of mMCP-1 and mMCP-2 was increased, as demonstrated by semiquantitative RT-PCR (18). There was no obvious variation in the transcription of mMCP-4 and mMCP-5, which was consistent with the immunohistochemical and ELISA results, which suggest that TGF-β1 regulates the increased expression of mMCP-1 and mMCP-2 (18).

There were substantial morphological differences between the two mBMMC populations. Control mBMMC grown in the absence of TGF-β1 had pseudopodia and the granules were vacuolated and less distinct. The addition of TGF-β1 was associated with a more compact mBMMC, lacking pseudopodia, and with densely stained granules of variable shape and size (18). A morphometric study was carried out to compare granule size and staining intensity using Leishman-stained cytosmears from mBMMC supplemented with TGF-β1/IL-9 or IL-9 alone, as described above. Morphological measurements were obtained from 24-bit colour digital images of mBMMC using Object Image 1.62n16 (38)*. rhTGF-β1 had a marked effect on mBMMC size and morphology (Table III). There was a significant reduction in mBMMC size in cultures maintained in

Table III
mBMMC Cultured in the Presence of TGF-β1 Show Altered Staining Properties as Assessed by Leishman-stained Cytosmears

	Median staining intensity (0–255)	Median variation in Staining Intensity (0–255)	Median area (μm^2)	Median Diameter (μm)	Median Roundness (0.0–1.0)
I/W/S	102±4	66±2	263.5±12.3	20±0.4	0.7059±0.0240
T/I/W/S	139±2	82±1	167.6±3.7	15.8±0.2	0.8115±0.0097
p value	<0.0001	<0.0001	<0.0001	<0.0001	<0.0001

Morphological analysis of mBMMC cultured in the presence or absence of TGF-β1. Cytokines are indicated as follows: T= TGF-β1; I = rmIL-9; W = WEHI-3B IL-3-rich supernatant; S = rrSCF. See text for details of analysis.

*24-bit colour digital images were acquired using a JVC TK-1280E charge couple device camera module attached to an Olympus BX-50 conventional compound microscope (Olympus Optical Co., London, U.K.). Live phase alternating line (PAL) video was digitized using an AV framegrabber mounted in a Power Macintosh 7100/80 AV (Apple Computers, Cupertino, U.S.A.), running Object Image 1.62n16 (Norbert Vischer, Universiteit van Amsterdam, The Netherlands) and Plug-in Digitiser v1.2.4 (Cyrus Daboo, Cyrusoft International, Cambridge, U.K.). Image acquisition and analysis were performed within Object Image 1.62n16. Object Image (38) is a public domain software package based on NIH Image (39) and can be downloaded via the internet from http://simon.bio.uva.nl/object-image.html

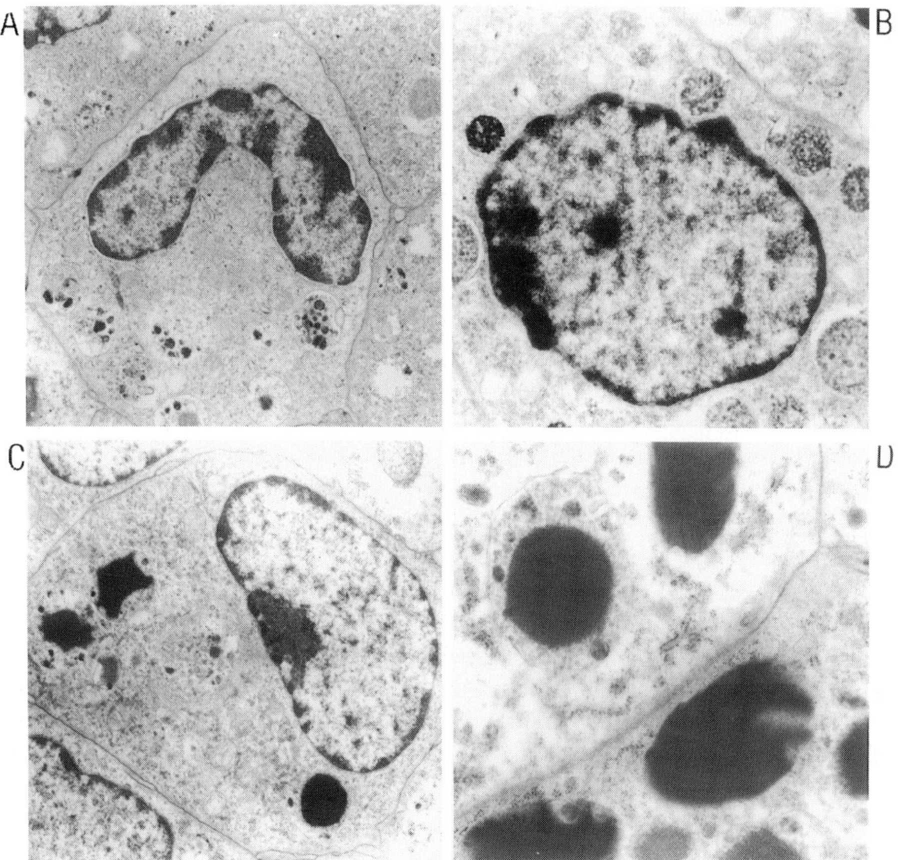

Fig. 3 Electron micrographs showing typical appearance of mBMMC cultured in the presence of WEHI/rrSCF (A) (original magnification: ×8500), WEHI/rrSCF/rmIL-9 (B) (original magnification: ×11,000) or WEHI/rrSCF/rmIL-9/TGF-β1 (original magnification: ×8550 (C) and original magnification: ×23,000 (D)). The granules of mBMMC supplemented with TGF-β1 are much more electron-dense.

TGF-β1 when compared to controls. mBMMC that were exposed to rhTGF-β1 had significantly higher roundness values, which was consistent with the observed reduction in the number of pseudopodia present on these cells. Granule size and staining intensity with Leishman's was also markedly increased in mBMMC exposed to rhTGF-β1, as indicated by significant increases in staining intensity and variability. Ultrastructurally, IMMC exposed to TGF-β1 contained densely stained homogeneous granules (Fig. 3C, D), whereas the granules of the control cells were vacuolated as described previously (Fig. 3A, B) (18, 40).

SCF and TGF-β1 Synergize in Regulating the Expression and Secretion of mMCP-1

The expression and secretion of mMCP-1 by different combinations of cytokines was explored by growing bone marrow cells for 1 week in WEHI/SCF (>80% BMMC, >90% viable) before transferring them at 5×10^5 cells ml^{-1} into 48-well plates. The cells were

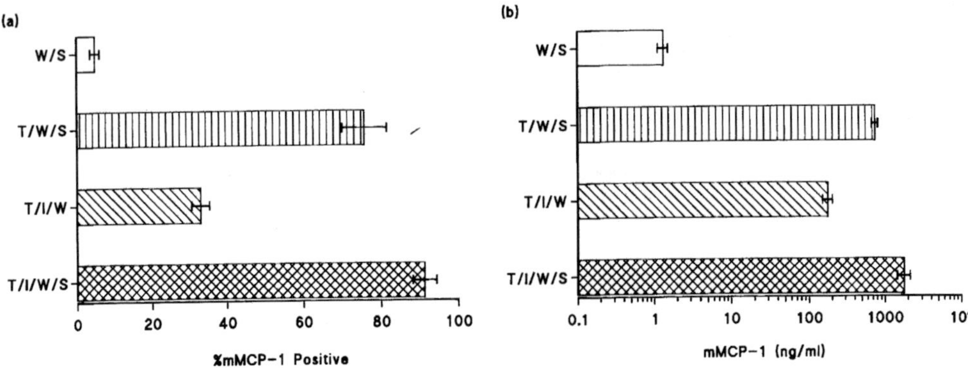

Fig. 4 TGF-β1 induces maximal mMCP-1 expression. mBMMC were grown in flasks for 1 week in WEHI/SCF before they were transferred at 5×10^5 ml^{-1} into 48-well plates and cultured for a further 4 days in the various cytokine combinations shown; T = TGF-β1; I = rmIL-9; W = WEHI-3B IL-3-rich supernatant; S = rrSCF. The graphs show the percentage of mMCP-1$^+$ mast cells as assessed by immunohistochemical staining of cytosmears by mAb RF 6.1 (a), and concentrations of mMCP-1 in culture supernatants (ng ml^{-1}) as assessed by ELISA using mAb RF 6.1 (b), measured 4 days after transfer.

cultured for a further 4 days in the cytokine combinations shown in Fig. 4. In the cytokine combinations supplemented with TGF-β1, the proportion of mMCP-1$^+$ mBMMC and the concentrations of mMCP-1 in the culture supernatants increased compared with those from WEHI/SCF controls (Fig. 4). Supplementation of mBMMC cultures with both IL-9 and TGF-β1, in the presence of WEHI/SCF, induced maximal expression and secretion of mMCP-1 (Fig. 4).

The more extensive experiments described by Miller *et al.* (18) showed that TGF-β1 in combination with WEH/SCF/IL-9 promotes maximal cell growth in association with mMCP-1 expression. Whereas mBMMC in WEHI/SCF showed a 4-fold increase in numbers of mBMMC by day 4, mMCP-1 expression ranged from minimal to absent (18). The combination of TGF-β1/rmIL-9/WEHI/SCF resulted in a 3-fold increase in numbers of mBMMC; conversely, supplementation with TGF-β1/SCF alone was associated with very poor cell viability and a drop in cell numbers (18) as reported previously (41).

Expression and Extracellular Release of mMCP-1 Requires the Continuous Presence of TGF-β1

In vivo, IMMC close to or within the mucosal epithelium invariably express mMCP-1 (6), and the level of expression can vary depending on the stimulus (7). It was important to determine, therefore, whether mBMMC expressing mMCP-1 after the addition of TGF-β1 continue to store and release this chymase when TGF-β1 is withdrawn from the culture.

Bone marrow cells grown in the presence of WEHI/rmIL-9/rrSCF and TGF-β1 for 7 days were transferred in quadruplicate to 48-well plates and cultured in varying combinations of cytokines, as shown in Fig. 5. Withdrawal of TGF-β1 resulted in a fall in the concentrations of mMCP-1 in the supernatants and in the cell pellets 4 days later (Fig. 5). Cell growth was greatest in the presence of WEHI/rrSCF, but this combination was associated with a substantial drop in the percentage of mMCP-1$^+$ cells (18) and in the intra- and extracellular levels of mMCP-1 (Fig. 5). These data show that TGF-β1 maintains the expression and extracellular release of mMCP-1.

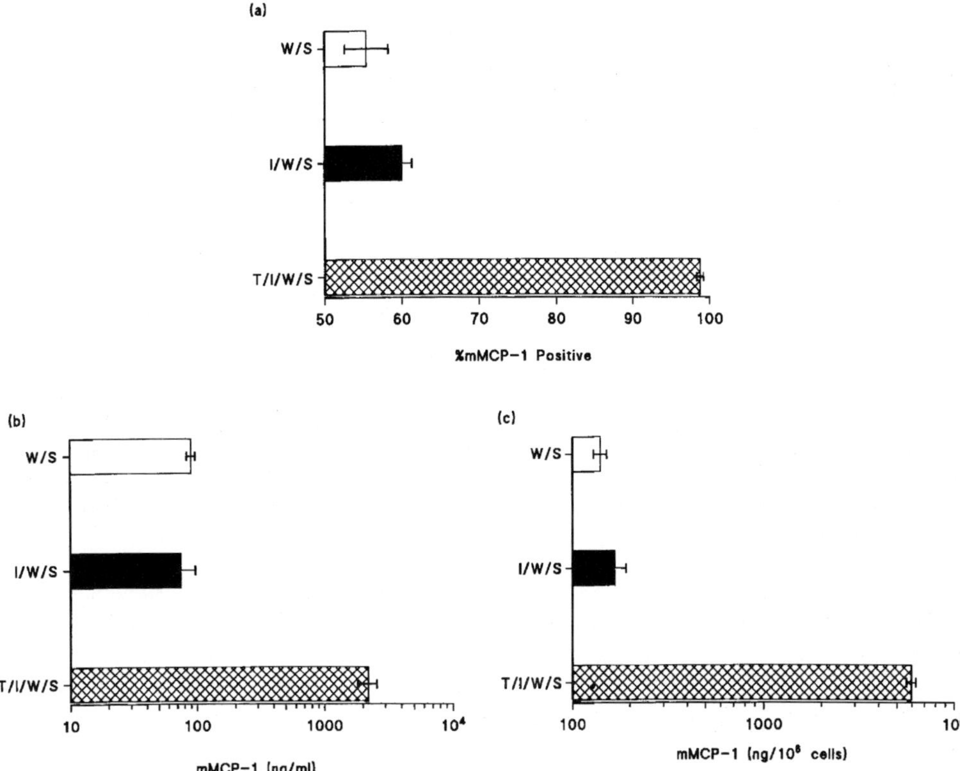

Fig. 5 Effect of withdrawal of TGF-β1 on mMCP-1 expression. mBMMC were grown in the presence of WEHI/rmIL-9/rrSCF and TGF-β1 (1 ng ml^{-1}) for 7 days. These cultures yielded 15.2×10^5 mBMMC per ml, 88% viable, and 99.6% MCP-1^+, with 547 ng mMCP-1 per ml of supernatant. Cells were transferred in quadruplicate at 5×10^5 per ml to 48-well plates and cultured in varying combinations of cytokines as shown; T = TGF-β1; I = rmIL-9; W = WEHI-3B IL-3-rich supernatant; S = rrSCF. The graphs show percentage of mMCP-1^+ mast cells as assessed by immunohistochemical staining of cytosmears by mAb RF 6.1 (a), concentrations of mMCP-1 in culture supernatants (ng ml^{-1}) (b) and cell pellets (ng per 10^5 cells) (c), as assessed by ELISA using mAb RF 6.1, measured 4 days after transfer into cytokine combinations.

A Role for Endogenous TGF-β1 in the Expression of mMCP-1

Whereas IL-9 induces increased expression of mMCP-1 and mMCP-2 in mBMMC (16, 17), our results show that TGF-β1 is a much more potent stimulus than IL-9 for the expression of mMCP-1 (18). One theory was that IL-9 was increasing the endogenous secretion of active TGF-β1 and/or the processing of latent TGF-β1 (18) and that low levels of this cytokine were responsible for the suboptimal mMCP-1 expression in IL-9-induced cultures. To test this hypothesis, mBMMC were grown in WEHI/SCF for 7 days before transferring them in quadruplicate to 48-well plates supplemented with TGF-β1 (100 pg ml^{-1})/rmIL-9 (5 ng ml^{-1})/WEHI (15%)/rrSCF (50 ng ml^{-1}), or with IL-9/SCF or WEHI/SCF at the same concentrations. Included in these cultures were chicken anti-TGF-β1 antibody (1μg ml^{-1}) or control chicken IgG antibody (1μg ml^{-1}). At 48 h mMCP-1 was detected in 5.1 ± 0.3% of the cells supplemented with IL-9/SCF to which chicken IgG had been added (Fig. 6) However, the addition of anti-TGF-β1 antibody suppressed

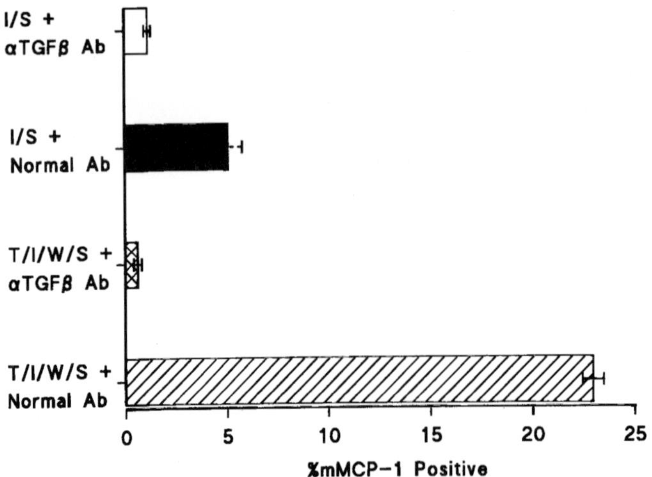

Fig. 6 Effect of anti-TGF-β1 antibody (Ab) on mMCP-1 expression. mBMMC were grown in WEHI/SCF for 7 days before transferring them in quadruplicate to 48-well plates supplemented with TGF-β1 (100 pg ml^{-1})/rmIL-9 (5 ng ml^{-1})/WEHI (15%)/rrSCF (50 ng ml^{-1}), or with IL-9/SCF or WEHI/SCF at the same concentrations. Cytokines: T = TGF-β1; I = rmIL-9; W = WEHI-3B IL-3-rich supernatant; S = rrSCF. Included in these cultures were chicken anti-TGF-β antibody (1 μg ml^{-1}), control chicken IgG (normal Ab) (1 μg ml^{-1}) as indicated. The graph shows the percentage of mMCP-1^{+} mast cells as assessed by immunohistochemical staining of cytosmears by mAb RF 6.1 measured 2 days after transfer into cytokines and antibodies.

the intracellular staining for mMCP-1 in the cultures supplemented with IL-9 (1.1 ± 0.3%; $p<0.03$; Mann–Whitney) and TGF-β1 (0.6 ± 0.2% v. 23 ± 0.5% in the control; $p<0.03$; Mann–Whitney) (Fig. 6). It also suppressed the extracellular release of mMCP-1 in the cultures supplemented with TGF-β1 (18).

The transcription of granule proteases was analysed in mBMMC supplemented with WEHI/SCF, with TGF-β1 (100 pg ml^{-1})/IL-9/WEHI/SCF, or with IL-9/WEHI/SCF in the presence of anti-TGF-β antibody or of control IgG. Substantial suppression of transcription of mMCP-1 and mMCP-2 was induced by anti-TGF-β antibody, but transcripts for mMCP-4 and mMCP-5 were apparently unaffected (18). Interestingly, transcripts for TGF-β1 were detected in all cell pellets regardless of treatment (18). This suggests that IL-9-induced mMCP-1 expression is mediated via autostimulation with TGF-β1 and that the mechanism is post-transcriptional.

DISCUSSION

Two main areas of our research have been addressed in this review. The first focused on the function of the abundant β-chymase mMCP-1 *in vivo*, which is being addressed by homologous gene targeting. The second examined the mechanisms regulating the expression of mMCP-1. An interesting preliminary finding in the mMCP-1 –/– mice was that there was a small but significant effect on worm survival (Table I) and this appeared to be associated with more severe pathology of the mucosa in the –/– mice. This finding requires confirmation in mice with congenic backgrounds and using microdissection techniques, studies which are currently under way. Infection with *N. brasiliensis* was associated with different IMMC kinetics in the –/– group of mice when compared with the +/+ controls in two separate experiments (Table I) (9); the possible reasons for this are discussed below.

Light microscopy suggested that IMMC in –/– mice have smaller, less intensely toluidine blue-positive granules which often did not stain for esterase. Ultrastructural analysis established that the classical paracrystalline structures, frequently described in IMMC and clearly present in IMMC from +/+ mice, were absent from –/– IMMC (9). The granules retained toluidine blue-positive staining, indicating that proteoglycans were still present, but the altered granule matrices detected ultrastructurally (Fig. 1) (9) suggest that there were profound biochemical changes associated with the depletion of mMCP-1. Since expression of mMCP-2, -4 and -5 in IMMC appeared to be normal, the depletion of esterase activity in the –/– IMMC was somewhat surprising (9). It may be that mMCP-2 has little esterase activity and that the concentrations of mMCP-4 and mMCP-5 are too low to generate strong esterase activity. Alternatively, the proteases in IMMC granules apart from mMCP-1 may simply be in low abundance. A third possibility, given the structural differences in the granules, is that the stacking of the proteases on the glycosaminoglycan side-chains (42) is disrupted and that the proteases are either non-functional or are catabolized in the granules.

The preliminary data on mucosal pathology indicate that, in addition to augmented mast cell hyperplasia, villus atrophy/crypt hyperplasia is more severe in *N. brasiliensis*-infected mMCP-1 –/– mice than +/+ controls. We are currently seeking to confirm these studies in congenic mice on a Balb/C background, with different gastrointestinal nematode species and using better-developed microdissection techniques (29). Should further studies in congenic mMCP-1 –/– and +/+ BALB/c mice confirm these observations, there will be several mechanisms of mMCP-1 *in vivo* that can be addressed experimentally.

The most interesting of these is the role of mMCP-1 in promoting the permeability of the jejunal epithelium. If a major function of the rodent β-chymases in IMMC is to increase epithelial permeability in the immediate environs of each IMMC, this would provide a highly regulated and localized mechanism for permitting the escape of potentially damaging accumulations of fluid from beneath the epithelium itself. *Ex vivo* perfusion studies, as well as *in vitro* epithelial cell monolayer experiments, show that rMCP-2 induces permeability without causing detectable epithelial damage (22, 23). Indeed, recent freeze-fracture studies of duodenal and ileal epithelial tight junctions in experimentally induced colitis in the rat have elegantly demonstrated that the transmembrane protein occludin is disrupted and irregular (43). These morphological alterations to the tight junction were associated with greatly increased permeability to lanthanum (43). It is interesting that the occludin in MDCK epithelial cell monolayers was disrupted after basolateral but not apical exposure to rMCP-2 (23), and it is possible that the β-chymases released in the immediate vicinity of intraepithelial mast cells can, similarly, disrupt the occludin in the tight junctions. Whether this a direct effect of the chymase on chymotrypsin-sensitive sites of the occludin molecule or, more likely, via a protease-sensitive receptor-mediated process within the epithelial cells themselves, remains to be seen.

There are alternative or even parallel mechanisms to consider which may contribute to more severe mucosal pathology in mMCP-1 –/– mice when compared with +/+ controls. The first of these is the substantially increased numbers of IMMC in the epithelium in –/– mice when compared with controls (Table II). It is possible that the resultant, more abundant, chemokines, cytokines and lipid and other inflammatory mediators, cause changes in the viability/turnover of the epithelium. A third possibility is that the absence of mMCP-1 results in reduced activation of the metalloproteinase family of pro-collagenases and that this limits the remodelling of the epithelium relative to the basement membrane that separates the epithelium from the lamina propria. It is probable that a combination of each of these various mechanisms could result in more severe pathology in nematode-infected –/– mice.

The mechanisms underlying mast cell heterogeneity, particularly with regard to the apparent tissue specificity of granule chymase expression (30, 44), are not well defined. mBMMC have been widely used to explore the mechanisms of heterogeneity. They were originally considered to be IMMC-like (40), although it is now clear that they lack the granule structure seen *in vivo*, and do not express the IMMC-specific chymase mMCP-1 (30). The present results show clearly that it is possible to induce the expression of high levels of mMCP-1 in mBMMC by the addition of TGF-β1 to the cultures.

The focus of this *in vitro* study was the expression of mMCP-1 in mBMMC since this is the only chymase that is expressed uniquely at mucosal surfaces (7, 45), and mechanisms that significantly upregulate expression *in vitro* are also likely to operate *in vivo*. The present data strongly suggest that TGF-β1 is a potent stimulus for the expression of mMCP-1, and that mBMMC cultured for several days in this cytokine not only express and release microgram quantities of this chymase, but that the cells themselves are more heavily granulated and the granules are more homogeneous than has been described previously for mBMMC (44).

Lantz and Huff (46) generated 35% mBMMC within 7 days after supplementing bone marrow cells with either WEHI/SCF or WEHI/IL-9/SCF. Differentiation of mBMMC was usually greater than 80% with both combinations of cytokines by day 7 in our studies (18). The results are in agreement with previous studies showing that WEHI/SCF was associated with barely detectable expression of mMCP-1, but that the presence of IL-9 was associated with upregulated expression of mMCP-1 (16). Importantly, there was mBMMC heterogeneity, as described previously, where a high proportion of cells did not express this chymase (15). However, the addition of TGF-β1 resulted in a very rapid increase in the expression of mMCP-1 in >95% of mBMMC, in increased transcription of mMCP-2 and in substantial release of mMCP-1 into the supernatant (18).

Expression and release of mMCP-1 were dependent on the concentration of TGF-β1 in the culture and diminished when TGF-β1 was withdrawn (18). The reduced mMCP-1 expression observed on addition of anti-TGF-β antibody suggests that mMCP-1 induction is through a TGF-β-dependent pathway. It is likely that TGF-β1 is responsible for the mucosal mast cell-specific expression as well as the IgE-independent release of mMCP-1. Such a mechanism could explain why, in normal mice, there is systemic secretion of mMCP-1, and why such high systemic levels of this protease are evident at very early stages of intestinal nematode infection, before specific IgE responses can be detected (3).

The readily induced expression of mMCP-1 *in vitro*, which can be measured in culture supernatants using sensitive immunoassays (7), provides a novel system in which to analyse TGF-β receptor expression and signal transduction pathways, and the transcriptional or post-transcriptional mechanisms operating on the mMCP-1 gene. For example, the relative contributions of TGF-β receptors I and II (TβI and TβII) to the expression of mMCP-1 and possible mechanisms of mRNA stabilization, which is reported to occur for mMCP-2 in the presence of exogenous IL-10 (17), can both be explored using this culture system. Our assumption is that IL-9 and IL-10 both induce autostimulation via endogenously secreted TGF-β and, since IL-9 has no obvious effect on TGF-β1 transcription, it is probable that post-translational activation of latent TGF-β1, as has been well described for macrophages and monocytes (34), is involved. This is unlikely to be the only role for IL-9 since, in mice transgenic for overexpression of IL-9, there is mucosal mast cell hyperplasia and substantial levels of mMCP-1 in the blood (14). It is probable that IL-9 synergizes with TGF-β1 and SCF in promoting the maturation of IMMC and the expression and secretion of mMCP-1.

These new data suggest that the expression and release of mMCP-1 is important in the regulation of mast cell hyperplasia, the expulsion of intestinal nematodes and the development of mucosal pathology. The fact that it appears to be highly TGF-β1-dependent extends the range of functions for this important cytokine at mucosal surfaces. Future work should address the localization of the site of expression of TGF-β1 in parasitized gut and its regulation during helminth infection in other allergic responses for which mucosal mast cell hyperplasia has been reported.

ACKNOWLEDGEMENTS

Supported by grant no. 050065 from the Wellcome Trust and grant no. 15/S 10130 from the British Biotechnology and Biology Council. We are grateful to Andrew Read's group, Division of Cell and Population Biology, for supplying *S. ratti* larvae.

REFERENCES

1. Perdue, M. H. and McKay, D. M. Mucosal regulators and immunopathophysiology. *Curr. Opin. Gastroenterol.* **12**:591, 1996.
2. Catto-Smith, A. G. and Ripper, J. L. Mucosal mast cells and developmental changes in gastric absorption. *Am. J. Physiol.* **268**:G121, 1995.
3. Miller, H. R. P. Mucosal mast cells and the allergic response against nematode parasites. *Vet. Immunol. Immunopathol.* **54**:331, 1996.
4. Miller, H. R. P., Huntley, J. F. and Newlands, G. F. J. Mast cell chymases in helminthosis and hypersensitivity. In *Mast Cell Proteases in Immunology and Biology* (Caughey, G. H., ed.), p. 203. Marcel Dekker, New York, 1995.
5. Wastling, J. M., Scudamore, C. L., Thornton, E. M., Newlands, G. F. J. and Miller, H. R. P. Constitutive expression of mouse mast cell protease-1 in normal BALB/c mice and its up-regulation during intestinal nematode infection. *Immunology* **90**:308, 1997.
6. Miller H. R. P., Huntley, J. F., Newlands, G. F. J., Mackeller, A., Lammas, D. and Wakelin, D. Granule proteinases define mast cell heterogeneity in the serosa and the gastrointestinal mucosa of the mouse. *Immunology* **65**:559, 1988.
7. Scudamore, C. L., McMillan, L., Thornton, E. M., Wright, S. H., Newlands, G. F. J. and Miller, H. R. P. Mast cell heterogeneity in the gastrointestinal tract: variable expression of mouse mast cell protease-1 (mMCP-1) in intraepithelial mucosal mast cells in nematode infected and normal BALB/c mice. *Am. J. Pathol.* **150**:1661, 1997.
8. Miller, H. R. P., Huntley, J. F., Newlands, G. F. J. and Irvine, J. Granule chymases and the characterisation of mast cell phenotype and function in rat and mouse. *Monogr. Allergy* **27**:1, 1990.
9. Wastling, H. M., Knight, P., Ure, J., Wright, S., Thornton, E. M., Scudamore, C. L., Mason, J., Smith, A. and Miller, H. R. P. Histochemical and ultrastructural modification of mucosal mast cell granules in parasitized mice lacking the β-chymase, mouse mast cell protease-1. *Am. J. Pathol.* **153**:491, 1998.
10. Galli, S. J., Geissler, E. N. and Zsebo, K. M. The *kit*-ligand, stem cell factor. *Adv. Immunol.* **55**:1, 1994.
11. Klimpel, G. R., Langley, K. E., Wypych, J., Abrams, J. S., Chopra, A. K. and Niesel, D. W. A role for

stem cell factor (SCF): c-*kit* interaction(s) in the intestinal tract response to *Salmonella typhimurium* infection. *J. Exp. Med.* **184**:271, 1996.

12. Abe, T. and Nawa, Y. Worm expulsion and mucosal mast cell response induced by repetitive IL-3 administration in *Strongyloides ratti*-infected nude mice. *Immunology* **63**:181, 1988.
13. Madden, K. B., Urban, J. F., Ziltener, H. J., Schrader, J. W., Finkelman, F. D. and Katona, I. M. Antibodies to IL-3 and IL-4 suppress helminth-induced intestinal mastocytosis. *J. Immunol.* **147**:1387, 1991.
14. Faulkner, H., Humphreys, M., Renauld, J. C., Van Snick, J. and Grencis, R. Interleukin-9 is involved in host protective immunity to intestinal nematode infection. *Eur. J. Immunol.* **27**:2536, 1997.
15. Newlands, G. F. J., Lammas, D. A., Huntley, J. F., MacKellar, A., Wakelin, D. and Miller, H. R. P. Heterogeneity of murine bone marrow-derived mast cells: analysis of their proteinase content. *Immunology* **72**:434, 1991.
16. Eklund, K. K., Ghildyal, N., Austen, K. F. and Stevens, R. L. Induction by IL-9 and suppression by IL-3 and IL-4 of the levels of chromosome 14-derived transcripts that encode late-expressed mouse mast cell proteases. *J. Immunol.* **151**:4266, 1993.
17. Xia, Z., Ghidyal, N., Austen, K. F. and Stevens, R. L. Post-transcriptional regulation of chymase expression in mast cells. A cytokine-dependent mechanism for controlling the expression of granule neutral proteases of haematopoietic cells. *J. Biol. Chem.* **271**:8747, 1996.
18. Miller, H. R. P., Wright, S. H., Knight, P. A. and Thornton, E. M. A novel function for TGF-β1: upregulation of the expression and the IgE-independent extracellular release of a mucosal mast cell granule-specific β-chymase, mouse mast cell protease-1. *Blood* **93**:3473, 1999.
19. Gurish, M. F., Nadeau, J. H., Johnson, K. R., McNeil, H. P., Grattan, K. M., Austen, K. F. and Stevens, R. L. A closely linked complex of mouse mast cell-specific chymase genes on chromosome 14. *J. Biol. Chem.* **268**:11372, 1993.
20. Crowle, P. K. and Phillips, D. E. Characteristics of mast cells in Chediak-Higashi mice: light and electron microscopic studies of connective tissue and mucosal mast cells. *Exp. Cell Biol.* **51**:130, 1983.
21. Friend, D. S., Ghildyal, N., Austen, K. F., Gurish, M. F., Matsumoto, R. and Stevens, R. L. Mast cells that reside at different locations in the jejunum of mice infected with *trichinella spiralis* exhibit sequential changes in their granule ultrastructre and chymase phenotype. *J. Cell. Biol.* **135**:279, 1996.
22. Scudamore, C. L., Thornton, E. M., McMillan, L., Newlands, G. F. J. and Miller, H. R. P. Release of the mucosal mast cell granule chymase, rat mast cell protease-II, during anaphylaxis is associated with the rapid development of paracellular permeability to macromolecules in rat jejunum. *J. Exp. Med.* **182**:1871, 1995.
23. Scudamore, C. L., Jepson, M. A., Hirst, B. H. and Miller, H. R. The rat mucosal mast cell chymase, RMCP-II, alters epithelial cell monolayer permeability in association with altered distribution of the tight junction proteins ZO-1 and occludin. *Eur. J. Cell. Biol.* **75**:321, 1998.
24. Stankiewicz, M., Pernathaner, W., Cabaj, W., Jonas, W. E., Douch, P. G. C., Bisset, S. A., Rabel, B., Pfeffer, A. and Green, R. S. Immunisation of sheep against parasitic nematodes leads to elevated levels of globule leukocytes in the small intestine lumen. *Int. J. Parasitol.* **25**:389, 1995.
25. Nawa, Y., Ishikawa, N., Tsuchiya, K., Horii, Y., Abe, T., Khan, A., Shi, B., Itoh, H., Ide, H. and Uchiyama, F. Selective effector mechanisms for the expulsion of intestinal helminths. *Parasite Immunol.* **16**:333, 1994.
26. Finkelman, F. D., Madden, K. B., Cheever, A. W., Katona, I. M., Morris, S. C., Gately, M. K., Hubbard, B. R., Gause, W. C. and Urban, J. F. Jr. Effects of interleukin 12 on immune responses and host protection in mice infected with intestinal nematode parasites. *J. Exp. Med.* **179**:1563, 1994.
27. Miller, H. R. P. *Vet. Immunol. Immunopathol.* **7(1)**:99, 1984.
28. Miller, H. R. P. Mast cells in the gastrointestinal tract. In Immuno Pharmacology of Mast Cells and Basophils (Foreman, J., ed.), p. 197, Academic Press, New York, 1993.
29. Lawrence, C. E., Paterson, J. C. M., Higgins, L. M., MacDonald, T. T., Kennedy, M. W. and Garside, P. IL-4 regulated enteropathy in an intestinal nematode infection. *Eur. J. Immunol.* **28**:2672, 1998.
30. Metcalfe, D. D., Baram, D. and Mekori, Y. A. Mast cells. *Physiol. Rev.* **77**:1033, 1997.
31. Hunt, J. E. and Stevens, R. L. Mouse mast cell proteases. In *Biological and Molecular Aspects of Mast Cell and Basophil Differentiation and Function* (Kitamura, Y., Yamamota, S., Galli, S. J. and Greaves, M. W. eds), p. 149. Raven Press, New York, 1995.
32. Friend, D. S., Ghildyal, N., Gurish, M. F., Hunt, J. Hu, X., Austen, K. F. and Stevens, R. L. Reversible expression of tryptases and chymases in the jejunal mast cells of mice infected with *Trichinella spiralis*. *J. Immunol.* **160**:5537, 1998.

33. Koyama, S. and Podolsky, D. K. Differential expression of transforming growth factors α and β in rat intestinal epithelial cells. *J. Clin. Invest.* **83**:1768, 1989.
34. Letterio, J. J. and Roberts, A. B. Regulation of immune responses by TGF-β. *Annu. Rev. Immunol.* **16**:137, 1998.
35. Smith, T. J., Ducharme, L. A., Shaw, S. K., Parker, C. M., Brenner, M. B., Kilshaw, P. J. and Weis, J. H. Murine M290 expression modulated by mast cell activation. *Immunity* **1**:393, 1994.
36. Wright, S. H., Knight, P. A., Brown, J., Thornton, E. M., Kilshaw, P. J. and Miller, H. R. P. TGF-β_1 mediates early expression of αE and the β chymase, mouse mast cell protease-1, in mouse bone marrow derived mast cells. (Submitted.)
37. Rumsaeng, V., Vliagoftis, H., Oh, C. K. and Metcalfe, D. D. Lymphotactin gene expression in mast cells following $Fc_{\epsilon}RI$ aggregation-modulation by TGF-β, IL-4, dexamethasone and cyclosporin A. *J. Immunol.* **158**:1353, 1997.
38. Vischer, N. O. E., Huls, P. G. and Woldringh, C. L. Object Image: an interactive image analysis program using structured point collection. *Binary* **6**:160, 1994.
39. Rasband, W. S. and Bright, D. S. NIH Image: A public domain image processing program for the Macintosh. *Microbeam Anal.* **4**:137, 1995.
40. Sredni, B., Friedman, M. M., Bland, C. E. and Metcalfe, D. D. Ultrastructural, biochemical, and functional characteristic of histamine-containing cells cloned from mouse bone marrow: tentative identification as mucosal mast cells. *J. Immunol.* **131**:915, 1983.
41. Mekori, Y. A. and Metcalfe, D. D. Transforming growth factor-β prevents stem cell factor-mediated rescue of mast cells from apoptosis after IL-3 deprivation. *J. Immunol.* **153**:2194, 1994.
42. Sali, A., Matsumoto, R., McNeil, P., Karplus, M. and Stevens, R. L. Three-dimensional models of four mouse mast cell chymases: identification of proteoglycan binding regions and protease-specific antigenic epitopes. *J. Biol. Chem.* **268**:9023, 1993.
43. Fries, W., Mazzon, E., Squarzoni, S., Martin, A., Martines, D., Micali, A., Sturniolo, G. C., Citi, S. and Longo, G. Experimental colitis increases small intestine permeability in the rat. *Lab. Invest.* **79**:49, 1999.
44. Galli, S. J. New insights into 'the riddle of the mast cell'. Microenvironmental regulation of mast cell development and phenotypic heterogeneity. *Lab. Invest.* **62**:5, 1990.
45. Huntley, J. F., Gooden, C., Newlands, G. F. J., MacKellar, A., Lammas, D. A., Wakelin, D., Woodbury, R. G. and Miller, H. R. P. Distribution of intestinal mast cell proteinase in blood and tissues of normal and *Trichinella*-infected mice. *Parasite Immunol.* **12**:85, 1990.
46. Lantz, C. S. and Huff, T. F. Differential responsiveness of purified mouse c-kit$^+$ mast cells and their progenitors to IL-3 and stem cell factor. *J. Immunol.* **155**:4024, 1995.

CHAPTER 18

Structure and Function of Human Chymase

NORMAN M. SCHECHTER,[1] PEDRO JOSÉ BARBOSA PEREIRA[2] and STEFAN STROBL[3]*

[1]*Department of Dermatology, University of Pennsylvania, Philadelphia, Pennsylvania, U.S.A.,*
[2]*Departamento de Biologia Molecular y Celular, Instituto de Biologia Molecular de Barcelona (IBMB) Jordi Girona, Barcelona, Spain, and*
[3]*Abteilung für Strukturforschung, Max-Planck-Institut für Biochemie, Planegg-Martinsried, Germany*

INTRODUCTION

Human chymase is a chymotrypsin-like protease found in mast cells. This review summarizes the genetic, enzymatic and biochemical properties of chymase. Also reviewed in detail is the X-ray crystal structure of recombinant human chymase (rHC) complexed to an irreversible peptide–chloromethylketone (CMK) inhibitor (1). The presence of the inhibitor is used to define the active site and extended substrate-binding site of chymase. Examination of the X-ray crystal structure reveals key interactions that may have important roles in determining the substrate specificity of human chymase.

GENERAL PROPERTIES

Chymase Is a Constituent of a Subpopulation of Human Mast Cells

Mast cells are potent inflammatory cells found in connective and mucosal tissues throughout the body (2, 3). They are noted for the large number of secretory granules that populate the cytoplasm. Within these secretory granules are stored many pre-formed mediators of biological importance (2, 3). The major protein constituents of mast cell granules are an assortment of proteolytic enzymes (4–7). Human chymase is one of these proteases (8–10). Chymase expression is limited to the subpopulation of mast cells termed MC_{TC} (4). These mast cells predominate in connective tissues (4). The concentration of human chymase within mast cells has been estimated at 4.5 pg per mast cell (5).

* Corresponding author.

MAST CELLS AND BASOPHILS
ISBN 0-12-473335-2

Genetic and Biochemical Properties

In contrast to many other species, human mast cells express only one chymase (Table I). The same enzyme has been isolated from skin, heart and tonsil tissues (11–14), and the single gene coding for the protease has been identified (15–17). As shown in Table I, human chymase exhibits a high degree of amino acid sequence homology with chymases from other animals, clearly making it a member of the chymase subfamily of serine proteases, which contains over 15 members at present. Within the chymase family, human chymase shows most identity with baboon chymase (97%) and least identity with mouse mast cell protease 2 (54%). It is 74% identical to rat chymase 2 (rat mast cell protease 2), the only other chymase for which there is a crystal structure.

The gene for human chymase has been mapped to chromosome 14 (18). The locus is near to the gene for cathepsin G, a serine protease expressed by neutrophils, monocytes and mast cells. Also near are the genes for several granzymes expressed by cytotoxic T lymphocytes. This chromosomal location places chymase within a cluster of genes for

TABLE I
Residues Important to the Extended Substrate-binding Site of Chymases and Related Proteases

Enzyme	Homology	Residues					
		40	41	143	175	189	192
Human chymase*	100†	Lys	Phe	Arg	Asp	Ser	Lys
Baboon chymase	97	Lys	Ser	Arg	Tyr	Ser	Lys
Dog chymase	82	Ala	Ser	Lys	Asp	Ser	Lys
Rat mast cell protease 1	61	Ala	Thr	Gln	His	Ser	Lys
Rat mast cell protease 2	59	Val	Ile	Lys	Tyr	Ala	Met
Rat mast cell protease 3		Ser	Ala	Arg	Ser	Asn	Lys
Mouse mast cell protease 1	58	Asp	Arg	Lys	Tyr	Thr	Met
Mouse mast cell protease 2	54	Asp	Arg	Lys	Asp	Ser	Gln
Mouse mast cell protease 4	65	Ala	Thr	Arg	Asp	Ser	Lys
Mouse mast cell protease 5	75	Ser	Ala	Arg	Ser	Asn	Lys
Mouse mast cell protease 9		Ala	Ile	Arg	His	Ser	Met
Mongolian gerbil chymase 1		Ala	Ser	Arg	His	Ser	Lys
Mongolian gerbil chymase 2		Ser	Thr	Lys	Asn	Asn	Lys
Sheep mast cell protease 1‡		Arg	Tyr	Ser	Lys	Asn	Ser
Sheep mast cell protease 3		Ile	Arg	Leu	Lys	Asp	Leu
Hamster chymase 1	63	Ala	Ser	Arg	His	Ser	Lys
Hamster chymase 2	70	Ala	Ser	Arg	Asp	Asn	Lys
Ancestral chymase`	77	Ala	Arg	Arg	Asp	Ser	Lys
Human cathepsin G	52	Ser	Arg	Arg	Asp	Ala	Lys
Bovine chymotrypsin	32	His	Phe	Leu	Lys	Ser	Met

* Assignments are based on sequence homologies between human chymase, rat mast cell protease 2, cathepsin G and chymotrypsin, as indicated by comparison of structures (1). In chymases position 40 and 41 are the residues immediately prior to the conserved Cys42; position 143 is the residue immediately after the conserved sequence Gly140–Trp141–Gly142; position 175 is 5 residues after the conserved Cys168, position 189 and 192 are located 6 and 3 residues prior to Ser 195, the catalytic serine.

† Homologies and sequence information based on published data (15, 22, 68, 71–77).

‡ Sheep mast cell protease 1 and 3 show more homology to cathepsin G than to chymases (76).

serine proteases that are specifically expressed by haematopoietic cells of different lineages.

The molecular mass of human chymase based on amino acid composition is approximately 25 kDa. On SDS gels, the protease migrates as a broad band centred at 30 kDa. The broad banding pattern is due to glycosylation (7, 19). Chymase is a basic protein, binds heparin through electrostatic interactions, and is likely stored in mast cell granules complexed to heparin-containing proteoglycans (8, 10, 20). A discrete binding site for heparin has not been identified. Analysis of the chymase structure finds patches or clusters of positive charge on the surface of the protein which may mediate the interaction with heparin (1, 21, 22).

Chymase is synthesized by the mast cell as a prepro-enzyme. The pro-enzyme contains a two-residue extension (Gly–Glu) at the N-terminus which is susceptible to removal by the thiol protease DPPI, a dipeptidyl aminopeptidase found in inflammatory cells (23, 24). Similar to the activation of the pancreatic serine proteases, full catalytic function is achieved after removal of the pro-peptide (19, 23, 24). The enzymatically active form of chymase is packaged within secretory granules. Thus degranulation of mast cells releases the catalytically active enzyme.

Recombinant Human Chymase

Several laboratories have reported the production of enzymatically active recombinant human chymase (19, 23–26). The most efficient system for production of the enzyme is the baculovirus–insect cell system. In this system the protease, which is usually expressed in an inactive form, is secreted into the media. Chymase has been expressed as the natural proenzyme which was activated by the thiol protease DPPI (23, 24). Chymase also has been expressed as a fusion protein linked to ubiquitin, an 8-kDa protein (19, 26). Separating the two proteins and replacing the native two-residue pro-peptide was the cleavage site sequence recognized by the proteinase enterokinase. Enterokinase converts trypsinogen to trypsin in the gut and displays a high degree of specificity for the sequence Asp–Asp–Asp–Asp–Lys–X (27). Activation of chymase from the fusion protein by enterokinase is highly efficient, showing that the natural propeptide is not necessary for the proper folding of the protease.

Recombinant chymase has been evaluated with respect to substrate specificity, interaction with inhibitors and heparin binding. Significant deviation from the characteristics of the native enzyme has not been observed. The X-ray crystal structure of recombinant chymase produced in baculovirus-infected insect cells revealed carbohydrate at the two consensus glycosylation sites, suggesting that both sites are available for glycosylation *in vivo* (1).

Substrates

Human chymase hydrolyses model peptide substrates (28), as well as biological peptides and proteins with specificity characteristic of a chymotrypsin-like protease (see below). That is, cleavage occurs on the carboxyl side of large aromatic and aliphatic amino acids (Fig. 1). Consistent with its substrate specificity, the protein protease inhibitors effective toward human chymase are those that display specificity for chymotrypsin (29–32). The 189 position in the amino acid sequence of serine proteases (numbering is based on homology with bovine chymotrypsin) is an important residue because it forms the base of

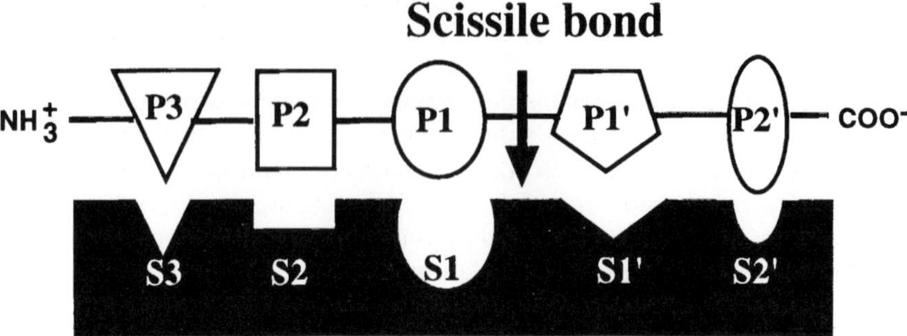

Fig. 1 Schematic representation of the extended substrate-binding site of a serine protease. Objects labelled P signify the amino acid residues of a peptide substrate where the bond between the P1 and P1′ position is the scissile bond. The area shaded in black represents the surface of the protease. The complementary surfaces on the protease marked S and S′ represent the subsites on the protease that interact with the corresponding P and P′ residues of the substrate. In a serine protease, the S1 subsite is a cavity that is considered the primary substrate-binding site. The structure of the S1 subsite determines whether a serine protease displays chymotrypsin-like, trypsin-like or elastase-like activity (59, 60). Cleavage of the scissile bond is mediated by a triad system composed of Ser195–His57–Asp102 (chymotrypsin numbering), which is not shown.

the primary specificity pocket (S1 pocket; see Fig. 1). In chymase this residue is a Ser as found in bovine pancreatic chymotrypsin (1).

Since its initial discovery and description, numerous *in vitro* studies have shown that chymase is capable of activating and inactivating many biological mediators by limited proteolysis at selected sites. Chymase produces vasoactive angiotensin II by cleavage of angiotensin I at a Phe–His bond (12, 33, 34), active interleukin-1β (IL-1β) by hydrolysis of the cell-secreted latent form at a Tyr–Val bond (35), active endothelin by cleavage of big endothelin at a Tyr–Gly bond (36, 37), mature collagen by hydrolysis of the globular pro-domain at a Leu–Ser bond (38), and active interstitial collagenase by cleavage of the secreted latent enzyme at a Leu–Thr bond (39). In addition, chymase has been shown to degrade basement membrane (40, 41), inactivate the thrombin receptor by cleavage of its extracellular domain (42), release membrane-bound stem cell factor by cleavage of a Phe–Met bond in the membrane anchoring domain (43) and stimulate secretion of mucus from serous cells (44). Dog chymase, a close relative (Table I), has been shown to activate latent gelatinase (45), a metalloproteinase which may play a key role in cancer metastasis. Lastly, rat chymase has been shown to degrade low-density lipoprotein (LDL), a finding implying that chymases may have a role in the development of atherosclerosis (46).

Many of the above reactions occur with remarkable selectivity and efficiency, suggesting that the substrate-binding site of human chymase is highly specialized. The properties determining the recognition of substrates by chymase are discussed in the next sections.

Regulation by Physiological Inhibitors

Chymase is susceptible to several inhibitors found in human plasma and mucosal secretions (e.g. bronchial and salivary fluids). In plasma, α_1-antichymotrypsin, α_1-proteinase inhibitor and α_2-macroglobulin irreversibly inactivate chymase (29, 31).

α_2-Macroglobulin is the primary inhibitor demonstrating an association rate constant ($5 \times 10^6\ M^{-1}\ s^{-1}$) that is two to three orders of magnitude higher than those of the other inhibitors.

Of the protease inhibitors in mucosal secretions, chymase is susceptible to inhibition by secretory leukocyte proteinase inhibitor (SLPI), a member of the mucus protease inhibitor family. The effectiveness of SLPI is dependent upon interaction with heparin, which induces a conformational change in the inhibitor that enhances inhibition (47). Mucus family inhibitors form reversible high-affinity complexes with proteases, and therefore their effectiveness is dependent on both an equilibrium dissociation constant and association rate constant. The equilibrium dissociation and association rate constants measured for the chymase–SLPI interaction in the presence of heparin were 3.5 nM and $2.5 \times 10^5\ M^{-1}\ s^{-1}$, respectively (30). In the absence of heparin the equilibrium constant increased to 50 μM. In bronchial fluids, where the concentration of SLPI is 5 μM (100–1000 times greater than the equilibrium constant value), chymase should be effectively inhibited in a stoichiometric manner (47).

Based on the concentration of inhibitors in plasma and bronchial fluids and the rate constants reported above, plasma would inactivate chymase with a half-life of approximately 0.05 sec, and bronchial fluid would inactivate chymase with a half-life of 0.5 sec (30). Because mast cells are present in tissues where the concentration of inhibitors is less than in fluids, the above half-lives should be considered as maximal rates of inhibition. Estimates of inhibition rates also are complicated by the possibility that chymase is secreted from mast cells complexed to heparin-containing proteoglycans. As part of a proteoglycan complex, chymase may not be accessible to physiological inhibitors which are relatively large proteins. Therefore the rate-determining step of inhibition may be the dissociation of the proteoglycan complex rather than the binding of the inhibitor. The stability of proteoglycan complexes from human mast cells and the accessibility of chymase present in these complexes has been difficult to establish with certainty (48, 49).

X-RAY CRYSTAL STRUCTURE OF rHC

The X-ray crystal structure of human chymase (HC) has been determined independently by two laboratories (1, 50). The first structure was established with rHC expressed in a specifically engineered *Bacillus subtilis*. For crystallization chymase was inhibited with phenylmethanesulphonyl fluoride (PMSF–rHC). The second structure was determined with rHC produced in a baculovirus–insect cell system. For crystallization chymase was inhibited with the peptide inhibitor succinyl–Ala–Ala–Pro–Phe–CMK (peptide–CMK–rHC).

The α carbon backbone structures of both human chymases are similar to each other and to that of rat chymase 2 (1, 50, 51). The general folding pattern of the polypeptide chain is similar to bovine pancreatic chymotrypsin, the archetype of the serine protease family (Plate I). Chymase is organized into two domains, each composed of a six-stranded β barrel structure and associated loops. The only substantial helix observed is a four turn–helix found at the C-terminus. The substrate binding cleft is at the interface of the two domains. The two glycosylation sites of human chymase are located on opposite sides of the molecule somewhat above and below the plane joining both domains. Glycosylation of chymase should not directly interfere with the binding of peptide substrates.

A significant difference in the two human chymase structures is found at the active site. The difference is primarily due to the interaction of the protease with the different inhibitors used for inactivation. In the PMSF–rHC structure, the sulphonyl sulphur is covalently bound to the γ oxygen of Ser195 and the sulphonyl oxygens are hydrogen-bonded to the backbone nitrogen of Gly193 and one of the imidazole nitrogens of His57. These interactions produce rotation of the His57 side-chain from its normal orientation in the catalytic triad (50). Additionally, the phenylmethane group of the inhibitor appears at the mouth of the S1 pocket instead of buried within the pocket.

In contrast to PMSF–rHC, the peptide–CMK–rHC structure demonstrated more typical interactions at the active site. The residues of the catalytic triad are correctly aligned and the side-chain of the P1 Phe residue is buried in the S1 pocket (1). As depicted in Fig. 2, the peptide–CMK inhibitor formed two covalent bonds with the protease. These bonds were between the P1 carbonyl carbon (originally that forming the ketone moiety) and the γ oxygen of Ser195, and between the methylene group (originally the chloromethyl group) and a nitrogen on the imidazole ring of His57. As a result of the first reaction, the carbonyl carbon of the P1 residue exhibited tetrahedral geometry with the formed oxyanion being stabilized by the NH groups of Gly193 and Ser195. These amide NH groups form the oxyanion hole of serine proteases. The oxyanion hole helps to stabilize substrates in the transition state (52–54). Thus the inhibitor appears to be bound to the protease in a conformation resembling a transition state analogue.

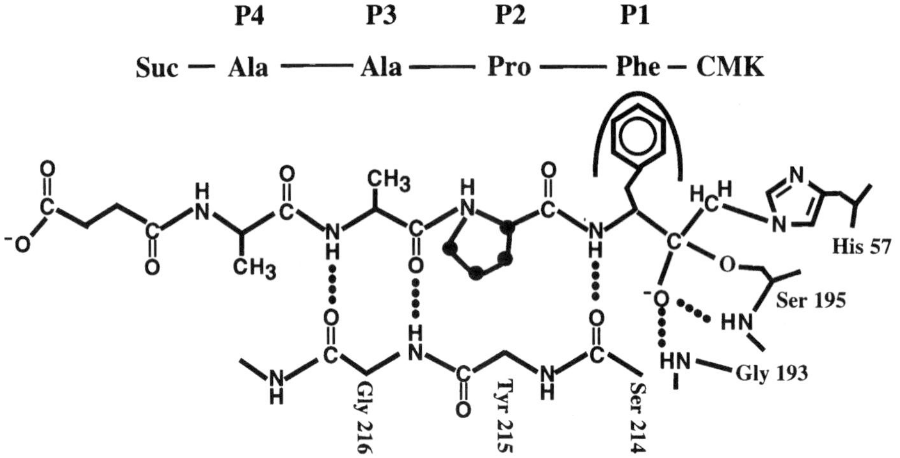

Fig. 2 Depiction of classical hydrogen and covalent bonds between inhibitor and chymase present in the peptide–CMK–rHC structure. Bonds were indicated from the closeness of groups in peptide–CMK–rHC X-ray crystal structure. Dotted lines indicate backbone hydrogen bonds between inhibitor and protease. Covalent bonds are solid lines between the γ oxygen of Ser195 and carbonyl carbon of the P1 residue, and the His57 imidazole ring and methylene group of the inhibitor. Stabilizing the oxyanion formed upon reaction of the γ oxygen of Ser195 with the carbonyl group of the P1 residue are the amide NH groups of Ser195 and Gly193.

Also stabilizing the inhibitor in the active site were three classical hydrogen bonds between the peptide backbones of the inhibitor and protease (Fig. 2). One was between the amide nitrogen of the P1 Phe residue and the carbonyl oxygen of Ser214. The other two were between the amide nitrogen and carbonyl oxygen of the P3 Ala residue and the complementary atoms of Gly216. These backbone interactions are observed in other serine protease–inhibitor structures, and function to align the reactive loops of inhibitors in the binding cleft (52–54). Similar interactions are hypothesized to occur with substrates.

THE EXTENDED SUBSTRATE-BINDING SITE OF HUMAN CHYMASE

The extended substrate-binding site is the region of the protease that interacts with amino acid residues flanking the scissile bond. It usually takes the form of a cleft in the protease capable of interacting with six to nine residues of a protein or peptide substrate (52, 53). As shown in the schematic representation in Fig. 1, residues of the substrate are designated as P or P′ depending upon the side of the scissile bond they appear (55). In an analogous manner, the sites on the protease that interact with the substrate are designated S or S′. In the peptide–CMK–rHC structure, the covalent and backbone interactions discussed above suggest that the peptide chain of the inhibitor is aligned in a manner that defines the structure, organization and side-chain preference of the subsites forming the extended substrate-binding site of chymase. Interactions between the S1–S4 subsites of the protease and the P1–P4 residues of the inhibitor are directly visualized. Information on the structure of the primed side was obtained by extrapolating the path of the polypeptide chain into this region and by modelling the association between chymase and a potent protein inhibitor.

S–P Interactions

In Plate IIA, a space-filling model of the peptide–CMK–rHC structure is shown with the bound inhibitor made transparent by the computer. The residues, except Ser195 (purple), that are important to the extended substrate-binding site are presented in CPK colours and are numbered according to homology with chymotrypsin (1). The rest of the protein is in green. Ser195 is at the centre of the structure, and the area below Ser195 is the opening of the S1 subsite. Of particular interest, the extended substrate-binding site contains a large number of charged residues. Within a relatively small area immediately to the right of Ser195 is a concentration of positive charge produced by Lys40, Lys192 and Arg143. The concentration of positive charge in this area appears to be a property exclusive to human and baboon chymase as shown in Table I. On the left side of the extended substrate-binding site is a negatively charged region composed of Asp175 and Ser97. At the bottom of the S1 pocket opening is the hydroxyl group of Ser218 and the positively charged residue, Arg217.

In Plate IIB, the P1 Phe residue of the peptide–CMK inhibitor is shown in orange. The phenylalanyl side-chain is buried sideways in the channel that produces the S1 pocket. The pocket is large, containing buried water, and can accommodate easily a Tyr or Trp residue. Surrounding the phenylalanyl side-chain at the opening to the S1 pocket is Lys192, Phe191, Ala220 (located behind Ser 218 and not visible in this view), Ser218, Gly216 and Tyr215.

In Plate IIC, the P2–P4 residues are shown. The P2 Pro and P4 Ala residues are shown in orange and the P3 Ala is shown in yellow; the P labelling is placed on the side-chains of each residue. The pyrrolidine ring of the Pro residue fits nicely into a V-shaped S2 subsite made by the side-chains of His57 and Leu99. Situated in this groove, the pyrrolidine ring kinks the polypeptide chain to position the P3 residue directly in front of Gly216, allowing for the production of two key backbone hydrogen bonds with this residue as discussed above (see Fig. 2). The methyl side-chain of the P3 Ala residue (sphere with label P3) is positioned over the S1 pocket. The methyl group is making close contact (4–4.5Å) with the methylene carbons of Lys192 and Ser218 and the phenylalanyl ring of the P1 Phe residue. The P4 Ala residue sits in a shallow hydrophobic groove made by the side-chains of Leu99, Tyr215 and Phe173. There are no backbone contacts stabilizing the P4 Ala residue of the inhibitor. The methyl side-chain of the P4 residue (sphere with label P4) is pointing towards the negatively charged region made by the carboxyl group of Asp175 and the carbonyl oxygen of Ser97. The succinyl moiety was not resolved in the crystal structure, thus there is no information on the S5 subsite.

Charge preferences of the S3 and S4 subsites

The P3–S3 and P4–S4 interactions defined by the peptide–CMK–rHC structure suggest that charge preferences may be associated with the S3 and S4 subsites of human chymase. The S3 subsite appears to overlap the S1 subsite, as indicated by the orientation of the side-chain of the P3 residue over the opening to the S1 pocket. In this orientation the side-chain of the P3 residue is close to the side-chain of Lys192. The nearness of the two side-chains suggests that electrostatic interactions could occur if the P3 residue was a charged amino acid. Because Lys192 is positively charged, an anionic P3 residue (Glu and Asp) would be preferred over a cationic residue (Lys and Arg). Interestingly, the opposite charge preference may be a factor determining the specificity of the S4 subsite. Although a distance of approximately 6.5 Å is found between the methyl group of the P4 Ala residue and the negatively charged region produced by Asp175 and Ser97, anionic and cationic amino acids have longer side-chains and could make contact with this region. Thus chymase may prefer cationic over anionic residues at the P4 position.

Biochemical evidence supporting these inferred charge preferences was indicated in a study mapping the extended substrate-binding site of human chymase with a series of peptide nitroanilide substrates (28). In addition, the presence of a Glu residue in the P3 position has been found to increase the effectiveness of peptide inhibitors of human chymase (56, 57).

Lys192 and effects of pH on activity

The analysis of human chymase hydrolysis of the substrate succinyl–Ala–Ala–Pro–Phe–pNA as a function of pH demonstrated a stair-step profile. Progressive increases in catalytic activity were observed over the pH range of 5.5–8.0 and over the alkaline pH range of 8–11 (58). Enhancement of catalytic activity over the high pH range is novel for a serine protease. The catalytic activity of a serine protease is dependent on the basicity of His57 of the catalytic triad. Because the pK_a for deprotonation of a His residue is 7.0, serine proteases typically achieve maximal catalytic activity around pH 8.0–8.5 (59, 60). The increased activity of chymase at high pH suggests that in addition to His57, residues with pK_a values in the alkaline range play a role in catalysis.

The identification of Lys192 as a residue contributing to both the S1 and S3 subsites of human chymase suggests a possible mechanism for the pH enhancement. Lys residues are positively charged at neutral pH and become neutral over the alkaline pH range, exhibiting a pK_a for the ε-amino group of approximately 10.5. Neutralization of the amino group of Lys192 by high pH would reduce its need to be solvated by water, thereby enhancing its ability to interact with groups of the substrate buried in the active site region. This enhanced interaction may produce more efficient catalysis. Conceivably a substrate with a negatively charged residue at P3 would not show the high pH enhancement because favourable electrostatic interactions in the neutral pH range would obviate the need for solvation of Lys192 by water.

S´–P´ Interactions

The presence of the CMK inhibitor made it possible to extrapolate the path of a polypeptide chain through the primed side of the extended substrate-binding site. Further definition of the primed subsites was obtained by docking to the active site the protein protease inhibitor, Bowman–Birk inhibitor, from soybean (61, 62). The interaction of this inhibitor with chymase demonstrates a strong equilibrium dissociation constant (K_d = 50 pM), suggesting highly complementary binding to the unprimed and primed sites of the extended substrate-binding site (32).

The notable features of the primed side are the protrusion of the side-chain of Phe41 into the binding cleft (Plates I, II), and the concentration of positive charge in the region of the S1´ and S2´ subsites. The protrusion of the phenylalanyl group appears to constrict the size of binding cleft opening at approximately the location of the S3´ subsite. This constriction may make it difficult for substrates with polypeptide chains extended beyond P2´ to bind, especially if the residues have large side-chains. Also appearing to constrict the primed side of the binding cleft are the positively charged residues Arg143 and Lys40, which are located on either side of Phe41. The reactive loop of inhibitors belonging to the Bowman–Birk inhibitor family have unusually short primed sides due to a sharp turn in the peptide chain mediated by a Pro residue at P3´. Such a structure may help to explain the strong interaction of this inhibitor with human chymase.

The 41 position in serine proteases is located next to a highly conserved disulphide bond formed between Cys42 and Cys58. This disulphide helps to form the substrate-binding cleft of the protease. Both human neutrophil elastase and chymotrypsin have a Phe at position 41 and a His at position 40. In contrast to chymase, the crystal structures of these proteases show both side-chains integrated into the protease body (Plate I). The difference in the orientation of residues 40–41 in chymase and chymotrypsin appears related to a three-residue insertion after position 35 in the chymotrypsin structure.

The cationic character of the primed side region suggests that the S1´ or S2´ subsites of human chymase might favour substrates and inhibitors with negatively charged residues at the complementary positions. Such a preference appears to be supported in the design of peptide and synthetic inhibitors of human chymase (56, 63). Negatively charged residues placed at positions to interact with the S1´ and S2´ subsites of chymase improve the strength of inhibition markedly.

MODELLING OF THE HUMAN CHYMASE–ANGIOTENSIN I INTERACTION

Dipeptidyl Carboxypeptidase-like Specificity

Cleavage of angiotensin I, the ten-residue peptide shown in Fig. 3, at the Phe8–His9 bond produces angiotensin II, a potent vasoconstrictor. The primary enzyme responsible for producing this cleavage is angiotensin-converting enzyme (ACE), a metalloproteinase with dipeptidyl carboxypeptidase specificity. Although serine proteases are considered better endopeptidases than exopeptidases (64), the presence of a Phe residue at position 8 of the sequence suggests that angiotensin I also may be a substrate for chymotrypsin-like

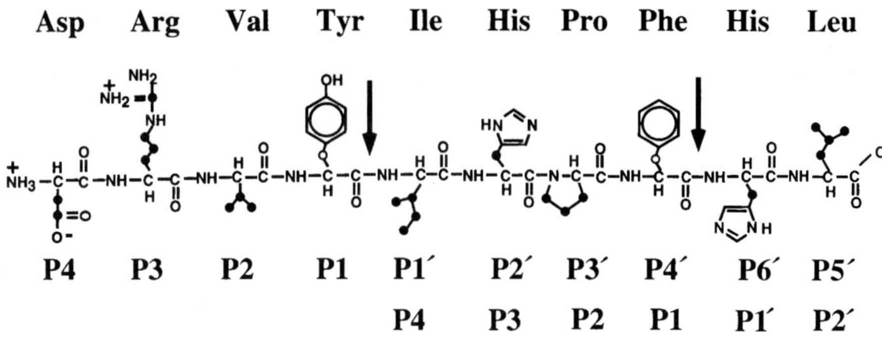

Asp Arg Val Tyr Ile His Pro Phe His Leu

P4 P3 P2 P1 P1′ P2′ P3′ P4′ P6′ P5′

P4 P3 P2 P1 P1′ P2′

Fig. 3 Structure of angiotensin I. The amino acid sequence of angiotensin I is above the structure. Arrows point to the two cleavage sites that may be recognized by a chymotrypsin-like protease. The designations below the structure are the substrate positions of residues if the scissile bond is at Tyr4–Ile5 or Phe8–His9.

proteases. Evaluation of a series of chymotrypsin-like proteases showed that human chymase was markedly better than chymotrypsin, cathepsin G and rat chymase 1 (rat mast cell protease I) in converting angiotensin I to II (33, 34, 65). The kinetic constants for the conversion were somewhat better than those determined for ACE (12, 34). Also of importance was the absence of chymase hydrolysis of the Tyr4–Ile5 bond, a second potential site for a chymotrypsin-like protease (Fig. 3). This bond is a degradative site that was cleaved by chymotrypsin and rat chymase (34, 65, 66). The selectivity and efficiency of Phe8–His9 bond cleavage suggests that human chymase displays an unusual dipeptidyl carboxypeptidase-like specificity. Computer modelling of chymase–angiotensin I interaction using the peptide–CMK–rHC crystal structure discussed next provides insight into this reaction specificity (1).

Interaction at the Site Producing Angiotensin II

Angiotensin I was docked to the active site of human chymase utilizing the backbone structure of the bound peptide–CMK inhibitor to orient the P1–P4 residues and the backbone structure of the Bowman–Birk protease inhibitor to orient the P1′–P2′ residues (1). Residues beyond P4 were not included because of the absence of structural information. The major interactions obtained from the modelling study are shown in Plate III.

On the unprimed side, the Phe and Pro residues dock as described for the peptide–CMK inhibitor. To dock the P3 His residue, the side-chain was positioned over the S1 pocket between Lys192 and Ser218. A hydrogen bond between a nitrogen on the imidazole ring and the hydroxyl group of Ser218 may help to stabilize the His residue in this position. The P4 Ile residue was accommodated well by the shallow pocket that forms the S4 subsite.

On the primed side, the side-chain of the P1′ His residue was positioned between His57 and Lys40. A hydrogen bond between a nitrogen of the imidazole ring and the carbonyl oxygen of His57 may help in the stabilization of this residue. The side-chain of the P2′ Leu residue was positioned pointing towards Arg143, thereby leaving the free carboxylate facing Phe41. In this position, the C-terminal P2′ Leu residue can be stabilized by several electrostatic interactions. The amide nitrogen and carbonyl oxygen of Phe41 are in a position to form hydrogen bonds with the carboxylate oxygens and amide nitrogen of the P2´ Leu residue, respectively. Also and perhaps key to the stabilization of the free carboxylate is the amino group of Lys40. The side-chain of Lys40 is flexible enough to rotate toward the free carboxyl group, positioning its amino group to help in charge neutralization.

The most striking finding from the modelling study was the ability of the S2´ subsite of chymase to stabilize a free carboxyl group through electrostatic interactions involving the residues Lys40 and Phe41. This may explain, in part, the unusual dipeptidyl carboxypeptidase-like activity of human chymase revealed in the cleavage of angiotensin I at the Phe8–His9 bond. The role of Phe41 and Lys40 in stabilizing a free carboxyl group is supported, to a degree, by biochemical studies. Truncating angiotensin I by removal of a single residue at the C-terminus markedly reduces the efficiency of hydrolysis, while extending angiotensin I at the C-terminus reduces the efficiency of hydrolysis by as much as 60% (66, 67).

Interaction at the Site Producing Degradation

Human chymase hydrolyses the Tyr4–IIe5 bond of angiotensin I extremely slowly regardless of the form of angiotensin (I or II) used as substrate. Chymotrypsin, on the other hand, hydrolyses the Tyr4–Ile5 bond and the Phe8–His9 bond relatively fast and with comparable efficiency. Thus, the lack of recognition of this bond is a property intrinsic to chymase.

Based on the properties of the extended substrate-binding site discussed so far many factors may contribute to the extremely low efficiency of hydrolysis of the Tyr4–Ile5 bond by human chymase. The Tyr4–Ile5 bond is located near the centre of the polypeptide (Fig. 3); therefore to bind at this site chymase must interact with residues beyond the P2´ position. The residues occupying P3´ and P4´ sites are relatively bulky with hydrophobic side-chains and may destabilize binding in the region due to steric clashes with Phe41. On the unprimed side virtually every interaction from P2–P4 appears suboptimal. The P2 residue is not a Pro, thus the chain is not held rigidly in a conformation optimal for alignment of the P1 and P3 residues in the active site. Although the P3 (Arg) and P4 (Asp) residues are charged, the polarity of each residue is the opposite of that preferred at the corresponding subsites of the enzyme. Thus, poor interactions on both sides of the Tyr4–Ile5 bond appear to contribute to the very slow cleavage of this site by human chymase.

In support of the above observations are kinetic studies with peptides duplicating the sequence of angiotensin I around the Tyr4–Ile5 bond (66). These studies implicate multiple poor interactions at S4–S2´ as the cause for the poor hydrolysis of the Tyr4–Ile5 bond by human chymase. A major improvement in catalysis was observed when an Ile residue was substituted for the Asp residue at P4. This result supports a role for charge in determining P4–S4 interactions.

Other Chymases

Chymases are a highly homologous family of proteases that may exhibit common features with respect to the hydrolysis of substrates like angiotensin I. Analysis of the peptide–CMK–rHC crystal structure identifies specific residues in the extended substrate-binding site of human chymase that may play a key role in angiotensin I hydrolysis. Residues 40 and 41 may function to stabilize the free carboxyl group at the P2′ position of angiotensin I, and to sterically prevent chymase from interacting with the degradative site at Tyr4–Ile5. Further inhibiting hydrolysis of the Try4–Ile5 bond are the charge preferences of the S3 and S4 subsites produced by Lys192 and Asp175. The crystal structure of rat chymase 2 (51) shows that residues 40 and 41 are oriented similarly to their counterpart residues in human chymase. Thus the backbone interactions between the 41 residue and the P2′ residue of a substrate likely are possible in all chymases (Plate I). Of the chymases identified to date (Table I), human chymase is the only member of the family with a Phe residue at position 41, and human and baboon chymase are the only members with a positively charged residue at the 40 position.

Information on angiotensin I hydrolysis has been obtained for rat chymase 1 (65, 68), monkey chymase (69), hamster chymase 1 (70) and a reconstructed ancestral chymase (68). Monkey chymase, which has a Lys at position 40 and a Ser at postion 41, selectively recognizes the angiotensin II forming site. The efficiency of conversion (k_{cat}/K_M value) is lower than that of human chymase. Ancestral chymase, which has an Ala at 40 and Arg at 41, produces angiotensin II with high efficiency. Although these results indicate that a positive charge at the 40 or 41 position is beneficial for angiotensin II production, hamster chymase 1, which has an Ala at position 40 and Ser at position 41, is also an efficient producer of angiotensin II. However, hamster chymase is not a pure converter, and recognizes the degradative site, albeit at a lower efficiency than the forming site. Rat chymase 1, which has an Ala at position 40 and a Thr at position 41, primarily recognizes the degradative site.

The four chymases discussed above have a Lys residue at the 192 position like human chymase, but only ancestral chymase has an Asp at the 175 position. The other chymases have a Tyr or His at the 175 position. Thus comparison by residue position shows no definitive trends among chymases with respect to angiotensin I conversion as developed in this review. Understanding species differences among chymases for angiotensin conversion will be a complex matter.

SUMMARY

Human chymase has the ability to interact with many important biological peptides and proteins in a selective manner. Although structurally similar to chymotrypsin and other chymotrypsin-like proteases, the active site of human chymase appears highly tailored to recognize substrates with a specific charge arrangement. The S2′, S3 and S4 subsites of human chymase all appear to exhibit a charge preference. The S3 and S4 subsites exhibit opposite charge preferences with the S3 preferring anionic and the S4 preferring cationic side-chains. The S2′ subsite appears not only capable of stabilizing a negatively charged side-chain via Arg143, but also a free carboxyl group via Lys40 and backbone hydrogen bonds made by Phe41. Another feature that may define chymase specificity is the constriction of the substrate binding cleft at the S3′ subsite. This constriction produced by Phe41 may limit P3′ residues to amino acids with relatively small side chains.

All these features indicate that human chymase is a highly focused protease. Perhaps this focus is the reason why human chymase expression is limited to a selected population of human mast cells. Although angiotensin I is hydrolysed efficiently by human chymase, a more optimal substrate seems possible. This review provides insight into the amino acid sequence that such substrates may possess.

ACKNOWLEDGEMENTS

We thank Michael Plotnick and Trevor Selwood for helpful discussions.

REFERENCES

1. Pereira, P. J. B., Wang, Z.-M., Rubin, H., Huber, R., Bode, W., Schechter, N. M. and Strobl, S. The 2.2 Å crystal structure of human chymase in complex with succinyl-Ala-Ala-Pro-Phe-chloromethylketone: structural explanation for its dipeptidyl carboxypeptidase specificity. *J. Mol. Biol.* **286**:163–173, 1999.
2. Costa, J. J., Weller, P. F. and Galli, S. J. The cells of the allergic response: mast cells, basophils, and eosinophils. *JAMA* **278**:1815–1822, 1997.
3. Metcalfe, D. D., Baram, D. and Mekori, Y. A. Mast cells. *Physiol. Rev.* **77**:1033–1079, 1997.
4. Irani, A.-M. A., Schechter, N. M., Craig, S. S., DeBlois, G. and Schwartz, L. B. Two types of human mast cells that have distinct neutral protease compositions. *Proc. Natl. Acad. Sci. USA* **83**:4464–4468, 1986.
5. Schwartz, L. B., Irani, A.-M. A., Roller, K., Castells, M. C. and Schechter, N. M. Quantitation of histamine, tryptase and chymase in dispersed human T and TC mast cells. *J. Immunol.* **138**:2611–2615, 1987.
6. Goldstein, S. M., Kaempfer, C. E., Kealey, J. T. and Wintroub, B. U. Mast cell carboxypeptidase: purification and characterization. *J. Clin. Invest.* **83**:1630–1636, 1989.
7. Schechter, N. M., Irani, A.-M. A., Sprows, J. L., Abernethy, J., Wintroub, B. and Schwartz, L. B. Identification of a cathepsin G-like proteinase in the MCTC type of human mast cell. *J. Immunol.* **145**:2652–2661, 1990.
8. Sayama, S., Iozzo, R. V., Lazarus, G. S. and Schechter, N. M. Human skin chymotrypsin-like proteinase chymase: subcellular localization to mast cell granules and interaction with heparin and other glycosaminoglycans. *J. Biol. Chem.* **262**:6808–6815, 1987.
9. Craig, S. S., Schechter, N. M. and Schwartz, L. B. Ultrastructural analysis of human T and TC mast cells identified by immunoelectron microscopy. *Lab. Invest.* **58**:682–691, 1988.
10. Whitaker-Menezes, D., Schechter, N. M. and Murphy, G. F. Serine proteinases are regionally segregated within mast cells granules. *Lab. Invest.* **72**:34–41, 1995.
11. Schechter, N. M., Choi, J. K., Slavin, D. A., Deresienski, D. T., Sayama, S., Dong, G., Lavker, R. M., Proud, D. and Lazarus, G. S. Identification of a chymotrypsin-like proteinase in human mast cells. *J. Biol. Chem.* **137**:962–970, 1986.
12. Urata, H., Kinoshita, A., Misono, K. S., Bumpus, M. and Husain, A. Identification of a highly specific chymase as the major angiotensin II-forming enzyme in the human heart. *J. Biol. Chem.* **263**:22348–22357, 1990.
13. Sukenaga, Y., Kido, H., Neki, A., Enomoto, M., Ishida, K., Takagi, K. and Katunuma, N. Purification and molecular cloning of chymase from human tonsils. *FEBS Lett.* **323**:119–122, 1993.
14. McEuen, A. R., Gaca, M. D. A., Buckley, M. G., He, S. H., Gore, M. G. and Walls, A. F. Two distinct forms of human mast cell chymase. Differences in affinity for heparin and in distribution in skin, heart, and other tissues. *Eur. J. Biochem.* **256**:461–470, 1998.
15. Caughey, G. H., Zerweck, E. H. and Vanderslice, P. Structure, chromosomal assignment, and deduced amino acid sequence of a human gene for mast cell chymase. *J. Biol. Chem.* **266**:12956–12963, 1991.
16. Urata, H., Kinoshita, A., Perez, D. M., Misono, K. S., Bumpus, F. M., Graham, R. M. and Husain, A. Cloning of the gene and cDNA for human heart chymase. *J. Biol. Chem.* **266**:17173–17179, 1991.
17. Schechter, N. M., Wang, Z.-M., Blatcher, R. W., Lessin, S. R., Lazarus, G. S. and Rubin, H. Determination of primary structures for human skin chymase and cathepsin G from cutaneous mast cells of urticaria pigmentosa lesions. *J. Immunol.* **152**:4062–4069, 1994.
18. Caughey, G. H., Schaumberg, T. H., Zerweck, E. H., Butterfield, J. H., Hanson, R. D., Silverman, G. A. and Ley, T. J. The human mast cell chymase gene (CMA1): mapping to the cathepsin G/granzyme gene cluster and lineage-restricted expression. *Genomics* **15**:614–620, 1993.
19. Wang, Z.-M., Rubin, H. and Schechter, N. M. Production of active recombinant human chymase from a construct containing the enterokinase cleavage site of trypsinogen in place of the native propeptide sequence. *Biol. Chem.* **376**:681–684, 1995.

20. Goldstein, S. M., Leong, J., Schwartz, L. B. and Cooke, D. Protease composition of exocytosed human skin mast cell protease-proteoglycan complexes. *J. Immunol.* **148**:2475–2482, 1992.
21. Trong, H. L., Parmelee, D. C., Walsh, K. A., Neurath, H. and Woodbury, R. G. Amino acid sequence of rat mast cell protease I (chymase). *Biochemistry* **26**:6988–6994, 1987.
22. Sali, A., Matsumoto, R., McNeil, H. P., Karplus, M. and Stevens, R. L. Three-dimensional models of four mouse mast cell chymases: identification of the proteoglycan regions and protease-specific antigenic epitopes. *J. Biol. Chem.* **268**:9023–9034, 1993.
23. Murakami, M., Karnik, S. S. and Hussain, A. Human prochymase activation: a novel role for heparin in zymogen processing. *J. Biol. Chem.* **270**:2218–2223, 1995.
24. McEuen, A. R., Ashworth, D. M. and Walls, A. F. The conversion of recombinant human mast cell prochymase to enzymatically active chymase by dipeptidyl peptidase I is inhibited by heparin and histamine. *Eur. J. Biochem.* **253**:300–308, 1998.
25. McGrath, M. E., Osawa, E., Barnes, M. G., Clark, J. M., Mortara, K. D. and Schmidt, B. F. Production of crystallizable human chymase from a *Bacillus subtilis* system. *FEBS Lett.* **413**:486–488, 1997.
26. Wang, Z. M., Walter, M., Selwood, T., Rubin, H. and Schechter, N. M. Recombinant expression of human mast cell proteases chymase and tryptase. *Biol. Chem.* **379**:167–174, 1998.
27. Anderson, L. E., Walsh, K. A. and Neurath, H. Bovine enterokinase. Purification, specificity, and some molecular properties. *Biochemistry* **16**:3354–3360, 1977.
28. Powers, J. C., Tanaka, T., Harper, J. W., Minematsu, Y., Barker, L., Lincoln, D., Crumley, K. V., Fraki, J. E., Schechter, N. M., Lazarus, G. L., Nakajima, K., Nakashino, K., Neurath, H. and Woodbury, R. G. Mammalian chymotrypsin-like enzymes comparative reactivities of rat mast cell proteases, human and dog chymases, and human cathepsin G with peptide 4-nitroanilide substrates and with peptide chloromethyl ketone and sulfonyl fluoride inhibitors. *Biochemistry* **24**:2048–2058, 1985.
29. Schechter, N. M., Sprows, J. L., Schoenberger, O. L., Lazarus, G. S., Cooperman, B. S. and Rubin, H. Reaction of human skin chymotrypsin-like proteinase chymase with plasma proteinase inhibitors. *J. Biol. Chem.* **264**:21308–21315, 1989.
30. Walter, M., Plotnick, M. and Schechter, N. M. Inhibition of human mast cell chymase by secretory leukocyte proteinase inhibitor (SLPI): enhancement of the interaction by heparin. *Arch Biochem. Biophys.* **327**:81–88, 1996.
31. Walter, M., Sutton, R. M. and Schechter, N. M. Highly efficient inhibition of human chymase by alpha(2)-macroglobuin. *Arch. Biochem. Biophys*, (in press).
32. Ware, J. H., Wan, X. S., Rubin, H., Schechter, N. M. and Kennedy, A. R. Soybean Bowman–Birk protease inhibitor is a highly effective inhibitor of human mast cell chymase. *Arch. Biochem. Biophys.* **344**:133–138, 1997.
33. Reilly, C. F., Tewksbury, D. A., Schechter, N. M. and Travis, J. Rapid conversion of angiotensin I to angiotensin II by neutrophil and mast cell proteinases. *J. Biol. Chem.* **257**:8619–8622, 1982.
34. Wintroub, B. U., Schechter, N. M., Lazarus, G. S., Kaempfer, C. E. and Schwartz, L. B. Angiotensin I conversion by human and rat chymotryptic proteinases. *J. Invest. Dermatol.* **83**:336–339, 1984.
35. Mizutani, H., Schechter, N. M., Lazarus, G. S., Black, R. A. and Kupper, T. S. Rapid and specific conversion of precursor interleukin 1β (IL-1β) to an active IL-1 species by human mast cell chymase. *J. Exp. Med.* **174**:821–825, 1991.
36. Kido, H., Nakano, A., Okishima, N., Wakabayashi, H., Kishi, F., Nakaya, Y., Yoshizumi, M. and Tamaki, T. Human chymase, and enzyme forming novel bioactive 31-amino acid length endothelins. *Biol. Chem.* **379**:885–891, 1998.
37. Takai, S., Shiota, N., Jin, D. and Miyazaki, M. Chymase processes big-endothelin-2 to endothelin-2-(1-31) that induces contractile responses in the isolated monkey trachea. *Eur. J. Pharmacol.* **358**:229–230, 1998.
38. Kofford, M. W., Schwartz, L. B., Schechter, N. M., Yager, D. R., Diegelmann, R. F. and Graham, M. F. Cleavage of type I procollagen by human mast cell chymase initiates collagen fibril formation and generates a unique carboxyl-terminal propeptide. *J. Biol. Chem.* **272**:7127–7131, 1997.
39. Saarinen, J., Kalkkinen, N., Welgus, H. G. and Kovanen, P. T. Activation of human interstitial procollagenase through direct cleavage of the Leu83-Thr bond by mast cell chymase. *J. Biol. Chem.* **269**:18134–18140, 1994.
40. Briggaman, R. A, Schechter, N. M., Fraki, J. and Lazarus, G. S. Degradation of the epidermal–dermal junction by proteolytic enzymes from human skin and human polymorphonuclear leukocytes. *J. Exp. Med.* **160**:1027–1042, 1984.
41. He, S. H. and Walls, A. F. The induction of a prolonged increase in microvascular permeability by human mast cell chymase. *Eur. J. Pharmacol.* **352**:91–98, 1998.

42. Schechter, N. M., Brass, L. F., Lavker, R. M. and Jensen, P. J. Reaction of mast cell proteases tryptase and chymase with protease activated receptors (PARs) on keratinocytes and fibroblasts. *J. Cell. Physiol.* **176**:365–373, 1998.
43. Longley, J. B., Tyrrell, L., Ma, Y., Williams, D. A., Halaban, R., Langley, K., Lu, H. S. and Schechter, N. M. Chymase cleavage of stem cell factor yields a bioactive, soluble product. *Proc. Natl. Acad. Sci. USA* **94**:9017–9021, 1997.
44. Sommerhoff, C. P., Caughey, G. H., Finkbeiner, W. E., Lazarus, S. C., Basbaum, C. B. and Nadel, J. A. Mast cell chymase: a potent secretagogue for airway gland serous cells. *J. Immunol.* **142**:2450–2456, 1989.
45. Fang, K. C., Raymond, W. W., Blount, J. L. and Caughey, G. H. Dog mast cell alpha-chymase activates progelatinase B by cleaving the Phe(88)–Gln(89) and Phe(91)–Glu(92) bonds of the catalytic domain. *J. Biol. Chem.* **272**:25628–25635, 1997.
46. Kokkonen, J. O., Lindstedt, K. A. and Kovanen, P. T. Role of mast cell proteases and proteoglycans in lipoprotein metabolism. *Clin. Allergy Immunol.* **6**:257–285, 1995.
47. Faller, B., Yves, M., Gerard, D. and Bieth, J. G. Heparin-induced conformational change and activation of mucus proteinase inhibitor. *Biochemistry* **31**:8285–8290, 1992.
48. Kokkonen, J. O., Saarinen, J. and Kovanen, P. T. Angiotensin II formation in the human heart: an ACE or non-ACE-mediated pathway? *Ann. Med.* **30 (Suppl 1)**:9–13, 1998.
49. Takai, S., Shiota, N., Jin, D. and Miyazaki, M. Functional role of chymase in angiotensin II formation in human vascular tissue. *J. Cardiovasc. Pharmacol.* **32**:826–833, 1998.
50. McGrath, M. E., Mirzadegan, T. and Schmidt, B. F. Crystal structure of phenylmethanesulfonyl fluoride-treated human chymase at 1.9 Å. *Biochemistry* **36**:14318–14324, 1997.
51. Remington, S. J., Woodbury, R. G., Reynolds, R. A., Matthews, B. W. and Neurath, H. The structure of rat mast cell proteinase II at 1.9 Å resolution. *Biochemistry* **27**:8097–8105, 1988.
52. Bode, W., Meyer, E. J. and Powers, J. C. Human leukocyte and porcine pancreatic elastase: X-ray crystal structures, mechanism, substrate specificity, and mechanism-based inhibitors. *Biochemistry* **28**:1951–1963, 1989.
53. Perona, J. J. and Craik, C. S. Structural basis of substrate specificity in the serine proteases. *Protein Sci.* **4**:337–359, 1995.
54. Lesk, A. M. and Fordham, W. D. Conservation and variability in the structures of serine proteinases of the chymotrypsin family. *J. Mol. Biol.* **258**:501–537, 1996.
55. Schechter, I. and Berger, A. C. On the size of the active site in proteases I. Papain. *Biochem. Biophys. Res. Commun.* **27**:157–162, 1967.
56. Bastos, M., Maeji, N. J. and Abeles, R. H. Inhibitors of human heart chymase based on a peptide library. *Proc. Natl. Acad. Sci. USA* **92**:6738–6742, 1995.
57. Eda, M., Ashimori, A., Akahoshi, F., Yoshimura, T., Inoue, Y., Fukaya, C., Nakajima, M., Fukuyama, H., Imada, T. and Nakamura, N. Peptidyl human heart chymase inhibitors. 2. Discovery of highly selective difluoromethylene ketone derivatives with Glu at P3 site. *Bioorg. Med. Chem. Lett.* **8**:919–924, 1998.
58. Schechter, N. M., Plotnick, M., Selwood, T., Walter, M. and Rubin, H. Diverse effects of pH on the inhibition of human chymase by serpins. *J. Biol. Chem.* **272**:24499–24507, 1997.
59. Shotton, D. The molecular architecture of the serine proteinases. In *Proceedings of the International Research Conference* (Fritz, H. and Tscheche, H., eds), pp. 47–55. Walter de Gruyter, Berlin, 1971.
60. Polgar, L. *Mechanism of Protease Action*. CRC Press, Boca Raton, FL, 1989.
61. Lin, G., Bode, W., Huber, R., Chi, C. and Engh, R. A. The 0.25-nm structure of the Bowman–Birk-type inhibitor from mung bean in ternary complex with porcine trypsin. *Eur. J. Biochem.* **212**:549–555, 1993.
62. Prakash, B., Selvaraj, S., Murthy, M. R. N., Sreerama, Y. N., Rao, D. R. and Gowda, L. R. Analysis of the amino acid sequences of plant Bowman-Birk inhibitors. *J. Mol. Evol.* **42**:560–569, 1996.
63. Eda, M., Ashimori, A., Akahoshi, F., Yoshimura, T., Inoue, Y., Fukaya, C., Nakajima, M., Fukuyama, H., Imada, T. and Nakamura, N. Peptidyl human heart chymase inhibitors. 1. Synthesis and inhibitory activity of difluoromethylene ketone derivatives bearing P′ binding subsites. *Bioorg. Med. Chem. Lett.* **8**:913–918, 1998.
64. Barrett, A. J. The many forms and functions of cellular proteinases. *Fed. Proc.* **39**:9–14, 1980.
65. Trong, H. L., Neurath, H. and Woodbury, R. G. Substrate specificity of the chymotrypsin-like protease in secretory granules isolated from rat mast cells. *Proc. Natl. Acad. Sci. USA* **84**:364–367, 1987.
66. Sanker, S., Chandrasekharan, U. M., Wilk, D., Glynias, M. J., Karnik, S. S. and Husain, A. Distinct multisite synergistic interactions determine substrate specificities of human chymase and rat chymase-1 for angiotensin. *J. Biol. Chem.* **272**:2963–2968, 1997.
67. Kinoshita, A., Urata, H., Bumpus, F. M. and Husain, A. Multiple determinants for the high substrate

specificity of an angiotensin II-forming chymase from human heart. *J. Biol. Chem.* **266**:19192–19197, 1991.

68. Chandrasekharan, U. M., Sanker, S., Glynias, M. J., Karnik, S. S. and Husain, A. Angiotensin II-forming activity in a reconstructed ancestral chymase. *Science* **271**:502–505, 1996.
69. Takai, S., Shiota, N., Kobayashi, S., Matsumura, E. and Miyazaki, M. Induction of chymase that forms angiotensin II in monkey atherosclerotic aorta. *FEBS Lett.* **412**:86–90, 1997.
70. Takai, S., Shiota, N., Yamamoto, D., Okunishi, H. and Miyazaki, M. Purification and characterization of angiotensin II-generating chymase from hamster cheek pouch. *Life Sci.* **58**:591–597, 1996.
71. Huang, R. and Hellman, L. Genes for mast cell serine protease and their molecular evolution. *Immunogenetics* **40**:397–414, 1994.
72. Ide, H., Itoh, H., Tomita, M., Murakumo, Y., Kobayashi, T., Maruyama, H., Osada, Y. and Nawa, Y. Cloning of the cDNA encoding a novel rat mast-cell proteinase, rMCP-3, and its expression in comparison with other rat mast-cell proteinases. *Biochem. J.* **311**:675–680, 1995.
73. Itoh, H., Murakumo, Y., Tomita, M., Ide, H., Kobayashi, T., Maruyawa, H., Horii, Y. and Nawa, Y. Cloning of the cDNAs for mast-cell chymases from the jejunum of Mongolian gerbils, *Meriones unguiculatus*, and their sequence similarities with chymases expressed in the connective-tissue mast cells of mice and rats. *Biochem. J.* **314**:923–929, 1996.
74. Hunt, J. E., Friend, D. S., Gurish, M. F., Feyfant, E., Sali, A., Huang, C. F., Ghildyal, N., Stechschulte, S., Austen, K. F. and Stevens, R. L. Mouse mast cell protease 9, a novel member of the chromosome 14 family of serine proteases that is selectively expressed in uterine mast cells. *J. Biol. Chem.* **272**:29158–29166, 1997.
75. Shiota, N., Fukamizu, A., Takai, S., Okunishi, H., Murakami, K. and Mizuo, M. Activation of angiotensin II-forming chymase in the cardiomyopathic hamster heart. *J. Hypertens.* **15**:431–440, 1997.
76. McAleese, S. M., Pemberton, A. D., McGrath, M. E., Huntley, J. F. and Miller, H. R. P. Sheep mast-cell proteinases-1 and -3: cDNA cloning, primary structure and molecular modelling of the enzymes and further studies on substrate specificity. *Biochem. J.* **333**:801–809, 1998.
77. Shiota, N., Fukamizu, A., Okunishi, H., Takai, S., Murakami, K. and Miyazaki, M. Cloning of the gene and cDNA for hamster chymase 2, and expression of chymase 1, chymase 2 and angiotensin-converting enzyme in the terminal stage of cardiomyopathic hearts. *Biochem. J.* **333**:417–424, 1998.

CHAPTER 19

Structure and Function of Human Mast Cell Tryptase

ANDREW F. WALLS

Immunopharmacology Group, Southampton General Hospital, U.K.

INTRODUCTION

The serine protease tryptase (EC 3.4.21.59) was first identified in mast cells some 40 years ago using histochemical substrates for tryptic substrates (1). For much of the time that has elapsed since, the study of this unique mast cell enzyme has been largely neglected. This is paradoxical, as tryptase must represent the most abundant constituent of human mast cells. Some 10 pg per cell has been detected in mast cells in the lung and up to 35 pg in skin mast cells (2). Understanding of mast cell function has been dominated by considerations of this cell as a major source of histamine (approximately 1 or 2 pg stored per cell), or its ability to generate prostaglandin D_2 or leukotriene C_4, and, more recently, for the potential to secrete various inflammatory cytokines (3, 4). Over the past decade, however, tryptase has become established as an important marker for human mast cells, and there has been an accumulation of evidence that this major mast cell product represents a key mediator of inflammation and a promising target for therapeutic intervention in inflammatory disease.

DISTRIBUTION AND SECRETION IN DISEASE

The advent of monoclonal antibodies specific for mast cell tryptase allowed the cellular distribution of this enzyme to be determined more accurately than was possible using histochemical substrates. Studies with a diverse range of human tissues have indicated that tryptase is present almost exclusively in mast cells (5, 6). Tryptase has not been demonstrated convincingly in any other cell type except for the basophil. The quantities of tryptase present in basophils, however, are very much less than those in mast cells (7),

MAST CELLS AND BASOPHILS
ISBN 0-12-473335-2

and immunocytochemical staining procedures with tryptase-specific antibodies frequently fail to detect basophils (5).

Tryptase has come to be considered a discriminating marker for mast cells. Other proteases detected in this cell type tend to be present in smaller amounts and appear to be restricted to distinct subpopulations of cells. Tryptase-containing cells in which chymase is also present (designated MC_{TC}) predominate in connective tissue sites, whereas those without chymase (MC_T) are particularly plentiful in mucosal tissues (8). The categorization of mast cell heterogeneity in this way has proved useful in establishing the concept that mast cell populations and hence their functions are altered in disease. It has been suggested that MC_T cells may depend on intact T lymphocyte function for their growth and survival, as reduced numbers of this subset have been reported in the gastrointestinal mucosa of AIDS patients (9). Increased numbers of the MC_T subpopulation do appear to be associated with the inflammatory changes in seasonal allergic rhinitis (10), atopic dermatitis (11), vernal conjunctivitis (12), scleroderma (13) and rheumatoid arthritis (14), although this selective expansion may also be a feature in osteoarthritis, a condition in which signs of inflammation are frequently absent (15).

Carboxypeptidase (16) and cathepsin G (17) have both been localized to MC_{TC} cells. Moreover, in biopsy tissue from the upper and lower airways of allergic subjects, it has been found that interleukin 4 (IL-4) was present in a greater proportion of MC_{TC} than MC_T cells; whereas IL-5 and IL-6 were present almost exclusively in the MC_T subset (18). This would suggest that the two subpopulations defined could fulfil a range of different functions. The possibility that there may be some mast cells that contain chymase but not tryptase has been raised by certain studies (19, 20). However, it remains to be determined to what extent mast cells wholly deficient in tryptase may exist. The ability to demonstrate the so-called MC_C phenotype by immunocytochemistry will depend on the sensitivity of the detection system, and it has been found that the proportion of such cells can be affected to a large extent by the length of the period for which tissue sections are incubated with the tryptase-specific antibody (21).

As tryptase is an abundant secretory product of the mast cell and is restricted almost entirely to this cell type, its measurement in biological fluids can provide compelling evidence for mast cell activation. Moreover, unlike other mast cell products, such as histamine and prostaglandin D_2 which are rapidly metabolized *in vivo*, tryptase is relatively stable, at least in an immunoreactive form, and it is cleared quite slowly from the circulation. Concentrations of tryptase reported in various body fluids are shown in Table I.

The measurement of tryptase in serum or plasma has proved particularly helpful as a diagnostic aid in cases of systemic anaphylaxis and in mastocytosis, as an elevation in blood tryptase levels is rare except in these conditions (22, 23). Following experimentally induced anaphylactic shock with a bee sting, tryptase levels in the circulation peak after 1 or 2 h, and then decline under apparent first-order kinetics with a half-life of about 2 h (24). Tryptase assays have proved useful as forensic tools in cases of fatal anaphylaxis, although, for reasons that remain unclear, tryptase levels are much less likely to be elevated in cases of allergic food reaction than in reactions provoked by the parenteral injection of the allergen (e.g. with venoms or drugs) (25). The recent demonstration that tryptase levels may be increased in the saliva of subjects undergoing food-induced anaphylaxis has recently suggested an alternative approach in the diagnosis of anaphylaxis provoked by ingested allergens (26).

Measurements of tryptase in bronchoalveolar lavage (BAL) fluid indicate that quite

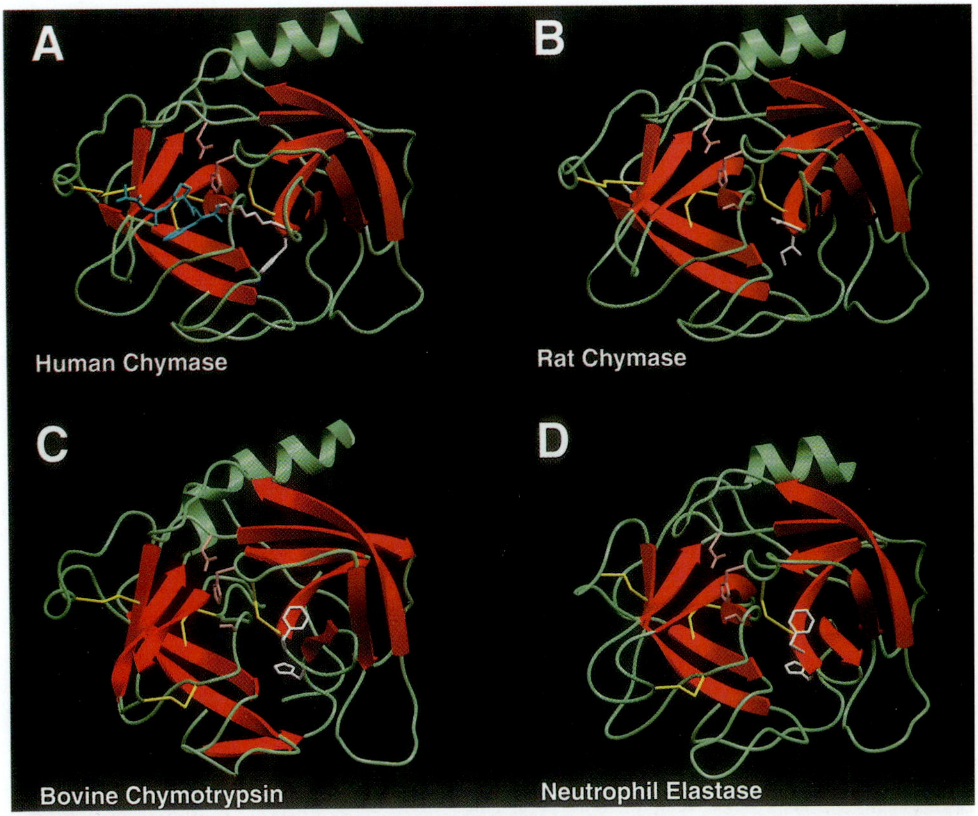

A

Human Chymase

B

Rat Chymase

C

Bovine Chymotrypsin

D

Neutrophil Elastase

Plate I Structure of human chymase (A) compared to other serine proteases (B–D). β sheet structures are shown in red and α helices and loops are shown in green. Residues presented as pink wireframe structures are Ser195, His57 and Asp102. Side-chains of the residues at the 40 and 41 positions of each protease are shown as white wireframe structures (Arg and Phe in human chymase, Val and Ile in rat chymase, and His and Phe in chymotrypsin and elastase). Blue wireframe structure in chymase is the bound inhibitor, succinyl–Ala–Ala–Pro–Phe–CMK. Disulphide bonds are depicted as yellow wireframe structures. All structures are available at the Protein Data Bank of Brookhaven National Laboratory. All numbering of proteases is according to homology with bovine chymotrypsin (1).

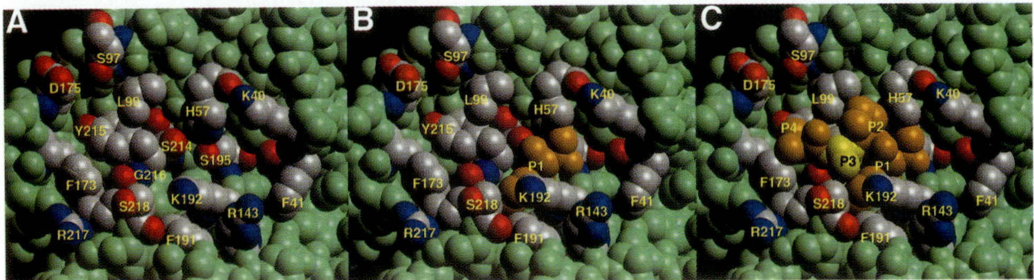

Plate II Space-filling model of the peptide–CMK–rHC structure showing residues forming the extended substrate binding site of human chymase as defined by the bound peptide–CMK inhibitor. Important residues of the extended substrate binding site are presented in CPK colours (carbons gray, oxygens red, nitrogens light blue) and are labelled according to chymotrypsin numbering. The peptide inhibitor is shown in orange, except for the P3 residue which is shown in yellow. P1–P4 labels were placed on spheres representing the side-chain of each residue. (A) Chymase structure with inhibitor made transparent by the computer. (B) Structure showing the binding of the P1 Phe residue. (C) Structure with bound inhibitor completely visible.

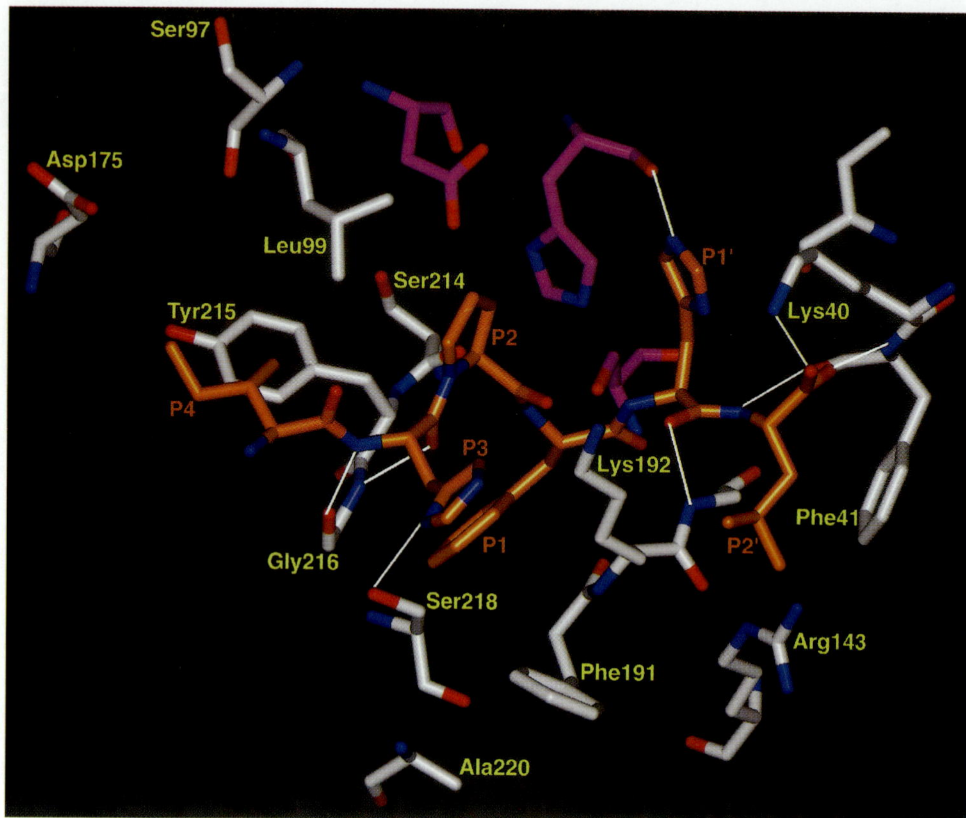

Plate III Computer docking of angiotensin I to peptide–CMK–rHC structure. The angiotensin I sequence around the Phe8–His9 peptide bond (P4–P2′), as shown in Fig. 3, was docked to human chymase. Angiotensin I is the wireframe structure shown in orange except for nitrogens and oxygens coloured in dark blue and red, respectively. Residues of the protease are presented as wireframe structures with CPK colouring (see Plate II). Predicted hydrogen and electrostatic bonds are indicated by white lines.

TABLE I
Tryptase Levels Reported in Biological Fluids

Condition	Fluid	Tryptase concentration (ng ml^{-1})
Anaphylaxis	Serum/plasma	2–500
Mastocytosis	Serum/plasma	2–500
Normal	Serum/plasma	2–20
Atopic asthma		
Baseline	Bronchoalveolar lavage	<2.0
Following allergen	Bronchoalveolar lavage	Up to 30
Atopic rhinitis		
Baseline	Nasal lavage	<2.0
Following allergen	Nasal lavage	Up to 50*
Atopic conjunctivitis		
Baseline	Tears	<40–150*
Following allergen	Tears	Up to 1200*
Normal	Tears	<40*
Urticaria		
Baseline	Skin blister fluid	<5.0*
Following allergen	Skin blister fluid	Up to 700*
Rheumatoid arthritis	Synovial	0.5–17
Spondyloarthritis	Synovial	0.5–32
Osteoarthritis	Synovial	0.5–12
Normal	Synovial	?
Cystitis	Dialysed urine	0.7–2.8
Normal	Dialysed urine	3.0–4.5*

Data relate to studies employing immunoassays with antibodies that detect both α and β forms of tryptase, or where indicated (*) with an immunoassay that detects predominantly β-tryptase.

high concentrations of tryptase may be reached in the lining fluid of the airways in atopic asthma. Even during asymptomatic periods, mild asthmatic subjects can have appreciably higher levels of tryptase in BAL fluid than that in non-asthmatic controls (27–29). It would thus appear that mast cell degranulation is a continuous process in asthmatic airways. BAL fluid tryptase levels may be decreased following a course of inhaled steroids, which is doubtless associated with the concomitant decline in mast cell numbers in the airways (30). Treatment with the β_2-adrenoceptor agonist salmeterol, however, leaves both tryptase levels and the mast cell numbers unaltered (31).

Dramatic elevations in BAL fluid tryptase concentration may be provoked within minutes of introducing allergen into the airways (32, 33). Twenty-four h following allergen challenge, an increase in tryptase levels has not been observed (34). Bronchoconstriction induced by local challenge with hypertonic saline (35) or by exercise (36, 37) is not associated with increases in BAL fluid tryptase concentrations.

The measurement of tryptase in BAL fluid has provided useful information on the potential roles of mast cells in bronchial asthma, but elevations in BAL fluid are by no means restricted to allergic airways disease. Increased BAL fluid tryptase concentrations have been reported in patients with interstitial lung disease or with bronchial carcinoma (38), and even in subjects who smoke (39).

The secretion of tryptase into various body fluids has been demonstrated following local allergen challenge in sensitized subjects. These include nasal lavage fluid from rhinitic subjects who have had allergen introduced into the nose (40–42), skin blister fluid from subjects in whom experimentally induced blisters are injected with allergen (43) and tears from sufferers of allergic conjunctivitis who have had allergen instilled into the eye (44). A role for mast cells in rheumatoid and seronegative spondyloarthritis, and even in osteoarthritis, has been suggested by the detection of tryptase in synovial fluid (45). Low levels of tryptase have been detected also in the urine of patients with cystitis (46).

STRUCTURE AND PHYSICOCHEMICAL PROPERTIES

The structure of tryptase is highly unusual. Alone among the serine proteases, this enzyme is a tetramer with a molecular mass of about 130 kDa, and with variably glycosylated subunits of some 28–38 kDa (5, 47–49). Two cDNA molecules for tryptase have been derived from a genomic library prepared from lung mast cells (50, 51) and independently at the same time three, designated I, II and III, were derived from a skin library (52). Molecular masses for the catalytic portions of each of these tryptase forms have been estimated to be approximately 27.5 kDa, and there are pre-pro regions of 3.0 kDa. β-Tryptase is identical to tryptase II, and exhibits some 98% identity with the catalytic portions of tryptases I and III and 100% identity with the pre-pro regions. For this reason the nomenclature may be unified and tryptases I, II and III have been renamed βI, βII and βIII. The α-tryptase gene exhibits some 90% identity with the catalytic portions of the β-tryptases, and 87% with the pre-pro regions.

Functionally active forms of both α- and β-tryptases have been obtained from insect and COS cell expression systems (53–58). An early suggestion that α-tryptase may be secreted from mast cells in the inactive pre-enzyme form (53) now seems unlikely, although α-tryptase has a more restricted substrate specificity than β-tryptase (54). Monoclonal antibodies prepared against tryptase generally appear to react equally well with recombinant α- and β-tryptases in Western blots (22, 45), although one antibody has been found that has a greater affinity for β-tryptase than for α-tryptase (22). Differences in cell staining patterns have not been observed when these different antibodies have been employed in immunocytochemistry with human tissues, strongly suggesting that β-tryptase forms are present in the secretory granules of all mast cells. It has been proposed that α-tryptase is not stored in the granules at all but is released constitutively from cells. Comparisons of findings with immunoassays with the different monoclonal antibodies have indicated it is the β form that predominates in serum from patients with anaphylaxis (22). However, in normal subjects and in patients with mastocytosis, α-tryptase appears to be the major form (22). In synovial fluid from patients with rheumatoid or osteoarthritis, it has been deduced that most tryptase detected is the α form, though relative proportions can differ between subjects (45). There is a need for further information on the biological significance, but α-tryptase could be considered a clinical marker of mast cell hyperplasia and β-tryptase a marker for mast cell activation.

The crystal structure of a preparation of tryptase purified from human lung tissue has now been reported (59). This was found to have the sequence of βII-tryptase. The four subunits were arranged in a flat ring, with the catalytic sites within a central pore. Heparin, with which tryptase is co-stored in the granules and co-released on activation, is likely to bind to elongated positively charged patches noted to span adjacent pairs of monomers.

NON-HUMAN TRYPTASES

When tryptase was first identified, it was reported that this unique tryptic enzyme could be detected in the mast cells of man and the dog, but not in those of the rat, rabbit, guinea pig or mouse (60). Subsequent investigations employing more sensitive histochemical substrates now suggest that tryptase is present in the mast cells of all mammalian species. Nonetheless, tryptases from different species exhibit major differences in their physicochemical and enzymatic properties, their distribution and almost certainly in the functions they fulfil.

Tryptase has been purified from tissues of the rat (65, 66), guinea pig (64), cow (141), sheep (142) and monkey (67). In the rat (68) and the mouse (69, 70), cDNA sequences have been derived for two quite distinct tryptases (frequently termed rat or mouse mast cell proteases 6 and 7). A single sequence has been derived also for dog tryptase (71) and for sheep tryptase (142). With the exception of rat tryptase, all of the tryptases so far characterized bind to heparin, and their enzymatic activity is not altered by endogenous protease inhibitors, including α_1-antitrypsin and α_2-macroglobulin. Purified rat tryptase is unusual in that it fails to bind heparin under physiological conditions, and it can be inhibited by α_1-antitrypsin and several other protease inhibitors which are without effect on human tryptase, including soybean trypsin inhibitor, antipain and aprotinin (65, 66).

As is the case with their human counterpart, tryptases of the rat, mouse, sheep and monkey are tetrameric, and have a molecular mass of 120–140 kDa. In contrast, bovine tryptase has a molecular mass of about 360 kDa (141) and guinea pig tryptase is a massive structure of some 860 kDa (64). All tryptases do appear to be composed of equally sized glycosylated subunits of 30–40 kDa and it will be interesting to know the nature of their association in cow and guinea pig tryptases.

Tryptase activity has been detected by enzyme histochemistry in the mast cells of the rat, mouse, guinea pig, gerbil, dog, cow and monkey (1, 60, 61, 64, 72–74). Such studies have tended to indicate that tryptase is present exclusively in mast cells. There have been few attempts to determine the proportions of mast cells expressing tryptase in non-human species. However, the application of a rat tryptase-specific antibody in immunohistochemistry has indicated that in this species many mast cells may be deficient in this protease (62). It was reported that tryptase is restricted to a subpopulation of connective tissue mast cells with marked differences in the proportions in different tissue compartments. In mice it has been found that expression of tryptases in mast cells may be altered in disease (63). *Trichinella* infection can lead to a reversible loss in tryptase from gut mast cells, and a redistribution of the tryptase-containing subset.

On account of the major differences in tryptases from different species, in the present review the main focus will be on tryptase from human mast cells. Experimental models with non-human species can yield useful insights into the pathobiological roles of mast cell proteases, but the extent of the variation that is emerging does call for caution when considering the relevance of findings to human disease.

REGULATION OF TRYPTASE ACTIVITY

Tryptase is stored within mast cell granules in a fully active state (60), although it is unlikely that this represents the primary site of action of this enzyme. With various chromogenic substrates, tryptase has an optimal pH of 7–8 (75) and has relatively little activity in the acidic conditions of the mast cell granule (76). Moreover, tryptase activity can be inhibited by histamine at the concentrations that are likely to be found in the secretory granule (77). Histamine can act as reversible inhibitor of tryptase, and induce allosteric behaviour.

Tryptase has been reported to cleave fibrinogen most efficiently at a pH of 6.5, indicating possible roles for this enzyme in acidic conditions (78). However, the main actions of tryptase are likely to be in the neutral pH conditions of the extracellular fluid following mast cell degranulation. Unusually for a serine protease, no endogenous inhibitors of tryptase have been identified in man. Perhaps on account of the inaccessibility of the catalytic sites of tryptase, large molecular weight circulating inhibitors such as α_1-antitrypsin, α_2-macroglobulin and C1 esterase inhibitor are without inhibitory effect. Similarly, various inhibitors isolated from tissues such as low molecular weight elastase inhibitor and bikunin do not inhibit tryptase (47, 48, 79–82), although some studies have suggested that there may be conditions under which secretory leukoprotease inhibitor (SLPI) may inhibit tryptase (83). Intriguingly, an inhibitor of the medicinal leech can act as an inhibitor of tryptase (84).

In the absence of an effective inhibitor of tryptase, factors affecting enzymatic stability may provide a more important means of regulating activity. Tryptase is secreted in a complex with proteoglycans of about 200 kDa (85), and this interaction appears to be important for the maintenance of activity. Purified tryptase is highly unstable, appearing to convert spontaneously and irreversibly from the active tetramer to inactive monomers (76, 86, 87), but the speed of this process is greatly reduced in the presence of heparin or other proteoglycans (88). Proteins that can compete with tryptase for heparin, such as antithrombin III (79), neutrophil lactoferrin (89) and myeloperoxidase (90), can effectively diminish the activity of tryptase, presumably by destabilizing tryptase–proteoglycan complexes.

PEPTIDE SUBSTRATES

Tryptase, like trypsin, will preferentially cleave peptide or ester bonds on the carboxyl side of basic amino acids (91). Unlike trypsin, tryptase is highly selective in the substrates it will cleave, and the list of defined substrates for this enzyme is quite limited (Table II). The location of the catalytic sites within the central pore of tetrameric tryptase is likely to restrict access to large molecular weight substrates.

The neuropeptides vasoactive intestinal peptide (VIP), peptide histidine methionine (PHM) and calcitonin gene-related peptide (CGRP) possess appropriate tryptic cleavage sites, are relatively small and are very efficiently cleaved by tryptase (92, 143). They also fulfil the necessary criterion for a potential natural substrate for tryptase of being present in the immediate vicinity of degranulating mast cells. There has long been evidence that mast cells are sited in close apposition to peptidergic nerves and that there is a close functional association (94, 95).

Table II
Actions of Human Tryptase on Peptide and Protein Substrates

Substrate	Action of tryptase
VIP	Degradation
PHM	Degradation
CGRP	Degradation
Fibrinogen	Inactivation
Kininogens	Generation of kinins
Fibronectin	Cleavage
Type VI collagen	Cleavage
Pro-stromelysin	Activation
Pro-urokinase	Activation
Pre-kallikrein	Activation
PAR-2	Activation

VIP, a peptide of 28 amino acids, is a potent bronchodilator, a powerful vasodilator and can inhibit glandular secretion. PHM, which is encoded by the same gene as VIP, has similarities in primary structure and biological actions. Tryptase may well play a role in the normal physiological control of these neuropeptides, but increased tryptase secretion in disease could provoke neurogenic inflammation. Purified dog tryptase, like its human counterpart, can degrade VIP (96) and can reverse smooth muscle relaxation induced by VIP in isolated tracheal rings from the ferret (97). In keeping with this observation, inhibitors of tryptase and other proteases have been found to potentiate the actions of VIP in tissues of the guinea pig and man (98–100). The cleavage of CGRP and its inactivation by tryptase may provide a rare example of this protease having an anti-inflammatory action. It has been demonstrated that the potent and long-acting vasodilator actions of this 37 amino acid neuropeptide may be abolished by experimental activation of mast cells *in vivo* (101). Similarities in inhibition profiles found for CGRP degradation by tryptase and lysates of purified mast cells suggest that tryptase may be the major enzyme of human mast cells responsible (93).

PROTEIN SUBSTRATES

Although relatively few protein substrates have been identified for tryptase, their cleavage *in vivo* could be important in processes of inflammation and remodelling. Tryptase purified from human tissues has been demonstrated in several studies to be effective at inactivating fibrinogen as a clottable substrate for thrombin (78, 102, 104). Following mast cell degranulation, tryptase could thus serve to limit the clotting reacting and facilitate the free movement of soluble mediators in the immediate vicinity. The finding that tryptase-induced fibrinogen cleavage occurs optimally at a pH which is slightly acidic (78) suggests that the actions of tryptase as an anticoagulant would be most marked in conditions of tissue acidity. Interestingly, a recent study with recombinant tryptases has suggested that it is the β form of tryptase alone that is able to cleave fibrinogen and that α-tryptase lacks this property (55).

Despite some initial reports to the contrary, tryptase has now become recognized as a kininogenase. The ability to generate kinins from both low (93, 105) and high molecular weight kininogen (105, 106) could contribute in an important way to the increased kinin production described following mast cell degranulation *in vivo*. Moreover, tryptase can activate prekallikrein (106) and, in cooperation with neutrophil elastase, can generate bradykinin with an efficiency similar to that of plasma kallikrein (107).

A role for tryptase in tissue remodelling is strongly suggested by the ability to activate the pre-enzyme forms of stromelysin (matrix metalloprotease 3) (108) and urinary plasminogen activator (109). Both of these enzymes have been implicated as having a major contribution in tissue degradation. In addition, tryptase can cleave 72-kDa gelatinase (matrix metalloprotease 2) and fibronectin (110), and type VI collagen (111).

ACTIONS ON CELLS AND TISSUES

Measurements of tryptase concentrations in various biological fluids indicate that cells in the vicinity of degranulating mast cells will be exposed to some very high concentrations of tryptase. The potential consequences for understanding the role of tryptase in inflammation (Fig. 1) and tissue remodelling (Fig. 2) are only just beginning to be appreciated. The behaviour of many cell types may be modulated in the presence of tryptase.

Glenner and Cohen's early prediction that tryptase could have a role in mast cell activation (112) has received support in recent years with the demonstration that purified human tryptase can provoke histamine release from mast cells dispersed from human tonsils (113) and from guinea pig lung and skin (114). Tryptase could thus provide an amplification signal in allergic disease, with tryptase from activated mast cells

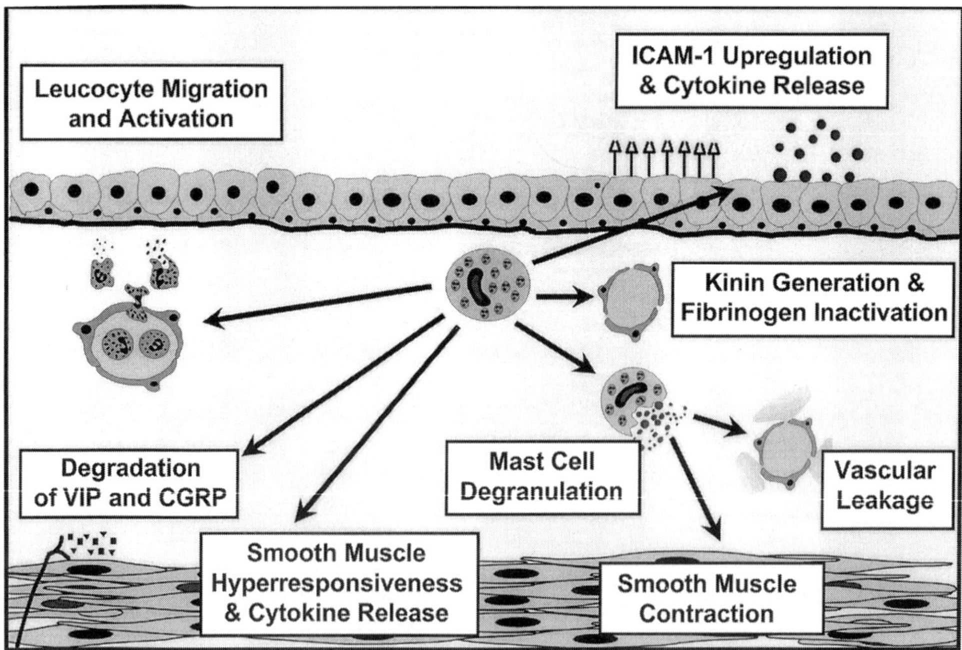

Fig. 1 Potential roles of human mast cell tryptase in acute inflammation.

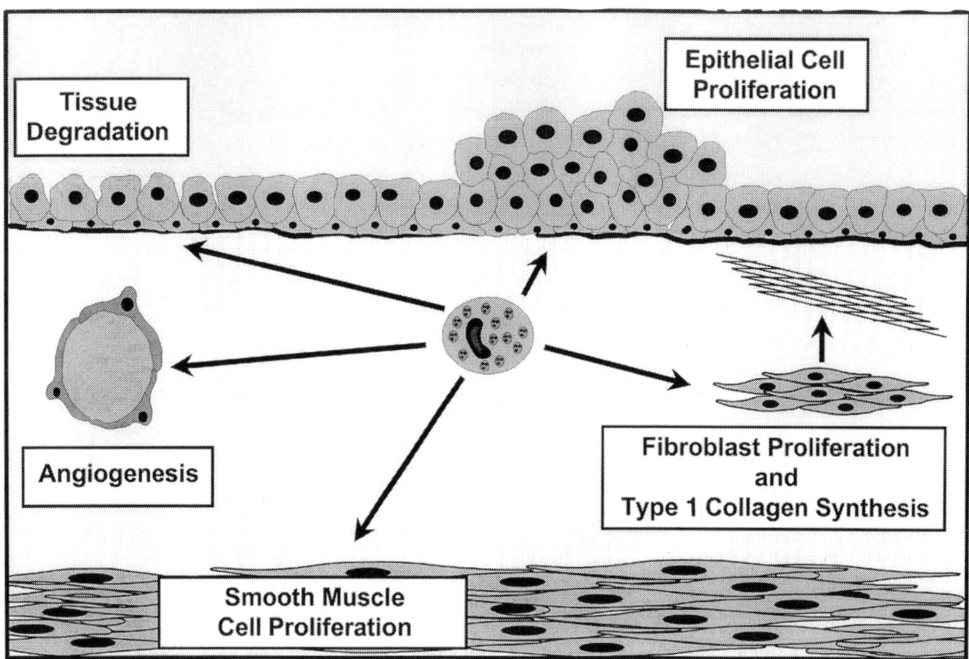

Fig. 2 Potential roles of human mast cell tryptase in tissue remodelling.

stimulating further mast cell degranulation. The ability of human tryptase to induce microvascular leakage in guinea pig and sheep skin (114, 115) and bronchoconstriction in sheep airways (116) appears to be dependent on the stimulation of mast cell activation in these animal models, as these responses may be blocked by antihistamine pre-treatment. Tryptase by itself fails to induce the contraction of human airways tissues *in vitro*, although it can induce hyper-responsiveness to histamine, an alteration that has been linked with a changed distribution of mast cells in the tissues (117).

Injection of human tryptase into the skin of guinea pigs or the peritoneum of mice has been found to provoke within 6 h a massive neutrophilia and eosinophilia (118). Selective increases in eosinophil numbers were observed in the mouse model when tryptase was co-injected with histamine, suggesting that *in vivo* the effects of tryptase will be modified by the actions of other mast cell mediators. Co-injection of heparin at certain concentrations was also able to alter the composition of the cellular infiltrate (118), although co-injection of chymase, which is itself a potent stimulus for cell accumulation in this system, has not been found to affect the response to tryptase (119).

Tryptase can act directly on human peripheral blood eosinophils and neutrophils *in vitro*, provoking chemotaxis and shape change responses and stimulating secretion of eosinophil cationic protein from eosinophils (120). A stimulus for inflammatory cell infiltration may be provided also by the release of chemoattractants from various other cells on which tryptase may act. Thus tryptase has been found to provoke the release of IL-8, IL-6 and granulocyte–macrophage colony-stimulating factor (GM-CSF) from epithelial cell lines and to upregulate expression of mRNA for these cytokines (121, 122). Incubation of tryptase with primary cultures of endothelial cells has also been found to stimulate IL-8 secretion, and mRNA for IL-8 and for IL-1β can be upregulated (123). Some evidence that such responses may be of biological relevance has been provided by the demonstration that supernatants from tryptase-treated endothelial cells can promote neutrophil migration *in vitro* (124).

Tryptase is emerging as a potent growth factor for a number of cell types. A potential role for tryptase in angiogenesis has been indicated by a report that tryptase can stimulate the proliferation of human dermal microvascular endothelial cells and promote vascular tube formation in culture (125). Tryptase is a mitogen also for human umbilical vein endothelial cells but fails to induce cell proliferation responses (123), possibly reflecting heterogeneity in responsiveness to tryptase in endothelial cell populations.

The potent actions of tryptase as a growth factor for fibroblasts (126–128) has focused attention on the potential contribution of this protease in fibrosis. Tryptase can act also as a chemoattractant for fibroblasts *in vitro* (128). Of particular interest is the observation that tryptase can stimulate the synthesis and release of collagen from fibroblasts in culture, as well as provoking secretion of collagenase (127, 128). The ability of tryptase to induce the proliferation of airway smooth muscle cells (129) could be of relevance in conditions such as bronchial asthma, in which smooth muscle cell hyperplasia is a feature. As a growth factor for epithelial cells (121), tryptase could have a role in tissue repair processes.

CELLULAR RECEPTORS FOR TRYPTASE

Precise mechanisms whereby tryptase can modulate cell function remain to be determined. In every case reported so far, it has proved possible to inhibit the actions of tryptase on cells using protease inhibitors. Nevertheless, it seems likely that tryptase is able to interact with cells in different ways. Some of the actions of tryptase on cells may be indirect and mediated by the generation or activation of extracellular or pericellular mediators. Alternatively, tryptase could engage specific receptors.

A family of protease-activated receptors (PARs) has been identified. These G protein-coupled receptors share the same basic structure, and comprise seven transmembrane domains with three extracellular and three intracellular loops (Fig. 3). Proteolytic cleavage of the extracellular N-terminus at a certain point will expose a 'tethered ligand'. The binding of the terminal six amino acids of the tethered ligand to the second extracellular loop triggers cell signalling. The first PAR to be cloned, now termed PAR-1, was characterized originally as a receptor for thrombin (130). Subsequently PAR-2, a receptor activated by trypsin, was discovered, and more recently PAR-3 and PAR-4, both of which may be activated by thrombin and trypsin (131). In certain recombinant expression systems tryptase has been reported to be able to activate PAR-1, though the biological significance of this relatively weak reaction is open to question (132). More important could be the findings that PAR-2 can be activated by purified lung tryptase (132) and by recombinant α- and β-tryptases (133).

There is as yet no selective antagonist for PAR-2 that could be used to explore the extent to which the actions of tryptase on cells could be mediated through this receptor. However, experimental activation of PAR-2 with proteases or with peptides corresponding to the tethered ligand-binding region has been found to provoke some effects reminiscent of those that have been evoked with tryptase. Thus, for example, as found with tryptase (114, 118), injection of PAR-2 agonists into a rodent model can induce an acute inflammatory response with persistent oedema and granulocyte infiltration (134). There are some similarities in patterns of cytokine release from

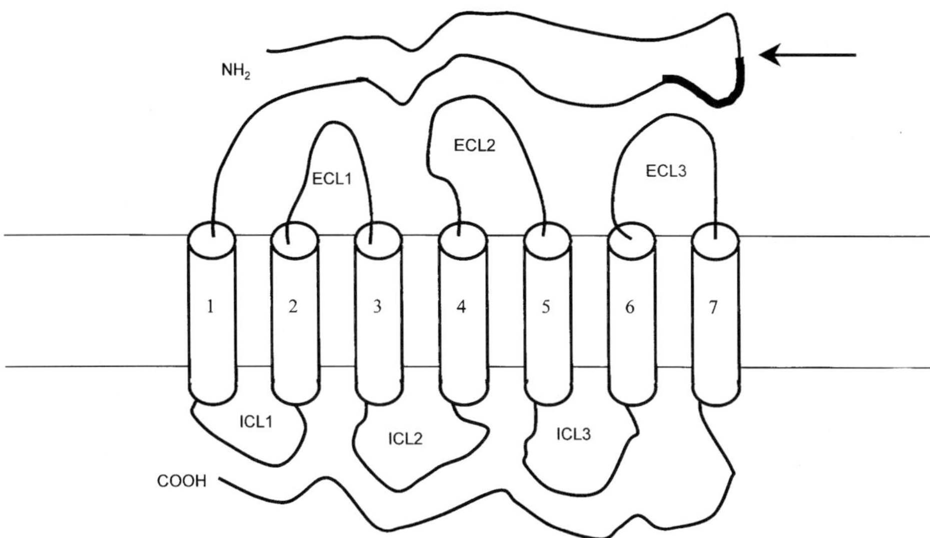

Fig. 3 Schematic diagram of a protease-activated receptor (PAR). The seven transmembrane domains are numbered, as are the three extracellular loops (ECL) and three intracellular loops (ICL). In the case of PAR-2, tryptase or trypsin can cleave the extracellular N-terminus at a point (arrowed) that would expose a 'tethered ligand' which binds to a region of ECL2 to initiate cell signalling.

epithelial cells treated with tryptase or PAR-2 agonists (122). However, some important discrepancies in findings are emerging. While PAR-2 activation can induce HUVEC proliferation (135), this has not been observed with tryptase (123). Perhaps more strikingly, exposure of airway smooth muscle tissue to tryptase can enhance contractile responses (117), but relaxant effects have been described for PAR-2 activation (136). To date there have been few direct comparisons of the effects of tryptase and PAR-2 agonists under the same experimental conditions, but the actions of tryptase cannot be explained simply in terms of the activation of PAR-2.

TRYPTASE AS A THERAPEUTIC TARGET

As a potent mediator of inflammation and tissue remodelling, tryptase has come to attract attention as a new target for drug development. The range of conditions in which an elevation in tryptase concentration has been detected in biological fluids (Table I) suggests applications in a number of therapeutic areas, although there has been particular interest to date in the potential of inhibitors of tryptase as anti-asthma agents.

All actions of tryptase on cells and tissues described to date have been dependent on an intact catalytic site. However, there are still very few reports on the ability of tryptase inhibitors to modulate processes in which the primary stimulus is not provided simply by administration of purified tryptase. A particular difficulty in this field is posed by the extent of the differences between human tryptases and their counterparts in experimental animals. Nevertheless, administration of tryptase inhibitors in a sheep model of asthma has had effects wholly consistent with our understanding of the mediator roles of tryptase. In naturally sensitized allergic sheep challenged with allergen, the synthetic tryptase inhibitors APC-366 and BABIM have been reported to significantly reduce both early- and late-phase bronchoconstriction, and to block allergen-induced hyper-responsiveness (137). Treatment with APC-366 was associated also with reductions in eosinophil infiltration and in microvascular leakage in the lungs. In the same sheep model, improvement in lung function has been reported also following administration of the tryptase destabilizer lactoferrin (89) and with the broad spectrum inhibitor SLPI, which can inhibit tryptase under certain conditions (138). SLPI has also been found to inhibit airway hyper-responsiveness in a sensitized guinea pig model (139).

A phase II clinical trial with inhaled APC-366 in 16 mild to moderate asthmatics has provided some support for the concept that tryptase inhibitors could have a role in the treatment of allergic disease (140). Late-phase bronchoconstriction to allergen was significantly reduced following administration of APC-366 and there was a trend for protection against early-phase bronchoconstriction, although allergen-induced hyper-responsiveness was unaffected. While the clinical trial findings with this first-generation tryptase inhibitor may not have matched those with the animal models, they are nonetheless encouraging. Tryptase inhibitors deserve further study as anti-inflammatory agents, and they could be particularly valuable if they can prevent or reverse the processes of tissue remodelling and fibrosis in which tryptase is implicated. Developments in this area seem set to provide further important information on the function of this major mast cell product.

ACKNOWLEDGEMENTS

Financial support from the National Asthma Campaign, the Medical Research Council and Action Research, U.K., is gratefully acknowledged.

REFERENCES

1. Lagunoff, D. and Benditt, E. P. Proteolytic enzymes of mast cells. *Ann. N.Y. Acad. Sci.* **103**:184, 1963.
2. Schwartz, L. B., Irani, A. A., Roller, K., Castells, M. C. and Schechter, N. M. Quantitation of histamine, tryptase and chymase in dispersed human T and TC mast cells. *J. Immunol.* **138**:2611–2615, 1987.
3. Bradding, P., Walls, A. F. and Church, M. K. Mast cells and basophils: their role in initiating and maintaining inflammatory responses. In *Immunology of the Respiratory System* (Holgate, S. T., ed.), pp. 53–84. Academic Press, London, 1995.
4. Church, M. K., Holgate, S. T., Shute, J. K., Walls, A. F. and Sampson, A. P. Mast cell derived mediators. In *Allergy: Principles and Practice*, 5th edn. (Middleton, E., Reed, C. E., Ellis, E., Adkinson, N. F., Yunginger, J. W. and Busse, W. W., eds), pp. 146–167. Mosby, St. Louis, 1998.
5. Walls, A. F., Bennett, A. R., McBride, H. M., Glennie, M. J., Holgate, S. T. and Church, M. K. Production and characterisation of monoclonal antibodies specific for human mast cell tryptase. *Clin. Exp. Allergy* **20**:581–589, 1990.
6. Craig, S. S., De Blois, G. and Schwartz, L. B. Mast cells in human keloid, small intestine and lung by an immunoperoxidase technique using a murine antibody against tryptase. *Am. J. Pathol.* **124**:427–435, 1986.
7. Castells, M. C., Irani, A. A. and Schwartz, L. B. Evaluation of human peripheral blood leukocytes for mast cell tryptase. *J. Immunol.* **138**:2184–2189, 1987.
8. Irani, A. A. Tissue and developmental variation of protease expression in human mast cells. In *Mast Cell Proteases in Immunology and Biology* (Caughey, G. H., ed.), pp. 127–143. Marcel Dekker, New York, 1995.
9. Irani, A. A., Craig, S. S., DeBlois, G., Elson, C. O., Schechter, N. M. and Schwartz, L. B. Deficiency of

the tryptase-positive, chymase-negative mast cell type in gastrointestinal mucosa of patients with defective lymphocyte function. *J. Immunol.* **138**:4381–4386, 1987.

10. Bentley, A. M., Jacobson, M. R., Cumberworth, V. *et al.* Immunohistology of the nasal mucosa in seasonal allergic rhinitis: increases in activated eosinophils and epithelial mast cells. *J. Allergy Clin. Immunol.* **89**:877–883, 1992.
11. Irani, A. A., Sampson, H. A. and Schwartz, L. B. Mast cells in atopic dermatitis. *Allergy* **44**:31–34, 1989.
12. Irani, A. A., Butrus, S. I., Tabbara, K. F. and Schwartz, L. B. Human conjunctival mast cells: distribution of MCT and MCTC in vernal conjunctivitis and giant papillary conjunctivitis. *J. Allergy Clin. Immunol.* **86**:34–40, 1990.
13. Irani, A. A., Gruber, B. L., Kaufman, L. D., Kahaleh, M. B. and Schwartz, L. B. Mast cell changes in scleroderma: presence of MCT cells in the skin and evidence of mast cell activation. *Arthritis Rheum.* **35**:933–939, 1992.
14. Tetlow, L. C. and Woolley, D. E. Distribution, activation and tryptase/chymase phenotype of mast cells in the rheumatoid lesion. *Ann. Rheum. Dis.* **54**:549–555, 1995.
15. Buckley, M. G., Gallagher, P. J. and Walls, A. F. Mast cell subpopulations in the synovial tissue of patients with osteoarthritis: selective increase in numbers of tryptase-positive, chymase-negative mast cells. *J. Pathol.* **186**:67–74, 1998.
16. Irani, A. A., Goldstein, S. M., Wintroub, B. U., Bradford, T. and Schwartz, L. B. Human mast cell carboxypeptidase. Selective localisation to MCTC cells. *J. Immunol.* **147**:247–253, 1991.
17. Schechter, N. M., Irani, A. A., Sprows, J. L., Abernethy, J. Wintroub, B. and Schwartz, L. B. Identification of a cathepsin G-like proteinase in the MCTC type of human mast cell. *J. Immunol.* **145**:2652–2661, 1990.
18. Bradding, P., Okayama, Y., Howarth, P. H., Church, M. G. and Holgate, S. T. Heterogeneity of human mast cells based on cytokine content. *J. Immunol.* **155**:297–307, 1995.
19. Weidner, N. and Austen, K. F. Heterogeneity of mast cells at multiple body sites. Fluorescent determination of avidin binding and immunofluorescent determination of chymase, tryptase and carboxypeptidase content. *Pathol. Res. Pract.* **189**:156–162, 1993.
20. Kleinjan, A., Godthelp, T., Blom, H. M. and Fokkens, W. J. Fixation with Carnoy's fluid reduces the number of chymase-positive mast cells: not all chymase-positive mast cells are also positive for tryptase. *Allergy* **51**:614–620, 1996.
21. Beil, W. J., Schultz, M., McEuen, A. R., Buckley, M. G. and Walls, A. F. Number, fixation properties, dye binding and protease expression of duodenal mast cells: comparisons between healthy subjects and patients with gastritis or Crohn's disease. *Histochem. J.* **29**:1–15, 1997.
22. Schwartz, L. B., Sakai, K., Bradford, T. R., Ren, S., Zweiman, B., Worobec, A. S. and Metcalfe, D. D. The α form of human tryptase is the predominant form present in blood at baseline in normal subjects and is elevated in those with systemic mastocytosis. *J. Clin. Invest.* **96**:2702–2710, 1995.
23. Schwartz, L. B., Metcalfe, D. D., Miller, J. S., Earl, H. and Sullivan, T. Tryptase levels as an indicator of mast cell activation in systemic anaphylaxis and mastocytosis. *N. Engl. J. Med.* **316**:1622–1626, 1987.
24. Schwartz, L. B., Yunginger, J. W., Miller, J., Bokhari, R. and Dull, D. Time course of appearance and disappearance of human mast cell tryptase in the circulation after anaphylaxis. *J. Clin. Invest.* **83**:1551–1555, 1989.
25. Yunginger, J. Y., Nelson, D. R., Squillace, D. L. *et al.* Laboratory investigation of deaths due to anaphylaxis. *J. Forensic Sci.* **36**:857, 1991.
26. Buckley, M. G., McEuen, A. R., Kilburn, S. A., Soilleux, S., Hourihane. J. O'B., Warner, J. O. and Walls, A. F. Elevated concentrations of salivary tryptase in food-induced allergic reactions. *Immunology* **98 Suppl. 1**:189, 1999.
27. Walls, A. F., Djukanović, R., Walters, C., Brander, M. L., Makker, H. K., Roberts, J. A., Gratziou, C., Carroll, M. P., Frew, A. J., Pritsinas, K., Lampe, F. C., Howarth, P. H. and Holgate, S. T. Mast cell, eosinophil and neutrophil activation in the lungs of asthmatic patients. *Am. J. Respir. Crit. Care. Med.* **151**:A38, 1995.
28. Broide, D. H., Gleich, G. J., Cuomo, A. J., Coburn, D. A., Federman, E. C., Schwartz, L. B. and Wasserman, S. I. Evidence of ongoing mast cell and eosinophil degranulation in symptomatic asthma airway. *J. Allergy Clin. Immunol.* **88**:637–648, 1991.
29. Jarjour, N. N., Calhoun, W. J., Schwartz, L. B. and Busse, W. W. Elevated bronchoalveolar lavage fluid histamine levels in allergic asthmatics are associated with increased airway obstruction. *Am. Rev. Respir. Dis.* **144**:83–87, 1991.

30. Djukanović, R., Wilson, J. W., Britten, K. M., Wilson, S. J., Walls, A. F., Roche, W. R., Howarth, P. H. and Holgate, S. T. Effect of an inhaled corticosteroid on airway inflammation and symptoms in asthma. *Am. Rev. Respir. Dis.* **145**:669–674, 1992.
31. Roberts, J. A., Bradding, P., Britten, K. M., Walls, A. F., Walters, C., Wilson, S., Gratziou, C., Holgate, S. T. and Howarth, P. H. The effect of long-acting inhaled β2 agonist salmeterol xinafoate on indices of airway inflammation in asthma. *Eur. Respir. J.* **14**:275–282, 1999.
32. Wenzel, S. E., Fowler, A. A. and Schwartz, L. B. Activation of pulmonary mast cells by bronchoalveolar allergen challenge. *In vivo* release of histamine and tryptase in atopic subjects with and without asthma. *Am. Rev. Respir. Dis.* **137**:1002–1008, 1988.
33. Salmonsson, P., Gronneberg, R., Gilljam, H., Andersson, O., Billing. B., Enander, I., Alkner, U. and Persson, C. G. A. Bronchial exudation of bulk plasma at allergen challenge in allergic asthma. *Am. Rev. Respir. Dis.* **146**:1535–1542, 1992.
34. Frew, A. J., St Pierre, J., Teran, L. M., Trefillieff, A., Madden, J., Peroni, D., Bodey, K. M., Walls, A. F., Howarth, P. H., Carroll, M. P. and Holgate, S. T. Cellular and mediator responses twenty four hours after local endobronchial allergen challenge of asthmatic airways. *J. Allergy Clin. Immunol.* **98**:133–143, 1996.
35. Makker, H. K., Walls, A. F., Goulding, D., Montefort, S., Varley, J. J., Carroll, M., Howarth, P. H. and Holgate, S. T. Local airway challenge with hypertonic saline in asthma: Effect on airway size, mast cell mediator release and epithelial structure. *Am. Rev. Respir. Crit. Care Med.* **149**:1012–1019, 1994.
36. Broide, D. H., Eisman, S., Ramsdell, J. W., Ferguson, P., Schwartz, L. B. and Wasserman, S. I. Airway levels of mast cell-derived mediators in exercise-induced asthma. *Am. Rev. Respir. Dis.* **141**:563–568, 1990.
37. Jarjour, N. N., Calhoun, W. J., Stevens, C. A. and Salisbury, S. M. Exercise induced asthma is not associated with mast cell activation or airway inflammation. *J. Allergy Clin. Immunol.* **89**:60–68, 1992.
38. Walls, A. F., Bennett, A. R., Godfrey, R. C., Holgate, S. T. and Church, M. K. Mast cell tryptase and histamine concentrations in bronchoalveolar lavage fluid from patients with interstitial lung disease. *Clin. Sci.* **81**:183–188, 1991.
39. Kalenderian, R., Raju, L., Roth, W., Schwartz, L. B., Gruber, B. and Janoff, A. Elevated histamine and tryptase levels in smokers' bronchoalveolar lavage fluid. Do lung mast cells contribute to smokers' emphysema? *Chest* **94**:119–123, 1988.
40. Castells, M. and Schwartz, L. B. Tryptase levels in nasal-lavage fluid as an indicator of the immediate allergic response. *J. Allergy Clin. Immunol.* **82**:348–355, 1988.
41. Juliusson, S., Holmberg, K., Baumgarten, C. R., Olsson, M., Enander, I. and Pipkorn, U. Tryptase in nasal lavage fluid after local allergen challenge. Relationship to histamine levels and TAME-esterase activity. *Allergy* **46**:459–465, 1991.
42. Proud, D., Bailey, G. S., Naclerio, R. M., Reynolds, C. J., Cruz, A. A., Egglestone, P. A., Lichtenstein, L. M. and Togias, A. G. Tryptase and histamine as markers to evaluate mast cell activation during the responses to nasal challenge with allergen, cold, dry air, and hyperosmolar solutions. *J. Allergy Clin. Immunol.* **89**:1098–1110, 1992.
43. Shalit, M., Schwartz, L. B., Golzar, N., Von Allman, C., Valenzano, M., Fleekop, P., Atkins, P. C. and Zweiman, B. Release of tryptase *in vivo* after prolonged cutaneous challenge with allergen in humans. *J. Immunol.* **141**:821–826, 1988.
44. Befus, S. I., Ochsner, K. I., Abelson, M. B. and Schwartz, L. B. The level of tryptase in human tears. An indicator of activation of human mast cells. *Ophthalmology* **97**:1678–1683, 1990.
45. Buckley, M. G., Walters, C., Wong, W. M., Cawley, M. I. D., Ren, S., Schwartz, L. B. and Walls, A. F. Mast cell activation in arthritis: detection of α and β tryptase, histamine and eosinophil cationic protein in synovial fluid. *Clin. Sci.* **93**:363–370, 1997.
46. Boucher, W., El-Mansoury, M., Pang, X., Sant, G. R. and Theoharides, T. C. Elevated mast cell tryptase in the urine of patients with interstitial cystitis. *Br. J. Urol.* **76**:94–100, 1995.
47. Schwartz, L. B., Lewis, R. A. and Austen, K. F. Tryptase from human pulmonary mast cells. Purification and characterisation. *J. Biol. Chem.* **256**:11939–11943, 1981.
48. Smith, T. J., Hougland, M. W. and Johnson, D. A. Human lung tryptase. Purification and characterisation. *J. Biol. Chem.* **259**:11046–11051, 1984.
49. Benyon, C. R, Enciso, J. A. and Befus, A. D. Analysis of human skin mast cell proteins by two-dimensional gel electrophoresis. Identification of tryptase as a sialylated glycoprotein. *J. Immunol.* **151**:2699–2706, 1993.
50. Miller, J. S., Moxley, G. and Schwartz, L. B. Cloning and characterisation of a second complementary

DNA for human tryptase. *J. Clin. Invest.* **86**:864–870, 1990.

51. Miller, J. S., Westin, E. H. and Schwartz, L. B. Cloning and characterisation of complementary DNA for human tryptase. *J. Clin. Invest.* **84**:1188–1195, 1989.
52. Vanderslice, P., Ballinger, S. M., Tam, E. K., Goldstein, S. M., Craik, C. S. and Caughey, G. H. Human mast cell tryptase: multiple cDNAs and genes reveal a multigene serine protease family. *Proc. Natl. Acad. Sci. USA* **87**:3811–3815, 1990.
53. Sakai, K., Ren, S. and Schwartz, L. B. A novel heparin-dependent processing pathway for human tryptase: autocatalysis followed by activation by dipeptidyl peptidase I. *J. Clin. Invest.* **97**:988–995, 1996.
54. Mirza, H., Schmidt, V. A., Derian, C. K., Jesty, J. and Bahon, W. F. Mitogenic responses mediated through the protease-activated receptor-2 are induced by expressed forms of mast cell α- or β-tryptases. *Blood* **90**:3914–3922, 1997.
55. Huang, C., Li, L., Krilis, S. A., Chanasyk, K., Tang, Y., Li, Z., Hunt, J. E. and Stevens, R. L. Human tryptases α and β/II are functionally distinct due, in part, to a single amino acid difference in one of the surface loops that forms the substrate-binding cleft. *J. Biol. Chem.* **274**:19670–19676, 1999.
56. Wang, Z. W., Walter, M., Selwood, T., Rubin, H. and Schechter, N. M. Recombinant expression of human mast cell proteases chymase and tryptase. *Biol. Chem.* **379**:167–174, 1998.
57. Niles, A. L., Maffitt, M., Haak-Frendscho, M., Wheeless, C. J. and Johnson, D. A. Recombinant human mast cell tryptase: stable expression in *Pichia pastris* and purification of fully active enzyme. *Biotechnol. Appl. Biochem.* **28**:125–131, 1998.
58. Chan, H., Elrod, K., Numarof, R. P., Sideris, S. and Clark, J. M. Expression and characterisation of recombinant mast cell tryptase. *Protein Express. Purif* **15**:251–257, 1999.
59. Pereira, P. J., Bergner, A., Macedo Ribeiro, S., Huber, R., Matschiner, G., Fritz, H., Sommerhof, C. P. and Bode, W. Human β-tryptase is a ring-like tetramer with active sites facing a central pore. *Nature* **392**:306–311, 1998.
60. Glenner, G. G. and Cohen, L. A. Histochemical demonstration of a species-specific trypsin-like enzyme in mast cells. *Nature* **185**:846–847, 1960.
61. Valchanov, K. P. and Proctor, G. B. Enzyme histochemistry of tryptase in the stomach mucosal mast cells of the mouse. *J. Histochem. Cytochem.* **47**:617–622, 1999.
62. Chan, Z., Irani, A. A., Bradford, T. R., Craig, S. S., Newlands, G., Miller, H., Huff, T., Simmons, W. H. and Schwartz, L. B. Localisation of rat tryptase to a subset of the connective tissue type of mast cell. *J. Histochem. Cytochem.* **41**:961, 1993.
63. Friend, D. S., Ghildyal, N., Gurish, M. F., Hunt, J., Hu, X., Austen, K. F. and Stevens, R. L. Reversible expression of tryptases and chymases in the jejunal mast cells of mice infected with *Trichinella spiralis*. *J. Immunol.* **160**:5537–5545, 1998.
64. McEuen, A. R., He, S., Brander, M. L. and Walls, A. F. Guinea pig lung tryptase. Localisation to mast cells and characterisation of the partially purified enzyme. *Biochem. Pharmacol.* **52**:331–340, 1996.
65. Katunuma, N. and Kido, H. Recent advances in research on tryptases and endogenous tryptase inhibitors. In *Neutral Proteases of Mast Cells* (Schwartz, L. B., ed.), pp. 51–66. Karger, Basel, 1990.
66. Lagunoff, D., Rickard, A. and Marquart, C. Rat mast cell tryptase. *Arch. Biochem. Biophys.* **291**:52–58, 1991.
67. Robinson, T. L. and Muller, D. K. Purification and characterisation of cynomolgus monkey tryptase. *Comp. Biochem. Physiol.* **118B**:783–792, 1997.
68. Lutzelschwab, C., Pejler, G., Aveskogh, M. and Hellman, L. Secretory granule proteases in rat mast cells. Cloning of 10 different serine proteases and a carboxypeptidase A from various rat mast cell populations. *J. Exp. Med.* **185**:13–29, 1997.
69. Reynolds, D. S., Gurley, D. S., Austen, K. F. and Serafin, W. E. Cloning of the cDNA and gene of mouse mast cell protease-6. Transcription by progenitor mast cells and mast cells of the connective tissue subclass. *J. Biol. Chem.* **266**:3845–3853, 1991.
70. McNeil, H. P., Reynolds, D. S., Schiller, V., Ghildyal, N., Gurley, D. S., Austen, K. F. and Stevens, R. L. Isolation, characterisation and transcription of the gene encoding mouse mast cell protease 7. *Proc. Natl. Acad. Sci. USA* **89**:11174–11178, 1992.
71. Vanderslice, P., Craik, C. S., Nadel, J. A. and Caughey, G. H. Molecular cloning of a dog mast cell tryptase and a related protease. Structural evidence of a unique mode of serine protease activation. *Biochemistry* **28**:41–48,1989.
72. Chiu, H. and Lagunoff, D. Histochemical comparison of vertebrate mast cells. *Histochem. J.* **4**:134–135, 1972.

73. Nawa, Y., Horiü Okada, M. and Arizono, N. Histochemical and cytological characterisations of mucosal and connective tissue mast cells of Mongolian gerbils (*Meriones unguiculatus*). *Int. Arch. Allergy Immunol.* **104**:249–254, 1994.
74. Welle, M. M., Proske, S. M., Harvima, I. T. and Schechter, N. M. Demonstration of tryptase in bovine cutaneous and tumor mast cells. *J. Histochem. Cytochem.* **43**:1139–1144, 1995.
75. Tanaka ,T., McRae, B. J., Cho, K., Cook, R., Fraki, J. E., Johnson, D. A. and Powers, J. C. Mammalian tissue trypsin-like enzymes. Comparative reactivities of human skin tryptase, human lung tryptase, and bovine trypsin with peptide 4 nitroanilide and thioester substrates. *J. Biol. Chem.* **258**:3552–3557, 1983.
76. Schechter, N. M., Eng, G. Y. and McCaslin, D. R. Human skin tryptase: kinetic characterisation of its spontaneous inactivation. *Biochemistry* **32**:2617–2625, 1993.
77. Alter, S. C and Schwartz, L. B. Effect of histamine and divalent cations on the activity and stability of tryptase from human mast cells. *Biochim. Biophys. Acta* **991**:426–430, 1989.
78. Ren, S., Lawson, A. E., Carr, M., Baumgarten, C. M. and Schwartz, L. B. Human tryptase fibrinogenolysis is optimal at acidic pH and generates anticoagulant fragments in the presence of the anti-tryptase monoclonal antibody B12. *J. Immunol.* **159**:3540–3548, 1997.
79. Alter, S. C., Kramps, J. A., Janoff, A. and Schwartz, L. B. Interactions of human mast cell tryptase with biological protease inhibitors. *Arch. Biochem. Biophys.* **276**:26–31, 1990.
80. Hochstrasser, K., Gebhard, W., Albrecht, G,. Rasp, G. and Kastenbauer, E. Interaction of human mast cell tryptase and chymase with low-molecular-mass serine proteinase inhibitors from the human respiratory tract. *Eur. Arch. Otorhinolaryngol.* **249**:455–458, 1993.
81. Sommerhoff, C. P., Söllner, C., Mentele, R., Piechottka, G. P., Auserswald, E. A. and Fritz, H. A Kazal-type inhibitor of human mast cell tryptase: isolation from the medicinal leech *Hirudo medicinalis*, characterisation and sequence analysis. *Biol. Chem. Hoppe-Seyler* **375**:685–694, 1994.
82. Kido, H. and Katunuma, N. Control of tryptase and chymase activity by protease inhibitors. In *The Biology of Mast Cell Proteases* (Caughey, G. H., ed.), pp. 145–167. Marcel Dekker, New York, 1995.
83. Robinson, T., Delavia, K., Harris, P., Lindall, D., Gundel, R. and Muller, D. Secretory leukoprotease as an inhibitor of mast cell tryptase. *Am. J. Respir. Crit. Care Med.* **153**:A399, 1996.
84. Sommerhoff, C. P., Sollner, C., Mentele, R., Piechottka, G. P., Auerswald, E. A. and Fritz, H. A Kazal-type inhibitor of human mast cell tryptase: isolation from the medicinal leech *Hirudo medicinalis*, characterisation and sequence analysis. *Biol. Chem. Hoppe-Seyler* **375**:685–694, 1994.
85. Goldstein, S. M., Leong, J., Schwartz, L. B. and Cooke, D. Protease composition of exocytosed human skin mast cell protease–proteoglycan complexes: tryptase resides in a complex distinct from chymase and carboxypeptidase. *J. Immunol.* **148**:2475–2482, 1992.
86. Schwartz, L. B. and Bradford, T. R. Regulation of tryptase from human lung mast cells by heparin. Stabilisation of the active tetramer. *J. Biol. Chem.* **261**:7372–7379, 1986.
87. Schwartz, L. B., Bradford, T. R., Lee, D. C. and Chlebowski, J. F. Immunologic and physicochemical evidence for conformation changes occurring on conversion of human mast cell tryptase from active tetramer to inactive monomer. Production of monoclonal antibodies recognising active tryptase. *J. Immunol.* **144**:2304–2311, 1990.
88. Alter, S. C., Metcalfe, D. D., Bradford, T. R. and Schwartz, L. B. Stabilisation of human mast cell tryptase: effects of enzyme concentration, ionic strength and the structure and negative charge density of polysaccharides. *Biochem. J.* **248**:821–827, 1987.
89. Elrod, K. C., Moore, W. R., Abraham, W. M. and Tanaka, R. D. Lactoferrin, a potent tryptase inhibitor, abolishes late phase airway responses in allergic sheep. *Am. J. Respir. Crit. Care. Med.* **156**:375–381, 1997.
90. Cregan, L., Elrod, K. C., Putnam, D. and Moore, W. R. Neutrophil myeloperoxidase is a potent and selective inhibitor of mast cell tryptase. *Arch. Biochem. Biophys.* **366**:125–130, 1999.
91. Tanaka, T., McRae, B. J., Cho, K., Cook, R., Fraki, J. E., Johnson, D. A. and Powers, J. C. Mammalian tissue trypsin-like enzymes. Comparative reactivities of human skin tryptase, human lung tryptase, and bovine trypsin with peptide 4-nitroaniclide and thioester substrates. *J. Biol. Chem.* **258**:13552–13557, 1983.
92. Walls, A. F., Brain, S. D., Jose, P. J., Desai, A., Hawkings, E., Church, M. K. and Williams, T. J. Human mast cell tryptase attenuates the vasodilator activity of calcitonin gene-related peptide (CGRP). *Biochem. Pharmacol.* **43**:1243–1248, 1992.
93. Proud, D., Siekierski, E. S. and Bailey, G. S. Identification of human lung mast cell kininogenase as tryptase and relevance of tryptase kininogenase activity. *Biochem. Pharmacol.* **37**:1473–1480, 1988.

94. Bienenstock, J., Perdue, M., Blennerhassett, M., Stead, R., Kakuta, N., Sestini, P., Vancheri, C. and Marshall, J. Inflammatory cells and the epithelium. Mast cell/nerve interactions in the lung *in vitro* and *in vivo*. *Am. Rev. Respir. Dis.* **138**:S31–34, 1988.
95. Hislop, A. A., Wharton, J., Allen, K. M., Polak, J. M. and Haworth, S. G. Immunohistochemical localisation of peptide-containing nerves in human airways: age related changes. *Am. J. Respir. Cell Mol. Biol.* **3**:191–198, 1990.
96. Caughey, G. H., Leidig, F., Viro, N. F. and Nadel, J. A. Substance P and vasoactive intestinal peptide degradation by mast cell tryptase and chymase. *J. Pharmacol. Exp. Ther.* **244**:133–137, 1988.
97. Franconi, G. M., Graf, P. D., Lazarus, S. C., Nadel, J. A. and Caughey, G. H. Mast cell tryptase and chymase reverse airway smooth muscle relaxation induced by vasoactive intestinal peptide in the ferret. *J. Pharmacol. Exp. Ther.* **248**:947–951, 1989.
98. Tam, E. K., Franconi, G. M., Nadel, J. A. and Caughey, G. H. Protease inhibitors potentiate smooth muscle relaxation induced by vasoactive intestinal peptide in isolated human bronchi. *Am. J. Respir. Cell Mol. Biol.* **2**:449–452,1990.
99. Thompson, D. C., Diamond, L. and Altiere, R. J. Enzymatic modulation of vasoactive intestinal peptide and nonadrenergic noncholinergic inhibitory responses in guinea pig tracheae. *Am. Rev. Respir. Dis.* **142**:1119–1123, 1990.
100. Hulsmann, A. R., Jongejan, R. C., Raatgeep, H. R., Stijnen, T., Bonta, I. L., Kerrebijn, K. F. and De Jongste, J. C. Epithelium removal and peptide inhibition enhance relaxation of human airways to vasoactive intestinal peptide. *Am. Rev. Respir. Dis.* **147**:1483–1486, 1993.
101. Brain, S. D. and Williams, T. J. Substance P regulates the vasodilator activity of calcitonin gene-related peptide. *Nature* **335**:73–75, 1988.
102. Schwartz, L. B., Bradford, T. R., Littman, B. H. and Wintroub, B. U. The fibrinogenolytic activity of purified tryptase from human lung mast cells. *J. Immunol.* **135**:2762–2767, 1985.
103. Ren, S., Lawson, A. E., Carr, M., Baumgarten, C. and Schwartz, L. B. Human tryptase fibrinogenolysis is optimal at acid pH and generates anticoagulant fragments in the presence of the anti-tryptase monoclonal antibody. *Br. J. Immunol.* **159**:3540–3548, 1997.
104. Thomas. V. A., Wheeless, C. J., Stack, M. S. and Johnson, D. A. Human mast cell tryptase fibrinogenolysis: kinetics, anticoagulation mechanism and cell adhesion disruption. *Biochemistry* **37**:2291–2298, 1998.
105. Walls, A. F., Bennett, A. R., Suieras-Diaz, J. and Olsson, H. The kininogenase activity of human mast cell tryptase. *Biochem. Soc. Trans.* **20**:260S, 1992.
106. Imamura, T., Dubin, A., Moore, W., Tanaka, R. and Travis, J. Induction of vascular permeability enhancement by human tryptase: dependence on activation of prekallikrein and direct release of bradykinin from kininogens. *Lab. Invest.* **74**:861–870, 1996.
107. Kozik, A., Moore, R. B., Potempa, J., Imamura, T., Rapala-Kozik, M. and Travis, J. A novel mechanism for bradykinin production at inflammatory sites. Diverse effects of a mixture of neutrophil elastase and mast cell tryptase versus tissue and plasma kallikreins on native and oxidised kininogens. *J. Biol. Chem.* **273**:33224–33229, 1998.
108. Gruber, B. L., Marchese, M. J., Suziki, K., Schwartz, L. B., Okada, Y., Nagase, H. and Ramamurthy, N. S. Synovial procollagenase activation by human mast cell tryptase. Dependence upon matrix metalloproteinase 3 activation. *J. Clin. Invest.* **84**:1657–1662, 1989.
109. Stack, M. S. and Johnson, D. A. Human mast cell tryptase activates single-chain urinary-type plasminogen activator (pro-urokinase). *J. Biol. Chem.* **269**:9416–9419, 1994.
110. Lohi, J., Harvima, I. and Keski-Oja, J. Pericellular substrates of human mast cell tryptase: 72,000 dalton gelatinase and fibronectin. *J. Cell. Biochem.* **50**:337–349, 1992.
111. Kielty, C. M., Lees, M., Shuttleworth, C. A. and Woolley, D. Catabolism of intact type VI collagen microfibrils: susceptibility to degradation by serine proteases. *Biochem. Biophys. Res. Commun.* **191**:1230–1236, 1993.
112. Glenner, G. G. and Cohen, L. A. Histochemical demonstration of a species specific trypsin-like enzyme in mast cells. *Nature* **185**:846–847, 1960.
113. He, S., Gaça, M. D. A. and Walls, A. F. A role for tryptase in the activation of human mast cells: modulation of histamine release by tryptase and inhibitors of tryptase. *J. Pharmacol. Exp. Ther.* **286**:289–297, 1998.
114. He, S. and Walls, A. F. Human mast cell tryptase: a stimulus of microvascular leakage and mast cell activation. *Eur. J. Pharmacol.* **328**:89–97, 1997.
115. Molinari, J. F., Moore, W. R., Clark, J., Tanaka, R., Butterfield, J. H. and Abraham, W. M. Role of

tryptase in immediate cutaneous responses in allergic sheep. *J. Appl. Physiol.* **79**:1966–1973, 1995.
116. Molinari, J. F., Scuri, M., Moore, W. R., Clark, J., Tanaka, R. and Abraham, W. M. Inhaled tryptase causes bronchoconstriction in sheep via histamine release. *Am. J. Respir. Crit. Care Med.* **154**:649–653, 1996.
117. Berger, P., Compton, S. J., Molimard, M., Walls, A. F., N'Guyen, C., Marthan, R. and Tunon de Lara, J. M. Mast cell tryptase as a mediator of hyperresponsiveness in human isolated bronchi. *Clin. Exp. Allergy* **29**:804–812, 1999.
118. He, S., Peng, Q. and Walls, A. F. Potent induction of a neutrophil and eosinophil-rich infiltrate *in vivo* by human mast cell tryptase: Selective enhancement of eosinophil recruitment by histamine. *J. Immunol.* **159**:6216–6225, 1997.
119. He, S. and Walls, A. F. Human mast cell chymase induces the accumulation of neutrophils, eosinophils and other inflammatory cells *in vivo. Br. J. Pharmacol.* **125**:1491–1500, 1998.
120. Walls, A. F., He, S., Teran, L., Buckley, M. G., Jung, K.-S., Holgate, S. T., Shute, J. K. and Cairns, J. A. Granulocyte recruitment by human mast cell tryptase. *Int. Arch. Allergy Immunol.* **107**:372–373, 1995.
121. Cairns, J. A. and Walls, A. F. Mast cell tryptase is a mitogen for epithelial cells. Stimulation of IL-8 production and intercellular adhesion molecule-1 expression. *J. Immunol.* **156**:275–283, 1996.
122. Perng, D.-W., Leir, S.-H., Compton, S. J., Lackie, P. M., Holgate, S. T. and Walls, A. F. Mast cell tryptase stimulates cytokine synthesis and secretion from bronchial epithelial cells: a role for protease activated receptor 2 (PAR-2). *Am. J. Respir. Crit. Care Med.* **159**:A336, 1999.
123. Compton, S. J., Cairns, J. A., Holgate, S. T. and Walls, A. F. The role of mast cell tryptase in regulating endothelial cell proliferation, cytokine release and adhesion molecule expression: tryptase induces expression of mRNA for IL-1β and IL-8 and stimulates the selective release of IL-8 from human umbilical vein endothelial cells. *J. Immunol.* **161**:1939–1946, 1998.
124. Compton, S. J., Cairns, J. A., Holgate, S. T. and Walls, A. F. Interaction of human mast cell tryptase with endothelial cells to stimulate inflammatory cell recruitment. *Int. Arch. Allergy Immunol.* **118**:200–205, 1999.
125. Blair, R. J., Meng, H., Marchese, M. J., Ren, S., Schwartz, L. B. and Tonnesen, M. G. Human mast cells stimulate vascular tube formation: tryptase is a novel potent angiogenic factor. *J. Clin. Invest.* **99**:2691–2700, 1997.
126. Hartmann, T., Ruoss, S. J., Raymond, W. W., Seuwen, K. and Caughey, G. H. Human tryptase as a potent, cell-specific mitogen: role of signalling pathways in synergistic responses. *Am. J. Physiol.* **262**:L528–534, 1992.
127. Cairns, J. A. and Walls, A. F. Mast cell tryptase stimulates the synthesis of type I collagen in human lung fibroblasts. *J. Clin. Invest.* **99**:1313–1321, 1997.
128. Gruber, B. L., Kew, R. R., Jelaska, A., Marchese, M. J., Garlick, J., Ren, S., Schwartz, L. B. and Korn, J. H. Human mast cells activate fibroblasts. *J. Immunol.* **158**:2310–2317, 1997.
129. Thabrew, H., Cairns, J. A. and Walls, A. F. Mast cell tryptase is a growth factor for human airway smooth muscle. *J. Allergy Clin. Immunol.* **97**:969, 1996.
130. Vu, T.-K., Hung, D. T., Wheaton, V. I. and Coughlin, S. R. Molecular cloning of a functional thrombin receptor reveals a novel proteolytic mechanism of receptor activation. *Cell* **64**:1057–1068, 1991.
131. Déry, O. and Bunnett, N. W. Proteinase activated receptors: a growing family of heptahelical receptors for thrombin, trypsin and tryptase. *Biochem. Soc. Trans.* **27**:246–254, 1999.
132. Molino, M., Barnathan, E. S., Numerof, R., Clark, J., Dreyer, M., Cumashi, A., Hoxie, J. A., Schechter, N., Woolkalis, M. and Brass, I. F. Interactions of mast cell tryptase with thrombin receptors and PAR-2. *J. Biol. Chem.* **272**:4043–4049, 1997.
133. Mirza, H., Schmidt, V. A., Derian, C. K., Jesty, J. and Bahou, W. F. Mitogenic responses mediated through the proteinase activated receptor-2 are induced by expressed forms of mast cell α- or β-tryptases. *Blood* **90**:3914–3922, 1997.
134. Vergnolle, N., Hollenberg, M. D., Sharkey, K. A. and Wallace, J. L. Characterisation of the inflammatory response to protease-activated receptor-2 (PAR-2)-activating peptides in the rat paw. *Br. J. Pharmacol.* **127**:1083–1090, 1999.
135. Mirza, H., Yatsula, V. and Bahou, W. F. The proteinase activated receptor 2 (PAR-2) mediates mitogenic responses in human vascular endothelial cells. *J. Clin. Invest.* **97**:1705–1714, 1996.
136. Cocks, T. M., Fong, B., Chow, J. M., Anderson, G. P., Frauman, A. G., Goldie, R. G., Henry, P. J., Carr, M. J., Hamilton, J. R. and Moffat, J. D. A protective role for protease activated receptors in the airways. *Nature* **398**:156–160, 1999.
137. Clark, J. M., Abraham, W. M., Fishman, C. E., Forteza, R., Ahmed, A., Cortes, A., Warne, R. L., Moore,

W. R. and Tanaka, R. D. Tryptase inhibitors block allergen induced airway and inflammatory responses in allergic sheep. *Am. J. Respir. Crit. Care Med.* **152**:20760–20783, 1995.

138. Fath, M. A., Wu, X., Hilemam, R. E., Linhardt, R. J., Kashem, M. A., Nelson, R. M., Wright, C. D. and Abraham, W. M. Interaction of secretory leukoprotease inhibitor with heparin inhibits proteases involved in asthma. *J. Biol. Chem.* **273**:13563–13569, 1998.
139. Havill, A. M., Middleton, S., Lyons, D. and Wright, C. Secretory leukocyte protease inhibitor (SLPI) prevents the development of airway hyperresponsiveness (AWHR) in allergen challenged guinea pigs. *Am. J. Respir. Crit. Care Med.* **155**:A654, 1997.
140. Krishna, T. K., Chauham, A. J., Little, L., Sampson, K., Mant, T. G. K., Hawksworth, R., Djukanović, R., Lee, T. H. and Holgate, S. T. Effect of inhaled APC-366 on allergen-induced bronchoconstriction and airway hyperresponsiveness to histamine in atopic asthmatics. *Am. J. Respir. Crit. Care Med.* **157**:A456, 1998.
141. Fiorucci, L., Erba, F. and Ascoli, F. Bovine tryptase: purification and characterisation. *Biol. Chem. Hoppe-Seyler* **373**:483–90, 1992.
142. Pemberton, A. D., McAleese, S. M., Huntley, J. F., Collie, D. D. S., Sandamore, C. L., McEwen, A. R., Walls, A. F. and Miller, H. R. P. cDNA sequence of sheep mast cell tryptase and the differential expression of tryptase and sheep mast cell proteinase-1 in lung, dermis and gastrointestinal tract. *Clin. Exp. Allergy.* (In press.) 2000.
143. Tam, E. K. and Caughey, G. H. Degradation of airway neuropeptides by human lung tryptase. *Am. Rev. Respir. Cell Mol. Biol.* **3**:27–32, 1990.

SECTION SIX

NERVE–MAST CELL INTERACTIONS: PHYSIOLOGY AND PATHOLOGY

CHAPTER 20

Mast Cell–Nerve Interactions: Possible Significance of Nerve Growth Factor

JOHN BIENENSTOCK

Departments of Medicine, Pathology and Molecular Medicine, McMaster University, Hamilton, Canada

My first interest in a possible relationship between mast cells and nerves was inspired by the work of Gil Castro in his seminal paper on what he subsequently termed integrative physiology (1, 2). In essence, he proposed that the various structural elements in the mucosa and submucosa of the gastrointestinal tract must be linked to explain the mass of data that was emerging from the work of physiologists, endocrinologists, anatomists and immunologists as they explored different aspects of normal and altered intestinal homeostasis. We ourselves had begun to explore the functional heterogeneity of mucosal mast cells as opposed to those found in connective tissue, and had begun to show significant differences in selective action of a variety of secretogogues in mast cell mediator release from these two cell types (3, 4). Our interest was further aroused by the observations of Shanahan, who showed that, out of a variety of neurotransmitters tested, including endorphins, the only one that seemed to have secretory activity upon intestinal mucosal mast cells was substance P (SP) (5). It was only later, when we began to explore the literature widely, that we began to suspect that nerve growth factor (NGF) might be playing a role in mast cell physiology, a thought first documented by the experiments of Aloe and Levi-Montalcini (6). These latter investigators showed that injections of NGF into neonatal rats produced significant mast cell hyperplasia.

In this chapter I will attempt to summarize much of the information that defines and characterizes mast cell–nerve interactions, and the increasingly central role that NGF is playing in these observations. While there is now incontrovertible evidence for purposeful interactions between mast cells and nerves (7, 8), and that they form an important unit involved in the regulation of tissue homeostasis, there is also beginning to be evidence that many other cell types may be involved in similar bi-directional communication pathways between nervous and immune systems (9, 10). These pathways are now being widely explored, and form the evidence for neuro-immune interactions on a broad scale throughout the body. Recent confocal scanning microscopic studies in the

MAST CELLS AND BASOPHILS
ISBN 0-12-473335-2

skin of primates, rodents and humans have extended these observations and shown that unmyelinated axons are associated with both dermal mast cells and Langerhans cells (11).

EVIDENCE FOR MAST CELL–NERVE INTERACTIONS

Anatomical Observations

In a careful morphometric study we showed that mast cells and nerves were closely and invariably approximated in rat intestinal villi (12). The nerves associated with mast cells contained SP and calcitonin gene-related peptide (CGRP). We subsequently went on to show that the association was not random and occurred also in the human intestine (13). Others confirmed these studies and extended them to show similar approximations in rat intestine, between eosinophils, B cells and nerves, in addition to those of mast cells (14). Similar observations have now identified such close interactions between mast cells and nerves in a variety of different tissues in many species. These now include skin (11), intestine (12, 15), urinary bladder (16) and several other tissues (17).

Ex Vivo Physiological Studies

There have been a number of experiments in this category which show mast cell–nerve communication. Most of them have described evidence for antigen–mast cell–nerve–epithelial cell interactions in classic Ussing chamber experiments in animals sensitized to make IgE antibodies against specific antigens (7, 18, 19). The first such observations were described in guinea pig intestine using lactoglobulin as antigen (20). Many such experiments have been performed since. A limited number of experiments have been performed with rat trachea and have largely reproduced findings from experiments with intestine (21). Conclusive proof that mast cells are involved in this homeostatic unit and dependent upon specific antigen interaction in sensitized animals was shown by Perdue *et al.* using w/w^V animals and their littermates (22). These animals constitutively lack mast cells, as a result of a genetic deficiency of the receptor for c-kit, and can be reconstituted generally by injection of bone marrow, or by local injection of cultured bone marrow-derived mast cells into particular tissue sites. Using these techniques Perdue *et al.* showed conclusively that 80–90% of antigen reactivity in the Ussing chamber was dependent on mast cells, and could be reconstituted by mast cell repopulation. Importantly, these experiments also showed that electrical field stimulation, which engenders nerve activation without effect upon mast cells themselves, was also able to induce changes in short circuit current generation that were dependent upon the presence of mast cells. These experiments thus conclusively showed that mast cells were crucial, both for the antigen-specific effects that were observed, but that they were also involved in bi-directional communication with nerves to effect these responses. Subsequent experiments in the guinea pig showed that acetylcholine was specifically released in the same time course as short circuit current generation upon exposure to specific antigen (23).

Other evidence for mast cell–nerve interaction comes from experiments on the colonic submucous nervous plexus during exposure to β-lactoglobulin in guinea pigs sensitized to milk early in life. Antigen had complex effects on excitatory post-synaptic potentials, which included differing and opposite effects of histamine on pre-synaptic and post-synaptic activity mediated through different histamine receptors (24). These types of results have been reviewed (18) and parallel in many ways those obtained by Weinreich (25, 26). In these latter experiments antigen induced long-term potentiation of nicotinic synaptic transmission in guinea pig superior cervical ganglia in sensitized animals. In addition, this work showed antigen-induced synaptic plasticity in sympathetic ganglia, which was clearly due, at least in part, to sensitization of mast cells since this occurred in both actively and passively sensitized animals.

These experiments can be summarized as demonstrating bi-directional communication between mast cells and nerves in the intestine. Many mast cell mediators in the form of histamine, serotonin and products of arachidonic acid metabolism are involved, as well as neurotransmitters including acetylcholine, SP, CGRP, platelet-activating factor. Pharmacological blockade experiments using relatively selective mast cell inhibitors such as doxantrazole, which specifically inhibits intestinal mucosal mast cells, and a sodium channel blocker (tetrodotoxin) which specifically inhibits nerve conduction, also demonstrate that communication between mast cells, nerves and epithelium is essential to allow antigen-specific and -dependent chloride ion secretion by epithelium to occur.

It is worth noting that the nature and extent to which cytokines are involved in this communication pathway has yet to receive the attention it deserves, and little definitive work has been done in this area (27).

Nerve (Electrical) Stimulation

If mast cells are in direct communication with nerves, electrical stimulation should produce some evidence of mast cell secretion. However, upon casual reading of the literature, this issue appears to be controversial (28). In balance there is excellent evidence to support the conclusion that electrical stimulation of nerves will, depending upon its nature, duration, timing of impulses, etc., either cause ultrastructural changes in associated mast cell granules (29, 30) or actual degranulation (31). This has been best shown in the dura mater, stomach and urinary bladder.

Many of the experiments that have shown mast cell degranulation upon electrical stimulation have also shown that these effects were inhibited by atropine or prior treatment with capsaicin. This suggests that SP and acetylcholine are involved. While there appears to be clear-cut evidence for the former, the involvement of the latter is less clear, at least in terms of direct degranulation effects upon the mast cell (32, 33). The involvement of cholinergic nerves in stimulation of mast cells appears to be indisputable; what remains unclear, however, is how, and under what circumstances, acetylcholine directly activates mast cells.

Perhaps some of the most striking evidence for nerve–mast cell interaction and the effects of electrical stimulation come from *in vivo* experiments. Miura *et al.* (34) showed, in cats sensitized to ascaris, that inhalation of ascaris antigens caused increased bronchial resistance and an expected increase in plasma histamine levels. However, in animals pre-treated with cholinergic and adrenergic blockers, bilateral electrical stimulation of both vagal nerves in the neck, prior to antigen inhalation, caused complete inhibition both of the generation of bronchial resistance and elevation of plasma histamine. This unique set of experiments showed that a non-adrenergic, non-cholinergic (NANC) system is present in the cat which is able to inhibit mast cell degranulation induced by antigen. The nature of the neurotransmitters involved in this inhibitory system has yet to be characterized. Furthermore, these types of experiments have yet to be performed in other tissues.

In Vitro

Since we had shown in morphological experiments the close association between mast cells and nerves, we co-cultured these using both superior cervical ganglion and dorsal root ganglion neurites and RBL-2H3 as a source of intestinal-like mast cells (35–37). We have also less frequently used rat peritoneal mast cells. In these experiments we showed that there was a selective association between mast cells and nerves. Once associations formed, they were present for up to 120 h in culture without the formation of specific synapse-like structures (38). The contacts between membranes were less than 20 nm and dense-core vesicles were found in the neurites approximating to mast cells. In turn, RBL granules were located in close proximity to the area of membrane contact. Changes in electrical properties of mast cell membranes upon contact were also noted (39), and RBL stopped dividing upon contact and increased their mediator content and number of granules. Mast cells displayed both tropic and trophic activity towards neurites, but the exact nature of either of these signals has not been ascertained, although it is tempting to assume that NGF, synthesized by the mast cells (40) is contributing at least to the trophic effect.

Our most recent experiments looking at co-culture of mast cells and superior cervical ganglion neurites have shown direct communication, using calcium-binding dyes to indicate evidence for activation and signalling (41). In these confocal scanning microscopy experiments, dose-dependent signalling with initial activation of neurites and subsequent calcium increases in the accompanying RBL cells were shown with scorpion venom and bradykinin, neither of which had any direct effect on the mast cells alone. This neurite signalling was dependent upon SP and the expression of an NK1 receptor upon the RBL surface. Such receptors have been shown before on RBL, but never on mast cells themselves (42). That this communication could be bi-directional was demonstrated with an antibody to the IgE receptor, which showed initial RBL activation followed after a brief lag period by neurite activation of the associated nerve fibre.

Since the whole role of SP has been questioned as far as mast cell activation is concerned, mostly because of the lack of any clear-cut expression of neurokine receptors on mast cells other than mast cell lines or RBL, we conducted a series of patch clamping experiments on the effects of SP on RBL and mast cells (43). At micromolar concentrations of SP, electrical effects and degranulation occurred. At picomolar concentrations, no effect on the mast cells was observed. After a rest period of 25 min or so and then re-stimulation of the cell with subthreshold picomolar concentrations of SP, they responded after a long lag period. Significant numbers of mast cells even degranulated under these conditions. We have also shown that this priming effect occurs with picomolar concentrations of SP followed by subthreshold amounts of anti-IgE, which by itself had no degranulating effect. Thus SP can be added to the list of substances known to prime mast and other cells for subsequent degranulation by known agonists, with synergistic effect. This has been shown before for basophils by Bischoff and Dahinden with a large number of growth factors, including NGF, and agonists, including C5a (44).

In Vivo

A number of experiments have demonstrated mast cell–nerve interactions *in vivo*. Extensive work on interstitial cystitis by Theoharides has demonstrated many of the aspects of nerve mast cell associations outlined previously. These include morphological associations, and elevated mast cell tryptase in patients with the disease, as well as various pharmacological effects of carbachol, oestrogen and capsaicin in experimental models (45). The human condition is characterized by chronic inflammation and fibrosis of the urinary bladder.

In experiments designed to investigate the mechanisms of the intestinal effects of *Clostridium difficile* toxin A, an inflammatory enterotoxin, and *Cholera* toxin, a non-inflammatory enterotoxin, evidence was obtained that the former involved mast cells and capsaicin-sensitive sensory afferent neurons, whereas the latter did not (46).

These investigators have gone on to look at the effect of stress on intestinal tissue and have made some striking observations (47–49). In short, stress will cause activation and secretion by goblet cells of mucin with discharge into the gastrointestinal lumen, increased secretion of fluid by epithelial cells, activation of mast cells and also enteric nerves, especially those synthesizing acetylcholine and SP. In addition, inhibitors of neurotensin as well as corticotropin-releasing factor (CRF) were able to inhibit this whole process. An inhibitor of CRF prevented the intestinal effects of stress. Intracerebral administration of CRF had the same effect as peripheral injection, suggesting an important central role for this molecule in the process. CRF appears to have direct mast cell degranulating capabilities (29). Thus, again, mast cells play a central role involved in a stress-induced set of events culminating in intestinal physiological alterations.

This effect of stress, namely the activation and degranulation of mast cells, has not just been observed in the intestine, but has also been shown in the urinary bladder and dura mater in the brain (29, 30). All of these effects are mimicked by CRF and blocked by CRF inhibition.

As yet unpublished, results of Santos and Perdue have found that three different types of stress have similar effects on intestinal permeability to small and large molecules, which are all inhibitable by the α-helical CRF inhibitor and reproduced *in vitro* by CRF itself. These effects are extremely complex, involve a variety of different nerves and ganglionic neurons but confirm and extend the findings described above.

The involvement of the brain in central control of mast cell activation is raised by the central and peripheral effects of CRF. Indeed, previous work has implicated the brain in psychological or Pavlovian conditioning of mast cell degranulation (50). In these experiments, rats were conditioned by exposure to flashing lights and noise and coupled with injections of antigen which caused mild anaphylactic reactions. Subsequent exposure of animals to the conditioning stimuli alone was sufficient to promote intestinal mast cell degranulation which was characterized by rat mast cell protease II release, as well as cyclical nervous discharges, characteristic of intestinal allergic responses, which were blocked by tetrodotoxin. Whether these latter effects were mediated by CRF, as appears likely now, through nerve transmission, or both, has not yet been explored.

What is clear, however, is that mast cells are regulatable by both peripheral and central nervous systems, through local and even emotional processes. Therefore these processes must be taken into account in attempting to explain physiological and disease mechanisms.

POSSIBLE ROLE OF NGF

From the foregoing descriptions of mast cell–nerve interactions it is now possible to begin to piece together a plausible hypothesis that involves NGF.

NGF has pleiotropic activities (Table I). In addition to its nerve growth promoting activity, it may have active mastopoietic effects in rats and mice, in part through its

TABLE I
Non-neuronal Effects of Nerve Growth Factor

- Promotes wound healing, prevents carrageenin-induced inflammation (58)
- Promotes growth, differentiation, proliferation and survival of B cells (57, 70)
- Enhances IgG4 production (69)
- Prevents apoptosis in mast cells, eosinophils and neutrophils (54–56)
- Strong mastopoietic effect in rodents, partly through mast cell degranulation (6, 51)
- Primes basophils for agonist action (e.g. C5a) (44)
- Enhances human haematopoietic colony growth of basophils and eosinophils; synergistic with GM-CSF and IL-5 (52, 53)

degranulating effect on mast cells, and in part through probable T cell activation (51). It acts as a basophil and eosinophil colony-stimulating factor in the human in concert with GM-CSF and IL-5 (52, 53). It seems to have an anti-apoptotic effect on mast cells (54), neutrophils (55), eosinophils (56) and B cells (57), primes basophils for subsequent secretagogue action (44) and promotes wound healing (58). It is released into the circulation in both experimental animals (59) and man (60) upon stress of a variety of sorts, and causes long-lasting hyperalgesia after as little as a single injection in rodents (Fig. 1) (61, 62). Transgenic animals which overexpress NGF in the lung, show bronchial smooth muscle hyper-reactivity (63).

NGF is synthesized by a variety of cells, which include keratinocytes (64), mast cells (40), eosinophils (65), T cells (especially Th2) (66), B cells (57) and macrophages, as well as Langerhans cells (67) and fibroblasts (68). NGF appears to have a selective action on the synthesis of IgG4 by human B cells (69), acts as an endocrine survival factor for B cells (67) and promotes B cell growth and differentiation (70).

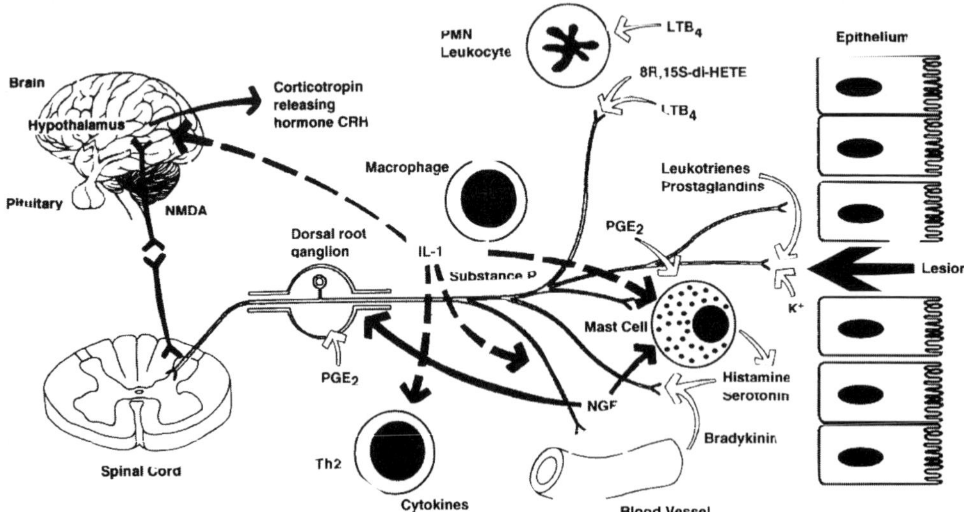

Fig. 1 This figure attempts to represent schematically the multiple and complex pathways involved in visceral hyperalgesia. It can be seen that NGF is involved or acts upon many different cells and systems within these pathways. The data used to develop this figure were drawn from a significant number of different publications, but the reader is referred to *Bueno et al.* (80) for one of the best reviews of the subject.

While NGF appears to be increased in the circulation in a variety of inflammatory and autoimmune conditions, it most consistently appears to be elevated in the circulation of patients with multiple allergic diseases, including asthma (71). We have recently been involved in a study of patients with allergic rhinitis (72), which has shown that NGF is present in nasal lavage of such patients following specific antigen stimulation within 10 min of challenge. Polymerase chain reaction (PCR) of nasal scrapings has shown the presence of mRNA for NGF, and cultured human airway epithelial cells may synthesize and release NGF (73). Renz and colleagues have shown that neutralizing antibody to NGF, instilled into the nasal cavity of mice that had been pre-sensitized to specific antigen, prevented the responsiveness of lung tissue to antigen, and were able to show release of NGF after antigen provocation in segmental lavage of human asthmatic lungs (74).

At first sight, the role of NGF in inflammation appears to be opposed to the notion that it may have beneficial effects since it causes the activation of many processes including the degranulation of mast cells and eosinophils. On the other hand there are the observations of the beneficial effects of NGF in the carrageenin model of inflammation (61), corneal ulcer healing (75) and wound healing in diabetic mice (52).

We have recently made an observation that may help to begin to resolve these discrepancies (76). While NGF promotes the synthesis and release by rat peritoneal mast cells of IL-6, at the same time it suppresses the synthesis of TNF-α through the autocoid release of PGE_2. Thus it may inhibit the inflammatory process by its discriminatory effect on cytokine synthesis.

The effects of stress in many systems are complex multifactorial processes which are still not understood. For example, it is becoming clearer that acute and chronic stress may have widely differing, and even opposite, effects on the expression of delayed hypersensitivity reactions (77). Thus, while CRF appears to have the properties of mediating potentially noxious effects in the intestine through mast cell and nervous involvement, this may not be the end of the story. Indeed, intracerebral CRF injections can actually inhibit the onset of one of the models of experimental colitis, induced by hapten, invoking a protective effect for the molecule (78).

In this regard, it is interesting to reflect upon the role of NGF which has been shown to be elevated in the circulation of fighting mice (59) and human parachutists (60). Since sialadenectomy prevented the rise of NGF in serum consequent to wounding of the skin, in the experiments referred to above (58), it seems likely that the rise of NGF in the circulation is at least in part due to release from the submandibular glands into the circulation. NGF, similar to CRF, may have equal and opposite effects *in vivo* so it seems reasonable to begin to explore under what circumstances NGF may be protective and even beneficial to the host.

Thus there is emerging a complex picture of a set of interactions between cells of the immune system and their interaction with the nervous system. NGF may play a central role in this communication which we are only just beginning to understand. The selective stimuli which may discriminate between the synthesis of NGF and other molecules such as cytokines and classic mediators of inflammation are largely unknown. However, nerve growth is a constant accompaniment to any inflammatory event (79). Mast cells, which make NGF, as well as eosinophils which have recently been shown to be capable of such synthesis (65), are involved in a number of inflammatory responses, including those of an allergic nature. In addition, these cell types appear to be involved in the normal healing process in which repair occurs following inflammation. As such they may be involved in both repair processes, as well as the promotion of disease upon continuing stimulation, such as may occur through nerve stimulation or as a result of stress. The fact that stress can activate physiological processes and disturb normal homeostasis in the intestine, brain and the urinary bladder suggests more mechanisms whereby neuroendocrine–immune interactions may be occurring. As we explore these systems further, the full complexity of these interactions becomes more evident, but it becomes harder to ignore them in attempting to understand the mechanisms which promote health, and may offer new therapeutic approaches to the healing or prevention of disease.

ACKNOWLEDGEMENT

Support by the Medical Research Council of Canada for the work conducted by the author in this study is gratefully acknowledged.

REFERENCES

1. Castro, G. A. Immunological regulation of epithelial function. *Am. J. Physiol.* **243**:G321–G329, 1982.
2. Castro, G. A. and Arntzen, C. J. Immunophysiology of the gut: a research frontier for integrative studies of the common mucosal immune system. *Am. J. Physiol.* **265**:G599–G610, 1993.
3. Befus, A. D., Pearce, F. L., Gauldie, J., Horsewood, P. and Bienenstock, J. Mucosal mast cells. I. Isolation and functional characteristics of rat intestinal mast cells. *J. Immunol.* **128**:2475–2480, 1982.
4. Pearce, F. L., Befus, A. D., Gauldie, J. and Bienenstock, J. Mucosal mast cells. II. Effects of anti-allergic compounds on histamine secretion by isolated intestinal mast cells. *J. Immunol.* **128**:2481–2486, 1982.
5. Shanahan, F., Denburg, J. A., Fox, J., Bienenstock, J. and Befus, D. Mast cell heterogeneity: effects of neuroenteric peptides on histamine release. *J. Immunol.* **135**:1331–1337, 1985.
6. Aloe, L. and Levi-Montalcini, R. Mast cells increase in tissues of neonatal rats injected with the nerve growth factor. *Brain Res.* **133**:358–366, 1977.
7. McKay, D. M. and Bienenstock, J. The interaction between mast cells and nerves in the gastrointestinal tract. *Immunol. Today* **15**, 533–538, 1994.
8. Perdue, M. H. and McKay, D. M. Integrative immunophysiology in the intestinal mucosa. *Am. J. Physiol.* **267**:G151–G165, 1994.
9. Smith, E. M., Harbour-McMenamin, D. and Blalock, J. E. Lymphocyte production of endorphins and endorphin-mediated immunoregulatory activity. *J. Immunol.* **135**:779s–782s, 1985.
10. Pascual, D.W., Stanisz, A. M., Bienenstock, J. and Bost, K. L. Neural intervention in mucosal immunity. In *Mucosal Immunology*, 2nd edn (Ogra, P., Mestecky, J., Lamm, M., Strober, W., Bienenstock, J. and McGhee, J., eds), pp. 631–642. Academic Press, San Diego, 1999.
11. Egan, C. L., Viglione-Schneck, M. J., Walsh, L. J., Green, B., Trojanowski, J. Q., Whitaker-Menezes, D. and Murphy, G. F. Characterization of unmyelinated axons uniting epidermal and dermal immune cells in primate and murine skin. *J. Cutan. Pathol.* **25**:20–29, 1998.
12. Stead, R. H., Tomioka, M., Quinonez, G., Simon, G. T., Felten, S. Y. and Bienenstock, J. Intestinal mucosal mast cells in normal and nematode-infected rat intestines are in intimate contact with peptidergic nerves. *Proc. Natl. Acad. Sci. USA* **84**:2975–2979, 1987.
13. Stead, R. H., Dixon, M. F., Bramwell, N. H., Riddell, R. H. and Bienenstock, J. Mast cells are closely apposed to nerves in the human gastrointestinal mucosa. *Gastroenterology* **97**:575–585, 1989.
14. Arizono, N., Matsuda, H., Hattori, T., Kojima, Y., Maeda, T. and Galli, S. J. Anatomical variation in mast cell nerve associations in the rat small intestine, heart, lung, and skin. Similarities of distances between neural processes and mast cells, eosinophils, or plasma cells in jejunal lamina propria. *Lab. Invest.* **62**:626–634, 1990.
15. Yonei, Y., Oda, M. and Nakamura, M. Evidence for direct interaction between the cholinergic nerve and mast cells in rat colonic mucosa. An electron microscope cytochemical and autoradiographic study. *J. Clin. Electron Microsc.* **18**:560–561, 1985.
16. Letourneau, R., Pang, X., Sant, G. R. and Theoharides, T. C. Intragranular activation of bladder mast

cells and their association with nerve processes in interstitial cystitis. *Br. J. Urol.* **77**:41–54, 1996.

17. Heine, H. and Forster, F. J. Histophysiology of mast cells in skin and other organs. *Arch. Dermatol. Res.* **253**:225–228, 1975.
18. Cooke, H. J. Neuroimmune signaling in regulation of intestinal ion transport. *Am. J. Physiol.* **266**:G167–G178, 1994.
19. Bienenstock, J. Cellular communication networks. Implications for our understanding of gastrointestinal physiology. *Ann. N.Y. Acad. Sci.* **664**:1–9, 1992.
20. Baird, A. W. and Cuthbert, A. W. Neuronal involvement in type 1 hypersensitivity reactions in gut epithelia. *Br. J. Pharmacol.* **92**:647–655, 1987.
21. Sestini, P., Bienenstock, J., Crowe, S. E., Marshall, J. S., Stead, R. H., Kakuta, Y. and Perdue, M. H. Ion transport in rat tracheal epithelium *in vitro*. Role of capsaicin-sensitive nerves in allergic reactions. *Am. Rev. Respir. Dis.* **141**:393–397, 1990.
22. Perdue, M. H., Masson, S., Wershil, B. K. and Galli, S. J. Role of mast cells in ion transport abnormalities associated with intestinal anaphylaxis. Correction of the diminished secretory response in genetically mast cell-deficient W/Wv mice by bone marrow transplantation. *J. Clin. Invest.* **87**:687–693, 1991.
23. Javed, N. H., Wang, Y. Z. and Cooke, H. J. Neuroimmune interactions: role for cholinergic neurons in intestinal anaphylaxis. *Am. J. Physiol.* **263**:G847–G852, 1992.
24. Frieling, T., Cooke, H. J. and Wood, J. D. Neuroimmune communication in the submucous plexus of guinea pig colon after sensitization to milk antigen. *Am. J. Physiol.* **267**:G1087–G1093, 1994.
25. Weinreich, D. Cellular mechanisms of inflammatory mediators acting on vagal sensory nerve excitability. *Pulmon. Pharmacol.* **8**:173–179, 1995.
26. Weinreich, D., Undem, B. J., Taylor, G. and Barry, M. F. Antigen-induced long-term potentiation of nicotinic synaptic transmission in the superior cervical ganglion of the guinea pig. *J. Neurophysiol.* **73**:2004–2016, 1995.
27. Miller Jonakait, G. and Hart, R. P. Immune cytokine regulation of sympathetic ganglion response to injury. *Neuroimmunomodulation* **2**:240, 1995.
28. Baraniuk, J. N., Kowalski, M. L. and Kaliner, M. A. Relationships between permeable vessels, nerves, and mast cells in rat cutaneous neurogenic inflammation. *J. Appl. Physiol.* **68**:2305–2311, 1990.
29. Theoharides, T. C., Singh, L. K., Boucher, W., Pang, W., Letourneau, R., Webster, E. and Chrousos, G. Corticotropin-releasing hormone induces skin mast cell degranulation and increased vascular permeability, a possible explanation for its proinflammatory effects. *Endocrinology* **139**:403–413, 1998.
30. Dimitriadou, V., Buzzi, M. G., Moskowitz, M. A. and Theoharides, T. C. Trigeminal sensory fiber stimulation induces morphological changes reflecting secretion in rat dura mater mast cells. *Neuroscience* **44**:97–112, 1991.
31. Gottwald, T. P., Hewlett, B. R., Lhotak, S. and Stead, R. H. Electrical stimulation of the vagus nerve modulates the histamine content of mast cells in the rat jejunal mucosa. *Neuroreport* **7**:313–317, 1995.
32. Masini, E., Fantozzi, R., Blandina, P., Brunelleschi, S. and Mannaioni, P. F. The riddle of cholinergic histamine release from mast cells. *Prog. Med. Chem.* **22**: 267–291, 1985.
33. Masini, E., Fantozzi, R., Conti, A., Blandina, P., Brunelleschi, S. and Mannaioni, P. F. Immunological modulation of cholinergic histamine release in isolated rat mast cells. *Agents Actions* **16**:152–154, 1985.
34. Miura, M., Inoue, H., Ichinose, M., Kimura, K., Katsumata, U. and Takishima, T. Effect of nonadrenergic noncholinergic inhibitory nerve stimulation on the allergic reaction in cat airways. *Am. Rev. Respir. Dis.* **141**:29–32, 1990.
35. Blennerhassett, M. G. and Bienenstock, J. Sympathetic nerve contact causes maturation of mast cells *in vitro*. *J. Neurobiol.* **35**:173–182, 1998.
36. Blennerhassett, M. G., Janiszewski, J. and Bienenstock, J. Sympathetic nerve contact alters membrane resistance of cells of the RBL-2H3 mucosal mast cell line. *Am. J. Respir. Cell Mol. Biol.* **6**:504–509, 1992.
37. Blennerhassett, M. G. and Bienenstock, J. Apparent innervation of rat basophilic leukaemia (RBL-2H3) cells by sympathetic neurons *in vitro*. *Neurosci. Lett.* **120**: 50–54, 1990.
38. Blennerhassett, M. G., Tomioka, M. and Bienenstock, J. Formation of contacts between mast cells and sympathetic neurons *in vitro*. *Cell Tissue Res.* **265**:121–128, 1991.
39. Janiszewski, J., Bienenstock, J. and Blennerhassett, M. G. Substance P induces whole cell current transients in RBL-2H3 cells. *Am. J. Physiol.* **263**:C736–C742, 1992.
40. Leon, A., Buriani, A., Dal, T. R., Fabris, M., Romanello, S., Aloe, L. and Levi-Montalcini, R. Mast cells synthesize, store, and release nerve growth factor. *Proc. Natl. Acad. Sci. USA* **91**:3739–3743, 1994.
41. Suzuki, R., Furuno, T., McKay, D. M., Wolvers, D., Teshima, R., Nakanishi, M. and Bienenstock, J.

Direct neurite-mast cell (RBL) communication *in vitro* occurs via the neuropeptide substance P. *J. Immunol.* (in press) 1999.

42. Cooke, H. J., Fox, P., Alferes, L., Fox, C. C. and Wolfe, S. A. J. Presence of NK1 receptors on a mucosal-like mast cell line, RBL-2H3 cells. *Can. J. Physiol. Pharmacol.* **76**:188–193, 1998.
43. Janiszewski, J., Bienenstock, J. and Blennerhassett, M. G. Picomolar doses of substance P trigger electrical responses in mast cells without degranulation. *Am. J. Physiol.* **267**:C138–C145, 1994.
44. Bischoff, S. C. and Dahinden, C. A. Effect of nerve growth factor on the release of inflammatory mediators by mature human basophils. *Blood* **79**:2662–2669, 1992.
45. Theoharides, T. C., Pang, X., Letourneau, R. and Sant, G. R. Interstitial cystitis: a neuro-immunoendocrine disorder. *Ann. N.Y. Acad. Sci.* **840**:619–634, 1998.
46. Castagliuolo, I., LaMont, J. T., Letourneau, R., Kelly, C., O'Keane, J. C., Jaffer, A., Theoharides, T. C. and Pothoulakis, C. Neuronal involvement in the intestinal effects of *Clostridium difficile* toxin A and *Vibrio cholerae* enterotoxin in rat ileum. *Gastroenterology* **107**:657–665, 1994.
47. Castagliuolo, I., Wershil, B. K., Karalis, K., Pasha, A., Nikulasson, S. T. and Pothoulakis, C. Colonic mucin release in response to immobilization stress is mast cell dependent. *Am. J. Physiol.* **274**:G1094–G1100, 1998.
48. Pothoulakis, C., Castagliuolo, I. and Leeman, S. E. Neuroimmune mechanisms of intestinal responses to stress. Role of corticotropin-releasing factor and neurotensin. *Ann. N.Y. Acad. Sci.* **840**:635–648, 1998.
49. Castagliuolo, I., Leeman, S. E., Bartolak-Suki, E., Nikulasson, S., Qiu, B., Carraway, R. E. and Pothoulakis, C. A neurotensin antagonist, SR 48692, inhibits colonic responses to immobilization stress in rats. *Proc. Natl. Acad. Sci. USA* **93**:12611–12615, 1996.
50. MacQueen, G., Marshall, J., Perdue, M., Siegel, S. and Bienenstock, J. Pavlovian conditioning of rat mucosal mast cells to secrete rat mast cell protease II. *Science* **243**:83–85, 1989.
51. Marshall, J. S., Stead, R. H., McSharry, C., Nielsen, L. and Bienenstock, J. The role of mast cell degranulation products in mast cell hyperplasia. I. Mechanism of action of nerve growth factor. *J. Immunol.* **144**:1886–1892, 1990.
52. Matsuda, H., Coughlin, M. D., Bienenstock, J. and Denburg, J. A. Nerve growth factor promotes human hemopoietic colony growth and differentiation. *Proc. Natl. Acad. Sci. USA* **85**:6508–6512, 1988.
53. Tsuda, T., Wong, D., Dolovich, J., Bienenstock, J., Marshall, J. and Denburg, J. A. Synergistic effects of nerve growth factor and granulocyte–macrophage colony-stimulating factor on human basophilic cell differentiation. *Blood* **77**:971–979, 1991.
54. Kawamoto, K., Okada, T., Kannan, Y., Ushio, H., Matsumoto, M. and Matsuda, H. Nerve growth factor prevents apoptosis of rat peritoneal mast cells through the trk proto-oncogene receptor. *Blood* **86**:4638–4644, 1995.
55. Kannan, Y., Usami, K., Okada, M., Shimizu, S. and Matsuda, H. Nerve growth factor suppresses apoptosis of murine neutrophils. *Biochem. Biophys. Res. Commun.* **186**: 1050–1056, 1992.
56. Hamada, A., Watanabe, N., Ohtomo, H. and Matsuda, H. Nerve growth factor enhances survival and cytotoxic activity of human eosinophils. *Br. J. Haematol.* **93**:299–302, 1996.
57. Torcia, M., Bracci-Laudiero, L., Lucibello, M., Nencioni, L., Labardi, D., Rubartelli, A., Cozzolino, F., Aloe, L. and Garaci, E. Nerve growth factor is an autocrine survival factor for memory B lymphocytes. *Cell* **85**:345–356, 1996.
58. Matsuda, H., Koyama, H., Sato, H., Sawada, J., Itakura, A., Tanaka, A., Matsumoto, M., Konno, K., Ushio, H. and Matsuda, K. Role of nerve growth factor in cutaneous wound healing: accelerating effects in normal and healing-impaired diabetic mice. *J. Exp. Med.* **187**:297–306, 1998.
59. Aloe, L., Alleva, E., Bohm, A. and Levi-Montalcini, R. Aggressive behavior induces release of nerve growth factor from mouse salivary gland into the bloodstream. *Proc. Natl. Acad. Sci. USA.* **83**:6184–6187, 1986.
60. Aloe, L., Bracci-Laudiero, L., Alleva, E., Lambiase, A., Micera, A. and Tirassa, P. Emotional stress induced by parachute jumping enhances blood nerve growth factor levels and the distribution of nerve growth factor receptors in lymphocytes. *Proc. Natl. Acad. Sci. USA.* **91**:10440–10444, 1994.
61. Banks, B. E., Vernon, C. A. and Warner, J. A. Nerve growth factor has anti-inflammatory activity in the rat hindpaw oedema test. *Neurosci. Lett.* **47**:41–45, 1984.
62. Woolf, C. J., Allchorne, A., Safieh-Garabedian, B. and Poole, S. Cytokines, nerve growth factor and inflammatory hyperalgesia: the contribution of tumour necrosis factor alpha. *Br. J. Pharmacol.* **121**:417–424, 1997.
63. Hoyle, G. W., Graham, R. M., Finkelstein, J. B., Nguyen, K. P., Gozal, D. and Friedman, M. Hyperinnervation of the airways in transgenic mice overexpressing nerve growth factor. *Am. J. Respir.*

Cell Mol. Biol. **18**:149–157, 1998.

64. Tron, V. A., Coughlin, M. D., Jang, D. E., Stanisz, J. and Sauder, D. N. Expression and modulation of nerve growth factor in murine keratinocytes (PAM 212). *J. Clin. Invest.* **85**:1085–1089, 1990.
65. Solomon, A., Aloe, L., Pe'er, J., Frucht-Pery, J., Bonini, S. and Levi-Schaffer, F. Nerve growth factor is preformed in and activates human peripheral blood eosinophils. *J. Allergy Clin. Immunol.* **102**:454–460, 1998.
66. Lambiase, A., Bracci-Laudiero, L., Bonini, S., Starace, G., D'Elios, M. M., De, C. M. and Aloe, L. Human CD4+ T cell clones produce and release nerve growth factor and express high-affinity nerve growth factor receptors. *J. Allergy Clin. Immunol.* **100**:408–414, 1997.
67. Torii, H., Yan, Z., Hosoi, J. and Granstein, R. D. Expression of neurotrophic factors and neuropeptide receptors by Langerhans cells and the Langerhans cell-like cell line XS52: further support for a functional relationship between Langerhans cells and epidermal nerves. *J. Invest. Dermatol.* **109**:586–591, 1997.
68. Young, M., Ofer, J., Blanchard, M. H., Asdourian, H., Amos, H. and Arnason, B. G. W. Secretion of a nerve growth factor by primary chick fibroblast cultures. *Science* **187**:361–362, 1974.
69. Kimata, H., Yoshida, A., Ishioka, C., Kusunoki, T., Hosoi, S. and Mikawa, H. Nerve growth factor specifically induces human IgG4 production. *Eur. J. Immunol.* **21**:137–141, 1991.
70. Otten, U., Ehrhard, P. and Peck, R. Nerve growth factor induces growth and differentiation of human B lymphocytes. *Proc. Natl. Acad. Sci. USA* **86**:10059–10063, 1989.
71. Bonini, S., Lambiase, A., Angelucci, F., Magrini, L., Manni, L. and Aloe, L. Circulating nerve growth factor levels are increased in humans with allergic diseases and asthma. *Proc. Natl. Acad. Sci. USA.* **93**:10955–10960, 1996.
72. Sanico, A. M., Stanisz, A. M., Gleeson, T. D., Bora, S., Proud, D., Bienenstock, J, Koliatsos, V. E., and Togias, A. Nerve growth factor expression and release is dysregulated in allergic inflammatory disease of the upper airways. (unpublished).
73. Fox, A., Barnes, P. and Belvisi, M. Release of nerve growth factor from human airway epithelial cells. *Am. J. Respir. Crit. Care Med.* **155**:A157, 1997.
74. Virchow, J. C., Julius, P., Lommatzsch, M., Luttmann, W., Renz, H. and Braun, A. Neurotrophins are increased in bronchoalveolar lavage fluid after segmental allergen provocation. *Am. J. Respir. Crit. Care Med.* **158**:2002–2005, 1998.
75. Lambiase, A., Rama, P., Bonini, S., Caprioglio, G. and Aloe, L. Topical treatment with nerve growth factor for corneal neurotrophic ulcers. *N. Engl. J. Med.* **338**:1174–1180, 1998.
76. Marshall, J. S., Gomi, K., Blennerhassett, M. G. and Bienenstock, J. Nerve growth factor modifies the expression of inflammatory cytokines by mast cells via a prostanoid-dependent mechanism. *J. Immunol.* **162**:4271–4276, 1999.
77. Dhabhar, F. S. and McEwen, B. S. Enhancing versus suppressive effects of stress hormones on skin immune function. *Proc. Natl. Acad. Sci. USA* **96**:1059–1064, 1999.
78. Million, M., Tache, Y. and Anton, P. Susceptibility of Lewis and Fischer rats to stress-induced worsening of TNB-colitis: protective role of brain CRF. *Am. J. Physiol.* **276**:G1027–G1036, 1999.
79. Stead, R. H., Kosecka-Janiszewska, U., Oestreicher, A. B., Dixon, M. F. and Bienenstock, J. Remodeling of B-50 (GAP-43)- and NSE-immunoreactive mucosal nerves in the intestines of rats infected with *Nippostrongylus brasiliensis. J. Neurosci.* **11**:3809–3821, 1991.
80. Bueno, L., Fioramonti, J., Delvaux, M. and Frexinos, J. Mediators and pharmacology of visceral sensitivity: from basic to clinical investigations. *Gastroenterology* **112**:1714–1743, 1997.

CHAPTER 21

Nerve Growth Factor, Mast Cells and Allergic Inflammation

LUIGI ALOE,[1] ALESSANDRA MICERA[1] and SERGIO BONINI[2]*

[1]Institute of Neurobiology, CNR, Rome, and
[2]Department of Allergy and Clinical Immunology, Second University of Naples, Italy

Nerve growth factor (NGF), originally discovered for its neurotrophic activity on peripheral sensory and sympathetic neurons, is now known as a molecule exerting a variety of multidirectional effects on non-neuronal cells. The starting point of this NGF property began in 1977 when it was shown that injection of NGF in neonatal rats increased the number of mast cells in various tissues. Subsequent studies carried out in different laboratories confirmed and extended these findings by showing that NGF enhances human basophilic/mast cell colony, induces phenotypic changes and that these cells produce NGF. More recently it has been shown that NGF promotes growth survival and functional properties of T and B cell-mediated immune responses. Evidence of a possible functional role of NGF within the immune system is suggested by results obtained in animal models and in humans showing that NGF is present in the bloodstream and that its levels undergo significant changes in inflammatory and autoimmune diseases and in allergic status. In this brief account, evidence supporting the hypothesis of a NGF role is presented and discussed.

NERVE GROWTH FACTOR: EARLY DISCOVERY AND EMERGING DATA

NGF is a polypeptide, which plays an important role on differentiation, growth and survival of central and peripheral nervous system neurons. It was discovered in the 1950s (1, 2) and represents the first isolated and best characterized member of a growing family of neurotrophins, which includes brain-derived neurotrophic factor (BDNF) and neurotrophins 3/5 (3). The term NGF was introduced more than 40 years ago to describe an effect on the developing sensory and sympathetic neurons of birds and mammals (1). NGF is a target-derived factor responsible for the survival and phenotypic maintenance

* Corresponding author.

MAST CELLS AND BASOPHILS
ISBN 0-12-473335-2

of specific sets of peripheral and central neurons during development and maturation, which continue to depend on NGF for survival into adulthood. The biological activities of NGF are mediated by the binding to two receptors: trkA, a tyrosine kinase receptor and p75, a low-affinity receptor, which can also bind the other neurotrophins (3). Studies published in recent years demonstrated that the NGF acts not only on cells of the peripheral nervous system but also on cells of the central nervous system (CNS) and on cells of the endocrine and immune systems (1, 4). Physiologically relevant quantities of NGF are produced and released by a variety of non-neuronal and neuronal cells in birds and mammals. However, the largest amount of this factor is produced in the salivary gland (SG) of adult male mice, which remains so far the best available source of NGF. Structurally, the NGF isolated and purified from male mice SG (2.5S) is a dimer of two identical subunits linked together by non-covalent bounds and with a molecular mass of about 26 kDa (5). The amino acid sequence and primary structure of this neurotrophin have been characterized (6), and indicate that NGF is a highly conserved molecule that shares a great homology within different species (7). Molecular studies have revealed that the gene for NGF is located on the proximal short arm of chromosome 1 (8, 9). The functional significance of the large amounts of NGF observed in mouse SG is not fully understood, though recent studies reported by our and other laboratories have demonstrated that the NGF produced in the SG is released into the bloodstream following intraspecific aggressive behaviour and acts on nerve cells and chromaffin cells, as well as on immune cells such as mast cells (MCs) and lymphocytes (10–15). Cumulatively, the latter observation led to the hypothesis that salivary NGF may play a functional role immune cells.

NGF AND THE NERVOUS SYSTEM

There is considerable evidence that the constitutive levels of neurotrophins in general and NGF in particular upregulate after neurological insults. The available data support the hypothesis that the increase of NGF synthesis after neurotrauma contributes to restoration mechanisms such as neurite elongation, synaptogenesis and neural reorganization (1–3). Within the nervous system, NGF functions as a retrograde trophic messenger between target tissues and their innervating cells. NGF and NGFmRNA have been detected in peripheral tissues of various mammalian species, whereas the NGF receptor (NGF-R) is localized on the corresponding sensory and sympathetic ganglia. Structural, biochemical and molecular analysis has shown the presence of two different types of NGF-R: p75 acting as a low-affinity NGF-R ($k_D = 10^{-9}$ M), and p140trk, a high-affinity NGF-R ($k_D = 10^{-11}$ M) (3, 16–18). NGF exerts its neurotrophic activity in response to ligand–receptor binding, and by internalization and retrograde transport as NGF/NGF-R complex to the neuronal cell body. The functional role of NGF on nerve cells is supported by numerous observations indicating that daily injection of purified NGF into neonatal rodents results in an increase in volume of sympathetic ganglia and hypertrophy of sensory ganglia. Exogenous administration of NGF enhances the synthesis and accumulation of noradrenergic and cholinergic neurotransmitters in NGF target cells (19–21), while removal of circulating NGF through administration of NGF antibodies results in death of sympathetic neurons, leading to what has been known for years as immunosympathectomy (22). In the last 10 years several studies carried out both *in vitro* and *in vivo* also revealed that NGF is produced in the CNS, that cholinergic basal forebrain neurons (CBFN) bear NGF-R, and that these neurons are highly receptive to the action of NGF (2, 23). NGF and its receptors are transported from cortex and hippocampus to CBFN where it exerts a trophic action. A functional role of NGF on cells of the CNS is supported by observations that exogenous administration of NGF in rodents increases cholinergic markers in the CNS while preventing damage induced in these neurons by surgical or chemical insults (24–26).

NGF AND THE ENDOCRINE SYSTEM

In addition to its well-known roles in the survival, development and differentiation of nerve cells, NGF is also known to take part in the regulation of specific neuroendocrine functions (27–30). It was shown that both NGF and NGFmRNA increase in the hypothalamus following stressful events (11, 31), whereas the injection of NGF antibodies into rat fetuses produces a marked neuroendocrine deficit in post-natal life, including reduction of peripheral sympathetic and sensory ganglia (30, 32). Deleterious effects of NGF deprivation during fetal life have also been reported in rats and guinea pigs (30, 32), causing loss of body weight, sensory deficits and high mortality, probably associated with neuroendocrine and immune alterations. It has also been shown that exogenous NGF administration acts on the hypothalamic–pituitary–adrenal axis by influencing the release of hypothalamic hormone (27). Moreover, intracerebral injection of purified NGF can modulate peripheral immune responses (33), thus implying that NGF participate in the complex network of neuroimmunoendrocrine interaction.

NGF AND THE IMMUNE SYSTEM

A variety of experimental studies published in the last few years have demonstrated that *in vivo* administration of NGF into neonatal rats causes a widespread increase in the size and number of MCs in several peripheral tissues (14), and that exposure to NGF antibodies results in a cytological alteration and reduction of MC numbers (12). The hypothesis of a possible functional significance of NGF on growth and differentiation of MCs is supported by evidence that these cells, similar to nerve cells, bear NGF-R and that NGF induces degranulation and histamine release from MCs (34). These findings, along with the observation that cytokines are known to affect NGF synthesis, while MCs produce a variety of biological mediators, including NGF, suggest the existence of an autocrine/paracrine feedback loop regulating MC–NGF interaction. Besides MCs, other cells of the immune system lineage express cell-surface receptors for NGF and are responsive to the action of NGF (15). It is known that NGF is implicated in the modulation of lymphocyte activities such as proliferation and/or differentiation (35–37). Moreover, treatment of young rats with NGF prior to and after immunization with sheep erythrocytes results in enhancement of T lymphocyte-dependent antibody synthesis (38). As illustrated in Fig. 1, rat MCs express trkA NGF-R, suggesting that these cells are highly receptive to NGF. Studies carried out in different laboratories have shown that NGF enhances lymphocyte proliferation in both B and T cell populations and stimulates production of IgM, IgA and IgG (39–42); also that it is capable of inducing high-affinity interleukin-2 (Il-2) receptors in human peripheral blood mononuclear cells (35) and of promoting human haematopoietic colony growth and differentiation (41, 42). Moreover,

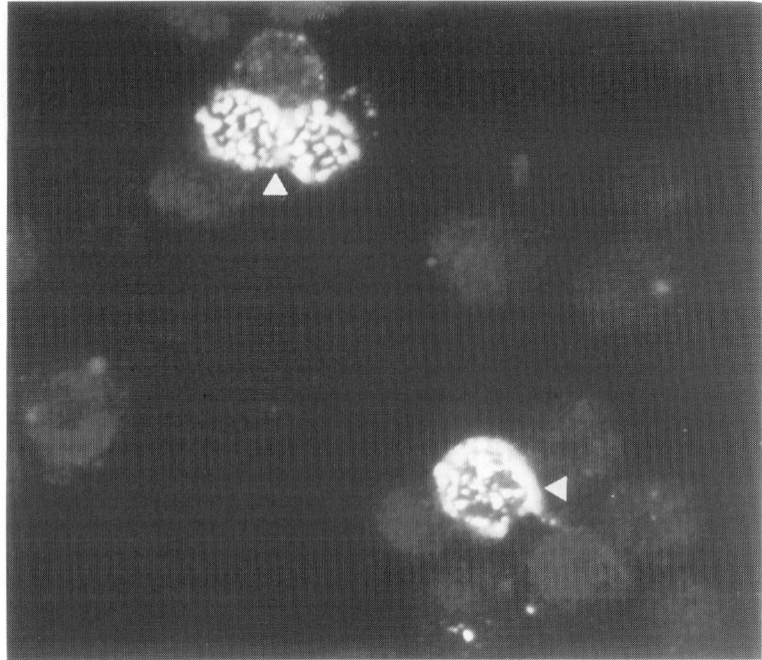

Fig. 1 Confocal photomicrograph of rat peritoneal mast cells (arrows) expressing high-affinity NGF-R. Original magnification: ×620.

NGF is chemotactic for neutrophilic leukocytes both *in vitro* and *in vivo* (43, 44).

The hypothesis that NGF plays a role on cells of the immune system is supported by several other studies published from different laboratories demonstrating that spleen mononuclear cells, thymocytes, human T and B lymphocytes and monocytes are responsive to the action of NGF (1, 15, 39). NGF also acts as a colony-stimulating factor for human and murine myeloid progenitor cells (45, 46), induces mediator release from basophils, eosinophils and neutrophils (47–49), acts as a chemoattractant for polymorphonuclear leukocytes (43, 44) and enhances neutrophil phagocytosis (50). Moreover, studies on mature lymphoid cells demonstrated that NGF has a dose-dependent proliferative effect on both B and T cells and causes differentiation of B cells into immunoglobulin-secreting plasma cells (39, 40). More recently it was also shown that lymphocytes are able to produce NGF (51) and that NGF is implicated in the regulation of survival in memory B lymphocytes (36), suggesting that NGF is regulated through autocrine mechanisms.

NGF AND MAST CELLS

The fact that MCs produce NGF, that NGF accumulates at the site of inflammation and that IL-1β and tumour necrosis factor-α (TNF-α) are potent inducers of NGF secretion (52–55) led to the hypothesis of a functional link between cytokines and NGF in certain inflammatory diseases. MCs and their biologically active mediators are known to initiate and modulate a variety of important inflammatory physiological and pathophysiological events (56–59). MCs are distributed in all tissues and are more numerously localized in cutaneous tissues, around blood vessels and within the lung, spleen, kidney, heart, peripheral nerves and brain. They are associated classically with hypersensibility reactions, involving interaction of allergens with cell-fixed IgE, and respond functionally, through specific receptors, to numerous biologically active compounds derived from both central and peripheral nerve cells, suggesting that MCs are not only constitutive parts of homeostatic events. Structural studies indicated that MCs and nerve cells are closely associated with peripheral nerve fibres and that when they are exposed to physiological amounts of neuroactive molecules they are capable of synthesizing and releasing a variety of biological mediators, including NGF. The first evidence that MCs are receptive to NGF showed that exogenous administration of highly purified NGF into newborn rats induces a marked increase in the number and size of MCs in peripheral tissues (14), including the iris, as illustrated in Fig. 2.

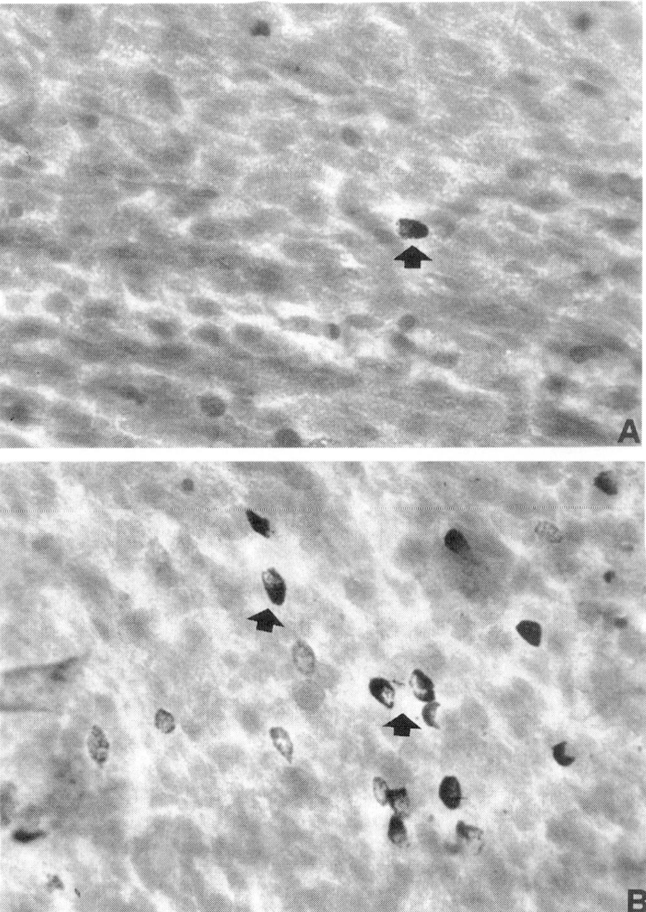

Fig. 2 Iris of 20-day-old rats injected subcutaneously for 8 consecutive days with 1 μg g^{-1} body weight of purified NGF (B) as compared to control (A). As indicated with arrows, NGF treatment induces an increase of mast cells in the iris tissue (B). Toluidine blue stain. Original magnification: ×250.

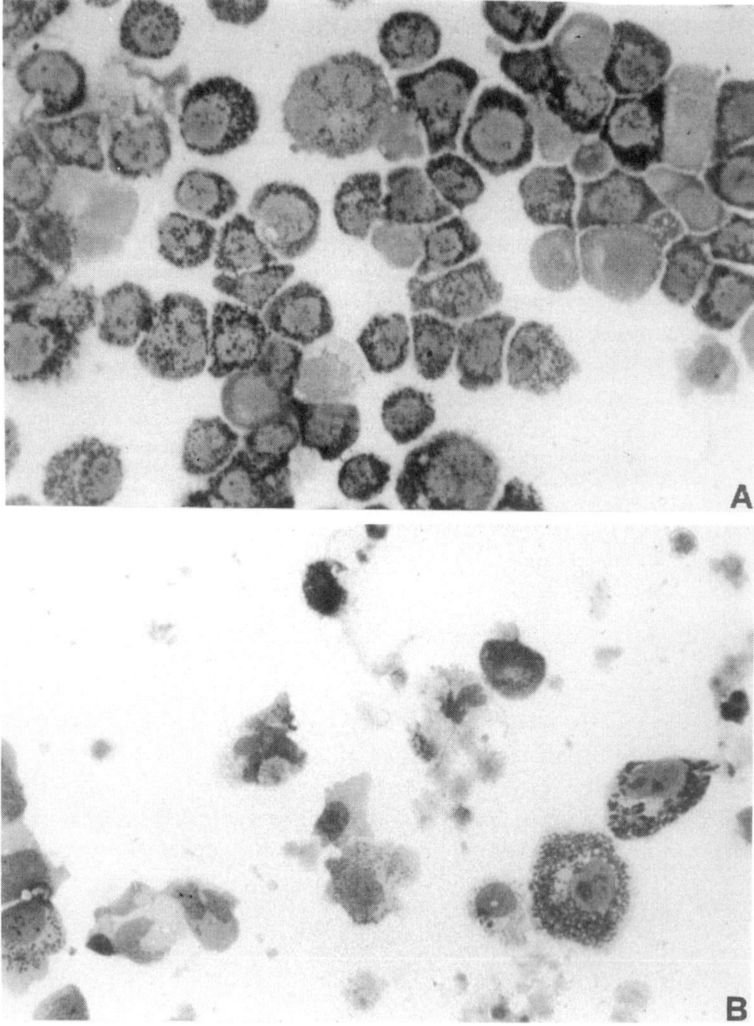

Fig. 3 Spleen cells of newborn rats cultured for 40 days in a semisolid medium in the presence (A) or absence (B) of 50 ng ml^{-1} NGF. Under these conditions NGF induces phenotypic transformation of spleen cells into mast cells. May–Grunwald–Giemsa stain. Original magnification: ×600.

Subsequent *in vitro* analysis demonstrated that dissociated spleen cells of neonatal rats cultured for 40 days in the presence of physiological amounts of NGF acquire morphological characteristics similar to those of fully differentiated MCs (Fig. 3).

These observations were confirmed and extended in later years by others who showed that these cells bear NGF-R, differentiate when exposed to physiological amounts of NGF *in vitro* (60–67) and that primed spleen cells transplanted into the brain ventricles of donor rodents are phenotypically converted into MCs (65). Additional evidence that NGF is implicated in survival of MCs was obtained after the observation that injection of NGF antibodies in developing rodents induces MC death (12). It is worth mentioning, however, that NGF is not the only neurotrophic factor acting on MCs, since recent findings indicate that other neurotrophins such as BDNF and ciliary neurotrophic factor

(CNTF) seem to be able to induce histamine release from these cells (34). How NGF synthesis and release are regulated by MCs and the pathophysiological relevance of NGF–MC–cytokine interactions are largely unknown. Certainly, a better understanding of this interaction may be important for the identification of mechanisms implicated in the development of certain neuroimmune pathologies. An important question raised by these findings is whether NGF–MC interaction is associated with pro-inflammatory action and whether this effect is deleterious for the tissues. Although a variety of observations support the hypothesis of a pro-inflammatory role of MCs, there is also evidence that the release of NGF by MCs and/or during inflammation may not necessarily be cytotoxic. Thus, the possibility that under certain circumstances this interaction may have a protective role on damaged cells, and be implicated in adaptive responses after noxious stimuli, cannot be excluded.

NGF AND INFLAMMATION

There are numerous morphological, biochemical and molecular studies indicating that NGF is involved in inflammatory responses. These include the evidence that NGF induces MC proliferation and degranulation (14, 34, 54, 55, 63), promotes growth and differentiation of lymphocytes (39), and that the amount of circulating NGF levels is elevated in several inflammatory and autoimmune disorders (15). There is growing evidence, however, supporting the hypothesis that the presence of NGF during inflammation may be associated with remodelling events. One of the first indications that NGF was implicated in inflammatory responses was an observation reported by Levi-Montalcini in 1960, that experimentally induced granulomas produce and release NGF (68). The demonstration that NGF not only stimulates MC proliferation of immature MCs, but also promotes MC degranulation of fully differentiated MCs (14, 34), provided additional evidence of an involvement of NGF in these events. More recently, we have also reported that liver granulomas induced in rodents by parasitic infection are characterized by elevated NGF levels and a marked accumulation of MCs (69), and that autoimmune inflammatory disorders display altered circulating local levels of NGF (15). It is known that inflammation, a localized reaction to injury or infection, is mainly associated with pain and sensory hypersensitivity, and concomitant phenotypic changes in sensory neurons. Thus, NGF seems to contribute to inflammatory pain hypersensitivity through an action in the periphery and by upregulation of peptides such as substance P which are neuromodulators released from C-fibre central terminal within the spinal cord (70). A link between NGF and inflammatory responses is also suggested by the evidence that NGF treatment increases substance P and calcitonin gene-related peptide expression in adult sensory neurons (15), two peptides which are involved in local plasma leakage, MC degranulation, and vasodilatation, causing hyperaemia, infiltration of leukocytes and other effects typical of neurogenic inflammation (19, 71). Despite these observations, the question as to whether NGF triggers or exacerbates the inflammatory responses needs to be fully demonstrated. Since NGF play a crucial role in growth and regeneration of nerve cells and is implicated in memory B cells and in tissue repair, it is highly possible that NGF is involved in tissue remodelling. Evidence that NGF exerts an anti-inflammatory activity is also available (72–74).

NGF AND ALLERGIC RESPONSES

Since MCs play a crucial role in allergic responses and since NGF is a powerful promoter of MC degranulation, the question was asked whether allergic response causes change in the basal NGF levels. The first indication that NGF released by MCs might be implicated in allergy came in 1992 during a study of the effect of psychological stress induced in a young soldier by parachute jumping. This study revealed that the plasma of a young soldier suffering from allergy contained a high level of NGF (13). However, the first evidence supporting this hypothesis was demonstrated by studying patients affected by vernal keratoconjunctivitis (VKC) (75) and later with subjects affected by various forms of human allergies (76). The previous findings, obtained in collaboration with Bonini and Lambiase (Department of Ophthalmology, University of Rome 'Tor Vergata'), showed that the concentration of NGF increases in the plasma of patients affected by VKC and that this increase correlates with the histopathology and immunopathology of the disease (75). A positive correlation between NGF levels and the number of MCs was observed by examining biopsy specimens of tarsal and bulbar conjunctiva of these patients. Concomitant studies revealing that human $CD4^+$ T cell clones express high-affinity NGF receptors (77) and that eosinophils produce and release NGF (78) strengthened the hypothesis of a possible functional link between NGF and allergic responses. The results of these studies showed that VKC patients with the highest plasma concentration of NGF also displayed the highest presence of SP, suggesting a close relationship between NGF,

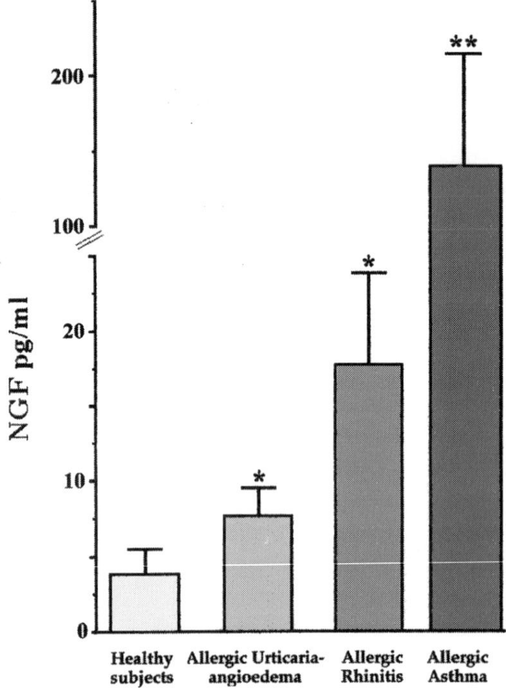

Fig. 4 Presence of NGF in serum of patients with various forms of allergy compared with the level in healthy subjects. As indicated, the amount of NGF increased significantly *($p<0.05$); **($p<0.01$).

SP, MCs and eosinophils (76). Subsequent studies carried out in collaboration with Bonini (Department of Allergology and Clinical Immunology, Second University of Naples) provided the crucial evidence that circulating NGF levels are enhanced in humans with allergic diseases and asthma (76). These studies revealed that NGF increases in patients with asthma, rhinoconjunctivitis and urticaria–angioedema compared to controls (55, 76), while patients affected with more than one allergic diseases express higher NGF values than those with a single disease (see Fig. 4). The fact that NGF levels correlated with total IgE antibody titre further supports the hypothesis of a link between NGF and allergic responses.

A key question raised by these observations is whether, in allergic responses, activated MCs or other pro-inflammatory cells produce the circulating NGF. Eosinophils are the predominant inflammatory cells of the late phase and chronic development of allergic inflammation, which, along with MCs and Th2 cells, play a central role in the pathogenesis of allergic diseases. Collaborative studies with Levi-Shaffer (Hebrew University of Jerusalem) indicate that NGF can induce circulating human eosinophils to release erythropoietin (EPO), an important pro-inflammatory mediator. This release does not necessarily seem to be linked to a cytotoxic effect of NGF on these cells, since no significant increase in the mortality rate of eosinophils was observed (78, 79). These studies also demonstrated that eosinophils contain variable levels of NGF protein and NGFmRNA, suggesting that eosinophils have the ability to produce and store NGF. Using an animal model of allergy it has been recently demonstrated that allergic airway inflammation in adult rodents is accompanied by enhanced local NGF production (79). In order to gain additional information regarding the role of overexpression of circulating NGF and allergic responses and better understand mechanisms of NGF–MC interaction, we have recently investigated the effect of highly purified NGF on the behaviour of MCs in nasal epithelia and lungs. As illustrated in Fig. 5A, dose–response studies revealed that topical application of NGF in the nasal cavity of adult mice induces degranulation of

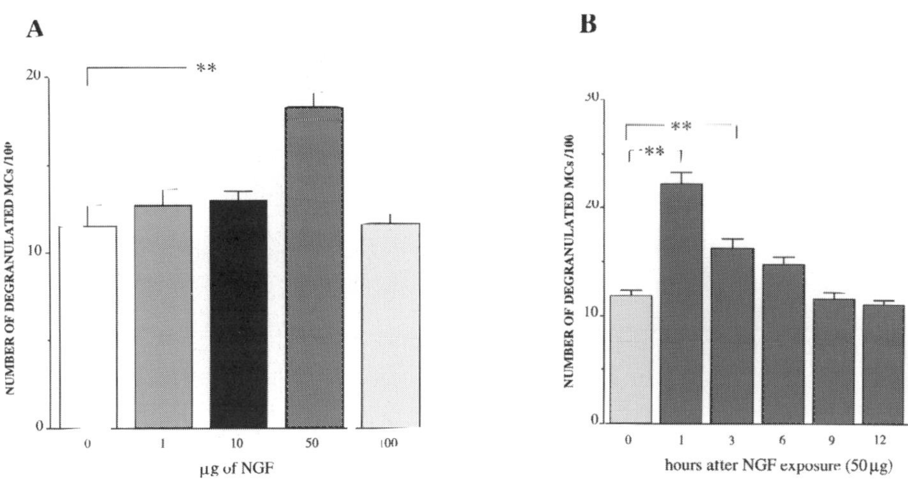

Fig. 5 (A) Dose–response data for the effect of topical application of NGF in the nasal cavity of 40-day-old rats (n=7) on mast cell degranulation. This study indicated that NGF is able to stimulate mast cell degranulation at a dose of 50 μg. (B) Time–course data on mast cell degranulation for topical application of 50 μg of NGF in the nasal cavity of 40-day-old rats (n=7). This study indicated that the major effect of NGF on mast cell degranulation occurs 1 h after exposure to NGF **($p<0.01$).

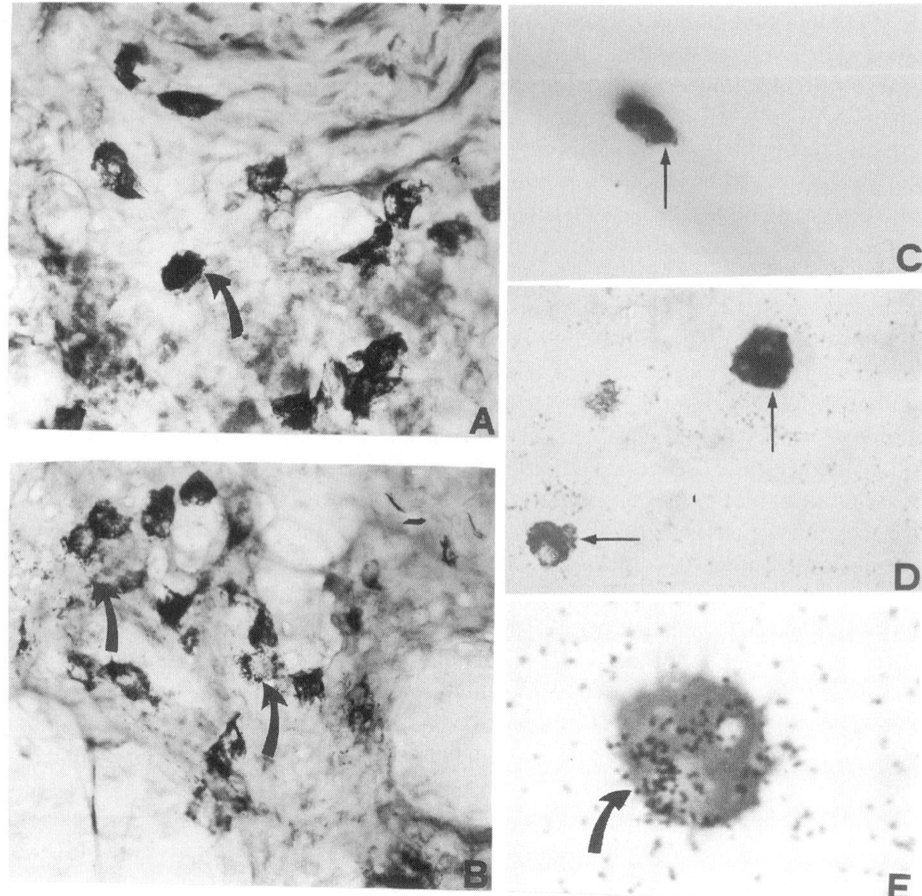

Fig. 6 Photomicrographs illustrating morphological characterization of mast cell–NGF relationship. Mast cells in nasal tissue of rats before (A) and after (B) NGF exposure in the nasal cavity. See degranulated mast cells in B (arrows). NGF-radiolabelled cell (arrows) localized in the thalamic area of rats in the proximity of a blood vessel (C) and near the habenula (D). (E) Rat peritoneal mast cell *in vitro* exposed to radiolabelled NGF for 39 min. Note a marked accumulation of iodinated NGF, indicating the ability of mast cells to take up NGF. Original magnifications: A, B, ×400; C, ×480; D, ×600; E, ×920.

MCs at a concentration of 50 μg and that this effect occurs within the first few hours (Fig. 5B). No changes in MC distribution and/or MC degranulation were observed in lung MCs (unpublished).

The evidence that NGF induces MC degranulation when it is present in the nasal cavity is shown in Fig. 6, which also illustrates the ability of MCs localized in the CNS (C, D) and in the peritoneum (E) to take up NGF.

BEHAVIOURAL INFLUENCE ON NGF LEVELS AND ALLERGIC RESPONSES

We have previously reported that altered behavioural activities induce changes in blood NGF circulation and in MC distribution in various tissues (15). We have also shown that disruption of maternal–infant interaction during early post-natal life can, most probably through MC–NGF interaction, significantly influence the susceptibility of developing neuroinflammatory diseases in adult life. For example, increased numbers of MCs have been found in mouse brain following chronic subordinate stress and courtship, intraspecific fighting and in rodents exposed to a novel odour (80). Whether the release of NGF by MCs or other cells has a functional role, however, is not known. There are growing clinical indications that overly fearful youngsters are at risk of later emotional distress and allergic disorders, and that there is a close relationship between psychological factors and allergy (81). Additionally, a possible correlation between behavioural alterations and development of allergic is suggested by recent observations that psychogenic stimuli induced by asthma, eczema and migraine have been associated with MC activation (80). Finally, patients displaying anxious, depressive or stress-related behaviour seem to suffer more frequently from allergic diseases compared to healthy subjects (81). As psychological factors can significantly influence the allergic conditions and the constitutive concentration of circulating NGF, one may wonder whether a functional link exists between certain psychosocially stressful events, MC activation and NGF release. The above evidence and the observations that MCs and their mediators are functionally linked with the nervous system led to the hypothesis that the release of NGF following MC degranulation could function as a general alerting signal utilized by the organism in settings of stress and anxiety to ‘prime’ the immune system towards external noxious perturbation. The fact that NGF is able to regulate synthesis and/or release of pro-inflammatory neuropeptides in a variety of cells of both nervous and immune systems provides additional evidence that NGF may be important in the development of neuroimmune regulation and a key factor in behaviourally related allergic events.

POSSIBLE FUTURE DIRECTIONS

We have briefly presented past and recent data supporting the hypothesis that, in addition to its role as a neurotrophic regulator, NGF displays survival and differentiative actions on cells of the immune system involved in inflammatory responses. We have also presented and discussed current knowledge about the role of NGF in MC growth and functions. The available data on NGF–MC interaction led to the observation that allergic responses in humans are also characterized by the release into the bloodstream of NGF and that Th2 cells, which play a pivotal role in allergy, are NGF-responsive and NGF-producing cells. Nonetheless, key questions that remain to be further investigated are whether NGF is involved either in preventing or reducing the neuronal damage and whether NGF plays a functional role in the pathogenesis of immune-related responses. Because inflammatory and allergic responses are mainly associated with MC activation and release of a variety of biological mediators including NGF, it has been suggested that NGF is involved in the inflammatory response. However, the hypothesis of a pro-inflammatory role of NGF requires further and more appropriate studies, since it is known that not only of NGF, but also MCs might be associated with reparative and remodelling action (72, 73, 82). The fact that exogenous administration of highly purified NGF in normal rodents does not induce inflammation supports the latter hypothesis. Future studies, both on animal models and in humans, will tell whether NGF is harmful or helpful in allergic and/or inflammatory disorders. Thus, 50 years after its discovery as a neurotrophic factor (83) and over 20 years after the report of its effect on MCs (14), NGF appears as a molecule that may contribute to understanding some important mechanisms involved in immune and allergic responses.

ACKNOWLEDGEMENTS

The authors dedicate this paper to the fiftieth anniversary of NGF discovery and the ninetieth birthday of Rita Levi-Montalcini. This study has been supported by the project 'Multiple Sclerosis' from Istituto Superiore di Sanità, Rome, Italy. We thank E. Vigneti, from the Institute of Cellular Biology, CNR, Rome, for the confocal photomicrograph.

REFERENCES

1. Levi-Montalcini, R. The nerve growth factor 35 years later. *Science* **237**:1154–1162, 1987.
2. Thoenen, H., Bandtlow, C. and Heumann, R. The physiological function of nerve growth factor in the central nervous system: comparison with the periphery. *Rev. Physiol. Biochem.* **109**:145–178, 1987.
3. Barde, Y. A. The nerve growth factor family. *Prog. Growth Factor Res.* **2**:237–248, 1990.
4. Levi-Montalcini, R., Aloe, L. and Alleva, E. A role for nerve growth factor in nervous, endocrine and immune system. *Prog. Neuroendocrinimmunol.* **3**:1–10, 1990.
5. Bocchini, G. and Angeletti, P. U. The nerve growth factor: purification as a 30,000 molecular-weight protein. *Proc. Natl. Acad. Sci. USA* **64**:787–794, 1996.
6. Hogue-Angeletti, R. and Bradshaw, R. A. Nerve growth factor from mouse submaxillary gland; amino acid sequence. *Proc. Natl. Acad. Sci. USA* **68**:2417–2420, 1971.
7. Goetz, R. and Schartl, M. The conservation of neurotrophic factors during vertebrate evolution. *Comp. Biochem. Physiol.* **108**:1–10, 1994.
8. Francke, U., De Martinville, B., Coussen, L. and Ullrich, A. The human gene for the beta subunit of nerve growth factor is located on the proximal short harm of chromosome 1. *Science* **222:**1248–1251, 1983.
9. Scott, J., Selby, M., Urdea, M. *et al.* Isolation of nucleotide of a cDNA encoding the precursor of mouse nerve growth factor. *Nature* **302**:538–540, 1983.
10. Aloe, L., Alleva, E., Böhm, A. *et al.* Aggressive behaviour induces release of nerve growth factor from mouse salivary gland into the bloodstream. *Proc. Natl. Acad. Sci. USA* **83**:6184–6187, 1986.
11. Aloe, L., Alleva, E. and De Simone, R. Changes of NGF level in mouse hypothalamus following intermale aggressive behavior: biological and immunohistochemical evidence. *Behav. Brain Res.* **39**:53–61, 1990.
12. Aloe, L. The effect of nerve growth factor and its antibody on mast cells *in vivo*. *J. Neuroimmunol.* **18**:1–12, 1988.
13. Aloe, L., Bracci-Laudiero, L., Alleva, E. *et al.* Emotional stress induced by parachute jumping changes blood nerve growth factor levels and the distribution of nerve growth factor receptors in lymphocytes. *Proc. Natl. Acad. Sci. USA* **91**:10440–10444, 1994.
14. Aloe, L. and Levi-Montalcini, R. Mast cells increase in tissues of neonatal rats injected with the nerve growth factor. *Brain Res.* **133**:358–366, 1977.
15. Aloe, L., Skaper, S. D., Leon, A. *et al.* Nerve growth factor and autoimmune diseases. *Autoimmunity* **19**:141–150, 1994.
16. Barbacid, M. Nerve growth factor: a tale of two receptors. Oncogene **8**:2033–2042, 1993.
17. Meakin, S. O. and Shooter, E. M. The nerve growth factor family of receptors. *Trends Neurosci.* **15**:323–331, 1992.
18. Snider, W. D. and Johnson, E. M. Neurotrophic molecules. *Ann. Neurol.* 26:489–506, 1989.
19. Lindsay, R. M. and Harmar, A. J. Nerve growth factor regulates expression of neuropeptide genes in adult sensory neurones. *Nature* **337**:362–364, 1989.
20. Donnerer, J., Schuligoi, R. and Stein, C. Increased content and transport of substance P and calcitonin gene-related peptide in sensory nerves innervating inflamed tissue: evidence for a regulatory function of nerve growth factor *in vivo*. *Neuroscience* **49**:693–698, 1992.
21. Lewin, G. R., Ritter, A. M. and Mendell, L. M. Nerve growth factor-induced hyperalgesia in the neonatal and adult rat. *J. Neurosci.* **13:**2136–2148, 1993.

22. Levi-Montalcini, R. and Angeletti, P. U. Immunosympathectomy. *Pharmacol. Rev.* **18**:619–628, 1966.
23. Whittemore, S. R. and Seiger, A. The expression, localization and functional significance of β-nerve growth factor in the central nervous system. *Brain Res.* **12**:439–464, 1987.
24. Fisher, W., Wiktorin, K., Bjorklund, A. *et al.* Amelioration of cholinergic neuron atrophy and spatial memory impairment in aged rats by nerve growth factor. *Nature* **329**:65–68, 1987.
25. Chen, K S. and Gage, F. H. Somatic gene transfer of NGF to the aged brain: behavioral and morphological amelioration. *J. Neurosci.* **15**:2819–2825, 1995.
26. Liberini, P. and Cuello, A. C. Effects of nerve growth factor in primate models of neurodegeneration. *Neurol. Rev. Neurosci.* **5**:89–104, 1994.
27. Otten, U., Baumann, J. B. and Girard, J. Stimulation of the pituitary–adrenocortical axis by nerve growth factor. *Nature* **282**:413–414, 1979.
28. Lahtinen, T., Soinila, S. and Lakshmanan, J. Biological demonstration of nerve growth factor in the rat pituitary gland. *Neuroscience* **30**:165–170, 1989.
29. Aloe, L. Adrenalectomy decreases nerve growth factor in young adult rat hippocampus. *Proc. Natl. Acad. Sci. USA* **86**:5636–5640, 1989.
30. Aloe, L., Cozzari, C., Calissano, P. *et al.* Somatic and behavioral postnatal effects of fetal injections of nerve growth factor antibodies in the rat. *Nature* **291**:413–415, 1981.
31. Spillantini, M. G., Aloe, L., Alleva, E. *et al.* Nerve growth factor mRNA and protein increase in hypothalamus in a mouse model of aggression. *Proc. Natl. Acad. Sci. USA* **86**:8555–8559, 1989.
32. Johnson, E. M., Osborne, P. A., Rydel, R. E. *et al.* Characterization of the effects of autoimmune nerve growth factor deprivation in the developing guinea pig. *Neuroscience* **8**:631–642, 1983.
33. Sacerdote, P., Manfredi, B., Aloe, L. *et al.* Centrally injected nerve growth factor modulates peripheral immune responses in the rat. *Neuroendocrinology* **64**:274–279, 1996.
34. Bruni, A., Bigon, E., Boarato, E. *et al.* Interaction between nerve growth factor and lysophosphatidylserine on rat peritoneal mast cells. *FEBS Lett.* **138**:190–192, 1982.
35. Thorpe, L. W., Werrbach-Perez, K. and Perez-Polo, J. R. Effects of nerve growth factor expression on interleukin-2 receptors on cultured human lymphocytes. *Ann. N. Y. Acad. Sci.* **496**:310–311, 1987.
36. Torcia, M., Bracci-Laudiero, L., Lucibello, M. *et al.* Nerve growth factor is an autocrine survival factor for memory B lymphocytes. *Cell* **85**:345–356, 1996.
37. Thorpe, L. W., Stach, R. W., Hashim, G. A. *et al.* Receptors for nerve growth factor on rat spleen mononuclear cells. *J. Neurosci. Res.* **17**:128–134, 1987.
38. Manning, P. T., Russel, G. H., Simmons, B. *et al.* Protection from guanethidine-induced neuronal destruction by nerve growth factor: effects of NGF on immune function. *Brain Res.* **340**:61–69, 1985.
39. Otten, U., Ehrhard, P. and Peck, R. Nerve growth factor induces growth and differentiation of human B lymphocytes. *Proc. Natl. Acad. Sci. USA* **86**:10059–10063, 1989.
40. Kimata, H., Yoshida, A., Ishioka, C. *et al.* Nerve growth factor specifically induces human IgG4 production. *Eur. J. Immunol.* **21**:137–141, 1991.
41. Brodic, C. and Gelfand, E. W. Functional nerve growth factor receptors on human B lymphocytes. Interaction with IL-2. *J. Immunol.* **148**:3492–3497, 1992.
42. Brodie, C., Oshiba, A., Renz, H. *et al.* Nerve growth factor and anti-CD40 provide opposite signals for the production of IgE in interleukin-4-treated lymphocytes. *Eur. J. Immunol.* **26**:171–178, 1996.
43. Gee, A. P., Boyle, M. D. P., Munger, K. L. *et al.* Nerve growth factor: stimulation of polymorphonuclear leukocyte chemotaxis *in vitro*. *Proc. Natl. Acad. Sci. USA* **80**:7215–7218, 1983.
44. Boyle, M. D. P., Lawman, M. J. P., Gee, A. P. *et al.* Nerve growth factor: a chemotactic factor for polymorphonuclear leukocytes *in vivo*. *J. Immunol.* **134**:564–568, 1985.
45. Matsuda, H., Coughlin, M. D., Bienenstock, J. *et al.* Nerve growth factor promotes human hemapoietic colony growth and differentiation. *Proc. Natl. Acad. Sci. USA* **85**:6508–6512,1988.
46. Kannan, Y., Matsuda, H., Ushio, H. *et al.* Murine granulocyte–macrophage and mast cell colony formation promoted by nerve growth factor. *Int. Arch. Allergy Immunol.* **102**:362–367, 1993.
47. Tsuda, T., Wong, D., Dolovich, J. *et al.* Synergistic effects of nerve growth factor and granulocyte–macrophage colony-stimulating factor on human basophilic cell differentiation. *Blood* **77**:971–979, 1991.
48. Bischoff, S. C. and Dahinden, C. A. Effect of nerve growth factor on the release of inflammatory mediators by mature human basophils. *Blood* **79**:2662–2669, 1992.
49. Takafuji, S., Bischoff, S. C., De Weck, A. L. *et al.* Opposing effects of tumor necrosis factor-α and nerve growth factor upon leukotriene C4 production by human eosinophils triggered with *N*-formyl-methionyl-leucyl-phenylalanine. *Eur. J. Immunol.* **22**:969–974, 1992.
50. Kannan, Y., Ushio, H., Koyama, H. *et al.* Nerve growth factor enhances survival, phagocytosis, and

superoxide production of murine neutrophils. *Blood* **77**:1320–1325, 1991.

51. Santambrogio, L., Benedetti, M., Chao, M. V. *et al.* Nerve growth factor production by lymphocytes. *J. Immunol.* **153**:4488–4492, 1994.
52. Gadient, R. A., Cron, K. C. and Otten, U. Interleukin-1β and tumor necrosis factor-α synergistically stimulate nerve growth factor (NGF) release from cultured rat astrocytes. *Neurosci. Lett.* **117**:335–340, 1990.
53. Lindholm, D., Heumann, R., Meyer, M. *et al.* Interleukin-1β regulates synthesis of nerve growth factor in non-neuronal cells of rat sciatic nerve. *Nature* **330**:658–659, 1987.
54. Marshall, J. S., Stead, R. H., McSharry, C. *et al.* The role of mast cell degranulation products in mast cell hyperplasia. I. Mechanism of action of nerve growth factor. *J. Immunol.* **144**:1886–1892, 1990.
55. Aloe, L., Bracci-Laudiero, L., Bonini, S. *et al.* The expanding role of nerve growth factor: from neurotrophic activity to immunologic disease. *Allergy* **52**:883–894, 1997.
56. Galli, S. J., Gordon, J. R. and Wershil, B. K. Cytokine production by mast cells and basophils. *Curr. Opin. Immunol.* **3**:865–872, 1991.
57. Purcell, W. M. and Atterwill, C. K. Mast cells in neuroimmune function: neurotoxicological and neuropharmacological perspectives. *Neurochem. Res* **20**:521–532, 1995.
58. Johnson, D. and Krenger, W. Interactions of mast cells with the nervous system: recent advances. *Neurochem. Res.* **9**:939–951, 1992.
59. Galli, S. J., Dvorak, A. M. and Dvorak, H. F. Basophils and mast cells: morphologic insights into their biology, secretory patterns and function. *Prog. Allergy* **34**:1–141, 1984.
60. Bohm, A. and Aloe, L. Nerve growth factor enhances precocious differentiation and numerical increase in mast cells in cultures of rat spenocytes. *Accad. Naz. Lincei Ser. VIII* **LXXX**:1–6v, 1986.
61. Kawamoto, K., Okada, T., Kannan, Y. *et al.* Nerve growth factor prevents apoptosis of rat peritoneal mast cells through the trk proto-oncogene receptor. *Blood* **86**:4638–4644, 1995.
62. Jippo, T., Ushio, H., Hirota, S. *et al.* Poor response of cultured mast cells derived from *mi/mi* mutant mice to nerve growth factor. *Blood* **84**:2977–2983, 1994.
63. Horigome, K., Bullock, E. D. and Johnson, E. M. Effects of nerve growth factor on rat peritoneal mast cells. *J. Biol. Chem.* **269**:2695–2702, 1994.
64. Leon, A., Buriani, A., Dal Toso, R. *et al.* Mast cells synthesize, store, and release nerve growth factor. *Proc. Natl. Acad. Sci. USA* **91**:3739–3743, 1994.
65. Aloe, L. and De Simone, R. NGF primed spleen cells injected in brain of developing rats differentiate into mast cells. *Int. J. Dev. Neurosci.* **7**:565–573, 1989.
66. Tam, S. Y., Tsai, M., Yamaguchi, M. *et al.* Expression of functional TrkA receptor tyrosine kinase in the HMC-1 human mast cell line and in human mast cells. *Blood* **90**:1807–1820, 1997.
67. Nilsson, G., Forsberg-Nilsson, K., Xiang, Z. *et al.* Human mast cells express functional trkA and are a source of nerve growth factor. *Eur. J. Immunol.* **27**:2295–2301, 1997.
68. Levi-Montalcini, R. and Angeletti, P. U. Biological properties of nerve growth promoting protein and its antiserum. In: *Regional Neurochemistry. The Fourth International Neurochemical Symposium* (Keti, S. S. and Elkes, J., eds). Pergamon Press, New York, 1960.
69. Aloe, L. and Fiore, M. Neuroinflammatory implications of *Schistosoma mansoni* infection: new information from the mouse model. *Parasitol. Today* **14**:314–316, 1998.
70. Millan, M. J. The induction of pain: an integrative review. *Prog. Neurobiol.* **57**:1–164, 1999.
71. Donnerer, J., Schuligoi, R. and Stain, C. Increased content and transport of substance P and calcitonine gene related peptide in sensory nerves innervating inflamed tissues: evidence for a regulatory function of NGF *in vitro*. *Neuroscience* **49**:693–698, 1992.
72. Li, A. K. C., Koroly, M. J., Schattenkerk, M. E. *et al.* Nerve growth factor: acceleration of the rate of wound-healing in mice. *Proc. Natl. Acad. Sci. USA* **77**:4379–4381, 1980.
73. Banks, B. E. C., Vernon, C. A. and Warner, J. A. Nerve growth factor has anti-inflammatory activity in the rat hindpaw oedema test. *Neurosci. Lett.* **47**:41–45, 1984.
74. Matzuda, H., Koyama, H., Sato, H. *et al.* Role of the nerve growth factor in cutaneous wound healing: accelerating effects in normal and healing-impaired diabetic mice. *J. Exp. Med.* **187**:297–306, 1998.
75. Lambiase, A., Bonini, S. T., Bonini, S. E. *et al.* Increased plasma levels of nerve growth factor in vernal keratoconjunctivitis and relationship to conjunctival mast cells. *Inv. Ophthalmol. Vis. Sci.* **36**:2127–2132, 1995.
76. Bonini Se., Lambiase, A., St. Bonini, F. *et al.* Circulating nerve growth factor levels are increased in human with allergic diseases and asthma. *Proc. Natl. Acad. Sci. USA* **93**:10955–10960, 1996.
77. Lambiase, A., Bracci-Laudiero, L., Bonini Se. *et al.* Human CD 4+ T-clones produce and release nerve growth factor and express TrkA. *J. Allergy Clin. Immunol.* **100:**408–414, 1997.

78. Solomon, A., Aloe, L., Pe'er, J. *et al.* Nerve growth factor is preformed in and activates human peripheral blood eosinophils. *J. Allergy Clin. Immunol.* **102**:454–460, 1998.
79. Braun, A., Appel, E., Baruch, R. *et al.* Role of nerve growth factor in a mouse model of allergic airways inflammation and asthma. *Eur. J. Immunol.* **28**:3240–3251, 1998.
80. Theoharides, T. C., Spanor, C., Pang, X. *et al.* Stress induces intracranial mast cell degranulation: corticotropin releasing hormone mediate effect. *Endocrinology* **136**:5745–5750, 1995.
81. Michel, F. B. Physiology of allergic patients. *Allergy* **49**:28–30, 1994.
82. Lu, X. and Richardson, P. M. Inflammation near the nerve cell body enhances axonal regeneration. *J. Neurosci.* **11**:972–978, 1991.
83. Levi-Montalcini, R. and Hamburger, V. Selective growth stimulating effects of mouse sarcoma on the sensory and sympathetic nervous system of the chick embryo. *J. Exp. Zool.* **116**:321, 1951.

CHAPTER 22

Interactions Between Neurotrophins and Mast Cells

BEATE M. HENZ,[*1] BARBARA HERMES[2] and PIA WELKER[1]*

[1]*Department of Dermatology, Charité, Humboldt-University, and*
[2]*Krankenhaus Neukölln, Berlin, Germany*

DEFINITION AND BASIC ASPECTS OF NEUROTROPHINS

Neurotrophins (NTs) encompass a family of structurally and functionally related polypeptides with up to 50% amino acid sequence homology. During development as well as in adult life, NTs induce neuronal proliferation, differentiation, survival and recovery from insult. Neurotrophic molecules have also been shown to regulate and prevent degeneration of cholinergic neurons and to maintain functional aspects of these and also peptidergic neurons (1–4). Furthermore, there is increasing evidence that NTs exert numerous biological effects on cells involved in the immune response and that they play a role in autoimmune, allergic and neurodegenerative diseases, as well as in a broad spectrum of inflammatory reactions, including such diverging processes as wound healing and psoriasis (5–14). Apart from neuronal tissue, NTs are produced in various organs and tissues by different cell types, including fibroblasts, myelomonocytic cells, lymphocytes, keratinocytes (11, 15–19) and, as will be discussed below in more detail, also by mast cells.

Based on their structure and receptor interactions, neurotrophic growth factors are classified into three major groups (see Table I). The first group of NTs as such includes nerve growth factor (NGF), brain-derived neurotrophic factor (BDNF) and neurotrophins NT-3 to NT-7. The family of glia cell line-derived neurotrophic factor (GDNF) encompasses two additional factors, neuroturin (NTN) and persephin (PSP). Only two members of the neurokine family have been identified so far, ciliary neurotrophic factor (CNTF) and leukaemia inhibitory factor (LIF) (4). The latter is identical to cholinergic differentiation factor (CNDF).

Epidermal growth factor (EGF), heparin-binding neurite-promoting factor (HBNF), insulin-like growth factor (IGF-2), α and β fibroblast growth factor (FGF), transforming

* Corresponding author.

ISBN 0-12-473335-2

TABLE I
Currently Known Neurotrophins and Their Receptors

Groups of neurotrophic factors	Receptors
A. Neurotrophins	
NGF	p75, TrkA
BDNF	p75, TrkB
NT-3	p75, TrkA, B, C
NT-4	p75, TrkB
B. GDNF group	
GDNF	RET/GRFα-1, -2 complex
NTN	RET/GRFα-1, -2 complex
PSP	?, GRFα-4
C. Neurokines	
CNTF	CNTFα/gp130/LIFR-β complex
LIF	Gp 130/LIFR-β complex
D. Diverse other factors with neurotrophic activities	
EGF, HBNF, IGF-2, αFGF, βFGF, TGF-β, PDGF, NSE, activin A	Specific receptors for the respective molecules

For abbreviations, see text. ? = unknown.

growth factor (TGF)-β, platelet-derived growth factor (PDGF), neuron-specific enolase (NSE) and activin A also have neurotrophic activities, but are not grouped with the NT families because of different structural features and receptor interactions.

For the specific neurotrophic factors (Table I), several distinct receptor proteins have been identified. Biological activities of the NTs are mediated by either of two distinct cellular receptors, a 140-kDa molecule (p140) and a 75-kDa (p75) molecule, which, in contrast to previous concepts, bind both with similar affinities. The 140-kDa receptor belongs to the group of receptor tyrosine kinases and is encoded by the tyrosine receptor kinase (trk)A proto-oncogene with specificity for NGF. A structural homologue TrkB (66% amino acid sequence identity) specifically binds NT-3, NT-4 and BDNF, and TrkC (68% amino acid sequence identity) binds only NT-3. For all three Trk receptors, different isoforms have been described which are produced by alternative splicing of the specific mRNA and are differentially expressed in neuronal and non-neuronal tissues.

Cells found to express the p140 Trk belong primarily to the central nervous system, but include also human monocytes, B and T lymphocytes, and keratinocytes. In contrast, the p75 receptor binds all NTs and is expressed by many cell types.

Receptors for the GDNF family include the newly discovered GFR receptors with the common signalling receptor subunit RET, while the neurokine family binds to two separate receptor molecules: a gp130/LIFR-β complex and a CNTFRα/gp130/LIFR-β complex (4, 20–22) (Table I). Taken together, there is thus considerable receptor sharing and diversity of binding for the different neurotrophic factors. However, very little is known so far about the specific biological or pathological role of the different receptors.

MAST CELLS AS A SOURCE OF NEUROTROPHIC FACTORS

A close proximity of mast cells and nerve fibres was noted more than a century ago by Paul Ehrlich, the discoverer of mast cells, and this was confirmed for the central and peripheral nervous system by several groups (23–25). Even before the latter publications, we were able to show a direct contact between nerves and mast cells at the ultrastructural level in the skin (26). On the basis of these observations, it is thus a distinct possibility that mast cells might contribute to nerve growth, differentiation and function by producing NTs, particularly in view of their ability to produce a broad spectrum of cytokines and growth factors (27–29). Recent studies have confirmed this possibility. Thus, rat peritoneal mast cells (RPMC) were shown to express NGF mRNA, an anti-NGF antibody stained the vesicular compartment of purified RPMC and of mesenchymal mast cells in histological sections, and supernatants of cultured RPMC contained NGF (30). RPMC were shown by another group to also produce the neurotrophic factor LIF (31). Rat basophilic leukaemia cells (RBL) were furthermore reported to stimulate PC12 (rat phaeochromocytoma cells) neurite outgrowths by producing NGF, with some contribution also from RBL cell-derived IL-6 (32).

Evidence that human mast cells are able to produce NTs as well comes primarily from studies of the human mast cell line HMC-1. Culture supernatants of these cells were shown to stimulate the growth of cultured sensory ganglia, and this effect could be blocked by an anti-NGF-antibody (33). Furthermore, HMC-1 cells expressed mRNA for NGF, BDNF and NT-3, and mast cells cultured from umbilical cord precursors NGF and BDNF (33, 34).

EFFECTS OF NEUROTROPHINS ON MAST CELLS

Induction of Secretory Functions

Compared to the paucity of data on NT production, the effects of NTs and particularly NGF on mast cells have been studied more extensively. Such data were generated mostly in animal models. Thus, injections of NGF into rat paw induced local mast cell degranulation (35). Furthermore, NGF caused *in vitro* histamine release from RPMC in the presence of phosphatidyl serine (36, 37), and in this same setting IL-6 was produced and TNF-α secretion inhibited via prostanoid-dependent pathways (38). Also in RPMC, NGF stimulated the delayed phase of PGD_2 generation in dependence of cyclo-oxygenase-2 (39) and induced the mRNA expression of IL-3, IL-4, IL-10, TNF-α and GM-CSF (40). In an interesting study which addressed the question of the relative effects of different NTs, rat brain mast cells were found to respond more to BDNF than to CNTF, followed by NGF, whereas RPMC responded only to NGF and not at all to the other two NTs studied (41). Taken together, these studies provide an as yet incomplete picture of the effect of NTs on murine mast cell function which may be improved once a thorough analysis of the involved signalling pathways is available.

In human mast cells, NGF has until now only been studied with respect to histamine release from human placental mast cells (42). Histamine secretion occurred under similar *in vitro* conditions as described for RPMC (36), i.e. in the presence of phosphatidyl serine (42). In view of the fact that mast cells can also produce NTs such as NGF (see above), an autocrine loop might be operative during the NGF-dependent induction of mast cell secretory activity.

Effects of Neurotrophins on Murine Mast Cell Growth, Differentiation and Survival

More than two decades ago, injections of NGF into neonatal rat were shown to enhance the numbers of local mast cells (43). Matsuda *et al.* (44) reported subsequently that, in murine bone marrow cells, NGF induces the *in vitro* development of mast cells. This was confirmed in other studies (45) and, furthermore, NGF was shown to enhance IL-3-dependent *in vitro* mast cell growth, an effect that was not observed in p75-deficient mi/mi mice (46). At about the same time, another group failed to demonstrate any effect of NGF on mast cell development in a fibroblast-dependent murine bone marrow culture system (47). Different *in vitro* culture conditions are most likely responsible for these divergent results.

As shown before for nerve cells, NGF is also a survival factor for mast cells, an effect that has been proposed to be mediated secondarily via increased cytokine production from murine mast cells (48). NGF also increased bcl-2 expression in RPMC, but not that of other genes of the bcl family (40). When NGF was compared to the mast cell growth factor SCF (stem cell factor), both factors were shown to prevent DNA fragmentation and loss of microvilli, but, unlike SCF, NGF was considered to have no proliferative effects on the cells, an assumption based on the failure of NGF to induce RPMC to pass from the G_0/G_1 to the S/G_2M phase of the cell cycle (49).

EVIDENCE THAT NGF IS ALSO A HUMAN MAST CELL GROWTH FACTOR

While the majority of data suggest that NGF is a differentiation factor for murine mast cells, the question as to whether NGF can also act as a human mast cell growth factor has been unsolved until recently. In an older study, enhanced histamine production was reported with cultures of human umbilical cord blood cells, but the cellular source of histamine, whether mast cells or basophils, was not clarified (50). In another study employing murine NGF, at a concentration of 100 ng ml^{-1}, this molecule failed to induce mast cell-specific tryptase in two different human bone marrow cultures, whereas SCF was clearly effective, and the combination of both factors even markedly reduced SCF-induced tryptase upregulation (51). On the other hand, there has been a general consensus that NGF functions as a human basophil growth factor (summarized in refs 52, 53).

Our own group has recently tried to clarify this question while studying fibroblast (FS) and keratinocyte supernatants (KS) for their mast cell diffentiating effects. We compared the activities of these crude cell supernatants to those of SCF which is currently considered to be the only human mast cell growth factor (reviewed in ref. 54). As a model of *in vitro* mast cell growth and differentiation, we used cultures of adherent peripheral blood mononuclear cells as a source of mast cell precursors and FS or KS as mast cell conditioning factor (55–58). In contrast to the cell supernatants, SCF was unable to induce mast cell differentiation in the absence of serum or additional haematopoietic growth factors, in agreement with observations regarding SCF function in other haematopoietic cell culture systems (54).

In another mast cell differentiation model employing immature HMC-1 cells, FS

TABLE II
Percentage Increase in Mast Cell Markers Compared with Normal Culture Medium, After a 10-Day Culture of HMC-1 Cells

Mast cell marker	Addition to baseline culture conditions	
	FS, 30%	NGF, 10 ng ml^{-1}
Tryptase activity	142	91
Histamine	66	50
$Fc_{\varepsilon}RI\alpha$	263	236

Note the marked increase in intracellular tryptase activity and histamine contents, and of cells reactive for antibodies against $Fc_{\varepsilon}RI\alpha$ on immunohistochemistry (means of $n = 3$ different experiments; increases were significant in each case at $p<0.01$). For details of the experiment, see refs 59, 61.

induced a clear increase of intracellular histamine and tryptase as well as $Fc_{\varepsilon}RI\alpha$ expression during a 10-day culture at both the mRNA and the protein level (Table II) (59). HMC-1 cells failed, however, to respond in a similar fashion to SCF (59), probably due to the well-established functional mutations of the SCF receptor, a protein coding by the proto-oncogene c-kit, in these cells (60). On the basis of these findings, we reasoned that fibroblasts and keratinocytes might produce mast cell growth factors other than SCF.

In search of such molecules, we first embarked on an analysis of the mast cell differentiating activity in FS, using the increase of intracellular tryptase as a simple screening model. On column fractionation, two peaks of such activity could be identified, one at a molecular mass between 40 and 60 kDa, the other at <10 kDa (59).

Next, we examined known growth factors from fibroblasts and keratinocytes with a molecular mass close to that of the high molecular mass tryptase-inducing activity. When TGF-α, βFGF and PDGF were tested, no significant effects were observed regarding an enhanced production of mast cell markers, whereas NGF had basically the same effect as unfractionated FS and KS (Table II) (61).

TABLE III
Effect of Different Mast Cell Growth Stimuli on the Percentage Increase of Tryptase Immunoreactive Cells and Intracellular Tryptase Activity During a 2-Week Culture of Human Cord Blood Mononuclear Cells, Compared to Controls (means from $n = 12$ experiments)

Stimulus of mast cell growth	Tryptase, % increase	
	Immunoreactive cells	Enzyme activity*
FS, 30%	131	52
NGF, 10 ng ml^{-1}	169	96
SCF, 50 ng ml^{-1}	192	116
NGF + SCF†	231	215

* Determined photometrically by enzymatic cleavage of a specific substrate (57).
† Concentrations as for individual factors.

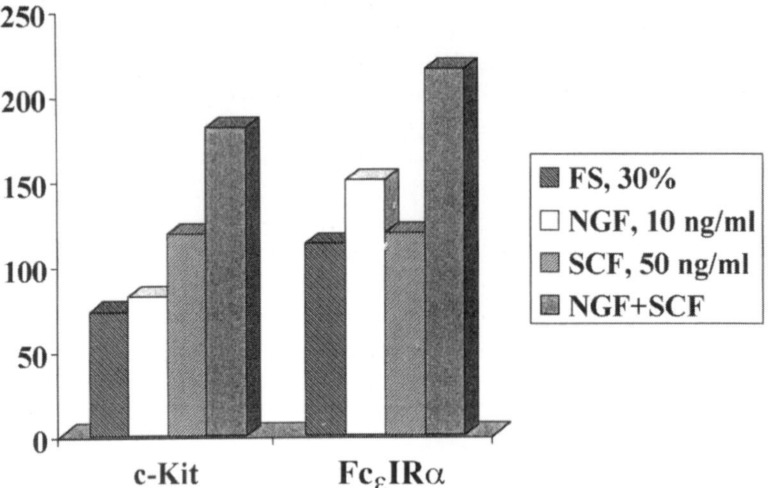

Fig. 1 Expression of c-Kit and $Fc_{\varepsilon}RI\alpha$ in cord blood-derived mast cells induced by different mast cell growth-promoting media. Data are expressed as percentage increase of immunoreactive cells compared to cells kept in control media.

A mast cell differentiating activity of NGF could subsequently also be demonstrated in cultures of mast cells from cord blood precursors. As shown in Table III, NGF caused a marked upregulation of total extractable tryptase enzymatic activity and of numbers of cells with tryptase immunoreactivity already after 2 weeks of culture. Furthermore, the numbers of cells expressing c-Kit and $Fc_{\varepsilon}RI\alpha$ increased as well (Fig. 1). As can be seen, at an NGF concentration of 10 ng ml^{-1}, the induction of mast cell-surface markers and of tryptase was comparable to that of SCF at 50 ng ml^{-1}. Combinations of both stimuli were most effective, although there were no synergisms and not even additive effects, with the exception of tryptase activity (Table III, Fig. 1). This finding was supported by the observation that intracellular immunocytochemical staining of tryptase was more intense within individual cells when both stimuli had been added in combination to the cultures (not shown). These data suggest thus that there is not only an increased recruitment of cells with mast cell characteristics like tryptase, c-Kit and $Fc_{\varepsilon}RI\alpha$ in the presence of both SCF and NGF, but also an increased tryptase synthesis in individual cells bearing mast cell characteristics.

The neosynthesis of mast cell markers in cord blood cultures in the presence of NGF was confirmed at the mRNA level. Only minor quantities of $Fc_{\varepsilon}RI\alpha$ and c-kit were detected prior to culture, while the expression of both molecules was markedly upregulated and tryptase was *de novo* expressed after 3 weeks of culture. After an additional 2 weeks, chymase mRNA was detectable for the first time as well (not shown).

When these data had been obtained, a crucial issue still remained unsettled, namely whether indeed mast cell or only basophil differentiation had been induced since basophils share a number of properties with mast cells, such as the expression of the high-affinity IgE receptor $Fc_{\varepsilon}RI\alpha$ and intracellular contents of histamine (27, 52). This question was resolved using two different approaches. Basophils differ from mast cells in that they express at most low levels of c-Kit and α-tryptase and lack mast cell-specific β-tryptase (62–65). Using restriction analysis in order to differentiate between α- and β-tryptase (63, 65) with the enzymes *Fok*I and *Pst*I, we found fragment sizes compatible with a mixture of cDNA for α- and β-tryptase from cells cultured with either rhSCF or rhNGF. This co-expression of both α- and β-tryptase has been described before for human cord blood-derived lung and cutaneous mast cells and is not found in basophils (63, 65). The second approach involved the use of the basophil-specific monoclonal antibody 2D7 (66), kindly provided by Dr L.B. Schwartz. None of the cord blood-derived cells showed immunoreactivity prior to, during and at the end of culture with NGF, whereas basophils in normal blood, which served as positive controls, were strongly reactive. Taken together, these data provide convincing evidence that NGF promotes not only the maturation of already differentiated mast cells, such as leukaemic HMC-1 cells, but that it can also induce human mast cell differentiation from normal precursors contained in cord blood. Since the effects were comparable to SCF, NGF must be viewed as another human mast cell growth factor. However, its relevance in different pathological conditions remains to be determined.

NEUROTROPHIN RECEPTORS ON MAST CELLS

In order to examine whether the biological effects of NGF on mast cells, as discussed above, are specific and receptor-mediated, several groups have studied murine and human mast cell NT receptor expression. In RPMC, Kawamoto *et al.* (49) showed that only the TrkA and not the p75 NGF receptor was detectable by flow cytometry, using specific antibodies. Their findings were supported by the inhibition of TrkA phosphorylation and NGF-induced apoptosis in the same cells, using the tyrosine kinase inhibitors herbimycin A and K-252a (49). On the other hand, the p75 receptor seems to play a role during combined IL-3- and NGF-induced mast cell differentiation in normal mice since, in p75-deficient mi/mi mice, NGF was functionally inactive (46).

Analysis of the NGF receptors in human mast cells has been performed primarily in HMC-1 cells. Thus, TrkA expression was demonstrated at the mRNA, protein and functional level, using polymerase chain reaction (PCR), Western blot analysis of cell lysates, flow cytometry and immunoblotting for assessment of tyrosine phosphorylation (33). Similar data were obtained at our laboratory, using reverse transcriptase (RT)-PCR, flow cytometry and immunocytochemistry (61). Another group looked at the NT receptor subtypes and found TrkA as well as truncated TrkB and TrkC mRNA in HMC-1 cells (34). In SCF-induced cultured human cord blood mast cells, the TrkA mRNA was demonstrated as well (33).

A search for the expression of the p75 receptor in HMC-1 cells was, however, fruitless (33, 34), except for a minor band at the mRNA level which was detected in our laboratory (61). The receptor could, however, not be demonstrated at the protein level, in agreement with findings of the other groups (33, 34). Cultured immature human mast cells also lacked p75 mRNA (33).

Our observations with mature human skin mast cells differ from these findings. Thus, we could demonstrate the p75 receptor on 70% of avidin-positive mast cells of normal skin, using double-staining methods, namely a combination of avidin fluorescence (specific for mature connective tissue mast cells) and immunohistochemistry, using an anti-NGF receptor antibody (67). Interestingly, this reactivity was no longer detectable in avidin-positive mast cells in human scar tissue, suggesting a receptor downregulation under pathological conditions. Taken together, these data also show that mature, resting human mast cells exhibit a different NGF receptor profile than immature cells like

HMC-1 cells and cultured mast cells. It remains to be shown whether this differential expression of the NGF receptors on cells at different levels of maturity also has functional implications.

CLINICAL FINDINGS SUGGESTING MAST CELL AND NEUROTROPHIN ACTIVITY

Almost all data described so far were obtained in experimental set-ups. The question therefore arises as to whether NTs play a role in mast cell-associated diseases or in the maintenance of normal tissue homeostasis.

It is striking that increased mast cell numbers and increased expression of NTs have been described in a number of inflammatory, autoimmunological and neurodegenerative conditions (7–11, 14) (Table IV).

TABLE IV
Pathological Conditions with Elevated Tissue Mast Cell Numbers as well as Increased Expression or Circulating levels of Neurotrophins

- Allergic diseases, including asthma and vernal conjunctivitis
- Wound healing
- Systemic sclerosis
- Psoriasis
- Rheumatoid arthritis
- Idiopathic sensory urinary urgency
- Multiple sclerosis
- Peripheral neuropathies

Only very recently, *Bonini et al.* summarized their own data, showing a close relationship between NGF plasma levels and numbers of mast cells infiltrating the conjunctiva (14). In human cutaneous scars, we have recently also shown a striking increase of avidin-negative mast cells, with a concomitant decrease of avidin-positive, supposedly resting, mast cells (68). Since NGF accelerates wound healing (12, 14), we also examined the tissue for expression of the p75 NGF receptor and found a marked increase of p75 immunoreactivity in the entire dermal inflammatory infiltrate whereas, paradoxically, the receptor was downmodulated on avidin-positive mast cells (12, 67, 68).

In accordance with these observations, increased mast cell numbers as well as NGF immunoreactivity have been described in other types of pathological fibrosing disorders, such as in systemic sclerosis (69). In psoriasis where dermal mast cells (70) and epidermal NGF expression are also increased (71), basal epidermal TrkA immunostaining is decreased in both lesional and non-lesional skin, compared to normal skin, and p75 staining of nerve fibres is also markedly lost in lesional psoriatic skin (72). UVB irradiation, which is used to treat psoriasis and other types of dermatoses, also causes a decreased expression of both NGF receptors, but only in the epidermis (72). Rheumatoid arthritis and idiopathic sensory urinary urgency are two additional conditions where a local mast cell increase as well as NGF levels have been detected in the joint fluid and the tissue, respectively (73, 74).

Much evidence for a role of mast cells and NTs has also come from experimental and clinical research in neurodegenerative diseases. Thus, neurotoxic effects of mast cells have been described on astrocytes via the action of TNF-α, with subsequent generation of nitric oxide (75). In experimental allergic neuritis and autoimmune encephalomyelitis, increased and partly degranulated mast cells have been observed in inflammed nervous tissue (10). In patients with multiple sclerosis, increased numbers of mast cells were noted in the involved tissue, NGF and mast cell-specific tryptase were increased in the cerebrospinal fluid, and an increased BDNF reactivity was found in the diseased tissue (76–78). A differential expression of NTs and their receptors was reported in peripheral neuropathies (6), a condition in which mast cell numbers are also increased (79). NTs and mast cells are also discussed in the context of other neurodegenerative diseases, such as acute myelogenous leukaemia, Alzheimer's and Parkinson's disease, and treatment with NGF as well as NGF antibodies has been tried, but has so far been hampered by NGF-induced toxicities, particularly hyperalgesia and inflammation (reviewed in refs 7, 41, 80, 81).

Apart from these attempts to clarify the role of mast cells and NTs in pathological conditions, little is known about their role in normal tissue homeostasis. Exceptions are recent studies from our group, showing that mast cell numbers fluctuate during the murine hair cycle, that NGF enhances keratinocyte proliferation and that BDNF and NT-4 mRNA expression peak during follicle regression (82–85).

MAST CELLS AND NEUROTROPHINS: OPEN ISSUES

The accumulating experimental and clinical evidence regarding the association of mast cells and NTs, as summarized here (see also Fig. 2), leaves no doubt as to the potential significance of their inter-relationship regarding, on the one hand, the effect of NTs on mast cells and, on the other, the activities mediated by the production of these molecules by mast cells.

Major questions as to the significance of these interactions in normal physiology and pathology, however, remain unanswered. Thus, it is unclear whether mast cell-derived

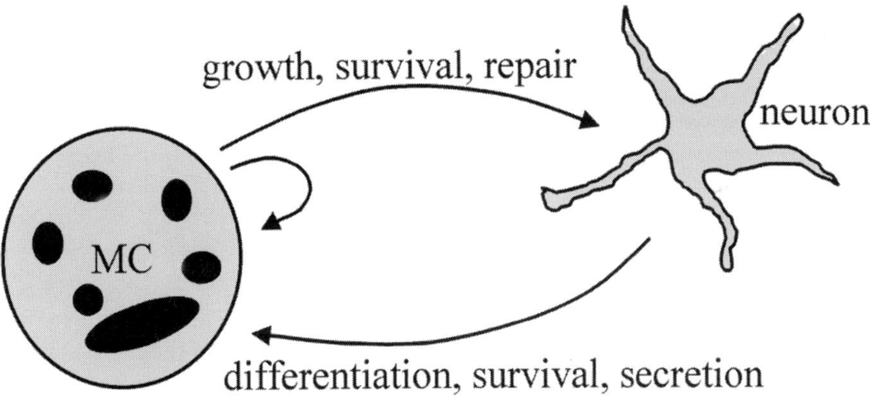

Fig. 2 Potential functions of neurotrophins released from mast cells (MC) and neurons on the other cell type. The small arrow indicates a possible autocrine loop of neurotrophin secretion from mast cells.

NTs are able to affect mast cell growth or function also in humans and under physiological conditions, or whether the effects of these molecules in tissue pathology derive predominantly from nerves, other resident tissue cells or infiltrating inflammatory cells. It remains also to be clarified whether mast cells residing in different organs and microenvironments or at different stages of maturation make different NTs, and which of the latter particularly affect the function of these different types of mast cells. Conversely, the role of mast cell-derived NTs in the growth, survival and repair of neurons needs to be elucidated.

Finally, the potentially beneficial role of mast cells and NTs has to be weighed against the potentially neurotoxic functions of mast cells and the pro-inflammatory effects of NTs. Since target cells of these molecules have the ability to modulate their receptor profile, thus warding off harmful or enhancing useful effects of NTs depending on their specific needs, NT receptors will always have to be studied as well when trying to unravel the puzzle of NT–mast cell interactions, their role in health and disease, and the possible targeting of this potentially complex interplay to the treatment of specific pathological conditions.

ACKNOWLEDGEMENT

Part of the work reported here was supported by a grant from the German Research Foundation.

REFERENCES

1. Ulrich, A., Gray, A. and Berman, C. Human beta-nerve growth factor gene sequence highly homologous to that of mouse. *Nature* **303**:821–825, 1983.
2. Aloe, L., Bracci-Laudiero, L., Bonini, S. and Manni, L. The expanding role of nerve growth factor: from neurotrophic activity to immunological diseases. *Allergy* **52**:883–894, 1997.
3. Levi-Montalcini, R., Skaper, S. D., Dal Toso, R., Tetrelli, L. and Leon, A. Nerve growth factor: from neurotrophin to neurokine. *Trends Neurosci.* **19**:1130–1132, 1996.
4. Ibanez, C. F. Emerging themes in structural biology of neurotrophic factors. *Trends Neurosci.* **21**:438–444, 1998.
5. Bothwell, M. Neurotrophin function in skin. *J. Invest. Dermatol. Symp. Proc.* **2**:27–30, 1997.
6. Sobue, G., Yamamoto, M., Doyu, M., Li, M., Yasasuda, T. and Mitsuma, T. Expression of mRNAs for neurotrophins (NGF, BDNF, NT-3) and their receptors (p75NGFR, trk, trkB, trkC) in human peripheral neuropathies. *Neurochem. Res.* **23**:821–829, 1998.
7. Ebadi, M., Bashir, R. M., Heidrick, M. L., Hamada, F. M., Rafaey, H. E., Hamed, A., Helal, G., Baxi, M. D., Cerutid, D. R. and Lassi, N. K. Neurotrophins and their receptors in nerve injury and repair. *Neurochem. Int.* **30**:347–374, 1997.
8. Merrill, J. E. and Benveniste, E. N. Cytokines in inflammatory brain lesions: helpful and harmful. *Trends Neurosci.* **19**:331–338, 1996.
9. Skaper, S. D., Facci, L., Romanello, S. and Leon, A. Mast cell activation causes delayed neurodegeneration in mixed hippocampal cultures via the nitric oxide pathway. *J. Neurochem.* **66**:1157–1166, 1996.
10. Aloe, L., Skaper, S. D., Leon, A. and Levi-Montalcini, R. Nerve growth factor and autoimmune diseases. *Autoimmunity* **19**:141–150, 1994.
11. Kerschensteiner, M., Gallmeier, E., Behrens, L., Leal, V. V., Misgeld, T., Klinkert, W. E., Kolbeck, R., Hoppe, E., Oropeza-Wekerle, R. L., Bartke, I., Stadelmann, C., Lassmann, H., Wekerle, H. and Hohlfeld, R. Activated human T cells, B cells, and monocytes produce brain-derived neurotrophic factor *in vitro* and in inflammatory brain lesions: a neuroprotective role of inflammation? *J. Exp. Med.* **189**:865–870, 1999.

12. Matsuda, H., Koyama, H., Sato, H., Sawada, J., Itakura, A., Tanaka, A., Matsumoto, M., Konno, K., Ushio, H. and Masuda, K. Role of nerve growth factor in cutaneous wound healing: accelerating effects in normal and healing-impaired diabetic mice. *J. Exp. Med.* **187**:297–306, 1998.
13. Bull, H. A., Leslie, T. A., Chopra, S. and Dowd, P. M. Expression of nerve growth factor receptors in cutaneous inflammation. *Br. J. Dermatol.* **139**:776–783, 1998.
14. Bonini, S., Lambiase, A., Levi-Schaffer, F. and Aloe, L. Nerve growth factor: an important molecule in allergic inflammation and tissue remodelling. *Int. Arch. Allergy Immunol.* **118**:159–162, 1999.
15. Calamandrei, G. and Alleva, E. Neuronal growth factors, neurotrophins and memory deficiency. *Behav. Brain Res.* **66**:129–132, 1995.
16. Pincelli, C., Sevignani, C., Manfredini, R., Grande, A., Fantini, F., Bracci-Laudiero, L., Aloe, L., Ferrari, S., Cossarizza, A. and Giannetti, A. Expression and function of nerve growth factor and nerve growth factor receptor on cultured keratinocytes. *J. Invest. Dermatol.* **103**:13–18, 1994.
17. Cartwright, M., Mikheev, A. M. and Heinrich, G. Expression of neutrotrophic genes in human fibroblasts: differential regulation of the brain-derived neutrotrophic factor gene. *Int. J. Dev. Neurosci.* **12**:685–693, 1994.
18. Patrick, C. W. Jr, Kukreti, S. and McIntire, L. V. Quantitative effects of peripheral monocytes and nerve growth factor on CNS neural morphometric outgrowth parameters *in vitro*. *Exp. Neurol.* **138**:277–285, 1996.
19. Santambrogio, L., Benedetti, M., Chao, M. V., Muzaffar, R., Kulig, K., Gabellini, N. and Hochwald, G. Nerve growth factor production by lymphocytes. *J. Immunol.* **153**:4488–4495, 1994.
20. Raffioni, S., Bradshaw, R. A. and Buxser, S. E. The receptors for nerve growth factor and other neurotrophins. *Annu. Rev. Biochem.* **62**:823–850, 1993.
21. Klein, R., Jing, S., Nanduri, U., O'Rouke, E. and Barbacid, M. The trk proto-oncogene encodes a receptor for nerve growth factor. *Cell* **65**:189–197, 1991.
22. Ehrhard, P. B., Ganter, U., Stalder, A., Bauer, J. and Otten, U. Expression of functional trk protooncogene in human monocytes. *Proc. Natl. Acad. Sci. USA* **90**:5423–5427, 1993.
23. Stead, R. H., Tomioka, M., Quinonez, G., Simon, G. T., Felten, S. Y. and Bienenstock, J. Intestinal mucosal mast cells in normal and nematode infected intestines are in intimate contact with peptidergic nerves. *Proc. Natl. Acad. Sci. USA* **84**:2975–2979, 1987.
24. Liang, Y., Marcusson, J. A., Jacobi, H. H., Haak-Frendscho, M. and Johansson, O. Histamine-containing mast cells and their relationship to NGF-immunoreactive nerves in prurigo nodularis: a reappraisal. *J. Cutan. Pathol.* **25**:189–198, 1998.
25. Williams, R. M., Berthoud, H. R. and Stead, R. H. Vagal afferent nerve fibres contact mast cells in rat small intestinal mucosa. *Neuroimmunomodulation* **4**:266–270, 1997.
26. Wiesner-Menzel, L., Schulz, B., Vakilzadeh, F. and Czarnetzki, B. M. Electron microscopical evidence for a direct contact between nerve fibers and mast cells. *Acta. Derm. Venereol.* **61**:465–469, 1981.
27. Metcalfe, D. D., Baram, D. and Mekori, Y. A. Mast cells. *Physiol. Rev.* **77**:1033–1079, 1997.
28. Möller, A., Henz, B. M., Grützkau, A., Lippert, U., Schwarz, T., Aragane, Y. and Krüger-Krasagakes, S. Comparative cytokine gene expression, regulation and release by human mast cells. *Immunology* **93**:289–295, 1998.
29. Artuc, M., Hermes, B., Steckelings, M. U., Grützkau, A. and Henz, B. M. Mast cells and their mediators in wound healing – active participants or innocent bystanders? *Exp. Dermatol.* **8**:1–16, 1999.
30. Leon, A., Buriani, A., Dal-Toso, R., Fabris, M., Romanello, S., Aloe, L. and Levi-Montalcini, R. Mast cell synthesize, store and release nerve growth factor. *Proc. Natl. Acad. Sci. USA* **91**:3739–3743, 1994.
31. Marshall, J. S., Gauldie, L., Nielsen, L. and Bienenstock, J. Leukemia inhibitory factor production by rat mast cells. *Eur. J. Immunol.* **23**:2116–2120, 1993.
32. Suzuki, M., Furuno, T., Teshima, R., Sawada, J. and Nakanishi, M. Soluble factors from rat basophilic leukemia (RBL-2H3) cells stimulated cooperatively the neurite outgrowth of PC12 cells. *Biol. Pharm. Bull.* **21**:1267–1270, 1998.
33. Nilsson, G., Forsberg-Nilsson, K., Xiang, Z., Hallbook, F., Nilsson, K. and Metcalfe, D. D. Human mast cells express functional TrkA and are a source of nerve growth factor. *Eur. J. Immunol.* **27**:2295–2301, 1997.
34. Tam, S. Y., Tsai, M., Yamaguchi, M., Yano, K., Butterfield, J. H. and Galli, S. J. Expression of functional TrkA receptor tyrosine kinase in the HMC-1 human mast cell line and in human mast cells. *Blood* **90**:1807–1820, 1997.
35. Tal, M. and Liberman, R. Local injection of nerve growth factor (NGF) triggers degranulation of mast cells in rat paw. *Neurosci. Lett.* **221**:129–133, 1997.

36. Bruni, A., Bignon, E., Boarato, E., Leon, A. and Toffano, G. Interaction between nerve growth factor and lysophosphatidylserine on rat peritoneal mast cells. *FEBS Lett.* **138**:190–192, 1982.
37. Pearce, F. L. and Thompson, H. L. Some characteristics of histamine secretion from rat peritoneal mast cells stimulated with nerve growth factor. *J. Physiol.* **372**:379–393, 1986.
38. Marshall, J S., Gomi, K., Blennerhassett, M. G. and Bienenstock, J. Nerve growth factor modifies the expression of inflammatory cytokines by mast cells via prostanoid-dependent mechanism. *J. Immunol.* **162**:4271–4276, 1999.
39. Tada, K., Murakami, M., Kambe, T. and Kudo, I. Induction of cyclooxygenase-2 by secretory phospholipases A2 in nerve growth factor-stimulated rat serosal mast cells is facilitated by interaction with fibroblasts and mediated by a mechanism independent of their enzymatic functions. *J. Immunol.* **161**:5008–5015, 1998.
40. Bullock, E. D. and Johnson, E. M. Jr. Nerve growth factor induces the expression of certain cytokine genes and bcl-2 in mast cells: potential role in survival promotion. *J. Biol. Chem.* **271**:27500–27508, 1996.
41. Purcell, W. M., Westgate, C. and Atterwill, C. K. Rat brain cells: An *in vitro* paradigm for assessing the toxic effects of neurotrophic therapeutics. *NeuroToxicology* **17**:845–850, 1996.
42. Purcell, W. M. and Atterwill, C. K. Human placental mast cells as an *in vitro* model system in aspects of neuro-immunotoxicity testing. *Hum. Exp. Toxicol.* **13**:429–433, 1994.
43. Aloe, L. and Levi-Montalcini, R. Mast cells increase in tissues of neonatal rats injected with the nerve growth factor. *Brain Res.* **133**:358–366, 1977.
44. Matsuda, H., Kannan, Y., Ushio, H., Kiso, Y., Kanemoto, T., Susuki, H. and Kitamura, Y. Nerve growth factor induces development of connective tissue-type mast cells *in vitro* from murine bone marrow cells. *J. Exp. Med.* **174**:7–14, 1991.
45. Kannan, Y., Matsuda, H., Ushio, H., Kawamoto, K. and Shimada, Y. Murine granulocyte–macrophage and mast cell colony formation promoted by nerve growth factor. *Int. Arch. Allergy Immunol.* **101**:362–367, 1993.
46. Jippo, T., Ushio, H., Hirota, S., Mizuno, H., Yamatodani, A., Nomura, S., Matsuda, H. and Kitamura, Y. Poor response of cultured mast cells derived from mi/Mi mutant mice to nerve growth factor. *Blood* **84**:2977–2983, 1994.
47. Nakamura, K., Tanaka, T., Morita, E., Kameyoshi, Y. and Yamamoto, S. Enhancement of fibroblast-dependent mast cell growth in mice by conditioned medium of keratinocyte-derived squamous cell carcinoma cells. *Arch. Dermatol. Res.* **28**:91–96, 1994.
48. Horigome, K., Bullock, E. D. and Johansson, E. M. Jr. Effects of nerve growth factor on rat peritoneal mast cells. Survival promotion and immediate early gene induction. *J. Biol. Chem.* **269**:2695–2702, 1994.
49. Kawamoto, K., Okada, T., Kannan, Y., Ushio, H., Matsumoto, M. and Matsuda, H. Nerve growth factor prevents apoptosis of rat peritoneal mast cells through the trk proto-oncogene receptor. *Blood* **86**:4638–4644, 1995.
50. Richard, A., McColl, S. R. and Pelletier, G. Interleukin 4 and nerve growth factor can act as co-factor for interleukin-3-induced histamine production in human umbilical cord blood cell in serum-free culture. *Br. J. Haematol.* **81**:6–11, 1992.
51. Valent, P., Spanböchl, E., Sperr, W. R., Sillaber, C., Zsebo, K. M., Agis, H., Strobl, H., Geissler, K., Bettelheim, P. and Lechner, K. Induction of differentiation of human mast cells from bone marrow and peripheral blood mononuclear cells by recombinant human stem cell factor/kit-ligand in long-term culture. *Blood* **80**:2237–2245, 1992.
52. Denburg, J.A. Differentiation of human basophils and mast cells. In: *Human Basophils and Mast Cells: Biological Aspects* (Marone, G., ed.), pp. 49–71. Karger, Basle, 1995.
53. Valent, P. Cytokines involved in growth and differentiation of human basophils and mast cells. *Exp. Dermatol.* **4**:255–259, 1995.
54. Grabbe, J., Welker, P., Dippel, E. and Czarnetzki, B. M. Stem cell factor, a novel cutaneous growth factor for mast cells and melanocytes. *Arch. Dermatol. Res.* **287**:78–84, 1994.
55. Czarnetzki, B. M., Krüger, G. and Sterry, W. *In vitro* generation of mast cell-like cells from human peripheral mononuclear phagocytes. *Int. Arch. Allergy Appl. Immunol.* **71**:161–167, 1983.
56. Czarnetzki, B. M., Figdor, C. G., Kolde, G., Vroom, T., Aalberse, R. and de Vries, J. E. Development of human connective tissue mast cells from purified blood monocytes. *Immunology* **51**:549–554, 1984.
57. Grabbe, J., Welker, P., Möller, A., Dippel, E., Ashmann, L. F. and Czarnetzki, B. M. Comparative cytokine release from human monocytes, monocyte-derived immature mast cells and a human mast cell

line (HMC1). *J. Invest. Dermatol.* **103**:504–508, 1994.

58. Grabbe, J., Welker, P., Rosenbach, T., Nürnberg, W., Krüger-Krasagakes, S., Artuc, M., Fiebiger, E. and Henz, B. M. Release of stem cell factor (SCF) from human keratinocytes (HaCaT) is increased in differentiating versus proliferating cells. *J. Invest. Dermatol.* **107**:219–224, 1996.
59. Grabbe, J., Welker, P., Grützkau, A., Dippel, E., Nürnberg, W., Zuberbier, T. and Henz, B. M. Induction of human leukemic mast cell differentiation by fibroblast supernatants, but not by stem cell factor. *Scand. J. Immunol.* **47**:324–331, 1998.
60. Furitsu, T., Tsujima, T., Tono, T., Ikeda, H., Kitayama, H., Koshimizu, U., Sugahara, H., Butterfield, J. H., Ashman, L. K., Kanayama, Y., Matsuzawa, Y., Kitamura, Y. and Kanehara, Y. Identification of mutations in the coding sequence of the proto-oncogene c-kit in a human mast cell leukemia cell line causing ligand-independent activation of c-kit product. *J. Clin. Invest.* **99**:1736–1744, 1993.
61. Welker, P., Grabbe, J., Grützkau, A. and Henz, B. M. Effects of NGF and other fibroblast-derived growth factors on immature human mast cells (HMC-1). *Immunology* **94**:310–317, 1998.
62. Schwartz, L. B. and Kepley, C. Development of markers for human basophils and mast cells. *J. Allergy Clin. Immunol.* **94**:1231–1240, 1994.
63. Xia, H. Z., Kepley, C. L., Sakai, K., Chelliah, L., Irani, A. and Schwartz, L. B. Quantitation of tryptase, chymase, Fc epsilon RI alpha and Fc epsilon RI gamma mRNAs in human mast cells and basophils by competitive reverse transcription–polymerase chain reaction. *J. Immunol.* **154**:5472–5480, 1995.
64. Mitsui, H., Furitsu, T., Dvorak, A. M., Irani, A. A., Schwartz, L. B., Inagaki, M., Takei, M., Ishizaka, K., Zsebo, K. M., Gillis, S. and Ishizaka, T. Development of human mast cells from umbilical cord blood cells by recombinant human and murine c-Kit ligand. *Proc. Natl. Acad. Sci. USA* **90**:735–739, 1993.
65. Nilsson, G., Blom, T., Harvima, I., Kusche-Gullberg, M. and Nilsson, K. Stem cell factor-dependent human cord blood derived mast cells express α- and β-tryptase, heparin and chondroitin sulphate. *Immunology* **88**:308–314, 1996.
66. Kepley, C. L., Craig, S. S. and Schwartz, L. B. Identification and partial characterization of a unique marker for human basophils. *J. Immunol.* **154**:6548–6555, 1994.
67. Hermes, B., Feldmann-Böddeker, I., Krüger-Krasagakes, S. and Henz, B. M. Mast cells express NGF- and GM-CSF-receptors. Comparison of normal versus scar tissue. *Arch. Dermatol. Res.* **289**:40, 1997.
68. Algermissen, B., Hermes, B., Feldmann-Böddeker, I., Bauer, F. and Henz, B. M. Mast cell chymase and tryptase during tissue turnover: analysis on *in vitro* mitogenesis of fibroblasts and keratinocytes and alterations in cutaneous scars. *Exp. Dermatol.* **8**:193–198, 1999.
69. Tuveri, M. A., Passiu, G., Mathieu, A. and Aloe, L. Nerve growth factor and mast cell distribution in the skin of patients with systemic sclerosis. *Clin. Exp. Rheumatol.* **11**:319–322, 1993.
70. Algermissen, B., Bauer, F., Schadendorf, D. and Czarnetzki, B. M. Analysis of mast cell subpopulations (MC_T, MC_{TC}) in cutaneous inflammation using novel enzyme-histochemical staining techniques. *Exp. Dermatol.* **3**:290–297, 1994.
71. Raychaudhuri, S. P., Jiang, W. Y. and Farber, E. M. Psoriatic keratinocytes express high levels of nerve growth factor. *Acta. Derm. Venereol.* **78**:84–86, 1998.
72. Bull, H. A., Leslie, T. A., Chopra, S. and Dowd, P. M. Expression of nerve growth factor receptors in cutaneous inflammation. *Br. J. Dermatol.* **139**:776–783, 1998.
73. Aloe, L., Tuveri, M. A., Carcassi, U. and Levi-Montalcini, R. Nerve growth factor in the synovial fluid of patients with chronic arthritis. *Arthritis Rheum.* **35**:351–355, 1992.
74. Lowe, E. M., Anand, P., Tereghi, G., Williams-Chestmut, R. E. and Osborne, J. L. Increased nerve growth factor levels in the urinary bladder of women with idiopathic sensory urgency and interstitial cystitis. *Br. J. Urol.* **79**:572–577, 1997.
75. Skaper, S. D., Facci, L., Romanello, S. and Leon, A. Mast cell activation causes delayed neurodegeneration in mixed hippocampal cultures via the nitric oxide pathway. *J. Neurochem.* **66**:1157–1166, 1996.
76. Theoharides, T. C. Mast cells: the immune gate to the brain. *Life Sci* **46**:607, 1990.
77. Rozniecki, J. J., Hauser, S. L., Stein, M., Lincoln, R. and Theoharides, T. C. Elevated mast cell tryptase in cerebrospinal fluid of multiple sclerosis patients. *Ann. Neurol.* **37**:63–66, 1995.
78. Kerschensteiner, M., Gallmeier, E., Behrens, L., Leal, V. V., Misgeld, T., Klinkert, W. E., Kolbeck, R., Hoppe, E., Oropeza-Wekerle, R. L., Bartke, I., Stadelmann, C., Lassmann, H., Wekerle, H. and Hohlfeld, R. Activated human T cells, B cells, and monocytes produce brain-derived neurotrophic factor *in vitro* and in inflammatory brain lesions: a neuroprotective role of inflammation? *J. Exp. Med.* **189**:865–870, 1999.
79. Theoharides, T. C. The mast cell: a neuroimmunoendocrine master player. *Int. J. Tissue React.* **18**:1–21, 1996.

80. Petty, B. G., Cornblath, D. R., Adornato, B. T., Chaudhry, V., Flexner, C., Wachsman, M., Sinicropi, D., Burton, L. E. and Peroutka, S. J. The effect of systemically administered recombinant human nerve growth factor in healthy human subjects. *Ann. Neurol.* **36**:244–246, 1994.
81. Olson, L. The coming of age of the GDNF family and its receptors: gene delivery in a rat Parkinson model may have clinical implications. *Trends Neurosci.* **20**:277–279, 1997.
82. Paus, R., Maurer, M., Slominski, A. and Czarnetzki, B. M. Mast cell involvement in murine hair growth. *Dev. Biol.* **163**:230–240, 1994.
83. Paus, R., Lüftl, M. and Czarnetzki, B. M. Nerve growth factor modulates keratinocyte proliferation in murine skin organ culture. *Br. J. Dermatol.* **130**:174–180, 1994.
84. Botchkarev, V. A., Botchkareva, N. V., Welker, P., Metz, M., Lewin, G. R., Subramaniam, A., Bulfone-Paus, S., Hagen, E., Braun, A., Lommatzsch, M., Renz, H. and Paus, R. A new role for neurotrophin-3: involvement of brain-derived neurotrophic factor and neurotrophin-4 in hair cycle control. *FASEB J.* **13**:395–410, 1999.
85. Botchkarev, V. A., Welker, P., Albers, K. M., Botchkareva, N. V., Metz, M., Lewin, G. R., Bulfone-Paus, S., Peters, E. M., Lindner, G. and Paus, R. A new role for neurotrophin-3: involvement in the regulation of hair follicle regression (catagen). *Am. J. Pathol.* **153**:785–799, 1998.

CHAPTER 23

Modulation of Peripheral Neurotransmission Associated with Mast Cell Activation

BRADLEY J. UNDEM, RADHIKA KAJEKAR and ALLEN C. MYERS*

Johns Hopkins Asthma and Allergy Center, Baltimore, Maryland, U.S.A.

Antigenic activation of tissue mast cells is an early event in a complex pathway leading to the pathophysiology of allergic disease. A consideration of the various symptoms of allergy, including itching, coughing, sneezing, abnormalities in gastrointestinal motility, bronchoconstriction and excessive airway secretions, reveals that alterations in the peripheral nervous system play the central role in transducing mast cell activation to allergic symptomatology. This justifies the rather large literature published on various aspects of mast cell–nerve interactions. The vast majority of the literature on mast cell–nerve interactions pertains to the anatomical and morphological association between mast cells and nerves. Indeed, in virtually all tissues of the body, mast cells are found in close proximity to nerve fibres. Another body of literature describes the various mechanisms by which nerve stimulation may alter the function of tissue mast cells. Neither of these aspects of nerve–mast cell interactions will be reviewed here. The interested reader can find these topics covered in various excellent reviews on mast cell–nerve interactions (1–3). Rather, this chapter will provide an overview of the surprisingly sparse literature that deals specifically with the issue of how mast cell activation may alter peripheral neurotransmission.

The study of mast cell–nerve interactions, or for that matter any cell–cell interaction, requires that the appropriate cellular architecture is maintained. Thus, information on the regulation of neuronal function by mast cell activation can only be gained from isolated tissue. This raises an important caveat: although there are relatively selective methods by which mast cells can be stimulated, none of them are specific. Moreover, a change in neuronal function following mast cell activation within a given tissue should not be interpreted as a direct mast cell–nerve interaction. Perhaps more often is the case where mast cell activation leads to processes that indirectly, through stimulation of intermediate cells, alter neuronal function. These points are implicit in the following overview of the

* Corresponding author.

MAST CELLS AND BASOPHILS
ISBN 0-12-473335-2

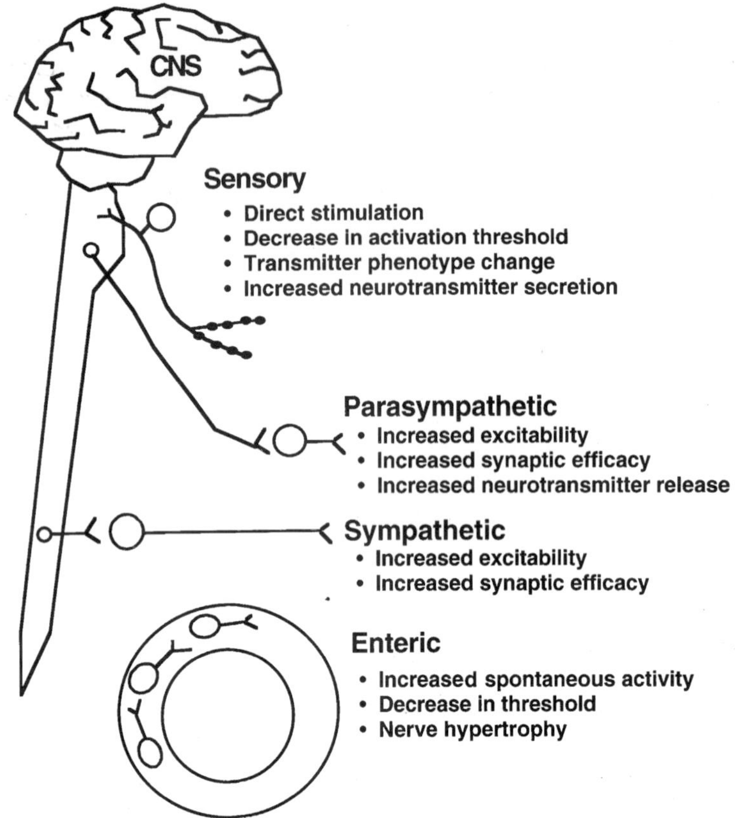

Fig. 1 Schematic representation of peripheral sensory, enteric, sympathetic and parasympathetic neural pathways associated with mast cell activation. See text for details.

interactions between mast cell activation and sensory, enteric, sympathetic and parasympathetic neurotransmission (Fig. 1).

SENSORY (AFFERENT) NERVOUS SYSTEM

The most self-evident neuronal response to mast cell activation is stimulation of sensory nerves. This is especially the case when mast cells are activated by a specific allergen. Allergen provocation leads to intense itching, coughing and sneezing. These reflexes are triggered by stimulation of primary afferent nerves in the skin and airways, respectively (4, 5). It is very likely that mast cell activation in visceral tissues will stimulate visceral sensory nerves as well. Stimulation of visceral sensory nerves often fails to lead to an overt response, but can cause large changes in physiology via the activation of various reflex pathways.

The mechanisms and mediators involved in mast cell-mediated sensory nerve activation have not been completely worked out. Histamine, via histamine H_1 receptor activation, is a prototypical chemical activator of sensory nerves. Not surprisingly, histamine mimics the ability of mast cell activation to cause cutaneous and airway reflexes (4, 5). Histamine receptor antagonists, however, are only marginally effective at controlling allergen-induced sensory stimulation. This is because of the myriad receptors and ion channels on sensory nerves that are also likely involved in mast cell-mediated afferent neuromodulation. With respect to the mechanism of modulation, the literature, as discussed below, has revealed three general modes of regulation. Mast cell activation can lead to: overt electrophysiological changes and increases in action potential discharge; increased neuropeptide gene expression and synthesis in the sensory neuronal soma; and increased action potential-dependent neuropeptide secretion from afferent nerve endings (Fig.1).

Allergen challenge *in vivo* has been shown to lead to action potential discharge in pulmonary afferent nerves (6–9). The electrophysiological basis for this process is not known. In the vagal afferent system, at least a component of the afferent discharge in response to allergen challenge is likely due to secondary effects of bronchial smooth muscle contraction and vascular effects. An electrophysiological study of vagal sensory neurons within the nodose ganglia revealed that activation of the resident mast cells in the ganglia by antigen challenge did not lead to overt activation of the neurons. Rather, mediators were released that increased the *excitability* of the neuron (10). For example, histamine released from mast cells caused an inhibition of a resting or leak-type potassium current that resulted in a slight membrane depolarization and an increased responsiveness of the neurons to other stimuli. Eicosanoids such as leukotriene C_4, prostacyclin and prostaglandin D_2 released as a consequence of antigen challenge selectively inhibit a potassium current that is involved in setting the refractory state of the neurons (10, 11). These direct electrophysiological studies on sensory cell bodies in ganglia suggest that mast cell activation is more effective at altering the electrical excitability of the sensory nerve than at overtly evoking action potential formation. Supporting this speculation are studies of airway afferent nerve endings *in vitro* which demonstrate that antigen challenge does not cause action potential discharge in a subset of sensory nerves, but significantly potentiates their response to mechanical stimuli (12). A 10-min exposure to specific antigen, in airways from actively or passively sensitized animals, leads to a 3-fold reduction in the amount of force required to excite the mechanoceptors in the guinea pig trachea/bronchial wall. The mediators that cause this change in excitability have not been defined, but are unlikely to be histamine or eicosanoids (12).

Recent studies in guinea pigs have demonstrated that specific antigen challenge to the airways leads to a significant elevation in the amount of sensory neuropeptides (substance P, neurokinin A and calcitonin gene-related peptide) in the pulmonary tissues (13). Upon further investigation it was demonstrated that this effect was due, at least in part, to a stimulation of neuropeptide precursor gene transcription in sensory vagal ganglia (13).

A potentially important twist on these findings is that allergen challenge leads to substance P production in a subset of sensory fibres that are typically devoid of neuropeptides. In other words, allergen challenge leads to a phenotypic switch in the sensory neuropeptide innervation of the airways. In the airways, and other tissues, sensory neuropeptides such as substance P are found primarily in neurons of small diameter that project thin unmyelinated fibres to the tissue of innervation (14, 15). Due to their size and lack of myelination these fibres conduct action potentials relatively slowly ($<1\ m\ s^{-1}$) and are referred to as C-fibres. The sensory C-fibres in the airways are difficult to stimulate physiologically, but respond directly to stimuli such as bradykinin, capsaicin

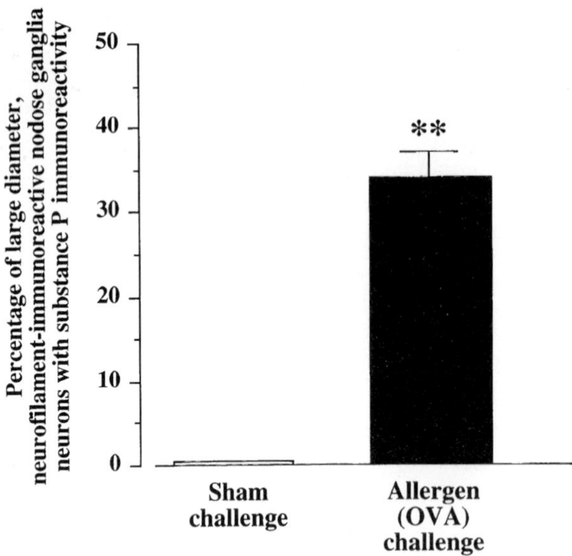

Fig. 2 Effect of allergen inhalation on substance P immunoreactivity in guinea pig vagal sensory ganglion neurons. The nodose ganglia of guinea pigs consists of small (~10 mm) diameter neurons that are substance P-positive, and large diameter (~40 mm) neurons that are neurofilament-immunoreactive but negative for substance P (see ref. 15). Within 24 h after allergen inhalation, 30% of the neurofilament-positive, large diameter population of neurons were induced to express substance P immunoreactivity. Values are mean ± SEM, n = 5. **p<0.01.

and decreases in pH (15–17). These so-called nociceptive(-like) fibres can thus be recruited during inflammatory processes. Other sensory fibres innervating the airways are larger diameter neurons that project myelinated fibres to the airways. These faster conducting A-fibres are devoid of neuropeptides, are positively stained with antibodies to neurofilament, are more sensitive to mechanical stimulation and are important in the regulation of physiological processes such as respiration (slowly adapting stretch receptors), or similar to C-fibres in responding to 'irritants' and initiating defensive reflexes (rapidly adapting 'irritant' receptors) (see refs 6–9). With this brief background in mind, it is interesting to note that allergen challenge of guinea pig airways leads to substantial elevations in preprotachykinin gene expression (13), and consequently substance P production, in large diameter fast conducting A-type sensory neurons innervating the airways (see Fig. 2). If the induced neuropeptides are transported to the peripheral and central ending of the sensory nerves, this would obviously significantly affect the sensory physiology both peripherally via axon reflexes, and, perhaps more importantly, centrally via alterations in synaptic neurotransmission in the brainstem.

The phenotypic change in neuropeptide-producing sensory nerves is observed within 1 day of allergen inhalation (13). We have noted this effect in both actively and passively sensitized guinea pigs (unpublished observation). Although not proven, it would seem likely that airway mast cell activation is an early step in the sequence of events leading to elevations in neuropeptide production by sensory cells. A 'mediator' that is known to increase preprotachykinin gene expression in sensory nerves is nerve growth factor (18). When nerve growth factor was instilled into the guinea pig trachea, it mimicked precisely the effect of allergen challenge in stimulating neuropeptide production in large, fast conducting sensory neurons (19). Nerve growth factor is known to be produced by mast cells (20–22), and is elevated in human airways immediately following allergen provocation (23), but the hypothesis that mast cell-derived nerve growth factor is involved in antigen-induced neuropeptide production in the airways has not yet been directly addressed.

In addition to increasing the excitability of sensory nerves, and increasing neuropeptide production, mast cell activation is also associated with an augmentation of neuropeptide secretion. In the guinea pig isolated airway preparation, specific allergen (ovalbumin), at a concentration that caused only a threshold level of mediator release and muscle contraction, profoundly elevated the tachykinergic contraction of the airway in response to electrical nerve stimulation (24). This effect was studied in the isolated trachea and thus was independent of potential effects on neuropeptide synthesis (the cell bodies in the vagal ganglia were removed from the preparation). Two mast cell-associated mediators that appear to play an important role in this response are histamine and Cys-leukotrienes (24–26).

The studies demonstrating that allergen challenge and mast cell activation can increase electrical excitability, neuropeptide production and neuropeptide secretion can explain many of the physiological responses to allergen challenge. In the airways, ion transport across the epithelium is enhanced upon allergen challenge, and a component of this response is secondary to neuropeptide release from sensory nerves (27). Similarly, a component of antigen-induced bronchial smooth muscle contraction is mediated via central reflexes stimulated by afferent nerve activation (6). It is likely that elevations in the 'responsiveness' of the primary afferent nervous system occurs in human tissues in which mast cells are chronically stimulated. Indeed, increases in local axon- and central-reflex responses are observed in the upper airways of seasonally allergic subjects when studied in their relevant allergic season (28). This may explain the beneficial effects of afferent nerve desensitization therapy in allergic rhinitis (29).

ENTERIC NERVOUS SYSTEM

Intestinal anaphylaxis is associated with alterations in intestinal motility and secretions. Numerous studies have led to the hypothesis that both motility and secretory changes associated with antigenic activation of mast cells are due to modulation of the enteric nervous system (see refs 30–32).

The strongest physiological evidence of mast cell-mediated modulation of nervous activity is found in an elegant study that demonstrated that antigen-induced increases in ion transport (short circuit current) in mouse intestine was inhibited by the neurotoxin tetrodotoxin (33). When the experiments were repeated in mast cell-deficient mice, the response on ion transport was reduced by about 50%, and, relevant to this discussion, this residual response was unaffected by tetrodotoxin. Reconstitution of the mast cell-deficient mice with mast cells restored the neuronal component of the antigen challenge, thus directly demonstrating, *in vivo*, a physiological interaction between mast cells and the nervous system.

Direct single cell electrophysiological studies on submucosal neurons also reveal a role of the enteric nervous system in intestinal anaphylaxis. Immediately upon antigen challenge the submucosal neurons become more spontaneously active, and are more excitable to stimulants such as acetylcholine (34). Antigen-induced activation of enteric nerves has also been noted by monitoring neurotransmitter release and Fos expression following antigen challenge (35). Through the recruitment of the enteric nervous system, alterations in gastrointestinal motility have been noted at sites remote from the antigen challenge. This requires an intact enteric nervous system, but seems to occur independent of central reflex control (36). The mediators that appear to be associated with antigen-induced activation of enteric nerves include histamine and eicosanoids. Mast cell tryptase may also be involved in mast cell-mediated recruitment of enteric nerve activity. Tryptase increases calcium flux in guinea pig myenteric neurons via the stimulation of proteinase-activated receptors (PARs) (37).

As with the sensory nerves, antigenic activation of mast cells may also cause neuroplastic effects in the gut. For example, histological and morphological studies have led to the hypothesis that mast cell activation leads to hypertrophy of enteric nerve trunks in patients suffering from Hirschsprung's disease (38). It is further suggested by these investigators that mast cell-associated nerve growth factor contributes to this process.

SYMPATHETIC NERVOUS SYSTEM

Autonomic ganglia are sites at which information (action potentials) arising from the central nervous system (CNS) is transmitted to the periphery via synaptic neurotransmission. The information from the CNS can be amplified, inhibited (filtered) or, in the case of a simple relay, left unaltered. Mechanisms that increase or decrease the efficacy of synaptic transmission will therefore have substantive effects on organ and vascular physiology. Mast cell activation is one mechanism by which synaptic transmission can be modified in mammals.

The mast cell content of guinea pig superior cervical and stellate sympathetic ganglia is similar to that of lung tissue (39). Mast cells within the sympathetic ganglia can be immunologically activated following either active or passive sensitization protocols. Activation of ganglion mast cells is associated with the release of mast cell mediators including histamine, proteolytic enzymes, prostaglandin D_2, and leukotrienes. As with other tissues, these primary mediators can then act on other cell types, leading to a cascade of mediator release within the ganglion (39). Simple electrophysiological studies in the isolated sympathetic ganglia have demonstrated that, in addition to releasing mediators, mast cell activation is associated with a striking increase in synaptic efficacy (40).

The simplest monitor of synaptic efficacy is the compound action potential. When an impulse is stimulated in the pre-ganglionic nerve trunk, a compound action potential can be recorded in the post-ganglionic nerve trunk. At an individual cell level, the pre-ganglionic action potential leads to acetylcholine release into the synaptic clef of a post-ganglionic neuron. This causes an excitatory post-synaptic potential (EPSP). If the EPSP is of sufficient magnitude, the threshold is reached for action potential generation. The action potential then propagates down the post-ganglionic nerve fibre to the tissue of innervation. If the EPSP is not of sufficient magnitude, no such action potential is elicited and neurotransmission is effectively aborted. By placing a recording electrode along the post-ganglionic nerve trunk (containing most post-ganglionic nerve fibres) a compound action potential can be recorded per each impulse applied to the pre-ganglionic nerve trunk (containing thousands of pre-ganglionic nerve fibres). The amplitude of the post-ganglionic compound action potential is proportional to the number of post-ganglionic nerves in which pre-ganglionic stimulation successfully led to an action potential. Using this technique, it was noted that, within 5 min of applying the sensitizing antigen to a guinea pig isolated superior cervical ganglion, the compound action potential nearly doubled in amplitude (40). This effect persists for over 2 h after the antigen has been washed from the tissue. This response was referred to as 'antigen-induced long-term potentiation' of sympathetic synaptic activity (41). It has been extensively studied in various guinea pig sympathetic ganglia both at the whole ganglion level as well as at the single neuron level using intracellular microelectrodes (41–43). The studies at the level of the single neurons reveal that there are multiple electrophysiological events occurring, likely involving several chemical mediators. Direct communication between mast cells and sympathetic neurons has been observed in elegant experiments using co-culture techniques (45). As in intact ganglia, these studies demonstrate that activation of mast cells causes a decrease in the membrane resistance of the sympathetic neuron and a depolarization of the membrane potential (45).

The extent to which mast cells within sympathetic ganglia are activated in humans is not known, nor is the potential consequence of such activation. Nevertheless, it is tempting to speculate that changes in sympathetic neurotransmission contribute to the physiology of anaphylactic responses where the relevant antigen is found in the circulation. With respect to anaphylaxis, it is interesting to note that, in studies in which the post-ganglionic nerve trunk leaving the superior cervical ganglia is severed, there is a marked reduction in antigen-induced late-phase pulmonary inflammation (46). The mechanism of this response has not been completely defined, but it appears that the relevant post-ganglionic fibres are those innervating the submandibular glands (47).

PARASYMPATHETIC NERVOUS SYSTEM

There have been relatively few studies on the role of mast cell activation on parasympathetic nerve activity. In contrast to sympathetic neurons that reside in large, easily accessible ganglia, parasympathetic neurons reside in very small ganglia located within or near the tissue of innervation (48). In the lower airways, the parasympathetic ganglia typically contain fewer than 20 neurons. These ganglia, as discussed above with sympathetic ganglia, are integration sites between the CNS and the organ of innervation. Elegant studies carried out *in vivo* in cats revealed that the synapses in airway parasympathetic ganglia effectively filter much of the pre-ganglionic input arising from the CNS (49). That is, many of the EPSPs evoked by action potentials propagated down the pre-ganglionic nerve fibre do not reach activation threshold for action potential generation in the post-ganglionic fibre. When this occurs the transmission between the CNS and the tissue of innervation is terminated. Thus, any process that increases synaptic efficacy will therefore lead to generalized increases in parasympathetic tone at the level of the effector cells (glands, airway smooth muscle, bronchial circulation, etc.) in the airways.

Surrounding airway parasympathetic ganglia are mast cells that degranulate upon allergen challenge (48). Associated with this mast cell degranulation in guinea pig airways is the depolarization of the membrane potential of principal neurons within the airway parasympathetic ganglia (48). A component of the membrane depolarization is due to histamine acting to inhibit a resting potassium current (50). This increases the responsiveness of the post-ganglionic neuron and thus increases synaptic efficacy. Histamine may have opposing effects within the airway ganglia as well, via the activation of histamine H_3 receptors (50). Antigenic activation of mast cells in the airways is also associated with decreases in 'accommodation' of the neurons (48). The antigen-induced decrease in accommodation results in the neurons generating more action potentials during prolonged excitatory membrane potentials. For example, when the parasympathetic neurons are stimulated with a 500-msec suprathreshold depolarization, only one or two action potentials are elicited. After antigen challenge this same neuron responds to the same stimulus with up to 25 action potentials. This antigen-induced decrease in accommodation is long-lasting and may be observed for over an hour after the antigen has been removed from the superfusate solution. Pharmacological and biophysical studies have suggested that the antigen-induced change in accommodation is secondary to the production of prostaglandin D_2 and the modulation of specific calcium and potassium currents (48, 51). Although antigen challenge leads to several ionic events at the level of the airway ganglia, the net effect is an increase in the efficacy of synaptic transmission (52). This, therefore, is a mechanism by which mast cell activation can control parasympathetic tone within a tissue.

Mast cell activation may also serve to modulate neurotransmission between the parasympathetic nerve and effector cell. When synaptic input sufficiently depolarized the ganglion neuron to action potential threshold, the action potential propagates down post-ganglionic axons where it invades transmitter-laden varicosities, causing voltage-dependent transmitter secretion into the so-called neuro-effector junction. The prototypical parasympathetic neurotransmitter is acetylcholine. In the pulmonary system, antigen challenge has been associated with an augmentation of acetylcholine release and potentiation of cholinergic contraction of airway smooth muscle (53). This effect may be indirect via the recruitment of other inflammatory cells that, in turn, release substances that inhibit the negative feedback inhibition caused by acetylcholine muscarinic M_2 receptor activation (53). In most tissues, vasoactive intestinal peptide (VIP) is also a neurotransmitter localized within post-ganglionic parasympathetic nerve terminals. Mast cell activation can serve to modulate VIP transmission by releasing tryptase that efficiently degrades VIP, and thereby inhibits the effect of VIP on effector cells (44).

SUMMARY

Mast cells are anatomically associated with all types of peripheral nerves. Most of the studies showing mast cell-mediated neuromodulation are based on isolated tissue or whole animal experiments where the nervous system is monitored before and after antigenic stimulation of mast cells. The complexity of the tissues makes it imprudent to conclude firmly that the neuromodulation studied is necessarily a direct consequence of mast cell activation. Nevertheless, these studies reveal that, at least associated with the selective activation of tissue mast cells, the neurophysiology of primary afferent (sensory) nerves, enteric nerves, sympathetic nerves and parasympathetic nerves is dramatically altered (Table 1). At the risk of overgeneralizing, it appears that typically the neuronal consequence of mast cell activation is one of increased function. This is true regardless of the nerve type studied. The mediators and ionic mechanism of mast cell-associated neural modulations are complex and strictly dependent on the neuronal system and effect studied. Histamine is effective in activating sensory and enteric nerves and can modulate synaptic transmission in autonomic ganglia. Other mast cell mediators including prostaglandin D_2, leukotrienes, tryptase and nerve growth factor have also been implicated in the neuromodulation associated with allergy.

Whereas this review has focused attention on the potential mechanisms by which the peripheral nervous system transduces antigen challenge into the symptoms of allergy and anaphylaxis, it should not go unmentioned that mast cells may be activated by other physiological stimuli. This fact, considered with the anatomical association of mast cells and nerves, and the profound influence that mast cell mediators have on neuronal activity, leads to speculation that mast cell–nerve interaction may be more than a mechanism by which the symptoms of allergy are triggered. For example, the stimulation of the enteric nervous system by mast cell activation may play an important role in mast cell-mediated host defence. Mast cell–nerve interactions have also been interpreted as important in neuronal survival and tissue repair mechanisms following injury (54, 55). Whether regarded from the physiological or pathophysiological vantage point, the literature on the biology of mast cell–nerve interactions amply provides for provocative conjecture. Unfortunately, until more research is completed on this fascinating topic, conjecture remains the appropriate noun.

TABLE I
Summary of Neuromodulatory Effects Associated with Mast Cell Activation

Nerve type	Effect associated with mast cell activation	Reference(s)
Sensory	Action potential discharge	6–9
	Membrane depolarization	11
	Decreased input impedance	11
	Inhibition of refractory period	11
	Increased mechanical sensitivity	12
	Increased neuropeptide production	13
	Potentiation of neuropeptide secretion	23
Enteric	Cholinergic nerve activation	32
	Increased submucosal nerve excitability	34
	Increased Fos expression	35
	Increased enteric reflex activity	36
	Neuronal hypertrophy	38
Sympathetic	Membrane depolarization	42–43
	Decreased input impedance	42–43
	Increased synaptic efficacy	39–43
Parasympathetic	Membrane depolarization	48, 50
	Increased input impedance	48, 50
	Decreased accommodation	48
	Increased synaptic efficacy	52
	Increased acetylcholine release	53
	Increased VIP degradation	44

REFERENCES

1. Williams, R. M., Bienenstock, J. and Stead, R. H. Mast cells: the neuroimmune connection. *Chem. Immunol.* **61**:208–235, 1995.
2. Bienenstock, J., MacQueen, G., Sestini, P., Marshall, J. S., Stead, R. H. and Perdue, M. H. Mast cell/nerve interactions *in vitro* and *in vivo*. *Am. Rev. Respir. Dis.* **143**:S55–S58, 1991.
3. Purcell, W. M. and Atterwill, C. K. Mast cells in neuroimmune function: neurotoxicological and neuropharmacological perspectives. *Neurochem. Res.* **20**:521–532, 1995.
4. Greaves, M. W. and Wall, P. D. Pathophysiology of itching. *Lancet* **348**:938–940, 1996.
5. Chang, A. B. Cough, cough receptors, and asthma in children. *Pediatr. Pulmonol.* **28**:59–70, 1999.
6. Mills, J. E. and Widdicombe, J. G. Role of the vagus nerves in anaphylaxis and histamine-induced bronchoconstriction in guinea pigs. *Br. J. Pharmacol.* **39**:724–731, 1970.
7. Mills, J. E., Sellick, H. and Widdicombe, J. G. Activity of lung irritant receptors in pulmonary microembolism, anaphylaxis and drug-induced bronchoconstrictions. *J. Physiol.* **203**:337–357, 1969.
8. Bergren, D. R., Myers, D. L. and Mohrman, M. Activity of rapidly-adapting receptors to histamine and antigen challenge before and after sodium cromoglycate. *Arch. Int. Pharmacodyn. Ther.* **273**:88–99, 1985.
9. Coleridge, H. M., Coleridge, J. C. G. and Schultz, H. D. Afferent pathways involved in reflex regulation of airway smooth muscle. *Pharmacol. Ther.* **42**:1–63, 1989.
10. Undem, B. J. and Weinreich, D. Electrophysiological properties and chemosensitivity of guinea pig nodose ganglion neurons *in vitro*. *J. Auton. Nerv. Syst.* **44**:17–34, 1993.
11. Undem, B. J., Hubbard, W. and Weinreich, D. Immunologically induced neuromodulation of guinea pig nodose ganglion neurons. *J. Auton. Nerv. Syst.* **44**:35–44, 1993.
12. Riccio, M. M., Myers, A. C. and Undem, B. J. Immunomodulation of afferent neurons in guinea-pig isolated airway. *J. Physiol.* **491**:499–509, 1996.
13. Fischer, A., McGregor, G. P., Saria, A., Philippin, B. and Kummer, W. Induction of tachykinin gene and peptide expression in guinea pig nodose primary afferent neurons by allergic airway inflammation. *J. Clin. Invest.* **98**:2284–2291, 1996.
14. Kummer, W., Fischer, A., Kurkowski, R. and Heym, C. The sensory and sympathetic innervation of guinea-pig lung and trachea as studied by retrograde neuronal tracing and double-labelling immunohistochemistry. *Neuroscience* **49**:715–737, 1992.
15. Riccio, M., Kummer, W., Myers, A. C. and Undem, B. J. Interganglionic segregation of airway vagal afferent fiber phenotypes. *J. Physiol.* **496**:521–530, 1996.
16. Kajekar, R., Proud, D., Meeker, S. N., Myers, A. C. and Undem, B. J. Characteristics of vagal afferent subtypes stimulated by bradykinin in guinea pig trachea. *J. Pharmacol. Exp. Ther.* **289**:682–687, 1999.
17. Fox, A. J., Barnes, P. J., Urban, L. and Dray, A. Effects of capsazepine against capsaicin- and proton-evoked excitation of single airway C-fibers and vagus nerve from the guinea-pig. *Neuroscience* **67**:741–752, 1995.
18. Levi-Montalcini, R., Skaper, S. D., DalToso, R., Petrelli, L. and Leon, A. Nerve growth factor: from neurotrophin to neurokine. *Trends Neurosci.* **19**:514–520, 1996.
19. Hunter, D. D. and Undem, B. J. Nerve growth factor causes a phenotypic change in substance P containing vagal neurons innervating guinea pig trachea. *Am. J. Respir. Crit. Care Med.* 2000 (in press).
20. Leon, A., Buriani, A., Toso, R. D., *et al.* Mast cells synthesize, store, and release nerve growth factor. *Proc. Natl. Acad. Sci. USA* **91**:3739–3743, 1994.
21. Tam, S. Y., Tsai, M., Yamaguchi, M., Yano, K., Butterfield, J. H. and Galli, S. J. Expression of functional TrkA receptor tyrosine kinase in the HMC-1 human mast cell line and in human mast cells. *Blood* **90:**1807–1820, 1997.
22. Nilsson, G., Forsberg-Nilsson, K., Xiang, Z., Hallbook, F., Nilsson, K. and Metcalfe, D. D. Human mast cells express functional TrkA and are a source of nerve growth factor. *Eur. J. Immunol.* **27**:2295–2301, 1997.
23. Sanico, A. M., Koliatsos, V. E., Stanisz, A. M., Bienenstock, J. and Togias, A. Neural hyperresponsiveness and nerve growth factor in allergic rhinitis. *Int. Arch. Allergy Immunol.* **118**:154–158, 1999.
24. Ellis, J. L. and Undem, B. J. Antigen-induced enhancement of noncholinergic contractile responses to vagus nerve and electrical field stimulation in guinea pig isolated trachea. *J. Pharmacol. Exp. Ther.* **262**:646–653, 1992.

25. Ellis, J. L. and Undem, B. J. Role of peptidoleukotrienes in capsaicin-sensitive sensory fibre-mediated responses in guinea-pig airways. *J. Physiol.* **436**:469–484, 1991.
26. McAlexander, M. A. and Undem, B. J. Inhibition of 5-lipoxygenase diminishes neurally evoked tachykinergic contraction of guinea pig isolated airway. *J. Pharmacol. Exp. Ther.* **285**:602–607, 1998.
27. Sestini, P., Bienenstock, J., Crowe, S. E., *et al.* Ion transport in rat tracheal epithelium *in vitro*. Role of capsaicin-sensitive nerves in allergic reactions. *Am. Rev. Respir. Dis.* **141**:393–397, 1990.
28. Riccio, M. M. and Proud, D. Evidence that enhanced nasal reactivity to bradykinin in symptomatic allergic individuals is mediated by neural reflexes. *J. Allergy Clin. Invest.* **97**:1252–1263, 1996.
29. Stjarne, P., Rinder, J., Heden-Blomquist, E., Cardell, L. O. and Lundberg, J. Capsaicin desensitization of the nasal mucosa reduces symptoms upon allergen challenge in patients with allergic rhinitis. *Acta. Otolaryngol.* **118**:235–239, 1998.
30. McKay, D. M. and Bienenstock, J. The interaction between mast cells and nerves in the gastrointestinal tract. *Immunol. Today* **15**:533–538, 1994.
31. Cooke, H. J. Neuro-modulation of ion secretion by inflammatory mediators. *Ann. N. Y. Acad. Sci.* **664**:346–352, 1992.
32. Javed, N. H., Wang, Y. Z. and Cooke, H. J. Neuroimmune interactions: role for cholinergic neurons in intestinal anaphylaxis. *Am. J. Physiol.* **263**:G847–G852, 1992.
33. Perdue, M. H., Masson, S., Wershil, B. K. and Galli, S. J. Role of mast cells in ion transport abnormalities associated with intestinal anaphylaxis. *J. Clin. Invest.* **87**:687–693, 1991.
34. Frieling, T., Palmer, J. M., Cooke, H. J. and Wood, J. D. Neuroimmune communication in the submucous plexus of guinea pig colon after infection with *Trichinella spiralis. Gastroenterology* **107**:1602–1609, 1994.
35. Miampamba, M., Tan, D. T., Oliver, M. R., Sharkey, K. A. and Scott, R. B. Intestinal anaphylaxis induces Fos immunoreactivity in myenteric plexus of rat small intestine. *Am. J. Physiol.* **272**:G181–G189, 1997.
36. Scott, R. B., Tan, D. T., Miampamba, M. and Sharkey, K. A. Anaphylaxis-induced alterations in intestinal motility: role of extrinsic neural pathways. *Am. J. Physiol.* **275**:G812–G821, 1998.
37. Corvera, C. U., Dery, O., McConalogue, K., *et al.* Thrombin and mast cell tryptase regulate guinea-pig myenteric neurons through proteinase-activated receptors-1 and -2. *J. Physiol.* **517**:741–756, 1999.
38. Kobayashi, H., Yamataka, A., Fujimoto, T., Lane, G. J. and Miyano, T. Mast cells and gut nerve development: implications for Hirschsprung's disease and intestinal neuronal dysplasia. *J. Pediatr. Surg.* **34**:543–548, 1999.
39. Undem, B. J., Hubbard, W. C., Christian, E. P. and Weinreich, D. Mast cells in the guinea pig superior cervical ganglion: a functional and histological assessment. *J. Auton. Nerv. Syst.* **30**:75–88, 1990.
40. Weinreich, D. and Undem, B. J. Immunological regulation of synaptic transmission in isolated guinea pig autonomic ganglia. *J. Clin. Invest.* **79**:1529–1532, 1987.
41. Weinreich, D., Undem, B. J., Taylor, G. and Barry, M. F. Antigen-induced long-term potentiation of nicotinic synaptic transmission in the superior cervical ganglion of the guinea pig. *J. Neurophysiol.* **73**:2004–2016, 1995.
42. Christian, E. P., Undem, B. J. and Weinreich, D. Endogenous histamine excites neurones in the guinea pig superior cervical ganglion *in vitro*. *J. Physiol.* **409**:297–312, 1989.
43. Cavalcante de Albuquerque, A. A., Leal-Cardoso, J. H. and Weinreich, D. Antigen-induced synaptic plasticity in sympathetic ganglia from actively and passively sensitized guinea-pigs. *J. Auton. Nerv. Syst.* **61**:139–144, 1996.
44. Franconi, G., Graf, P., Lazarus, S., Nadel, J. A. and Caughey, G. Mast cell tryptase and chymase reverse airway smooth muscle relaxation induced by vasoactive intestinal peptide in ferret. *J. Pharmacol. Exp. Ther.* **248**:947–951, 1989.
45. Janiszewski, J., Bienenstock, J. and Blennerhassett, M. G. Activation of rat peritoneal mast cells in coculture with sympathetic neurons alters neuronal physiology. *Brain Behav. Immun.* **4**:139–150, 1990.
46. Waddell, S. C., Davison, J. S., Befus, A. D. and Mathison, R. D. Role for the cervical sympathetic trunk in regulating anaphylactic and endotoxin shock. *J. Manipulative Physiol. Ther.* **15**:10–15, 1992.
47. Mathison, R., Davison, J. S. and Befus, A. D. Neuroendocrine regulation of inflammation and tissue repair by submandibular gland factors. *Immunol. Today* **15**:527–532, 1994.
48. Myers, A. C., Undem, B. J. and Weinreich, D. Influence of antigen on membrane properties of guinea pig bronchial ganglion neurons. *J. Appl. Physiol.* **71**:970–976, 1991.
49. Mitchell, R. A., Herbert, D. A., Baker, D. G. and Basbaum, C. B. *In vivo* activity of tracheal parasympathetic ganglion cells innervating tracheal smooth muscle. *Brain Res.* **437**:157–160, 1987.
50. Myers, A. C. and Undem, B. J. Antigen depolarizes guinea pig bronchial parasympathetic ganglion

neurons by activation of histamine H1 receptors. *Am. J. Physiol.* **268**:L879–L884, 1995.

51. Myers, A. C. Ca^{2+} and K^+ currents regulate accommodation and firing frequency in guinea pig bronchial ganglion neurons. *Am. J. Physiol.* **275**:L357–L364, 1998.
52. Undem, B. J., Myers, A. C. and Weinreich, D. Antigen-induced modulation of autonomic and sensory neurons *in vitro*. *Int. Arch. Allergy Appl. Immunol.* **94**:319–324, 1991.
53. Fryer, A. D. and Wills-Karp, M. Dysfunction of M_2-muscarinic receptors in pulmonary parasympathetic nerves after antigen challenge. *J. Appl. Physiol.* **71**:2255–2261, 1991.
54. Gottwald, T., Coerper, S., Schaffer, M., Koveker, G. and Stead, R. H. The mast cell–nerve axis in wound healing: a hypothesis. *Wound Repair Regen.* **6**:8–20, 1998.
55. Murphy, P. G., Borthwick, L. S., Johnston, R. S., Kuchel, G. and Richardson, P. M. Nature of the retrograde signal from injured nerves that induces interleukin-6 mRNA in neurons. *J. Neurosci.* **19**:3791–3800, 1999.

CHAPTER 24

Regulation of Gastrointestinal Mucin Production by Nerve–Mast Cell Interactions

CHARALABOS POTHOULAKIS

Division of Gastroenterology, Beth Israel Deaconess Medical Center, Harvard Medical School, Boston, Massachusetts, U.S.A.

INTRODUCTION

Mucins are high molecular weight glycoproteins secreted from specialized epithelial cells, called goblet cells, into the gastrointestinal lumen by exocytosis. It is well established that mucins participate in many physiological and pathophysiological gastrointestinal processes, including lubrication of mucosa surfaces, maintenance of the intestinal diffusion barrier, and defence against viruses, bacteria and bacterial toxins and proteases (1). Mucins have been also involved in the pathophysiology of intestinal inflammation (2). These mucin properties are largely dependent on the formation of a gel-like continuous extracellular matrix that covers the gastrointestinal mucosa. Increased mucin secretion occurs in response to several stimuli, including chemical irritation, bacterial toxins and pro-inflammatory cytokines such as interleukin-1 (IL-1) (3, 4). Several studies have shown that communications between the nervous, endocrine and immune networks play an important role in the regulation of mucin secretion from goblet cells. Neuropeptides are able to stimulate release of mucin from goblet cells. Recent studies from our laboratory showed that acute immobilization stress in rats and mice resulted in colonic mucin and prostaglandin E_2 (PGE_2) secretion and that the peptides corticotropin-releasing factor (CRF) and neurotensin (NT) are involved in these stress responses by interacting with colonic mucosal mast cells. Taken together, these results indicate that mucin release during non-traumatic stress involves communication between mast cells and neuropeptides.

A functional interaction between mast cells and nerves in the regulation of mucin secretion may be also important in the pathophysiology of irritable bowel syndrome (IBS) in humans. The aetiology of IBS appears to involve dysregulation of the pathways communicating signals between the central and the enteric nervous systems. Because stress has been linked to the pathophysiology of IBS and neuropeptides and their

MAST CELLS AND BASOPHILS
ISBN 0-12-473335-2

receptors participate in the responses to stress, considerable attention has been also given to the mechanism(s) by which neuropeptides communicate signals in the intestine. This chapter will discuss some of these mechanisms and address their potential importance for IBS.

NEURONAL MEDIATION OF MUCIN SECRETION

It is well established that nerves can control mucin secretion from goblet cells in the intestine. For example, Specian and Neutra (5) showed that the parasympathomimetic compound pilocarpine stimulated release of mucin from rat intestine. Phillips *et al.* (6) showed that application of electrical field stimulation to full-thickness rat distal ileum or descending colon mounted in Ussing chambers resulted in mucin secretion from crypt goblet cells. Interestingly, pre-incubation of rat ileum with either tetrodotoxin or atropin diminished goblet cell release in response to electric field stimulation, suggesting that cholinergic nerves regulate mucin secretion from goblet cells. Recent results from Plaisancie *et al.* (7) confirmed and extended these observations. These investigators showed that intra-arterial administration of the cholinergic agonist betanechol to isolated perfused rat colon resulted in a dose-dependent increase in mucin secretion. In another study, Phillips *et al.* (8) showed that exposure of isolated epithelial cells to the cholinergic secretagogue carbachol resulted in significant mucin exocytosis from secretory granules of crypt goblet cells. These results indicate that the mechanism of regulation of mucin secretion by cholinergic nerves in the gut involves a direct action of cholinergic secretagogues to crypt epithelial cells to elicit mucin secretion. Evidence also indicates that, in addition to cholinergic nerves, sensory nerves can also modulate intestinal mucin secretion *in vivo.* Laporte *et al.* (9) showed that intravenous administration of the neurotoxin capsaicin, which targets primary sensory afferent neurons, increased mucin secretion from rat duodenum and that substance P mimicked part of this effect. Along the same lines, Moore *et al.* (10) showed that desensitization of primary sensory neurons by neonatal capsaicin treatment diminished cholera toxin-induced mucin secretion in rat small intestine, while atropine only partially reduced this response.

Subsequent studies identified several specific neurotransmitters able to stimulate goblet cell release. Studies using the isolated rat perfused model showed that intra-arterial administration of vasoactive intestinal peptide (VIP), bombesin, peptide YY, or serotonin caused a substantial increase in mucin discharge, whereas, under the same conditions, calcitonin gene-related peptide had no significant effect (7, 11). However, these studies did not show whether neuropeptides cause mucin secretion by acting directly on goblet cells or indirectly via stimulation of other cells of the intestinal mucosa.

EFFECTS OF STRESS ON COLONIC MUCIN AND PROSTAGLANDIN RELEASE

Since, as discussed above, intestinal nerves are involved in the regulation of intestinal mucin secretion, and a large number of patients with IBS suffer from mucin diarrhoea (12), we sought to determine the effect of acute stress in colonic mucin secretion using the rat restraint model of stress. Our results demonstrated that 30 min of restraint stress of rats stimulated release of colonic mucin from goblet cells (13). Histological examination of colonic tissues revealed that surface goblet cells were the likely source of mucin in

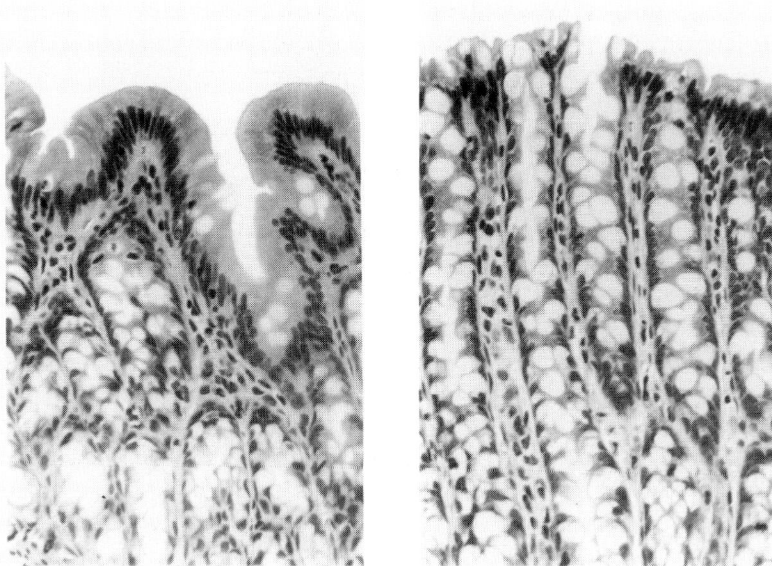

Fig. 1 Immobilization stress in rats causes colonic goblet cell depletion of mucin. Formalin-fixed colonic tissue sections obtained from immobilized (30 min) or non-immobilized (control) rats were stained with haematoxylin and eosin and Alcian blue. Colonic section from a control rat (right) shows several mucin-containing goblet cells in the superficial epithelium and crypts. Colonic section obtained from a rat exposed to immobilization stress (left) shows significant disappearance of mucin-containing cells in the superficial mucosal layer. (Reproduced, with permission, from *Am. J. Physiol.* **271**:G884–G892, 1996.)

these experiments since a significant number of mucin-containing surface goblet cells in the rat colon were absent following immobilization stress, whereas crypt goblet cells were intact (Fig. 1) (13). Pre-treatment of rats with the ganglionic blocker hexamethonium, the muscarinic cholinergic antagonist atropine, or the adrenergic blocker bretylium blocked stress-mediated colonic mucin secretion (13), indicating that parasympathetic and sympathetic nerves mediate these responses to stress. Interestingly, Phillips *et al.* previously showed that both crypt and villus intestinal goblet cells were able to secrete mucin in response to the cholinergic agent carbachol (8), indicating that different mechanisms for mucin secretion are responsible in our study (13) and the study of Phillips *et al.* (8).

Because mucin secretion is prostaglandin-dependent (14), we also evaluated changes in the mRNA expression of the prostaglandin synthesis rate-limiting enzyme prostaglandin H synthase/cyclo-oxygenase (COX) and in colonic PGE_2 levels after immobilization stress. Our results showed a dramatic increase in the colonic levels of the inducible form of COX, COX-2 mRNA, while the expression of the constitutive form COX-1 did not change (13). Restraint stress also increased colonic PGE_2 release which was inhibited by hexamethonium, while administration of the prostaglandin synthesis inhibitor indomethacin attenuated colonic mucin secretion in response to stress (13). Thus, PGE_2 release in response to immobilization stress is neuronally mediated, accompanied by an increase in the expression of the transcription factor COX-2 and is functionally linked to goblet cell degranulation and mucin secretion. These data are in keeping with previous findings showing that vascular perfusion of isolated rat colon with dimethyl-PGE_2 results in mucin discharge (7), and that dimethyl-PGE_2 can stimulate secretion of mucin glycoproteins from a human colonic adenocarcinoma cell line (15).

ROLE OF CORTICOTROPIN-RELEASING FACTOR IN STRESS-INDUCED COLONIC MUCIN SECRETION

Many studies have underlined the significance of the peptide corticotropin-releasing factor (CRF), a major regulatory peptide of the hypothalamic–pituitary–adrenal (HPA) axis, in the responses to stress. CRF is produced in the hypothalamus and other areas of the brain during stress and leads eventually to the release of corticosteroids (16). It is well documented that CRF participates in several stress-related colonic functions, including ion secretion and permeability (17), and colonic motility and fecal pellet output (18, 19). Since our results demonstrated increased colonic mucin secretion in response to stress, we next determined whether CRF is involved in this response. Our results showed that intravenous injection of the CRF antagonist α helical $CRF_{(9-14)}$ prior to immobilization stress inhibited mucin and PGE_2 release caused by stress (13). Furthermore, intravenous or intracerebroventricular injection of CRF itself in freely moving rats mimicked the effects of immobilization stress, since both PGE_2 and mucin secretion were increased in colonic explants from these animals (13). These results indicate that both central and peripheral CRF are involved in the mediation of stress-induced colonic responses. An important question arising from these results is whether peripheral receptors for CRF in the rat colon are involved in these responses. CRF acts via specific receptors as the central mediator of the HPA axis stress response (20). Two main receptor subtypes (type 1 and type 2) have been identified for CRF (20, 21). *In situ* hybridization studies showed distribution of CRF receptor subtype 2 mRNA in the submucosal layer and over cells at the base of the villi in the mouse duodenum (22). In a preliminary report, Lembo *et al.* (23) identified both CRF receptor subtypes in human colon by *in situ* hybridization. However, whether one or both CRF receptor subtypes in the colon mediate CRF-related responses remains to be elucidated.

NEUROTENSIN IS AN IMPORTANT MEDIATOR OF COLONIC STRESS RESPONSES

Neurotensin (NT), a 13-amino acid peptide originally isolated from bovine hypothalamus (24), is an important neurotransmitter in the central and peripheral nervous system (25, 26). In the gastrointestinal tract, which represents one of the most abundant sources for NT, this peptide influences multiple gastrointestinal functions, including cell growth, motility, chloride secretion and colonic inflammation (27, 28, 29). In humans and rodents the effects of NT are mediated by specific high- and low-affinity receptors belonging to the seven transmembrane domain G protein-coupled receptor family (30, 31). An important advance in the NT research field has been the identification and cloning of the human NTR1 from a colonic adenocarcinoma (HT-29) cell line (30). NT binds specifically to plasma membranes from these cells (31) and NT exposure stimulates mucin secretion (32) via mechanism that involves release of Ca^{2+} from intracellular stores (33).

Based on these considerations we next examined whether NT is also involved in PGE_2 and mucin release from rat colon following stress. We found a significant decrease in stress-induced colonic mucin and PGE_2 release and goblet cell degranulation in animals pre-treated with the NT receptor antagonist SR48692 15 min prior to immobilization stress (34). Although the cellular target(s) of the NT receptor antagonist in our experiments was not identified, our results demonstrated that the high-affinity NT receptor is expressed in the rat colonic mucosa, including colonic epithelial cells, and its expression is decreased after immobilization stress compared to colonic tissues from non-immobilized animals (34). Interestingly, immobilization stress in rats also caused increased NT blood levels as early as 10 min after restraint stress (35). These results (34, 35) indicate that NT and its high-affinity receptor are important in colonic goblet cell release following immobilization stress.

Since central and peripheral CRF was found to mediate stress-induced colonic responses, we next examined whether NT is involved in the CRF-mediated effects in rat colon. Our results showed that pre-treatment of rats with the NT receptor antagonist SR48692 prior to CRF injection inhibited colonic mucin and PGE_2 release as well as colonic mast cell degranulation induced by intravenous administration of CRF (36). These results indicate that CRF may mediate mucin secretion and PGE_2 release during experimental stress via a NT-dependent pathway and that the high-affinity receptor for NT participates in these CRF-induced colonic responses.

MAST CELLS AS A MAJOR LINK OF STRESS-MEDIATED MUCIN SECRETION

A large body of evidence indicates that mast cells play an important role in several intestinal functions, including intestinal transport, motility, permeability to ions and gut inflammation (37). These effects are thought to occur via mast cell mediators acting either directly on the epithelium and/or indirectly via intestinal nerves. The importance of the cross-talk between mast cells and nerves in the mediation of gastrointestinal functions has been also underscored by the close anatomical proximity of these two cell types in the intestinal tract (38). Several early studies suggested that mast cells may participate in the regulation of mucin secretion in the lung and the intestine. For example, addition of antigen to IgE-sensitized human bronchial airways in culture caused increased mucus glycoprotein production (39). Along the same lines, selective histamine agonists increased mucus secretion (40), suggesting the possibility that anaphylaxis of airways results in increased mucus release partly through histamine stimulation. Several groups have also demonstrated release of mucus during intestinal anaphylaxis in man. Gray and Walzer (41) injected serum from a donor allergic to peanuts into the rectal mucosa of a non-sensitive individual. Following ingestion of peanuts by the recipient there was rapid appearance of hyperaemia and mucus secretion at the site of the injection (41). Infection with *Nippostrongylus brasiliensis* is accompanied by degranulation of intestinal mucosal mast cells and goblet cell release (42, 43). Walker *et al.* (44) showed that immune complexes or anaphylaxis may stimulate rat intestinal goblet cells to secrete mucin.

Because mediators derived from mast cells may stimulate mucus glycoprotein release and since early studies suggested that central nervous system events are linked to activation of intestinal mast cells (45), we examined the role of mast cells in mucin release during acute immobilization stress. Our results showed that 30 min of immobilization stress to rats caused colonic mucosal mast cell activation, as shown by the increased levels of the specific mucosal mast cell product rat mast cell protease II

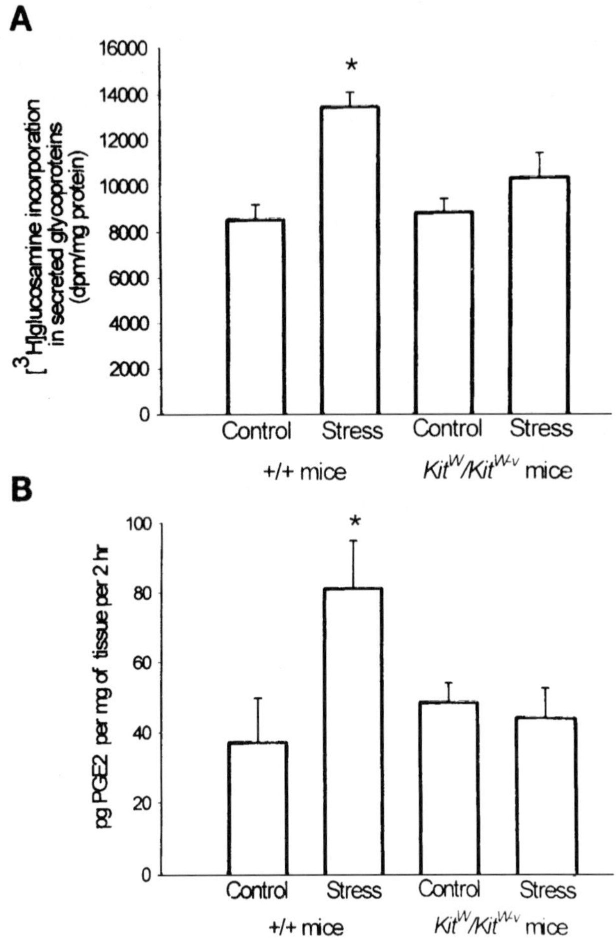

Fig. 2 Mast cell-deficient *Kit*^{W}/*Kit*$^{W\text{-}v}$ mice demonstrate reduced colonic mucin and PGE_2 release following immobilization stress. Mice were immobilized but control mice moved freely in their cages. After 30 min, mice were sacrificed, colon was removed and colonic explants were cultured. (A) Explants were cultured in medium containing [^{3}H]glucosamine and incubated at 37°C for 18 h. Mucin release was measured by incorporation of [^{3}H]glucosamine into TCA/PTA precipitates of culture supernatants. (B) Colonic explants were placed in culture dishes and incubated (37°C for 2 h) in modified Kreb's buffer. PGE_2 levels were measured in aliquots of the supernatants by an immunoenzymatic assay. Each bar represents the mean ± SEM of 5–13 different experiments, each with duplicate determinations. *$p<0.05$ versus all other groups. (Reproduced, with permission, from *Am. J. Physiol.* **274**:G1094–G1100, 1998.)

(RMCPII) (13). Also, 30 min of immobilization of rats also caused microscopic changes in ileal mast cells, evidenced by the disappearance of electron-dense granules of mast cells in ileal tissue sections (46). Pharmacological evidence also indicates that colonic mucin and PGE_2 secretion in response to stress is also mast cell-dependent. Thus, pretreatment of rats with the mast cell inhibitor lodoxamide inhibited stress-mediated mucin and PGE_2 release following immobilization stress (13). To directly assess the contribution of mast cells in colonic mucin secretion we compared colonic mucin and PGE_2 secretion in response to stress in mice that genetically lack mast cells. In previous studies these mice were found to respond less to several intestinal stimuli, including short circuit current changes following antigenic stimulation (47), enterotoxin-mediated ileal secretion and inflammation (48) and substance P-induced intestinal secretion (49). We found that mast cell-deficient mice secrete considerably less colonic mucin and PGE_2 in response to stress than normal mice do (50) (Fig. 2). Since mast cell-deficient mice, in addition to their mast cell deficiency, also exhibit profound anaemia and low numbers of the interstitial cells of Cajal (51, 52), we also examined colonic responses to stress in mast cell-deficient mice that had been selectively reconstituted with mast cells obtained from their congenic controls. Selectively reconstituted mast cell-deficient mice showed colonic mucin secretion and PGE_2 release similar to those of normal mice (50), providing further evidence for participation of mast cells in these stress-mediated colonic functions. Additional experiments using this animal model showed that, following immobilization stress, mast cell-deficient mice had similarly increased corticosterone levels to those of normal mice (50), suggesting that mast cell deficiency does not affect mobilization of the HPA axis following stress.

Our observation that mast cells are involved in several intestinal functions in animal models of stress is not surprising. Santos *et al.* showed that cold pain stress in humans causes release of the mast cell mediators tryptase, histamine and prostaglandin D_2 into the duodenum and increased jejunal water secretion, indicating that mast cells may be important in stress-induced intestinal dysfunction (53). It is also well established that stress activates mast cells in non-intestinal organs. For example, Theoharides *et al.* (54) showed that immobilization stress induces degranulation of mast cells in the dura mater of rats, and Spanos *et al.* (55) demonstrated, by light and electron microscopy, activation of 70% of bladder mast cells following 30 min of restraint stress. Acute immobilization stress also results in skin mast cell degranulation (56) and proliferation and degranulation of testis mast cells (57). Furthermore, isolation stress increases hypothalamic histamine content (58), and Pavlovian conditioning in rats causes activation of mucosal mast cells in the intestine (45). These results demonstrate a very close association between stressful conditions and mast cell activation, which may be important in several pathophysiological, stress-related processes.

CRF AND NT AS MEDIATORS OF INTESTINAL MAST CELL ACTIVATION DURING STRESS

The next question was to identify the neuropeptides involved in the communication between the nervous system and mast cells in these stress effects. Based on our observation that CRF mediates colonic mucin and PGE_2 release (13), and since CRF is a major peptide in the responses to stress, we examined its role in stress-induced mucosal mast cell activation. Our results showed that systemic administration of the peptide CRF antagonist α helical $CRF_{(9-14)}$ before immobilization stress significantly reduced release of the enzyme rat mast cell protease II from rat colon following immobilization stress (13). Similarly, injection of CRF either peripherally or centrally in freely moving rats also caused colonic mast cell activation (13). Administration of the CRF antagonist α helical $CRF_{(9-14)}$ before restraint stress also inhibited stress-induced colonic ion secretion and ion permeability, and peripheral administration of CRF itself reproduced these effects (17). Interestingly, CRF-induced ion permeability changes in rats were also diminished by administration of the mast cell inhibitor doxantrazole (17). Similarly, Gue *et al.* (59) demonstrated that CRF enhances abdominal contractions following experimental rectal distension by a mechanism involving central release of CRF and colonic mast cells. Taken together, these results indicate that CRF, either directly or indirectly, plays a significant role in colonic mast cell activation and may represent a major link whereby events in the central nervous system trigger mast cell degranulation in peripheral organs. Although the mechanism of activation of intestinal mast cells in response to CRF is not entirely clear, several reports also indicate that CRF and mast cell interactions are involved in other stress-mediated pathophysiological conditions such as activation of intracranial mast cells (54) and skin mast cells (60).

Several studies also documented the ability of the peptide NT to stimulate mast cell activation. Castagliuolo *et al.* (34) showed that colonic mast cell activation in response to immobilization stress was inhibited by antagonism of the high-affinity NT receptor. This NT-dependent effect could be the result of a direct interaction between NT and mast cells, since NT is able to directly activate mast cells *in vitro* to release mediators via a receptor-dependent mechanism (61, 62). A NT–mast cell interaction has also been suggested in stress-induced cardiac and bladder mast cell degranulation (63, 64).

CLINICAL IMPLICATIONS

To date no therapy has been convincingly proven to be effective in the treatment of most patients with stress-related intestinal conditions, including IBS. Since CRF and its receptors have been implicated in the pathophysiology of stress-related colonic dysfunction in animal models of stress, some laboratories examined its possible role in humans with IBS. In a recent study Fukudo *et al.* (65) examined the effect of intravenous administration of CRF on intestinal motility in humans and determined whether patients with IBS have a different response to CRF. Their results showed that CRF induced significantly higher increases in colonic motility in IBS patients and these were associated with higher ACTH levels in IBS patients compared with control patients (65). A likely source of CRF during pathophysiological conditions is the colon. For example, CRF-positive cells are present in the human colon and CRF has been localized in human colonic enterochromaffin cells (66) and mucosal macrophages (67). Interestingly, CRF-containing cells are also increased in the colon of patients with inflammatory bowel disease (67).

NT antagonism may represent an interesting possibility for treating some groups of patients with stress-related colonic dysfunction. Based on our results indicating that NT and its receptor are involved in stress-mediated colonic responses, we examined whether NT and NT receptor mRNA levels were altered in the colon of patients with IBS. Our results showed that colonic mucosal NT mRNA levels were approximately 2.5-fold higher in IBS patients than in control subjects (68), indicating the possible involvement of this receptor in IBS. We also investigated the ability of NT to cause secretion in human colonic mucosa and examined the mechanism of this response. We found that luminal administration of NT resulted in increased chloride secretion (69). This response was inhibited by the neuronal blocker tetrodotoxin, and the prostaglandin synthesis inhibitor indomethacin, indicating that the secretory effects of NT involve mucosal nerves and prostaglandins (69). Interestingly, NT receptor expression was also significantly elevated in the colon of patients with *Clostridium difficile*-associated pseudomembranous colitis (70), and administration of a NT receptor antagonist to rats inhibited colitis mediated by toxin A of *Clostridium difficile* (29). Thus, NT and its receptor appear to be involved in the mediation of several colonic responses in animals and humans, including chloride secretion, colonic inflammation, mast cell degranulation, and mucin and prostaglandin release in response to stress, and may represent a target peptide for treatment of these conditions.

REFERENCES

1. Neutra, M. R. and Forstner, J. F. Gastrointestinal mucus: synthesis, secretion, and function. In: *Physiology of the Gastrointestinal Tract* (Johnson, L. R., ed,), pp. 975–1009. Raven Press, New York, 1992.
2. Buisine, M. P., Desreumaux, P., Debailleul, V., Gambiez, L., Geboes, K., Ectprs, N., Delescaunt, M. P., Degand, P., Aubert, J. P., Colombel, J. F. and Porchet, N. Abnormalities in mucin gene expression in Crohn's disease. *Inflamm. Bowel Dis.* **5**:24–32, 1999.
3. Moore, B. A., Sharkey, K. A. and Mantle, M. Role of 5-HT in cholera toxin-induced mucin secretion in the rat small intestine. *Am. J. Physiol.* **270**:G1001–G1009, 1996.
4. Cohan, V. L., Scott, A. L., Dinarello, C. A. and Prendergast, R. A. Interleukin-1 is a mucus secretagogue. *Cell Immunol.* **136**:425–434, 1991.
5. Specian, R. D. and Neutra, M. R. Regulation of intestinal goblet cell secretion. I. Role of parasympathetic stimulation. *Am. J. Physiol.* **242**:G370–G379, 1982.
6. Phillips, T. E., Phillips, T. H. and Neutra, M. R. Regulation of intestinal goblet cell secretion. IV. Electrical field stimulation *in vitro*. *Am. J. Physiol.* **247**:G682–G687, 1984.
7. Plaisancie, P., Barcelo, A., Moro, F., Claustre, J., Chayvialle, J.-A., and Cuber, J.-A. Effects of neurotransmitters, gut hormones, and inflammatory mediators on mucus discharge in rat colon. *Am. J. Physiol.* **275**:G1073–G1088, 1998.
8. Phillips, T. E., Phillips, T. H. and Neutra, M. R. Regulation of intestinal goblet cell secretion. III. Isolated intestinal epithelium. *Am. J. Physiol.* **247**:G674–G681, 1984.
9. Laporte, J. L., Dauge-Geffroy, M. C., Chariot, J., Roze, C. and Potet, F. Sensory fibers sensitive to capsaicin can modulate secretion of the duodenal mucus. A morphometric study in rats. *Gastroenterol. Clin. Biol.* **17**:535–541, 1993.
10. Moore, B. A., Sharkey, K. A. and Mantle, M. Neural mediation of cholera toxin-induced mucin secretion in the rat small intestine. *Am. J. Physiol.* **265**:G1050–G1056, 1993.
11. Plaisancie, P., Bosshard, A., Meslin, J. C. and Cuber, J. C. Colonic mucin discharge by a cholinergic agonist, prostaglandins, and peptide YY in the isolated vascularly perfused rat colon. *Digestion* **58**:168–175, 1997.
12. Thompson, W. G., Creed, F., Drossman, D. A., Heaton, K. W. and Mazzacca, G. Functional bowel disorders and chronic functional abdominal pain. *Gastroenterol. Int.* **5**:75–91, 1992.
13. Castagliuolo, I., LaMont, J. T., Qiu, B., Fleming, S. M., Bhaskar, K. R., Nikulasson, S. T., Kornetsky, C. and Pothoulakis, C. Acute stress causes mucin release from rat colon: role of corticotropin releasing factor and mast cells. *Am. J. Physiol.* **271**:G884–G892, 1996.
14. Seidler, U. and Sewing, K. F. R. Ca^{2+}-dependent and -independent secretagogue action on gastric mucus secretion in rabbit mucosal explants. *Am. J. Physiol.* **256**:G739–G746, 1989.
15. Phillips, T. E., Stanley, C. M. and Wilson, J. The effect of 16,16-dimethyl prostaglandin E2 on proliferation of an intestinal goblet cell line and its synthesis and secretion of mucin glycoproteins. *Prostaglandins Leukot. Essent. Fatty Acids* **48**:423–428, 1993.
16. Lightman, S. L. and Young, W. S. Influence of steroids on the hypothalamic corticotropin releasing factor and pre-proenkephalin mRNA responses to stress. *Proc. Natl. Acad. Sci. USA* **86**:4306–4310, 1989.
17. Santos, J., Saunders, P. R., Hanssen, N. P., Yang, P. C., Yates, D., Groot, J. A. and Perdue, M. Corticotropin-releasing hormone mimics stress-induced colonic epithelial pathophysiology in the rat. *Am. J. Physiol.* **277**:G391–G399, 1999.
18. Williams, C. L., Peterson, J. L., Villar, R. G. and Burks, T. F. Corticotropin-releasing factor directly mediates colonic responses to stress. *Am. J. Physiol.* **253**:G582–G586, 1987.
19. Monnikes, H., Schmidt, B. G. and Taché, Y. Psychological stress-induced accelerated colonic transit in rats involves hypothalamic corticotropin-releasing factor. *Gastroenterology* **104**:716–723, 1993.
20. Chalmers, D T., Lovemberg, T. W., Grigoriadis, D. E., Behan, D. P. and De Souza, E B. Corticotropin-

releasing factor receptors: from biology to drug design. *Trends Pharmacol. Sci.* **17**:166–172, 1996.
21. Dieterich, K. D., Lehnert, H. and De Souza, E. Corticotropin-releasing factor receptors: an overview. *Exp. Clin. Endocrinol. Diabetes* **105**:65–82, 1997.
22. Lovemberg, T. W., Liaw, C. W., Grigoriadis, D. E., Clevenger, W., Chalmers, D. T., De Souza, E. B. and Oltersdorf, T. Cloning and characterization of a functionally distinct corticotropin-releasing factor receptor subtype from rat brain. *Proc. Natl. Acad. Sci. USA* **92**:836–840, 1995.
23. Lembo, T., Chalmers, D., De Souza, E. B., Bernstein, C. N., Tache, Y. and Mayer, E. A. Localization of novel corticotropin-releasing factor receptor (CRF2) mRNA in human colon. *Gastroenterology* **110**:A1093, 1996.
24. Carraway, R. E. and Leeman, S. E. The isolation of a new hypotensive peptide from bovine hypothalamus. *J. Biol. Chem.* **248**:6854–6861, 1973.
25. Nemeroff, C. B., Bissette, G., Manberg, P. J., Osbahr, A. J. III, Breese, G. A. and Prange, A. J. Jr. Neurotensin-induced hypothermia: evidence for an interaction with dopaminergic systems and the hypothalamic–pituitary–thyroid axis. *Brain Res.* **195**:69–84, 1980.
26. Kasckow, J. and Nemeroff, C. B. The neurobiology of neurotensin: focus on neurotensin–dopamine interactions. *Regul. Pept.* **36**:153–164, 1991.
27. Walsh, J. H. Gastrointestinal hormones. In: Physiology of the Gastrointestinal Tract, 2nd edn (Johnson, L. R., ed.) pp. 181–253. Raven Press, New York, 1992.
28. Kachur, J. F., Miller, R. J., Field, M. and Rivier, J. Neurohumoral control of ileal electrolyte transport II. Neurotensin and substance P. *J. Pharmacol. Exp. Ther.* **220**: 456–483, 1982.
29. Castagliuolo, I., Wang, C.-C., Valenick, L., Pasha, A., Nikulasson, S., Carraway, R. E. and Pothoulakis, C. Neurotensin is a proinflammatory peptide in colonic inflammation. *J. Clin. Invest.* **103**:843–849, 1999.
30. Vita, N., Laurent, P., Lefort, S., Chalon, P., Dumont, X., Kaghad, M., Gully, D., Lefur, G., Ferrara, P. and Caput, D. Cloning and expression of a complementary DNA encoding a high affinity human neurotensin receptor. *FEBS Lett.* **317**: 139–142, 1993.
31. Mazella, J., Botto, J. M., Guillemare, E., Coppola, T., Sarret, P. and Vincent, J. P. Structure, functional expression and cerebral localization of the levocabastine-sensitive neurotensin/neuromedin N receptor from mouse brain. *J. Neurosci.* **16**:5613–5620, 1996.
32. Augeron, C., Voisin, T., Maoret, J. J., Berthon, B., Laburthe, M. and Labiosse, C. L. Neurotensin and neuromedin N stimulate mucin output from human goblet cells (C1.16E) via neurotensin receptors. *Am. J. Physiol.* **262**:G470–G476, 1992.
33. Bou-Hanna, C., Berthon, B., Combettes, L., Claret, M. and Laboisse, C. L. Role of calcium in carbachol- and neurotensin-induced mucin exocytosis in a human colonic goblet cell line and cross-talk with the cyclic AMP pathway. *Biochem. J.* **299**:579–585, 1994.
34. Castagliuolo, I., Leeman, S. E., Bortolac-Suki, E., Nikulasson, S. T., Qiu, B. S., Carraway, R. E. and Pothoulakis, C. A neurotensin antagonist, SR-48692, inhibits colonic responses to immobilization stress in rats. *Proc. Natl. Acad. Sci. USA* **93**:12611–12615, 1996.
35. Pothoulakis, C., Castagliuolo, I. and Leeman, S. E. Neuroimmune mechanisms of intestinal responses to stress. *Ann, N. Y. Acad. Sci.* **840**:635–648, 1998.
36. Castagliuolo, I., Qiu, B. S., Leeman, S. E., LaMont, J. T. and Pothoulakis, C. Neurotensin and intestinal mast cells mediate colonic mucin and prostaglandin E2 release caused by corticotropin-releasing factor. *Gastroenterology* **110**:1062, 1996.
37. Perdue, M. H. and McKay, D. M. Integrative immunophysiology in the intestinal mucosa. *Am. J. Physiol.* **267**:G151–G165, 1994.
38. Stead, R. H., Dixon, R. F., Bramwell, N. H., Ridell, R. H. and Bienenstock, J. Mast cells are closely apposed to nerves in the human gastrointestinal mucosa. *Gastroenterology* **97**:575–585, 1989.
39. Shelhamer, J. H., Marom, Z. and Kalinger, M. Immunologic and neuropharmacologic stimulation of mucus glycoprotein release from human airways *in vitro*. *J. Clin. Invest.* **66**:1400–1412, 1980.
40. Boat, T. F. and Kleinerman, J. I. Human respiratory tract secretions: effect of cholinergic and adrenergic agents on *in vitro* release of protein and mucus glycoprotein. *Chest* **67**:S325–S332, 1975.
41. Gray, I. and Waltzer, M. Studies in mucus membrane hypersensitiveness. III. The allergic reaction of the passively sensitized rectal mucus membrane. *Am. J. Dig. Dis.* **4**:707, 1938.
42. King, S. J., Miller, H. R., Woodbury, R. G. and Newlands, G. F. Gut mucosal mast cells in *Nippostrongylus*-primed rats are the major source of secreted rat mast cell protease II following systemic anaphylaxis. *Eur. J. Immunol.* **16**:151–155, 1986.
43. Koninkx, J. F., Mirck, M. H., Hendriks, H. G., Mouwen, J. M. and van Dijk, J E. *Nippostrongylus*

brasiliensis: histochemical changes in the composition of mucins in goblet cells during infection in rats. *Exp. Parasitol.* **65**:84–90, 1988.
44. Walker, W. A., Wu, M. and Bloch, K. J. Stimulation by immune complexes of mucus release from goblet cells in the rat small intestine. *Science* **4**:707–710, 1977.
45. MacQueen, G., Marshall, J., Perdue, M., Siegel, S. and Bienenstock, J. Pavlovian conditioning of rat mucosal mast cells to secrete rat mast cell protease II. *Science* **243**:83–85, 1989.
46. Theoharides, T. C., Letourneau, R., Patra, P., Pang, X., Boucher, W. and Mompoint, C. Stress-induced rat intestinal mast cell intragranular activation and inhibitory effect of sulfated proteoglycans. *Dig. Dis. Sci.* **44**:87S–93S, 1999.
47. Perdue, M. H., Masson, S., Wershil, B. K. and Galli, S. J. Role of mast cells in ion transport abnormalities associated with intestinal anaphylaxis. Correction of the diminished secretory response in genetically mast cell-deficient W/W^v mice by bone marrow transplantation. *J. Clin. Invest.* **87**:687–693, 1991.
48. Wershil, B., Castagliuolo, I. and Pothoulakis, C. Mast cell involvement in *Clostridium difficile* toxin A-induced intestinal fluid secretion and neutrophil recruitment in mice. *Gastroenterology* **114**:956–964, 1998.
49. Wang, L., Stanisz, A. M., Wershil, B K., Galli, S. J. and Perdue, M. H. Substance P induces ion secretion in mouse small intestine through effects on enteric nerves and mast cells. *Am. J. Physiol.* **269**:G85–G92, 1995.
50. Castagliuolo, I., Wershil, B. K., Karalis, K., Pasha, A., Nikulasson, S. T. and Pothoulakis, C. Colonic mucin release in response to immobilization stress is mast cell-dependent. *Am. J. Physiol.* **274**:G1094–G1100, 1998.
51. Galli, S. J., Zsebo, K. M. and Geissler, E N. The kit ligand, stem cell factor. *Adv. Immunol.* **55**:1–96, 1994.
52. Huizinga, L., Thuneberg, L., Kluppel, M., Malysz, L., Mikkelsen, H. B. and Bernstein, A. Kit/W gene required for interstitial cells of Cajal and for intestinal pacemaker activity. *Nature* **373**:347–349, 1995.
53. Santos, J., Saperas, E., Nogueiras, C., Mourelle, M., Antolin, M., Cadahia, A. and Malagelada, J. R. Release of mast cells mediators into the jejunum by cold pain stress in humans. *Gastroenterology* **114**:640–648, 1998.
54. Theoharides, T C., Spanos, C., Pang, X., Alferes, L., Ligris, K., Letourneau, R., Rozniecki, J J., Webster, E. and Chrousos, G. P. Stress-induced intracranial mast cell degranulation: a corticotropin-releasing hormone-mediated effect. *Endocrinology* **136**:5745–5750, 1995.
55. Spanos, C., Pang, X., Ligris, K., Letourneau, R., Alferes, L., Alexakos, N., Sant, G. R. and Theoharides, T. C. Stress-induced bladder mast cell activation: implications for interstitial cystitis. *J. Urol.* **157**:669–672, 1997.
56. Singh, L. K., Pang, X., Alexakos, G., Letourneau, R. and Theoharides, T. C. Acute immobilization stress triggers skin mast cell degranulation via corticotropin-releasing hormone, neurotensin, and substance P: a link to neurogenic skin disorders. *Brain Behav. Immun.* **13**:225–239, 1999.
57. Tuncel, N., Gurer, F., Aral, E., Uzuner, K., Aydin, Y. and Baycu, C. The effect of vasoactive intestinal peptide on mast invasion/degranulation in testicular interstitium of immobilized + cold stressed and beta-endorphin-treated rats. *Peptides* **17**:817–821, 1996.
58. Bugajski, A. J., Chlap, Z., Gadek-Michalska, A. and Bugajski, A. Effect of isolation stress on brain mast cells and brain histamine levels in rats. *Agents Actions* **41**:C75–C76, 1994.
59. Gue, M., Rio-Lacheze, C., Eutamene, H., Theodorou, V., Fioramonti, J. and Bueno, L. Stress-induced visceral hypersensitivity to rectal distension in rats: role of CRF and mast cells. *Neurogastroenterol. Motil.* **9**:271–279, 1997.
60. Theoharides, T. C., Singh, L. K., Boucher, W., Pang, X., Letourneau, R., Webster, E. and Chrousos, G. Corticotropin-releasing hormone induces skin mast cell degranulation and increased vascular permeability, a possible explanation for its proinflammatory effects. *Endocrinology* **139**:403–413, 1998.
61. Feldberg, R. S., Cochrane, D. E., Carraway, R. E., Brown, E., Sawyer, R., Hartunian, M. and Wentworth, D. Evidence for a neurotensin receptor in rat serosal mast cells. *Inflamm. Res.* **47**:245–250, 1998.
62. Barrocas, A. M., Cochrane, D. E., Carraway, R. E. and Feldberg, R. S. Neurotensin stimulation of mast cell secretion is receptor-mediated, pertussis-toxin sensitive and requires activation of phospholipase C. *Immunopharmacology* **41**:131–137, 1999.
63. Pang, X., Alexacos, N., Letourneau, R., Seretakis, D., Gao, W., Boucher, W. and Cochrane, D. E. A neurotensin receptor antagonist inhibits acute immobilization stress-induced cardiac mast cell degranulation, a corticotropin-releasing hormone-dependent process. *J. Pharmacol. Exp. Ther.* **287**:307–314, 1998.

64. Alexacos, N., Pang, X., Boucher, W., Cochrane, D. E., Sant, G. R. and Theoharides, T. C. Neurotensin mediates rat bladder mast cell degranulation triggered by acute psychological stress. *Urology* **53**:1035–1040, 1999.
65. Fukudo, S., Nomura, T. and Hongo, M. Impact of corticotropin-releasing hormone on gastrointestinal motility and adrenocorticotropic hormone in normal controls and patients with irritable bowel syndrome. *Gut* **42**:845–849, 1998.
66. Kawahito, Y., Sano, H., Kawata, M., Yuri, K., Mukai, S., Yamamura, Y., Kato, H., Chrousos, G. P., Wilder, R. L. and Kondo, M. Local secretion of corticotropin-releasing hormone by enterochromaffin cells in human colon. *Gastroenterology* **106**:859–865, 1994.
67. Kawahito, Y., Sano, H., Mukai, S. Asai, K., Kimura, S., Yamamura, Y., Kato, H., Chrousos, G. P., Wilder, R. L. and Kondo, M. Corticotropin-releasing hormone in colonic mucosa in patients with ulcerative colitis. *Gut* **37**:544–551, 1995.
68. Castagliuolo, I., Lembo, T., Pasha, A. and Pothoulakis, C. Increased neurotensin receptor mRNA in colonic mucosa of patients with irritable bowel disease. *Gastroenterology* **114**:A357, 1998.
69. Riegler, M., Castagliuolo, I., Matthews, J. B., Wlk, M., Sogukoglu, T., Wenzl, E. and Pothoulakis, C. Neurotensin induces chloride secretion in human colonic mucosa through an adenosine-dependent pathway. *Gastroenterology* **114**:A1174, 1998.
70. Pasha, A., Wang, C., Nikulasson, S., Pothoulakis, C. and Castagliuolo, I. Increased neurotensin receptor mRNA in the colonic mucosa during *Clostridium difficile* pseudomembranous colitis in humans. *Gastroenterology* **114**:A1058, 1998.

SECTION SEVEN

MAST CELLS AND BASOPHILS IN HOMEOSTASIS AND HOST DEFENCE

CHAPTER 25

Mast Cell–Enterobacteria Interactions during Infection

SOMAN N. ABRAHAM*[1] and RAVI MALAVIYA[2]

[1]*Department of Pathology and Microbiology, Duke University Medical Center, Durham, North Carolina, and* [2]*Hughes Institute, St Paul, Minnesota, MN 55113, U.S.A.*

INTRODUCTION

Although mast cells were identified over 150 years ago, their physiological role in the body has remained a mystery. They are the primary effector cell mediating the pathophysiology of allergic diseases, which is attributable to the presence of high-affinity receptors for IgE on their cell-surface membranes and their intrinsic capacity to release significant amounts of pro-inflammatory mediators (1). Mast cells also contribute to a number of chronic inflammatory conditions, including stress-induced intestinal ulceration, rheumatoid arthritis, interstitial cystitis, scleroderma, psoriasis, neurofibromatosis and Crohn's diseases. In spite of these deleterious properties, mast cells have been preserved through evolution even among the lowest orders of animals, indicating that these cells must be serving a valuable function in the body.

At any given time, the host is challenged by a wide range of potentially infectious agents. Indeed, the most common cause of mortality and morbidity in man and animals is infectious diseases. The location of mast cells in the body ensures early contact with any invading pathogen, consistent with a role in immune surveillance. Indeed, mast cells are found preferentially in the skin, mucosal surfaces and around blood vessels, which correspond with portals of entry for many microorganisms (2, 3). Based on their active role in allergic disease, it is clear that mast cells have the intrinsic capacity to mobilize a rapid and vigorous inflammatory response in the host. That mast cells contribute to certain aspects of the adaptive immune response to pathogens, especially through IgE-mediated responses to certain parasitic infections, is well established (4, 5). Recently, several groups, including ours, have reported that mast cells play a critical and more central role in the host's immune response to infectious agents. These studies reveal mast cell involvement in various aspects of the immediate innate immune response to various

* Corresponding author.

MAST CELLS AND BASOPHILS
ISBN 0-12-473335-2

infectious agents (6–10). Cumulatively, they suggest that one physiological role for mast cells is to modulate protective host responses to various infectious agents. Here we review some of our studies relating to the role of mast cells in mediating bacterial clearance *in vivo* and some of our recent efforts at elucidating the molecular basis for mast cell recognition of bacteria. For these studies we have used enterobacterial strains of bacteria, primarily *Escherichia coli* and *Klebsiella pneumoniae.* These bacteria are opportunistic pathogens accounting for a large proportion of infections in immunocompromised and aged patients.

PROTECTIVE ROLE OF MAST CELLS AGAINST BACTERIAL INFECTION

The availability of mutant mice virtually lacking mast cells has made it possible to evaluate the specific contribution of the mast cell to the host's defence against enterobacterial challenge. One such mast cell-deficient mutant is the WBB6F1$^-$ W/W^v mouse, which has defective c-kit proteins, the receptor for stem cell factor (SCF). Through many elegant studies, Galli's laboratory has demonstrated the value of this mouse model system for analysing mast cell function (11). By quantitating differences in biological responses between mast cell-deficient mice (W/W^v) and their congenic mast cell-sufficient WBB6F1$^-$+/+ (+/+) controls and then analysing the responses in W/W^v mice that have been selectively reconstituted with cultured primary mast cells (W/W^v+MC), they have been able to define the specific *in vivo* contributions of mast cells to a variety of inflammatory reactions. We sought to utilize this system to define the role of mast cells in host defence against bacterial infections by comparing the susceptibility of genetically mast cell-deficient mice and their +/+ littermate controls following intraperitoneal challenge by mouse-virulent enterobacteria. We found that mast cell-deficient W/W^v mice experienced as much as 80% mortality compared with no mortality for the wild-type +/+ mice (7). Furthermore, W/W^v+MC mice exhibited the same resistance to infection as that exhibited by wild-type mice (7). This confirmed that the observed difference in susceptibility to bacterial infection was solely due to mast cells and not to other abnormalities that might exist in these mice. When we compared the extent of bacterial clearance in the three groups of mice, we noticed that clearance in W/W^v mice was at least 30-fold less effective than in mast cell-sufficient, +/+ or W/W^v+MC, mice (7). Remarkably, there was a 5-fold decrease in neutrophil numbers in the peritoneal cavities of W/W^v mice compared to both +/+ and W/W^v+MC mice, indicating that the neutrophil response to bacteria was impaired in the mast cell-deficient mice, which could account for their inability to effectively clear bacteria (7).

Mast cells have previously been shown to be a critical source of neutrophil chemoattractants following immune-complex injury and allergic inflammation (12, 13). Of the multiple chemoattractants that can potentially be released by the mast cell, tumour necrosis factor-α (TNF-α) is of special interest because of the discovery by Galli and coworkers that mast cells have the unique capacity to store pre-synthesized TNF-α and thus are able to release this cytokine immediately following activation (11, 14). Of note, TNF-α potentiates neutrophil bactericidal properties in addition to facilitating neutrophil extravasation through endothelial walls by triggering endothelial cell expression of various cell adhesion molecules (11). We detected a burst in extracellular TNF-α levels immediately preceding the influx of neutrophils to the site of bacterial instillation in the mast cell-sufficient mice, a phenomenon that was not detected in mast cell-deficient mice

(7). Injection of mast cell-sufficient mice with a monoclonal antibody directed at TNF-α but not another monoclonal directed at interleukin-1β (IL-1β), another mast cell pro-inflammatory cytokine, blocked up to 70% of the neutrophil response, confirming that mast cell-derived TNF-α plays a significant role in recruiting neutrophils to sites of bacterial infection (7). Working independently, Echtenacher and colleagues used a mast cell-deficiency model to reach a similar conclusion (6). These investigators showed that reconstitution of W/W^v mice with mast cells prevented death from surgically induced bacterial peritonitis. This effect was abolished by the administration of antibodies to TNF-α. Interestingly, the direct injection of TNF-α into W/W^v mice was also protective against enterobacterial infection, albeit only within a limited range of cytokine concentrations (6).

Recently, we have determined that TNF-α is not the only neutrophil chemoattractant produced by mast cells upon bacterial challenge. We found that the leukotriene LTB_4 is also a potent neutrophil chemoattractant produced by mast cells during bacterial infection. Leukotrienes (LT) are a family of biologically active compounds that are produced from arachidonic acid in a multistep process via activation of the 5-lipoxygenase pathway. Compound A63162, a hydroxamic acid derivative, is an orally active compound that selectively inhibits LT formation *in vitro* as well as *in vivo* in rodents (13). Three groups of mice (+/+, W/W^v and W/W^v+MC mice) were pre-treated with A63162 and challenged intraperitoneally with enterobacteria. Two hours later, we determined that the neutrophil influx in A63162-treated +/+ mice or W/W^v+MC mice was significantly decreased compared to untreated controls (Fig. 1a). Furthermore, the neutrophil response in A63162-treated +/+ mice appeared close to that of control W/W^v mice (Fig. 1a). A63162 did not have a significant effect on the neutrophil response of W/W^v mice (Fig. 1a), suggesting that mast cells were the source of the chemoattractant. The A63162-treated +/+ or W/W^v+MC mice also revealed a reduced level of bacterial clearance compared to untreated controls, indicating, once again, a correlation between neutrophil influx and bacterial clearance in the peritoneum (Fig. 1b). Our data strongly suggest that LT released from peritoneal mast cells following activation by enterobacteria is important for the influx of neutrophils and the resulting clearance of bacteria. For direct evidence that bacteria-activated mast cells can release LT, we examined cultured bone marrow-derived mast cells *in vitro* for LT release after 1 h exposure to enterobacteria. Limited amounts of LTB_4 were detected in the extracellular medium (Fig. 2), confirming that enterobacteria can induce the release of this potent chemoattractant from a variety of host cells (15, 16). Because LTB_4 is readily metabolized, its level in this assay represented only recently generated LTB_4. Therefore, we assayed the medium for LTC_4 that is secreted concomitantly with LTB_4, but which is markedly more stable. In Fig. 2, appreciable release of LTC_4 from bacteria-activated mast cells is indicated. It is noteworthy that LTC_4 is a vasoactive factor which can increase microvascular permeability. Taken together, these findings indicate that mast cells release LT in response to bacteria and that this LT plays an important role in modulating neutrophil influx and bacterial clearance at sites of infection.

MAST CELL PHAGOCYTOSIS OF BACTERIA

Although the phagocytic capability of mast cells has been known for some time (17), the physiological significance and implications of this mast cell property have not been

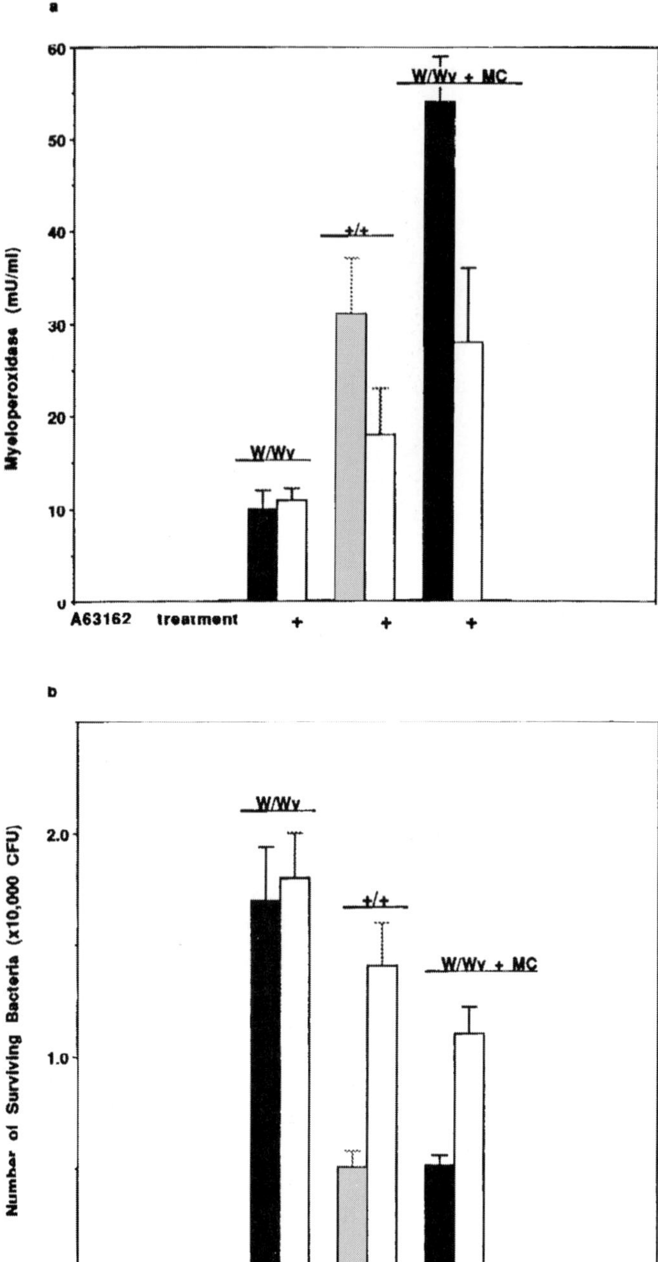

Fig. 1 Effect of an LT synthesis inhibitor on neutrophil recruitment and bacterial clearance. Neutrophil influx (a) and bacterial clearance (b) in the peritoneal cavities of A63162 pre-treated or control W/W^{V}, +/+ and W/W^{V}+MC mice were estimated 2 h after intraperitoneal bacterial challenge. A63162 treatment involved introducing 100 mg kg^{-1} (p.o.) of the compound 30 min before bacterial challenge. Neutrophils were quantitated employing a standard myeloperoxidase assay and bacterial viability was determined by standard colony counts on agar plates. Data are expressed as mean ± SEM, *n*=3.

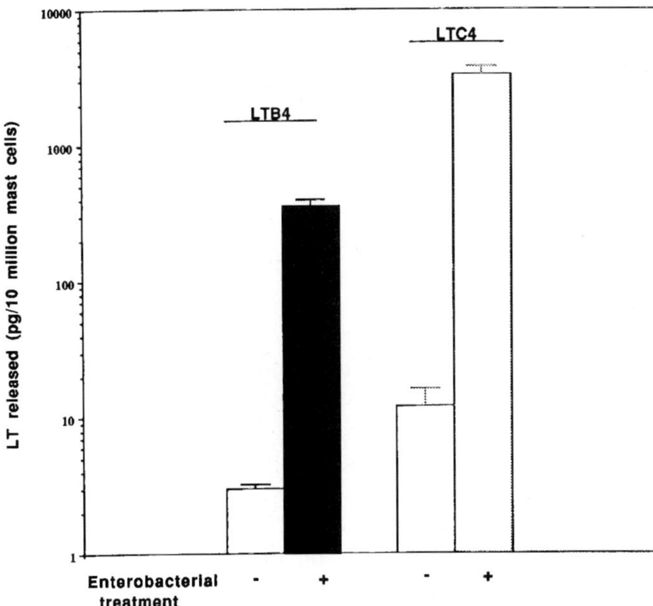

Fig. 2 *In vitro* LT release by mast cells following exposure to enterobacteria. Mast cells (1×10^6 ml^{-1}) were incubated with enterobacteria (1×10^7) for 1 h. LTB_4 and LTC_4 were estimated in cell-free supernatant before and after exposure to bacteria using an enzyme-linked immunoassay. Data are expressed as mean ± SEM, n=3.

adequately examined. Sher and colleagues were one of the first groups to investigate the mast cell phagocytosis of bacteria and reported that mast cells could bind and phagocytose *Salmonella* species (18). In light of the intrinsic invasive nature of *Salmonella*, it is not clear how much of this process was mediated by the mast cell. We demonstrated that classically non-invasive enterobacteria, including strains of *Escherichia coli*, *Enterobacter cloacae* and *K. pneumoniae*, were phagocytosed by mast cells in the presence of serum and subsequently killed (19). Viability assays indicated that, after 1 h of incubation, the amount of surviving bacteria had been lowered by 40%. In contrast, when the same bacteria were incubated with mouse 3T3 fibroblasts, viability increased by 60%, reflecting bacterial growth during the incubation period (19). Traditional phagocytes such as neutrophils and macrophages kill bacteria through a combination of oxidative and non-oxidative killing systems. The oxidative systems involve generation of toxic oxygen radicals and the non-oxidative systems involve acidification of phagocytic vacuoles and the fusion of lysosomal granules to the vacuole. Both mechanisms of bactericidal activity appear to be present in mast cells (19).

MOLECULAR BASIS FOR MAST CELL–ENTEROBACTERIA RECOGNITION

The Enterobacterial Component Responsible for Binding and Activating Mast Cells

Like cells that are traditionally known to mediate the innate immune responses to microbial infection, mast cells are able to discern and bind a variety of infectious agents even in the absence of specific antibodies to the pathogen. Mast cells can also be activated at a distance by invading bacteria through the release of toxins. This has been demonstrated with cholera toxin (20) and by cell wall components such as lipopolysaccharides (LPS) (21). Host-derived proteins generated during bacterial infection which could potentially activate mast cells include the by-products of complement activation, C3a and C5a, and the subfragments of fibrinogen and fibronectin that are generated following cleavage by plasmin (22, 23). Other mast cell activators such as granule-associated cationic polypeptides may originate from inflammatory cells such as macrophages, neutrophils and eosinophils following their activation by agonists including the bacterial peptide fMLP (24, 25).

In naïve hosts, mast cells exhibit two basic mechanisms of microbial recognition: opsonin-dependent and opsonin-independent recognition. The former requires soluble host components such as complement to first opsonize the microorganism before their recognition by the mast cell. For example, microbes coated with the iC3b fragment of complement will be readily recognized by the CR3 receptor on the mast cell membrane (18, 26). The critical role of complement system in mast cell recognition of bacteria comes from the recent finding that inflammatory responses to enterobacteria in complement-deficient mice were significantly reduced compared to those in wild-type mice (8). In opsonin-independent interactions, specific receptors on mast cells and complementary ligands on the bacterial cell surface are thought to be involved. In *in vitro* experiments involving monolayers of cultured mouse bone marrow-derived mast cells and various Gram-negative and Gram-positive bacteria, we found that mast cells possess a striking capacity to bind many bacteria even in the absence of host-derived opsonins (19) (unpublished findings). Among the Gram-negative bacteria tested, mast cells bound effectively to *E. coli*, *K. pneumoniae*, *Salmonella typhimurium* and *Helicobacter pylori*. Among Gram-positive bacteria, mast cells bound extremely well to *Staphylococcus aureus* and *Streptococcus faecalis* but exhibited considerably less affinity for *Streptococcus pyogenes* (unpublished findings). The molecular basis for many of these interactions is not known, but it is conceivable that ‘pattern recognition receptors’, molecules that display binding specificity for structural patterns displayed by cell-surface molecules common to many microorganisms (e.g. LPS), on the mast cell membrane are somehow involved (27).

Electron microscopy has revealed that mast cell binding to enteric bacteria correlated with expression by these organisms of hair-like appendages of attachment called fimbriae on the cell surface. Since the most commonly expressed fimbriae on enteric bacteria are the type 1 fimbriae, which are characterized by their capability to mediate mannose-inhibitable binding reactions, we examined the ability of 100 mM of D-mannose or its analogue, α-methyl-D-mannopyranoside, to block the binding interaction between bacteria and mast cells. We found that the binding of mast cells to *E. coli*, *K. pneumoniae*, *Enterobacter cloacae* and *Serratia marcescens* strains is mannose-sensitive (19), indicating the involvement of type 1 fimbriae in these binding interactions. At this time, it is unclear whether bacteria or mast cells play a more active role in the binding interaction since the intensity and mannose sensitivity of binding between dead bacteria and dead mast cells are comparable to that seen with their viable counterparts (unpublished). Type 1 fimbriae are heteropolymeric appendages 1–2 μm long and 7 nm thick. They are composed of a major subunit, FimA, and at least three minor subunits, including FimH, a mannose-binding lectin. We have determined that FimH is typically 29 kDa and is presented preferentially at the tips of the fimbriae (28, 29). Because of the peritrichous

TABLE I
Binding of FimH-expressing Bacteria and FimH-coated Beads to Mast Cells

	Properties	Number of adherent particles per 50 mast cells	% inhibition by D-mannose
Bacteria			
E. coli ORN103	Non-fimbriated/FimH$^-$	96 ± 12	2
E. coli ORN103(pSH2)	Fimbriated/FimH$^+$	570 ± 9	75
E. coli ORN103(pUT2002)	Fimbriated/FimH$^-$	82 ± 14	4
Beads			
FimH-coated		970 ± 23	65
FimA-coated		195 ± 18	12

The results are expressed as mean ± SE. $n = 3$.

and radial arrangement of the fimbriae, FimH is likely to be the first component on intact bacteria to make contact with host cells. To further confirm the potency of the interactions mediated by bacterial type 1 fimbriae, we transformed *E. coli* ORN103, a non-adhesive and non-fimbriated laboratory K12 strain, with pSH2, a plasmid encoding all the genes necessary for the expression of fully functional type 1 fimbriae (28). We found that, in contrast to *E. coli* ORN103, the transformed type 1 fimbriated *E. coli* strain ORN103(pSH2) mediated a high level of binding to mast cell monolayers (Table I). The binding of several *E. coli* ORN103(pSH2) to a mast cell is shown in Fig. 3. To demonstrate the specific role of FimH in the binding interaction, we examined the

Fig. 3 Scanning electron micrograph of a mast cell with several adherent FimH-expressing bacteria.

binding mediated by *E. coli* ORN103(pUT2002), a mutant created by specifically knocking-out the *fimH* gene at the distal end of the *fim* gene cluster in *E. coli* ORN103(pSH2) (28). The binding mediated by the fimbriated FimH$^-$ mutant was comparable to that mediated by the non-fimbriated ORN103 strain (Table I). To directly demonstrate the adhesive role of FimH, we coated inert beads with recombinant *E. coli* FimH and examined their ability to bind mast cell monolayers. We found that, compared to beads coated with bovine serum albumin, FimH-coated beads mediated a remarkably high level of binding (Table I) (30). Moreover, this binding could be inhibited by mannose (Table I). Finally, to prove that FimH, in the absence of all other bacterial components, was capable of binding and activating mast cells, we instilled beads coated with either *E. coli* FimH or *E. coli* FimA, the major and non-adhesive subunit, into the peritoneal cavities of mice. Compared to FimA-coated beads, FimH beads evoked at least an 8-fold higher level of histamine release from the peritoneal mast cells (Fig. 4) (30).

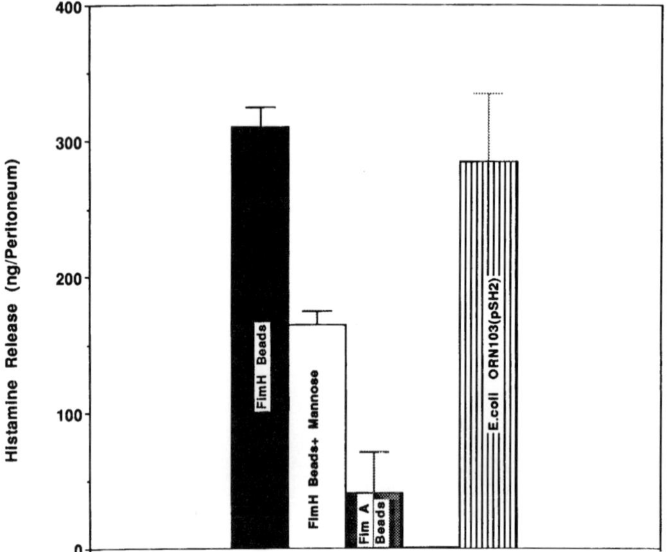

Fig. 4 Histamine release triggered by FimH- and FimA-coated beads in the mouse peritoneum. For comparison, histamine release by the injection of FimH-expressing *E. coli* ORN103(pSH2) is also shown. The bacteria or beads were injected intraperitoneally, and 60 min later the peritoneal fluid was assayed for histamine by a fluorometric method. n=3–5.

Thus, the component on enterobacteria largely responsible for binding and activating mast cells is the minor fimbrial component, FimH. Since, the binding between FimH-expressing bacteria or FimH-coated beads and mast cells is mannose-sensitive, the complementary 'receptor' on mast cell membrane is predictably a mannose-containing moiety. It is noteworthy that expression of type 1 fimbriae by enterobacteria plays a critical role in their ability to successfully colonize various mucosal epithelia and to resist the flushing actions of mucosal secretions (31). While the binding interactions between FimH-expressing bacteria and mast cells appear to largely benefit the host, it is not inconceivable that, in immunocompromised individuals, this interaction may be co-opted by these typically opportunistic pathogens for their own benefit.

The Mast Cell Moiety Responsible for Binding FimH and Triggering the Mast Cell Response

Because of its relevance in identifying the molecular events leading to mast cell responses to enterobacteria, we sought to identify the mast cell receptor for bacterial FimH. Our strategy is outlined in Fig. 5. We chose to isolate the FimH receptor from the

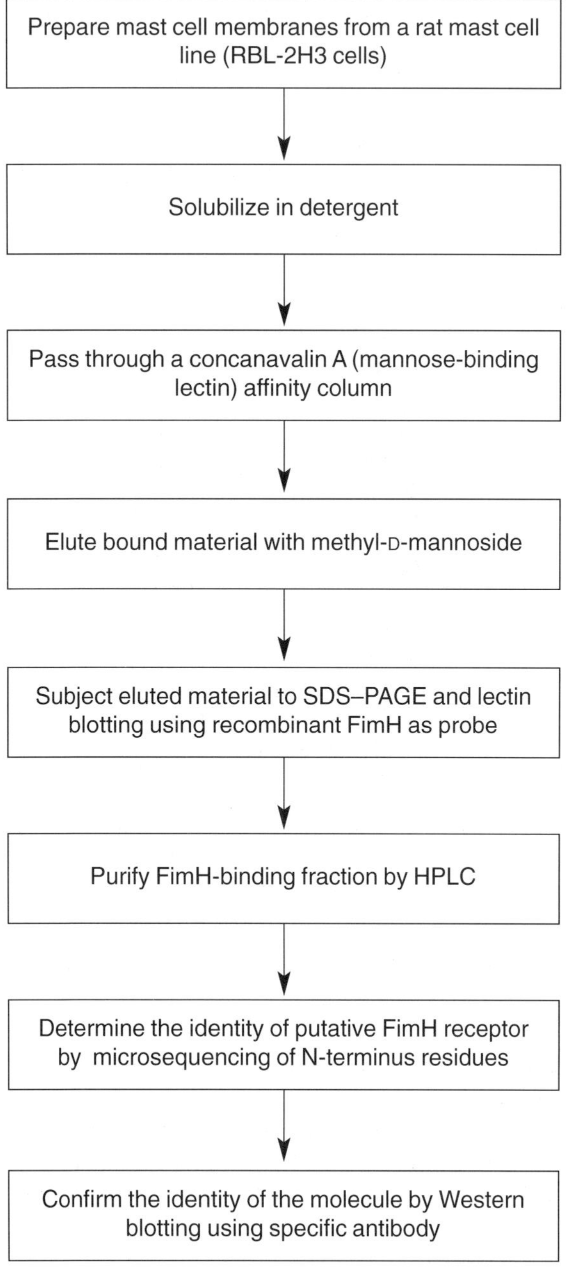

Fig. 5 Strategy used to isolate and identify the FimH receptor from mast cells.

rat mast cell line RBL-2H3 because, unlike bone marrow-derived cultured mast cells or mouse peritoneal mast cells, these cells could be readily cultured generating large numbers of cells. We reasoned that, because the FimH receptor contained mannose, we could use the mannose-binding lectin concanavalin A (Con A) to enrich for mannose-containing molecules from the mast cell membrane preparations. Pooled batch cultures of approximately 10^{11} cells were prepared and a Triton X-100-soluble membrane fraction (100,000 × *g* fraction) was obtained. This membrane fraction was passed through a Sepharose–Con A affinity column to isolate candidate mannosylated compounds. α-Methyl-D-mannopyranoside 100 mM was employed to elute all material bound via their mannose residues. The resulting eluate was dialysed to remove α-methyl-D-mannopyranoside and was then concentrated and subjected to SDS–PAGE. In order to identify the putative FimH-recognizing moiety among the mannoside eluted material, we electrophoretically transferred the material after SDS–PAGE onto PVDF membranes. The immobilized material was then exposed to ^{125}I labelled recombinant *E. coli* FimH. The FimH probe specifically bound to a 45-kDa band only in the absence of α-methyl-D-mannopyranoside (32). Furthermore, when the blot was exposed to FimH-expressing *E. coli* ORN103(pSH2) and mutant FimH-deficient *E. coli* ORN103(pUT2002), only the former bound to the 45-kDa band (32). The binding reaction of *E. coli* ORN103(pSH2) could be inhibited by 100 mM α-methyl-D-mannopyranoside. These data indicate that *E. coli* FimH binds specifically to a 45-kDa rat mast cell membrane component via mannose residues.

To determine the identity of the 45-kDa band, we purified the protein from the Con A eluted fraction to homogeneity and then subjected it to microsequencing. A sequence of 12 amino acid residues in the N-terminus were identified as 100% homologous to rat CD48. CD48 is a glycosylphosphatidylinositol (GPI)-linked molecule that has been reported to be present primarily on cells of haematopoietic lineage (33). Since GPI-linked moieties are cleaved off the surface of cells with phospholipase C (PLC), we incubated bone marrow-derived mouse mast cells with increasing amounts of PLC prior to exposure to FimH-expressing *E. coli.* PLC pre-treatment of mast cells was found to inhibit bacterial binding in a dose-dependent fashion (32). More direct evidence implicating CD48 as the putative *E. coli* FimH receptor on rodent mast cells comes from the observation that pre-treatment of these mast cells with antibodies to CD48 inhibited the adherence of FimH-expressing *E. coli* in a dose-dependent fashion, whereas antibodies to CD117 (c-kit), a well-known mast cell membrane marker, did not (32). We also wanted to know whether expression of CD48 on cells that do not normally express this molecule could promote binding of FimH-expressing *E. coli.* We stably transfected Chinese hamster ovary (CHO) cells with the full-length cDNA encoding CD48. We then examined the transfectants for their capacity to bind FimH-expressing *E. coli.* The association of bacteria with these CD48-expressing cells was at least 4-fold higher than the number associated with CHO cells transfected with control cDNA (32). Thus, CD48 molecules on transfected CHO cells are functional as FimH receptors. Taken together, the data show that the putative FimH receptor on mast cells is a GPI-linked moiety, CD48.

We have already shown that CD48-specific antibody inhibits FimH-mediated bacterial binding to mast cells (32). We have also reported previously that the early mast cell TNF-α response to bacteria plays a critical role in triggering the innate immune response (7). We sought to demonstrate the role of CD48 in eliciting the mast cell TNF-α response to FimH-expressing bacteria after (1) exposing mast cell surface CD48 to CD48-specific antibody and (2) removing CD48 from mast cell surface with PLC. Antibody to CD48, but not antibody to CD117, blocked mast cell TNF-α release in a dose-dependent manner (32). Further confirmation of the critical role of CD48 in the mast cell TNF-α response comes from the finding that pre-treatment of bone marrow-derived mast cells with increasing concentrations of PLC significantly reduced the mast cells' capacity to release TNF-α following exposure to FimH-expressing *E. coli* (32). Taken together, these observations provide definitive evidence that the mast cell TNF-α response to FimH-expressing bacteria is mediated by CD48 molecules present on the mast cell surface. Although there are likely to be other mannosylated moieties on the mast cell membrane capable of binding FimH-expressing *E. coli*, these studies show that CD48 is the biologically relevant receptor.

MAST CELL ACTIVATION VIA CD48

CD48 was the first physiologically relevant receptor to be identified on mast cells for an infectious agent. CD48 has been referred to as BCM1 in mice, OX45 in rats and Blast-1 in humans (32). The expression of CD48 is restricted to cells of haematopoietic lineage, particularly lymphocytes, monocytes and mast cells. CD48 was discovered as a cell-surface molecule expressed by human B lymphocytes in response to Epstein–Barr virus (EBV) infections, but its physiological role in the body is still unclear. The involvement of CD48 in bacterial recognition and in triggering TNF-α release in inflammatory cells represents a novel function for this molecule. Rodent CD48 is glycosylated, and the primary structure reveals several potential sites of glycosylation (34). The importance of the sugar moiety on CD48 and specifically its mannose residues is indicated by the observation that binding of FimH-expressing *E. coli* to mast cells is mannose-sensitive.

CD48 joins a growing class of GPI-anchored cell-surface molecules that serve as receptors for microbes and their toxins. Some notable examples of these receptors include CD14 for LPS, Thy-1 for aerolysin and CD55 for echovirus, and group B coxsackieviruses (35, 36). Engagement of these receptors by microbes or their products have been shown to trigger cellular responses (35, 37). How CD48 and other GPI-anchored receptors, which are only linked to the exoplasmic leaflet of the lipid bilayer of the plasma membrane, actually transduce intracellular signals is not clear. Many GPI-anchored moieties, including CD48, are typically found in special 'glycolipid-enriched microdomains' in the plasma membranes of cells (38). These microdomains are rich in signalling molecules such as the heterotrimeric GTP-binding proteins which can potentially mediate signal transduction from GPI-anchored proteins. G proteins are involved in many signal transduction pathways, including stimulation of adenylate cyclase, regulation of Ca^{2+} channels, stimulation of phospholipase A_2 and stimulation of phosphoinositol 3-kinase or phospholipase C (39). A recent immunochemical study has shown that CD48 in lymphocytes is physically associated with GTP-binding proteins (39). Thus, engagement of CD48 could potentially trigger a mast cell response via a signalling pathway involving GTP-binding proteins. Engagement of CD48 has also been shown to activate Src family member tyrosine kinases, which are also important effectors of signal transduction found associated with glycolipid-enriched microdomains of the plasma membrane (40).

INTERACTIONS OF HUMAN MAST CELLS WITH ENTEROBACTERIA

Because of the remarkable heterogeneity in morphology and responsiveness to agonists of mast cells from different animal species and even among cells from different sites in the same animal, it cannot be readily assumed that the response of human mast cells to enterobacteria is similar to that of their rodent counterparts. Recently, we examined the interactions of cord blood-derived human mast cells with *E. coli* ORN103(pSH2) as well as type 1 fimbriated strains of *K. pneumoniae* and *Citrobacter freundii.* We found that all the enteric bacteria bound avidly to mast cells in a mannose-inhibitable fashion, indicating the critical contribution of FimH (41). Although human mast cells express CD48, we have, as yet, not determined if this molecule serves as the FimH receptor on these cells. Nevertheless, the mast cells evoked a strong TNF-α response to these FimH-expressing enterobacteria. Indeed, the amount of TNF-α evoked in 6 h by *E. coli* ORN103(pSH2) was comparable to the amounts of TNF-α evoked during the same time period by the well-known agonists phorbol myristate acetate and calcium ionophore (41). Recent studies employing the human mast cell line HMC-1 5C6 has also revealed that mast cell activation by *E. coli* ORN103(pSH2) involves protein kinase C (42).

FINAL THOUGHTS

Mast cells can play a key role in modulating the early inflammatory response to enterobacterial infection. During infection, mast cells recognize enterobacteria and respond to these organisms by releasing neutrophil chemoattractants such as TNF-α and LT. We showed that the mast cell TNF-α and LT responses were crucial for bacterial clearance and for the survival of mice when challenged by potentially lethal doses of enterobacteria. A diagrammatic depiction of the mast cell's role is presented in Fig. 6. Although there are potentially many ways that mast cells can be activated during bacterial infection, we have identified an opsonin-independent mechanism involving the direct binding of enterobacteria to mast cell plasma membrane. The specific molecules that are involved are FimH, a mannose-binding lectin presented on the distal tips of bacterial fimbriae, and CD48 (at least in rodent mast cells) a molecule that is anchored to the mast cell plasma membrane via a GPI moiety. Because many potentially pathogenic enterobacteria express FimH, mast cells have the intrinsic capacity to recognize and bind a wide range of enterobacteria. The coupling of FimH with CD48 results in mast cell activation, leading to the release of several pro-inflammatory mediators which can potentially play a key role in determining the nature and intensity of the early inflammatory response to the infecting bacteria. It should be emphasized, however, that while this mode of recognition between mast cells and FimH-expressing enterobacteria exists, its utility depends to a large extent on the immune status of the host. Because of the preponderance of enterobacteria amongst our endogenous microflora, most healthy individuals have relatively high levels of enterobacteria-specific antibodies in the circulation and in mucosal secretions. Thus, invading enterobacteria are liable to be opsonized first with specific antibody or by complement fragments prior to encountering any mast cells. In which case, the interactions between the opsonized bacteria and mast cells may involve IgG or complement receptors on the mast cell rather than CD48. At any given time, the interactions between FimH-expressing bacteria and mast cells *in vivo* are

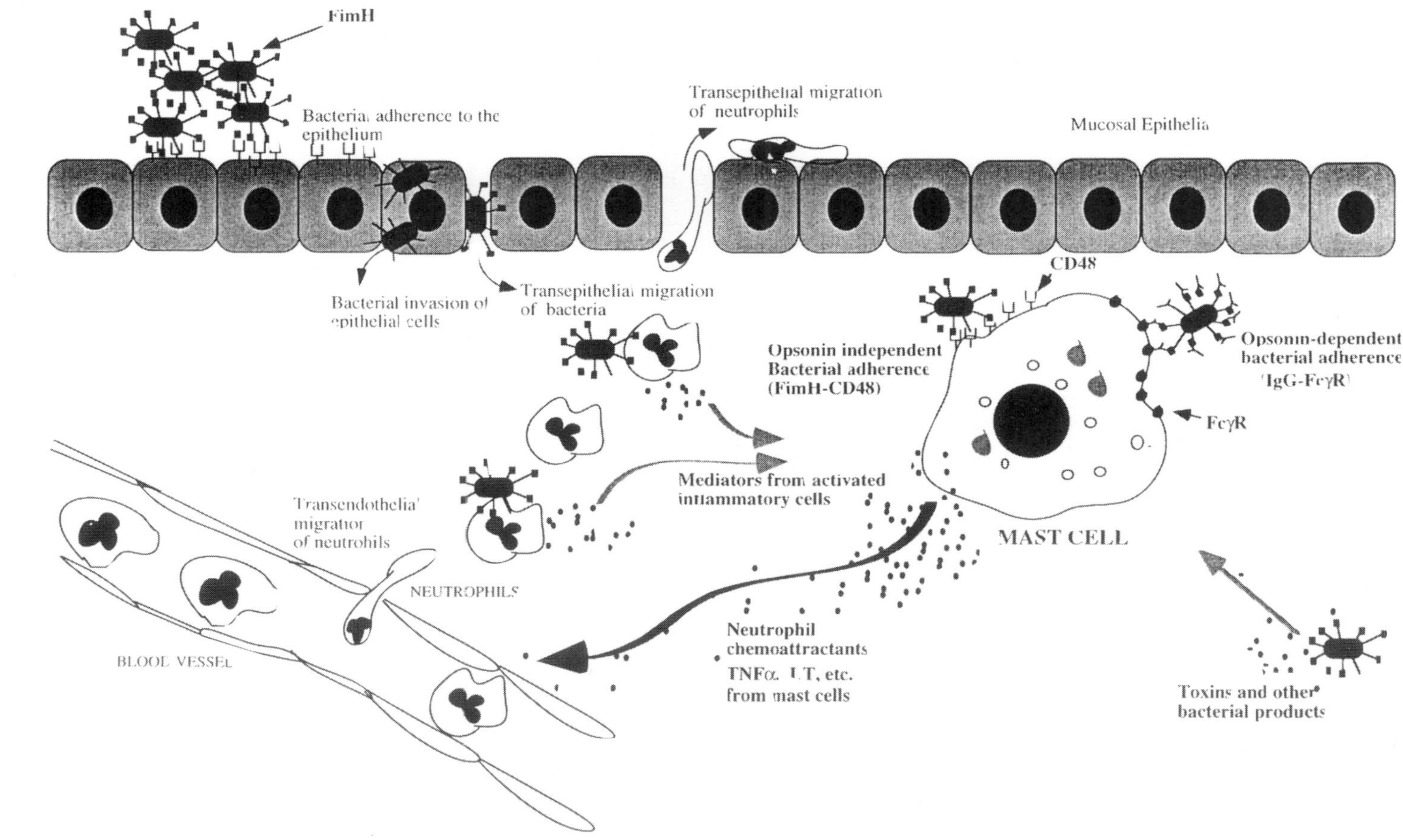

Fig. 6 Mast cell–bacteria interactions in infection.

likely to involve both opsonin-dependent and opsonin-independent interactions. In healthy hosts, the opsonin-dependent interactions are likely to predominate whereas, in naïve or immunocompromised patients, the opsonin-independent interactions such as that mediated between FimH and CD48 are likely to predominate. Although both types of interactions can result in mast cell activation, the range and amounts of mediator release are likely to be different since distinct mast cell receptors are involved. Nevertheless, the capacity of mast cells to recognize and respond to enterobacteria even in the absence of opsonins is consistent with its proposed role in contributing to the host's innate immune system.

ACKNOWLEDGEMENTS

We are grateful to James McLachlan and Matthew Duncan for critical review of the manuscript. This work was supported in part from research grants from the NIH (AI 35678 and DK 50814).

REFERENCES

1. Marone, G., Casolaro, V., Patella, V., Florio, G. and Triggiani, M. Molecular and cellular biology of mast cells and basophils. *Int. Arch. Allergy Immunol.* **114**:207–217, 1997.
2. Abraham, S. N. and Arock, M. Mast cells and basophils in innate immunity. *Semin. Immunol.* **10**:373–381, 1998.
3. Abraham, S. N. and Malaviya, R. Mast cells in infection and immunity. *Infect. Immun.* **65**:3501–3508, 1997.
4. Levy, D. A. and Frondoza, C. Immunity to intestinal parasites: role of mast cells and goblet cells. *Fed. Proc.* **42**:1750–1755, 1983.
5. Miller, H. R. Mucosal mast cells and the allergic response against nematode parasites. *Vet. Immunol. Immunopathol.* **54**:331–336, 1996.
6. Echtenacher, B., Mannel, D. N. and Hultner, L. Critical protective role of mast cells in a model of acute septic peritonitis. *Nature* **381**:75–77, 1996.
7. Malaviya, R., Ikeda, T., Ross, E. and Abraham, S. N. Mast cell modulation of neutrophil influx and bacterial clearance at sites of infection through TNF-alpha. *Nature* **381**:77–80, 1996.
8. Prodeus, A. P., Zhou, X., Maurer, M., Galli, S. J. and Carroll, M. C. Impaired mast cell-dependent natural immunity in complement C3-deficient mice. *Nature* **390**:172–175, 1997.
9. Rosenkranz, A. R., Coxon, A., Maurer, M., Gurish, M. F., Austen, K. F., Friend, D. S., Galli, S. J. and Mayadas, T. N. Impaired mast cell development and innate immunity in Mac-1 (CD11b/CD18, CR3)-deficient mice. *J. Immunol.* **161**:6463–6467, 1998.
10. Talkington, J. and Nickell, S. P. *Borrelia burgdorferi* spirochetes induce mast cell activation and cytokine release. *Infect. Immun.* **67**:1107–1115, 1999.
11. Galli, S. J. and Wershil, B. K. The two faces of the mast cell. *Nature* **381**:21–22, 1996.
12. Lukacs, N. W., Strieter, R. M., Chensue, S. W. and Kunkel, S. L. Activation and regulation of chemokines in allergic airway inflammation. *J. Leukoc. Biol.* **59**:13–17, 1996.
13. Zhang, Y., Ramos, B. F. and Jakschik, B. A. Neutrophil recruitment by tumor necrosis factor from mast cells in immune complex peritonitis. *Science* **258**:1957–1959, 1992.
14. Galli, S. J., Gordon, J. R. and Wershil, B. K. Cytokine production by mast cells and basophils. *Curr. Opin. Immunol.* **3**:865–872, 1991.
15. Bremm, K. D., Brom, H. J., Alouf, J. E., Konig, W., Spur, B., Crea, A. and Peters, W. Generation of leukotrienes from human granulocytes by alveolysin from *Bacillus alvei*. *Infect. Immun.* **44**:188–193, 1984.
16. Konig, W., Scheffer, J., Bremm, K. D., Hacker, J. and Goebel, W. Role of bacterial adherence and toxin production from *Escherichia coli* on leukotriene generation from human polymorphonuclear

granulocytes. *Int. Arch. Allergy Appl. Immunol.* **77**:118–120, 1985.

17. Padawer, J. Uptake of colloidal thorium dioxide by mast cells. *J. Cell. Biol.* **40**:747–760, 1969.
18. Sher, A., Hein, A., Moser, G. and Caulfield, J. P. Complement receptors promote the phagocytosis of bacteria by rat peritoneal mast cells. *Lab. Invest.* **41**:490–499, 1979.
19. Malaviya, R., Ross, E. A., MacGregor, J. I., Ikeda, T., Little, J. R., Jakschik, B. A. and Abraham, S. N. Mast cell phagocytosis of FimH-expressing enterobacteria. *J. Immunol.* **152**:1907–1914, 1994.
20. Leal-Berumen, I., Snider, D. P., Barajas-Lopez, C. and Marshall, J. S. Cholera toxin increases IL-6 synthesis and decreases TNF-alpha production by rat peritoneal mast cells. *J. Immunol.* **156**:316–321, 1996.
21. Leal-Berumen, I., Conlon, P. and Marshall, J. S. IL-6 production by rat peritoneal mast cells is not necessarily preceded by histamine release and can be induced by bacterial lipopolysaccharide. *J. Immunol.* **152**:5468–5476, 1994.
22. Galli, S. J. New concepts about the mast cell. *N. Engl. J. Med.* **328**:257–265, 1993.
23. Wojtecka-Lukasik, E. and Maslinski, S. Fibronectin and fibrinogen degradation products stimulate PMN-leukocyte and mast cell degranulation. *J. Physiol. Pharmacol.* **43**:173–181, 1992.
24. Zheutlin, L. M., Ackerman, S. J., Gleich, G. J. and Thomas, L. L. Stimulation of basophil and rat mast cell histamine release by eosinophil granule-derived cationic proteins. *J. Immunol.* **133**:2180–2185, 1984.
25. Fantozzi, R., Brunelleschi, S., Rubino, A., Tarli, S., Masini, E. and Mannaioni, P. F. FMLP-activated neutrophils evoke histamine release from mast cells. *Agents Actions* **18**:155–158, 1986.
26. Sher, A. and McIntyre, S. L. Receptors for C3 on rat peritoneal mast cells. *J. Immunol.* **119**:722–725, 1977.
27. Medzhitov, R. and Janeway, C. A. Jr. Innate immunity: impact on the adaptive immune response. *Curr. Opin. Immunol.* **9**:4–9, 1997.
28. Abraham, S. N., Goguen, J. D., Sun, D., Klemm, P. and Beachey, E. H. Identification of two ancillary subunits of *Escherichia coli* type 1 fimbriae by using antibodies against synthetic oligopeptides of fim gene products. *J. Bacteriol.* **169**:5530–5536, 1987.
29. Abraham, S. N., Sun, D., Dale, J. B. and Beachey, E. H. Conservation of the D-mannose-adhesion protein among type 1 fimbriated members of the family Enterobacteriaceae. *Nature* **336**:682–684, 1988.
30. Malaviya, R., Ross, E., Jakschik, B. A. and Abraham, S. N. Mast cell degranulation induced by type 1 fimbriated *Escherichia coli* in mice. *J. Clin. Invest.* **93**:1645–1653, 1994.
31. Thankavel, K., Madison, B., Ikeda, T., Malaviya, R., Shah, A. H., Arumugam, P. M. and Abraham, S. N. Localization of a domain in the FimH adhesin of *Escherichia coli* type 1 fimbriae capable of receptor recognition and use of a domain-specific antibody to confer protection against experimental urinary tract infection. *J. Clin. Invest.* **100**:1123–1136, 1997.
32. Malaviya, R., Gao, Z., Thankavel, K., Merwe, P. A. and Abraham, S. N. The mast cell tumor necrosis factor alpha response to FimH-expressing *Escherichia coli* is mediated by the glycosylphosphatidylinositol-anchored molecule CD48. *Proc. Natl. Acad. Sci. USA* **96**:8110–8115, 1999.
33. van der Merwe, P. A., Barclay, A. N., Mason, D. W., Davies, E. A., Morgan, B. P., Tone, M., Krishnam, A. K., Ianelli, C. and Davis, S. J. Human cell-adhesion molecule CD2 binds CD58 (LFA-3) with a very low affinity and an extremely fast dissociation rate but does not bind CD48 or CD59. *Biochemistry* **33**:10149–10160, 1994.
34. Killeen, N., Moessner, R., Arvieux, J., Willis, A. and Williams, A. F. The MRC OX-45 antigen of rat leukocytes and endothelium is in a subset of the immunoglobulin superfamily with CD2, LFA-3 and carcinoembryonic antigens. *EMBO J.* **7**:3087–3091, 1988.
35. Clarkson, N. A., Kaufman, R., Lublin, D. M., Ward, T., Pipkin, P. A., Minor, P. D., Evans, D. J. and Almond, J. W. Characterization of the echovirus 7 receptor: domains of CD55 critical for virus binding. *J. Virol.* **69**:5497–5501, 1995.
36. Fenton, M. J. and Golenbock, D. T. LPS-binding proteins and receptors. *J. Leukoc. Biol.* **64**:25–32, 1998.
37. Baorto, D. M., Gao, Z., Malaviya, R., Dustin, M. L., van der Merwe, A., Lublin, D. M. and Abraham, S. N. Survival of FimH-expressing enterobacteria in macrophages relies on glycolipid traffic. *Nature* **389**:636–639, 1997.
38. Simons, K. and Ikonen, E. Functional rafts in cell membranes. *Nature* **387**:569–572, 1997.
39. Solomon, K. R., Rudd, C. E. and Finberg, R. W. The association between glycosylphosphatidylinositol-anchored proteins and heterotrimeric G protein alpha subunits in lymphocytes. *Proc. Natl. Acad. Sci. USA* **93**:6053–6058, 1996.

40. Cebecauer, M., Cerny, J. and Horejsi, V. Incorporation of leucocyte GPI-anchored proteins and protein tyrosine kinases into lipid-rich membrane domains of COS-7 cells. *Biochem. Biophys. Res. Commun.* **243**:706–710, 1998.
41. Arock, M., Ross, E., Lai-Kuen, R., Averlant, G., Gao, Z. and Abraham, S. N. Phagocytic and tumor necrosis factor alpha response of human mast cells following exposure to gram-negative and gram-positive bacteria. *Infect. Immun.* **66**:6030–6034, 1998.
42. Lin, T-J, Gao Z., Arock M. and Abraham, S.N. Internalization of FimH$^+$ *Escherichia coli* by the human mast cell line HMC-1 5C6 involves protein kinase C. *J. Leukoc. Biol.* **66**:1031–1038, 1999.

CHAPTER 26

Human Mast Cells and Basophils in Immune Responses to Infectious Agents

*VINCENZO PATELLA, GIOVANNI FLORIO, ALFONSO ORIENTE, GIUSEPPE SPADARO, VIRGINIA FORTE, ARTURO GENOVESE and GIANNI MARONE**

Division of Clinical Immunology and Allergy, University of Naples Federico II, Naples, Italy

INTRODUCTION

Mast cells and basophils, the only cells that express high-affinity receptors for IgE ($Fc_{\varepsilon}RI$) and synthesize histamine, are widely recognized as primary effector cells in allergic disorders (1–3). $Fc_{\varepsilon}RI^+$ cells exert a fundamental role in the pathophysiology of allergic diseases through the elaboration and release of a myriad of pro-inflammatory and immunoregulatory molecules and the expression of a wide spectrum of surface receptors for cytokines and chemokines (4–7). Our group has demonstrated that $Fc_{\varepsilon}RI^+$ cells are involved in the pathophysiology of cardiovascular diseases (5, 8–10), rheumatic disorders (11–13), and chronic inflammatory diseases (11, 14).

Infectious diseases are the most common cause of mortality and morbidity in man and animals. Mast cells are widely distributed in all vascularized tissues. Moreover, mature basophils and precursors of mast cells in the circulatory system ensure early contact with bacterial and viral pathogens. Furthermore, the strategic location of $Fc_{\varepsilon}RI^+$ cells at the host–environment interface and their ability to release inflammatory mediators, cytokines and chemokines suggest that these cells are implicated in bacterial and viral infections. In addition, considering that mast cells and basophils have been preserved through evolution (15), $Fc_{\varepsilon}RI^+$ cells are probably fundamental for natural and acquired immunity.

$Fc_{\varepsilon}RI^+$ cells play a role in the host's immune response to parasites (16–18). In particular, mast cells are required for the expression of IgE dependent host defence against parasites (19). Data obtained in experimental models have indicated that mast cells are also central to the host's immune response to various bacterial agents (20–23). In contrast, the involvement, if any, of $Fc_{\varepsilon}RI^+$ cells in viral infections in man remains to be established.

* Corresponding author.

ISBN 0-12-473335-2

Mast cells tend to accumulate at sites of chronic bacterial infections and can phagocytose various bacterial agents in the absence of opsonizing antibodies (24). Moreover, bacterial products can activate human $Fc_{\varepsilon}RI^{+}$ cells to release pro-inflammatory mediators and cytokines (20, 21, 25). Finally, viral infections can directly or indirectly activate human $Fc_{\varepsilon}RI^{+}$ cells (26–29).

MAST CELLS IN HOST DEFENCE AGAINST BACTERIAL INFECTIONS

Various Gram-negative and Gram-positive bacteria possess the capacity to bind, albeit to varying degrees, to rodent mast cells. Among Gram-negative bacteria, mast cells bind to *Escherichia coli*, *Klebsiella pneumoniae*, *Salmonella typhimurium* and *Helicobacter pylori* (24); among Gram-positive bacteria, they bind to *Staphylococcus aureus* and *Streptococcus faecalis*. *Escherichia coli* adheres to mast cells through fimbriae type 1 (22, 23). The adhesive interaction is mediated by CD48, a mannosylated glycoprotein on mast cells, and FimH, a fimbrial protein on the bacteria. This adhesive interaction triggers mast cell internalization of bacteria and concomitant secretion of several mast cell mediators. Host-derived proteins are also an important cause of mast cell and basophil activation. Bacterial lipopolysaccharide (LPS) has also been reported to induce histamine release from basophils through complement activation (30). The bacterial formyl-containing tripeptide (FLMP) can also induce mediator release through interaction with a specific receptor present on human basophils (21, 31).

Metchnikoff was probably the first to suggest that mast cells have a phagocytic function and might thereby contribute to host defence (32). This property was confirmed almost 80 years later (33). More recently, Malaviya and co-workers demonstrated that mast cells can indeed phagocytose FimH-expressing enteric bacteria, leading to the killing of the bacteria within acidic vacuoles and through the release of superoxide anions. Moreover, it has been demonstrated that rodent mast cells are capable of processing bacterial antigens through a phagocyte route for class I MHC presentation (34). The latter observation extends the range of antigen-processing cells and suggests that mast cells may have a previously unrecognized role in the induction of specific immune response to bacteria.

The clinical significance of these *in vitro* observations emerges from experiments in which congenitally mast cell-deficient W/W^{v} mice and normal +/+ littermates were used to analyse the role of mast cells in models of acute septic peritonitis. These studies showed that mast cells and mast cell-derived tumour necrosis factor-α (TNF-α) protected against acute bacterial peritonitis (35, 36). Moreover, there is a reduction of mast cell-dependent natural immunity against acute septic peritonitis in C_4–C_3-deficient mice (37). Finally, Galli and co-workers have demonstrated that treatment with stem cell factor (SCF), the most important cytokine for mast cell development and activation (5, 38, 39), can enhance innate immunity in a model of acute bacterial peritonitis (40).

These findings indicate that a lack of mast cells, or a deficit in other components of innate defence mechanisms (e.g. TNF-α, complement), can result in impaired natural immunity to bacterial infections. Moreover, manipulation with SCF to increase mast cell numbers can enhance mast cell function and improve the ability to express innate immunity (40).

If the results of these elegant studies in different experimental models prove to be valid in humans, it can be envisioned that mast cells might play a significant and complex role in the immune response to bacterial infections.

MAST CELLS AND BASOPHILS IN *HELICOBACTER PYLORI* INFECTION

Helicobacter pylori is a Gram-negative, spiral-shaped bacterium that has established its ecological niche in the human stomach where it induces an inflammatory response. Differences in virulence of *H. pylori* isolates could account for different clinical outcomes of infection. There is evidence of interactions between *H. pylori* and $Fc_{\varepsilon}RI^{+}$. As mentioned above, *H. pylori* can bind to rodent mast cells (24). This bacterium appears to induce a specific IgE immune response in patients with chronic gastritis, and *H. pylori* extracts can cause histamine release from basophils obtained from these patients (41). Studies with rat peritoneal mast cells exposed to *H. pylori* extracts *in vitro* have yielded conflicting results (42–44). However, *in vivo* exposure to *H. pylori* appears to elicit perivascular mast cell degranulation (45) and gastric mast cell hyperplasia (46). Patients with duodenal ulcer and *H. pylori* infection have higher concentrations of duodenal histamine content compared with controls (47). Mast cell density is greater in the mucosa of patients with gastritis with or without *H. pylori* infection (48). More recently, *H. pylori* extracts have been reported to induce non-cytotoxic Ca^{2+}-dependent degranulation of rat mast cells (49). It is not known how *H. pylori* strains $CagA^{+}/VacA^{+}$ and Cag A^{-}/Vac A^{-} affect mast cells isolated from human tissues (gastric, skin, etc.). This information is of paramount importance to answer key questions regarding the possible involvement of $Fc_{\varepsilon}RI^{+}$ in gastric inflammation associated with *H. pylori* infection.

LEUKOTRIENES IN THE DEFENCE AGAINST BACTERIAL AND VIRAL INFECTIONS

Leukotrienes (LTs) are lipid mediators derived from arachidonic acid by the action of 5-lipoxygenase (5-LO) (50). Leukotriene B_4 (LTB_4) is a potent effector of neutrophil chemotaxis and activation (51, 52), while the cysteinyl LTs, C_4, D_4 and E_4, augment vascular permeability and smooth muscle tone (53, 54). Immunologically challenged human mast cells and basophils synthesize LTC_4 (55, 56), while neutrophils and macrophages preferentially synthesize LTB_4 (57, 58). Little is known about the relevance, if any, of endogenous LTs in mediating the host response to bacterial and viral infections in humans.

LT-deficient mice manifest increased lethality from *K. pneumoniae* in association with decreased alveolar macrophage phagocytic and bacterial activities (59). This increased susceptibility can be overcome by supplying exogenous LTB_4 to alveolar macrophages. It has been reported that mast cells are essential for the production of LTs involved in the neutrophil recruitment in immune complex peritonitis (60).

Recent evidence suggests that an intact LT-generating system and LTB_4 receptors play a role in host defence against viral infections. HIV-1-infected patients have a deficit in alveolar macrophage 5-LO metabolism (61). It has also been demonstrated that the LTB_4 receptor (BLTR) is a novel type of HIV-1 co-receptor, along with the previously identified chemokine receptors (CCR5 and CXCR4) (62). CCR5 and CXCR4 are the major HIV-1 co-receptors and their interaction with M- and T-tropic virus strains, respectively, is well established (63–68). Dual-tropic isolates use both receptors (66). During disease progression, the viruses expand their receptor usage to include secondary co-receptors, such as BLTR, providing the viruses with new cell populations to target.

However, the importance of an intact LT-generating system and BLTR for host defence to viral infections in humans remains to be established. This aspect appears of great clinical relevance since 5-LO inhibitors are now used in patients with bronchial asthma (69).

ACTIVATION OF HUMAN BASOPHILS AND MAST CELLS BY PROTEIN A

Staphylococcus aureus remains a versatile and dangerous pathogen in humans (70). The frequency of both community-acquired and hospital-acquired staphylococcal infections have increased steadily, with little change in overall mortality (70). Treatment of these infections has become more difficult because of the emergence of multidrug-resistant strains.

Staphylococci produce various surface proteins (e.g. protein A) and secreted proteins (e.g. enterotoxin B, α-toxin, TSST-1). These bacterial products may facilitate the spread of infection to adjoining tissues, although their complex role in the pathogenesis of disease is not completely defined. Protein A is a 45-kDa bacterial membrane protein, produced by the majority of clinical isolates of *S. aureus* (70). Protein A has anti-phagocytic properties that are based on its ability to bind the Fc portion of immunoglobulins. Protein A binds to Fc_γ, a constant region of human IgG subclasses 1, 2 and 4 (71). It has also been established that protein A also reacts with a structure located in the Fab region of immunoglobulin which is shared by human IgM, IgG, IgA and IgE (72, 73). Unlike rabbit IgG, which reacts with protein A through the Fc_γ region alone, human IgG can react with protein A via both the Fab and Fc_γ regions, and the protein A-binding Fab region is found on a proportion (15–50%) of human IgE, IgM and IgA (72, 73). Structural analyses of protein A have identified five extramembrane regions, termed domains E, D, A, B and C (74–77). Each domain consists of 58–61 amino acids that are highly homologous in sequence. Each domain possesses immunoglobulin-binding capacity (76, 77). Domain B expresses Fc_γ-binding activity, but has been reported to express little or no Fab-binding capacity (78, 79). In contrast, domain D of protein A has the capacity to interact with the Fab region of human immunoglobulin (80).

In collaboration with Sergio Romagnani we have investigated the capacity of *S. aureus* Cowan 1 and *S. aureus* Wood 46 to induce mediator release from human basophils *in vitro* (20, 81). *S. aureus* Cowan 1 (10^5–10^7 ml^{-1}), which synthesizes protein A, stimulated the release of histamine from basophils, whereas *S. aureus* Wood 46 (10^5–10^7 ml^{-1}), which does not synthesize protein A, did not induce histamine secretion. Soluble protein A (10^{-1} to 30 μg ml^{-1}) also induced histamine secretion from human basophils (Fig. 1).

We have also investigated the mechanism by which protein A activates basophils to release mediators. It has been demonstrated that hyper-iodination of protein A destroys over 90% of the original Fc_γ reactivity without altering the Fab-binding site (82). This procedure did not alter the ability of the protein to induce histamine release from basophils. The stimulating effect of protein A was dose-dependently inhibited by pre-incubation with human polyclonal IgG (0.3–100 μg ml^{-1}) and a human monoclonal IgM (0.3–100 μg ml^{-1}) which have F(ab′)–protein A reactivity. In contrast, rabbit IgG, which possesses only Fc–protein A reactivity, and a protein A-unreactive human monoclonal

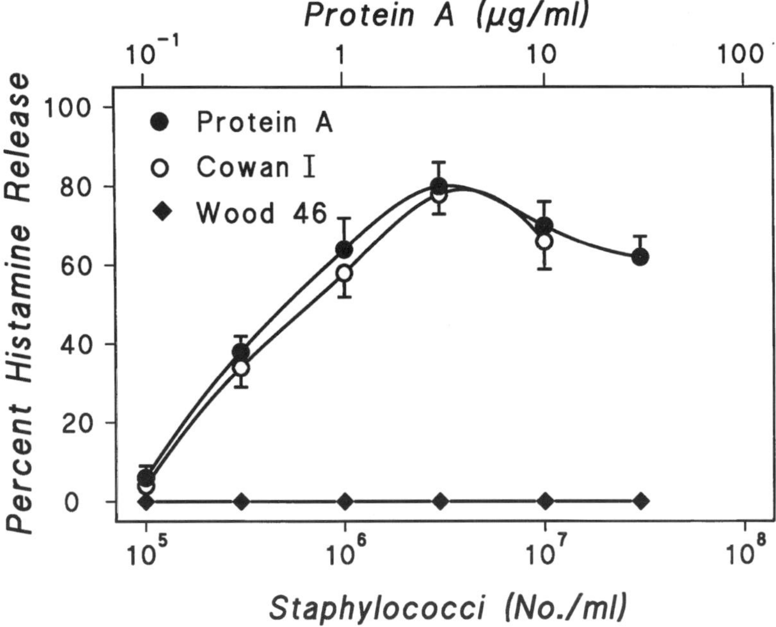

Fig. 1 Effect of protein A, *Staphylococcus aureus* Cowan 1 and *S. aureus* Wood 46 on histamine secretion from human basophils. Each point represents the mean ± SEM of four experiments. Error bars are not shown when details are too small.

IgM did not inhibit protein A activity. These results suggested that staphylococci synthesizing protein A and soluble protein A induce mediator release from human basophils through the interaction with the alternative (Fab) binding site. This hypothesis was supported by a series of desensitization experiments. Pre-incubation with either protein A or anti-IgE resulted in complete desensitization to subsequent challenge with the homologous stimulus. Protein A and anti-IgE induced cross-desensitization to the heterologous stimulus. Moreover, basophils from which IgE had been dissociated by brief exposure to lactic acid (21) no longer released histamine in response to anti-IgE and protein A. When basophils from which IgE had been dissociated were incubated with human polyclonal IgE, they regained their ability to induce histamine release in response to protein A and anti-IgE. In contrast, two monoclonal IgEs that do not bind to protein A did not restore the basophils' responsiveness to protein A. Furthermore, there was complete cross-desensitization between soluble protein A and *S. aureus* Cowan 1, whereas cells desensitized to *S. aureus* Wood 46 released normally with protein A and *S. aureus* Cowan 1. These results demonstrate that protein A and *S. aureus* Cowan 1 activate mediator release from human basophils by interacting with the $F(ab')_2$ region of IgE present on the cell surface.

It has been demonstrated that the ability of Fab to bind to protein A is a functional marker for immunoglobulin encoded by the largest human V_H gene family, V_H3 (83–86). Furthermore, protein A binding was demonstrated with a large proportion of V_H3 IgM but only with a lower proportion of the V_H3 IgA and V_H3 IgG fragments tested (83, 87). These findings suggest that specificity for protein A is encoded by the germ line sequences of many of the commonly expressed V_H3 genes. The Fab site to which protein A binds involves V_H family-specific residues, which have been demonstrated to reside outside the antigen-binding site (88, 89). The association of V_H3 H chains with specificity for protein A is analogous to the ability of certain T cell receptor (TCR) V_γ molecules to bind bacterial superantigens (90). Therefore, protein A, which is a potent polyclonal activator of human B cells (91), is now considered an immunoglobulin superantigen (87, 92).

To investigate further the structural basis of the interaction between IgE and protein A, we have evaluated the effect of human immunoglobulins of different specificities to modulate the releasing activity of protein A on basophils. In these experiments, one of which is illustrated in Fig. 2, we found that monoclonal IgM V_H3^+ blocked the histamine-releasing activity of protein A, whereas monoclonal IgM V_H6^+ did not modify the histamine-releasing activity. These results indicate that human monoclonal IgM V_H3^+ acts as a competitive antagonist of IgE bound to basophils at the level of IgE binding site on protein A. Our results are consistent with the hypothesis that protein A-synthesizing staphylococci and soluble protein A, acting as an immunoglobulin superantigen, interact with V_H3^+ IgE to induce the release of mediators from human basophils.

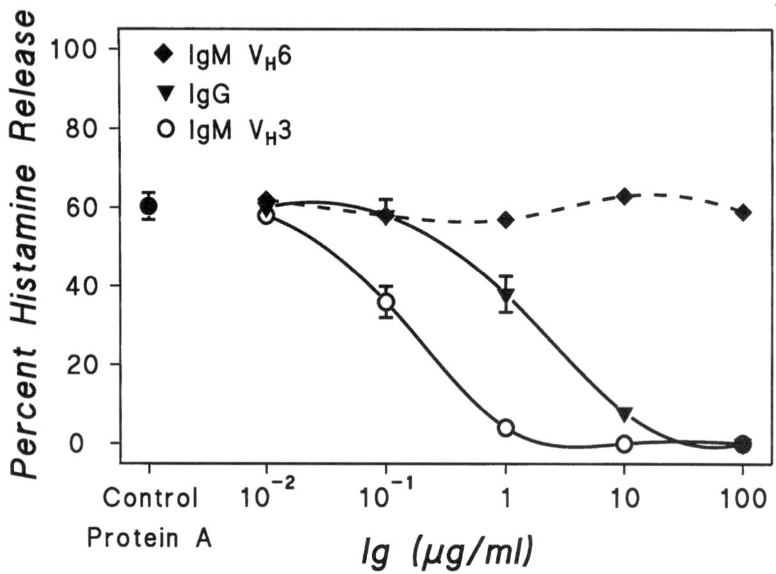

Fig. 2 Effect of preincubation (10 min, 37°C) of protein A with monoclonal IgM V_H6, monoclonal IgM V_H3 or polyclonal IgG on the activation of human basophils. Each point represents the mean ± SEM of triplicate determinations from typical experiments. The percentage histamine release (Control) in the absence of pre-incubation with immunoglobulins (Ig) is indicated (●).

ACTIVATION OF HUMAN BASOPHILS AND MAST CELLS BY PROTEIN L

The anaerobic bacterial strain *Peptostreptococcus magnus* expresses a cell wall protein that possesses a high affinity for the L chains of human immunoglobulin (93) and is thus a candidate for an immunoglobulin-binding protein with broad binding specificity (93). The protein has been isolated and designated 'protein L' (94). This protein binds specifically to a different region of human immunoglobulin, compared to protein A and protein G. Protein A binds to the Fc_γ region of human IgG subclasses 1, 2 and 4 (71), and also reacts with a structure located in the $F(ab')_2$ region of immunoglobulin that is shared by human IgM, IgE, IgG and IgA (72, 73). Protein G purified from streptococci binds with high affinity to all human IgG subclasses (95, 96) and thus has a broader IgG-binding capacity than protein A. In contrast to proteins A and G, which both bind to the $C\gamma2$–$C\gamma3$ interface region of human immunoglobulin, protein L reacts with the L chains of immunoglobulin, which means that the molecule has affinity for all classes of immunoglobulin (94). Protein L binds predominantly to L chains of the κ type (94) and there is evidence that the expression of protein L is correlated to peptostreptococcal virulence (97).

In collaboration with Lars Björk we found that *Peptostreptococcus magnus* strain 312 (10^6–10^8 ml^{-1}), which synthesizes a protein that binds to κ L chains of human immunoglobulin (protein L), stimulated the release of histamine from human basophils *in vitro* (98). *P. magnus* strain 644, which does not synthesize protein L, did not induce histamine secretion. Soluble protein L (3×10^{-2} to 3 μg ml^{-1}) also induced histamine release from human basophils (Fig. 3). The characteristics of the release reaction were similar to those of anti-IgE: it was Ca^{2+}- and temperature-dependent, optimal release occurring at 37°C in the presence of 1.0 mM extracellular Ca^{2+}. There was an excellent correlation ($r = 0.82$; $p < 0.001$) between the maximal percentage histamine release induced by protein L and that induced by anti-IgE, as well as between protein L and protein A from *S. aureus* ($r = 0.52$; $p < 0.01$). Pre-incubation of basophils with either protein L or anti-IgE resulted in complete cross-desensitization to a subsequent challenge with the heterologous stimulus. IgE purified from myeloma patients P.S. and P.P. (λ chains) blocked anti-IgE-induced histamine release but failed to block the histamine-releasing activity of protein L. In contrast, IgE purified from myeloma patient A.D.Z. (κ chains) blocked both anti-IgE- and protein L-induced releases, whereas human polyclonal IgG selectively blocked protein L-induced secretion (Fig. 4).

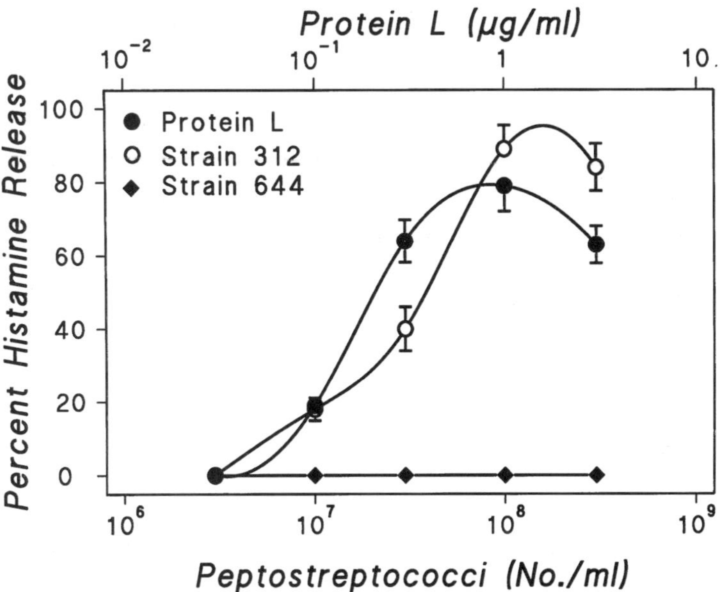

Fig. 3 Effect of increasing concentrations of protein L and *Peptostreptococcus magnus* strains 312 and 644 on histamine secretion from human basophils. Each point represents the mean ± SEM of five experiments. Error bars are not shown when details are too small.

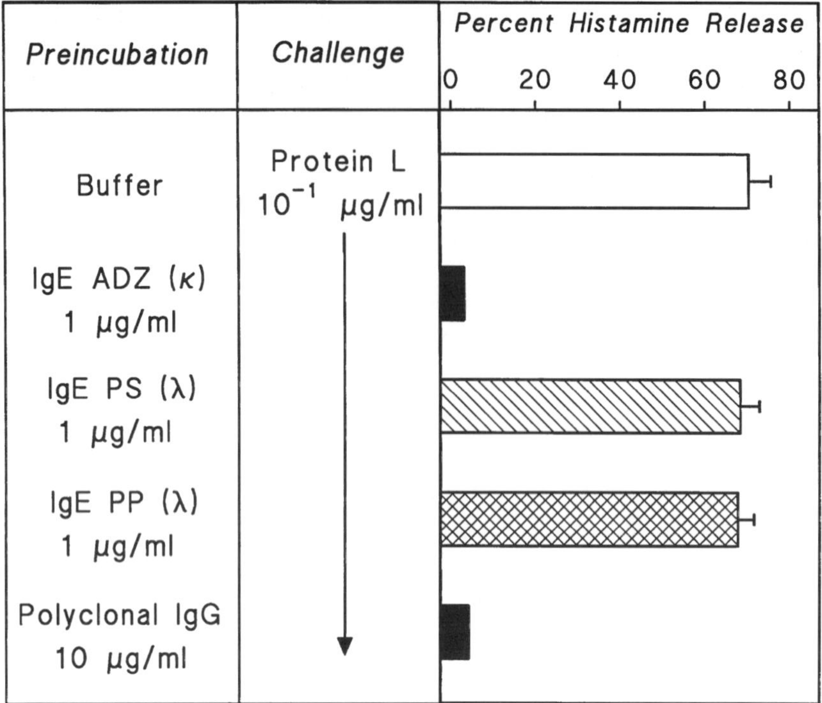

Fig. 4 Effect of pre-incubation of protein L with monoclonal IgE or polyclonal IgG on histamine release from human basophils. Protein L (0.1 μg ml^{-1}) was pre-incubated (15 min, 37°C) with human monoclonal IgE from patients A.D.Z. (κ chains), P.S. (λ chains) and P.P. (λ chains), or with human polyclonal IgG. Leukocytes were then added, and the incubation was continued for an additional 45 min at 37°C. Each bar represents the mean ± SEM from triplicate incubations of the cells from the same donor.

In these experiments we found that protein L acted as a complete secretagogue; i.e. it activated basophils to release cysteinyl leukotriene C_4 (LTC_4) as well as histamine. Protein L (10^{-1} to 3 μg ml^{-1}) also induced the release of pre-formed (histamine) and *de novo* synthesized mediators (LTC_4 and/or PGD_2) from mast cells isolated from lung parenchyma and skin tissues. The *in vivo* relevance of these findings is supported by the observation that intradermal injections of protein L (0.01–10 μg ml^{-1}) in non-allergic subjects caused a dose-dependent wheal and flare reaction.

In conclusion, protein L, which exerts a potent activating effect on human basophils and mast cells *in vitro* and *in vivo* by interacting with κ L chains of the IgE isotype, should be considered a novel immunoglobulin superantigen. These observations appear also of clinical relevance because there is evidence that the expression of protein L is correlated to peptostreptococcal virulence (98).

ACTIVATION OF HUMAN BASOPHILS BY PEPSTATIN A

Pepstatin A is a natural pentapeptide that has been isolated from culture filtrates of various species of actinomycetes (99) and which plays an important role in several lung diseases, including hypersensitivity pneumonitis (100–102). Its amino acid sequence, *N*-isovaleryl-L-valyl-L-valyl-AHMHA-L-alanyl-AHMHA, includes the unusual amino acid 4-amino-hydroxy-6-methylheptanoic acid (AHMHA) (103). This pentapeptide is a potent chemoattractant from human neutrophils, monocytes and eosinophils (104). Pepstatin A also induces the release of lysosomal enzymes and the generation of superoxide anion (O_2^-) from human neutrophils (105). These activities are similar to those of the formylated tripeptide, f-Met–peptide (FMLP), which binds to a cell-surface receptor present on human neutrophils and basophils (106), and it has been suggested that pepstatin A competes with [^{3}H]FMLP for binding to human polymorphonuclear cells (107).

FMLP induce histamine release from human basophils by activating a specific cell-surface receptor independent of both $Fc_\varepsilon RI$ and C5aR (8, 13). We have investigated the mechanisms by which pepstatin A induces histamine secretion from human basophils (21, 108, 109). The characteristics of this reaction were similar to those of FMLP-induced histamine release: pepstatin A-induced release required Ca^{2+}, and the release reaction was complete within 2 min at 22 or 37°C but did not occur at 4°C. There was excellent correlation ($r = 0.93$; $p < 0.001$) between the maximal histamine release induced by pepstatin A and FMLP, but there was no relationship with the capacity of basophils to release with anti-IgE or the Ca^{2+} ionophore A23187. Release by pepstatin A was reversibly inhibited by two non-releasing analogues of FMLP, CBZ-Phe–Met and Boc-Met–Leu–Phe. Boc-Met–Leu–Phe competitively inhibited the effect of both FMLP and pepstatin A on histamine release from basophils and the dissociation constant (K_D) for the Boc-Met–Leu–Phe-receptor complex in both conditions was approximately 10^{-6} M. Furthermore, there was complete cross-desensitization between pepstatin A and FMLP, whereas cells desensitized to pepstatin A released normally with anti-IgE and vice versa.

It is therefore likely that the natural pentapeptide pepstatin A, synthesized by various species of actinomycetes, induces histamine release from human basophils by activating a cell-surface receptor(s), was also activated by the synthetic tripeptide FMLP (108). The specific cell-surface receptor activated by FMLP and pepstatin A on basophils is independent of the $Fc_\varepsilon RI$ (Fig. 5).

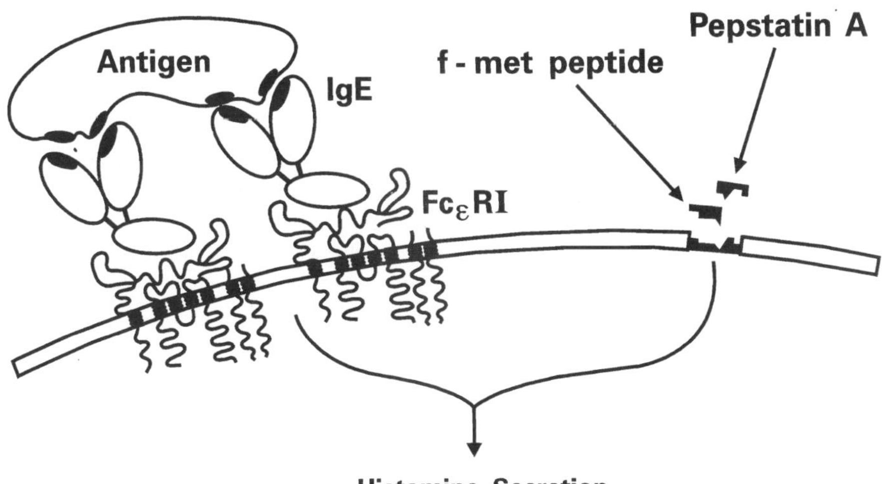

Fig. 5 Schematic representation of the $Fc_\varepsilon RI$–IgE network and the receptor of FMLP and pepstatin A. FMLP and pepstatin A are structurally distinct molecules, but activate the same surface receptor on human basophils which is distinct from the $Fc_\varepsilon RI$ for IgE. The engagement of $Fc_\varepsilon RI$ by antigen or anti-$Fc_\varepsilon RI\alpha$ or FMLP receptor by pepstatin A or FMLP leads to mediator secretion from human basophils.

ENDOGENOUS SUPERALLERGEN PROTEIN Fv IN VIRAL HEPATITIS

Protein Fv, formerly referred to as 'protein F' (110), is a novel immunoglobulin-binding protein that is present in normal liver and is largely released in the digestive tract and in biological fluids during viral hepatitis (111). Protein Fv is an acidic trypsin-resistant sialoprotein (pHi 4.0) with an apparent molecular mass of 175 kDa, that dissociates into two monomers under dissociative conditions. Binding of protein Fv to immunoglobulin from various mammalian and non-mammalian species has been characterized (112). Protein Fv binds the variable domain of H chains, irrespective of immunoglobulin class, subclass and L chain type (113).

Protein Fv is released in the digestive lumen of a significant percentage of patients suffering from B (~35%) and C (~43%) acute and chronic viral hepatitis (111). However, its pathophysiological role in humans is still largely unknown. Acute and chronic viral hepatitis are associated with a wide spectrum of hepatic and extrahepatic tissue injuries, not directly related to the cytotoxic effects of HBV and HCV infection. These include a variety of skin rashes and cutaneous and systemic vasculitis (114–116). Urticarial reactions involving skin mast cells and their vasoactive and pro-inflammatory mediators occur in about 5% of patients with viral hepatitis (117–120). Mast cells and basophils, through the secretion of vasoactive mediators, also play a role in cutaneous and systemic vasculitis (55, 121), a feature frequently associated with concomitant HCV infection (122). Moreover, human mast cells and basophils, through the release of pro-inflammatory mediators (histamine, tryptase, etc.) and the elaboration of various cytokines (IL-4, IL-13, TNF-α, etc.) (4, 29, 123–126), stimulate fibroblast proliferation (127–129) and collagen synthesis (130–132), which could contribute to the fibrotic response in chronic liver diseases (133).

In collaboration with Jean-Pierre Bouvet we have investigated the effects of protein Fv isolated from different patients with viral hepatitis on the release of histamine and LTC_4 from purified basophils (27). Protein Fv (1–10 ng ml^{-1}) concentration-dependently induced histamine and LTC_4 release from basophils. Protein Fv adsorbed with protein A–Sepharose coated with polyclonal IgG did not induce mediator secretion. The characteristics of the release reaction induced by protein Fv were similar to those of anti-IgE. There was an excellent correlation ($r_s = 0.83$; $p < 0.001$) between the maximal percentage histamine release induced by protein Fv and that induced by anti-IgE from basophils. Pre-incubation of basophils with either protein Fv or anti-IgE resulted in complete cross-desensitization to subsequent challenge with heterologous stimulus. Basophils from which IgE had been dissociated by brief exposure to lactic acid no longer released histamine in response to anti-IgE and protein Fv. A monoclonal IgE purified from a myeloma patient (patient A.D.Z.) blocked both anti IgE- and protein Fv-induced releases, whereas human polyclonal IgG selectively blocked protein Fv-induced secretion.

Protein Fv also induced the release of pre-formed (histamine and tryptase) and *de novo* synthesized mediators (LTC_4 and/or PGD_2) from mast cells purified from human lung parenchyma and skin tissues. There was a significant correlation between the maximal percentage histamine release induced by protein Fv and anti-IgE from skin mast cells. There was also an excellent correlation between histamine and tryptase release caused by protein Fv from both lung and skin mast cells. Thus, we established that protein Fv acts as a novel activator of human basophils and mast cells by interacting with IgE.

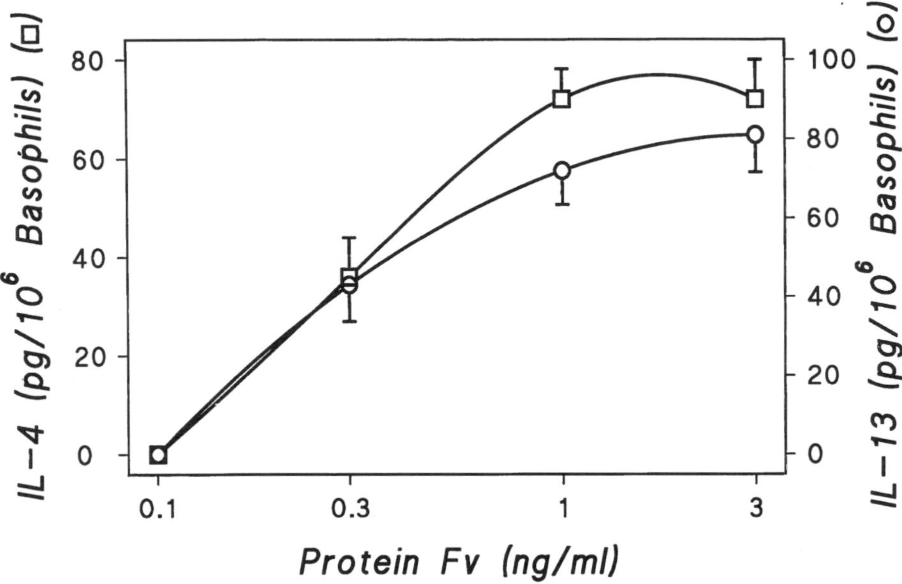

Fig. 6 Effects of various concentrations of protein Fv on IL-4 and IL-13 release from purified (>98%) human basophils. Basophils were incubated for 4 h at 37°C for IL-4 release and for 18 h at 37°C for IL-13 release. Values are expressed as the mean ± SEM from triplicate incubations of the cells from the same donor. Error bars are not shown when details are too small.

More recently we investigated the mechanism whereby protein Fv induces mediator release from basophils and mast cells and evaluated whether it also induces cytokine (IL-4 and IL-13) synthesis in basophils. We found that protein Fv is a potent stimulus for IL-4 and IL-13 release from purified basophils (26). Figure 6 illustrates the results of a typical experiment showing that protein Fv (0.1–3 ng ml^{-1}) concentration-dependently induced IL-14 and IL-13 release from enriched preparation (> 95%) of human basophils. Histamine and IL-4 secretion from basophils activated by protein Fv was significantly correlated ($r_s = 0.70$; $p < 0.001$). Although there was also a correlation between the maximum protein Fv- and anti-IgE-induced IL-4 release from basophils (26), the average half-time for protein Fv-induced IL-4 release was higher than for histamine release. IL-4 mRNA, constitutively present in basophils, was increased after stimulation by protein Fv and was inhibited by cyclosporin A and tacrolimus. Removal of IgE from basophils abolished the release of IL-4 in response to protein Fv.

We have also investigated the mechanism by which protein Fv induces cytokine release from human basophils. To this end, three human V_H3^+ monoclonal IgMs concentration-dependently inhibited protein Fv-induced secretion of IL-4 and histamine from basophils and of histamine from lung mast cells. In contrast, V_H6^+ monoclonal IgM did not inhibit the release of IL-4 and histamine induced by protein Fv from $Fc_\varepsilon RI^+$ cells. These results indicate that protein Fv acts as an endogenous superallergen interacting with the V_H3 domain of IgE to induce the synthesis and release of IL-4 and IL-13 from human basophils.

ACTIVATION OF HUMAN BASOPHILS AND MAST CELLS BY HIV-1 GLYCOPROTEIN gp120

There is now compelling evidence that serum IgE levels are increased in some patients with HIV-1 infection (134–137). More importantly, elevated IgE levels in HIV-1-infected children and in adults have been associated with progression of HIV-1 disease (138, 139). Thus, an increased IgE level appears to be a marker of poor prognosis in some patients with HIV-1 infection (138, 139).

Clerici and colleagues (140, 141) suggested that, during the early stages of HIV-1 infection, there is a switch from 'Th1-like' towards a 'Th2-like' pattern of cytokine production. However, in early studies, an overall shift in the cytokine pattern towards the Th2 subset was not detected in lymph nodes or in T cell clones of HIV-1-infected individuals (142, 143). Maggi and colleagues found a depletion of $CD4^+$ Th2-type cells in the advanced phases of HIV-1-infection, and that HIV-1 replicates preferentially in Th2 rather than in Th1 clones (142). Subsequent reports have added to the controversy (144–148). The apparently conflicting results could be due to technical reasons, the production of Th2-like cytokines by cell types other than lymphocytes or stimulation by specific viral superantigens. In addition, most studies have focused on IL-4, whereas recent data show that other cytokines such as IL-13 are critical for Th2 cell polarization (149–152).

Immunologically challenged human basophils synthesize a restricted profile of cytokines (IL-4 and IL-13) (26, 153–155), which are critical for Th2 cell polarization (149–152). In addition, human mast cells synthesize IL-4 and IL-13 (156, 157). Moreover, HIV antigens induce histamine release from basophils (158) and histamine blood levels are increased in HIV-1-infected children (159).

The entry of HIV into host cells is mediated by interaction of the viral envelope glycoprotein, gp120, with the CD4 glycoprotein (160) and chemokine receptors (CCR5 and CXCR4) on the surface of $CD4^+$ cells (65, 67, 68). Recent evidence indicates that HIV-1 gp120 is a new member of the immunoglobulin superantigen family (92, 161). Immunoglobulin V_H3 gene products are the ligand for gp120 (162) and this interaction might explain the superantigen activation of B lymphocytes in patients with AIDS (163).

We have recently found that nanomolar concentrations of HIV-1 gp120 from different clades induce the release of IL-4 from human peripheral blood basophils purified from healthy individuals who were seronegative for antibodies to HIV-1 and HIV-2 (Fig. 7). Gp120 simulated the release of IL-4 and IL-13 in parallel to the secretion of histamine from basophils. In contrast, interferon-γ (IFN-γ) mRNA was not detectable in any of the basophil preparations stimulated with gp120 (29).

We have also evaluated the mechanism whereby gp120 activates basophils purified from healthy individuals. Removal of IgE by brief exposure of basophils to lactic acid completely blocks the effect exerted by gp120 on histamine and cytokine release from basophils. These data are compatible with the hypothesis that gp120 activates $Fc_\varepsilon RI^+$ cells through the interaction with IgE bound on basophils.

Incubation of gp120 with three preparations of monoclonal IgM (M3, M11 and LAN), which possess the V_H3 domain (6, 26), concentration-dependently inhibited the effect of gp120 on IL-4 and IL-13 synthesis. In contrast, a monoclonal IgM (M14), which possesses a V_H6 domain, had no effect. Thus, binding to the V_H3 domain inhibits the interaction of gp120 with IgE bound to $Fc_\varepsilon RI$ on basophils.

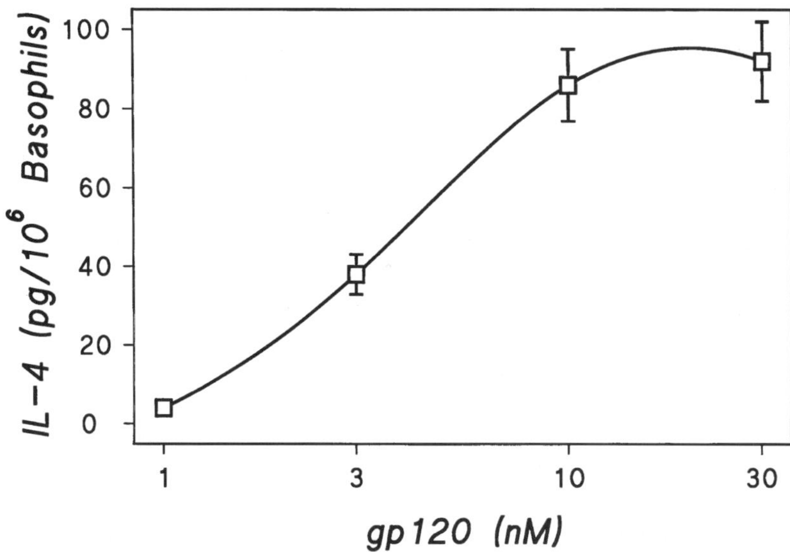

Fig. 7 Effect of various concentrations of $gp120_{MN}$ on IL-4 release from purified (>98%) human basophils. Basophils were pre-incubated (15 min, 37°C) with rhIL-3 (10 ng ml^{-1}) and then challenged with gp120 for 4 h at 37°C. Values are expressed as the mean ± SEM from triplicate incubations of the cells from the same donor. Error bars are not shown when details are too small.

Our results provide the first demonstration that gp120 triggers the release of two critical cytokines (IL-4 and IL-13) for Th2 polarization from human $Fc_{\varepsilon}RI^{+}$ cells, thus acting as a potent viral superantigen. This novel finding appears to be of clinical relevance. In fact, it suggests that, during the early phase of HIV infection associated with high levels of viraemia (164), basophils exposed to virus-bound or shed gp120 (165) might represent the initial source of IL-4 and IL-13 thereby favouring a shift from Th0 towards a Th2 phenotype. In advanced HIV-1 infection, when $CD4^{+}$ T cells are decreased, $Fc_{\varepsilon}RI^{+}$ cells might also represent a significant source of Th2-like cytokines. Therefore, basophils and mast cells might be a relevant source of cytokines contributing to the polarization of $CD4^{+}$ cells toward Th2 cells during HIV-1 infection.

These findings might be relevant also from a quantitative viewpoint. Th2 cells represent 0.2–2% of $CD4^{+}$ cells (166), whereas basophils represent 1% of peripheral blood leukocytes (1, 2). Although viral antigens interact with individual Th clones, viral superantigen gp120 can produce a rapid and massive activation of basophils via $V_{H}3^{+}$ IgE. Because the $V_{H}3$ family is the largest in the human repertoire (approximately 50%) (161), it is likely that shed or virus-bound gp120 interacts with a high frequency with $V_{H}3^{+}$ IgE bound to basophils of normal or early-infected individuals. Finally, the levels of IL-4 produced by human lymphocytes are about 10–20% of those generated by immunologically challenged basophils (167).

Thus, basophils may play an important role since they are capable of producing IL-4 and IL-13 in a restricted manner without synthesizing Th1-type cytokines (e.g. IFN-γ).

In conclusion, we have provided the first evidence that gp120 can induce the release of IL-4 and IL-13 from human $Fc_{\varepsilon}RI^{+}$ cells, which might be a novel source of Th2 cytokines contributing to the dysregulation of the immune system in HIV-1 infection.

THE 'THREE FACES' OF MAST CELLS AND BASOPHILS IN THE IMMUNE RESPONSE TO INFECTIOUS AGENTS

The results accumulated over the recent years suggest that $Fc_{\varepsilon}RI^{+}$ cells represent a central component of host defence against bacterial and viral infections. Increasing evidence suggests that mast cells and basophils might be involved in at least three distinct mechanisms of immunological responses against bacteria and viruses (Fig. 8). Mast cell and basophil mediators can be released from $Fc_{\varepsilon}RI^{+}$ cells by immunoglobulin-independent mechanisms. This aspect of natural immunity might involve direct activation of specific receptors (e.g. CD48) with fimbrie of bacteria, activation by bacterial products of specific membrane receptors (FMLP) on basophils or the activation of complement. All these distinct aspects can participate in natural immunity against bacteria and perhaps viruses, but may also induce non-immunological tissue damage.

A second mechanism, previously unrecognized, but assuming increasing importance, involves the activation of human basophils and mast cells by various immunoglobulin superantigens. Several bacterial and viral superantigens have been recently identified and

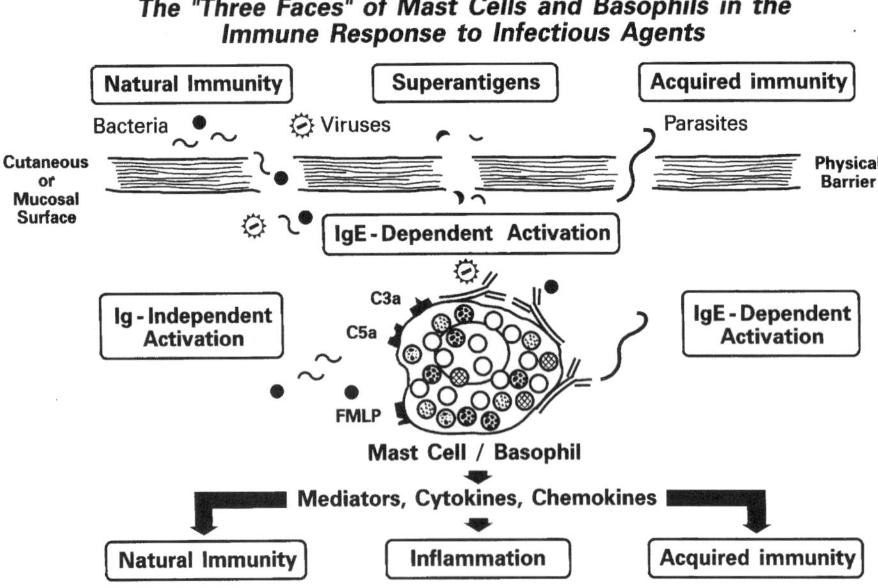

Fig. 8 The 'three faces' of mast cells and basophils in the immune response to infectious agents. Mast cells and basophils can contribute to host defence against microbial infection by at least three distinct mechanisms. Bacterial products (e.g. FMLP, pepstatin A) can participate in natural immunity by activating specific membrane receptors leading to the release of pro-inflammatory mediators. Bacteria can also induce complement activation leading to the formation of anaphylatoxins (C3a, C4a, C5a) and C3b which can activate specific membrane receptors. Mast cells and basophils can also take part in acquired immune responses against parasites through the interaction with specific IgE bound to $Fc_{\varepsilon}RI$. A third and novel mechanism by which mast cells and basophils participate in host defence is mediated by the interaction of bacterial (e.g. protein A, protein L) and viral products (e.g. gp120) with IgE bound to $Fc_{\varepsilon}RI$. The latter mechanism involves the interaction of bacterial and viral products with IgE in an immunoglobulin superantigenic mechanism.

characterized. Protein A of *S. aureus* and protein L of *P. magnus* are potent immunoglobulin superantigens interacting with distinct regions of IgE bound to $Fc_{\varepsilon}RI^{+}$ cells. Viral infections can also directly or indirectly act on $Fc_{\varepsilon}RI^{+}$ cells. For instance protein Fv, induced *in vivo* by HBV and HBC can act as an endogenous superantigen interacting with $V_{H}3$ IgE bound to basophils and mast cells. Recently we have demonstrated that HIV-1 gp120 can stimulate histamine and cytokine release from human basophils and mast cells isolated from non-infected individuals. Therefore, during early and advanced HIV-1 infection, basophils and mast cells may be a source of relevant cytokines (e.g. IL-4 and IL-13) that contribute to immunological dysregulation in HIV infection. The latter finding highlights the importance of a specific viral superantigen, gp120, acting on cell types other than lymphocytes, in the production of Th2-like cytokines.

Finally, there is ample evidence that $Fc_{\varepsilon}RI^{+}$ cells represent a central component of host defence against parasite infections.

Taken together, these findings highlight a novel and intricate universe of multifactorial interactions in which mast cells and basophils might have a prominent position in host defence against a variety of bacterial and viral infections.

ACKNOWLEDGEMENTS

This work was supported by grants from the National Research Council (C.N.R.) (Targeted Project Biotechnology No. 98.00085.PF31 and 99.00401.PF49) and the Ministry of Health (ISS - AIDS Project No. 40B.64) (Rome, Italy). The authors thank Lina Tagliaferri for her secretarial assistance on this manuscript.

REFERENCES

1. Marone, G. *Human Basophils and Mast Cells. I. Biological Aspects.* Karger, Basel, 1995.
2. Marone, G. *Human Basophils and Mast Cells. II. Clinical Aspects.* Karger, Basel, 1995.
3. Marone, G., Casolaro, V., Patella, V., Florio, G. and Triggiani, M. Molecular and cellular biology of mast cells and basophils. *Int. Arch. Allergy Immunol.* **114**:207–217, 1997.
4. Gauchat, J. F., Henchoz, S., Mazzel, G., Aubry, J. P., Brunner, T., Blasey, H., Life, P., Talabot, D., Flores-Romo, L., Thompson, J., Kishi, K., Butterfield, J., Dahinden, C. and Bonnefoy, J. Y. Induction of human IgE synthesis in B cells by mast cells and basophils. *Nature* **365**:340–343, 1993.
5. Patella, V., Marinò, I., Arbustini, E., Lamparter-Schummert, B., Verga, L., Adt, M. and Marone, G. Stem cell factor in mast cells and increased mast cell density in idiopathic and ischemic cardiomyopathy. *Circulation* **97**:971–978, 1998.
6. de Paulis, A., Minopoli, G., Del Piaz, F., Pucci, P., Russo, T. and Marone, G. Novel autocrine and paracrine loops of the stem cell factor/chymase network. *Int. Arch. Allergy Immunol.* **118**:422–425, 1999.
7. Romagnani, P., de Paulis, A., Beltrame, C., Annunziato, F., Dente, V., Maggi, E., Romagnani, S. and Marone, G. Tryptase–chymase double-positive human mast cells express the eotaxin receptor (CCR3) and are attracted by CCR3-binding chemokines. *Am. J. Pathol.* **155**:1195–1204, 1999.
8. Patella, V., Marinò, I., Lampärter, B., Arbustini, E., Adt, M. and Marone, G. Human heart mast cells: isolation, purification, ultrastructure, and immunologic characterization. *J. Immunol.* **154**:2855–2865, 1995.
9. Marone, G., de Crescenzo, G., Adt, M., Patella, V., Arbustini, E. and Genovese, A. Immunological characterization and functional importance of human heart mast cells. *Immunopharmacology* **31**:1–18, 1995.
10. Marone, G., de Crescenzo, G., Florio, G., Granata, F., Dente, V. and Genovese, A. Immunological

modulation of human cardiac mast cells. *Neurochem. Res.* **24**:1195–1202, 1999.

11. de Paulis, A., Marinò, I., Ciccarelli, A., de Crescenzo, G., Concardi, M., Verga, L., Arbustini, E. and Marone, G. Human synovial mast cells. I. Ultrastructural *in situ* and *in vitro* immunologic characterization. *Arthritis Rheum.* **39**:1222–1233, 1996.
12. de Paulis, A., Ciccarelli, A., Marinò, I., de Crescenzo, G., Marinò, D. and Marone, G. Human synovial mast cells. II. Heterogeneity of the pharmacologic effects of antiinflammatory and immunosuppressive drugs. *Arthritis Rheum.* **40**:469–478, 1997.
13. Marone, G. Mast cells in rheumatic disorders: mastermind or workhorse? *Clin. Exp. Rheumatol.* **16**:245–249, 1998.
14. de Paulis, A., Valentini, G., Spadaro, G., Lupoli, S., Tirri, G. and Marone, G. Human basophil releasability. VIII. Increased basophil releasability in patients with scleroderma. *Arthritis Rheum.* **34**:1289–1296, 1991.
15. Chieffi Baccari, G., de Paulis, A., Di Matteo, L., Gentile, M., Marone, G. and Minucci, S. *In situ* characterization of mast cells in the frog *Rana esculenta*. *Cell Tissue Res.* **292**:151–162, 1998.
16. Brown, S. J., Galli, S. J., Gleich, G. J. and Askenase, P. W. Ablation of immunity to *Amblyomma americanum* by anti-basophil serum: cooperation between basophils and eosinophils in expression of immunity to ectoparasites (ticks) in guinea pigs. *J. Immunol.* **129**:790–796, 1982.
17. Hussain, R., Hamilton, R. G., Kumaraswami, V., Adkinson, N. F. Jr and Ottesen, E. A. IgE responses in human filariasis. I. Quantitation of filaria-specific IgE. *J. Immunol.* **127**:1623–1629, 1981.
18. Lantz, C. S., Boesiger, J., Song, C. H., Mach, N., Kobayashi, T., Mulligan, R. C., Nawa, Y., Dranoff, G. and Galli, S. J. Role for interleukin-3 in mast cell and basophil development and in immunity to parasites. *Nature* **392**:90–93, 1998.
19. Matsuda, H., Watanabe, N., Kiso, Y., Hirota, S., Ushio, H., Kannan, Y., Azuma, M., Koyama, H. and Kitamura, Y. Necessity of IgE antibodies and mast cells for manifestation of resistance against larval *Haemaphysalis tongicornis* ticks in mice. *J. Immunol.* **144**:259–262, 1990.
20. Marone, G., Poto, S., Petracca, R., Triggiani, M., De Lutio di Castelguidone, E. and Condorelli, M. Activation of human basophils by staphylococcal protein A. I. The role of cyclic AMP, arachidonic acid metabolites, microtubules and microfilaments. *Clin. Exp. Immunol.* **50**:661–668, 1982.
21. Marone, G., Columbo, M., Soppelsa, L. and Condorelli, M. The mechanism of basophil histamine release induced by pepstatin A. *J. Immunol.* **133**:1542–1546, 1984.
22. Malaviya, R., Ross, E. A., MacGregor, J. I., Ikeda, T., Little, J. R., Jakschik, B. A. and Abraham, S. N. Mast cell phagocytosis of FimH-expressing Enterobacteria. *J. Immunol.* **152**:1907–1914, 1994.
23. Malaviya, R., Ross, E. A., Jakschik, B. A. and Abraham, S. N. Mast cell degranulation induced by type 1 fimbriated *Escherichia coli* in mice. *J. Clin. Invest.* **93**:1645–1653, 1994.
24. Abraham, S. N. and Arock, M. Mast cells and basophils in innate immunity. *Semin. Immunol.* **10**:373–381, 1998.
25. O'Flaherty, J. T., Showell, H. J., Kreutzer, D. L., Ward, P. A. and Becker, E. L. Inhibition of *in vivo* and *in vitro* neutrophil responses to chemotactic factors by a competitive antagonist. *J. Immunol.* **120**:1326–1332, 1978.
26. Patella, V., Bouvet, J. P. and Marone, G. Protein Fv produced during viral hepatitis is a novel activator of human basophils and mast cells. *J. Immunol.* **151**:5685–5698, 1993.
27. Patella, V., Giuliano, A., Bouvet, J. P. and Marone, G. Endogenous superallergen protein Fv induces IL-4 secretion from human $Fc_{\varepsilon}RI^{+}$ cells through interaction with the V_H3 region of IgE. *J. Immunol.* **161**:5647–5655, 1998.
28. Patella, V., Giuliano, A., Florio, G., Bouvet, J. P. and Marone, G. Endogenous superallergen protein Fv interacts with the V_H3 region of IgE to induce cytokine secretion from human basophils. *Int. Arch. Allergy Immunol.* **118**:197–199, 1999.
29. Patella, V., Florio, G., Petraroli, A. and Marone, G. HIV-1 gp 120 induces IL-4 and IL-13 release from human $Fc\varepsilon RI^{+}$ cells through interaction with the V_H3 region of IgE. *J. Immunol.* **164**:589–595, 2000.
30. Norn, S., Jensen, C., Dahl, B. T., Skov, P. S., Baek, L., Permin, H., Jarlov, J. O. and Sorensen, H. Endotoxins release histamine by complement activation and potentiate bacteria-induced histamine release. *Agents Actions* **18**:149–152, 1986.
31. Cirillo, R., Triggiani, M., Siri, L., Ciccarelli, A., Pettit, G. R., Condorelli, M. and Marone, G. Cyclosporin A rapidly inhibits mediator release from human basophils presumably by interacting with cyclophilin. *J. Immunol.* **144**:3891–3897, 1990.
32. Metchnikoff, E. *Leçons sur la Pathologie Comparée de l'Inflammation*. Masson, Paris, 1892.
33. Padawer, J. Uptake of colloidal thorium dioxide by mast cells. *J. Cell Biol.* **40**:747–760, 1969.

34. Malaviya, R., Twesten, N. J., Ross, E. A., Abraham, S. N. and Pfeifer, J. D. Mast cells process bacterial ags through a phagocytic route for class I MHC presentation to T cells. *J. Immunol.* **156**:1490–1496, 1996.
35. Echtenacher, B., Männel, D. N. and Hültner, L. Critical protective role of mast cells in a model of acute septic peritonitis. *Nature* **381**:75–77, 1996.
36. Malaviya, R., Ikeda, T., Twesten, N. J., Ross, E. A. and Abraham, S. N. Mast cell modulation of neutrophil influx and bacterial clearance at sites of infections through TNF-α. *Nature* **381**:77–80, 1996.
37. Prodeus, A. P., Zhou, X., Maurer, M., Galli, S. J. and Carroll, M. C. Impaired mast cell-dependent natural immunity in complement C3-deficient mice. *Nature* **390**:172–175, 1997.
38. de Paulis, A., Ciccarelli, A., Cirillo, R., de Crescenzo, G., Columbo, M. and Marone, G. Modulation of human lung mast cell function by the c-*kit* receptor ligand. *Int. Arch. Allergy Immunol.* **99**:326–329, 1992.
39. de Paulis, A., Ciccarelli, A., de Crescenzo, G., Cirillo, R., Patella, V. and Marone, G. Cyclosporin H is a potent and selective competitive antagonist of human basophil activation by FMLP. *J. Allergy Clin. Immunol.* **98**:152–164, 1996.
40. Maurer, M., Echtenacher, B., Hültner, L., Kollias, G., Männel, D. N., Langley, K. E. and Galli, S. J. The c-*kit* ligand, stem cell factor, can enhance innate immunity through effects on mast cells. *J. Exp. Med.* **188**:2343–2348, 1998.
41. Aceti, A., Celestino, D., Caferro, M., Casale, V., Citarda, F., Conti, E. M., Grassi, A., Grilli, A., Pennica, A. and Sciarretta, F. Basophil-bound and serum immunoglobulin E directed against *Helicobacter pylori* in patients with chronic gastritis. *Gastroenterology* **101**:131–137, 1991.
42. Bechi, P., Dei, R., Di Bello, M. G. and Masini, E. *Helicobacter pylori* potentiates histamine release from serosal rat mast cells *in vitro*. *Dig. Dis. Sci.* **38**:944–949, 1993.
43. Masini, E., Bechi, P., Dei, R., Di Bello, M. G. and Sacchi, T. B. *Helicobacter pylori* potentiates histamine release from rat serosal mast cells induced by bile acids. *Dig. Dis. Sci.* **39**:1493–1500, 1994.
44. Lutton, D. A., Bamford, K. B., O'Loughlin, B. and Ennis, M. Modulatory action of *Helicobacter pylori* on histamine release from mast cells and basophils *in vitro*. *J. Med. Microbiol.* **42**:386–393, 1995.
45. Kurose, I., Granger, D. N., Evans, D. J. Jr, Evans, D. G., Graham, D. Y., Miyasaka M., Anderson, D. C., Wolf, R. E., Cepinskas, G. and Kvietys, P. R. *Helicobacter pylori*-induced microvascular protein leakage in rats: role of neutrophils, mast cells, and platelets. *Gastroenterology* **107**:70–79, 1994.
46. van Doorn, N. E., van Rees, E. P., Namavar, F., Ghiara, P., Vandenbroucke-Grauls, C. M. and de Graaff, J. The inflammatory response in CD1 mice shortly after infection with a CagA+/VacA+ *Helicobacter pylori* strain. *Clin. Exp. Immunol.* **115**:421–427, 1999.
47. Bechi, P., Romagnoli, P., Bacci, S., Dei, R., Amorosi, A., Cianchi, F. and Masini, E. *Helicobacter pylori* and duodenal ulcer: evidence for a histamine pathways-involving link. *Am. J. Gastroenterol.* **91**:2338–2343, 1996.
48. Nakajima, S., Krishnan, B., Ota, H., Segura, A. M., Hattori, T., Graham, D. Y. and Genta, R. M. Mast cell involvement in gastritis with or without *Helicobacter pylori* infection. *Gastroenterology* **113**:746–754, 1997.
49. Yamamoto, J., Watanabe, S., Hirose, M., Osada, T., Ra, C. and Sato, N. Role of mast cells as a trigger of inflammation in *Helicobacter pylori* infection. *J. Physiol. Pharmacol.* **50**:17–23, 1999.
50. Triggiani, M., Oriente, A., de Crescenzo, G. and Marone, G. Metabolism of lipid mediators in human basophils and mast cells. In *Human Basophils and Mast Cells. I. Biological Aspects* (Marone, G., ed.), pp. 135–147. Karger, Basel, 1995.
51. Ford-Hutchinson, A. W., Bray, M. A., Doig, M. V., Shipley, M. E. and Smith, M. J. H. Leukotriene B, a potent chemokinetic and aggregating substance released from polymorphonuclear leukocytes. *Nature* **286**:264–265, 1980.
52. Goldman, D. W., Gifford, L. A., Olson, D. M. and Goetzl, E. J. Transduction by leukotriene B_4 receptors of increases in cytosolic calcium in human polymorphonuclear leukocytes. *J. Immunol.* **135**:525–530, 1985.
53. Dahlén, S. E., Björk, J., Hedqvist, P., Arfors, K. E., Hammarström, S., Lindgren, J. A. and Samuelsson, B. Leukotrienes promote plasma leakage and leukocyte adhesion in postcapillary venules: *in vivo* effects with relevance to the acute inflammatory response. *Proc. Natl. Acad. Sci. USA* **78**:3887–3891, 1981.
54. Vigorito, C., Giordano, A., Cirillo, R., Genovese, A., Rengo, F. and Marone, G. Metabolic and hemodynamic effects of peptide leukotriene C_4 and D_4 in man. *Int. J. Clin. Lab. Res.* **27**:178–184, 1997.
55. Marone, G., Columbo, M., Triggiani, M., Cirillo, R., Genovese, A. and Formisano, S. Inhibition of IgE-

mediated release of histamine and peptide leukotriene from human basophils and mast cells by forskolin. *Biochem. Pharmacol.* **36**:13–20, 1987.

56. Columbo, M., Taglialatela, M., Warner, J. A., MacGlashan, D. W. Jr, Yasumoto, T., Annunziato, L. and Marone, G. Maitotoxin, a novel activator of mediator release from human basophils, induces large increases in cytosolic calcium resulting in histamine, but not leukotriene C_4 release. *J. Pharmacol. Exp. Ther.* **263**:979–986, 1992.
57. Triggiani, M., Oriente, A. and Marone, G. Differential roles for triglyceride and phospholipid pools of arachidonic acid in human lung macrophages. *J. Immunol.* **152**:1394–1403, 1994.
58. Triggiani, M., De Marino, V., Sofia, M., Faraone, S., Ambrosio, G., Carratù, L. and Marone, G. Characterization of platelet-activating factor acetylhydrolase in human bronchoalveolar lavage. *Am. J. Respir. Crit. Care Med.* **156**:94–100, 1997.
59. Bailie, M. B., Standiford, T. J., Laichalk, L. L., Caffey, M. J., Strieter, R. and Peters-Golden, M. Leukotriene-deficient mice manifest enhanced lethality from *Klebsiella pneumonie* in association with decreased alveolar macrophage phagocytic and bactericidal activities. *J. Immunol.* **157**:5221–5224, 1996.
60. Ramos, B. F., Zhang, Y., Qureshi, R. and Jakschik, B. A. Mast cells are critical for the production of leukotrienes responsible for neutrophil recruitment in immune complex-induced peritonitis in mice. *J. Immunol.* **147**:1636–1641, 1991.
61. Coffey, M., Phare, S., Kazanjian, P. and Peters-Golden, M. 5-Lipoxygenase metabolism in alveolar macrophages from subjects infected with the human immunodeficiency virus. *J. Immunol.* **157**:393–399, 1996.
62. Owman, C., Garzino-Demo, A., Cocchi, F., Popovic, M., Sabirsh, A. and Gallo, R. C. The leukotriene B_4 receptor functions as a novel type of coreceptor mediating entry of primary HIV-1 isolates into CD4-positive cells. *Proc. Natl. Acad. Sci. USA* **95**:9530–9534, 1998.
63. Alkhatib, G., Combadiere, C., Broder, C. C., Feng, Y., Kennedy, P. E., Murphy, P. M. and Berger, E. A. CC-CKR-5: a RANTES, MIP-1α, MIP-1β receptor as a fusion cofactor for macrophage-tropic HIV-1. *Science* **272**:1955–1958, 1996.
64. Choe, H., Farzan, M., Sun, Y., Sullivan, N., Rollins, B., Ponath, P. D., Wu, L., Mackay, C. R., LaRosa, G., Newman, W., Gerard, N., Gerard, C. and Sodroski, J. The beta-chemokine receptors CCR3 and CCR5 facilitate infection by primary HIV-1 isolates. *Cell* **85**:1135–1148, 1996.
65. Deng, H. K., Liu, R., Ellmeier, W., Choe, S., Unutmaz, D., Burkhart, M., Di Marzio, P., Marmon, S., Sutton, R. E., Hill, C. M., Davis, C. B., Peiper, S. C., Schall, T. J., Littman, D. R. and Landau, N. R. Identification of major co-receptor for primary isolates of HIV-1. *Nature* **381**:661–666, 1996.
66. Doranz, B. J., Rucker, J., Yi, Y., Smyth, R. J., Samson, M., Peiper, S. C., Parmentier, M., Collman, R. G. and Doms, R. W. A dual-tropic primary HIV-1 isolate that uses fusin and the beta-chemokine receptors CKR-5, CKR-3, and CKR-2b as fusion cofactors. *Cell* **85**:1149–1158, 1996.
67. Dragic, T., Litwin, V., Allaway, G. P., Martin, S. R., Huang, Y., Nagashima, K. A., Cayanan, C., Maddon, P. J., Koup, R. A., Moore, J. P. and Paxton, W. A. HIV-1 entry into $CD4^+$ cells is mediated by the chemokine receptor CC-CKR-5. *Nature* **381**:667–673, 1996.
68. Feng, Y., Broder, C. C., Kennedy, P. E. and Berger, E. A. HIV-entry cofactors: functional cDNA cloning of a seven-transmembrane, G protein-coupled receptor. *Science* **272**:872–877, 1996.
69. Holgate, S. T., Bradding, P. and Sampson, A. P. Leukotriene antagonists and synthesis inhibitors: new directions in asthma therapy. *J. Allergy Clin. Immunol.* **98**:1–13, 1996.
70. Lowy, F. D. *Staphylococcus aureus* infections. *N. Engl. J. Med.* **339**:520–532, 1998.
71. Forsgren, A. and Sjöquist, J. 'Protein A' from *S. aureus*. I. Pseudoimmune reaction with human γ-globulin. *J. Immunol.* **99**:822–827, 1966.
72. Inganäs, M., Johansson, S. G. O. and Bennich, H. H. Interaction of human polyclonal IgE and IgG from different species with protein A from *Staphylococcus aureus*: demonstration of protein A-reactive sites located in the Fab'_2 fragment of human IgG. *Scand. J. Immunol.* **12**:23–31, 1980.
73. Inganäs, M. Comparison of mechanisms of interaction between protein A from *Staphylococcus aureus* and human monoclonal IgG, IgA and IgM in relation to the classical Fcγ and the alternative $F(ab')_2$ protein interactions. *Scand. J. Immunol.* **13**:343–352, 1981.
74. Sjodahl, J. Structural studies on the four repetitive Fc-binding regions in protein A from *Staphylococcus aureus*. *Eur. J. Biochem.* **78**:471–479, 1977.
75. Lofdahl, S., Guss, B., Uhlen, M., Philipson, L. and Lindberg, M. Gene for staphylococcal protein A. *Proc. Natl. Acad. Sci. USA* **80**:697–701, 1983.
76. Uhlen, M., Guss, B., Nilsson, B., Gatenbeck, S., Philipson, L. and Lindberg, M. Complete sequence of

the staphylococcal gene encoding protein A. A gene evolved through multiple duplications. *J. Biol. Chem.* **259**:1695–1702, 1984.

77. Moks, T., Abrahmsen, L., Nilsson, B., Hellman, U., Sjoquist, J. and Uhlen, M. Staphylococcal protein A consists of five IgG-binding domains. *Eur. J. Biochem.* **156**:637–642, 1986.
78. Ibrahim, S. Immunoglobulin binding specificities of the homology regions (domains) of protein A. *Scand. J. Immunol.* **38**:368–374, 1993.
79. Ljungberg, U. K., Jansson, B., Niss, U., Nilsson, R., Sandberg. B. E. and Nilsson, B. The interaction between different domains of staphylococcal protein A and human polyclonal IgG, IgA, IgM and $F(ab')_2$: separation of affinity from specificity. *Mol. Immunol.* **30**:1279–1285, 1993.
80. Roben, P. W., Salem, A. N. and Silverman, G. J. V_H3 family antibodies bind domain D of staphylococcal protein A. *J. Immunol.* **154**:6437–6445, 1995.
81. Marone, G., Tamburini, M., Giudizi, M. G., Biagiotti, R., Almerigogna, F. and Romagnani, S. Mechanism of activation of human basophils by *Staphylococcus aureus* Cowan 1. *Infect. Immun.* **55**:803–809, 1987.
82. Sjöholm, I., Bierken, A. and Sjöquist, J. Protein A from *Staphylococcus aureus.* XIV. The effect of nitration of protein A with tetranitromethane and subsequent reduction. *J. Immunol.* **110**:1562–1569, 1973.
83. Sasso, E. H., Silverman, G. J. and Mannik, M. Human IgM molecules that bind staphylococcal protein A contain VHIII H chains. *J. Immunol.* **142**:2778–2783, 1989.
84. Hillson, J. L., Karr, N. S., Oppliger, I. R., Mannik, M. and Sasso, E. H. The structural basis of germline-encoded V_H3 immunoglobulin binding to staphylococcal protein A. *J. Exp. Med.* **178**:331–336, 1993.
85. Sasano, M., Burton, D. R. and Silverman, G. J. Molecular selection of human antibodies with an unconventional bacterial B cell antigen. *J. Immunol.* **151**:5822–5839, 1993.
86. Silverman, G. J., Sasano, M. and Wormsley, S. B. Age-associated changes in binding of human B lymphocytes to a V_H3-restricted unconventional bacterial antigen. *J. Immunol.* **151**:5840–5855, 1993.
87. Sasso, E. H., Silverman, G. J. and Mannik, M. Human IgA and IgG $F(ab')_2$ that bind to staphylococcal protein A belong to the V_HIII subgroup. *J. Immunol.* **147**:1877–1883, 1991.
88. Capra, J. D. and Kehoe, J. M. Variable region sequences of five human immunoglobulin heavy chains in the V_HIII subgroup: definitive identification of four heavy chain hypervariable regions. *Proc. Natl. Acad. Sci. USA* **71**:845–848, 1974.
89. Schroeder, H. W., Hillson, J. L. and Perlmutter, R. M. Structure and evolution of mammalian V_H families. *Int. Immunol.* **2**:41–50, 1990.
90. Herman, A., Kappler, J. W., Marrack, P. and Pullen, A. M. Superantigens: mechanism of T-cell stimulation and role in immune responses. *Annu. Rev. Immunol.* **9**:745–772, 1991.
91. Romagnani, S., Giudizi, M. G., Almerigogna, F. and Ricci, M. Interaction of staphylococcal protein A with membrane components of IgM- and/or IgD-bearing lymphocytes from human tonsil. *J. Immunol.* **124**:1620–1626, 1980.
92. Silverman, G. J. B-cell superantigens. *Immunol. Today* **18**:379–386, 1997.
93. Myhre, E. B. and Erntell, M. A non-immune interaction between the light chain of human immunoglobulin and a surface component of a *Peptococcus magnus* strain. *Mol. Immunol.* **22**:879–885, 1985.
94. Björk, L. Protein L: a novel bacterial cell wall protein with affinity for Ig L chains. *J. Immunol.* **140**:1194–1197, 1988.
95. Björk, L. and Kronvall, G. Purification and some properties of streptococcal protein G, a novel IgG-binding reagent. *J. Immunol.* **133**:969–974, 1984.
96. Reis, K., Ayoub, E. M. and Boyle, M. D. P. Streptococcal Fc receptors. I. Isolation and characterization of the receptor from a group C streptococcus. *J. Immunol.* **132**:3091–3097, 1984.
97. Kastern, W., Holst, E., Nielsen, E., Sjöbring, U. and Björk, L. Protein L: a bacterial immunoglobulin-binding protein and a possible virulence determinant. *Infect. Immun.* **58**:1217–1222, 1990.
98. Patella, V., Casolaro, V., Björk, L. and Marone, G. Protein L: a bacterial Ig-binding protein that activates human basophils and mast cells. *J. Immunol.* **145**:3054–3061, 1990.
99. Umezawa, H., Aoyagi, T., Morishima, H., Matsuzaki, M., Hamada, M. and Takeuchi, T. Pepstatin, a new pepsin inhibitor produced by actinomycetes. *J. Antibiot.* **23**:259–262, 1970.
100. Lacey, J. Thermophilic actinomycetes: characteristics and identification. *J. Allergy Clin. Immunol.* **61**:231–237, 1976.
101. Patterson, R., Roberts, M., Roberts, R. C., Emanuel, D. A. and Fink, J. N. Antibodies of different immunoglobulin classes against antigens causing farmer's lung. *Am. Rev. Respir. Dis.* **114**:315–324, 1976.

102. Grouchow, H. W., Hoffmann, R. G., Marx, J. J. Jr, Emanuel, D. A. and Rimm, A. A. Precipitating antibodies to farmer's lung antigens in a Wisconsin farming population. *Am. Rev. Respir. Dis.* **124**:411–415, 1981.
103. Morishima, H., Takita, T., Aoyagi, T., Takeuchi, T. and Umezawa, H. The structure of pepstatin. *J. Antibiot.* **23**:263–265, 1970.
104. Ackerman, S. K. and Douglas, S. D. Pepstatin A: a human leukocyte chemoattractant. *Clin. Immunol. Immunopathol.* **14**:244–250, 1979.
105. Smith, R. J., Bowman, B. J. and Iden, S. S. Pepstatin A-induced degranulation and superoxide anion generation by human neutrophils. *Clin. Immunol. Immunopathol.* **22**:83–93, 1982.
106. Schmitt, M., Painter, R. G., Jesaitis, A. J., Preissner, K., Sklar, L. A. and Cochrane, C. G. Photoaffinity labeling of the *N*-formyl peptide receptor binding site of intact human polymorphonuclear leukocytes. *J. Biol. Chem.* **258**:649–654, 1983.
107. Nelson, R. D., Ackerman, S. K., Fiegel, V. D., Bauman, M. P. and Douglas, S. D. Cytotaxin receptors of neutrophils: evidence that F-methionyl peptides and pepstatin share a common receptor. *Infect. Immun.* **26**:996–999, 1979.
108. Marone, G., Columbo, M., Tamburini, M. and Romagnani, S. Activation of human basophils by bacterial products. *Int. Arch. Allergy Appl. Immun.* **77**:213–215, 1985.
109. de Paulis, A., Minopoli, G., Arbustini, E., de Crescenzo, G., Dal Piaz, F., Pucci, P., Russo, T. and Marone, G. Stem cell factor is localized in, released from, and cleaved by human mast cells. *J. Immunol.* **163**:2799–2808, 1999.
110. Bouvet, J. P. and Pillot, J. Letter to the editor. *J. Immunol.* **145**:3944, 1990.
111. Bouvet, J. P., Pirès, R., Lunel-Fabiani, F., Crescenzo-Chaigne, B., Maillard, P., Valla, D., Opolon, P. and Pillot, J. Protein F: a novel F(ab)-binding factor, present in normal liver, and largely released in the digestive tract during hepatitis. *J. Immunol.* **145**:1176–1180, 1990.
112. Bouvet, J. P., Pirès, R., Charlemagne, J., Pillot, J. and Iscaki, S. Non-immune binding of human protein Fv to immunoglobulins from various mammalian and non-mammalian species. *Scand. J. Immunol.* **34**:491–496, 1991.
113. Bouvet, J. P., Pirès, R., Quan, C., Iscaki, S. and Pillot, J. Non-immune VH-binding specificity of human protein Fv. *Scand. J. Immunol.* **33**:381–386, 1991.
114. Shumaker, J. B., Goldfinger, S. E., Alpert, E. and Isselbacher, K. J. Arthritis and rash. Clues to anicteric viral hepatitis. *Arch. Intern. Med.* **133**:483–485, 1974.
115. Weiss, T. D., Tsai, C. C., Baldassare, A. R. and Zuckner, J. Skin lesions in viral hepatitis. Histologic and immunofluorescent findings. *Am. J. Med.* **64**:269–273, 1978.
116. Popp, J. W., Harrist, T. J., Dienstag, J. L., Bhan, A. K., Wands, J. R., LaMont, J. T. and Mihm, M. C. Cutaneous vasculitis associated with acute and chronic hepatitis. *Arch. Intern. Med.* **141**:623–629, 1981.
117. Ljunggren, B. and Möller, H. Hepatitis presenting as transient urticaria. *Acta Derm. Venereol.* **51**:295–297, 1971.
118. Lockshin, N. A. and Hurley, H. Urticaria as a sign of viral hepatitis. *Arch. Dermatol.* **105**:570–571, 1972.
119. Segool, R. A., Lejtenyi, C. and Taussig, L. M. Articular and cutaneous prodromal manifestations of viral hepatitis. *J. Pediatr.* **87**:709–712, 1975.
120. Vaida, G. A., Goldman, M. A. and Bloch, K. J. Testing for hepatitis B virus in patients with chronic urticaria and angioedema. *J. Allergy Clin. Immunol.* **72**:193–198, 1983.
121. Miadonna, A., Tedeschi, A., Leggieri, E., Invernizzi, F. and Zanussi, C. Basophil activation in idiopathic mixed cryoglobulinemia. *J. Clin. Immunol.* **2**:309–313, 1982.
122. Agnello, V., Chung, R. T. and Kaplan, L. M. A role for hepatitis C virus infection in type II cryoglobulinemia. *N. Engl. J. Med.* **327**:1490–1495, 1992.
123. Walsh, L. J., Trinchieri, G., Waldorf, H. A., Whitaker, D. and Murphy, G. F. Human dermal mast cells contain and release tumor necrosis factor alpha, which induces endothelial leukocyte adhesion molecule 1. *Proc. Natl. Acad. Sci. USA* **88**:4220–4224, 1991.
124. Moller, A., Lippert, U., Lessmann, D., Kolde, G., Hamann, K., Welker, P., Schadendorf, D., Rosenbach, T., Luger, T. and Czarnetzki, B. M. Human mast cells produce IL-8. *J. Immunol.* **151**:3261–3266, 1993.
125. Rumsaeng, V., Cruikshank, W. W., Foster, B., Prussin, C., Kirshenbaum, A. S., Davis, T. A., Kornfeld, H., Center, D. M. and Metcalfe, D. D. Human mast cells produce the $CD4^+$ T lymphocyte chemoattractant factor, IL-16. *J. Immunol.* **159**:2904–2910, 1997.
126. Yano, K., Yamaguchi, M., de Mora, F., Lantz, C. S., Butterfield, J. H., Costa, J. J. and Galli, S. J.

Production of macrophage inflammatory protein-1α by human mast cells: increased anti-IgE-dependent secretion after IgE-dependent enhancement of mast cell IgE-binding ability. *Lab. Invest.* **77**:185–193, 1997.

127. Baud, L., Perez, J., Denis, M. and Ardaillou, R. Modulation of fibroblast proliferation by sulfidopeptide leukotrienes: effect of indomethacin. *J. Immunol.* **138**:1190–1195, 1987.
128. Ruoss, S. J., Hartmann, T. and Caughey, G. H. Mast cell tryptase is a mitogen for cultured fibroblasts. *J. Clin. Invest.* **88**:493–499, 1991.
129. Hartmann, T., Ruoss, S. J., Raymond, W. W., Seuwen, K. and Caughey, G. H. Human tryptase as a potent, cell-specific mitogen: role of signaling pathways in synergistic responses. *Am. J. Physiol.* **262**:L528–L534, 1992.
130. Hatamochi, A., Fujiwara, K. and Ueki, H. Effects of histamine on collagen synthesis by cultured fibroblasts derived from guinea pig skin. *Arch. Dermatol. Res.* **277**:60–64, 1985.
131. Cairns, J. A. and Walls, A. F. Mast cell tryptase stimulates the synthesis of type I collagen in human lung fibroblasts. *J. Clin. Invest.* **99**:1313–1321, 1997.
132. Gruber, B. L., Kew, R. R., Jelaska, A., Marchese, M. J., Garlick, J., Ren, S., Schwartz, L. B. and Korn, J. H. Human mast cells activate fibroblasts. Tryptase is a fibrogenic factor stimulating collagen messenger ribonucleic acid synthesis and fibroblast chemotaxis. *J. Immunol.* **158**:2310–2317, 1997.
133. Biagini, G. and Ballardini, G. Liver fibrosis and extracellular matrix. *J. Hepatol.* **8**:115–124, 1989.
134. Paganelli, R., Scala, E., Mezzaroma, I., Pinter, E., D'Offizi, G., Fanales-Belasio, E., Rosso, R. M., Ansotegui, I. J., Pandolfi, F. and Aiuti, F. Immunologic aspects of hyperimmunoglobulinemia E-like syndrome in patients with AIDS. *J. Allergy Clin. Immunol.* **95**:995–1003, 1995.
135. Shor-Posner, G., Miguez-Burbano, M. J., Lu, Y., Feaster, D., Fletcher, M. A., Sauberlich, H. and Baum, M. K. Elevated IgE level in relationship to nutritional status and immune parameters in early human immunodeficiency virus-1 disease. *J. Allergy Clin. Immunol.* **95**:886–892, 1995.
136. Koutsonikolis, A., Nelson, R. P. Jr, Fernandez-Caldas, E., Brigino, E. N., Seleznick, M., Good, R. A. and Lockey, R. F. Serum total and specific IgE levels in children infected with human immunodeficiency virus. *J. Allergy Clin. Immunol.* **97**:692–697, 1996.
137. Secord, E. A., Kleiner, G. I., Auci, D. L., Smith-Norowitz, T., Chice, S., Finkielstein, A., Nowakowski, M., Fikrig, S. and Durkin, H. G. IgE against HIV proteins in clinically healthy children with HIV disease. *J. Allergy Clin. Immunol.* **98**:979–984, 1996.
138. Israël-Biet, D., Labrousse, F., Tourani, J. M., Sors, H., Andrieu, J. M. and Even, P. Elevation of IgE in HIV-infected subjects: a marker of poor prognosis. *J. Allergy Clin. Immunol.* **89**:68–75, 1992.
139. Viganò, A., Principi, N., Crupi, L., Onorato, J., Zuccotti, G. V. and Salvaggio, A. Elevation of IgE in HIV-infected children and its correlation with the progression of disease. *J. Allergy Clin. Immunol.* **95**:627–632, 1995.
140. Clerici, M. and Shearer, G. M. A Th1–Th2 switch is a critical step in the etiology of HIV infection. *Immunol. Today* **14**:107–111, 1993.
141. Clerici, M., Hakim, F. T., Venzon, D. J., Blatt, S., Hendrix, C. W., Wynn, T. A. and Shearer, G. M. Changes in interleukin-2 and interleukin-4 production in asymptomatic, human immunodeficiency virus-seropositive individuals. *J. Clin. Invest.* **91**:759–765, 1993.
142. Graziosi, C., Pantaleo, G., Gantt, K. R., Fortin, J. P., Demarest, J. F., Cohen, O. J., Sékaly, R. P. and Fauci, A. S. Lack of evidence for the dichotomy of T_H1 and T_H2 predominance in HIV-infected individuals. *Science* **265**:248–252, 1994.
143. Maggi, E., Mazzetti, M., Ravina, A., Annunziato, F., De Carli, M., Piccinni, M. P., Manetti, R., Carbonari, M., Pesce, A. M., Del Prete, G. and Romagnani, S. Ability of HIV to promote a T_H1 to T_H0 shift and to replicate preferentially in T_H2 and T_H0 cells. *Science* **265**:244–248, 1994.
144. Barcellini, W., Rizzardi, G. P., Borghi, M. O., Fain, C., Lazzarin, A. and Meroni, P. L. T_H1 and T_H2 cytokine production by peripheral blood mononuclear cells from HIV-infected patients. *AIDS* **8**:757–762, 1994.
145. Meyaard, L., Otto, S. A., Keet, I. P. M., Van Lier, R. A. W. and Miedema, F. Changes in cytokine secretion patterns of $CD4^+$ T-cell clones in human immunodeficiency virus infection. *Blood* **84**:4262–4268, 1994.
146. Meroni, L., Trabattoni, D., Balotta, C., Riva, C., Gori, A., Moroni, M., Villa, M. L., Clerici, M. and Galli, M. Evidence for type 2 cytokine production and lymphocyte activation in the early phases of HIV-1 infection. *AIDS* **10**:23–30, 1996.
147. Fakoya, A., Matear, P. M., Filley, E., Rook, G. A. W., Stanford, J., Gilson, R. J. C., Beecham, N., Weller, I. V. D. and Vyakarnam, A. HIV infection alters the production of both type 1 and 2 cytokines but does

not induce a polarized type 1 or 2 state. *AIDS* **11**:1445–1452, 1997.

148. Klein, S. A., Dobmeyer, J. M., Dobmeyer, T. S., Pape, M., Ottmann, O. G., Helm, E. B., Hoelzer, D. and Rossol, R. Demonstration of the T_H1 to T_H2 cytokine shift during the course of HIV-1 infection using cytoplasmic cytokine detection on single cell level by flow cytometry. *AIDS* **11**:1111–1118, 1997.
149. Emson, C. L., Bell, S. E., Jones, A., Wisden, W. and McKenzie, A. N. J. Interleukin (IL)- 4-independent induction of immunoglobulin (Ig)E, and perturbation of T cell development in transgenic mice expressing IL-13. *J. Exp. Med.* **188**:399–404, 1998.
150. Grünig, G., Warnock, M., Wakil, A. E., Venkayya, R., Brombacher, F., Rennick, D. M., Sheppard, D., Mohrs, M., Donaldson, D. D., Locksley, R. M. and Corry, D. B. Requirement for IL-13 independently of IL-4 in experimental asthma. *Science* **282**:2261–2263, 1998.
151. McKenzie, G. J., Emson, C. L., Bell, S. E., Anderson, S., Fallon, P., Zurawski, G., Murray, R., Grencis, R. and McKenzie, A. N. Impaired development of T_H2 cells in IL-13-deficient mice. *Immunity* **9**:423–432, 1998.
152. Wills-Karp, M., Luyimbazi, J., Xu, X., Schofield, B., Neben, T. Y., Karp, C. L. and Donaldson, D. D. Interleukin-13: central mediator of allergic asthma. *Science* **282**:2258–2261, 1998.
153. Brunner, T., Heusser, C. H. and Dahinden, C. A. Human peripheral blood basophils primed by interleukin 3 (IL-3) produce IL-4 in response to immunoglobulin E receptor stimulation. *J. Exp. Med.* **177**:605–611, 1993.
154. MacGlashan, D. Jr, White, J. M., Huang, S. K., Ono, S. J., Schroeder, J. T. and Lichtenstein, L. M. Secretion of IL-4 from human basophils: the relationship between IL-4 mRNA and protein in resting and stimulated basophils. *J. Immunol.* **152**:3006–3016, 1994.
155. Ochensberger, B., Daepp, G. C., Rihs, S. and Dahinden, C. A. Human blood basophils produce interleukin-13 in response to IgE-receptor-dependent and -independent activation. *Blood* **88**:3028–3037, 1996.
156. Bradding, P., Feather, I. H., Howarth, P. H., Mueller, R., Roberts, J. A., Britten, K., Bews, J. P. A., Hunt, T. C., Okayama, Y., Heusser, C. H., Bullock, G. R., Church, M. K. and Holgate, S. T. Interleukin 4 is localized to and released by human mast cells. *J. Exp. Med.* **176**:1381–1386, 1992.
157. Burd, P. R., Thompson, W. C., Max, E. E. and Mills, F. C. Activated mast cells produce interleukin 13. *J. Exp. Med.* **181**:1373–1380, 1995.
158. Pedersen, M., Nielsen, C. M. and Permin, H. HIV antigen-induced release of histamine from basophils from HIV infected patients. *Allergy* **46**:2006–2012, 1991.
159. Burtin, C., Blanche, S., Galoppin, L., Merval, R., Griscelli, C. and Scheinmann, P. Blood histamine levels in HIV-1-infected infants and children. *Int. Arch. Allergy Appl. Immunol.* **91**:142–144, 1990.
160. Kwong, P. D., Wyatt, R., Robinson, J., Sweet, R. W., Sodroski, J. and Hendrickson, W. A. Structure of an HIV gp 120 envelope glycoprotein in complex with the CD4 receptor and a neutralizing human antibody. *Nature* **393**:648–659, 1998.
161. Karray, S. and Zouali, M. Identification of the B cell superantigen-binding site of HIV-1 gp 120. *Proc. Natl. Acad. Sci. USA* **94**:1356–1360, 1997.
162. Berberian, L., Goodglick, L., Kipps, T. J. and Braun, J. Immunoglobulin V_H3 gene products: natural ligands for HIV gp 120. *Science* **261**:1588–1591, 1993.
163. Schnittman, S. M., Lane, H. C., Higgins, S. E., Folks, T. and Fauci, A. S. Direct polyclonal activation of human B lymphocytes by the acquired immune deficiency syndrome virus. *Science* **233**:1084–1086, 1986.
164. Fauci, A. S. Host factors and the pathogenesis of HIV-induced disease. *Nature* **384**:529–534, 1996.
165. Gelderblom, H. R., Hausmann, E. H. S., Özel, M., Pauli, G. and Koch, M. A. Fine structure of human immunodeficiency virus (HIV) and immunolocalization of structural proteins. *Virology* **156**:171–176, 1987.
166. Andrew, D. P., Chang, M., McNinch, J., Wathen, S. T., Rihanek, M., Tseng, J., Spellberg, J. P. and Elias, C. G. III. STCP-1 (MDC) CC chemokine acts specifically on chronically activated Th2 lymphocytes and is produced by monocytes on stimulation with Th2 cytokines IL-4 and IL-13. *J. Immunol.* **161**:5027–5038, 1998.
167. Schroeder, J. T., Howard, B. P., Jenkens, M. K., Kagey-Sobotka, A., Lichtenstein, L. M. and MacGlashan, D. W. Jr. IL-4 secretion and histamine release by human basophils are differentially regulated by protein kinase C activation. *J. Leukoc. Biol.* **63**:692–698, 1998.

CHAPTER 27

Differential Regulation of Mast Cell Mediator Secretion by Bacterial Products

TONG-JUN LIN, FU-GANG ZHU and JEAN S. MARSHALL*

Departments of Pathology and Microbiology & Immunology, Dalhousie University, Halifax, Canada

INTRODUCTION

Mast cells have critical roles in a number of physiological and pathological events, ranging from type 1 hypersensitivity and chronic inflammation to host defence against pathogens. The diverse roles of mast cells in these processes depend upon the selective expression and secretion of specific mast cell mediators. These include well-characterized granule-associated products, lipid mediators and a wide range of cytokines and chemokines (1–3). Although mast cells were discovered more than 100 years ago and have been implicated in parasite infection for many decades, only recently has their critical role in host defence been fully recognized (4, 5). Despite several studies of the consequences of mast cell function in infection (6, 7, 125) we still have relatively little understanding of the signals that induce mast cell activation and the mechanisms by which mast cells act selectively in the innate immune response without inducing the physiological consequences associated with allergic disease. Recent studies by several laboratories, including our own, have begun to examine these issues, particularly by reference to the known bacteria-derived activators of other cell types in the front line of host defence, such as macrophages and epithelial cells.

RECEPTORS FOR BACTERIA ON MAST CELLS

The microorganism-induced host response is the reflection of a complex contest between two highly evolved systems. During pathogen infection, the responses observed can be a strategy employed by the host cells to combat the foreign pathogen, or a mechanism initiated and orchestrated by the pathogen for its own survival. It is

* Corresponding author.

MAST CELLS AND BASOPHILS
ISBN 0-12-473335-2

increasingly apparent that many pathogens have developed the ability to interact with host cell receptor molecules (surface or intracellular) to induce their own entry and survival and perturb host cell function. The mast cell surface is replete with many molecules such as CD48, CD55 and β_1 integrins, which potentially serve as receptors for different bacterial ligands.

In general, the initial interaction between bacteria and host cells occurs when bacterial surface molecules bind to host cell receptors. In the case of *Pseudomonas aeruginosa*, for example, the bacterial ligands responsible for the binding to host cells include pilus, alginate, surface-associated exo-enzyme S and lipopolysaccharide (LPS) (8, 9). Among them, the type 4 pilus accounts for about 90% of the adherence capability of *P. aeruginosa* to human pneumocyte cells and is responsible for more than 90% of the virulence in mice (8, 10). The mammalian receptor for pili of *P. aeruginosa* has been identified as the carbohydrate moiety of the glycosphingolipids asialo-G_{M1}, asialo-G_{M2}, lactosyl ceramide and ceramide trihexoside (8). Although, *P. aeruginosa* significantly stimulates mast cell degranulation (11) and induces intracellular calcium mobilization (Lin and Marshall, unpublished), the role of type 4 pili and other virulence factors in these effects remains to be determined. In *Escherichia coli*, FimH, a protein located in the tip of type 1 pili, has been demonstrated as the ligand responsible for binding to rodent (12, 12a) and human mast cells (13, 14). Because FimH binds to the mannosyl residues of glycoprotein on the host cell surface, the identity of FimH receptors on host cells varies. This receptor was demonstrated as CD48 in rodent mast cells and macrophages (15), CD11/CD18 and NCA-50 in human blood leukocytes and epithelial cells (16, 17). However, the FimH receptor on human mast cells remains to be determined experimentally. Several other cell-surface proteins such as β_1 integrins, E-cadherin and CD66 have been identified as the receptors for the invasin of *Yersinia* (18), internalin of *Listeria monocytogenes* (19) and opacity proteins of *Neisseria gonorrhoeae* (20), respectively, in other cell types. Interaction of these bacterial ligands with their corresponding host cell receptors induced profound protein tyrosine phosphorylation in human neutrophils (20) and epithelial cells (21).

The surface molecules of host cells are not always employed by bacteria as receptors. Some organisms such as enteropathogenic *E. coli* transfer the bacterial protein (Tir) into epithelial cells as the receptor for bacterial ligand intimin (22). Moreover, surface protein expression on host cells can be significantly modulated upon bacterial activation. For example, CD28 expression by rodent mast cells is significantly induced by bacterial products such as LPS of *Salmonella typhosa* and the outer surface protein A lipoprotein from *Borrelia burgdorferi* (23). Similarly, bacterial infection activates T cells (24), macrophages (25) and endothelial cells and induces various phenotypic and functional changes such as increased surface expression of E-selectin and the adhesion molecules ICAM-1 and VCAM-1. Tashiro *et al.* (26) demonstrated that stimulation of CD28 induces tumour necrosis factor-α (TNF-α) secretion and enhances $Fc_{\varepsilon}RI$-induced secretion of TNF-α in mouse mast cells. Thus, it is likely that during bacterial infection mast cells undergo phenotypic changes to enhance their capability to secrete cytokines to promote host defence against pathogens (4).

SELECTIVE MAST CELL RESPONSES TO SPECIFIC BACTERIAL PRODUCTS

As a consequence of the closely regulated signalling and receptor mechanisms present on immune effector cells, many bacterial products will selectively activate cells to selectively produce a particular profile of cytokines and other mediators or to alter the expression of receptors important to immune regulation. Mast cells provide very good examples of such selective activation, and a number of bacterial products have been demonstrated to have unique and potent effects on mast cell function.

Mast Cell Response to Bacterial Superantigens

The term 'superantigen' has been used to describe a group of microbial antigens that differ from conventional antigens and share several properties which provide the basis by which these molecules fulfil the general requirements for consideration as a superantigen (27). Superantigens are presented by class II major histocompatibility complex (MHC) molecules through interaction with conserved amino acid residues that are on the outside of the peptide-binding cleft as compared with the conventional antigens which are presented in MHC peptide-binding groove (27). The recognition of superantigen by T cell receptors (TCR) almost entirely depends on the variable domain of the TCR β chain (Vβ) with little regard for the other diversity components such as Dβ, Jβ, Vα, Jα (27, 28). Thus, superantigen is capable of interacting with a large population (up to 30%) of the T cell repertoire as compared with conventional antigen with a less than 1 in 1000 responding frequency (27). Superantigens can be broadly divided into two classes: the minor lymphocyte-stimulating (Mls) antigens encoded by endogenous retroviral genes (endogenous superantigens), and the exogenous superantigens represented by a growing list of bacterial, mycoplasmal and viral proteins (27). Although mast cells contain transcripts predicted to encode Mls-1a, their biological relevance remains unclear (29). In contrast, a role for mast cells in bacterial superantigen-mediated pathogenesis has long been suspected (30, 31). In mice, an intracutaneous injection of staphylococcal enterotoxin B (SEB), one of the best studied bacterial superantigens, elicits a strong inflammatory response, including mast cell degranulation, vasodilation and leukocyte infiltration (32). *In vitro*, SEB reduces interleukin-4 (IL-4) and TNF-α mRNA expression and inhibits IL-4 protein production in the human mast cell line HMC-1 (33). In RBL cells, Komisar *et al.* (34) were not able to demonstrate MHC class II expression; however, they reported preliminary data that SEB induces serotonin secretion from these cells. Thus, they proposed new receptors for SEB other than MHC class II on mast cells. These preliminary findings require further study. Olenick *et al.* (35) have shown that SEB is able to specifically bind to rat skin mast cells and rat mast cell line (RBL cells). Recently, Dimitriadou *et al.* (36) demonstrated that the receptor for bacterial superantigen on mast cells is MHC class II, similar to that observed on other cells. Binding of staphylococcal enterotoxin A (SEA) to MHC class II on human mast cells has been demonstrated to elicit significant ultrastructural changes (36). Recognition of SEA-bound MHC class II molecules on HMC-1 mast cells by the T cell receptor of K25 cells, an SEA-specific murine T cell hybridoma, triggers significant IL-2 secretion by these T cell hybridomas (36). Thus, mast cells are able to specifically recognize bacterial superantigens and participate in the innate and acquired immunity through altering mediator secretion and antigen presentation to T cells.

LPS Activation of Mast Cells

Bacterial lipopolysacharride (LPS) can be an important signal for mobilizing the immune response and is thought to be critical in inducing much of the pathology associated with septicaemia due to infection with Gram-negative bacteria. Immunologists have studied the role of this family of molecules in activating monocytes and macrophages for many years. However, mast cells have been recognized as an LPS target only relatively recently. Early work by our group demonstrated that highly purified rat peritoneal mast cells can be activated by high doses of LPS (1–5 $\mu g\ ml^{-1}$) to produce IL-6 without concurrent degranulation (37). This finding was of interest because it opened up the possibility that a range of other pathogen-associated stimuli might be able to induce mast cells to selectively produce cytokines even if they had been demonstrated not to induce mast cell degranulation. Further studies, by us and others, demonstrated selective cytokine secretion by LPS-treated mouse bone marrow mast cells (mBMMC) (38, 39) and the production of TNF-α by LPS-activated rat peritoneal mast cells (40, 41). More recent data (see Chapter 34) have demonstrated that human intestinal mast cells will also respond to LPS activation and selectively produce the cytokine IL-8. Given these data it is tempting to consider LPS as a broad inducer of mast cell cytokine expression without degranulation. However, LPS does not induce the production of a number of other cytokines from rodent mast cells including granulocyte–macrophage colony-stimulating factor (GM-CSF) and interferon-γ (IFN-γ) under conditions in which other stimuli such as IgE-mediated activation (GM-CSF) or IL-12 treatment (IFN-γ) were effective. There is also no evidence to suggest that LPS can induce lipid mediator production from mast cells. Rat peritoneal mast cells produce predominately the prostaglandin PGD_2 following IgE-mediated activation. LPS treatment for up to 18 h does not induce significant production of this mediator. Recently, we have shown that 2.5S nerve growth factor can induce the production of PGE_2 from rat peritoneal mast cells and mBMMC. LPS was also found to be ineffective in inducing production of this prostanoid (42).

The mechanism through which LPS activates mast cells remains unclear. Generally much larger amounts of LPS are required to activate mast cells *in vitro* than are required to activate macrophages. Mast cells express little or no surface CD14, suggesting that one or more of a number of putative alternative LPS receptor systems may be employed by the mast cell. Obvious candidates include the β_2 integrin family (43) and the macrophage scavenger receptors (44) both of which have been shown to bind and induce cellular functions in response to LPS in other cell types.

In vitro there is little evidence to suggest that LPS can induce mast cell degranulation; in fact, all the studies to date suggest that LPS does not directly induce the release of histamine or β-hexosaminidase from mast cells. However, *in vivo* studies have demonstrated that administration of LPS can be associated with the release of pre-formed mast cell mediators, especially proteases. Brown *et al.* (45) have shown that administration of LPS induced an increase in plasma RMCPII which could be partially blocked by pre-treatment of the animals with doxantrazole. Similarly, Suzuki *et al.* have demonstrated that short-term treatment of hamsters with LPS induces release of a chymase-like activity from perivascular mast cells (46). It is not clear whether these effects of LPS are direct or indirect, via the production of other mediators by LPS-responsive cells within the mast cell microenvironment.

One early study has suggested that mast cells are not critical for an effective TNF-α response to LPS since both mast cell-deficient (w/w^v) mice and mast cell-containing controls gave a similar response to low-dose LPS challenge (47). However, the role of mast cells in the *in vivo* cytokine response to LPS is still under active investigation. It is likely that the mast cell response is most important in local tissue sites where there are fewer macrophages, such as the skin, or when the LPS concentrations are high, such as in sepsis. The type of cytokine response induced in mast cells by LPS is likely to be very dependent on the presence of other mediators. For example, both the TNF-α and IL-6 responses to LPS by rodent mast cells are profoundly inhibited in the presence of IL-10 (40); corticosteroids can also inhibit such responses very effectively. In contrast, treatment of mast cells with LPS in the presence of low doses of PGE_2 leads to a synergistic enhancement of mast cell IL-6 responses, while the TNF-α response is reduced compared with LPS treatment alone (see later).

LPS effects on mast cells do not appear to be limited to mediator production; several reports have indicated a role for LPS in enhancing the expression of surface molecules (48). One of the first reports of LPS action on mast cells was the demonstration that LPS treatment of mBMMC leads to an increase in MHC class II expression by these cells (23). More recently, expression of the co-receptor CD28 was described on murine cultured mast cells and peritoneal mast cells. CD28 expression was markedly upregulated by LPS pre-treatment of the cells (23). The combined effects of LPS inducing cytokines necessary for mobilization of immune effector cells and the upregulation of receptor molecules important for potential antigen presentation and co-activation may be important in the development of an effective immune response to Gram-negative bacteria. Both further *in vivo* studies and a detailed evaluation of human mast cell responses to LPS will be necessary in order to understand the importance of this response in disease and host defence.

Mast Cell Response to *Clostridium difficile* Toxins

Enterotoxin A and cytotoxin B are the major virulence factors of *Clostridium difficile*, an organism which causes antibiotic-associated diarrhoea and pseudomembranous colitis (49). Toxin A elicits an acute inflammatory response such as neutrophil infiltration, fluid secretion and increased permeability when injected into animal intestine (50, 51). The receptor for toxin A has been identified as a galactose- and *N*-acetylglucosamine-containing glycoprotein coupled to a G protein (52). Several studies suggest that mast cells play an important role in the pathophysiology of toxin A. *In vivo* administration of toxin A into rat ileal loops mediates mast cell activation, as evidenced by the secretion of rat mast cell protease II (RMCPII), a specific mucosal mast cell mediator (50, 51). Exposure of rat colon to toxin A or human anti-galactose IgG caused mast cell activation, fluid secretion and increased permeability (53). Mast cell degranulation occurs as early as 15 min after exposure to toxin A (54). Moreover, treatment of animals with the H_1 histamine antagonist lodoxamide, with histaminase, or with the anti-inflammatory agent ketotifen reduced the inflammation and secretory responses caused by toxin A (54–56). Castagliulo *et al.* (51) demonstrated that neurotensin together with substance P are involved in the toxin A-induced mast cell activation in rat. Pre-treatment of rats with the neurotensin receptor antagonist SR-48,692 inhibits toxin A-induced changes in colonic secretion and mucosal permeability and histological damage. Mast cell degranulation in colonic explants after exposure to toxin A or neurotensin is inhibited by SR-48,692. Moreover, the substance P receptor antagonist CP-96,345 significantly inhibits neurotensin-induced mast cell degranulation in colonic explants (51). Thus, toxin A-mediated mast cell activation *in vivo* likely involves both direct and indirect mechanisms. The toxin A-mediated inflammatory response in the intestine involves several cell types, such as epithelial cells, neutrophils, mast cells and lymphocytes (57). The specific role of the mast cells was recently addressed in a mast cell-deficient mouse model. Wershil *et al.* (58) demonstrated that the intestinal fluid secretion and neutrophil infiltration that are induced by intraluminal exposure of small intestinal loops to toxin A occurs by mechanisms that are partially mast cell-dependent. *In vitro*, the direct effects of toxin A on mast cell activation and survival have been examined (59). Toxin A at 10 ng ml^{-1} or 1 μg ml^{-1} for 4 h stimulates rat peritoneal mast cells to release TNF-α without inducing significant histamine release and nitric oxide production. TNF-α has been implicated in intestinal inflammatory responses such as neutrophil infiltration and activation, and necrosis of intestinal epithelial cells (60). Thus, toxin A-mediated TNF-α release from mast cells likely contributes to the early inflammatory response during *C. difficile* infection. However, longer exposure (24 h) of mast cells to toxin A caused chromatin condensation, cytoplasmic blebbing, appearance of apoptotic-like vesicles and DNA fragmentation (59). Thus, toxin A-mediated impaired mast cell function and survival could hamper the capacity of these cells to counteract the infection, thus prolonging the pathogenic process during *C. difficile* infection.

Unlike cholera toxin and pertussis toxin which target heterotrimeric G proteins ('big' GTP-binding proteins), *C. difficile* toxins A and B, using UDP-glucose as co-substrate, monoglucosylate Rho proteins, a subfamily of small GTPases (61, 62). Toxin A-catalysed glucosylation has been demonstrated in mast cells (62). Proteins glucosylated by toxin A have been identified as RhoA, Rac1 and Cdc42Hs (62). Similarly, toxin B mediates RhoA and Cdc42 gluocosylation in mast cells (63). Rho proteins participate in a number of signalling pathways, including Fc receptor-mediated kinase cascade, and are involved in a variety of cellular events such as phagocytosis, cytoskeleton rearrangement, apoptosis and proliferation. In mast cells, degranulation (63) and Fc receptor-mediated phagocytosis (64) can be inhibited by toxin B, an effect likely mediated by this glucosylation of Rho proteins. In human monocytes, toxin B strongly stimulates IL-1, TNF-α and IL-6 production (65). Moreover, toxin B and LPS showed synergistic interactions in the stimulation of IL-1, TNF-α and IL-6 production (65). The effects of toxin B on cytokine production by mast cells have not yet been studied.

Cholera Toxin Activation of Mast Cells

Cholera toxin (CTX) has been widely examined as a potential mucosal adjuvant (66) and is known to activate mast cells through elevation of intracellular cAMP levels. We have demonstrated that treatment of freshly isolated rat peritoneal mast cells with CTX will induce a dose-dependent increase in IL-6 and decrease in TNF-α production (41). Given the role of IL-6 as an inducer of plasma cell differentiation and the acute-phase response with a predominantly anti-inflammatory role, such a pattern of cytokine modulation may limit local inflammation while aiding in the generation of effective immunity.

In vivo CTX treatment at the time of oral antigen administration has been shown to enhance an anaphylactic response to later antigen challenge (67). While much of this effect may be due to increased IgE antibody production, the possibility that changes in mast cell responses or cytokine expression may play a role also needs to be considered. Since CTX has both profound effects on the local cytokine microenvironment and direct effects on cellular activities, understanding the outcome may be complex. The importance of mast cells to the intestinal secretory response to CTX has been directly examined using mast cell-deficient mice (68). Interestingly, tissues from w/w^v or Sl/Sl^d mast cell-deficient animals failed to give an intestinal secretory response to luminal CTX challenge. However, reconstitution of the local mast cell population of w/w^v mice with bone marrow cells failed to restore tissue responsiveness to cholera toxin, suggesting that another SCF/c-kit-dependent cell type or signalling pathway may be responsible for the initial observations. It has been suggested that agents which elevate cAMP will lead to increased mast cell development based on mouse *in vitro* studies (69), and cholera toxin itself has been shown to promote the factor-dependent growth of murine mast cell progenitors (70). However, direct examination of rectal biopsy tissues suggests that infection with *Cholera* does not have such an effect in the human. Similar numbers and distribution of mast cells were observed in subjects who had cholera as in normal controls (71). There is currently little work which directly examines the effects of CTX on other aspects of human mast cell regulation and function.

Mast Cell Responses to Bacterial DNA

Specific bacterial DNA sequences are known to be potent activators of some immune effector cells (72–74). One such set of sequences are unmethylated CpG motifs, which are found in much higher concentration on the genome of bacteria than in mammalian systems (73, 75). Native DNA or oligonucleotides containing these sequences are known to activate human and rodent macrophages (76), natural killer (NK) cells and B cells (73); however, mammalian DNA and synthetic DNA containing no unmethylated CpG motifs are not able to stimulate these effector cells (77). It is likely that these sequences are an example of structures that are recognized by the innate immune response and which provide critical inducing signals to the immune system. Given the location of mast cells, their role in antibacterial responses and as sentinal cells, it is necessary to consider the possibility that mast cells might be activated by CpG motifs or other bacterial sequences to produce pro-inflammatory cytokines or other signals related to immune activation.

We have recently demonstrated (Zhu and Marshall, unpublished) that mBMMC selectively produce the cytokines TNF-α and IL-6 following activation with either native bacterial DNA derived from *E. coli* or with unmethylated CpG motif-containing oligonucleotides. The oligonucleotides used in this study were all 20-mers and shown to be endotoxin-free. It was notable that 20-mer oligonucleotides in which, by a single base pair change, the -C-G- portion of the motif sequence was replaced by a -G-C- did not induce any significant cytokine response when examined in parallel experiments. Treatment of bacterial DNA with methylase blocked the mast cell cytokine response, indicating a need for a methylated sequence for mast cell activation. Other cytokines such as GM-CSF and IFN-γ were not induced by CpG-ODN (oligodeoxynucleotide) treatment. An investigation into the ability of CpG-ODN to induce release of pre-formed mast cell mediators suggested that CpG-ODN treatment did not induce the short-term release of either histamine or β-hexosaminidase. The ability of CpG sequences to induce TNF-α and IL-6 production from mast cells was dose-dependent, and the amount of TNF-α and IL-6 produced correlated closely with the number of CpG motifs included within a 20-mer sequence.

The mechanisms by which CpG-ODN exert their effects on mast cells and other immune effector cells are still unclear. It has been suggested that there may be specific cell-surface receptors for CpG motifs (78, 79) or that CpG-ODN are taken up via a pinocytotic mechanism (80) and interact with transcription factors such as NF-κB within the nucleus (81). There is also recent evidence that CpG DNA can induce MAP (mitogen-activated protein) kinase activity (82). Our observations would support the latter hypothesis, since binding studies suggest that the uptake of Texas red-labelled CpG-ODN cannot be blocked by a large excess of unlabelled CpG DNA. The patterns of Texas red-labelled CpG-ODN and control GpC-ODN uptake observed in BMMC by confocal microscopy are very similar and consistent with a rapid pinocytotic uptake (see Fig. 1). This suggests that mast cells do not discriminate between bacterial and control DNA sequences at the level of uptake but are nevertheless able to mount a specific and selective cytokine response. These observations are in keeping with the early prediction, based on information from other cell types, that the major signalling mechanisms mediating the stimulatory effects of bacterial and synthetic DNA occur following their uptake (83).

Fig. 1 Typical photographs of confocal microscopic examination of mBMMC after incubation for 6 h at 37°C with 2 μm Texas red-conjugated CpG-ODN. The flourescence intensity and intracellular distribution vary from cell to cell. The majority of cells displayed peripheral patches of flourescence, and a small number showed a strong but heterogeneous distribution of flourescence throughout the whole cell. A very similar uptake pattern was observed in mBMMC incubated with Texas red-conjugated control ODN in a parallel experiment.

COMBINED SIGNALS ALTER AND ENHANCE MAST CELL RESPONSES

When considering mast cells within the context of a complex immune and inflammatory response to injury or infection, the established systems to predict the responses of mast cells to single bacterial products are clearly insufficient. *In vivo* studies such as the landmark papers describing the importance of mast cells in bacterial host defence (4, 84, 125), and others described above using w/w^v mice, allow for an evaluation of the importance of mast cells but do not easily lend themselves to developing a full understanding of the signalling and regulatory mechanisms involved.

Studies in our laboratory examining the simple model of LPS-mediated activation of mast cells suggest that the presence of other mediators can have a profound effect on the observed response. PGE_2 is usually considered to be an anti-inflammatory mediator and has been demonstrated to inhibit both the early- and late-phase response in exercise-induced and allergen-induced asthma (85, 86). Short-term products of mast cell degranulation, including TNF-α and histamine, have been shown to act on stromal cells such as fibroblasts to induce the production of PGE_2. If mast cells are treated with a combination of LPS and PGE_2 the IL-6 response of the cells is synergistically enhanced while the TNF response to LPS treatment is inhibited. Recent data suggest that the overall effect of IL-6 may be anti-inflammatory (87). Therefore, we could interpret these data to suggest that the presence of PGE_2 might convert a predominantly pro-inflammatory mast cell cytokine response to LPS to one that is predominantly anti-inflammatory. In this regard we should also consider the potential effects on effective host defence of a number of bacterial products which directly induce PGE_2. The possibility that early mast cell mediators could trigger a PGE_2 response capable of regulating the later cytokine cascade by mast cells, and other cell types, may need to be considered in the development of therapeutic strategies, which have usually focused on preventing mast cell granule release (see Fig. 2).

One potential approach to delineating the combined effects of many bacterial signals is to examine the consequences of mast cells interacting with whole bacteria, which provide

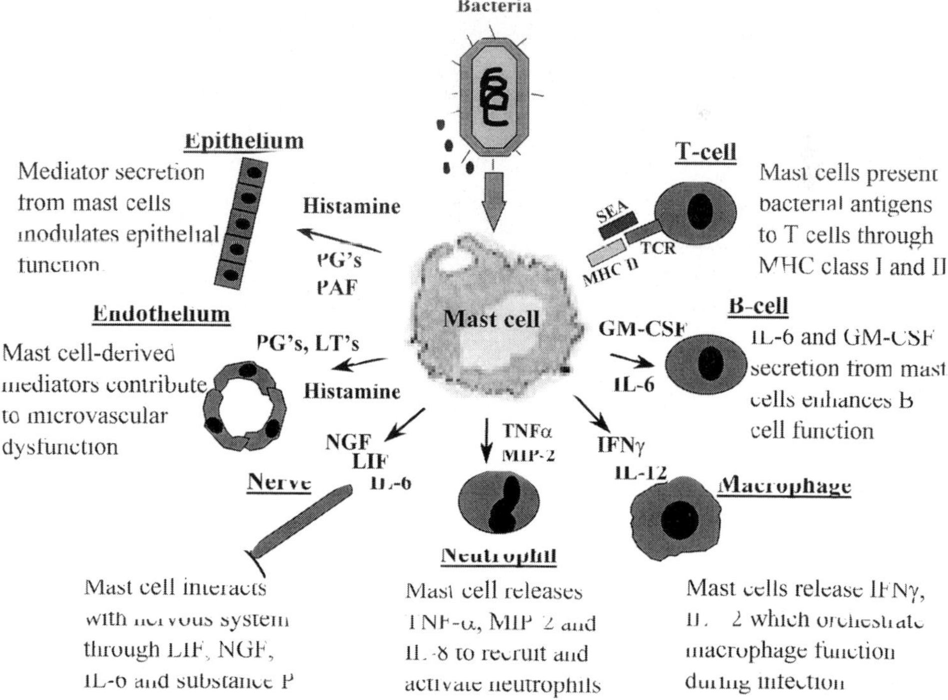

Fig. 2 Role of mast cells in innate and acquired immunity. The diverse roles of mast cells during bacterial infection depend on the selective expression and secretion of specific mast cell mediators in local tissue.

TABLE I
Bacterial Products Capable of Modulating Mast Cell Function

Bacterial components	Effects on mast cells		Reference(s)
Secreted products	**Degranulation**	**Other effects**	
Clostridium difficile toxin A *in vitro*	–	↑TNF-α	59
Clostridium difficile toxin A *in vivo*	+		50, 58
Clostridium difficile toxin B	–	↓phagocytosis	63, 64
Cholera toxin	–	↑IL-6, ↓TNF-α	41
Pertussis toxin	–		107
Staphylococcal enterotoxin A	?	Structural changes	36
Staphylococcal enterotoxin B	–	↓IL-4, ↓TNF-α	32–34
Leukotoxin of *Pasteurella haemolytica*	+		108
Toxin STb of *Escherichia coli*	+		109
d-Toxin of *Staphylococcus aureus*	?	Permeabilization	110
Haemolysins			
Vibro vulnificus	+		111
Escherichia coli	+		112
Serratia marcescens	+		112
Aeromonas hydrophilia	+		112
Listeria monocytogenes	+		112
Bacterial wall components			
Fimbriae			
Escherichia coli	+	↑TNF-α	7, 13
Klebsiella pneumoniae	+	↑TNF-α	7, 13
Polysaccharide			
Group A streptococci	+		113
Lipopolysaccharide			
Escherichia coli	–	↑IL-6, THF-α	37
Salmonella typhosa	?	↑CD28 expression	23
Fusobacterium nucleatum	+		114
Bacteroides oralis	+		114
Veillonella parvula	+		114
Lipid A-associated protein	+		115
Protein A of *Borrelia burgdorferi*	?	↑CD28 expression	23
Protein L of *Peptostreptococcus magnus*	+	↑LTC_4, PGD_2	116
Peptidoglycan of *Staphylococcus aureus*	+		117
Nuclear products			
Bacterial DNA	–	↑IL-6, THF-α	118
Uncharacterized components			
Extract of *Helicobacter pylori*	+		119
Whole bacteria			
Gram-negative			
Aeromonas hydrophilia	+		112
Borrelia burgdorferi	+	↑TNF-α	120
Citrobacter freundii	?	↑TNF-α	13
Enterobacter cloacae	+		121
Escherichia coli	+	↑TNF-α	7, 13
Haemophilus influenzae	+		122
Helicobacter pylori	–		123
Klebsiella oxytoca	+		121
Klebsiella pneumoniae	+	↑TNF-α	13, 121
Proteus vulgaris	+		121
Pseudomonas aeruginosa	+		11
Serratia marcescens	+		112
Vibrio vulnificus	+	↑LDH	124
Gram-positive			
Listeria monocytogenes	+		112
Staphylococcus aureus	+	↑TNF-α	13, 122
Staphylococcus epidermidis	+		121
Staphylococcus faecium	?	↑TNF-α	13

+ stimulation; – inhibition; ? not tested

multiple interactions capable of mast cell activation (see Table I). While the use of live bacteria has certain advantages in terms of comparison with disease conditions, the rapid growth of bacteria and subsequent effects of non-physiologically high levels of bacterial products and cell debris make evaluation of the long-term mediator response of mast cells difficult. Therefore, recent studies in our laboratory have examined the ability of fixed *Staphylococcus aureus* Cowan strain 1 (SAC) bacteria to activate mBMMC. Such treatment does not induce the short-term release of mast cell mediators such as histamine and β-hexosaminidase. We predicted that such mast cell activation might lead to a TNF-α response, as has been reported for other bacterial systems. However, we also examined the production of other cytokines not traditionally associated with acute inflammation such as IL-10 and IL-4. We observed substantial, selective, dose-dependent increases in the production of a number of cytokines over a 24-h time period (see Fig. 3).

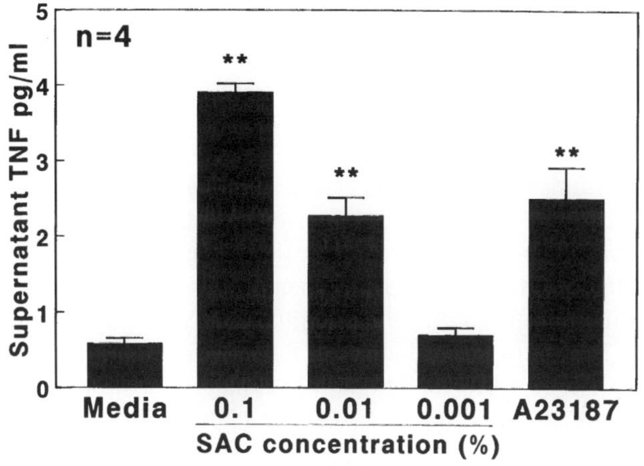

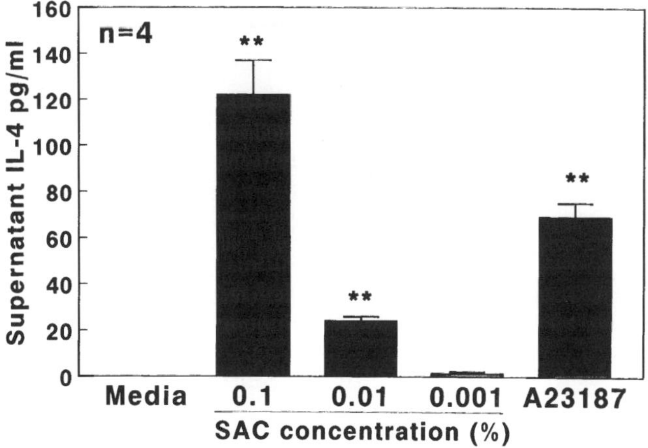

Fig. 3 Cytokine (TNF, IL-4) production in mBMMC in response to fixed *Staphylococcus aureus* Cowan 1 (SAC). mBMMC at 1×10^6/ml were incubated with different concentrations of SAC at 37°C for 24 h, followed by measurement of cytokines in supernatants using ELISA methods. ** $p < 0.01$.

BACTERIA AND BACTERIAL PRODUCT-MEDIATED SIGNALLING EVENTS IN MAST CELLS

When interacting with the host, some pathogens use a 'long-distance' approach via the production of toxins or noxious virulence factors, which exert effects without cell–cell contact. Others rely on the intimate interplay between bacteria and host cells through cell–cell contact or microenvironmental effects. Whatever the approach, the signalling cascades of the host cell usually mediate alterations in function. We still have little knowledge of the details of these interactions in mast cells but can learn a great deal from the work that has been done in other cell systems.

The role of intracellular calcium

Calcium is an important element in signal transduction and is involved in a variety of cellular functions such as degranulation (exocytosis), phagocytosis (endocytosis) and cell movement (chemotaxis). The fact that histamine secretion from mast cells can be significantly stimulated by purified glycolipid (heat-stable haemolysin) of *P. aeruginosa* (11) and calcium ionophore-mediated histamine release can be altered by *P. aeruginosa* (88) suggests that intracellular calcium in mast cells is likely one of the signalling targets altered by this bacterium. Other bacteria, such as *Salmonella typhimurium* (89) and enteropathogenic *E. coli* (90), also modulate intracellular calcium levels during bacterial invasion. *Pseudomonas* pyocyanine-mediated calcium increase appears to result from release of intracellular stores because it is independent of extracellular calcium and was completely abolished by thapsigargin (91). Mobilization of intracellular calcium stored in the endoplasmic reticulum is mediated by the second messenger inositol (1,4,5)-trisphosphate (IP_3), derived from membrane phosphatidylinositol bisphosphate (PIP_2) through the action of phospholipase C (PLC) (92). Foubister *et al.* (93) have demonstrated that *E. coli* infection triggers the release of IP_3 in host cells, leading to the hypothesis that PLC was activated by bacteria under these conditions. In response to this, a more recent study demonstrated that *E. coli* infection induces tyrosine phosphorylation of PLC-γ_1, suggesting the activation of this kinase (94).

The importance of PKC-mediated pathways

The action of PLC on PIP_2 generates also the second-messenger molecule diacylglycerol (DAG), which is involved in the activation of protein kinase C (PKC). PKC is a family of protein serine/threonine kinases which are involved in a range of cellular functions such as cytokine production and phagocytosis. In epithelial cells *E. coli* (95) or *Legionella pneumophila* (96) infection induces PKC activation. However, the exact role of PKC in these infections remains to be determined. Given the significant roles of PKC in the modulation of receptor-mediated phagocytosis in macrophages (97, 98) and the unique effects of phorbol 12-myristate 13-acetate (PMA), a PKC activator, on the cytoskeletal rearrangement in mast cells (99), it is likely that PKC may modulate bacterial internalization by mast cells. Indeed, a recent study (14) demonstrated that internalization of fimbrial expressing *E. coli* by human mast cells was significantly stimulated by PKC activation (short-term PMA treatment) and dramatically decreased by PKC inhibitors (Ro 31-8220 and staurosporine) or PKC depletion (long-term PMA treatment). Moreover, bacterial internalization was accompanied by significant expression of PKC-β1 and δ. Fluorescence microscopy demonstrated accumulation of PKC-β_1 on internalized bacteria. These data suggest that in mast cells PKC activation may be a critical signalling event in bacterial internalization. The involvement of PKC in bacterial internalization together with the calcium mobilization after bacterial challenge suggest an important role for the PLC-mediated signalling pathway in the mast cell response to bacterial infection.

Tyrosine phosphorylation events in mast cell–bacteria interactions

Modification of protein activity by tyrosine phosphorylation is of critical importance for the development and function of eukaryotic cells. The extent of protein tyrosine phosphorylation in eukaryotic cells is precisely regulated by the concerted actions of protein tyrosine kinases (PTKases) and protein tyrosine phosphatases (PTPases). During bacterial infection a variety of proteins are activated by tyrosine phosphorylation as part of the strategy of host defence. Probably the best characterized tyrosine kinases activated following bacteria infection are those of the MAP kinase family, including extracellular signalling-regulated kinases 1 and 2 (ERK1/2), the p46 and p54 isoforms of c-Jun N-terminal kinase (JNK) and p38 (100, 101). After activation, these kinases translocate to the nucleus and activate the transcription factor AP-1 (100, 102). Bacteria and bacterial products also activate the transcription factor NF-κB through activation of the IκBα kinase and subsequent phosphorylation and degradation of the NF-κB inhibitor IκBα (90). In epithelial cells, Li *et al.* (103) have demonstrated that *P. aeruginosa* activates a c-Src–Ras–MEK1/2–MAPK–pp90rsk signalling pathway that leads to activation of nuclear factor NF-κB. The activation of AP-1 or NF-κB results in the production of various cytokines such as TNF-α and IL-8 (100). *Salmonella*-mediated activation of MAP kinase pathways leading to the cytokine production, such as IL-8, requires both AP-1 and NF-κB in epithelial cells (100). Although human and rodent mast cells respond to various Gram-negative and Gram-positive bacteria and bacterial products to produce cytokines such as TNF-α and IL-6 (13, 37), the underlying signal transduction pathways have not been defined. Recently, it was demonstrated that internalization of FimH-expressing *E. coli* by human mast cell line HMC-1 was not affected by protein tyrosine kinase inhibitors such as genistein and PP1, but was significantly inhibited by PKC inhibitors (14). However, bacterial internalization and cytokine production may involve distinct mechanisms, and the roles of protein tyrosine kinases and PKC in bacteria-mediated cytokine production by mast cells remain to be determined.

A range of bacterial pathogens, such as *P. aeruginosa*, *Legionella*, *E. coli*, *Salmonella* and *Yersinia*, can exploit and subvert the tyrosine kinase function in a variety of host cells as part of their strategy for survival through adhesion (ligand–receptor binding) or secretion of virulent proteins (18, 21, 104). For example, once bacteria bind to host cells, a set of virulent bacterial proteins such as Yops (YopJ, YopH) of *Yersinia* (105), SptP and SipA of *Salmonella* (106), are translocated into the cytoplasm of the host cells at the point of contact. Some of these bacterial proteins are tyrosine phosphatases such as YopH and SptP, which interrupt host cell function through tyrosine dephosphorylation (105, 106). Palmer *et al.* (105) demonstrated that TNF-α production by macrophages in response to *Yersinia* infection was dramatically inhibited by YopJ through downregulation of MAP kinases p38 and JNK. Similar bacteria–host cell interactions have not yet been examined in mast cells although mast cell-derived TNF-α has been convincingly demonstrated as the critical mediator in host defence against bacterial infection (4, 5).

CONCLUSIONS

Interactions between mast cells and bacteria are critical to host defence under some circumstances. However, the interactions are complex and we need to consider that the responses evoked may be beneficial to either the host or the pathogen. Understanding mast cell–pathogen interactions is of obvious importance given the location and role of mast cells as early activators of effective immunity and regulators of the tissue-specific response. The nature of the bacterial components that activate mast cells, the signalling mechanisms used and the co-ordinate effects of multiple mediator responses need to be considered in order for a good understanding of mast cell–bacterial interactions to be developed. Understanding the mechanisms by which bacterial products and other stimuli selectively activate mast cells to produce specific cytokine profiles without degranulation may provide new opportunities to locally modify mast cell responses or the cytokine microenvironment in tissue-specific disease. A particular emphasis is needed in understanding the responses in human mast cells, which may be very different from those in rodent cells which have provided the vast majority of the available literature. We are at an early stage in this field of study and can learn a great deal from the responses observed in other cell types, particularly macrophages. However, the mast cell appears to have uniquely selective responses and a unique role in local innate immunity, the study of which provides many exciting new opportunities.

REFERENCES

1. Galli, S. J. New insights into 'the riddle of the mast cells': microenvironmental regulation of mast cells development and phenotypic heterogeneity. *Lab. Invest.* **62**:5, 1990.
2. Schwartz, L. and Huff, T. Biology of mast cells and basophils. In: *Allergy Principles and Practice:* 4th edn (Middleton, E. Jr., Reed, C. E., Ellis, E. F., Adkinson, N. F. Jr., Yunginger, J. W. and Busse, W. W., eds), pp. 135–168. Mosby, St Louis, MO, 1993.
3. MacDonald, A. J., Wills, F. L., Lin, T. J. and Befus, A. D. The role of mast cells in inflammation and homeostasis. In: *Allergic Disease: Basic and Clinical Aspects* (Denburg, J., ed.), pp. 103–125. Humana Press, Totowa, NJ, 1997.
4. Echtenacher, B., Mannel, D. N. and Hultner, L. Critical protective role of mast cells in a model of acute septic peritonitis. *Nature* **381**:75, 1996.
5. Galli, S. J. and Wershil, B. K. The two faces of the mast cell. *Nature* **381**:21, 1996.
6. Galli, S. J., Maurer, M. and Lantz, C. S. Mast cells as sentinels of innate immunity. *Curr. Opin. Immunol.* **11**:53, 1999.
7. Abraham, S. N. and Malaviya, R. Mast cells in infection and immunity. *Infect. Immun.* **65**:3501, 1999.
8. Hahn, H. P. The type-4 pilus is the major virulence-associated adhesion of *Pseudomonas aeruginosa* – a review. *Gene* **192**:99, 1997.
9. Gupta, S. K., Berk, R. S., Masinick, S. and Hazlett, L. D. Pili and lipopolysaccharide of *Pseudomonas aeruginosa* bind to the glycolipid asialo GM1. *Infect. Immun.* **62**:4572, 1994.
10. Chi, E., Mehl, T., Nunn, D. and Lory, S. Interaction of *Pseudomonas aeruginosa* with A549 pneumocyte cells. *Infect. Immun.* **59**:822, 1991.
11. Bergmann, U., Scheffer, J., Koller, M., Schonfeld, W., Erbs, G., Muller, F. E. and Konig, W. Introduction of inflammatory mediators (histamine and leukotrienes) from rat peritoneal mast cells and human granulocytes by *Pseudomonas aeruginosa* strains from burn patients. *Infect. Immun.* **57**:2187, 1989.
12. Malaviya, R., Ross, E., Jakschjik, B. A. and Abraham, S. N. Mast cell degranulation induced by type 1 fimbriated *Escherichia coli. J. Clin. Invest.* **93**:1645–1653, 1994.

12a. Malaviya, R., Ross, E. A., MacGregor, J. I., Ikeda, T., Little, J. R., Jakschik, B. A. and Abraham, S. N. Mast cell phagocytosis of FimH-expressing enterobacteria. *J. Immunol.* **152**:1907, 1994.
13. Arock, M., Ross, E., Lai-Kuen, E., Averlant, G., Gao, Z. and Abraham, S. N. Phagocytic and tumor necrosis factor alpha response of human mast cells following exposure to gram-negative and gram-positive bacteria. *Infect. Immun.* **66**:6030, 1998.
14. Lin, T. J., Gao, Z., Arock, M. and Abraham, S. N. Internalization of $FimH^+$ *Escherichia coli* by the human mast cell line (HMC-1 5C6) involves protein kinase C. *J. Leukocyte Biol.* (in press).
15. Baorto, D. M., Gao, Z., Malaviya, R., Dustin, M. L., van der Merwe, A., Lublin, D. M. and Abraham, S. N. Survival of FimH-expressing enterobacteria in macrophages relies on glycolipid traffic. *Nature* **389**:636, 1997.
16. Gbarah, A., Gahmberg, C. G., Ofek, I., Jacobi, U. and Sharon, N. Identification of the leukocyte adhesion molecules CD11 and CD18 as receptors for type 1-fimbriated (mannosespecific) *Escherichia coli. Infect. Immun.* **59**:4524, 1991.
17. Sauter, S. L., Rutherfurd, S. M., Wagener, C., Shively, J. E. and Hefta, S. A. Indentification of the specific oligosaccharide sites recognized by type 1 fimbriae from *Escherichia coli* on nonspecific cross-reacting antigen, a CD66 cluster granulocyte glycoprotein. *J. Biol. Chem.* **268**:15510, 1993.
18. Bliska, J. B., Galan, J. E. and Falkow, S. Signal transduction in the mammalian cell during bacterial attachment and entry. *Cell* **73**:903, 1993.
19. Mengaud, J., Ohayon, H., Gounon, P., Mege, R.-M. and Cossart, P. E-Cadherin is the receptor for internalin, a surface protein required for entry of *L. monocytogenes* into epithelial cells. *Cell* **84**:923, 1996.
20. Hauck, C. R., Meyer, T. F., Lang, F. and Gulbins, E. CD66-mediated phagocytosis of Opa52 *Neisseria gonorrhoeae* requires a Src-like tyrosine kinase-and Rac 1-dependent signaling pathway. *EMBO J.* **17**:443, 1998.
21. Bliska, J. B. and Falkow, S. The role of host tyrosine phosphorylation in bacterial pathogenesis. *Trends Genet.* **9**:85, 1993.
22. Kenny, B., DeVinney, R., Stein, M., Reinscheid, D. J., Frey, E. A. and Finlay, B. B. Enteropathogenic *E. coli* (EPEC) transfers its receptor for intimate adherence into mammalian cells. *Cell* **91**:511, 1997.
23. Marietta, E. V., Weis, J. J. and Weis, J. H. CD28 expression by mouse mast cells is modulated by lipopolysaccharide and outer surface protein A lipoprotein from *Borrelia burgdorferi. J. Immunol.* **159**:2840, 1997.
24. Busch, D. H., Pilip, I. M., Vijh, S. and Pamer, E. G. Coordinate regulation of complex T cell populations responding to bacterial infection. *Immunity.* **8**:353, 1998.
25. Schaible, U. E., Collins, H. L. and Kaufmann, S. E. Confrontation between intracellular bacteria and the immune system. *Adv. Immunol.* **71**:267, 1999.
26. Tashiro, M., Kawakami, Y., Abe, R., Han, W., Hata, D., Sugie, K., Yao, L., and Kawakami, T. Increased secretion of TNF-alpha by costimulation of mast cells via CD28 and Fc epsilon RI. *J. Immunol.* **158**:2382, 1997.
27. Kotzin, B. L., Leung, D. Y. M., Kappler, J. and Marrack, P. Superantigens and their potential role in human disease. *Adv. Immunol.* **54**:99, 1999.
28. Misfeldt, M. L. Microbial ‘superantigens’. *Infect. Immun.* **58**:2409, 1990.
29. Frandji, P., Tkaczyk, C., Oskeritzian, C., David, B., Desaymard, C. and Mecheri, S. Exogenous and endogenous antigens are differentially presented by mast cells to CD4+ T lymphocytes. *Eur. J. Immunol.* **26**:2517, 1996.
30. Scheuber, P. H., Golecki, J. R., Kickhofen, B., Scheel, D., Beck, G. and Hammer, D. K. Skin reactivity of unsensitized monkeys upon challenge with staphylcoccal enterotoxin B: a new approach for investigating the site of toxin action. *Infect. Immun.* **50**:869, 1985.
31. Bamberger, U., Scheuber, P. H., Sailer, K. B., Bartsch, K., Hartmann, A., Beck, G. and Hammer, D. K. Anti-idiotypic antibodies that inhibit immediate-type skin reactions in unsensitized monkeys on challenge with staphylococcal enterotoxin. *Proc. Natl. Acad. Sci. USA* **83**:7054, 1986.
32. Saloga, J., Leung, D. Y., Reardon, C., Giorno, R., Born, W. and Gelfand, E. W. Cutaneous exposure to the superantigen staphylococcal entertoxin B elicits a T-cell dependent inflammatory response. *J. Invest. Dermatol.* **106**:982, 1996.
33. Ackerman, L., Pelkonen, J. and Harvima, I. T. Staphylococcal enterotoxin B inhibits the production of interleukin-4 in a human mast cell line HMC-1. *Immunology* **94**:247, 1998.
34. Komisar, J., Rivera, J., Vega, A. and Tseng, J. Effects of staphylococcal enterotoxin B on rodent mast cells. *Infect. Immun.* **60**:2969–2975, 1992.

35. Olenick, J. G., Nauman, R. K. and Jett, M. Staphylococcal enterotoxin B: immunolabeling and visualization of target cells. *J. Histochem. Cytochem.* **39**:373–377, 1991.
36. Dimitriadou, V., Mecheri, S., Koutsilieris, M., Fraser, W., Al-Daccak, R. and Mourad, W. Expression of functional major histocompatibility complex class II molecules on HMC-1 human mast cells. *J. Leukoc. Biol.* **64**:791, 1998.
37. Leal-Berumen, I., Conlon, P. and Marshall, J. S. IL-6 production by rat peritoneal mast cells is not necessarily preceded by histamine release and can be induced by bacterial lipopolysaccharide. *J. Immunol.* **152**:5468, 1994.
38. Gagari, E., Tsai, M., Lantz, C. S., Fox, L. G. and Galli, S. J. Differential release of mast cell interleukin-6 via c-kit. *Blood* **89**:2654, 1997.
39. Gomi, K., Zhu, F. G. and Marshall, J. S. PGE2 enhances IgE mediated GM-CSF production by murine mast cells. *FASEB J.* **12**:A895, 1998 (abstract).
40. Marshall, J. S., Leal-Berumen, I., Nielsen, L., Glibetic, M. and Jordana, M. Interleukin (IL)-10 inhibits long-term IL-6 production but not preformed mediator release from rat peritoneal mast cells. *J. Clin. Invest.* **97**:1122, 1996.
41. Leal-Berumen, I., Snider, D. P., Barajas-Lopez, C. and Marshall, J. S. Cholera toxin increases IL-6 synthesis and decreases TNF-alpha production by rat peritoneal mast cells. *J. Immunol.* **156**:316, 1996.
42. Marshall, J. S., Gomi, K., Blennerhassett, M. G. and Bienenstock, J. Nerve growth factor modifies the expression of inflammatory cytokines by mast cells via a prostanoid-dependent mechanism. *J. Immunol.* **162**:4271, 1999.
43. Fenton, M. J. and Golenbock, D.T. LPS binding proteins and receptors. *J. Leukoc. Biol.* **64**:25, 1998.
44. Hawarth, R., Platt, N., Keshav, S., Hughes, D., Darley, E., Suzuk, H., Kurihara, Y., Kodama, T. and Gordon, S. The macrophage scavenger receptor typeA is expressed by activated macrophages and protects the host against lethal endotoxic shock. *J. Exp. Med.* **186**:1431, 1997.
45. Brown, J. F., Chafee, K. A. and Tepperman, B. L. Role of mast cells, neutrophils and nitric oxide in endotoxin induced damage to the neonatal rat colon. *Br. J. Pharmacol.* **123**:31, 1998.
46. Suzuki, H., Caughey, G. H., Gao, X. P. and Rubinstein, I. Mast cell chymase-like protease(s) modulates *Escherichia coli* lipopolysaccharide-induced vasomotor dysfunction in skeletal muscle *in vivo*. *J. Pharmacol. Exp. Ther.* **284**:1156, 1998.
47. Gatti, S., Faggioni, R., Sironi, M. and Ghezzi, P. Mast cells do not contribute to the rapid appearance of TNF in the serum of LPS treated mice: a study using mast cell deficient mice. *Int. J. Immunopharmacol.* **15**:551, 1993.
48. Frandji, P., Oskeritzian, C., Cacaraci, F., Lapeyre, J., Peronet, R., David, B., Guillet, J. G. and Mecheri, S. Antigen-dependent stimulation by bone marrow-derived mast cells of MHC class II-restricted T cell hybridoma. *J. Immunol.* **151**:6318, 1993.
49. Bartlett, J. G. *Clostridium difficile*: history of its role as an enteric pathogen and the current state of knowledge about the organism. *Clin. Infect. Dis.* **18 (Suppl. 4)**:S265, 1994.
50. Pothoulakis, C., Karmeli, F., Kelly, C. P., Eliakim, R., Joshi, M. A., O'Keane, J. C., Castagliuolo, I., LaMont, J. T. and Rachmilewitz, D. Ketotifen inhibits *Clostridium difficile* toxin A-induced enteritis in rat ileum. *Gastroenterology* **105**:701, 1993.
51. Castagliuolo, I., LaMont, J. T., Letourneau, R., Kelly, C. P., O'Keane, C. J., Jaffer, A., Theorides, T. C. and Pothoulakis, C. Neuronal involvement in the intestinal effects of *Clostridium difficile* toxin A and *Vibrio cholera* enterotoxin in rat ileum. *Gastroenterology* **107**:657, 1994.
52. Pothoulakis, C., LaMont, J. T., Eglow, R., Gao, N., Rubins, J. B., Theoharides, T. C. and Dickey, B. F. Characterization of rabbit ileal receptors for *Clostridium difficile* toxin A: evidence for a receptor-coupled G protein. *J. Clin. Invest.* **88**:119, 1991.
53. Pothoulakis, C., Galili, U., Castagliuolo, I., Kelly, C. P., Nikulasson, S., Dudeja, P. K., Brasitus, T. A. and LaMont, J. T. A human antibody binds to alpha-galatose receptors and mimics the effects of *Clostridium difficile* toxin A in rat colon. *Gastroenterology* **110**:1704, 1996.
54. Pothoulakis, C., Castagliuolo, I., LaMont, J. T., Jaffer, A., O'Keane, J. C., Snider, R. M. and Leeman, S.E. CP-96,345, a substance P antagonist, inhibits rat intestinal responses to *Clostridium difficile* toxin A but not cholera toxin. *Proc. Natl. Acad. Sci. USA* **91**:947, 1994.
55. Kurose, I., Pothoulakis, C., LaMont, J. T., Anderson, D. C., Paulson, J. C., Miyasaka, M., Wolf, R. and Granger, D. N. *Clostridium difficile* toxin A-induced microvascular dysfunction. Role of histamine. *J. Clin. Invest.* **94**:1919, 1994.
56. Eliakim, R., Karmeli, F., Okon, E. and Radaszkiewicz, T. Ketoifen effectively prevents mucosal damage in experimental colitis. *Gut* **33**:1498, 1992.

57. Fiorentini, C. and Thelestam, M. *Clostridium difficile* toxin A and its effects on cells. *Toxicon* **29**:543, 1991.
58. Wershil, B. K., Castagliuolo, I. and Pothoulakis, C. Direct evidence of mast cell involvement in *Clostridium difficile* toxin A-induced enteritis in mice. *Gastroenterology* **114**:956, 1998.
59. Calderon, G. M., Torres-Lopez, J., Lin, T. J., Chavez, B., Hernandez, M., Munoz, O., Befus, A. D. and Enciso, J. A. Effects of toxin A from *Clostridium difficile* on mast cell activation and survival. *Infect. Immun.* **66**:2755, 1998.
60. Sartor, R. B. Cytokines in intestinal inflammation: pathophysiological and clinical considerations. *Gastroenterology* **106**:533, 1994.
61. Just, I., Selzer, J., Wilm, M., von Eichel-Streiber, C., Mann, M. and Aktories, K. Glucosylation of Rho proteins by *Clostridium difficile* toxin B. *Nature* **375**:500, 1995.
62. Just, I., Wilm, M., Selzer, J., Rex, G., von Eichel-Streiber, C., Mann, M. and Aktories, K. The enterotoxin from *Clostridium difficile* (ToxA) monoglucosylates the Rho proteins. *J. Biol. Chem.* **270**:13932, 1995.
63. Prepens, U., Just, I., von Eichel-Streiber, C., and Aktories, K. Inhibition of FcεRI-mediated activation of rat basophilic leukemia cells by *Clostridium difficile* toxin B (monoglucosyltransferase). *J. Biol. Chem.* **271**:7324, 1996.
64. Massol, P., Montcourrier, P., Guillemot, J. C. and Chavrier, P. Fc receptor-mediated phagocytosis requires CDC42 and Rac1. *EMBO J.* **17**:6219, 1998.
65. Flegel, W. A., Muller, F., Daubener, W., Fischer, H. G., Hadding, U. and Northoff, H. Cytokine response by human monocytes to *Clostridium difficile* toxin A and toxin B. *Infect. Immun.* **59**:3659, 1991.
66. Snider, D. P. The mucosal adjuvant properties of ADP-ribosylating bacterial enterotoxins. *Crit. Rev. Immunol.* **15**:317, 1995.
67. Snider, D. P., Marshall, J. S., Perdue, M. H. and Liang, H. Production of IgE antibody and allergic sensitization of intestinal and peripheral tissues after oral immunization with protein antigen and cholera toxin. *J. Immunol.* **153**:647, 1994.
68. Klimpel, G. R., Chopra, A. K., Langley, K. E., Wypych, J., Annable, C. A., Kaiserlian, D., Ernst, P. B. and Peterson, J. W. A role for stem cell factor and c-kit in the murine intestinal tract secretory response to cholera toxin. *J. Exp. Med.* **182**:1931, 1995.
69. Hu, Z. Q., Asano, K., Seki, H. and Shimamura, T. An essential role of prostaglandin E on mouse mast cell induction. *J. Immunol.* **155**:2134, 1995.
70. Saito, H., Sanai,Y., and Nagai, Y. Cholera toxin enhances factor dependent colony growth of murine mast cell progenitors. *Exp. Hematol.* **13**:261, 1985.
71. Pulimood, A. B., Mathan, M. M. and Mathan, V. I. Quantitative and ultrastructural analysis of rectal mucosal mast cells in acute infectious diarrhea. *Dig. Dis. Sci.* **43**:2111, 1998.
72. Krieg, A. M. An innate immune defense mechanism based on the recognition of CpG motifs in microbial DNA. *J. Lab. Clin. Med.* **128**:128, 1996.
73. Krieg, A. M., Yi, A. K., Matson, S., Waldschmidt, T. J., Bishop, G. A., Teasdale, R., Koretzky, G. A. and Klinman, D.M. CpG motifs in bacterial DNA trigger direct B-cell activation. *Nature* **374**:546, 1995.
74. Pisetsky, D. The immunologic properties of DNA. *J. Immunol.* **412**, 1997.
75. Tokunaga, T., Yano, O., Kuramoto, E., Kimura, Y., Yamamoto, T., Kataoka, T. and Yamamoto, S. Synthetic oligonucleotides with particular base sequences from the cDNA encoding proteins of *Mycobacterium bovis* BCG induce interferons and activate natural killer cells. *Microbiol. Immunol.* **36**:55, 1992.
76. Sparwasser, D., Miethke, T., Lipford, G., Borschert, K., Hacker, H., Heeg, K. and Wagner, H. Bacterial DNA causes septic shock. *Nature* **386**:336, 1997.
77. Lipford, G., Bauer, M., Blank, C., Reiter, R., Wagner, H. and Heeg, K. CpG-containing synthetic oligonucleotides promote B and cytotoxic T cell responses to protein antigen: a new class of vaccine adjuvants. *Eur. J. Immunol.* **27**:2340, 1997.
78. Benimetskaya, L., Loike, J. D., Khaled, Z., Loike, G., Silverstein, S. C., Cao, L., El-Khoury, J., Cai, T. Q. and Stein, C. A. Mac-1 CD11b/CD18 is an oligodeoxynucleotide binding protein. *Nat. Med.* **3**:414, 1997.
79. Bennett, R. M., Cornell, K. A., Merritt, M. J., Bakke, A.C., Mourich, D. and Hefenieder, S. H. Idiotypic mimicry of a cell surface DNA receptor: evidence for anti-DNA antibodies being a subset of anti-anti-DNA receptor antibodies. *Clin. Exp. Immunol.* **90**:428, 1992.
80. Stein, C. A., Tonkinson, J. L., Zhang, L. M., Yakubov, L., Gervasoni, J., Taub, R. and Rotenberg, S. A. Dynamics of the internalization of phosphodiesteroligodeoxynucleotides in HL60 cells. *Biochemistry* **32**:4855, 1993.

81. Yi, A.K. and Krieg, A. M. CpG DNA rescue from anti-IgM-induced WEHI-231 B lymphoma apoptosis via modulation of I kappa B alpha and I kappa B beta and sustained activation of nuclear factor-kappa B/c-Rel. *J. Immunol.* **160**:1240, 1998.
82. Yi, A.K. and Krieg, A. M. Rapid induction of mitogen-activated protein kinases by immune stimulatory CpG DNA. *J. Immunol.* **161**:4493, 1998.
83. Pisetsky, D., Reich, C., Crowley, S. D. and Halpern, M. D. Immunological properties of bacterial DNA. *Ann. N.Y. Acad. Sci.* **772**:152, 1995.
84. Prodeus, A. P., Zhou, X., Maurer, M., Galli, S. J. and Carroll, M. C. Impaired mast cell-dependent natural immunity in complement C3 deficient mice. *Nature* **390**:172, 1997.
85. Melillo, E., Wooley, K. L., Manning, P. J. and O'Byrne, P. Effect of inhaled PGE2 on exercise induced bronchoconstriction in asthmatic subjects. *Am. J. Respir. Crit.Care Med.* **149**:1138, 1994.
86. Pavord, I. D., Wong, C. S., Williams, J. and Tattersfield, A. E. Effect of inhaled prostaglandin E2 on allergen induced asthma. *Am. Rev. Respir. Dis.* **148**:87, 1993.
87. Xing, Z., Gauldie, J., Cox, G., Baumann, H., Jordana, M., Lei, X. F. and Achong, M. K. IL-6 is an anti-inflammatory cytokine required for controlling local or systemic acute inflammatory responses. *J. Clin. Invest.* **101**:311, 1998.
88. Friedl, P. and Konig, W. Effects of mucoid and non-mucoid *Pseudomonas aeruginosa* isolates from cystic fibrosis patients on inflammatory mediator release from human polymorphonuclear granulocytes and rat mast cells. *Immunology* **76**:86, 1992.
89. Pace, J., Hayman, M. J. and Galan, J. E. Signal transduction and invasion of epithelial cells by *S. typhimurium. Cell* **72**:505, 1993.
90. Baldwin, T. J., Ward, W., Aitken, A., Knutton, S. and Williams, P. H. Elevation of intracellular free calcium levels in Hep-2 cells infected with enteropathogenic *Escherichia coli. Infect. Immun.* **59**:1599–1604, 1991.
91. Denning, G. M., Railsback, M. A., Rasmussen, G. T., Cox, C. D. and Britigan, B. F. *Pseudomonas* pyocyanine alters calcium signaling in human airway epithelial cells. *Am. J. Physiol.* **274**:L893, 1998.
92. Nishizuka, Y. Intracellular signaling by hydrolysis of phospholipids and activation of protein kinase C. *Science* **258**:607, 1992.
93. Foubister, V., Rosenshine, I. and Finlay, B.B. A diarrheal pathogen, enteropathogenic *Escherichia coli* (EPEC), trigger a flux of inositol phosphates in infected epithelial cells. *J. Exp. Med.* **179**:993, 1994.
94. Kenny, B. and Finlay, B. B. Intimin-dependent binding of enteropathogenic *Escherichia coli* to host cells triggers novel signaling events, including tyrosine phosphorylation of phospholipase C-γ1. *Infect. Immun.* **65**:2528, 1997.
95. Crane, J. K. and Oh, J. S. Activation of host cell protein kinase C by enteropathogenic *Escherichia coli. Infect. Immun.* **65**:3285, 1997.
96. Coxon, P. Y., Summersgill, J. T., Ramirez, J. A. and Miller, R. D. Signal transduction during *Legionella pneumophila* entry into human monocytes. *Infect. Immun.* **66**:2905, 1998.
97. Zheleznyak, A. and Brown, E. J. Immunoglobulin-mediated phagocytosis by human monocytes requires protein kinase C activation. *J. Biol. Chem.* **267**:12042, 1992.
98. Panetti, T. S., Wilcox, S. A., Horzempa, C. and McKeown-Longo, P. J. $\alpha_v B_5$ Integrin receptor-mediated endocytosis of vitronectin is protein kinase C-dependent. *J. Biol. Chem.* **270**:18593, 1995.
99. Ludowyke, R. I., Kawasugi, K. and French, P. W. PMA and calcium inophore induce myosin and F-actin rearrangement during histamine secretion from RBL-2H3 cells. *Cell Motil. Cytoskeleton* **29**:354, 1994.
100. Hobbie, S., Chen, L. M., Davis, R. J. and Galan, J. E. Involvement of mitogen-activated kinase pathways in the nuclear responses and cytokine production induced by *Salmonella typhimurium* in cultured intestinal epithelial cells. *J. Immunol.* **159**:5550, 1997.
101. Tang, P., Sutherland, C. L., Gold, M. R. and Finlay, B. B. *Listeria monocytogenes* invasion of epithelial cells requires the MEK-1/ERK-2 mitogen-activated protein kinase pathway. *Infect. Immun.* **66**:1106, 1998.
102. Witmarsh, A.J. and Davis, R. J. Transcription factor AP-1 regulation by mitogen-activated protein kinase signal transduction pathways. *J. Mol. Med.* **74**:589, 1996.
103. Li, J.-D., Feng, W., Gallup, M., Kim, J.-H., Gum, J., Kim, Y. and Basbaum, C. Activation of NF-κB via Src-dependent Ras–MAPK–pp90rsk pathway is required for *Pseudomonas aeruginosa*-induced mucin overproduction in epithelial cells. *Proc. Natl. Acad. Sci. USA* **95**:5718, 1998.
104. Finlay, B.B. and Cossart, P. Exploitation of mammalian host cell functions by bacterial pathogens. *Science* **276**:718, 1997.

105. Palmer, L. E., Hobbie, S., Galan, J. E. and Bliska, J.B. YopJ of *Yersinia pseudotuberculosis* is required for the inhibition of macrophage TNF production and downregulation of the MAP kinases p38 and JNK. *Mol. Microbiol.* **27**:953, 1998.
106. Fu, Y. and Galan, J. E. The *Salmonella typhimurium* tyosine phosphatase SptP is translocated into host cells and disrupts the actin cytoskeleton. *Mol. Microbiol.* **27**:358, 1998.
107. Saito H., Okajima, F., Molski, T. F. P., Sha'afi, R. I., Ui, M. and Ishizaka, T. Effects of ADP-ribosylation of GTP-binding protein by pertussis toxin on immunoglobulin E-dependent and -independent histamine release from mast cells and basophils. *J. Immunol.* **138**:3827, 1987.
108. Adusu, T. E., Conlon, P. D., Shewen, P. E. and Black, W. D. *Pasteurella haemolytica* leukotoxin induces histamine release from bovine pulmonary mast cells. *Can. J. Vet. Res.* **58**:1–5, 1994.
109. Harville, B. A. and Dreyfus, L. A. Release of serotonin from RBL-2H3 cells by the *Escherichia coli* peptide toxin STb. *Peptides* **17**:363–366, 1996.
110. Hohman, R. J. Aggregation of IgE receptors induces degranulation in rat basophilic leukemia cells permeabilized with alpha-toxin from *Staphylococcus aureus. Proc. Natl. Acad. Sci. USA* **85**:1624–1628, 1988.
111. Yamanaka, H., Sugiyama, K., Furuta, H., Miyoshi, S. and Shinoda, S. Cytolytic action of *Vibrio vulnificus* haemolysin on mast cells from rat peritoneal cavity. *J. Med. Microbiol.* **32**:39–43, 1990.
112. Scheffer, J., Konig, W., Braun, V. and Goebel, W. Comparison of four hemolysin-producing organisms (*Escherichia coli, Serratia marcescens, Aeromonas hydrophila* and *Listeria monocytogenes*) for release of inflammatory mediators from various cells. *J. Clin. Microbiol.* **26**:544–551, 1988.
113. Dalldorf, F. G., Anderle, S. K., Brown, R. R., Schwab, J. H. *et al.* Mast cell activation by group A streptococcal polysaccharide in the rat and its role in experimental arthritis. *Am. J. Pathol.* **132**:258–264, 1988.
114. Nygren, H. and Dahlen, G. Complement-dependent histamine release from rat peritoneal mast cells, induced by lipopolysaccharides from *Bacteroides oralis, Fusobacterium nucleatum* and *Veillonella parvula. J. Oral Pathol.* **10**:87–94, 1981.
115. Morrison, D. C. and Betz, S. J. Chemical and biological properties of a protein-rich fraction of bacterial lipopolysaccharides. II. The *in vitro* rat peritoneal mast cell response. *J. Immunol.* **119**:1790–1795, 1977.
116. Patella, V., Casolaro, V., Bjorck, L. and Marone, G. Protein L.A. bacterial Ig-binding protein that activates human basophils and mast cells. *J. Immunol.* **145**:3054–3061, 1990.
117. Espersen, F., Jarlov, J. O., Jensen, C., Skov, P. S. and Norn, S. *Staphylococcus aureus* peptidoglycan induces histamine release from basophil human leukocytes *in vitro. Infect. Immun.* **46**:710–714, 1984.
118. Zhu, F. G., Gomi, K. and Marshall, J. S. Short term and long term cytokine release by mouse bone marrow mast cells and the differentiated KU-812 cell line is inhibited by Brefeldin A. *J. Immunol.* (in press).
119. Yamamoto, J., Watanabe, S., Hirose, M., Osada, T., Ra, C. and Sata, N. Role of mast cells as a trigger of inflammation in *Helicobacter pylori* infection. *J. Physiol. Pharmacol.* **50**:17–23, 1999.
120. Talkington, J. and Nickell, S. P. *Borrelia burgdorferi* spirochetes induce mast cell activation and cytokine release. *Infect. Immun.* **67**:1107–1115, 1999.
121. Church, M. K., Norn, S., Pao, G. J. and Holgate, S. T. Non-IgE-dependent bacteria-induced histamine release from human lung and tonsillar mast cells. *Clin. Allergy* **17**:341–353, 1987.
122. Clementsen, P., Larsen, F. O., Milman, N., Skov, P. S. and Norn, S. *Haemophilus influenzae* release histamine and enhance histamine release from human bronchoalveolar cells. Examination of patients with chronic bronchitis and controls. *APMIS* **103**:806–812, 1995.
123. Lutton, D. A., Bamford, K. B., O'Loughlin, B. and Ennis, M. Modulatory action of *Helicobacter pylori* on histamine release from mast cells and basophils *in vitro. J. Med. Microbiol.* **42**:386–393, 1995.
124. Kim, J. S., Chae, M. R., Chang, K., Park, K. H., Rho, H. W., Park, B. H., Park, J. W. and Kim, H. R. Cytotoxicity of *Vibrio vulnificus* cytolysin on rat peritoneal mast cells. *Microbiol. Immunol.* **42**:837–843, 1998.
125. Malaviya, R., Ikeda, T., Ross, E. and Abraham, S. N. Mest cell modulation of neutrophil influx and bacterial clearance at sites of infection through TNF-α. *Nature* **381**:77–80, 1996.

CHAPTER 28

The Role of Mast Cells, Basophils and Interleukin-3 (IL-3) in Immune Responses to Parasites: Studies with Mast Cell- and IL-3-Deficient Mice

CHRIS S. LANTZ[1] *and STEPHEN J. GALLI*[*2]

[1]*Department of Biology, James Madison University, Harrisonburg, Virginia, and*
[2]*Department of Pathology, Stanford University Medical Center, Stanford, California, U.S.A.*

INTRODUCTION

Several lines of evidence support the widely held concept that mast cells and basophils can participate in host defence against parasites. Infection with helminthic parasites is associated with increased levels of parasite-specific IgE and non-specific IgE, as well as with mast cell hyperplasia and/or blood basophilia (1, 2). Certain worm antigens can induce the degranulation of mast cells obtained from parasite-infected animals, and some mast cell- or basophil-derived mediators can have toxic effects on these parasites (1, 2). Finally, some mast cell- or basophil-derived mediators, including histamine and serotonin, have physiological effects on vascular permeability, intestinal ion and mucous secretion, and gut motility, which might in turn enhance local expressions of host defence against parasites (1, 2).

Accordingly, it has been hypothesized that mast cell and basophil sensitization by parasite-specific IgE (or, in mice, IgG1) antibodies, followed by mast cell and basophil degranulation and mediator secretion in response to exposure to the parasite's antigens, can promote the expulsion of the parasites (1, 2). On the other hand, one might argue that the enhancement of local blood flow and vascular permeability, by increasing the delivery of blood-borne nutrients, can promote the survival and/or fecundity of parasites. By this reasoning, the effects of mediator production at sites of parasitic infection by resident mast cells or recruited basophils may, at least in certain circumstances, benefit the parasite more than the host.

So, how can one decide, in the context of a specific parasite–host interaction, to what extent a particular potential element of host defence (e.g. mast cells, basophils, individual cytokines, etc.) tips the balance of the elicited response in favour of the parasite or the

---

* Corresponding author.

MAST CELLS AND BASOPHILS
ISBN 0-12-473335-2

host? One approach is to analyse the expression of the response in animals that differ solely in lacking or having the cell or cytokine in question. In this chapter we focus primarily on the results of experiments performed to analyse the expression of host resistance to parasite infections in mice that genetically lack mast cells, interleukin-3 (IL-3), or both elements of the host response.

STUDIES OF PARASITE INFECTION IN GENETICALLY MAST CELL-DEFICIENT MICE

Because their c-*kit* mutations result in markedly reduced stem cell factor (SCF)/c-kit signalling (3, 4), $Kit^{W}/Kit^{W\text{-}v}$ mice are ordinarily profoundly mast cell-deficient (5). We have utilized mast cell-deficient $Kit^{W}/Kit^{W\text{-}v}$ mice that have been locally or systematically reconstituted by adoptive transfer of *in vitro* derived cultured mast cells ('mast cell knockin mice') as a model to search for differences in the expression of biological responses in anatomical sites, or in whole animals, that differ only in that one site, or one set of animals, contains mast cells, whereas the control sites or animals remain profoundly mast cell-deficient (6–8). This general approach is summarized in Table I.

TABLE I
General Scheme for Investigating Mouse Mast Cell Function *In Vivo*

1. Search for quantitative differences in the expression of biological responses in genetically mast cell-deficient WBB6F$_1$-$Kit^{W}/Kit^{W\text{-}v}$ and WCB6F$_1$-$Mgf^{Sl}/Mgf^{Sl\text{-}d}$ mice and the congenic normal (+/+) mice.
   Note: Such studies should include appropriate histological analysis, since certain biological responses can result in the appearance of increased numbers of mast cells in the tissues of $Kit^{W}/Kit^{W\text{-}v}$ (but not $Mgf^{Sl}/Mgf^{Sl\text{-}d}$) mice (8, 9, 78).
2. Compare the responses in $Kit^{W}/Kit^{W\text{-}v}$ mice and in $Kit^{W}/Kit^{W\text{-}v}$ mice that have received bone marrow transplantation from congenic +/+ mice.
   Note: This determines whether the response that is abnormally expressed in $Kit^{W}/Kit^{W\text{-}v}$ mice is influenced by mast cells or other cells derived from haematopoietic precursors.
3. Analyse the response in $Kit^{W}/Kit^{W\text{-}v}$ mice that have been *selectively* reconstituted with mast cells ('mast cell knockin mice').
   Note: This determines whether the response that is abnormally expressed in $Kit^{W}/Kit^{W\text{-}v}$ mice has a mast cell-dependent component.
4. Define the mechanism(s) by which mast cells contribute the response.
   Note: One can, for example, compare the responses in $Kit^{W}/Kit^{W\text{-}v}$ mice that have been reconstituted with wild-type mast cells or mast cells that express genetically determined alterations in the production or structure of mediators or other factors that might influence the expression of the response.

Modified from ref. 79.

It is important to emphasize that the c-*kit* mutations in $Kit^{W}/Kit^{W\text{-}v}$ mice result in a constellation of abnormalities in addition to their profound mast cell deficiency. For example, $Kit^{W}/Kit^{W\text{-}v}$ mice are mildly anaemic, lack germ cells and cutaneous melanocytes (reviewed in 7, 9) and also virtually lack the interstitial cells of Cajal that confer electrical pacemaker activity in the intestines (10–12). Accordingly, if differences in host response against parasites (or in other biological responses) are detected in $Kit^{W}/Kit^{W\text{-}v}$ vs. wild-type mice, one cannot be sure that these differences are mast cell-dependent unless the abnormalities can be at least partially normalized in Kit^{W}/Kit^{W-v} mice that have been selectively repaired of their mast cell deficiency (see Table I).

Another point regarding the mast cell deficiency of Kit^{W}/Kit^{W-v} mice and Mgf^{Sl}/Mgf^{Sl-d} mice (which are mast cell-deficient because they do not produce the membrane-associated form of SCF, also known as kit ligand, steel factor, or mast cell growth factor (MGF); reviewed in refs 9, 13) also should be emphasized. Although both Kit^{W}/Kit^{W-v} and Mgf^{Sl}/Mgf^{Sl-d} mice are profoundly mast cell-deficient (ordinarily <0.5% the wild-type number of mast cells in the dermis; none in the respiratory tract, stomach, and intestine, peritoneal cavity and most other sites), they do contain mast cell progenitors (reviewed in refs 7, 9). This is readily demonstrable *in vivo* (e.g. when bone marrow cells of Kit^{W}/Kit^{W-v} mice are maintained in IL-3 (6, 14, 15), but also may be revealed under certain circumstances *in vivo* (see below). Current evidence indicates that Kit^{W}/Kit^{W-v} mice are much more likely than Mgf^{Sl}/Mgf^{Sl-d} mice to develop recognizable (i.e. cytoplasmic granule-containing) mast cells during biological responses *in vivo*. Thus, Kit^{W}/Kit^{W-v} mice develop small numbers of 'mucosal-type' mast cells in the intestines at sites of infections with certain parasites (16–18), probably via an IL-3-driven but relatively c-kit/SCF-independent pathway of mast cell development (see below and ref. 18). Kit^{W}/Kit^{W-v} mice can also develop heparin-containing dermal mast cells at sites of certain types of chronic, cutaneous inflammation (19, 20) or after IL-3 injection (21). By contrast, the same types of cutaneous inflammation do not induce dermal mast cell development in Mgf^{Sl}/Mgf^{Sl-d} mice (19, 20). Indeed, the only procedure that has been demonstrated to induce mast cell development in Mgf^{Sl}/Mgf^{Sl-d} mice is the long-term *in vivo* administration of SCF (22, 23). Another important distinction between the two types of genetically mast cell-deficient mice is that the adult Mgf^{Sl}/Mgf^{Sl-d} mice appear to contain far fewer (or no) mast cell progenitors in the skin or other peripheral tissues (24).

There are two practical implications of these findings that must be kept in mind when genetically mast cell-deficient mice are used for analysis of the potential roles of mast cells in biological responses *in vivo*. First, appropriate histological studies should be performed to assess whether the biological response under investigation might have induced the development of recognizable mast cells at the site of the reaction (Table I). This is more likely to occur in Kit^{W}/Kit^{W-v} mice than in Mgf^{Sl}/Mgf^{Sl-d} mice, and is more likely to occur in chronic or persistent biological responses, particularly those associated with high levels of IL-3 production, than in more acute reactions. Second, one must acknowledge the theoretical possibility, in studies with Kit^{W}/Kit^{W-v} mice, that small numbers of mast cell progenitors (that lack those features usually used to identify cells in the mast cell lineage, such as prominent cytoplasmic granules), or the very small numbers of recognizable 'mature' mast cells that are present in some tissues in these mice, may provide important functions during certain biological responses (8, 9). Based on this consideration, a lack of a difference in the expression of a particular biological response in Kit^{W}/Kit^{W-v} vs. +/+ mice does not formally prove that mast cells make little or no unique contribution to the response, but only that the mast cell-dependent contribution in question (if it exists) cannot be detected in this model system.

Nevertheless, much interesting information about the roles of mast cells in biological responses already has been derived from studies employing mast cell-reconstituted Kit^{W}/Kit^{W-v} mice. This work has identified three patterns of mast cell involvement in biological responses (reviewed in refs 8, 9). In some reactions, mast cells appear to have an essential role, in that the responses are not detectably expressed in Kit^{W}/Kit^{W-v} or Mgf^{Sl}/Mgf^{Sl-d} mast cell-deficient mice. In other responses, mast cells appear to regulate the intensity or kinetics of the response, but the reactions can be detectably expressed in the absence of mature mast cells. In yet other responses, no specific mast cell-dependent contribution has yet been identified. For example, this approach has established that virtually all of the inflammation associated with IgE-dependent reactions elicited in mouse skin (25, 26) or stomach (27) is mast cell-dependent, and that mast cells are responsible for the bronchial hyper-responsiveness to methacholine observed after anti-IgE challenge in mice (28). Essentially, all of the inflammation produced by intradermal injection of substance P is also mast cell-dependent (29, 30), whereas mast cells can significantly augment, but are not essential for, the inflammation induced by phorbol 12-myristate 13-acetate (PMA) (31) or immune complexes (32). By contrast, most analyses have detected no impairment of the expression of T cell-mediated contact hypersensitivity reactions in the skin of mast cell-deficient mice (8, 33).

For a variety of reasons, it has been difficult to employ $Kit^{W}/Kit^{W\text{-}v}$ mice to define precisely the roles of mast cells in host defence against parasites. For example, studies of *Trichinella spiralis* or *Strongyloides ratti* infections, as well as some experiments with the roundworm *Nippostrongylus brasiliensis*, showed that the duration of these experimental parasitic infections was prolonged in mast cell-deficient mice when compared with the results obtained in the wild-type (+/+) animals (34). However, the impairment of immunity in mast cell-deficient mice was never as severe as in athymic nude mice and, in each instance, the mast cell-deficient mice eventually were able to resolve the infection (34). Moreover, the successful elimination of parasites in the absence (or virtual absence) of a specific IgE response has also been reported (35, 36). Thus, several lines of evidence, including that derived from studies of $Kit^{W}/Kit^{W\text{-}v}$ mice, indicate that mucosal mast cell hyperplasia and activation may contribute to host defence against certain helminthic infections, but that mast cells may not ordinarily represent an essential or 'unique' component of these immune responses.

Indeed, in several studies, primary host responses against *N. brasiliensis* were essentially indistinguishable in $Kit^{W}/Kit^{W\text{-}v}$ vs. wild-type mice (34). And even in those settings in which parasite rejection was prolonged in $Kit^{W}/Kit^{W\text{-}v}$ vs. +/+ mice (as in primary infections with *S. ratti*, attempts to employ mast cell reconstitution of $Kit^{W}/Kit^{W\text{-}v}$ mice to 'normalize' their responses to the parasite were unsuccessful (8, 34, 37). This may have reflected abnormalities in the anatomical location and/or phenotype of the adoptively transferred mast cells in the intestines of the $Kit^{W}/Kit^{W\text{-}v}$ recipients, as opposed to the native mast cells in the intestines of the wild-type mice (8, 34, 37).

In certain settings, mast cell-dependent processes may contribute to the pathology associated with host reactions to parasites without influencing the effectiveness of the immune response to that organism. For example, studies in $Kit^{W}/Kit^{W\text{-}v}$, wild-type and mast cell knockin mice show that mast cells can contribute significantly to the extent and the persistence of the local cutaneous inflammation, including the ulceration, associated with experimental infections with *Leishmania major* (38). On the other hand, mast cells had no detectable effect on the ultimate course of the infection. It should be noted, however, that the method for quantifying viable *L. major* in this setting probably would not have detected even a 2-fold difference in the numbers of the organisms (38).

The most compelling evidence for a significant role for mast cells or basophils in host defence against parasites is in immune responses to ectoparasites, such as ticks, not in host responses to nematodes. However, the relative importance of basophils or mast cells as effectors in immune reactions to ticks may vary according to the species of host and the species of tick. Studies utilizing mast cell knockin mice have demonstrated that IgE and mast cells are essential for the expression of immune resistance to the cutaneous feeding of larval *Haemaphysalis longicornis* ticks; such immune resistance is measured as reduced numbers of larval ticks that successfully feed, as well as reduced weights of the engorged ticks (39).

In contrast, basophils may be more important than mast cells in immune resistance to the feeding of the larvae of a different species of Ixodid tick (*Dermacentor variabilis*) in mice. Thus, $Kit^{W}/Kit^{W\text{-}v}$ mice did not exhibit a significant impairment in their ability to express immune resistance to the feeding of larval *D. variabilis* ticks, and electron microscopic examination indicated that the tick feeding sites contained many basophils, both in $Kit^{W}/Kit^{W\text{-}v}$ and wild-type mice (40). Studies with heterologous anti-basophil and anti-eosinophil antisera suggest that both basophils and eosinophils are required for immune resistance to the feeding of larval *Amblyomma americanum* ticks in guinea pigs (41). Finally, there has even been a single report of a patient who lacked basophils and eosinophils (and expressed an IgA deficiency), who suffered from severe scabies (42). These and other findings support the concept that resident mast cells and recruited basophils may have similar, overlapping, or complementary functions in immune responses to ectoparasites, worms, and perhaps other parasites, with the relative contributions of each cell type varying according to the type of parasite, species of host animal, or other factors (43).

ASSESSING THE ROLE OF IL-3 IN HOST RESPONSES TO NEMATODES USING IL-3 –/– MICE

It has long been proposed that IL-3, which can be derived from T cells and other sources, can represent an important link between the immune and haematopoietic systems, and that IL-3 may be especially important in expanding mast cell and basophil populations during host responses to parasites (44–46). To test these hypotheses, and to explore the importance of IL-3 in host resistance to certain nematodes, we utilized IL-3 –/– mice. First, mice that lacked IL-3 were produced using gene targeting in embryonic stem cells (47). IL-3 –/– mice are healthy and, unlike $Kit^{W}/Kit^{W\text{-}v}$ mice, are fertile. Moreover, like mice which carry an inactivating mutation in the α chain of the heterodimeric IL-3 receptor (48) or which lack both IL-3 and the common β subunit of the receptors for IL-3, IL-5 and granulocyte–macrophage colony-stimulating factor (GM-CSF) (49), IL-3 –/– mice exhibit no detectable abnormalities in multiple aspects of haematopoiesis *in vitro* or *in vivo* (47). Thus, in comparison to mice with mutations resulting in impaired c-kit or SCF expression or function, the phenotype of IL-3 –/– mice was remarkably normal.

However, we found that IL-3 –/– mice did exhibit abnormalities in mast cell development *in vitro* and *in vivo*. In accord with previous work indicating that exogenous IL-3 can augment SCF-dependent mast cell development *in vitro* (9, 50–55), we found that SCF induced fewer mast cells to develop *in vitro* in suspension cultures of bone marrow cells derived from IL-3 –/– mice as opposed to IL-3 +/+ mice (18). By contrast, substantially higher, and essentially equivalent, numbers of mast cells developed when bone marrow cells from either IL-3 –/– or IL-3 +/+ mice were maintained *in vitro* in exogenous SCF plus IL-3 (18).

Our cell culture studies thus showed that endogenous IL-3 can enhance, but is not required for, mast cell development from bone marrow progenitors in the presence of exogenous SCF *in vitro*. To assess the role of IL-3 in mast cell development *in vivo*, we quantified mast cells in the tissues of IL-3 –/– vs. wild-type mice at baseline or after 21 daily subcutaneous injections of recombinant rat SCF (rrSCF, at 100 μg kg^{-1} day^{-1}) or of vehicle alone (18). The results of these experiments showed that endogenous IL-3 is not essential for the development of mast cells under physiological conditions *in vivo*. Indeed, in all sites examined, levels of tissue mast cells at baseline in adult IL-3 –/– mice were very similar to those in the corresponding sites in IL-3 +/+ mice. Moreover, in contrast to our observations in the *in vitro* system, we found that endogenous IL-3 was not required for rrSCF-induced mast cell hyperplasia *in vivo*. In fact, in certain tissues (the trachea, forestomach and spleen), mast cell levels after rrSCF treatment were significantly *greater* (by up to 140%) in IL-3 –/– mice than in the corresponding wild-type mice (18).

To assess the role of IL-3 in the augmented development of basophils and mast cells that is observed during parasite infections, we quantified tissue mast cells and bone marrow and blood basophils, and assessed the expression of parasite immunity, in IL-3 –/– and corresponding wild-type mice that had been infected with the intestinal nematode, *Strongyloides venezuelensis* (*S.v.*). *S.v.* is a naturally occurring parasite of murine rodents that is rejected by a T cell-dependent immune response which is associated with extensive mast cell hyperplasia in the intestinal mucosa (17, 37). In three separate experiments, we found that IL-3 –/– mice that had been inoculated with 2000 *S.v.* third-stage infective larvae (L_3), in comparison to the corresponding wild-type mice, exhibited both significantly delayed expulsion of the adult worms and a significantly prolonged (by 2–3 days, depending on BALB/c or C57BL/6 strain background) production of the parasite's eggs.

In addition, the *S.v.*-infected IL-3 –/– mice exhibited striking abnormalities in their basophil and mast cell responses to the parasite. First, while baseline percentages of bone marrow basophils were essentially identical in IL-3 –/– vs. wild-type mice, *S.v.* infection induced a significant increase in bone marrow basophil levels in the wild-type mice but not in the IL-3 –/– mice. These findings confirm the hypothesis, which had been based largely on analyses of the effects of recombinant IL-3 (56–58), that endogenous IL-3 can function to expand basophil populations *in vivo*. However, they also show that IL-3 is not required for baseline levels of bone marrow basophil production in mice.

Second, we found that endogenous IL-3 was required for a substantial proportion (~76%), but not all, of the increases in mast cell numbers which occurred in C57BL/6 mice near the major site of *S.v.* infection, the jejunum. IL-3 appeared to make a lesser contribution to the increases in jejunal mast cell numbers (~50%) during *S.v.* infection in mice on the BALB/c background (18), perhaps because of strain-dependent differences in levels of other cytokines which can influence mast cell development in mice. Based primarily on the results of *in vitro* analyses, as well as a limited number of *in vivo* experiments, cytokines that may promote the development of certain mast cell populations in mice include IL-4 (59, 60), IL-6 (61), IL-9 (62, 63), IL-10 (64), tumour necrosis factor-α (TNF-α) (61) and nerve growth factor (NGF) (65–67). However, IL-3 was required for essentially all of the increases in mast cells which developed in the ileum or spleen of *S.v.*-infected C57BL/6 mice or BALB/c mice (18).

In subsequent studies with IL-3 –/– and +/+ mice, we found that IL-3 was not required for maintaining baseline levels of blood basophils in mice, but was necessary for virtually all of the substantial blood basophilia that occurred during infection with the nematodes *S.v.* or *N. brasiliensis* (68).

KIT^{W}/*KIT*$^{W\text{-}v}$, IL-3 –/– MICE EXHIBIT A PROFOUND IMPAIRMENT OF BASOPHILIA, MUCOSAL MAST CELL DEVELOPMENT AND IMMUNITY DURING INFECTION WITH *STRONGYLOIDES VENEZUELENSIS*

Previous work showed that host immunity to *S.v.* is even more impaired in *Kit*^{W}/*Kit*$^{W\text{-}v}$ mice (17) than in IL-3 –/– mice. SCF/c-kit signalling is more important in the development of mast cells than basophils. For example, mature human basophils (69) or mouse bone marrow basophils (70, 71) exhibit low or undetectable levels of c-kit surface expression, and *Kit*^{W}/*Kit*$^{W\text{-}v}$ mice have been reported to have normal levels of blood basophils (72). Moreover, IL-3 can induce mast cell development in *Kit*^{W}/*Kit*$^{W\text{-}v}$ mice (21), which has been thought to account, at least in part, for the modest numbers of mast cells which develop in the intestines of these animals during infections with some parasites, including *Trichinella spiralis* (16) and *S.v.* (17). Finally, studies with neutralizing antibodies to SCF indicate that adequate SCF/c-kit signalling is required for the intestinal mast cell hyperplasia induced by *Trichinella spiralis* infection in mice, as well as for the expression of normal immunity to this helminth (73).

Taken together, these and other findings suggested that interactions between SCF- and IL-3-dependent signalling mechanisms might importantly influence mast cell development, and immunity to certain parasites, in mice. However, parasite egg production during a primary infection with *N. brasiliensis* was significantly *less* in c-*kit* mutant *Ws*/*Ws* mast cell-deficient rats than the corresponding wild-type rats (74). Moreover, normal rats that were treated with an anti-SCF antibody, as compared to rats which had been treated with a control antibody preparation, exhibited both significantly diminished intestinal mucosal mast cell hyperplasia and significantly *diminished* parasite egg production during primary infection with *N. brasiliensis* (75). These findings raised the possibility that, under some circumstances, SCF- and/or IL-3-dependent mucosal mast cell hyperplasia may have consequences that are more advantageous to the parasite than to the host.

To examine *S.v.* infection in mice which both have markedly impaired c-kit function and cannot make IL-3, we produced *Kit*^{W}/*Kit*$^{W\text{-}v}$, IL-3 –/– mice (18). Adult *Kit*^{W}/*Kit*$^{W\text{-}v}$, IL-3 –/– mice were clinically healthy and resembled *Kit*^{W}/*Kit*$^{W\text{-}v}$, IL-3 +/+ mice in haematocrit and percentage of bone marrow basophils at baseline (Fig. 1d). However, *Kit*^{W}/*Kit*$^{W\text{-}v}$, IL-3 –/– mice exhibited a much more profound defect in their ability to reject *S.v.* than did either *Kit*^{W}/*Kit*$^{W\text{-}v}$, IL-3 +/+ mice (Fig. 1a,b) or *Kit* +/+, IL-3 –/– mice (data not shown).

Kit^{W}/*Kit*$^{W\text{-}v}$, IL-3 –/– mice, unlike *Kit*^{W}/*Kit*$^{W\text{-}v}$, IL-3 +/+ or wild-type mice, also exhibited little or no enhancement of bone marrow basophil production during *S.v.* infection (Fig. 1d). By contrast, *Kit*^{W}/*Kit*$^{W\text{-}v}$, IL-3 +/+ mice developed a significant enhancement of bone marrow basophil levels during the course of *S.v.* infection, findings that are in accord with previous work indicating that c-*kit* mutant *Ws*/*Ws* mast cell-deficient rats exhibit little or no impairment of the blood basophilia that is induced in rats by infection with the parasite *N. brasiliensis* (76). On the other hand, we found that levels of bone marrow basophils were ~50% lower at baseline, and ~35% lower at the time of parasite expulsion, in *Kit*^{W}/*Kit*$^{W\text{-}v}$, IL-3 +/+ mice than in wild-type mice (Fig. 1d).

S.v.-infected *Kit*^{W}/*Kit*$^{W\text{-}v}$, IL-3 –/– mice exhibited levels of histologically detectable mast cells in the jejunum, ileum and spleen that were even more profoundly reduced (vs. the corresponding levels in wild-type mice) than those in the tissues of *S.v.*-infected

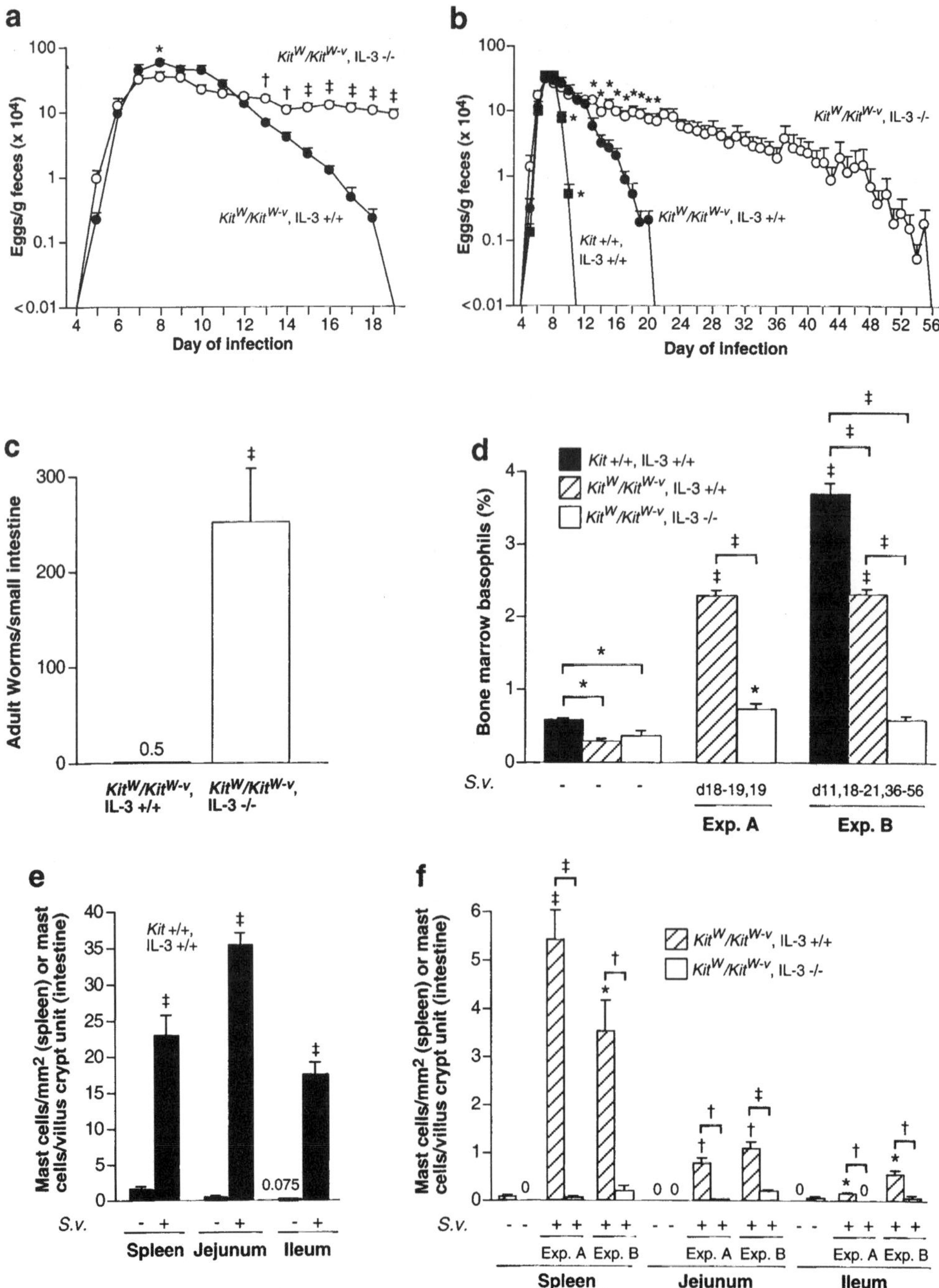
a
Eggs/g feces (x 10^4)
Day of infection
Kit^W/Kit^W-v, IL-3 -/-
Kit^W/Kit^W-v, IL-3 +/+
b
Eggs/g feces (x 10^4)
Day of infection
Kit^W/Kit^W-v, IL-3 -/-
Kit^W/Kit^W-v, IL-3 +/+
Kit +/+, IL-3 +/+
c
Adult Worms/small intestine
0.5
Kit^W/Kit^W-v, IL-3 +/+
Kit^W/Kit^W-v, IL-3 -/-
d
Bone marrow basophils (%)
Kit +/+, IL-3 +/+
Kit^W/Kit^W-v, IL-3 +/+
Kit^W/Kit^W-v, IL-3 -/-
S.v.
d18-19,19
d11,18-21,36-56
Exp. A
Exp. B
e
Mast cells/mm2 (spleen) or mast cells/villus crypt unit (intestine)
Kit +/+, IL-3 +/+
0.075
S.v.
Spleen Jejunum Ileum
f
Mast cells/mm2 (spleen) or mast cells/villus crypt unit (intestine)
Kit^W/Kit^W-v, IL-3 +/+
Kit^W/Kit^W-v, IL-3 -/-
S.v.
Exp. A Exp. B
Spleen Jejunum Ileum

Kit^W/Kit^{W-v}, IL-3 +/+ mice (Fig. 1e,f). Indeed, while levels of mast cells in the spleen, jejunum or ileum of *S.v.*-infected Kit^W/Kit^{W-v}, IL-3 +/+ mice were ~23%, 3.5% or 3.2% those in the corresponding tissues of *S.v.*-infected wild-type mice, mast cells were virtually absent from the corresponding tissues of *S.v.*-infected Kit^W/Kit^{W-v}, IL-3 –/– mice (Fig. 1e,f).

The findings in IL-3 –/– or Kit^W/Kit^{W-v}, IL-3 –/– mice demonstrate that IL-3 importantly contributes to the mast cell hyperplasia and enhanced bone marrow and blood basophil development observed in mice during *S.v.* infection, as well as to the expression of normal host immunity to this nematode. Our data also indicate that IL-3 and SCF may express overlapping and/or synergistic roles in maintaining an adequate immune response to this parasite. However, it should be emphasized that our findings do not prove that the abnormalities in mast cell and basophil development in IL-3 –/– or Kit^W/Kit^{W-v}, IL-3 –/– mice are the sole basis for (or, necessarily, even significantly contribute to) the impaired immunity to *S.v.* expressed by these mice. The expression of contact hypersensitivity reactions (but not T cell-dependent immunity to tumour cells) is moderately reduced in IL-3 –/– mice (47), indicating that IL-3 –/– mice may express defects in some T cell-dependent responses that are not due solely to problems with mast cell or basophil mobilization or function. In addition, Kit^W/Kit^{W-v} mice virtually lack interstitial cells of Cajal, which generate gut electrical pacemaker activity (10–12), and can exhibit reduced numbers of γδ T cells in the gastrointestinal tract (77). Thus, it is possible that abnormalities in addition to those affecting their mast cell and basophil responses contributed to the delayed resolution of *S.v.* infections in IL-3 –/– or Kit^W/Kit^{W-v}, IL-3 –/– mice.

SUMMARY

Host–parasite interactions are exceedingly complex, reflecting long periods of co-evolution. Th2 cell-associated immune responses, which often include both the development of a blood basophilia and increased numbers of mast cells at the site of infection, are widely regarded as important in the expression of host defence against

---

Fig. 1 Markedly defective responses to *Strongyloides venezuelensis* (*S.v.*) infection in Kit^W/Kit^{W-v}, IL-3 –/– mice. Kinetics of *S.v.* egg production in (a) nine Kit^W/Kit^{W-v}, IL-3 +/+ mast cell-deficient mice (five males, four females) vs. six Kit^W/Kit^{W-v}, IL-3 –/– mice (three males, three females) (all killed on day 18 or 19 of infection) and (b) in groups of six male WBB6F$_1$-*Kit* +/+, IL-3 +/+ wild-type mice, WBB6F$_1$-Kit^W/Kit^{W-v}, IL-3 +/+ mice or Kit^W/Kit^{W-v}, IL-3 –/– mice inoculated with 2000 *S.v.* L$_3$ (all the mice were killed on the day of clearance of infection). All data are mean ± SEM. $*p < 0.05$, $\dagger p < 0.001$, $\ddagger p < 0.0001$ vs. corresponding values for WBB6F$_1$-Kit^W/Kit^{W-v}, IL-3 +/+ mice. In (b) the day parasite egg production fell to zero was day 11 for all the wild-type mice, day 18 (two mice) and day 21 (four mice) for Kit^W/Kit^{W-v}, by IL-3 +/+ mice, and day 36 (one mouse), day 38 (two mice), day 48 (one mouse) and day 56 (two mice) for Kit^W/Kit^{W-v}, IL-3 –/– mice. Similar results were obtained in another experiment. (c) Numbers of adult *S.v.* per small intestine on day 18 or 19 of infection in the mice shown in experiment (a). $\ddagger p < 0.0001$ vs. values for Kit^W/Kit^{W-v}, IL-3 +/+ mice. (d) Percentage of bone marrow basophils at baseline ($n = 3$–7 mice) or at various days after infection in the mice shown in (a) (Exp. A) and (b) (Exp. B). $*p<0.05$, $\ddagger p < 0.0001$ vs. corresponding baseline values for mice of the same genotype or vs. values indicated by the bracket. (e,f) Numbers of mast cells in the spleen, proximal jejunum, and ileum at baseline ($n = 3$ or 4 mice) or at various days after *S.v.* infection in the mice shown in (a,b). $*p < 0\ 05$, $\dagger p < 0.001$, $\ddagger p < 0.0001$ vs. corresponding values for uninfected mice of the same genotype or vs. values indicated by the bracket. (a–f) All data are mean ±SEM, except that only mean values are shown for very low values. (Reproduced, with permission, from *Nature* **392**:90–93, 1998.)

parasites. However, certain aspects of such immune responses, such as enhanced vascular permeability at sites of mast cell or basophil mediator release, may have effects that actually can promote parasite survival, development, or fecundity. Accordingly, it may be very difficult to assess the specific contributions of mast cells or other individual potential elements of host defence in parasite–host interactions unless one can examine the expression of the response in the presence and virtual absence of the element in question.

Studies in mast cell-reconstituted genetically mast cell-deficient $Kit^{W}/Kit^{W\text{-}v}$ mice indicate that mast cells can be essential for immune resistance to the feeding of larvae of one species of Ixodid tick, *Haemaphysalis longicornis* (39). By contrast, mast cells are not essential for expression of immunity to the feeding of the larvae of a second species of Ixodid tick, *Dermacentor variabilis* (40). In the case of *D. variabilis* in mice, or *Amblyomma americanum* in guinea pigs (41), the basophil may represent an important effector cell of immune resistance to the feeding of the larval ticks.

Analyses of nematode infections in $Kit^{W}/Kit^{W\text{-}v}$ mast cell-deficient mice indicate that, for *N. brasiliensis*, minimal or no impairment of the primary expulsion response can be detected; by contrast, for nematodes such as *S. ratti* or *S. venezuelensis*, the $Kit^{W}/Kit^{W\text{-}v}$ mice exhibit expulsion responses that clearly are impaired compared to those in the wild-type mice (reviewed in refs 34, 37). Yet, even for *S. ratti* or *S. venezuelensis*, the $Kit^{W}/Kit^{W\text{-}v}$ mice eventually clear the infection.

The expulsion of *S.v.* is impaired much more severely in $Kit^{W}/Kit^{W\text{-}v}$ mice that also lack IL-3 than in $Kit^{W}/Kit^{W\text{-}v}$, IL-3 +/+ mice, or wild-type mice (18). $Kit^{W}/Kit^{W\text{-}v}$, IL-3 –/– mice not only virtually lack bone marrow or blood basophilia during *S.v.* infection (as do IL-3 –/– mice that are wild-type for c-*kit*), but also remain profoundly mast cell-deficient during the course of the infection (18, 68). Thus, their impaired resistance to *S.v.* infection might reflect their inability to mount either mast cell- or basophil-dependent responses to the infection. However, $Kit^{W}/Kit^{W\text{-}v}$, IL-3 –/– mice may have mast cell- or basophil-independent problems that also contribute to their markedly diminished resistance to infection with *S.v.* (18). In other words, findings in IL-3 –/– or $Kit^{W}/Kit^{W\text{-}v}$, IL-3 –/– mice are consistent with the hypothesis that mast cells and basophils express overlapping, but important, roles in host resistance to *S.v.* infection, but they do not unequivocally confirm this hypothesis.

In normal mice, infection with the nematode *N. brasiliensis*, like infections with *S.v.*, induces increased numbers of mucosal mast cells in the intestine, and a bone marrow and blood basophilia. However, preliminary evidence indicates that neither IL-3 –/– nor $Kit^{W}/Kit^{W\text{-}v}$, IL-3 –/– mice exhibit any detectable impairment of their ability to expel a primary infection with *N. brasiliensis* (18) (Lantz and Galli, unpublished data). These findings strongly support other lines of evidence (34, 37) indicating that mast cells and basophils do not critically contribute to host defence against this particular nematode in mice. And, as has been noted above, the results of certain experiments in rats even raise the possibility that mast cells have activities that favour the fecundity of *N. brasiliensis* (74, 75).

In summary, one of the clearest conclusions to draw from the studies in mice and rats would appear to be that the appearance of large numbers of mast cells, or the development of blood basophilia, during the course of an immune response to a parasite cannot be taken as evidence that mast cells and/or basophils necessarily contribute importantly to host resistance to the parasite. Indeed, work with mast cell knockin mice indicates that mast cells can contribute to the extent and persistence of the cutaneous pathology associated with *L. major* infections without detectably influencing the ultimate course of the infection (38). In certain contexts, such as immune resistance to the cutaneous feeding of larval *H. longicornis* ticks, mast cells do appear to make an important contribution to host defence. In other settings, such as resistance to the feeding of other species of ticks, or the expulsion of certain nematodes (such as *S.v.*), mast cells and basophils may express overlapping but important effector functions. Yet, even in those instances in which evidence for a protective role for mast cells and/or basophils appears to be strongest, the mechanisms by which those cells mediate such functions largely remain to be determined.

ACKNOWLEDGEMENTS

Some of the work reviewed here was supported by United States Public Health Service grants (Al-23990, CA-72074 and Al-33372), the Beth Israel Hospital Pathology Foundation, Inc. and AMGEN Inc. S.J.G. performs research funded by, and consults for, AMGEN Inc., under terms that are in accord with Beth Israel Deaconess Medical Center, Harvard Medical School, and Stanford University conflict-of-interest policies.

REFERENCES

1. Askenase, P. W. Immunopathology of parasitic diseases: involvement of basophils and mast cells. *Springer Semin. Immunopathol.* **2**:417–442, 1980.
2. Jarrett, E. E. E. and Miller, H. R. P. Production and activities of IgE in helminth infections. *Prog. Allergy* **31**:178–233, 1982.
3. Nocka, K., Tan, J. C., Chiu, E. *et al.* Molecular bases of dominant negative and loss of function mutations at the murine c-*kit*/white spotting locus: W^{37}, W^{v}, W^{41}, and W. *EMBO J.* **9**:1805–1813, 1990.
4. Hayashi, S., Kunisada, T., Ogawa, M., Yamaguchi, K. and Nishikawa, S. Exon skipping by mutation of an authentic splice site of c-*kit* gene in *W/W* mouse. *Nucleic Acids Res.* **19**:1267–1271, 1991.
5. Kitamura, Y., Go, S. and Hatanaka, K. Decrease of mast cells in W/W^{v} mice and their increase by bone marrow transplantation. *Blood* **52**:447–452, 1978.
6. Nakano, T., Sonoda, T., Hayashi, C. *et al.* Fate of bone marrow-derived cultured mast cells after intracutaneous, intraperitoneal, and intravenous transfer into genetically mast cell-deficient W/W^{v} mice. Evidence that cultured mast cells can give rise to both connective tissue type and mucosal mast cells. *J. Exp. Med.* **162**:1025–1043, 1985.
7. Galli, S. J. and Kitamura, Y. Animal models of human disease. Genetically mast-cell-deficient W/W^{v} and Sl/Sl^{d} mice: their value for the analysis of the roles of mast cells in biological responses *in vivo*. *Am. J. Pathol.* **127**:191–198, 1987.
8. Galli, S. J., Geissler, E. N., Wershil, B. K. *et al.* Insights into mast cell development and function derived from analyses of mice carrying mutations at beige, *W*/c-*kit* or *Sl/SCF* (c-*kit* ligand) loci. In: *The Role of the Mast Cell in Health and Disease* (Kaliner, M. A. and Metcalfe, D. D., eds), pp. 129–202. Marcel Dekker, New York, 1992.
9. Galli, S. J., Zsebo, K. M. and Geissler, E. N. The kit ligand, stem cell factor. *Adv. Immunol.* **55**: 1–96, 1994.
10. Maeda, H., Yamagata, A., Nishikawa, S. *et al.* Requirement of c-*kit* for development of intestinal pacemaker system. *Development* **116**:369–375, 1992.
11. Ward, S., Burns, A., Torihashi, S. and Sanders, K. Mutation in the proto-oncogene c-*kit* blocks development of interstitial cells and electrical rhythm in murine intestine. *J. Physiol.* **480**:91–97, 1994.
12. Huizinga, J. D., Thuneberg, L., Kluppel, M., Malysz, J., Mikkelsen, H. B. and Bernstein, A. *W/Kit* gene required for interstitial cells of Cajal and for intestinal pacemaker activity. *Nature* **373**:347–349, 1995.
13. Kitamura, Y. and Go, S. Decreased production of mast cells in Sl/Sl^{d} anemic mice. *Blood* **53**:492–497, 1979.

14. Yung, Y.-P. and Moore, M. A. S. Long-term *in vitro* culture of murine mast cells. III. Discrimination of mast cell growth-factor and granulocyte CSF. *J. Immunol.* **129**:1256–1261, 1982.
15. Suda, T., Suda, J., Spicer, S. S. and Ogawa, M. Proliferation and differentiation in culture of mast cell progenitors derived from mast cell-deficient mice of genotype W/W^v. *J. Cell. Physiol.* **122**:187–192, 1985.
16. Alizadeh, H. and Murrell, K. D. The intestinal mast cell response to *Trichinella spiralis* infection in mast cell-deficient W/W^v mice. *J. Parasitol.* **70**:767–773, 1984.
17. Khan, A. I., Horii, Y., Tiuria, R., Sato, Y. and Nawa, Y. Mucosal mast cells and the expulsive mechanisms of mice against *Strongyloides venezuelensis. Int. J. Parasitol.* **23**:551–555, 1993.
18. Lantz, C. S., Boesiger, J., Song, C. H. *et al.* Role for interleukin-3 in mast cell and basophil development and parasite immunity. *Nature* **392**:90–93, 1998.
19. Galli, S. J., Arizono, N., Murakami, T., Dvorak, A. M. and Fox, J. G. Development of large numbers of mast cells at sites of idiopathic chronic dermatitis in genetically mast cell-deficient $WBB6F_1$-W/W^v mice. *Blood* **69**:1661–1666, 1987.
20. Gordon, J. R. and Galli, S. J. Phorbol 12-myristate 13-acetate-induced development of functionally active mast cells in W/W^v but not Sl/Sl^d genetically mast cell-deficient mice. *Blood* **75**:1637–1645, 1990.
21. Ody, C., Kindler, V. and Vassalli, P. Interleukin 3 perfusion in W/W^v mice allows the development of macroscopic hematopoietic spleen colonies and restores cutaneous mast cell numbers. *J. Exp. Med.* **172**:403–406, 1990.
22. Zsebo, K. M., Williams, D. A., Geissler, E. N. *et al.* Stem cell factor (SCF) is encoded at the *Sl* locus of the mouse and is the ligand for the c-*kit* tyrosine kinase receptor. *Cell* **63**:213–224, 1990.
23. Ando, A., Martin, T. R. and Galli, S. J. Effects of chronic treatment with the c-*kit* ligand, stem cell factor, on immunoglobulin E-dependent anaphylaxis in mice: genetically mast cell-deficient Sl/Sl^d mice acquire anaphylactic responsiveness, but the congenic normal mice do not exhibit augmented responses. *J. Clin. Invest.* **92**:1639–1649, 1993.
24. Tsai, M., Shih, L., Newlands, G. F. J. *et al.* The rat c-*kit* ligand, stem cell factor, induces the development of connective tissue-type and mucosal mast cells *in vivo.* Analysis by anatomical distribution, histochemistry, and protease phenotype. *J. Exp. Med.* **174**:125–131, 1991.
25. Wershil, B. K., Mekori, Y. A., Murakami, T. and Galli, S. J. ^{125}I-fibrin deposition in IgE-dependent immediate hypersensitivity reactions in mouse skin. Demonstration of the role of mast cells using genetically mast cell-deficient mice locally reconstituted with cultured mast cells. *J. Immunol.* **139**:2605–2614, 1987.
26. Wershil, B. K., Wang, Z.-S., Gordon, J. R. and Galli, S. J. Recruitment of neutrophils during IgE-dependent cutaneous late phase responses in the mouse is mast cell dependent: partial inhibition of the reaction with antiserum against tumor necrosis factor-alpha. *J. Clin. Invest.* **87**:446–453, 1991.
27. Wershil, B. K., Furuta, G. T., Wang, Z.-S. and Galli, S. J. Mast cell-dependent neutrophil and mononuclear cell recruitment in immunoglobulin E-induced gastric reaction in mice. *Gastroenterology* **110**:1452–1490, 1996.
28. Martin, T. R., Takeishi, T., Katz, H. R. *et al.* Mast cell activation enhances airway responsiveness to methacholine in the mouse. *J. Clin. Invest.* **9**:1176–1182, 1993.
29. Matsuda, H., Kawakita, K., Kiso, Y. *et al.* Substance P induces granulocyte infiltration through degranulation of mast cells. *J. Immunol.* **142**:927–931, 1989.
30. Yano, H., Wershil, B. K., Arizono, N. and Galli, S. J. Substance P-induced augmentation of cutaneous vascular permeability and granulocyte infiltration in mice is mast cell dependent. *J. Clin. Invest.* **84**:1276–1286, 1989.
31. Wershil, B. K., Murakami, T. and Galli, S. J. Mast cell-dependent amplification of an immunologically nonspecific inflammatory response. Mast cells are required for the full expression of cutaneous acute inflammation induced by phorbol 12-myristate 13-acetate. *J. Immunol.* **140**:2356–2360, 1988.
32. Ramos, B. F., Qureshi, R., Olsen, K. M. and Jakschik, B. A. The importance of mast cells for the neutrophil influx in immune complex-induced peritonitis in mice. *J. Immunol.* **145**:1868–1873, 1990.
33. Galli, S. J. and Hammel, I. Unequivocal delayed hypersensitivity in mast cell-deficient and beige mice. *Science* **226**:710–713, 1984.
34. Reed, N. D. Function and regulation of mast cells in parasite infections. In: *Mast Cell and Basophil Differentiation and Function in Health and Disease* (Galli, S. J. and Austen, K. F., eds), pp. 205–215. Raven Press, New York, 1989.
35. Jacobson, R. H., Reed, N. D. and Manning, D. D. Expulsion of *Nippostrongylus brasiliensis* from mice lacking antibody production potential. *Immunology* **32**:867–874, 1977.

36. Watanabe, N., Katakura, K., Kobayashi, A. *et al.* Protective immunity and eosinophilia in IgE deficient SJA/9 mice infected with *Nippostrongylus brasiliensis* and *Trichinella spiralis. Proc. Natl. Acad. Sci. USA* **85**:4460–4462, 1988.
37. Nawa, Y., Ishikawa, N., Tsuchiya, K. *et al.* Selective effector mechanisms for the expulsion of intestinal helminths. *Parasite Immunol.* **16**:333–338, 1994.
38. Wershil, B. K., Theodos, C. M., Galli, S. J. and Titus, R. G. Mast cells augment lesion size and persistence during experimental *Leishmania major* infection in the mouse. *J. Immunol.* **152**:4563–4571, 1994.
39. Matsuda, H., Watanabe, N., Kiso, Y. *et al.* Necessity of IgE antibodies and mast cells for manifestation of resistance against larval *Haemaphysalis longicornis* ticks in mice. *J. Immunol.* **144**:259–262, 1990.
40. Steeves, E. B. and Allen, J. R. Basophils in skin reactions of mast cell-deficient mice infested with *Dermacentor variabilis. Int. J. Parasitol.* **20**:655–667, 1990.
41. Brown, S. J., Galli, S. J., Gleich, G. J. *et al.* Ablation of immunity to *Amblyomma americanum* by anti-basophil serum: cooperation between basophils and eosinophils in expression of immunity to ectoparasites (ticks) in guinea pigs. *J. Immunol.* **129**:790–796, 1982.
42. Juhlin, L. and Michaelsson, G. A new syndrome characterized by absence of eosinophils and basophils. *Lancet* **i**:1233–1235, 1977.
43. Galli, S. J. and Askenase, P. W. Cutaneous basophil hypersensitivity. In: *The Reticuloendothelial System: A Comprehensive Treatise* (Abramoff, P., Phillips, S. M. and Escobar, M. R., eds), pp. 321–369. Plenum Press, New York, 1986.
44. Ihle, J. N., Keller, J., Orsozlan, S. *et al.* Biological properties of homogeneous interleukin 3. I. Demonstration of WEHI-3 growth factor activity, mast cell growth factor activity, P cell stimulating factor activity and histamine producing factor activity. *J. Immunol.* **131**:282–287, 1983.
45. Ihle, J. N. and Weinstein, Y. Immunological regulation of hematopoietic/lymphoid stem cell differentiation by interleukin 3. *Adv. Immunol.* **39**:1–50, 1986.
46. Ihle, J. N. Interleukin-3 and hematopoiesis. *Chem. Immunol.* **51**:63–106, 1992.
47. Mach, N., Lantz, C. S., Galli, S. J. *et al.* Involvement of interleukin-3 in delayed-type hypersensitivity. *Blood* **91**:778–783, 1998.
48. Ichihara, M., Hara, T., Takagi, M., Cho, L. C., Gorman, D. M. and Miyajima, A. Impaired interleukin-3 (IL-3) response of the A/J mouse is caused by a branch point deletion in the IL-3 receptor α subunit gene. *EMBO J.* **14**:939–950, 1995.
49. Nishinakamura, R., Miyajima, A., Mee, P. J., Tybulewicz, V. W. and Murray, R. Hematopoiesis in mice lacking the entire granulocyte–macrophage colony-stimulating factor/interleukin-3/interleukin-5 functions. *Blood* **88**:2458–2464, 1996.
50. Tsuji, K., Zsebo, K. M. and Ogawa, M. Murine mast cell colony formation supported by IL-3, IL-4, and recombinant rat stem cell factor, ligand for c-*kit*. *J. Cell. Physiol.* **148**:362–369, 1991.
51. Gurish, M. F., Ghildyal, N., McNeil, H. P., Austen, K. F., Gillis, S. and Stevens, R. L. Differential expression of secretory granule proteases in mouse mast cells exposed to interleukin 3 and c-*kit* ligand. *J. Exp. Med.* **175**:1003–1012, 1992.
52. Haig, D. M., Huntley, J. F., Mackellar, A. *et al.* Effects of stem cell factor (kit-ligand) and interleukin-3 on the growth and serine proteinase expression of rat bone marrow derived or serosal mast cells. *Blood* **83**:72–83, 1994.
53. Lantz, C. S. and Huff, T. F. Differential responsiveness of purified mouse c-*kit*+ mast cells and their progenitors to IL-3 and stem cell factor. *J. Immunol.* **155**:4024–4029, 1995.
54. Rennick, D., Hunte, B., Holland, G. and Thompson-Snipes, L. Cofactors are essential for stem cell factor dependent growth and maturation of mast cell progenitors: comparative effects of interleukin-3 (IL-3), IL-4, IL-10, and fibroblasts. *Blood* **85**:57–65, 1995.
55. Rodewald, H.-R., Dessing, M., Dvorak, A. M. and Galli, S. J. Identification of a committed precursor for the mast cell lineage. *Science* **271**:818–822, 1996.
56. Donahue, R. E., Seehra, J., Metzger, M. *et al.* Human IL-3 and GM-CSF act synergistically in stimulating hematopoiesis in primates. *Science* **241**:1820–1823, 1988.
57. Saito, H., Hatake, K., Dvorak, A. M. *et al.* Selective differentiation and proliferation of hematopoietic cells induced by recombinant human interleukins. *Proc. Natl. Acad. Sci. USA* **85**:2288–2292, 1988.
58. Mayer, P., Valent, P., Schmidt, G., Liehl, E. and Bettelheim, P. The *in vivo* effects of recombinant human interleukin-3: demonstration of basophil differentiation factor, histamine-producing activity, and priming of GM-CSF-responsive progenitors in nonhuman primates. *Blood* **74**:613–621, 1989.
59. Mosmann, T. R., Bond, M. W., Coffman, R. L., Ohara, J. and Paul, W. E. T-cell and mast cell lines respond to B-cell stimulatory factor 1. *Proc. Natl. Acad. Sci. USA* **83**:5654–5658, 1986.

60. Smith, C. A. and Rennick, D. M. Characterization of a murine lymphokine distinct from interleukin 2 and interleukin 3 (IL-3) possessing a T-cell growth factor activity and a mast-cell growth factor activity that synergizes with IL-3. *Proc. Natl. Acad. Sci. USA* **83**:1857–1861, 1986.
61. Hu, Z.-Q., Kobayashi, K., Zenda, N. and Shimamura, T. Tumor necrosis factor-α- and interleukin-6 triggered mast cell development from mouse spleen cells. *Blood* **89**:526–533, 1997.
62. Hültner, L., Druez, C., Moeller, J. *et al.* Mast cell growth-enhancing activity (MEA) is structurally related and functionally identical to the novel mouse T cell growth factor P40/TCGFIII (interleukin 9). *Eur. J. Immunol.* **20**:1413–1416, 1990.
63. Khalil, R. M., Luz, A., Mailhammer, R. *et al. Schistosoma mansoni* infection in mice augments the capacity for interleukin 3 (IL-3) and IL-9 production and concurrently enlarges progenitor pools for mast cells and granulocytes–macrophages. *Infect. Immun.* **64**:4960–4966, 1996.
64. Thompson-Snipes, L., Dhar, V., Bond, M. W., Mosmann, T. R., Moore, K. W. and Rennick, D. Interleukin-10: a novel stimulatory factor for mast cells and their progenitors. *J. Exp. Med.* **173**:507–510, 1991.
65. Aloe, L. and Levi-Montalcini, R. Mast cells increase in tissues of neonatal rats injected with the nerve growth factor. *Brain Res.* **133**:358–366, 1977.
66. Aloe, L. The effect of nerve growth factor and its antibody on mast cells *in vivo. J. Neuroimmunol.* **18**:1–12, 1988.
67. Matsuda, H., Kannan, Y., Ushio, H. *et al.* Nerve growth factor induces development of connective tissue-type mast cells *in vitro* from murine bone marrow cells. *J. Exp. Med.* **174**:714, 1991.
68. Lantz, C. S., Song, C. H., Dranoff, G. and Galli, S. J. Interleukin-3 (IL-3) is required for blood basophilia, but not for increased basophil IL-4 production, in response to parasite infection in mice. *FASEB J.* **13**:A235 (abstract 255.18), 1999.
69. Columbo, M., Horowitz, E. M., Botana, L. M. *et al.* The human recombinant c-*kit* receptor ligand, rhSCF, induces mediator release from human cutaneous mast cells and enhances IgE dependent mediator release from both skin mast cells and peripheral blood basophils. *J. Immunol.* **149**:599–608, 1992.
70. Dvorak, A. M., Seder, R. A., Paul, W. E., Morgan, E. S. and Galli, S. J. Effects of interleukin-3 with or without the c-*kit* ligand, stem cell factor, on the survival and cytoplasmic granule formation of mouse basophils and mast cells *in vitro. Am. J. Pathol.* **144**:160–170, 1994.
71. Lantz, C. S., Yamaguchi, M., Oettgen, H. C. *et al.* IgE regulates mouse basophil FcεRI expression *in vivo. J. Immunol.* **158**:2517–2521, 1997.
72. Jacoby, W., Cammarata, P. V., Findlay, S. and Pincus, S. H. Anaphylaxis in mast cell-deficient mice. *J. Invest. Dermatol.* **83**:302–304, 1984.
73. Donaldson, L. E., Schmitt, E., Huntley, J. F., Newlands, G. F. J. and Grencis, R. K. A critical role of stem cell factor and c-*kit* in host protective immunity to an intestinal helminth. *Int. Immunol.* **8**:559–567, 1996.
74. Arizono, N., Kasugai, T., Yamada, M. *et al.* Infection of *Nippostrongylus brasiliensis* induces development of mucosal-type but not connective tissue-type mast cells in genetically mast cell-deficient *Ws/Ws* rats. *Blood* **81**:2572–2578, 1993.
75. Newlands, G. F. J., Miller, H. R. P., MacKellar, A. and Galli, S. J. Stem cell factor contributes to intestinal mucosal mast cell hyperplasia in rats infected with *Nippostrongylus brasiliensis* or *Trichinella spiralis*, but anti-stem cell factor treatment decreases parasite egg production during *N. brasiliensis* infection. *Blood* **86**:1968–1976, 1995.
76. Kasugai, T., Okada, M., Morimoto, M. *et al.* Infection of *Nippostrongylus brasiliensis* induces normal increases of basophils in mast cell-deficient *Ws/Ws* rats with a small deletion at the kinase domain of c-*kit*. *Blood* **81**:2521–2529, 1993.
77. Puddington, L., Olson, S. and Lefrancois, L. Interactions between stem cell factor and c-*kit* are required for intestinal immune system homeostasis. *Immunity* **1**:733–739, 1994.
78. Galli, S. J. New concepts about the mast cell. *N. Engl. J. Med.* **328**:257–265, 1993.
79. Galli, S. J. New insights into 'the riddle of mast cells': microenvironmental regulation of mast cell development and phenotypic heterogeneity. *Lab. Invest.* **62**:5–33, 1990.

SECTION EIGHT

MAST CELLS IN ATHEROSCLEROSIS AND CARDIAC DISEASES

CHAPTER 29

Human Heart Mast Cells: Immunological Characterization *In Situ* and *In Vitro*

GIANNI MARONE,[1] GENNARO DE CRESCENZO,[1] VINCENZO PATELLA,[1] FRANCESCOPAOLO GRANATA,[1] LAURA VERGA,[2] ELOISA ARBUSTINI[2] and ARTURO GENOVESE[1]*

[1]*Division of Clinical Immunology and Allergy, University of Naples Federico II, Naples and* [2]*Department of Pathology, University of Pavia, Pavia, Italy*

INTRODUCTION

Human mast cells have long been considered as mainly inflammatory cells, but through the elaboration of various cytokines (1–5) and chemokines (3, 6, 7) they play a more complex role than was originally thought (8–10). Their physiological roles have remained largely elusive, although increasing evidence suggests that they are involved in defence against host invasion by parasites (11) and bacterial and viral infections (12–19). These cells and their mediators are pivotal in several allergic, inflammatory (9, 10, 20) and chronic disorders affecting various organs (21–24).

Mast cells have been identified in heart tissue in animals (25) and in man (26–33), and in normal and atherosclerotic human arterial intima (34–36). However, their role and that of their mediators in the cardiovascular system is just starting to be elucidated (37). Difficulties in obtaining samples of human heart tissue have impeded the isolation of cardiac mast cells for *in vitro* studies. However, the increasing practice of heart transplantation has provided material for the isolation and purification of these cells from heart tissue.

We have systematically approached the question of clarifying the possible roles of human heart mast cells (HHMC) and their mediators in health and disease (38–40). We examined by electron microscopy heart tissue from patients undergoing heart transplantation and from subjects who had died in accidents. In addition, we developed an original technique for the isolation and partial purification of HHMC (38–40). Here we briefly review evidence supporting the involvement of HHMC in different pathophysiological conditions.

* Corresponding author.

MAST CELLS AND BASOPHILS
ISBN 0-12-473335-2

ISOLATION OF HHMC AND THEIR CHARACTERIZATION *IN SITU*

We examined sections of more than 50 hearts from patients undergoing heart transplantation, mostly for dilated cardiomyopathy, and from subjects without cardiovascular diseases who had died in accidents. Mast cells, stained with toluidine blue, were mainly found around blood vessels and between myocardial fibres in all sections of all hearts. These cells were frequently close to blood vessels, suggesting that circulating antigens, autoantibodies (e.g. anti-IgE, anti-$Fc_{\varepsilon}RI$), drugs (e.g. general anaesthetics, protamine) or agents used during diagnostic procedures (e.g. radiocontrast media) can reach the perivascular HHMC. Activated mast cells can in turn release vasoactive substances (e.g. histamine, cysteinyl leukotrienes, platelet-activating factor (PAF), etc.) that affect blood vessels (Fig. 1). HHMC are also found in close contact with sarcolemma (Fig. 2). Mast cells are also identified in the human arterial intima, particularly at the site of coronary atheroma (Fig. 3). Table I summarizes the strategic localization of mast cells in human heart tissue.

Cytoplasmic granules of HHMC displayed diverse ultrastructural patterns: homogeneously dense, combined homogeneously dense/scrolls, scrolls and crystals (38).

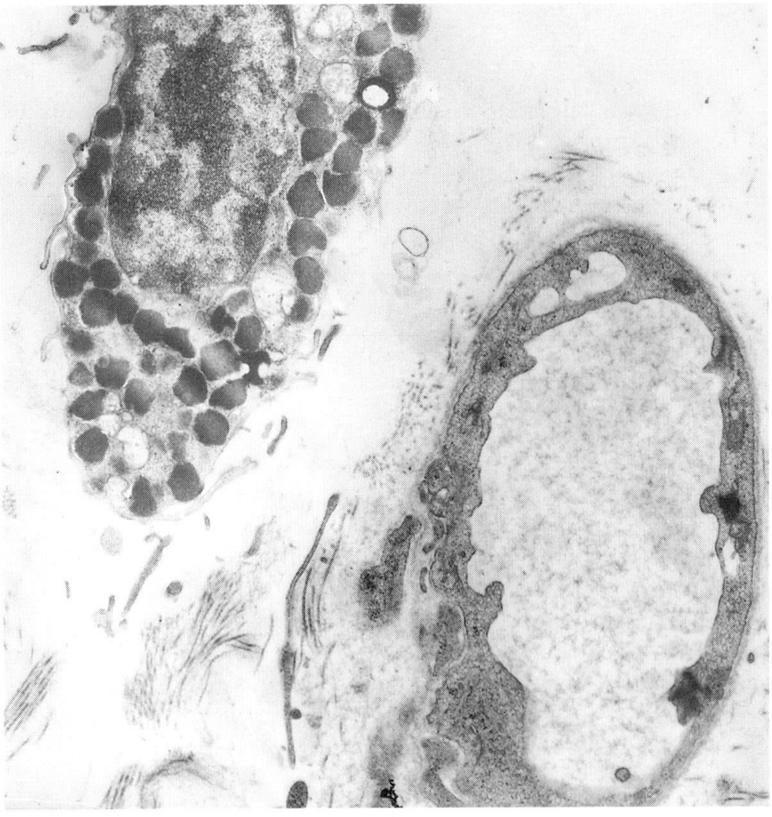

Fig. 1 Electron micrograph of human cardiac mast cell in close contact with the capillary vessel wall (uranyl acetate and lead citrate). Original magnification: 9000×.

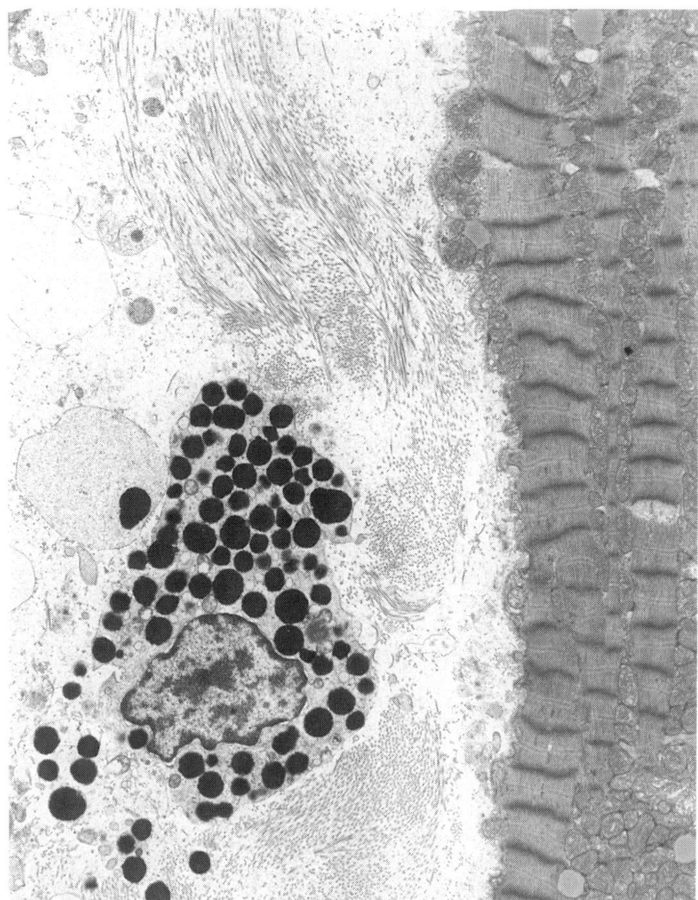

Fig. 2 Electron micrograph of human cardiac mast cell in close contact with the myocyte sarcolemma. Some granules are released, whereas others are in the process of being released from the mast cell. The mast cell is surrounded by collagen fibers (uranyl acetate and lead citrate). Original magnification: 9000×.

TABLE I
Location of Mast Cells in Human Heart Tissue

Interstitial mast cells	Close proximity to myocytes Increased density in ischaemic and idiopathic cardiomyopathy
Perivascular and adventitial mast cells	Always found in the adventitia Increased density in coronary spasm and thrombosis Few mast cells in the arterial media
Intimal mast cells	Present in the human arterial intima, at the site of atherogenesis Preferentially located in the shoulder region of atheromas Increased density of activated mast cells in human coronary atheroma

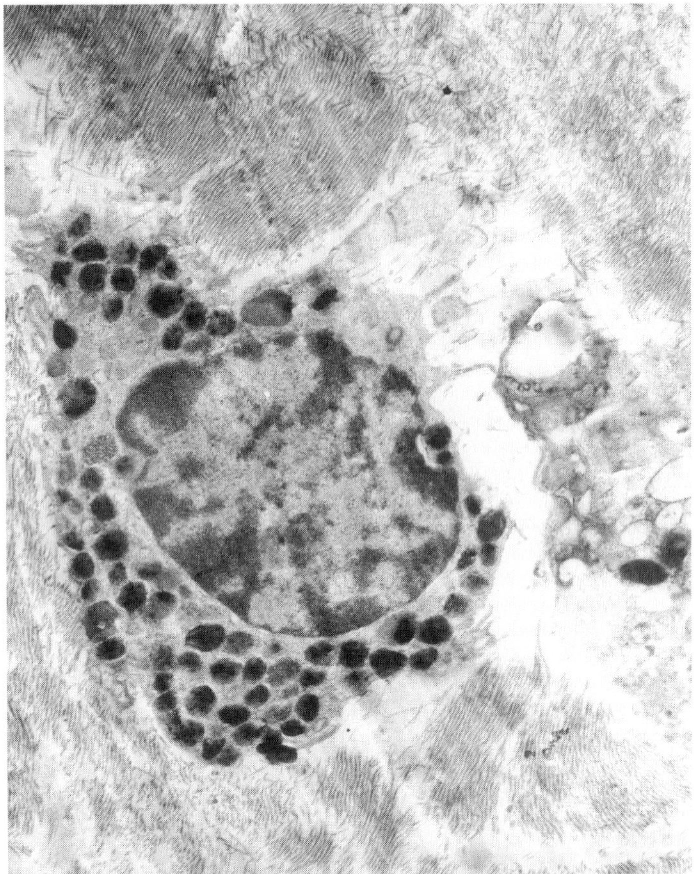

Fig. 3 Electron micrograph of human cardiac mast cell in coronary atheroma. Note that the mast cell is adjacent to a foam cell and surrounded by collagen fibres (uranyl acetate and lead citrate). Original magnification: 9000×.

Non-membrane-bound lipid bodies were present and frequently displayed focal lucent areas, as in gut and uterine mast cells (41, 42). A small percentage of HHMC (≈10%) showed partial degranulation, occurring through the extrusion of altered, membrane-free granules to the cell's exterior, in addition to the intracytoplasmic solubilization of granules and opening of multiple intracytoplasmic degranulation channels.

To characterize the HHMC it was necessary to isolate and partially purify these cells. In our laboratory we isolated and partially purified HHMC using hearts from patients undergoing heart transplantation and from victims of car accidents (38). Using a complex sequence of procedures (Percoll gradients, countercurrent elutriation, etc.) HHMC can be enriched from < 0.1% to ≈18%, which is remarkable considering the low percentage of mast cells in the starting suspension. Using this technique we examined various aspects of HHMC biology *in vitro* (38–40, 43, 44).

HISTAMINE, PROTEOLYTIC ENZYMES AND PROTEOGLYCANS SYNTHESIZED BY HHMC

In a first series of experiments we measured the histamine content of sections of human heart. The histamine content of left (≈5.4 μg g^{-1} wet tissue) and right ventricles (≈5.3 μg g^{-1} wet tissue) and the septum (≈5.6 μg g^{-1} wet tissue) was similar. These values are lower than in lung parenchyma (≈22.7 μg g^{-1} wet tissue). Our results agree with previous reports of no regional differences in the tissue concentrations of histamine in the left and right ventricle in the normal and pathological myocardium (45). The histamine content of isolated HHMC was ≈3.3 pg per cell, comparable to the content of lung parenchymal (≈3.7 pg per cell) and skin mast cells (≈3.8 pg per cell), but higher than human basophils (≈1.1 pg per cell). The fact that the mast cells isolated from different human anatomical sites did not really differ in histamine content suggests that the procedures used to isolate the different mast cells have little effect on their histamine content.

Human mast cells contain at least three main proteolytic enzymes: tryptase, chymase and carboxypeptidase (46, 47). These may account for 20–50% of the total cell protein in rat and human mast cells (48, 49). Human mast cells have been differentiated on the basis of their content of tryptase alone (MC_T) or tryptase together with chymase (MC_{TC}) (48). The vast majority of human mast cells (lung, skin, etc.) contain tryptase, although in different amounts, whereas peripheral blood basophils contain only very small amounts (< 0.05 pg per cell) (48). The mean tryptase content of HHMC (≈24 μg per 10^6 cells) is lower than skin mast cells (HSMC, ≈35 μg per 10^6 cells) and higher than lung mast cells (HLMC, ≈10 μg per 10^6 cells). More importantly, IgE-mediated activation of HHMC caused the release of tryptase in parallel to the secretion of histamine (38) (Fig. 4). Tryptase and chymase-positive cells were found in the intima of human normal and atherosclerotic aorta (35, 36).

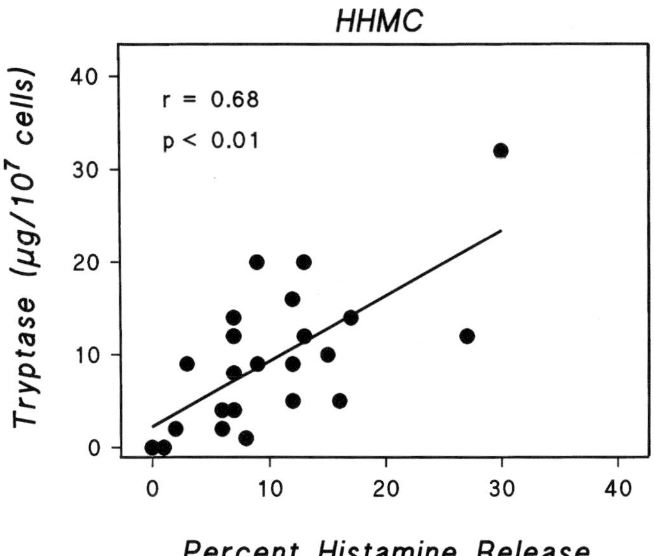

Fig. 4 Correlation between the maximum percentage of histamine release and tryptase secretion from mast cells isolated from human heart tissue, in response to anti-$Fc_{\varepsilon}RI$ challenge *in vitro*. Each point represents the mean of duplicate determinations.

Using a polyclonal anti-chymase antiserum (50) we detected chymase in all secretory granules in the vast majority of HHMC (>95%). The mean chymase content of HHMC (≈4.0 μg per 10^6 cells) is higher than HLMC. Therefore, HHMC contain tryptase as well as chymase. This is particularly important because the activity of human heart chymase can generate several biological compounds. For instance, chymase generates the vasoactive peptide angiotensin II from angiotensin I, acting as an angiotensin-converting enzyme (51). We recently demonstrated that supernatants of HHMC challenged *in vitro* with anti-IgE convert angiotensin I into angiotensin II (4), suggesting that chymase released from immunologically challenged HHMC could also play a role in the homeostatic control of blood pressure. This latter observation is particularly important because it might imply that activation of HHMC and perivascular mast cells and the release of chymase can influence cardiovascular functions through activation of the angiotensin system.

We have recently found that recombinant human chymase can cleave big endothelin 1 to form endothelin 1 (Marone *et al.*, unpublished observation). This is of particular interest because acute coronary syndromes are accompanied by vasoconstriction. Increased endothelin 1-like immunoreactivity has been found in atherosclerotic lesions associated with acute coronary syndromes (52).

Human mast cells from different anatomical sites differ in proteoglycan content. For example, HSMC mainly contain heparin, whereas HLMC contain heparin and chondroitin sulphate E (53, 54). In intestinal mast cells chondroitin sulphate E is a major proteoglycan (55), whereas basophils contain chondroitin sulphate A (56). No information is available at the moment on the types of proteoglycan(s) in HHMC. However, the heparin proteoglycan matrix of exocytosed granules of rodent mast cells binds low-density lipoprotein (LDL), whereas the neutral proteases of the granules (chymase and carboxypeptidase A) degrade the apolipoprotein B-100 (apoB-100) component of the bound LDL (57, 58). Chymase first hydrolyses apoB-100, then carboxypeptidase A cleaves the C-terminal amino acids. The proteolytic degradation of LDL by granule neutral proteases profoundly influences the interaction between LDL and the exocytosed granules, which are phagocytosed by macrophages then converted into foam cells in the subendothelial space of the arterial intima. These fascinating findings suggest that vascular mast cells, their proteolytic enzymes and proteoglycans could play a role in the pathogenesis of atherosclerosis.

PROSTAGLANDINS AND CYSTEINYL LEUKOTRIENES SYNTHESIZED *DE NOVO* BY HHMC

Anti-IgE cross-linking of isolated HHMC led to the *de novo* synthesis of approximately 18 ng of prostaglandin D_2 (PGD_2) per 10^6 cells through cyclo-oxygenase activity (38). PGD_2 is a potent coronary constrictor (57) and its *in vivo* release from HHMC can cause coronary vasoconstriction in man. Activation *in vitro* of HHMC with anti-IgE or anti-$Fc_\varepsilon RI$ induced *de novo* synthesis of the cysteinyl leukotriene C_4 (LTC_4) with a mean production of ≈18 ng per 10^6 cells.

The pathophysiological role of the eicosanoids *de novo* synthesized by HHMC is still uncertain. Cysteinyl leukotrienes may modulate fibroblast proliferation when the synthesis of prostaglandins is blocked (59), suggesting that the fibrotic process depends partly on the local balance between prostaglandins and leukotriene synthesis. Moreover, intravenous and intracoronary injection of LTC_4 and LTD_4 produces several cardiovascular and metabolic effects (60–62). Intravenous injection of LTD_4 in patients undergoing coronary angiography for diagnostic purpose increases coronary resistance and reduces coronary blood flow (Fig. 5). There are several intra- and extracardiac sources of cysteinyl leukotrienes in the human cardiovascular system (Table II). LTC_4 can be immunologically released by HHMC *in vitro* and possibly *in vivo*. In addition, LTC_4 can be generated by intercellular transfer of LTA_4 from neutrophils to endothelial cells to synthesize LTC_4 (63). Finally, during allergic reactions, leukotrienes synthesized by HLMC can reach the heart through the pulmonary vessels. In summary, the heart can be challenged twice. There is preliminary evidence that HHMC can also synthesize PAF; Table III summarizes the mediators pre-formed and newly synthesized by immunologically activated HHMC.

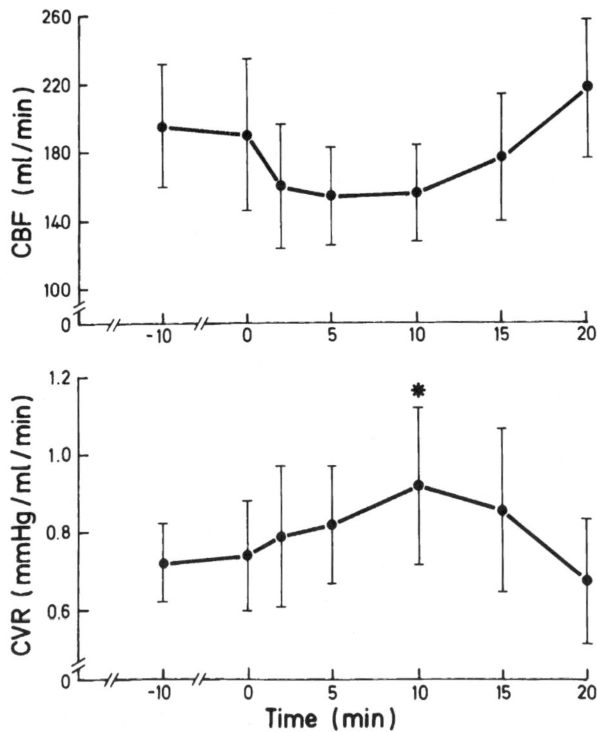

Fig. 5 Time course of the effect of intravenous LTC_4 on coronary blood flow (CBF) and coronary vascular resistance (CVR). $*p < 0.05$ (vs. time 0). (Reprinted, with permission, from *Int. J. Clin. Lab. Res.* **27**:178–184, 1997.)

TABLE II
Sources of Cysteinyl Leukotrienes in the Human Cardiovascular System

- *In situ* generation of LTC_4 by human heart mast cells
- *In situ* generation of LTC_4 by mast cells and macrophages in the coronary adventitia or circulating blood cells
- Intercellular transfer of LTA_4 from neutrophils to endothelial cells to synthesize LTC_4
- Cysteinyl leukotrienes synthesized by human lung mast cells reach the heart through the pulmonary vessels

TABLE III
Chemical Mediators Synthesized by Human Heart Mast Cells

Histamine	≈3 pg per cell
Tryptase	≈20 pg per cell
Chymase	≈4 pg per cell
Prostaglandin D_2	≈18 ng per 10^6 cells
Cysteinyl leukotriene C_4	≈18 ng per 10^6 cells
Platelet-activating factor	+

The arachidonic acid metabolism of HHMC provides additional examples of their unique biochemical properties. IgE cross-linking of human lung and gut mast cells induces the synthesis of approximately the same amount (≈60 ng per 10^6 cells) of PGD_2 and LTC_4 (21, 64). Uterine mast cells (HUMC) synthesize more PGD_2 (≈90 ng per 10^6 cells) than LTC_2 (≈45 ng per 10^6 cells) (42); HSMC prefer the cyclo-oxygenase pathway (PGD_2 ≈45 ng per 10^6 cells) with little metabolism through the 5-lipoxygenase pathway ($LTC_4 < 5$ ng per 10^6 cells) (64, 65). Immunological activation of HHMC leads to *de novo* synthesis of equal amounts of PGD_2 and LTC_4 (≈18 ng per 10^6 cells). Figure 6 compares the amounts of PGD_2 and LTC_4 immunologically released by isolated HHMC, HLMC, HSMC and HUMC. Therefore, HHMC differ quantitatively and qualitatively from other mast cells with respect to the arachidonic acid metabolism.

IMMUNOLOGICAL AND NON-IMMUNOLOGICAL STIMULI THAT ACTIVATE HHMC *IN VITRO*

Mast cells dispersed from human heart tissue can be immunologically activated by IgE- and non-IgE-mediated stimuli. IgE-mediated activation can be induced by antigen, anti-IgE and anti-$Fc_{\varepsilon}RI$ (38). Activation of HHMC with antigen or anti-IgE does not require passive sensitization with IgE, suggesting that enzymatic dispersion and the procedure to purify these cells had no deleterious effect on the surface structures required for IgE binding. The activation of HHMC by anti-IgE and by a monoclonal antibody against the α chain of $Fc_{\varepsilon}RI$ (anti-$Fc_{\varepsilon}RI\alpha$) may be clinically relevant. In fact, histamine releasing autoantibodies against IgE (anti-IgE) or $Fc_{\varepsilon}RI\alpha$ are present in the circulation of some patients with atopic dermatitis, chronic urticaria and intrinsic asthma (66–68).

Complement deposition has been documented in infarcted areas of human heart (69) and complement depletion or inhibition reduces tissue injury in myocardial ischaemia (70–72). There is also experimental evidence that C5a can cause several cardiovascular derangements, either directly or through the release of vasoactive mediators (73–75). It has also been demonstrated that cardiac mast cells are involved in C5a-induced myocardial ischaemia (76). Furthermore, complement activation and anaphylatoxin formation (C3a and C5a) occur during cardiac (77) and systemic anaphylaxis (78).

Incubation of isolated HHMC with recombinant human C5a caused rapid, dose-dependent release of histamine. Interestingly, rhC5a does not activate HLMC, whereas HSMC are responsive (79), suggesting that HHMC and HSMC are the only human mast cells so far isolated possessing C5a receptors. HHMC are also responsive to stem cell factor (SCF), a cytokine synthesized by fibroblasts and other cells (80), which activates a

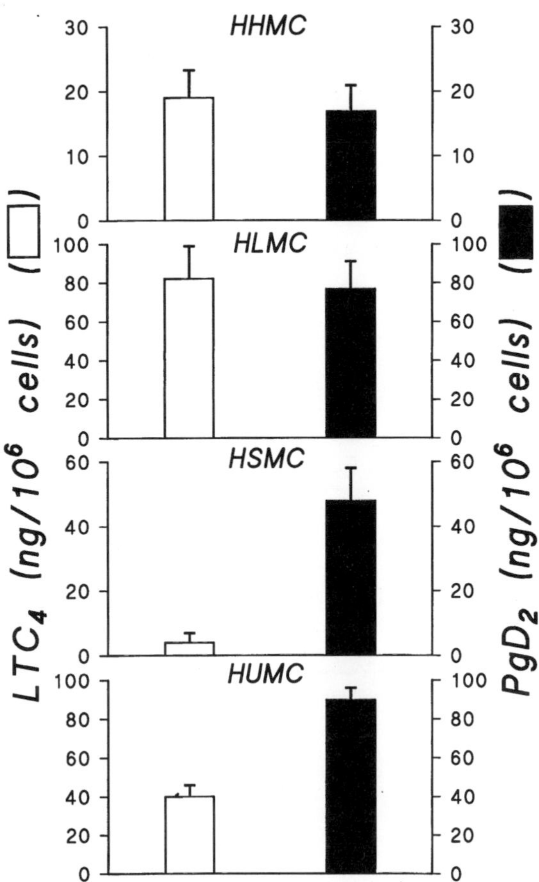

Fig. 6 Comparison of the effects of maximal stimulation with anti-IgE (3 μg ml^{-1}) on LTC_4 and PGD_2 release from mast cells isolated from human heart (HHMC), lung parenchyma (HLMC), skin tissue (HSMC) and uterine tissue (HUMC). Error bar indicates the mean ± SEM of ten experiments.

membrane receptor on HSMC and HLMC, inducing histamine release (81, 82).

Table IV summarizes the immunological stimuli that can activate mast cells isolated from human heart, lung and skin tissue.

A variety of non-immunological stimuli including the calcium ionophore A23187, compound 48/80, maitotoxin, protamine and several contrast media can activate HHMC *in vitro* (64, 83). Some are clinically relevant because they might explain certain of the adverse effects when these compounds are used for diagnostic or therapeutic purposes. For example, protamine, widely used to neutralize heparin, can induce histamine release from HHMC (84), and radiocontrast media, injected into the coronary arteries for diagnostic purposes, can activate HHMC *in vitro* (40). The presence of HHMC in coronary blood vessels and in human coronary atheroma (35, 36) suggests that high intracoronary doses of contrast media can induce mast cell activation and the *in vivo* release of vasoactive mediators. This may explain some of the cardiac effects of these agents, particularly in patients with underlying cardiovascular diseases (62, 85, 86).

Table V summarizes the wide variety of responses of mast cells isolated from different human tissues in response to non-immunological stimuli.

TABLE IV
Immunological Stimuli that Activate Mast Cells Isolated from Human Heart Tissue (HHMC), Lung Parenchyma (HLMC) and Skin (HSMC)

	HHMC	HLMC	HSMC
Antigen	+	+	+
Anti-IgE	+	+	+
Anti-$Fc_{\varepsilon}RI$	+	+	+
Protein Fv	+	+	+
Protein L	+	+	+
C5a	+	–	+
SCF	+	+	+
ECP	+	–	–
MBP	+	–	–

TABLE V
Non-immunological Stimuli that Activate Mast Cells Isolated from Human Heart Tissue (HHMC), Lung Parenchyma (HLMC) and Skin (HSMC)

	HHMC	HLMC	HSMC
A23187	+	+	+
48/80	+	–	+
Maitotoxin	+	+	+
Substance P	–	–	+
Protamine	+	–	+
Morphine	–	–	+
Radiocontrast Media	+	+	–

MAST CELLS AND REGULATION OF COAGULATION AND FIBRINOGENOLYSIS

Mast cells are located in strategic opposition to vascular cells and blood vessels. They contain and release a variety of enzymes and mediators to regulate coagulation and fibrino(geno)lysis at different levels. The involvement of cardiac mast cells in thrombus formation is suggested by the increase in their density in auricular thrombosis (87). Moreover, mast cell deficient W/W^{+} mice show an increased susceptibility to thrombogenic stimuli (88).

Mast cells are an important source of the anticoagulant heparin, and heparin-like anticoagulants may be increased in asthmatics (89) and in patients with mastocytosis (90). Mast cell chymase can cleave thrombin, a central player in the coagulation cascade (91). In addition, tryptase has anticoagulant activities, degrading fibrinogen (92) and high molecular weight kininogen (93).

Valent and colleagues demonstrated that HLMC constitutively express and release tissue-type plasminogen activator (tPA) without producing plasminogen activator inhibitors (PAI) (94). tPA activates the pro-enzyme plasminogen to form plasmin, a key enzyme in fibrinolysis. Although a number of different mast cells express and release tPA, the expression and release of tPA protein is apparently constitutive. The mast cells' potent pro-fibrinolytic effect is apparently due also to their lack of inhibitors (PAI). Valent's group also detected tPA in HHMC. The biological relevance of mast cell tPA is not yet known.

HHMC IN SYSTEMIC AND CARDIAC ANAPHYLAXIS AND IN ANAPHYLACTOID REACTIONS

Levi and collaborators provided compelling evidence that the heart is directly involved in anaphylaxis in the guinea pig (73, 95) through the release of chemical mediators from cardiac mast cells (96). We have shown that human heart tissue fragments immunologically challenged *in vitro* can release mediators from mast cells (32, 97). Cardiac involvement has also been documented in human anaphylaxis (78, 98, 99) and attributed to mediators originating from the lung and reaching the heart. However, the local release of vasoactive mediators by cardiac mast cells themselves may contribute to anaphylactic and anaphylactoid reactions. For instance, complement activation and C5a formation have been documented during systemic anaphylaxis in man (78). We have demonstrated that HHMC possess $Fc_{\varepsilon}RI$ and IgE bound to membrane surface and C5a receptors. Therefore, it is likely that IgE- and C5a-mediated activation of these cells is involved in systemic and cardiac anaphylaxis in man.

HHMC and their mediators are also very likely involved in anaphylactoid reactions involving the heart. HHMC can be directly activated by several agents used intravenously for therapeutic (general anaesthetics, protamine, etc.) or diagnostic purposes (radiocontrast media, etc.) that may cause anaphylactoid reactions (83, 100). Therefore, the release of vasoactive mediators from perivascular and interstitial cardiac mast cells might play a major part in anaphylactoid reactions to these agents.

HHMC AND ATHEROSCLEROSIS

Immunological and inflammatory mechanisms appear to play a role in the progression of atherosclerosis (101). Mast cells are present in the adventitia of all blood vessels (102, 103) and can be identified in the intima and subintima too. Mast cells and their mediators may well be involved in various stages of atherosclerosis.

Early studies suggested that mast cell density is proportional to the severity of atheroma (104). Clustered infiltration of mast cells has been described in the adventitia of the coronary arteries of patients with unstable angina who suffered coronary death (30, 105). Coronary arteries of patients with sudden cardiac death contain significantly higher concentrations of histamine than those of non-cardiac patients (106); the coronary vessels of cardiac patients are also hyper-responsive to histamine (86, 106). Therefore, vasoactive substances secreted by vascular mast cells (e.g. histamine, PGD_2, or cysteinyl leukotrienes) may have a role in the pathogenesis of coronary spasm.

There is increasing evidence that mast cells play some role in the early and late stages of atherogenesis. A link between mast cells and atherosclerosis was first suggested in

1953 by Costantinides (107). Mast cells and macrophages co-exist in the intima and adventitia where low-density lipoproteins (LDL) are oxidized in atherosclerosis (108). Elegant studies by Kovanen and collaborators (35, 36, 109–111) showed that stimulation of rat mast cells in the presence of LDL leads to their modification, with subsequent LDL uptake by macrophages (58, 112). The stimulated mast cell seems critical for LDL modification and macrophage uptake. The apolipoprotein B of LDL (apoB-100) binds to heparin released by mast cells and the neutral proteases, chymase and carboxypeptidase A, degrade apoB-100, whereupon the LDL particles on the remaining granule surface fuse. The granule remnants, laden with fused LDL particles, are phagocytosed by the macrophages with subsequent cholesterol ester accumulation and foam cell formation in the subendothelial space of the arterial intima. Heparin can form large, insoluble complexes with LDL, which are then taken up by macrophages through scavenger receptor-mediated phagocytosis, resulting in cholesterol accumulation (113).

Experimental studies have reported that mast cells also affect LDL oxidation, rendering it resistant to copper ion-mediated oxidation (114) through a mechanism of proteolytic degradation of LDL by chymase, leading to release of copper-containing apoB-100 peptide from the LDL. This allows the small peptides released to bind free copper ions with formation of redox-inactive copper ion complexes (114). Histamine released from mast cells can chelate copper ions, preventing oxidation of LDL. If these findings, mainly obtained in rodents, prove to be extendable to humans, activated mast cells may prevent cell-mediated oxidation of LDL, implying a potential preventive role for these cells in atherosclerosis.

Mast cell density appears to be high in human coronary atheroma, particularly in the shoulder regions (35, 36). Moreover, there were more partially degranulated mast cells than in the normal coronary intima. Finally, there were signs of mast cell degranulation (i.e. tryptase-positive granules) in the extracellular microenvironment of mast cells (35, 36, 110, 111). The same group of investigators also demonstrated the presence of mast cells with tumour necrosis factor-α (TNF-α)-containing secretory granules, particularly in the shoulder region of human coronary atheromas (115). The latter finding has now been extended by an experimental study showing that cardiac mast cells degranulate after myocardial ischaemia, releasing histamine and TNF-α (116). Mast cells might thus play an active role in inflammatory reactions of these rupture-prone areas of atheromas.

The following question then arises: under what conditions would mast cell activation occur in the arterial intima? Mast cells and basophils could be activated by any stimulus originating in the peripheral blood such as IgE-mediated mechanisms (8), or VLDL (117), or generated in the microenvironment by activated T lymphocytes (118), macrophages (119), fibroblasts (81, 82), and eosinophils (120) which participate in chronic inflammatory reactions. Thus, in active inflammatory areas of the intima, such as the shoulder region of coronary atheromas, containing T lymphocytes, macrophages, fibroblasts and eosinophils, the mast cells can be stimulated to degranulate and release mediators that participate in the various stages of atherogenesis.

Collagen comprises 40–60% of the protein and approximately 30% of the weight of human fibre atherosclerotic plaques. Its deposition in the arterial intima is responsible for the occlusive and irreversible nature of coronary arterial disease (121). Tryptase (122–126), cysteinyl leukotrienes (59), and histamine (127) are mitogens and co-mitogens for fibroblasts and can stimulate collagen accumulation. Their release from intimal and adventitial mast cells might therefore also be relevant to atherosclerosis through this mechanism.

Despite all these fascinating observations, additional studies are still needed to define the precise role of mast cells, whether atherogenic or anti-atherogenic, at different stages of the disease.

HHMC IN PATIENTS WITH MYOCARDITIS AND HYPER-EOSINOPHILIA

Mast cells have been described in experimental *Trypanosoma cruzi*-induced acute and chronic myocarditis (128, 129). Increased numbers of mast cells in close contact with myocytes were observed in endomyocardial biopsy specimens from patients with Chagas cardiomyopathy (130), and increased levels of histamine have been found in the heart of patients with chronic Chagas cardiomyopathy (45). Interestingly, the number of mast cells is markedly increased in the hearts of patients with eosinophilic myocarditis (29, 131). This is particularly relevant because eosinophil infiltration and the deposition of eosinophil granule protein (ECP and MBP) has been documented in these patients (132, 133).

We examined the effects of eosinophil granule proteins on the activation of HHMC and found that eosinophil cationic protein (ECP) and major basic protein (MBP), but not eosinophil peroxidase (EPO) and eosinophil-derived neurotoxin (EDN) caused rapid release of histamine and tryptase from HHMC. ECP and MBP also induced the arachidonic acid metabolism through both the cyclo-oxygenase and 5-lipoxygenase pathways (120, 134). Interestingly, ECP, MBP, EPO and EDN do not activate human mast cells isolated from skin (134, 135) and lung tissues (134). Our observation that ECP and MBP activate HHMC to release pro-inflammatory and vasoactive mediators suggests that eosinophil–mast cell interactions might be clinically relevant in patients with hyper-eosinophilia.

HHMC AND ANGIOGENESIS IN THE HEART

Stimulation of vascular growth may play a role in the hypertrophic heart, in wound healing and tumour angiogenesis. Mast cell density is increased in highly vascularized tumours (136) and, when tumours are implanted in mast cell-deficient mice (W/W^v), angiogenesis and subsequent tumour growth are retarded (137). Chronic inflammatory reactions are associated with neovascularization. It has been suggested that mast cell-derived mediators stimulate capillary endothelial cell migration. Mast cell granules and histamine induce endothelial proliferation (138); heparin stimulates migration of capillary endothelial cells (139) and protamine inhibits capillary formation in growing rat hearts (140) and several models of angiogenesis (141).

TNF-α, a mast cell-derived cytokine (1, 142), is a potent angiogenic molecule (143). Canine cardiac mast cells degranulate after myocardial ischaemia, releasing histamine and TNF-α (116) and mast cells in human coronary atheromas store and release TNF-α (115). These observations suggest that mast cells and their mediators present in human heart tissue play a role in angiogenesis, a process that might be important in the hypertrophic heart and after myocardial infarction.

HHMC AND MYOCARDIAL ISCHAEMIA

Several groups have reported that the density of mast cells in and around the coronary arteries is increased in patients with ischaemic heart disease (29, 30, 34). Low doses of histamine and cysteinyl leukotrienes secreted by mast cells may be important in causing coronary spasm in some patients with unstable angina (60–62). Another intriguing observation is that IgE levels are increased in patients with myocardial infarction and with unstable angina (144–146). It has consequently been suggested that IgE-mediated events may play a role in these conditions. Other potentially relevant stimuli are the complement-derived peptides. Complement activation and C5a-mediated events come into play during ischaemic myocardial injury. *In vitro* activation of complement by isolated human heart subcellular membranes has been described (147). *In vivo* C5a localizes in experimentally infarcted myocardium (70, 148, 149) and C3a and C5a cause cardiac dysfunction (73–75). *In vivo* activation of the complement system occurs in patients with myocardial infarction (150, 151), and the terminal C5–9 complement complex is deposited in infarcted areas of the human myocardium (69). Depletion or inhibition of complement reduces tissue injury in ischaemic infarction (70–72). The complement system is therefore an important mediator of the acute inflammatory response following myocardial ischaemia. We have demonstrated that C5a is a potent stimulus for the activation of HHMC and the release of vasoactive mediators, suggesting that complement activation and C5a formation might be important in ischaemic myocardial injury in man.

Recent findings indicate that mast cells influence myocardial ischaemia through the production of cytokines. In a canine model of myocardial ischaemia TNF-α but not interleukin IL-1β was constitutively expressed in mast cells (116, 152). Myocardial ischaemia was followed by histamine release and immunohistochemical evidence of mast cell degranulation and TNF-α secretion. TNF-α released from cardiac mast cells appeared to play a major role in inducing IL-6 in infiltrating mononuclear cells. These findings suggest that mast cells are critical in experimental myocardial injury. TNF-α-containing mast cells have been detected in the shoulder region of human coronary atheromas (115), the most frequent site of rupture (35, 36). In conclusion, recent studies indicate that some cardiac mast cells contain and release TNF-α, which may have pro-inflammatory effects in myocardial ischaemia.

HHMC IN IDIOPATHIC AND ISCHAEMIC CARDIOMYOPATHY

Fibrosis is characterized by an increase in fibrous components of the extracellular matrix and is a hallmark of cardiomyopathy (153). It is not known which cells are responsible for early changes in the fibrotic cascade in cardiomyopathy. Mast cells are involved in many types of inflammation and repair processes and are found in increased numbers in fibrotic tissue in skin (154), intestine (155) and lung (156, 157).

An association between increased mast cell density and endomyocardial fibrosis in Africans was first suggested by Fernex in 1968 (28). A role for HHMC has also been suggested in patients with hyper-eosinophilic syndrome (120, 158) and in dilated cardiomyopathy secondary to systemic sclerosis (159). We have compared the density of HHMC of patients with cardiomyopathy and subjects who died of non-cardiovascular causes (160). The heart tissue (left ventricle) from patients with dilated cardiomyopathy contained on average four times more mast cells than controls. There was a significant correlation between the histamine content of the heart and cardiac mast cell density, suggesting that these cells are the main, possibly the only, source of histamine in human heart tissue (5). Therefore, the histamine content of human hearts with cardiomyopathy was significantly higher than in normal controls.

We also examined the histamine secretion and release per gram of wet tissue from HHMC in both groups of subjects, induced by immunological (anti-IgE, anti-$Fc_{\varepsilon}RI$, SCF and C5a) and non-immunological stimuli (A23187 and 48/80). The release was always similar in the two groups. However, the absolute release of histamine was significantly higher in HHMC from cardiomyopathy patients than in those from controls (5).

Histamine, cysteinyl leukotrienes, and tryptase are mitogens and co-mitogens for human fibroblasts (59, 122–127) and stimulate collagen synthesis (126). SCF, a major product of human fibroblasts, is a fundamental growth factor for human mast cells (80, 161, 162). These findings suggest the possibility of a positive feedback between mast cells and fibroblasts in the fibrotic cascade leading to certain forms of cardiomyopathy.

Our findings of increased release of mediators in patients with cardiomyopathy might also have clinical relevance. Histamine is arrhythmogenic (85, 96) and causes coronary vasoconstriction in some patients with unstable angina (86). *In vivo* administration of LTC_4 and LTD_4 can increase coronary vascular resistance in man (60–62). Thus, the high cardiac mast cell density and the increased release of their vasoactive mediators in cardiomyopathy might contribute to certain cardiovascular features in these patients.

PERSPECTIVE AND CONCLUSIONS

Mast cells are present in normal and even more abundant in diseased human heart tissue. Within heart tissue mast cells lie between myocytes and in close contact with blood vessels. These cells are also found in the coronary adventitia and in the shoulder region of coronary atheromas. The density of HHMC is higher in some patients with myocarditis and idiopathic and ischaemic cardiomyopathy than in accident victims without cardiovascular diseases. More importantly, in some of these conditions there is *in situ* evidence of mast cell activation.

Immunological and ultrastructural characterization of HHMC *in situ* has provided novel information. Using the immunogold technique we showed that HHMC contain tryptase and chymase. Activation of isolated HHMC caused the release of these proteolytic enzymes that act on the angiotensin I/II and the big endothelin 1/endothelin 1 systems.

HHMC may contain several cytokines. Immunoelectron microscopy showed that SCF^{1-166} resides within their cytoplasmic secretory granules and can be immunologically released (163, 164). Interestingly, SCF^{1-166} can be cleaved to SCF^{1-159} and SCF^{1-149} by mast cell chymase, indicating a novel autocrine loop. TNF-α is found in secretory granules of mast cells in the shoulder region of human coronary atheromas (115). Thus HHMC express and secrete several cytokines that might play a prime role in triggering and maintaining inflammatory processes in the human heart.

We have developed an original technique to isolate and purify HHMC that can then be studied *in vitro*. Mast cells obtained with this procedure are viable and in response to immunological and non-immunological stimuli release preformed (histamine, tryptase and chymase) and newly generated mediators (PGD_2 and LTC_4). HHMC differ from those in other anatomical districts in that they are activated by specific immunological and non-immunological stimuli and with respect to the arachidonic acid metabolism, suggesting that the local microenvironment influences their phenotypic and biochemical characteristics.

Observations by other groups of investigators as well as our own findings suggest that HHMC have complex and significant roles in different pathophysiological conditions involving the cardiovascular system. Direct activation of HHMC by therapeutic (general anaesthetics, protamine) and diagnostic (radiocontrast media) substances injected intravenously explains some of the anaphylactoid reactions caused by these agents. HHMC possess $Fc_{\varepsilon}RI$ and IgE bound to the surface and C5a receptors, which could explain the involvement of these cells in systemic and cardiac anaphylaxis. Cardiac mast cells and those in human coronary arteries also play a role in the early and late stages of atherogenesis, during ischaemic myocardial injury and in different stages of dilated cardiomyopathies. Finally, HHMC and their mediators seem to play a prime role in the regulation of coagulation and fibrinogenolysis and in angiogenesis in the heart.

The immunological characterization of HHMC is in its infancy, but is already progressing rapidly. The *in vitro* isolation of HHMC and their *in situ* immunohistochemical and ultrastructural characterization will be of paramount importance to identify additional mediators synthesized and released, stimuli relevant to human pathophysiology and pharmacological agents that selectively affect the activation of these cells and their mediators. Drugs specifically acting on HHMC or their mediators may prove useful for patients suffering from different cardiovascular diseases.

ACKNOWLEDGEMENTS

This work was supported by grants from the National Research Council (CNR) (Targeted Project Biotechnology No. 99.00216.PF31 and No. 99.00401.PF49) and the MURST (Rome, Italy).

REFERENCES

1. Walsh, L. J., Trinchieri, G., Waldorf, H. A., Whitaker, D. and Murphy, G. F. Human dermal mast cells contain and release tumor necrosis factor α, which induces endothelial leukocyte adhesion molecule 1. *Proc. Natl. Acad. Sci. USA* **88**:4220–4224, 1991.
2. Bradding, P., Feather, I. H., Howarth, P. H., Mueller, R., Roberts, J. A., Britten, K., Bews, J. P. A., Hunt, T. C., Okayama, Y., Heusser, C. H., Bullock, G. R., Church, M. K. and Holgate, S. T. Interleukin 4 is localized to and released by human mast cells. *J. Exp. Med.* **176**:1381–1386, 1992.
3. Möller, A., Lippert, U., Lessmann, D., Kolde, G., Hamann, K., Welker, P., Schadendorf, D., Rosenbach, T., Luger, T. and Czarnetzki, B.M. Human mast cells produce IL-8. *J. Immunol.* **151**:3261–3266, 1993.
4. Marone, G., de Crescenzo, G., Patella, V. and Genovese A. Human cardiac mast cells and their role in severe allergic reactions. In: *Asthma and Allergic Diseases* (Marone, G., Austen, K. F., Holgate, S. T., Kay A. B. and Lichtenstein, L. M., eds), pp. 237–257. Academic Press, London, 1998.
5. Patella, V., Marinò, I., Arbustini, E., Lampärter-Schummert, B., Verga, L., Adt, M. and Marone G. Stem cell factor in mast cells and increased cardiac mast cell density in idiopathic and ischemic cardiomyopathy. *Circulation* **97**:971–978, 1998.
6. Rumsaeng, V., Cruikshank, W. W., Foster, B., Prussin, C., Kirshenbaum, A. S., Davis, T. A, Kornfeld, H., Center, D. M. and Metcalfe, D. D. Human mast cells produce the $CD4^+$ T lymphocyte chemoattractant factor, IL-16. *J. Immunol.* **159**:2904–2910, 1997.

7. Yano, K., Yamaguchi, M., de Mora, F., Lantz, C. S., Butterfield, J. H., Costa, J. J. and Galli, S. J. Production of macrophage inflammatory protein-1α by human mast cells: increased anti-IgE-dependent secretion after IgE-dependent enhancement of mast cell IgE-binding ability. *Lab. Invest.* **77**:185–193, 1997.
8. Marone, G., Casolaro, V., Cirillo, R., Stellato, C. and Genovese, A. Pathophysiology of human basophils and mast cells in allergic disorders. *Clin. Immunol. Immunopathol.* **50**:S24–S40, 1989.
9. Marone, G., ed. *Human Basophils and Mast Cells in Health and Disease. Biological Aspects.* Karger, Basel, 1995.
10. Marone, G., ed. *Human Basophils and Mast Cells in Health and Disease. Clinical Aspects.* Karger, Basel, 1995.
11. Galli, S. J. New concepts about the mast cell. *N. Engl. J. Med.* **328**:257–265, 1993.
12. Marone, G., Tamburini, M., Giudizi, M. G., Biagiotti, R., Almerigogna, F. and Romagnani, S. Mechanism of activation of human basophils by *Staphylococcus aureus* Cowan 1. *Infect. Immun.* **55**:803–809, 1987.
13. Echtenacher, B., Männel, D. N. and Hültner, L. Critical protective role of mast cells in a model of acute septic peritonitis. *Nature* **381**:75–77, 1996.
14. Malaviya, R., Ikeda, T., Ross, E. and Abraham, S. N. Mast cell modulation of neutrophil influx and bacterial clearance at sites of infection through TNF-α. *Nature* **381**:77–80, 1996.
15. Prodeus, A. P., Zhou, X., Maurer, M., Galli, S. J. and Carroll, M. C. Impaired mast cell-dependent natural immunity in complement C3-deficient mice. *Nature* **390**:172–175, 1997.
16. Patella, V., Casolaro, V., Björck, L. and Marone, G. Protein L: a bacterial Ig-binding protein that activates human basophils and mast cells. *J. Immunol.* **145**:3054–3061, 1990.
17. Patella, V., Bouvet, J.-P. and Marone, G. Protein Fv produced during viral hepatitis is a novel activator of human basophils and mast cells. *J. Immunol.* **151**:5685–5698, 1993.
18. Patella, V., Giuliano, A., Bouvet, J.-P. and Marone, G. Endogenous superallergen protein Fv induces IL-4 secretion from human $Fc_{\varepsilon}RI^{+}$ cells through interaction with the V_H3 region of IgE. *J. Immunol.* **161**:5647–5655, 1998.
19. Patella, V., Giuliano, A., Florio, G., Bouvet, J.-P. and Marone, G. Endogenous superallergen protein Fv interacts with the V_H3 region of IgE to induce cytokine secretion from human basophils. *Int. Arch. Allergy Immunol.* **118**:197–199, 1999.
20. Marone, G. Asthma: recent advances. *Immunol. Today* **19**:5–9, 1998.
21. de Paulis, A., Cirillo, R., Ciccarelli, A., de Crescenzo, G., Oriente, A. and Marone, G. Characterization of the anti-inflammatory effect of FK-506 on human mast cells. *J. Immunol.* **147**:4278–4285, 1991.
22. de Paulis, A., Valentini, G., Spadaro, G., Lupoli, S., Tirri, G. and Marone, G. Human basophil releasability. VIII. Increased basophil releasability in patients with scleroderma. *Arthritis Rheum.* **34**:1289–1296, 1991.
23. de Paulis, A., Ciccarelli, A., Marinò, I., de Crescenzo, G., Marinò, D. and Marone, G. Human synovial mast cells. II. Heterogeneity of the pharmacologic effects of antiinflammatory and immunosuppressive drugs. *Arthritis Rheum.* **40**:469–478, 1997.
24. Marone, G. Mast cells in rheumatic disorders: mastermind or workhorse? *Clin. Exp. Rheumatol.* **16**:245–249, 1998.
25. Ghanem, N. S., Assem, E. S. K., Leung, K. B. P. and Pearce, F. L. Cardiac and renal mast cells: morphology, distribution, fixation and staining properties in the guinea pig and preliminary comparison with human. *Agents Actions* **23**:223–226, 1988.
26. McGovern, V. J. Mast cells and their relationship to endothelial surfaces. *J. Pathol. Bacteriol.* **71**:1–6, 1956.
27. Pepler, W. J. and Meyer, B. J. Mast cells in the coronary arteries. A comparative study of South African Bantu and European hearts. *Arch. Pathol.* **71**:103–107, 1961.
28. Fernex, M. *The Mast-Cell System. Its Relationship to Atherosclerosis, Fibrosis and Eosinophils.* Williams & Wilkins, Baltimore, 1968.
29. Estensen, R. D. Eosinophilic myocarditis: a role for mast cells? *Arch. Pathol. Lab. Med.* **108**:358–359, 1984.
30. Forman, M. B., Oates, J. A., Robertson, D., Robertson, R. M., Roberts, L. J. and Virmani, R. Increased adventitial mast cells in a patient with coronary spasm. *N. Engl. J. Med.* **313**:1138–1141, 1985.
31. Dvorak, A. M. Mast-cell degranulation in human hearts. *N. Engl. J. Med.* **315**:969–970, 1986.
32. Marone, G., Triggiani, M., Cirillo, R., Giacummo, A., Hammarström, S. and Condorelli, M. IgE-mediated activation of human heart *in vitro*. *Agents Actions* **18**:194–196, 1986.

33. Assem, E. S. K. and Ghanem, N. S. Demonstration of IgE-sensitized mast cells in human heart and kidney. *Int. Arch. Allergy Appl. Immunol.* **87**:101–104, 1988.
34. Pouchlev, A., Youroukova, Z. and Kiprov, D. A study of changes in the number of mast cells in the human arterial wall during the stages of development of atherosclerosis. *J. Atheroscler. Res.* **6**:324–351, 1966.
35. Kaartinen, M., Penttilä, A. and Kovanen, P. T. Mast cells of two types differing in neutral protease composition in the human aortic intima. Demonstration of tryptase- and tryptase/chymase-containing mast cells in normal intimas, fatty streaks and the shoulder region of atheromas. *Arterioscler. Thromb.* **14**:966–972, 1994.
36. Kaartinen, M., Penttilä, A. and Kovanen, P. T. Accumulation of activated mast cells in the shoulder region of human coronary atheroma, the predilection site of atheromatous rupture. *Circulation* **90**:1669–1678, 1994.
37. Patella, V., Genovese, A. and Marone, G. What are human mast cells for? In: *Human Basophils and Mast Cells. Clinical Aspects* (Marone, G., ed.), pp. 171–186. Karger, Basel, 1995.
38. Patella, V., Marinò, I., Lampärter, B., Arbustini, E., Adt, M. and Marone, G. Human heart mast cells. Isolation, purification, ultrastructure and immunologic characterization. *J. Immunol.* **154**:2855–2865, 1995.
39. Patella, V., de Crescenzo, G., Ciccarelli, A., Marinò, I., Adt, M. and Marone, G. Human heart mast cells: a definitive case of mast cell heterogeneity. *Int. Arch. Allergy Immunol.* **106**:386–393, 1995.
40. Patella, V., Marinò, I., Lampärter, B., Genovese, A., Adt, M. and Marone, G. Immunologic and non-immunologic release of histamine and tryptase from human heart mast cells. *Inflamm. Res.* **44(S-I)**:S22–S23, 1995.
41. Dvorak, A. M., Schulman, E. S., Peters, S. P., MacGlashan, D. W. Jr, Newball, H. H., Schleimer, R. P. and Lichtenstein, L. M. Immunoglobulin E-mediated degranulation of isolated human lung mast cells. *Lab. Invest.* **53**:45–56, 1985.
42. Massey, W. A., Guo, C.-B., Dvorak, A. M., Hubbard, W. C., Bhagavan, B. S., Cohan, V. L., Warner, J. A., Kagey-Sobotka, A. and Lichtenstein, L. M. Human uterine mast cells. Isolation, purification, characterization, ultrastructure, and pharmacology. *J. Immunol.* **147**:1621–1627, 1991.
43. Marone, G., Patella, V., de Crescenzo, G. and Adt, M. Human heart mast cells in anaphylaxis and cardiovascular disease. *Int. Arch. Allergy Immunol.* **107**:72–75, 1995.
44. Marone, G., de Crescenzo, G., Adt, M., Patella, V., Arbustini, E. and Genovese, A. Immunological characterization and functional importance of human heart mast cells. *Immunopharmacology* **31**:1–18, 1995.
45. Correa-Araujo, R., Oliveira, J. S. M. and Ricciardi Cruz, A. Cardiac levels of norepinephrine, dopamine, serotonin and histamine in Chagas' disease. *Int. J. Cardiol.* **31**:329–336, 1991.
46. Irani, A. A., Schechter, N. M., Craig, S. S., DeBlois, G. and Schwartz, L. B. Two types of human mast cells that have distinct neutral protease compositions. *Proc. Natl. Acad. Sci. USA* **83**:4464–4468, 1986.
47. Schwartz, L. B. Mast cells and asthma. In: *Asthma and Allergic Diseases. Physiology, Immunopharmacology and Treatment* (Marone, G., Austen, F. K., Holgate, S. T., Kay A. B. and Lichtenstein, L. M., eds), pp. 223–235. Academic Press, London, 1998.
48. Schwartz, L. B., Lewis, R. A., Seldin, D. and Austen, K. F. Acid hydrolases and tryptase from secretory granules of dispersed human lung mast cells. *J. Immunol.* **126**:1290–1294, 1981.
49. Schwartz, L. B., Riedel, C., Caulfield, J. P., Wasserman, S. I. and Austen, K. F. Cell association of complex of chymase, heparin proteoglycan and protein after degranulation by rat mast cells. *J. Immunol.* **126**:2071–2078, 1981.
50. Schechter, N. M., Irani, A.-M., Sprows, J. L., Abenerthy, J., Wintroub, B. and Schwartz, L. B. Identification of a cathepsin G-like proteinase in the MC_{TC} type of human mast cell. *J. Immunol.* **145**:2652–2661, 1990.
51. Urata, H., Kinoshita, A., Misono, K. S., Bumpus, F. M. and Husain, A. Identification of highly specific chymase as the major angiotensin II-forming enzyme in the human heart. *J. Biol. Chem.* **265**:22348–22357, 1990.
52. Zeiher, A. M., Ihling, C., Pistorius, K., Schächinger, V. and Schaefer, H.-E. Increased tissue endothelin immunoreactivity in atherosclerotic lesions associated with acute coronary syndromes. *Lancet* **344**:1405–1406, 1994.
53. Metcalfe, D. D., Bland, C. E. and Wasserman, S. I. Biochemical and functional characterization of proteoglycans isolated from basophils of patients with chronic myelogenous leukemia. *J. Immunol.* **132**:1943–1950, 1984.

54. Thompson, H. L., Schulman, E. S. and Metcalfe, D. D. Identification of chondroitin sulfate E in human lung mast cells. *J. Immunol.* **140**:2708–2713, 1988.
55. Eliakim, R., Gilead, L., Ligmusky, M., Okon, E., Raachmilewitz, D. and Razin, E. Histamine and chondroitin sulphate E proteoglycans released by cultured human colonic mucosa: indication for possible presence of E mast cells. *Proc. Natl. Acad. Sci. USA* **83**:461–464, 1986.
56. Metcalfe, D. D., Lewis, R. A., Silbert, J. E., Rosenberg, R. D., Wasserman, S. I. and Austen, K. F. Isolation and characterization of heparin from human lung. *J. Clin. Invest.* **64**:1537–1543, 1979.
57. Hattori, Y. and Levi, R. Effect of PGD_2 on cardiac contractility: a negative inotropism secondary to coronary vasoconstriction conceals a primary positive inotropic action. *J. Pharmacol. Exp. Ther.* **237**:719–724, 1986.
58. Kokkonen, J. O. and Kovanen, P. T. Stimulation of mast cells leads to cholesterol accumulation in macrophages *in vitro* by a mast cell granule-mediated uptake of low density lipoprotein. *Proc. Natl. Acad. Sci. USA* **84**:2287–2291, 1987.
59. Baud, L., Perez, J., Denis, M. and Ardaillou, R. Modulation of fibroblast proliferation by sulfidopeptide leukotrienes: effect of indomethacin. *J. Immunol.* **138**:1190–1195, 1987.
60. Marone, G., Giordano, A., Cirillo, R., Triggiani, M. and Vigorito, C. Cardiovascular and metabolic effects of peptide leukotrienes in man. *Ann. N. Y. Acad. Sci.* **524**:321–333, 1988.
61. Vigorito, C., Marone, G. and Condorelli, M. Cardiovascular effects of histamine and sulfidopeptide leukotrienes. In: *Human Inflammatory Disease. Clinical Immunology. Vol. I* (Marone, G., Lichtenstein, L. M., Condorelli, M. and Fauci, A. S., eds), pp. 107–123. B. C. Decker, Toronto, 1988.
62. Vigorito, C., Giordano, A., Cirillo, R., Genovese, A., Rengo, F. and Marone, G. Metabolic and hemodynamic effects of peptide leukotriene C_4 and D_4 in man. *Int. J. Clin. Lab. Res.* **27**:178–184, 1997.
63. Feinmark, S. J. and Cannon, P. J. Endothelial cell leukotriene C_4 synthesis results from intercellular transfer of leukotriene A_4 synthesized by polymorphonuclear leukocytes. *J. Biol. Chem.* **261**:16466–16472, 1986.
64. Stellato, C., Cirillo, R., de Paulis, A., Casolaro, V., Patella, V., Mastronardi, P., Mazzarella, B. and Marone, G. Human basophil/mast cell releasability. IX. Heterogeneity of the effects of opioids on mediator release. *Anesthesiology* **77**:932–940, 1992.
65. Stellato, C., de Paulis, A., Ciccarelli, A., Cirillo, R., Patella, V., Casolaro, V. and Marone, G. Anti-inflammatory effect of cyclosporin A on human skin mast cells. *J. Invest. Dermatol.* **98**:800–804, 1992.
66. Marone, G., Casolaro, V., Paganelli, R. and Quinti, I. IgG anti-IgE from atopic dermatitis induces mediator release from basophils and mast cells. *J. Invest. Dermatol.* **93**:246–252, 1989.
67. Marone, G., Spadaro, G., Palumbo, C. and Condorelli, G. The anti-IgE/anti-FcεRIα autoantibody network in allergic and autoimmune diseases. *Clin. Exp. Allergy* **29**:17–27, 1999.
68. Hide, M., Francis, D. M., Grattan, C. E. H., Hakimi, J., Kochan, J. P. and Greaves, M. W. Autoantibodies against the high-affinity IgE receptor as a cause of histamine release in chronic urticaria. *N. Engl. J. Med.* **328**:1599–1604, 1993.
69. Schafer, H., Mathey, D., Hugo, F. and Bhakdi, S. Deposition of the terminal C5b–9 complement complex in infarcted areas of human myocardium. *J. Immunol.* **137**:1945–1949, 1986.
70. Crawford, M. H., Grover, F. L., Kolb, W. P., McMahan, C. A., O'Rourke, R. A., McManus, L. M. and Pinckard, R. N. Complement and neutrophil activation in the pathogenesis of ischemic myocardial injury. *Circulation* **78**:1449–1458, 1988.
71. Weisman, H. F., Bartow, T., Leppo, M. K., Marsh, H.C . Jr., Carson, G. R., Concino, M. F., Boyle, M. P., Roux, K. H., Weisfeldt, M. L. and Fearon, D. T. Soluble human complement receptor type 1: *in vivo* inhibitor of complement suppressing post-ischemic myocardial inflammation and necrosis. *Science* **249**:146–151, 1990.
72. Shandelya, S. M. L., Kuppusamy, P., Herskowitz, A., Weisfeldt, M. L. and Zweier, L. Soluble complement receptor type 1 inhibits the complement pathway and prevents contractile failure in the postischemic heart. Evidence that complement activation is required for neutrophil-mediated reperfusion injury. *Circulation* **88**:2812–2826, 1993.
73. Hachfeld del Balzo, U., Levi, R. and Polley, M. J. Cardiac dysfunction caused by purified human C3a anaphylatoxin. *Proc. Natl. Acad. Sci. USA* **82**:886–890, 1985.
74. del Balzo, U., Polley, M. J. and Levi, R. Activation of the third complement component (C3) and C3a generation in cardiac anaphylaxis: histamine release and associated inotropic and chronotropic effects. *J. Pharmacol. Exp. Ther.* **246**:911–916, 1988.
75. del Balzo, U., Sakuma, I. and Levi, R. Cardiac dysfunction caused by recombinant human C5a

anaphylatoxin: mediation by histamine, adenosine and cyclooxygenase arachidonate metabolites. *J. Pharmacol. Exp. Ther.* **253**:171–179, 1990.

76. Ito, B. R., Engler, R. L. and del Balzo, U. Role of cardiac mast cells in complement C5a-induced myocardial ischemia. *Am. J. Physiol.* **264**:H1346–H1354, 1993.
77. del Balzo, U., Polley, M. J. and Levi, R. Cardiac anaphylaxis. Complement activation as an amplification system. *Circ. Res.* **65**:847–857, 1989.
78. Smith, P. L., Kagey-Sobotka, A., Bleecker, E. R., Traystman, R., Kaplan, A. P., Gralnick, H., Valentine, M. D., Permutt, S. and Lichtenstein, L. M. Physiologic manifestations of human anaphylaxis. *J. Clin. Invest.* **66**:1072–1080, 1980.
79. Schulman, E. S., Post, T. J., Henson, P. M. and Giclas, P. Differential effects of the complement peptides, C5a and C5a des Arg on human basophil and lung mast cell histamine release. *J. Clin. Invest.* **81**:918–923, 1988.
80. Tsai, M., Shih, L. S., Newlands, G. F. J., Takeishi, T., Langley, K. E., Zsebo, K. M., Miller, H. R. P., Geissler, E. N. and Galli, S. J. The rat c-*kit* ligand, stem cell factor, induces the development of connective tissue-type and mucosal mast cells *in vivo*. Analysis by anatomical distribution, histochemistry, and protease phenotype. *J. Exp. Med.* **174**:125–131, 1991.
81. Columbo, M., Horowitz, E. M., Botana, L. M., MacGlashan, D. W. Jr, Bochner, B. S., Gillis, S., Zsebo, K. M., Galli, S. J. and Lichtenstein, L. M. The human recombinant c-*kit* receptor ligand, rhSCF, induces mediator release from human cutaneous mast cells and enhances IgE-dependent mediator release from both skin mast cells and peripheral blood basophils. *J. Immunol.* **149**:599–608, 1992.
82. de Paulis, A., Ciccarelli, A., Cirillo, R., de Crescenzo, G., Columbo, M. and Marone, G. Modulation of human lung mast cell function by the c-*kit* receptor ligand. *Int. Arch. Allergy Immunol.* **99**:326–329, 1992.
83. Stellato, C., de Crescenzo, G., Patella, V., Tatangelo, F., Mastronardi, P., Mazzarella, B. and Marone, G. Human basophil/mast cell releasability. XI. Heterogeneity of the effects of contrast media on mediator release. *J. Allergy Clin. Immunol.* **97**:838–850, 1996.
84. Patella, V., Ciccarelli, A., Lampärter-Schummert, B., de Paulis, A., Adt, M. and Marone, G. Heterogeneous effects of protamine on human mast cells and basophils. *Br. J. Anaesth.* **78**:724–730, 1997.
85. Vigorito, C., Russo, P., Picotti, G. B., Chiariello, M., Poto, S. and Marone, G. Cardiovascular effects of histamine infusion in man. *J. Cardiovasc. Pharmacol.* **5**:531–537, 1983.
86. Vigorito, C., Poto, S., Picotti, G. B., Triggiani, M. and Marone, G. Effect of activation of the H_1 receptor on coronary hemodynamics in man. *Circulation* **73**:1175–1182, 1986.
87. Bankl, H. C., Radaszkiewicz, T., Klappacher, G. W., Glogar, D., Sperr, W. R., Großschmidt, K., Bankl, H., Lechner, K. and Valent, P. Increase and redistribution of cardiac mast cells in auricular thrombosis. Possible role of kit ligand. *Circulation* **91**:275–283, 1995.
88. Kitamura, Y., Taguchi, T., Yokoyama, M., Inoue, M., Yamatodani, A., Asano, H., Koyama, T., Kanamaru, A., Hatanaka, K. and Wershil, B. K. Higher susceptibility of mast cell-deficient W/W^v mutant mice to brain thromboembolism and mortality caused by intravenous injection of India ink. *Am. J. Pathol.* **122**:469–480, 1986.
89. Lasser, E. D., Simon, R. A., Lyon, S. G., Hamblin, A. E. and Stein, R. Heparin-like anticoagulants in asthma. *Allergy* **42**:619–625, 1987.
90. Kauhanen, P., Kovanen, P. T., Reunala, T. and Lassila, R. Effects of skin mast cells on bleeding time and coagulation activation at the site of platelet plug formation. *Thromb. Haemost.* **79**:843–847, 1998.
91. Pejler, G., Söderström, K. and Karlström, A. Inactivation of thrombin by a complex between rat mast-cell protease I and heparin proteoglycan. *Biochem. J.* **299**:507–513, 1994.
92. Schwartz, L. B., Bradford, T. R., Littman, B. H. and Wintroub, B. U. The fibrinogenolytic activity of purified tryptase from human lung mast cells. *J. Immunol.* **135**:2762–2767, 1985.
93. Maier, M., Spragg, J. and Schwartz, L. B. Inactivation of human high molecular weight kininogen by mast cell tryptase. *J. Immunol.* **130**:2352–2356, 1983.
94. Sillaber, C., Baghestanian, M., Bevec, D., Willheim, M., Agis, H., Kapiotis, S., Füreder, W., Bankl, H. C., Kiener, H. P., Speiser, W., Binder, B. R., Lechner, K. and Valent, P. The mast cell as site of tissue-type plasminogen activator expression and fibrinolysis. *J. Immunol.* **162**:1032–1041, 1999.
95. Capurro, N. and Levi, R. The heart as a target organ in systemic allergic reactions: comparison of cardiac anaphylaxis *in vivo* and *in vitro*. *Circ. Res.* **36**:520–525, 1975.
96. Levi, R., Malm, J. R., Bowman, F. O. and Rosen, M. R. The arrhythmogenic actions of histamine on human atrial fibers. *Circ. Res.* **49**:545–550, 1981.

97. Marone, G., Triggiani, M., Cirillo, R., Vigorito, C., Genovese, A., Spampinato, N. and Condorelli, M. Chemical mediators and the human heart. *Prog. Biochem. Pharmacol.* **20**:38–54, 1985.
98. Bernreiter, M. Electrocardiogram of patient in anaphylactic shock. *JAMA* **170**:1628–1630, 1959.
99. Booth, B. H. and Patterson, R. Electrocardiographic changes during human anaphylaxis. *JAMA* **211**:627–631, 1970.
100. Stellato, C. and Marone, G. The role of basophils and mast cells in adverse reactions to drugs used during general anesthesia. In: *Human Basophils and Mast Cells in Health and Disease. Clinical Aspects* (Marone, G., ed.), pp. 108–131. Karger, Basel, 1995.
101. Ross, R. Atherosclerosis: an inflammatory disease. *N. Engl. J. Med.* **340**:115–126, 1999.
102. Hjelmman, G. Über das erste Auftreten der Mastzellen in einigen Geweben und Organen bei Homoembryonen mit Berücksichtigung der Zunahme dieser Zellen während der Embryonalentwicklung. *Soc. Sci. Fenn. Commun. Biol.* **13**:1–53, 1952.
103. Pollak, O. J. Mast cells in the circulatory system of man. *Circulation* **16**:1084–1089, 1957.
104. Pomerance, A. Peri-arterial mast cells in coronary atheroma and thrombosis. *J. Pathol. Bacteriol.* **76**:55–70, 1958.
105. Kolodgie, F. D., Virmani, R., Cornhill, J. F., Herderick, E. E. and Smialek, J. Increase in atherosclerosis and adventitial mast cells in cocaine abusers: an alternative mechanism of cocaine-associated coronary vasospasm and thrombosis. *J. Am. Coll. Cardiol.* **17**:1553–1560, 1991.
106. Kalsner, S. and Richards, R. Coronary arteries of cardiac patients are hyperreactive and contain stores of amines: a mechanism for coronary spasm. *Science* **223**:1435–1436, 1984.
107. Constantinides, P. Mast cells and susceptibility to experimental atherosclerosis. *Science* **117**:505–506, 1953.
108. Bloom, G.D. A short history of the mast cell. *Acta Otolaryngol. Suppl.* **414**:87–92, 1984.
109. Kovanen, P. T. The mast cell: a potential link between inflammation and cellular cholesterol deposition in atherogenesis. *Eur. Heart J.* **14 (Suppl K)**:105–117, 1993.
110. Kovanen, P. T. Role of mast cells in atherosclerosis. In: *Human Basophils and Mast Cells in Health and Disease. Vol. II. Clinical Aspects* (Marone, G., ed.), pp. 132–170. Karger, Basel, 1995.
111. Kovanen, P. T., Kaartinen, M. and Paavonen, T. Infiltrates of activated mast cells at the site of coronary atheromatous erosion or rupture in myocardial infarction. *Circulation* **92**:1084–1088, 1995.
112. Kokkonen, J. O. and Kovanen, P. T. Proteolytic enzymes of mast cell granules degrade low density lipoproteins and promote their granule-mediated uptake by macrophages *in vitro*. *J. Biol. Chem.* **264**:10749–10755, 1989.
113. Lindstedt, K. A., Kokkonen, J. O. and Kovanen, P. T. Soluble heparin proteoglycans released from stimulated mast cells induce uptake of low density lipoproteins by macrophages via scavenger receptor-mediated phagocytosis. *J. Lipid Res.* **33**:65–75, 1992.
114. Lindstedt, K. A., Kokkonen, J. O. and Kovanen, P. T. Inhibition of copper-mediated oxidation of LDL by rat serosal mast cells: a novel cellular protective mechanism involving proteolysis of the substrate under oxidative stress. *Arterioscler. Thromb.* **13**:23–32, 1993.
115. Kaartinen, M., Penttilä, A. and Kovanen, P. T. Mast cells in rupture-prone areas of human coronary atheromas produce and store TNF-α. *Circulation* **94**:2787–2792, 1996.
116. Frangogiannis, N. G., Lindsey, M. L., Michael, L. H., Youker, K. A., Bressler, R. B., Mendoza, L. H., Spengler, R. N., Smith, C. W. and Entman, M. L. Resident cardiac mast cells degranulate and release preformed TNF-α, initiating the cytokine cascade in experimental canine myocardial ischemia/reperfusion. *Circulation* **98**:699–710, 1998.
117. Gonen, B., O'Donnel, P., Post, T. J., Quinn, T. J. and Schulman, E. S. Very low density lipoproteins (VLDL) trigger the release of histamine from human basophils. *Biochim. Biophys. Acta* **917**:418–424, 1987.
118. Goetzl, E. J., Foster, D. W. and Payan, D. G. A basophil-activating factor from human T lymphocytes. *Immunology* **53**:227–234, 1984.
119. Schulman, E. S., Liu, M. C., Proud, D., MacGlashan, D. W. Jr., Lichtenstein, L. M. and Plaut, M. Human lung macrophages induce histamine release from basophils and mast cells. *Am. Rev. Respir. Dis.* **131**:230–235, 1985.
120. Patella, V., de Crescenzo, G., Marinò, I., Genovese, A., Adt, M., Gleich, G. J. and Marone, G. Eosinophil granule proteins activate human heart mast cells. *J. Immunol.* **157**:1219–1225, 1996.
121. McCullagh, K. A. and Balian, G. Collagen characterization and cell transformation in human atherosclerosis. *Nature* **258**:73–75, 1975.
122. Ruoss, S. J., Hartmann, T. and Caughey, G. H. Mast cell tryptase is a mitogen for cultured fibroblasts. *J.*

Clin. Invest. **88**:493–499, 1991.

123. Hartmann, T., Ruoss, S. J., Raymond, W. W., Seuwen, K. and Caughey, G. H. Human tryptase is a potent, cell-specific mitogen: role of signaling pathways in synergistic responses. *Am. J. Physiol.* **262**:L528–L534, 1992.
124. Cairns, J. A. and Walls, A. F. Mast cell tryptase is a potent mitogen for epithelial cells. Stimulation of IL-8 production and intercellular adhesion molecule-1 expression. *J. Immunol.* **156**:275–283, 1996.
125. Cairns, J. A. and Walls, A. F. Mast cell tryptase stimulates the synthesis of type I collagen in human lung fibroblasts. *J. Clin. Invest.* **99**:1313–1321, 1997.
126. Gruber, B. L., Kew, R. R., Jelaska, A., Marchese, M. J., Garlick, J., Ren, S., Schwartz, L. B. and Korn, J. H. Human mast cells activate fibroblasts. Tryptase is a fibrogenic factor stimulating collagen messenger ribonucleic acid synthesis and fibroblast chemotaxis. *J. Immunol.* **158**:2310–2317, 1997.
127. Hatamochi, A., Fujiwara, K. and Ueki, H. Effects of histamine on collagen synthesis by cultured fibroblasts derived from guinea pig skin. *Arch. Dermatol. Res.* **277**:60–64, 1985.
128. Ben Younes-Chennoufi, A., Said, G., Eisen, H., Durand, A. and Hontebeyrie-Joskowics, M. Cellular immunity to *Trypanosoma cruzi* is mediated by helper T cells ($CD4^+$). *Trans. R. Soc. Trop. Med. Hyg.* **82**:84–89, 1988.
129. Postan, M., Correa, R., Ferrans, V. J. and Tarleton, R. L. *In vitro* culture of cardiac mast cells from mice experimentally infected with *Trypanosoma cruzi. Int. Arch. Allergy Immunol.* **105**:251–257, 1994.
130. Laranja, F. S., Dias, E., Nogrega, G. and Miranda, A. Chagas' disease: a clinical, epidemiological and pathological study. *Circulation* **14**:1035–1060, 1956.
131. Hiruta, Y., Adachi, K., Okamoto, T., Fujiura, Y. and Toshima, H. Cardiac mast cells in myocardial diseases. *Kokyu To Junkan* **39**:1133–1138, 1991, (in Japanese).
132. Tai, P.-C., Spry, C. J. F., Olsen, E. G. J., Ackerman, S. J., Dunnette, S. and Gleich, G. J. Deposits of eosinophil granule proteins in cardiac tissues of patients with eosinophilic endomyocardial disease. *Lancet* **i**:643–647, 1987.
133. Desreumax, P., Janin, A., Dubucquoi, S., Copin, M.-C., Torpier, G., Capron, A., Capron, M. and Prin, L. Synthesis of interleukin-5 by activated eosinophils in patients with eosinophilic heart diseases. *Blood* **82**:1553–1560, 1993.
134. Patella, V., de Crescenzo, G., Marinò, I., Genovese, A., Adt, M. and Gleich, G. J. Eosinophil granule proteins are selective activators of human heart mast cells. *Int. Arch. Allergy Immunol.* **113**:200–202, 1997.
135. Okayama, Y., el-Lati, S. G., Leiferman, K. M. and Church, M. K. Eosinophil granule proteins inhibit substance P-induced histamine release from human skin mast cells. *J. Allergy Clin. Immunol.* **93**:900–909, 1994.
136. Glowacki, J. and Mulliken, J. B. Mast cells in hemangiomas and vascular malformations. *Pediatrics* **70**:48–51, 1982.
137. Dethlefsen, S. M., Matsura, N. and Zetter, B. R. Tumor growth and angiogenesis in wild type and mast cell deficient mice. *FASEB J.* **4(3)**:A623, 1990.
138. Marks, R. M., Roche, W. R., Czerniecki, M., Penny, R. and Nelson, D. S. Mast cell granules cause proliferation of human microvascular endothelial cells. *Lab. Invest.* **55**:289–294, 1986.
139. Azizkhan, R. G., Clifford Azizkhan, J., Zetter, B. R. and Folkman, J. Mast cell heparin stimulates migration of capillary endothelial cells *in vitro. J. Exp. Med.* **152**:931–944, 1980.
140. Rakusan, K. and Turek, Z. Protamine inhibits capillary formation in growing rat hearts. *Circ. Res.* **57**:393–398, 1985.
141. Taylor, S. and Folkman, J. Protamine is an inhibitor of angiogenesis. *Nature* **297**:307–312, 1982.
142. Gordon, J. R. and Galli, S. J. Mast cells as a source of both preformed and immunologically inducible TNF-α/cachectin. *Nature* **346**:274–276, 1990.
143. Folkman, J. and Brem, H. Angiogenesis and inflammation. In: *Inflammation: Basic Principles and Chemical Correlates*, Vol. 40 (Gallin, J. I., Goldstein I. M. and Snyderman, R., eds), pp. 821–839. Raven Press, New York, 1992.
144. Criqui, M. H., Lee, E. R., Hamburger, R. N., Klauber, M. R. and Coughlin, S. S. IgE and cardiovascular disease. Results from a population-based study. *Am. J. Med.* **82**:964–968, 1987.
145. Szczeklik, A., Sladek, K., Szczerba, A. and Dropinski, J. Serum immunoglobulin E response to myocardial infarction. *Circulation* **77**:1245–1249, 1988.
146. Korkmaz, M. E., Oto, A., Saraçlar, Y., Oram, E., Oram, A., Ugurlu, S., Karamehmetoglu, A. and Karaagaoglu, E. Levels of IgE in the serum of patients with coronary arterial disease. *Int. J. Cardiol.* **31**:199–204, 1991.

147. Giclas, P. C., Pinkard, R. N. and Olson, M. S. *In vitro* activation of complement by isolated human heart subcellular membranes. *J. Immunol.* **122**:146–151, 1979.
148. Pinckard, R. N., O'Rourke, R. A., Crawford, M. H., Grover, F. S., McManus, L. M., Ghidoni, J. J., Storrs, S. B. and Olson, M. S. Complement localization and mediation of ischemic injury in baboon myocardium. *J. Clin. Invest.* **66**:1050–1056, 1980.
149. Rossen, R. D., Michael, L. H., Kagiyama, A., Savage, H. E., Hanson, G., Reisberg, M. A., Moake, J. N., Kim, S. H., Self, D., Weakley, S., Giannini, E. and Entman, M. L. Mechanism of complement activation after coronary artery occlusion: evidence that myocardial ischemia in dogs causes release of constituents of myocardial subcellular origin that complex with human C1q *in vivo. Circ. Res.* **62**:572–584, 1988.
150. Pinckard, R. N., Olson, M. S., Giclas, P. C., Terry, R., Boyer, J. T. and O'Rourke, R. A. Consumption of classical complement components by heart subcellular membranes *in vitro* and in patients after acute myocardial infarction. *J. Clin. Invest.* **56**:740–750, 1975.
151. Langlois, P. F. and Grawryl, M. S. Detection of the terminal complement complex in patient plasma following acute myocardial infarction. *Atherosclerosis* **70**:95–105, 1988.
152. Frangogiannis, N. G., Perrard, J. L., Mendoza, L. H., Burns, A. R., Lindsey, M. L., Ballantyne, C. M., Michael, L. H., Wayne Smith, C. and Entman, M. L. Stem cell factor induction is associated with mast cell accumulation after canine myocardial ischemia and reperfusion. *Circulation* **98**:687–698, 1998.
153. Mukherjee, D. and Sen, S. Alteration of collagen phenotypes in ischemic cardiomyopathy. *J. Clin. Invest.* **88**:1141–1146, 1991.
154. Seibold, J. R., Giorno, R. C. and Claman, H. N. Dermal mast cells degranulation in systemic sclerosis. *Arthritis Rheum.* **33**:1702–1709, 1990.
155. Fox, C. C., Lazenby, A. J., Moore, W. C., Yardley, J. H., Bayless, T. M. and Licthenstein, L. M. Enhancement of human intestinal mast cell mediator release in active ulcerative colitis. *Gastroenterology* **99**:119–124, 1990.
156. Kawanami, O., Ferrans, V. J., Fulmer, J. D. and Crystal, R. G. Ultrastructure of pulmonary mast cells in patients with fibrotic lung disorders. *Lab. Invest.* **40**:717–734, 1979.
157. Broide, D. H., Smith, C. M. and Wasserman, S. I. Mast cells and pulmonary fibrosis. Identification of a histamine releasing factor in bronchoalveolar lavage fluid. *J. Immunol.* **145**:1838–1844, 1990.
158. Fauci, A. S., Harley, J. B., Roberts, W. C., Ferrans, V. J., Gralnic, H. R. and Björnson, B. H. NIH Conference: The idiopathic hypereosinophilic syndrome. Clinical, pathophysiologic and therapeutic considerations. *Ann. Intern. Med.* **87**:78–92, 1982.
159. Lichtbroun, A. S., Sandhaus, L. M., Giorno, R. C., Kim, H. and Seibold, J. R. Myocardial mast cells in systemic sclerosis. A report of three fatal cases. *Am. J. Med.* **89**:372–376, 1990.
160. Patella, V., de Crescenzo, G., Lampärter-Schummert, B., De Rosa, G., Adt, M. and Marone, G. Increased cardiac mast cell density and mediator release in patients with dilated cardiomyopathy. *Inflamm. Res.* **46(S-I)**:S31–S32, 1997.
161. Valent, P., Spanblöchl, E., Sperr, W. R., Sillaber, C., Zsebo, K. M., Agis, H., Strobl, H., Geissler, K., Bettelheim, P. and Lechner, K. Induction of differentiation of human mast cells from bone marrow and peripheral blood mononuclear cells by recombinant human stem cell factor/*kit* ligand in long-term culture. *Blood* **80**:2237–2245, 1992.
162. Saito, H., Sakaguchi, N., Matsumoto, K., Tsubaki, T., Numazaki, T., Ebisawa, M., Kobayashi, M., Ozawa, R., Yanagi, H., Akasawa, A. and Iikura, Y. Growth in methylcellulose of human mast cells in hematopoietic colonies stimulated by steel factor, a c-*kit* ligand. *Int. Arch. Allergy Immunol.* **103**:143–151, 1994.
163. de Paulis, A., Minopoli, G., Dal Piaz, F., Pucci, P., Russo, T. and Marone, G. Novel autocrine and paracrine loops of the stem cell factor/chymase network. *Int. Arch. Allergy Immunol.* **118**:422–425, 1999.
164. de Paulis, A., Minopoli, G., Arbustini, E., de Crescenzo, G., Dal Piaz, F., Pucci, P., Russo, T. and Marone, G. Stem cell factor is localized in, released from and cleaved by human mast cells. *J. Immunol.* **163**:2799–2808, 1999.

CHAPTER 30

Mast Cells in Atherosclerotic Human Coronary Arteries: Implications for Coronary Fatty Streak Formation, Plaque Ulceration and Control of Local Haemostatic Balance

PETRI T. KOVANEN

Wihuri Research Institute, Helsinki, Finland

INTRODUCTION

Atherothrombosis is a slowly progressing inflammatory disease of the innermost arterial layer, the intima: first fatty streaks appear, then atheromas or atherosclerotic plaques develop, and, finally, the fibrous cap of a plaque may erode or rupture and trigger formation of a thrombus on the newly exposed thrombogenic surface.

Immunohistochemical observations on atherosclerotic lesions in human coronary arteries have revealed that these lesions contain more mast cells than does the normal coronary intima. Most importantly, the fraction of degranulated mast cells in the lesions is increased. Biochemical and cell culture experiments with degranulated rat serosal mast cells have suggested several mechanisms which, if operative *in vivo*, would provide plausible mechanisms explaining how mast cells could influence the development of coronary atherothrombosis in man. First, the heparin proteoglycans and the neutral protease chymase in exocytosed mast cell granules can induce the formation of foam cells, the hallmarks of fatty streak lesions. Second, granule heparin proteoglycans can inhibit proliferation of cultured smooth muscle cells, and granule chymase can induce their apoptosis. Since smooth muscle cells in the fibrous cap of a plaque produce collagen and so provide tensile strength in the cap, their loss would weaken the cap and render it susceptible to rupture. Third, granule heparin proteoglycans effectively block platelet aggregation on collagen surfaces under high shear stress, an effect that would tend to prevent thrombotic occlusion of an affected coronary artery. Taken together, experimental work has unravelled several functions by which mast cells may either accelerate or retard the development of atherothrombosis.

Detection of these mast cell-dependent functions in the human coronary arterial wall and definition of their actual impact on the development of human coronary atherothrombosis is an exciting challenge for future studies.

MAST CELLS AND BASOPHILS
ISBN 0-12-473335-2

ATHEROSCLEROSIS IS A DISORDER OF LIPID METABOLISM IN THE ARTERIAL INTIMA WITH A STRONG INFLAMMATORY COMPONENT

Atherosclerosis is a disease of the innermost layer of the arterial wall, the intima (1). The intima is composed of an endothelial cell layer with underlying stromal tissue. The stromal cells of the intima are smooth muscle cells. The intima is separated from the medial layer by an internal elastic lamina. Even during fetal life, medial smooth muscle cells migrate through the intimal elastic layer into the subendothelial space, where they begin to divide and secrete the various components of the extracellular matrix: collagen, elastin and proteoglycans.

The common type of human coronary atherosclerosis is associated with high levels of plasma low-density lipoproteins (LDL) and may ultimately lead to angina pectoris and myocardial infarction. This type is characterized by local accumulations of LDL-derived cholesterol in the affected sites of the coronary intima (2, 3). Initially, cholesterol accumulates subendothelially in the intimal monocyte-derived macrophages. When the cytoplasm of the macrophages is filled with cholesteryl ester droplets, these cells appear 'foamy' in electron and light microscope sections, and are therefore called foam cells. The foam cells lie beneath the transparent endothelial cell layer, giving the inner surface of the vessel a yellowish appearance. Such yellow areas are called fatty streaks (4). Fatty streaks neither obstruct the arterial lumen nor cause local thrombus formation, and so are clinically silent. They are precursors of the true atherosclerotic lesions or atheromas, also called atherosclerotic plaques (5). In atheromas, cholesterol has accumulated, not only in foam cells, but also in the extracellular space of the deep layers of the intima, where it forms a soft extracellular lipid core.

The layer overlying the lipid core of the atheroma and separating it from the circulation is called the fibrous cap (5). This is characterized by a dense extracellular matrix, the major component of which is collagen. Advanced atherosclerotic plaques vary in their architecture, from a solid fibrous lesion with a small lipid core and a thick cap to a lipid-rich lesion with a large lipid core and a thin cap. An atheroma with a thick cap is essentially stable, whereas an atheroma with a thin cap is prone to rupture (6). Recent clinical and pathological studies of coronary arteries of patients suffering myocardial infarction have demonstrated that atheromas typically rupture in the shoulder region, an intimal area that is characterized by high circumferential stress. A photomicrograph of a coronary atheroma and a schematic representation showing the locations of the various regions of the atheroma are shown in Fig. 1.

As the lipid core grows, it pushes the cap toward the lumen of the artery, and finally would occlude the lumen. However, this tendency to occlude is counteracted by 'remodelling' of the atheroma, a process that allows maintenance of blood flow, the lesion expanding outwards (into the perivascular tissue) and the cap becoming thinner. But, as the lipid core grows and the cap thins, the stable fibrous lesion is converted into an unstable lipid-rich lesion. Even lesions with stable morphology can become thrombogenic when the surface of a thick cap is eroded – i.e. the endothelial cells are detached and subendothelial thrombogenic structures become exposed (7).

Data emerging from clinical, pathological and experimental studies on atherogenesis have revived the paradigm according to which atherosclerosis is an inflammatory disease (8). The key elements in the production of local inflammation in atherosclerotic lesions are inflammatory cells, notably macrophages, T lymphocytes and mast cells. The

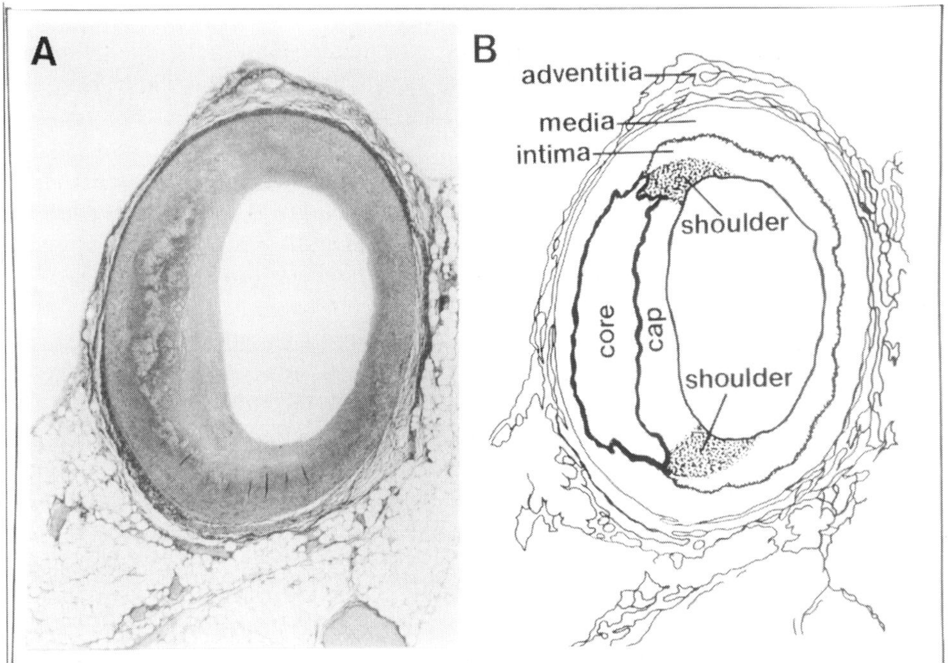

Fig. 1 The anatomy of an atheroma. (A) Light microscopic view of a cross-section of the origin of the left anterior descending coronary artery. (B) Schematic representation (shadow picture) showing the locations of the various regions of the atheroma. Note the normal architecture of the arterial wall opposite to the atheroma (eccentric lesion). Also note the shoulder regions, which are the predilection sites of rupture of the atheroma. This atheroma has a relatively thick cap, and, therefore, is not likely to undergo deep ulceration (rupture) into the lipid core. However, superficial erosions of the cap and of the shoulder regions are frequent events even in stable-appearing coronary atheromas, like the one shown. Original magnification: ×20. (Reproduced, with permission, from *Circulation* **90**:1669–1678, 1994.)

macrophages and T lymphocytes appear already in early childhood in the intima at atherosclerotic predilection sites of the arterial tree (9). In the coronary arterial tree, the areas prone to the development of atherosclerosis are the outer curvatures of bifurcations of the epicardial coronary arteries, where the intima is already thick at birth. No systematic studies have been performed on the timing of the appearance of mast cells in the atherosclerosis-prone areas of coronary arteries. The available evidence in human aortic samples suggests that very few if any mast cells are present in the intima of young people (10).

CIRCULATING MAST CELL PROGENITORS ALSO FIND THEIR WAY INTO THE ARTERIAL INTIMA

Mast cells originate in the bone marrow. They develop from circulating multilineage c-*kit*$^+$, CD34$^+$, Ly$^-$, CD14$^-$, CD17$^-$ haematopoietic progenitors, and, accordingly, differ from circulating monocytes (CD14$^+$) and blood basophils (CD17$^+$) (11). The morphologically indeterminate circulating progenitor cells migrate into various tissues, notably the various mucosal surfaces and the skin. The chemokine responsible for the migration of these precursors into peripheral tissues is thought to be the stem cell factor

(SCF), which is secreted by the stromal cells, such as the fibroblasts, of the tissues (12). In the tissues, the precursors are then converted into mature mast cells, the hallmark of which is their very high content of cytoplasmic granules (13). The granules have three main components: histamine, neutral proteases and proteoglycans.

Once appropriately activated, the mast cells degranulate, expelling some of their granules into the surrounding extracellular fluid, where histamine is released from the proteoglycans, diffuses away and exerts its various functions. A variable fraction of the proteoglycans is also released in soluble form. Another fraction of the proteoglycans remains bound to the neutral proteases in the form of protease–proteoglycan complexes. These residual complexes are called granule remnants (14).

Human mast cells can be divided phenotypically into two types, according to their content of neutral proteases: those containing tryptase and those containing both tryptase and chymase (15). Since all mast cells and only mast cells contain tryptase, we routinely stain sections of the tissue with a monoclonal antibody directed against tryptase in order to count all the mast cells. To define the phenotype of the mast cells, we also stain for chymase. Our studies have revealed that, in the human arterial intima (whether normal or atherosclerotic), a highly variable fraction of the cells contains chymase in addition to tryptase. Thus, in some subjects, all of the mast cells contain chymase, in other subjects no mast cells contain chymase, and in yet other subjects a fraction of the mast cells contain chymase. What factors could account for this great inter-individual variation in chymase expression? *In vitro* studies on human mast cell development have shown that, in culture systems containing mast cell precursors, addition of SCF induces strong expression of tryptase and only weak expression of chymase (16). But only if the system includes a stromal cell layer will the developing mast cells express chymase in significant amounts (17). Thus, it appears that, in addition to SCF, some co-factors produced by stromal cells are necessary for the expression of chymase. Our finding of the great inter-individual variation in chymase expression suggests differences in the production of such co-factors in the intimal stroma. Since the stromal environment of the intima is created by smooth muscle cells, we are left with the challenging question of whether these cells or other cells in the intima (endothelium, macrophages and T lymphocytes) express factors that may modify the phenotype of the mast cells.

IMMUNOHISTOCHEMICAL OBSERVATIONS ON MAST CELLS IN THE HUMAN CORONARY ARTERIAL INTIMA

Mast Cells in Normal Coronary Intima and in Coronary Fatty Streaks

In a systematic study on the density of mast cells in human coronary atherosclerosis, we found that in normal coronary intima and in fatty streaks, respectively, there were on average 1 and 5 mast cells per mm^2, the mast cells amounting to 0.1 and 0.9% of all nucleated cells (18). Thus, in the coronary arteries, the density of mast cells was 5-fold higher in areas where foam cells were also present than in areas without foam cells (Fig. 2). To investigate whether the coronary mast cells were activated, electron microscopy was used to detect signs of degranulation. This method allowed a clear distinction between intact, dark, electron-dense cytoplasmic granules with typical heterogeneous morphology, and light, electron-lucent granule remnants, either inside or outside the mast cells. The presence of the latter demonstrated that the granules had been involved in the degranulation process and had lost some of their contents. To study the proportions of

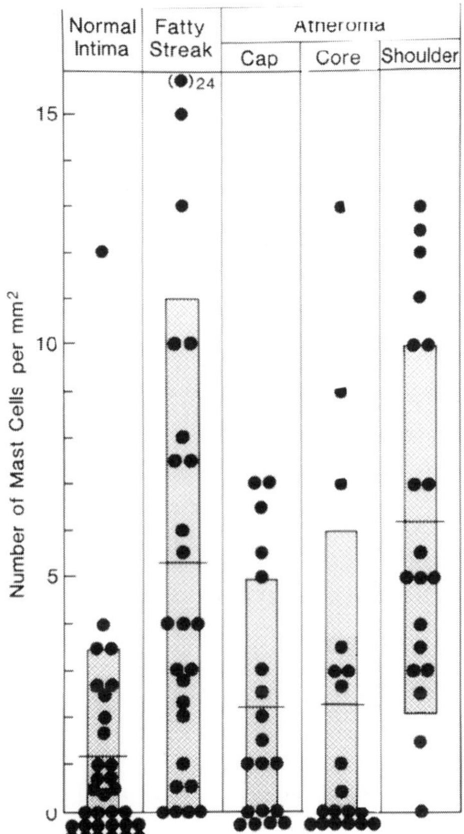

Fig. 2 Bar graph showing the densities of mast cells in normal and atherosclerotic coronary intimas. Both early atherosclerotic lesions (fatty streaks) and late atherosclerotic lesions (atheromas or plaques) were studied. In the atheromas, three distinct anatomical areas – the cap, core and shoulder regions – were analysed separately. Cross-sectional intimal areas were measured by planomorphometry, and the numbers of all tryptase-positive cells in these areas were counted. Individual data and means are shown. Bars shown are mean ± SD. The statistical significances of the differences between the atherosclerotic intimas and the normal intimas were as follows: fatty streaks vs. normal intimas, $p<0.0001$; caps vs. normal intimas, $p<0.05$; cores vs. normal intimas, $p<0.05$; shoulders vs. normal intimas, $p<0.0001$. (Reproduced, with permission, from *Circulation* **90**:1669–1678, 1994.)

mast cells that are degranulated, we turned to light microscopy and observed extracellular tryptase-positive granules at high magnification. Degranulated mast cells were found in every intimal sample, and, most importantly, they were more numerous in the intimal areas in which the atherosclerotic process had advanced. Thus, in the normal coronary intima, about every fifth mast cell, and in the fatty streaks about every other mast cell, showed signs of degranulation. The factors leading to mast cell degranulation in the arterial intima are unknown. However, we have observed that the degree of degranulation is always high in intimal areas with signs of inflammation, suggesting that the degranulating factors are of inflammatory nature, derived either from other intimal cells (endothelial cells, smooth muscle cells, T lymphocytes, macrophages) or from plasma.

Mast Cells in Coronary Atheromas

In coronary atheromas, we made the surprising observation that mast cells are distributed unevenly in a typical way: 6 per mm^2 in the shoulder region and 2 per mm^2 in both the cap and the core regions (Fig. 2) (18). It is noteworthy that the increase in the number of degranulated mast cells was especially pronounced in the shoulder region of atheromas, the predilection site for atheromatous rupture. In this region, the proportion of the mast cells that were degranulated was 85%, but in the normal intima it was only 18%. Moreover, in the shoulder region, the average ratio of mast cells to T lymphocytes was 1:4 (26%) and to macrophages it was 1:8 (12%).

Mast Cells in Eroded or Ruptured Coronary Atheromas

The most important mechanism underlying the sudden onset of coronary syndromes, including unstable angina and acute myocardial infarction, is erosion or rupture of a coronary atheroma (19, 20). The risk of atheromatous erosion or rupture appears to depend critically on the cellular and extracellular composition of the atheroma. We identified mast cells at the sites of atheromatous erosion or rupture in patients who had died of acute myocardial infarction (21). At the immediate site of erosion or rupture, there

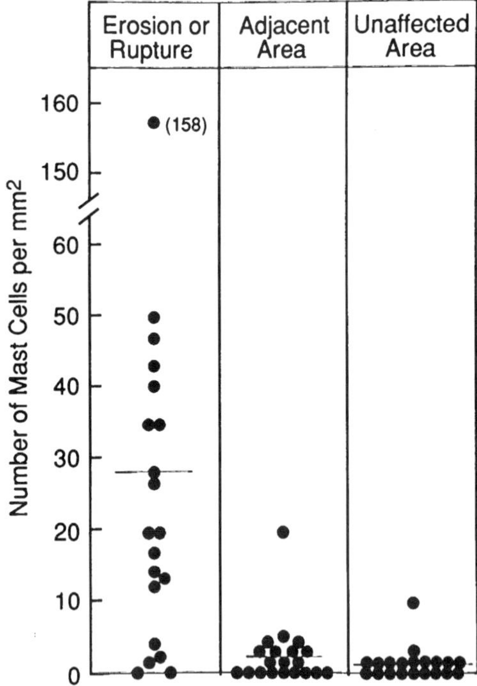

Fig. 3 Bar graph showing the densities of mast cells in specimens of infarct-related coronary arteries from patients who had died of acute myocardial infarction. Data are shown for each individual and means are given for mast cell densities at the site of erosion or rupture, in an adjacent area, and in an unaffected area. Each point represents one patient (n=20). The probability values for the differences between the eroded or ruptured sites and the adjacent or unaffected areas were both $p<0.05$. (Reproduced, with permission, from *Circulation* **92**:1084–1088, 1995.)

were, on average, about 25 mast cells per mm^2 (Fig. 3), the mast cells amounting to 6% of all nucleated cells; meanwhile, the mast cells amounted in the adjacent atheromatous area to 1%, and in the unaffected intimal area to only 0.1% (Table I). The proportions of the mast cells that were degranulated were 86% at the site of erosion or rupture, 63% in the adjacent atheromatous area and 27% in the unaffected intima. From these figures we could calculate that the density of degranulated mast cells was 200-fold higher at the eroded or ruptured site than in the unaffected intimal area. At the site of erosion or rupture, the numbers of T lymphocytes and macrophages were also increased, and the number of smooth muscle cells was decreased (Table I).

Table I

Percentages of Mast Cells, Macrophages, T Lymphocytes and Smooth Muscle Cells in Relation to the Total Number of Cells in Coronary Intimas of Patients who Died of Myocardial Infarction

Intimal area	Mast cells	Macrophages	T Lymphocytes	Smooth muscle cells	All nucleated cells
Erosion/rupture (*n*=20), %	6.1 (0–36)	51 (9–85)	7.1 (0–16)	32 (1–86)	100
Adjacent area (*n*=20), %	1.1 (0–7)	29 (1–67)	5.1 (0.1–22)	62 (6–96)	100
Unaffected area (*n*=20), %	0.1 (0–0.6)	7.1 (2–24)	1.1 (0.1–6)	88 (70–96)	100

Values (%) are means and ranges. Note that the predominant cell type at the erosion or rupture site was macrophages, the second most frequent cell type being smooth muscle cells, and T lymphocytes equalling the mast cells in number. At the immediate site of erosion or rupture, mast cells amounted to 6% of all nucleated cells, in the adjacent intimal areas they amounted to 1%, while in the unaffected intimal areas the proportion of mast cells was very low (0.1%). (Ref. 21: reproduced with permission)

Most recently, to define better the role of mast cells in the destabilization of a coronary atheroma, we studied coronary atherectomy specimens from patients with chronic stable angina, unstable angina and severe unstable angina, who had undergone directional coronary atherectomy (22). The coronary atherectomy specimens provided a unique source of plaque tissues because they made it possible to correlate features of plaque biology with the clinical status of the patient. Samples of culprit lesions were snap-frozen and analysed immunohistochemically. We found that the numbers of mast cells tend to increase with the clinical severity of the coronary syndromes. Thus, the average densities of mast cells in the lesions were 1, 2.5 and 5 per mm^2 in chronic stable, unstable and severe unstable angina, respectively. Moreover, the distribution of mast cells in these samples was highly uneven. In the clusters, the densities of mast cells typically varies between 30 and 50 cells per mm^2, the highest density observed being 100 mast cells per mm^2. We did not determine the degree of mast cell degranulation, as degranulation might have been caused artificially by the mechanical trauma associated with the surgical procedure. Finally, the numbers of T lymphocytes and macrophages also gradually increased from stable through unstable to very unstable angina pectoris. The study implied that the infiltrates of inflammatory cells are present in the lesions prior to erosion or rupture, rather than being part of an inflammatory response to such symptom-causing ulceration.

POTENTIAL MECHANISMS BY WHICH MAST CELLS MAY PLAY A ROLE IN ATHEROGENESIS

Mast Cells and Foam Cell Formation

Mast cells trigger cholesterol uptake by macrophages and smooth muscle cells and promote their conversion into foam cells in vitro

From the above quantitative studies on the distribution of mast cells and on the degree of their activation, a clear picture has emerged in which the numbers of degranulated mast cells in human fatty streaks and in the shoulder regions of atheromas are increased. The most characteristic feature of human fatty streaks is their high content of foam cells (9). Likewise, the growing edges of atheromas, the shoulder regions, contain numerous foam cells. This led to the suggestion that mast cells play an active role in the formation of foam cells in human fatty streaks and atheromas.

To gain insight into the mechanisms by which mast cells might participate in the formation of foam cells, we used cell culture methods, employing mast cells derived from the rat peritoneal cavity (23). These studies defined a tightly regulated sequence of events leading to foam cell formation. The mast cells were co-cultured with rat or mouse peritoneal macrophages in a medium to which LDL particles had been added, and were then stimulated to degranulate (either immunologically or non-immunologically). We found that the apolipoprotein (apo)B-100 component of the LDL particles was bound by the heparin proteoglycan component of the exocytosed granules (i.e. the granule remnants). The granule remnant-bound LDL was then proteolysed by granule remnant-bound chymase. When proteolysed, the LDL particles became unstable and fused into larger lipid droplets; thus the capacity of each granule remnant to bind and carry LDL was, on average, increased from a full load of 10,000 to a full load of 50,000 LDL particles. Finally, the granule remnants coated with fused LDL particles were phagocytosed by the co-cultured macrophages, the result being massive uptake of LDL by these cells, with ultimate accumulation of LDL-derived cholesterol as cytoplasmic cholesteryl ester droplets (i.e. formation of foam cells). Cultured rat aortic smooth muscle cells of synthetic phenotype (corresponding to the phenotype in atherosclerotic lesions) could also ingest such LDL-coated granule remnants, and so became filled with cholesterol (24). With the aid of immunoelectron microscopy, we have found evidence that, in the human arterial intima, exocytosed mast cell granules bind apoB-100-containing LDL, and that such granule remnants may be ingested by the intimal phagocytes (either macrophages or smooth muscle cells) (25).

Mast cells prevent cholesterol efflux from foam cells in vitro

One key factor in the prevention of atherosclerosis is considered to be the ability of high-density lipoproteins (HDL) to remove excess cholesterol from the cholesterol-loaded foam cells. This is the initial step in reverse cholesterol transport, along the pathway by which cholesterol is carried from the extrahepatic tissues, such as the arterial intima, back to the liver (26, 27). The major protein component of HDL is apolipoprotein (apo)A-I, which has an important biological function, playing the key role in the initiation of reverse cholesterol transport (28). To test the possible blocking effect of mast cells on the removal of cholesterol from macrophage foam cells, we have worked with a model system in which rat serosal mast cells are incubated with macrophage foam cells in a medium to which HDL_3 particles have been added (29). Provided the mast cells are not stimulated to degranulate, HDL_3 efficiently removes cholesterol from the foam cells. In sharp contrast, when the mast cells are stimulated, this initial step of reverse cholesterol transport is blocked and the cholesterol is not removed by the HDL_3 particles, the chymase of the exocytosed granules proteolysing the HDL_3 particles (i.e. degrading their apoA-I) and thereby reducing their ability to induce efflux of cholesterol from the foam cells.

The above experiments were conducted in a protein-free incubation medium. To elucidate the potential of exocytosed mast cell granules to modify HDL_3 proteolytically in the presence of its physiological inhibitors (e.g. α_1-antitrypsin), we incubated HDL_3 and cultured macrophage foam cells with mast cell granule remnants in samples of extracellular fluid obtained from human aortic intimas (30). We found that, even under these more stringent conditions, the granule remnants prevented the HDL_3 particles from effectively inducing cholesterol efflux from the cultured foam cells. In other experiments, we found that the granule remnants also prevented intimal fluid or human serum from inducing cholesterol efflux from the foam cells. Western blot analysis showed that the granule remnants had degraded apoA-I in the intimal fluid and the serum. The results showed that, despite the presence of physiological protease inhibitors, the residual chymase activity of the granule remnants was sufficient to degrade apoA-I and block cholesterol efflux. The other important observation was that the rapid efflux of cholesterol at low concentrations of apoA-I was chymase-sensitive, which, accordingly, was designated the 'protease-sensitive high-affinity cholesterol efflux-promoting component'. Since this high-affinity cholesterol efflux was impeded when only <5% of the apolipoproteins of HDL_3 were degraded, we deduced that the chymase-sensitive high-affinity component is a minor subfraction of the particles in the HDL_3 fraction that is particularly susceptible to proteolytic cleavage.

The HDLs are a highly heterogeneous family of particles. In recent years, much attention has been focused on the small fractions of lipid-poor discoidal HDL particles that exhibit electrophoretic pre-beta mobility (pre-β HDL), in contrast to the major spherical components of HDL, which exhibit alpha mobility (31). The pre-β HDL has been suggested to act as a shuttle, transporting cholesterol from the plasma membrane of foam cells to the fully lipidated mature HDL species. In these particles, the sole apolipoprotein is apoA-I, and its conformation appears to render it exquisitely sensitive to proteolytic cleavage (27). To test the hypothesis that granule remant chymase degrades this minor fraction, human HDL_3 and unfractionated serum were incubated with rat granule remants and analysed for their content of the pre-β HDL (32). The analysis revealed that, upon incubation of human HDL_3 or serum at 37°C, the discoidal pre-β_1 HDL particles contained in them, but not the spherical α-migrating particles, were rapidly depleted, and that this loss closely paralleled the loss of the ability of the HDL_3 or serum to induce cholesterol efflux from macrophage foam cells. In addition, these studies revealed that another species of small HDL particle, apoA-IV-containing particles, which are present in human blood plasma and actively induce cellular cholesterol efflux, were also depleted. Together, loss of the small apoA-I- and apoA-IV-containing lipoprotein particles is sufficient to explain the sensitivity of the high-affinity component of the cholesterol efflux-inducing ability of human serum and aortic intimal fluid to mast cell chymase.

Taken together, the cell culture experiments carried out with rodent mast cells suggest the possibility that stimulated mast cells, both by promoting the uptake of LDL and by inhibiting the release of cellular cholesterol, can accelerate foam cell formation in the

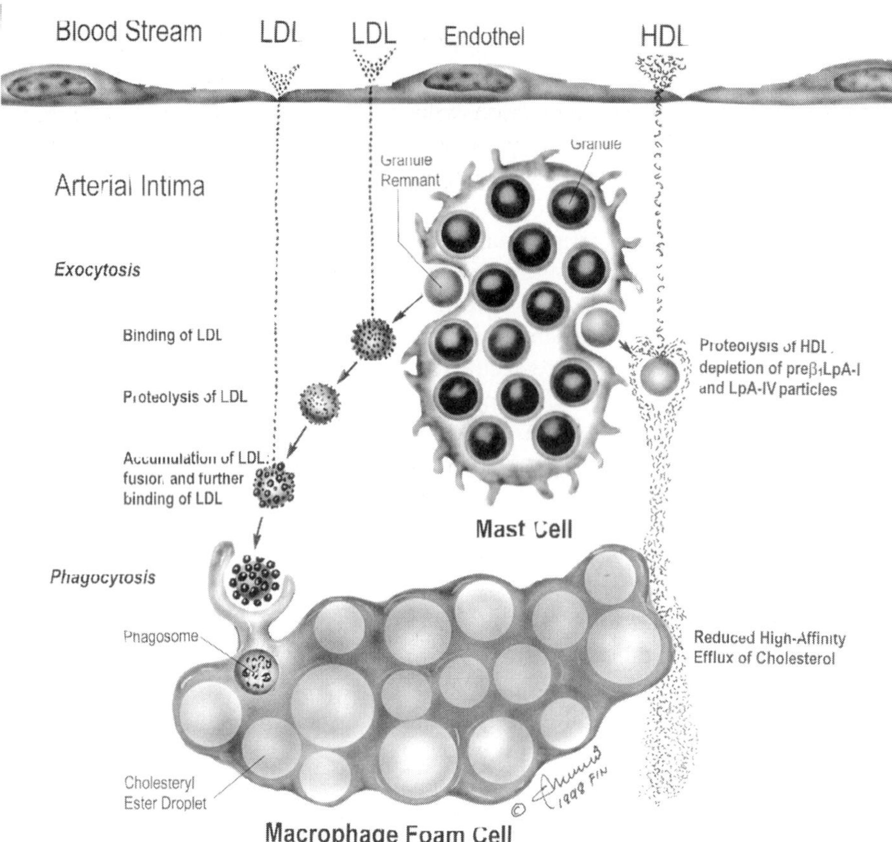

Fig. 4 The proposed dual action of exocytosed mast cell granules on lipoprotein metabolism in the two sequences of events by which a stimulated mast cell converts a macrophage into a foam cell. A degranulating mast cell is shown residing in the subendothelial space of the arterial intima. Left: An exocytosed mast cell granule (i.e. granule remnant) binds and proteolyses LDL particles, so that the particles become unstable and fuse on the remnant surface. The fusion of LDL particles allows new particles to be bound to the surface of the granule remnant, so increasing the maximal load that can be carried by the remnant. The granule remnant, with its load of fused LDL, is phagocytosed and degraded by the macrophage, the result being a macrophage foam cell with typical cytoplasmic cholesteryl ester droplets. Right: Simultaneously, another granule remnant proteolyses the apoA-I- and apoA-IV-containing subfractions of HDL, and so blocks the high-affinity component of the HDL-dependent efflux of cholesterol from the macrophage foam cell. This dual action of granule remnants on lipoprotein metabolism facilitates conversion of macrophages into foam cells and prevents their regression to normal macrophages devoid of cholesteryl ester droplets. (Modified from ref. 23.)

intimal areas in which mast cells, macrophages and smooth muscle cells co-exist (Fig. 4). Inhibition of cholesterol removal from foam cells would inevitably alter the delicate balance between cholesterol influx and efflux, and so contribute to the maintenance of foam cells. This, again, would tend to prevent regression of fatty streaks, even if influx of cholesterol were low.

Potential Roles of Mast Cells in the Erosion and Rupture of Coronary Atheromas

Novel functions for intimal mast cells other than those directly related to lipid metabolism are emerging. These functions relate to the concept that coronary atheromas may rupture because their fibrous caps have been weakened by the locally increased activity of the enzymes involved in digestion of the extracellular matrix (33). In the human arterial intima, as in other tissues, mast cells are the major local source of neutral proteases. This high storage capacity of proteases suggested to us that mast cells may contribute to the weakening of the atheromatous plaques. When stimulated, human mast cells release their neutral proteases, depending on phenotype, chymase and tryptase or tryptase alone, into the surrounding microenvironment. Both are secreted as active enzymes (not as pro-enzymes), and are capable of degrading the various components of the pericellular matrix, such as fibronectin (34, 35). In addition, they can effectively activate matrix metalloproteinases (MMPs). This is necessary since, in contrast to the mast cell-derived neutral proteases, the MMPs are synthesized and secreted as zymogens, i.e. as inactive pro-forms (pro-MMPs), and consequently, upon secretion, still have to be activated (36). One of these matrix-degrading enzymes is interstitial collagenase (MMP-1), also found in atherosclerotic lesions (37). We found that human skin chymase effectively activates pro-MMP-1 by cleaving the proenzyme at the Leu83-Thr84 position (38). Moreover, studies in other laboratories have demonstrated that tryptase can activate pro-stromelysin (pro-MMP-3) (39), which, in addition to being a powerful matrix-degrading enzyme, can activate other MMPs. Recently, Johnson *et al.* (40) incubated small pieces prepared from carotid endarterectomy specimens with the mast cell-degranulating agent 48/80. They observed a 2-fold increase in the activity of mast cell tryptase in the incubation medium, and an increase in MMP activity. This *in vitro* study showed that degranulation of mast cells in an atherosclerotic plaque may also activate the MMPs present in the lesion. Finally, granules of mast cells in rupture-prone areas of human coronary atheromas have been found to contain the potent pro-inflammatory cytokine tumour necrosis factor-α (TNF-α) (41). Importantly, TNF-α can induce macrophages to synthesize 92-kDa gelatinase (MMP-9), another member of the MMP family present in coronary atheromas (42). Interestingly, in coronary atherectomy specimens from patients with chronic stable angina, unstable angina and severe unstable angina, the numbers of TNF-α-positive mast cells and of MMP-9-positive macrophages were highest in the specimens from patients with the most severe symptoms (22). Taken together, the above *in vitro* findings and the observation that degranulated mast cells are present at the actual site of coronary rupture, point to the participation of an entirely new plaque-destabilizing factor in acute coronary syndromes – that is, of neutral proteases and pro-inflammatory cytokines released from mast cells in the vulnerable coronary plaques (Fig. 5).

In addition to the neutral proteases, the heparin contained in the cytoplasmic secretory granules may also participate in mast cell-induced weakening of the fibrous caps of coronary plaques. Mast cells are the major source of tissue heparin, and experimental studies have shown that the macromolecular heparin proteoglycans released from rat serosal mast cells can inhibit proliferation of rat aortic smooth muscle cells in culture (43). Moreover, exocytosed mast cell granules (chymase) can promote apoptosis of growth-arrested smooth muscle cells *in vitro*. Since smooth muscle cells are the only significant producers of the various components of the extracellular matrix of the fibrous cap, notably collagen, reduction in their number will also reduce matrix synthesis in the

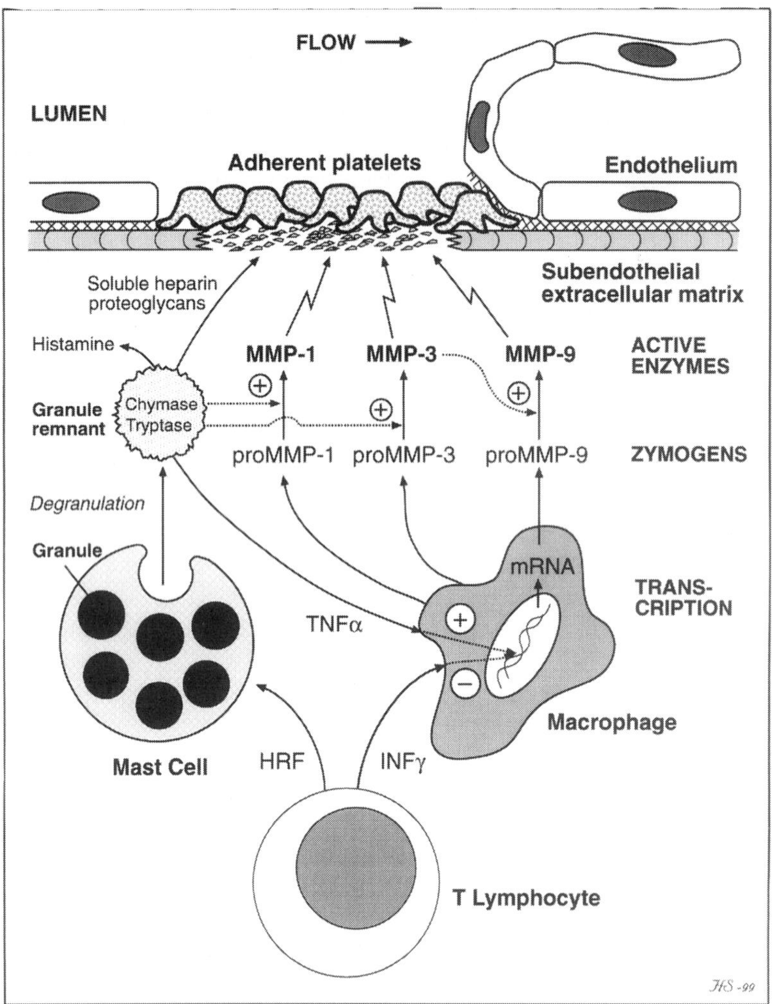

Fig. 5 Hypothetical model of mast cell-dependent erosion of a coronary plaque, with ensuing sealing of the ulcerated area by adherent thrombocytes. The vulnerable shoulder region of a coronary plaque is shown to contain inflammatory cells of three types: a mast cell, a T lymphocyte and a macrophage. The lymphocyte secretes a histamine-releasing factor (HRF) which stimulates the mast cell to degranulate. TNF-α is released from the exocytosed mast cell granule, and stimulates synthesis of pro-MMP-9 by the macrophage. INF-γ, which is secreted by the T lymphocyte, counteracts this stimulatory effect. The macrophage also synthesizes and secretes pro-MMP-1 and pro-MMP-3. The neutral protease chymase contained in the granule remnant activates pro-MMP-1, and the neutral protease tryptase activates the pro-MMP-3. The active MMP-3 can activate other pro-MMPs, including pro-MMP-9. Together, the activated MMPs can degrade various components of the extracellular matrix, such as collagen. Chymase and tryptase can degrade the components of the pericellular matrix, such as fibronectin. Degradation of the various subendothelial extracellular matrix components may cause detachment of endothelial cells and exposure of the thrombogenic surface. The heparin proteoglycans released by the mast cells strongly inhibit platelet–collagen interaction and platelet aggregation, but allow platelets to adhere on collagen. The clinical endpoint is an unnoticed (subclinical) acute coronary event, which would lead to a healing process without thrombotic occlusion of the lumen. In the lesion with dysfunctional or lost endothelium (low or absent nitric oxide production), the histamine released from mast cells tends to cause contraction of medial smooth muscle cells, and hence aggravates a coronary spasm. Spasm may cause sufficient lowering of blood flow to cause angina pectoris, but is not likely to cause total occlusion of the lumen.

cap. Thus, activated mast cells may not only induce degradation of collagen, but also reduce its production; both effects would weaken the cap of a coronary atheroma and so predispose to a greater risk of plaque rupture.

Coronary spasm contributes to the development of thrombotic complications of coronary atherosclerosis by reducing blood flow, and so increasing the tendency for thrombus growth. Histamine causes dilatation of normal coronary arteries, but paradoxically causes vasocontriction in ulcerated areas where the endothelium is damaged and has lost its opposing vasodilatory capacity (44). Thus, in atherosclerotic coronary segments with ruptured plaques, some of the histamine released from degranulated mast cells is likely to diffuse into the medial layer, where the smooth muscle may contract in response to the histamine. Similarly, mast cells in the outer layer of coronary arteries, the adventitia, have been suggested to contribute to coronary spasm in variant angina (45). Recently, we found increased numbers of degranulated mast cells around the small adventitial vessels, the vasa vasorum, in the adventitial layer backing ruptured coronary plaques in patients who had died of myocardial infarction (46). Since the vasa vasorum nourish the medial layer, this finding suggests that adventitial mast cells, by releasing histamine, which is then carried into the medial layer via the vasa vasorum, may also contribute to the development of thrombotic coronary artery occlusion by provoking coronary spasm. Finally, recent experimental data imply that activation of mast cells, with ensuing secretion of heparin proteoglycans, can locally attenuate the thrombogenicity of matrix collagen (47). Thus, the soluble macromolecular heparin proteoglycans exocytosed from stimulated rat serosal mast cells strongly inhibited platelet–collagen interactions. Since, under the conditions used, the heparin proteoglycans did not bind to collagen or to the resting platelets, they must have disturbed platelet–collagen interactions through other mechanisms. Importantly, the heparin proteoglycans inhibited platelet aggregation on, but not adhesion to, a collagen-coated surface at both low and high shear rates. Accordingly, glycoprotein Ia/IIa-mediated platelet adhesion was not blocked, but subsequent collagen-dependent platelet activation and aggregation were attenuated. Furthermore, the ability of heparin proteoglycans to bind to von Willebrand factor and to attenuate the von Willebrand factor-dependent glycoprotein IIb/IIIa-mediated platelet functions is likely to be important for the anti-thrombotic efficacy of the heparin proteoglycans in flowing blood. The results implied that the heparin proteoglycans attenuated transmission of the activation signal from glycoprotein Ia/IIa to glycoprotein IIb/IIIa, without directly interfering with glycoprotein IIb/IIIa.

The ability of the intact heparin proteoglycans derived from rat serosal mast cells (average molecular weight 750,000) to inhibit platelet function was greater than that of the heparin glycosaminoglycan chains released from the proteoglycans by treatment with alkali (average molecular weight 75,000), which, again, was greater than that of commercial high molecular weight heparin (average molecular weight 15,000) or commercial low molecular weight heparin (average molecular weight 5000). In structural analysis of the soluble macromolecular heparin proteoglycans, the composition of disaccharide units is typical of heparin (J.-P. Li, P. Kovanen, U. Lindahl, unpublished results). Taken together, these findings indicate that the most important factors contributing to the observed inhibition by heparin proteoglycans from rat serosal mast cells of collagen-dependent platelet aggregation in flowing blood at high shear rates are the large size of the heparin glycosaminoglycan chains and their attachment to a core protein (to create 'macromolecular' heparin). Provided that the mast cells in advanced coronary atherosclerotic lesions also possess this inhibitory function, they may exert a local physiological anti-thrombotic function by regulating haemostasis in the injured vascular wall, as recently observed in wounded human skin (48). An important additional point of control is the unique fibrinolytic potential of mast cells (49). It remains to be seen how much the mast cells contribute to the endogenous fibrinolytic potential leading to clot lysis and spontaneous reperfusion of the infarct-related occluded coronary artery often found in patients who have survived an acute myocardial infarction (50).

Taken together, all the evidence currently available suggests that coronary mast cells possess both pro-thrombotic and anti-thrombotic propensities, the former depending on protease-induced plaque destabilization with ensuing erosion or rupture, and the latter on heparin proteoglycan-dependent inhibition of platelet–collagen interaction (Fig. 5). How can we understand this dual role of mast cells in the development of atherothrombotic complications of coronary atherosclerosis? Much evidence has accumulated that plaque ulceration (erosion or rupture) frequently occurs in asymptomatic individuals (51). Consequently, most coronary thrombi do not produce clinical manifestations. Accordingly, most mural thrombi do not occlude the vessel, as they do not grow, but rather are transient phenomena. Such subclinical coronary thrombi will heal and, according to an old hypothesis, a healing mural thrombus, although clinically silent, may still contribute to plaque growth (52). Thus, the ulcerated area will be covered by endothelium, the small thrombus being incorporated into the plaque in a process called organization of a thrombus. Moreover, activated platelets will secrete potent fibrogenic mediators, such as platelet-derived growth factor and transforming growth factor-β, which together will induce growth of smooth cells with collagen synthesis (53).

Why do some plaque ulcerations lead to an occlusive thrombus that may cause myocardial infarction and sudden cardiac death, while others pass unnoticed? Libby *et al.* have recently elegantly summarized how the thrombotic potential of an atheroma may be regulated (54). Accordingly, the locally prevailing haemostatic and fibrinolytic balance probably proves decisive in determining the fate of a given disruption of an atheroma. Considering the evidence that has recently accumulated suggesting that mast cells may both trigger plaque ulceration and inhibit platelet aggregation and induce fibrinolysis, it is tempting to speculate that one factor that can regulate this delicate regional balance in the coronary atheroma is the mast cell.

CONCLUSION AND FUTURE DIRECTIONS

Systematic studies on mast cells in human coronary atherosclerotic lesions have demonstrated that the numbers of these cells are increased both in the foam cell-rich early lesions (fatty streaks) and in the rupture-prone advanced lesions (atheromas, also called atherosclerotic plaques). In the early lesions, the mast cells reside in areas where foam cells are also present. In the advanced lesions they are most numerous in the areas known to be prone to rupture. In these rupture-prone areas, the numbers of other inflammatory cells (T lymphocytes and macrophages) are also increased and the numbers of the matrix-producing smooth muscle cells are decreased.

Although few in number, the mast cells are unique in that, in the tissues, they are the major sources of neutral proteases, heparin proteoglycans and histamine. These considerations have prompted us to start a persistent search for mast cell-specific mechanisms in atherogenesis. Regarding the potential roles of mast cells in the foam cell-rich early lesions, the experimental results obtained with rat serosal mast cells served to identify a tightly regulated mechanism by which stimulated mast cells promote the formation of foam cells when co-cultured with rodent peritoneal macrophages or with rodent smooth muscle cells. The essential elements in this sequence of events are the macromolecular heparin proteoglycans and the neutral protease chymase.

Regarding the potential roles of mast cells in the fibrous caps in advanced atherosclerotic plaques, three fundamental observations were made *in vitro*. First, when rodent or human arterial smooth muscle cells are co-cultured with rat serosal mast cells, which are then stimulated to degranulate, the heparin proteoglycans released from the stimulated mast cells inhibit proliferation of smooth muscle cells, and the chymase released induces their apoptosis. Second, experiments *in vitro* have revealed that human mast cell chymase (and tryptase) can activate matrix-degrading metalloproteinases. Third, the mast cell-derived soluble heparin proteoglycans possess significant inhibitory capacity in platelet reactivity toward collagen. If operative *in vivo*, these three effects would tend to weaken the fibrous cap of an atherosclerotic plaque, and so predispose to plaque rupture, but would also tend to limit thrombus growth. Thus, mast cells may prove to be important in determining the fate of a given disruption of an atheroma.

The conceptual framework for the possible functions of mast cells in atherogenesis presents a compilation of data obtained by immunohistochemical techniques in human autopsy samples and experimental data obtained from animal cell culture systems and from biochemical experiments. To close the gap between human and animal data, it will be necessary to perform *in vitro* experiments using human mast cells. In addition, *in vivo* experiments will be necessary with various rodent models, notably knockout mice. Ultimately, we will be faced with the challenge of testing the actual function, either harmful or beneficial, of mast cells in human coronary disease. This could be done by stabilizing the coronary mast cells, or by specifically inhibiting the function of individual components derived from such cells.

ACKNOWLEDGEMENTS

The author wishes to thank Dr Riitta Lassila for fruitful discussions on the role of mast cells in arterial thrombosis, and Mrs Jean Margaret Perttunen for revision of the text.

REFERENCES

1. Stary, H. C. Evolution and progression of atherosclerotic lesions in coronary arteries of children and young adults. *Arteriosclerosis* **9**:I-19–I-32, 1989.
2. Smith, E. B. The relationship between plasma and tissue lipids in human atherosclerosis. *Adv. Lipid Res.* **12**:1–49, 1974.
3. Kovanen, P. T. Atheroma formation: defective control in the intimal round-trip of cholesterol. *Eur. Heart J.* **11 (Suppl. E)**:238–246, 1990.
4. McGill, H. C. Jr. Fatty streaks in the coronary arteries and aorta. *Lab. Invest.* **18**:560–564, 1968.
5. Stary, H. C., Chandler, A. B., Dinsmore, R. E. *et al.* A definition of advanced types of atherosclerotic lesions and a histological classification of atherosclerosis. A report from the Committee on Vascular Lesions of the Council on Arteriosclerosis, American Heart Association. *Circulation* **92**:1355–1374, 1995.
6. Richardson, P. D., Davies, M. J. and Born, G. V. Influence of plaque configuration and stress distribution on fissuring of coronary atherosclerotic plaques. *Lancet* **ii**:941–944, 1989.

7. Farb, A., Burke, A. P., Tang, A. L., *et al.* Coronary plaque erosion without rupture into a lipid core. A frequent cause of coronary thrombosis in sudden coronary death. *Circulation* **93**:1354–1363, 1996.
8. Ross, R. Atherosclerosis – an inflammatory disease. *N. Engl. J. Med.* **340**:115–126, 1999.
9. Stary, H. C., Chandler, A. B., Glagov, S. *et al.* A definition of initial, fatty streak, and intermediate lesions of atherosclerosis: a report from the Committee on Vascular Lesions of the Council on Atherosclerosis, American Heart Association. *Arterioscler. Thromb.* **14**:840–856, 1994.
10. Atkinson, J. B., Harlan, C. W., Harlan, G. C. and Virmani, R. The association of mast cells and atherosclerosis: a morphologic study of early atherosclerotic lesions in young people. *Hum. Pathol.* **25**:154–159, 1994.
11. Agis, H., Willheim, M., Sperr, W. R. *et al.* Monocytes do not make mast cells when cultured in the presence of SCF. Characterization of the circulating mast cell progenitor as c-kit$^+$, CD34$^+$, Ly$^-$, CD14$^-$, CD17$^-$, colony-forming cell. *J. Immunol.* **151**:4221–4227, 1993.
12. Galli, S. J., Iemura, A., Garlic, D. S., Gamba-Vitalo, C., Zsebo, K. M. and Andrews, R. G. Reversible expansion of primate mast cell populations *in vivo* by stem cell factor. *J. Clin. Invest.* **91**:148–152, 1993.
13. Schwartz, L. B. and Austen, K. F. Structure and function of the chemical mediators of mast cells. *Prog. Allergy* **34**:271–321, 1984.
14. Kovanen, P. T. Mast cell granule-mediated uptake of low density lipoproteins by macrophages: a novel carrier mechanism leading to the formation of foam cells. *Ann. Med.* **23**:551–559, 1991.
15. Irani, A. A., Schechter, N. M., Craig, S. S., DeBlois, G. and Schwartz, L. B. Two types of human mast cells that have distinct neutral protease compositions. *Proc. Natl. Acad. Sci. USA* **83**:4464–4468, 1986.
16. Irani, A. A., Nilsson, G., Miettinen, U. *et al.* Recombinant human stem cell factor stimulates differentiation of mast cells from dispersed human fetal liver cells. *Blood* **80**:3009–3021, 1992.
17. Irani, A. A., Craig, S. S., Nilsson, G., Ishizaka, T. and Schwartz, L. B. Characterization of human mast cells developed *in vitro* from fetal liver cells cocultured with murine 3T3 fibroblasts. *Immunology* **77**:136–143, 1992.
18. Kaartinen, M., Penttilä, A. and Kovanen, P. T. Accumulation of activated mast cells in the shoulder region of human coronary atheroma, the predilection site of atheromatous rupture. *Circulation* **90**:1669–1678, 1994.
19. Fuster, V., Badimon, L., Badimon, J. J. and Chesebro, J. H. The pathogenesis of coronary artery disease and the acute coronary syndromes. *N. Engl. J. Med.* **326**:242–250, 1992.
20. Falk, E. Why do plaques rupture? *Circulation* **86 (Suppl III)**:30–42, 1992.
21. Kovanen, P. T., Kaartinen, M. and Paavonen, T. Infiltrates of activated mast cells at the site of coronary atheromatous erosion or rupture in myocardial infarction. *Circulation* **92**:1084–1088, 1995.
22. Kaartinen, M., van der Wal, A. C., van der Loos, C. M., Piek, J. J., Koch, K. T., Becker, A. E. and Kovanen, P. T. Mast cell infiltration in acute coronary syndromes: implications for plaque rupture. *J. Am. Coll. Cardiol.* **32**:606–612, 1998.
23. Kovanen, P. T. Mast cells in human fatty streaks and atheromas: implications for intimal lipid accumulation. *Curr. Opin. Lipidol.* **7**:281–286, 1996.
24. Wang, Y., Lindstedt, K. A. and Kovanen, P. T. Mast cell granule remnants carry LDL into smooth muscle cells of the synthetic phenotype and induce their conversion into foam cells. *Arterioscler. Thromb. Vasc. Biol.* **15**:801–810, 1995.
25. Kaartinen, M., Penttilä, A. and Kovanen, P. T. Extracellular mast cell granules carry apolipoprotein B-100-containing lipoproteins into phagocytes in human arterial intima: functional coupling of exocytosis and phagocytosis in neighboring cells. *Arterioscler. Thromb. Vasc. Biol.* **15**:2047–2054, 1995.
26. Johnson, W. J., Mahlberg, F. H., Rothblat, G. H. and Phillips, M. S. Cholesterol transport between cells and high-density lipoproteins. *Biochim. Biophys. Acta.* **1085**:273–298, 1991.
27. Barter, P. High-density lipoproteins and reverse cholesterol transport. *Curr. Opin. Lipidol.* **4**:210–217, 1993.
28. Tailleux, A. and Fruchart, J.-C. HDL heterogeneity and atherosclerosis. *Crit. Rev. Clin. Lab. Sci.* **33**:163–202, 1996.
29. Lee, M., Lindstedt, L. K. and Kovanen, P. T. Mast cell-mediated inhibition of reverse cholesterol transport. *Arterioscler. Thromb.* **12**:1329–1335, 1992.
30. Lindstedt, L., Lee, M., Castro, G. R., Fruchart, J.-C. and Kovanen, P. T. Chymase in exocytosed rat mast cell granules effectively proteolyses apolipoprotein AI-containing lipoproteins, so reducing the cholesterol efflux-inducing ability of serum and aortic intimal fluid. *J. Clin. Invest.* **97**:2174–2182, 1996.
31. Von Eckardstein, A. Cholesterol efflux from macrophages and other cells. *Curr. Opin. Lipidol.* **7**:308–319, 1996.

32. Lee, M., von Eckardstein, A., Lindstedt, L., Assmann, G. and Kovanen, P. T. Depletion of Preβ_1LpA1 and LpA4 particles by mast cell chymase reduces cholesterol efflux from macrophage foam cells induced by plasma. *Arterioscler. Thromb. Vasc. Biol.* **19**:1066–1074, 1999.
33. Libby, P. Molecular bases of the acute coronary syndromes. *Circulation* **91**:2844–2850, 1995.
34. Vartio, T., Seppä, H. and Vaheri, A. Susceptibility of soluble and matrix fibronectin to degradation by tissue proteinases, mast cell chymase and cathepsin G. *J. Biol. Chem.* **256**:471–477, 1981.
35. Lohi, J., Harvima, I. and Keski-Oja, J. Pericellular substrates of human mast cell tryptase: 72000 dalton gelatinase and fibronectin. *J. Cell Biochem.* **50**:337–349, 1992.
36. Birkedal-Hansen, H., Moore, W. G. I., Bodeen, M. K. *et al.* Matrix metalloproteinases: A review. *Crit. Rev. Oral Biol. Med.* **4**:197–250, 1993.
37. Nikkari, S. T., O'Brien, K. D., Ferguson, M. *et al.* Interstitial collagenase (MMP-1) expression in human carotid atherosclerosis. *Circulation* **92**:1393–1398, 1995.
38. Saarinen, J., Kalkkinen, N., Welgus, H. G. and Kovanen, P. T. Activation of human interstitial procollagenase through direct cleavage of the Leu83–Thr84 bond by mast cell chymase. *J. Biol. Chem.* **269**:18134–18140, 1994.
39. Gruber, B. L., Marchese, M. J., Suzuki, K. *et al.* Synovial procollagenase activation by human mast cell tryptase: dependence upon matrix metalloproteinase 3 activation. *J. Clin. Invest.* **84**:1657–1662, 1989.
40. Johnson, J. L., Jackson, C. L., Angelini, G. D. and George, S. J. Activation of matrix-degrading metalloproteinases by mast cell proteinases in atherosclerotic plaques. *Arterioscler. Thromb. Vasc. Biol.* **18**:1707–1715, 1998.
41. Kaartinen, M., Penttilä, A. and Kovanen, P. T. Mast cells in rupture-prone areas of human coronary atheromas produce and store TNF-alpha. *Circulation* **94**:2787–2792, 1996.
42. Saren, P., Welgus, H. G. and Kovanen, P. T. TNF-alfa and IL-1beta selectively induce expression of 92-kDa gelatinase by human macrophages. *J. Immunol.* **157**:4159–4165, 1996.
43. Wang, Y. and Kovanen, P. T. Heparin proteoglycans released from rat serosal mast cells inhibit proliferation of rat aortic smooth muscle cells in culture. *Circ. Res.* **84**:74–83, 1999.
44. Kalsner, S. and Richards, R. Coronary arteries of cardiac patients are hyperreactive and contain stores of amines: a mechanism for coronary spasm. *Science* **223**:1435–1437, 1984.
45. Forman, M. B., Oates, J. A., Robertson, D., Robertson, R. M., Roberts, L. J. and Virmani, R. Increased adventitial mast cells in a patient with coronary spasm. *N. Engl. J. Med.* **313**:1138–1141, 1985.
46. Laine, P., Kaartinen, M., Penttilä, A., Panula, P., Paavonen, T. and Kovanen, P. T. Association between myocardial infarction and the mast cells in the adventitia of the infarct-related coronary artery. *Circulation* **99**:361–369, 1999.
47. Lassila, R., Lindstedt, K. and Kovanen, P. T. Native macromolecular heparin proteoglycans exocytosed from stimulated rat serosal mast cells strongly inhibit platelet-collagen interactions. *Arterioscler. Thromb. Vasc. Biol.* **17**:3578–3587, 1997.
48. Kauhanen, P., Kovanen, P. T., Reunala, T. and Lassila, R. Effects of skin mast cells on bleeding time and coagulation activation at the site of platelet plug formation. *Thromb. Haemost.* **79**:843–847, 1998.
49. Sillaber, C., Baghestanian, M., Bevec, D., Willheim, M., Agis, H., Kapiotis, S., Fureder, W., Bankl, H. C., Kiener, H. P., Speiser, W., Binder, B. R., Lechner, K. and Valent, P. The mast cell as site of tissue-type plasminogen activator expression and fibrinolysis. *J. Immunol.* 162:1032–1041, 1999.
50. Christian, T. F., Milavetz, J. J., Miller, T. D., Clements, I. P., Holmes, D. R. and Gibbons, R. J. Prevalence of spontaneous reperfusion and associated myocardial salvage in patients with acute myocardial infarction. *Am. Heart. J.* **135**:421–427, 1998.
51. Fuster, V. and Lewis, A. Conner Memorial Lecture. Mechanisms leading to myocardial infarction: insights from studies of vascular biology. *Circulation* **90**:2126–2146, 1994.
52. Duguid, J. B. Thrombosis as a factor in the pathogenesis of coronary atherosclerosis. *J. Pathol. Bacteriol.* **58**:207–212, 1946.
53. Ross, R. The pathogenesis of atherosclerosis: a perspective for 1990s. *Nature* **362**:801–809, 1993.
54. Libby, P., Mach, F., Schönbeck, U., Bourcier, T. and Aikawa, M. Regulation of the thrombotic potential of atheroma. *Thromb. Haemost.* **82**:736–741, 1999.

CHAPTER 31

Role of Mast Cells in Endogenous Fibrinolysis and Related (Patho)physiological Processes

PETER VALENT

Department of Internal Medicine I, Division of Hematology & Hemostaseology, University of Vienna, Austria

INTRODUCTION

Mast cells (MC) are multifunctional immune cells producing diverse vasoactive and proinflammatory mediator substances, such as histamine, tumour necrosis factor-α, prostaglandin D_2 and proteolytic enzymes (1–4). Most of these mediators are important regulators of vascular endothelial cells (EC) (5, 6). MC-derived enzymes have also been implicated in the biosynthesis (processing) or degradation of vasoactive hormones, including endothelin, ANF (atrial natriuretic factor), angiotensin, and VIP (vasoactive intestinal peptide) (7–9).

Accumulating evidence suggests that interactions between MC and EC contribute to a number of pathophysiological reactions *in vivo*. For example, mast cell-derived histamine regulates expression of selectins in EC with consecutive rolling of leukocytes. MC-derived TNF-α is involved in the rapid transmigration of neutrophils into inflammed tissues in experimental animals (10–12). Another well-established concept is that MC and their mediators contribute to the capillary leak and rapid oedema formation in IgE-dependent (or other MC-related) inflammatory responses. Likewise, it was found that the rapid IgE-mediated influx of ^{125}I-fibrinogen into skin tissue of mice is mast cell-dependent (13, 14). This influx is followed by local formation and transient deposition of fibrin.

In most inflammatory reactions, the altered tissue is capable of entering complete reconstitution (homeostasis) by virtue of several repair processes. A number of cells and mechanisms may contribute to the repair that follows oedema formation and fibrin deposition. Likewise, EC have been implicated in the regulation of fibrinolysis. Other cell types, such as the macrophage or smooth muscle cell, may also contribute to local tissue repair. However, little is known about the possible function of the MC as a repair cell (15, 16). The notion that MC express the repair-associated molecule heparin (17) as

MAST CELLS AND BASOPHILS
ISBN 0-12-473335-2

well as several proteases (18) raised the possibility that MC indeed can act as repair cells. The emerging concept that will be presented here is that MC contribute to endogenous fibrinolysis as part of the tissue repair following inflammatory or other local reactions accompanied by (transient) fibrin deposition. The concept is based on the observation that MC, in addition to heparin and other mediators and molecules, express enzymatically active tissue-type plasminogen activator (tPA) and urinary-type plasminogen activator (uPA) receptor (uPAR), but do not express inhibitors of tPA or uPA.

CONTRIBUTION OF MAST CELLS TO FIBRINOGEN EXTRAVASATION AND FIBRIN FORMATION

Following an IgE-dependent or -independent activation of MC, the release of vasoactive mediators (histamine, vascular endothelial growth factor (VEGF), prostaglandins, etc.) leads to an increase in capillary permeability. The increased vascular permeability leads to the translocation of plasma molecules and blood cells into the tissues, and the extravasation of fibrinogen is followed by the formation of fibrin (13, 14). However, in virtually all biological processes, fibrin deposition has to be transient and terminated by endogenous fibrinolysis. Indeed, even in case of recurring oedema formation, no permanent deposition of fibrin is seen in the skin of allergic patients.

MAST CELL-DEPENDENT FIBRINOLYSIS: ROLE OF TRYPTASE AND CHYMASE

Several previous observations have already raised the possibility that MC are involved in the process of endogenous fibrinolysis. Thus, the mast cell enzyme tryptase is capable of degrading fibrinogen (19) (Table I). Tryptase also activates pro-urokinase with consecutive fibrinolysis (20). The second proteolytic MC enzyme, chymase, is able to inactivate thrombin (21). Heparin, which is also stored in MC (17), prevents coagulation by acting as co-factor of anti-thrombin III. However, heparin has also been recognized as an important co-factor of tryptase (22) and tPA (23) (Table I). Histamine (produced by MC and basophils) reportedly decreases the antifibrinolytic activity of EC by increasing the cellular tPA/PAI (plasminogen activator inhibitor) ratio (24). However, this histamine effect is probably antagonized by MC-derived TNF-α, which strongly induces expression of PAI-1 in EC. Histamine may also act as a 'pro-thrombotic' agent by inducing platelet aggregation.

Table I
Mast Cell-derived Molecules Involved in Fibrino(geno)lysis

Molecule	Effect(s)
tPA	Induces plasminogen activation
uPAR	Binding uPA (required for uPA activity)
Tryptase	Induces fibrinogenolysis, activates pro-uPA
Heparin	Activates tPA and tryptase

THE MAST CELL AS SITE OF tPA EXPRESSION

The key enzyme of fibrinolysis, plasmin, is generated from plasminogen through the catalytic action of two enzymes: the tissue-type plasminogen activator tPA, and the urinary-type plasminogen activator (uPA, urokinase) (25, 26). Both tPA and uPA are considered to play a role in endogenous fibrinolysis. Thus, the functional disruption of the *tPA* gene in experimental animals (knockout mice) leads to impaired clot lysis and spontaneous fibrin deposition (27). Mice with a combined deficiency (*tPA* and *uPA* gene) suffer from extensive fibrin deposition with impaired organ function, occurrence of vascular thrombosis, loss of fertility, and reduced survival (27). The emerging concept is that endogenous tPA and uPA are involved in the degradation of locally generated fibrin. The exact distribution of tPA in the various cells in the microvasculature in health and disease, however, has not been defined.

Several previous and more recent data suggest that MC are capable of producing tPA. In a previous study, a mouse mast cell line was found to express the tPA protein (28). The cultured bone marrow derived interleukin-3 (IL-3)-dependent mouse mast cell also expresses tPA (unpublished observation). More recently, primary human tissue MC, as well as the human mast cell line HMC-1, were found to express tPA mRNA (29) (Table I). In addition, the tPA protein was detected in supernatants of isolated lung MC (29) as well as in tissue MC by *in situ* immunohistochemistry (29–31). HMC-1 cells were also found to contain and release detectable tPA (29). As assessed by Western blot analysis, the tPA protein in the HMC-1 supernatants exhibits a molecular mass of approximately 70 kDa, similar to recombinant tPA (Fig. 1). The second protagonist of fibrinolysis, uPA, was not

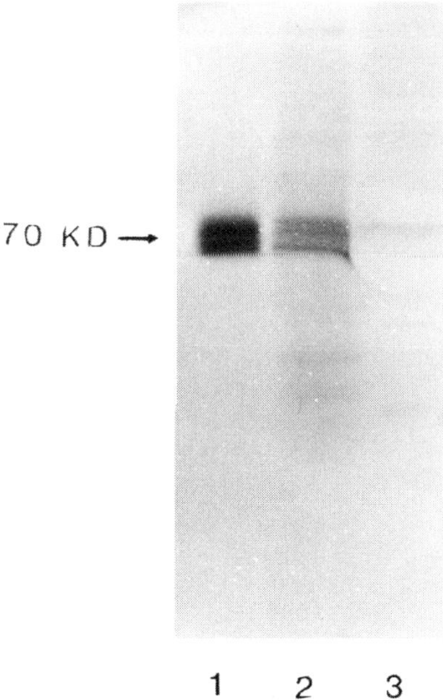

Fig. 1 Western blot analysis of supernatant of HMC-1 cells using an antibody against tPA (3-VPA). The HMC-1 supernatant (lane 2) was found to contain the 70-kDa form of tPA corresponding to recombinant CHO-derived tPA (lane 1). The second plasminogen activator, uPA (anti-uPA antibody: scuPA) was not detectable in HMC-1 supernatants by Western blotting (lane 3).

detectable in primary human MC or in the mast cell line HMC-1 (29). The constitutive expression of tPA mRNA in MC *in vitro* is associated with constant release and accumulation of the mature protein in cell supernatants (29). Whether such constant release of tPA from MC also occurs *in vivo*, is not known. Also, little is known about the regulation of expression and release of tPA in mast cells by cytokines or other factors. However, recent data suggest that stem cell factor (SCF) regulates both the expression and release of tPA in MC (29).

HUMAN MAST CELLS EXPRESS THE UROKINASE RECEPTOR (uPAR, CD87)

The activity of urokinase is dependent on ligand binding to a specific receptor, the urokinase receptor (uPAR, CD87). Recent data suggest that MC express detectable amounts of uPAR/CD87 on their surface (32) (Table I). Since MC do not express uPA, it is tempting to speculate that uPAR on MC interacts with uPA released from neighbouring cells in the microvasculature. This is of particular interest since uPA–uPAR interactions have also been implicated in the regulation of cell migration (chemotaxis). Indeed, urokinase was found to induce migration of human MC most probably by interacting with uPAR (32). Figure 2 shows the dose-dependent effect of uPA on migration of human lung mast cells.

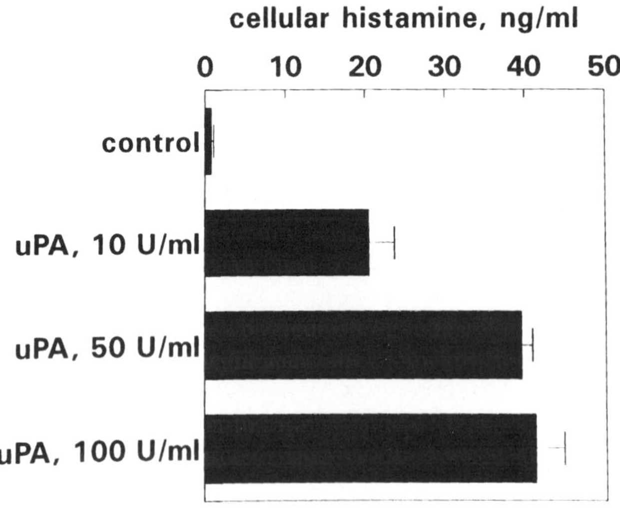

Fig. 2 Dose-dependent effect of uPA on migration of human lung mast cells. Mast cell migration was analysed in a chemotaxis chamber system. Various concentrations of uPA were applied as indicated. The amount of migrated mast cells in primary lung cell suspension was quantified by measuring cellular histamine levels (cells from equal volumes analysed) after migration. As can be seen, exposure to uPA was followed by a significant increase in cellular histamine (mast cells) compared to control.

FUNCTIONAL SIGNIFICANCE OF MC-tPA: MAST CELLS ARE 'PRO-FIBRINOLYTIC' CELLS

The cellular basis of endogenous fibrinolysis has been a matter of numerous speculations. EC apparently are a source of both tPA and uPA. Activated macrophages and smooth muscle cells are also capable of producing plasminogen activators. However, under diverse *in vitro* conditions, these cells do not exhibit fibrinolytic activity. By contrast, some tumour cells express tPA or uPA in excess over PAIs, so that they can trigger fibrinolysis. However, a physiological cell, which would serve as an exclusive and potent source of fibrinolytic activity, had not been identified before. The observation that MC not only express tPA but also trigger fibrinolysis *in vitro* (29) is remarkable. Thus, in contrast to all other cell types (EC, smooth muscle cells, fibroblasts, blood cells) human MC (lysates or supernatants) induced fibrinoclot dissolution *in vitro* (29). It thus appears that tPA-dependent fibrinolysis *in vitro* is largely confined to MC (comparing normal cells). The role of tPA in MC-dependent fibrinolysis was confirmed by the observation that the clot lysis effect of MC can be neutralized by anti-tPA antibody (29).

MOLECULAR BASIS OF MAST CELL DEPENDENT FIBRINOLYSIS: ROLE OF INHIBITORS (PAIs)

The profibrinolytic activities of tPA and uPA are counterbalanced by plasminogen activator inhibitors (PAIs) (33). Likewise, the clot lysis effect of MC (supernatants or lysates) can be neutralized by addition of inhibitory PAI-1 (29). In most 'physiological' cells that express tPA (EC, smooth muscle cells, macrophages, etc.), significant amounts of inhibitors (PAI-1 or PAI-2) have been detected. By contrast, however, primary human lung MC (as well as HMC-1) were found to lack detectable amounts of PAI-1, PAI-2 or PAI-3 (by Northern blotting and/or ELISA) (29). The absence of PAIs in MC is a reasonable explanation for the unique expression of huge amounts of active (uncomplexed) tPA in these cells, as opposed to other microenvironmental cells.

FIBRIN-INDEPENDENT PLASMINOGEN ACTIVATION

The effect of MC/HMC-1 supernatants on plasminogen activation is striking and different from recombinant tPA, or tPA purified from HMC-1 supernatant. Thus, while crude HMC-1 supernatant induced plasminogen conversion into plasmin in the absence of any accessory molecule, rtPA or purified HMC-1 tPA showed no significant plasminogen-activation, unless the reaction mixture was supplemented with either fibrin or heparin (29). Since MC express heparin (17), it is tempting to speculate that heparin (or similar proteoglycans) may be the responsible accessory molecule that acts as a co-factor of MC tPA.

CLINICAL SIGNIFICANCE

So far we do not have definitive evidence that the fibrinolytic effect of MC is of pathophysiological significance or has a protective (repair) function. However, several observations that would be in favour of our concept exist. Malone *et al.* (34) have reported on an inverse correlation between the number of MC and the amount of fibrin deposits in patients with rheumatoid arthritis. In addition, MC-dependent (rapid) fibrin deposition in experimental animals seems to resolve more rapidly than MC-independent fibrin deposition. Whether such correlations are indeed caused by the production and release of tPA in MC is currently under investigation. Another important model could be mastocytosis, a disease characterized by abnormal growth and accumulation of tissue MC in diverse organs. Interestingly, patients with severe forms of the disease (malignant mastocytosis, mast cell leukaemia) often suffer from coagulation disorders. In this regard it is noteworthy that signs of hyperfibrinolysis have recently been reported in a case of malignant mast cell disease (35). Whether indeed mast cell-derived tPA plays a role in the coagulation disorder observed in such patients remains to be elucidated.

MAST CELLS ACCUMULATE AT THE SITE OF THROMBOSIS

An important question seems whether MC are (also) involved in the repair following intravascular formation and deposition of fibrin (i.e. thrombosis). In fact, MC are a source of both tPA and heparin, molecules that are of pathophysiological significance (thrombosis) and can be used as specific therapy. Moreover, it is well known that endogenous fibrinolysis can occur in cases of thrombosis and (under anticoagulant therapy) can even result in a dissolution of the thrombotic mass. The mechanisms of endogenous thrombolysis, however, are not known. The possibility that MC are involved in the process of endogenous thrombolysis following a thromboembolic event is an attractive hypothesis, especially in the light of the current data (tPA expression in MC). For example, it is well established that MC-deficient mice have an increased incidence to develop fatal thromboembolic events after India ink injection compared with their normal littermates (36). We have recently observed an association between thromboembolic events and a local accumulation of mast cells. In particular, MC were found to increase in number and to redistribute to the upper endocardium in patients with auricular thrombosis (30, 37). A similar increase in MC numbers was also found in patients with deep venous thrombosis (38) and also in many other thromboembolic disorders (31). Interestingly, the MC accumulating at the site of thrombus formation were found to express less tPA protein compared with the non-redistributed MC in the same tissue sections (30). Whether this loss of mast cell tPA is due to release from MC, reduced synthesis or other mechanism(s) is not known.

SCF REGULATES EXPRESSION AND RELEASE OF tPA IN MAST CELLS

The stem cell factor (SCF), also termed mast cell growth factor (MGF), is a major regulator of human (and rodent) MC. Thus, SCF induces MC differentiation, as well as chemotaxis and mediator secretion in MC (39–44). The receptor for SCF (c-kit) is expressed in MC progenitors as well as on mature MC (45–47).

The possible role of SCF and c-kit in the repair functions of MC is the subject of ongoing research. One important aspect is that SCF is detectable in activated (isolated) EC as well as in local sites of thrombosis (where EC are considered to be in an activated state) (37). Furthermore, it was found that thrombin activates cultured EC to produce and secrete the soluble form of SCF which induces MC migration (48). Thus, EC-derived SCF may be capable of triggering a local accumulation of MC by inducing MC chemotaxis. Since SCF is also expressed in membrane-bound form in endothelial cells or other stroma cells, the localizing effect of SCF may also be due to a direct adhesive interaction between MC and local microenvironmental cells (49). However, SCF not only induces chemotaxis in human MC but also mediator secretion (41, 42). Notably, SCF promotes the release of tPA from MC (29). Moreover, SCF upregulates expression of uPAR in MC (32). It thus appears that SCF is critically involved in the regulation of expression and release of pro-fibrinolytic substances in MC.

CONCLUDING REMARKS AND FUTURE DIRECTIONS

So far MC have primarily been implicated in the 'effector phase' of allergic and other inflammatory reactions, notably in oedema formation, EC activation and accumulation of blood-derived leukocytes. The possibility that MC, in addition, fulfils important repair functions has not been dicussed in detail. The notion that MC express pro-fibrinolytic molecules and are able to induce fibrinolysis *in vitro*, points to a novel role of this cell in the tissue repair following intra- or extravascular fibrin formation. The emerging concept is that the same cell that activates EC and induces a significant capillary leak or pro-thrombotic condition in the (micro)vasculature, also contributes to local fibrinolysis as part of specific repair. Whether this concept holds true for the situation *in vivo* (various disease models) remains to be elucidated.

ACKNOWLEDGEMENT

Supported by Fonds zur Förderung der Wissenschaftlichen Forschung in Österreich, grant no. F005/01.

REFERENCES

1. Galli, S. J. Biology of disease: New insights into 'the 'riddle of the mast cells': microenvironmental regulation of mast cell development and phenotypic heterogeneity. *Lab. Invest.* **62**: 5–33, 1990.
2. Schwartz, L. B. The mast cell. In: *Allergy,* Vol. 1 (Kaplan, A. P. ed.), pp. 53–92. Churchill Livingstone, Edinburgh, 1985.
3. Serafin, W. E. and Austen, K. F. Mediators of immediate hypersensitivity reactions. *N. Engl. J. Med.* **317**: 30–34, 1987.
4. Burd, P. R., Rogers, H. W., Gordon, J. R., Martin, C. A., Jayaraman, S., Wilson, S. D., Dvorak, A. M., Galli, S. J. and Dorf, M. E. Interleukin 3-dependent and -independent mast cells stimulated with IgE and antigen express multiple cytokines. *J. Exp. Med.* **170**: 245–257, 1989.
5. Klein, L. M., Lavker, R. M., Matis, W. L. and Murphy, G. F. Degranulation of human mast cells induces an endothelial cell antigen central to leukocyte adhesion. *Proc. Natl. Acad. Sci. USA* **86**: 8972–8976, 1989.
6. Walsh, L. J., Trinchieri, G., Waldorf, H. A., Whitaker, D. and Murphy, G. F. Human dermal mast cells contain and release tumor necrosis factor α, which induces endothelial leucocyte adhesion molecule 1. *Proc. Natl. Acad. Sci. USA* **88**:4220–4224, 1991.
7. Wypij, D. M., Nochols, S. J., Novak, P. J., Stacy, D. L., Berman, J. and Wisman, S. J. Role of mast cell chymase in the extracellular processing of big-endothelin-1 into endothelin-1 in the perfused rat lung. *Biochem. Pharmacol.* **43**:845–853, 1992.
8. Urata, H., Boehm, K. D., Philip, A., Kinoshita, A., Gabrovsek, J., Bumpus, F. M. and Husain, A. Cellular localization and regional distribution of an angiotensin II-forming chymase in the heart. *J. Clin. Invest.* **91**:1269–1281, 1993.
9. Proctor, G. B., Chan, K. M., Garrett, J. R. and Smith, R. E. Proteinase activities in bovine atrium and the possible role of mast cell tryptase in the processing of atrial natriuretic factor (ANF). *Comp. Biochem. Physiol.* **99**:839–844, 1991.

10. Zhang, Y., Ramos, B. F. and Jakschik, B. A. Neutrophil recruitment by tumor necrosis factor from mast cells in immune complex peritonitis. *Science* **258**:1957–1959, 1992.
11. Yano, H., Wershil, B. K., Arizono, N. and Galli, S. J. Substance P-induced augmentation of cutaneous vascular permeability and granulocyte infiltration in mice is mast cell-dependent. *J. Clin. Invest.* **84**:1276–1286, 1989.
12. Malaviya, R., Ikeda, T., Ross, E. and Abraham, S. N. Mast cell modulation of neutrophil influx and bacterial clearance at sites of infection through TNF-α. *Nature* **381**:77–80, 1996.
13. Wershil, B. K., Mekori, Y. A., Murakami, T. and Galli, S. J. ^{125}I-fibrin deposition in IgE-dependent immediate hypersensitivity reactions in mouse skin. Demonstration of the role of mast cells using genetically mast cell-deficient mice locally reconstituted with cultured mast cells. *J. Immunol.* **139**:2605–2614, 1987.
14. Mekori, Y. A. and Galli, S. J. (125I)fibrin deposition occurs at both early and late intervals of IgE-dependent or contact sensitivity reactions elicited in mouse skin. Mast cell-dependent augmentation of fibrin deposition at early intervals in combined IgE-dependent and contact sensitivity reactions. *J. Immunol.* **145**:3719–3727, 1990.
15. Dvorak, A. M. Granule changes of human skin mast cells characteristic of piecemeal degranulation and association with recovery during wound healing *in situ*. *J. Leukoc. Biol.* **49**:197–210, 1991.
16. Norrby, K. Mast cells and *de novo* angiogenesis: angiogenic capability of individual mast cell mediators such as histamine, TNF, IL-8, and bFGF. *Inflamm. Res.* **46 (Suppl. 1)**:7–8, 1997.
17. Yurt, R. W., Leid, R. W., Austen, K. F. and Silbert, J. E. Native heparin from rat peritoneal mast cells. *J. Biol. Chem.* **252**:518–521, 1977.
18. Irani, A. A., Schechter, N. M., Craig, S. S., DeBlois, G. and Schwartz, L. B. Two human mast cell subsets with distinct neutral protease compositions. *Proc. Natl. Acad. Sci. USA* **83**:4464–4468, 1984.
19. Schwartz, L. B., Badford, T. R., Littman, B. H. and Wintroub, B. U. The fibrinogenolytic activity of purified tryptase from human lung mast cells. *J. Immunol.* **135**:2762–2767, 1985.
20. Stack, M. S. and Johnson, D. A. Human mast cell tryptase activates single chain urinary-type plasminogen activator (pro-urokinase). *J. Biol. Chem.* **269**:9416–9419, 1994.
21. Pejler, G. and Karlstrom, A. Thrombin is inactivated by mast cell secretory granule chymase. *J. Biol. Chem.* **268**:11817–11822, 1993.
22. Sakai, K., Ren, S. and Schwartz, L. B. A novel heparin-dependent processing pathway for human tryptase. Autocatalysis followed by activation with dipeptidyl peptidase I. *J. Clin. Invest.* **97**:895–896, 1996.
23. Stein, P. L., van-Zonneveld, A. J., Pannekoek, H. and Strickland, S. Structural domains of human tissue type plasminogen activator that confer stimulation by heparin. *J. Biol. Chem.* **264**:15441–15444, 1989.
24. Yamamoto, C., Kaji, T., Sakamoto, M., Kozuka, H. and Koizumi, F. Calcium regulation of tissue type plasminogen activator and plasminogen activator inhibitor-1 release from cultured human vascular endothelial cells. *Thromb. Res.* **74**:163–168, 1994.
25. Collen, D. and Lijnen, H. R. Basic and clinical aspects of fibrinolysis and thrombolysis. *Blood* **78**:3114–3124, 1991.
26. Vassalli, J. D., Sappino, A. P. and Belin, D. The plasminogen activator/plasmin system. *J. Clin. Invest.* **88**:1067–1072, 1991.
27. Carmeliet, P., Schoonjans, L., Kieckens, L., Ream, B., Degen, J., Bronson, R., De Vos, R., van den Oord, J. J., Collen, D. and Mulligan, R. C. Physiological consequences of loss of plasminogen activator gene function in mice. *Nature* **368**:419–424, 1994.
28. Bartholomew, J. S. and Woolley, D. E. Plasminogen activator release from cultured murine mast cells. *Biochem. Biophys. Res. Commun.* **153**:540–544, 1988.
29. Sillaber, C., Baghestanian, M., Bevec, D., Willheim, M., Agis, H., Kapiotis, S., Füreder, W., Bankl, H. C., Kiener, H., Speiser, W., Binder, B. R., Lechner, K. and Valent, P. Mast cells as site of tissue type plasminogen activator production and fibrinolysis. *J. Immunol.* **162**:1032–1041, 1999.
30. Bankl, H. C., Radaszkiewicz, T., Pikula, B., Baghestanian, M., Mherabi, M. R., Bankl, H., Lechner, K. and Valent, P. Expression of fibrinolytic antigens in redistributed cardiac mast cells in auricular thrombosis. *Hum. Pathol.* **28**:1283–1290, 1997.
31. Valent, P., Sillaber, C., Baghestanian, M., Bankl, H. C., Kiener, H. P., Lechner, K. and Binder, B. R. What have mast cells to do with edema formation, the consecutive repair, and fibrinolysis? *Int. Arch. Allergy Immunol.* **115**:2–18, 1998.
32. Sillaber, C., Baghestanian, M., Hofbauer, R., Virgolini, I., Bankl, H. C., Füreder, W., Agis, H., Willheim, M., Leimer, M., Scheiner, O., Binder, B. R., Kiener, H., Bevec, D., Fritsch, G., Majdic, O., Kress, H. G.,

Gadner, H., Lechner, K. and Valent, P. Molecular and functional characterization of the urokinase receptor on human mast cells. *J. Biol. Chem.* **272**:7824–7832, 1997.

33. Kruithof, E. K. O. Plasminogen activator inhibitors – a review. *Enzyme* **40**:113–121, 1988.
34. Malone, D. G., Wilder, R. L., Saavedra Delgado, A. M. and Metcalfe, D. D. Mast cell numbers in rheumatoid synovial tissues. Correlations with quantitative measures of lymphocytic infiltration and modulation by antiinflammatory therapy. *Arthritis Rheum.* **30**:130–137, 1987.
35. Wimazal, F., Sperr, W. R., Horny, H.-P., Carroll, V., Binder, B. R., Fonatsch, C., Walchshofer, S., Födinger, M., Schwarzinger, I., Samorapoompichit, P., Chott, A., Dvorak, A. M., Lechner, K. and Valent, P. Hyperfibrinolysis in a case of myelodysplastic syndrome with leukemic spread of mast cells. *Am. J. Hematol.* (in press).
36. Kitamura, Y., Taguchi, T., Yokoyama, M., Inoue, M., Yamatodani, A., Asano, H., Koyama, T., Kanamaru, A., Hatanaka, K. and Wershil, B. K. Higher susceptibility of mast-cell-deficient W/W^{v} mutant mice to brain thromboembolism and mortality caused by intravenous injection of India ink. *Am. J. Pathol.* **122**:469–480, 1986.
37. Bankl, H. C., Radaszkiewicz, T., Klappacher, G. W., Glogar, D., Sperr, W. R., Grossschmidt, K., Bankl, H., Lechner, K. and Valent, P. Increase and redistribution of cardiac mast cells in auricular thrombosis. Possible role of kit ligand. *Circulation* **91**:275–283, 1995.
38. Bankl, H. C., Radaszkiewicz, T., Baghestanian, M., Mherabi, M. R., Bankl, H., Kapiotis, S., Hofbauer, R., Lechner, K. and Valent, P. Mast cells are augmented in deep venous thrombosis and express a profibrinolytic phenotype. *Hum. Pathol.* **30**:188–194, 1999.
39. Irani, A. M., Nilsson, G., Miettinen, U., Craig, S. S., Ashman, L. K., Ishizaka, T., Zsebo, K. M. and Schwartz, L. B. Recombinant human stem cell factor stimulates differentiation of human mast cells from dispersed fetal liver cells. *Blood* **80**:3009–3016, 1992.
40. Mitsui, H., Furitsu, T., Dvorak, A. M., Irani, A. M., Schwartz, L. B., Inagaki, N., Takei, M., Ishizaka, K., Zsebo, K. M. and Gillis, S. Development of human mast cells from umbilical cord blood cells by recombinant human and murine c-kit ligand. *Proc. Natl. Acad. Sci. USA* **90**:735–739, 1993.
41. Valent, P., Spanblöchl, E., Sperr, W. R., Sillaber, C., Agis, H., Zsebo, K. M., Geissler, K., Bettelheim, P. and Lechner, K. Induction of differentiation of human mast cells from bone marrow and peripheral blood mononuclear cells by recombinant human stem cell factor (SCF)/kit ligand (KL) in long term culture. *Blood* **80**:2237–2245, 1992.
41. Bischoff, S. C. and Dahinden, C. A. c-kit ligand: a unique potentiator of mediator release by human lung mast cells. *J. Exp. Med.* **175**:237–244, 1992.
42. Sperr, W. R., Czerwenka, K., Mundigler, G., Müller, M. R., Semper, H., Klappacher, G., Glogar, D., Lechner, K. and Valent, P. Specific activation of human mast cells by the ligand for c-kit: comparison between lung-, uterus- and heart mast cells. *Int. Arch. Allergy Appl. Immunol.* **102**:170–175, 1993.
43. Nilsson, G., Butterfield, J. H., Nilsson, K. and Siegbahn, A. Stem cell factor is a chemotactic factor for human mast cells. *J. Immunol.* **153**:3717–3723, 1994.
44. Baghestanian, M., Hofbauer, R., Bevec, D., Agis, H., Willheim, M., Scheiner, O., Füreder, W., Müller, M. R., Kress, H. G., Lechner, K. and Valent, P. The c-kit ligand stem cell factor and anti-IgE promote expression of monocyte chemoattractant protein 1 (MCP-1) in human lung mast cells. *Blood* **90**:4438–4449, 1997.
45. Sillaber, C., Bevec, D., Ashman, L. K., Butterfield, J. H., Lechner, K., Maurer, D., Bettelheim, P. and Valent, P. IL-4 regulates c-kit gene product expression in human myeloid- and mast cell progenitors. *J. Immunol.* **147**:4224–4228, 1991.
46. Valent, P. The riddle of the mast cell: c-kit ligand as missing link? *Immunol. Today* **15**:111–114, 1994.
47. Valent, P. and Bettelheim, P. Cell surface structures on human basophils and mast cells: biochemical and functional characterization. *Adv. Immunol.* **52**:333–423, 1992.
48. Baghestanian, M., Hofbauer, R., Kress, H. G., Wojta, J., Fabry, A., Binder, B. R., Kaun, C., Müller, M. R., Mehrabi, M. R., Kapiotis, S., Sengoelge, G., Ghannadan, M., Lechner, K. and Valent, P. Thrombin augments vascular cell-dependent migration of human mast cells: role of MGF. *Thromb. Haemost.* **77**:577–584, 1997.
49. Adachi, S., Tsujimura, T., Jippo, T., Morimoto, M., Isozaki, K., Kasugai, T., Nomura, S. and Kitamura, Y. Inhibition of attachment between cultured mast cells and fibroblasts by phorbol 12-myristate 13-acetate and stem cell factor. *Exp. Hematol.* **23**:58–65, 1995.

CHAPTER 32

Mast Cells in Myocardial Ischaemia and Reperfusion

*NIKOLAOS G. FRANGOGIANNIS and MARK L. ENTMAN**

Section of Cardiovascular Sciences, The Methodist Hospital, and The DeBakey Heart Center, Department of Medicine, Baylor College of Medicine, Houston, Texas, U.S.A.

INTRODUCTION

Myocardial ischaemia and reperfusion is associated with an intense inflammatory reaction leading to healing and scar formation (1–4). Early descriptive studies of the inflammatory process associated with myocardial infarction by Mallory *et al.* (5) concluded that 'polymorphonuclear leukocytes are attracted and infiltrate around and into the necrotic muscle' and that 'the infiltration is much more active . . . in those portions adjacent to the uninvolved muscle'.

Coronary artery occlusion critically reduces the blood flow to the portion of the myocardium subserved, markedly impairing the energy metabolism, leading to cell death. Myocardial cell necrosis results in the release of subcellular membrane constituents, rich in mitochondria, which are capable of activating the complement cascade (6, 7). This represents the initial chemotactic event responsible for neutrophil influx in the ischaemic myocardium. An additional leukotactic factor appearing in the ischaemic myocardium with reperfusion, is interleukin-8 (IL-8) (8, 9), which participates in neutrophil-mediated myocardial injury by enhancing neutrophil adhesiveness and motility.

Subsequently, activated neutrophils adhere to the endothelium and transmigrate to the extravascular space. Leukocyte–endothelial interactions are regulated by a cascade of molecular steps that correspond to the morphological changes that accompany adhesion (10). Selectins (11, 12) appear to be critical to early neutrophil margination by promoting neutrophil 'rolling'. Although rolling appears to be a prerequisite for eventual firm adherence to blood vessels under conditions of flow, selectin-dependent leukocyte adhesion does not lead to firm adhesion unless another family of adhesion molecules is engaged: the integrins (13). Finally, the neutrophil transmigrates to the extravascular

* Corresponding author.

MAST CELLS AND BASOPHILS
ISBN 0-12-473335-2

space, a process that requires both a chemotactic stimulus and engagement of the β_2 integrin LFA-1 with the endothelial intercellular adhesion molecule ICAM-1 (14, 15). Neutrophils accumulating in the ischaemic and reperfused areas release proteolytic enzymes or reactive oxygen species, injuring surrounding myocytes. However, *in vivo*, these toxic products are almost exclusively secreted by adherent neutrophils. Recent experiments have indicated that neutrophil adhesion to cardiac myocytes is dependent on neutrophil integrin activation and on the induction of ICAM-1 on cardiac myocytes (16–18). ICAM-1 has a highly specific localization to ischaemic but viable myocardium, demarcating a 'border zone' susceptible to neutrophil-mediated injury (19). Myocyte ICAM-1 induction is dependent on a cytokine cascade leading to IL-6 expression (20) in mononuclear cells (21) and myocytes (22). Cardiac mast cells appear to have a significant role in initiating this cytokine cascade (21) and may also serve as important sources of fibrogenic factors during healing of a reperfused myocardial infarct (23, 24).

This chapter summarizes information derived from *in vivo* and *in vitro* experiments on the potential role of mast cells in initiating the cytokine cascade associated with myocardial ischaemia and reperfusion. We begin with a general description of the resident mast cell population in the normal heart, and analyse the potential role of mast cells as a source of cytokines in the ischaemic myocardium. Subsequently, we examine the role of mast cells and their products in the healing process.

MAST CELLS IN NORMAL HEARTS

Mast cells are multifunctional effector cells of the immune system, capable of producing a variety of vasoactive mediators, cytokines and growth factors (25–27). They are found resident in tissues throughout the body, often in association with structures such as blood vessels and nerves, and in proximity to surfaces that interface the external environment. Although mast cells share many characteristics, they do not represent a homogeneous population. Mast cell heterogeneity was established by the pioneering work of Enerback, who demonstrated a distinctive mucosal mast cell phenotype in the gastrointestinal tract of the rat (28–30). In rodents, the use of modified fixation helps to distinguish mast cell subpopulations: mucosal or 'atypical' mast cells are smaller in size and their granules may become resistant to metachromatic staining after routine fixation with formalin, whereas 'typical' or connective tissue mast cells contain large amounts of histamine and stain metachromatically regardless of fixation. By analogy to rodent mast cells, human mast cells also demonstrate evidence of heterogeneity (31). Human mast cell subpopulations are recognized by the presence of distinct protease expression patterns: MC_T cells have tryptase-containing cytoplasmic granules, whereas MC_{TC} cells express both tryptase and chymase (32). Different mast cell subtypes can also vary in mediator content and susceptibility to potential secretagogues.

The presence of mast cells has been established within human heart tissue. Mast cell involvement in the pathogenesis of coronary spasm (33), cardiomyopathy (34) atherosclerosis (35) and myocardial ischaemia (21) has been suggested. Human heart mast cells, obtained from tissue from patients undergoing heart transplantation, have been identified and characterized by their content of tryptase and chymase as MC_{TC}-type mast cells (36–38), capable of producing and releasing a variety of vasoactive mediators. Human heart mast cells released histamine in response to C5a, cross-linking of $Fc_{\varepsilon}RI$, recombinant human stem cell factor (SCF) and compound 48/80 (37). Recently, we

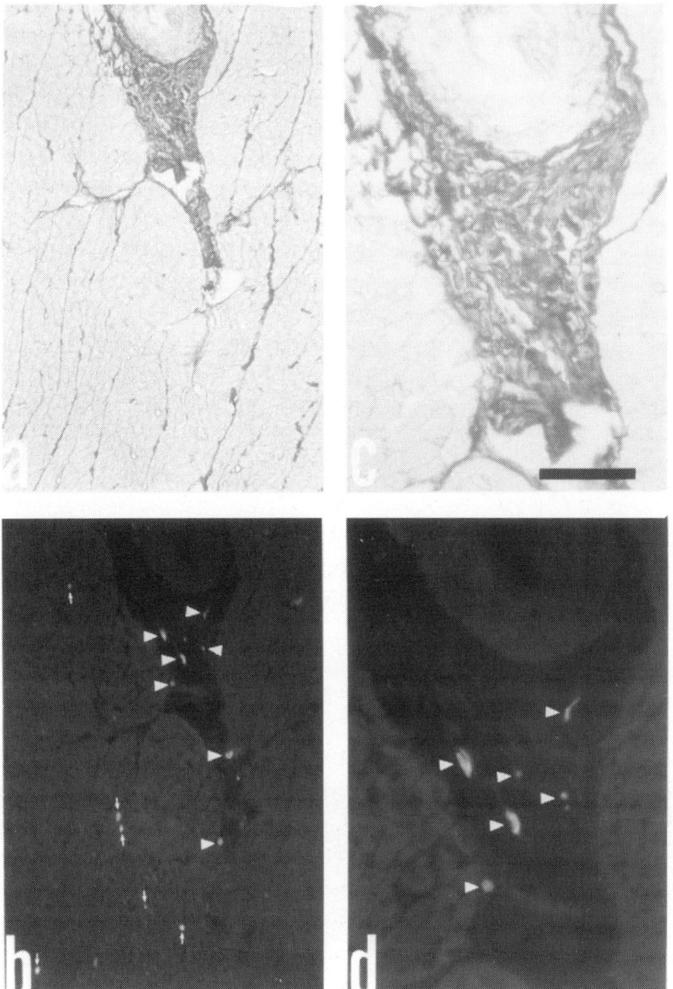

Fig. 1 Resident mast cells in the canine heart. (a, c) Control canine heart stained with picrosirius red. (b, d) Same section stained with FITC–avidin to identify resident mast cells. Note that the majority of cardiac mast cells are perivascular (arrowheads). A significant number of intramyocardial mast cells is also noted.

characterized the histochemical and morphological characteristics of mast cells in normal canine hearts (Fig. 1) (39). Our findings show that canine cardiac mast cells contain chymase and tryptase (Fig. 2) and suggest the presence of a formalin-sensitive population of mast cells in the canine heart (39).

THE MAST CELL AS A SOURCE OF CYTOKINES

Mast cells generate and release a wide variety of inflammatory mediators, including histamine, proteoglycans, neutral proteases and lipid-derived substances (23). Recently, mast cells have been recognized as an important source of pre-formed and newly synthesized cytokines, chemokines and growth factors (40–43). Gordon and Galli (40)

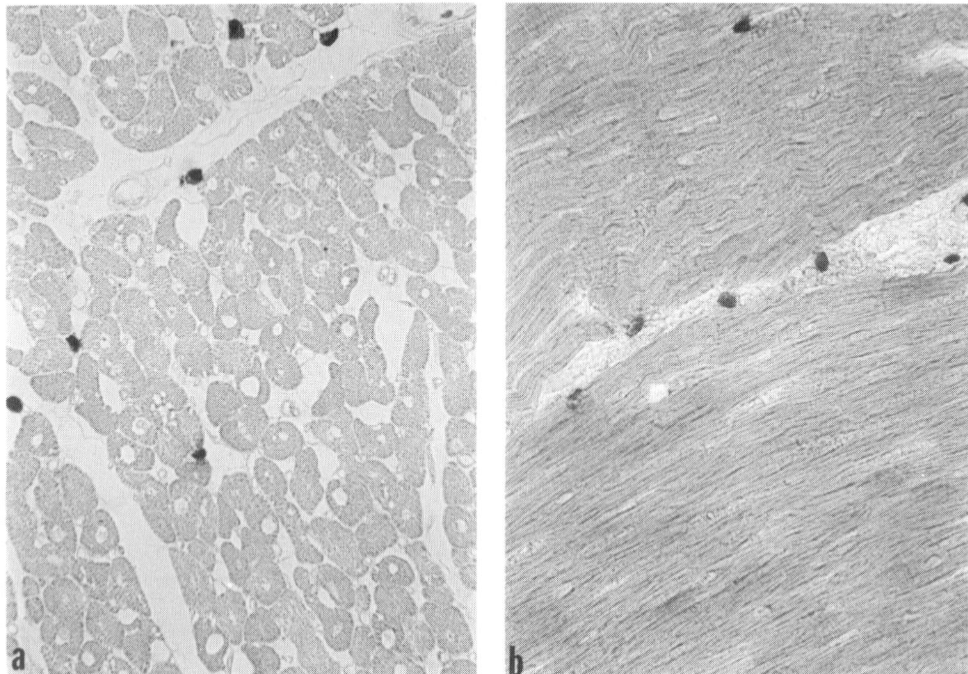

Fig. 2 Canine cardiac mast cells contain tryptase and chymase. (a) Histochemical staining for tryptase. (b) Staining for chymase.

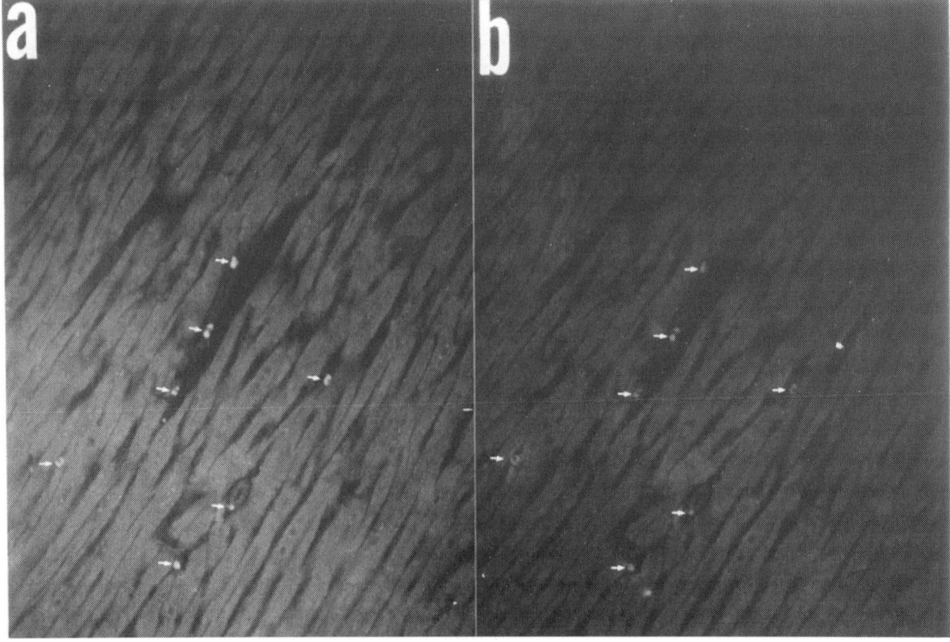

Fig. 3 Canine cardiac mast cells contain preformed TNF-α. Dual fluorescence combining FITC–avidin (a) to identify mast cells with immunohistochemical staining for TNF-α (b). (Reprinted, with permission, from *Circulation* **68**:699–710, 1998.)

identified mouse peritoneal mast cells as an important source of both preformed and immunologically induced tumour necrosis factor-α (TNF-α). Additional observations suggested that mast cells are capable of producing multiple cytokines and chemokines, such as interleukins (IL)-4, IL-5 (44), IL-8 (42), macrophage inflammatory proteins (MIP)-1α and MIP-1β, macrophage chemotactic protein (MCP)-1 and lymphotactin (45). Cytokine expression by different types of mast cells has not been adequately studied. However, Bradding *et al.* (46) indicated that bronchial mucosal mast cells are heterogeneous with respect to cytokine content, suggesting that the biological functions of MC_{TC} and MC_{T} differ as a result of their capacity to generate different cytokine profiles.

Cytokine expression by cardiac mast cells has recently been demonstrated. Mast cells in rupture-prone areas of human coronary atheromas were positive for TNF-α (35). Furthermore, our laboratory (21) has demonstrated constitutive expression of TNF-α in canine cardiac mast cells (Fig. 3).

MAST CELL DEGRANULATION FOLLOWING MYOCARDIAL ISCHAEMIA

Kanwar and Kubes (47) have demonstrated a significant contribution of oxidant-induced mast cell degranulation in granulocyte infiltration and tissue dysfunction associated with reperfusion of the ischaemic intestine. A possible role for cardiac mast cells in mediating injury was suggested in acute re-oxygenation injury in the isolated rat heart (48) and in a porcine model of C5a-mediated myocardial ischaemia (49).

Our studies (21, 23) indicated a role for mast cell mediators in initiating the cytokine cascade ultimately responsible for ICAM-1 induction in the reperfused canine myocardium. We used a canine model of myocardial ischaemia and reperfusion, developed in our laboratory (50), that allows collection of cardiac lymph from chronically instrumented animals, in which all inflammatory sequelae of the instrumentation surgery have dissipated.

The ability of post-ischaemic cardiac lymph to induce ICAM-1 in myocytes was neutralized by an antibody to IL-6 (20). In contrast, ICAM-1 expression in canine endothelial cells stimulated with post-ischaemic cardiac lymph was not neutralized by anti-IL-6 antibody (20). These observations suggested the presence of additional cytokine activity, appearing early following reperfusion of the ischaemic myocardium responsible for IL-6 induction and subsequent ICAM-1 upregulation.

The constitutive presence of TNF-α in mast cells in control canine hearts led us to postulate that mast cell-derived TNF-α may be released following myocardial ischaemia, representing the ‘upstream’ cytokine responsible for initiating the inflammatory cascade.

Our experiments demonstrated a rapid release of histamine (Fig. 4) and TNF-α (Fig. 5) bioactivity in the early post-ischaemic cardiac lymph. In addition, histochemical studies indicated mast cell degranulation in ischaemic, but not in control, sections of canine myocardium. These findings suggested rapid mast cell degranulation and mediator release following myocardial ischaemia. C5a, adenosine and reactive oxygen may represent the stimuli responsible for initiation of mast cell degranulation. Furthermore, *in vitro* experiments showed that early post-ischaemic cardiac lymph is capable of inducing IL-6 expression in canine mononuclear cells. Incubation with a neutralizing antibody to TNF-α in part inhibited IL-6 upregulation (21), suggesting an important role for TNF-α as the upstream cytokine inducer. These studies allowed us to refine our hypothesis

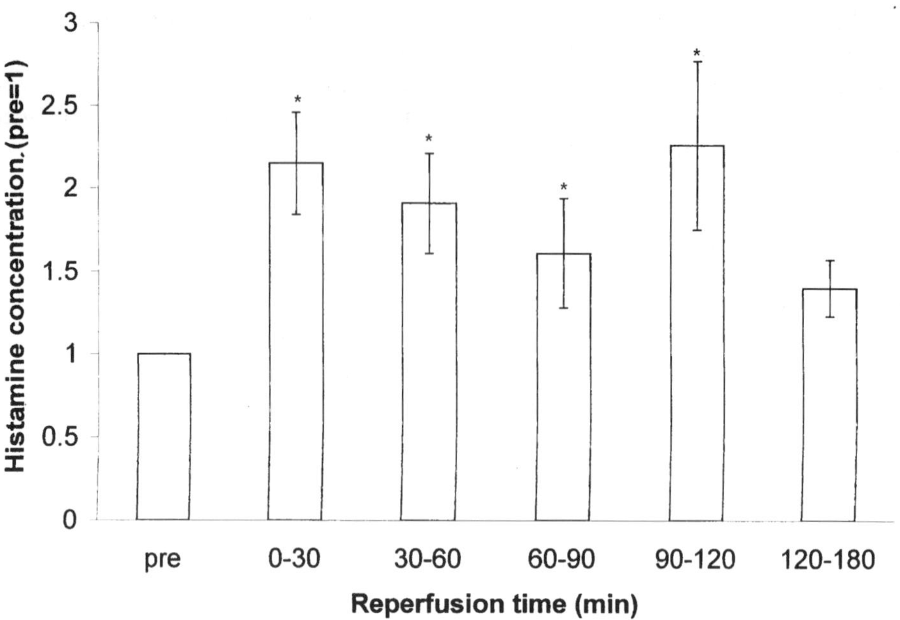

Fig. 4 Kinetics of histamine release in canine post-ischaemic cardiac lymph. A significant early increase in histamine concentration was noted (0–30 min, 2.15±0.3-fold increase; *p<0.05 vs. pre-ischaemic lymph). Numbers of experiments of ischaemia and reperfusion: n=8 for pre, 0–30 min and 30–60 min; n=6 for 60–120 min and 120–180 min. (Reprinted, with permission, from *Circulation* **68**:699–710, 198.)

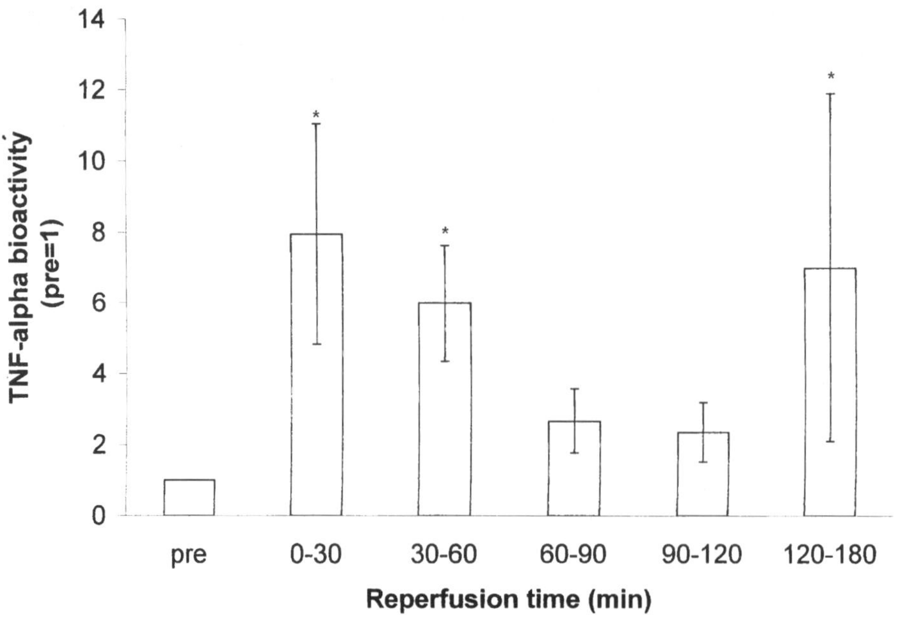

Fig. 5 Release of TNF-α bioactivity in canine cardiac lymph after myocardial ischaemia and reperfusion. A significant increase in TNF-α bioactivity was noted early after reperfusion (0–30 min, 7.94±3.1-fold increase; *p<0.05 vs. pre-ischaemic lymph, n=8). (Reprinted, with permission, from *Circulation* **8**:699–710, 1998.)

regarding the role of mast cells in myocardial ischaemia and reperfusion. Mast cell degranulation appears to be confined in the ischaemic area and results in rapid release of TNF-α, inducing IL-6 in infiltrating mononuclear cells. Histamine may also represent an important autacoid by stimulating surface expression of P-selectin from Weibel–Palade bodies and inducing leukocyte rolling (51).

MAST CELLS IN HEALING MYOCARDIAL INFARCTS

There is significant evidence that mast cells may participate in the fibrotic process. In systemic mastocytosis, a disease characterized by tissue accumulation of mast cells, significant fibrosis occurs in the bone marrow, spleen and liver (52). Furthermore, mast cell involvement in pulmonary fibrotic disorders (53, 54) and the chronic fibrotic process associated with progressive systemic sclerosis (55) has been suggested.

We postulated that mast cell numbers may increase in the healing phase of reperfused canine myocardial infarcts. Using staining with fluorescein isothiocyanate (FITC)-labelled avidin we demonstrated a striking accumulation of mast cells (Fig. 6) in areas of collagen deposition and cell proliferation. The increase in mast cell numbers was first noted after 72 h of reperfusion (23). Following 5–7 days of reperfusion mast cell numbers in fibrotic areas, in which myocytes were fully replaced by scar, were markedly higher

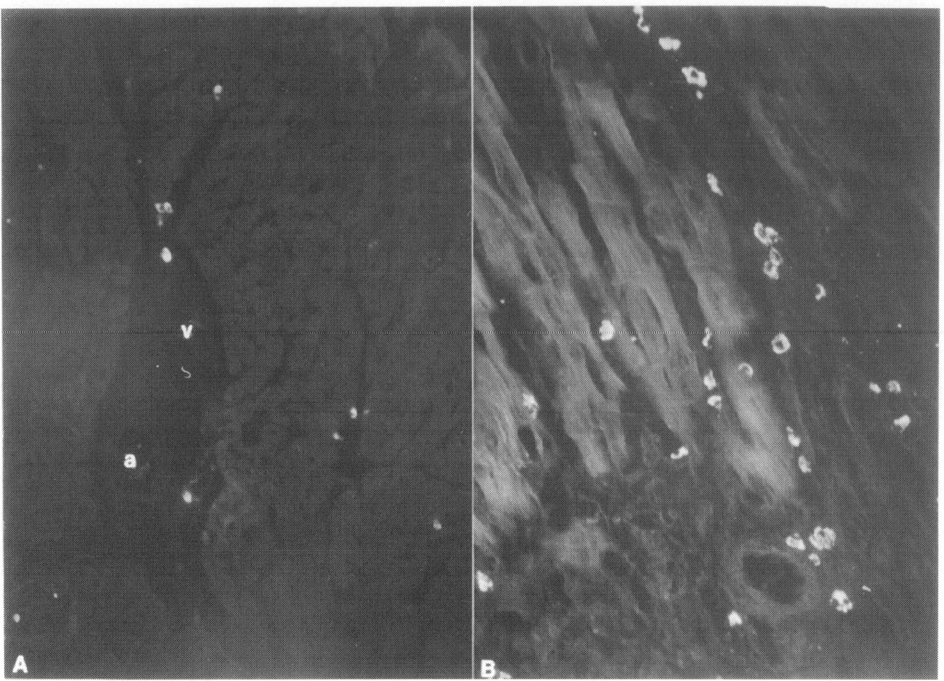

Fig. 6 Increase in mast cell number after myocardial ischaemia and reperfusion. (A) Control canine heart stained with FITC–avidin to identify mast cells, located predominantly along arterioles (a) and venules (v). (B) Ischaemic section of canine myocardium after 1 h of ischaemia and 7 days reperfusion stained with FITC–avidin. A striking accumulation of mast cells is noted in the healing area. (Reprinted, with permission, from *Circulation* **98**:687–698, 1998.)

than the numbers from areas of the same section showing intact myocardium (12.0±2.6-fold increase; $p<0.01$, $n=5$) (23). Partially fibrotic areas also had significantly higher mast cell numbers than non-fibrotic areas (6.1±2.2-fold increase; $p<0.05$, $n=5$). Utilizing immunohistochemistry for the proliferating cell nuclear antigen (PCNA), we identified significant numbers of proliferating cells in healing myocardial infarcts. Our studies showed an excellent correlation between the numbers of tryptase-positive mast cells and PCNA-positive proliferating cells in the same field, indicating that mast cells accumulated in areas of active cell proliferation. Interestingly, we failed to detect significant numbers of proliferating mast cells. Using dual immunohistochemical techniques, we identified the majority of the proliferating PCNA-positive cells as smooth muscle myosin negative myofibroblasts (Fig. 7). These phenotypically altered fibroblasts develop several ultrastructural and biochemical features of smooth muscle cells (56), including the expression of α-smooth muscle actin (57) and have been shown to be the main source of type I and type III pro-collagen mRNA in the infarcted heart (58).

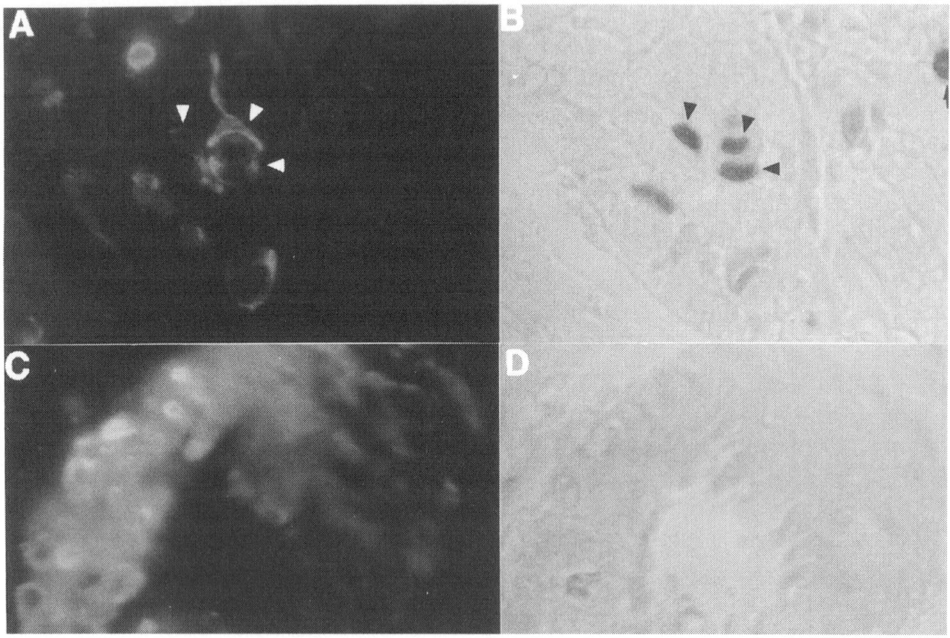

Fig. 7 Many proliferating cells in the healing myocardial scar are identified as α-smooth muscle actin-positive myofibroblasts. Sections from an experiment of 1 h of ischaemia and 5 days of reperfusion were stained for α-smooth muscle actin (A,C) and PCNA (B,D). Many α-smooth muscle actin-positive cells demonstrated evidence of proliferation, as indicated by PCNA expression (arrowheads). In contrast, α-smooth muscle actin-positive cells in control areas were predominantly located in the media of vessels and did not show PCNA expression. (Reprinted, with permission, from *Circulation* **98**:687–698, 1998.)

MAST CELL PRECURSORS INFILTRATE THE HEALING HEART

Our experiments failed to demonstrate significant numbers of proliferating mast cells in the healing heart. Although the contribution of mast cell proliferation cannot be ruled out, we suggest that chemotaxis of circulating mast cell precursors in the healing myocardium may be the predominant mechanism responsible for mast cell accumulation in the ischaemic myocardium. Mast cells originate from $CD34^+$ haematopoietic stem cells and circulate as immature precursors in the peripheral blood (59, 60). Recently, Rodewald *et al.* (61) identified a cell population in murine fetal blood that fulfils the criteria of progenitor mastocytes. It is defined by the phenotype Thy-1 (lo) c-kit (hi), expresses RNAs encoding mast cell-associated proteases, but lacks expression of the high-affinity IgE receptor (61). Using histochemical techniques, we identified in the healing canine myocardium, intravascular tryptase positive, FITC–avidin-positive cells that do not express the basophil marker CD11b (23). Furthermore, tryptase-positive metachromatic cells were found in the post-ischaemic cardiac lymph as early as 48 h post-reperfusion (23). These cells may represent immature mast cell precursors infiltrating the ischaemic and reperfused heart.

SCF INDUCTION FOLLOWING MYOCARDIAL ISCHAEMIA/REPERFUSION

The factors responsible for mast cell accumulation in areas of fibrosis remain to be defined. SCF is a potent mast cell chemoattractant (62, 63) that stimulates directional motility of both mucosal and connective tissue-type mast cells. In addition, several angiogenic factors, such as platelet-derived growth factor (PDGF) AB, VEGF (vascular endothelial growth factor) and bFGF (basic fibroblast growth factor) have been demonstrated to promote murine mast cell chemotaxis *in vitro* (64). However, SCF along with the anaphylatoxins C3a and C5a (65) are the only factors shown to induce migration of human mast cells. Subcutaneous administration of recombinant human SCF to baboons produced a marked expansion of the mast cell population, which was reversed when the cytokine was discontinued (66), providing the first direct evidence that a specific factor can regulate mast cell development *in vivo*. Our studies demonstrated significant upregulation of SCF mRNA expression in ischaemic segments of canine myocardium following 1 h of ischaemia and 72 h of reperfusion (23). At the same time point, an increase in mast cell numbers is noted in the healing myocardium. In addition to being a mast cell chemoattractant, SCF critically regulates the maturation and survival of mast cells by suppressing mast cell apoptosis (67), enhancing mast cell maturation (68) and inducing mast cell adhesion to fibronectin (69). Furthermore, SCF is capable of inducing substantial mast cell histamine release and can promote the functional activation of mast cells *in vivo* (70, 71). All these actions may be important in regulating mast cell growth and activity after myocardial ischaemia. Recent studies suggested that the ability of SCF to support mast cell differentiation is influenced by interactions with specific co-factors, such as IL-3, IL-4 and IL-10 (72). Experiments from our laboratory have demonstrated induction of IL-10 mRNA in the infarcted myocardium peaking at 72–96 h of reperfusion (4). IL-10 may be important in costimulating and sustaining SCF-dependent mast cell accumulation.

Recently, Patella *et al.* (34) demonstrated increased mast cell density and SCF expression in patients with idiopathic and ischaemic cardiomyopathy, suggesting that sustained mast cell hyperplasia in cardiomyopathic hearts may contribute to collagen accumulation and fibrosis.

MACROPHAGES ARE THE MAIN SOURCE OF SCF FOLLOWING MYOCARDIAL ISCHAEMIA

SCF can be produced by a variety of cells, including fibroblasts, (73) smooth muscle and endothelial cells (74, 75). Recently, bone marrow macrophages (76) and mast cells (77) were found to produce SCF. It seemed reasonable to hypothesize that cytokine-stimulated fibroblasts or endothelial cells may be the source of SCF in the ischaemic myocardium. However, immunohistochemical studies utilizing a monoclonal antibody to canine SCF (78) showed that SCF immunoreactivity in the healing myocardial scar was predominantly localized in a subset of macrophages identified with the macrophage specific antibody AM-3K (Fig. 8) (79). These findings suggest that macrophages may orchestrate the healing process, promoting mast cell accumulation in the ischaemic myocardium.

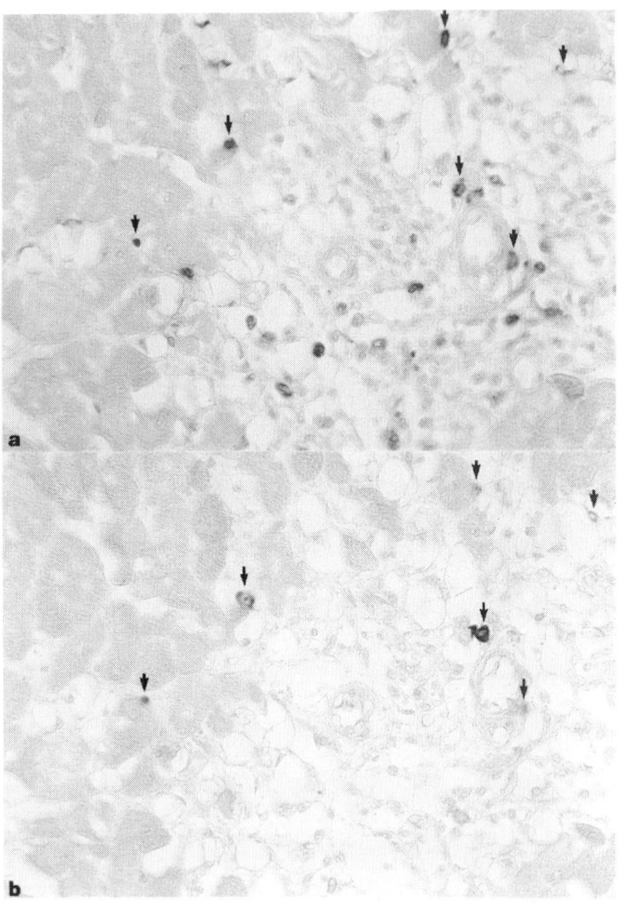

Fig. 8 SCF immunoreactivity in canine ischaemic myocardium. Serial sections from ischaemic areas of canine heart after 1 h of coronary occlusion and 5 days of reperfusion were immunostained for the macrophage-specific antibody AM-3K (a) and for SCF (b). A subset of macrophages (arrows) was shown to be the source of SCF immunoreactivity in the healing myocardium. (Reprinted, with permission, from *Circulation* **98**:687–698, 1998.)

POTENTIAL ROLE OF MAST CELLS IN HEALING

The potential role of mast cells in the healing process remains to be elucidated.

Mast cell degranulation products have been demonstrated to induce fibroblast proliferation (80). When activated mast cells were co-cultured with fibroblasts they were found to increase collagen synthesis and stimulate fibroblast proliferation (81), indicating a direct involvement of mast cells in the fibrotic process. Many mast cell-derived mediators may potentially influence fibroblast growth and function. Histamine has been shown to stimulate fibroblast growth and collagen synthesis *in vitro* (82, 83). Tryptase, the most abundant of the proteases found in mast cell granules, induces fibroblast proliferation (84), stimulates fibroblast chemotaxis and upregulates type I collagen production (85, 86). Furthermore, mast cells are important sources of transforming growth factor (TGF)-β (43), bFGF (87) and VEGF (88), factors that can regulate fibroblast growth, modulate extracellular matrix metabolism and stimulate angiogenesis. Mast cell-derived TGF-β may have a significant role in mediating myofibroblast α-smooth muscle actin expression in the healing scar (89). Finally, mast cells may influence healing and tissue remodelling by expressing gelatinases A and B (90), which are implicated in extracellular matrix degradation and angiogenesis.

THE MAST CELL IN MYOCARDIAL ISCHAEMIA/REPERFUSION: FRIEND OR FOE?

Mast cells are critical effector cells in immediate hypersensitivity reactions; however, their exact role in mediating other immunological or pathological responses remains unclear. We have presented evidence suggesting the importance of mast cells in myocardial ischaemia and reperfusion, initially as contributors in the initiation of the cytokine cascade. Resident cardiac mast cells, strategically located along venules, are an important source of TNF-α, histamine and other inflammatory mediators, which, when released may modulate endothelial function and mononuclear cell cytokine expression. During the healing phase of a myocardial infarction, significant mast cell accumulation is noted in areas of fibrosis and myofibroblast proliferation. Mast cells may now act as a source of growth factors and fibrogenic substances (such as TGF-β, bFGF, tryptase and VEGF), promoting collagen deposition and angiogenesis. Obviously, other cell types are capable of producing many of the mediators secreted by mast cells, complicating studies on the importance of mast cells in myocardial ischaemia.

Furthermore, mast cells are capable of producing a wide variety of substances, which may have diverse biological effects. It is therefore difficult to predict the net effect of mast cell activation in a particular biological response. Further experiments utilizing mast cell-deficient mice may provide more information in determining the contribution of mast cells and their mediators in myocardial ischaemia.

The intense inflammatory response associated with myocardial ischaemia and reperfusion may extend and complicate cardiac injury, but is also crucial for healing and scar formation. The catastrophic experience of the methylprednisolone trial in patients with acute myocardial infarction (91), which led to a significant increase in ventricular aneurysm formation and cardiac rupture, emphasizes the need for a better understanding of the cellular and molecular events associated with myocardial ischaemia. Interestingly, recent studies demonstrated that glucocorticoids decrease tissue mast cell number *in vivo* by reducing SCF production by resident cells (92). This effect may be an important factor explaining the deleterious effects of steroid administration in patients with myocardial infarction. Understanding of the role of the mast cell is important in order to design site-specific interventions that could mitigate inflammatory injury, without interfering with myocardial healing.

ACKNOWLEDGEMENT

This work was supported by NIH grant HL-42550.

REFERENCES

1. Hillis, L. D. and Braunwald, E. Myocardial ischemia. *N. Engl. J. Med.* **296**:1093, 1977.
2. Entman, M. L. and Smith, C. W. Postreperfusion inflammation: a model for reaction to injury in cardiovascular disease. *Cardiovasc. Res.* **28**:1301–1311, 1994.
3. Frangogiannis, N. G. and Entman, M. L. Role of inflammation following myocardial ischemia and reperfusion. In: *Textbook of Coronary Thrombosis and Thrombolysis* (Becker, R. C., ed.), pp. 569–584. Kluwer, Norwell, MA, 1997.
4. Frangogiannis, N. G., Youker, K. A., Rossen, R. D., Gwechenberger, M., Lindsey, M. L., Mendoza, L. H., Ballantyne, C. M., Smith, C. W. and Entman, M. L. Cytokines and the microcirculation in ischemia and reperfusion. *J. Mol. Cell. Cardiol.* **30**:2567–2576, 1998.
5. Mallory, G. K., White, P. D. and Salcedo-Salgar, J. The speed of healing of myocardial infarction. A study of the pathologic anatomy in seventy-two cases. *Am. Heart. J.* **18**:647, 1939.
6. Hill, J. H. and Ward, P. A. The phlogistic role of C3 leukotactic fragment in myocardial infarcts of rats. *J. Exp. Med.* **133**:885–900, 1971.
7. Pinckard, R. N., Olson, M. S., Giclas, P. C., Terry, R., Boyer, J. T. and O'Rourke, R. A. Consumption of classical complement components by heart subcellular membranes *in vitro* and in patients after acute myocardial infarction. *J. Clin. Invest.* **56**:740–750, 1975.
8. Kukielka, G. L., Smith, C. W., LaRosa, G. J., Manning, A. M., Mendoza, L. H., Hughes, B. J., Youker, K. A., Hawkins, H. K., Michael, L. H., Rot, A. and Entman, M. L. Interleukin-8 gene induction in the myocardium following ischemia and reperfusion *in vivo*. *J. Clin. Invest.* **95**:89–103, 1995.
9. Ivey, C. L., Williams, F. M., Collins, P. D., Jose, P. J. and Williams, T. J. Neutrophil chemoattractants generated in two phases during reperfusion of ischemic myocardium in the rabbit. *J. Clin. Invest.* **95**:2720–2728, 1995.
10. Adams, D. H. and Shaw, S. Leukocyte–endothelial interactions and regulation of leukocyte migration. *Lancet* **343**:831–836, 1994.
11. Tedder, T. F., Steeber, D. A., Chen, A. and Engel, P. The selectins: vascular adhesion molecules. *FASEB J.* **9**:866–873, 1995.
12. Lasky, L. A. Selectins: interpretors of cell-specific carbohydrate information during inflammation. *Science* **258**:964–969, 1992.
13. Luscinskas, F. W. and Lawler, J. Integrins as dynamic regulators of vascular function. *FASEB J.* **8**:929–938, 1994.
14. Albelda, S. M., Smith, C. W. and Ward, P. A. Adhesion molecules and inflammatory injury. *FASEB J.* **8**:504–512, 1994.
15. Jaeshke, H. and Smith, C. W. Cell adhesion and migration. III. Leukocyte adhesion and transmigration in the liver vasculature. *Am. J. Physiol.* **273**:G1169–1173, 1997.
16. Entman, M. L., Youker, K. A., Shoji, T., Kukielka, G. L., Shappell, S. B., Taylor, A. A. and Smith, C. W. Neutrophil-induced oxidative injury of cardiac myocytes: a compartmented system requiring CD11b/CD18-ICAM-1 adherence. *J. Clin. Invest.* **90**:1335–1345, 1992.
17. Entman, M. L., Youker, K. A., Shappell, S. B., Siegel, C., Rothlein, R., Dreyer, W. J., Schmalstieg, F. C. and Smith, C. W. Neutrophil adherence to isolated adult canine myocytes: evidence for a CD18-

dependent mechanism. *J. Clin. Invest.* **85**:1497–1506, 1990.

18. Smith, C. W., Entman, M. L., Lane, C. L., Beaudet, A. L. Ty, T. I., Youker, K. A., Hawkins, H. K. and Anderson, D. C. Adherence of neutrophils to canine cardiac myocytes *in vitro* is dependent on intercellular adhesion molecule-1. *J. Clin. Invest.* **88**:1216–1223, 1991.
19. Youker, K. A., Hawkins, H. K., Kukielka, G. L., Perrard, J. L., Michael, L. H., Ballantyne, C. M., Smith, C.W. and Entman, M. L. Molecular evidence for induction of intercellular adhesion molecule-1 in the viable border zone associated with ischemia–reperfusion injury of the dog heart. *Circulation* **89**;2736–2746, 1994.
20. Youker, K. A., Smith, C. W., Anderson, D. C., Miller, D., Michael, L. H., Rossen, R. D. and Entman, M. L. Neutrophil adherence to isolated adult cardiac myocytes: induction by cardiac lymph collected during ischemia and reperfusion. *J. Clin. Invest.* **89**:602–609, 1992.
21. Frangogiannis, N. G., Lindsey, M. L., Michael, L. H., Youker, K. A., Bressler, R. B., Mendoza, L. H., Spengler, R. N., Smith, C. W. and Entman, M. L. Resident cardiac mast cells degranulate and release preformed TNF-α, initiating the cytokine cascade in experimental canine myocardial ischemia/reperfusion. *Circulation* **98**:699–710, 1998.
22. Gwechenberger, M., Mendoza, L. H., Youker, K. A., Frangogiannis, N. G., Smith, C. W., Michael, L. H. and Entman, M. L. Cardiac myocytes produce Interleukin-6 in culture and in the viable border zone of reperfused infarctions. *Circulation* **99**:546–551, 1999.
23. Frangogiannis, N. G., Perrard, J. L., Mendoza, L. H., Burns, A. R., Lindsey, M. L., Ballantyne, C. M., Michael, L. H., Smith, C. W. and Entman, M. L. Stem cell factor induction is associated with mast cell accumulation after canine myocardial ischemia and reperfusion. *Circulation* **98**:687–698, 1998.
24. Frangogiannis, N. G., Perrard, J. L., Mendoza, L. H., Youker, K. A., Lindsey, M. L., Ballantyne, C. M., Michael, L. H., Smith, C. W. and Entman, M. L. Evolving role of the mast cell in myocardial ischemia and reperfusion. In: *Coronary Microcirculation During Ischaemia and Reperfusion (Alfred Benzon Symposium 41)* (Haunso, S., Aldershvile, J., Svendsen, J. H., eds), pp. 278–293. Munksgaard, Copenhagen, 1997.
25. Metcalfe, D. D., Baram, D. and Mekori, Y. A. Mast cells. *Physiol. Rev.* **77**:1033–1071, 1997.
26. Galli, S. J. The mast cell: a versatile effector cell for a challenging world. *Int. Arch. Allergy Immunol.* **113**:14–22, 1997.
27. Galli, S. J. New concepts about the mast cell. *N. Engl. J. Med.* **328**:257–265, 1993.
28. Enerback, L. Mast cells in rat gastrointestinal mucosa. 1. Effects of fixation. *Acta. Pathol. Microbiol. Scand.* **66**:289–302, 1966.
29. Enerback, L. Mast cells in rat gastrointestinal mucosa. 2. Dye-binding and metachromatic properties. *Acta. Pathol. Microbiol. Scand.* **66**:303–312, 1966.
30. Enerback, L., Miller, H. R. P. and Mayrhofer, G. Methods for the identification and characterization of mast cells by light microscopy. In: *Mast Cell Differentiation and Heterogeneity* (Befus, A. D., Bienenstock, J. and Denburg, J. A., eds), pp. 405–417. Raven Press, New York, 1986.
31. Miller, J. S. and Schwartz, L. B. Human mast cell proteases and mast cell heterogeneity. *Curr. Opin. Immunol.* **1**:637–642, 1989.
32. Irani, A. M. and Schwartz, L. B. Human mast cell heterogeneity. *Allergy Proc.* **15**:303–308, 1994.
33. Forman, M. B., Oates, J. A., Robertson, R. M., Roberts, L. J. and Virmani, R. Increased adventitial mast cells in a patient with coronary spasm. *N. Engl. J. Med.* **313**:1138–1141, 1985.
34. Patella, V., Marino, I., Arbustini, E., Lamparter-Schummert, B., Verga, L., Adt, M. and Marone, G. Stem cell factor in mast cells and increased mast cell density in idiopathic and ischemic cardiomyopathy. *Circulation* **97**:971–978, 1998.
35. Kaartinen, M., Penttila, A. and Kovanen, P. T. Mast cells in rupture-prone areas of human coronary atheromas produce and store TNF-alpha. *Circulation* **94**:2787–2792, 1996.
36. Sperr, W. R., Bankl, H. C., Mundigler, G., Klappacher, G., Grossschmidt, K., Agis, H., Simon, P., Laufer, P., Imhof, M., Radaszkiewicz, T., Glogar, D., Lechner, K. and Valent, P. The human cardiac mast cell: localization, isolation, phenotype, and functional characterization. *Blood* **84**:3876–3884, 1994.
37. Patella, V., Marino, I., Lamparter, B., Arbustini, E., Adt, M. and Marone, G. Human heart mast cells. Isolation, purification, ultrastructure, and immunologic characterization. *J. Immunol.* **154**:2855–2865, 1995.
38. Patella, V., deCrescenzo, G., Ciccarelli, A., Marino, I., Adt, M. and Marone, G. Human heart mast cells: a definitive case of mast cell heterogeneity. *Int. Arch. Allergy Immunol.* **106**:386–393, 1995.
39. Frangogiannis, N. G., Burns, A. R., Michael, L. H. and Entman, M. L. Histochemical and morphological characteristics of canine cardiac mast cells. *Histochem. J.* **31**:221–229, 1999.

40. Gordon, J. R. and Galli, S. J. Mast cells as a source of both preformed and immunologically inducible TNF-α/cachectin. *Nature* **346**:274–276, 1990.
41. Gordon, J. R., Burd, P. R. and Galli, S. J. Mast cells as a source of multifunctional cytokines. *Immunol. Today* **11**:458–464, 1990.
42. Moller, A., Lippert, U., Lessmann, D., Kolde, G., Hamann, K., Welker, P., Schadendorf, D., Rosenbach, T., Luger, T. and Czarnetzki, B. M. Human mast cells produce IL-8. *J. Immunol.* **151**:3261–3266, 1993.
43. Pennington, D. W., Lopez, A. R., Thomas, P. S., Peck, C. and Gold, W. M. Dog mastocytoma cells produce transforming growth factor beta 1. *J. Clin. Invest.* **90**:35–41, 1992.
44. Bradding, P., Feather, I. H., Wilson, S., Bardin, P. G., Heusser, C. H., Holgate, S. T. and Howarth, P. H. Immunolocalization of cytokines in the nasal mucosa of normal and perennial rhinitic subjects. The mast cell as a source of IL-4, IL-5, and IL-6 in human allergic mucosal inflammation. *J. Immunol.* **151**:3853–3865, 1993.
45. Rumsaeng, V., Vliagoftis, H., Oh, C. K. and Metcalfe, D. D. Lymphotactin gene expression in mast cells following Fc (epsilon) receptor I aggregation: modulation by TGF-beta, IL-4, dexamethasone, and cyclosporine A. *J. Immunol.* **158**:1353–1360, 1997.
46. Bradding, P., Okayama, Y., Howarth, P. H., Church, M. K. and Holgate, S. T. Heterogeneity of human mast cells based on cytokine content. *J. Immunol.* **155**:297–307, 1995.
47. Kanwar, S. and Kubes, P. Mast cells contribute to ischemia–reperfusion-induced granulocyte infiltration and intestinal dysfunction. *Am. J. Physiol.* **267(2 Pt 1)**:G316–321, 1994.
48. Keller, A. M., Clancy, R. M., Barr, M. L., Marboe, C. C. and Cannon, P. J. Acute reoxygenation injury in the isolated rat heart: role of resident cardiac mast cells. *Circ. Res.* **63**:1044–1052, 1988.
49. Ito, B. R., Engler, R. L. and delBalzo, U. Role of cardiac mast cells in complement C5a-induced myocardial ischemia. *Am. J. Physiol.* **264 (5 Pt 2)**:H1346–1354, 1993.
50. Michael, L. H., Lewis, R. M., Brandon, T. A. and Entman, M. L. Cardiac lymph from conscious dogs. *Am. J. Physiol.* **237**:H311–317, 1979.
51. Ley, K. Histamine can induce leukocyte rolling in rat mesenteric venules. *Am. J. Physiol.* **267**:H1017–1023, 1994.
52. Yam, L. T., Chan, C. H. and Li, C. Y. Hepatic involvement in systemic mast cell disease. *Am. J. Med.* **80**:819–826, 1986.
53. Inoue, Y., King, T. E. Jr, Tinkle, S. S., Dockstader, K. and Newman, L. S. Human mast cell fibroblast growth factor in pulmonary fibrotic disorders. *Am. J. Pathol.* **149**:2037–2054, 1996.
54. Pesci, A., Bertorelli, G., Gabrielli, M. and Olivieri, D. Mast cells in fibrotic lung disorders. *Chest* **103**:989–996, 1993.
55. Hawkins, R. A., Claman, H. N., Clark, R. A. and Steigerwald, J. C. Increased dermal mast cell populations in progressive systemic sclerosis: a link in chronic fibrosis? *Ann. Intern. Med.* **102**:182–186, 1985.
56. Gabbiani, G. Evolution and clinical implications of the myofibroblast concept. *Cardiovasc. Res.* **38**:545–548, 1998.
57. Willems, I. E., Havenith, M. G., De Mey, J. G. and Daemen, M. J. The α-smooth muscle actin-positive cells in healing human myocardial scars. *Am. J. Pathol.* **145**:865–875, 1994.
58. Cleutjens, J. P., Verluyten, M. J., Smiths, J. F. and Daemen, M. J. Collagen remodeling after myocardial infarction in the rat heart. *Am. J. Pathol.* **147**:325–338, 1995.
59. Agis, H., Willheim, M., Sperr, W. R., Wilfing, A., Kromer, E., Kabrna, E., Spanbochl, E., Strobl, H., Geissler, K., Spittler, A., Bolz-Nitulescu, G., Majdic, O., Lechner, K. and Valent, P. Monocytes do not make mast cells when cultured in the presence of SCF: characterization of the circulating mast cell progenitor as a c-kit+, CD34+, Ly–, CD14–, CD17–, colony forming cell. *J. Immunol.* **151**:4221–4227, 1993.
60. Rottem, M., Okada, T., Goff, J. P. and Metcalfe, D. D. Mast cells cultured from the peripheral blood of normal donors and patients with mastocytosis originate from a CD34+/Fc epsilon RI– cell population. *Blood* **84**:2489–2496, 1994.
61. Rodewald, H. R., Dessing, M., Dvorak, A. M. and Galli, S. J. Identification of a committed precursor for the mast cell lineage. *Science* **271**:818–822, 1996.
62. Galli, S. J., Zsebo, K. M. and Geissler, E. N. The kit ligand, stem cell factor. *Adv. Immunol.* **22**:1–96, 1994.
63. Meininger, C. J., Yano, H., Rottapel, R., Bernstein, A., Zsebo, K. M, and Zetter, B. R. The c-kit receptor ligand functions as a mast cell chemoattractant. *Blood* **79**:958–963, 1992.
64. Gruber, B. L., Marchese, M. J. and Kew, R. Angiogenic factors stimulate mast cell migration. *Blood*

86:2488–2493, 1995.
65. Hartmann, K., Henz, B. M., Kruger-Krasagakes, S., Kohl, J., Burger, R., Guhl, S., Haase, I., Lippert, U. and Zuberbier, T. C3a and C5a stimulate chemotaxis of human mast cells. *Blood* **89**:2863–2870, 1997.
66. Galli, S. J., Iemura, A., Garlick, D. S., Gamba-Vitalo, C., Zsebo, K. M. and Andrews, R. G. Reversible expansion of primate mast cell populations *in vivo* by stem cell factor. *J. Clin. Invest.* **91**:148–152, 1993.
67. Iemura, A., Tsai, M., Ando, A., Weshil, B. K. and Galli, S. J. The c-kit ligand, stem cell factor, promotes mast cell survival by suppressing apoptosis. *Am. J. Pathol.* **144**:321–328, 1994.
68. Tsai, M., Takeishi, T., Thompson, H., Langley, K. E., Zsebo, K. M., Metcalfe, D. D., Geissler, E. N. and Galli, S. J. Induction of mast cell proliferation, maturation and heparin synthesis by the rat c-kit ligand, stem cell factor. *Proc. Natl. Acad. Sci. USA* **88**:6382–6386, 1991.
69. Dastych, J. and Metcalfe, D. D. Stem cell factor induces mast cell adhesion to fibronectin. *J. Immunol.* **152**:213–219, 1994.
70. Lukacs, N. W., Kunkel, S. L., Strieter, R. M., Evanoff, H. L., Kunkel, R. G., Key, M. L. and Taub, D. D. The role of stem cell factor (c-kit ligand) and inflammatory cytokines in pulmonary mast cell activation. *Blood* **87**:2262–2268, 1996.
71. Costa, J. J., Demetri, G. D., Harrist, T. J., Dvorak, A. M., Hayes, D. F., Merica, E. A., Menchaca, D. M., Gringeri, A. J., Schwartz, L. B. and Galli, S. J. Recombinant human stem cell factor (kit ligand) promotes human mast cell and melanocyte hyperplasia and functional activation *in vivo*. *J. Exp. Med.* **183**:2681–2686, 1996.
72. Rennick, D., Hunte, B., Holland, G. and Thompson-Snipes, L. Cofactors are essential for stem cell factor-dependent growth and maturation of mast cell progenitors: comparative effects of interleukin-3 (IL-3), IL-4, IL-10, and fibroblasts. *Blood* **85**:57–65, 1995.
73. Linenberger, M. L., Jacobson, F. W., Bennett, L. G., Broudy, V. C., Martin, F. H. and Abkowitz, J. L. Stem cell factor production by human marrow stromal fibroblasts. *Exp. Hematol.* **23**:1104–1114, 1995.
74. Miyamoto, T., Sasaguri, Y., Sasaguri, T., Azakami, S., Yasukawa, H., Kato, S., Arima, N., Sugama, K. and Morimatsu, M. Expression of stem cell factor in human aortic endothelial and smooth muscle cells. *Atherosclerosis* **129**:207–213, 1997.
75. Koenig, A., Yakisan, E., Reuter, M., Huang, M., Sykora, K. W., Cobacioglu, S. and Welte, K. Differential regulation of stem cell factor mRNA expression in human endothelial cells by bacterial pathogens: an *in vitro* model of inflammation. *Blood* **83**:2836–2843, 1994.
76. Temeles, D. S., McGrath, H. E,. Kittler, L. W., Schadduck, R. K., Kister, V. K., Crittenden, R. B., Turner, B. L. and Quesenberry, P. J. Cytokine expression from bone marrow derived macrophages. *Exp. Hematol.* **21**:388–393, 1993.
77. Zhang, S., Anderson, D. F., Bradding, P., Coward, W. R., Baddeley, S. M., MacLeod, J. D., McGill, J. I., Church, M. K., Holgate, S. T. and Roche, W. R. Human mast cells express stem cell factor. *J. Pathol.* **186**:59–66, 1998.
78. Huss, R., Hong, D. S., Beckham, C., Storb, R., Broudy, V. C. and Deeg, H. J. Neutralization of stem cell factor *in vitro* and *in vivo*: dose dependent inhibition of stromal cells and induction of granulocyte/monocyte differentiation. *Ann. Hematol.* **72**:253–259, 1996.
79. Zeng, L., Takeya, M., Ling, X., Nagasaki, A. and Takahashi, K. Interspecies reactivities of anti-human macrophage monoclonal antibodies to various animal species. *J. Histochem. Cytochem.* **44**:845–853, 1996.
80. Pennington, D. W., Ruoss, S. J. and Gold, W. M. Dog mastocytoma cells secrete a growth factor for fibroblasts. *Am. J. Respir. Cell. Mol. Biol.* **6**:625–632, 1992.
81. Levi-Schaffer, F., and Rubinchick, E. Activated mast cells are fibrogenic for 3T3 fibroblasts. *J. Invest. Dermatol.* **104**:999–1003, 1995.
82. Russell, J. D., Russell, S. B. and Trupin, K. M. The effect of histamine on the growth of cultured fibroblasts isolated from normal and keloid tissue. *J. Cell. Physiol.* **93:**389–394, 1977.
83. Kupietzky, A. and Levi-Schaffer, F. The role of mast cell derived histamine in the closure of an *in vitro* wound. *Inflamm. Res.* **45**:176–180, 1996.
84. Ruoss, S. J., Hartmann, T. and Caughey, G. H. Mast cell tryptase is a mitogen for cultured fibroblasts. *J. Clin. Invest.* **88**:493–499, 1991.
85. Gruber, B. L., Kew, R. R., Jelaska, A., Marchese, M. J., Garlick, J., Ren, S., Schwartz, L. B. and Korn, J. H. Human mast cells activate fibroblasts: tryptase is a fibrogenic factor stimulating collagen messenger ribonucleic acid synthesis and fibroblast chemotaxis. *J. Immunol.* **158**:2310–2317, 1997.
86. Cairns, J. A. and Walls, A. F. Mast cell tryptase stimulates the synthesis of type I collagen in human lung fibroblasts. *J. Clin. Invest.* **99**:1313–1321, 1997.

87. Qu, Z., Liebler, J. M., Powers, M. R., Galey, T., Ahmadi, P., Huang, X. N., Ansel, J. C., Butterfield, J. H., Planck, S. R. and Rosenbaum, R. Mast cells are a major source of basic fibroblast growth factor in chronic inflammation and cutaneous hemangioma. *Am. J. Pathol.* **147**:564–573, 1995.
88. Boesiger, J., Tsai, M., Maurer, M., Yamaguchi, M., Brown, L. F., Claffey, K. P., Dvorak, H. F. and Galli, S. J. Mast cells can secrete vascular permeability factor/vascular endothelial cell growth factor and exhibit enhanced release after immunoglobulin E-dependent upregulation of Fc epsilon receptor I expression. *J. Exp. Med.* **188**:1135–1145, 1998.
89. Desmouliere, A., Geinoz, A., Gabbiani, F. and Gabbiani, G. Transforming growth factor beta-1 induces alpha smooth muscle actin expression in granulation tissue myofibroblasts and in quiescent and growing cultured fibroblasts. *J. Cell Biol.* **122**:103–111, 1993.
90. Fang, K. C., Wolters, P. J., Steinhoff, M., Bidgol, A., Blount, J. L. and Caughey, G. H. Mast cell expression of gelatinases A and B is regulated by kit ligand and TGF-beta. *J. Immunol.* **162**:5528–5535, 1999.
91. Roberts, R., DeMello, V. and Sobel, B. E. Deleterious effects of methylprednisolone in patients with myocardial infarction. *Circulation* **53 (Suppl. 1)**:204, 1976.
92. Finotto, S., Mekori, Y. A. and Metcalfe, D. D. Glucocorticoids decrease tissue mast cell number by reducing the production of the c-kit ligand, stem cell factor, by resident cells: *in vitro* and *in vivo* evidence in murine systems. *J. Clin. Invest.* **99**:1721–1728, 1997.

SECTION NINE

LEUKOTRIENES, CYTOKINES AND CHEMOKINES IN ALLERGIC DISORDERS

CHAPTER 33

Lipid Mediators from Mast Cells and Basophils in Allergic Diseases

MASSIMO TRIGGIANI, CRISTIANA PALUMBO, MARCO GENTILE, FRANCESCOPAOLO GRANATA, CECILIA CALABRESE and GIANNI MARONE*

Division of Clinical Immunology and Allergy, University of Naples Federico II, Naples, Italy

INTRODUCTION

Mast cells and basophils play a primary role in allergic and autoimmune diseases (1, 2). In addition, they contribute to defence against bacterial and parasitic infections and they are involved in the pathogenesis of some cardiovascular, neurological and gastrointestinal disorders (2, 3). The pathophysiological functions of mast cells and basophils are largely mediated by the production of soluble molecules including pre-formed mediators, *de novo* synthesized lipid mediators, cytokines, and chemokines (4–9).

Mast cells and basophils are major producers of lipid mediators among human inflammatory cells (10–12). These cells produce two classes of lipid mediators: the metabolites of arachidonic acid, collectively referred to as eicosanoids, and the group of 2-acetylated phospholipids, including platelet-activating factor (PAF) and related molecules (13, 14). The two classes of lipid mediators are biochemically related, since they both derive from phospholipids containing arachidonic acid (AA) esterified at the *sn*-2 position (2-arachidonoyl phospholipids) (15, 16). Mast cells and basophils contain large quantities of this common precursor of eicosanoids and PAF. Activation of the common biosynthetic pathway explains why the two classes of lipid mediators are often produced simultaneously by stimulated mast cells and basophils.

The initial step in the biosynthetic pathway of lipid mediators is the hydrolysis of AA from 2-arachidonoyl phospholipids by an enzyme of the phospholipase A_2 (PLA_2) family (17, 18). The molecules resulting from PLA_2 hydrolysis are free AA and a 2-lysoglycerophospholipid. AA may have one of several metabolic fates: it may be converted to eicosanoids by lipoxygenases and cyclo-oxygenases, released from the cell without further metabolism, or be re-acylated into other glycerolipid molecules as part of a complex process defined as 'glycerolipid remodelling' (19, 20). Lysophospholipids, in

* Corresponding author.

MAST CELLS AND BASOPHILS
ISBN 0-12-473335-2

turn, can be converted to PAF or to other 2-acetylated phospholipids, or they can be re-acylated with other fatty acids and re-incorporated into cell membranes (20–22). The metabolic fate of AA largely depends on the state of cellular activation and on the influence of other mediators, particularly cytokines, to which the cells are exposed. Conversion of AA to leukotrienes (LTs) is initiated by 5-lipoxygenase (5-LO), which in concert with 5-LO activating protein (FLAP) generates the unstable intermediate LTA_4 (23, 24). The addition of a cysteinyl residue to LTA_4 results in its conversion to LTC_4 by LTC_4 synthase (25). LTC_4 is the predominant leukotriene produced by human mast cells and basophils (5, 13, 26–28). Very little conversion of LTC_4 to the other cysteinyl leukotrienes (LTD_4 and LTE_4) occurs in these cells (29, 30). Two cyclo-oxygenases COX-1 and COX-2 are responsible for the formation of prostaglandins and thromboxanes (31). These two enzymes appear to be both constitutively expressed at least in murine mast cells, although the levels of COX-2 are strongly influenced by cytokines such as IL-3 and stem cell factor (32, 33).

The profile of eicosanoids produced by stimulated human mast cells and basophils is quite different. In general, mast cells produce both 5-LO and COX-1/COX-2-derived eicosanoids, primarily LTC_4 and prostaglandin D_2 (PGD_2) (29, 34–37). Human basophils produce only LTC_4 and no COX metabolite has been conclusively demonstrated in these cells (11, 30, 38, 39). Furthermore, qualitative and quantitative differences exist in the profile of eicosanoids produced by mast cells isolated from different tissues (i.e. lung, skin, small intestine and heart). Immunologically activated lung mast cells produce approximately equal amounts of LTC_4 and PGD_2 ($\approx$ 60 ng per 10^6 cells) (10, 29). Cardiac mast cells also produce both eicosanoids, but in smaller quantities ($\approx$ 20 ng per 10^6 cells) (36, 40). Skin mast cells produce a quantity of PGD_2 similar to that produced by lung mast cells, but they generate, at least *in vitro*, very little amounts of LTC_4 (34). Figure 1 shows the amount of LTC_4 produced by mast cells isolated from different tissues and

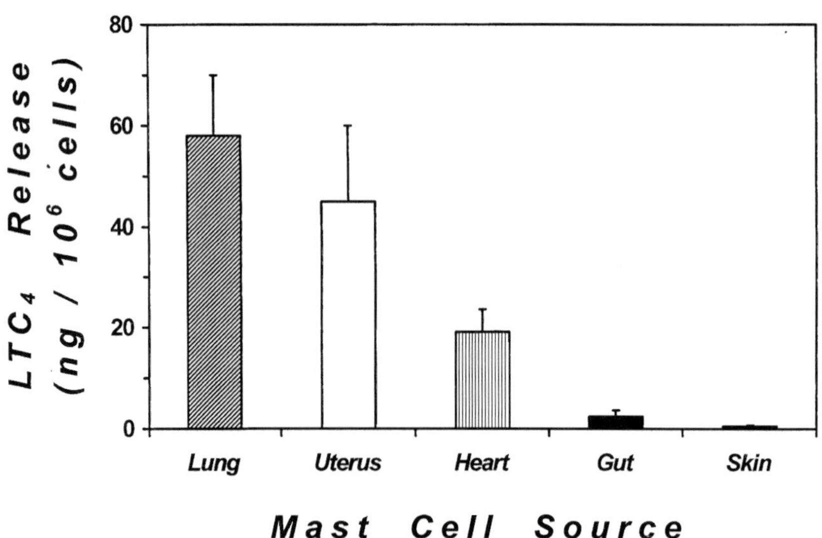

Fig. 1 Release of LTC_4 from human mast cells isolated from different sites. Mast cells were isolated and purified (80–95%) from surgical specimens of lung, uterus, heart, gut (small intestine) and skin. The cells were stimulated *in vitro* with a rabbit antibody anti-human IgE and the release of LTC_4 was determined by high-performance liquid chromatography and radioimmunoassay.

stimulated with an antibody anti-human IgE. These data suggest that lipid mediator synthesis is controlled by a complex series of biochemical events, including precursor accumulation, enzyme expression and substrate–enzyme coupling, and that these events are differentially regulated in basophils and in mast cells isolated from different sites.

In the last 5 years many aspects of the biochemistry and molecular biology of lipid mediators in mast cells and basophils have been elucidated. The identification of AA pools with specific functions and subcellular location, the characterization of novel eicosanoid-forming enzymes with distinct specificities and sites of action and, finally, the genetic control of lipid mediator synthesis are just some of the major breakthroughs in this area. Most of these findings have been reported in cells of non-human origin, but at least some of them have been confirmed in human cells.

ARACHIDONIC ACID POOLS IN HUMAN MAST CELLS AND BASOPHILS

Human lung mast cells contain approximately 8 nmol of AA per 10^6 cells (41). This amount of AA is 2–3-fold larger than that of other human inflammatory cells, including neutrophils, eosinophils, macrophages, monocytes and platelets (17). Skin and cardiac mast cells contain a similar amount of AA, whereas human basophils isolated from peripheral blood of normal donors contain approximately 3 nmol of AA per 10^6 cells (17, 42). In both mast cells and basophils, AA is esterified into different classes of glycerolipids including phosphatidylethanolamine (PE), phosphatidylinositol (PI), phosphatidylserine (PS), phosphatidylcholine (PC) and triglycerides (TG) (41). A unique feature of human mast cells is that the majority of AA (>60%) is esterified in the TG (17). This is in contrast with the distribution of AA in other inflammatory cells, including basophils, in which this fatty acid is predominantly incorporated into phospholipids, primarily PE and PC (41–43). The amount of AA esterified into TG in mast cells isolated from different sites is linearly correlated with the total cellular content of AA (44).

Studies in human macrophages and neutrophils have revealed that, in contrast to phospholipids, TG are not substrates for phospholipase A_2 (PLA_2), the major AA-mobilizing enzyme (45, 46). Therefore, the TG pool of AA cannot be used directly for eicosanoid synthesis (46). The abundance of TG-associated AA in mast cells prompted us to evaluate its role in the intracellular metabolism of AA. TG is the first pool in which exogenous AA is incorporated (45). From TG, AA is subsequently transferred to phospholipids (PC, PI and PE) (44). The transfer of AA from TG to phospholipids occurs at a slow rate in quiescent mast cells and it is responsible for the continuous remodelling of the glycerolipid pools of AA (47, 48). The remodelling of AA pools is important for maintaining a steady-state level of intracellular AA and to preserve the correct composition of membrane glycerolipids (48, 49).

Turnover of AA within glycerolipid pools is dramatically accelerated in stimulated mast cells (48, 50). In this case, activation of phospholipases leads to a massive mobilization of AA, predominantly from PE and PC. In the early phase (within minutes) of mast cell activation, AA is largely mobilized from phospholipids but there is no significant release of AA from TG, since this pool is not a substrate for phospholipases. The TG pool rather increases its AA content because this pool rapidly recaptures free AA before it is converted to eicosanoids (45, 47, 50). However, studies performed with mast cells in which the TG pool was selectively labelled with [^{3}H] AA, revealed that, 30–60 min after mast cell activation, there is an increase in AA transfer from TG to phospholipids (Triggiani *et al.*, unpublished observations). Taken together, these observations suggest that TG may represent a pool involved in the long-term control of intracellular levels of AA in resting and stimulated mast cells. In resting cells, TG function as a high-capacity storage pool, whereas in stimulated cells they act by controlling the quantities of AA to be converted to eicosanoids and by feeding back AA to phospholipid pools (41, 46, 51).

A series of studies have been performed to identify the subcellular location of the TG pool of AA. Lipid bodies are non-membrane-bound cytoplasmic organelles present in several types of human inflammatory cells (52). The number of lipid bodies, however, varies greatly from cell to cell (17, 52, 53). For example, mast cells contain a large number (> 15) of lipid bodies per cell, whereas quiescent basophils and eosinophils have fewer than 5 lipid bodies per cell (54, 55). Raising the intracellular content of AA results in the accumulation of AA in the TG pool and an increased number of lipid bodies (17). The increase in lipid bodies correlates linearly with the accumulation of AA into the TG pool (41). Although these data do not provide conclusive evidence for the subcellular location of the TG pool, they strongly suggest that lipid bodies may be the site of accumulation of this pool of AA. However, lipid bodies are not only the putative location of TG, they are also multifunctional organelles important to the overall metabolism of AA. Elegant studies by Dvorak, Weller and co-workers provided evidence that lipid bodies of mast cells and eosinophils contain most of the enzymes responsible for eicosanoid biosynthesis, including PLA_2, 5-LO, LTC_4 synthase, COX-1 and COX-2 (53, 56, 57). Although a functionally active metabolism of AA in isolated lipid bodies has still to be demonstrated, these findings suggest that these organelles possess the complete biochemical machinery necessary for eicosanoid synthesis.

As mentioned above, the quantity and distribution of AA in human basophils is different from that in mast cells. This is reflected also in the smaller number of lipid bodies in basophils (58). The distribution of AA within glycerolipid pools of human basophils more closely resembles that of neutrophils and eosinophils (large pools in phospholipids with little AA in TG) (17, 42). However, when neutrophils and eosinophils are recruited into an inflammatory area, such as in the airways of patients with adult respiratory distress syndrome (ARDS) (neutrophils) or with bronchial asthma (eosinophils), a large accumulation of AA in the TG pool occurs (17, 59). Whether such a redistribution of AA in the TG pool also takes place in basophils upon immunological activation *in vitro* or *in vivo* is unknown.

PHOSPHOLIPASE A_2 IN HUMAN MAST CELLS AND BASOPHILS

The initial step in the common biosynthetic pathway of eicosanoids and PAF is the hydrolysis of AA from phospholipids catalysed by enzymes of the PLA_2 family (15, 60). Several isoforms of PLA_2 have been described, each with a distinct primary sequence, cellular location and Ca^{2+} requirements (61, 62). These enzymes are currently divided into two major classes: secretory PLA_2s ($sPLA_2$s), and cytosolic PLA_2s ($cPLA_2$s). Secretory PLA_2s are low molecular mass (12–16 kDa) enzymes, containing several disulphide bridges; they are usually released and active in the extracellular environment where they find the millimolar concentrations of Ca^{2+} required for their activation (63). In contrast, $cPLA_2$ are high molecular mass (80–110 kDa) enzymes located in the cytosol of resting cells. Cytosolic PLA_2 are activated by micromolar concentrations of Ca^{2+}, such as those generally reached inside stimulated cells (group IV), or they are Ca^{2+}-independent (group VI) (64).

Human lung mast cells contain at least two PLA_2s: a group IV $cPLA_2$ and a recently characterized group IIA $sPLA_2$ (65). Both enzymes appear to be involved in the release of AA used for eicosanoid synthesis although their relative contribution is still unclear. Preliminary evidence suggest that human mast cells isolated from lung parenchyma also express mRNA and synthesize a group V secretory PLA_2 (Triggiani *et al.*, unpublished observation). Human basophils also contain both a group IV and a group IIA $sPLA_2$ (66). However, there is no conclusive evidence that $sPLA_2$ is effectively released from stimulated basophils at least in *in vitro* experiments (67).

Cytosolic PLA_2 in murine mast cells has a high specificity for AA at the *sn*-2 position of phospholipids and it recognizes as preferential substrate the 1-alkyl subclasses of PE and PC (17, 68). Activation of mast cells by immunological stimuli results in a rapid (within minutes) increase in $cPLA_2$ activity (69). The early activation of $cPLA_2$ requires, in addition to the intracellular Ca^{2+} elevation, its phosphorylation at Ser505 by ERK, a member of the MAP kinase pathway (66, 69). Activated $cPLA_2$ translocates to the nuclear membrane, to initiate AA hydrolysis. In addition to the activation of preformed $cPLA_2$, an increased expression of this enzyme has been demonstrated to occur during the late activation of murine mast cells (6–12 h after IgE-mediated activation) (70). The late expression of $cPLA_2$ is responsible for prostanoid synthesis detected 12–18 h after antigen challenge of murine mast cells *in vitro*.

A group IIA $sPLA_2$ is localized within the secretory granules (71) and is released during IgE-mediated activation of murine and human mast cells (72). The release of $sPLA_2$ is rapid and occurs simultaneously with the release of histamine (65). Convincing evidence has been obtained that $sPLA_2$ participates to AA mobilization in murine mast cells (73). However, the precise mechanism by which $sPLA_2$ hydrolyse AA has not yet been determined. Activation of $sPLA_2$ requires millimolar concentrations of Ca^{2+}, found only in the extracellular environment. It has been hypothesized that $sPLA_2$ would bind to the external surface of the mast cell to carry out the hydrolysis of AA from phospholipids of the outer leaflet of cell membrane. However, a number of observations do not fully confirm this model. For example, specific inhibitors of $sPLA_2$ do not completely suppress the release of AA in mast cells incubated with an exogenous $sPLA_2$ (74). Furthermore, a monoclonal antibody anti-$sPLA_2$, with blocking activity on the isolated enzyme, fails to inhibit AA release from human basophils challenged with anti-IgE (67). These results support the hypothesis that $sPLA_2$ may contribute to AA mobilization in activated mast cells and basophils by mechanisms alternative to phospholipid hydrolysis on the extracellular surface. The studies performed to test this hypothesis led to the identification and characterization of specific $sPLA_2$ receptors.

EXTRACELLULAR EFFECTS OF MAST CELL-DERIVED SECRETORY PLA_2

In the last 15 years a large body of evidence has accumulated on the presence of high levels of low molecular weight $sPLA_2$ in fluids retrieved from inflammatory sites (75). $sPLA_2$ has been detected in the synovial fluid of patients with rheumatoid arthritis (76, 77) and in the bronchoalveolar lavage (78) and nasal lavage (79) of patients with bronchial asthma and allergic rhinitis, respectively. In addition, elevated plasma levels of the enzyme were found in patients with septic shock (80), ARDS (75), bronchial asthma (65) and mastocytosis (81). Together these observations suggested that $sPLA_2$ is released in large quantities during systemic or local inflammation. Moreover, the identification of mast cells as a major cellular source of $sPLA_2$ released in upper and lower airways of allergic patients raised important questions on its role in diseases such as asthma and rhinitis. $sPLA_2$ released in the extracellular space during the inflammatory reaction can act enzymatically on phospholipids on the external surface of cells recruited in the inflamed sites (82). In addition, $sPLA_2$ may find its substrate directly in the extracellular space – e.g. the phospholipids of surfactant in the airways of asthmatic patients (83). In any event, $sPLA_2$ hydrolysis would result in the generation of lysophospholipids and of free fatty acids that, in most cases, consist of free AA. Extracellular free AA can then be taken up by infiltrating inflammatory cells (e.g. eosinophils), in the airways of asthmatics, and converted to bioactive eicosanoids (84). Lysophospholipids, in turn, are highly cytotoxic molecules that can alter epithelium homeostasis and surfactant properties (85). These effects of $sPLA_2$ are exclusively due to its enzymatic activity. However, studies performed by instilling $sPLA_2$ in the airways of experimental animals indicated that the pro-inflammatory activities of this enzyme could not be explained solely by its hydrolytic activity. For example, intratracheal administration of $sPLA_2$ results in severe alterations of the respiratory functions with impairment of gas exchange (86). Histological examination reveals that $sPLA_2$ induces extensive tissue damage in the lung and massive infiltration of neutrophils and eosinophils (86). These observations suggested that other mechanisms, unrelated to the enzymatic activity, were responsible for the pro-inflammatory effects of $sPLA_2$ in the lung. We have previously demonstrated that large quantities of $sPLA_2$ are released by human lung mast cells activated by immunological stimuli (65). Furthermore, we have shown that patients with bronchial asthma have an increased level of $sPLA_2$ in the bronchoalveolar lavage and that this $sPLA_2$ is biochemically and immunologically identical to that released from mast cells (78, 83). Murakami *et al.* first demonstrated the existence of specific membrane receptors for $sPLA_2$ on murine mast cells (87). Binding of $sPLA_2$ to this receptor induced the release of histamine and the production of eicosanoids (87). It was also shown that $sPLA_2$ could induce neutrophil degranulation (88), interstitial cell proliferation (89) and bronchoconstriction (90). All these effects of $sPLA_2$ were due to the binding of specific membrane receptors and not to the enzymatic activity. These findings provided additional mechanisms by which $sPLA_2$ may sustain inflammation in bronchial asthma and they indicated that $sPLA_2$ could exert its effects by interacting with specific and functionally active receptors.

We also hypothesized that a major cellular target of mast cell-derived $sPLA_2$ in the airways might be the alveolar macrophage. This hypothesis has been confirmed by showing that $sPLA_2$ induces the release of lysosomal enzymes and the production of IL-6 from human macrophages (65). These effects are mediated by the activation of two distinct receptors for $sPLA_2$ expressed on macrophages. Thus, $sPLA_2$ released from mast cells may significantly contribute to the pathogenesis of airway inflammation and tissue damage by promoting cytolytic enzyme secretion and pro-inflammatory cytokine production from macrophages. Other biological activities of $sPLA_2$ potentially relevant to the pathogenesis of bronchial asthma are reported in Fig. 2.

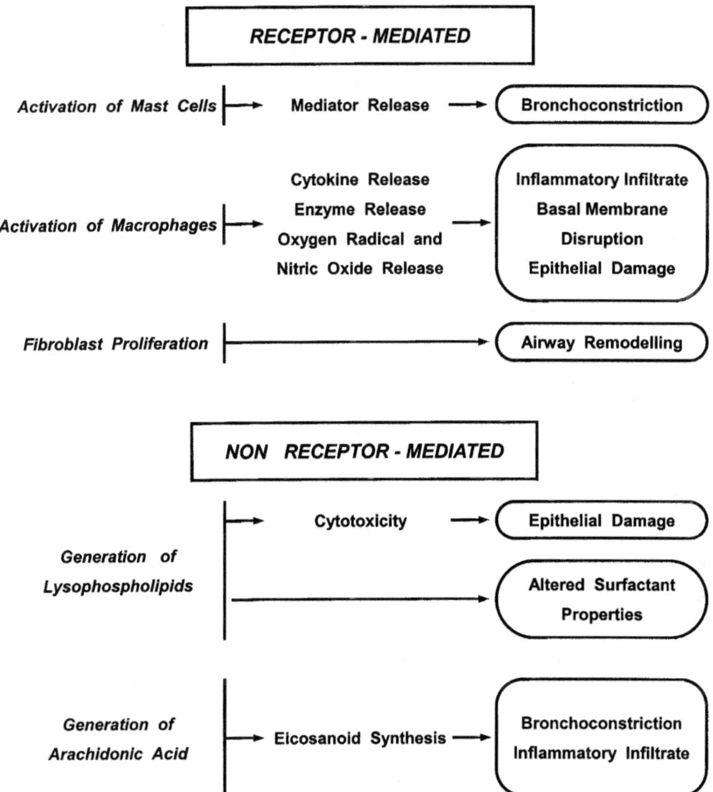

Fig. 2 Biological effects of mast cell-derived secretory PLA_2 ($sPLA_2$) potentially relevant to the pathogenesis of bronchial asthma. The effects of $sPLA_2$ can be either receptor-mediated or non-receptor-mediated. A functionally active receptor for $sPLA_2$ has been demonstrated, among other cells, on mast cells, macrophages and fibroblasts. Binding of $sPLA_2$ to the receptor on mast cells induces the release of histamine and LTD_4, two bronchoconstrictor mediators. Activation of macrophages results in exocytosis and release of oxygen radicals and nitric oxide, which are responsible for epithelial and basal membrane damage. $sPLA_2$ also induces the production of pro-inflammatory cytokines (IL-6 and TNF-α) from lung macrophages. Activation of $sPLA_2$ receptor on fibroblasts results in cell proliferation, a crucial event for airway remodelling in asthmatic patients. The non-receptor-mediated effects of $sPLA_2$ are related to its enzymatic activity and to the generation of lysophospholipids and arachidonic acid. Lysophospholipids are cytolytic molecules that can cause epithelial cell damage and alter the tensioactive properties of surfactant. Free extracellular arachidonic acid can be incorporated by inflammatory cells recruited in the airways (e.g. eosinophils) and converted to bronchoconstrictor and pro-inflammatory eicosanoids.

PAF AND RELATED PHOSPHOLIPIDS

PAF is a unique phospholipid characterized as 1-alkyl-2-acetyl-*sn*-glycero-3-phosphocholine (1-alkyl-2-acetyl-GPC) (91). Although physiological levels of PAF are involved in the homeostasis of kidney, brain and liver functions, this phospholipid has a primary role as a pro-inflammatory mediator (92, 93). PAF is a potent stimulus for degranulation, chemotaxis and mediator release from human platelets, neutrophils and eosinophils (94–96) and induces histamine and LTC_4 release from human basophils (95). In addition, PAF has been shown to promote oxygen radical generation and cytokine production in human monocytes and macrophages and to regulate IgE synthesis in human B lymphocytes (97). Administration of PAF *in vivo* results in a profound hypotension, bronchoconstriction, increased vascular permeability and lung injury, primarily due to intravascular platelet aggregation and neutrophil recruitment (97, 98). Recent evidence indicates that PAF exerts part of these effects by inducing $sPLA_2$ secretion from neutrophils (99). These observations, together with the detection of PAF in inflammatory areas, suggested a primary role for this phospholipid as a mediator of inflammatory reactions.

Several cell types have been shown to produce PAF after appropriate stimulation. These cells include neutrophils, eosinophils, basophils, monocytes, macrophages, mast cells and endothelial cells (14, 93, 94). Early studies have shown that PAF produced by several inflammatory cells is composed of different molecular species that differ in chain length and degree of unsaturation of the chain residue at the *sn*-1 position of the glycerol backbone (100). Naturally occurring PAF contains variable proportions of 16:0, 18:0 and 18:1 molecular species. All PAF molecular species possess similar biological activities although they display a different degree of potency (101, 102).

As already discussed the metabolism of PAF is closely linked to that of AA (18, 94). The lysophospholipid derived from the activities of $cPLA_2$ and $sPLA_2$ is subsequently acetylated at the *sn*-2 position by an acetyl transferase to form PAF, or one of the other 2-acetylated phospholipids (16, 103). 2-Acetylated phospholipids are a class of biologically active molecules that share with PAF the common feature of an acetate group at the *sn*-2 position of the glycerol backbone. Naturally occurring 2-acetylated phospholipids include 1-acyl-2-acetyl-GPC (AAGPC), 1-alkyl-1´-enyl-2-acetyl-GPC and 1-alkyl-1´-enyl-2-acetyl-sn-glycero-3-phosphoethanolamine (1-alkyl-1´-enyl-2-acetyl-GPE) (12, 14).

We have previously demonstrated that immunologically activated human lung mast cells produce large amounts of AAGPC (12). The amount of AAGPC and PAF produced by mast cells can be quite variable depending on experimental conditions because these two molecules are extensively catabolized *in vitro*; however, human lung mast cells stimulated *in vitro* with a monoclonal antibody anti-IgE synthesize approximately 6 pmoles of AAGPC per 10^6 cells and 2 pmoles of PAF per 10^6 cells (14). The molecular species of AAGPC and PAF synthesized by mast cells are very similar, the predominant being the 16:1 and, to a lesser extent, 18:0 and 18:1. Human lung mast cells also synthesize 1-alkyl-1´-enyl-2-acetyl-GPE (12). A predominant synthesis of AAGPC over that of PAF is not characteristic of mast cells but it can be demonstrated also in the human basophils and endothelial cells (14, 104).

The biological activities of the 2-acetylated phospholipids are still largely unknown. Most of these molecules, including AAGPC, are weak stimuli for human neutrophils (102, 105) and platelets (101). However, their activity can be potentiated more than 100-fold in cells primed with 5-HETE (106). A number of experimental observations also support the hypothesis that the 2-acetylated phospholipids can act either as primers or as antagonists of the pro-inflammatory activities of PAF depending on the cell and on the type of response examined. For example, AAGPC inhibits in a non-competitive fashion the intracellular calcium signal as well as the release of lysosomal enzymes in the neutrophil (105) and the release of histamine and LTC_4 from basophils induced by PAF (107). In contrast, priming with AAGPC significantly enhances the generation of superoxide radicals in neutrophils stimulated with PAF (102).

PAF exerts most of its biological effects at subnanomolar concentrations. Therefore, the rate of degradation of this molecule is crucial to limit its pro-inflammatory effects. The most efficient system to regulate PAF levels is its catabolism by acetylhydrolase, a specific enzyme that cleaves the acetate at the *sn*-2 position of PAF producing lyso-PAF, a virtually inactive metabolite (108). Acetylhydrolase exists in two forms: an intracellular cytosolic form widely distributed in inflammatory cells and an extracellular form present in large amounts in human plasma associated with low-density lipoproteins (LDL) (109). Interestingly, a genetic deficiency of the plasma form of acetylhydrolase has been associated with the occurrence of bronchial asthma (110). An extracellular form of acetylhydrolase different from the plasma enzyme is detectable in the bronchoalveolar lavage. Mast cells have the ability to release acetylhydrolase upon immunological activation *in vitro* (111); these cells may therefore be the source of acetylhydrolase in the airways (83, 112). The levels of acetylhydrolase in the bronchoalveolar lavage are reduced in patients with asthma, indicating that the catabolism of PAF may be impaired in the airways of these patients (83).

We have previously demonstrated that oxygen-derived free radicals can rapidly and irreversibly inactivate acetylhydrolase in plasma and bronchoalveolar lavage fluid (113). This effect can be prevented by the radical scavenger superoxide dismutase, but not by catalase, pointing to superoxide anion as the molecule primarily responsible for inactivation of acetylhydrolase. Many cells, including eosinophils, macrophages and mast cells, are able to generate oxygen free radicals in the airways of asthmatic patients (114, 115). Therefore, in these patients, the concurrent production of PAF and oxygen free radicals results in a local increase and prolongation of PAF pro-inflammatory activities.

CONCLUDING REMARKS

Lipid mediators are pivotal molecules in initiating and maintaining the inflammatory response in allergic diseases. The role of these mediators in bronchial asthma has been recently emphasized by the favourable clinical results obtained with leukotriene receptor antagonists in this disease (116).

Two cells primarily involved in allergic disorders – the mast cell and the basophil – are rich sources of lipid mediators when immunologically activated. Our view of the biochemistry and molecular biology of lipid mediators has greatly changed in the last 10 years. The overall picture has been made more complicated by the discovery of new enzymes, of various pools of AA with different specificities and subcellular locations and, finally, of the genetic control of lipid mediator synthesis and catabolism. The impact of these new findings on the pathogenesis of human diseases is currently under evaluation. It is reasonable to hypothesize, however, that understanding the complexity of the lipid mediator network will provide new targets for the pharmacological modulation of their effects in allergic diseases.

ACKNOWLEDGEMENTS

This work was supported by grants from the National Research Council (CNR) (Targeted Project Biotechnology No. 98.00085.PF31 and No. 99.00401.PF49) and the Ministry of Health (ISS-AIDS Project No. 40B.64) (Rome, Italy).

REFERENCES

1. Marone, G., Casolaro, V., Patella, V., Florio, G. and Triggiani, M. Molecular and cellular biology of mast cells and basophils. *Int. Arch. Allergy Immunol.* **114:** 207–217, 1997.
2. Marone, G. ed. *Human Basophils and Mast Cells. Biological Aspects.* Karger, Basel, 1995.
3. Marone, G. ed. *Human Basophils and Mast Cells. Clinical Aspects.* Karger, Basel, 1995.
4. Serafin, W. E. and Austen, K. F. Mediators of immediate hypersensitivity reactions. *N. Engl. J. Med.* **317**:30–34, 1987.
5. Triggiani, M., Casolaro, V., Genovese, A., Spadaro, G. and Marone, G. Heterogeneity of human FcεRI-bearing cells. *Ann. Allergy* **71**:133–138, 1993.
6. Bradding, P., Okayama, Y., Howarth, P. H., Church, M. K. and Holgate, S. T. Heterogeneity of human mast cells based on cytokine content. *J. Immunol.* **155**:297–307, 1995.
7. Patella, V., Marinò, I., Arbustini, E., Lamparter-Schummert, B., Verga, L., Adt, M. and Marone, G. Stem cell factor in mast cells and increased mast cell density in idiopathic and ischemic cardiomyopathy. *Circulation* **97**:971–978, 1998.
8. Patella, V., Giuliano, A., Bouvet, J. P. and Marone, G. Endogenous superallergen protein Fv induces IL-4 secretion from human $Fc_{\varepsilon}RI^{+}$ cells through interaction with the VH_3 region of IgE. *J. Immunol.* **161**:5647–5655, 1998.
9. de Paulis, A., Minopoli, G., Arbustini, E., de Crescenzo, G., Dal Piaz, F., Pucci, P., Russo, T. and Marone, G. Stem cell factor is localized in, released from, and cleaved by human mast cells. *J. Immunol.* **163**:2799–2808, 1999.
10. Peters, S. P., MacGlashan, D. W. Jr, Schulman, E. S., Schleimer, R. P., Hayes, E. C., Rokach, J., Adkinson, N. F. Jr and Lichtenstein, L. M. Arachidonic acid metabolism in purified human lung mast cells. *J. Immunol.* **132**:1972–1979, 1984.
11. MacGlashan, D. W. Jr, Peters, S. P., Warner, J. A. and Lichtenstein, L. M. Characteristics of human basophils sulfidopeptide leukotriene release: releasability defined as the ability of the basophil to respond to dimeric cross-links. *J. Immunol.* **136**:2231–2239, 1986.
12. Triggiani, M., Hubbard, W. C. and Chilton, F. H. Synthesis of 1-acyl-2-acetyl-*sn*-glycero-3-phosphocholine by an enriched preparation of human lung mast cells. *J. Immunol.* **144**:4773–4780, 1990.
13. Warner, J. A., Peters, S. P., Lichtenstein, L. M., Hubbard, W. C., Yancey, K. B., Stevenson, H. C., Miller, P. J. and MacGlashan, D. W. Jr. Differential release of mediators from human basophils: differences in arachidonic acid metabolism following activation by unrelated stimuli. *J. Leukoc. Biol.* **45**:558–566, 1989.
14. Triggiani, M., Schleimer, R. P., Warner, J. A. and Chilton, F. H. Differential synthesis of 1-acyl-2-acetyl-*sn*-glycero-3-phosphocholine and platelet-activating factor by human inflammatory cells. *J. Immunol.* **147**:660–666, 1991.
15. Chilton, F. H., Ellis, J. M., Olson, S. C. and Wykle, R. L. 1-*O*-Alkyl-2-arachidonoyl-*sn*-glycero-3-phosphocholine: a common source of platelet-activating factor and arachidonate in the human polymorphonuclear leukocytes. *J. Biol. Chem.* **259**:12014–12019, 1984.
16. Triggiani, M. and Chilton, F. H. Factors that influence the biosynthesis of 1-acyl-2-acetyl-*sn*-glycero-3-phosphocholine and platelet-activating factor. *Biochem. J.* **286**:497–503, 1992.
17. Triggiani, M., Oriente, A., Seeds, M. C., Bass, D. A., Marone, G. and Chilton, F. H. Migration of human inflammatory cells into the lung results in the remodeling of arachidonic acid into a triglyceride pool. *J. Exp. Med.* **182**:1181–1190, 1995.
18. Chilton, F. H., Fonteh, A. N., Surette, M. E., Triggiani, M. and Winkler, J. D. Control of arachidonate levels within inflammatory cells. *Biochim. Biophys. Acta* **1299**:1–15, 1996.
19. MacDonald, J. I. S. and Sprecher, H. Phospholipid fatty acid remodeling in mammalian cells. *Biochim. Biophys. Acta* **1084**:105–121, 1991.
20. Triggiani, M., D'Souza, D. M. and Chilton, F. H. Metabolism of 1-acyl-2-acetyl-*sn*-glycero-3-phosphocholine in the human neutrophil. *J. Biol. Chem.* **266**:6928–6935, 1991.
21. Waku, K. Origins and fates of fatty acyl-CoA esters. *Biochim. Biophys. Acta* **1124**:101–111, 1992.
22. Triggiani, M., De Marino, V., de Crescenzo, G. and Marone, G. Arachidonic acid remodeling in human inflammatory cells migrating to the lung *in vivo*. *Int. Arch. Allergy Immunol.* **113**:190–192, 1997.
23. Murphy, R. C., Hammarstrom, S. and Samuelsson, B. Leukotriene C: a slow reacting substance from murine mastocytoma cells. *Proc. Natl. Acad. Sci. USA* **76:**4275–4279, 1997.

24. Dahinden, C. A., Clancy, R. M., Gross, M., Chiller, J. M. and Hugli, T. E. Leukotriene C_4 production by murine mast cells: evidence of a role for extracellular leukotriene A_4. *Proc. Natl. Acad. Sci. USA* **82**:6632–6636, 1985.
25. Zhao, J.-L., Austen, K. F. and Lam, B. K. Prospective on the functional characteristics of human LTC_4 synthase. In: *Asthma and Allergic Diseases: Physiology, Immunopharmacology, and Treatment* (Marone, G., Austen, K. F., Holgate, S. T., Kay, A. B. and Lichtenstein, L. M. eds), pp. 3–12. Academic Press, London, 1998.
26. Schwartz, L. B. Mediators of human mast cells and human mast cell subsets. *Ann. Allergy* **58**:226–235, 1987.
27. Cirillo, R., Triggiani, M., Siri, L., Ciccarelli, A., Pettit, G. R., Condorelli, M. and Marone, G. Cyclosporin A rapidly inhibits mediator release from human basophils presumably by interacting with cyclophilin. *J. Immunol.* **144**:3891–3897, 1990.
28. Casolaro, V., Galeone, D., Giacummo, A., Sanduzzi, A., Melillo, G. and Marone, G. Human basophil/mast cell releasability. V. Functional comparison of cells obtained from peripheral blood, lung parenchyma, and bronchoalveolar lavage in asthmatics. *Am. Rev. Respir. Dis.* **139**:1375–1382, 1989.
29. Schleimer, R. P., MacGlashan, D. W. Jr, Peters, S. P., Pinckard, R. N., Adkinson, N. F. Jr and Lichtenstein, L. M. Characterization of inflammatory mediator release from purified human lung mast cells. *Am. Rev. Respir. Dis.* **133**:614–617, 1986.
30. Warner, J. A. and MacGlashan, D. W. Jr. Signal transduction events in human basophils. A comparative study of the role of protein kinase-C in basophils activated by anti-IgE antibody and formyl-methionyl-leucyl-phenylalanine. *J. Immunol.* **145**:1897–1905, 1990.
31. Bakhle, Y. S. and Botting, R. M. Cyclooxygenase-2 and its regulation in inflammation. *Mediat. Inflamm.* **5**:305–323, 1996.
32. Murakami, M., Matsumoto, R., Urade, Y., Austen, K. F. and Arm, J. P. c-Kit ligand mediated increased expression of cytosolic phospholipase A_2, prostaglandin endoperoxide synthase-1, and hematopoietic prostaglandin D_2 synthase and increased IgE-dependent prostaglandin D_2 generation in immature mouse mast cells. *J. Biol. Chem.* **270**:3239–3246, 1995.
33. Murakami, M., Tada, K., Nakajima, K. and Kudo, I. Cyclooxygenase-2-dependent delayed prostaglandin D_2 generation is initiated by nerve growth factor in rat peritoneal mast cells. Its augmentation by extracellular type II secretory phospholipase A_2. *J. Immunol.* **159**:439–446, 1997.
34. Lawrence, I. D., Warner, J. A., Cohan, V. L., Hubbard, W. C., Kagey-Sobotka, A. and Lichtenstein, L. M. Purification and characterization of human skin mast cells: evidence for human mast cell heterogeneity. *J. Immunol.* **139**:3062–3069, 1987.
34. Massey, W. A., Guo, C. B., Dvorak, A. M., Hubbard, W. C., Bhagavan, B. S., Cohan, V. L., Warner, J. A., Kagey-Sobotka, A. and Lichtenstein, L. M. Human uterine mast cells. Isolation, purification, characterization, ultrastructure, and pharmacology. *J. Immunol.* **147**:1621–1627, 1991.
36. Patella, V., Marinò, I., Lamparter, B., Arbustini, E., Adt, M. and Marone, G. Human heart mast cells. Isolation, purification, ultrastructure, and immunologic characterization. *J. Immunol.* **154**:2855–2865, 1995.
37. Bischoff, S. C., Schwengberg, S., Raab, R. and Manns, M. P. Functional properties of human intestinal mast cells cultured in a new culture system: enhancement of IgE receptor-dependent mediator release and response to stem cell factor. *J. Immunol.* **159**:5560–5567, 1997.
38. Warner, J. A., Peters, S. P., Lichtenstein, L. M., Hubbard, W., Yancey, K. B., Stevenson, H. C., Miller, P. J. and MacGlashan, D. W. Jr. Differential release of mediators from human basophils: differences in arachidonic acid metabolism following activation by unrelated stimuli. *J. Leukoc. Biol.* **45**:558–571, 1989.
39. Marone, G., Casolaro, V., Cirillo, R., Stellato, C. and Genovese, A. Pathophysiology of human basophils and mast cells in allergic disorders. *Clin. Immunol. Immunopathol.* **50**:524–540, 1989.
40. Marone, G., de Crescenzo, G., Patella, V. and Genovese, A. Human cardiac mast cells and their role in severe allergic reactions. In: *Asthma and Allergic Diseases. Physiology, Immunopharmacology and Treatment* (Marone, G., Austen, K. F., Holgate, S. T., Kay, A. B. and Lichtenstein, L. M., eds), pp. 237–259. Academic Press, London, 1998.
41. Triggiani, M., Oriente, A., de Crescenzo, G. and Marone, G. Metabolism of lipid mediators in human basophils and mast cells. In: *Human Basophils and Mast Cells. Biological Aspects* (Marone, G., ed.), pp. 135–147. Karger, Basel, 1995.
42. MacGlashan, D. W. Jr and Hubbard, W. C. IL-3 alters free arachidonic acid generation in C5a-stimulated human basophils. *J. Immunol.* **151**:6358–6369, 1993.

43. Triggiani, M., Oriente, A. Lipid mediator of basophils and mast cells. In: *Basophils, Mast Cells and Airway Epithelial Cells* (Marone, G., Pozzi, E. and Rennard, S. I. eds), pp. 43–55. Masson, Milan, 1993.
44. Triggiani, M., Oriente, A., de Crescenzo, G., Rossi, G. and Marone, G. Biochemical functions of a pool of arachidonic acid associated with triglycerides in human inflammatory cells. *Int. Arch. Allergy Immunol.* **107**:261–263, 1995.
45. Triggiani, M., Oriente, A. and Marone, G. Differential roles for triglyceride and phospholipid pool of arachidonic acid in human lung macrophages. *J. Immunol.* **152**:1394–1403, 1994.
46. Johnson, M. M., Vaughn, B., Triggiani, M., Swan, D. D., Fonteh, A. H. and Chilton, F. H. Role of arachidonoyl triglycerides within lipid bodies in eicosanoid formation by human polymorphonuclear cells. *Am. J. Respir. Cell Mol. Biol.* **21**:253–258, 1999.
47. Nakamura, T., Fonteh, A. N., Hubbard, W. C., Triggiani, M., Inagaki, N., Ishizaka, T. and Chilton, F. H. Arachidonic acid metabolism during antigen and ionophore activation of the mouse bone marrow derived mast cell. *Biochim. Biophys. Acta* **1085**:191–200, 1991.
48. Fonteh, A. N. and Chilton, F. H. Rapid remodeling of arachidonate from phosphatidylcholine to phosphatidylethanolamine pools during mast cell activation. *J. Immunol.* **148**:1784–1791, 1992.
49. Chilton, F. H. and Murphy, R. C. Remodelling of arachidonate-containing phosphoglycerides within the human neutrophil. *J. Biol. Chem.* **261**:7771–7777, 1986.
50. Fonteh, A. N. and Chilton, F. H. Mobilization of different arachidonate pools and their roles in the generation of leukotrienes and free arachidonic acid during immunologic activation of mast cells. *J. Immunol.* **150**:563–570, 1993.
51. Blank, M. L., Smith, Z. L. and Snyder, F. Contributing factors in the trafficking of [^{3}H]-arachidonate between phospholipids. *Biochim. Biophys. Acta* **1124**:262–272, 1992.
52. Dvorak, A. M. Similarities in the ultrastructural morphology and developmental and secretory mechanisms of human basophils and eosinophils. *J. Allergy Clin. Immunol.* **94**:1103–1134, 1994.
53. Weller, P. F. and Dvorak, A. Lipid bodies: intracellular sites for eicosanoid production. *J. Allergy Clin. Immunol.* **94**:1151–1156, 1994.
54. Dvorak, A. M., Dvorak, H. F., Peters, S. P., Schulman, E. S., MacGlashan, D. W. Jr, Pyne, K., Harvey, V. S., Galli, S. J. and Lichtenstein, L. M. Lipid bodies: cytoplasmic organelles important to arachidonate metabolism in macrophages and mast cells. *J. Immunol.* **131**:2965–2976, 1983.
55. Weller, P. F., Monahan-Earley, R. A., Dvorak, H. F. and Dvorak, A. M. Cytoplasmic lipid bodies of human eosinophils. Subcellular isolation and analysis of arachidonate incorporation. *Am. J. Pathol.* **138**:141–148, 1991.
56. Weller, P. F. and Dvorak, A. M. Arachidonic acid incorporation by cytoplasmic lipid bodies of human eosinophils. *Blood* **65**:1269–1274, 1985.
57. Bozza, P. T., Yu, W., Penrose, J. F., Morgan, E. S., Dvorak, A. M. and Weller, P. F. Eosinophil lipid bodies: specific, inducible intracellular sites for enhanced eicosanoid formation. *J. Exp. Med.* **186**:909–920, 1997.
58. Dvorak, A. M. Ultrastructural analysis of human mast cells and basophils. In: *Human Basophils and Mast Cells. Biological Aspects* (Marone, G., ed.), pp. 1–33. Karger, Basel, 1995.
59. Triggiani, M., Rossi, G. and Marone, G. Remodelling of arachidonic acid pools in inflammatory cells migrating into the lung. In: *Asthma and Allergic Diseases: Physiology Immunopharmacology and Treatment* (Marone, G., Austen, K. F., Holgate, S. T., Kay, A. B. and Lichtenstein, L. M., eds), pp. 133–141. Academic Press, London, 1998.
60. Dennis, E. A. Diversity of group types, regulation, and function of phospholipase A_2. *J. Biol. Chem.* **269**:13057–13060, 1994.
61. Kudo, I., Murakami, M., Shuntaro, H. and Inoue, K. Mammalian non-pancreatic phospholipases A_2. *Biochim. Biophys. Acta* **117**:217–231, 1993.
62. Mukherjee, A. B., Miele, L. and Pattabiraman, N. Phospholipase A_2 enzymes: regulation and physiological role. *Biochem. Pharmacol.* **48**:1–10, 1994.
63. Tischfield, J. A. A reassessment of the low molecular weight phospholipase A_2 gene family in mammals. *J. Biol. Chem.* **272**:17247–17250, 1997.
64. Dennis, E. A. The growing phospholipase A_2 superfamily of signal transduction enzymes. *Trends Biochem. Sci.* **22**:1–2, 1997.
65. De Marino, V., Gentile, M., Granata, F., Marone, G. and Triggiani, M. Secretory phospholipase A_2: a putative mediator of airway inflammation. *Int. Arch. Allergy Immunol.* **118**:200–201,1999.
66. Miura, K., Hubbard, W. C. and MacGlashan, D. W. Jr. Phosphorylation of cytosolic phospholipase A_2 by IL-3 is associated with increased free arachidonic acid generation and leukotriene C_4 release in

human basophils. *J. Allergy Clin. Immunol.* **102**:512–520, 1998.

67. Hundley, T. R., Marshall, L. A., Hubbard, W. C. and MacGlashan, D. W. Jr. Characteristics of arachidonic acid generation in human basophils: relationship between the effects of inhibitors of secretory phospholipase A_2 activity and leukotriene C_4 release. *J. Pharmacol. Exp. Ther.* **284**:847–857, 1998.
68. Kramer, R. M. and Sharp, J. D. Structure, function and regulation of Ca^{2+}-sensitive cytosolic phospholipase A_2 ($cPLA_2$). *FEBS Lett.* **410**:49–53, 1997.
69. Nakatami, Y., Murakami, M., Kudo, I. and Inoue, K. Dual regulation of cytosolic phospholipase A_2 in mast cells after cross-linking of Fcε-receptor. *J. Immunol.* **153**:796–802, 1994.
70. Fujishima, H., Sanchez Mejia, R. O., Bingham, C. O., Lam, B. K., Sapirstein, A., Bonventre, J. V., Austen, K. F. and Arm, J. P. Cytosolic phospholipase A_2 is essential for both the immediate and the delayed phases of eicosanoid generation in mouse bone marrow-derived mast cells. *Biochemistry* **96**:4803–4807, 1999.
71. Chock, S. P., Schmauder-Chock, E. A., Cordella-Miele, E., Miele, L. and Mukherjee, A. B. The localization of phospholipase A_2 in the secretory granule. *Biochem. J.* **300**:619–622, 1994.
72. Murakami, M., Kudo, I., Suwa, Y. and Inoue, K. Release of 14-kDa group II phospholipase A_2 from activated mast cells and its possible involvement in the regulation of the degranulation process. *Eur. J. Biochem.* **209**:257–265, 1992.
73. Fonteh, A. N., Bass, D. A., Marshall, L. A., Seeds, M., Samet, J. M. and Chilton, F. H. Evidence that secretory phospholipase A_2 plays a role in arachidonic acid release and eicosanoid biosynthesis by mast cells. *J. Immunol.* **152**:5438–5446, 1994.
74. Fonteh, A. N., Samet, J. M., Surette, M., Reed, W. and Chilton, F. H. Mechanism that accounts for the selective release of arachidonic acid from intact cells by secretory phospholipase A_2. *Biochim. Biophys. Acta* **1392**:253–266, 1998.
75. Vadas, P. and Pruzansky, W. Biology of disease. Role of secretory phospholipase A_2 in the pathobiology of disease. *Lab. Invest.* **55**:391–404, 1986.
76. Kramer, R. M., Kramer, R. M. Hession, C., Johansen, B., Hayes, G., McGray, P., Chow, E. P., Tizard, R. and Pepinsky, R. B. Structure and properties of a human non-pancreatic phospholipase A_2. *J. Biol. Chem.* **264**:5768–5775, 1988.
77. Seilhamer, J., Pruzansky, W., Vadas, P., Plant, S., Miller, J. A., Miller, J. A., Kloss, J. and Johnson, L. K. Cloning and recombinant expression of phospholipase A_2 present in rheumatoid arthritic synovial fluid. *J. Biol. Chem.* **264**:5335–5338, 1989.
78. Chilton, F. H., Averill, F. J., Hubbard, W. C., Fonteh, A. N., Triggiani, M. and Liu, M. C. Antigen-induced generation of lyso-phospholipids in human airways. *J. Exp. Med.* **183**:2235–2245, 1996.
79. Stadel, J. M., Hoyle, K., Naclerio, R. M., Roshak, A. and Chilton, F. H. Characterization of phospholipase A_2 from human nasal lavage. *Am. J. Respir. Cell Mol. Biol.* **11**:108–113, 1994.
80. Nakae, H., Endo, S., Inada, K., Yamashita, H., Yamada, Y., Takakuwa, T., Kasai, T., Ogawa, M. and Uchida, K. Plasma concentrations of type II phospholipase A_2, cytokines and eicosanoids in patients with burns. *Burns* **6**:422–426, 1995.
81. Bernton, E., Hershey, J., Franson, R., Engler, R., Schwartz, L., Metcalfe, D. and Worobec, A. Serum secretory phospholipase A_2 levels in patients with mastocytosis. *J. Allergy Clin. Immunol.* **97**:267 (abstract), 1996.
82. Fleisch, J. H., Armstrong, C. T., Roman, C. R., Mihelich, E. D., Spaethe, S. M., Jackson, W. T., Bobbitt, J. L., Draheim, S., Bach, N. J., Dillard, R. D., Martinelli, M., Fouts, R. and Snyder, D. W. Recombinant human secretory phospholipase A_2 released thromboxane from guinea pig bronchoalveolar lavage cells: *in vitro* and *ex vivo* evaluation of a novel secretory phospholipase A_2 inhibitor. *J. Pharmacol. Exp. Ther.* **278**:252–257, 1996.
83. Triggiani, M., De Marino, V., Sofia, M., Faraone, S., Ambrosio, G., Carratù, L. and Marone, G. Characterization of platelet-activating factor acetylhydrolase in human bronchoalveolar lavage. *Am. J. Respir. Crit. Care Med.* **156**:94–100, 1997.
84. Cowburn, A. S., Sladek, K., Soja, J., Adamek, L., Nizankowska, E., Szczeklik, A., Lam, B. K., Penrose, J. F., Austen, K. F., Holgate, S. T. and Sampson, A. P. Overexpression of leukotriene C_4 synthase in bronchial biopsies from patients with aspirin-intolerant asthma. *J. Clin. Invest.* **101**:834–846, 1998.
85. Weltzien, H. U. Cytolitic and membrane-perturbing properties of lysophosphatidylcholine. *Biochim. Biophys. Acta* **559**:259–287, 1979.
86. Edelson, J. D., Vadas, P., Villar, J., Mullen, J. B. M. and Pruzanski, W. Acute lung injury induced by phospholipase A_2. Structural and functional changes. *Am. Rev. Respir. Dis.* **143**:1102–1109, 1991.

87. Murakami, M., Hara, N., Kudo, I. and Inoue, K. Triggering of degranulation in mast cells by exogenous type II phospholipase A_2. *J. Immunol.* **151**:5675–5684, 1993.
88. Takasaki, J., Kawauchi, Y. and Masuho, Y. Synergistic effect of type II phospholipase A_2 and platelet-activating factor on Mac-1 surface expression and exocytosis of gelatinase granules in human neutrophils: evidence for the 5-lipoxygenase-dependent mechanism. *J. Immunol.* **160**:5066–5072, 1998.
89. Wada, A., Tojo, H., Toshiro, S., Fujiwara, Y., Kamada, T., Ueda, N. and Okamoto, M. Group II phospholipase A_2 as an autocrine growth factor mediating interleukin-1 action on mesangial cells. *Biochim. Biophys. Acta* **1345**:99–108, 1997.
90. Kanemasa, T., Arimura, A., Kishino, J., Ohtani, M. and Arita, H. Contraction of guinea pig lung parenchyma by pancreatic type phospholipase A_2 via its specific binding site. *FEBS Lett.* **303**:217–220, 1992.
91. Demopoulos, C. A., Pinckard, R. N. and Hanahan, D. J. Platelet-activating factor. Evidence for 1-*O*-alkyl-2-acetyl-*sn*-glyceryl-3-phosphorylcholine as the active component: a new class of lipid chemical mediators. *J. Biol. Chem.* **254**:9355–9358, 1979.
92. Spencer, D. A. An update on PAF. *Clin. Exp. Allergy* **22**:521–524, 1992.
93. Venable, M. E., Zimmermann, G. A., McIntyre, T. M. and Prescott, S. M. Platelet-activating factor: a phospholipid autacoid with diverse actions. *J. Lipid Res.* **34**:691–702, 1993.
94. Chilton, F. H. Platelet-activating factor: synthesis, metabolism, and relationship to arachidonate products. In: *Handbook of Inflammation. Vol. 6. Mediators of the Inflammatory Process* (Henson, P. M., Murphy, R. C., eds), pp. 77–112. Elsevier, Amsterdam, 1989.
95. Columbo, M., Casolaro, V., Warner, J. A., MacGlashan, D. W. Jr, Kagey-Sobotka, A. and Lichtenstein, L. M. The mechanism of mediator release from human basophils induced by platelet-activating factor. *J. Immunol.* **145**:3855–3861, 1990.
96. Henson, P. M., Barnes, P. J. and Banks-Schlegel, S. P. Platelet-activating factor: role in pulmonary injury and disfunction and blood abnormalities. *Am. Rev. Respir. Dis.* **145**:726–731, 1992.
97. Barnes, P. J., Chung, K. F. and Page, C. P. Platelet-activating factor as a mediator of allergic disease. *J. Allergy Clin. Immunol.* **83**:919–934, 1988.
98. Pinckard, R. N., Ludwig, J. C. and McManus, L. M. Platelet activating factors. In: *Inflammation: Basic Principles and Clinical Correlates* (Gallin, J. I., Goldstein, I. M. and Snyderman, R., eds), pp. 139–167. Raven Press, New York, 1988.
99. Tan, X. D., Wang, H., Gonzales-Crussi, F. X., Chang, H., Gonzales-Crussi, F. and Hsueh, W. Platelet-activating factor and endotoxin increase the enzyme activity and gene expression of type II phospholipase A_2 in the rat intestine. Role of polymorphonuclear leukocytes. *J. Immunol.* **156**:2985–2990, 1996.
100. Mueller, H. W., O'Flaherty, J. T. and Wykle, R. L. The molecular species distribution of platelet-activating factor synthesized by rabbit and human neutrophils. *J. Biol. Chem.* **259**:14554–14559, 1984.
101. Blank, M. L., Cress, E. A., Lee, T. C., Malone, B., Surles, J. R., Piantadosi, C., Hajdu, J. and Snyder, F. Structural features of platelet-activating factor (1-alkyl-2-acetyl-*sn*-glycero-3-phosphocholine) required for hypotensive and platelet serotonin responses. *Res. Commun. Chem. Pathol. Pharmacol.* **38**:3–20, 1982.
102. Pinckard, R. N., Showell, H. J., Castillo, R., Lear, C., Breslow, R,. McManus, L. M., Woodard, D. S. and Ludwig, J. C. Differential responsiveness of human neutrophils to the autocrine actions of 1-*O*-alkyl homologs and 1-acyl analogs of platelet-activating factor. *J. Immunol.* **148**:3528–3535, 1992.
103. Chilton, F. H., Cluzel, M. and Triggiani, M. Recent advances in our understanding of the biochemical interactions between platelet-activating factor and arachidonic acid. *Lipids* **26**:1021–1027, 1991.
104. Whatley, R. E., Clay, K. L., Chilton, F. H., Triggiani, M., Zimmerman, G. A., McIntyre, T. M. and Prescott, S. M. Relative amounts of 1-*O*-alkyl- and 1-acyl-2-acetyl-*sn*-glycero-3-phosphocholine in stimulated endothelial cells. *Prostaglandins* **43**:21–29, 1992.
105. Triggiani, M., Goldman, D. W. and Chilton, F. H. Biological effects of 1-acyl-2-acetyl-*sn*-glycero-3-phosphocholine in the human neutrophil. *Biochim. Biophys. Acta* **1084**:41–47, 1991.
106. O'Flaherty, J. T., Tessner, T., Greene, D., Redman, J. R. and Wykle, R. L. Comparison of 1-*O*-alkyl-, 1-*O*-alkyl-1´-enyl-, and 1-*O*-acyl-2-acetyl-*sn*-glycero-3-phosphoethanolamines and 3-phosphocholines as agonists of the platelet-activating factor family. *Biochim. Biophys. Acta* **1210**:209–216, 1994.
107. Columbo, M., Horowitz, E. M., Patella, V., Kagey-Sobotka, A., Chilton, F. H. and Lichtenstein, L. M. A comparative study of the effects of 1-acyl-2-acetyl-*sn*-glycero-3-phosphocholine and platelet-activating factor on histamine and leukotriene C_4 release from human leukocytes. *J. Allergy Clin.*

Immunol. **92**:325–333, 1993.

108. Farr, R. S., Wardlow, M. L., Cox, C. P., Meng, K. E. and Green, D. E. Human serum acid-labile factor is an acylhydrolase that inactivates platelet-activating factor. *Fed. Proc.* **42**:3120–3122, 1983.
109. Stafforini, D. M., Prescott, S. M., Zimmerman, G. A. and McIntyre, T. M. Platelet-activating factor acetylhydrolase activity in human tissues and blood cells. *Lipids* **26**:979–985, 1991.
110. Miwa, M., Miyake, T., Yamanaka, T., Sugatani, J., Suzuki, Y., Sakata, S., Araki, Y. and Matsumoto, M. Characterization of serum platelet-activating factor (PAF) acetylhydrolase: correlation between deficiency of serum PAF-acetylhydrolase and respiratory symptoms in asthmatics children. *J. Clin. Invest.* **82**:1983–1991, 1988.
111. Triggiani, M. and Chilton, F. H. Influence of immunologic activation and cellular fatty acid levels on the catabolism of platelet-activating factor within the murine mast cell (PT-18). *Biochim. Biophys. Acta* **1006**:41–51, 1989.
112. Shin, M. H., Averill, F. J., Hubbard, W. C., Chilton, F. H., Baroody, F. M., Liu, M. C. and Naclerio, R. M. Nasal allergen challenge generates 1-*O*-hexadecyl-2-lyso-*sn*-glycero-3-phosphocholine. *Am. Rev. Respir. Crit. Care Med.* **149**:660–666, 1994.
113. Ambrosio, G., Oriente, A., Napoli, C., Palumbo, G., Chiariello, P., Marone, G., Condorelli, M., Chiariello, M. and Triggiani, M. Oxygen radicals inhibit human plasma acetylhydrolase, the enzyme that catabolizes platelet-activating factor. *J. Clin. Invest.* **93**:2408–2416, 1994.
114. Sedgwick, J. B., Geiger, K. M. and Busse, W. W. Superoxide generation by hypodense eosinophils from patients with asthma. *Am. Rev. Respir. Dis.* **142**:120–125, 1990.
115. Jarjour, N. N., Busse, W. W. and Calhoun, W. J. Enhanced production of oxygen radicals in nocturnal asthma. *Am. Rev. Respir. Dis.* **146**:905–911, 1992.
116. O'Byrne, P. M., Israel, E. and Drazen, J. M. Antileukotrienes in the treatment of asthma. *Ann. Intern. Med.* **127**:472–480, 1997.

CHAPTER 34

Regulation and Function of Human Intestinal Mast Cells

STEPHAN C. BISCHOFF

Medical School of Hannover, Department of Gastroenterology and Hepatology, Hannover, Germany

INTRODUCTION

Localization of Mast Cells in the Gut

The human gastrointestinal tract contains numerous mast cells, which are located primarily in the lamina propria of the mucosa. Intra-epithelial mast cells are rare or absent in normal tissue, but the subepithelial lamina propria contains about 2–3% mast cells under normal conditions (1). In the submucosa, mast cell density is lower compared to the lamina propria (about 1%). The muscularis contains almost no mast cells (<0.1%). The serosa, which is a common source of mast cells in rodents and is therefore frequently used for *in vitro* studies ('peritoneal mast cells'), only contains small quantities of mast cells in humans. This mast cell distribution may be profoundly changed under pathological conditions. For example, in the course of fibrotic transformations, the number of mast cells in the lamina muscularis propria and in the submucosa may be sharply increased (2), whereas the lamina propria mast cell density remains largely unchanged (Table I). These data suggest that mast cells may be involved in the pathogenesis of tissue fibrosis in different tissues such as intestine, lung and liver (3–6). The factors regulating mast cell density locally are not yet clearly identified. A candidate that may cause the recruitment of mast cells is stem cell factor (SCF), which is produced by stromal cells (7).

Heterogeneity of Intestinal Mast Cells

Mast cells are subjected to a considerable heterogeneity, defined primarily by different staining properties of rodent mast cell subsets. Such subsets have been further defined by morphological, biochemical and functional criteria (8). Enerbäck and co-workers (9) were the first to describe two subsets of mast cells in the rat: the mucosal mast cell

MAST CELLS AND BASOPHILS
ISBN 0-12-473335-2

(MMC) located in the lamina propria, and the connective tissue mast cells (CTMC) located predominantly in the submucosa and serosa (Table II). CTMC were similar to mast cells found in other connective tissues (e.g. skin), whereas mast cells of the MMC type were found at all mucosal sites (e.g. respiratory tract). In humans, this classification

TABLE I
Mast Cell Density in Human Gut

Layer	Normal	Fibrosis/ strictures	Inflammatory bowel disease (with or without strictures)	Mastocytosis
Epithelium	–/+	–/+	–/+	+
Lamina propria	++	++	+++	++++
Submucosa	+	++	++	+++
Muscularis	–	+++	+	+
Serosa	–/+	++	–/+	+

– to ++++ indicate 'no' to 'very high' numbers of mast cells in tissue.

TABLE II
Properties of Rat Mast Cell Subtypes

Property	MMC	CTMC
Size (diameter)	~10 μm	~20 μm
Nucleus	1–2 lobes	Unilobed
Granule size	Few variable	Many uniform
Cell membrane microvilli	Few	Numerous
Fixation	Formalin-sensitive	Formalin-resistant
Thymus-dependent proliferation	Yes	No
Half-life	< 40 days	Longer
IgE	Surface and cytoplasmic	Surface

After *Ann. Allergy* **59**:44–53, 1987.

TABLE III
Localization of Human Mast Cell Subtypes

Localization	MC_T	MC_{TC}
Lung	90%	10%
Intestinal mucosa	80–100%	0–20%
Intestinal submucosa	20–40%	60–80%
Intestinal muscularis/serosa	–	100%
Skin	–	100%

could not be confirmed, although different mast cell types exist. Human mast cells are commonly classified according to their protease content (10, 11). Mast cells containing tryptase only (MC_T) resemble rodent MMC, whereas mast cells containing tryptase and chymase (MC_{TC}) show similarities to rodent CTMC. However, the distribution of MC_T and MC_{TC} between mucosa and submucosal layers is not as clear in humans as previously described in rats (Table III). Moreover, a third subgroup may exist in human GI tissue (1, 12), the chymase-positive, tryptase-negative mast cells (MC_C), but this could not be confirmed by other investigators (L. B. Schwartz, personal communication). Most importantly, it is unclear if the phenotype of mast cells, defined according to their protease content, is irreversible or may change depending on the environmental conditions of the cells. Finally, the functional significance of this classification needs to be elucidated. For example, the triggering agents may be different, since skin-derived MC_{TC} respond to IgE-independent agonists such as C5a or substance P, whereas MC_T of the lung do not (13). However, it is unknown whether these data could be transferred to the gut, where both phenotypes are located. Bradding *et al.* (14) suggested that the cytokine pattern produced by lung-derived mast cells differs between MC_T (interleukins IL-5 and IL-6) and MC_{TC} (IL-4), a finding that has to be confirmed for human intestinal mast cells. Such experiments are hampered by the fact that the different subsets cannot be separated *in vitro*, therefore limiting such experiments to immunocytochemical approaches known to be at risk of many pitfalls. The characterization of increasing numbers of proteases and protease isoforms may lead to a more sophisticated classification (15), but it remains mandatory that such classifications are related to different functional properties of the cells. In conclusion, it became evident that, not only interspecies differences in mast cells exist, but that mast cells from different sites within the same species may exhibit heterogeneity. Therefore, data obtained from studies performed with human lung, nasal or skin mast cells, and in particular studies performed in the rodent system, are not readily transferable to the human gastrointestinal tract. Therefore, we focus on human gastrointestinal mast cells, for which information is rare compared to that for other mast cells, because the gut is a remarkable reservoir of mast cells in the human body, which may be of clinical significance.

EFFECTOR FUNCTIONS OF HUMAN INTESTINAL MAST CELLS

Mast cells exert their effector functions almost exclusively by releasing bioactive molecules, either constitutively or upon stimulation. Apart from these humoral effects, which are summarized in Fig. 1, other effector functions such as cytotoxicity or phagocytosis have been hardly characterized. Very recently, Mécheri and David suggested that mast cells may also function as antigen-presenting cells since they express MHC molecules and accessory molecules and they can activate T cells in an antigen-specific manner (16). These data, which were obtained from experiments in mice, have to be confirmed for human mast cells.

Release of Granule Constituents: Histamine, Proteases and Proteoglycans

The histamine content of human intestinal mast cells has been estimated to be 1–3 pg per cell, but may vary depending on the mast cell subtype and the activation status (10, 17, 18). Histamine is the only bioactive amine found in human mast cells, whereas rodent

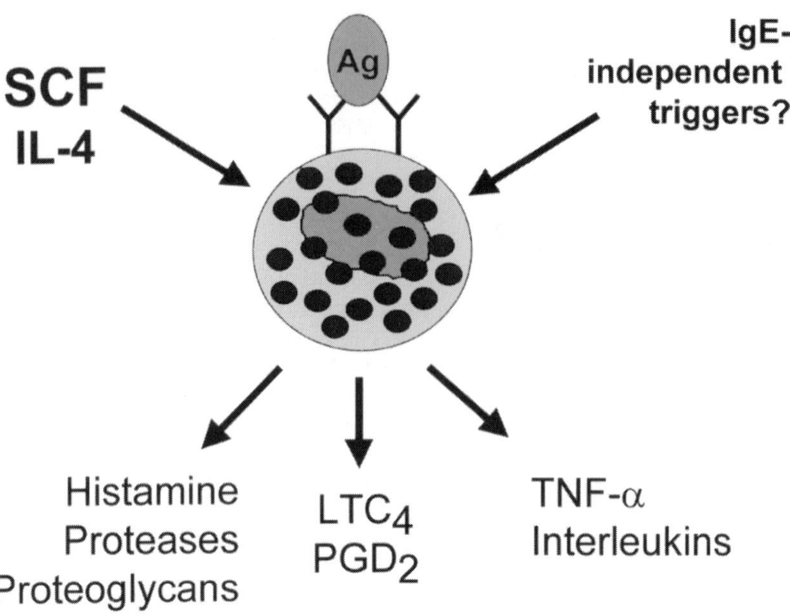

Fig. 1 Regulatory molecules and effector functions of human intestinal mast cells.

mast cells contain also serotonin and possibly other monamines. In humans, serotonin is found in platelets and in intestinal enterochromaffin cells, but has not been reported in normal intestinal mast cells. The multiple biological activities of histamine are mediated through three types of receptors (H_1–H_3) which are expressed on different cell types (19). Histamine is a regulatory mediator with particular relevance for the gastrointestinal tract, since it regulates epithelial functions such as secretion of mucus, hypochloric acid and electrolytes, muscle functions such as peristalsis, splanchnic blood flow and vascular permeability, leukocyte functions and transmitter secretion by neuronal cells (Table IV).

Apart from histamine, proteases and proteoglycans are found in human mast cell granules. The two main proteases are tryptase and chymase, for which different isoforms have been identified (11, 15). Because of the heterogeneity of mast cells expressing these proteases, they were used to define subsets of mast cells. The proteases have numerous biological effects by degrading other peptides such as fibrinogen, vasoactive intestinal peptide (VIP), neuropeptides and activating cascades such as complement, kalikrein–kinin and the angiotensin–aldosterone system. Moreover, chymase promotes mucus secretion in the intestine (Table IV).

Heparin and chondroitin sulphate E are the major proteoglycans found in human mast cells. In the rat it was suggested that only CTMC, but not MMC, contain heparin (20). However, it is unclear whether heparin production in humans could be attributed exclusively to one mast cell subtype. Apart from their well-known anticoagulatory effect, heparin and, albeit to a lesser extent, chrondroitin sulphate E have multiple regulatory effects on other bioactive molecules (e.g. by stabilizing cytokines or neutralizing their effects) (Table IV).

TABLE IV
Biological Effects of Mast Cell Granule Constituents

Histamine		
Via H_1 receptors	Epithelium:	Secretion of mucus ↑
	Muscles:	Constriction ↑ Peristalsis ↑
	Blood vessels:	Vasodilatation ↑ Permeability ↑
Via H_2 receptors	Epithelium:	Acid and chloride secretion ↑
		Permeability ↑
	Lymphocytes:	Cytotoxicity ↑
	Granulocytes:	Chemokinesis ↑ Receptors ↑
Via H_3 receptors	Endothelium:	PGI_2 release ↑
	Neuronal cells:	Acetylcholine↑ Histamine ↑
Proteases		
Tryptase	Degradation of fibrinogen, kininogen, neuropeptides and vasoactive intestinal peptide	
	Stimulation of fibroblast proliferation	
	Generation of C3a, activation of prostomelysin (MMP-3)	
	Enhancement of histamine effects	
	Regulation of endothelial functions	
Chymase	Degradation of basal membranes and substance P	
	Stimulation of mucus secretion in goblet cells	
	Conversion of angiotensin I to angiotensin II	
	Accumulation of neutrophils and eosinophils	
	Activation of IL-1β precursor	
Carboxypeptidase	Cleavage of angiotensin I (similar to zinc metallopeptidases)	
Proteoglycans		
Heparin	Binding and stabilization of histamine, proteases (in particular tryptase) and acidic hydrolases	
	Neutralization of major basic protein (MBP)	
	Enhancement of fibronectin binding to collagen	
	Stimulation of endothelial cell migration	
	Stabilization of cytokines; anticoagulating activity	
	Inhibition of complement and kallikrein activation	
Chondroitin sulphate	Type E (in mast cells), type A (in basophils)	
	Similar effects as heparin, but weaker	

Release of Eicosanoids: Pro-inflammatory and Regulatory Effects

Apart from the release of pre-formed mediators stored in the granules, human intestinal mast cells are capable of releasing *de novo* synthesized mediators such as leukotrienes and prostaglandins upon activation by ionomycin or IgE receptor cross-linking (17, 21, 22). In contrast to skin mast cells, human intestinal mast cells are not activated by compound 48/80 (own unpublished results). Sulphidoleukotrienes (LTC_4, LTD_4, LTE_4) are predominantly produced, whereas LTB_4 could be detected only at low levels. Moreover, human intestinal mast cells (in contrast to basophils) produce substantial amounts of prostaglandin D_2 (PGD_2), thought to play an important role in the late phase of allergic reactions at different mucosal sites (23; own unpublished results). Sulphidoleukotrienes and PGD_2 effects are similar to those of histamine (e.g. increase of vascular permeability, enhancement of peristalsis, etc.), but the mediators are about 100 times more potent on a molar basis (24). The release of eicosanoids by mast cells generally requires two signals. IgE receptor cross-linking by antigen (*in vivo,* in allergic patients) or by anti-IgE or anti-IgE receptor antibodies (*in vitro*) causes the release of histamine and other granule constituents but in most cases only marginal release of leukotrienes. However, after pre-incubation with appropriate cytokines such as SCF (mast cells) or IL-3, IL-5, granulocyte–macrophage colony-stimulating factor (GM-CSF) or nerve growth factor (NGF) (basophils) and subsequent stimulation by IgE receptor cross-linking, both human mast cells and human basophils release high amounts of eicosanoids (24–26). This phenomenon was termed 'priming' and seems to be a general regulatory principle occurring in different cell types and not restricted to IgE-dependent stimulation (27).

Release of Tumour Necrosis Factor: Role in Host Defence Against Bacteria and Sepsis

Histamine, leukotrienes and PGD_2 are classical pro-inflammatory mediators released by human intestinal mast cells. More recently, it was recognized that mast cells are capable of producing cytokines such as tumour necrosis factor-α (TNF-α). TNF-α immunoreactivity was found first in granules of rodent mast cells (28), and is released constitutively at low levels into supernatants. The production and release of TNF-α can be substantially enhanced upon activation of the cells by IgE receptor cross-linking. These findings could be confirmed for human skin, nasal and lung mast cells (29, 30). We could show that human intestinal mast cells also produce and release TNF-α both *in vitro* and *in vivo* (31). Interestingly, not only IgE receptor cross-linking, but also incubation of mast cells with particular strains of Gram-negative bacteria such as *Escherichia coli,* enhance TNF-α production. Studies in animal models indicate that mast cell-derived TNF-α may have a protective function against bacterial infection, since mortality from artificial bacterial peritonitis was substantially enhanced in mast cell-deficient mice compared with normal littermates. The effect of mast cell deficiency could be simulated by administration of neutralizing anti-TNF-α antibodies, which also enhanced the mortality. These findings strongly suggest that the protective effect of mast cells against bacterial infection is indeed due to their production of TNF-α that can be triggered directly by the bacteria (18, 32). The data could be confirmed in part by human studies showing that administration of anti-TNF-α antibodies in patients with bacterial sepsis does not improve the clinical course, as expected, but rather worsened it (33). On the other hand, there is increasing body of evidence that the permanent release of relatively high amounts of TNF-α is of crucial importance in the pathogenesis of chronic inflammatory diseases such as rheumatoid arthritis and Crohn's disease, and that blocking the cytokine using humanized anti-TNF-α antibodies improves the symptoms dramatically (34, 35). According to multiple *in vitro* and *in vivo* data, two cell types are the principal sources of TNF-α production in the human gut: monocytes and mast cells (31, 36). These data emphasize that mast cells may play an important role at the gastrointestinal barrier, for both host defence and pathogenesis of chronic inflammation.

Release of Interleukins and Growth Factors: Role in Immunoregulation and Neuroimmune Interactions

The significance of human mast cells as immunoregulatory cells was recognized by the observation that they can produce a large array of different cytokines, including interleukins, growth factors and chemokines. As reported before for TNF-α, the first discovery of cytokine production was made in rodent mast cells (37), but much of the data has since been confirmed for human mast cells (38). We examined cytokine production in purified human intestinal mast cells (purity 98–100%) by RT–PCR, RNase protection assay, FACS analysis (intracellular cytokine staining), Western blot and ELISA (own unpublished results). These experiments revealed that human intestinal mast cells produce different sets of cytokines depending on the kind of stimulation. Pro-inflammatory cytokines such as TNF-α, IL-6 and IL-8 are produced at low levels spontaneously without stimulation of the cells. Stimulation by IgE receptor cross-linking leads to an enhanced production of pro-inflammatory cytokines and *de novo* production of Th2 cytokines such as IL-3, IL-5, IL-10 and IL-13 (but not IL-4). Pre-incubation of the cells with IL-4 and subsequent stimulation by IgE receptor cross-linking causes a further enhancement of the production of Th2 cytokines but not of TNF-α or IL-6. Interestingly, IL-6 production is inhibited by IL-4 pre-incubation. Gram-negative bacteria, in contrast to IgE receptor cross-linking, do not induce the release of Th2 cytokines but enhance that of pro-inflammatory cytokines (39, 40). The findings, which are summarized in Fig. 2, may have important clinical implications. Whereas bacteria may drive mast cells to release exclusively pro-inflammatory cytokines such as TNF-α, IL-4 and IgE receptor cross-linking induce the almost selective release of Th2 cytokines. These cytokines may facilitate the specific immune response towards a distinct direction typical for allergic

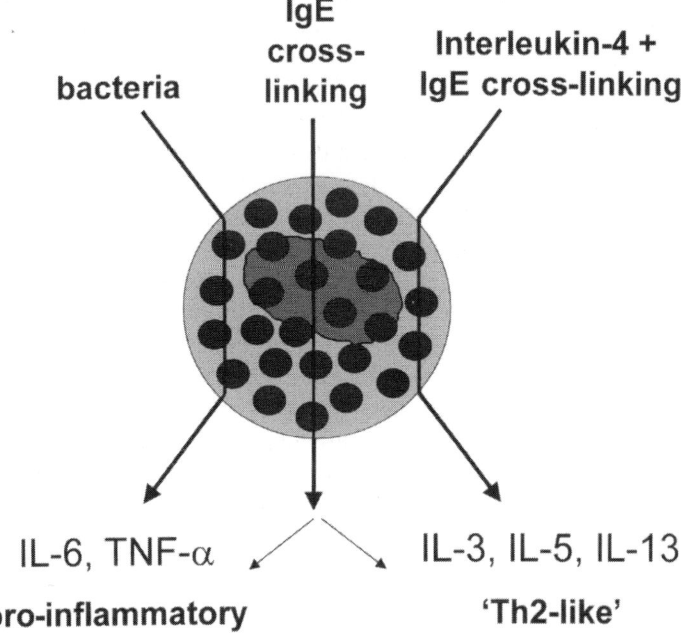

Fig. 2 Cytokine patterns released by human intestinal mast cells depending on the stimulus.

reactions (Th2 response, eosinophil recruitment) or bacterial infection (pro-inflammatory, neutrophil recruitment).

Nerve–mast cell interactions have for a long time been suggested to play a role at the gastrointestinal barrier, mostly based on morphological studies indicating a close relationship between mast cells and nerve endings (41–43). However, the mechanism of interactions on a molecular level has not yet been elucidated. It has been reported that substance P activates rodent mast cells and human skin mast cells for histamine release (13, 30), but these findings could not be confirmed for human lung and intestinal mast cells (44; own unpublished results). Moreover, other neuropeptides and gastrointestinal peptides such as VIP failed to activate human intestinal mast cells for mediator release. The findings were confirmed by the fact that human intestinal mast cells do not generally express the NK_1–NK_3 receptors (own unpublished results). On the other hand, mast cells may modulate nerve functions in the gut mucosa by different mediators such as histamine, prostaglandins, leukotrienes, cytokines and other mediators. Of particular interest in this respect is the finding that rodent mast cells are capable of producing NGF (45), but this was not confirmed for mature human mast cells so far. The role of NGF and related peptides as regulators of inflammatory effector cells is not yet fully elucidated. Whereas human basophils and eosinophils clearly respond to NGF, the results for mast cells are conflicting, possibly because of the heterogeneity of mast cells (26, 46, 47). In rodent mast cells, NGF induces IL-6 and PGE_2 production, and inhibits TNF-α release and mast cell apoptosis (48–51). These data were partly confirmed by *in vivo* studies (52). In man, the situation is less clear. The human mast cell line HMC-1, and possibly also human mast cells derived from particular origins and cultured with SCF, express the high-affinity NGF receptor trkA (53, 54). Data on human intestinal mast cells, however, are not available. Clearly, more studies are needed to clarify the possible functional interactions between intestinal mast cells and the enteric nervous system in man, which may be of particular relevance for a variety of gastrointestinal diseases such as chronic diarrhoea, constipation, motility disorders and chronic inflammation.

REGULATION OF HUMAN INTESTINAL MAST CELL FUNCTIONS

IgE Cross-linking

Cross-linking of cell surface-bound IgE by antigen or of IgE receptors by anti-IgE receptor antibodies is the only well-defined triggering event leading to degranulation, activation and mediator release in human intestinal mast cells. It induces the release of histamine, LTC_4, PGD_2 and TNF-α *in vitro* and, most likely, *in vivo* (10, 17, 55). The release is dose- and time-dependent and involves different signal transduction pathways causing tyrosine phosphorylation and intracellular calcium mobilization. *In vitro*, the release reaction is more pronounced if the cells have been stimulated with anti-IgE receptor antibodies such as mAb 29C6 directed against the α chain of $Fc_{\varepsilon}RI$ compared to stimulation with anti-IgE antibodies, which depends on the number of IgE receptors and the receptor occupancy with IgE. The role of IgG and IgA receptors for mast cell activation has to be elucidated. It was reported that anti-IgG antibodies may also, albeit less pronounced than anti-IgE antibodies, induce mediator release, which may be due to cross-linking of IgG receptors on cell surfaces or due to cross-linking of naturally occurring autoantibodies directed against immunoglobulins bound on cell surfaces (56, 57).

IgE-independent Triggering Agents

Several IgE-independent mast cell triggers have been described, such as C5a, substance P and other neuropeptides, adenosine, compound 48/80 and chemokines. In contrast to IgE receptor cross-linking, the efficacy of these agonists is subject to considerable variability, depending on the mast cell type (rodent versus human, skin versus mucosal, etc.). Using human intestinal mast cells freshly isolated from surgery specimens or cultured in the presence of cytokines, we found no mediator release at all in response to all these IgE-independent agonists, suggesting that human intestinal mast cells are less sensitive towards common triggering agents compared to mast cells of other sources (17). Apart from IgE receptor cross-linking, only two stimulation pathways have been described for human intestinal mast cells. Gram-negative bacteria may activate human intestinal mast cells, as mentioned before, although the exact mechanism is not clear (31). Moreover, SCF can activate human intestinal mast cells directly under particular conditions (21).

Mast Cell Growth Factors and Modulatory Cytokines

Since mast cells are a potent source of highly active mediators, their function must be carefully regulated. Apart from IgE receptor cross-linking, growth factors have been proposed to be involved in the regulation of mast cell function. The first mast cell growth factor was identified in 1990 as the ligand of the proto-oncogene c-kit and is now termed stem cell factor (SCF) or c-kit ligand (58). SCF is produced by many tissue cells such as fibroblasts, endothelial cells and stromal cells. It is an essential factor for mast cell development, since defects in the genes encoding for SCF or c-kit preventing translation into protein result in mast cell deficiency. We could show that SCF is not only involved in mast cell development but also of importance for the regulation of mature mast cell functions. In human lung mast cells, SCF primes the cells for enhanced IgE receptor-dependent mediator release (24) and survival (59). Similar observations were made for human intestinal mast cells, which can be maintained in culture for several months if the culture medium is supplemented with SCF (17, 21). Interestingly, mast cells cultured for a few days in the absence of SCF can be triggered for histamine and leukotriene release by SCF (100 ng ml^{-1}, incubation time 30 min), indicating that SCF acts not only as a modulator of mediator release but also as a triggering agent (21). Defects in the SCF/c-kit system also have clinical implications since they are thought to be the molecular basis of mastocytosis. In particular, mutations within the gene encoding for the c-kit tyrosine kinase, causing permanent phosphorylation or dimerization, were found in patients with systemic mastocytosis (60, 61).

For many years evidence has existed that, apart from SCF, other growth factors, namely T cell-derived cytokines, are involved in the regulation of human mast cell function (62). A number of interleukins (IL-3, IL-4, IL-9, IL-10) have been described as promoting mast cell growth in rodents, but failed to affect human mast cell functions (24, 62). It remained unclear which molecule mediates the known interaction between mast cells and the specific immune system. Recently, we and others demonstrated that IL-4 plays a crucial role in human mast cell development, survival and function (63–65). In contrast to SCF, IL-4 has virtually no effect on human mast cell functions, but it magnifies some effects of SCF. We found that the proliferation of isolated human intestinal mast cells could be enhanced by 200–500% if IL-4 and SCF are added to the culture medium compared to SCF supplementation alone. This dramatic effect occurs dose-dependently (ED_{50} = 50 pg ml^{-1}) and is most likely mediated through the IL-4 receptor expressed on human intestinal mast cells. Moreover, pre-incubation of the mast cells with IL-4 and SCF, compared to SCF alone, strongly enhances the mediator release reaction in response to IgE receptor cross-linking and polarizes the cytokine production to a set of Th2 cytokines such as IL-3, IL-5 and IL-10 (63; unpublished results). These findings indicate that IL-4 plays an important role in regulating human intestinal mast cell functions. SCF promotes mast cell development and enhances their functional activity, whereas IL-4 decreases the development of mast cells from progenitor cells, but induces final differentiation, increased IgE receptor expression and enhanced functional activity (64, 65). The effects are summarized in Fig. 3.

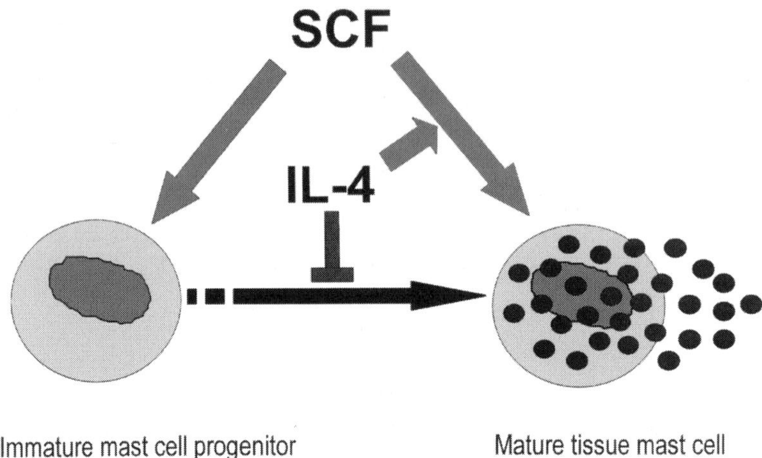

Fig. 3 Regulatory role of interleukin-4 (IL-4) and stem cell factor (SCF) on human mast cells. In contrast to SCF, IL-4 suppresses proliferation of immature mast cells, but IL-4 potentiates the agonistic effects of SCF on mature mast cells.

CLINICAL IMPLICATIONS

Mast Cells and Gastrointestinal Allergy

Whereas allergic reactions have been studied extensively in human skin and respiratory mucosa, little is known about the pathophysiology of intestinal hypersensitivity reactions, and most available information has come from animal studies (66–77). The mechanisms of intestinal allergic reactions towards food proteins and other antigens are poorly defined (69, 78–80). For example, it is largely unclear to what extent such reactions may be IgE-mediated and whether other mechanisms such as IgG immune complexes or T cell-mediated reactions may be involved (81–83). Reactions occurring within a few minutes after food allergen exposure and manifesting in the upper digestion tract (e.g. oral allergy syndrome or gastric symptoms) are thought to be mainly IgE-mediated, since most patients have positive skin reactivity and specific serum IgE against particular food antigens. This is not necessarily true in patients with lower gastrointestinal tract symptoms such as abdominal pain, diarrhoea, cramps and flatulence often starting with a delay of hours to days after ingestion of suspected food antigens. However, the possibility that local IgE production may occur in patients with food allergy having negative skin tests or RAST cannot be excluded (84). One of the most relevant food allergies often involving the gastrointestinal tract is cow's milk allergy, which occurs in both children and adults (85–91).

The relationship between increased intestinal permeability and food allergy is also a controversial issue (92). Although it is recognized that intestinal food allergy is generally accompanied by increased intestinal permeability, it remains unclear whether food allergy causes the increased permeability because of intestinal inflammation, or whether increased permeability, which may be genetically determined or caused by non-allergic bowel diseases, may subsequently lead to intestinal hypersensitivity towards luminal antigens such as bacterial proteins or food proteins, or even to extra-intestinal allergic disease (93–96).

Mast cells are recognized as crucial effector cells in intestinal anaphylaxis (77, 97–100). More than a decade ago it was clearly shown that intestinal mucosal mast cells degranulate upon allergen stimulation in sensitized rats (77). Also some human studies indicated an involvement of mast cells in hypersensitivity reactions of the gastrointestinal tract (101). In endoscopic mucosal provocation tests mast cell and histamine changes in the mucosa were reported (102–106). In particular, mast cells regulate the secretory response of the gut and thus diarrhoea, which is frequently observed in patients suffering from intestinal allergy (107, 108). The role of cytokines such as TNF-α produced by intestinal mast cells in the pathogenesis of intestinal allergic reactions is equivocal (109).

Many questions remain to be answered in the field of allergic enteropathy, including the most appropriate means of diagnosis and the basic pathological mechanisms involved in intestinal allergic reactions (110). Clinical manifestations of gastrointestinal allergy are variable and unspecific. They include symptoms such as abdominal pain, cramping, and diarrhoea, which also occur in other gastrointestinal disorders, particularly in inflammatory bowel disease or irritable bowel syndrome (66–68). Oral challenge procedures with blinded food antigens are now accepted as the 'gold standard' in diagnosing food allergy (111), although these tests have several practical problems, especially in patients with intestinal manifestations of food hypersensitivity (68). However, new test systems have been developed, such as measurement of IgE and eosinophil parameters in stool samples (84, 112, 113), or endoscopic provocation tests performed locally in the gastrointestinal mucosa (102, 105, 106, 114). These tests may improve the possibility of identifying patients with allergic enteropathy and, in addition, they can provide new insights into the pathophysiology of intestinal hypersensitivity reactions. There is increasing evidence that, consistent with findings in the skin and respiratory system, mast cells and eosinophils play a critical role in mediating intestinal allergic reactions, although the basic mechanisms of allergic enteropathy are still puzzling. Further investigations are necessary for a better understanding of gastrointestinal allergy which may allow the development of reliable diagnostic tests and new pharmacological agents for treatment. The clinical relevance of gastrointestinal allergic reactions to food is also largely unclear (115). In particular, it is unknown to what extent food allergy may be involved in gastrointestinal diseases such as irritable bowel syndrome (116–118), inflammatory bowel disease (68), coeliac disease (101, 119), eosinophilic gastroenteritis and other disease entities of unknown aetiology. On the other hand, there is clear evidence from many epidemiological studies that the prevalence of food allergy is rather high (about 4% in children, about 1–2% in adults), and that it is increasing (91, 120, 121). Therefore, and also because of the possible severity of symptoms (122), there is an urgent need to improve our knowledge on the pathological mechanisms and management of intestinal food allergy.

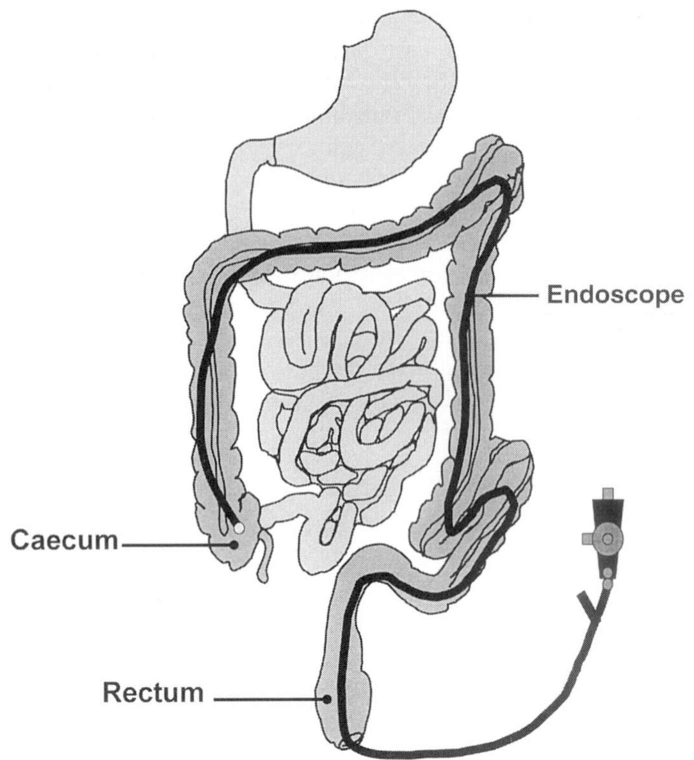

Fig. 4 Colonoscopic allergen provocation (COLAP) performed in the caecum using a flexible endoscope. For details, see text and Bischoff *et al.* (102).

Mast Cells and Infectious Diseases

It has been known for a decade that rodent mast cells are capable of producing TNF-α. This pro-inflammatory cytokine is stored in the granules and released upon activation by IgE receptor cross-linking (28). More recently, we and others have shown that this is also true for human mast cells, including human intestinal mast cells (29–31, 123, 124). We isolated human intestinal mast cells, purified them to homogeneity and found low levels of TNF-α in unstimulated mast cells. If stimulated by IgE receptor cross-linking, mast cells start to enhance TNF-α production and release (28). Most interestingly, we and others found that not only IgE receptor cross-linking but also Gram-negative bacteria such as particular *E. coli* strains are capable of enhancing TNF production in human intestinal mast cells (28, 125; and own unpublished results). The biological relevance of these findings may be that mast cells are involved in neutrophil recruitment which happens in the course of many inflammatory reactions such as allergy and peritonitis (109, 126, 127). The results fit with some very interesting observations using a mouse model of septic peritonitis (18, 32). Mortality from sepsis was strongly enhanced if the mice lacked mast cells (WW^V mice), indicating that mast cells are of importance for host defence against bacterial infection. The authors could show that this mast cell-dependent decrease of mortality was due to TNF-α release from mast cells, because the effect of mast cell deficiency could be simulated by administering neutralizing anti-TNF antibodies. These experiments provided evidence for the concept of the 'two faces of mast cells': a pro-inflammatory one important for allergic diseases, for example, and a protective one involved in host defence against bacterial infection (128). This concept has been indirectly confirmed to some extent for humans, since clinical studies showed that blocking TNF-α has no benefit for the outcome of bacterial sepsis, which gets even worse in some cases (33). This is in sharp contrast to the clinical findings in patients with chronic inflammatory diseases such as rheumatoid arthritis or Crohn's disease, in which the blockage of TNF derived from macrophages and mast cells by anti-TNF antibodies significantly improves the course of the disease (34, 35).

Of particular interest for the gastroenterologist are the possible interactions between gastric mast cells and *Helicobacter pylori.* These bacteria have been identified as major pathogens of chronic gastritis and ulcer disease. The role of mast cells in the pathogenesis of *Helicobacter pylori*-induced gastritis is not defined, but first studies suggest an interaction between mast cells and this type of bacteria (129, 130).

Whereas the involvement of mast cells in host defence against bacterial infections is

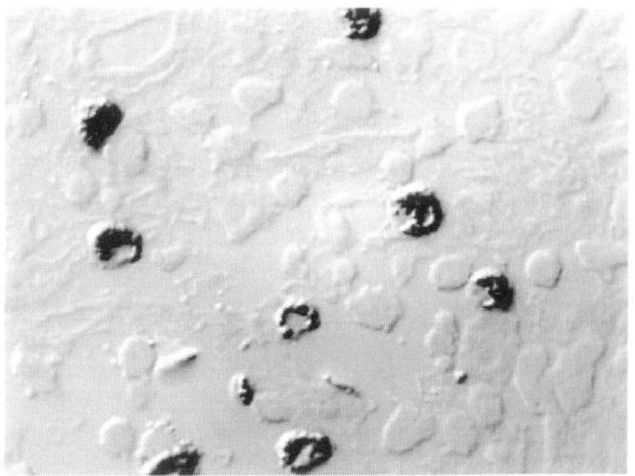

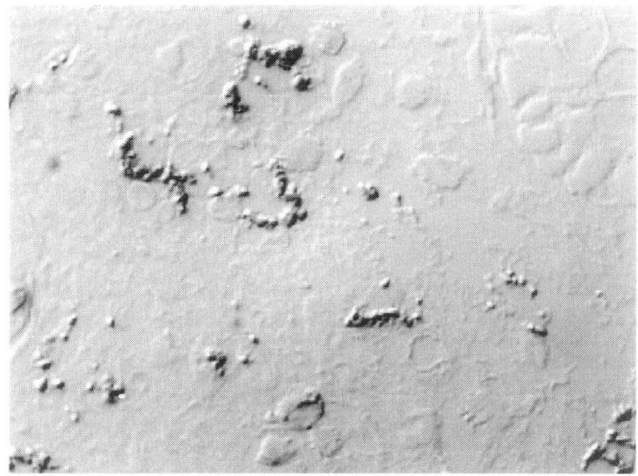

Fig. 5 Immunohistochemistry of human colonic tissue using an anti-histamine antibody. (Upper panel) Resting mast cells after administration of NaCl 0.9% (negative control solution) in the COLAP. (Lower panel) Activated mast cells after administration of antigen in a patient with food allergy.

rather new in mast cell research, great efforts have been made to elucidate the role of mast cells during parasitic infections, which generally involve the gastrointestinal tract (131–137). This topic has been repeatedly reviewed in detail elsewhere.

Mast Cells and Inflammatory Bowel Disease

Crohn's disease and ulcerative colitis are the two major forms of chronic inflammatory bowel disease of unclear aetiology. Multiple aetiological factors have been discussed, such as genetic factors and environmental factors leading to an abnormal immune reaction towards yet unknown antigens (138–141). Apart from food allergens (87, 142–147), bacteria-derived luminal antigens have been discussed as triggering molecules (148, 149). An impaired balance between protective factors promoting oral tolerance (IL-4, IL-10, TGF-β) and proinflammatory factors (TNF-α, IL-1, IL-6, IL-8) may cause the excessive inflammatory reaction and subsequently symptoms such as abdominal cramps, diarrhoea and rectal bleeding. It was hypothesized that Crohn's disease is related to a Th1-dominated immune response characterized by the enhanced release of IL-2, IL-12 and IFN-γ (interferon-γ) and tissue infiltration with monocytes and neutrophils, whereas ulcerative colitis is characterized by a Th2-dominated response involving IL-4, IL-5 and inflammatory cells such as eosinophils (138, 141). This paradigm was questioned by the observations of a French research group who found that the type of immune response in Crohn's disease is obviously dependent on the stage of disease. Long-standing Crohn's disease is indeed characterized by a Th1-type immune response, but early phases of Crohn's disease (onset phase or recurrence phase after surgery) show typical features of Th2-type immune response with increased levels of IL-5 and enhanced eosinophil numbers in the mucosa (150, 151). Therefore, it is not surprising that enhanced numbers and/or enhanced activation of mast cells has been repeatedly reported for both ulcerative colitis (104, 152–156) and Crohn's disease (2, 104, 157–162). In our own studies, we have confirmed and extended these results and found that mast cells become activated in the course of active Crohn's disease as well as active ulcerative colitis (1). Moreover, *in vitro* studies with mast cells isolated from human intestinal tissue indicate that the release of histamine, LTC_4 and TNF-α is more pronounced in cells derived from patients with inflammatory bowel disease compared to controls (17, 31, 154, 159, 161). Of particular interest is the observation that mast cells may act not only as pro-inflammatory effector cells but also as regulatory cells promoting wound healing after tissue destruction. For example, it has been shown that mast cells accumulate in particular at the demarcation line of active ulcerative colitis (155), and, more recently, Gelbman *et al.* showed that mast cells accumulate at sites of strictures in Crohn's disease (2). Since strictures may be viewed as an abnormal healing process, this finding further supports the hypothesis that mast cells are involved in tissue remodelling (2; own unpublished observations). The fact that mast cells may be involved in the pathogenesis of different phases of inflammatory bowel disease such as the active inflammatory phases and healing processes may be because of different mast cell subpopulations found in the gastrointestinal tissue. For example, mast cells in the mucosa (predominantly MC_T) are thought to exert mainly pro-inflammatory and immunomodulatory effects by releasing histamine, leukotrienes and cytokines, whereas mast cells in the deeper layers (predominantly MC_{TC}) may be specialized to interact with fibroblasts by releasing anti-inflammatory and tissue remodelling factors such as TGF-β and bFGF (basic fibroblast growth factor). This hypothesis needs to be confirmed by further studies.

Mast Cells and Other Gastrointestinal Diseases of Unclear Origin

Microscopic colitis and chronic idiopathic diarrhoea

The aetiology of chronic diarrhoea remains often obscure, despite extensive diagnostic tests which may all yield negative results (163). In some cases (about 1–10%), infectious diseases are excluded and endoscopic findings are completely negative, but histology reveals a so-called microscopic colitis, a disease entity consisting of at least two forms, named collagenous colitis, characterized by subepithelial deposition of type III collagen, and lymphocytic colitis, with numerous $CD8^+$ lymphocytes located in the lamina propria and the epithelium layer (164, 165). Some reports indicate that the number of intestinal mast cells is increased at sites of microscopic colitis, suggesting that mast cells are involved in the pathogenesis (166, 167). The mechanism of this mast cell activation is totally unclear. Possibly, autoimmune mechanisms may be anticipated, since it has been shown that, for example, gut epithelia proteins can trigger intestinal mast cells for mediator release in idiopathic inflammatory bowel disease (168).

Coeliac disease and eosinophilic gastroenteritis

Coeliac disease characterized by gluten hypersensitivity leading to a flat mucosa and subsequent malassimilation is accompanied by a marked infiltration of the gastrointestinal tissue with mast cells, suggesting a pathophysiological role of this cell type (169–171). Although it has been demonstrated that intestinal gliadin administration causes mast cell degranulation in patients with coeliac disease (172), this does not indicate that coeliac disease is equal to food allergy (119). Similarly, eosinophilic gastroenteritis, another form of idiopathic intestinal inflammation with a marked eosinophil infiltration of particular bowel layers, is frequently accompanied by mast cell accumulation (173, 174).

Gastrointestinal fibrosis

Evidence is now accumulating that indicates a role for mast cells in the pathogenesis of fibrosis at different sites such as lung, liver and other tissues (3, 4, 5, 7, 175, 176). It is tempting to speculate that this may be also true for the gastrointestinal tract, where mast cell accumulation has been found at sites of strictures occurring in Crohn's disease (2, 177). The mechanisms by which mast cells might stimulate fibroblasts and thus induce fibrosis have to be determined. Candidate mediators are bFGF, TGF-β and TNF-α, three cytokines known to be produced by human intestinal mast cells (2, 31, 178, 179; own unpublished observations).

Mastocytosis

Gastrointestinal mastocytosis is characterized by an accumulation of mast cells in the gastrointestinal tissue, which may occur in the course of systemic mastocytosis involving other organs such as bone marrow, liver, skin and different mucosal sites or isolated in the gastrointestinal tract. Patients with mastocytosis (adults or children) typically suffer from diarrhoea, abdominal pain and cramps and malassimilation (180–185). The basis of mastocytosis is most likely an alteration or hyperactivation of the proto-oncogene c-kit expressed on mast cell surfaces and acting as the receptor of SCF. In particular, the more aggressive forms of mastocytosis not restricted to the skin (urticaria pigmentosa) are associated with mutations in the c-kit gene, causing a ligand-independent permanent activation of the SCF receptor (60, 61, 186–189). Treatment of mastocytosis consists of symptomatic therapy with H_1 and H_2 antagonists, leukotriene antagonists and corticosteroids. The experimental therapy with IFN-α2b proved to be generally ineffective (190). In severe cases, with invasive mastocytosis leading to impairment of bone marrow function, treatment with chemotherapy must be considered (180, 182, 183).

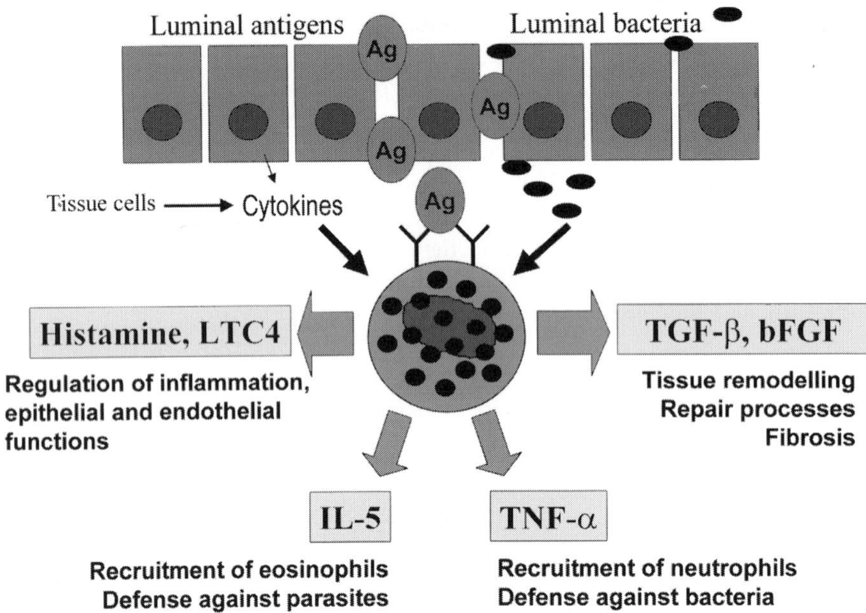

Fig. 6 Possible physiological functions of mast cells at the gastrointestinal barrier.

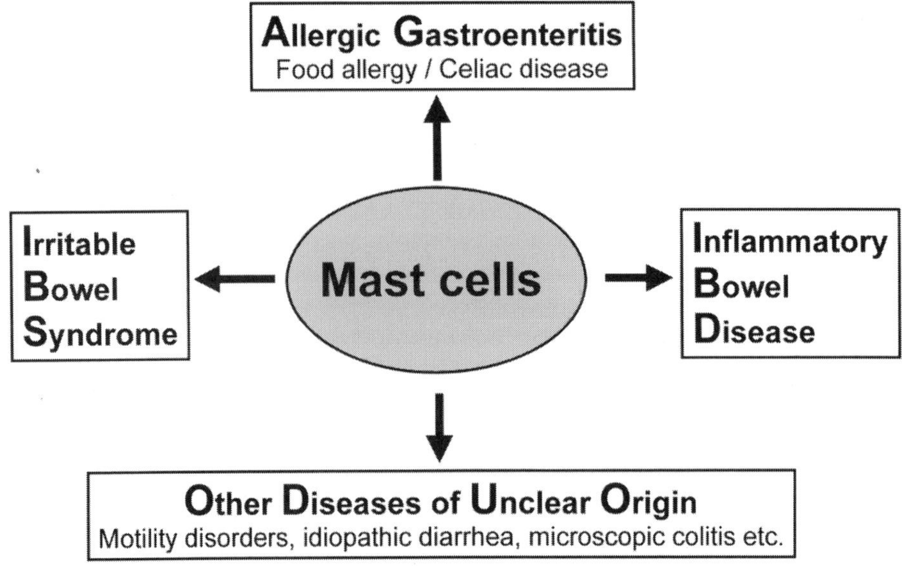

Fig. 7 Gastrointestinal disease with a pathophysiological role for mast cells has been suggested.

FUTURE DIRECTIONS

Because of the fact that the methods have been improved for the isolation and purification of human intestinal mast cells, our understanding of the biology of this cell type is increasing. Some findings previously described in rodent mast cells or human skin or lung mast cells have been confirmed for human intestinal mast cells, others not. Because of the well-known heterogeneity of mast cells, data are not easily transferable from one subtype to another. Both the physiological functions of mast cells at the gastrointestinal barrier (Fig. 6) as well as the pathophysiological functions in food allergy, inflammatory bowel disease, irritable bowel syndrome and other gastrointestinal diseases such as coeliac disease, eosinophilic gastroenteritis, microscopic colitis, etc. (Fig. 7) are increasingly being elucidated. Clearly, further studies are necessary to answer the many open questions on human intestinal mast cell functions. In particular, we are interested in the interaction with the local nervous system, their precise role in defence against bacteria and parasites, and their role in mediating allergy and autoimmune diseases. In particular, we need more information about the cross-talk between human intestinal mast cells and neighbouring cells, and about the extra- and intracellular signalling events regulating mast cell actions for successful pharmacological intervention.

REFERENCES

1. Bischoff, S. C., Wedemeyer, J., Herrmann, A., Meier, P. N., Trautwein, C., Cetin, Y., Maschek, H., Stolte, M., Gebel, M. and Manns, M. P. Quantitative assessment of intestinal eosinophils and mast cells in inflammatory bowel disease. *Histopathology* **28**:1–13, 1996.
2. Gelbmann, C. M., Mestermann, S., Gross, V., Kollinger, M., Scholmerich, J. and Falk, W. Strictures in Crohn's disease are characterized by an accumulation of mast cells colocalized with laminin but not with fibronectin or vitronectin. *Gut* **45**:210–217, 1999.
3. Armbrust, O. T., Batusic, D., Ringe, B. and Ramadori, G. Mast cells distribution in human liver disease and experimental rat liver fibrosis. Indications for mast cell participation in development of liver fibrosis. *J. Hepatol.* **26**:1042–1054, 1997.
4. Gruber, B. L. Mast cells: accessory cells which potentiate fibrosis. *Int. Rev. Immunol.* **12**:259–279, 1995.
5. Matsunaga, Y., Kawasaki, H. and Terada, T. Stromal mast cells and nerve fibers in various chronic liver diseases: relevance to hepatic fibrosis. *Am. J. Gastroenterol.* **94**:1923–1932, 1999.
6. Sugihara, A. Evaluation of role of mast cells in the development of liver fibrosis using mast cell-deficient rats and mice. *J. Hepatol.* **30**:859–867, 1999.
7. Gaca, M. D., Pickering, J. A., Arthur, M. J. and Benyon, R. C. Human and rat hepathic stellate cells produce stem cell factor: a possible mechanism for mast cell recruitment in liver fibrosis. *J. Hepatol.* **30**:850–858, 1999.
8. Adkins, F. M. Intestinal mucosal mast cells. *Ann. Allergy* **59**:44–53, 1987.
9. Ennerbäck, L. Mast cells in the rat gastrointestinal mucosa. I. Effects of fixation. *Acta Pathol. Microbiol. Scand.* **66**:289–302, 1966.
10. Befus, A. D., Dyck, N., Goodacre, R. and Bienenstock J. Mast cells from the human intestinal lamina propria. Isolation, histochemical subtypes, and functional characterization. *J. Immunol.* **138**:2604–2610, 1987.
11. Irani, A. A., Bradford, T. R., Kepley, C. L., Schechter, N. M. and Schwartz, L. B. Detection of MC_T and MC_{TC} types of human mast cells by immunohistochemistry using new monoclonal anti-tryptase and anti-chymase antibodies. *J. Histochem. Cytochem.* **37**:1509–1515, 1989.
12. Weidner, N. and Austen, K. F. Heterogeneity of mast cells at multiple body sites. *Pathol. Res. Pract.* **189**:156–162, 1993.
13. Church, M. K., Lowman, M. A., Robinson, C., Holgate, S. T. and Benyon, R. C. Interaction of

neuropeptides with human mast cells. *Int. Arch. Allergy Appl. Immunol.* **88**:70–78, 1989.
14. Bradding, P., Okayama, Y., Howarth, P. H., Church, M. K. and Holgate, S. T. Heterogeneity of human mast cells based on cytokine content. *J. Immunol.* **155**:297–307, 1995.
15. Stevens, R. L. and Austen, K. F. Recent advances in the cellular and molecular biology of mast cells. *Immunol. Today* **10**:381–386, 1989.
16. Mécheri, S. and David, B. Unravelling the mast cell dilemma: culprit or victim of its generosity? *Immunol. Today* **18**:212–215, 1997.
17. Bischoff, S. C., Schwengberg, S., Wordelmann, K., Weimann, A., Raab, R. and Manns, M. P. Effect of c-*kit* ligand, stem cell factor, on mediator release by human intestinal mast cells isolated from patients with inflammatory bowel disease and controls. *Gut* **38**:104–114, 1996.
18. Echtenacher, B., Mannel, D. N. and Hultner, L. Critical protective role of mast cells in a model of acute septic peritonitis. *Nature* **381**:75–77, 1996.
19. Metcalfe, D. D. Mast cell mediators with emphasis on intestinal mast cells. *Ann. Allergy* **53**:563–575, 1984.
20. Tas, J. and Berndsen, R. G. Does heparin occur in mucosal mast cells of the rat small intestine? *J. Histochem. Cytochem.* **25**:1058, 1977.
21. Bischoff, S. C., Schwengberg, S., Raab, R. and Manns, M. P. Functional properties of human intestinal mast cells cultured in a new culture system: enhancement of IgE receptor-dependent mediator release and response to stem cell factor. *J. Immunol.* **159**:5560–5567, 1997.
22. Bischoff, S. C., Sellge, G., Schwengberg, S., Lorentz, A. and Manns, M. P. Stem cell factor-dependent survival, proliferation and enhanced releasability of purified mature mast cells isolated from human intestinal tissue. *Int. Arch. Allergy Immunol.* **118**:104–107, 1999.
23. Naclerio, M., Proud, D., Togias, T. G., Adkinson, N. F., Meyers, D. A., Kagey-Sobotka, A., Plaut, M., Norman, P. S. and Lichtenstein, S. Inflammatory mediators in late antigen-induced rhinitis. *N. Engl. J. Med.* **313**:65–70, 1985.
24. Bischoff, S. C. and Dahinden, C. A. c-*kit* ligand: a unique potentiator of mediator release by human lung mast cells. *J. Exp. Med.* **175**:237–244, 1992.
25. Bischoff, S. C., Brunner, T., De-Weck, A. L. and Dahinden CA. Interleukin 5 modifies histamine release and leukotriene generation by human basophils in response to diverse agonists. *J. Exp. Med.* **172**:1577–1582, 1990.
26. Bischoff, S. C. and Dahinden, C. A. Effect of nerve growth factor on the release of inflammatory mediators by mature human basophils. *Blood* **79**:2662–2669, 1992.
27. Bischoff, S. C. and Dahinden, C. A. Cytokines and allergy. *Allergy Today* **24**:1–4, 1990.
28. Gordon, J. R. and Galli, S. J. Mast cells as a source of both preformed and immunologically inducible TNF-alpha/cachectin. *Nature* **346**:274–276, 1990.
29. Bradding, P., Mediwake, R., Feather, I. H., Madden, J., Church, M. K., Holgate, S. T. and Howarth, P. H. TNF alpha is localized to nasal mucosal mast cells and is released in acute allergic rhinitis. *Clin. Exp. Allergy* **25**:406–415, 1995.
30. Okayama, Y., Ono, Y., Nakazawa, T., Church, M. K. and Mori, M. Human skin mast cells produce TNF-alpha by substance P. *Int. Arch. Allergy Immunol.* **117 (Suppl. 1)**:48–51, 1998.
31. Bischoff, S. C., Lorentz, A., Schwengberg, S., Weier, G., Raab, R. and Manns, M. P. Mast cells are an important cellular source of tumour necrosis factor α in human intestinal tissue. *Gut* **44**:643–652, 1999.
32. Malaviya, R., Ikeda, T., Ross, E. and Abraham, S. N. Mast cell modulation of neutrophil influx and bacterial clearance at sites of infection through TNF-alpha. *Nature* **381**:77–80, 1996.
33. Fisher, C. J., Agosti, J. M., Opal, S. M., Lowry, S. F., Balk, R. A., Sadoff, R. A., Abraham, E., Schein, R. M. and Benjamin, E. Treatment of septic shock with the tumor necrosis factor receptor: Fc fusion protein. *N. Engl. J. Med.* **334**:1697–1702, 1996.
34. Moreland, L. W., Baumgartner, S. W., Schiff, M. H., Tindall, E. A., Fleischmann, R. M., Weaver, A. L., Ettlinger, R. E., Cohen, S., Koopman, W. J., Mohler, K., Widmer, M. B. and Blosch, C. M. Treatment of rheumatoid arthritis with a recombinant human tumor necrosis factor receptor (p75)-Fc fusion protein. *N. Engl. J. Med.* **337**:141–147, 1997.
35. Targan, S. R., Hanauer, S. B., Van Deventer, S. J. H., Mayer, L., Present, D. H., Braakman, T., DeWoody, K. L., Schaible, T. F. and Rutgeerts, P. J. A short-term study of chimeric monoclonal antibody cA2 to tumor necrosis factor α for Crohn's disease. *N. Engl. J. Med.* **337**:1029–1035, 1997.
36. Bazzoni, F. and Beutler, B. The tumor necrosis factor ligand and receptor families. *N. Engl. J. Med.* **334**:1717–1725, 1996.
37. Gordon, J. R., Burd, P. R. and Galli, S. J. Mast cells as a source of multifunctional cytokines. *Immunol.*

Today **11**:458–464, 1990.

38. Church, M. K. and Levi Schaffer, F. The human mast cell. *J. Allergy Clin. Immunol.* **99**:155–160, 1997.
39. Lorentz, A., Schwengberg, S., Mierke, C., Manns, M. P. and Bischoff, S. C. Human intestinal mast cells produce IL-5 *in vitro* upon IgE receptor cross-linking and *in vivo* in the course of intestinal inflammatory disease. *Eur. J. Immunol.* **29**:1496–1503, 1999.
40. Lorentz, A., Schwengberg, S., Sellge, S., Manns, M. P. and Bischoff, S. C. Human intestinal mast cells are capable of producing different cytokine profiles: Role of IgE receptor-crosslinking and IL-4. *J. Immunol.* **164**:43–48, 2000.
41. Bienenstock, J., MacQueen, G. A., Sestini, P., Marshall, J. S., Stead, R. H. and Perdue MH. Mast cell/ nerve cell interactions *in vitro* and *in vivo*. *Am. Rev. Respir. Dis.* **143**:S55–S58, 1991.
42. Stead, R. H., Dixon, M. F., Bramwell, N. H., Riddell, R. H. and Bienenstock, J. Mast cells are closely apposed to nerves in the human gastrointestinal mucosa. *Gastroenterology* **97**:575–585, 1989.
43. Williams, R. M., Bienenstock, J. and Stead, R. H. Mast cells: the neuroimmune connection. *Chem. Immunol.* **61**:208–235, 1995.
44. Miadonna, A., Leggieri, E., Tedeschi, A., Lorini, M., Froldi, M. and Zanussi, C. Study of the effect of some neuropeptides and endogenous opioid peptides on *in vitro* histamine release from human lung mast cells and peripheral blood basophils. *Agents Actions* **25**:11–16, 1988.
45. Leon, A., Buriani, A., Dal Toso, R., Fabris, M., Romanello, S., Aloe, L. and Levi-Montalcini, R. Mast cells synthesize, store, and release nerve growth factor. *Proc. Natl. Acad. Sci. USA* **91**:3739–3743, 1994.
46. Levi-Montalcini, R., Skaper, S. D., Dal Tosa, R., Petrelli, L. and Leon, A. Nerve growth factor: from neurotrophin to neurokine. *Trends Neurosci.* **19**:514–520, 1996.
47. Takafuji, S. and Bischoff, S. C. Opposing effects of tumor necrosis factor alpha and nerve growth factor upon leukotriene C4 production by human eosinophils triggered with *N*-formyl-methionyl-leucyl-phenylalanine. *Eur. J. Immunol.* **22**:969, 1992.
48. Horigome, K., Pryor, J. C., Bullock, E. D. and Johnson, E. M. J. Mediator release from mast cells by nerve growth factor. Neurotrophin specificity and receptor mediation. *J. Biol. Chem.* **268**:14881–14887, 1993.
49. Kawamoto, K., Okada, T., Kannan, Y., Ushio, H., Matsumoto, M. and Matsuda, H. Nerve growth factor prevents apoptosis of rat peritoneal mast cells through the trk proto-oncogene receptor. *Blood* **86**:4638–4644, 1995.
50. Marshall, J. S., Gomi, K., Blennerhassett, M. G. and Bienenstock, J. Nerve growth factor modifies the expression of inflammatory cytokines by mast cells via a prostanoid-dependent mechanism. *J. Immunol.* **162**:4271–4276, 1999.
51. Matsuda, H., Kannan, Y., Ushio, H., Kiso, Y., Kanemoto, T., Suzuki, H. and Kitamura, Y. Nerve growth factor induces development of connective tissue-type mast cells *in vitro* from murine bone marrow cells. *J. Exp. Med.* **174**:7–14, 1991.
52. Tal, M. and Liberman, R. Local injection of nerve growth factor (NGF) triggers degranulation of mast cells in rat paw. *Neurosci. Lett.* **221**:129–132, 1997.
53. Nilsson, G., Forsberg, N. K., Xiang, Z., Hallbook, F., Nilsson, K. and Metcalfe, D. D. Human mast cells express functional TrkA and are a source of nerve growth factor. *Eur. J. Immunol.* **27**:2295–2301, 1997.
54. Tam, S. Y., Tsai, M., Yamaguchi, M., Yano, K., Butterfield, J. H. and Galli, S. J. Expression of functional TrkA receptor tyrosine kinase in the HMC-1 human mast cell line and in human mast cells. *Blood* **90**:1807–1820, 1997.
55. Fox, C. C., Dvorak, A. M., Peters, S. P., Kagey, S. A. and Lichtenstein, L. M. Isolation and characterization of human intestinal mast cells. *J. Immunol.* **135**:483–491, 1985.
56. Beauvais, F., Hieblot, C. and Benveniste, J. Effect of sodium cromoglycate and nedocromil sodium on anti-IgE-induced and anti-IgG4-induced basophil degranulation. *Drugs* **37 (Suppl. 1)**:4–8, 1989.
57. Kaneko, M., Swanson, M. C., Gleich, G. J. and Kita, H. Allergen-specific IgG1 and IgG3 through Fc gamma RII induce eosinophil degranulation. *J. Clin. Invest.* **95**:2813–2821, 1995.
58. Galli, S. J., Tsai, M. and Wershil, B. K. The c-*kit* receptor, stem cell factor, and mast cells. What each is teaching us about the others. *Am. J. Pathol.* **142**:965–974, 1993.
59. Bischoff, S. C. and Dahinden, C. A. Effect of c-*kit* ligand on mediator release by human lung mast cells. *Int. Arch. Allergy Immunol.* **99**:319–322, 1992.
60. Longley, B. J., Metcalfe, D. D., Tharp, M., Wang, X, Tyrrell, L., Lu, S. Z., Heitjan, D. and Ma, Y. Activating and dominant inactivating c-*kit* catalytic domain mutations in distinct clinical forms of human mastocytosis. *Proc. Natl. Acad. Sci. USA* **96**:1609–1614, 1999.
61. Nagata, H., Okada, T., Worobec, A. S., Semere, T. and Metcalfe, D. D. c-*kit* mutation in a population of

patients with mastocytosis. *Int. Arch. Allergy Immunol.* **113**:184–186, 1997.

62. Irani, A. M., Craig, S. S., DeBlois, G., Elson, C. O., Schechter, N. M. and Schwartz, L. B. Deficiency of the tryptase-positive, chymase-negative mast cell type in gastrointestinal mucosa of patients with defective T lymphocyte function. *J. Immunol.* **138**:4381–4386, 1987.
63. Bischoff, S. C., Sellge, G., Lorentz, A., Sebald, W., Raab, R. and Manns, M. P. IL-4 enhances proliferation and mediator release in mature human mast cells. *Proc. Natl. Acad. Sci. USA* **96**:8080–8085, 1999.
64. Toru, H., Eguchi, M., Matsumoto, R., Yanagida, M., Yata, J. and Nakahata, T. Interleukin-4 promotes the development of tryptase and chymase double-positive human mast cells accompanied by cell maturation. *Blood* **91**:187–195, 1998.
65. Toru, H., Ra, C., Nonoyama, S., Suzuki, K., Yata, J. and Nakahata, T. Induction of the high-affinity IgE receptor (Fc epsilon RI) on human mast cells by IL-4. *Int. Immunol.* **8**:1367–1373, 1996.
66. Ahmed, T. and Fuchs, G. J. Gastrointestinal allergy to food: a review. *J. Diarrhoeal Dis. Res.* **15**:211–223, 1997.
67. Aiuti, F. and Paganelli, R. Food allergy and gastrointestinal diseases. *Ann. Allergy* **51**:275–280, 1983.
68. Bischoff, S. C., Herrmann, A., Mayer, J. and Manns, M. P. Food allergy in patients with gastrointestinal disease. *Monogr. Allergy* **32**:130–142, 1996.
69. Crowe, S. E. and Perdue, M. H. Gastrointestinal food hypersensitivity: Basic mechanisms of pathophysiology. *Gastroenterology* **103**:1075–1095, 1992.
70. Crowe, S. E., Sestini, P. and Perdue, M. H. Allergic reactions of rat jejunal mucosa. Ion transport responses to luminal antigen and inflammatory mediators. *Gastroenterology* **99**:74–82, 1990.
71. Curtis, G. H., Patrick, M. K., Catto-Smith, A. G. and Gall, D. G. Intestinal anaphylaxis in the rat. Effect of chronic antigen exposure. *Gastroenterology* **98**:1558–1566, 1990.
72. De-Weck, A. L. Pathophysiologic mechanisms of allergic and pseudo-allergic reactions to foods, food additives and drugs. *Ann. Allergy* **53**:583–586, 1984.
73. Dharmsathaphorn, K. and Barrett, K. E. Allergic reactions alter gastric function. *Gastroenterology* **99**:1184–1185, 1990.
74. Fargeas, M. J., Theodourou, V., Fioramonti, J. and Bueno, L. Relationship between mast cell degranulation and jejunal myoelectric alterations in intestinal anaphylaxis in rats. *Gastroenterology* **102**:157–162, 1992.
75. Finn, R. Food allergy – fact or fiction: a review. *J. R. Soc. Med.* **85**:560–564, 1992.
76. Metcalfe, D. D. Food allergy. *Prim. Care* **25**:819–829, 1998.
77. Patrick, M. K., Dunn, I. J., Buret, A., Miller, H. R., Huntley, J. F., Gibson, S. and Gall, D. G. Mast cell protease release and mucosal ultrastructure during intestinal anaphylaxis in the rat. *Gastroenterology* **94**:1–9, 1988.
78. Knutson, T. W., Bengtsson, U., Dannaeus, A., Ahlstedt, S., Stalenheim, G., Hallgren, R. and Knutson, L. Intestinal reactivity in allergic and nonallergic patients: an approach to determine the complexity of the mucosal reaction. *J. Allergy Clin. Immunol.* **91**:553–559, 1993.
79. Metcalfe, D. D. Immune mechanisms in food allergy. *Clin. Exp. Allergy* **21 (Suppl. 1)**:321–324, 1991.
80. Shigematsu, T., Miura, S., Hirokawa, M., Hokari, R., Higuchi, H., Tsuzuki, Y., Kimura, H., Nakatsumi, R. C., Serizawa, H., Saito, H. and Ishii, H. Endothelins promote egg albumin-induced intestinal anaphylaxis in rats. *Gastroenterology* **115**:348–356, 1998.
81. Barnes, R. M. IgG and IgA antibodies to dietary antigens in food allergy and intolerance. *Clin. Exp. Allergy* **25 (Suppl. 1)**:7–9, 1995.
82. Halpern, G. M. and Scott, J. R. Non-IgE antibody mediated mechanisms in food allergy. *Ann. Allergy* **58**:14–27, 1987.
83. Kemeny, D. M., Urbanek, R., Amlot, P. L., Ciclitira, P. J., Richards, D. and Lessof, M. H. Sub-class of IgG in allergic disease. I. IgG sub-class antibodies in immediate and non-immediate food allergy. *Clin. Allergy* **16**:571–581, 1986.
84. Kolmannskog, S. and Haneberg, B. Immunoglobin E in faeces from children with allergy. Evidence of local production of IgE in the gut. *Int. Arch. Allergy Appl. Immunol.* **76**:133–137, 1985.
85. Host, A., Jacobsen, H. P., Halken, S. and Holmenlund, D. The natural history of cow's milk protein allergy/intolerance. *Eur. J. Clin. Nutr.* **49 (Suppl. 1)**:S13–S18, 1995.
86. Iacono, G., Carroccio, A., Cavataio, F., Montalto, G., Kazmierska, I., Lorello, D., Soresi, M. and Notarbatolo, A. Gastroesophageal reflux and cow's milk allergy in infants: a prospective study. *J. Allergy Clin. Immunol.* **97**:822–827, 1996.
87. Jenkins, H. R., Pincott, J. R., Soothill, J. F., Milla, P. J. and Harries, J. T. Food allergy: the major cause of

infantile colitis. *Arch. Dis. Child.* **59**:326–329, 1984.
88. Keller, K. M., Burgin Wolff, A., Menger, H., Lippold, R., Wirth, S. and Baumann, W. IgG, IgA, and IgE antibodies to cow milk proteins in an allergy prevention study. *Adv. Exp. Med. Biol.* **310**:467–473, 1991.
89. Levy, F. S., Bircher, A. J. and Gebbers, J. O. Adult onset of cow's milk protein allergy with small-intestinal mucosal IgE mast cells. *Allergy* **51**:417–420, 1996.
90. Li, X. M., Schofield, B. H., Huang, C. K., Kleiner, G. I. and Sampson, H. A. A murine model of IgE-mediated cow's milk hypersensitivity. *J. Allergy Clin. Immunol.* **103**:206–214, 1999.
91. Strobel, S. Epidemiology of food sensitivity in childhood – with special reference to cow's milk allergy in infancy. *Monogr. Allergy* **31**:1993.
92. Walker, W. A. Pathophysiology of intestinal uptake and absorption of antigens in food allergy. *Ann. Allergy* **59**:7–15, 1987.
93. Andre, C., Andre, F., Colin, L. and Cavagna, S. Measurement of intestinal permeability to mannitol and lactulose as a means of diagnosing food allergy and evaluating therapeutic effectiveness of disodium cromoglycate. *Ann. Allergy* **59**:127–130, 1987.
94. DuMont, G. C. L., Beach, R. C. and Menzies, I. S. Gastrointestional permeability in food-allergic eczematous children. *Clin. Allergy* **14**:55–59, 1984.
95. Fählt-Magnusson, K., Kjellman, N.-I. M., Odelram, H., Sundqvist, T. and Magnusson, K.-E. Gastrointestinal permeability in children with cow's milk allergy: effect of milk challenge and sodium cromoglycate as assessed with polyethylenglycols (PEG 400 and PEG 1000). *Clin. Allergy* **16**:543–551, 1986.
96. Turner, M. W., Boulton, P. and Strobel, S. Experimental intestinal hypersensitivity. Effect of four anti-allergic drugs on protein uptake, permeability to sugars and mucosal mast-cell activation. *Clin. Exp. Allergy* **25**:448–455, 1995.
97. Barrett, K. E. and Metcalfe, D. D. The mucosal mast cell and its role in gastrointestinal allergic diseases. *Clin. Rev. Allergy* **2**:39–53, 1984.
98. Bischoff, S. C. Mucosal allergy: role of mast cells and eosinophil granulocytes in the gut. *Baillières Clin. Gastroenterol.* **10**:443–459, 1996.
99. Gui, X. Y. Mast cells: a possible link between psychological stress, enteric infection, food allergy and gut hypersensitivity in the irritable bowel syndrome. *J. Gastroenterol. Hepatol.* **13**:980–989, 1998.
100. Saavedra-Delgado, A. M. and Metcalf, D. D. The gastrointestinal mast cell in food allergy. *Ann. Allergy* **51**:185–189, 1983.
101. Kosnai, I., Kuitunen, P., Savilahti, E. and Sipponen, P. Mast cells and eosinophils in the jejunal mucosa of patients with intestinal cow's milk allergy and celiac disease of childhood. *J. Pediatr. Gastroenterol. Nutr.* **3**:368–372, 1984.
102. Bischoff, S. C., Mayer, J., Wedemeyer, J., Meier, P. N., Zeck, K. G., Wedi, B., Kapp, A., Cetin, Y., Gebel, M. and Manns, M. P. Colonoscopic allergen provocation (COLAP): a new diagnostic approach for gastrointestinal food allergy. *Gut* **40**:745–753, 1997.
103. Pollard, H. M. and Stuart, G. J. Experimental reproduction of gastric allergy in human beings with controlled observations on the mucosa. *J. Allergy* **13**:467–473, 1942.
104. Raithel, M., Matek, M., Baenkler, H. W., Jorde, W. and Hahn, E. G. Mucosal histamine content and histamine secretion in Crohn's disease, ulcerative colitis and allergic enteropathy. *Int. Arch. Allergy Immunol.* **108**:127–133, 1995.
105. Reimann, H. J. and Lewin, J. Gastric mucosal reactions in patients with food allergy. *Am. J. Gastroenterol.* **83**:1212–1219, 1988.
106. Reimann, H. J., Ring, J., Ultsch, B. and Wendt, P. Intragastral provocation under endoscopic control (IPEC) in food allergy: mast cell and histamine changes in gastric mucosa. *Clin. Allergy* **15**:195–202, 1985.
107. Crowe, S. E., Luthra, G. K. and Perdue, M. H. Mast cell mediated ion transport in intestine from patients with and without inflammatory bowel disease. *Gut* **41**:785–792, 1997.
108. Perdue, M. H., Masson, S., Wershil, B. K. and Galli, S. J. Role of mast cells in ion transport abnormalities associated with intestinal anaphylaxis. Correction of the diminished secretory response in genetically mast cell-deficient W/Wv mice by bone marrow transplantation. *J. Clin. Invest.* **87**:687–693, 1991.
109. Furuta, G. T., Schmidt-Choudhury, A. and Wang, M. Y. Mast cell-dependent tumor necrosis factor α production participates in allergic gastric inflammation in mice. *Gastroenterology* **113**:1560–1569, 1997.
110. Freed, D. L. J. Laboratory diagnosis of food intolerance. In: *Food Allergy and Intolerance* (Brostoff, J.

and Challacombe, S. J., eds), pp 873–897. W. B. Saunders, London, 1987.

111. Sampson, H. A. Immunologically mediated food allergy: the importance of food challenge procedures. *Ann. Allergy* **60**:262–269, 1988.
112. Bischoff, S. C., Grabowsky, J. and Manns, M. P. Quantification of inflammatory mediators in stool samples of patients with inflammatory bowel disorders and controls. *Dig. Dis. Sci.* **42**:394–403, 1997.
113. Kosa, L., Kereki, E. and Borzsonyl, L. Copro-eosinophil cationic protein (ECP) in food allergy. *Allergy* **51**:964–966, 1996.
114. Bagnato, G. F., Di Cesare, E., Caruso, R. A., Gulli, S., Cugliari, A. and Morabito Lo Prete, A. Gastric mucosal mast cells in atopic subjects. *Allergy* **50**:322–327, 1995.
115. Shanahan, F. Food allergy: fact, fiction, and fatality. *Gastroenterology* **104**:1229–1231, 1993.
116. Jones, V. A., McLaughlan, P., Shorthouse, M., Workman, E. and Hunter, J. O. Food intolerance: a major factor in the pathogenesis of irritable bowel syndrome. *Lancet* **ii**:1115–1117, 1982.
117. Nanda, R., James, R., Smith, H., Dudley, C. R. K. and Jewell, D. P. Food intolerance and the irritable bowel syndrome. *Gut* **30**:1099–1104, 1989.
118. Zwetchkenbaum, J. F. and Burakoff, R. Food allergy and the irritable bowel syndrome. *Am. J. Gastroenterol.* **83**:901–904, 1988.
119. Hed, J. Coeliac disease: a food allergy? *Monogr. Allergy* **32**:204–210, 1996.
120. Sampson, H. A. Epidemiology of food allergy. *Pediatr. Allergy Immunol.* **7**:42–50, 1996.
121. Young, E., Stoneham, M. D., Petruckevitch, A., Barton, J. and Rona, R. A population study of food intolerance. *Lancet* **343**:1127–1130, 1994.
122. Sampson, H. A. Fatal and near fatal anaphylactic reactions to food in children and adolescents. *N. Engl. J. Med.* **327**:380–384, 1992.
123. Benyon, R. C., Bissonnette, E. Y. and Befus, A. D. Tumor necrosis factor-alpha dependent cytotoxicity of human skin mast cells is enhanced by anti-IgE antibodies. *J. Immunol.* **147**:2253–2258, 1991.
124. Ohkawara, Y., Yamauchi, K., Tanno, Y., Tamura, G., Ohtani, H., Nagura, H., Ohkuda, K. and Takishima, T. Human lung mast cells and pulmonary macrophages produce tumor necrosis factor-alpha in sensitized lung tissue after IgE receptor triggering. *Am. J. Respir. Cell Mol. Biol.* **7**:385–392, 1992.
125. Arock, M., Ross, E., Lai, K. R., Averlant, G., Gao, Z. and Abraham, S. N. Phagocytic and tumor necrosis factor alpha response of human mast cells following exposure to gram-negative and gram-positive bacteria. *Infect. Immun.* **66**:6030–6034, 1998.
126. Wershil, B. K., Wang, Z. S., Gordon, J. R. and Galli, S. J. Recruitment of neutrophils during IgE-dependent cutaneous late phase reactions in the mouse is mast cell-dependent. Partial inhibition of the reaction with antiserum against tumor necrosis factor-alpha. *J. Clin. Invest.* **87**:446–453, 1991.
127. Zhang, Y., Ramos, B. and Jakschik, B. A. Neutrophil recruitment by tumor necrosis factor from mast cells in immune complex peritonitis. *Science* **258**:1957–1959, 1957.
128. Galli, S. J. and Wershil, B. K. The two faces of the mast cell. *Nature* **381**:21–22, 1996.
129. Nakajima, S., Krishnan, B., Ota, H., Segura, A. M., Hattori, T., Graham, D. Y. and Genta, R. M. Mast cell involvement in gastritis with or without *Helicobacter pylori* infection. *Gastroenterology* **113**:746–754, 1997.
130. Yamamoto, J., Watanabe, S., Hirose, M., Osada, T., Ra, C. and Sato, N. Role of mast cells as a trigger of inflammation in *Helicobacter pylori* infection. *J. Physiol. Pharmacol.* **50**:17–23, 1999.
131. Alizadeh, H. and Murrell, K. D. The intestinal mast cell response to *Trichinella spiralis* infection in mast cell-deficient w/wv mice. *J. Parasitol.* **70**:767–773, 1984.
132. Arizono, N., Kasugai, T., Yamada, M., Okada, M., Morimoto, M., Tei, H., Newlands, G. F. J. and Miller, H. R. P. Infection of *Nippostrongylus brasiliensis* induces development of mucosal-type but not connective tissue-type mast cells in genetically mast cell-deficient Ws/Ws rats. *Blood* **81**:2572–2578, 1993.
133. Newlands, G. F., Miller, H. R., MacKellar, A. and Galli, S. J. Stem cell factor contributes to intestinal mucosal mast cell hyperplasia in rats infected with *Nippostrongylus brasiliensis* or *Trichinella spiralis*, but anti-stem cell factor treatment decreases parasite egg production during *N. brasiliensis* infection. *Blood* **86**:1968–1976, 1995.
134. Ruitenberg, E. J. and Elgersma, A. Absence of intestinal mast cell response in congenitally athymic mice during *Trichinella spiralis* infection. *Nature* **264**:258–260, 1976.
135. Shaikh, N., Rivera, J., Hewlett, B. R., Stead, R. H., Zhu, F. G. and Marshall, J. S. Mast cell Fc epsilonRI expression in the rat intestinal mucosa and tongue is enhanced during *Nippostrongylus brasiliensis* infection and can be up-regulated by *in vivo* administration of IgE. *J. Immunol.* **158**:3805–3812, 1997.
136. Wastling, J. M., Scudamore, C. L., Thornton, E. M., Newlands, G. F. and Miller, H. R. Constitutive

expression of mouse mast cell protease-1 in normal BALB/c mice and its up-regulation during intestinal nematode infection. *Immunology* **90**:308–313, 1997.

137. Wershil, B. K., Theodos, C. M., Galli, S. J. and Titus, R. G. Mast cells augment lesion size and persistence during experimental *Leishmania major* infection in the mouse. *J. Immunol.* **152**:4563–4571, 1994.
138. Fiocchi, C. Inflammatory bowel disease. Etiology and pathogenesis. *Gastroenterology* **115**:182–205, 1998.
139. Lowes, J. R. and Jewel, D. P. The immunology of inflammatory bowel disease. *Springer Semin. Immunopathol.* **12**:251–268, 1990.
140. Podolsky, D. K. Inflammatory bowel disease. *N. Engl. J. Med.* **325**:928–937, 1991.
141. Sartor, R. B. Cytokines in intestinal inflammation: pathophysiological and clinical considerations. *Gastroenterology* **106**:533–539, 1994.
142. Ballegaard, M., Bjergstrom, A., Bronndum, S., Hylander, E. and Jensen, L. Self-reported food intolerance in chronic inflammatory bowel disease. *Scand. J. Gastroenterol.* **32**:569–571, 1997.
143. Farah, D. A., Calder, I., Benson, L. and Mackenzie, J. F. Specific food intolerances: its place as a cause of gastrointestinal symptoms. *Gut* **26**:164–168, 1985.
144. Jones, V. A., Dickinson, R. .J., Workman, E., Wilson, A. J., Freeman, A. H. and Hunter, J. O. Crohn's disease: maintenance of remission by diet. *Lancet* **ii**:177–180, 1985.
145. Knoflach, P., Park, B. H., Cunningham, R., Weiser, M. M. and Albini, B. Serum antibodies to cow's milk proteins in ulcerative colitis and Crohn's disease. *Gastroenterology* **92**:479–485, 1987.
146. Pearson, M., Teahon, K., Jonathan Levi, A. and Bjarnason, I. Food intolerance and Crohn's disease. *Gut* **34**:783–787, 1993.
147. Truelove, S. C. Ulcerative colitis provoked by milk. *Br. Med. J.* **21**:154–160, 1961.
148. Duchmann, R., Kaiser, I., Hermann, E., Mayet, W., Ewe, K. and Meyer zum Büschenfelde, K. H. Tolerance exists towards resident intestinal flora but is broken in active inflammatory bowel disease (IBD). *Clin. Exp. Immunol.* **102**:448–455, 1995.
149. Maluenda, C., Phillips, A. D., Briddon, A. and Walker-Smith, J. A. Quantitative analysis of small intestinal mucosa in cow's milk-sensitive enteropathy. *J. Pediatr. Gastroenterol. Nutr.* **3**:349–356, 1984.
150. Desreumaux, P., Brandt, E., Gambiez, L., Emilie, D., Geboes, K., Klein, O., Ectors, N., Cortot, A., Capron, M. and Colombel, J. F. Distinct cytokine patterns in early and chronic ileal lesions of Crohn's disease. *Gastroenterology* **113**:118–126, 1997.
151. Dubucquoi, S., Janin, A., Klein, O., Desreumaux, P., Quandalle, P., Cortot, A., Capron, A. and Colombel, J. F. Activated eosinophils and interleukin 5 expression in early recurrence of Crohn's disease. *Gut* **37**:242–246, 1995.
152. Balazs, M., Illyes, G. and Vadasz, G. Mast cells in ulcerative colitis. Quantitative and ultrastructural studies. *Virchows Arch. B Cell Pathol. Mol. Pathol.* **57**:353–360, 1989.
153. Benfield, G. F., Bryan, R. and Crocker, J. Lamina propria eosinophils and mast cells in ulcerative colitis: comparison between Asians and Caucasians. *J. Clin. Pathol.* **43**:27–31, 1990.
154. Fox, C. C., Lazenby, A. J., Moore, W. C., Yardley, J. H., Bayless, T. M. and Lichtenstein, L. M. Enhancement of human intestinal mast cell mediator release in active ulcerative colitis. *Gastroenterology* **99**:119–124, 1990.
155. King, T., Biddle, W., Bhatia, P., Moore, J. and Miner, P. B. Colonic mucosal mast cell distribution at line demarcation of active ulcerative colitis. *Dig. Dis. Sci.* **37**:490–495, 1992.
156. Sarin, S. K., Malhotra, V., Sen Gupta, S., Karol, A., Gaur, S. K. and Anand, B. S. Significance of eosinophil and mast cell counts in rectal mucosa in ulcerative colitis – a prospective controlled study. *Dig. Dis. Sci.* **32**:363–367, 1987.
157. Araki,Y., Kakegawa, T. and Stadil, F. Mast cells and histamine release in Crohn's disease. *Kurume Med. J.* **40**:93–99, 1993.
158. Beil, W. J., Schulz, M., McEuen, A. R., Buckley, M. G. and Walls, A. F. Number, fixation properties, dye-binding and protease expression of duodenal mast cells: comparisons between healthy subjects and patients with gastritis or Crohn's disease. *Histochem. J.* **29**:759–773, 1997.
159. Casellas, F., Guarner, F., Antolin, M., Rodriguez, R., Slas, A. and Malagelada, J. R. Abnormal leukotriene C4 release by unaffected jejunal mucosa in patients with inactive Crohn's disease. *Gut* **35**:517–522, 1994.
160. Dvorak, A. M., Monahan, R. A., Osage, J. E. and Dickersin, G. R. Crohn's disease: transmission electron microscopic studies. II. Immunologic inflammatory response. Alterations of mast cells,

basophils, eosinophils, and the microvasculature. *Hum. Pathol.* **11**:606–619, 1980.

161. Knutson, L., Ahrenstedt, O., Odlind, B. and Hallgren, R. The jejunal secretion of histamine is increased in active Crohn's disease. *Gastroenterology* **98**:849–854, 1990.
162. Sanderson, I. R., Leung, K. B., Pearce, F. L. and Walker, S. J. Lamina propria mast cells in biopsies from children with Crohn's disease. *J. Clin. Pathol.* **39**:279–283, 1986.
163. Afzalpurkar, R. G., Schiller, L. R., Little, K. H., Santangelo, W. C. and Fordtran, J. S. The self-limited nature of chronic idiopathic diarrhea. *N. Engl. J. Med.* **327**:1849–1852, 1992.
164. Flejou, J. F., Grimaud, J. A., Molas, G., Baviera, E. and Potet, F. Collagenous colitis. Ultrastructural study and collagen immunotyping of four cases. *Arch. Pathol. Lab. Med.* **108**:977–982, 1984.
165. Giardiello, F. M., Lazenby, A. J., Bayless, T. M., Levine, E. J., Bias, W. B., Ladenson, P. W., Hutcheon, D. F., Derevjanik, N. L. and Yardley, J. H. Lymphocytic (microscopic) colitis – clinicopathologic study of 18 patients and comparison to collagenous colitis. *Dig. Dis. Sci.* **34**:1730–1738, 1989.
166. Baum, C. A., Bhatia, P. and Miner, P. B. Increased colonic mucosal mast cells associated with severe watery diarrhea and microscopic colitis. *Dig. Dis. Sci.* **34**:1462–1465, 1989.
167. Molas, G. J., Flejou, J. F. and Potet, F. Microscopic colitis, collagenous colitis, and mast cells. *Dig. Dis. Sci.* **35**:920–921, 1990.
168. Fox, C. C., Lichtenstein, L. M. and Roche, J. K. Intestinal mast cell responses in idiopathic inflammatory bowel disease. Histamine release from human intestinal mast cells in response to gut epithelial proteins. *Dig. Dis. Sci.* **38**:1105–1112, 1993.
169. Dollberg, L., Gurevitz, M. and Freier, S. Gastrointestinal mast cells in health, and in coeliac disease and other conditions. *Arch. Dis. Child.* **55**:702–705, 1980.
170. Marsh, M. N. and Hinde, J. Inflammatory component of celiac sprue mucosa. I. Mast cells, basophils, and eosinophils. *Gastroenterology* **89**:92–101, 1985.
171. Strobel, S., Busuttil, A. and Ferguson, A. Human intestinal mucosal mast cells: expanded population in untreated coeliac disease. *Gut* **24**:222–227, 1983.
172. Horvath, K., Simon, K., Horn, G. and Bodanszky, H. Mast cell degranulation after a single dose of gliadin in the jejunum of patients with coeliac disease. *Acta Paediatr. Hung.* **27**:311–316, 1986.
173. Clouse, R. E., Alpers, D. H., Hockenbery, D. M. and DeSchryver-Kecskemeti, K. Pericrypt eosinophilic enterocolitis and chronic diarrhea. *Gastroenterology* **103**:168–176, 1992.
174. Lee, C., Changchien, C., Chen, P., Lin, D., Sheen, I., Wang, C., Tai, D., Sheen-Chen, S., Chen, W. and Wu, C. Eosinophilic gastroenteritis: 10 years experience. *Am. J. Gastroenterol.* **88**:70–74, 1993.
175. Broide, D. H., Smith, C. M. and Wasserman, S. I. Mast cells and pulmonary fibrosis. Identification of histamine releasing factor in bronchoalveolar lavage fluid. *J. Immunol.* **145**:1838–1844, 1990.
176. Claman, H. N. Mast cells, T cells and abnormal fibrosis. *Immunol. Today* **6**:192–195, 1985.
177. Stead, R. H., Franks, A. J., Goldsmith, C. H., Bienenstock, J. and Dixon, M. F. Mast cells, nerves and fibrosis in the appendix: a morphological assessment. *J. Pathol.* **161**:209–219, 1990.
178. Gordon, J. R. and Galli, S. J. Promotion of mouse fibroblast collagen gene expression by mast cells stimulated via the Fc epsilon RI. Role for mast cell-derived transforming growth factor beta and tumor necrosis factor alpha. *J. Exp. Med.* **180**:2027–2037, 1994.
179. Kendall, J. C., Li, X. H., Galli, S. J. and Gordon, J. R. Promotion of mouse fibroblast proliferation by IgE-dependent activation of mouse mast cells: role for mast cell tumor necrosis factor-alpha and transforming growth factor-beta 1. *J. Allergy Clin. Immunol.* **99**:113–123, 1997.
180. Austen, K. F. Systemic mastocytosis. *N. Engl. J. Med.* **326**:639–640, 1992.
181. Avila, N. A., Ling, A., Worobec, A. S., Mican, J. M. and Metcalfe, D. D. Systemic mastocytosis: CT and US features of abdominal manifestations. *Radiology* **202**:367–372, 1997.
182. Cherner, J. A., Jensen, R. T., Dubois, A., O'Doriso, T. M. and Gardner, J. D. Gastrointestinal dysfunction in systemic mastocytosis. A prospective study. *Gastroenterology* **95**:657–667, 1988.
183. Kettelhut, B. V. and Metcalfe, D. D. Pediatric mastocytosis. *Ann. Allergy* **73**:197–202, 1994.
184. Mican, J. M., Di Bisceglie, A. M., Fong, T. L., Travis, W. D., Kleiner, D. E., Baker, B. and Metcalfe, D. D. Hepatic involvement in mastocytosis: Clinicopathologic correlations in 41 cases. *Hepatology* **22**:1163–1170, 1995.
185. Miner, P. B. J. The role of the mast cell in clinical gastrointestinal disease with special reference to systemic mastocytosis. *J. Invest. Dermatol.* **96**:40S–43S, 1991.
186. Longley, B. J., Tyrrell, L., Lu, S. Z., Ma, Y. S., Langley, K., Ding, T. G., Duffy, T., Jacobs, P., Tang, L. H. and Modlin, I. Somatic c-*kit* activating mutation in urticaria pigmentosa and aggressive mastocytosis: establishment of clonality in a human mast cell neoplasm. *Nat. Genet.* **12**:312–314, 1996.
187. Nagata, H., Worobec, A. S., Oh, C. K., Chowdhury, B. A., Tannenbaum, S., Suzuki, Y. and Metcalfe, D.

D. Identification of a point mutation in the catalytic domain of the protooncogene c-*kit* in peripheral blood mononuclear cells of patients who have mastocytosis with an associated hematologic disorder. *Proc. Natl. Acad. Sci. USA* **92**:10560–10564, 1995.

188. Worobec, A. S., Akin, C., Scott, L. M. and Metcalfe, D. D. Cytogenetic abnormalities and their lack of relationship to the Asp816Val c-*kit* mutation in the pathogenesis of mastocytosis. *J. Allergy Clin. Immunol.* **102**:523–524, 1998.
189. Worobec, A. S., Semere, T., Nagata, H. and Metcalfe, D. D. Clinical correlates of the presence of the Asp816Val c-*kit* mutation in the peripheral blood mononuclear cells of patients with mastocytosis. *Cancer* **83**:2120–2129, 1998.
190. Worobec, A. S., Kirshenbaum, A. S., Schwartz, L. B. and Metcalfe, D. D. Treatment of three patients with systemic mastocytosis with interferon alpha-2b. *Leuk. Lymphoma* **22**:501–508, 1996.

CHAPTER 35

Differential Regulation of Basophil Functions by Chemokines

CLEMENS A. DAHINDEN

Institute of Immunology and Allergology, Inselspital, University Hospital Bern, Bern, Switzerland

BASOPHIL PHENOTYPE AND FUNCTIONS

Phenotype

Although basophils share some properties with tissue mast cells, such as basophilic staining, histamine content and the expression of high levels of the high-affinity IgE receptor ($Fc_{\varepsilon}RI$), they are actually specialized blood granulocytes most closely related to eosinophils. Basophils and eosinophils develop along the myeloid lineage, and both derive from a common committed eosinophil/basophil progenitor under the control of the same growth factors, in particular interleukins IL-3, IL-5 and granulocyte–macrophage colony-stimulating factor (GM-CSF). Both cell types express an almost identical set of CD markers, including receptors for inflammatory mediators and cytokines, indicating that the control of their emigration into inflammatory sites and their activation is similar and mediated by the same factors. It is likely, therefore, that basophils participate in most, if not all, inflammatory diseases involving eosinophils. Indeed, basophils have been found to accumulate in allergic late-phase reactions, and their number appears to correlate with the severity of asthma (1, 2). However, basophils also express a number of receptors normally not found on eosinophils, such as $Fc_{\varepsilon}RI$ or CCR2 (see below), indicating that basophils are involved in certain inflammatory processes lacking eosinophils.

Basophil Functions

Chemotaxis

Like other blood leukocytes, basophils have to migrate into the inflamed tissue along a chemotactic gradient to fulfil their function. All well-defined basophil chemotaxins interact with seven-transmembrane G protein-coupled receptors of the chemotactic factor receptor subfamily. The cellular responses following activation of chemotactic factor

MAST CELLS AND BASOPHILS
ISBN 0-12-473335-2

receptors (e.g. induction of shape change, transient elevation of intracellular calcium concentrations, activation of integrin receptors, cell polarization, chemotaxis) have been most extensively studied in neutrophils, but the responses of basophils seem very similar.

Exocytosis

The most extensively studied function of basophils is the release of histamine, which occurs rapidly within minutes through exocytosis of the pre-formed mediators stored in the basophilic granules. The basophil is the only blood leukocyte type containing significant amounts of histamine, and exocytosis of basophils can, thus, be assessed also in mixed leukocyte cultures. Histamine is released after $Fc_{\varepsilon}RI$ cross-linking and in response to several exogenous and endogenous humoral or cell-derived mediators belonging to the chemotactic agonist family, including certain chemokines.

Leukotriene formation

The basophil has a particularly high capacity to produce the lipid mediator leukotriene C_4 (LTC_4). Although basophils represent only a very minor fraction in mixed leukocytes, they are the only relevant source of LTC_4 when cells are activated by physiological stimuli (3, 4). LTC_4 is formed upon $Fc_{\varepsilon}RI$ cross-linking (e.g. by allergen) (5). Among the chemotactic agonists, the bacterial leader sequence analogue fMLP (*N*-formyl-Met–Leu–Phe) is the only known ligand of a G protein-coupled receptor capable of directly inducing LTC_4 formation (5). For endogenous agonists, including chemokines, two sequential signals (an appropriate 'priming' growth factor, IL-3, IL-5, GM-CSF or NGF (nerve growth factor), followed by a chemotactic agonist) are needed to induce lipid mediator formation (6–10). In primed basophils LTC_4 is formed in large amounts within seconds after activation of a G protein-coupled receptor in a transient burst which is complete in about 2 min (3, 10).

Cytokine expression

More recent studies from several laboratories showed that basophils are capable of expressing and secreting cytokines (11–16). Basophils are rather unique as they express large amounts of IL-4 and IL-13 in a very restricted manner without any Th1 cytokines (e.g.interferon-γ (INF-γ), LT, IL-2) and even without most pro-inflammatory cytokines (e.g. tumour necrosis factor-α (TNF-α), IL-1, IL-6) (15). This indicates that basophils are not simply effector cells of inflammation but also play a key immunoregulatory role by skewing immune responses towards the Th2 type (17). IL-4 and IL-13 expression are induced most efficiently by $Fc_{\varepsilon}RI$ cross-linking (11–13) or by C5a in synergy with IL-3 (15, 16). Interestingly, IgE-independent cytokine expression by basophils in response to combined stimulation with IL-3 and C5a is accompanied by a more continuous late phase of LTC_4 formation occurring 4–18 h after activation (15). In contrast to the rapid burst of LTC_4 formation, which is only observed if the growth factor is added a few minutes before the chemotactic agonist, the induction of cytokine expression and of this late phase of LTC_4 formation does not depend on the sequence of the stimuli, but rather requires their continuous presence.

Basophils also express the chemokine IL-8 (unpublished data) and the macrophage inflammatory protein MIP-1α (18) upon IgE receptor cross-linking. However, the biological meaning of these observations is questionable, since other cells produce much higher quantities of IL-8 in response to diverse stimuli. Furthermore, our own unpublished studies indicate that basophils are not an important source of other chemokines. In particular, basophils do not seem to produce their own major chemokine chemoattractants and/or activators of mediator release, such as eotaxin or MCP-1, under conditions which promote the expression of large amounts of IL-4 and IL-13. Thus, in contrast to neutrophils which amplify their recruitment by the release of IL-8, basophils do not directly promote the accumulation of further basophils and eosinophils. Nevertheless, IL-4 and IL-13 are emerging as the major regulators of the production of chemokines with selectivity for the leukocyte types found at sites of allergic inflammation. IL-4 (in synergy with TNF-α) is a most potent stimulus for the expression of eotaxin in connective tissue fibroblasts (19). IL-4 and IL-13 also upregulate the expression of adhesion molecules (e.g. vascular cell adhesion molecule, VCAM) that are important for the emigration of basophils, eosinophils and Th2 cells (20). Thus, IL-4 and IL-13 produced by basophils in response to antigen-dependent and -independent activation may strongly amplify allergic inflammation indirectly by inducing appropriate chemokines in tissue cells.

Regulators of Basophil Functions

The major ligands regulating the effector functions of basophils can be divided into three groups based on the class of receptors with which they interact, the signal-transduction pathways involved and the cellular responses that are induced: (1) allergens or anti-IgE cross-linking the $Fc_{\varepsilon}RI$; (2) growth factors interacting with receptors of the cytokine or tyrosine kinase receptor families; and (3) ligands interacting with seven-transmembrane G protein-coupled receptors.

Priming cytokines

Of particular importance for the basophil response to chemokines and other agonists is the priming effect of certain growth factors which enhance all effector functions, such as chemotaxis, mediator release and cytokine expression, and which can even qualitatively alter the response of the cell (by priming for leukotriene formation and inducing responsiveness to otherwise ineffective agonists) (3–8, 21, 22). The most effective cytokines are IL-3, IL-5, GM-CSF and NGF, which prime basophils with identical efficacies but different potencies (in the order IL-3>NGF>IL-5, GM-CSF) (10). Among these cytokines, IL-3, IL-5 and NGF are relatively specific for basophils and/or eosinophils and are thus more important in allergic inflammation, while GM-CSF is a more pleiotrophic pro-inflammatory cytokine affecting all myeloid cell types. Thus, the profile of cytokines produced in an immune response is one important component controlling the type of effector cells that are attracted and activated in a site of inflammation.

Consistent with these bioactivities, human basophils express high levels of the α chain of IL-3 receptors, lower levels of α chains of receptors for IL-5 and GM-CSF and intermediate levels of the common β chain (23). Basophils also express the proto-oncogene TRK-A, the high-affinity NGF receptor, but not the low-affinity NGF receptor and other neurotropin receptors (TRK-B, TRK-C) (24). The potent priming and modulatory effects of NGF on human basophils indicates an important link between the nervous system and stress responses and immunodeviation towards a Th2 type in allergic inflammation.

Basophil agonists

Basophils can be attracted and/or activated by ligands interacting with G protein-coupled receptors. Basophils appear to express a particularly broad spectrum of such chemoattractant receptors. Apart from chemokine receptors, they also express receptors for more pleiotrophic chemotactic factors, such as receptors for C5a, C3a, fMLP and platelet-activating factor (PAF) (7, 23), and most probably also receptors for LTB_4 and 5-oxo-ETE (unpublished observations). Thus, basophils have the potential to respond to a very wide variety of inflammatory stimuli. However, apart from certain chemokines discussed below, only C5a and fMLP are direct inducers of exocytosis in normal blood basophils. All other agonists need priming by one of the growth factors discussed above for an effective response.

CHEMOKINES

Structure and Biology of Chemokines

After the discovery of the first chemokine, IL-8, more than 12 years ago, an increasing number of other members of the chemokine family have been identified in rapid succession. There are now over 40 known human chemokines and 16 chemokine receptors. Chemokines represent a ligand superfamily of more or less homologous proteins of about 10 kDa with similar secondary and tertiary structure. Based on the amino acid motif after the first cysteine, the chemokines can be divided in two major families, the CXC and the CC chemokine family, and two other chemokine types, the C chemokine lymphotactin lacking one disulphide bridge and the CXXXC chemokine fractalkine with three amino acids between the first two cysteines. Many chemokines of the CXC family are chemoattractants of neutrophil granulocytes and are important in acute inflammatory responses to infection and tissue injury, while most CC chemokines act on monocytes, different lymphocyte subsets, basophils and eosinophils, and are thus major mediators of more chronic inflammatory processes. Chemokines can also be classified according to whether they are induced in inflammation or constitutively expressed in lymphoid organs and tissue. The latter chemokines, which can belong to both the CXC and the CC group, have been discovered more recently and regulate the complex traffic of different leukocyte types in normal immune surveillance.

All chemokine receptors identified so far belong to the superfamily of the ‘serpentine’ seven-transmembrane G protein-coupled receptors. Among this family they form a subfamily with homologies among themselves and also with other receptors for chemotactic factors such as the C5a and C3a receptors, and receptors for fMLP and LTB_4. Most chemokine receptors, like other chemoattractant receptors, are pertussis toxin-sensitive, indicating coupling to $G_{\alpha i}$ proteins and similarity in signalling pathways.

Activation of Basophils by Chemokines

CXC chemokines and receptors

The first chemokine shown to activate basophils was IL-8 (21), despite the fact that IL-8 primarily attracts neutrophil granulocytes. However, IL-8 is a relatively weak basophil chemoattractant, and induces significant mediator release only in basophils primed by IL-3, IL-5, GM-CSF or NGF (6, 21, 22). Binding studies demonstrated that basophils constitutively express considerable numbers of IL-8 receptors (about one-fourth of the level on neutrophils), and competition experiments in leukaemic basophils indicated that the predominant IL-8 receptor is of the IL-8 receptor A (CXCR1) subtype (25). This could be confirmed by flow cytometry using monoclonal antibodies against CXCR1 and CXCR2, demonstrating prominent expression of CXCR1, as well as lower, but still considerable, levels of CXCR2 (23). Interestingly, the ligands with selectivity for CXCR2 (e.g. GRO family, ENA-78, NAP-2) are very poor basophil chemoattractants, and do not induce mediator release at reasonable concentrations even in IL-3-primed cells (25). The reason why CXCR2 is inefficiently coupled to cellular activation in basophils is unknown. Constitutive expression of CXCR1 and CXCR2 distinguishes basophils from the related eosinophils, which normally do not express IL-8 receptors, but upregulate CXCR2 in hyper-eosinophilic conditions *in vivo* or upon culture with IL-5 *in vitro* (26). Consistent with their main function in Th1-type immune responses (27, 28), basophils do not respond to INF-γ-induced CXC chemokines (e.g. IP-10, MIG) and thus do not express CXCR3 (unpublished data). However, basophils express CXCR4, which is also found on most other cell types. Indeed, the CXCR4 ligand SDF has some chemotactic activity at higher concentrations (unpublished observations), but whether chemotaxis is the major function of SDF on basophils is, in our opinion, rather questionable.

CC chemokines and receptors

More recent studies clearly showed that the chemokines most important in allergy belong to the CC family. Monocyte chemotactic protein MCP-1 was the first chemokine to be identified as a potent inducer of exocytosis and mediator release in basophils (29–31), and RANTES as an efficient chemoattractant of eosinophils and basophils (32–34), indicating an important role of CC chemokines in the selective attraction and activation of effector cells in allergic inflammation. Further studies showed that MCP-3 is both a potent agonist of basophil mediator release and a chemoattractant of basophils and eosinophils, and thus shares the bioactivities of MCP-1 and RANTES (35, 36). Functional studies and cross-desensitization of calcium transients indicated the presence of a shared MCP-1/MCP-3 receptor on basophils and monocytes, and of a unique shared RANTES/MCP-3 receptor selectively expressed by basophils and eosinophils (32, 35, 37). Further studies resulted in the identification of an increasing number of chemokines activating basophils and/or eosinophils and resulted in the identification of the MCP receptor CCR2 and the eotaxin receptor CCR3, which appear to be the key chemokine receptors regulating the effector functions of human basophils.

The MCP receptor, CCR2, is expressed by basophils but not eosinophils (38) and is activated by MCP-1, -2, -3 and -4 (29–41). The eotaxin-receptor, CCR3, is selectively expressed by basophils (38), eosinophils (42–44) and polarized Th2 lymphocytes (45–47), and is activated by a particularly broad range of CC chemokines. These ligands include the CCR3-selective chemokines eotaxin (48, 49) and eotaxin-2 (50), the more promiscuous CC chemokines, MCP-4 (38–41), MCP-3 (35, 36), MCP-2 (51), RANTES (32, 33), and even a viral chemokine, vMIP-II (52). Basophils also express the CC chemokine receptor, CCR1, consistent with the chemotactic activity of MIP-1α on this cell type. Basophils, however, do not respond to MIP-1β and accordingly do not stain with monoclonal antibodies against CCR5. The lack of expression of CCR5 in basophil is another indication of their 'Th2-like' phenotype, since CCR5, like CXCR3, is preferentially expressed by T cells secreting a Th1 cytokine profile (28). Finally, we have no indication that basophils express CCR4, although CCR4 has been cloned from a cDNA library of a human basophilic cell line and this group reported that normal basophils express this receptor, as indicated by RT–PCR (53). These studies also showed that MCP-1, RANTES and MIP-1α are the ligands of CCR4. CCR4 was therefore viewed as the major chemokine receptor mediating the effects of these chemokines on basophils (54). It is now clear, however, that MDC and TARC are the physiological high-affinity ligands of CCR4 (55, 56). Furthermore, in our hands, no CCR4 mRNA was detectable by northern blotting in highly purified basophils, and basophils could not be activated by MDC or TARC, at least with regards to chemotaxis or mediator release (unpublished results).

Basophils express particularly high levels of both CCR2 and CCR3. In unfractionated blood, leukocytes, basophils and eosinophils are the only cell types that constitutively express high levels of CCR3 mRNA, as demonstrated by northern blotting, and a high level of CCR3 protein, as evidenced by flow cytometry (38). In all other leukocytes, staining with anti-CCR3 mAb is minimal or absent, giving further support to a key role of CCR3 agonists in the selective attraction of these effectors in allergic inflammation. In marked contrast to eosinophils, basophils express CCR2 mRNA at levels similar to CCR3 mRNA, which exceed by far that found in monocytes and even that in IL-2-activated T cells (Fig. 1a). More recent studies with a CCR2-specific monoclonal antibody indeed showed a selective bright staining of basophils, demonstrating that the basophil is the only leukocyte type with a constitutively high level of expression of this chemokine receptor (Fig. 1a). These observations may also indicate a role for basophil-derived mediators in pathological conditions involving monocytes and/or activated T cells without the participation of eosinophils, such as the formation of arteriosclerotic plaques.

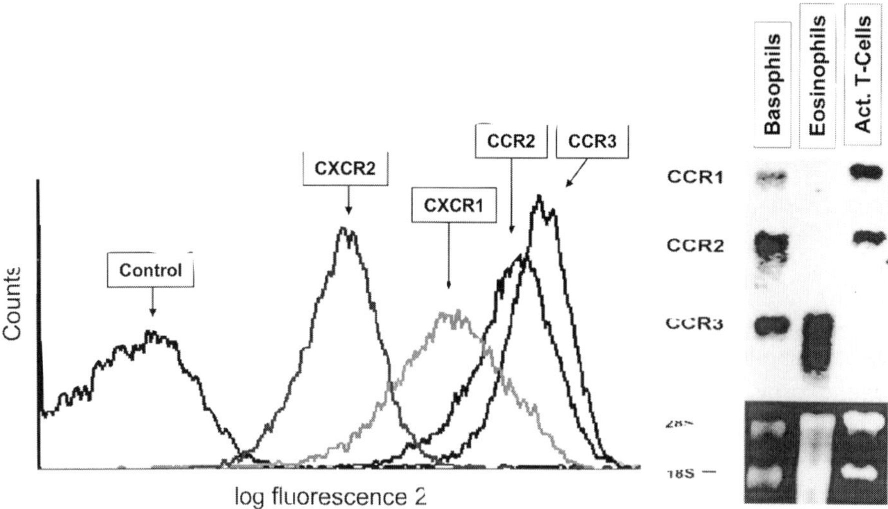

Fig. 1(a) Chemokine receptors expressed on basophils. Left: receptor densities assessed by flow cytometry. Right: mRNA by northern blotting. Note that approximately five times as much RNA of eosinophils and IL-2-activated lymphocytes was used for the comparison.

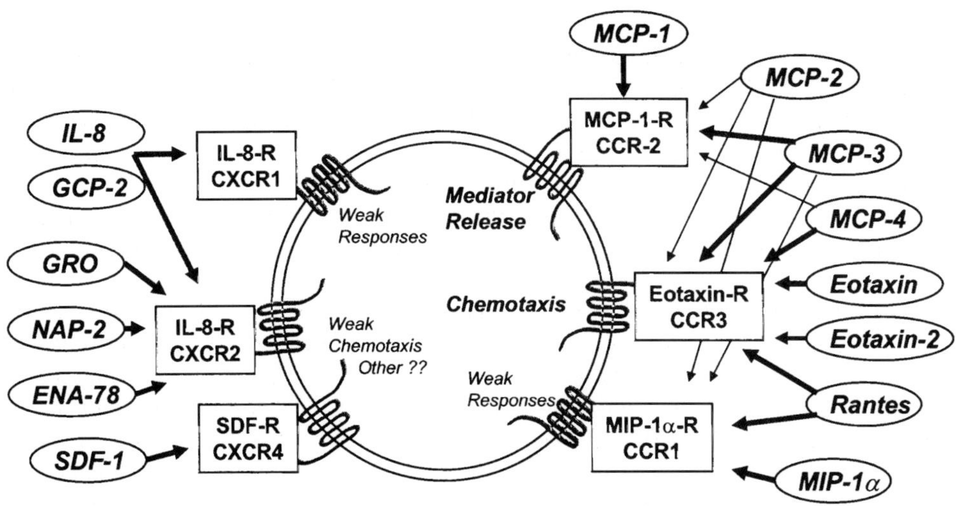

Fig. 1(b) Summary of chemokine ligands, receptors and functions in human basophils.

Differential Regulation of Basophil Functions

Despite the fact that CCR2 and CCR3 are expressed by basophils at similarly high densities, the two closely related chemokine receptors appear to mediate quite distinct effector functions, as evidenced by the analysis of the bioactivity profile of different CC chemokines, the use of receptor blocking antibodies and receptor desensitization studies (Fig. 1).

All the ligands for CCR3 are potent chemoattractants for basophils (and eosinophils), but induce significant mediator release only in primed basophils, unless they are also agonists for CCR2. The conclusion that CCR3 is the major receptor mediating the attraction of basophils by diverse CC chemokines was confirmed using blocking antibodies to CCR3, which abrogate the chemotactic activity of the CCR3-selective chemokines eotaxin and eotaxin-2, as well as that of MCP-4 and RANTES, which interact also with CCR2 and CCR1/CCR5, respectively (38). The redundancy of the chemokine system, and in particular the large number of chemokines interacting with CCR3, makes it difficult to estimate the importance of an individual chemokine in allergic inflammation. Thus, targeting a single chemokine, such as eotaxin, as a means to treat allergic disease may not be sufficiently protective. However, the fact that CCR3 appears to be responsible for the chemotactic activity of the most efficient CC chemokines together with the restricted expression pattern of this receptor on basophils, eosinophils and activated Th2-type lymphocytes, makes this receptor a particularly promising drug target for inhibiting the co-ordinated attraction of these major effector cells in allergic inflammation.

By contrast, CCR2 seems to mediate predominantly mediator release but only weak migratory responses (29, 32, 38). CCR2 is the only chemokine receptor capable of inducing strong and consistent release responses in blood basophils without the need of priming the cell with a growth factor such as IL-3. This conclusion is based on the fact that MCP-1, a selective CCR2 agonist, is a most potent inducer of histamine release with minimal chemotactic activity (28–30, 34). The importance of CCR2 in promoting mediator release in response to more promiscuous CC chemokines had to be examined more indirectly by desensitization experiments, since an efficient blocking antibody of this receptor has not yet been generated. These studies showed that pre-treatment of basophils with MCP-1 before stimulation with IL-3 followed by MCP-2, MCP-3 or MCP-4 completely blocked leukotriene generation, indicating that these other MCP homologues promote mediator release primarily by activating CCR2 (38). However, the efficacy of MCP-1 and MCP-3 to induce basophil degranulation is clearly superior to that of MCP-2 and MCP-4, despite identical displacement curves of MCP-3 and MCP-4 in equilibrium binding to CCR2 transfectants (38). This indicates partially separate requirements for binding and activation, consistent with models of distinct binding and activation domains for the interaction of larger peptides with G protein-coupled receptors.

As mentioned above, other chemokines interacting with CXCR1 or CCR1 consistently activate only basophils primed by an appropriate growth factor such as IL-3, and the ligands of CXCR2, such as NAP-2 and GRO, are even ineffective in primed cells, despite the fact that CXCR2 is expressed at considerable density (Fig. 1).

Another important function of basophils is the expression of IL-4 and IL-13, key cytokines in Th2-type immune responses (11–17). We found that IL-4/IL-13 expression is not only induced by IgE receptor cross-linking but also by chemotactic agonists C5a, when acting in synergy with IL-3 (15, 16). Since basophils express several chemokine receptors, of which CCR2 and CCR3 are found at similarly high densities as the C5a receptor (23), and since chemotactic agonists seem to utilize similar signal transduction pathways, it was reasonable to expect that at least some chemokines may also regulate IL-4 and IL-13 expression. However, with the exception of a minor effect of MCP-1, the most potent CCR2 agonist, none of the chemokines of the CXC and CC families influenced cytokine expression. Similar observations were made for the 'late phase' of leukotriene formation which occurs in parallel with IL-4/IL-13 secretion 4–18 h after combined stimulation with IL-3 and C5a. Interestingly fMLP was moderately active at inducing cytokine production, despite the fact that fMLP receptors are expressed at clearly lower levels than CCR2 or CCR3 (23). These observations further indicate that chemoattractant receptors mediate quite distinct cellular functions unrelated to their receptor density (summarized in Fig. 1b and Table I).

The reason why the closely related receptors for chemokines and other chemotactic agonists differ in their capacity to promote the various effector functions of basophils remain to be determined. These G protein-coupled receptors seem to utilize very similar signal transduction pathways. At least the early signalling events, such as the kinetic and amplitude of the rise in intracellular calcium concentration, cannot explain the differences in the cellular responses. The only exception is the more prolonged calcium transient in response to fMLP, which results in still elevated calcium levels when the MAP kinase pathway is activated. This may explain the production of leukotrienes in response to fMLP alone without the need of priming the cells with IL-3, since cytosolic phospholipase A_2 needs both phosphorylation by mitogen-activated protein (MAP) kinase and elevated calcium for optimal activity. However, the calcium transients following activation of CCR2 and CCR3, which primarily promote exocytosis and chemotaxis, respectively, are identical. To further complicate the issue, the CCR2 agonist MCP-1 is an effective chemoattractant of monocytes, indicating that the cellular background also affects the function of chemokine receptors.

In conclusion, we find that basophils express a particularly broad range of receptors for

TABLE I
Function and Receptor Expression for Chemotactic Agonists on Human Basophils

Agonist	Receptor	Receptor expression	Chemotaxis	Exocytosis	Lipid Mediator Formation			IL-4 expression	
					–IL-3 unprimed (burst)	+IL-3 primed (burst)	+IL-3 (late phase)	–IL-3	+IL-3
C5a	C5aR	+++	+++	+++	–	+++	+++	–	+++
fMLP	fMLPR	+	n.d.	+++	++	+++	++	–	++
PAF	PAFR	++	n.d.	(+)	–	+	–	–	–
C3a	C3aR	n.d.	n.d.	(+)	–	+	–	–	–
MCP-1	CCR2	+++	(+)	++	–	+++	(+)	–	(+)
Eotaxin	CCR3	+++	+++	(+)	–	+	–	–	–
Mip-1α	CCR1	+	+	(+)	–	+	–	–	–
IL-8	CXCR1	++	+	(+)	–	+	–	–	–
Nap-2	CXCR2	++	(+)	–	–	–	–	–	–
SDF	CXCR4	++	+	–	–	–	–	–	–

Number of '+' signs indicates the expression level or efficacy of the cellular response. (+) indicates that the response is weak and only observed in cells of select donors, or after priming the basophil with IL-3. n.d., not determined.

chemokines and chemotactic agonists. However, each receptor appears to mediate a quite distinct profile of the different basophil effector functions. This may allow the fine-tuning of cellular responses of basophils at inflammatory cites and control their functions as inflammatory or immunoregulatory cells.

ACKNOWLEDGEMENT

The work of the author is supported by the Swiss National Science Foundation.

REFERENCES

1. Maruyama, N., Tamura, G., Aizawa, T., *et al.* Accumulation of basophils and their chemotactic activity in the airways during natural narrowing in asthmatic individuals. *Am. J. Respir. Crit. Care Med.* **150**:1086–1093, 1994.
2. Koshino, T., Aria, Y., Miyamoto, Y., *et al.* Mast cell and basophil number in the airway correlate with the bronchial responsiveness of asthmatics. *Int. Arch. Allergy Immunol.* **107**:378–379, 1995.
3. Kurimoto, Y., de Weck, A. L. and Dahinden, C. A. Interleukin 3-dependent mediator release in basophils triggered by C5a. *J. Exp. Med.* **170**:467–479, 1989.
4. Takafuji, S., Bischoff, S. C., de Weck, A. L., *et al.* Interleukin 3 and interleukin 5 prime normal human eosinophils to produce leukotriene C4 in response to soluble agonists. *J. Immunol.* **147**:3855–3861, 1991.
5. Kurimoto, Y., de Weck, A. L. and Dahinden, C. A. The effect of interleukin 3 upon IgE-dependent and IgE-independent basophil degranulation and leukotriene generation. *Eur. J. Immunol.* **21**:361–368, 1991.
6. Bischoff, S. C., Brunner, T., de Weck, A. L., *et al.* Interleukin 5 modifies histamine release and leukotriene generation by human basophils in response to diverse agonists. *J. Exp. Med.* **172**:1577–1582, 1990.

7. Bischoff, S. C., de Weck, A. L. and Dahinden, C. A. Interleukin-3 and granulocyte/macrophage-colony-stimulating factor render human basophils responsive to low concentrations of complement component C3a. *Proc. Natl. Acad. Sci. USA* **87**:6813–6817, 1990.
8. Brunner, T., de Weck, A. L. and Dahinden, C. A. Platelet-activating factor induces mediator release by human basophils primed with IL-3, granulocyte-macrophage colony-stimulating factor, or IL-5. *J. Immunol.* **147**:237–242, 1991.
9. Bischoff, S. C. and Dahinden, C. A. Effect of nerve growth factor on the release of inflammatory mediators by mature human basophils. *Blood* **79**:2662–2669, 1992.
10. Bürgi, B., Brunner, T. and Dahinden, C. A. The degradation product of the C5a anaphylatoxin $C5a_{desarg}$ retains basophil-activating properties. *Eur. J. Immunol.* **24**:1583–1589, 1994.
11. Brunner, T., Heusser, T. H. and Dahinden, C. A. Human peripheral blood basophils primed by interleukin 3 (IL-3) produce IL-4 in response to immunoglobulin E receptor stimulation. *J. Exp. Med.* **177**:605–611, 1994.
12. MacGlashan, D. W. Jr., White, J. M., Huang, S. K., *et al.* Secretion of IL-4 from human basophils. *J. Immunol.* **152**:3006–3015, 1994.
13. Schroeder, J. T., MacGlashan, D. W Jr., Kagey-Sobotka, A,. *et al.* IgE-dependent IL-4 secretion by human basophils. The relationship between cytokine production and histamine release in mixed leukocyte cultures. *J. Immunol.* **153**:1808–1817, 1994.
14. Ochensberger, B., Rihs, S., Brunner, T., *et al.* IgE-independent interleukin-4 expression and induction of a late phase of leukotriene C4 formation in human blood basophils. *Blood* **86**:4039–4049, 1995.
15. Ochensberger, B., Daepp, G. C., Rhis, S., *et al.* Human blood basophils produce interleukin-13 in response to IgE-receptor-dependent and -independent activation. *Blood* **88**:3028–3037, 1996.
16. Li, H., Sim, T. C. and Alam, R. IL-13 released by and localized in human basophils. *J. Immunol.* **156**:4833–4838, 1996.
17. Seder, R. A. and Paul, W. E. Acquisition of lymphokine-producing phenotype by $CD4^+$ T cells. *Annu. Rev. Immunol.* **12**:635–673, 1994.
18. Li, H., Sim, T. C., Grant, J. A., *et al.* The production of macrophage inflammatory protein-1 alpha by human basophils. *J. Immunol.* **157**:1207–1212, 1996.
19. Mochizuki, M., Bartels, J., Mallet, A. L., *et al.* IL-4 induces eotaxin: a possible mechanism of selective eosinophil recruitment in helminth infection and atopy. *J. Immunol.* **160**:60–68, 1998.
20. Bochner, B., Klunk, D., Sterbinsky, S., *et al.* IL-13 selectively induces vascular cell adhesion molecule-1 expression in human endothelial cells. *J. Immunol.* **154**:799–803, 1995.
21. Dahinden, C. A., Kurimoto, Y., de Weck, A. L., *et al.* The neutrophil-activating peptide NAF/NAP-1 induces histamine and leukotriene relese by interleukin-3 primed basophils. *J. Exp. Med.* **170**:1787–1792, 1989.
22. Dahinden, C. A., Bischoff, S. C., Brunner, T., *et al.* Regulation of mediator release by human basophils: importance of the sequence and time of addition in the combined action of different agonists. *Int. Arch. Allergy Appl. Immunol.* **94**:161–164, 1991.
23. Ochensberger, B., Tassera, L., Bifare, D., *et al.* Regulation of cytokine expression and leukotriene formation in human basophils by growth factors, chemokines and chemotactic agonists. *Eur. J. Immunol.* **29**:11–22, 1999.
24. Bürgi, B., Otten, U. H., Ochensberger, B., *et al.* Basophil priming by neurotrophic factors: activation through the TRK receptor. *J. Immunol.* **157**:5582–5588, 1996.
25. Krieger, M., Brunner, T., Bischoff, S. C., *et al.* Activation of human basophils through the interleukin 8 receptor. *J. Immunol.* **8**:2662–2667, 1992.
26. Heath, H., Qin, S., Rao, P., *et al.* Chemokine receptor usage by human eosinophils. The importance of CCR3 demonstrated using an antagonistic monoclonal antibody. *J. Clin. Invest.* **99**:178–184, 1997.
27. Loetscher, M., Gerber, B., Loetscher, P., *et al.* Chemokine receptor specific for IP10 and mig: structure, function, and expression in activated T-lymphocytes. *J. Exp. Med.* **184**:963–969, 1996.
28. Qin, S., Rottman, J. B., Myers, P., *et al.* The chemokine receptors CXCR3 and CCR5 mark subsets of T cells associated with certain inflammatory reactions. *J. Clin. Invest.* **101**:746–754, 1998.
29. Bischoff, S. C., Krieger, M., Brunner, T., *et al.* Monocyte chemotactic protein-1 is a potent activator of human basophils. *J. Exp. Med.* **175**:1271–1275, 1992.
30. Kuna, P., Reddigari, S. R., Rucinski, D., *et al.* Monocyte chemotactic and activating factor is a potent histamine-releasing factor for human basophils. *J. Exp. Med.* **175**:489–493, 1992.
31. Alam, R., Lett-Brown, M. A., Forsythe, P. A., *et al.* Monocyte chemotactic and activating factor is a potent histamine releasing factor in basophils. *J. Clin. Invest.* **89**:723–728, 1992.
32. Rot, A., Krieger, M., Brunner, T., *et al.* Rantes and MIP-1α induce the migration and activation of normal

human eosinophil granulocytes. *J. Exp. Med.* **176**:1489–1495, 1992.
33. Kameyoshi, Y., Dorschner, A., Mallet, A. L., *et al.* Cytokine RANTES released by thrombin-stimulated platelets is a potent attractant for human eosinophils. *J. Exp. Med.* **176**:587–592, 1992.
34. Bischoff, S. C., Krieger, M., Brunner, T., *et al.* RANTES and related chemokines activate human basophil granulocytes through different G protein-coupled receptors. *Eur. J. Immunol.* **23**:761–767, 1993.
35. Dahinden, C. A., Geiser, T., Brunner, T., *et al.* Monocyte chemotactic protein 3 is a most effective basophil- and eosinophil-activating chemokine. *J. Exp. Med.* **179**:751–756, 1994.
36. Proost, P., Wuyts, A. and Van Damme, J. Human monocyte chemotactic proteins-2 and -3: structural and functional comparison with MCP-1. *J. Leukoc. Biol.* **59**:67–74, 1996.
37. Baggiolini, M. and Dahinden, C. A. CC chemokines in allergic inflammation. *Immunol. Today* **15**:127–133, 1994.
38. Uguccioni, M. G., Mackay, C., Ochensberger, B., *et al.* High expression of the chemokine receptor CCR3 in human blood basophils. *J. Clin. Invest.* **100**:1137–1143, 1997.
39. Uguccioni, M. G., Loetscher, P., Forssmann, U., *et al.* Monocyte chemotactic protein 4 (MCP-4), a novel structural and functional analogue of MCP-3 and eotaxin. *J. Exp. Med.* **183**:2379–2384, 1996.
40. Garcia-Zepeda, E. A., Combadiere, C., Rothenberg, M. E., *et al.* Human monocyte chemoattractant protein (MCP)-4 is a novel CC chemokine with activities on monocytes, eosinophils, and basophils induced in allergic and nonallergic inflammation that signals through the CC chemokine receptors (CCR)-2 and -3. *J. Immunol.* **157**:5613–5626, 1996.
41. Stellato, C., Collins, P., Ponath, P. D., *et al.* Production of the novel C-C chemokine MCP-4 by airway cells and comparison of its biological activity to other C-C chemokines. *J. Clin. Invest.* **99**:926–936, 1997.
42. Kitaura, M., Nakajima, T., Imai, T., *et al.* Molecular cloning of human eotaxin, an eosinophil-selective CC chemokine, and identification of a specific eosinophil eotaxin receptor, CC chemokine receptor 3. *J. Biol. Chem.* **271**:7725–7730, 1996.
43. Ponath, P. D., Qin, S., Post, T. W., *et al.* Molecular cloning and characterization of a human eotaxin receptor expressed selectively on eosinophils. *J. Exp. Med.* **183**:2437–2448, 1996.
44. Daugherty, B. L., Siciliano, S. J., De Martino, J. A., *et al.* Cloning, expression, and characterization of the human eosinophil eotaxin receptor. *J. Exp. Med.* **183**:2349–2354, 1996.
45. Sallusto, F., Mackay, C. R. and Lanzavecchia, A. Selective expression of the eotaxin receptor CCR3 by human T helper 2 cells. *Science* **277**:2005–2007, 1997.
46. Sallusto, F., Lenig, D., Mackay, C. R., *et al.* Flexible programs of chemokine receptor expression on human polarized T helper 1 and 2 lymphocytes. *J. Exp. Med.* **187**:875–883, 1998.
47. Bonecchi, R., Bianchi, G., Bordignon, P. P., *et al.* Differential expression of chemokine receptors and chemotactic responsiveness of type 1 T helper cells (Th1s) and Th2s. *J. Exp. Med.* **187**:129–134, 1998.
48. Garcia-Zepeda, E. A., Rothenberg, M. E., Ownbey, R. T., *et al.* Human eotaxin is a specific chemoattractant for eosinophil cells and provides a new mechanism to explain tissue eosinophilia. *Nat. Med.* **2**:449–456, 1996.
49. Ponath, P. D., Qin, S., Ringler, D. J., *et al.* Cloning of the human eosinophil chemoattractant, eotaxin. Expression, receptor binding, and functional properties suggest a mechanism for the selective recruitment of eosinophils. *J. Clin. Invest.* **97**:604–612, 1996.
50. Forssmann, U., Uguccioni, M., Loetscher, P., *et al.* Eotaxin-2, a novel CC chemokine that is selective for the chemokine receptor CCR3, and acts like eotaxin on human eosinophil and basophil leukocytes. *J. Exp. Med.* **185**:2171–2176, 1997.
51. Weber, M., Uguccioni, M., Ochensberger, B., *et al.* The monocyte chemotactic protein MCP-2 activates human basophil and eosinophil leukocytes similar to MCP-3. *J. Immunol.* **154**:4166–4172, 1995.
52. Boshoff, C., Endo, Y., Collins, P. D. *et al.* Angiogenic and HIV-inhibitory functions of KSHV-encoded chemokines. *Science* **278**:290–294, 1997.
53. Power, C. A., Meyer, A., Nemeth, K., *et al.* Molecular cloning and functional expression of a novel CC chemokine receptor cDNA from a human basophilic cell line. *J. Biol. Chem.* **270**:19495–19500, 1995.
54. Murphy, P. M. The molecular biology of leukocyte chemoattractant receptors. *Annu. Rev. Immunol.* **12**:593–633, 1994.
55. Imai, T., Chantry, D., Raport, C. J., *et al.* Macrophage-derived chemokine is a functional ligand for the CC chemokine receptor 4. *J. Biol. Chem.* **273**:1764–1768, 1998.
56. Imai, T., Baba, M., Nishimura, M., *et al.* The T cell-directed CC chemokine TARC is a highly specific biological ligand for CC chemokine receptor 4. *J. Biol. Chem.* **272**:15036–15042, 1997.

CHAPTER 36

Chemokine Receptors on Human Mast Cells

PAOLA ROMAGNANI,*[1] AMATO DE PAULIS,[3] CHIARA BELTRAME,[2] GIANNI MARONE[3] and SERGIO ROMAGNANI[2]

[1]*Department of Pathophysiology, Endocrinology Unit,*
[2]*Department of Internal Medicine, Section of Clinical Immunology, Allergy and Respiratory Diseases, University of Florence, and*
[3]*Division of Clinical Immunology and Allergy, University of Naples Federico II, Naples, Italy*

CHEMOKINES

The attraction of leukocytes to tissues is a basic mechanism for inflammation and the host response to infections. This process is controlled by a large number of chemotactic cytokines, called chemokines (1–5). Chemokines not only cause chemotactic migration of leukocytes, but also have other functions, being involved in angiogenesis, collagen production and the proliferation of haematopoietic precursors.

Chemokine Structure and Production

Chemokines are 8–10-kDa proteins with 20–70% homology in amino acid sequences. Based on the relative position of their cysteine residues, at least four families of chemokines have been identified, the CXC, CC, C and CX3C families, two of which (CXC and CC) have been extensively characterized. The CXC, or α-chemokines, and the CC, or β-chemokines, appear to be the largest families. In the α-chemokines, one amino acid (X) separates the first two cysteine residues (cysteine–X–cysteine, or CXC), whereas in the β-chemokines the first two cysteine residues are adjacent to each other (cysteine–cysteine, or CC). The C chemokine lymphotactin has only two cysteines in the mature protein, and the CX3C chemokine fractalkine has three amino acids separating the first two cysteines. So far, about 50 chemokines have been identified in humans (Table I).

As mentioned above, the main function of chemokines is chemotaxis for leukocytes. Schematically, CXC (or α-) chemokines containing the glutamic acid–leucine–arginine sequence near the N-terminal (preceding the CXC sequence) are chemotactic for polymorphonuclear neutrophils, whereas those not containing the sequence act on T and

* Corresponding author.

MAST CELLS AND BASOPHILS
ISBN 0-12-473335-2

Table I
Chemokine Families and Chemokines

Structure	Chemokine
CXC (α)	IL-8
	GRO (α, β, γ)
	ENA-78
	NAP-2
	GCP-2
	PF-4
	Mig
	IP-10
	I-TAC
	SDF-1α/β
	BCA-1
CC (β)	MIP-1α
	MIP-1β
	RANTES
	MDC/STCP-1
	TARC
	TECK
	MIP-3α/LARC/Exodus
	MIP-3β/ELC/SCL/Exodus-2/6Ckine/TCA-4
	Eotaxin 1, 2, 3
	MCP-1, 2, 3, 4, 5
	DC-CK-1/PARC/MIP-4/AMAC-1
	HCC-1
	HCC-4/NCC-4/ILINK/LEC/LMC
	MIP-5/HCC-2/Lkn-1
	MPIF-1/CKβ8
C (γ)	Lymphotactin
CX3C (δ)	Fractalkine

B lymphocytes. The CC (or β-) chemokines, in general, do not act on neutrophils but exert their action on multiple leukocyte subtypes, including monocytes, polymorphonuclear eosinophils and basophils, lymphocytes, dendritic cells (DC) and natural killer (NK) cells with variable selectivity (1–3). Chemokines are redundant in their action on target cells; no chemokine is uniquely active on one leukocyte population, and usually a given leukocyte population has receptors for, and responds to, different molecules. Of note, mononuclear phagocytes, the most evolutionary ancient cell type involved in natural immunity, are responsive to the majority of chemokines.

The different cell types can produce chemokines according to two main modalities. Some chemokines are produced constitutively by specialized cells and tissues (e.g. MDC), or in a less restricted way (e.g. SDF-1). However, the great majority of chemokines are produced following cell activation. Of note, multiple chemokines are produced in a redundant way by a single cell (polyspeirism). ‘Polyspeirism’, the ability to produce more than one chemokine, is a function particularly of cells involved in natural immunity, such as mononuclear phagocytes and endothelial cells (EC). The redundancy of chemokines, as well as their polyspeirism, have recently been attributed to the need for robust outputs of these chemotactic cytokine networks in order to avoid their failure, even in the presence of genetic or epigenetic alterations (6).

Chemokine Functions

Chemokines provide the directional cues for the movements of leukocytes in development and homeostasis during inflammatory processes; they also play an important role in both angiogenesis and tumour growth.

Role in leukocyte development and homeostasis

During their development and differentiation, leukocytes move through different compartments. Although the pathways, the role of adhesion molecules in the recognition of homing sites and several highly effective chemokines are known, the regulation of this intricate cellular trafficking is still partially unknown (7). A number of critical points, however, are being clarified. Some chemokines have dual effects on haematopoiesis depending on the maturity of the progenitors being treated. MIP-1α, MIP-1β, GRO-β and GRO-γ enhance the formation of granulocyte–macrophage colony-forming unit (GM-CFU) from unfractionated bone marrow (BM), but only in the presence of M- or GM-colony-stimulating factor (CSF) (reviewed in ref. 1). In contrast, other chemokines suppress the proliferation of more immature progenitors (e.g. CFU-S, CFU-A, CFU-GM, CFU-GEMM and BFU-E). This effect occurs directly on the progenitor cell because the suppression is more complete on purified $CD34^+$ cells (reviewed in ref. 1).

Another chemokine with potential involvement in leukocyte maturation and in other homeostatic functions is SDF-1, which was originally described as a growth factor for B lymphocyte precursors (8). Mice lacking the SDF-1 gene have severely impaired lymphopoiesis and abnormally low numbers of B cells and myeloid BM precursors (9). SDF-1 is chemotactic for pro- and pre-B cells, which depend on stromal cell contact for growth and differentiation, but not for mature forms (10). SDF-1 also induces chemotaxis of $CD34^+$ cells of different lineages. Thus, SDF-1 may be involved in directing progenitor B cells into the microenvironment of stromal cells, where growth and differentiation factors are released (10). More generally, they may be involved in directing progenitor cells into the appropriate maturation sites in the BM and in supporting the colonization of the BM haematopoietic precursors during embryogenesis (11). Murine SDF-1 also attracts resting T cells *in vitro* and *in vivo*, usually with high efficacy (11). Human SDF-1 is chemotactic for T cells, monocytes and neutrophils (11, 12).

Of particular interest is a group of CC chemokines, including TARC, MDC, ELC, SLC, LARC and DC-CK1, which, except for LARC, are expressed constitutively at high levels in the thymus and in other lymphoid tissues. Recently, human thymus MDC expression was localized to the medullary thymic epithelial cells (TEC) (13), and MDC-attracted CCR4-expressing murine cells were found to be double-positive $CD4^+CD^{dull}$ thymocytes (13). This suggests that MDC produced by medullary TEC is responsible for the migration of cortical thymocytes which have passed positive selection to the medullary areas where they are subjected to negative selection. By contrast, ELC, which attracts more mature thymocytes, was found to be produced by human TEC mainly localized around the thymic vessels (P. Romagnani *et al.*, unpublished results), suggesting its possible role in favouring the passage of mature thymocytes into the circulation.

Other chemokines allow the physiological homing of different cell types of the immune system in lymph nodes and in other lymphoid tissues. DK-CK1 is produced by DC of germinal centres and T cell areas of secondary lymphoid organs and is chemotactic for naive T cells, suggesting a role in the initiation of the immune response (14). TECK is produced by thymic DC and is chemotactic for murine macrophages, DC and thymocytes (15). Another example of such a housekeeping function of chemokines is provided by BCA-1/BLC, a chemokine expressed in lymphoid tissues and selective for B lymphocytes (16, 17). Disruption of the CXCR5 gene, the putative BCA-1/BLC receptor, which is highly expressed in B lymphocytes (18), leads to a loss of inguinal lymph nodes and the defective formation of primary follicles and germinal centres in the spleen and Peyer's patches. Receptor-deficient B lymphocytes enter the T cell areas of these tissues, but fail to migrate into B cell areas. These results suggest the existence of chemokines that direct homing of lymphocytes and other cell types into specific anatomical areas and regulate the development of functional lymphoid tissues.

Role in inflammation

Leukocyte extravasation from the blood into inflamed tissues is a regulated multistep process consisting of a series of co-ordinated interactions between leukocytes and EC (19). Several families of molecular regulators, such as selectins, integrins and chemokines themselves, control different aspects of this process. Chemokines provide the signals that convert the low-affinity, selectin-mediated interaction into the higher-affinity, integrin-mediated interaction that leads to extravasation of leukocytes (3). The dramatic increase in the production of chemokines during inflammation results in the selective recruitment of leukocytes into inflamed tissue. Chemokines have been detected during inflammation in most organs, including the skin, brain, joints, meninges, lungs, blood vessels, kidney and gastrointestinal tract. They have also been identified in many types of cells during inflammation in these organs, suggesting that most, if not all, cells can secrete chemokines, given the appropriate stimulus. The main stimuli for chemokine production during inflammation are bacterial products, such as lipopolysaccharide (LPS), and viral infections, but also early pro-inflammatory cytokines, such as interleukin-1 (IL-1) and tumour necrosis factor-α (TNF-α), and cytokines produced by Th1 or Th2 specific effectors, such as interferon-γ (IFN-γ) or IL-4, respectively (20–22). During the inflammatory processes, chemokines do not act alone in recruiting inflammatory cells; rather, they collaborate with other cytokines in modulating leukocyte responsiveness, as well as the expression of particular adhesion molecules, thus providing a certain degree of specificity to, and controlling the type of, inflammatory response. For example, in many acute disease processes, such as bacterial pneumonia and adult respiratory distress syndrome (ARDS), there is a massive influx of neutrophils into lung tissue, and in the bronchoalveolar lavage fluid of the same patients the concentration of a potent neutrophil chemoattractant, such as IL-8, is strongly increased (23). By contrast, IP-10 and MCP-1 are increased in the cerebrospinal fluid of patients with viral meningitis, tuerculoid leprosy and sarcoidosis, where mononuclear cell infiltration is prevalent (24, 25). Finally, in allergic disorders, such as asthma, rhinitis and atopic dermatitis, production of RANTES, MCP-3, and particularly that of ETXs, is thought to be responsible for the selective accumulation and/or activation of eosinophils (26), basophils (27), Th2 cells (28) and MC (29, 30) (see below).

Role in angiogenesis

Several CXC chemokines have been found to play an important role in angiogenesis *in vivo* and on EC *in vitro*. For example, IL-8, ENA-78, GCP2, GRO-α, GRO-β and GRO-γ have been reported to be chemotactic for EC and are also angiogenic in the rat cornea neovascularization assay (31, 32). By contrast, PF4, IP-10 and Mig are not only inactive chemoattractants and non-angiogenic themselves, but they inhibit the angiogenic effects of the other CXC chemokines (31, 33). One model for this angiostatic property would be interference with activation of growth factor receptors on EC. In fact, PF4 binds to full-length vascular endothelial growth factor (VEGF) and prevents its heparin-dependent interaction with its receptor (34). However, PF4 also inhibits the mitogenic effects of a truncated VEGF, whose receptor binding is not heparin-dependent and is not prevented by PF4. This is a strong argument for angiostatic chemokines exerting their effects through their own cognate receptors, and affecting events at post-growth factor receptor activation stage (1). Recently, we have indeed observed that IP-10 inhibits the proliferation of cultured EC. More importantly, the expression of its receptor (CXCR3) during the S phase of the mitotic cell cycle of these cells was detected by flow cytometry. By using immunohistochemistry, CXCR3 expression was also observed *in vivo* in microvascular EC (P. Romagnani *et al.*, unpublished).

Because of their ability to modulate neovascularization, chemokines can also promote or inhibit the development of tumour growth and metastasis (32). Several lung carcinomas have marginally increased levels of IL-8, and the angiogenic activity present in extracts from these tumours is almost entirely caused by IL-8 (35). Moreover, administration of anti-IL-8 antibody to SCID mice bearing a human IL-8-expressing lung cancer xenograft has provided beneficial effects (32). MCP-1, -2 and -3 have been isolated from glioma and osteosarcoma (36). Many other chemokines are produced by tumour cells *in vitro* and are present in human tumours (21, 22, 32). Recently, we have found high IP-10 expression in kidney tumours with a low degree of malignancy, but little, if any, in those with a high degree of malignancy (P. Romagnani, unpublished). However, the overall role of chemokines in tumour biology is still unclear, in as much as tumour-associated leukocytes may either stimulate or inhibit tumour growth.

Chemokine Receptors

Chemokines induce cell migration and activation by binding to specific G protein-coupled cell-surface receptors possessing a seven-transmembrane domain on target cells (1–3, 5, 6). Ten human CC chemokine receptors (CCR1–CCR10), five CXC chemokine receptors (CXCR1–CXCR5) and one CX3C (CX3CR1) receptor have been identified (Table II). Chemokine receptors are expressed on different types of leukocytes, some being restricted to certain cells, whereas others are more widely expressed not only on leukocytes, but also on non-haematopoietic cells. Although most chemokine receptors bind more than one chemokine, in general CC receptors bind only CC chemokines and CXC receptors bind only CXC chemokines (1–3, 5, 6). In addition, some chemokine receptors are constitutively expressed on some cells, whereas they are inducible on others. For example, CCR1 and CCR2 are constitutively present on monocytes, but are expressed also on lymphocytes after IL-2 stimulation (37). In contrast, the expression of other chemokine receptors is restricted to a cell state of activation and differentiation. For example, CCR5 is expressed on activated T cells showing the Th1 cytokine profile (38), whereas CCR4 is preferentially expressed on activated T cells showing a Th0/Th2 profile of cytokine production (39).

Table II
Chemokine Receptors

Receptor	Ligand(s)	Cellular distribution
CCR1	MIP-1α; RANTES; MCP-3	Mono; T; NK; imm. DC; Neu
CCR2 B/A	MCP-1–4	Mono; act. T; act. NK
CCR3	Eotaxin 1, 2, 3; MCP-3; RANTES	Eo; Ba; MC; Th2 (poor)
CCR4	MDC, TARC	Med. thym.; Th2; Tc2; Mono
CCR5	RANTES; MIP-1α; MIP-1β	Mono; Th1; Tc1; imm. DC
CCR6	MIP-3α/LARC/Exodus	T; imm. DC; (CD34)
CCR7	ELC/SLC/MIP-3β	Thym; T; Mono; mat. DC
CCR8	I-309; TARC	Mono; Th2; Tc2
CCR9,CCR10	Eotaxin; MIP-1α, β; RANTES; MCP-1–4	Mono; T
CXCR1	IL-8; GCP-2	Neu
CXCR2	IL-8; GRO-α, β, γ; NAP-2	Neu
CXCR3	IP-10; MIG; ITAC	Th1; Tc1; mesangial cells; EC of small arteries
CXCR4	SDF-1	Many cell types
CXCR5	BCA-1	B
CX3CR1	Fractalkine	Mono; NK; T

Mono, monocytes; imm. DC, immature dendritic cells; neu, neutrophils; act., activated; Eo, eosinophils; Ba, basophils; MC, mast cells; med. thym., medullary thyocytes; mat. DC, mature dendritic cells.

Utilization of chemokine receptors by infectious agents

There are several intriguing examples of infectious agents that utilize chemokines or chemokine receptors for their entry into the cells or even encode chemokine receptor-like molecules. The most famous example of the first condition is HIV-1 infection. M-tropic HIV-1 strains utilize CCR5 as co-receptor for their cell entry, whereas T cell line tropic HIV-1 strains utilize CXCR4 (reviewed in ref. 40). This is the reason why the two main types of HIV-1 strains have been re-named R5- and X4-tropic, respectively. Recently, we have shown that Th2 cells support the expression of X4-tropic HIV-1 strains better than Th1 cells, because IL-4 upregulates, whereas IFN-γ downregulates, CXCR4 expression (41). Moreover, HIV-infected individuals showing high CXCR4 $CD4^+$ T cell expression exhibited a higher frequency of X4-tropic highly aggressive HIV-1 isolates than those with lower CXCR4 $CD4^+$ T cell expression (R. Manetti, unpublished). Surprisingly, however, although the R5-tropic strain showed higher entry into Th1 cells than into Th2 cells because of their higher CCR5 co-receptor expression, even R5-tropic HIV-1 strains showed higher spreading among Th2 cells, due to the lower ability of these latter to produce RANTES, MIP-1α and MIP-1β (42), the β-chemokines that inhibit the entry of

R5-tropic strains into T cells (43). Of note, a substantial proportion of certain human populations carry a 32 deletion variant of the CCR5 gene in a heterozygous or homozygous form, which results in increased resistance to HIV (44–46). Although CCR5 and CXCR4 are the major co-receptors for HIV-1 cell entry, chemokine receptors such as CCR2b, CCR3, CCR8 and CCR9, as well as other not yet well-characterized receptors, can also be utilized by this virus. Another well-known example of chemokine receptor utilization by infectious agents is DARC, which allows the entry of *Plasmodium vivax* into red blood cells (47, 48).

Some infectious agents can also encode chemokines or chemokine receptor-like molecules. For example, EC3F3 from *Herpesvirus saimiri* predicts a heptahelical membrane protein with a sequence similar to CXCR1 and CXCR2, and the transfected recombinant protein can bind and signal in response to IL-8, GRO-α and NAP-2 (49). The US28 open reading frame from human cytomegalovirus is highly similar to CCR1, and the transfected protein can bind a wide range of CC, but not CXC, chemokines (50, 51). Finally, the Kaposi's sarcoma-associated herpesvirus 8 encodes for a viral inflammatory protein II (vMIP-II), which is an agonist for CCR8, a chemokine receptor preferentially expressed by Th2 and Tc2 cells (52).

Signal transduction pathways

Chemokine receptors, like other members of the family of G protein-coupled receptors, are functionally linked to phospholipases through G proteins (37, 53). Many chemokine-induced signalling events are inhibited by *Bordetella pertussis* toxin, suggesting that chemokine receptors are associated with PTx-sensitive GTP-binding proteins. Receptor triggering leads to a cascade of cellular activation, including activation of phosphatidylinositol-specific phospholipase C, protein kinase C, small GTPases, Src-related tyrosine kinases, phosphatidylinositol-3-OH kinases and protein kinase B (20, 54, 55). Phospholipase C delivers two second messengers, inositol (1,4,5)-trisphosphate, which releases Ca^{2+} from intracellular stores leading to a transient rise of cytosolic Ca^{2+} concentration, and diacylglycerol, which activates protein kinase C. Mobilization of Ca^{2+} is essential for granule release and superoxide production, but is not required for the cytoskeletal rearrangements leading to shape change (20). Phosphatidylinositol-3-OH kinases can be activated by the βγ subunit of G proteins, small GTPases or Src-related tyrosine kinases (54). Small GTPases regulate cytoskeletal rearrangements involved in adhesion and chemotaxis (56, 57), mediate activation of phospholipase D and are involved in the assembly of the superoxide-forming oxidase (56). Signalling by other chemokine receptors has not been investigated in a comparable manner, and virtually no information is available about chemokine receptors that mediate homeostatic rather than inflammatory functions of chemokines, for which different signal transduction mechanisms can be assumed (2, 6). Chemokine receptor signalling also activates small GTP-binding proteins of the Ras and Rho families (57). Rho proteins are involved in cell motility through regulation of actin-dependent processes, such as membrane ruffling, pseudopod formation and assembly of focal adhesion complexes. Recent findings have shown that both CC and CXC chemokines activate PYk2/RAFTK tyrosine kinase and the Ras/Raf/MAP kinase pathways (58–60), and inhibitors of MAP kinases reduce the chemotactic response to MCP-1 (59). Thus, chemokine receptors activate multiple intracellular signalling pathways that regulate the intracellular machinery necessary to propel the cell in its chosen direction (6).

MAST CELLS

The expression of chemokine receptors on mast cells (MC) has not been investigated as in other cells, such as polymorphonuclear leukocytes, macrophages, DC and T cells. This is mainly because of the remarkable difficulty in handling MC in humans, and because the pathways of differentiation and subtyping, which have aroused much controversy and confusion, are still unclear.

Biological Significance of MC

MC are found in almost all major organs and tissues of the body and it has been estimated that, if all MC in the body were assembled in one organ, their mass would equal that of the spleen (61). MC, however, can be found in large numbers wherever the body is in contact with its environment (e.g. the skin, the respiratory system and the digestive tract). This selective accumulation at tissue sites where foreign material attempts to invade the host suggests that MC are among the first cells to initiate defence mechanisms. MC are one of the major effector cells in the pathogenesis of immediate-type hypersensitivity (allergic) reactions, but also in a number of non-allergic immune disorders, as well as in normal physiological processes. Activation of MC, either by IgE-dependent or IgE-independent agonists leads to the release of a variety of pre-formed and newly synthesized mediators, which then act on distinct effector cells to produce the symptoms of allergy and anaphylaxis (62, 63). More recently, it has been established that MC are also able to contribute to the chronic inflammatory events of allergic diseases via the secretion of cytokines. Thus, these cells have the capacity to make major contributions to both acute and chronic elements of pulmonary disease (64).

However, MC play a major role not only in allergic diseases but also in a number of non-allergic immune reactions. An increased number of MC can be found in scar tissue, especially in keloids, in callus tissues, in osteoporosis-linked diseases and in various neuropathies. Furthermore, they have been observed to increase in different diseases, such as interstitial pneumonia, ulcerative colitis, intestinal helminthosis, ectodermal parasitosis and in particular skin disorders, such as atopic dermatitis, psoriasis, bullous pemphigoid, scleroderma, lichen planus, wound healing and cancer (64). Apart from the above-mentioned pathological conditions, MC also play a role in normal physiological processes. The mast cells mediators, histamine and heparin, have been shown to enhance vascularization and EC proliferation (65, 66).

MC Development

The different phases of MC differentiation are still not fully known. They were initially suggested to be derived from T cells, fibroblasts or macrophages (67–69). At present, however, the general consensus suggests that MC originate from pluripotent haematopoietic stem cells in the BM (64). MC leave the BM and circulate in the blood only as progenitors, and it is not until they enter the tissues that they undergo their terminal differentiation into mature MC. Experiments using cells from CFU injected into mouse skin demonstrated that MC originate from cells less differentiated than those precursors committed to either the neutrophil–macrophage or erythroid lineage (70). Whereas most descendants of pluripotent stem cells, such as erythrocytes, neutrophils, basophils or eosinophils, do not leave the haematopoietic tissue until their differentiation is completed, MC invade the connective or the mucosal tissue as morphologically unidentifiable MC precursors (MCPr). Only at tissue level do they differentiate into phenotypically identifiable MC (71–73). The *in vitro* studies performed so far have allowed the identification of some MC growth factors that give rise to MC proliferation from their precursors. They include IL-3, IL-4, stem cell factor (SCF), IL-9 and IL-10 (74–76). More recently, the development of homogeneously pure MCPr from uncommitted mouse BM cells using a triad of SCF, IL-6 and IL-10 was described (77). The more recent data, based on the identification and characterization of chemokine receptors on both MCPr and mature MC, will probably yield further information and enable additional hypotheses on the mechanisms responsible for MC development (see below).

Mast Cell Heterogeneity

MC subtypes in rodents

A still highly controversial issue in MC physiology is their heterogeneity, which is based on histochemical, biochemical and functional criteria and is strictly connected with their origin. On the basis of their different fixative and staining properties, two MC subtypes in rodents have been distinguished. First, there are the atypical T cell-dependent mucosal MC (MMC), which are mainly found in the mucosa of the gastrointestinal tract and in the lamina propria of the respiratory tract. The T cell-dependent proliferation of MMC has been shown to be mediated by IL-3, IL-4, IL-9 and IL-10 (78). Second, there are T cell-independent connective tissue MC (CTMC), which occur mostly in the submucosa of the gastrointestinal tract, in the skin and in the peritoneum. Fibroblast-derived factors (e.g. SCF) appear to mediate the development of this subtype (78). In addition to their different growth factor dependencies, MMC and CTMC can be distinguished by size, proteoglycan and histamine content, and by the expression of different MC-specific serine proteases, which, according to their cleavage specificity, have been classified as chymases or tryptases. Members of the mouse chymase family include MMCP-1, MMCP-2, MMCP-4A, MMCP-4B, MMCP-L and MMCP-5. The two MC tryptases that have been characterized in mice are MMCP-6 and MMCP-7 (64).

MC subtypes in humans

In humans, the strict classification of MC into MMC and CTMC has not been possible. However, even to a minor extent, it is possible to define MC subtypes in humans according to different staining and fixation properties (79). Since the extensive investigations of Irani *et al.* (80, 81), the content of MC-specific proteases chymase and tryptase as the main criterion of differentiation for human MC has been established, and MC containing only tryptase (MC_T) can be distinguished from MC containing both tryptase and chymase (MC_{TC}). Tryptase is present in all MC identified by classic histochemistry, where metachromasia occurs with acid dyes, and it is therefore considered as a selective marker for human MC (82). Chymase positivity was mainly localized to human skin MC by immunofluorescence, but it was usually lost in formalin-fixed tissues. Weidner and Austen discovered a third MC type containing only chymase but no tryptase (MC_C), prevalently present in the submucosa of the gastrointestinal tract and in lymph nodes (83). However, the existence of this MC subtype is still controversial.

CHEMOKINE RECEPTORS ON MC

The presence of chemokine receptors on MC was first investigated by assessing the ability of different chemokines to induce histamine release by human MC in comparison to basophils. No histamine release was observed by lung, uterus or tonsil MC after activation with MCAF, IL-8, MIP-1α, MIP-1β, GRO, IP-10 and RANTES, whereas MCAF caused direct and dose-dependent histamine release by human basophils (84). Accordingly, no effects of RANTES, MCP-1, MCP-2, MCP-3, MIP-1α or MIP-1β on histamine release or changes in the cytosolic free calcium level by the human MC line HMC-1 were observed (85). However, MC haptotactic responses by unstimulated MC was found in response to both MCP-1 and RANTES, whereas IgE-activated MC showed significant migration to MCP-1, RANTES, PF4 and MIP-1α (86).

More recently, the expression of both types of IL-8 receptors (CXCR1 and CXCR2) were demonstrated by polymerase chain reaction (PCR) and flow cytometry on the human leukaemic MC line HMC-1. This finding was supported by a dose-dependent rise of cytosolic free calcium concentration and significant migratory response induced by both IL-8 and its homologues MGSA and NAP-2 (87). Using the RNAase protection assay, CXCR2 mRNA, but not CXCR1 mRNA, expression was detected in HMC-1 cells. The activity of this receptor was further explored using IL-8, which was found to induce dose-dependent chemotactic response in both HMC-1 cells and *in vitro* cultured human umbilical cord blood (UCB)-derived MC (88). On the other hand, MC recruitment was observed after RANTES injection *in vivo* in both rats and mice (89, 90), suggesting the existence of at least one of the different RANTES receptors (CCR1, CCR3, CCR4, CCR5, DARC) (1, 2, 4, 6) on rodent MC.

CCR3 is the Central Chemokine Receptor of Human MC

Two independently performed studies have recently demonstrated that CCR3, the receptor which binds ETX 1, 2 and 3, RANTES and MCP-2, 3 and 4 is the central chemokine receptor in human MC (29, 30). CCR3 was found to be expressed in human eosinophils (26, 78, 91), basophils (27) and Th2 cells (28, 92). Based on these findings, it has been suggested that the pathophysiologically relevant leukocytes share CCR3 and can be recruited concomitantly to sites of allergic inflammation by the same chemokines (2).

Ochi *et al.* (29) have assessed the expression of different chemokine receptors in human MCPr derived *in vitro* from UCB and during their differentiation into fully mature MC obtained by SCF and IL-6 as comitogenic cytokines, and with IL-10 added to suppress monocyte development. MCPr were found to express CCR3, CXCR4, CCR5 and CXCR2. These receptors each mediated intracellular calcium flux in human MCPr in response to their ligands (ETX, SDF-1, MIP-1α and IL-8, respectively), and both ETX and SDF-1 promoted chemotaxis in a dose-dependent fashion (29).

In our study, by using both immunohistochemistry and flow cytometry, we were unable to detect CCR3 expression on Th2 cells both *in vitro* and *in vivo*, in apparent discrepancy with the data reported by Sallusto *et al.* (28) and by Gerber *et al.* (92). However, by using the same antibody (7B11), we found CCR3 expression in circulating basophils and, at tissue level, not only in eosinophils, but also in a remarkable proportion of tryptase-positive cells. Although ETX did not induce histamine release by enriched

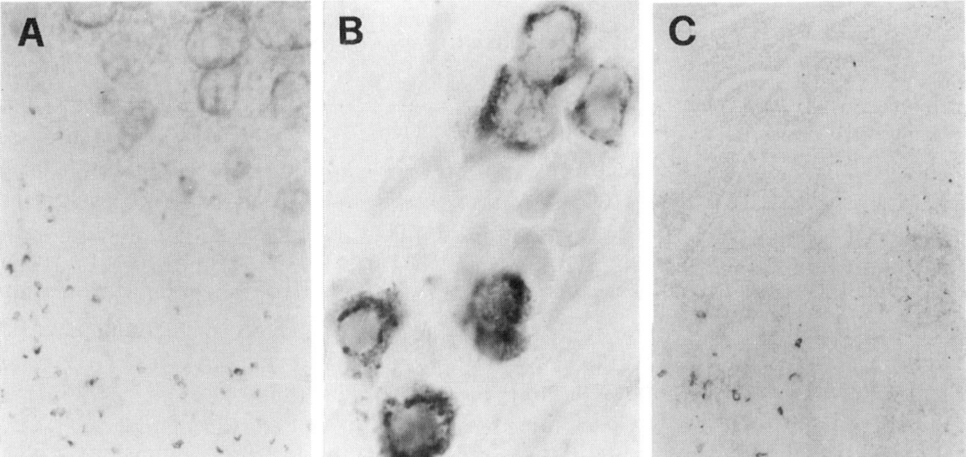

Fig. 1 Detection of CCR3-expressing cells in different human tissues. (A) High numbers of CCR3$^+$ cells (original magnification: ×100) in the intestinal submucosa of normal gut. (B) CCR3$^+$ cells at high-power magnification (original magnification: ×1000) in the intestinal submucosa. (C) A few CCR3$^+$ cells (original magnification: ×100) in the interstitium from normal lung. Immunohistochemical analysis was performed according to the technique described by Romagnani *et al.* (30).

suspensions of human lung MC, both ETX and RANTES exerted chemotactic activity on the same cells. The tissue distribution of CCR3$^+$ MC, as well as the co-expression of CCR3 and chymase, was investigated. Of note, the numbers of CCR3$^+$ MC were higher in the skin and in the intestinal submucosa than in intestinal mucosa and lungs (Fig. 1). More importantly, CCR3 expression was virtually a property of MC_{TC}, since virtually all CCR3$^+$ MC also showed chymase expression (30), suggesting that migration and/or persistence of this MC subset into non-inflamed and inflamed tissues may largely depend on its CCR3 expression.

Possible Role of CCR3 and Other Chemokine Receptors in MC Pathophysiology

The above-mentioned data provide evidence that chemokine receptors are expressed in human MC and that chemokines play a critical role in their development and/or tissue homing. Some chemokine receptors, such as CXCR4, CCR5 and CXCR2, are expressed only in MCPr and not in mature MC, suggesting the role of their ligands (SDF-1, RANTES and/or MIP-1α, β) in the distribution of MCPr under basal conditions, as well as their recruitment to diverse sites of inflammation. By contrast, CCR3 was found to be expressed on both MCPr derived from UCB progenitors *in vitro* and was the only chemokine receptor detectable in both *in vitro* differentiated mature MC and in MC present *in vivo* in normal, as well as inflamed, tissues. The observation that ETX elicited chemotaxis on 4-week-old MCPr but not on 9-week-old mature MC, where it caused sustained calcium flux, suggests that it may have functions for stationary MC within tissues that are distinct from its action on bloodborne MCPr (29). Thus, according to the model suggested by Ochi *et al.* (29), the transit of MCPr from the circulation to various sites within the tissues may be regulated by their expression of CXCR2, CCR3, CXCR4 and CCR5 and the local availability of the corresponding respective ligands. SDF-1 may act via CXCR4 at stromal level, MIP-1α via CCR5 at lung level and IL-8 via CXCR2 at skin level. On the other hand, interaction of ETX with CCR3, which is expressed throughout all phases of MC development, may provide an important stimulus for MCPr movement toward mucosal surfaces, as well as the basic mechanism for their tissue retention, which is compatible with the proposed critical role for ETX in allergic inflammation (29). The important role of ETX in MC development is in agreement with the results of experiments recently performed in mice, showing that ETX acts synergistically with SCF to accelerate the differentiation of embryonic MCPr via its interaction with CCR3. CCR3 was indeed found to be expressed at the level of fetal liver and yolk sac, and ETX in combination with SCF or fibroblasts alone promoted the differentiation of fetal MCPr into differentiated MC (93). Of note, the BM-derived MC co-cultured with SCF and fibroblasts phenotypically resembled CTMC, as revealed by their expression of mMCP-4 (94). Thus, ETX which is produced in high amounts by connective tissue, may be of particular importance in favouring not only the development of MC, but also their preferential homing into connective tissues. Of note, a significant increase in ETX has been observed in response to SCF production during fibroblast–MC interaction, and fibroblasts have been found to produce ETX in response to IL-4 stimulation (95), suggesting an amplification circuit among IL-4, fibroblasts, ETX and CTMC.

The results of our experiments, showing prevalent CCR3 expression in MC present in the skin and in the intestinal submucosa rather than in intestinal mucosa or in lungs, as well as the demonstration that *in vivo* CCR3 is virtually expressed only in human MC_{TC} (30), is consistent with the possibility that ETX–CCR3 interactions, not only favour MC development but also play an important role in the recruitment of the MC_{TC} subtype showing preferential homing at connective tissue level (Fig. 2).

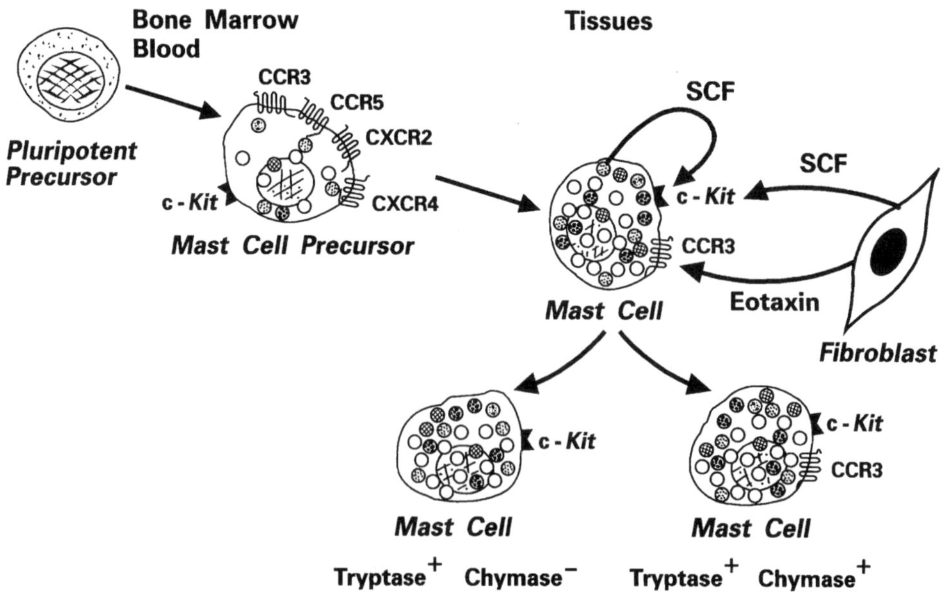

Fig. 2 Hypothetical view of the possible role of chemokine receptors, particularly CCR3, in the differentiation and/or homing of human MC. See text for explanations.

CONCLUDING REMARKS

Recent data indicate that human MC can express several chemokine receptors during their differentiation *in vitro* to mature MC, whereas CCR3 remains the unique chemokine receptor present in mature MC generated *in vitro*, and is easily detectable in a proportion of human MC present *in vivo* in different tissues. Of note, the number of $CCR3^+$ MC was higher in tissues showing elevated proportions of double-positive MC_{TC} and appeared to be characteristic of these latter rather than of MC showing only tryptase, but not chymase, expression. The reason why CCR3 is primarily expressed by MC_{TC}, which predominate in connective tissues rather than in the mucosa (64, 81, 83), is unknown, and the mechanisms that regulate the *in situ* differentiation of human MC are largely unclear.

Some years ago, Furitsu *et al.* (96) showed that prolonged co-culture of human UCB nucleated cells with skin fibroblasts results in the development of mature MC, of which the great majority were MC_{TC}, suggesting that fibroblasts not only facilitate the differentiation of MC precursors to mature MC, but also contribute to the determination of the MC_{TC} phenotype. Of note, murine MC which develop in response to the combined action of SCF and eotaxin exhibit the CTMC phenotype (94). Since eotaxin is constitutively expressed even in 'non-immune' tissues, such as intestine, skin and mammary gland (97), where the great majority of MC are MC_{TC} (64, 81, 83), it is tempting to speculate that ETX through its interaction with CCR3 plays an important role, not only in the preferential differentiation of MC precursors into the MC_{TC} phenotype, but also favours their migration to the connective tissues, as well as the maintenance of their differentiation pathway and/or survival. Furthermore, the demonstration that MC express CCR3 and therefore can respond to eotaxin and to other CCR3-interacting cytokines, may account for their upregulation in tissues known to be sites of allergic reactions such as the airways, where they usually associate with CCR3-expressing basophils and eosinophils. Accordingly, high ETX expression has recently been observed in the epithelium and submucosa of bronchial biopsies from patients with atopic asthma (98). Thus, the study of chemokine receptors on MC may open new avenues, not only for the identification of mechanisms involved in their differentiation and homing, but possibly also for a better understanding of their pathophysiological significance.

ACKNOWLEDGEMENT

The experiments reported in this paper have been performed with grants provided by ISS projects to S.R. and G.M.

REFERENCES

1. Rollins, B. J. Chemokines. *Blood* **90**:909–928, 1998.
2. Baggiolini, M. Chemokines and leukocyte traffic. *Nature* **392**:565–568, 1998.
3. Luster, A. D. Chemokines – chemotactic cytokines that mediate inflammation. *N. Engl. J. Med.* **338**:436–445, 1998.
4. Ward, S.G., Bacon, K., and Westwick, J. Chemokines and T lymphocytes: more than an attraction. *Immunity* **9**:1–11, 1998.

5. Mantovani, A., Allavena, P., Vecchi, A. and Sozzani, S. Chemokines and chemokine receptors during activation and deactivation of monocytes and dendritic cells and in amplification of Th1 versus Th2 responses. *Int. J. Clin. Lab. Res.* **28**:77–82, 1998.
6. Mantovani, A. The chemokine system: redundancy for robust outputs. *Immunol. Today* **20**:254–257, 1999.
7. Butcher, E. C. and Picker, L. Lymphocyte homing and homeostasis. *Science* **272**:60–66, 1996.
8. Nagasawa, T., Kikutani, H. and Kishimoto, T. Molecular cloning and structure of a pre B-cell growth factor. *Proc. Natl. Acad. Sci. USA* **91**:2305–2309, 1994.
9. Nagasawa, T., Hirota, S., Tachibana, K., Takakura, N., Nishikawa, S., Kitamura, Y., Yoshida, N., Kikutani, H. and Kishimoto, T. Defects of B-cell lymphopoiesis and bone marrow myelopoiesis in mice lacking the CXC chemokine PBSF/SDF-1. *Nature* **382**:635–638, 1996.
10. D'Apuzzo, M., Rolink, A., Loetscher, M., Hoxie, J. A., Clark-Lewis, I., Melchers, F., Baggiolini, M. and Moser, B. The chemokine SDF-1, stromal-cell derived factor 1, attracts early stage B cell precursors via the chemokine receptor CXCR4. *Eur. J. Immunol.* **27**:1788–1793, 1997.
11. Bleul, C. C., Farzan, M., Choe, H., Parolin, C., Clark-Lewis, I., Sodroski, J. and Springer, T. A. The lymphocyte chemoattractant SDF-1 is a ligand for LESTR/fusin and blocks HIV-1 entry. *Nature* **382**:829–833, 1996.
12. Oberlin, E. Amara, A., Bachelerie, F., Bessia, C., Virelizier, J.L., Arenzana-Seisdedos, F., Schwartz, O., Heard, J. M., Clark-Lewis, I., Legler, D. F., Loetscher, M., Baggiolini, M. and Moser, B. The lymphocyte chemoattractant SDF-1 is a ligand for LESTR/fusin and prevents infection by T-cell-line-adapted HIV-1. *Nature* **382**:833–835, 1996.
13. Chantry D., Romagnani P., Epp, A., Romagnani, S. and Gray, P. A. Macrophage-derived chemokine is localized to thymic medullary epithelial cells and is a chemoattractant for $CD3^+$, $CD4^+$, CD8 low thymocytes. *Blood* **94**:1890–1898, 1999.
14. Feng, Y., Broder, C. C., Kennedy, P. E. and Berger, E. A. HIV-1 entry cofactor: functional cDNA cloning of a seven-transmembrane, G protein-coupled receptor. *Science* **272**:872–877, 1996.
15. Vicari, A. P. Figueroa, D. J., Hedrick, J. A., Foster, J. S., Singh, K. P., Meonon, S., Copeland, N. G., Gilbert, D. J., Jenkins, N. A., Bacon, K. B. and Zlotnik, A. TECK: a novel CC chemokine specifically expressed by thymic dendritic cells and potentially involved in T cell development. *Immunity* **7**:291–301, 1997.
16. Legler, D. F., Loetscher, M., Roos, R. S., Clark-Lewis, I., Baggiolini, M., Moser, B. B-cell attracting chemokine 1, a human CXC chemokine expressed in lymphoid tissues, selectively attracts B lymphocytes via BLR1/CXCR5. *J. Exp. Med.* **187**:655–660, 1998.
17. Gunn, M. D., Ngo, V. N., Ansel, K. M., Ekland, E. H., Cyster, J. G. and Williams, L. T. A B-cell homing chemokine made in lymphoid follicles activated Burkitt's lymphoma receptor-1. *Nature* **391**:799–803, 1998.
18. Forster, R. *et al.* A putative chemokine receptor BLR1, directs B cell migration to defined lymphoid organs and specific anatomic compartments of the spleen. *Cell* **87**:1037–1047, 1996.
19. Springer, T. A. Traffic signals for lymphocyte recirculation and leukocyte emigration: the multistep paradigm. *Cell* **76**:301–314, 1994.
20. Baggiolini, M., Dewald, B. and Moser, B. Interleukin-8 and related chemotactic chemokines – CXC and CC chemokines. *Adv. Immunol.* **55**:97–179, 1994.
21. Garcia-Zepeda, E. A., Rothenberg, M. E., Ownbey, R. T., Celestin, J., Leder, P., Luster, A. D. Human eotaxin is a specific chemoattractant for eosinophil cell and provides a new mechanism to explain tissue eosinophilia. *Nat. Med.* **2**:449–456, 1996.
22. Garcia-Zepeda, E. A., Combadiere, C., Rothenberg, M. E., Sarafi, M. N., Lavigne, F., Hamid, Q., Murphy, P. M. and Luster A. D. Human monocyte chemoattractant protein (MCP)-4 is a novel CC chemokine with activities on monocytes, eosinophils and basophils induced in allergic and non-allergic inflammation that signals through the CC chemokine receptor (CCR)-2 and -3. *J. Immunol.* **157**:5613–5626, 1996.
23. Chollet-Martin, S., Montravers, P., Gilbert, C., Elbim, C., Desmonts, J. M., Fagon, J. Y. and Gougerot-Pocidalo, M. A. High levels of interleukin-8 in the blood and alveolar spaces of patients with pneumonia and adult respiratory distress syndrome. *Infect. Immun.* **61**:4453–4459, 1993.
24. Lahrtz, F., Piali, L., Nadal, D., Pfister, H. W., Spanaus, K. S., Baggiolini, M. and Fontana, A. Chemotactic activity on mononuclear cells in the cerebrospinal fluid of patients with viral meningitis is mediated by interferon-gamma inducible protein-10 and monocyte chemotactic protein-1. *Eur. J. Immunol.* **27**:2484–2489, 1997.

25. Gottlieb, A. B., Luster, A. D., Posnett, D. N. and Carter, D. M. Detection of a gamma interferon-induced protein IP-10 in psoriatic plaques. *J. Exp. Med.* **168**:941–948, 1988.
26. Ponath, P. D., Qin, S. X., Post, T. W., Wang, J., Wu, L., Gerard, N. P., Newman, W., Gerard, C. and Mackay, C. R. Molecular cloning and characterization of a human eotaxin receptor expressed selectively on eosinophils. *J. Exp. Med.* **183**:2437–2448, 1997.
27. Uguccioni, M., Mackay, C. R., Ochensberger, B., Loetscher, P., Rhis, S., LaRosa, G. J., Rao, P., Ponath, P. D., Baggiolini, M. and Dahinden, C. A. High expression of the chemokine receptor CCR3 in human blood basophils. Role in activation by eotaxin, MCP-4 and other chemokines. *J. Clin. Invest.* **100**:1137–1143, 1997.
28. Sallusto, F., Mackay, C. R. and Lanzavecchia, A. Selective expression of the eotaxin receptor CCR3 by human T helper 2 cells. *Science* **277**:2005–2007, 1997.
29. Ochi, H., Hirani, W. M., Yuan, Q., Friend, D., Austen, F. K. and Boyce, J. A. T helper cell type 2 cytokine-mediated comitogenic responses and CCR3 expression during differentiation of human mast cells *in vitro*. *J. Exp. Med.* **190**:267–280, 1999.
30. Romagnani, P., De Paulis, A., Beltrame, C., Annunziato, F., Dente, V., Maggi, E., Romagnani, S. and Marone, G. Tryptase-chymase double positive human mast cells express the eotaxin receptor (CCR3) and are attracted by CCR3-binding chemokines. *Am. J. Pathol.* **155**:1195–1204, 1999.
31. Strieter, R. M., Polverini, P. J., Kunkel, S. L., Arenberg, D. A., Burdick, M. D., Kasper, J., Dzuiba, J., Van, D. J., Walz, A., Marriott, D., Chan, S. Y., Roczniak, S. and Shanafelt, A. B. *J. Biol. Chem.* **270**:27348–27357, 1995.
32. Strieter, R. M., Polverini, P. J., Arenberg, D. A., Walz, A., Opdenakker, G., Van, D. J. and Kunkel, S. L. Role of C-X-C chemokines as regulators of angiogenesis in lung cancer. *J. Leukoc. Biol.* **57**:752–762, 1995.
33. Angiolillo, A. L., Sgadari, C., Taub, D. D., Liao, F., Farber, J. M., Maheshwari, S., Kleinman, H. K., Reaman, G. H. and Tosato, G. Human interferon-inducible protein 10 is a potent inhibitor of angiogenesis *in vivo*. *J. Exp. Med.* **182**:155–162, 1995.
34. Gengrinovitch, S., Greenberg, S. M., Cohen, T., Gitay, G. H., Rockwell, P., Maione, T. E., Levi, B. Z. and Neufeld, G. Platelet factor-4 inhibits the mitogenic activity of VEGF121 and VEGF165 using several concurrent mechanisms. *J. Biol. Chem.* **270**:15059–15065, 1995.
35. Smith, D. R., Polverini, P. J., Kunkel, S. L., Orringer, M. B., Whyte, R. I., Burdick, M. D., Wilke, C. A. and Strieter, R. M. Inhibition of interleukin-8 attenuates angiogenesis in bronchogenic carcinoma. *J. Exp. Med.* **179**:1409–1415, 1994.
36. Yoshimura, T., Robinson, E. A., Tanaka, S., Appella, E., Kuratsu, J. and Leonard, E. L. Purification and aminoacid analysis of two human glioma-derived monocyte chemoattractants. *J. Exp. Med.* **169**:1449–1459, 1989.
37. Loetscher, P., Seitz, M., Baggiolini, M. and Moser, B. Interleukin-2 regulates CC chemokine receptor expression and chemotactic responsiveness in T lymphocytes. *J. Exp. Med.* **184**:569–577, 1996.
38. Loetscher, P., Uguccioni, M., Bordoli, L., Baggiolini, M., Moser, B., Chizzolini, C. and Dayer, J. M. CCR5 is characteristic for Th1 lymphocytes. *Nature* **391**:344–345, 1998.
39. Bonecchi, R., Bianchi, G., Panina-Bordignon, P., D'Ambrosio, D., Lang, R., Borsatti, A., Sozzani, S., Allavena, P., Gray, P. A., Mantovani, A. and Sinigaglia, F. Differential expression of chemokine receptors and chemotactic responsiveness of type 1 T helper cells (Th1s) and Th2s. *J. Exp. Med.* **187**:129–134, 1998.
40. Fauci, A. S. Host factors and the pathogenesis of HIV-induced disease. *Nature* **384**:529–533, 1996.
41. Galli, G., Annunziato, F., Mavilia, C., Romagnani, P., Cosmi, L., Manetti, R., Pupilli, C., Maggi, E. and Romagnani, S. Enhanced HIV expression during Th2-oriented responses explained by the opposite regulatory effect of IL-4 and IFN-γ on fusin/CXCR4. *Eur. J. Immunol.* **28**:3280–3290, 1998.
42. Annunziato, F., Galli, G., Nappi, F., Cosmi, L., Manetti, R., Maggi, E., Ensoli, B. and Romagnani, S. Limited expression of R5-tropic HIV-1 in CCR5-positive type-1 ploarized T cells explained by their ability to produce rantes, MIP-1α, and MIP-1β. *Blood* **95**:1167–1174, 2000.
43. Cocchi, F., DeVico, A. L., Garzino-Derno, A., Arya, S. K., Gallo, R. C. and Lusso, P. Identification of RANTES, MIP-1α and MIP-1β as the major HIV-suppressive factors produced by CD8$^+$ T cells. *Science* **270**:1811–1815, 1995.
44. Samson, M., Libert, F., Doranz, B. J., Rucker, J., Liesnard, C., Farber, C. M., Saragosti, S., Lapoumeroulie, C., Cognaux, J., Forceille, C., Muyldermans, G., Verhofstede, C., Burtonboy, G., Georges, M., Imai, T., Rana, S., Yi, Y., Smith, R. J., Collman, R. G., Doms, R. W., Vassart, G. and Parmentier, M. Resistance to HIV-1 infection in caucasian individuals bearing mutant alleles of the CCR-

5 chemokine receptor gene. *Nature* **382**:722–725, 1996.
45. Dean, M., Carrington, M., Winkler, C., Huttley, G. A., Smith, M. W., Allikmets, R., Goedert, J. J., Buchbinder, S. P., Vittinghoff, E., Gomperts, E., Donfield, S., Vlahov, D., Kaslow, R., Saah, A., Rinaldo, C., Detels, R. and O'Brien, S. J. Genetic restriction of HIV-1 infection and progression to AIDS by a deletion allele of the CKR5 structural gene. Hemophilia Growth and Development Study, Multicenter Hemophilia Cohort Study, San Francisco City Cohort, Alive Study. *Science* **273**:1856–1862, 1996.
46. Liu, R., Paxton, W. A., Choe, S. Ceradini, D., Martin, S. R., Horuk, R., MacDonald, M. E., Stuhlmann, H., Koup, R. A. and Landau, N. R. Homozygous defect in HIV-1 coreceptor accounts for resistance of some multiply-exposed individuals to HIV-1 infection. *Cell* **86**:367–377, 1996.
47. Neote, K., Darbonne, W., Ogez, J., Oruk, R. and Schall, T. J. Identification of a promiscuous inflammatory peptide receptor on the surface of red blood cells. *J. Biol. Chem.* **268**:12247–12249, 1993.
48. Horuk, R., Chitnis, C. E., Darbonne, W. C., Colby, T. J., Rybicki, A., Hadley, T. J. and Miller, L. H. A receptor for the malarial parasite *Plasmodium vivax*: the erythrocyte chemokine receptor. *Science* **261**:1182–1184, 1993.
49. Ahuja, S. K. and Murphy, P. M. Molecular piracy of mammalian interleukin-8 receptor type B by *Herpesvirus saimiri. J. Biol. Chem.* **270**:20691, 1993.
50. Neote, K., Di Gregorio, D., Mak, J. Y., Horuk, R. and Schall, T. J. Molecular cloning, functional expression, and signaling characteristics of a C-C chemokine receptor. *Cell* **72**:415–425, 1993.
51. Gao, J. L. and Murphy, P. M. Human cytomegalovirus open reading frame US28 encodes a functional beta chemokine receptor. *J. Biol. Chem.* **268**:28539–28542, 1994.
52. Sozzani, S., Luini, W., Bianchi, G., Allavena, P., Wells, T. N., Napolitano, M., Bernardini, G., Vecchi, A., D'Ambrosio, D., Mazzeo, D., Sinigaglia, F., Santoni, A., Maggi, E., Romagnani, S. and Mantovani, A. The viral chemokine macrophage inflammatory protein-II is a selective Th2 chemoattractant. *Blood* **92**:4036–4039, 1998.
53. Murphy, P. M. The molecular biology of leukocyte chemoattractant receptors. *Annu. Rev. Immunol.* **12**:593–633, 1994.
54. Baggiolini, M., Dewald, B. and Moser, B. Human chemokines: an update. *Annu. Rev. Immunol.* **15**:675–705, 1997.
55. Tilton, B., Andjelkovic, M., Didichenko, S. A., Hemmings, B. A. and Thelen, M. G-protein coupled receptors and Fcγ-receptors mediate activation of Akt/protein kinase B in human phagocytes. *J. Biol. Chem.* **272**:28096–28101, 1997.
56. Bokoch, G. M. Chemoattractant signaling and leukocyte activation. *Blood* **86**:1649–1660, 1995.
57. Laudanna, C., Campbell, J. J. and Butcher, E. C. Role of Rho in chemoattractant-activated leukocyte adhesion through integrins. *Science* **271**:981–983, 1996.
58. Kuziel, W. A., Morgan, S. J., Dawson, T. C., Griffin, S., Smithies, O., Ley, K. and Maeda, N. Severe reduction in leukocyte adhesion and monocyte extravasation in mice deficient in CC chemokine receptor 2. *Proc. Natl. Acad. Sci. USA* **94**:12053–12058, 1997.
59. Kurihara, T., Warr, G., Loy, J. and Bravo, R. Defects in macrophage recruitment and host defense in mice lacking the CCR2 chemokine receptor. *J. Exp. Med.* **186**:1757–1762, 1997.
60. Boring, L., Gosling, J., Chensue, S. W., Kunkel S. L., Farese R. V. Jr, Broxmeyer H. E. and Charo I. F. Impaired monocyte migration and reduced type 1 (Th1) cytokine responses in C-C chemokine receptor 2 knockout mice. *J. Clin. Invest.* **100**:2552–2561, 1997.
61. Caughey, G. H. The structure and airway biology of mast cell proteinases. *Am. J. Respir. Cell Mol. Biol.* **4**:387–394, 1991.
62. Benyon, C. The human skin mast cell. *Clin. Exp. Allergy* **19**:375–387, 1989.
63. Huntley, J. F. Mast cells and basophils: a review of their heterogeneity and function. *J. Comp. Pathol.* **107**:349–372, 1992.
64. Welle, M. Development, significance, and heterogeneity of mast cells with particular regard to the mast cell-specific proteases, chymase and tryptase. *J. Leukoc. Biol.* **61**:233–245, 1997.
65. Folkman, J. How is blood vessel growth regulated in normal and neoplastic tissue? *Cancer Res.* **46**:467–473, 1986.
66. Norrby, K., Jacobsson, A. and Sorbo, J. Mast cell-mediated angiogenesis: a novel experimental model using the rat mesentery. *Virchows Arch. Ser. B* **52**:195–206, 1986.
67. Burnet, F. M. The probable relationship of some or all mast cells to the T-cell system. *Cell. Immunol.* **30**:358–360, 1977.
68. Zucker-Franklin, D., Grusky, G., Hirayama, N. and Schnipper, E. The presence of mast cell precursors in rat peripheral blood. *Blood* **58**:544–551, 1981.

69. Czarnetzky, B. M., Sterry, W., Bazin, H. and Kalveram, K. J. Evidence that tissue mast cells derive from mononuclear phagocytes. *Int. Arch. Allergy Appl. Immunol.* **67**:44–48, 1982.
70. Sonoda, T., Kitamura, Y., Haku, Y., Hara, H. and Mori, K. J. Mast-cell precursors in various haemopoietic colonies of mice produced *in vivo* and *in vitro*. *Br. J. Haematol.* **53**:611–620, 1983.
71. Hatanaka, K., Kitamura, Y. and Nishimune, Y. Local development of mast cell from bone marrow-derived precursors in the skin of mice. *Blood* **53**:142–147, 1979.
72. Hayashi, C., Sonoda, T., Nakano, T., Nakayama, H. and Kitamura, Y. Mast-cell precursors in the skin of mouse embryos and their deficiency in embryos of Sl/Sld genotype. *Dev. Biol.* **109**:234–241, 1985.
73. Kitamura, Y. and Fujita, J. Regulation of mast cell differentiation. *BioEssays* **10**:193–196, 1989.
74. Schrader, J. W. The *in vitro* production and cloning of the P cell, a bone marrow-derived null cell that expresses H-2 and Ia-antigens, has mast cell granules, and is regulated by a factor released by activated T cells. *J. Immunol.* **126**:452–458, 1981.
75. Ihle, J. H., Keller, J., Oroszlan, S., Henderson, L. E., Copeland, T. D., Fitch, F., Prystowsky, M. B., Goldwasser, E., Schrader, J. W., Palaszynski, E., Dy, M. and Lebel, B. Biological properties of homogeneous interleukin-3. I. Demonstration of WEHI-3 growth factor activity, mast cell growth factor activity, P cell-stimulating factor activity, colony-stimulating factor activity, and histamine-producing cell-stimulating factor activity. *J. Immunol.* **131**:282–287, 1983.
76. Thompson-Snipes, L., Dhar, V., Bond, M. W., Mosmann, T. R., Moore, K. W. and Rennkk, D. M. Interleukin-10: a novel stimulatory factor for mast cells and their progenitors. *J. Exp. Med.* **173**:507–510, 1991.
77. Yuan, Q., Gurish, F., Friend, D. S., Austen, K. F. and Boyce, J.A. Generation of a novel stem cell factor-dependent mast cell progenitor. *J. Immunol.* **161**:5143–5146, 1998.
78. Kitamura, M., Nakajima, T., Imai, T., Harada, S., Combadiere, C., Tiffany, H. L., Murphy, P. M. and Yoshie, O. Molecular cloning of human eotaxin, an eosinophil-selective CC chemokine and identification of a specific eosinophil eotaxin receptor. *J. Biol. Chem.* **271**:7725–7730, 1996.
79. Strobel, S., Miller, H. R. P. and Ferguson, A. Human intestinal mucosal mast cells: evaluation of fixation and staining techniques. *J. Clin. Pathol.* **34**:851–858, 1981.
80. Irani, A.-M., Schechter, N. M., Craig, S. S., DeBlois, G. and Schwartz, L. B. Two types of human mast cells that have distinct neutral protease composition. *Proc. Natl. Acad. Sci. USA* **83**:4464–4468, 1986.
81. Irani, A.-M. and Schwartz, L. B. Human mast cell heterogeneity. *Allergy Proc.* **15**:303–308, 1994.
82. Castells, M. C., Irani, A.-M., and Schwartz, M. A. Evaluation of human peripheral blood leukocytes from mast cell tryptase. *J. Immunol.* **138**:2184–2189, 1987.
83. Weidner, N. and Austen, K. F. Heterogeneity of mast cells at multiple body sites. Fluorescent determination of avidin binding and immunofluorescent determination of chymase, tryptase, and carboxypeptidase content. *Pathol. Res. Pract.* **189**:156–162, 1993.
84. Fureder, W., Agis, H., Semper, H., Keil, F., Maier, U., Muller, M. R., Czerwenka, K., Hofler, H., Lechner, K. and Valent, P. Differential response of human basophils and mast cells to recombinant chemokines. *Ann. Hematol.* **70**:251–258, 1995.
85. Hartmann, K., Beiglbock, F., Czarnetzki, B. M. and Zuberbier, T. Effect of CC chemokines on mediator release from human skin mast cells and basophils. *Int. Arch. Allergy Immunol.* **108**:224–230, 1995.
86. Taub, D., Dastych, J., Inamura, U., Upton, J., Kelvin, D., Metcalfe, D. and Oppenheim J. Bone marrow-derived murine mast cells migrate, but do not degranulate, in response to chemokines. *J. Immunol.* **154**:2393–2402, 1995.
87. Lippert, U., Artuc, M., Grutzkau, A., Moller, A., Kenderessy-Szabo, A., Schadendorf, D., Norgauer, J., Hartmann, K., Schweitzer-Stenner, R., Zuberbier, T., Henz, B. M. and Kruger-Krasagakes, S. Expression and functional activity of the IL-8 receptor type CXCR1 and CXCR2 on human mast cells. *J. Immunol.* **161**:2600–2608, 1998.
88. Nilsson, G., Mikovits, J. A., Metcalfe, D. D. and Taub, D. D. Mast cell migratory response to interleukin-8 is mediated through interaction with chemokine receptor CXCR2/interleukin-8RB. *Blood* **93**:2791–2797, 1999.
89. Conti, P., Reale, M., Barbacane, R.C., Felaco, M., Grilli, A. and Theoharides, T. C. Mast cell recruitment after subcutaneous injection of RANTES in the sole of the rat paw. *Br. J. Haematol.* **103**:798–803, 1998.
90. Conti, P., Reale, M., Barbacane, R. C., Letourneau, R. and Theoharides, T. C. Intramuscular injection of hRANTES causes mast cell recruitment and increased transcription of histidine decarboxylase in mice: lack of effects in genetically mast cell-deficient W/WV mice. *FASEB J.* **12**:1693–1700, 1998.
91. Daugherty, B. L., Siciliano, S. J., DeMartino, J. A., Malkowitz, L., Sirotina, A. and Springer, M. S. Cloning, expression, and characterization of the human eosinophil eotaxin receptor. *J. Exp. Med.*

183:2349–2354, 1996.

92. Gerber, B. O., Zanni, M. P., Uguccioni, M., Loescher, P., Mackay, C. R., Pichler, W. J., Yawalkar, N., Baggiolini, M. and Moser, B. Functional expression of the eotaxin receptor CCR3 in T lymphocytes co-localizing with eosinophils. *Curr. Biol.* **7**:836–843, 1997.
93. Quackenbush, E. J., Wershil, B. K., Aguirre, V. and Gutierrez-Ramos, J.-C. Eotaxin modulates myelopoiesis and mast cell development from embryonic hematopoietic progenitors. *Blood* **92**:1887–1897, 1998.
94. Jippo, T., Tsujino, K., Kim, H.-M., Kim, D.-K., Lee, Y.-M., Nawa, Y. and Kitamura, Y. Expression of mast-cell-specific proteases in tissues of mice studied by *in situ* hybridization. *Am. J. Pathol.* **150**:1373–1382, 1997.
95. Hogaboam, C., Kunkel, S. L., Strieter, R. M., Taub, D. D., Lincoln, P., Standiford, T. J. and Lukacs, N. W. Novel role of transmembrane SCF for mast cell activation and eotaxin production in mast cell–fibroblast interactions. *J. Immunol.* **160**:6166–6171, 1998.
96. Furitsu, T., Saito, H., Dvorak, A. M., Schwartz, L. B., Irani, A.-M., Burdick, J. F., Ishizaka, K. and Ishizaka, T. Development of human mast cells *in vitro*. *Proc. Natl. Acad. Sci. USA* **86**:10039–10043, 1989.
97. Rothenberg, M. E., Luster, A. D. and Leder, P. Murine eotaxin: an eosinophil chemoattractant inducible in endothelial cells and in interleukin-4-induced tumor suppression. *Proc. Natl. Acad. Sci. USA* **92**:8960–8964, 1995.
98. Ying, S., Robinson, D. S., Meng, Q., Rottman, J., Kennedy, R., Ringler, D. J., Mackay, C. R., Daugherty, B. L., Springer, M. S., Durham, S. R., Williams, T. J. and Kay, A. B. Enhanced expression of eotaxin and CCR3 mRNA and protein in atopic asthma. Association with airways hyperresponsiveness and predominant colocalization of eotaxin mRNA to bronchial epithelial and endothelial cells. *Eur. J. Immunol.* **27**:3507–3516, 1997.

CHAPTER 37

Release and Cleavage of Stem Cell Factor by Human Mast Cells

AMATO DE PAULIS,[1] GIUSEPPINA MINOPOLI,[2] ELOISA ARBUSTINI,[4] GENNARO DE CRESCENZO,[1] FABRIZIO DAL PIAZ,[3] PIERO PUCCI,[3] TOMMASO RUSSO[2] and GIANNI MARONE[1]*

[1]*Division of Clinical Immunology and Allergy,*
[2]*Department of Biochemistry and Medical Biotechnology,*
[3]*International Mass Spectrometry Facility Center, University of Naples Federico II, Naples, and*
[4]*Department of Pathology, University of Pavia, Pavia, Italy*

INTRODUCTION

Stem cell factor (SCF) is the most important cytokine in the regulation of human mast cell growth, proliferation (1–5) and functions *in vitro* (6–10) and *in vivo* (11). SCF is produced by fibroblasts (12, 13), stromal cells (14, 15), keratinocytes (16, 17) and endothelial cells (15, 17–19). It can also be produced by neuroblastoma cells (20), and by some tumour cell lines (21). SCF binds the c-*kit* receptor (c-*kit*R), which is a member of the receptor family for the platelet-derived growth factor (PDGFR) and the macrophage colony-stimulating factor (CSF-1R) (22, 23). The c-*kit* product is selectively expressed on rodent (24–30) and human mast cells (1, 2, 4, 6, 7), on melanocytes (31) and on a small fraction of progenitor cells present in bone marrow (2, 3), fetal liver (1), and cord blood mononuclear cells (4). SCF acts synergistically with other haematopoietic growth factors to stimulate the growth and differentiation of a variety of progenitors including human mast cell progenitors (1, 2, 5).

The gene encoding SCF resides at the *steel* locus (*Sl*) on human chromosome 12 (32). The *Sl* gene encodes a primary translation product of 248 amino acids with a leader sequence, followed by three domains: extracellular, transmembrane and cytoplasmic (12, 33). This protein contains a proteolytic cleavage site encoded by the exon 6 sequence, and post-translational processing at this site leads to the secretion of a 165-amino acid, biologically active protein (34). An alternatively spliced cDNA form encodes a smaller 220-amino acid polypeptide that lacks the exon 6 sequence, including the proteolytic cleavage sites, and hence results in a membrane-bound protein (12).

There is still some controversy concerning the release of various cytokines from human mast cells immunologically activated *in vitro*. Although it was demonstrated that murine mast cell lines produce a number of cytokines (35, 36), including interleukin-4

* Corresponding author.

MAST CELLS AND BASOPHILS
ISBN 0-12-473335-2

(IL-4), which resembled the profile of those made by activated Th2 lymphocytes, those findings have been difficult to duplicate in *in vitro* studies using mast cells isolated from human tissues. Church, Holgate and their co-workers reported that human mast cells contain and release IL-4 (37); however, in other studies neither IL-4 mRNA nor protein has been observed, even though functionality has been demonstrated by IgE-dependent histamine release. This failure to observe significant cytokine generation by human mast cells may reflect the action of biochemical circuits in which these cytokines can be rapidly degraded by chymase and other mast cell proteases.

Using immunogold staining of human heart tissue, we provided the first evidence that the secretory granules of cardiac mast cells (HHMC) store SCF (9). Recently, Longley *et al.* (38) demonstrated that human mast cell chymase, a chymotrypsin-like protease (39), also present in the secretory granules of human mast cells (8, 10, 39, 40), cleaves SCF at the peptide bond between Phe158 and Met159 (38), which are encoded by exon 6 of the SCF gene (34). More recently, constitutive synthesis of SCF mRNA was demonstrated in human mast cells by reverse transcriptase–polymerase chain reaction (RT–PCR) (41).

In this brief review we discuss our recent findings that the secretory granules of human mast cells store SCF and we analyse the potential role of chymase, a chymotrypsin-like protease, also present in the secretory granules of human mast cells, in the cleavage process of SCF.

STEM CELL FACTOR EXPRESSION IN HUMAN MAST CELLS

SCF in Skin Mast Cells of Patients with Systemic Mastocytosis

We have previously demonstrated the presence of SCF in the secretory granules of HHMC from patients with idiopathic and ischaemic cardiomyopathy (9). By using RT–PCR, Roche and his collaborators demonstrated that human mast cells isolated from skin and lung tissues constitutively express SCF mRNA (41). Dilated cardiomyopathies (9) and systemic mastocytosis (16, 17, 42) are characterized by local or systemic mast cell hyperplasia. More recently, we extended the previous observations by demonstrating the presence of SCF in secretory granules of mast cells from patients with mastocytosis and in human lung mast cells (HLMC) (43).

In our study, skin tissue from four patients with mastocytosis and eight preparations of human lung parenchyma from patients undergoing thoracotomy were studied by electron microscopy and the immunogold technique. The ultrastructural localization of SCF in human skin mast cells (HSMC) was examined by the immunogold procedure using the mAb 7H6 against region 79–97 of human SCF (44). Gold particles were present throughout the secretory granules of HSMC but not in the perigranular cytoplasm (Fig. 1). Gold particles were also found throughout the secretory granules of mast cells from the skin tissue of individuals undergoing cosmetic surgery. Similar results were obtained with rabbit and sheep polyclonal antibodies against multiple epitopes of SCF and a different mAb (hkl-12) against region 150–154 of SCF (17). Human skin mast cells incubated with a murine myeloma against an irrelevant antigen at concentrations similar to those of the anti-SCF antibody showed no gold particles over secretory granules.

These findings suggest that SCF is present *in situ* in secretory granules of HSMC. Similar results were obtained in preparations of human lung parenchyma examined for SCF expression in HLMC.

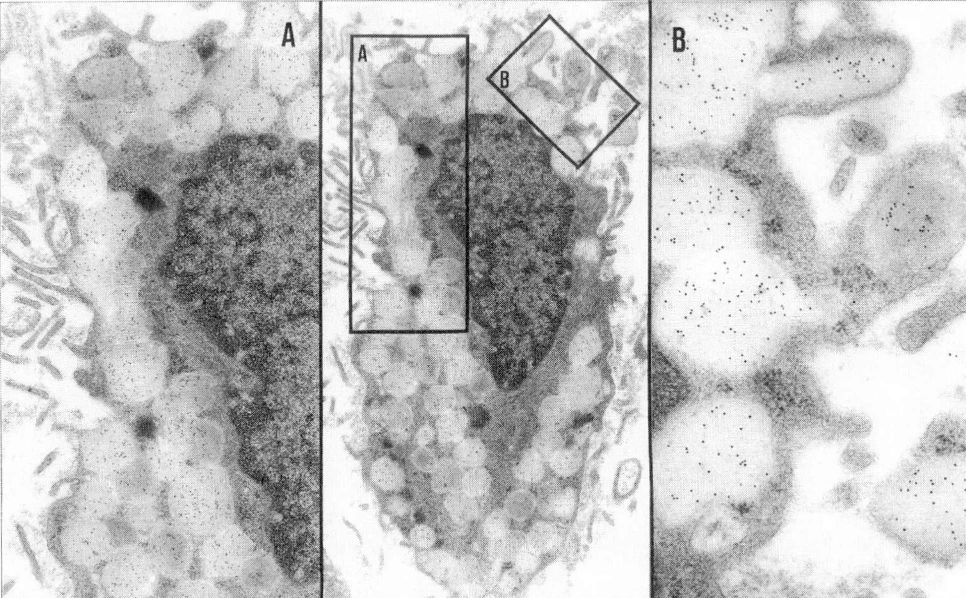

Fig. 1 Immunogold staining for SCF of a human skin mast cell from a patient with systemic mastocytosis. Sections were stained for SCF (mAb 7H6 anti-human SCF). SCF-locating gold particles are present over all secretory granules of the mast cell (uranyl acetate and lead citrate stained; original magnification, ×20,300). Inset A shows the early phases of exocytosis at higher magnification. Secretory granules are still in the cell, but the cell membrane is partly lost at the outer surface of the cell. Inset B shows, at high magnification, that extruded secretory granules retain the specific immunocytochemical localization of SCF.

Measurement of SCF Immunoreactivity (iSCF) in Human Lung Mast Cells

A significant difference was found between mast cells and circulating basophils, the only two cells known to express $Fc_\varepsilon RI$ and to synthesize histamine (25, 26, 45), with respect to SCF. In fact, iSCF was undetectable in cell lysates of purified basophils and could not be identified by the immunogold technique.

In contrast, iSCF was detected in cell lysates of highly purified preparations of HLMC. In a series of seven experiments the concentration of iSCF in lysates of HLMC was ~50.2 pg per 10^6 cells. iSCF was undetectable in lysates of five preparations of purified basophils. Interestingly, SCF has marginal effects on human basophils (7). Thus, these findings confirm and extend the many immunological and biochemical differences between human basophils and mast cells (45) also in terms of cytokines synthesized and released (46–51).

Kinetics of Histamine and iSCF Release from Immunologically Activated HLMC

After demonstrating the presence of SCF in HLMC granules, we devised experiments to evaluate whether this cytokine could be immunologically released together with other mediators. To this end, mast cells isolated and purified from lung parenchyma (HLMC) were challenged *in vitro* with an optimal concentration (1 μg ml^{-1}) of anti-IgE to compare the release of histamine and iSCF. The histamine release induced by maximal stimulation with anti-IgE was complete within 1 min. The kinetics reached a plateau after 1 min and remained unchanged for 60–120 min. In contrast, iSCF concentration showed a peak after 3–15 min, and progressively declined between 30 and 120 min. Similar results were obtained in experiments in which purified HLMC were stimulated *in vitro* with anti-$Fc_{\varepsilon}RI$.

These data are compatible with the hypothesis that iSCF, present in human mast cells, can be rapidly released by immunologically challenged HLMC, along with the preformed mediator histamine. Histamine is stable in the supernatants of anti-IgE-activated HLMC, while iSCF declines rapidly. This might be due to prompt internalization of SCF bound to its cognate receptor, c-*kit* (52), or to rapid hydrolysis by proteolytic enzymes, suggesting that the immunological activation of human mast cells also results in enzymatic activity capable of degrading iSCF. In recent years we have provided evidence in support of the latter hypothesis (43).

STEM CELL FACTOR–PROTEASE INTERACTIONS

Mast Cell Proteases

Proteases are the major secretory products of the mast cell on a weight basis. In humans, mast cells contain, in the secretory granules, a cumulative total of up to 60 pg of proteolytic enzymes (53). This compares with just 1–2 pg of histamine. Several specific mast cell proteases have been purified and their cDNA sequences derived: tryptase, chymase and carboxypeptidase. These proteases have been implicated in defence against helminthic parasites, allergic reactions, cardiovascular diseases and chronic inflammatory diseases (8, 54).

Chymase, a monomeric serine protease of 30 kDa (55), is stored in the same secretory granules as those that contain tryptase (56), but it is released from mast cells in a macromolecular complex distinct from that of tryptase (57). Chymase is stored in a complex together with proteoglycans. Like tryptase, chymase is present in a catalytically active form in the granules (57), but, in the acidic conditions therein, appears to have relatively little activity (58). Chymase is secreted in a fully active form (59). The contribution of bound proteoglycans to the regulation of chymase activity remains unclear. Chymase is inhibited by the circulating inhibitors α_1-anti-chymotrypsin and α_1-anti-trypsin (55, 60).

Detailed analyses of human, rodent and dog chymases substrate specificities have revealed an extended substrate binding site with a preference for phenylalanine at residue P_1 and for hydrophobic amino acid at residues P_2 and P_3 (61, 62). A number of small, naturally occurring peptides with potential cleavage sites fulfilling these criteria are cleaved by chymase. Among these, angiotensin I is efficiently hydrolysed at the Phe8–Phe9 bond, forming the vasoconstrictor and cardiac inotrope angiotensin II (55, 62). Chymase also selectively cleaves big-ET-1, -2 and -3 at their Tyr31–Gly32 bonds, and produces novel bioactive 31-amino acid length ET^{1-31}, without any further degradation products (63). Chymase has been implicated in the induction of microvascular leakage and neutrophil accumulation, the stimulation of mucus secretion, and the activation or inactivation of cytokines (58).

Longley *et al.* (38) demonstrated that recombinant human SCF ($rhSCF^{1-166}$) is rapidly cleaved to SCF^{1-159} and to a septapeptide by human mast cell chymase. We have confirmed their findings by showing that chymase selectively and rapidly cleaves SCF^{1-166} into two components (43).

We submitted rhSCF$^{1-166}$ to limited proteolysis by recombinant human chymase. To identify protease-sensitive sites within the rhSCF$^{1-166}$ molecule, enzymatic hydrolysis was performed under strictly controlled conditions to ensure maximal stability of the protein conformation and to address protease action toward specific sites, as previously reported (64). rhSCF$^{1-166}$ was incubated with human chymase in an appropriate enzyme-to-substrate ratio (E:S) and the process was monitored on a time-course basis by sampling the incubation mixture at intervals and analysing samples by HPLC. Fragments released from the recombinant protein were identified by electron spray mass spectrometry leading to the assignment of cleavage sites.

The HPLC profile of the sample at 3 h of incubation of rhSCF$^{1-166}$ with chymase using an E:S of 1:500 (w/w) shows that chymase displays greater cleavage selectivity, only two fragments being generated by proteolysis. Mass spectral analysis of the major fraction showed two components, whose mass values were determined as 17,977.1 ± 0.6 Da and 18,656.9 ± 0.9 Da (Fig. 2). The main component was identified as peptide 1–159 and the minor species corresponded to the intact rhSCF$^{1-166}$ still present in the incubation mixture. Analysis of the minor fraction showed the complementary peptide 160–166 (mass value 697.4 ± 0.1 Da). No further cleavages were observed later, even when proteolysis was continued for 24 h. These results indicate that chymase very specifically cleaves rhSCF$^{1-166}$ at Phe159.

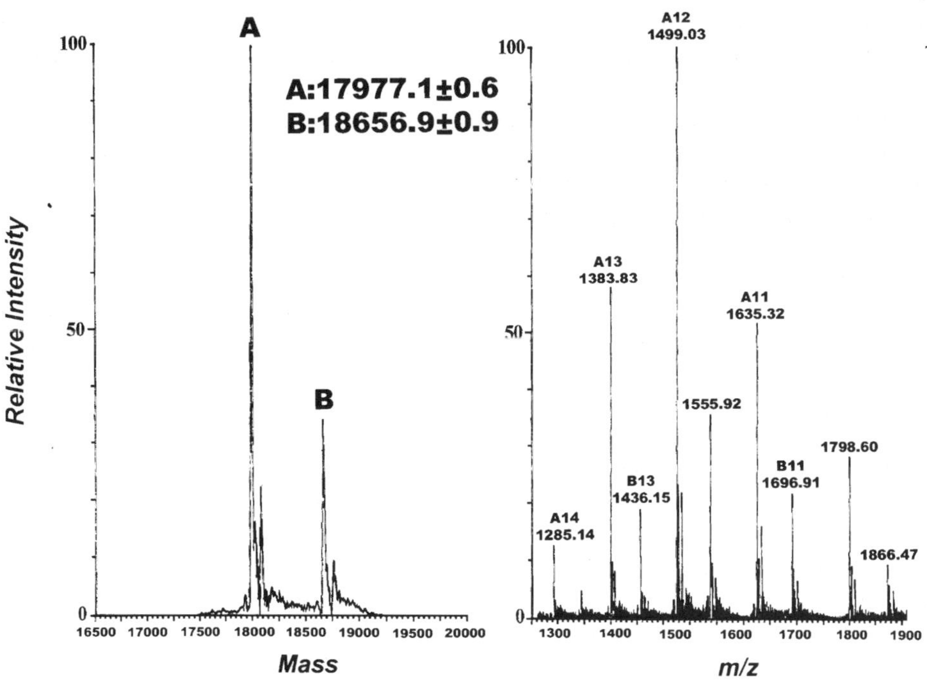

Fig. 2 Mass-spectrometric analysis of the major product of digestion of SCF$^{1-166}$ by human chymase. The figure shows the multiply charged ions (right panels) and the transformed (left panel) mass spectra. The major species (A) has a molecular mass of 17,977.1 ± 0.6 Da, corresponding to fragment 1–159 (theoretical mass 17,976.6 Da) originating from a single proteolytic event at Phe159. The smaller component (B) represents the residual undigested SCF$^{1-166}$.

Kinetics of the Interaction Between SCF^{1-166} and Immunologically Activated HLMC

Human mast cells contain several proteases, such as chymase (39, 40), tryptase (8, 10, 40), carboxypeptidase (57, 65), and cathepsin G (66). Experiments performed using chymase indicated that, although chymase very specifically cleaves $rhSCF^{1-166}$ at Phe159, there are further cleavage sites within SCF^{1-166}. By Western blot, using the mAb 7H6, we evaluated by HPLC whether immunologically activated HLMC cleave $rhSCF^{1-166}$. After 30 min at 37°C, exogenous $rhSCF^{1-166}$ added to a suspension of immunologically activated HLMC was converted to at least two different forms, with molecular mass corresponding to SCF^{1-159} and SCF^{1-144}, respectively. To exclude the possibility that the disappearance of $SCF^{1-166}$was due to its internalization in HLMC (52), we performed a similar experiment incubating SCF^{1-166} with cell-free supernatants of immunologically activated HLMC. Also in this case, SCF^{1-166} was rapidly cleaved to at least two forms, one with a molecular mass compatible with SCF^{1-159}. These results indicate that the immunological secretion of proteases from HLMC leads to the formation of several SCF species, suggesting there may be several cleavage sites within SCF^{1-166}. An important control was performed with anti-IgE-activated basophils. It is well established that human basophils contain extremely low concentrations of proteases (40). Incubation of SCF with immunologically activated basophils did not cleave the cytokine.

These results indicate that, besides the native form of SCF^{1-166}, at least two others (SCF^{1-159} and SCF^{1-144}) may be formed *in vivo*. Thus, chymase might exert specific enzymatic activity selectively acting at Phe159, and other as yet unknown protease(s) may cleave SCF, leading to the formation of several SCF species. Whatever the findings, it appears that SCF, chymase and other proteolytic enzymes present in human mast cells participate in a complex biochemical system similar to the angiotensin I–ACE–angiotensin II–angiotensin III system (55, 62). Interestingly, chymase in human skin (39, 40) and heart mast cells (8) exerts angiotensin-converting enzyme (ACE) activity, cleaving angiotensin I to angiotensin II (67).

BIOLOGICAL EFFECTS OF SCF^{1-166} AND SCF^{1-159} ON HLMC AND HSMC

The results reported above indicate that chymase rapidly and specifically cleaves SCF^{1-166} at a site encoded within exon 6 of the SCF gene, leading to the formation of SCF^{1-159} and of a C-terminal septapeptide (34, 38). To investigate the possible biological role of these three peptides, we evaluated the effects of $rhSCF^{1-166}$, SCF^{1-159} and the septapeptide on HLMC and HSMC. In a series of experiments, we found that SCF^{1-166} and SCF^{1-159} induced the release of histamine from HLMC in a similar manner, whereas the small cleavage product, the C-terminal septapeptide had no activating effect whatsoever. SCF^{1-166} and SCF^{1-159} also enhanced the release of histamine from HLMC induced by anti-IgE. The septapeptide had no such effect. Results were similar when $rhSCF^{1-166}$, SCF^{1-159} and the septapeptide were examined on HSMC alone or in combination with anti-IgE.

Figure 3 shows the effects of SCF^{1-166}, SCF^{1-159} and the septapeptide $SCF^{160-166}$ on HLMC chemotaxis. SCF^{1-166} and SCF^{1-159} concentration-dependently induced the chemotaxis of HLMC. The C-terminal septapeptide had no effect. SCF^{1-166} and SCF^{1-159} are equally active on human mast cells. The biological activities of the third form, SCF^{1-144}, generated by incubation of SCF^{1-166} with HLMC or their supernatants, remain to be determined.

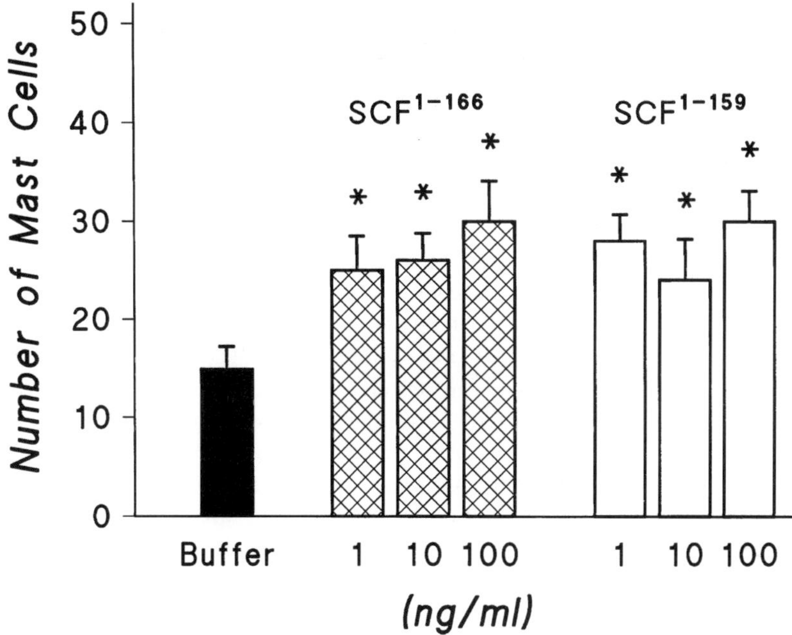

Fig. 3 Effect of increasing concentrations of SCF^{1-166} and SCF^{1-159} on HLMC chemotaxis. HLMC (50,000 per well) in the assay buffer were allowed to migrate toward the indicated concentrations of SCF^{1-166} and SCF^{1-159} for 3 h at 37°C in a humidified incubator (5% CO_2). Values are the mean ± SEM. $*p < 0.01$ compared with control.

CONCLUSIONS

We provided the first evidence that the secretory granules of human cardiac mast cells store SCF (9). This novel observation was extended and confirmed showing that SCF can be found in several tissues (skin and lung) from patients affected by diverse disorders (43). These findings led to the suggestion that SCF, present in the secretory granules of mast cells, might represent an autocrine factor sustaining the characteristic mast cell hyperplasia in allergic patients and in patients with mastocytosis. In addition, we have found that SCF can be immunologically released by isolated and purified mast cells, and that chymase and other proteolytic enzymes, also present in the secretory granules of human mast cells, can cleave SCF. The complexity of the *in vivo* interactions between several proteolytic mast cell enzymes and SCF can be easily envisioned. In fact, although chymase very specifically and rapidly cleaves $rhSCF^{1-166}$ at Phe159, there are other putative cleavage sites within SCF^{1-166}. Thus, another fascinating finding emerging from these studies is that, besides the native form of SCF^{1-166}, several species of SCF may be formed *in vivo*, and chymase selectively acts at Phe159 whereas other proteases may further cleave SCF. Thus, the immunological activation of human mast cells could lead to the concomitant release of SCF and of SCF-digesting proteases. These novel results highlight the complexity of the autocrine loops and negative feedbacks involving human mast cells and their metabolic products.

A new finding of these studies is that mast cell proteases can degrade mast cell-derived cytokine. Previous studies have provided contrasting results about the production of cytokines by human mast cells. Several cytokines have been detected by immunocytochemical techniques (48, 49, 51) and at the level of mRNA (41, 47, 50). However, some technical difficulties were observed in measuring cytokine release from isolated and purified human mast cells at the protein level (68). For example, constitutive synthesis of SCF mRNA was demonstrated in human lung and skin mast cells by RT–PCR, but iSCF was apparently undetectable in supernatants of anti-IgE-activated HLMC (41). Our results show that chymase and probably other mast cell-associated proteases can rapidly and efficiently cleave SCF.

The *in vivo* significance of the low amounts of SCF immunologically released from human mast cells remains to be established. We presume that they can exert local autocrine and paracrine functions. Our results also highlight the complexity of the autocrine loops and negative feedbacks involving human mast cells. These cells not only synthesize (41), store and release the autocrine factor SCF acting on c-*kit*R, but also elaborate several proteases that *in vivo* might modulate the biological effects of cytokines.

Our results also emphasize the importance of investigating the biological interactions between enzymes and substrates not only *in vitro*, but also in physiological and pathological conditions.

ACKNOWLEDGEMENTS

This work was supported by grants from the CNR (Target Project Biotechnology No. 98.00085.PF31 and No.99.00216.PF31), Ministero della Sanità–Istituto Superiore Sanità (AIDS Project 1998 No. 40B.64), MURST (Rome, Italy) and Associazione Italiana per la Ricerca sul Cancro.

REFERENCES

1. Irani, A.-M. A., Nilsson, G., Miettinen, U., Craig, S. S., Ashman, L. K., Ishizaka, T., Zsebo, K. M. and Schwartz, L. B. Recombinant human stem cell factor stimulates differentiation of mast cells from dispersed human fetal liver cells. *Blood* **80**:3009–3021, 1992.
2. Kirshenbaum, A. S., Goff, J. P., Kessler, S. W., Mican, J. M., Zsebo, K. M. and Metcalfe, D. D. Effect of IL-3 and stem cell factor on the appearance of human basophils and mast cells from $CD34^+$ pluripotent progenitor cells. *J. Immunol.* **148**:772–777, 1992.
3. Valent, P., Spanblöchl, E., Sperr, W. R., Sillaber, C., Zsebo, K. M., Agis, H., Strobl, H. , Geissler, K., Bettelheim, P. and Lechner, K. Induction of differentiation of human mast cells from bone marrow and peripheral blood mononuclear cells by recombinant human stem cell factor/*kit*-ligand in long-term culture. *Blood* **80**:2237–2245, 1992.
4. Mitsui, H., Furitsu, T., Dvorak, A. M., Irani, A.-M. A., Schwartz, L. B., Inagaki, N., Takei, M., Ishizaka, K., Zsebo, K. M., Gillis, S. and Ishizaka, T. Development of human mast cells from umbilical cord blood cells by recombinant human and murine c-*kit* ligand. *Proc. Natl. Acad. Sci. USA* **90**:735–739, 1993.
5. Saito, H., Ebisawa, M., Tachimoto, H., Shichijo, M., Fukagawa, K., Matsumoto, K., Iikura, Y., Awaji, T., Tsujimoto, G., Yanagida, M., Uzumaki, H., Takahashi, G., Tsuji, K. and Nakahata, T. Selective growth of human mast cells induced by *steel* factor, IL-6, and prostaglandin E_2 from cord blood mononuclear cells. *J. Immunol.* **157**:343–350, 1996.
6. Bischoff, S. C. and Dahinden, C. A. c-*kit* ligand: a unique potentiator of mediator release by human lung mast cells. *J. Exp. Med.* **175**:237–244, 1992.
7. Columbo, M., Horowitz, E. M., Botana, L. M., MacGlashan, D. W. Jr, Bochner, B. S., Gillis, S., Zsebo, K. M., Galli, S. J. and Lichtenstein, L. M. The human recombinant c-*kit* receptor ligand, rhSCF, induces mediator release from human cutaneous mast cells and enhances IgE-dependent mediator release from

both skin mast cells and peripheral blood basophils. *J. Immunol.* **149**:599–608, 1992.

8. Patella, V., Marinò, I., Lampärter, B., Arbustini, E., Adt, M. and Marone, G. Human heart mast cells. Isolation, purification, ultrastructure, and immunologic characterization. *J. Immunol.* **154**:2855–2865, 1995.
9. Patella, V., Marinò, I., Arbustini, E., Lampärter-Schummert, B., Verga, L., Adt, M. and Marone, G. Stem cell factor in mast cells and increased mast cell density in idiopathic and ischemic cardiomyopathy. *Circulation* **97**:971–978, 1998.
10. de Paulis, A., Marinò, I., Ciccarelli, A., de Crescenzo, G., Concardi, M., Verga, L., Arbustini, E. and Marone, G. Human synovial mast cells. I. Ultrastructural *in situ* and *in vitro* immunologic characterization. *Arthritis Rheum.* **39**:1222–1233, 1996.
11. Costa, J. J., Demetry, G. D., Harrist, T. J., Dvorak, A. M., Hayes, D. F., Merica, E. A., Menchaca, D. M., Gringeri, A. J., Schwartz, L. B. and Galli, S. J. Recombinant human stem cell factor (*kit* ligand) promotes human mast cell and melanocyte hyperplasia and functional activation *in vivo*. *J. Exp. Med.* **183**:2681–2686, 1996.
12. Flanagan, J. G., Chan, D. C. and Leder, P. Transmembrane form of the *kit* ligand growth factor is determined by alternative splicing and is missing in the Sl^{d} mutant. *Cell* **64**:1025–1035, 1991.
13. Huang, E. J., Nocka, K. H., Buck, J. and Besmer, P. Differential expression and processing of two cell associated forms of the *kit*-ligand: KL-1 and KL-2. *Mol. Biol. Cell* **3**:349–362, 1992.
14. Toksoz, D., Zsebo, K. M., Smith, K. A., Hu, S., Brankow, D., Suggs, S. V., Martin, F. H. and Williams, D. A. Support of human hematopoiesis in long-term bone marrow cultures by murine stromal cells selectively expressing the membrane-bound and secreted forms of the human homolog of the *steel* gene product, stem cell factor. *Proc. Natl. Acad. Sci. USA* **89**:7350–7354, 1992.
15. Heinrich, M. C., Dooley, D. C., Freed, A. C., Band, L., Hoatlin, M. E., Keeble, W. W., Peters, S. T., Silvey, K. V., Ey, F. S., Kabat, D., Maziarz, R. T. and Bagby, G. C. Jr. Constitutive expression of *steel* factor gene by human stromal cells. *Blood* **82**:771–783, 1993.
16. Longley, B. J. Jr, Morganroth, G. S., Tyrrell, L., Ding, T. G., Anderson, D. M., Williams, D. E. and Halaban, R. Altered metabolism of mast-cell growth factor (c-*kit* ligand) in cutaneous mastocytosis. *N. Engl. J. Med.* **328**:1302–1307, 1993.
17. Hamann, K., Haas, N., Grabbe, J. and Czarnetzki, B. M. Expression of stem cell factor in cutaneous mastocytosis. *Br. J. Dermatol.* **133**:203–208, 1995.
18. Broudy, V. C., Kovach, N. L., Bennett, L. G., Lin, N., Jacobsen, F. W. and Kidd, P. G. Human umbilical vein endothelial cells display high-affinity c-*kit* receptors and produce a soluble form of the c-*kit* receptor. *Blood* **83**:2145–2152, 1994.
19. Weiss, R. R., Whitaker-Menezes, D., Longley, J., Bender, J. and Murphy, G. F. Human dermal endothelial cells express membrane-associated mast cell growth factor. *J. Invest. Dermatol.* **104**:101–106, 1995.
20. Beck, D., Gross, N., Beretta Brognara, C. and Perruisseau, G. Expression of stem cell factor and its receptor by human neuroblastoma cells and tumors. *Blood* **86**:3132–3138, 1995.
21. Turner, A. M., Zsebo, K. M., Martin, F., Jacobsen, F. W., Bennett, L. G. and Broudy, V. C. Nonhematopoietic tumor cell lines express stem cell factor and display c-*kit* receptors. *Blood* **80**:374–381, 1992.
22. Yarden, Y., Kuang, W. J., Yang-Feng, T., Coussens, L., Munemitsu, S., Dull, T. J., Chen, E., Schlessinger, J., Francke, U. and Ullrich, A. Human proto-oncogene c-*kit*: a new cell surface receptor tyrosine kinase for an unidentified ligand. *EMBO J.* **6**:3341–3351, 1987.
23. Qiu, F. H., Ray, P., Brown, K., Barker, P. E., Jhanwar, S., Ruddle, F. H. and Besmer, P. Primary structure of c-*kit*: relationship with the CSF-1/PDGF receptor kinase family: oncogenic activation of v-*kit* involves deletion of extracellular domain and C terminus. *EMBO J.* **7**:1003–1011, 1988.
24. Zsebo, K. M., Wypych, J., McNiece, I. K., Lu, H. S., Smith, K. A., Karkare, S. B., Sachdev, R. K., Yuschenkoff, V. N., Birkett, N. C., Williams, L. R., Satyagal, V. N., Tung, W., Bosselman, R. A., Mendiaz, E. A. and Langley, K. E. Identification, purification, and biological characterization of hematopoietic stem cell factor from buffalo rat liver-conditioned medium. *Cell* **63**:195–201, 1990.
25. Tsai, M., Shih, L.-S., Newlands, G. F. J., Takeishi, T., Langley, K. E., Zsebo, K. M., Miller, H. R. P., Geissler, E. N. and Galli, S. J. The rat c-*kit* ligand, stem cell factor, induces the development of connective tissue-type and mucosal mast cells *in vivo*. Analysis by anatomical distribution, histochemistry, and protease phenotype. *J. Exp. Med.* **174**:125–131, 1991.
26. Tsai, M., Takeishi, T., Thompson, H., Langley, K. E., Zsebo, K. M., Metcalfe, D. D., Geissler, E. N. and Galli, S. J. Induction of mast cell proliferation, maturation, and heparin synthesis by the rat c-*kit* ligand,

stem cell factor. *Proc. Natl. Acad. Sci. USA* **88**:6382–6386, 1991.

27. Gurish, M. F., Ghildyal, N., McNeil, H. P., Austen, K. F., Gillis, S. and Stevens, R. L. Differential expression of secretory granule proteases in mouse mast cells exposed to IL-3 and c-*kit* ligand. *J. Exp. Med.* **175**:1003–1012, 1992.
28. Meininger, C. J., Yano, H., Rottapel, R., Bernstein, A., Zsebo, K. M. and Zetter, B. R. The c-*kit* receptor ligand functions as a mast cell chemoattractant. *Blood* **79**:958–963, 1992.
29. Wershil, B. K., Tsai, M., Geissler, E. N., Zsebo, K. M. and Galli, S. J. The rat c-*kit* ligand, stem cell factor, induces c-*kit* receptor-dependent mouse mast cell activation *in vivo*. Evidence that signaling through the c-*kit* receptor can induce expression of cellular function. *J. Exp. Med.* **175**:245–255, 1992.
30. Dvorak, A. M., Seder, R. A., Paul, W. E., Morgan, E. S. and Galli, S. J. Effects of interleukin-3 with or without the c-*kit* ligand, stem cell factor, on the survival and cytoplasmic granule formation of mouse basophils and mast cells *in vitro*. *Am. J. Pathol.* **144**:160–170, 1994.
31. Funasaka, Y., Boulton, T., Cobb, M., Yarden, Y., Fan, B., Lyman, S. D., Williams, D. E., Anderson, D. M., Zakut, R., Mishima, Y. and Halaban, R. c-*kit*-kinase induces a cascade of protein tyrosine phosphorylation in normal human melanocytes in response to mast cell growth factor and stimulates mitogen-activated protein kinase but is down-regulated in melanomas. *Mol. Biol. Cell.* **3**:197–209, 1992.
32. Geissler, E. N., Liao, M., Brook, J. D., Martin, F. H., Zsebo, K. M., Housman, D. E. and Galli, S. J. Stem cell factor (SCF), a novel hematopoietic growth factor and ligand for c-*kit* tyrosine kinase receptor, maps on human chromosome 12 between 12q14.3 and 12qter. *Somat. Cell. Mol. Genet.* **17**:207–214, 1991.
33. Brannan, C. I., Lyman, S. D., Williams, D. E., Eisenman, J. D., Anderson, D. M., Cosman, D., Bedell, M. A., Jenkins, N. A. and Copeland, N. G. Steel-Dickie mutation encodes a c-*kit* ligand lacking transmembrane and cytoplasmic domains. *Proc. Natl. Acad. Sci. USA* **88**:4671–4674, 1991.
34. Zsebo, K. M., Williams, D. A., Geissler, E. N., Broudy, V. C., Martin, F. H., Atkis, H. L., Hsu, R. Y., Birkett, N. C., Okino, K. H., Murdock, D. C., Jacobsen, F. W., Langley, K. E., Smith, K. A., Takeishi, T., Cattanach, B. M., Galli, S. J. and Suggs, S. V. Stem cell factor (SCF) is encoded at the *Sl* locus of the mouse and is the ligand for the c-*kit* tyrosine kinase receptor. *Cell* **63**:213–224, 1990b.
35. Plaut, M., Pierce, J. H., Watson, C. J., Hanley-Hyde, J., Nordan, R. P. and Paul, W. E. Mast cell lines produce lymphokines in response to cross-linkage of $Fc_{\varepsilon}RI$ or to calcium ionophores. *Nature* **339**:64–67, 1989.
36. Wodnar-Filipowicz, A., Heusser, C. H. and Moroni, C. Production of the haematopoietic growth factors GM-CSF and interleukin-3 by mast cells in response to IgE receptor-mediated activation. *Nature* **339**:150–152, 1989.
37. Bradding, P., Feather, I. H., Howarth, P. H., Mueller, R., Roberts, J. A., Britten, K., Bews, J. P., Hunt, T. C., Okayama, Y., Heusser, C. H., Bullock, G. R., Church, M. K. and Holgate, S. T. Interleukin 4 is localized to and released by human mast cells. *J. Exp. Med.* **176**:1381–1386, 1992.
38. Longley, B. J. Jr, Tyrrel, L., Ma, Y., Williams, D. A., Halaban, R., Langley, K., Lu, H. S. and Schechter, N. M. Chymase cleavage of stem cell factor yields a bioactive, soluble product. *Proc. Natl. Acad. Sci. USA* **94**:9017–9021, 1997.
39. Schechter, N. M., Choi, J. K., Slavin, D. A., Deresienski, D. T., Sayama, S., Dong, G., Lavker, R. M., Proud, D. and Lazarus, G. S. Identification of a chymotrypsin-like proteinase in human mast cells. *J. Immunol.* **137**:962–970, 1986.
40. Irani, A. A., Schechter, N. M., Craig, S. S., DeBlois, G. and Schwartz, L. B. Two types of human mast cells that have distinct neutral protease composition. *Proc. Natl. Acad. Sci. USA* **83**:4464–4468, 1986.
41. Zhang, S., Anderson, D. F., Bradding, P., Coward, W. R., Baddeley, S. M., MacLeod, J. D. A., McGill, J. I., Church, M. K., Holgate, S. T. and Roche, W. R. Human mast cells express stem cell factor. *J. Pathol.* **186**:59–66, 1998.
42. Metcalfe, D. D. Classification and diagnosis of mastocytosis: current status. *J. Invest. Dermatol.* **96**:2S–4S, 1991.
43. de Paulis, A., Minopoli, G., Arbustini, E., de Crescenzo, G., Dal Piaz, F., Pucci, P., Russo, T. and Marone, G. Stem cell factor is localized in, released from, and cleaved by human mast cells. *J. Immunol.* **163**:2799–2808, 1999.
44. Mendiaz, E. A., Chang, D. G., Boone, T. C., Grant, J. R., Wypych, J., Aguero, B., Egrie, J. C. and Langley, K. E. Epitope mapping and immunoneutralization of recombinant human stem-cell factor. *Eur. J. Biochem.* **239**:842–849, 1996.
45. Marone, G., Casolaro, V., Patella, V., Florio, G. and Triggiani, M. Molecular and cellular biology of mast cells and basophils. *Int. Arch. Allergy Immunol.* **114**:207–217, 1997.
46. Brunner, T., Heusser, C. H. and Dahinden, C. A. Human peripheral blood basophils primed by

interleukin 3 (IL-3) produce IL-4 in response to immunoglobulin E receptor stimulation. *J. Exp. Med.* **177**:605–611, 1993.

47. Möller, A., Lippert, U., Lessmann, D., Kolde, G., Hamann, K., Welker, P., Schadendorf, D., Rosenbach, T., Luger, T. and Czarnetzki, B. M. Human mast cells produce IL-8. *J. Immunol.* **151**:3261–3266, 1993.
48. Bradding, R., Roberts, J. A., Britten, K. M., Montefort, S., Djukanovic, R., Mueller, R., Heusser, C. H., Howarth, P. H. and Holgate, S. T. Interleukin-4, -5, and -6 and tumor necrosis factor-α in normal and asthmatic airways: evidence for the human mast cell as a source of these cytokines. *Am. J. Respir. Cell Mol. Biol.* **10**:471–480, 1994.
49. Bradding, P., Okayama, Y., Howarth, P. H., Church, M. K. and Holgate, S. T. Heterogeneity of human mast cells based on cytokine content. *J. Immunol.* **155**:297–307, 1995.
50. MacGlashan, D. Jr, White, J. M., Huang, S.-K., Ono, S. J., Schroeder, J. T. and Lichtenstein, L. M. Secretion of IL-4 from human basophils. The relationship between IL-4 mRNA and protein in resting and stimulated basophils. *J. Immunol.* **152**:3006–3016, 1994.
51. Kaartinen, M., Penttilä, A. and Kovanen, P. T. Mast cells in rupture-prone areas of human coronary atheromas produce and store TNF-α. *Circulation* **94**:2787–2792, 1996.
52. Shimizu, Y., Ashman, L. K., Du, Z. and Schwartz, L. B. Internalization of *kit* together with stem cell factor on human fetal liver-derived mast cells: new protein and RNA synthesis are required for reappearance of *kit*. *J. Immunol.* **156**:3443–3449, 1996.
53. Walls, A. F. The roles of neutral proteases in asthma and rhinitis. In: *Asthma and Rhinitis* (Busse, W. W. and Holgate, S. T., eds), pp. 801–824. Blackwell Science, Boston, 1994.
54. Walls, A. F., Shute, J. K., Warner, J., Cairns, J. A., Howarth, P. H. and Holgate, S. T. Mast cell and eosinophil proteases in the chronic allergic inflammatory diseases. In: *Asthma and Allergic Diseases: Physiology, Immunopharmacology and Treatment* (Marone, G., Austen, K. F., Holgate, S. T., Kay, A. B. and Lichtenstein, L. M., eds), pp. 119–131. Academic Press, London, 1998.
55. Urata, H., Kinoshita, A., Misono, K. S., Bumpus, F. M. and Husain, A. Identification of a highly specific chymase as the major angiotensin II-forming enzyme in the human heart. *J. Biol. Chem.* **265**:22348–22357, 1990.
56. Craig, S. S., Schechter, N. M. and Schwartz, L. B. Ultrastructural analysis of human T and TC mast cells identified by immunoelectron microscopy. *Lab. Invest.* **58**:683–691, 1988.
57. Goldstein, S. M., Leong, J., Schwartz, L. B. and Cooke, D. Protease composition of exocytosed human skin mast cells protease-proteoglycan complexes. Tryptase resides in a complex distinct from chymase and carboxypeptidase. *J. Immunol.* **148**:2475–2482, 1992.
58. McEuen, A. R., Gaça, M. D. A., Buckley, M. G., He, S., Gore, M. G. and Walls, A. F. Two distinct forms of human mast cell chymase: differences in affinity for heparin and in distribution in skin, heart and other tissues. *Eur. J. Biochem.* **256**:461–470, 1998.
59. Huntley, J. F., Newlands, G. F., Gibson, S., Ferguson, A. and Miller, H. R. Histochemical demonstration of chymotrypsin-like serine esterases in mucosal mast cells in four species including man. *J. Clin. Pathol.* **38**:375–384, 1985.
60. Schechter, N. M., Sprows, J. L., Shoenberger, O. L., Lazarus, G. S., Cooperman, B. S. and Rubin, H. Reaction of human skin chymotrypsin-like proteinase chymase with plasma proteinase inhibitors. *J. Biol. Chem.* **264**:21308–21315, 1989.
61. Powers, J. C., Tanaka, T., Harper, J. W., Minematsu, Y., Barker, L., Lincoln, D., Crumley, K. V., Fraki, J. E., Schechter, N. M. and Lazarus, G. G. Mammalian chymotrypsin-like enzymes. Comparative reactivities of rat mast cell proteases, human and dog skin chymases, and human cathepsin G with peptide 4-nitroanilide substrates and with peptide chloromethyl ketone and sulphonyl fluoride inhibitors. *Biochemistry* **24**:2048–2058, 1985.
62. Kinoshita, A., Urata, H., Bumpus, F. M. and Husain, A. Multiple determinants for the high substrate specificity of an angiotensin II-forming chymase from the human heart. *J. Biol. Chem.* **266**:19192–19197, 1991.
63. Kido, H., Nakano, A., Okishima, N., Wakabayashi, H., Kishi, F., Nakaya, Y., Yoshizumi, M. and Tamaki, T. Human chymase, an enzyme forming novel bioactive 31-amino acid length endothelins. *Biol. Chem.* **379**:885–891, 1998.
64. Zappacosta, F., Pessi, A., Bianchi, E., Venturini, S., Sollazzo, M., Tramontano, A., Marino, G. and Pucci, P. Probing the tertiary structure of proteins by limited proteolysis and mass spectrometry: the case of Minibody. *Protein Sci.* **5**:802–813, 1996.
65. Goldstein, S. M., Kaempfer, C. E., Proud, D., Schwartz, L. B., Irani, A.-M. and Wintroub, B. U. Detection and partial characterization of a human mast cell carboxypeptidase. *J. Immunol.*

139:2724–2729, 1987.

66. Schechter, N. M., Irani, A.-M. A., Sprows, J. L., Abernethy, J., Wintroub, B. and Schwartz, L. B. Identification of a cathepsin G-like proteinase in the MC_{TC} type of human mast cell. *J. Immunol.* **145**:2652–2661, 1990.
67. Marone, G., de Crescenzo, G., Patella, V. and Genovese, A. Human cardiac mast cells and their role in severe allergic reactions. In: *Asthma and Allergic Diseases: Physiology, Immunopharmacology and Treatment* (Marone, G., Austen, K. F., Holgate, S. T., Kay, A. B. and Lichtenstein, L. M., eds), pp. 237–257. Academic Press, London, 1998.
68. Redrup, A. C., MacGlashan, D. W., Lichtenstein, L. M., Kagey-Sobotka, A. and Schroeder, J. T. Minimal TNF-α production following activation of purified human lung mast cells. *J. Allergy Clin. Immunol.* **99**:S87, 1997.

CHAPTER 38

SCF-Induced Chemokine Production, Allergic Airway Inflammation and Airway Hyper-reactivity

NICHOLAS W. LUKACS, SANDRA H. P. OLIVEIRA and CORY M. HOGABOAM*

University of Michigan Medical School, Department of Pathology, Ann Arbor, Michigan, U.S.A.

INTRODUCTION

One of the underlying mechanisms that separate allergic reactions from other types of inflammatory responses is mast cell activation (1–4). Mast cells appear to be the primary cell initially activated during allergic events, and more recent evidence has demonstrated that they likely contribute to the responses of chronic disease development. Thus, controlling mast cell activation may be critical in effectively attenuating allergic responses. In addition to IgE-mediated mechanisms, it appears that c-kit ligand or stem cell factor (SCF) can directly induce mast cell activation as well as augment the IgE-mediated response (5–9). The prolonged activation of local airway mast cell populations by SCF after initial IgE-mediated events may play a significant role in persistent activation leading to late-phase responses (10–12). Airway inflammation and hyper-reactivity characterize these late-phase responses. SCF plays a central role in mast cell biology via its ability to initiate and augment activational events, leading to the production of various inflammatory mediators, including histamine and chemokines. The overall concept that SCF may play a role in disease progression has been relatively understudied.

SCF is not only an important haematopoietic factor that drives terminal differentiation of mast cells; it has also been shown to have other important roles in regulating mast cell biology, such as survival, activation and degranulation of mature mast cells. SCF binds to its surface receptor, c-kit, which is a member of the receptor tyrosine kinase family (6, 7). Endogenous SCF occurs in both transmembrane and soluble forms that differ depending upon whether a specific enzymatic cleavage site is present (13). The tissue distribution of the two forms of SCF has not been thoroughly investigated and the relative importance of either form during haematopoiesis or disease is not known. The SCF-dependent function

* Corresponding author.

MAST CELLS AND BASOPHILS
ISBN 0-12-473335-2

may vary depending on whether the c-kit$^+$ cell encounters SCF in its transmembrane or soluble form. The regulation of SCF in cell populations appears to occur by a combination of post-transcriptional and post-translational mechanisms. At the gene level, alternative transcripts are produced, while different protein products appear to be regulated in a tissue-specific manner. In addition to inducing mast cell activation, the membrane form appears to have the ability to allow adhesion of mast cells to cellular surface (14, 15). SCF receptor, c-kit, is found on haematopoietic progenitor cells, melanocytes, germ cells, eosinophils and some lymphocytes; it is highly expressed on mast cells, with little or no expression on basophils. Growth of mast cells from Sl/Sl^d (Sl/d) mutant mice (deficient in membrane SCF, but readily release a soluble form of SCF) can be induced by fibroblasts (16), suggesting that peripheral cell populations have a role in SCF production and long-term mast cell survival. Since the Sl/Sl^d mice have few or no mast cells and are often anaemic, it is apparent that the membrane form of SCF is required for specific haematopoietic functions. These latter observations may be very pertinent to the relationship of SCF and mast cell survival as well as the activation and degranulation of peripheral mast cells during disease progression. In addition, SCF has also been shown to have an activating effect on eosinophils. These observations may identify SCF as a central molecule in allergic inflammation. In this chapter the function of SCF during allergic events *in vivo* and during mast cell and eosinophil activation *in vitro* will be examined.

SCF-INDUCED MAST CELL ACTIVATION, INFLAMMATION AND AIRWAY HYPER-REACTIVITY

The function of mast cells during the progression of allergic responses has primarily been identified at the initiation of the response via IgE-mediated mechanisms. The ability of mast cells to release pre-formed mediators and quickly synthesize acute mediators, such as leukotrienes, that can influence airway physiology has been clearly established (1, 17, 18). Recent evidence has suggested that locally produced SCF can have prolonged effects on mast cells and allow them to participate in the later stages of the response. Direct effects of SCF can induce mast cells to degranulate and release histamine as well as other inflammatory factors, including leukotrienes (1, 5, 18, 19). SCF can greatly enhance antigen-specific IgE-mediated degranulation (10, 12, 20, 21) and has been shown to directly stimulate mast cell activation in human bronchi and induce smooth muscle cell contraction (22). In addition, these latter studies also demonstrated a release of arachidonic acid mediators, leukotrienes and prostaglandins, from the SCF-stimulated bronchus. In recent studies, significant increases in SCF levels have been detected in the lungs of allergic mice after allergen challenge, while the neutralization of SCF *in vivo* altered the inflammatory response by decreasing the eosinophil accumulation within and around the airways (23). Not surprisingly, when SCF was neutralized, a significant decrease in histamine levels was also observed, suggesting that SCF was functioning via continued mast cell activation. The association of SCF with eosinophils is quite striking and offers an interesting perspective on the role and relationship of SCF with mast cells and eosinophils. Although SCF is not directly chemotactic for eosinophils, seminal studies on the effect of SCF on eosinophils have demonstrated increased adhesion of eosinophils to activated endothelium via the ability of SCF to enhance the avidity of VLA-4 (very late antigen) to its ligands VCAM-1 (vascular cell adhesion molecule) and/or fibronectin (24). It is logical to assume that SCF has other activating effects on eosinophils other than augmentation of adhesion events. Unpublished studies have demonstrated the ability of SCF to induce significant induction of chemokines, such as RANTES and MCP-1 (monocyte chemotactic protein), from eosinophils, suggesting the ability to augment the inflammatory response during allergic reactions (Oliveira and Lukacs, unpublished). The ability of SCF to promote such responses locally within the lung may have detrimental effects on the system, including the exacerbation of chronic inflammation and degradation of lung function.

The major pathophysiological event that occurs during asthma is induction of airway reactivity during the late-phase response (25–28). The fact that c-kit ligand or SCF can directly induce mast cell activation, as well as augment the IgE-mediated responses, suggests that this cytokine may play a role during disease progression. Interestingly, results suggest that elevated histamine levels are maintained in the airways of allergic mice until 6–8 h post-allergen challenge and are at least partially dependent upon locally generated SCF (23). Thus, prolonged activation of local airway mast cell populations by SCF after initial IgE-mediated events may play a significant role in persistent activation leading to a late-phase response. In a recent study, multiple experimental protocols were used to identify SCF as a molecule involved in the induction of airway hyper-reactivity (29). Neutralization of local SCF in the airway using intratracheal exposure to SCF-specific antibody at the time of allergen challenge, significantly attenuated the induction of airway hyper-reactivity. Furthermore, addition of recombinant SCF along with allergen exacerbated and prolonged the airway hyper-reactivity responses. These responses were identified as being mediated by local mast cell populations using SCF-deficient (Sld; mast cell-deficient) mice. Moreover, when SCF alone was administered into the airways of normal mice it induced a hyper-reactive response that was prolonged up to 24 h (Fig. 1). In additional studies it appeared that SCF-induced airway changes

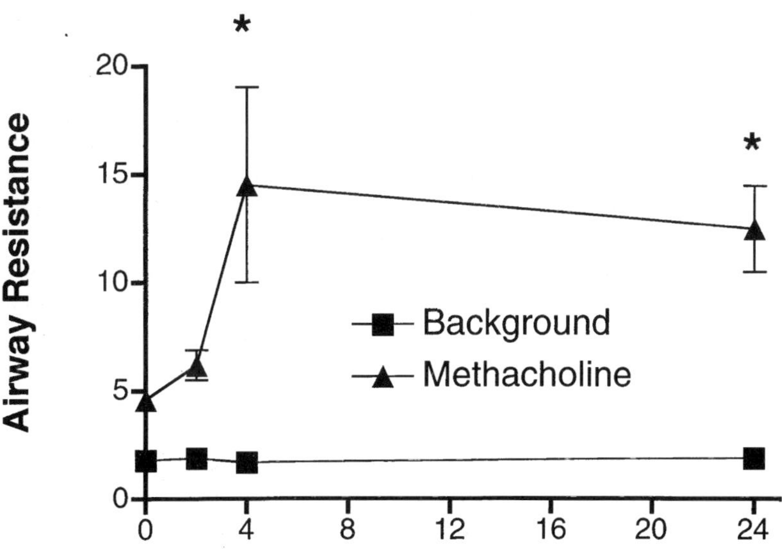

Fig. 1 SCF induces airway hyper-reactivity in normal mice. SCF (100ng) was injected into the airways of non-allergic, normal mice and subsequently challenge at various time points post-installation. Each point represents the mean ± SE of five mice per group. $*p < 0.05$.

were dependent upon leukotriene production. Pre-treatment of allergic mice with specific inhibitors to leukotriene LTD_4 prior to SCF instillation reduced the hyper-reactive response (unpublished data). Thus, SCF may have both direct and indirect roles in mediating airway inflammation and hyper-reactivity by enhancing eosinophil accumulation in allergic responses and via leukotriene release into the airway. Altogether these data suggest that SCF may play an important role in the initiation, progression and chronicity of allergic asthmatic responses. Furthermore, since SCF can directly induce these responses in the absence of IgE, SCF production alone may be a driving force for exacerbation of asthmatic responses in non-allergic conditions, such as viral infections.

REGULATION OF SCF PRODUCTION

A number of cell populations have now been identified that express SCF, including fibroblasts, endothelial cells, neural crest cells, melanocytes, hepatocytes, smooth muscle cells, mast cells, macrophages and epithelial cells as well as a number of tumour cell lines (8, 30, 31). Although it appears that SCF can be expressed in a cleavable and a non-cleavable truncated form of the protein, the predominant species seems to be cleavable in most cell populations. Both the soluble (cleavable) and transmembrane forms (non-cleavable) appear to be functional, and the specific action of SCF may depend upon whether the protein is encountered in its soluble or transmembrane form of the protein. Production and release of SCF are regulated at several levels, including transcriptional and translational as well as by surface cleavage of the protein by specific enzymes. Because SCF can be found on the surface of cells, a significant reservoir of SCF may be maintained constitutively until an appropriate inflammatory event occurs that allows the expression of enzymes, such as chymase (32, 33). SCF can then be cleaved from the surface of the cells and can interact with c-kit receptor on specific cell populations. The regulation of the different forms of SCF has not been thoroughly studied nor has the specific function of soluble versus membrane-associated SCF. Significant levels of SCF do appear to be overexpressed in disease tissue, such as in nasal polyps, and can be regulated by glucocorticoid treatment in asthmatic tissue (34–37). However, there is a paucity of data relating to the expression of SCF during allergic disease progression.

In recent studies the expression of SCF in various populations has begun to be elucidated. The ability of local cell populations within the lung and airway to produce SCF may be very important for exacerbation and prolonged activation of mast cells during asthmatic responses in the lung. Initial studies demonstrated that alveolar macrophages produce significant amounts of SCF in response to tumour necrosis factor (TNF) and interleukin-4 (IL-4), two cytokines strongly associated with asthmatic responses (23). Several laboratories have now shown that fibroblasts can indeed express mRNA for SCF, however, the protein data is quite variable (19, 38–40). This latter issue is likely due to the fact that both SCF protein species, soluble and transmembrane, are initially expressed in a membrane form and differ only by the fact that one can be enzymatically cleaved from the surface and one can not. In order to properly evaluate the level of SCF protein in fibroblasts (as well as other structural cell populations), which apparently do not produce the proper enzymes to cleave SCF, the protein levels should be evaluated from isolated and dispersed membranes. In studies examining pulmonary fibroblast production of SCF both IL-4 and TNF were again found to increase the expression of SCF (Fig. 2), however, the SCF was completely membrane-associated in

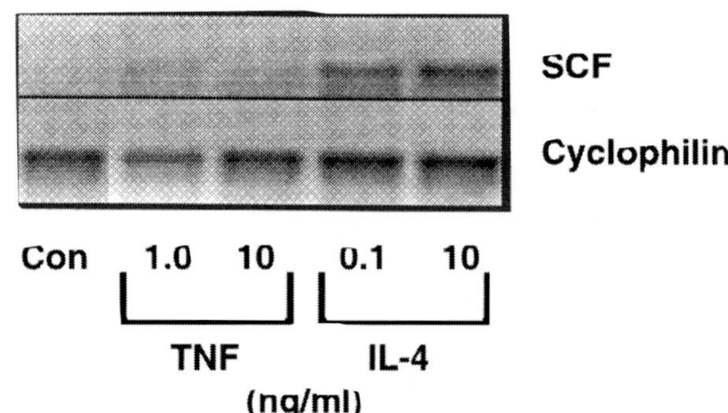

Fig. 2 Pulmonary fibroblasts express mRNA for SCF after TNF and IL-4 stimulation. Isolated fibroblast cultures were stimulated for 4–6 h, mRNA was isolated and RT–PCR analysis was used to examine the expression of SCF. Data are representative of three repeat experiments.

pure cultures. In contrast, activated alveolar macrophage populations apparently make the proper enzymes, as nearly all of the SCF produced from these cells is soluble and very little remains membrane-bound (unpublished observations). Because SCF is active whether it is in its membrane or soluble form, both of these cell populations can directly impact on the tissue mast cell populations within the lung and further activate system. We have also evaluated SCF production on other cell populations, including airway epithelial cells, tracheal smooth muscle cells and even on hepatocyte cell lines. The results from all of these structural cell populations demonstrate that SCF expression is primarily membrane-associated. The presence of SCF on these cell populations likely indicates that this cytokine is a relevant molecule that may be involved in disease progression and/or tissue integrity, such as protection from oxidative damage, proliferation factor, etc. (41–48). Future studies should examine the consequence of having SCF expressed by these various populations.

INITIATION OF CYTOKINE/CHEMOKINE PRODUCTION BY SCF IN MAST CELLS AND EOSINOPHILS

In addition to releasing early response mediators, mast cells also have the ability to produce and release cytokines and chemokines. Factors such as LTB_4 and PAF (platelet-activating factor) have non-specific chemotactic activity for leukocytes and can be released during mast cell degranulation, contributing to the late phase inflammatory response (49–51). Since neutralization of SCF alters eosinophil accumulation in chronic responses, specific chemokines have been examined that can be directly induced by SCF. Recent studies have identified mRNA for both CXC (IL-8) and CC (MIP-1α (macrophage inflammatory protein), MCP-1 and RANTES) chemokines in human mast cell populations (52). SCF has also been shown to augment MCP-1 and MIP-1 production from mast cells (20, 53). Furthermore, direct immunohistochemical staining techniques have identified the presence of similar chemokine proteins in murine airway-derived mast cell populations (54, 55). The production and release of these chemotactic factors may provide an initial mechanism for the recruitment and activation of particular leukocyte subsets during allergic inflammation and the prolonged production of these factors may help maintain the chronic influx of cells. The role of mast cells as only immediate-early-phase reactive cells has begun to be re-evaluated as investigators realize that mast cells may play a significant role in chronic inflammation. The chemokines, unlike PAF and LTB_4, preferentially recruit specific leukocyte populations. In particular, several CC (β) chemokines have potent mononuclear cell and eosinophil chemotactic activity and the release of these chemokines from mast cells during allergic responses may constitute an important mechanism for long-term recruitment of cell populations that lead to continued damage of the lung (56–59).

Since mast cells have now been shown to produce significant levels of multiple chemokines, such as MCP-1 and MIP-1α, multiple laboratories are now examining the activational requirements for their production. In a series of unpublished experiments using immunogold staining techniques the chemokines were found to be localized to the cytoplasmic regions and not to the granules of the mast cells. These studies would be consistent with the notion that mast cells produce chemokines predominantly in a *de novo* fashion and do not store large amounts of these mediators. In subsequent studies, our laboratory has continued to examine the time course of production and the contribution of specific factors on mast cell-derived chemokine production. Similar to the findings of others, we have been able to demonstrate that SCF drives significant levels of chemokine by itself and can further upregulate IgE-mediated production and release of the chemokines. Consistent with the above observation of *de novo* synthesis, the chemokine release peaked at 6–18 h post-stimulation with both IgE- and SCF-mediated production of chemokines. The data demonstrated that both SCF and IgE have the ability to promote the induction of chemokines and that, as previous observed (20), SCF augments IgE-mediated induction (Table I). As mentioned previously, the ability of soluble SCF to directly induce chemokine production without IgE-mediated mechanisms indicates that mast cells may play an important role in non-allergic responses.

Table I
SCF-induced Chemokine Production from Mast Cells

Chemokine	SCF	IgE	SCF + IgE
RANTES	++	–	+
Eotaxin	–	–	–
MIP-1α	+	+++	+++
MCP-1	+	++	+++

The activation of mast cells by SCF likely induces specific signal transduction pathways that may be common to other receptor-coupled responses. In particular, in comparison to IgE-mediated activation of mast cells most of the pathways of activation appear to be similar, including activation of protein tyrosine kinase, PI_3 kinase, MAP kinase, JNK, p38-MAP kinase and phospholipase D (12, 21, 47, 60–66), as well as activation of specific JAK–Stat proteins (64, 67–69). These pathways correlate well with the above data demonstrating that SCF can induce chemokine production. Recent studies have associated production of specific chemokines with signal transduction pathways involving MAP kinases (70, 71), thus closely associating the responses with one another. Preliminary data in our laboratory indicate that inhibitors of some of these pathways, such as protein tyrosine kinase and MAP kinase, block chemokine production from SCF-stimulated mast cells. Interestingly, IgE-mediated pathways also appear to utilize PKC (protein kinase C) pathways whereas SCF does not. However, these two mediator pathways synergize to increase the activation of other downstream proteins, such as Ras and Raf-1, leading to ERK protein activation (66), further suggesting that other differences in the way that IgE and SCF signal the mast cell may exist (12). Specific immune modulators differentially regulate common pathways of activation; IgE-mediated activation of JNK and p38-MAP kinase is inhibited by cyclosporin A and FK506, whereas these same pathways are not inhibited by cyclosporin A or FK506 after SCF-induced mast cell activation (12). These latter observations lead to interesting concepts that may be applicable to allergic diseases, such as asthma, where patients become increasingly resistant to specific inhibitors. Although the causative cell populations are still becoming activated during chronic asthmatic responses, the mediators, IgE versus SCF, may be changing as the disease progresses. These concepts may help explain how mast cells can participate in chronic allergic responses.

STRUCTURAL CELL-DERIVED SCF CONTRIBUTES TO DISEASE PROGRESSION: DIFFERENTIAL ACTIVATION OF MAST CELLS BY SOLUBLE VERSUS MEMBRANE SCF

The function of SCF in the peripheral system at the present time is unclear. It appears to perform several functions not related to haematopoietic mechanisms in multiple cell populations. An interesting view of a possible function of SCF comes from tumour cell research where, it appears that many tumour cell populations express SCF and its receptor (72–79). These observations are thought to be important for tumour cell growth and/or protection from lethal immune function (41–43, 45, 46). Even mast cells themselves have been observed to produce and cleave different forms of SCF (80). The function of SCF may ultimately depend upon whether a cell encounters SCF in a membrane or soluble form of the molecule. Over the past couple of years our laboratory has attempted to begin addressing the differential activation of c-kit$^+$ cell populations, mast cells and eosinophils in particular, with either soluble or 'membrane' SCF. This is an important issue since most of the SCF within the tissue appears to be membrane-bound and not solubilized (unpublished data). This potential 'reservoir' of SCF would initially be encountered in a membrane form and much of it cleaved from the surface of cells once the proper enzymes were released. In addition, the engagement of cells via soluble or membrane SCF may both change the type or strength of signal that is induced and therefore affect the activational pathway of the cells. In initial experiments we were interested in whether soluble versus bound SCF altered the production of chemokines from mast cell populations. These studies were performed using isolated populations cultured onto SCF-coated plates to simulate the membrane form of SCF. Bound SCF appears to induce eotaxin and RANTES production from mast cells, whereas soluble SCF induced production of only RANTES, MIP-1α and MCP-1. In contrast, MIP-1α and MCP-1 production was not increased by the plate-bound SCF, demonstrating a specificity of the interaction for eotaxin and RANTES production. The expression of these chemokines, in particular, from mast cells may specifically aid in the accumulation of eosinophils within the airway. Altogether, these data demonstrate that plate-bound SCF induces chemokine production from airway-derived mast cells, which may lead to increased eosinophilic inflammation within the airway. Co-culture of mast cells with lung fibroblasts induces significant histamine release (81–83). A significant decrease in histamine release can be observed by pre-treatment of the fibroblasts with anti-SCF antibody or by incubation of mast cells in a transwell set-up that separates them from the fibroblasts (84). This further supports the concept that membrane-associated SCF from fibroblasts is responsible for the activation. Interestingly, when similar assays were set up with alveolar macrophages the histamine release could not be decreased in transwell separated compared to co-cultured cell populations, suggesting that soluble SCF produced by macrophages could pass through the transwell membrane and activate the mast cells. The co-culture of mast cells with fibroblasts also demonstrated an SCF-dependent increase in eotaxin production that could be blocked by pre-treating the fibroblasts with anti-SCF or separating the cell populations by a transwell filter. In contrast, alveolar macrophage populations co-cultured with mast cells demonstrate no ability to drive eotaxin production consistent with the observation that soluble SCF does not induce chemokine production (unpublished data). These results suggest that SCF production from pulmonary cell populations contribute to the activation of surrounding mast cells within the upper airway. The fact that the two pulmonary cell populations, macrophages and fibroblasts, may initially express different forms of SCF, soluble or transmembrane, respectively, and have differential effects on mast cell-derived eotaxin and RANTES, but not histamine release, suggests differential activation cascades during allergic inflammation *in vivo*. The elucidation of the mechanisms involved in mast cell activation by membrane versus soluble SCF, with or without IgE stimulation, will provide novel information on the expression and role of SCF during allergic inflammation.

The above issues of mast cell activation by SCF expressed on structural cells may be relevant in several disease scenarios. The induction of chronic disorders leading to end-stage disease with associated fibrotic episodes, including recent data in asthma, may be of particular interest. Results over the past several years have clearly demonstrated a relationship between myofibroblast proliferation in fibrotic diseases with mast cell hyperplasia within the affected tissue (85–88). Although it appears that mast cells are not causative for fibrosis, the activation of mast cells may increase the speed of progression and/or severity of the fibrotic event (89–92). In addition, eosinophil accumulation within the fibrotic tissue also has strong associations with the severity and/or intensity of the response. Thus, the interaction between fibroblasts and mast cells leading to specific chemokine production and eosinophil accumulation may represent a relevant scenario for the progression of events that lead to end-stage fibrotic disease.

SUMMARY

The production of SCF during disease may play multiple roles in prolonging and intensifying the inflammatory response. One aspect of these mechanisms is represented in Fig. 3. The introduction of specific antigen (allergen) leads to the production of TNF and IL-4, both of which have the ability to drive SCF production in local macrophage and fibroblast populations. The production of SCF can directly and indirectly (via augmentation of IgE-mediated responses) activate mast cells, inducing histamine and leukotriene release (soluble and membrane SCF) and activation of chemokine production (membrane only). Subsequently, increased CC chemokine production leads to increased

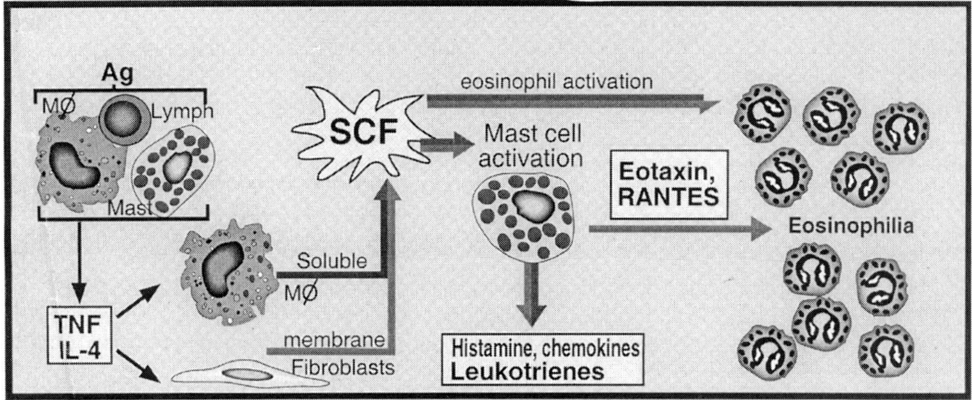

Fig. 3 The role of SCF in allergic responses.

leukocyte accumulation through the specific recruitment of mononuclear cells and eosinophils. The soluble form of SCF induces a level of mast cell activation associated with leukotriene release with immediate effects on the pathophysiology of the airway, whereas membrane-associated SCF activates the mast cell for RANTES and eotaxin production leading to eosinophil accumulation associated with long-term chronic damage to the airway. Thus, both forms of SCF appear to induce mechanisms that contribute to local airway damage and clinical disease. The overall concept of the role of SCF in exacerbation of disease may apply to multiple inflammatory disorders and may represent a suitable therapeutic target, especially for chronic allergic disorders.

REFERENCES

1. Wasserman, S. I. Mast cells and airway inflammation in asthma. *Am. J. Respir. Crit. Care. Med.* **150**:S39, 1994.
2. Miller, H. R. Mucosal mast cells and the allergic response against nematode parasites. *Vet. Immunol. Immunopathol.* **54**:331, 1996.
3. de Pater-Huijsen, F. L., Pompen, M., Jansen, H. M. and Out, T. A. Products from mast cells influence T lymphocyte proliferation and cytokine production – relevant to allergic asthma? *Immunol. Lett.* **57**:47, 1997.
4. Rossi, G. L. and Olivieri, D. Does the mast cell still have a key role in asthma? *Chest* **112**:523, 1997.
5. Columbo, M., Horowitz, E. M., Botana, L. M., MacGlashan, D. W. Jr, Bochner, B. S., Gillis, S., Zsebo, K. M., Galli, S. J. and Lichtenstein, L. M. The human recombinant c-kit receptor ligand, rhSCF, induces mediator release from human cutaneous mast cells and enhances IgE-dependent mediator release from both skin mast cells and peripheral blood basophils. *J. Immunol.* **149**:599, 1992.
6. Galli, S. J., Tsai, M. and Wershil, B. K. The c-kit receptor, stem cell factor, and mast cells. What each is teaching us about the others. *Am. J. Pathol.* **142**:965, 1993.
7. Galli, S. J. and Hammel, I. Mast cell and basophil development. *Curr. Opin. Hematol.* **1**:33, 1994.
8. Galli, S. J., Zsebo, K. M. and Geissler, E. N. The kit ligand, stem cell factor. *Adv. Immunol.* **55**:1, 1994.
9. Costa, J. J., Demetri, G. D., Harrist, T. J., Dvorak, A. M., Hayes, D. F., Merica, E. A., Menchaca, D. M., Gringeri, A. J., Schwartz, L. B. and Galli, S. J. Recombinant human stem cell factor (kit ligand) promotes human mast cell and melanocyte hyperplasia and functional activation *in vivo*. *J. Exp. Med.* **183**:2681, 1996.
10. Lin, T. J., Bissonnette, E. Y., Hirsh, A. and Befus, A. D. Stem cell factor potentiates histamine secretion by multiple mechanisms, but does not affect tumour necrosis factor-alpha release from rat mast cells. *Immunology* **89**:301, 1996.

11. Lukacs, N. W., Kunkel, S. L., Strieter, R. M., Evanoff, H. L., Kunkel, R. G., Key, M. L. and Taub, D. D. The role of stem cell factor (c-kit ligand) and inflammatory cytokines in pulmonary mast cell activation. *Blood* **87**:2262, 1996.
12. Ishizuka, T., Kawasome, H., Terada, N., Takeda, K., Gerwins, P., Keller, G. M., Johnson, G. L. and Gelfand, E. W. Stem cell factor augments Fc epsilon RI-mediated TNF-alpha production and stimulates MAP kinases via a different pathway in MC/9 mast cells. *J. Immunol.* **161**:3624, 1998.
13. Huang, E. J., Nocka, K. H., Buck, J. and Besmer, P. Differential expression and processing of two cell associated forms of the kit-ligand: KL-1 and KL-2. *Mol. Biol. Cell* **3**:349, 1992.
14. Dastych, J. and Metcalfe, D. D. Stem cell factor induces mast cell adhesion to fibronectin. *J. Immunol.* **152**:213, 1994.
15. Kinashi, T. and Springer, T. A. Steel factor and c-kit regulate cell–matrix adhesion. *Blood* **83**:1033, 1994.
16. Jippo-Kanemoto, T., Adachi, S., Ebi, Y., Matsuda, H., Kasugai, T., Nishikawa, S. and Kitamura, Y. BALB/3T3 fibroblast-conditioned medium attracts cultured mast cells derived from W/W but not from mi/mi mutant mice, both of which are deficient in mast cells. *Blood* **80**:1933, 1992.
17. O'Byrne, P. M. Airway inflammation and asthma. *Aliment. Pharmacol. Ther.* **10 (Suppl. 2)**:18, 1996.
18. Harris, R. R., Komater, V. A., Marett, R. A., Wilcox, D. M. and Bell, R. L. Effect of mast cell deficiency and leukotriene inhibition on the influx of eosinophils induced by eotaxin. *J. Leukoc. Biol.* **62**:688, 1997.
19. Zhang, S., Howarth, P. H., Roche, W. R., Wasserman, S. I., Bischoff, S. C. and Dahinden, C. A. Cytokine production by cell cultures from bronchial subepithelial myofibroblasts/Mast cells and airway inflammation in asthma c-kit ligand: a unique potentiator of mediator release by human lung mast cells. *J. Pathol.* **180**:95, 1996.
20. Baghestanian, M., Hofbauer, R., Kiener, H. P., Bankl, H. C., Wimazal, F., Willheim, M., Scheiner, O., Fureder, W., Muller, M. R., Bevec, D., Lechner, K. and Valent, P. The c-kit ligand stem cell factor and anti-IgE promote expression of monocyte chemoattractant protein-1 in human lung mast cells. *Blood* **90**:4438, 1997.
21. Ishizuka, T., Chayama, K., Takeda, K., Hamelmann, E., Terada, N., Keller, G. M., Johnson, G. L. and Gelfand, E. W. Mitogen-activated protein kinase activation through Fc epsilon receptor I and stem cell factor receptor is differentially regulated by phosphatidylinositol 3-kinase and calcineurin in mouse bone marrow-derived mast cells. *J. Immunol.* **162**:2087, 1999.
22. Undem, B. J., Lichtenstein, L. M., Hubbard, W. C., Meeker, S. and Ellis, J. L. Recombinant stem cell factor-induced mast cell activation and smooth muscle contraction in human bronchi. *Am. J. Respir. Cell Mol. Biol.* **11**:646, 1994.
23. Lukacs, N. W., Strieter, R. M., Lincoln, P. M., Brownell, E., Pullen, D. M., Schock, H. J., Chensue, S. W., Taub, D. D. and Kunkel, S. L. Stem cell factor (c-kit ligand) influences eosinophil recruitment and histamine levels in allergic airway inflammation. *J. Immunol.* **156**:3945, 1996.
24. Yuan, Q., Austen, K. F., Friend, D. S., Heidtman, M. and Boyce, J. A. Human peripheral blood eosinophils express a functional c-kit receptor for stem cell factor that stimulates very late antigen 4 (VLA-4)-mediated cell adhesion to fibronectin and vascular cell adhesion molecule 1 (VCAM-1). *J. Exp. Med.* **186**:313, 1997.
25. FitzGerald, J. M. and Macklem, P. Fatal asthma. *Annu. Rev. Med.* **47**:161, 1996.
26. Backman, K. S., Greenberger, P. A. and Patterson, R. Airways obstruction in patients with long-term asthma consistent with 'irreversible asthma'. *Chest* **112**:1234, 1997.
27. Manian, P. Genetics of asthma: a review. *Chest* **112**:1397, 1997.
28. Meza, C. and Gershwin, M. E. Why is asthma becoming more of a problem? *Curr. Opin. Pulmon. Med.* **3**:6, 1997.
29. Campbell, E., Hogaboam, C., Lincoln, P. and Lukacs, N. W. Stem cell factor-induced airway hyperreactivity in allergic and normal mice. *Am. J. Pathol.* **154**:1259, 1999.
30. Hamel, W. and Westphal, M. The road less travelled: c-kit and stem cell factor. *J. Neuro-oncol.* **35**:327, 1997.
31. Sieber-Blum, M. and Zhang, J. M. Growth factor action in neural crest cell diversification. *J. Anat.* **191**:493, 1997.
32. Longley, B. J., Tyrrell, L., Ma, Y., Williams, D. A., Halaban, R., Langley, K., Lu, H. S. and Schechter, N. M. Chymase cleavage of stem cell factor yields a bioactive, soluble product. *Proc. Natl. Acad. Sci. USA* **94**:9017, 1997.
33. de Paulis, A., Minopoli, G., Dal Piaz, F., Pucci, P., Russo, T. and Marone, G. Novel autocrine and paracrine loops of the stem cell factor/chymase network. *Int. Arch. Allergy Immunol.* **118**:422, 1999.
34. Samet, J. M., Fasano, M. B., Fonteh, A. N. and Chilton, F. H. Selective induction of prostaglandin G/H

synthase I by stem cell factor and dexamethasone in mast cells. *J. Biol. Chem.* **270**:8044, 1995.

35. Finotto, S., Mekori, Y. A. and Metcalfe, D. D. Glucocorticoids decrease tissue mast cell number by reducing the production of the c-kit ligand, stem cell factor, by resident cells: *in vitro* and *in vivo* evidence in murine systems. *J. Clin. Invest.* **99**:1721, 1997.
36. Kim, Y. K., Nakagawa, N., Nakano, K., Sulakvelidze, I., Dolovich, J. and Denburg, J. Stem cell factor in nasal polyposis and allergic rhinitis: increased expression by structural cells is suppressed by *in vivo* topical corticosteroids. *J. Allergy Clin. Immunol.* **100**:389, 1997.
37. Kassel, O., Schmidlin, F., Duvernelle, C., de Blay, F. and Frossard, N. Up- and down-regulation by glucocorticoids of the constitutive expression of the mast cell growth factor stem cell factor by human lung fibroblasts in culture. *Mol. Pharmacol.* **54**:1073,1998.
38. Hogaboam, C., Kunkel, S. L., Strieter, R. M., Taub, D. D., Lincoln, P., Standiford, T. J. and Lukacs, N. W. Novel role of transmembrane SCF for mast cell activation and eotaxin production in mast cell-fibroblast interactions. *J. Immunol.* **160**:6166, 1998.
39. Linenberger, M. L., Jacobson, F. W., Bennett, L. G., Broudy, V. C., Martin, F. H. and Abkowitz, J. L. Stem cell factor production by human marrow stromal fibroblasts. *Exp. Hematol.* **23**:1104, 1995.
40. Fireman, E., Kivity, S., Shahar, I., Reshef, T. and Mekori, Y. A. Secretion of stem cell factor by alveolar fibroblasts in interstitial lung diseases. *Immunol. Lett.* **67**:229, 1999.
41. Leigh, B. R., Webb, S., Hancock, S. L. and Knox, S. J. Stem cell factor enhances the survival of irradiated human bone marrow maintained in SCID mice. *Stem Cells* **12**:430, 1994.
42. Neta, R., Oppenheim, J. J., Wang, J. M., Snapper, C. M., Moorman, M. A. and Dubois, C. M. Synergy of IL-1 and stem cell factor in radioprotection of mice is associated with IL-1 up-regulation of mRNA and protein expression for c- kit on bone marrow cells. *J. Immunol.* **153**:1536, 1994.
43. Leigh, B. R., Khan, W., Hancock, S. L. and Knox, S. J. Stem cell factor enhances the survival of murine intestinal stem cells after photon irradiation. *Radiat. Res.* **142**:12, 1995.
44. McNiece, I. K. and Briddell, R. A. Stem cell factor. *J. Leukoc. Biol.* **58**:14, 1995.
45. Shui, C., Khan, W. B., Leigh, B. R., Turner, A. M., Wilder, R. B. and Knox, S. J. Effects of stem cell factor on the growth and radiation survival of tumor cells. *Cancer Res.* **55**:3431, 1995.
46. Neta, R. Modulation with cytokines of radiation injury: suggested mechanisms of action. *Environ. Health Perspect.* **105 (Suppl. 6)**:1463, 1997.
47. Krystal, G. W., DeBerry, C. S., Linnekin, D. and Litz, J. Lck associates with and is activated by Kit in a small cell lung cancer cell line: inhibition of SCF-mediated growth by the Src family kinase inhibitor PP1. *Cancer Res.* **58**:4660, 1998.
48. Huang, H. M., Li, J. C., Hsieh, Y. C., Yang-Yen, H. F. and Yen, J. J. Optimal proliferation of a hematopoietic progenitor cell line requires either costimulation with stem cell factor or increase of receptor expression that can be replaced by overexpression of Bcl-2. *Blood* **93**:2569, 1999.
49. Negro, J. M., Miralles, J. C., Ortiz, J. L., Funes, E. and Garcia, A. Leukotrienes and their antagonists in allergic disorders. *Allergol. Immunopathol.* **25**:104, 1997.
50. O'Byrne, P. M. Leukotrienes in the pathogenesis of asthma. *Chest* **111**:27S, 1997.
51. Wenzel, S. E. Arachidonic acid metabolites: mediators of inflammation in asthma. *Pharmacotherapy* **17**:3S, 1997.
52. Selvan, R. S., Butterfield, J. H. and Krangel, M. S. Expression of multiple chemokine genes by a human mast cell leukemia. *J. Biol. Chem.* **269**:13893, 1994.
53. Yano, K., Yamaguchi, M., de Mora, F., Lantz, C. S., Butterfield, J. H., Costa, J. J. and Galli, S. J. Production of macrophage inflammatory protein-1alpha by human mast cells: increased anti-IgE-dependent secretion after IgE-dependent enhancement of mast cell IgE-binding ability. *Lab. Invest.* **77**:185, 1997.
54. Lukacs, N. W., Strieter, R. M., Chensue, S. W. and Kunkel, S. L. Activation and regulation of chemokines in allergic airway inflammation. *J. Leukoc. Biol.* **59**:13, 1996.
55. Lukacs, N. W., Hogaboam, C. M., Kunkel, S. L., Chensue, S. W., Burdick, M. D., Evanoff, H. L. and Strieter, R. M. Mast cells produce ENA-78, which can function as a potent neutrophil chemoattractant during allergic airway inflammation. *J. Leukoc. Biol.* **63**:746, 1998.
56. Lukacs, N. W., Strieter, R. M., Standiford, T. J. and Kunkel, S. L. Characterization of chemokine function in animal models of diseases. *Methods* **10**:158, 1996.
57. Taub, D. D. Chemokine–leukocyte interactions. The voodoo that they do so well. *Cytokine Growth Factor Rev.* **7**:355, 1996.
58. Baggiolini, M., Dewald, B. and Moser, B. Human chemokines: an update. *Annu. Rev. Immunol.* **15**:675, 1997.

59. Moser, B., Loetscher, M., Piali, L. and Loetscher, P. Lymphocyte responses to chemokines. *Int. Rev. Immunol.* **16**:323, 1998.
60. Koike, T., Hirai, K., Morita, Y. and Nozawa, Y. Stem cell factor-induced signal transduction in rat mast cells. Activation of phospholipase D but not phosphoinositide-specific phospholipase C in c-kit receptor stimulation. *J. Immunol.* **151**:359, 1993.
61. Vogel, W., Lammers, R., Huang, J. and Ullrich, A. Activation of a phosphotyrosine phosphatase by tyrosine phosphorylation. *Science* **259**:1611, 1993.
62. Blume-Jensen, P., Ronnstrand, L., Gout, I., Waterfield, M. D. and Heldin, C. H. Modulation of Kit/stem cell factor receptor-induced signaling by protein kinase C. *J. Biol. Chem.* **269**:21793, 1994.
63. Linnekin, D., Keller, J. R., Ferris, D. K., Mou, S. M., Broudy, V. and Longo, D. L. Stem cell factor induces phosphorylation of a 200 kDa protein which associates with c-kit. *Growth Factors* **12**:57, 1995.
64. Linnekin, D., Mou, S., Deberry, C. S., Weiler, S. R., Keller, J. R., Ruscetti, F. W. and Longo, D. L. Stem cell factor, the JAK–STAT pathway and signal transduction. *Leuk. Lymphoma* **27**:439, 1997.
65. Pearson, M. A., O'Farrell, A. M., Dexter, T. M., Whetton, A. D., Owen-Lynch, P. J. and Heyworth, C. M. Investigation of the molecular mechanisms underlying growth factor synergy: the role of ERK 2 activation in synergy. *Growth Factors* **15**:293, 1998.
66. Sui, X., Krantz, S. B., You, M. and Zhao, Z. Synergistic activation of MAP kinase (ERK1/2) by erythropoietin and stem cell factor is essential for expanded erythropoiesis. *Blood* **92**:1142, 1998.
67. Linnekin, D., Weiler, S. R., Mou, S., DeBerry, C. S., Keller, J. R., Ruscetti, F. W., Ferris, D. K. and Longo, D. L. JAK2 is constitutively associated with c-Kit and is phosphorylated in response to stem cell factor. *Acta Haematol.* **95**:224, 1996.
68. Weiler, S. R., Mou, S., DeBerry, C. S., Keller, J. R., Ruscetti, F. W., Ferris, D. K., Longo, D. L. and Linnekin, D. JAK2 is associated with the c-kit proto-oncogene product and is phosphorylated in response to stem cell factor. *Blood* **87**:3688, 1996.
69. Gregory, R. C., Jiang, N., Todokoro, K., Crouse, J., Pacifici, R. E. and Wojchowski, D. M. Erythropoietin receptor and STAT5-specific pathways promote SKT6 cell hemoglobinization. *Blood* **92**:1104, 1998.
70. Sano, H., Nakagawa, N., Chiba, R., Kurasawa, K., Saito, Y. and Iwamoto, I. Cross-linking of intercellular adhesion molecule-1 induces interleukin-8 and RANTES production through the activation of MAP kinases in human vascular endothelial cells. *Biochem. Biophys. Res. Commun.* **250**:694, 1998.
71. Feoktistov, I., Goldstein, A. E. and Biaggioni, I. Role of p38 mitogen-activated protein kinase and extracellular signal-regulated protein kinase kinase in adenosine A2B receptor-mediated interleukin-8 production in human mast cells. *Mol. Pharmacol.* **55**:726, 1999.
72. Oka, M., Hirose, K., Iizuka, N., Aoyagi, K., Yamamoto, K., Abe, T., Hazama, S. and Suzuki, T. Cytokine mRNA expression patterns in human esophageal cancer cell lines. *J. Interferon Cytokine Res.* **15**:1005, 1995.
73. Papadimitriou, C. A., Topp, M. S., Serve, H., Oelmann, E., Koenigsmann, M., Maurer, J., Oberberg, D., Reufi, B., Thiel, E. and Berdel, W. E. Recombinant human stem cell factor does exert minor stimulation of growth in small cell lung cancer and melanoma cell lines. *Eur. J. Cancer* **31A**:2371, 1995.
74. Krystal, G. W., Hines, S. J. and Organ, C. P. Autocrine growth of small cell lung cancer mediated by coexpression of c-kit and stem cell factor. *Cancer Res.* **56**:370, 1996.
75. Shinohara, A., Shimizu, E., Takada, M. and Sone, S. Lack of c-mpl proto-oncogene transcripts and growth-stimulatory effects of thrombopoietin on human small cell lung cancer cell lines. *Oncology* **53**:426, 1996.
76. Bellone, G., Silvestri, S., Artusio, E., Tibaudi, D., Turletti, A., Geuna, M., Giachino, C., Valente, G., Emanuelli, G. and Rodeck, U. Growth stimulation of colorectal carcinoma cells via the c-kit receptor is inhibited by TGF-beta 1. *J. Cell Physiol.* **172**:1, 1997.
77. Krystal, G. W., Carlson, P. and Litz, J. Induction of apoptosis and inhibition of small cell lung cancer growth by the quinoxaline tyrphostins. *Cancer Res.* **57**:2203, 1997.
78. Yamamoto, T., Katayama, I. and Nishioka, K. Expression of stem cell factor in basal cell carcinoma. *Br. J. Dermatol.* **137**:709, 1997.
79. Pietsch, T., Nicotra, M. R., Fraioli, R., Wolf, H. K., Mottolese, M. and Natali, P. G. Expression of the c-Kit receptor and its ligand SCF in non-small-cell lung carcinomas. *Int. J. Cancer* **75**:171, 1998.
80. de Paulis, A., Minopole, G., Arbustini, E., de Crescenzok, G., Dal Piaz, F., Pucci, P., Russo, T. and Marone, G. Stem cell factor is localized in, released from and cleaved by human mast cells. *J. Immunol.* (in press).
81. Levi-Schaffer, F., Austen, K. F., Gravallese, P. M. and Stevens, R. L. Coculture of interleukin 3-dependent mouse mast cells with fibroblasts results in a phenotypic change of the mast cells. *Proc. Natl.*

Acad. Sci. USA **83**:6485, 1986. Published erratum appears in *Proc. Natl. Acad. Sci. USA* **83**:7805, 1986.

82. Levi-Schaffer, F., Austen, K. F., Caulfield, J. P., Hein, A., Gravallese, P. M. and Stevens, R. L. Co-culture of human lung-derived mast cells with mouse 3T3 fibroblasts: morphology and IgE-mediated release of histamine, prostaglandin D2, and leukotrienes. *J. Immunol.* **139**:494, 1987.
83. Levi-Schaffer, F., Dayton, E. T., Austen, K. F., Hein, A., Caulfield, J. P., Gravallese, P. M., Liu, F. T. and Stevens, R. L. Mouse bone marrow-derived mast cells cocultured with fibroblasts. Morphology and stimulation-induced release of histamine, leukotriene B4, leukotriene C4, and prostaglandin D2. *J. Immunol.* **139**:3431, 1987.
84. Hogaboam, C. M., Lukacs, N. W., Chensue, S. W., Strieter, R. M. and Kunkel, S. L. Monocyte chemoattractant protein-1 synthesis by murine lung fibroblasts modulates CD4+ T cell activation. *J. Immunol.* **160**:4606, 1998.
85. Hebda, P. A., Collins, M. A. and Tharp, M. D. Mast cell and myofibroblast in wound healing. *Dermatol. Clin.* **11**:685, 1993.
86. Gruber, B. L. Mast cells: accessory cells which potentiate fibrosis. *Int. Rev. Immunol.* **12**:259, 1995.
87. Levi-Schaffer, F. and Rubinchik, E. Mast cell role in fibrotic diseases. *Isr. J. Med. Sci.* **31**:450, 1995.
88. Levi-Schaffer, F. and Weg, V, B. Mast cells, eosinophils and fibrosis. *Clin. Exp. Allergy* **27 (Suppl. 1)**:64, 1997.
89. Galli, S. J., Wershil, B. K. and Mekori, Y. A. Analysis of mast cell function in biological responses not involving IgE. *Int. Arch. Allergy Appl. Immunol.* **82**:269, 1987.
90. Mori, H., Kawada, K., Zhang, P., Uesugi, Y., Sakamoto, O. and Koda, A. Bleomycin-induced pulmonary fibrosis in genetically mast cell-deficient WBB6F1-W/Wv mice and mechanism of the suppressive effect of tranilast, an antiallergic drug inhibiting mediator release from mast cells, on fibrosis. *Int. Arch. Allergy Appl. Immunol.* **95**:195, 1991.
91. Chyczewski, L., Debek, W., Chyczewska, E., Debek, K. and Bankowski, E. Morphology of lung mast cells in rats treated with bleomycin. *Exp. Toxicol. Pathol.* **48**:515, 1996.
92. Okazaki, T., Hirota, S., Xu, Z. D., Maeyama, K., Nakama, A., Kawano, S., Hori, M. and Kitamura, Y. Increase of mast cells in the liver and lung may be associated with but not a cause of fibrosis: demonstration using mast cell-deficient Ws/Ws rats. *Lab. Invest.* **78**:1431, 1998.

SECTION TEN

CLINICAL IMPLICATIONS OF HISTAMINE H_1 RECEPTOR ANTAGONISTS

CHAPTER 39

Pharmacological Profile of Mizolastine, a Novel Histamine H_1 Receptor Antagonist

N. SELVE,[3] Ph. PICHAT,[1] J. GOLDHILL,[1] H. DEPOORTERE[2] and S. ARBILLA[1]*

Synthélabo Recherche, [1]Rueil Malmaison and [2]Bagneux, France, and [3]Schwarz Pharma, Monheim, Germany

HISTAMINE AND ITS RECEPTOR SUBTYPES

Histamine has been known as an endogenous biogenic amine for nearly a century (1). The first histamine receptor antagonists were synthesized in the 1930s and were shown to be effective in protecting against bronchospasm produced in guinea pigs by anaphylaxis or administration of histamine (2). However, histamine was not conclusively associated with tissue mast cells until 1953 (3).

Histamine antagonists were shown to reduce the effects of histamine on many tissues, notably vascular and extravascular smooth muscle (e.g. guinea pig ileum), but it became evident that some of the effects of histamine were not antagonized by these so-called antagonists; for example, histamine-stimulated gastric secretion was shown to be unaffected by three different antihistaminic compounds (4).

The presence of more than one histamine receptor was suggested by Ash and Schild (5), who initially noted that the classic antihistamine mepyramine could block histamine-induced contractions of guinea pig ileum but not histamine-induced gastric acid secretion. The presence of a histamine H_2 receptor was later confirmed by Black (6), who introduced burimamide, the first effective H_2 antagonist. Thus, the actions of histamine are mediated through two distinct receptors defined pharmacologically by the actions of their respective agonists and antagonists. Histamine H_1 receptor-mediated activities are stimulated by the H_1 agonists 2-methylhistamine, 2-(2-pyridyl)ethylamine and 2-(2-thiazolyl)ethylamine and are inhibited by 'classical' H_1 antihistamines such as diphenhydramine, chlorpheniramine and mepyramine. Histamine H_2 receptor-mediated activities are stimulated by the H_2 agonists 4-methylhistamine and dimaprit (6) and are inhibited by the H_2 receptor antagonists cimetidine, burimamide and ranitidine. A large number of more potent and selective H_2 receptor antagonists have since been developed

* Corresponding author.

MAST CELLS AND BASOPHILS
ISBN 0-12-473335-2

(7). Further quantitative investigations of the antagonist potency of burimamide on other histamine-mediated responses contributed to the definition and classification of a further histamine receptor, subsequently termed the histamine H_3 receptor (8).

This third histamine receptor was also defined by a functional assay. Histamine was found to inhibit its own synthesis and release in rat cerebral cortical slices, and the effects of H_1 and H_2 receptor agonists and antagonists indicated a distinct receptor (8).

Some of the major operational characteristics used to define the nature of the histamine receptor in the different tissue responses are described in Table I.

TABLE I
Characterization of Histamine Receptors

Receptor subtype	Location	Main responses
Histamine H_1	Most smooth muscle, endothelial cells, adrenal medulla, heart, CNS	Smooth muscle contraction, stimulation of NO formation, endothelial cell contraction, increased vascular permeability, stimulation of hormone release, negative inotropism, depolarization (block of potassium current leak)
Histamine H_2	Gastric parietal cells, vascular smooth muscle, suppressor T cells, neutrophils, CNS, heart, uterus (rat)	Stimulation of gastric acid secretion, smooth muscle relaxation, stimulation of adenylyl cyclase, positive chronotropic and inotropic effects on cardiac muscle, decreased firing rate
Histamine H_3	CNS, peripheral nerve endings (heart, lung, gastrointestinal tract), endothelium, enterochromaffin cells	Inhibition of neurotransmitter release (histamine autoreceptors and heteroreceptors of pre-synaptic location), inhibition of gastric acid secretion (dog), increase in smooth muscle voltage-dependent Ca^{2+} current

Adapted from ref. 29.

SPECIFICITY AND SELECTIVITY OF MIZOLASTINE FOR H_1 HISTAMINE RECEPTOR SUBTYPES

Mizolastine is a novel histamine H_1 receptor antagonist (for structure of mizolastine, see Fig. 1). The compound interacts with histamine H_1 receptors in receptor binding studies (using [^{3}H]pyrilamine as a ligand). The displacement by mizolastine and reference antihistamine compounds of [^{3}H]pyrilamine binding to membranes from guinea pig cerebellum is shown in Table II. Mizolastine inhibited the binding of [^{3}H]pyrilamine with an IC_{50} of 47 nmol l^{-1}, similar to that of astemizole, whereas terfenadine, cetirizine and loratadine were relatively weak displacers.

The selectivity of mizolastine for the histamine H_1 receptor was determined by the pharmacological characteristics of [^{3}H]mizolastine binding sites in the guinea pig cerebellum. Among the classical antihistamines tested, pyrilamine and chlorpheniramine were the most potent inhibitors of [^{3}H]mizolastine; loratadine was less active. Ligands for

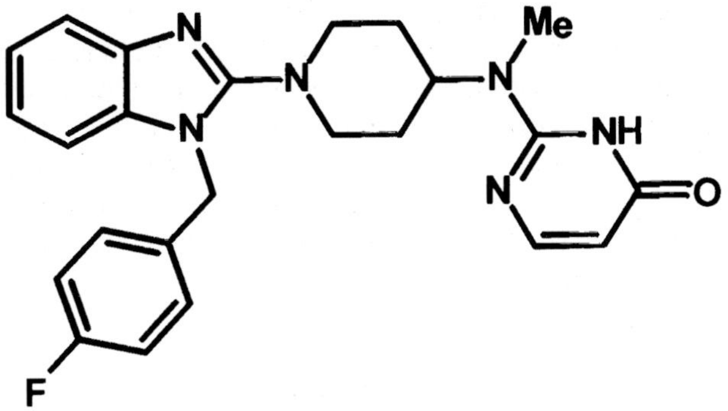

Fig. 1 Structure of mizolastine.

TABLE II
Histamine H_1 Receptor Binding Affinity of Mizolastine: Potency to Displace [^{3}H] Pyrilamine Binding to Membranes from Guinea Pig Cerebellum

Compound	IC_{50} (nmol l^{-1})
Mizolastine	47
Astemizole	52
Terfenadine	450
Cetirizine	470
Loratadine	1000

Adapted from ref. 9.

TABLE III
Pharmacological Characteristics of [^{3}H] Mizolastine Binding Sites in Guinea Pig Cerebellum

Compound	K_i (nmol l^{-1})
H_1 receptor ligands	
Pyrilamine	1
Chloropheniramine	6.4
Loratadine	50
H_2 receptor ligands	
Cimetidine	> 100,000
Dimaprit	> 100,000
H_3 receptor ligands	
Burimamide	> 100,000

Adapted from ref. 9.

the H_2 histamine receptor, such as dimaprit and cimetidine, and burimamide, a ligand for the histamine H_3 receptor, did not displace the binding of [^{3}H]mizolastine (Table III).

When evaluated in radioligand-binding assays for other neurotransmitter receptors and recognition sites, mizolastine displayed a very low (micromolar) affinity for serotonin (5-HT_{1A}, 5-HT_{1B}, 5-HT_{1C}, 5-HT_{1D}, 5-HT_2 and 5-HT_3), dopamine (D_1 and D_2), α_1- and α_2-adrenergic, adenosine and muscarinic cholinergic receptors. This compound is also devoid of noticeable affinity for the serotonin and norepinephrine transporters in rat brain and vas deferens membranes, respectively. Moreover, mizolastine does not inhibit, up to a concentration of 10 μmol l^{-1}, the binding of [^{3}H]nitrendipine or [^{3}H]batrachotoxin to the Ca^{2+} and Na^+ channels, respectively, or the binding of [^{3}H]flumazenil to W_1 or W_2 (benzodiazepine) modulatory sites of the $GABA_A$ receptor (9). Mizolastine also shows high specificity for the histamine H_1 receptor.

ANTIHISTAMINERGIC EFFECTS OF MIZOLASTINE

Mizolastine antagonizes the effects of histamine in a number of experimental models. Thus, it prevents histamine-induced paw oedema in the rat and mouse, the increase in capillary permeability induced by intradermal histamine injection in the rat and histamine-induced bronchoconstriction in the guinea pig (10).

Mizolastine has been shown to inhibit histamine-mediated reactions such as histamine-induced oedema, bronchoconstriction and capillary permeability. Mizolastine is 1.3- to >100-fold more potent than reference antihistamines in a range of different models and species.

Inhibition of Histamine-induced Oedema

Rat paw oedema was inhibited dose-dependently by oral mizolastine. This antihistaminic effect of mizolastine was superior to that of terfenadine and similar to that of astemizole and loratadine (Table IV) (10).

TABLE IV
Antihistaminergic Properties of Mizolastine in Various Animal Tests

Test	ED_{50} (mg kg^{-1} p.o.)			Ratio of activity (*b*/*a*)
	Mizolastine (*a*)	Reference drug (*b*)		
Rat paw oedema, induced by intraplantar injection of histamine (200 μg per 0.1 ml) *	0.50	Astemizole	0.8	1.6
		Loratadine	0.9	1.8
		Cetirizine	2.7	5.4
		Terfenadine	8.0	16.0
Skin oedema, induced by intradermal injection of histamine (0.5 μg per 0.1 ml) in the dog†	0.07	Terfenadine	7.5	> 100
Bronchoconstriction, induced by histamine (3 μg kg^{-1} i.v.) in the anaesthetized guinea pig*	0.03	Cetirizine	0.04	1.3
		Loratadine	0.20	6.7

* Taken from ref. 10.
† Taken from ref. 11.

Mizolastine also dose-dependently inhibited histamine-induced skin oedema in the dog, with a peak effect seen 2 h after treatment and a 100-fold higher potency compared to terfenadine (Table IV) (10, 11).

Inhibition of Histamine-induced Bronchoconstriction

Intravenous mizolastine is a potent inhibitor of histamine-induced bronchoconstriction in the guinea pig Konzett–Rössler test. Oral mizolastine was more potent than loratadine and equipotent to cetirizine in this test (Table IV) (10).

Inhibition of Histamine-induced Capillary Permeability

Significant inhibition of histamine-induced capillary permeability by mizolastine was found following oral treatment in rats and following intravenous treatment in guinea pigs. Oral mizolastine was as effective as astemizole, but much more effective than terfenadine or loratadine; intravenous mizolastine was three times more effective than astemizole or terfenadine (Table V). Mizolastine and astemizole inhibited histamine-induced tracheal and bronchopermeability, whereas terfenadine only inhibited the former (10).

TABLE V
Effects of Mizolastine on Vascular Permeability Induced by Histamine

Test	ED_{50} (mg kg^{-1})	
	Mizolastine	Reference drugs
Cutaneous permeability in the rat (histamine, 8μg per 0.1 ml)	< 0.3 p.o.	Astemizole < 0.3 p.o. Terfenadine < 2.5 p.o. Loratadine < 0.45 p.o.
Vascular permeability in guinea pig bronchi (histamine 50 μg kg^{-1} i.v.)	0.03 i.v.	Astemizole 0.09 i.v.

Taken from ref. 10.

ALLERGIC REACTION

The main function of the immune system is to maintain the integrity of a given organism, in particular by destroying invading organisms such as viruses or bacteria. There are four different parts of the immune system, which co-operatively supply, complete and reinforce each other; subdivision into unspecific (primary) versus specific and cellular versus humoral mechanisms is usually made.

The unspecific complement activation response and the specific immunoglobulin response at the humoral level (synthesis of antibodies) work together with unspecific macrophage/granulocyte activation and the specific T cell response at the cellular level. Antibodies can be produced in response to a vast range of substances (antigens). Antigens that cause allergic diseases (including urticaria, rhinoconjunctivitis, asthma and anaphylaxis) are called allergens. Most allergens are derived from natural organic sources (e.g. house dust mites, pollens, mould spores, insect and animal danders) as well as from foodstuffs and drugs (e.g. penicillin). The antibodies responsible for allergic diseases are immunoglobulins of the E class (IgE) (11a, 12).

Allergic reactions require prior exposure (i.e. pre-sensitization) to allergens, and several exposures may be required before the allergic reaction is initiated. The 'first' exposure (which may extend over weeks, months or even years) induces the production of IgE anti-allergen antibodies by B lymphocytes following antigen presentation by local immune competent cells. IgE, a plasma protein, binds to the surface of mast cells via their Fc_ε-receptor-1 (Fc_ε-R1). Once enough IgE antibody is present on the mast cells, the individual is sensitized to that allergen. Subsequent exposure to this allergen activates mast cells, resulting in an allergy, or type 1 hypersensitivity reaction, which is characterized by an initial release of histamine from these mast cells. In addition, spasmogenic eicosanoids (see below) can be produced. Other cells, macrophages or platelets (both of which have IgE receptors) and polymorphs are sometimes involved, and other mediators such as platelet-activating factor (PAF) can be produced. The effects may be localized to the nose (hay fever), the bronchial tree (the initial phase of asthma), the skin (urticaria) or the gastrointestinal tract. In some cases the reaction is more generalized and produces anaphylactic shock.

ANTI-ANAPHYLACTIC PROPERTIES OF MIZOLASTINE

Mizolastine shows anti-allergic activity, inhibits anaphylactic reactions, prevents lethal shock, inhibits clinical signs of bronchospasm and inhibits release of histamine from activated mast cells in animals.

Inhibition of Passive Cutaneous Anaphylaxis

Oral mizolastine inhibited passive cutaneous anaphylaxis in the rat. This effect was more potent than that of terfenadine and loratadine, and 2-fold weaker than that of astemizole (Table VI) (10).

TABLE VI
Effects of Mizolastine on Models of Anaphylaxis

Test	ED_{50} (mg kg^{-1} p.o.)		
	Mizolastine	Reference drugs	
Passive cutaneous anaphylaxis in the rat (ovalbumin challenge)	1.6	Astemizole	0.8
		Loratadine	4.8
		Terfenadine	10.0
Lethal shock in the rat (compound 48/80-induced)	0.07	Astemizole	0.07
		Terfenadine	1.3
		Loratadine	4.8
		Cetirizine	> 10.0

Taken from ref. 10.

Inhibition of Lethal Shock

Oral mizolastine inhibited compound 48/80-induced lethal shock in the rat. This effect was equipotent to that of astemizole, but 140, 70 and 20 times superior to cetirizine, loratadine and terfenadine, respectively (Table VI) (10).

Inhibition of Bronchospasm

Intravenous mizolastine at 0.05–0.2 mg kg^{-1}, prevented increased lung resistance (R_{max}) and decreased compliance preceding anaphylactic death in actively ovalbumin-sensitized guinea pigs. At 0.1 mg kg^{-1}, mizolastine provided almost complete protection. In addition, mizolastine significantly reduced alveolar distension and relative muscle thickness at doses of 0.05 and 0.1 mg kg^{-1} i.v., respectively. Cetirizine and terfenadine were less effective than mizolastine in terms of potency and efficacy (Table VII).

TABLE VII
Protective Effects of Mizolastine on Bronchopulmonary Hyper-reactivity in Ovalbumin-sensitive Guinea Pig

Test	Minimal effective dose (mg kg^{-1} p.o.)			Ratio of activity (*b*/*a*)
	Mizolastine (*a*)	Reference drug (*b*)		
Anaphylactic bronchospasm, induced by injection of ovalbumin (1 mg kg^{-1} i.v.)	0.03	Cetirizine	0.1	3
		Terfenadine	3.0	100

Taken from ref. 10.

Inhibition of Histamine Release

Mizolastine, administered orally, significantly and dose-dependently reduces histamine release from peritoneal mast cells in passively sensitized rats (Fig. 2) as well as from bronchial mast cells in actively sensitized guinea pigs (Fig. 3) (10), suggesting long-term preventive efficacy following mizolastine treatment. Loratadine p.o. and cromoglycate i.v. were less effective.

IMMUNE SYSTEM-MEDIATED INFLAMMATORY REACTION IN ALLERGY

Allergic diseases are a consequence of immune system-mediated inflammatory reactions. Mast cells have cytoplasmic granules that contain pre-formed inflammatory mediators bound to a proteoglycan matrix. When a sensitized individual is exposed to the sensitizing allergen, it is absorbed and rapidly combined with, thereby activating, IgE molecules on the surface of the mast cells. Activated mast cells degranulate (Fig. 4) (i.e. their granules are expelled into the surrounding environment) and the pre-formed mediators dissolve (13). The extracellular mediators rapidly induce an acute response, the allergic (or immediate hypersensitivity) reaction (12).

Pro-inflammatory mediators newly generated and secreted by activated mast cells

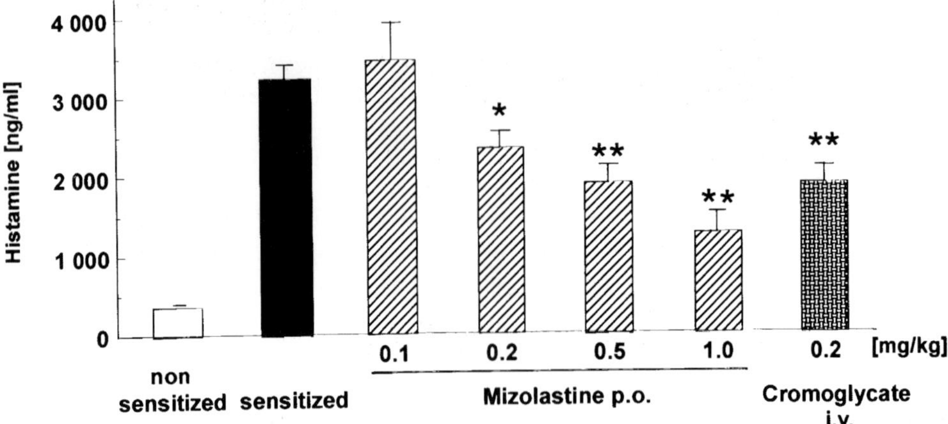

Fig. 2 Effects of mizolastine on the anaphylactic release of histamine from peritoneal mast cells in passively sensitized rats, compared with intravenous cromoglycate. $^*p < 0.05$, $^{**}p < 0.01$. (Reproduced from *Arzneim.-Forsch. Drug Res.* **45**:559–568, 1995.)

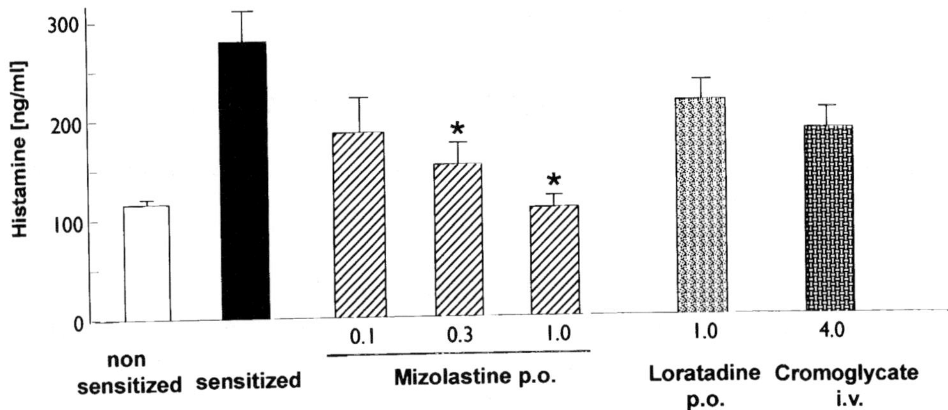

Fig. 3 Effects of oral mizolastine on the anaphylactic release of histamine from bronchial mast cells in actively sensitized guinea pigs: comparison with oral loratadine and intravenous sodium cromoglycate. $^*p < 0.05$. (Reproduced from *Arzneim.-Forsch. Drug Res.* **45**:559–568, 1995.)

include leukotrienes (LTs) derived from the 5-lipoxygenase (5-LO) pathway (i.e. LTB_4, LTC_4 and LTD_4, which have various physiological roles, including vasoconstriction, and induce bronchoconstriction, chemotaxis and chemokinesis), prostaglandins (PGs) and thromboxane derived from the cyclo-oxygenase pathway (which induce contraction of the bronchomuscles, aggregation of platelets and vasodilation) and PAF (which induces platelet activation). PGs, LTs and thromboxane are generated from the liberation of membrane-derived arachidonic acid (AA) products.

Together with the pre-formed mediators, these newly generated mediators create a transient local inflammatory reaction due to the migration of leukocytes (neutrophils, eosinophils and basophils) and plasma (causing oedema) from venules near the site of allergen entry (e.g. the nose or the skin).

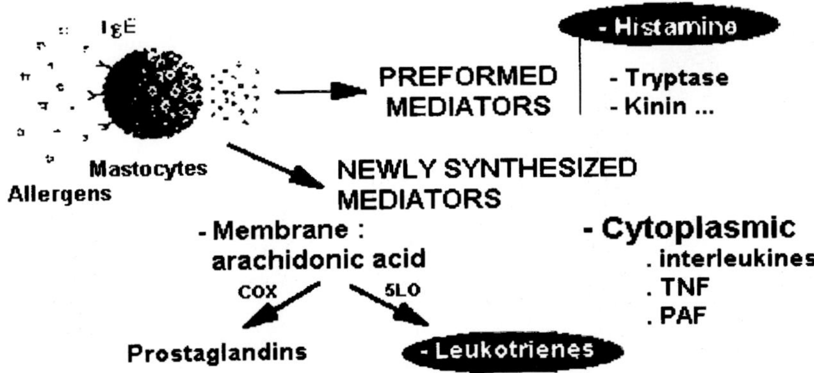

Fig. 4 Simplified representation of activated mast cells degranulation during inflammatory reaction in allergy.

Chemotactic factors also induce changes in leukocytes that allow them to move through the submucosal space. All these changes are transient and resolve when the allergen is no longer present. However, prolonged contact with an allergen, as in allergic rhinitis and asthma, results in chronic inflammation (12–14).

ANTI-ALLERGIC EFFECTS OF MIZOLASTINE

Mizolastine shows additional anti-allergic effects, such as inhibition of neutrophil migration and inhibition of AA-induced inflammation.

Inhibition of Neutrophil Migration

Ovalbumin challenge of sensitized guinea pigs increases the number and percentage of eosinophils and neutrophils in bronchoalveolar lavage fluid. Under histamine H_1 receptor blockade by pyrilamine (1 mg kg^{-1}, i.p.), mizolastine showed clear inhibition of eosinophil and significant inhibition of neutrophil cell infiltration into the bronchoalveolar lavage in actively ovalbumin sensitized guinea pigs (Fig. 5)(15). A dose-dependent effect of oral mizolastine (0.03–3 mg kg^{-1}) administered 1 h before and 8 h after intranasal challenge with ovalbumin at 1 mg per nostril was observed. Under these particular circumstances the antihistaminergic activity of mizolastine is unlikely to be responsible for this additional anti-allergic effect. Therefore, mizolastine is considered to act via another, additional mode of action here.

Inhibition of AA-induced Cutaneous Inflammation

At antihistaminic doses, mizolastine inhibited the cutaneous inflammatory reaction induced by AA in the rat (16). The anti-inflammatory effect of mizolastine was quick in onset and long-lasting. At a dose of comparable anti-inflammatory effect, dexamethasone showed long-lasting activity within the same time frame as mizolastine (Fig. 6). The reference antihistamines, loratadine and terfenadine, did not show any *in vivo* anti-

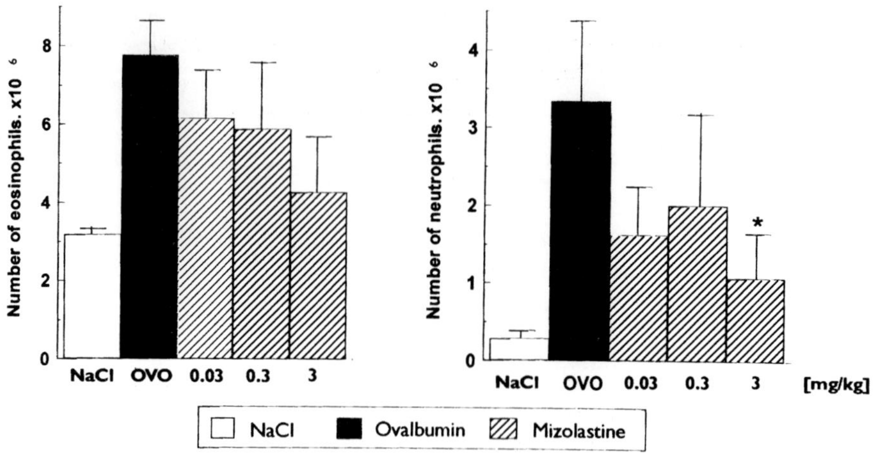

Fig. 5 Effect of orally administered mizolastine on cell migration into the bronchoalveolar lavage fluid in actively sensitized guinea pigs pre-treated with pyrilamine (1 mg kg^{-1}). $*p < 0.05$. (Adapted from ref. 15.)

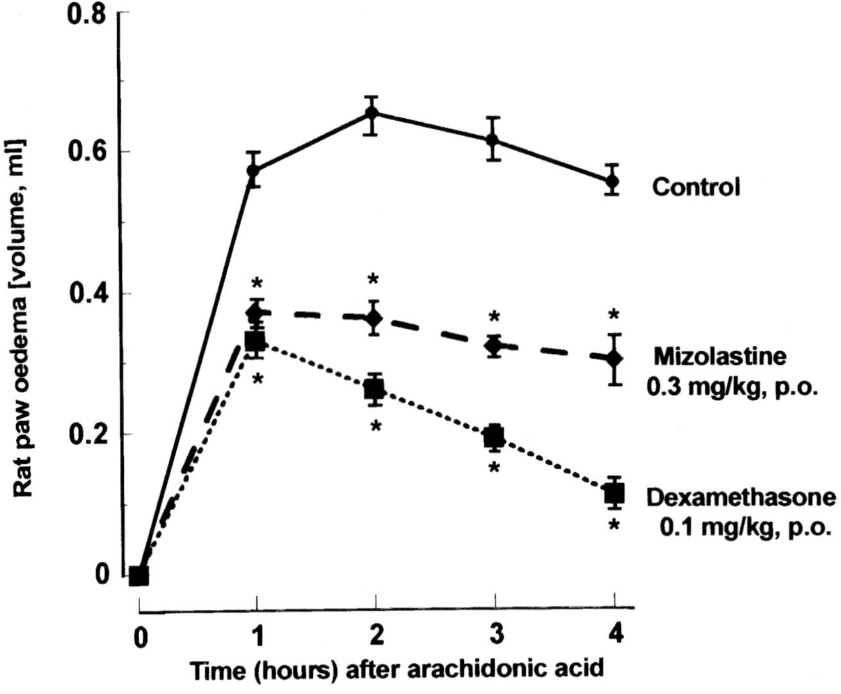

Fig. 6 Effects of mizolastine and dexamethasone (2 h pre-treatment) on rat paw inflammation induced by arachidonic acid (0.3%, 0.1 ml). * $p < 0.001$. (Reproduced from *Arzneim.-Forsch. Drug Res.* **48** (**Suppl.1**): 173–178, 1998.)

TABLE VIII
Effects of Mizolastine and Reference Compounds on Rat Paw Oedema Induced by Either Arachidonic Acid or Histamine

Drug	Histamine (200 μg per 0.1 ml) ED_{50} (mg kg^{-1} p.o.)	Arachidonic acid (0.3%, 0.1 ml) MED* (mg kg^{-1} p.o.)
Mizolastine	0.5	0.1
Loratadine	0.9	> 30
Terfenadine	8.0	> 30

* Minimal effective dose.
Taken from refs 10, 16.

inflammatory effects on AA-induced rat paw oedema up to 30 mg kg^{-1}, p.o. (16), although both compounds are active in inhibiting histamine-induced oedema (Table VIII) (10). Therefore, only mizolastine showed additional non-antihistaminic related properties among these antihistamines. Mizolastine inhibited 5-lipoxygenase activity in guinea pig peritoneal polymorphonuclear cells (IC_{50} = 3.7 μmol l^{-1}) *in vitro* (17).

In addition, the *in vivo* anti-inflammatory effect of mizolastine is probably related to inhibition of 5-LO. In fact, derived carrageenin-induced rat paw oedema, which is mainly mediated by cyclo-oxygenase metabolites (18, 19), was inhibited by the reference cyclo-oxygenase inhibitor ibuprofen, but was unaffected by mizolastine even at doses up to 50 mg kg^{-1}, p.o. (16).

Therefore, the anti-inflammatory effect of mizolastine is different to that of classical NSAIDs (non-steroidal anti-inflammatory drugs) and also differs from that of corticosteroids.

Inhibition of Visceral Afferent Sensitivity and Inflammation in Experimental Colitis

The anti-inflammatory potency of mizolastine was additionally investigated in a model of subchronic colonic inflammation. Oral mizolastine (0.03, 0.3, 3.0 mg kg^{-1}) dose-dependently reduced the various pathophysiological changes assessed in this model (Fig. 7). In contrast, terfenadine not only had no effect on trinitrobenzene sulphonic acid (TNBS)-induced pathophysiological changes, but in fact worsened some of the parameters tested (20).

OTHER PHARMACOLOGICAL EFFECTS OF MIZOLASTINE

Lack of Sedative Potential

Antihistamines are the most widely prescribed drugs for the treatment of allergic diseases, but a major limitation of the earlier H_1 receptor antagonists is their liability for inducing a variety of undesirable side-effects; for example, they often cause drowsiness and impair performance (21).

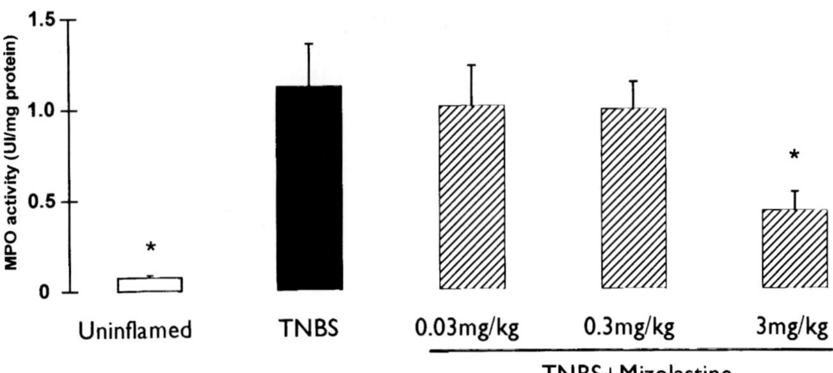

Fig. 7 Effect of mizolastine on TNBS-induced neutrophil infiltration in rat colon measured as an increase in myeloperoxydase activity (MPO). * $p < 0.05$ versus TNBS. (Adapted from *Arzneim.-Forsch. Drug Res.* **48 (Suppl. 1)**:179–184, 1998.)

The histamine synthesis inhibitor, α-fluoromethylhistidine, and the earlier H_1 receptor antagonists pyrilamine, diphenhydramine and chlorpheniramine reduce wakefulness and increase slow-wake sleep in rats, cats and dogs (22–24).

Recently, a new generation of H_1 receptor antagonists (astemizole, cetirizine, loratadine, terfenadine) has been developed and classified as non-sedating antihistamines owing to the fact that they do not interfere with CNS functions and, in addition, poorly cross the blood–brain barrier (24–26). Nevertheless, conflicting results have been reported concerning the central effects of some of these new antihistamines.

The pharmaco-EEG profiles of mizolastine and reference compounds have been characterized in chronically implanted rabbits as well as on the sleep–wake cycle in freely moving implanted rats.

EEG in Rodents

Mizolastine has no sedative potential on the EEG in rodents as increasing doses of mizolastine (3–10 mg kg^{-1}, i.v.) did not cause sedation in rabbits chronically implanted with cortical electrodes. No sedation was found with astemizole 1–10 mg kg^{-1} i.v., whereas loratadine 1–3 mg kg^{-1} i.v. and terfenadine 1–10 mg kg^{-1} i.v. induced sleep episodes that were characterized by increases in the 'activity' parameter of 20–50% and 20–70%, respectively (27).

Sleep–wake Cycle

Mizolastine at 10 mg kg^{-1} i.p. had no effect on the sleep–wake cycle in rats during the light period. During the dark period, the active period of rats, neither mizolastine nor cetirizine significantly modified the slow-wave sleep pattern, whereas astemizole, loratadine and terfenadine significantly increased the total duration of slow-wave sleep (27).

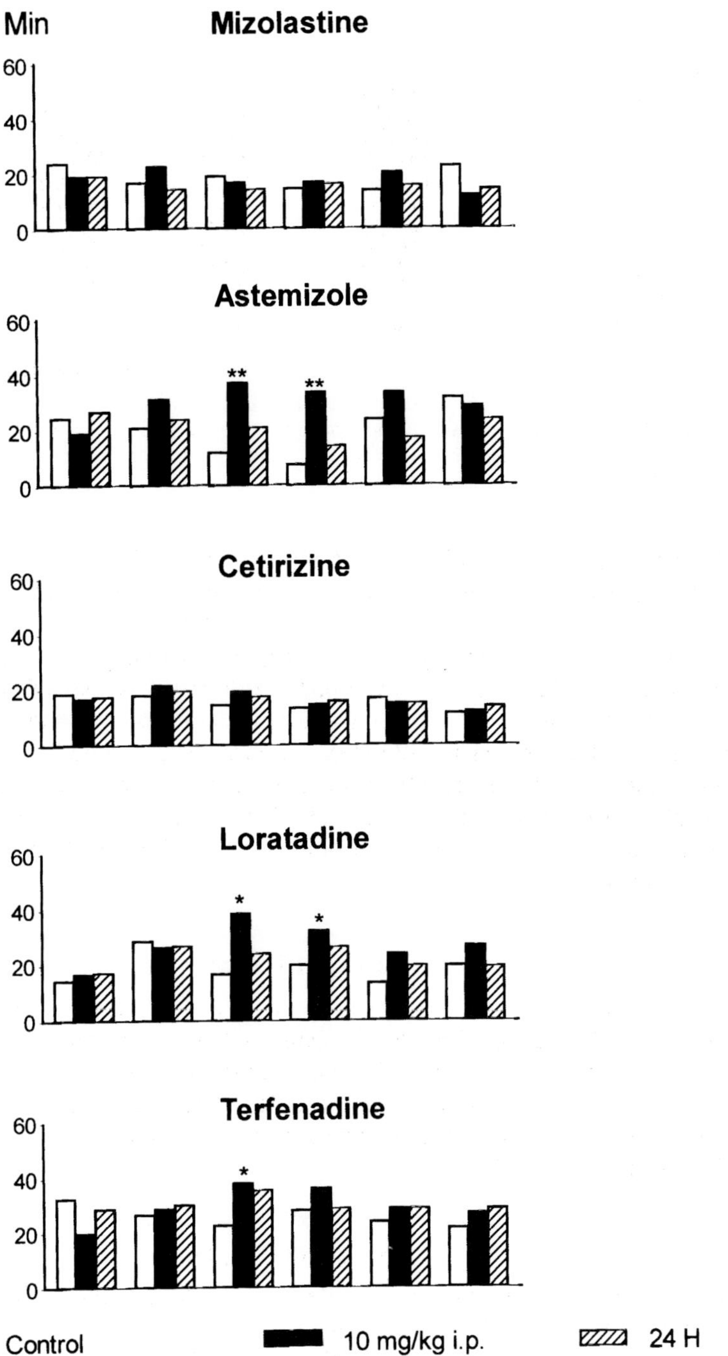

Fig. 8 Effect of various antihistamines (10 mg kg^{-1} i.p.) on the 6-h analysis of slow-wave sleep in rats recorded during the dark period. * $p < 0.05$, ** $p < 0.01$. (Reproduced from *Neuropsychobiology* **32**:214–221, 1995.)

Slow-wave Sleep

Hourly analysis of slow-wave sleep during the 6-h dark period in rats detected clear-cut differences between various antihistaminergics (Fig. 8).

At 10 mg kg^{-1} i.p. neither mizolastine nor cetirizine showed any effect on slow-wave sleeping time. In contrast, astemizole, loratadine and terfenadine significantly and reversibly enhanced the slow-wave sleeping time in at least one of the 6 h periods (27).

Lack of Anticholinergic Effects

The *in vitro* specificity (see below) can be additionally demonstrated *in vivo*. Mizolastine, even at a dose 60-fold higher than the antihistaminic dose did not change acetylcholine-induced decrease of blood pressure in rats, but significantly normalized histamine-induced reduction of blood pressure (9).

Lack of Effect on Cardiovascular Parameters at Doses Higher than Antihistaminic Doses in Different Species

In the rat, oral mizolastine at 30 mg kg^{-1}, i.e. 60-fold higher than the antihistaminic dose in this species, modify neither mean arterial pressure nor heart rate compared to placebo. In pithed rats, again oral mizolastine at 30 mg kg^{-1} had no effect on mean arterial pressure or heart rate compared to placebo, and failed to modify the increases in arterial blood pressure and heart rate evoked by exogenously administered noradrenaline, or serotonin, or electrical stimulation of the spinal cord (9).

At doses >1600 times the antihistaminic dose in guinea pig, mizolastine had no effect on the electrocardiogram (ECG) of conscious animals, while astemizole and terfenadine induced ECG changes (RR and QT interval) at doses less than 100 times and less than 600 times the antihistaminic dose, respectively (28).

CONCLUSION

Mizolastine is a potent specific and selective H_1 antihistaminergic *in vitro* and *in vivo*. It inhibits histamine release from activated mast cells and shows additional unique anti-allergic properties at similar doses *in vivo*. At antihistaminic and supra-antihistaminic doses, mizolastine is not sedative and does not show any anticholinergic effects.

Mizolastine at low doses exhibits a powerful and long-lasting anti-inflammatory action against AA-induced paw oedema. Based on the existing experimental data, the mechanism of the anti-inflammatory effect of mizolastine appears to involve the 5-LO pathway, and would exclude direct antagonism of histamine H_1 receptors as terfenadine, loratadine and pyrilamine were ineffective. In the animal models described so far, mizolastine reduces the infiltration of neutrophils and reduces inflammatory reactions.

Therefore, mizolastine may have an anti-inflammatory activity, which, combined with its antihistamine properties, could be valuable for the treatment of allergic inflammation.

ACKNOWLEDGEMENT

The authors wish to thank John Alexander for his invaluable input in the preparation of this manuscript.

REFERENCES

1. Dale, H. H. and Laidlaw, P. P. The physiological action of β-imidazolylethylamine. *J. Physiol.* **41**:318–344, 1911.
2. Bovet, D. and Staub, A. Action protectrice des éthers phénoliques au cours de l'intoxication histaminique. *C. R. Seances Soc. Biol. Fil.* **124**:547–549, 1936.
3. Riley, J. F. and West, D. B. Histamine and tissue mast cells. *J. Physiol.* **120**:528–537, 1953.
4. Ashford, C. A., Heller, H. and Smart, G. A. The action of histamine on hydrochloric acid and pepsin secretion in man. *Br. J. Pharmacol.* **4**:153–161, 1949.
5. Ash, A. S. F. and Schild, H. O. Receptors mediating some actions of histamine. *Br. J. Pharmacol.* **27**:427–439, 1966.
6. Black, J. W., Duncan, W. A. M., Durant, C. J., Ganellin, C. R. and Parsons, E. M. Definition and antagonism of histamine H_2-receptors. *Nature* **236**:385–390, 1972.
7. Cooper, D. G., Young, R. C., Durant, G. J. and Ganellin, C. R. Histamine receptors. In: *Comprehensive Medicinal Chemistry*, Vol. 3. (Emmett, J. C., ed.), pp. 323–421. Pergamon Press, Oxford, 1990.
8. Arrang, J. M., Garbarg, M. and Schwartz, J. C. Auto-inhibition of brain histamine release mediated by a novel class (H_3) of histamine receptor. *Nature* **302**:1–5, 1983.
9. Benavides, J., Schoemaker, H., Dana, C., Claustre, Y., Delahaye, M., Prouteau, M., *et al. In vivo* and *in vitro* interaction of the novel selective histamine H_1 receptor antagonist mizolastine with H_1 receptors in the rodent. *Arzneim.-Forsch. Drug Res.* **45**:551–558, 1995.
10. Levrier, J., Duval, D., Prouteau, M., Voltz, C., Berry, C., Lloyd, K. G., *et al.* Anti-anaphylactic activity of the novel selective histamine H_1 receptor antagonist mizolastine in the rodent. *Arzneim.-Forsch. Drug Res.* **45**:559–568, 1995.
11. Angel, I., Pichat, P. and Arbilla, S. Powerful anti-histamine properties of mizolastine in cutaneous oedema in the dog. *Allergy* **51 (Suppl. 31)**:P583, 1996.
11a. Holgate, S. T. and Church, M. K. Atopic dermatitis and contact dermatitis. Pathophysiology. In: *Allergy*, pp. 23.5–23.6, Gower Medical, London, 1993.
12. Janeway, C. A. and Travers, P. *Immunobiology: The Immune System in Health and Disease*, 2nd edn. London, Current Biology, 1996.
13. Church, M. K. and Levi-Schaffer, F. Updates on cells and cytokines. The human mast cell. *J. Allergy Clin. Immunol.* **99**:155–160, 1997.
14. Rajakulasingam, K. and Howarth, P. H. Relation of disordered airway function to the treatment of rhinitis. In: *Asthma and Rhinitis* (Busse, W. W. and Holgate, S. T., eds), pp. 1247–1254. Blackwell Scientific, Boston, 1995.
15. Vargaftig, B. B. Mechanisms of experimental bronchopulmonary hyperresponsiveness as related to eosinophils. In: *New Developments in the Therapy of Allergic Disorders and Asthma*, Vol. 6 (Langer, S. Z., Church, M. K., Vargartig, B. B. and Nicosia, S., eds), pp. 27–32. Karger, Basel, 1993.
16. Pichat, P., Angel, I. and Arbilla, S. Anti-inflammatory properties of mizolastine after oral administration on arachidonic acid-induced cutaneous reaction in the rat. *Arzneim.-Forsch. Drug Res.* **48 (Suppl. 1)**:173–178, 1998.
17. Sudo, K., Nagai, K. and Yamada, N. Inhibitory effect of mizolastine on 5-lipoxygenase. *Jpn. Pharmacol. Ther.* **26**:155–157, 1998.
18. Boughton-Smith, N. K., Deakin, A. M., Follenfant, R. L., Whittle, B. J. R. and Garland, L. G. Role of oxygen radicals and arachidonic acid metabolites in the reverse passive Arthus reaction and carrageenin paw oedema in the rat. *Br. J. Pharmacol.* **110**:896–902, 1993.
19. Winter, C. A. and Risley, E. A. Carrageenan-induced oedema in the hindpaw of rats as an assay for anti-inflammatory drugs. *Proc. Soc. Exp. Biol. Med.* **11**:544–547, 1962.
20. Goldhill, J., Pichat, P., Roome, N., Angel, I. and Arbilla, S. Effect of mizolastine on visceral sensory afferent sensitivity and inflammation during experimental colitis. *Arzneim.-Forsch. Drug Res.* **48 (Suppl. 1)**:179–184, 1998.
21. Nicholson, A. N. and Stone, B. M. Antihistamines: impaired performance and the dependency to sleep. *Eur. J. Clin. Pharmacol.* **30**:27–32, 1986.
22. Lin, J. S., Sakai, K. and Jouvet, M. Rôle des systèmes histaminergiques hypothalamiques dans la régulation des états de vigilance chez le chat. *CR Acad. Sci. III* **303**:469–474, 1986.
23. Kiyono, S., Seo, M. L., Shibagaki, M., Watanab, T., Maeyama, K. and Wada, H. Effects of α-fluoromethylhistamine on sleep–waking parameters in rats. *Physiol. Behav.* **34**:615–617, 1985.

24. Wauquier, A., Van Den Broeck, W. A. E., Awouters, F. and Janssen, P. A. J. A comparison between astemizole and other antihistamines on sleep–wakefulness cycles in dogs. *Neuropharmacology* **20**:853–859, 1981.
25. Marzanatti, M., Monopoli, A., Trampus, M. and Ongini, E. Effects of non-sedating histamine H_1-antagonists on EEG activity and behavior in the cat. *Pharmacol. Biochem. Behav.* **32**:861–866, 1989.
26. Weiner, M. Sedation and antihistaminics. *Arzneim.-Forsch. Drug Res.* **32**:1193–1195, 1982.
27. Depoortere, H., Decobert, M., Granger, P. and Françon, D. Mizolastine, a novel selective histamine H_1 receptor antagonist: lack of sedative potential on the EEG in the rodent. *Neuropsychobiology* **32**:214–221, 1995.
28. Ando, K., Watanabe, Y. and Ishibashi, A. Effects of mizolastine on electrocardiogram in non-anesthetized guinea-pigs. *Pharmacometrics* **55 (Suppl. 4)**:107–112, 1998.
29. Hill, S. J., Ganellin ,C. R., Timmerman, H., Schwartz, J.-C., Shankley, N. P., Young, J. M., Schunack, W., Levi, R. and Haas, H. International Union of Pharmacology. XIII. Classification of histamine receptors. *Pharmacol. Rev.* **49**:253–278, 1997.

CHAPTER 40

Adenosine-Induced Enhancement of Mast Cell Mediator Release in Asthma and COPD

STEPHEN T. HOLGATE,*[1] RICCARDO POLOSA,[1] WILLIAM R. COWARD[2] and MARTIN K. CHURCH[2]

[1]*Adult Respiratory and Molecular Sciences Research and*
[2]*Allergy and Inflammation Divisions, Southampton General Hospital, Southampton, U.K.*

INTRODUCTION

It has long been recognized that mediator release from human mast cells contributes to the airflow limitation and accompanying symptoms of asthma. In active disease, immunohistochemical and altered structural analysis of submucosal and epithelial mast cells reveals that many of them are actively degranulating (1, 2). This is reflected in increased concentrations of such mast cell mediators as histamine, prostaglandin (PG) D_2, tryptase and leukotriene (LT) D_4 in bronchial wash and bronchial alveolar lavage (BAL) fluid (3). Provocation with inhaled allergen or following exercise are the two stimuli most widely quoted to enhance mast cell mediator release and results in an early bronchoconstrictor response representing the composite actions of histamine, PGD_2 and LTD_4.

The role of the mast cell in the pathogenesis of chronic obstructive pulmonary disease (COPD) is more speculative. Increased levels of histamine have been found in the sputum of patients with obstructive bronchitis (4), and Postma *et al.* have reported an increase of the $N\tau$-methyl metabolite of histamine in the urine of patients with chronic airflow limitation (5). As reported for chronic asthma (6, 7), Grasshoff *et al.* (8) have shown that larger numbers of mast cells are present in the bronchiolar epithelium in COPD when compared to the airways of smokers without airway obstruction. Lamb and Lumsden (9) have also reported a greater number of mast cells in the respiratory epithelium of the distal airways of smokers compared to non-smokers. Thus, in the active 'inflammatory' phase of COPD, mediator secretion from mast cells may contribute to the airway pathology.

Mast cells in asthma appear to be hyper-responsive to a number of stimuli, a feature which may account for hyper-responsiveness observed with a variety of 'indirect'

---

* Corresponding author.

MAST CELLS AND BASOPHILS
ISBN 0-12-473335-2

stimuli. In addition to allergen and exercise, both hyper- and hypo-tonic stimuli cause acute bronchoconstriction which is susceptible to attenuation with H_1 antihistamines (10). Other indirect stimuli that may mediate bronchoconstriction in part through mast cell activation include inhaled benzalkonium chloride, sulphur dioxide and substance P. It is in this context that adenosine, a naturally occurring purine nucleoside, has aroused interest as a selective enhancer of mast cell mediator release in asthma and related disorders.

Adenosine is increased in inflammatory conditions of the airways. Increased levels have been reported in BAL fluid from sensitized rabbits following antigen provocation (11). High concentrations of adenosine have also been recovered from BAL fluid of subjects with asthma and chronic bronchitis when compared to normal individuals (12). All cells contain adenosine and adenine nucleotides. Extracellular adenosine is predominantly derived from the 5′-nucleotidase cleavage of adenosine 5′-monophosphate (AMP) from many pathways of intermediary metabolism. Cellular production of adenosine is greatly enhanced under conditions of local hypoxia, as may occur in inflammation when a large number of infiltrating inflammatory cells are competing for a limited oxygen supply. Adenosine is also released from activated mast cells following allergen challenge (13), from neutrophils (14) and, as has been known for a long time, platelets. Once generated, adenosine has the capacity to elicit a variety of cellular responses which are mediated through specific purinoceptors (15).

THE BRONCHOCONSTRICTOR EFFECT OF ADENOSINE IN ASTHMA

In 1983 we first reported that inhaled adenosine causes dose-related bronchoconstriction in patients with both allergic and non-allergic asthma, which could not be reproduced by the related purine nucleosides guanosine and inosine (16) but could be produced by inhaled AMP and ADP (17), presumably via 5′-nucleotidase degradation to adenosine. Since AMP is rapidly converted to adenosine and is more soluble than adenosine in aqueous solution, it has replaced adenosine as the most frequently used purine nucleoside bronchoprovocant.

Evidence that mast cell-derived mediators are involved in the bronchoconstrictor response to inhaled adenosine is derived from a number of observations.

(1) Adenosine and related purine nucleosides, acting through A_2 receptors on the surface of human lung mast cells (18, 19) or circulating basophils (20), enhance IgE-dependent histamine and eicosanoid release. At high concentrations, adenosine is inhibitory, stimulating adenylate cyclase to raise intracellular levels of cAMP and by augmenting intracellular methyltransferase pathways. Recently, Forsythe *et al.* (21) have shown that adenosine is able to directly stimulate histamine release from normal human mast cells obtained by BAL even though inhaled adenosine, in similar subjects, had no airway effects (16). In human airways isolated from an asthmatic with birch pollen asthma, but not normal airways, adenosine elicited a contractile response that could be effectively antagonized by an antihistamine and a cysteinyl LT_1 receptor blocker (22). Our own studies comparing the effects of adenosine analogues on BAL mast cells reveal a clear increased responsiveness if the mast cells are derived from asthmatic compared to non-asthmatic airways (Fig. 1).

(2) Bronchoprovocation provoked by AMP could be effectively inhibited by the mast cell stabilizing agents sodium cromoglycate and nedocromil sodium (23). In addition, the

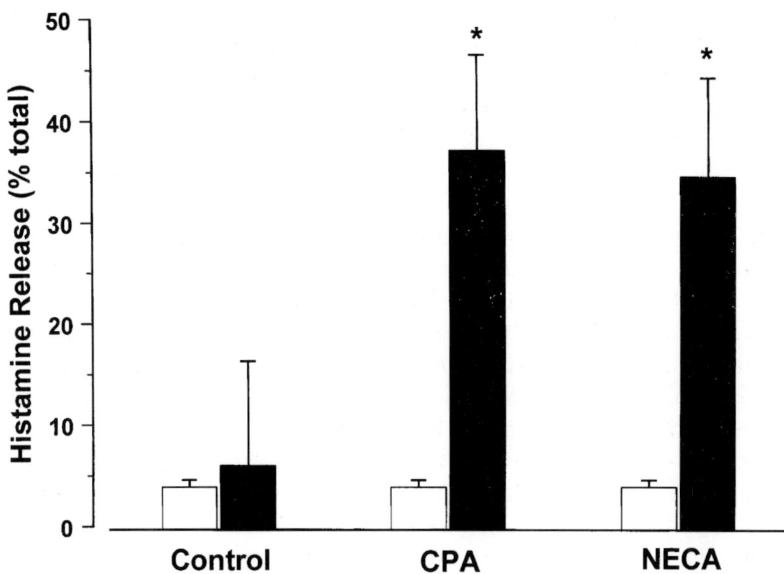

Fig. 1 Effect of adenosine analogues on histamine release from BAL mast cells. The adenosine A_1 agonist N^6-cyclopentyladenosine (CPA) and the $A_{1/2}$ agonist 5′*N*-ethylcarboxamideadenosine (NECA) were investigated for their ability to induce histamine release from mast cells obtained by BAL of non-atopic non-asthmatic (empty columns) and atopic asthmatic subjects (fiiled columns). Mast cells were sensitized for 16 h with human myeloma IgE (3 μg ml^{-1}) and then incubated for 10 min with 100 nM of the above agonists. Results are expressed as mean ± SEM for four experiments. Asterisks denote a significant difference ($p<0.05$) between the groups.

β_2-agonist salbutamol, which is a potent mast cell inhibitor (24), attenuates AMP-induced bronchoconstriction to a greater extent than bronchoprovocation provoked by the smooth muscle agonist methacholine (25).

(3) Direct instillation of AMP into asthmatic bronchi (26) or into the nose of patients with rhinitis (27) produced significant increases in the concentration of histamine and tryptase in lavage fluid. Inhaled bronchial provocation with AMP also produces a small, but significant, increase in circulating histamine levels (28).

(4) Pre-medication of asthmatic patients with H_1 histamine receptor antagonists (29, 30), cyclo-oxygenase inhibitors (31–33) or a 5-lipoxygenase inhibitor (5-lo) (34) markedly suppressed the acute bronchoconstrictor response to inhaled AMP.

Taken together, these studies provide overwhelming evidence that adenosine is able to enhance airway mast cell mediator release in patients with asthma and rhinitis. The capacity of adenosine to influence mast cell function requires the presence of a low-level second stimulus which may be either immunological or non-immunological. We suggest that this nucleoside elicits mediator release by interacting with cytokine-primed 'mast cells' on the surface of inflamed airways (35). Thus, purine-induced bronchoconstriction in asthmatics might well depend on the state of airway mast cell priming and, as such, could be a useful test for this *in vivo* (36). Bronchoconstriction provoked by inhaled AMP has a greater predictability than methacholine for the diagnosis of asthma (37).

Repeated inhalation of AMP causes a progressive lack of response to the purine nucleoside that lasts for 6–8 h (38). Cross-refractoriness exists between bronchoconstriction caused by inhaled AMP with bradykinin (39) and exercise (40), suggesting utilization of a common pathway, possibly the induction of PGE_2 release, although, in the bradykinin study, functional antagonism can be excluded because responsiveness to histamine was maintained. An alternative explanation is that adenosine receptors are important in setting the sensitivity for bradykinin- and exercise-induced bronchoconstriction.

Other anti-asthma drugs that are able to influence the asthmatic airway response to adenosine are β_2-agonists and inhaled corticosteroids. Repeated administration of β_2-agonists leads to progressive loss of β_2-adrenoceptor responsiveness, which is reflected to a greater extent on the airway response to inhaled AMP than it is to methacholine. This suggests that β_2-receptors on mast cells are more susceptible to downregulation than those on airway smooth muscle (41). Regular treatment with inhaled corticosteroids also results in a progressive loss of the airway response to inhaled AMP (42). This has proved useful as an efficacy measure for topical corticosteroid action (43), presumably by influencing the cytokines (e.g. stem cell factor) that support the survival of airway mast cells because regular treatment with inhaled corticosteroids results in a progressive loss of mast cells from asthmatic epithelium and submucosa (44). Loss of mast cells might also explain the marked effect that allergen avoidance at altitude has on AMP responsiveness when compared to methacholine responsiveness in severely asthmatic children (45).

THE AIRWAY RESPONSE TO ADENOSINE IN COPD

In addition to asthma, inhaled AMP has also been shown to cause bronchoconstriction in certain patients with smoking-induced airway obstruction. Airway inflammation has a particular role to play in AMP-induced bronchoconstriction in patients with COPD, since responsiveness is observed to a greater extent in current smokers than in ex-smokers, despite comparable responses to inhaled methacholine (46). However, different from asthma, treatment with inhaled corticosteroids failed to decrease hyper-responsiveness to AMP in patients with COPD (47). However, mast cells, but not cholinergic nerves, are implicated in AMP-induced bronchoconstriction in COPD since the H_1 histamine antagonist terfenadine was inhibitory, whereas the muscarinic antagonist ipatropium bromide was not (48). In a further study, Rutgers *et al.* have shown that the airway response to AMP in COPD appears to parallel the level of sputum eosinophilia and, therefore, as in the case in asthma, may reflect the underlying inflammatory response rather than a direct effect on smooth muscle (49).

ADENOSINE RECEPTORS ON MAST CELLS

Marquardt *et al.* were the first to report that adenosine, while ineffective alone, was able to potentiate histamine release induced by anti-IgE, concanavalin A, compound 48/80 and the calcium ionophore A23187 in rat peritoneal mast cells (50). The mechanisms that mediated the potentiating effect of adenosine and related nucleosides on these cells was unclear. Stimulation of adenylate cyclase by adenosine could be effectively blocked by 8-phenyl-theophylline but potentiation of histamine release was not, suggesting that the two effects were mediated by different adenosine receptors (51). A series of studies on isolated human lung mast cells and leukaemic basophils suggested that adenosine-induced enhancement of mediator release was linked to increased intracellular cAMP levels through activation of adenylate cyclase (18, 20). Further *in vitro* studies using N^6-ethylcarboxyamideadenosine (NECA) as the agonist demonstrated preferential mediator enhancement from human mast cells when compared to a variety of other adenosine analogues and supported the idea that adenosine was augmenting mediator release from human mast cells via A_2 receptors (18–20). Such a mechanism would explain the preferential inhibitory effect of theophylline on adenosine – when compared to methacholine-induced bronchoconstriction in patients with asthma (52, 53). However, this was questioned when it was shown that enprofilline, a cAMP phosphodiesterase inhibitor-lacking adenosine receptor antagonism, also exhibited preferential inhibitory activity against adenosine-induced bronchoconstriction and was efficacious as a therapy for asthma (54, 55).

There is growing evidence that, in rat and human mast cells, adenosine-induced enhancement of mediator release is affected through the A_{2B} receptor subtype, which is quite distinct from the A_{2A} subtype linked to adenylate cyclase (35, 55). In support of this, the failure of the A_{2A} selective agonist CGS-21680 to enhance mediator release from rodent mast cells, in contrast to NECA, suggests that the latter agent is acting preferentially on the A_{2B} receptor subtype (56). Recently, it has also been shown that enprofilline, while not being an antagonist at the A_{2A} receptor, is equipotent to theophylline as an A_{2B} receptor antagonist of NECA-induced activation of adenylate cyclase (57, 58). The dissociation constant of the antagonist–receptor complex of 7 μm is in close agreement with therapeutic plasma levels of enprofilline (5–25 $\mu mol\ l^{-1}$) (59).

Adenosine A_{2B} receptors have been shown to activate the human mast cell line HMC-1 (55). HMC-1 cells were derived from a patient with mast cell leukaemia and have a neutral protease content similar to that of human lung mast cells. They co-express both A_{2A} and A_{2B} receptors, which are coupled to adenylate cyclase through G_s proteins. However, only A_{2B} receptors activate HMC-1 cells, as reflected by the release of interleukin-8 using NECA but not with the selective A_{2A} agonist CGS21680 (55). cAMP appeared not to be involved in this process since neither forskolin nor 8-bromo-cAMP influenced IL-8 secretion. Currently it is thought that A_{2B}-enhanced mast cell activation is mediated by coupling of the receptor to phospholipase C and increased intracellular mobilization of calcium (57). These pathways also involve a G protein but this is insensitive to both cholera and pertussis toxin and belongs to the G_q family. As with human lung mast cells and basophils, activation of A_{2B} receptors on HMC-1 cells also potentiates mast cell secretion excited by a second stimulus (35).

In contrast to rat and human lung mast cells, adenosine-induced enhancement of mediator release in mouse mast cells appears to be under the control of A_3 receptors. *In situ* hybridization and reverse-transcriptase–polymerase chain reaction have failed to detect transcripts for A_3 receptors in human lung mast cells.

CONCLUDING COMMENTS

There is now ample evidence to indicate that mast cells in asthma and COPD are under the control of extracellular adenosine. The mechanisms whereby this purine nucleoside augments mediator release through the A_{2B} receptor subtype involving phosphatidyl-inositol activation and increased intracellular calcium mobilization requires elucidation. Based on the known clinical efficacy of xanthines, including enprofilline and theophylline in asthma, the A_{2B} receptor provides a future potential therapeutic target for diminishing mast cell responsiveness in this disease. Although more potent receptor antagonists for the A_{2B} receptor have been characterized, such as 1,3-dipropyl-8-sulphophenylxanthine (DPSPX) (58) and 1,3-dipropyl-8-cyclopentylxanthine (DPCPX) (59), they are relatively weak and non-selective. The potential role of A_{2B} receptors in augmenting mast cell activation and in the pathogenesis of asthma and the inflammatory component of COPD creates a novel approach to the treatment of these common disorders.

Finally, what could be the underlying reason why A_{2B} receptor function is upregulated on human airway mast cells in asthma? Possible mechanisms include upregulation of the receptors themselves, increased efficiency of G_q coupling once the receptor is activated, or a reduction in the threshold of mast cell activation by any stimulus (e.g. hypertonicity, allergen). Another intriguing possibility is loss of a negative regulatory pathway of mast cell activation. Mast cells contain both A_{2A} and A_{2B} receptors, the former serving an inhibitory function, the latter excitatory. In asthma it has long been known that β_2-receptor adenylate cyclase activity is reduced (β_2-adrenergic resistance) (60) and that further downregulation of β_2-adrenergic receptors on airway mast cells is especially sensitive to repeated administration of an inhaled β_2-agonist (61). Since loss of β_2-

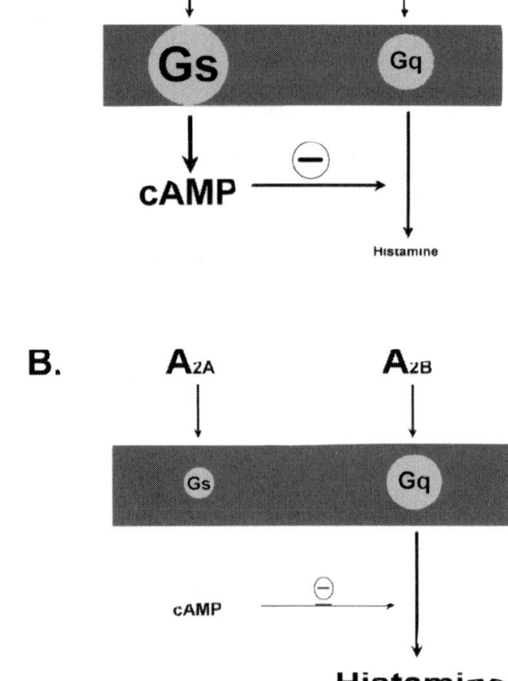

Fig. 2 Possible mechanism by which adenosine releases histamine in asthma. (A) The A_{2A} effects of adenosine lead to the generation of large quantities of intracellular cAMP which downregulates the biochemical pathways leading to histamine release. (B) Possible scenario in asthma where downregulation of G_s function leads to a reduction of cAMP generation and a reduced negative modulation of histamine release.

adrenoceptor function is in part mediated by reduced G_s–adenylate cyclase coupling, this would also have the effect of reducing the availability of adenylate cyclase coupling to A_{2A} receptors but not affect the function of A_{2B} receptors. The net result would be an overall increase in A_{2B}-receptor linked PIP_3 formation and mast cell activation (Fig. 2). In support of this hypothesis, rat mast cells, which appear to have relatively poor β-adrenoceptor coupling (62), release large amounts of histamine on stimulation with adenosine together with allergen (50).

REFERENCES

1. Church, M. K., Holgate, S. T., Shute, J. K., Walls, A. F. and Sampson, A. P. Mast cell-derived mediators. In: *Allergy: Principles and Practice*, Vol. 1 (Middleton, E., Reed, C. E., Elliot, E. F., Adkinson, N. F., Yunginger, J. W. and Busse, W. W., eds), pp. 146–167. Mosby, 1998.
2. Church, M. K. and Levi-Schaffer, F. The human mast cell. *J. Allergy Clin. Immunol.* **99**:155–160, 1997.
3. Casale, T., Wood, D., Richardson, H. *et al.* Elevated bronchoalveolar lavage fluid histamine levels in allergic asthmatics are associated with methacholine bronchial hyperresponsiveness. *J. Clin. Invest.* **79**:1197–1203, 1987.
4. Turnbull, L. W., Turnbull, L. S., Croft, J. and Kay, A. B. Variations in chemical mediators of hypersensitivity in the sputum of chronic bronchitics: correlation with peak expiratory flow. *Lancet* **2**:184–186, 1978.
5. Postma, D. S., Keyzer, J. J., Koëter, G. H., Sluiter, H. J. and De Vries, K. Influence of the parasympathetic and sympathetic nervous system on nocturnal bronchial obstruction. *Clin. Sci.* **69**:251–258, 1985.
6. Koshino, T., Arai, Y., Miyamoto, Y. *et al.* Airway basophil and mast cell density in patients with bronchial asthma and relationship to bronchial hyperresponsiveness. *J. Asthma* **33**:89–95, 1996.
7. Humbert, M., Grant, J. A., Taborda-Barata, L. *et al.* High-affinity IgE receptor (FcεR1)-bearing cells in bronchial biopsies from atopic and non-atopic asthma. *Am. J. Respir. Crit. Care Med.* **153**:1931–1937, 1996.
8. Grashoff, W. F. H., Sont, J. K., Krieken, J. H. J. M. v., Hiemstra, P. S., Stolk, J. and Sterk, P. J. Increased numbers of mast cells and macrophages in peripheral airway epithelium in smokers with airways obstruction. *Am. J. Pathol.* **151**:1785–1790, 1997.
9. Lamb, D. and Lumsden, A. Intra-epithelial mast cells in human airway epithelium: evidence for smoking-induced changes in their frequency. *Thorax* **37**:334–342, 1982.
10. Holgate, S. T. Experimental models in asthma. Antihistamines: back to the future. *Clin. Exp. Allergy* **29 (Suppl. 3)**:82–86, 1999.
11. Ali, S., Mustafa, S. J., Driver, A. G. and Metzger, W. J. Release of adenosine in bronchoalveolar lavage fluid following allergen bronchial provocation in allergic rabbits. *Am. Rev. Respir. Dis.* **143**:A417, 1991.
12. Driver, A. G., Kukoly, C. A., Ali, S. and Mustafa, S. J. Adenosine in bronchoalveolar lavage fluid in asthma. *Am. Rev. Respir. Dis.* **148**:91–97, 1993.
13. Marquardt, D. L., Gruber, H. E. and Wasserman, S. T. Adenosine released from stimulated mast cells. *Proc. Natl. Acad. Sci. USA* **81**:6192–6196, 1984.
14. Madara, J. L., Papapoff, T. W., Gillece-Castro, B., Colgan, S. P., Parkos, C. A., Delp, C. *et al.* 5′-AMP is the neutrophil-derived paracrine factor that elicits chloride secretion from T84 intestinal epithelial cell monolayers. *J. Clin. Invest.* **91**:2320–2325, 1993.
15. Gordon, J. L. Extracellular ATP: effects, sources and fate. *Biochem. J.* **233**:309–319, 1986.
16. Cushley, M. J., Tattersfield, A. E. and Holgate, S. T. Inhaled adenosine and guanosine on airway resistance in normal and asthmatic subjects. *Br. J. Clin. Pharmacol.* **15**:161–165, 1983.
17. Mann, J. S., Holgate, S. T., Renwick, A. G. and Cushley, M. J. Airway effects of purine nucleosides and nucleotides and release with bronchial provocation in asthma. *J. Appl. Physiol.* **61**:1667–1676, 1986.
18. Hughes, P. J., Holgate, S. T. and Church, M. K. Adenosine inhibits and potentiates IgE-dependent histamine release from human lung mast cells by A_2-purinoceptor mediated mechanism. *Biochem. Pharmacol.* **33**:3847–3852, 1984.
19. Peachell, P. T., Colombo, M., Kagey-Sobotka, A., Lichtenstein, L. M. and Marone, G. Adenosine potentiate mediator release from human lung mast cells. *Am. Rev. Respir. Dis.* **138**:1143–1151, 1988.

20. Church, M. K., Holgate, S. T. and Hughes, P. J. Adenosine inhibits and potentiates IgE-dependent histamine release from human basophils by an A2-receptor mediated mechanism. *Br. J. Pharmacol.* **80**:719–726, 1983.
21. Forsythe, P., McGarvey, L. P. A., Heaney, L. G., MacMahon, J. and Ennis, M. Adenosine stimulates human pulmonary mast cells. *Thorax* **50 (Suppl. 2)**:P132, 1995.
22. Björk, T., Gustafsson, L. E. and Dahlen, S. E. Isolated bronchi from asthmatics are hyperresponsive to adenosine, which apparently acts indirectly by liberation of leukotrienes and histamine. *Am. Rev. Respir. Dis.* **145**:1087–1091, 1992.
23. Phillips, G. D., Scott, V. L., Richard, R. and Holgate, S. T. Effect of nedocromil sodium and sodium cromoglycate against bronchoconstriction induced by inhaled adenosine 5′-monophosphate. *Eur. Respir. J.* **2**:210–217, 1989.
24. Church, M. K. and Hiroi, J. Inhibition of IgE-dependent histamine release from human dispersed lung mast cells by anti-allergic drugs and salbutamol. *Br. J. Pharmacol.* **90**:421–429, 1987.
25. Phillips, G. D., Finnerty, J. P. and Holgate, S. T. Comparative protective effect of the inhaled β_2-agonist salbutamol (albuterol) on bronchoconstriction provoked by histamine, methacholine, adenosine 5′-monophosphate in asthma. *J. Allergy Clin. Immunol.* **85**:755–762, 1990.
26. Polosa, R., Ng, W. H., Crimi, N., Vancheri, C., Holgate, S. T., Church, M. K. and Mistretta, A. Release of mast-cell-derived mediators after endobronchial adenosine challenge in asthma. *Am. J. Resp. Crit. Care Med.* **151**:624–629, 1995.
27. Polosa, R., Crimi, N., Raccuglia, D., Pagano, C., Walters, C., Church, M. K. *et al.* Histamine changes to nasal provocation with AMP and GMP in rhinitic and normal subjects. *J. Allergy Clin. Immunol.* **93**:329 (abstract), 1994.
28. Phillips, G. D., Ng, R., Church, M. K. and Holgate, S. T. The response of plasma histamine to bronchoprovocation with methacholine, adenosine 5′-monophosphate, and allergen in atopic non-asthmatic subjects. *Am. Rev. Respir. Dis.* **141**:9–13, 1990.
29. Phillips, G. D., Rafferty, P., Beasley, C. R. W. and Holgate, S. T. The effect of oral terfenadine on the bronchoconstrictor response to inhaled histamine and adenosine 5′-monophosphate in non-atopic asthma. *Thorax* **42**:939–945, 1987.
30. Rafferty, P., Beasley, C. R. and Holgate, S. T. The contribution of histamine to bronchoconstriction produced by inhaled allergen and adenosine 5′-monophosphate in asthma. *Am. Rev. Respir. Dis.* **136**:369–373, 1987.
31. Crimi, N., Palermo, F., Polosa, R., Oliveri, R., Maccarrone, C., Palermo, B. *et al.* Effect of indomethacin on adenosine-induced bronchoconstriction. *J. Allergy Clin. Immunol.* **83**:921–925, 1989.
32. Phillips, G. D. and Holgate, S. T. The effect of oral terfenadine alone and in combination with flurbiprofen on the bronchoconstrictor response to inhaled adenosine 5′-monophosphate in non-atopic asthma. *Am. Rev. Respir. Dis.* **139**:463–469, 1989.
33. Crimi, N., Polosa, R., Magri, S., Prosperini, G., Milazzo, L. V., Santonocito, G. *et al.* Inhaled lysine-acetylsalicylate (L-ASA) attenuates the bronchoconstrictor response to AMP in asthmatic subjects. *Eur. Respir. J.* **8**:905–912, 1995.
34. Van Schoor, J., Joos, G. F., Kips, J. C., Drajesk, J. F., Carpentier, P. J. and Pauwels, R. A. The effect of BAT0761, a novel 5-lipoxygenase inhibitor, on exercise- and adenosine-induced bronchoconstriction in asthmatic subjects. *Am. J. Respir. Crit. Care Med.* **155**:875–880, 1997.
35. Feoktistov, I., Polosa, R., Holgate, S. T. and Biaggioni, I. Adenosine A_{2B} receptors: a novel therapeutic target in asthma? *Trends Pharmacol. Sci.* **19**:148-151, 1998.
36. Polosa, R. and Holgate, S. T. Adenosine bronchoprovocation: a promising marker of allergic inflammation in asthma? *Thorax* **52**:919–923, 1997.
37. Wilkinson, J., Shipley, E., Smith, S., Schreiber, J., Morton, N. E. and Holgate, S. T. The improved ability of adenosine 5′-monophosphate over histamine bronchial provocation to predict asthma. *Eur. Respir. J.* (submitted).
38. Daxun, Z., Rafferty, P., Richard, R., Summerell, S. and Holgate, S. T. Airway refractoriness to adenosine-5′-monophosphate after repeated inhalation by atopic non-asthmatic subjects. *J. Allergy Clin. Immunol.* **87**:373–378, 1989.
39. Polosa, R., Rajakulasingham, K., Church, M. K. and Holgate, S. T. Repeated inhalation of bradykinin attenuates adenosine 5′-monophosphate (AMP)-induced bronchoconstriction in asthmatic airways. *Eur. Respir. J.* **5**:700–706, 1992.
40. Finnerty, J. P., Polosa, R. and Holgate, S. T. Repeated exposure of asthmatic airways to inhaled adenosine 5′-monophosphate attenuates bronchoconstriction provoked by exercise. *J. Allergy Clin. Immunol.* **86**:353–359, 1990.

41. O'Connor, B. J., Aikman, S. and Barnes, P. J. Tolerance to the non-bronchodilator effects of inhaled beta-agonists in asthma. *N. Engl. J. Med.* **327**:1204–1208, 1992.
42. O'Connor, B. J., Ridge, S. M., Barnes, P. J. and Fuller, R. W. Greater effect of inhaled budesonide on AMP-induced than on sodium metabisulphite-induced bronchoconstriction in asthma. *Am. Rev. Respir. Dis.* **146**:560–564, 1992.
43. Doull, I. J. M., Sandall, D., Smith, S., Schreiber, J., O'Toole, S., Freezer, N. J. and Holgate, S. T. The differential inhibitory effect of regular inhaled corticosteroids on airway responsiveness to adenosine 5′-monophosphate, methacholine and bradykinin in symptomatic children with recurrent wheeze. *Pediatr. Pulmonol.* **23**:404–411, 1997.
44. Djukanović, R., Wilson, J. W., Britten, K. M., Wilson, S. J., Walls, A. F., Roche, W. R., Howarth, P. H. and Holgate, S. T. The effect on inhaled corticosteroid on airway inflammation and symptoms of asthma. *Am. Rev. Respir. Dis.* **145**:669–674, 1992.
45. Van Velzen, E., van den Bos, J.-W., Benckhuijsen, J. A. W., van Essel, T., de Bruijn, R. and Aalbers, R. Effect of allergen avoidance at high altitude on direct and indirect bronchial hyperresponsiveness and markers of inflammation in children with allergic asthma. *Thorax* **51**:582–584, 1996.
46. Oosterhoff, Y., de Jong, J. W., Jansen, M. A. M., Koëter, G. H. and Postma, D. S. Airway responsiveness to adenosine 5′-monophosphate in chronic obstructive pulmonary disease is determined by smoking. *Am. Rev. Respir. Dis.* **147**:553–558, 1993.
47. Rutgers, S. R., Koëter, G. H., van der Mark, T. W. and Postma, D. S. Short-term treatment with budesonide does not improve hyperresponsiveness to adenosine 5′-monophosphate in COPD. *Am. J. Respir. Crit. Care Med.* **157**:880–886, 1998.
48. Rutgers, S. R., Koëter, G. H., van der Mark, T. W. and Postma, D. S. Protective effect of oral terfenadine and not inhaled ipratropium bromide on adenosine-5′-monophosphate-induced bronchoconstriction in patients with COPD. *Clin. Exp. Allergy* **29**:1287–1292, 1999.
49. Rutgers, S. R., Timens, W., Tzanakis, N., Kauffman, H. F., van der Mark, W., Koëter, G. H. and Postma, D. S. Airway inflammation and hyperresponsiveness to adenosine 5′-monophosphate in COPD (submitted for publication).
50. Marquardt, D. L., Parker, C. W. and Sullivan, T. J. Potentiation of mast cell mediator release by adenosine. *J. Immunol.* **120**:871–878, 1978.
51. Church, M. K., Hughes, P. J. and Vardey, C. J. Studies on the receptor mediating cyclic AMP-independent enhancement by adenosine of IgE-dependent mediator release from rat mast cells. *Br. J. Pharmacol.* **87**:233–242, 1986.
52. Mann, J. S. and Holgate, S. T. Specific antagonism of adenosine-induced bronchoconstriction in asthma by oral theophylline. *Br. J. Clin. Pharmacol.* **19**:685–692, 1985.
53. Clarke, H., Cushley, M. J., Persson, C. G. and Holgate, S. T. The protective effects of intravenous theophylline and enprofylline against histamine- and adenosine 5′-monophosphate-provoked bronchoconstriction: implications for the mechanisms of action of xanthine derivatives in asthma. *J. Respir. Pharmacol.* **2**:147–154, 1989.
54. Persson, C. G. A., Andersson, K. E. and Kjellin, G. Effects of enprofylline and theophylline may show the role of adenosine. *Life Sci.* **38**:1057–1072, 1986.
55. Feoktistov, I. and Biaggioni, I. Adenosine 2_{2B} receptors evoke IL-8 secretion in human mast cells – an enprofylline-sensitive mechanism with implications for asthma. *J. Clin. Invest.* **96**:1979-1986, 1995.
56. Marquardt, D. L., Walker, L. L. and Heinemann, S. Cloning of two adenosine receptor subtypes from mouse bone marrow-derived mast cells. *J. Immunol.* **152**:4508–4515, 1994.
57. Marquardt, D. L. and Walker, L. L. Inhibition of protein-kinase-a fails to alter mast-cell adenosine responsiveness. *Agents Actions* **43**:7–12, 1994.
58. Feoktistov, I. and Biaggioni, I. Characterization of adenosine receptors in human erythroleukemia- cells and platelets – further evidence for heterogeneity of adenosine-a2 receptor subtypes. *Mol. Pharmacol.* **43**: 909–914, 1993.
59. Brackett, L. E. and Daly, J. W. Functional-characterization of the a(2B) adenosine receptor in NIH 3T3 fibroblasts. *Biochem. Pharmacol.* **47**:801–814, 1994.
60. Tashkin, D. P., Connolly, M. E., Deutch, R. I. *et al.* Subsensitisation of beta-adrenoceptors in airways and lymphocytes of healthy and asthmatic subjects. *Am. Rev. Respir. Dis.* **125**:185–193, 1982.
61. Sears, M. R. The short and long term effects of β_2 agonists. In: *Asthma: Physiology, Immunopharmacology and Treatment* (Holgate, S. T., Austen, K. F., Lichtenstein, L. M. and Kay, A. B., eds), pp. 359–374. Academic Press, London, 1993.
62. Ishizaka, T. and Ishizaka, K. Activation of mast cells for mediator release through IgE receptors. *Progr. Allergy* **34**:188–235, 1984.

CHAPTER 41

Neural Activation and Inflammation in Allergic Rhinitis

ALVIN M. SANICO[*1] and ALKIS TOGIAS[1,2]*

[1]*Department of Medicine, Division of Clinical Immunology and*
[2]*Department of Medicine, Division of Respiratory and Critical Care Medicine,*
Johns Hopkins Asthma & Allergy Center, Baltimore, Maryland, U.S.A.

INTRODUCTION

Allergic rhinitis is a common chronic condition associated with a high socio-economic cost (1, 2). The two cardinal features of this disease are inflammation and hyper-responsiveness. Nasal mucosal inflammation is manifested by leukocyte infiltration, plasma extravasation and chronic structural changes in the epithelium and submucosa. Hyper-responsiveness, on the other hand, is suggested by an increased susceptibility to environmental irritants (3).

The relationship between inflammation and hyper-responsiveness in allergic airways disease has not been fully elucidated. There is increasing evidence that these elements may actually be linked through the peripheral nervous system. This chapter reviews findings demonstrating that nerve activation can lead to augmentation of allergic inflammation, and that allergic inflammation can lead to neurally mediated hyper-responsiveness. In closing, the possible involvement of neurotrophic factors and their cellular sources, such as mast cells, will be addressed.

The nervous system has been implicated as an integral component in the pathophysiology of various nasal diseases since the early 19th century (4). Imbalance of the autonomic nerves, with sympathetic underactivity (5) or parasympathetic overactivity (6), was initially cited as a cause of nasal mucosal swelling and rhinorrhea, respectively. More recently, the significance of other nerve fibres that are non-adrenergic and non-cholinergic has been recognized (7). For example, nerves containing neuropeptides such as the tachykinin substance P have been localized along gland acini and blood vessel walls within the nasal mucosa (8), strongly suggesting their regulatory role in the secretory and vascular functions of the nose. Studies indicate that these tachykinergic fibres may participate in the process of neurogenic inflammation, whereby nerve activation causes several inflammatory changes.

* Corresponding author.

MAST CELLS AND BASOPHILS
ISBN 0-12-473335-2

NEUROGENIC INFLAMMATION

The term neurogenic inflammation was originally used to describe the occurrence of dermal plasma extravasation upon application of various irritants (9). Similar observations have subsequently been attributed to neuropeptides that are released from sensory nerve endings upon stimulation (10, 11). Nerve-mediated inflammatory changes have since been demonstrated in other organ systems, including the respiratory tract in various animal models. In humans, *in vivo* studies on airway inflammation have been facilitated by the technique of nasal provocation, whereby stimuli can be applied directly and lavage samples can be collected repeatedly for evaluation (12, 13). For neural stimulation, an agent that has been extensively used is capsaicin, the pungent component of hot pepper that activates afferent nerves comprised mostly of unmyelinated C-fibres and some myelinated Aδ-fibres (14). The recently cloned receptor for capsaicin is a non-selective cation channel that is also involved in heat sensation (15). The occurrence of neurogenic inflammation in humans can, therefore, be examined by determining whether nasal provocation with capsaicin induces inflammatory changes such as leukocyte influx and plasma extravasation.

Nerve-mediated Leukocyte Influx

Nasal provocation with capsaicin has been previously shown to induce transient symptoms of local burning sensation, nasal congestion and rhinorrhoea (16–18). Two groups of investigators have independently shown that capsaicin nasal challenge causes significant increases in the leukocyte content of nasal lavage fluids collected at various time points up to 4 h post-challenge (19, 20). Subsequently, this phenomenon of capsaicin-induced leukocyte influx was shown to be dose-dependent. In a study on subjects with active allergic rhinitis, nasal provocation with 1, 10 and 100 μg of capsaicin was administered by metered pump sprays in a double-blind, randomized, cross-over manner on three occasions separated by 1 week. Nasal lavage fluids were collected before and 30 min, 1 h and 4 h after capsaicin challenge. Nasal provocation with 1 μg of capsaicin did not cause any significant change in the leukocyte counts of nasal lavage fluids at any time point. On the other hand, 10 μg of capsaicin produced significant increases in the leukocyte content of nasal lavage fluids at 1 h and 4 h, while the 100-μg dose produced significant increases at 30 min, 1 h and 4 h post-challenge. This appears to be a non-selective inflammatory cellular response, as parallel increases in neutrophils, eosinophils and mononuclear cells were observed (21). A similarly rapid onset of leukocyte recruitment has been previously demonstrated in the rat trachea, wherein adherence of neutrophils to vascular endothelium was found within 5 min of capsaicin administration. In that study, transendothelial migration of leukocytes was facilitated by inhibiting neutral endopeptidase, an enzyme that degrades neuropeptides (22). This finding supports the notion that capsaicin-induced influx of inflammatory cells is mediated by tachykinins such as substance P.

The release of substance P-like immunoreactivity following administration of capsaicin has been demonstrated in isolated animal airways (23–25). *In vitro*, substance P has been shown to induce activation and chemotaxis of rodent (26, 27) and human leukocytes (28–30). *In vivo*, nasal provocation with exogenous substance P induced an increase in the leukocyte content of nasal lavage fluids of subjects with allergic rhinitis

(31, 32). In the rat trachea, endothelial adhesion of neutrophils and eosinophils triggered by either capsaicin or substance P was shown to be mediated by neurokinin 1 (NK_1) receptors. In addition, capsaicin and substance P were also found to cause NK_1 receptor-mediated plasma extravasation (33).

Nerve-mediated Plasma Extravasation

In animal models, plasma extravasation can be demonstrated by the leakage of intravenously injected blue dye or of macromolecular tracers into surrounding tissue. Several investigators have shown this effect following administration of capsaicin in the rodent nasal mucosa (34, 35) and lower airways (10, 36–41).

In human airways, increased vascular permeability and resultant plasma leakage can be assessed by measuring serum proteins such as albumin and fibrinogen in lavage fluids. In a comparative study, nasal provocation with capsaicin significantly increased the albumin levels of nasal lavage fluids in subjects with active allergic rhinitis. This effect was not found in subjects with non-allergic rhinitis or in healthy individuals (84). Capsaicin-induced plasma extravasation has also been shown to be dose-dependent. Nasal provocation with capsaicin at a dose of 10 μg or 100 μg, but not 1 μg, induced significant increases in the albumin levels of nasal lavage fluids collected 30 min post-challenge. The plasma leakage following capsaicin challenge is unlikely due to any long-lasting disruption of the vascular wall, as the effect was no longer observed 1 h or 4 h post-challenge (21) (Fig. 1). Nasal provocation with capsaicin has similarly been shown to increase the nasal lavage levels of fibrinogen, which is a much larger serum protein than albumin. These effects cannot be attributable to protein secretion from glands, as the capsaicin-induced increases in albumin and fibrinogen were not attenuated by

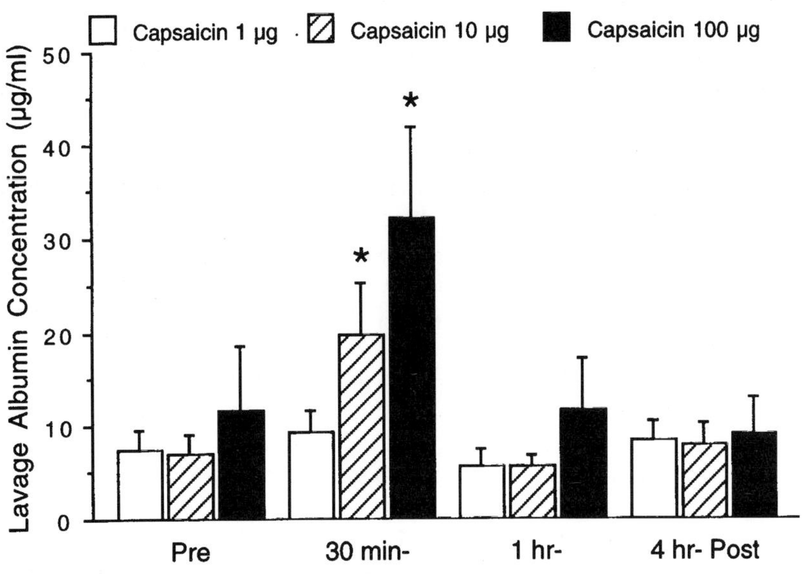

Fig. 1 Capsaicin-induced plasma extravasation. Nasal provocation with capsaicin in a group of 10 subjects with active allergic rhinitis demonstrated a dose-dependent increase in the concentration of albumin in nasal lavage fluids collected 30 min post-challenge (*$p = 0.005$, pre- versus post-challenge). This effect was acute and transient, with no evidence of plasma protein leakage at later time points. (Adapted from ref. 21.)

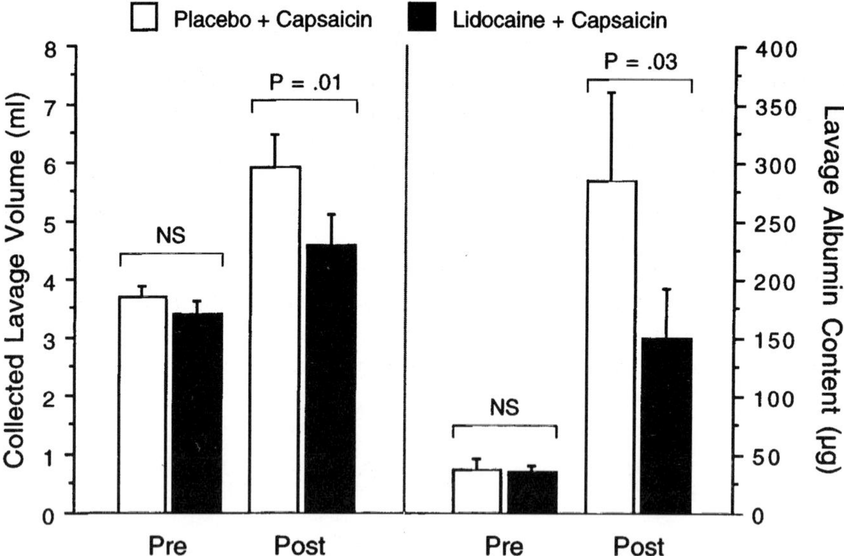

Fig. 2 Effect of nerve blockade on capsaicin-induced secretions and plasma extravasation. Nasal provocation with capsaicin in a group of 10 subjects with active allergic rhinitis resulted in a significant increase in the volume of collected nasal lavage fluids (left panel) and of their albumin content (right panel). Compared to sham pre-treatment, application of a local anaesthetic significantly reduced both capsaicin-induced nasal secretions and plasma extravasation (*p* values shown; NS, = not significant). (Adapted from ref. 43)

anticholinergic inhibition of gland activity (43). These findings provide further evidence that locally applied capsaicin can cause leakage of plasma proteins, a reflection of increased vascular permeability.

That capsaicin-induced plasma extravasation is truly nerve-mediated was shown by a study that employed nerve blockade with a local anaesthetic. In a randomized, cross-over manner, the nasal mucosae of subjects with active allergic rhinitis were pre-treated with either topical lidocaine or placebo 5 min before capsaicin nasal spray challenge. Pre-treatment with lidocaine significantly attenuated the capsaicin-induced increase in the volume of nasal secretions as well as in the albumin content of lavage fluids (43) (Fig. 2). These findings indicate that capsaicin-induced plasma extravasation is definitely associated with nerve activation. Again, this may involve tachykinin release, as application of exogenous substance P has been shown to significantly increase the levels of albumin in nasal lavage fluids of subjects with allergic rhinitis (31, 32). The fact that neurogenic plasma leakage was found by comparative studies only in subjects with allergic rhinitis suggests increased nerve-mediated responsiveness in this disease.

NEURAL HYPER-RESPONSIVENESS

Another indication of enhanced nasal responsiveness of allergic individuals is their increased susceptibility to develop symptoms upon exposure to environmental irritants (3). Exacerbation of allergic rhinitis is typically triggered not only by aeroallergens but also by non-allergenic elements such as tobacco smoke and cold dry air (44). This exaggerated symptomatic responsiveness appears to be neurally mediated, as shown by studies on nerve-mediated reactions, namely the sneezing reflex and the secretory reflex.

Exaggerated Sneezing Reflex

Sternutation or sneezing is a teleologically protective reflex triggered by mechanical or chemical stimulation of afferent nerves in the nasal mucosa (45, 46). During an allergic reaction, the biochemical likely responsible for the induction of sneezing is histamine. This mediator, which is released from mast cells and basophils upon allergen exposure (47), has been shown to produce several nerve-mediated nasal responses (48). Nasal provocation with exogenous histamine causes acute and dose-dependent sneezing in both allergic rhinitis and healthy subjects. However, in the former group of subjects, the threshold for sneezing is significantly decreased and the magnitude of response is considerably increased (52) (Fig. 3). A protective reflex such as sneezing, therefore, can become pathologically exaggerated in the setting of allergic inflammation.

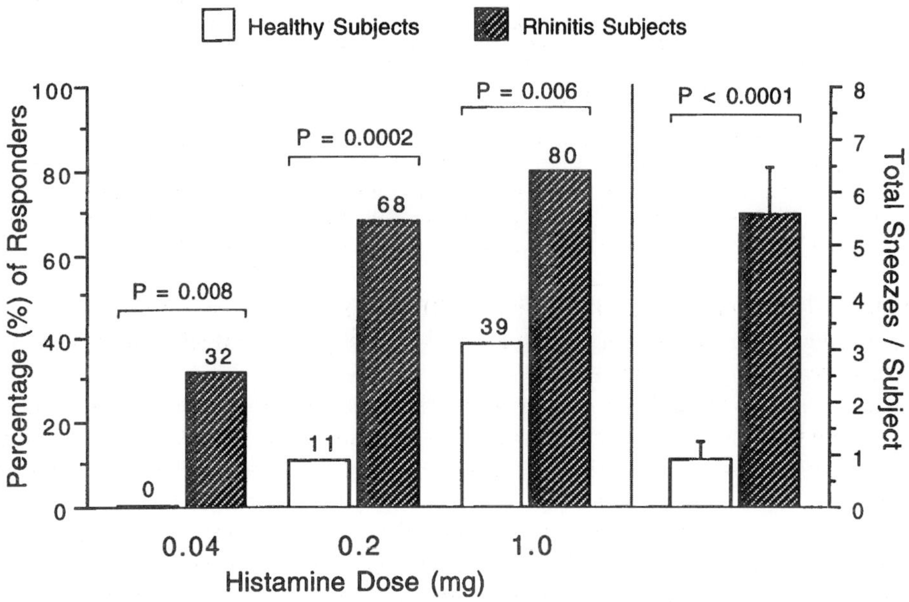

Fig. 3 Hyper-responsiveness in allergic rhinitis. Nasal provocation with histamine in 18 healthy subjects and 25 subjects with allergic rhinitis induced sneezing in a dose-dependent fashion in both groups. However, the threshold and magnitude of response were significantly decreased and increased, respectively, in subjects with allergic rhinitis compared to their healthy counterparts (*p* values shown). (Adapted from ref. 52.)

Exaggerated Secretory Reflex

Another natural protective nasal response to irritant exposure is the secretory reflex. Nasal secretions can be quantified by measuring markers of glandular activation such as lysozyme, lactoferrin and mucus glycoproteins in nasal fluids, the volume of returned lavage samples, or the weight of secretions absorbed by filter paper. Nasal provocation with capsaicin induces an acute and dose-dependent increase in the glandular product lysozyme in subjects with active allergic rhinitis (21). Separate studies have shown that capsaicin similarly causes a significant increase in the volume and weight of nasal fluids collected post-challenge (18, 43). A number of other stimuli have been shown to induce nasal secretions and rhinorrhoea, including hyperosmolar saline, cold dry air, histamine, bradykinin and methacholine (42, 48–50). The nasal secretory response to most, but not all, of these stimuli has been shown to be markedly exaggerated in subjects with allergic rhinitis. Interestingly, the nasal response to those stimuli that induce greater secretions in allergic rhinitis appears to be nerve-mediated.

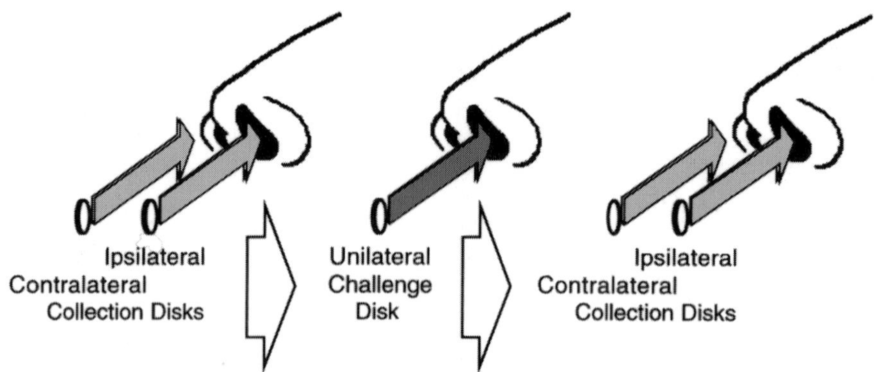

Fig. 4 Method of evaluating the nasonasal secretory reflex. Pre-weighed disks punched out from filter paper cards are applied onto the left and right septal mucosae for 30 sec to absorb secretions, before and after the challenge material is administered onto one nostril. The difference between the wet and dry weights of the disk represents the secretion weight at the time of collection.

One method of demonstrating nerve-mediated secretory response is through the so-called nasonasal reflex, whereby application of stimuli onto one nostril causes increased secretions not only at the ipsilateral side but also at the contralateral unchallenged nostril (51). In this type of experiment, pre-weighed filter paper collection disks are applied onto the left and right septal mucosae for 30 sec to absorb secretions, before and after the challenge material is administered onto one nostril (Fig. 4). The difference between the wet and dry weights of the collection disks represents the secretion weight. Studies applying this method have shown that capsaicin, hyperosmolar saline, cold dry air, histamine and bradykinin cause significant increases in ipsilateral and contralateral secretion weights, from pre- to post-challenge. In contrast, the direct gland activator methacholine was found to cause significantly increased secretions only ipsilaterally and not contralaterally (48). A comparative study has shown no significant difference between allergic rhinitis and healthy control subjects in the secretory response to methacholine (42). On the other hand, the nasonasal secretory reflex induced by capsaicin, hyperosmolar saline, cold dry air, histamine, and bradykinin was found to be significantly greater in subjects with allergic rhinitis compared to healthy individuals (42, 50, 52–54) (Fig. 5). Furthermore, the nerve-mediated secretory response to bradykinin was significantly greater during, compared to outside, the pollen season among subjects with seasonal allergic rhinitis (50). These findings suggest that the presence of allergic inflammation is a predisposing factor for enhanced neural activation. Another method of demonstrating that the nasal secretory response is neurally mediated is through nerve blockade with a local anaesthetic. Various studies have shown that pre-treatment with lidocaine significantly decreased the nasonasal secretory response to capsaicin, hyperosmolar saline, cold dry air and histamine, but not to methacholine (42, 43, 48, 49).

A more specific approach in the assessment of nerve involvement is through neural

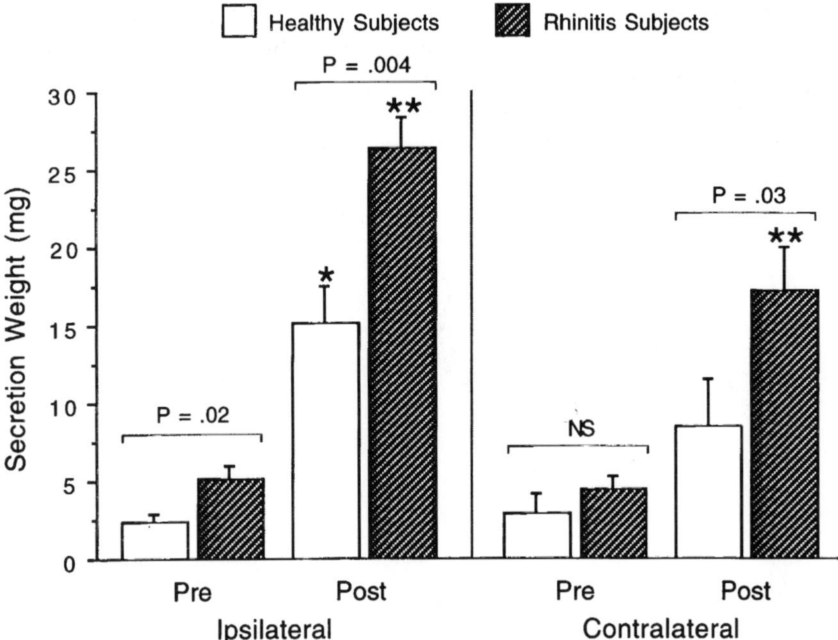

Fig. 5 Exaggerated secretory reflex in allergic rhinitis. Unilateral nasal provocation with hyperosmolar saline in 10 healthy subjects and 23 subjects with allergic rhinitis induced significantly increased secretions 1 min post-challenge, ipsilaterally for both groups and contralaterally only for the rhinitis group ($*p = 0.005$ and $**p < 0.0001$, pre- versus post-challenge). Group comparisons showed significantly increased secretory responsiveness in subjects with allergic rhinitis compared to healthy individuals (p values shown). (Adapted from ref. 42.)

desensitization with repeated application of capsaicin. This method has been previously shown to significantly reduce neural responsiveness in a reversible manner (17, 55, 56), presumably through decreased afferent transmission (57) and/or depletion of neuropeptides (58–60). This form of intervention has been previously reported to effectively reduce rhinitis symptoms (56, 58, 61–63). Repeated application of capsaicin at the site of challenge was shown to attenuate the bilateral secretory response to unilateral nasal provocation with hyperosmolar saline in subjects with allergic rhinitis. As expected, the secretory response to direct gland stimulation with methacholine was not affected by capsaicin desensitization (42).

It has become clear, therefore, that nasal hyper-responsiveness exists in the setting of allergic rhinitis, and that this can be attributable to neural elements. Nerve involvement may also hold true for bronchial hyper-responsiveness in the setting of asthma. Several studies have shown that neural stimuli such as histamine, bradykinin and hyperosmolar saline induce significantly greater bronchoconstriction in asthmatics compared to healthy control subjects (64–66). Furthermore, pre-treatment with lidocaine was shown to attenuate the bronchoconstrictive effect of hyperosmolar saline (64). In-depth studies on the interaction between nerve activation and inflammation in the lower airways, however, are precluded mainly by methodological limitations. This highlights the importance of nasal studies in allowing studies towards greater understanding of human airway pathophysiology.

NERVE GROWTH FACTOR

To further advance the understanding of neural hyper-responsiveness in allergic airways disease, recent studies have focused on identifying factors that may provide the molecular basis of this abnormality. One element that may help elucidate the effective relationship between inflammation and increased nerve activation is the prototypical neurotrophin, nerve growth factor (NGF). This factor was discovered by Rita Levi-Montalcini in the early 1950s (67), for which she eventually received the Nobel Prize for medicine. NGF may serve as an important link in the pathophysiology of allergic rhinitis because it is produced by cells that actively participate in allergic inflammation and it effects changes that can lead to enhanced neural reactivity.

Nerve Growth Factor Sources

There is evidence that NGF is synthesized, stored and released by mast cells (68, 69), eosinophils (70), $CD4^+$ T cells (71, 72) and B cells (73). Of note, these cells play an active role in allergic inflammatory disease of the airways. Mast cells are particularly interesting as they have been found in close apposition with neuropeptide-containing nerve endings in peripheral tissue (74). Levels of NGF have been reported to be elevated in the serum of patients with various allergic disorders (75). It is conceivable that NGF may, at least in part, be responsible for the increased neural responsiveness that characterizes allergic airways disease.

Nerve Growth Factor Effects

Nerve growth factor (NGF), which is an essential constituent in peripheral nerve development (76), effects multiple changes (77) that can lead to enhanced neural responsiveness. This is exemplified by transgenic mice overexpressing NGF that develop bronchial hyper-innervation. These mice have been found to exhibit hyper-responsiveness to inhalational challenge with capsaicin (78). NGF also upregulates neuropeptides such as substance P (79) that can be antidromically released upon nerve activation. This effect may provide an explanation for the finding of increased bronchial responsiveness to histamine after NGF administration in guinea pigs. In that study, the NGF-induced hyper-responsiveness was completely blocked by an NK_1 receptor antagonist (80).

In humans, recent findings also support the role of NGF in allergic airways disease. NGF has been detected in nasal fluids by Western immunoblotting and by enzyme-linked immunosorbent assay. The mean baseline level of NGF in nasal lavage fluids obtained from subjects with active allergic rhinitis was found to be 11-fold higher than in healthy individuals. Nasal provocation with allergen caused a significant acute increase in NGF levels of nasal lavage fluids in subjects with allergic rhinitis, but not in healthy control subjects. The levels of NGF in nasal fluids were not affected by control challenges with either the vehicle or histamine (81). These findings demonstrate that NGF release occurs as part of allergic reactions. In the lower airways, segmental allergen challenge in asthmatics was shown to cause a significant increase in the NGF levels of bronchoalveolar lavage fluids at a later time point (82). These observations are consistent with the knowledge of cellular sources of NGF, and of how these inflammatory cells can be activated by allergen stimulation. The rapid release of NGF in allergic rhinitis suggests, in particular, participation of mast cells. Previous studies have shown significant increases in the nasal lavage levels of mast cell mediators such as tryptase and histamine within minutes of allergen nasal challenge (83).

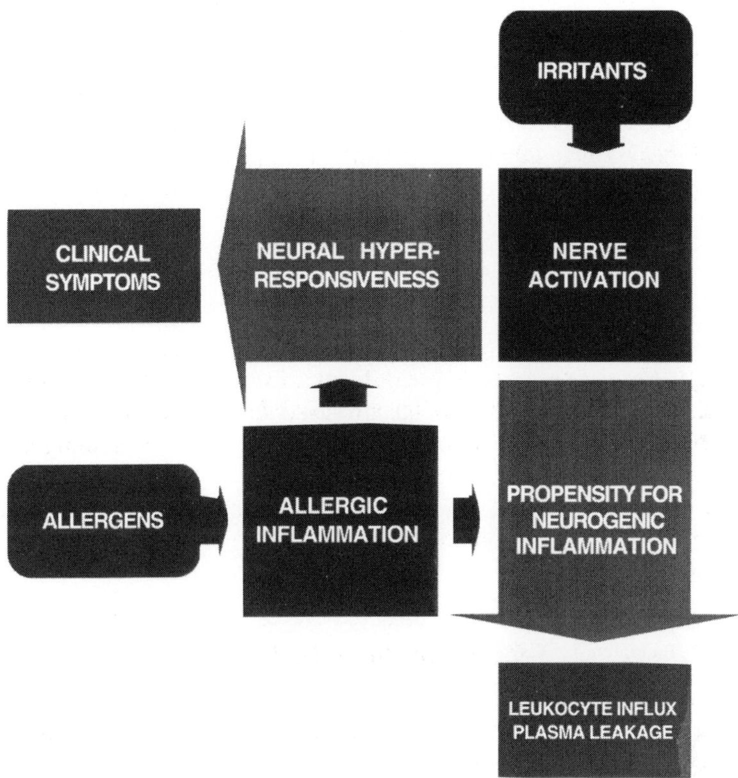

Fig. 6 Summary of our working hypothesis on the interaction between inflammation and neural activation in allergic rhinitis. Exposure to allergens causes allergic inflammation which, in turn, leads to a propensity for neurogenic inflammation and to neural hyper-responsiveness. In this setting, irritants may activate tachykinergic sensory nerves, leading to neurogenic inflammation manifested by leukocyte influx and plasma extravasation. These changes may interplay with, and magnify, allergic inflammation. In the presence of neural hyper-responsiveness, irritants can trigger exaggerated clinical symptoms such as sneezing and rhinorrhoea. We propose that the neurotrophin NGF mediates the various interactions between inflammation and nerve activation in this complex disease.

SUMMARY AND CONCLUSION

The interaction between inflammation and neural activation in allergic rhinitis is summarized in Fig. 6. Chronic exposure to aeroallergens results in allergic inflammation which, in turn, leads to a propensity for neurogenic inflammation and to neural hyper-responsiveness. In this setting, irritants may activate neuropeptide-containing sensory nerves along the nasal mucosa. The resultant release of tachykinins can cause neurogenic inflammation manifested by leukocyte influx and plasma extravasation. This development may interplay with, and result in augmentation of, allergic inflammatory changes. The presence of neural hyper-responsiveness increases susceptibility to non-allergenic irritants and causes an exaggeration of the sneezing and secretory reflexes. These enhanced responses can also be induced by products of allergic reactions such as histamine and bradykinin, which have sensorineural stimulatory properties. The development of neural hyper-responsiveness may be mediated by NGF. The levels of this neurotrophin, which can be released from various sources such as mast cells in the course of an allergic reaction, have been found to be significantly increased in subjects with allergic rhinitis. Overall, the role of sensory nerves in allergic airway disease appears to be central in the induction of both mucosal inflammation and symptomatic hyperresponsiveness. Further understanding of this role and its molecular basis may provide novel tools for therapeutic intervention.

REFERENCES

1. Nathan, R., Meltzer, E., Selner, J. and Storms, W. Prevalence of allergic rhinitis in the United States. *J. Allergy Clin. Immunol.* **99**:808–814, 1997.
2. Storms, W., Meltzer, E., Nathan, R. and Selner, J. The economic impact of allergic rhinitis. *J. Allergy Clin. Immunol.* **99 (Suppl.)**:820–824, 1997.
3. Cummings, K. M., Zaki, A. and Markello, S. Variation in sensitivity to environmental tobacco smoke among adult non-smokers. *Int. J. Epidemiol.* **20**:121–125, 1991.
4. Cloquet, J. *Osphresiologie, ou Traite des Odeurs, du Sens et des Organes de l'Olfaction,* 2 edn. Mequignon-Marvis, Paris, 1821.
5. Fowler, E. Jr. Unilateral vasomotor rhinitis due to interference with the cervical sympathetic system. *Arch. Otolaryngol.* **37**:710–712, 1943.
6. Malcomson, K. The vasomotor activities of the nasal mucous membrane. *J. Laryngol. Otol.* **73**:73–98, 1959.
7. Anggard, A. Vasomotor rhinitis – pathophysiological aspects. *J. Rhinol.* **17**:31–35, 1979.
8. Baraniuk, J., Lundgren, J., Okayama, M., Goff, J., Mullol, J., Merida, M., Shelhamer, J. and Kaliner, M. Substance P and neurokinin A in human nasal mucosa. *Am. J. Respir. Cell Mol. Biol.* **4**:228–236, 1991.
9. Jancso, N., Jancso-Gabor, A. and Szolcsanyi, J. Direct evidence for neurogenic inflammation and its prevention by denervation and by pretreatment with capsaicin. *Br. J. Pharmacol. Chemother.* **31**:138–151, 1967.
10. Brokaw, J., Hillenbrand, C., White, G. and McDonald, D. Mechanism of tachyphylaxis associated with neurogenic plasma extravasation in the rat trachea. *Am. Rev. Respir. Dis.* **141**:1434–1440, 1990.
11. Pernow, B. Role of tachykinins in neurogenic inflammation. *J. Immunol.* **135**:812–815, 1985.
12. Persson, C., Svensson, C., Greiff, L., Andersson, M. and Wollmer, P. The use of the nose to study the inflammatory response of the respiratory tract. *Thorax* **47**:993–1000, 1992.
13. Togias, A., Naclerio, R., Proud, D., Baumgarten, C., Peters, S., Warner, J., Kagey-Sobotka, A., Adkinson, N., Norman, P. and Lichtenstein, L. Mediator release during nasal provocation: a model to investigate the pathophysiology of rhinitis. *Am. J. Med.* **79 (Suppl. 6A)**:26–33, 1986.
14. Holzer, P. Capsaicin: cellular targets, mechanisms of action, and selectivity for thin sensory neurons. *Pharmacol. Rev.* **43**:143–201, 1991.
15. Caterina, M., Schumachert, M., Tominaga, M., Rosen, T., Levine, J. and Julius, D. The capsaicin receptor: a heat-activated ion channel in the pain pathway. *Nature* **389**:816–824, 1997.
16. Bascom, R., Kagey-Sobotka, A. and Proud, D. Effect of intranasal capsaicin on symptoms and mediator release. *J. Pharmacol. Exp. Ther.* **259**:1323–1327, 1991.
17. Geppetti, P., Fusco, B., Marabini, S., Maggi, C., Fanciullacci, M. and Sicuteri, F. Secretion, pain and sneezing induced by the application of capsaicin to the nasal mucosa in man. *Br. J. Pharmacol.* **93**:509–514, 1988.
18. Philip, G., Baroody, F., Proud, D., Naclerio, R. and Togias, A. The human nasal response to capsaicin. *J. Allergy Clin. Immunol.* **94**:1035–1045, 1994.
19. Philip, G., Sanico, A. and Togias, A. Inflammatory cellular influx follows capsaicin nasal challenge. *Am. J. Respir. Crit. Care Med.* **153**:1222–1229, 1996.
20. Roche, N., Lurie, S., Authier, S. and Dusser, D. Nasal response to capsaicin in patients with allergic rhinitis and in healthy volunteers: effect of colchicine. *Am. J. Respir. Crit. Care Med.* **151**:1151–1158, 1995.

21. Sanico, A., Atsuta, S., Proud, D. and Togias, A. Dose-dependent effects of capsaicin nasal challenge: *in vivo* evidence of human airway neurogenic inflammation. *J. Allergy Clin. Immunol.* **100**:632–641, 1997.
22. Umeno, E., Nadel, J. and McDonald, D. Neurogenic inflammation of the rat trachea: fate of neutrophils that adhere to venules. *J. Appl. Physiol.* **69**:2131–2136, 1990.
23. Martins, M., Shore, S. and Drazen, J. Capsaicin-induced release of tachykinins: effects of enzyme inhibitors. *J. Appl. Physiol.* **70**:1950–1956, 1991.
24. Saria, A., Martling, C., Elvar, Z., Theodorsson-Norheim, E., Gamse, R. and Lundberg, J. Release of multiple tachykinins from capsaicin-sensitive sensory nerves in the lung by bradykinin, histamine, dimethylphenyl piperazinium, and vagal nerve stimulation. *Am. Rev. Respir. Dis.* **137**:1330–1335, 1988.
25. Saria, A., Theodorsson-Norheim, E., Gamse, R. and Lundberg, J. Release of substance P- and substance K-like immunoreactivities from the isolated perfused guinea-pig lung. *Eur. J. Pharmacol.* **106**:207–208, 1984.
26. Helme, R., Eglezos, A. and Hosking, C. Substance P induces chemotaxis of neutrophils in normal and capsaicin-treated rats. *Immunol. Cell Biol.* **65**:267–269, 1987.
27. Roch-Arveiller, M., Regoli, D., Chanaud, B., Lenoir, M., Muntaner, O., Stralzko, S. and Giroud, J. Tachykinins: effects on motility and metabolism of rat polymorphonuclear cells. *Pharmacology* **33**:266–273, 1986.
28. Carolan, E. and Casale, T. Effects of neuropeptides on neutrophil migration through noncellular and endothelial barriers. *J. Allergy Clin. Immunol.* **92**:589–598, 1993.
29. Kolasinski, S., Haines, K., Siegel, E., Cronstein, B. and Abramson, S. Neuropeptides and inflammation: a somatostatin analog as a selective antagonist of neutrophil activation by substance P. *Arthritis Rheum.* **35**:369–375, 1992.
30. Wiedermann, C., Wiedermann, F., Apperl, A., Kieselbach, G., Gonwalinka, G. and Braunsteiner, H. *In vitro* human polymorphonuclear leukocyte chemokinesis and human monocyte chemotaxis are different activities of aminoterminal and carboxyterminal substance P. *Naunyn-Schmiedeberg's Arch. Pharmacol.* **340**:185–190, 1989.
31. Braunstein, G., Buvry, A., Lacronique, J., Desjardins, N. and Frossard, N. Do nasal mast cells release histamine on stimulation with substance P in allergic rhinitis? *Clin. Exp. Allergy* **24**:922–929, 1994.
32. Braunstein, G., Fajac, I., Lacronique, J. and Frossard, N. Clinical and inflammatory responses to exogenous tachykinins in allergic rhinitis. *Am. Rev. Respir. Dis.* **144**:630–635, 1991.
33. Baluk, P., Bertrand, C., Geppetti, P., McDonald, D. M. and Nadel, J. A. NK1 receptors mediate leukocyte adhesion in neurogenic inflammation in the rat trachea. *Am. J. Physiol.* **268**:L263–269, 1995.
34. Asakura, K., Shirasaki, H., Narita, S., Kojima, T. and Kataura, A. Study on the dye leakage response of nasal mucosa following topical, capsaicin challenge in guinea pigs. *Acta Otolaryngol.* **112**:545–551, 1992.
35. Lundblad, L., Saria, A., Lundberg, J. and Anggard, A. Increased vascular permeability in rat nasal mucosa induced by substance P and stimulation of capsaicin-sensitive trigeminal neurons. *Acta Otolarnygol.* **96**:479–484, 1983.
36. Erjefalt, I. and Persson, C. Anti-asthma drugs and capsaicin-induced microvascular effects in lower airways. *Agents Actions* **16**:9–10, 1985.
37. Erjefalt, I. and Persson, C. Pharmacologic control of plasma exudation into tracheobronchial airways. *Am. Rev. Respir. Dis.* **143**:1008–1014, 1991.
38. Lundberg, J. and Saria, A. Capsaicin-sensitive vagal neurons involved in control of vascular permeability in rat trachea. *Acta Physiol. Scand.* **115**:521–523, 1982.
39. Lundberg, J. and Saria, A. Capsaicin-induced desensitization of airway mucosa to cigarette smoke, mechanical and chemical irritants. *Nature* **302**:251–253, 1983.
40. Piedimonte, G., Pickles, R., Lehmann, J., McCarty, D., Costa, D. and Boucher, R. Replication-deficient adenoviral vector for gene transfer potentiates airway neurogenic inflammation. *Am. J. Respir. Cell Mol. Biol.* **16**:250–258, 1997.
41. Saria, A., Lundberg, J., Skofitsch, G. and Lembeck, F. Vascular protein leakage in various tissues induced by substance P, capsaicin, bradykinin, serotonin, histamine and by antigen challenge. *Naunyn-Schmiedeberg's Arch. Pharmacol.* **324**:212–218,1983.
42. Sanico, A. M., Philip, G., Lai, G. and Togias, A. Hyperosmolar saline induces reflex nasal secretions, evincing neural hyperresponsiveness in allergic rhinitis. *J. Appl. Physiol.* **86**:1202–1210, 1999.
43. Sanico, A., Atsuta, S., Proud, D. and Togias, A. Plasma extravasation through neuronal stimulation in

human nasal mucosa. *J. Appl. Physiol.* **84**:537–543, 1998.
44. Diemer, F. B., Sanico, A. M., Horowitz, E. and Togias, A. Non-allergenic inhalant triggers in seasonal and perennial allergic rhinitis. *J. Allergy Clin. Immunol.* **103**:S2, 1999.
45. Lucier, G. E. and Egizii, R. Characterization of cat nasal afferents and brain stem neurones receiving ethmoidal input. *Exp. Neurol.* **103**:83–89, 1989.
46. Wallois, F., Macron, J. M., Jounieaux, V. and Duron, B. Trigeminal afferences implied in the triggering or inhibition of sneezing in cats. *Neurosci. Lett.* **122**:145–147, 1991.
47. Naclerio, R., Proud, D., Togias, A., Adkinson, N., Meyers, D., Kagey-Sobotka, A. M. P., Norman, P. and Lichtenstein, L. Inflammatory mediators in late antigen-induced rhinitis. *N. Engl. J. Med.* **313**:65–70, 1985.
48. Baroody, F., Wagenmann, M. and Naclerio, R. Comparison of the secretory response of the nasal mucosa to methacholine and histamine. *J. Appl. Physiol.* **74**:2661–2671, 1993.
49. Philip, G., Jankowski, R., Baroody, F., Naclerio, R. and Togias, A. Reflex activation of nasal secretion by unilateral inhalation of cold dry air. *Am. Rev. Respir. Dis.* **148**:1616–1622, 1993.
50. Riccio, M. and Proud, D. Evidence that enhanced nasal reactivity to bradykinin in patients with symptomatic allergy is mediated by neural reflexes. *J. Allergy Clin. Immunol.* **97**:1252–1263, 1996.
51. Sanico, A. and Togias, A. Neuronal dysfunction and its role in rhinitis. Paper presented at the proceedings of the ICACI XVI/EAACI 97, 1997.
52. Sanico, A. M., Koliatsos, V. E., Stanisz, A. M., Bienenstock, J. and Togias, A. Neural hyperresponsiveness and nerve growth factor in allergic rhinitis. *Int. Arch. Allergy Immunol.* **118**:153–158, 1999.
53. Togias, A., Proud, D., Kagey-Sobotka, A., Lichtenstein, L. and Naclerio, R. Cold dry air and histamine induce more potent responses in perennial rhinitics compared to normal individuals. *J. Allergy Clin. Immunol.* **87**:148, 1991.
54. Van Wijk, R. and Dieges, P. Comparison of nasal responsiveness to histamine, methacholine and phentolamine in allergic rhinitis patients and controls. *Clin. Allergy* **17**:563–570, 1987.
55. Craft, R. M. and Porreca, F. Treatment parameters of desensitization to capsaicin. *Life Sci.* **51**:1767–1775, 1992.
56. Stjarne, P., Lundblad, L., Anggard, A. and Lundberg, J. Local capsaicin treatment of the nasal mucosa reduces symptoms in patients with nonallergic nasal hyperreactivity. *Am. J. Rhinol.* **5**:145–151, 1991.
57. Beydoun, A., Dyke, D., Morrow, T. and Casey, K. Topical capsaicin selectively attenuates heat pain and A delta fiber-mediated laser-evoked potentials. *Pain* **65**:189–196, 1996.
58. Lacroix, J., Buvelot, J., Polla, B. and Lundberg, J. Improvement of symptoms of non-allergic chronic rhinitis by local treatment with capsaicin. *Clin. Exp. Allergy* **21**:595–700, 1991.
59. Petersson, G., Malm, M., Ekman, R. and Hakanson, R. Capsaicin evokes secretion of nasal fluid and depletes substance P and calcitonin gene-related peptide from the nasal mucosa in the rat. *Br. J. Pharmacol.* **98**:930–936, 1989.
60. Zhang, R., Jiang, D. and Li, Z. Experimental study on blocking agent of substance P nerves in the treatment of allergic rhinitis. *Chin. J. Otorhinolaryngol.* **29**:282–285, 1994.
61. Blom, H. M., Van Rijswijk, J. B., Garrelds, I. M., Mulder, P. G. H., Timmermans, T. and Gerth Van Wijk, R. Intranasal capsaicin is efficacious in non-allergic, non-infectious perennial rhinitis. A placebo-controlled study. *Clin. Exp. Allergy* **27**:796–801, 1997.
62. Filiaci, F., Zambetti, G., Ciofalo, A., Luce, M., Masieri, S. and Lovecchio, A. Local treatment of aspecific nasal hyperreactivity with capsaicin. *Allergol. Immunopathol.* **22**:264–268, 1994.
63. Marabini, S., Ciabatti, P., Polli, G., Fusco, B. and Geppetti, P. Beneficial effects of intranasal applications of capsaicin in patients with vasomotor rhinitis. *Eur. Arch. Otorhinolaryngol.* **248**:191–194, 1991.
64. Makker, H. and Holgate, S. The contribution of neurogenic reflexes to hypertonic saline-induced bronchoconstriction in asthma. *J. Allergy Clin. Immunol.* 92, 82–88, 1993.
65. Rosiman, G., Lacronique, J., Desmazes-Duffeu, N., Carre, C., Cae, A. and Dusser, D. Airway responsiveness to bradykinin is related to eosinophilic inflammation in asthma. *Am. J. Respir. Crit. Care Med.* **153**:381–390, 1996.
66. Woolcock, A., Salome, C. and Yan, K. The slope of the dose–response curve to histamine in asthmatic and normal subjects. *Am. Rev. Respir. Dis.* **130**:71–75, 1984.
67. Levi-Montalcini, R. and Hamburger, V. Selective growth stimulating effects of mouse sarcoma on the sensory and sympathetic nervous system of the chick embryo. *J. Exp. Zool.* **116**:321–324, 1951.
68. Leon, A., Buriani, A., Dal Toso, R., Fabris, M., Romanello, S., Aloe, L. and Levi-Montalcini, R. Mast

cells synthesize, store, and release nerve growth factor. *Proc. Natl. Acad. Sci. USA* **91**:3739–3743, 1994.

69. Nilsson, G., Forsberg-Nilsson, K., Xiang, Z., Hallbook, F., Nilsson, K. and Metcalfe, D. Human mast cells express functional TrkA and are a source of nerve growth factor. *Eur. J. Immunol.* **27**:2295–2301, 1997.
70. Solomon, A., Aloe, L., Pe'er, J., Frucht-Pery, J., Bonini, S., Bonini, S. and Levi-Schaffer, F. Nerve growth factor is preformed in and activates human peripheral blood eosinophils. *J. Allergy Clin. Immunol.* **102**:454–460, 1998.
71. Ehrhard, P., Erb, P., Graumann, U. and Otten, U. Expression of nerve growth factor and nerve growth factor receptor tyrosine kinase Trk in activated CD4-positive T-cell clones. *Proc. Natl. Acad. Sci. USA* **90**:10984–10988, 1993.
72. Lambiase, A., Bracci-Laudiero, L., Bonini, S., Bonini, S., Starace, G., D'Elios, M., De Carli, M. and Aloe, L. Human CD4+ T cell clones produce and release nerve growth factor and express high-affinity nerve growth factor receptors. *J. Allergy Clin. Immunol.* **100**:408–414, 1997.
73. Torcia, M., Bracci-Laudiero, L., Lucibello, M., Nencioni, L., Labardi, D., Rubartelli, A., Cozzolino, F., Aloe, L. and Garaci, E. Nerve growth factor is an autocrine survival factor for memory B lymphocytes. *Cell* **85**:345–356, 1996.
74. Bienenstock, J., Tomioka, M., Matsuda, H., Stead, R., Quinonez, G., Simon, G., Coughlin, M. and Denburg, J. The role of mast cells in inflammatory processes: evidence for nerve/mast cell interactions. *Int. Arch. Allergy Appl. Immunol.* **82**:238–243, 1987.
75. Bonini, S., Lambiase, A., Bonini, S., Angelucci, F., Magrini, L., Manni, L. and Aloe, L. Circulating nerve growth factor levels are increased in humans with allergic diseases and asthma. *Proc. Natl. Acad. Sci. USA* **93**:10955–10960, 1996.
76. Crowley, C., Spencer, S., Nishimura, M., Chen, K., Pitts-Meek, K., Armanini, M., Ling, L., McMahon, L., Shelton, S., Levinson, A. and Phillips, H. Mice lacking nerve growth factor display perinatal loss of sensory and sympathetic neurons yet develop basal forebrain cholinergic neurons. *Cell* **76**:1001–1011, 1994.
77. Snider, W. and Wright, D. Neurotrophins cause a new sensation. *Neuron* **16**:229–232, 1996.
78. Hoyle, G., Graham, R., Finkelstein, J., Nguyen, K.-P., Gozal, D. and Friedman, M. Hyperinnervation of the airways in transgenic mice overexpressing nerve growth factor. *Am. J. Respir. Cell Mol. Biol.* **18**:149–157, 1998.
79. Donnerer, J., Schuligoi, R. and Stein, C. Increased content and transport of substance P and calcitonin gene related peptide in sensory nerves innervating inflamed tissue: evidence for a regulatory function of nerve growth factor *in vivo*. *Neuroscience* **49**:693–698, 1992.
80. de Vries, A., Dessing, M. C., Engels, F., Henricks, P. A. J. and Nijkamp, F. P. Nerve growth factor induces a neurokinin-1 receptor-mediated airway hyperresponsiveness in guinea pigs. *Am. J. Respir. Crit. Care Med.* **159**:1541–1544, 1999.
81. Sanico, A. M., Stanisz, A., Gleeson, T. D., Bora, S., Proud, D., Bienenstock, J., Koliatsos, V. and Togias, A. Dysregulation of nerve growth factor expression and release in allergic inflammatory disease of the upper airways. *Am. J. Respir. Crit. Care Med.* In press, 2000.
82. Virchow, J., Julius, P., Lommatzsch, M., Luttmann, W., Renz, H. and Braun, A. Neurotrophins are increased in bronchoalveolar lavage fluid after segmental allergen provocation. *Am. J. Respir. Crit. Care Med.* **158**:2002–2005, 1998.
83. Naclerio, R., Togias, A., Proud, D., Adkinson, N., Kagey-Sobotka, A., Plaut, M., Norman, P. and Lichtenstein, L. Inflammatory mediators in nasal secretions during early and late reactions. *J. Allergy Clin. Immunol.* **73**:148, 1984.
84. Sanico, A., Philip, G., Proud, D., Naclerio, R. and Togias, A. Comparison of nasal mucosal responsiveness to neuronal stimulation in nonallergic and allergic rhinitis: effects of capsaicin nasal challenge. *Clin. Exp. Allergy* **28**:92–100, 1998.

CHAPTER 42

Antihistaminic and Anti-inflammatory Effects of Mizolastine

MASSIMO TRIGGIANI, CRISTIANA PALUMBO, MARCO GENTILE, FRANCESCOPAOLO GRANATA and GIANNI MARONE*

Division of Clinical Immunology and Allergy, University of Naples Federico II, Naples, Italy

INTRODUCTION

Histamine is a central mediator in the pathogenesis of allergic and inflammatory disorders (1, 2). Its biological effects are mediated by the activation of specific receptors indicated as H_1, H_2 and H_3 (3–5). H_1 is the receptor predominantly involved in the clinical and biochemical manifestations of allergic diseases, including bronchial asthma (6), urticaria (7) and rhinitis (8). H_1 receptor antagonists have been available for the treatment of allergic disorders for over 60 years (9). However, the first-generation antihistamines were characterized by such untoward events as sedation, cardiac arrhythmias and anticholinergic effects (10, 11). The second generation of this class of drugs has been considerably improved particularly with regard to specificity for the H_1 receptor and reduced ability to cross the blood–brain barrier and induce somnolence and impaired psychomotor performance (12, 13). Furthermore, even high doses of most of the second-generation antihistamines do not affect cardiac repolarization and do not cause clinically relevant tachyarrhythmias (14).

Mizolastine, one of the newest H_1 receptor antagonists, is highly selective for the H_1 receptor and does not exert anticholinergic, anti-adrenergic or antiserotoninergic activity (15). Clinical data have shown that mizolastine is highly effective in the treatment of urticaria and rhinitis (16, 17). However, studies in both humans and animals indicated that, in addition to the antihistamine effect, mizolastine also possesses broad anti-inflammatory activities that may contribute to its clinical efficacy.

* Corresponding author.

MAST CELLS AND BASOPHILS
ISBN 0-12-473335-2

PHARMACOKINETICS OF MIZOLASTINE

Absorption of mizolastine after oral administration is rapid in healthy young adults (18, 19). The median time to peak concentration is 1.5 h and the mean plasma peak concentration is 276 ng ml^{-1} after a single administration of 10 mg (20). Intra-individual variability of mizolastine absorption is low; the coefficient of variation of the area under the curve is 9.6% (15, 20). Bioavailability is approximately 65% and is not significantly affected by concomitant ingestion of food or alcohol (21). However, at steady-state, plasma concentrations correlate with biological activity (22).

Mizolastine is extensively metabolized, although no active metabolites have been reported. [^{14}C]Mizolastine is largely excreted in the faeces (84–95%), with less than 0.5% of an administered dose being excreted unchanged in the urine (23). Glucuronidation and sulphation, the main metabolic pathway, account for 65% of an administered mizolastine dose. Minor metabolic pathways include oxidation of the benzimidazole ring, *N*-dealkylation by removal of the *p*-fluorobenzyl ring, and hydroxylation followed by methylation of the pyrymidinone ring (24). In *in vitro* studies, cytochrome P450 3A4 and cytochrome P450 2A6 have been found to be mainly responsible for oxidation (24).

Mizolastine elimination is described by an open two-compartment model with distribution and terminal half-life values of 2 and 13 h, respectively (19, 20). Given the latter value, mizolastine can be administered in a single daily dose (25). After regular daily oral administration for up to 2 weeks, steady-state is reached by the third day and no accumulation is observed (26).

CLINICAL EFFICACY OF MIZOLASTINE

A number of clinical studies have demonstrated the efficacy of mizolastine in the treatment of allergic rhinitis and chronic urticaria and the drug is now licensed in several countries for the treatment of these two conditions (16, 17). Initial studies have shown that a single oral administration of mizolastine effectively inhibits the wheal and flare response to intradermal injection of histamine (16, 22, 27). In a double-blind, cross-over study, mizolastine was more effective than loratadine and equally effective as cetirizine (16). Using the same clinical model, it was shown that subsensitivity did not develop during continuous treatment with mizolastine for up to 8 weeks (28).

A double-blind, randomized, placebo-controlled, parallel-group study was conducted in 56 patients with chronic idiopathic urticaria (17). Mizolastine was administered as a single daily dose of 10 mg orally for 4 weeks. Treatment with mizolastine significantly improved the urticaria symptoms, assessed by subjective and clinical scores, as compared to placebo. Mizolastine was well tolerated and somnolence was the only major side-effect (recorded in 7% of patients) (17).

The efficacy of mizolastine in seasonal or perennial allergic rhinitis has been evaluated in several studies. A multicentre, double-blind, placebo-controlled study was performed in 494 patients with seasonal allergic rhinitis (29). This study showed that mizolastine significantly improved symptom score, nasal score and ocular score compared with placebo. Initial improvement was recorded as early as after 2 days of treatment, and a full therapeutic effect was achieved after the first week, indicating that the pharmacological effect of mizolastine is very rapid (29). In addition, this study established that the optimal dose of mizolastine is 10 mg/day, no further advantage being obtained with higher doses (29).

Another study compared the effect of mizolastine (10 mg) and loratadine (10 mg) in 68 patients with perennial allergic rhinitis (30). After 4 weeks of treatment, mizolastine was as effective as loratadine, and induced approximately a 60%, 75% and 65% decrease in nasal score, ocular score and global total score, respectively (30). Safety was satisfactory with both mizolastine and loratadine. These results have recently been extended in a double-blind, placebo-controlled study of 257 patients suffering from perennial allergic rhinitis for more than 10 years (31). In addition to the symptom improvement recorded in the previous studies, mizolastine significantly ameliorated nasal blockade and induced a significant reduction in rhinoscopy findings. Since these events have been attributed to the release of histamine as well as to the local production of cysteinyl leukotrienes, the beneficial effect of mizolastine was related, at least in part, to its inhibitory effect on 5-lipoxygenase (see below) (31, 32).

A large trial attempted to determine whether the onset of hay fever symptoms could be delayed in patients known to suffer seasonal allergic rhinitis symptoms if mizolastine was given before the pollen season (33). This double-blind study involved 342 patients, allocated randomly to once-daily 10 mg mizolastine (n=115), once-daily 120 mg terfenadine (n=116), or placebo (n=111). All patients started treatment on 1 May, before the onset of the grass pollen season. The prophylactic effect of test drugs was determined according to their ability to delay the time to the first hay fever attack of the season. The first attack was defined by the occurrence of one of the following events: use of rescue medication, study withdrawal because of treatment failure, or total diary symptom score over 18. Active treatments prolonged the time to the first attack by approximately 1 week (mizolastine 55 days, terfenadine 57 days) in comparison with placebo (50 days) (survival curve analysis: log rank test, p=0.01; Wilcoxon test, p=0.03). Tolerability was satisfactory and comparable between groups.

Taken together, these data indicate that mizolastine can be used safely and effectively to treat symptoms of chronic urticaria and of both perennial and seasonal allergic rhinitis. Mizolastine is at least as effective as other second-generation antihistamines and it can be also used to delay the onset of seasonal rhinitis.

ANTI-INFLAMMATORY EFFECTS OF MIZOLASTINE

Since early studies in experimental animals, it was evident that mizolastine possessed broad anti-inflammatory activities that were apparently unrelated to its antihistaminic properties. For example, mizolastine inhibited the passive cutaneous anaphylactic reaction in guinea pigs, rats and mice (34). Furthermore, mizolastine protected rats from lethal shock induced by compound 48/80 and prevented bronchospasm induced by systemic injection of platelet-activating factor and cysteinyl leukotriene D_4 (LTD_4) (34). In isolated mast cells obtained by peritoneal or bronchoalveolar lavage of sensitized rats, mizolastine inhibited histamine release induced by antigen (34).

Mizolastine was not only effective in experimental models of allergic reactions but also in other models of inflammation. Oral mizolastine significantly reduced nociception, macroscopic and microscopic intestinal damage, intestinal tissue weight and myeloperoxidase activity in a rat model of inflammatory bowel disease (35). Interestingly, other antihistamines such as terfenadine were devoid of such activity.

These effects were attributed to a reduction in the generation of arachidonic acid metabolites from resident intestinal cells and from infiltrating inflammatory cells (mainly neutrophils). This hypothesis was confirmed by studies showing that mizolastine inhibited in a dose-dependent manner the time-course of rat paw inflammation induced by local injection of arachidonic acid (36). Experiments in disrupted cells evidenced that mizolastine effectively inhibited 5-lipoxygenase, but not cyclo-oxygenase, activity (36). Therefore, mizolastine is able to inhibit, at least *in vitro*, the production of both cysteinyl leukotrienes (LTC_4, LTD_4 and LTE_4) and LTB_4. Together with the inhibition of histamine release, these findings suggest that mizolastine might interfere with the production of two major classes of mediators produced by mast cells and involved in allergic disorders (histamine and cysteinyl leukotrienes).

The above effects of mizolastine were not related to its antihistaminic activity. However, it is worth mentioning other anti-inflammatory properties of mizolastine that are primarily due to the blockade of H_1 receptors on cells involved in the long-term regulation of inflammatory and immune responses. In fact, a body of experimental evidence indicates that histamine may possess inflammatory and immunomodulating activities. In addition to the well-described effects on vascular and respiratory smooth muscle cells and endothelial cells, histamine influences other effector cells of the immune response. For example, histamine may regulate the function of T lymphocytes (37) and NK (natural killer) cells (38) by inducing the production of immunomodulatory cytokines (39–44). Furthermore, histamine is able to increase chemotaxis and mediator release from human eosinophils in response to tryptase (45, 46).

We have recently shown that physiologically relevant concentrations of histamine induce exocytosis and cytokine production from human macrophages purified from lung parenchyma (47). These effects of histamine are mediated by activation of the H_1 receptor expressed on macrophages (48). In these cells, histamine induces the secretion of the lysosomal enzyme β-glucuronidase and the expression and release of interleukin-6 (IL-6) and tumour necrosis factor-α (TNF-α). Both events are Ca^{2+}-dependent and associated with an increase in cytosolic Ca^{2+}. These observations raised important questions about the role of histamine as a mediator involved in the long-term regulation of inflammatory and immune responses. Macrophages are cells involved in antigen presentation, T lymphocyte proliferation and recruitment of inflammatory cells in the airways (49). The ability of histamine to promote the release of the major cytokines produced by macrophages suggests that it may play an important role in the chronic airway inflammation found in patients with bronchial asthma. We have recently studied the effect of pharmacological concentrations of mizolastine on the activation of human lung macrophages induced by histamine. Figure 1 shows that mizolastine efficiently inhibits by up to 80% the release of β-glucuronidase induced by histamine from human macrophages with an IC_{50} of approximately 2 μM. In parallel experiments, mizolastine inhibited the production of IL-6 from human macrophages stimulated with histamine (data not shown). While these observations reinforce the potency of mizolastine as an H_1 receptor antagonist (50), they lead to the hypothesis that the drug may interfere with the production of mediators involved in the regulation of the immune response and tissue remodelling. This hypothesis deserves further experimental and clinical studies and, should it be confirmed, will open new pharmacological perspectives for the long-term use of mizolastine in the treatment of allergic disorders.

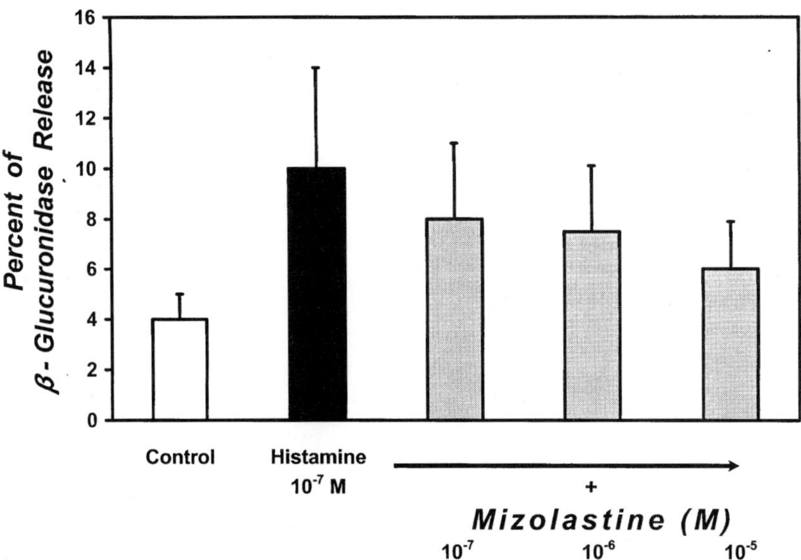

Fig. 1 Effect of increasing concentrations of mizolastine on the release of β-glucuronidase from human lung macrophages induced by histamine. Human macrophages were isolated and purified from the lung parenchyma of patients undergoing thoracic surgery. The macrophages were pre-incubated (15 min, 37°C) with the indicated concentrations of mizolastine and subsequently incubated (120 min, 37°C) with histamine (10^{-7} M). At the end of the incubation, the release of β-glucuronidase was determined and expressed as a percentage of the total cellular content of the enzyme.

CONCLUDING REMARKS

Mizolastine is a well-tolerated, highly specific H_1 receptor antagonist, useful in the treatment of allergic rhinitis and chronic urticaria (17, 29, 30, 50). The reduced incidence of somnolence, psychomotor impairment and anticholinergic and pro-arrhythmogenic effects makes this drug suitable for long-term use in patients with allergic diseases (50–52). However, a full evaluation of the pharmacological profile of mizolastine must take into account its anti-inflammatory properties and whether or not they are related to H_1 receptor antagonism. A complete characterization of the anti-inflammatory effects of mizolastine *in vitro* and *in vivo* may offer a stronger rationale for its chronic use in allergic patients. The results of *in vitro* experiments suggest the importance of a careful evaluation of mizolastine as an inhibitor of mediator release from mast cells (histamine and cysteinyl leukotrienes) and macrophages (remodelling enzymes and pro-inflammatory cytokines). If confirmed *in vivo*, these effects may contribute to extending the pharmacological profile of mizolastine in the prevention and treatment of allergic disorders.

REFERENCES

1. White, M. V. The role of histamine in allergic diseases. *J. Allergy Clin. Immunol.* **86**:599–605, 1990.
2. Marone, G. Milestones in the biology and pharmacology of H_1-receptor antagonists. *Allergy* **52 (Suppl. 34)**:7–13, 1997.

3. Gantz, I., Schaffer, M., Del Valle, J., Logsdon, C., Campbell, V., Uhler, M. and Yamada, T. Molecular cloning of a gene encoding the histamine H_2 receptor. *Proc. Natl. Acad. Sci. USA* **88**:429–433, 1991.
4. Marone, G. and Genovese, A., eds. Specific and novel aspects of histamine H_1-receptor antagonists. *Allergy* **52 (Suppl. 34)**, 1997.
5. Lovenberg, T. W., Roland, B. L., Wilson, S. J., Jiang, X., Pyati, J., Huvar, A., Jackson, M. R. and Erlander, M. G. Cloning and functional expression of the human histamine H_3 receptor. *Mol. Pharmacol.* **55**:1101–1107, 1999.
6. Casale, T. B., Wood, D. and Richerson, H. B. Elevated bronchoalveolar lavage fluid histamine levels in allergic asthmatics are associated with metacholine bronchial hyperresponsiveness. *J. Clin. Invest.* **79**:1197–1203, 1987.
7. Monroe, E. W. Nonsedating H_1 antihistamines in chronic urticaria. *Ann. Allergy* **71**:585–591, 1993.
8. Ash, A. S. F. and Shild, H. O. Receptors mediating some actions of histamine. *Br. J. Clin. Pharmacol.* **27**:427–439, 1966.
9. Simons, F. E. R. The pharmacology and use of H_1-receptor antagonist drugs. *N. Engl. J. Med.* **330**:1663–1670, 1994.
10. Woodward, J. K. Pharmacology of antihistamines. *J. Allergy Clin. Immunol.* **86**:606–612, 1990.
11. Vigorito, C., Poto, S., Picotti, G. B., Triggiani, M. and Marone, G. Effect of activation of the H_1 receptor on coronary hemodynamics in man. *Circulation* **73**:1175–1182, 1986.
12. Slater, J. W., Zechnich, A. D. and Haxby, D. G. Second generation antihistamines. *Drugs* **57**:31–47, 1999.
13. Woosley, R. L. Cardiac actions of antihistamines. *Annu. Rev. Pharmacol. Toxicol.* **36**:233–252, 1996.
14. Yap, Y. G. and Camm, A. J. The current cardiac safety situation with antihistamines. *Clin. Exp. Allergy* **29**:15–24, 1999.
15. Simons, F. E. R. Mizolastine: antihistaminic activity from preclinical data to clinical evaluation. *Clin. Exp. Allergy* **29 (Supp. 1)**:3–9, 1999.
16. Rosenzweig, P., Caplain, H., Chaufour, S., Ulliac, N., Cabanis, M. J. and Thebault, J. J. Comparative wheal and flare study of mizolastine vs terfenadine, cetirizine, loratadine and placebo in healthy volunteers. *Br. J. Clin. Pharmacol.* **40**:459–465, 1995.
17. Brostoff, J., Fitzharris, P., Dunmore, C., Theron, M. and Blondin, P. Efficacy of mizolastine, a new antihistamine, compared with placebo in the treatment of chronic idiopathic urticaria. *Allergy* **51**:320–325, 1996.
18. Pinquier, J. L., Caplain, H., Cabains, M. J., Dubruc, C., Stalla-Bourdillon, A. and Rosenzweig, P. Inhibition of histamine-induced skin weal and flare after 5 days of mizolastine. *J. Clin. Pharmacol.* **36**:72–78, 1996.
19. Mesnil, F., Dubruc, C., Mentre, F., Huet, S., Mallet, A. and Thenot, J. P. Pharmacokinetic analysis of mizolastine in healthy young volunteers after single oral and intravenous doses: noncompartmental approach and compartmental modelling. *J. Pharmacokinet. Biopharm.* **25**:125–147, 1997.
20. Rosenzweig, P., Thebault, J. J., Caplain, H., Dubruc, C., Bianchetti, G., Fuseau, E. and Morselli, P. L. Pharmacodynamics and pharmacokinetics of mizolastine (SL 85.0324), a new nonsedative H_1 antihistamine. *Ann. Allergy* **69**:135–139, 1992.
21. Ascalone, V., Guinebault, P. and Rouchouse, A. Determination of mizolastine, a new antihistaminic drug, in human plasma by liquid–liquid extraction, solid-phase extraction, and column-switching techniques in combination with high-performance liquid chromatography. *J. Chromatogr.* **619**:275–284, 1993.
22. Chosidow, O., Dubruc, C., Danjou, P., Fuseau, E., Espagne, E., Bianchetti, G., Thenot, J. P., Herson, S., Rosenzweig, P. and Revuz, J. Plasma and skin suction–blister–fluid pharmacokinetics and time course of the effects of oral mizolastine. *Eur. J. Clin. Pharmacol.* **50**:327–333, 1996.
23. Malavasi, B., Locatelli, M. and Ascalone, V. Coupled liquid–liquid extraction and column switching LC for the determination of a new antihistaminic H_1 drug in human urine. *Chromatographia* **36**:337–342, 1993.
24. Synthelabo Group. *Mizolastine Core Scientific Brochure, Medical and Regulatory Affairs Direction*, 1998.
25. Arbilla, S., Schoemaker, H., Scatton, B., Benavides, J., Claustre, Y., Proteau, M., Delaye, M., Pimouie, C., Graham, D., Manoury, P., Lloyd, K. G., Wick, A. and Langer, S. Z. SL 85.0324: a novel benzimidazole derivative with selective histamine H_1-receptor antagonist properties. *Eur. J. Clin. Pharmacol.* **183**:222, 1990 (abstract).
26. Mesnil, F., Mentre, F., Dubruc, C., Thenot, J. P. and Mallet, A. Population pharmacokinetic analysis of mizolastine and validation from sparse data on patients using the nonparametric maximum likelihood method. *J. Pharmacokinet. Biopharm.* **26**:133–161, 1998.

27. Pinquier, J. L., Caplain, H., Cabanis, M. J., Dubruc, C., Stalla-Bourdillon, A. and Rosenzweig, P. Inhibition of histamine-induced skin wheal and flare after 5 days of mizolastine. *J. Clin. Pharmacol.* **36**:72–78, 1996.
28. Bousquet, J., Chanal, I., Murrieta, M. and Stalla-Bourdillon, A. Lack of subsensitivity to mizolastine over 8-week treatment. *Allergy* **51**:251–256, 1996.
29. Leynadier, F., Bousquet, J., Murrieta, M., Attali, P. and the Rhinase Study Group. Efficacy and safety of mizolastine in seasonal allergic rhinitis. *Ann. Allergy Asthma Immunol.* **76**:163–168, 1996.
30. Bellioni, P., Catalano, B., Cervellera, G., Filiaci, F., Mira, E. and Carraro, A. Comparison of mizolastine with loratadine in the treatment of perennial allergic rhinitis. *Rhinology* **34**:101–104, 1996.
31. Bachert, C., Brostoff, J., Scadding, G. K., Tasman, J., Stalla-Bourdillon, A., Murrieta, M. and the RIPERAN Study Group. Mizolastine therapy also has an effect on nasal blockade in perennial allergic rhinoconjunctivitis. *Allergy* **53**:969–975, 1998.
32. Bousquet, J. Rapid symptom relief in rhinitis. *Clin. Exp. Allergy* **29**:25–29, 1999.
33. Stern, M., Blondin-Ertzbischoff, P., Murrieta-Aguttes, M., Hardwicke, C., Emmerson, E. B. M. and Judd, M. S. Rapid and sustained efficacy of mizolastine 10 mg once daily in seasonal allergic rhinitis. *J. Int. Med. Res.* **26**:292–303, 1998.
34. Levrier, J., Duval, D., Prouteau, M., Voltz, C., Berry, N., Lloyd, G. and Scatton, B. Anti-anaphylactic activity of the novel selective histamine H_1 receptor antagonist mizolastine in the rodent. *Arzneim.-Forsch. Drug Res.* **45**:559–568, 1995.
35. Goldhill, J., Pichat, P., Roome, N., Angel, I. and Arbilla, S. Effect of mizolastine on visceral sensory afferent sensitivity and inflammation during experimental colitis. *Arzneim.-Forsch. Drug Res.* **48**:179–184, 1998.
36. Pichat, P., Angel, I. and Arbilla, S. Anti-inflammatory properties of mizolastine after oral administration on arachidonic acid-induced cutaneous reaction in the rat. *Arzneim.-Forsch. Drug Res.* **48**:173–178, 1998.
37. Cameron, W., Doyle, K. and Rocklin, R. E. Histamine type 1 (H_1) receptor radioligand binding studies on normal T cell subset, B cells, and monocytes. *J. Immunol.* **136**:2116–2120, 1986.
38. Hill, S. J. Distribution, properties, and functional characteristics of three classes of histamine receptor. *Pharmacol. Rev.* **42**:45–83, 1990.
39. Beer, D. J., Matloff, S. M. and Rocklin, R. E. The influence of histamine on immune and inflammatory responses. *Adv. Immunol.* **35**:209–268, 1984.
40. Dohlsten, M., Sjogren, H. O. and Carlsson, R. Histamine acts directly on human T cells to inhibit interleukin-2 and interferon-γ production. *Cell. Immunol.* **109**:65–74, 1987.
41. Vannier, E., Miller, L. C. and Dinarello, C. A. Histamine suppresses gene expression and synthesis of tumor necrosis factor α via histamine H_2 receptor. *J. Exp. Med.* **174**:281–284, 1991.
42. Vannier, E. and Dinarello, C. A. Histamine enhances interleukin (IL)-1-induced IL-6 gene expression and protein synthesis via H_2 receptors in peripheral blood mononuclear cells. *J. Biol. Chem.* **269**:9952–9956, 1994.
43. Takizawa, H., Ohtoshi, T., Kikutani, T., Okazaki, H., Akiyama, N., Sato, M., Shoji, S. and Ito, S. Histamine activates bronchial epithelial cells to release inflammatory cytokines *in vitro*. *Int. Arch. Allergy Immunol.* **108**:260–267, 1995.
44. van der Pouw Kraan, T. C. T. M., Snijders, A., Boeije, L. C. M., de Groot, E. R., Alewijnse, A. E., Leurs, R. and Aarden, L. A. Histamine inhibits the production of interleukin-12 through interaction with H_2 receptors. *J. Clin. Invest.* **102**:1866–1873, 1998.
45. Raible, D. G., Schulman, E. S., Dimuzio, J., Cardillo, R. and Post, T. J. Mast cell mediators prostaglandin D_2 and histamine activate human eosinophils. *J. Immunol.* **148**:3536–3542, 1992.
46. He, S., Peng, Q. and Walls, A. F. Potent induction of a neutrophil and eosinophil-rich infiltrate *in vivo* by human mast cell tryptase: selective enhancement of eosinophil recruitment by histamine. *J. Immunol.* **159**:6216–6225, 1997.
47. Triggiani, M., Gentile, M., Palumbo, C., Granata, F. and Marone, G. Histamine induces IL-6 and lysosomal enzyme release from human lung macrophages. *Allergy* **54 (Suppl. 52)**:8, 1999.
48. Cluzel, M., Liu, M. C., Goldman, D. G., Undem, B. J. and Lichtenstein, L. M. Histamine acting on a histamine type 1 (H_1) receptor increases β-glucuronidase release from human lung macrophages. *Am. J. Respir. Cell Mol. Biol.* **3**:603–609, 1990.
49. Benavides, J., Schoemaker, H., Dana, C., Claustre, Y., Delahaye, M., Prouteau, M., Manoury, P., Allen, J., Scatton, B., Langer, S. Z. and Arbilla, S. *In vivo* and *in vitro* interaction of the novel selective histamine H_1 receptor antagonist mizolastine with H_1 receptors in the rodent. *Arzneim.-Forsch. Drug Res.* **45**:551–558, 1995.

50. Danjou, P., Molinier, P., Berlin, I., Patat, A., Rosenzweig, P. and Morselli, L. Assessment of the anticholinergic effect of the new antihistamine mizolastine in healthy subjects. *Br. J. Clin. Pharmacol.* **34**:328–331, 1992.
51. Kerr, J. S., Dunmore, C. and Hindmarch, I. The psychomotor and cognitive effects of a new antihistamine, mizolastine, compared to terfenadine, triprolidine and placebo in healthy volunteers. *Eur. J. Clin. Pharmacol.* **47**:331–335, 1994.
52. Patat, A., Perault, M. C., Vandel, B., Ulliac, N., Zieleniuk, I. and Rosenzweig, P. Lack of interaction between a new antihistamine, mizolastine, and lorazepam on psychomotor performance and memory in healthy volunteers. *Br. J. Clin. Pharmacol.* **39**:31–38, 1995.

CHAPTER 43

Electrophysiological and Molecular Basis for the Adverse Cardiovascular Effects of Histamine H_1 Receptor Antagonists

MAURIZIO TAGLIALATELA, ANNA PANNACCIONE, PASQUALINA CASTALDO, GIOVANNA GIORGIO and LUCIO ANNUNZIATO*

Section of Pharmacology, Department of Neuroscience and Behavioral Science, School of Medicine, University of Naples Federico II, Naples, Italy*

INTRODUCTION

The discovery of histamine, the description of its role in physiological and pathological responses, along with the development of pharmacological tools designed to block these effects, represent some of the greatest achievements in modern physiology, medicinal chemistry, and pharmacology. The history of these achievements is highlighted by the names of some of the greatest pharmacologists of the first half of this century (Nobel Laureates including Ehrlich, Richet, Bovet and Dale, have substantially contributed to this progress) and by the extensive co-operation among disciplines that were still in their infancy at that time. Although over 60 years have passed since the first drug capable of antagonizing the biological effects of histamine was described (1), the recent history of antihistamines is far from being immune from breakthrough discoveries and innovative findings. In fact, the development of the so-called 'second-generation antihistamines' in the last 20 years has clearly provided the physician with improved therapeutical tools, including acrivastine, astemizole, azelastine, cetirizine, ebastine, emedastine, fexofenadine, ketotifen, levocarbastine, loratadine, mizolastine, oxatomide, and terfenadine (just to mention those widely used presently or that have been utilized more in the recent past). In fact, these novel compounds are characterized by improved H_1 selectivity, absence of sedation and, possibly, anti-allergic properties distinct from their antihistaminic activity (2). However, even more recently, adverse cardiovascular effects associated with the use of some congeners belonging to this therapeutic class (particularly terfenadine and astemizole) have been reported, and a major concern over the therapeutic selection of antihistamines now deals with their potentially severe arrhythmogenic properties (3). This chapter attempts to review the recent advances in the understanding of the pathogenesis and aetiology of the cardiotoxic actions of some

* Corresponding author.

MAST CELLS AND BASOPHILS
ISBN 0-12-473335-2

second-generation antihistamines, with particular emphasis on their molecular actions at the level of ion channels controlling the cardiac action potential. In addition, the chapter will focus on the recent concerns over the potential adverse cardiac effects of first-generation H_1 receptor blockers. The modern developments in the history of antihistamines are also characterized by contributions from disciplines and techniques not traditionally related to immunology, such as molecular genetics and cellular electrophysiology, and it seems possible to anticipate that the merging of such different disciplines will provide drugs with still improved therapeutic efficacy and safety for the clinical management of allergic diseases.

CARDIOTOXICITY OF SECOND-GENERATION ANTIHISTAMINES

Introduction

Although histamine itself has important cardiovascular effects and the cardioactive properties of 'classical' antihistamines have been known for decades, the issue of a possible association between the use of antihistamines and cardiac arrhythmias has only received considerable attention in the last decade. In fact, an increasing number of reports have appeared in the recent literature showing the occurrence of prolongation of the QT interval, 'torsade de pointes' ventricular arrhythmias, syncope, and cardiac arrests either in patients taking the recommended doses of terfenadine or astemizole, or in patients with intentional or accidental overdoses of these two second-generation antihistamines (4–11). In 1990, after a review by the Food and Drug Administration (FDA) of 25 case reports of associations between cardiac arrhythmia and second-generation antihistamine use, labelling changes were introduced several times for terfenadine, and a prominent box cautioning against its use under specific settings was included in 1992. Furthermore, in an effort to inform the medical and patient communities about the serious and potentially fatal adverse cardiac effects associated with the inappropriate use of terfenadine, several 'Dear Health Care Professional' letters were sent in 1990, 1992 and 1996 to physicians and pharmacists in the USA warning them of the risks associated with the use of terfenadine (12, 13). After these letters, the number of reported cases describing such rare though potentially fatal arrhythmia in association with terfenadine or astemizole use grew to over 200 (14); therefore, in consideration of studies also showing co-prescription and co-dispensing of medications contraindicated with terfenadine continued to occurr (15), terfenadine has now been withdrawn from the market in several countries. Similar procedures are also being started for astemizole (16).

Factors Predisposing to Cardiotoxicity by Second-Generation Antihistamines

Under standard therapeutic settings and conventional doses, terfenadine can be considered a pro-drug since it undergoes hepatic first-pass extraction and extensive metabolization into an acid metabolite by the CYP3A4 isoenzyme of the P450 oxidative pathway (17). Therefore, in most normal subjects, its levels in the plasma are below the detection limits ($\approx$10 ng ml^{-1}). Careful evaluation of the clinical circumstances under which terfenadine elicits its serious cardiotoxic potential has revealed that these adverse effects were directly related to increased plasma levels of terfenadine itself. The case of a 39-year-old woman who experienced several syncopes and 'torsade de pointes' ventricular arrhythmias while taking the standard therapeutic dose of terfenadine

(120 mg per day for 12 days) when she started self-medication with the well-known inhibitor of the CYP3A4 P450 enzyme ketoconazole (200 mg twice a day for 2 days), in particular pointed towards impaired liver metabolism as one of the main predisposing factors for cardiotoxicity by terfenadine (5). In fact, most of the cases initially reviewed by the FDA throughout April 1992 involved the concomitant use of inhibitors of the CYP3A4 P450 enzyme, such as ketoconazole, itraconazole, or macrolide antibiotics (14). Patients at risk of developing arrhythmias also included those with impaired liver function (cirrhosis, ethanol abuse). The crucial role played by the hepatic metabolism in the expression of the cardiotoxic potential of the parent molecule, besides representing an important risk factor for patients taking these second-generation antihistamines, also requires the careful evaluation of potential cardiotoxicity of the main metabolic products generated by hepatic metabolism. As indicated, terfenadine is metabolized to an acid metabolite called fexofenadine or terfenadine carboxylate. This compound retains the histamine H_1 receptor antagonist and non-sedative properties of the parent compound, but does not affect the cardiac QT interval and does not undergo hepatic metabolism (18). Therefore, as terfenadine was being withdrawn from the market, fexofenadine began to be commercialized. Astemizole also undergoes extensive first-pass metabolism, its main metabolite being desmethylastemizole. Yet following the description of QT prolongation and 'torsade de pointes' in a patient with undetectable (>0.5 ng ml^{-1}) serum concentrations of astemizole but with higher concentrations (7.7 ng ml^{-1}) of desmethylastemizole, the cardiac safety of this metabolite has been questioned (19).

In addition to impaired liver metabolism of the parent molecule, other clinical conditions of pre-existing cardiac dysfunction, such as congenital QT prolongation (see below), ischaemic heart disease, and congestive heart failure, or electrolyte imbalance, such as hypokalaemia and hypomagnesaemia, may precipitate the arrhythmic episodes induced by second-generation antihistamines (2).

HERG K^+ Channels as Targets for the Adverse Cardiovascular Effects of Second-Generation Antihistamines

The co-ordinated contraction of the cardiac muscle is the result of the sequential opening of several classes of ion channels in different cells. Blockade of specific ion channels may exert both pro-arrhythmic and anti-arrhythmic effects, depending on the anatomical site of action, the characteristics of the cardiac rhythm, in addition to several other concomitant factors. The duration of the cardiac action potential is controlled by a fine equilibrium between inward and outward currents (20). As already described, most of the arrhythmic episodes occurring in predisposed patients upon taking terfenadine and astemizole were polymorphic ventricular tachycardias with 'torsade de pointes' ventricular fibrillation. This ventricular arrhythmia, as first described by Dessertenne (21), consists of a progressive modification of the amplitude and polarity of the QRS complexes on the surface ECG, which appear to be twisting around an imaginary isoelectric baseline (22). 'Torsade de pointes' generally occur in the setting of a marked bradycardia with prolongation of the QT interval (generally >440 ms). Several groups have observed the existence of a strict parallelism between the propensity of a drug to cause 'torsade de pointes' and its ability to induce specific and characteristic abnormalities of the cardiac repolarization process in experimental models such as the canine cardiac Purkinje fibres, defined as 'early after-depolarizations' (EADs) (Fig. 1). In fact: (1) both conditions are potentiated by low extracellular potassium concentrations

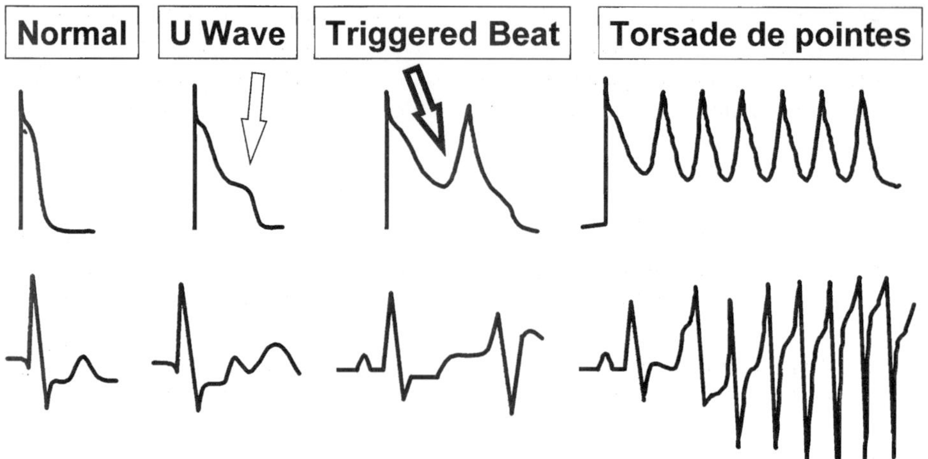

Fig. 1 Schematic representation of the temporal relationship between action potentials recorded in the conductive system (top row) and the corresponding electrocardiogram (bottom row). The thin arrow shows an abnormality of terminal repolarization (EAD; see text for details); the bolded arrow shows a triggered beat resulting from such EADs. Torsade de pointes may represent a 'run' of such triggered activity. (Modified from ref. 23.)

($[K^+]_e$) and are inhibited by high extracellular $[K^+]_e$; (2) both conditions are potentiated by low pacing rates and are inhibited by high pacing rates; (3) the initial triggered beat arises late during repolarization of a markedly prolonged action potential, while the initial beat of the 'torsade de pointes' arises late during repolarization following a markedly prolonged QT interval (23). Therefore, it seems plausible that similar mechanisms mediating EADs are also involved in the genesis of 'torsade de pointes'. Thus, understanding the mechanisms involved in cardiac action potential prolongation leading to EADs is of crucial importance in order to reveal the pathogenesis of the 'torsade de pointes' arrhythmic episodes. In most clinical circumstances, the action potential prolongation which results in EADs is consequent to the blockade of the repolarizing currents mainly carried by cardiac K^+ channels, although additional mechanisms, such as an increased depolarizing current via voltage-dependent Na^+ or Ca^{2+} channels, have also been called into play (24, 25).

Under physiological conditions, several classes of K^+ currents shape the action potential in cardiac cells (26): the transient outward current (I_{to}), the delayed rectifier repolarizing current made up of both rapid (I_{Kr}) and slow (I_{Ks}) components (27), and the inward rectifier I_{K1} current, which participates in the final phases of repolarization. Other K^+ channels are also active during the action potential plateau in selective heart regions such as the acetylcholine-activated K^+ channel ($I_{K(Ach)}$) in the atria and the Purkinje fibres, or, under specific pathological conditions, such as the ATP-dependent K^+ channel ($I_{K(ATP)}$), the Na^+-dependent K^+ current ($I_{K(Na)}$) and the fatty-acid activated K^+ current ($I_{K(FA)}$) (20). The recent cloning of most of the genes encoding for K^+ channels has revealed that their functional diversity is matched by an extraordinary degree of structural heterogeneity. Thus they can be classified on the basis of several criteria: putative transmembrane arrangement (six-transmembrane domains versus two- or four-transmembrane domains), pharmacological selectivity (to drugs such as tetra-ethylammonium and 4-aminopyridine, or toxins), permeation and rectification properties (inward rectifiers versus outward rectifiers), as well as their gating mechanisms (discriminating those activated by changes in transmembrane voltage from those activated by changes in the concentration of specific ligands, generally on the cytoplasmic side of the membrane). The last criterion has been utilized to generate the classifications shown in Fig. 2.

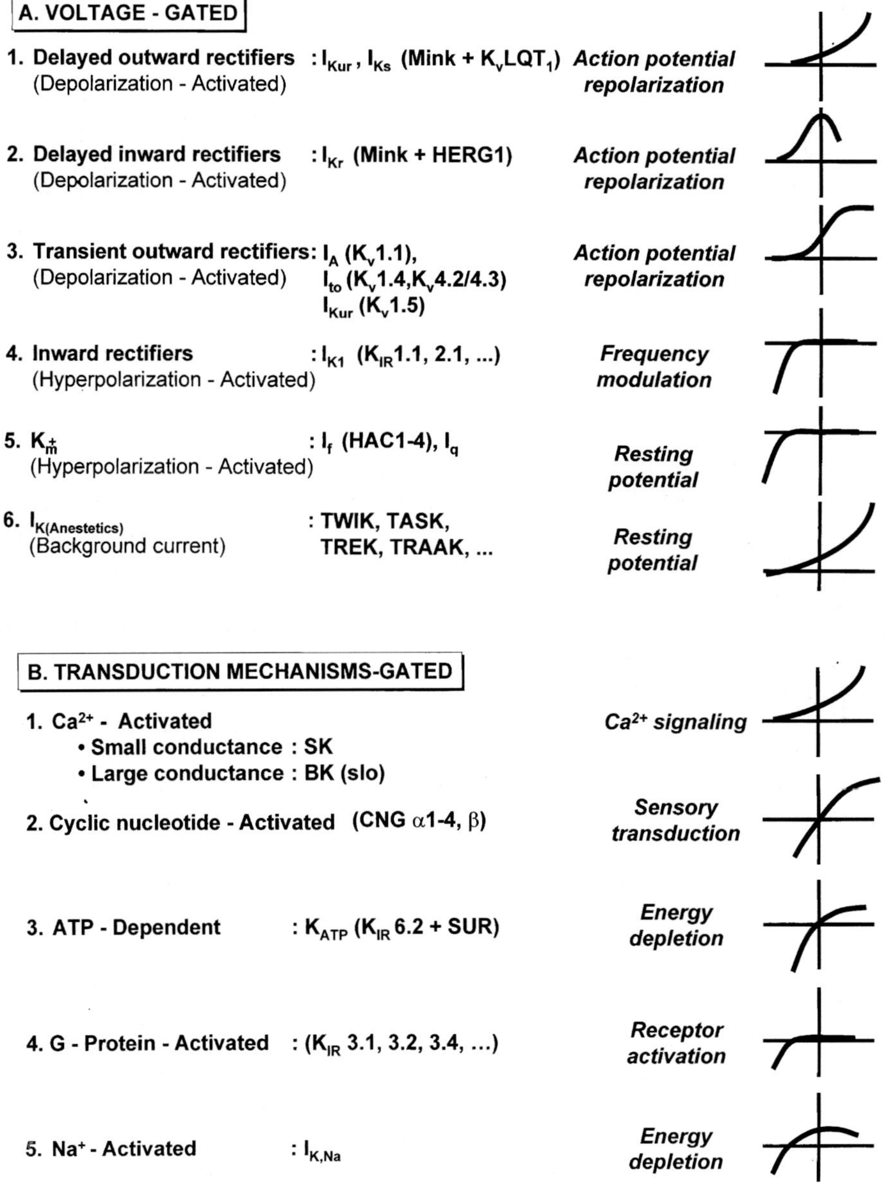

Fig. 2 Functional classification of K^+ channels: (A) voltage-gated channels; (B) transduction mechanisms-gated channels. In parentheses are shown the current nomenclature for the genes encoding each of these currents, when available. At the right of each panel a schematic representation of the current–voltage relationship for each channel subtype is shown.

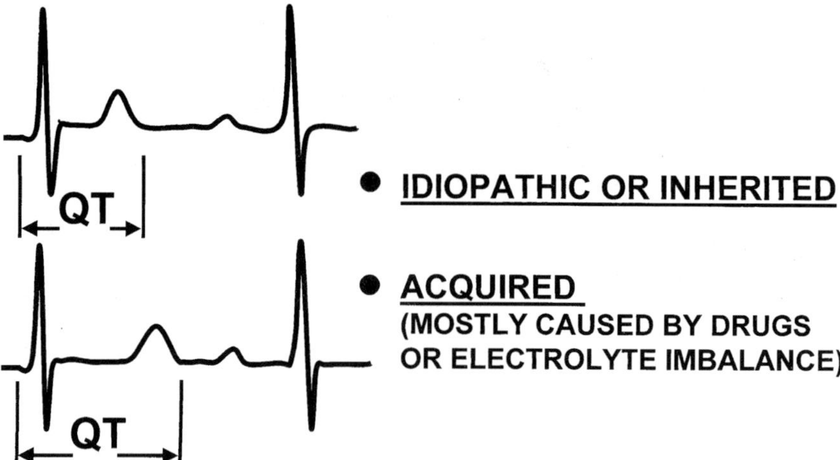

Fig. 3 Schematic representation of an ECG recording showing QT prolongation. Such prolongation may be of inheritable causes (idiopathic or inherited) or acquired by drug administration or electrolyte imbalance.

The molecular basis for most of these currents has now been identified, and a major contribution to the field has come from the study of the molecular genetics of the long QT syndrome, a life-threatening genetically transmitted disease characterized by a marked prolongation of the QT interval on the ECG and by frequent episodes of syncope or cardiac arrest usually occurring during conditions of psychological or physical stress (28). Similar to the previously described acquired LQTS induced by cardiotoxic second-generation antihistamines (Fig. 3), these syncopal episodes are caused by 'torsade de pointes' ventricular arrhythmias that often degenerate into ventricular fibrillation and

LQTS	Chromos.	Current affected	Clone	Mechanism of disease
1	11 (p15.5)	I_{Ks}	K_vLQT1	Decreased K^+ current
2	7 (q35-36)	I_{Kr}	hERG1	Decreased K^+ current
3	3 (p21-24)	I_{Na}	SCN5A	Increased Na^+ current
4	4 (q25-27)	?	?	?
5	21 (q22.1-22.2)	I_{Ks},I_{Kr}	MinK	Decreased K^+ current
6	21 (q22.1-22.2)	I_{Ks},I_{Kr}	MiRP1	Decreased K^+ current

Fig. 4 Genetic heterogeneity of congenital LQTS. The chromosomal location, the cardiac current affected, the gene involved and the proposed pathogenetic mechanism of each form are shown for the known forms (see text for details).

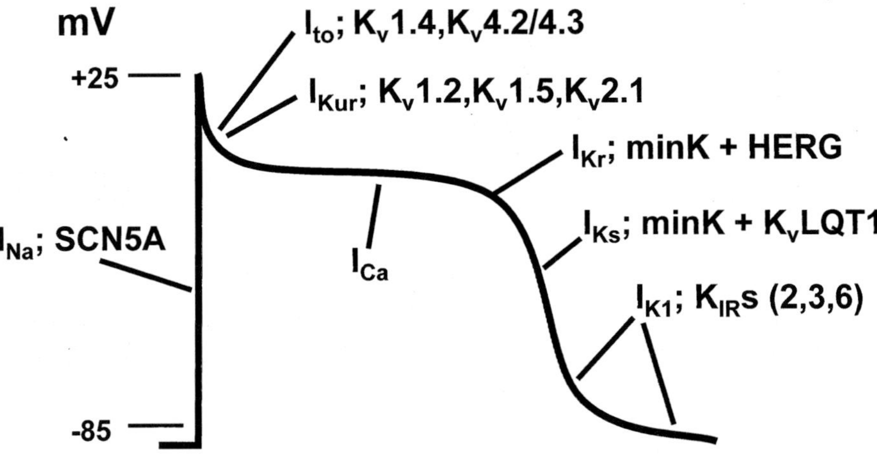

Fig. 5 Cardiac K^+ channels and genes. (Modified from *Trends Cardiovasc. Med.* **7**:118–124, 1997.)

eventually result in the sudden death of the affected patient (29). In the early 1990s, linkage studies revealed that this syndrome was genetically heterogeneous; several genetic defects have in fact been recognized in the last 4 years and associated to the arrhythmic phenotype. In particular, six genetic loci linked to LQTS have been described (Fig. 4); five of the genes of these loci have already been identified and mutations in these five genes have been associated to LQTS (30–34). Two of these genes (HERG and K_vLQT1) encode for the main subunits of cardiac K^+ channels, one (SCN5A) for a Na^+ channel subunit, and the last two (MinK and MiRP1) for accessory subunits regulating K^+ channel function. These studies, which gave a tremendous boost to the study of the molecular basis of drug-induced acquired LQTS, also allowed to establish a clear functional role for these genes encoding for cardiac ion channels. As a result of these achievements, the genes underlying each of the cardiac ionic currents (and particularly K^+ currents) are now known (Fig. 5). In particular, HERG [an acronym for 'human ether-a-gogo-related gene', since the HERG cDNA was cloned by homology to another gene, EAG or ether-a-gogo gene, which derived its nomenclature from the phenotype of the *Drosophila* mutants involving this gene, which displayed spontaneous repetitive firing of action potentials in the motor axons of the larval neuromuscular junction (35)] was soon recognized to encode for the main subunit of the rapid component of the ventricular repolarizing current I_{Kr} (36); more recently, it has been proposed that the association MinK + HERG reproduces the biophysical and pharmacological properties of the native I_{Kr} (37). More than 50 mutations in HERG have been described in the last 4 years in families affected by LQTS, leading to the hypothesis that these mutations, by impairing the rapid component of the ventricular repolarizing current I_{Kr}, may cause delayed repolarization, which, as already stated, represents a triggering mechanism for EADs and 'torsade de pointes'.

Constitutively or Heterologously Expressed HERG K^+ Channels as Cellular Models for the Prediction of Cardiotoxic Effects of Second-Generation Antihistamines

The crucial role played by HERG K^+ channels in the control of cardiac repolarization has prompted several studies that attempt to correlate the *in vivo* propensity of a drug to cause EADs and 'torsade de pointes' with its inhibitory action at the level of HERG K^+ channels *in vitro*. Given the complexity of isolating I_{Kr} in cardiac tissue, where several K^+ currents are simultaneously expressed in the same cell type (26), these studies have utilized several cellular models in which HERG K^+ currents could be properly resolved. These models include either cells in which HERG K^+ channels are constitutively present and easily recorded, such as SH-SY5Y human neuroblastoma (38), primary or clonal hypophyseal cells (39), the mouse tumour atrial myocytes AT-1 (40) or cells in which the heterologous expression of the channel can be achieved either by cDNA or cRNA microinjection techniques, such as *Xenopus* oocytes (41), or by transient or permanent transfection techniques, such as in human embryonic HEK 293 cells (42). Although each of these cellular models has specific advantages and drawbacks, a complete picture of the effect of a drug on a particular class of ion channels generally emerges upon comparison and careful evaluation of the results obtained in each of these different cellular models.

Using the oocyte expression system, both terfenadine (43) and astemizole (44) were found to effectively block HERG K^+ channels with IC_{50} values in the nanomolar range. Thus, the concentrations of terfenadine and astemizole required to produce such a pharmacological effect were clearly within the range found in those clinical situations (hepatic diseases; co-administration of itraconazole, ketoconazole, or macrolide antibiotics; see above) where patients underwent cardiotoxic manifestation by these two second-generation antihistamines (14, 45). The value of these *in vitro* studies in predicting the potential cardiotoxicity of compounds such as second-generation antihistamines is also shown by the results obtained with the main metabolic products of terfenadine and astemizole, (i.e. fexofenadine and desmethylastemizole). In fact, fexofenadine, which appears to lack the cardiotoxic potential of terfenadine (18), was shown to be ineffective in blocking HERG K^+ channels (43). On the other hand, both of the astemizole metabolites desmethylastemizole, which has been associated in clinical settings with the occurrence of 'torsade de pointes' (19), and norastemizole, display potent inhibitory actions on this class of K^+ channels (19, 46). Nevertheless, it should be pointed out that recognizing the cardiotoxic potential of terfenadine and astemizole has required several years of study and its prescription worldwide. Therefore a definitive answer on the safety of other second-generation antihistamines, and in particular of those introduced more recently, requires a longer period of observation and, possibly, several pharmacovigilance studies. In this respect, it should be pointed out that the cardiac safety of fexofenadine has recently been the subject of debate (47).

The important cardiotoxic effects exerted by two of the most widely used second-generation antihistamines have prompted speculation as to whether other molecules belonging to this therapeutic class might also possess similar pharmacological properties (2, 48, 49). The availability of these different *in vitro* models in order to evaluate the possible interference of H_1 receptor blockers with HERG K^+ channels, a pharmacological action which, by delaying action potential repolarization is directly related to their ability to exert cardiotoxic actions, has recently allowed comparative studies to be performed among the four second-generation antihistamines terfenadine, astemizole, loratadine and cetirizine (50). While the results obtained with terfenadine and astemizole in this study confirmed previously obtained results, the piperazine derivative cetirizine was shown to be devoid of HERG-blocking ability both upon the heterologous expression of this channel in *Xenopus* oocytes or in HEK-293 cells, as well as in cells constitutively expressing HERG K^+ channels. This lack of effect was observed for concentrations of cetirizine (1–30 μM) that are comparable to those achieved in the plasma of normal subjects (1–5 μM) after the administration of doses 2–6 times higher than the commonly recommended daily therapeutic dose (51). Although it is not always possible to formulate predictions on the *in vivo* pharmacological profile of a drug based only on *in vitro* data, this observation seems to suggest that 'torsade de pointes' are not likely to occur during conventional therapy with cetirizine. This conclusion seems to also be confirmed by the fact that cetirizine did not display significant prolongation of the QT interval in experimental animals (52) or humans (51), and that no study has yet appeared in the literature reporting cardiac arrhythmias or QT prolongation associated with its use (2, 53). In a more recent set of experiments, Carmeliet has studied the ability of cetirizine, as compared to terfenadine, to interfere with the rapid and slow components of the delayed rectifier K^+ current in guinea pig ventricular myocytes (54). The results obtained also show that both I_{Kr} and I_{Ks} in these cells were quite insensitive to cetirizine, although a certain degree of blockade could be detected at the very large doses of 0.1 mM and 1 mM, with I_{Kr} being slightly more sensitive than I_{Ks}. The IC_{50} values obtained for cetirizine were at least 10^3-fold higher than those obtained with terfenadine in the same study. *In vitro* data on the piperidinic second-generation antihistamine loratadine demonstrate that this compound was unable to block cloned HERG K^+ channels expressed in *Xenopus* oocytes (up to 10 μM) (50). These results are in agreement with those obtained in guinea pig ventricular myocytes (up to 3 μM) (55); however, higher concentrations of loratadine blocked both heterologously and constitutively expressed HERG K^+ channels (50, 56). Given that the concentrations of loratadine required to block HERG K^+ channels (3–30 μM) were at least 30 times higher than those achieved in the plasma during conventional therapy [after a single 40-mg dose the C_{max} of loratadine did not exceed 0.1 μM (57)], these results also might explain the lack of cardiac side-effects associated with its use in humans (58, 59) and experimental animals (60). In fact, it should be considered that, in a recent pharmacosurveillance study in which the risk profile for heart rhythm disorders and cardiac deaths was determined for some of the most common non-sedating antihistamines, cetirizine and loratadine displayed the lowest adverse drug reaction report rate per million of defined daily doses sold (61). A recent study (85) performed in human cardiac myocytes at physiological temperature and $[K^+]_e$, shows that Loratadine possesses I_{Kr}-blocking abilities similar to those of terfenadine, underlining the importance of a careful and critical evaluation of data obtained in pre-clinical studies before attempting to extrapolate their results into speculations on the clinical safety of specific compounds.

Another second-generation antihistamine available in several countries is ebastine, which bears considerable structural similarities to terfenadine. *In vivo* studies performed in laboratory animals (guinea pigs) have revealed that this compound might exert QT-prolonging effects (52), although on a smaller scale than terfenadine and clearly apparent only upon concomitant administration of ketoconazole. These results have been a matter of debate since, in a separate study, ketoconazole itself, when administered alone in the same experimental paradigm, caused an increase in the QT interval (62); nevertheless, *in vitro* studies revealed that ebastine was able to block both HERG K^+ channels heterologously expressed in *Xenopus* oocytes and native I_{Kr} in ventricular guinea pig myocytes (55). Interestingly, in both *in vivo* and *in vitro* models, the liver metabolic product of ebastine, carebastine, failed to show any effect. Although the structural similarities between terfenadine and ebastine should not be overlooked (63), the possible clinical relevance of these studies remains obscure.

Potential Cardiotoxicity of First-generation Antihistamines

The large number of studies demonstrating the cardiotoxicity of some second-generation antihistamines, as well as the similarities of the chemical structures of some older and new H_1-receptor blockers, has prompted the re-examination of possible adverse cardiac effects of older first-generation molecules which are generally considered to be free from cardiac toxicity and are still widely sold in most Western countries as 'over-the-counter' (OTC) or prescribed medications. Since these drugs are readily available to the public and are frequently implicated in accidental or intentional poisoning, this reassessment seems of particular pharmacological relevance. The issue of the potential cardiotoxicity of older antihistamines has also been raised by recent studies that have appeared in the literature focusing on the cardiac actions of these antiallergic drugs. In particular, using the patch-clamp electrophysiological technique, Khalifa *et al.* (64) reported that diphenhydramine blocked the repolarizing K^+ currents in guinea pig ventricular myocytes, an effect also shared by chlorpheniramine and pyrilamine (65); furthermore, the same study also showed the ability of therapeutic doses of diphenhydramine to lengthen cardiac repolarization *in vivo* in normal volunteers and in patients undergoing angioplasty (64). The ability of diphenhydramine, as well as of other 'conventional' antihistamines, to interfere with cardiac repolarization has also been confirmed *in vitro* by another study performed in feline hearts (66). In this model, the EC_{50} for the ability of these older molecules to delay cardiac repolarization ranged from 5 to 20 μM and, surprisingly, did not differ significantly from those of the two well-known cardiotoxic second-generation antihistamines astemizole (3.9 μM) and terfenadine (14.1 μM). A review of the available literature on clinical cases describing a possible association between the administration of older antihistamines and cardiac adverse effects shows that both diphenhydramine (67–69) and the cetirizine metabolic precursor hydroxyzine (70, 71), display the ability to induce QT prolongation when taken in larger than therapeutic doses, in some cases leading to the occurrence of cardiac arrhythmias of the 'torsade de pointes' type. The concentration of diphenhydramine and hydroxyzine found in the blood of intoxicated individuals (range 20–200 μM) were much higher than those considered to be within the standard therapeutic range (≈ 0.2–2 μM). The rare cardiovascular adverse effects observed with older antihistamines have often been interpreted as a consequence of their antimuscarinic properties; however, although this pharmacological property should not be underestimated, the antimuscarinic actions of older antihistamines seem to not always be directly related to their cardiac toxicity. In the case report of Clark and Vance (67), despite the massive intoxication with 30 25-mg diphenhydramine tablets, physical examinations showed only minimal anticholinergic effects. Similarly, Magera *et al.* (70), in their case report of a 13-month-old child intoxicated by 20–25 capsules (25 mg each) of hydroxyzine, noted that the classical autonomic peripheral symptoms of anticholinergic toxicity were absent.

Prompted by these observations, we have recently studied the *in vitro* blockade by hydroxyzine and diphenhydramine of HERG K^+ channels both heterologously expressed in *Xenopus* oocytes and constitutively present in SH-SY5Y cells (72). The results obtained suggest that both compounds exerted an inhibitory action on this class of K^+ channels at concentrations similar to those found to block I_{Kr} in guinea pig ventricular myocytes (64) (the IC_{50} values for HERG K^+ channel blockade in *Xenopus* oocytes were 10.7 and 21.5 μM, for hydroxyzine and diphenhydramine, respectively), and, more importantly, within the plasma concentration range found in patients undergoing poisoning with these first-generation antihistamines. These experimental and clinical data support the idea that older H_1 receptor antagonists possessing HERG-blocking ability could also exert cardiotoxic manifestations under specific clinical settings, as can occur with some of the second-generation antihistamines. This conclusion seems to find support in the results of a recent pharmaco-epidemiological study showing that the incidence of life-threatening ventricular arrhythmic events and cardiac arrests were more frequent in patients receiving OTC antihistamines, dominated by DPH (diphenhydramine) prescriptions, than in those receiving terfenadine (73). However, identifying the predisposing factors to such adverse reactions is complicated by the relatively little knowledge available on the pharmacokinetic profile of such compounds, and in particular on their metabolism. The CYP2D6 isoenzyme of cytochrome P450 seems to be involved in the metabolism of hydroxyzine and diphenhydramine (74). Genetic polymorphism of this isoenzyme is known to exist, and predisposition to adverse drug reaction is known to occur in individuals who are poor metabolizers or who make concomitant use of inhibitors of the CYP2D6 (75). Furthermore, older H_1 receptor antagonists inhibit their own metabolism by interfering with the CYP2D6; the IC_{50} values for diphenhydramine and hydroxyzine inhibition of this isoenzyme are 11 μM and 4 μM, respectively (74), a concentration range comparable to that achieved in the peripheral tissues by these compounds during accidental or intentional poisoning.

Interference of Second-Generation Antihistamines with Other Cardiac Ionic Channels

Beside their interference with I_{Kr}, second-generation antihistamines have been shown to block several other cardiac ionic currents, and these effects have also been correlated to their propensity to induce adverse cardiovascular effects. In particular, terfenadine has been shown to block the K^+ channels encoded by the $K_v1.5$ gene (76), which underlie I_{Kur} in cardiac myocytes. Both terfenadine enantiomers possess this pharmacological property. Ebastine, on the other hand, showed a much more reduced potency when compared to terfenadine (77). More interestingly, neither carebastine nor fexofenadine possessed any inhibitory action on these cardiac K^+ channels (78). Both loratadine (79) and its main liver metabolite descarboethoxy-loratadine (80) caused a certain degree of inhibition on the K^+ channels encoded by $K_v1.5$. Recently, Crumb (81) showed that loratadine caused a rate-dependent blockade of I_{to} in human atrial myocytes, an effect also observed at the low 'therapeutic' concentration of 10 nM. Crumb speculates that such a pharmacological action might provide a mechanism for the rare incidence of supraventricular arrhythmias reported with the use of this antihistamine (48).

Beside K^+ channels, voltage-gated Na^+ and Ca^{2+} channels have also been implicated in the cardiac actions of some second-generation antihistamines. In fact, terfenadine has been shown to block both Na^+ and Ca^{2+} channels in guinea pig ventricular myocytes (82, 83) in a concentration range ($\approx$ 100 nM) similar to that exerting inhibitory actions on the repolarizing K^+ currents. Interestingly, terfenadine bears a structural resemblance to the Ca^{2+} antagonists of the diphenylalkylamine class. In fact, verapamil has been shown to induce, at low concentrations, a prolongation of the cardiac action potential, suggesting a possible interaction with cardiac repolarizing currents. This hypothesis has recently been confirmed by studies demonstrating the ability of this Ca^{2+}-entry blocker to potently (IC_{50} of 143 nM) inhibit HERG K^+ channels. Interestingly, diltiazem was 100-fold less effective in this experimental model, whereas nifedipine was completely ineffective (84).

CONCLUSIONS

The debate over the possible adverse cardiac effects of second-generation antihistamines, following the reports of potentially fatal cardiac arrhythmias in predisposed patients or in intoxicated individuals, is a remarkable example of how molecular pharmacology studies can effectively match modern pharmacovigilance. The rapidly expanding knowledge of the molecular mechanisms underlying such adverse effects has made pre-clinical tests available (both *in vivo* and *in vitro*) which seem to display an appropriate predictive efficacy. In view of these considerations, the use of safer drugs to treat non-life-threatening diseases appears mandatory. Contributions from several disciplines have poured into this field, with the aim of reducing the cardiotoxic risk of widely used drugs such as antihistamines, as well as to improve the basic knowledge of the molecular mechanisms controlling action potential repolarization. It seems clear that the impact of such expanding knowledge will transcend the field of antihistamines, to affect other pharmacological classes, such as antipsychotics and antidepressants, which are known to possess proclivities to induce similar cardiovascular side-effects.

ACKNOWLEDGEMENTS

The authors are indebted to Dr M.T. Keating (Salt Lake City, UT, USA) for HERG cDNA. The study was supported by the following grants: Telethon 1058 to M.T., National Research Council (CNR) 97.04512.CT04, 97.01230.PF49 and 98.03149.CT04 to M.T., CNR 95.02857.CT04, 98.01048.CT04 and 98.00062.PF31 (PS Biotecnologie 5%) to L.A., and MURST 60% and 40% to L.A., and grants from the Regione Campania (P.O.P. and Legge 41) to L.A.

REFERENCES

1. Fourneau, E. and Bovet, D. Recherches sur l'action sympathicolytique d'un nouveau dérivé du dioxane. *Arch. Int. Pharmacodyn. Ther.* **46**:178–191, 1933.
2. Woosley, R. L. Cardiac actions of antihistamines. *Annu. Rev. Pharmacol. Toxicol.* **36**:233–252, 1996.
3. Slater, J. W., Zechnich, A. D. and Haxby, D. G. Second-generation antihistamines: a comparative review. *Drugs* **57**:31–47, 1999.
4. Davies, A. J., Harindra, V., McEwan, A. and Ghose, R. R. Cardiotoxic effect with convulsions in terfenadine overdose. *Br. Med. J.* **298**:325, 1989.
5. Monahan, B. P., Ferguson, C. L., Killeavy, E. S., Lloyd, B. K., Troy, J. and Cantilena, L. R. Jr. Torsades de pointes occurring in association with terfenadine use. *J. Am. Med. Assoc.* **264**:2788–2790, 1990.
6. Craft, T. M. Torsade de pointes after astemizole overdose. *Br. Med. J.* **292**:660, 1986.
7. Snook, J., Boothman-Burrell, D., Watkins, J. and Colin-Jones, D. Torsade de pointes ventricular tachycardia associated with astemizole overdose. *Br. J. Clin. Pract.* **42**:257–259, 1988.
8. Simons, F. E., Kesselman, M. S., Giddins, N. G., Pelech, A. N. and Simons, K. J. Astemizole-induced torsade de pointes. *Lancet* **ii**:624, 1988.

9. Bishop, R. O. and Gaudry, P. L. Prolonged Q–T interval following astemizole overdose. *Arch. Emerg. Med.* **6**:63–65, 1989.
10. Leor, J., Harman, M., Rabinowitz, B. and Mozes, B. Giant U waves and associated ventricular tachycardia complicating astemizole overdose: successful therapy with intravenous magnesium. *Am. J. Med.* **91**:94–97, 1991.
11. Hoppu, K., Tikanoja, T., Tapanainen, P., Remes, M., Saarenpaa-Heikkila, O. and Kouvalainen, K. Accidental astemizole overdose in young children. *Lancet* **338**:538–540, 1991.
12. Marion Merrel Dow, Inc. *Important Drug Warning.* Marion Merrel Dow, Cincinnati, OH, 6 August 1990.
13. FDA Federal Register. **62**:1889–1892, 1997, (docket no. 96N-1512).
14. Honig, P. and Baranjuk, J. N. In: *Histamine and H_1-Receptor Antagonists in Allergic Diseases* (Estelle, F. and Simons, R., eds), pp. 383–412. Marcel Dekker, New York, 1996.
15. Cavuto, N. J., Woosley, R. L. and Sale, M. Pharmacies and prevention of potentially fatal drug interactions. *J. Am. Med. Assoc.* **275**:1086–1087, 1996.
16. Anonymous. *FDA Med. Bull.* **22**:2–3, 1992.
17. Garteiz, D. A., Hook, R. H., Walker, B. J. and Okerholm, R. A. Pharmacokinetics and biotransformation studies of terfenadine in man. *Arzneimittelforschung* **32**:1185–1190, 1982.
18. Markham, A. and Wagstaff, A. J. Fexofenadine. *Drugs* **55**:269–274, 1998.
19. Vorperian, V. R., Zhou, Z., Mohammad, S., Hoon, T. J., Studenik, C. and January, C. T. Torsade de pointes with an antihistamine metabolite: potassium channel blockade with desmethylastemizole. *J. Am. Coll. Cardiol.* **28**:1556–1561, 1996.
20. Carmeliet, E. Mechanisms and control of repolarization. *Eur. Heart J.* **14**:3–13, 1993.
21. Dessertenne, F. La tachycardie ventriculaire a deux foyers opposes variables. *Arch. Mal. Coeur* **95**:263–272, 1966.
22. Genovese, A. and Spadaro, G. Highlights in cardiovascular effects of histamine and H_1-receptor antagonists. *Allergy* **52 (Suppl. 34)**:67–78, 1997.
23. Roden, D. M. Early after-depolarization and torsade de pointes: implications for the control of arrhythmias by prolonging repolarization. *Eur. Heart J.* **14**:56–61, 1993.
24. el-Sherif, N., Zeiler, R. H., Craelius, W., Gough, W. B. and Henkin, R. QTU prolongation and polymorphic ventricular tachyarrhythmias due to bradycardia-dependent early afterdepolarizations. Afterdepolarizations and ventricular arrhythmias. *Circ. Res.* **63**:286–305, 1988.
25. January, C. T., Riddle, J. M. and, Salata, J. J. A model for early afterdepolarizations: induction with the Ca^{2+} channel agonist Bay K 8644. *Circ. Res.* **62**:563–571, 1988.
26. Barry, D. M. and Nerbonne, J. M. Myocardial potassium channels: electrophysiological and molecular diversity. *Annu. Rev. Physiol.* **58**:363–394, 1996.
27. Sanguinetti, M. C. and Jurkiewicz, N. K. Two components of cardiac delayed rectifier K^+ current: differential sensitivity to block by class III antiarrhythmic agents. *J. Gen. Physiol.* **96**:195–215, 1990.
28. Schwartz, P. J., Locati, E. H., Napolitano, C. and Priori, S. G. In: *Cardiac Electrophysiology. From Cell to Bedside*, 2nd edn (Zipes, D. P. and Jalife, J., eds), pp. 788–811. W. B. Saunders, Philadelphia, 1995.
29. Napolitano, C., Priori, S. G. and Schwartz, P. J. Torsade de pointes. Mechanisms and management. *Drugs* **47**:51–65, 1994.
30. Wang, Q., Curran, M. E., Splawski, I., Burn, T. C., Millholland, J. M., VanRaay, T. J., Shen, J., Timothy, K. W., Vincent, G. M., de Jager, T., Schwartz, P. J., Toubin, J. A., Moss, A. J., Atkinson, D. L., Landes, G. M., Connors, T. D. and Keating, M. T. Positional cloning of a novel potassium channel gene: KVLQT1 mutations cause cardiac arrhythmias. *Nat. Genet.* **12**:17–23, 1996.
31. Curran, M. E., Splawski, I., Timothy, K. W., Vincent, G. M., Green, E. D. and Keating, M. T. A molecular basis for cardiac arrhythmia: HERG mutations cause long QT syndrome. *Cell* **80**:795–803, 1995.
32. Wang, Q., Shen, J., Splawski, I., Atkinson, D., Li, Z., Robinson, J. L., Moss, A. J., Towbin, J. A. and Keating, M. T. SCN5A mutations associated with an inherited cardiac arrhythmia, long QT syndrome. *Cell* **80**:805–811, 1995.
33. Splawski, I., Tristani-Firouzi, M., Lehmann, M. H., Sanguinetti, M. C. and Keating, M. T. Mutations in the hminK gene cause long QT syndrome and suppress IKs function. *Nat. Genet.* **17**:338–340, 1997.
34. Abbott, G. W., Sesti, F., Splawski, I., Buck, M. E., Lehmann, M. H., Timothy, K. W., Keating, M. T. and Goldstein, S. A. MiRP1 forms I_{Kr} potassium channels with HERG and is associated with cardiac arrhythmia. *Cell* **97**:175–187, 1999.
35. Warmke, J., Drysdale, R. and Ganetzky, B. A distinct potassium channel polypeptide encoded by the *Drosophila* eag locus. *Science* **252**:1560–1562, 1991.

36. Sanguinetti, M. C., Jiang, C., Curran, M. E. and Keating, M. T. A mechanistic link between an inherited and acquired cardiac arrhythmia: HERG encodes the I_{Kr} potassium channel. *Cell* **81**:299–307, 1995.
37. McDonald, T. V., Yu, Z., Ming, Z., Palma, E., Meyers, M. B., Wang, K. W., Goldstein, S. A. and Fishman, G. I. A minK–HERG complex regulates the cardiac potassium current I(Kr). *Nature* **388**:289–292, 1997.
38. Bianchi, L., Wible, B., Arcangeli, A., Taglialatela, M., Morra, F., Castaldo, P., Crociani, O., Rosati, B., Faravelli, L., Olivotto, M. and Wanke, E. herg encodes a K^+ current highly conserved in tumors of different histogenesis: a selective advantage for cancer cells? *Cancer Res.* **58**:815–822,1998.
39. Schafer, R., Wulfsen, I., Behrens, S., Weinsberg, F., Bauer, C. K. and Schwarz, J. R. The erg-like potassium current in rat lactotrophs. *J. Physiol.* **518**:401–416, 1999.
40. Yang, T., Snyders, D. J. and Roden, D. M. Ibutilide, a methanesulfonanilide antiarrhythmic, is a potent blocker of the rapidly activating delayed rectifier K^+ current (IKr) in AT-1 cells. Concentration-, time-, voltage-, and use-dependent effects. *Circulation* **91**:1799–1806,1995.
41. Taglialatela, M., Castaldo, P., Iossa, S., Pannaccione, A., Fresi, A., Ficker, E. and Annunziato, L. Regulation of the human ether-a-gogo related gene (HERG) K^+ channels by reactive oxygen species. *Proc. Natl. Acad. Sci. USA* **94**:11698–11703, 1997.
42. Zhou, Z., Gong, Q., Ye, B., Fan, Z., Makielski, J. C., Robertson, G. A. and January, C. T. Properties of HERG channels stably expressed in HEK 293 cells studied at physiological temperature. *Biophys. J.* **74**:230–241, 1998.
43. Roy, M.-L., Dumaine, R. and Brown, A. M. HERG, a primary human ventricular target of the non sedating antihistamine terfenadine. *Circulation* **94**:817–823, 1996.
44. Suessbrich, H., Waldegger, S., Lang, F. and Busch, A. E. Blockade of HERG channels expressed in *Xenopus* oocytes by the histamine receptor antagonists terfenadine and astemizole. *FEBS Lett.* **385**:77–80, 1996.
45. Sorkin, E. M. and Heel, R. C. Terfenadine: a review of its pharmacodynamic properties and therapeutic efficacy. *Drugs* **29**:34–56, 1985.
46. Zhou, Z., Vorperian, V. R., Gong, Q., Zhang, S. and January, C. T. Block of HERG potassium channels by the antihistamine astemizole and its metabolites desmethylastemizole and norastemizole. *J. Cardiovasc. Electrophysiol.* **10**:836–843, 1999.
47. Giraud, T. QT lengthening and arrhythmias associated with fexofenadine. *Lancet* **353**:2072–2073, 1999.
48. Good, A. P., Rockwood, R. and Schad, P. Loratadine and ventricular tachycardia. *Am. J. Cardiol.* **74**:1240–1245, 1994.
49. DuBuske, L. M. Second-generation antihistamines: the risk of ventricular arrhythmias. *Clin. Ther.* **21**:281–294, 1999.
50. Taglialatela, M., Pannaccione, A., Castaldo, P., Giorgio, G., Zhou, Z., January, C. T., Genovese, A., Marone, G. and Annunziato, L. The molecular basis for the lack of HERG K^+ channels block-related cardiotoxicity by the H_1 receptor blocker cetirizine as compared to other second-generation antihistamines. *Mol. Pharmacol.* **54**: 113–121, 1998.
51. Sale, M. E., Barbery, J. T., Woosley, R. L., Edwards, D., Yeh, J., Thakker, K. and Chung, M. The electrocardiographic effects of cetirizine in normal subjects. *Clin. Pharmacol. Ther.* **56**:295–301, 1994.
52. Hey, J. A., del Prado, M., Sherwood, J., Kreutner, W. and Egan, R. W. Comparative analysis of the cardiotoxicity proclivities of second generation antihistamines in an experimental model predictive of adverse clinical ECG effects. *Arznem.-Forsch. Drug Res.* **46**:153–156, 1996.
53. Coulie, P., Delaere, A., De Vos, C., Donnelly, F. and Rihoux, J. P. Non-sedating antihistamines and cardiac arrhythmias. *Lancet* **351**:451, 1998.
54. Carmeliet, E. Effects of cetirizine on the delayed K^+ currents in cardiac cells: comparison with terfenadine. *Br. J. Pharmacol.* **124**:663–668, 1998.
55. Ko, C. M., Ducic, I., Fan, J., Shuba, Y. M. and Morad, M. Suppression of mammalian K^+ channel family by Ebastine. *J. Pharmacol. Exp. Ther.* **281**:233–244, 1997.
56. Lacerda, A. E., Roy, M.-L., Lewis, E. W. and Rampe, D. Interaction of the nonsedating antihistamine loratadine with a Kv1.5 type potassium channel cloned from human heart. *Mol. Pharmacol.* **52**:314–322, 1997.
57. Haria, M., Fitton, A. and Peters, D. H. Loratadine. A reappraisal of its pharmacodynamic properties and therapeutic use in allergic disorders. *Drugs* **48**:617–638, 1994.
58. Brannan, M., Reidenberg, P., Radwanski, E., Shneyer, L., Lin, C. *et al.* Loratadine administered concomitantly with erythromycin: pharmacokinetic and electrocardiographic evaluations. *Clin. Pharmacol. Ther.* **58**:269–278, 1995.

59. Woosley, R. L. and Darrow, W. R. Analysis of potential adverse drug reactions – a case of mistaken identity. *Am. J. Cardiol.* **74**:208–209, 1994.
60. Hey, J. A., del Prado, M., Egan, R. W., Sherwood, J. and Kreutner, W. Loratadine produces antihistamine activity without adverse CNS, ECG or cardiovascular effects in guinea pigs. Comparative studies with terfenadine and sedating antihistamines. *Int. Arch. Allergy Immunol.* **107**:418–419, 1995.
61. Lindquist, M. and Edwards, I. R. Risk of non-sedating antihistamines. *Lancet* **349**:1322, 1997.
62. Gras, J., Llenas, J., Palacios, J. M. and Roberts, D. J. The role of ketoconazole in the QTc interval prolonging effects of H_1-antihistamines in a guinea-pig model of arrhythmogenicity. *Br. J. Pharmacol.* **119**:187–188, 1996.
63. Roberts, D. J. Towards the optimal antihistamine: studies with ebastine. *Inflamm. Res.* **74**:S36–S37, 1998.
64. Khalifa, M., Drolet, B., Daleau, P., Lefez, C., Gilbert, M., Plante, S., O'Hara, G. E., Gleeton, O., Hamelin, B. A. and Turgeon, J. Block of potassium currents in guinea pig ventricular myocytes and lengthening of cardiac repolarization in man by the histamine H_1 receptor antagonist diphenhydramine. *J. Pharmacol. Exp. Ther.* **288**:858–865, 1999.
65. Salata, J. J., Jurkiewicz, N. K., Wallace, A. A., Stupienski, R. F. 3rd, Guinosso, P. J. Jr and Lynch, J. J. Jr. Cardiac electrophysiological actions of the histamine H_1-receptor antagonists astemizole and terfenadine compared with chlorpheniramine and pyrilamine. *Circ. Res.* **76**:110–119, 1995.
66. Wang, W. X., Ebert, S. N., Liu, X. K., Chen, Y. W., Drici, M. D. and Woosley, R. L. 'Conventional' antihistamines slow cardiac repolarization in isolated perfused (Langendorff) feline hearts. *J. Cardiovasc. Pharmacol.* **32**:123–128, 1998.
67. Clark, R. F. and Vance, M. V. Massive diphenhydramine poisoning resulting in a wide-complex tachycardia: successful treatment with sodium bicarbonate. *Ann. Emerg. Med.* **21**:318–321, 1992.
68. Hestand, H. E. and Teske, D. W. Diphenhydramine hydrochloride intoxication. *J. Pediatr.* **90**:1017–1018, 1977.
69. Zareba, W., Moss, A. J., Rosero, S. Z., Hajj-Ali, R., Konecki, J. and Andrews, M. Electrocardiographic findings in patients with diphenhydramine overdose. *Am. J. Cardiol.* **80**:1168–1173, 1997.
70. Magera, B. E., Betlach, C. J., Sweatt, A. P. and Derrick, C. W. Jr. Hydroxyzine intoxication in a 13-month-old child. *Pediatrics* **67**:280–283, 1981.
71. Hollister, L. E. Hydroxyzine hydrochloride: possible adverse cardiac interactions. *Psychopharmacol. Commun.* **1**:61–65, 1975.
72. Taglialatela, M., Timmermann, H. and Annunziato, L. Cardiotoxic potential and CNS effects of first-generation antihistamines. *Trends Pharm. Sci.* **21**:52–56, 2000.
73. Pratt, C. M., Hertz, R. P., Ellis, B. E., Crowell, S. P., Louv, W. and Moyé, L. Risk of developing life-threatening ventricular arrhythmia associated with terfenadine in comparison with over-the-counter antihistamines, ibuprofen and clemastine. *Am. J. Cardiol.* **73**:346–352, 1994.
74. Hamelin, B. A., Bouayad, A., Drolet, B., Gravel, A. and Turgeon, J. *In vitro* characterization of cytochrome P450 2D6 inhibition by classic histamine H1 receptor antagonists. *Drug Metab. Dispos.* **26**:536–539, 1998.
75. Lennard, M. S. Genetically determined adverse drug reactions involving metabolism. *Drug Saf.* **9**:60–77, 1993.
76. Crumb, W. J. Jr, Wible, B., Arnold, D. J., Payne, J. P. and Brown, A. M. Blockade of multiple human cardiac potassium currents by the antihistamine terfenadine: possible mechanism for terfenadine-associated cardiotoxicity. *Mol. Pharmacol.* **47**:181–190, 1995.
77. Yang, T., Prakash, C., Roden, D. M. and Snyders, D. J. Mechanism of block of a human cardiac potassium channel by terfenadine racemate and enantiomers. *Br. J. Pharmacol.* **115**:267–274, 1995.
78. Valenzuela, C., Delpon, E., Franqueza, L., Gay, P., Vicente, J. and Tamargo, J. Comparative effects of nonsedating histamine H1 receptor antagonists, ebastine and terfenadine, on human Kv1.5 channels. *Eur. J. Pharmacol.* **326**:257–263, 1997.
79. Delpon, E., Valenzuela, C., Gay, P., Franqueza, L., Snyders, D. J. and Tamargo, J. Block of human cardiac Kv1.5 channels by loratadine: voltage-, time- and use-dependent block at concentrations above therapeutic levels. *Cardiovasc. Res.* **35**:341–350, 1997.
80. Caballero, R., Delpon, E., Valenzuela, C., Longobardo, M., Franqueza, L. and Tamargo, J. Effect of descarboethoxyloratadine, the major metabolite of loratadine, on the human cardiac potassium channel Kv1.5. *Br. J. Pharmacol.* **122**:796–798, 1997.
81. Crumb, W. J. Rate-dependent blockade of a potassium current in human atrium by the antihistamine loratadine. *Br. J. Pharmacol.* **126**:575–580, 1999.

82. Ming, Z. and Nordin, C. Terfenadine blocks time-dependent Ca^{2+}, Na^{+}, and K^{+} channels in guinea pig ventricular myocytes. *J. Cardiovasc. Pharmacol.* **26**:761–769, 1995.
83. Liu, S., Melchert, R. B. and Kennedy, R. H. Inhibition of L-type Ca^{2+} channel current in rat ventricular myocytes by terfenadine. *Circ. Res.* **81**:202–210, 1997.
84. Zhang, S., Zhou, Z., Gong, Q., Makielski, J. C. and January, C. T. Mechanism of block and identification of the verapamil binding domain to HERG potassium channels. *Circ. Res.* **84**:989–998, 1999.
85. Crumb, W. Loratadine blockade of K^{+} channels in human heart: comparison with Terfenadine under physiological conditions. *JPET* **292**:261–264, 2000.

Index

B

D

E

F

G

H

J

N

O

P

R

S

T